Lai-Massey Cipher Designs

Jorge Nakahara Jr.

Lai-Massey Cipher Designs

History, Design Criteria and Cryptanalysis

Springer

Jorge Nakahara Jr.
São Paulo, Brazil

Additional material for this book can be downloaded from http://extras.springer.com.

ISBN 978-3-030-09827-8 ISBN 978-3-319-68273-0 (eBook)
https://doi.org/10.1007/978-3-319-68273-0

This Springer imprint is published by the registered company Springer Nature Switzerland AG
The registered company address is: Gewerbestrasse 11, 6330 Cham, Switzerland

This book is dedicated to my parents
Jorge Nakahara and Misao Nakahara.
In memoriam

Preface

This book presents the first comprehensive account of the history, design and analysis of cryptographic block ciphers following the Lai-Massey design paradigm, which started with the PES and IDEA ciphers in 1990/1991.

The fact that these ciphers were originally designed by X. Lai and J.L. Massey led to the terminology *Lai-Massey cipher designs* to denote these and other ciphers that follow in one way or another the design guidelines set by Lai and Massey for PES and IDEA.

Other ciphers discussed in this book include: the MESH cipher family (including MESH-64, MESH-96, MESH-128, MESH-64(8) and MESH-128(8)), RIDEA, WIDEA-n, YBC, the FOX/IDEA-NXT cipher family, REESSE3+, IDEA* and Bel-T.

Lai-Massey cipher designs differ structurally from Feistel Network ciphers such as the DES, and from Substitution-Permutation Network (SPN) ciphers such as the AES. Lai-Massey ciphers have their own unique features such as: (i) complete text diffusion in a single round; (ii) they usually repeat a rather strong round function a small number of times, instead of repeating a comparatively weak round function a large number of times; (iii) originally, only three group operations are used as building blocks, such as bitwise exclusive-or, modular addition and modular multiplication (in a finite field); (iv) originally, they did not employ S-boxes nor MDS codes to achieve optimal confusion and diffusion, respectively.

This book consolidates research and cryptanalysis results in the publicly-available literature on Lai-Massey ciphers covering a 28-year period from 1990 to 2017. An extensive bibliographic research has been performed to collect the most relevant conference papers, journal articles, PhD and MSc theses and technical reports dealing with mathematical and computational aspects of Lai-Massey ciphers. Therefore, this book may serve as a useful resource for students, researchers and practitioners willing to understand, design and deploy ciphers based on the Lai-Massey design paradigm.

The attacks listed in this book cover both published, new, updated and revised papers. The attacks include: Differential Cryptanalysis, Linear Cryptanalysis, Square/Multiset, Impossible Differential, Truncated Differential, Biryukov-Demirci, Demirci, Slide, Advanced Slide, Boomerang, Biclique, Differential-Linear, Key-Dependent, BDK, Meet-in-the-Middle and Related-Key attacks.

Although we tried to be as accurate as possible, any errors, mistakes and inconsistencies are the author's sole responsibility. Any comments, corrections or suggestions for improvement are welcome. Please, forward them to

jorge.nakahara@springer.com

Many people helped and supported this book project over the years. To mention a few: Bart Preneel, Vincent Rijmen, Joos Vandewalle, Christian Cachin, Pela Noé, Beverley Ford, Alfred Hofmann, Ronan Nugent and Xuejia Lai.

The author appreciates the kind permission to include parts of papers, articles and theses that have been published (and are copyrighted) by Springer, by the Katholieke Universiteit Leuven (KUL), by the International Association for Cryptologic Research (IACR) or by other institutions, scientific societies or companies.

Some data concerning attack details of some Lai-Massey ciphers could not fit in this book due to space limitations. Please, follow the instructions on the book's webpage to access this information:

http://www.springer.com/978-3-319-68272-3

Finally, I would like to dedicate this book to my family, without whom nothing would have ever been accomplished at all.

São Paulo, Brazil

Jorge Nakahara Jr.
January 2018

List of Acronyms

ACPC	Adaptively-Chosen Plaintext and Ciphertext
AES	Advanced Encryption Standard
AMX	Addition, Multiplication, eXclusive-or
ARX	Addition, bitwise-Rotation, eXclusive-or
ASR	All-Subkey Recovery
AX	Addition-eXclusive-or (layer)
BDK	Biham-Dunkelman-Keller
CBC	Cipher Block Chaining (mode)
CC	Chosen Ciphertext
CCACP	Chosen-Ciphertext Adaptively-Chosen Plaintext
CFB	Cipher FeedBack (mode)
CP	Chosen Plaintext
CPACC	Chosen-Plaintext Adaptively-Chosen Ciphertext
CPCC	Chosen-Plaintext (non-adaptively)-Chosen Ciphertext
CO	Ciphertext Only
CPU	Central Processing Unit
CTR	Counter (mode)
DES	Data Encryption Standard
DFR	Distinguish-From-Random
ECB	Electronic Code Book (mode)
ES	Exhaustive Search
EKS	Exhaustive Key Search
GCD	Greatest Common Divisor
HW	Hamming Weight
IBA	Impossible-Boomerang Attack
IDEA	International Data Encryption Algorithm
IND-KPA	Indistinguishable under Known-Plaintext Attack
IND-CPA	Indistinguishable under Chosen-Plaintext Attack
IND-CCA1	Indistinguishable under non-adaptive Chosen-Ciphertext Attack
IND-CCA2	Indistinguishable under adaptive Chosen-Ciphertext Attack
io	inverse orthomorphism (function)
IPES	Improved Proposed Encryption Standard
IV	Initial Value(s)
KK	Known Key
KM	Key-Mixing (layer)
KP	Known Plaintext
KR	Key Recovery

KW	Key Whitening (layer)
KSA	Key Schedule Algorithm
LCM	Least Common Multiple
LFSR	Linear Feedback Shift Register
lsb	Least Significant Bit
MA	Memory Access or Multiplication-Addition (layer)
MA-box	Multiplication-Addition box
MAC	Message Authentication Code
MDS	Maximum Distance Separable (code)
MITM	Meet-in-the-Middle
msb	Most Significant Bit
NESSIE	New European Schemes for Signature, Integrity and Encryption
NIST	(US) National Institute of Standards and Technology
NLFSR	NonLinear Feedback Shift Register
OFB	Output FeedBack (mode)
OPP	Offline Pre-Processing
or	orthomorphism function
OT	Output Transformation
OTP	One-Time Pad
PES	Proposed Encryption Standard
PGP	Pretty Good Privacy
PRF	PseudoRandom Function
PRP	PseudoRandom Permutation
RK-CP	Related-Key Chosen-Plaintext
RK-KP	Related-Key Known-Plaintext
S-box	Substitution box
SPN	Substitution Permutation Network
SPRP	Strong PseudoRandom Permutation
TDC	Truncated Differential Cryptanalysis
TEA	Tiny Encryption Algorithm
TLU	Table LookUp
TMDTO	Time-Memory-Data Trade-Off
TMTO	Time-Memory Trade-Off
UD	Unicity Distance
WCU	whole codebook used
WKC	Weak-Key Class

List of Symbols

$\mathrm{lsb}_i(x)$	the i-th least significant bit of x
$\mathrm{msb}_i(x)$	the i-th most significant bit of x
$x \ll y$	the value of x left-shifted by y bits
$x \gg y$	the value of x right-shifter by y bits
$x \lll y$	the value of x left-rotated by y bits
$x \ggg y$	the value of x right-rotated by y bits
$\oplus$	bitwise exclusive-or operation
$\boxplus$	modular addition in $\mathbb{Z}_{2^w}$, $w \in \{2, 4, 8, 16\}$
$\boxminus$	modular subtraction in $\mathbb{Z}_{2^w}$, $w \in \{2, 4, 8, 16\}$
$\odot$	modular multiplication in $\mathrm{GF}(2^w + 1)$, $w \in \{2, 4, 8, 16\}$
$\boxdot$	modular division in $\mathrm{GF}(2^w + 1)$, $w \in \{2, 4, 8, 16\}$
$\mathbb{Z}_{2^n}$	the ring of integer modulo 2^n
$\mathrm{GF}(2^{16} + 1)$	the finite field of $2^{16} + 1$ elements and such that $0 \equiv 2^{16}$
$\mathrm{GF}(2^8 + 1)$	the finite field of $2^8 + 1$ elements and such that $0 \equiv 2^8$
$\mathrm{GF}(2^n)$	the finite field of 2^n elements i.e. $\mathrm{GF}(2)[x]/p(x)$ where $p(x)$ is an irreducible polynomial in $\mathrm{GF}(2)$ of degree n
subscript x	denotes an hexadecimal value e.g. $1a_x = 26$
subscript 2	denotes a binary value e.g. $1100_2 = 12$
1^m	denotes a sequence of m bits 1
0^t	denotes a sequence of t bits 0
$\lvert x \rvert$	the magnitude, length or the absolute value of x
$\mathbb{Z}_m^t$	the cartesian product of $\mathbb{Z}_m$ with itself t times: $\mathbb{Z}_m \times \mathbb{Z}_m \times \ldots \times \mathbb{Z}_m$
Λ-set	a multiset of 2^m n-bit text blocks

Contents

1 Introduction 1
1.1 Symmetric and Asymmetric Ciphers 1
1.2 Iterated (or Product) Ciphers 5
1.3 Symmetric Cryptosystems 11
1.4 Entropy 13
1.5 Confusion and Diffusion 14
1.6 Security Assumptions and Threat Models 15
1.7 PRP and SPRP 17
1.8 Modes of Operation 18
1.9 Unconditional Security 23
1.10 The Origins of the MESH Ciphers 24
References 28

2 Lai-Massey Block Ciphers 37
2.1 The PES Block Cipher 37
2.1.1 Encryption and Decryption Frameworks 40
2.1.2 Key Schedule Algorithm 41
2.2 The IDEA Block Cipher 44
2.2.1 Encryption and Decryption Frameworks 45
2.2.2 Key Schedule Algorithm 47
2.3 The MESH Block Ciphers 49
2.3.1 Encryption and Decryption Frameworks of MESH-64 51
2.3.2 Key Schedule Algorithm of MESH-64 55
2.3.3 Encryption and Decryption Frameworks of MESH-96 58
2.3.4 Key Schedule Algorithm of MESH-96 61
2.3.5 Encryption and Decryption Frameworks of MESH-128 64
2.3.6 Key Schedule Algorithm of MESH-128 68
2.3.7 Encryption and Decryption Frameworks of MESH-64(8) 71
2.3.8 Key Schedule Algorithm of MESH-64(8) 73
2.3.9 Encryption and Decryption Frameworks of MESH-128(8) 77

2.3.10 Key Schedule Algorithm of MESH-128(8) 80
2.4 The RIDEA Block Cipher 83
2.4.1 Encryption and Decryption Frameworks 83
2.4.2 Key Schedule Algorithm 84
2.5 The WIDEA-n Block Ciphers 85
2.5.1 Encryption and Decryption Frameworks 86
2.5.2 Key Schedule Algorithms 89
2.6 The FOX/IDEA-NXT Block Ciphers 90
2.6.1 Encryption and Decryption Frameworks 91
2.6.2 Key Schedule Algorithms 94
2.7 The REESSE3+ Block Cipher 96
2.7.1 Encryption and Decryption Frameworks 96
2.7.2 Key Schedule Algorithm 99
2.8 The IDEA* Block Cipher 100
2.8.1 Encryption and Decryption Frameworks 100
2.8.2 Key Schedule Algorithm 103
2.9 The Yi's Block Cipher 106
2.9.1 Encryption and Decryption Frameworks 106
2.9.2 Key Schedule Algorithm 108
2.10 The Bel-T Block Cipher 109
2.10.1 Encryption and Decryption Frameworks 110
2.10.2 Key Schedule Algorithm 111
References 112

3 Attacks 117
3.1 Exhaustive Search (Brute Force) Attack 117
3.2 Dictionary Attack 120
3.3 Birthday-Paradox Attacks 121
3.3.1 Generalized Birthday Paradox Attack 124
3.4 Time-Memory Trade-Off Attacks 124
3.4.1 Hellman's Attack 127
3.4.2 Time/Memory/Data Trade-Off Attacks 130
3.5 Differential Cryptanalysis 131
3.5.1 DC of PES Under Weak-Key Assumptions 145
3.5.2 DC of IDEA Under Weak-Key Assumptions 149
3.5.3 DC of MESH-64 Under Weak-Key Assumptions 154
3.5.4 DC of MESH-96 Under Weak-Key Assumptions 156
3.5.5 DC of MESH-128 Under Weak-Key Assumptions 163
3.5.6 DC of MESH-64(8) Under Weak-Key Assumptions 184
3.5.7 DC of MESH-128(8) Under Weak-Key Assumptions 206
3.5.8 DC of WIDEA-n Under Weak-Key Assumptions 218
3.5.9 DC of RIDEA Under Weak-Key Assumptions 230
3.5.10 DC of REESSE3+ Under Weak-Key Assumptions 236
3.5.11 DC of IDEA* Without Weak-Key Assumptions 253
3.5.12 DC of YBC Without Weak-Key Assumptions 254

3.6 Truncated Differential Cryptanalysis 256
3.6.1 TDC of IDEA 257
3.6.2 TDC of MESH-64 265
3.6.3 TDC of MESH-96 269
3.6.4 TDC of MESH-128 272
3.6.5 TDC of WIDEA-n 275
3.7 Multiplicative Differential Analysis 280
3.8 Impossible-Differential Cryptanalysis 282
3.8.1 ID Analysis of IDEA 283
3.8.2 ID Analysis of MESH-64 289
3.8.3 ID Analysis of MESH-96 291
3.8.4 ID Analysis of MESH-128 296
3.8.5 ID Analysis of MESH-64(8) 303
3.8.6 ID Analysis of MESH-128(8) 306
3.8.7 ID Analysis of FOX/IDEA-NXT 310
3.8.8 ID Analysis of REESSE3+ 315
3.8.9 ID Analysis of IDEA* 318
3.9 Slide Attacks 319
3.9.1 Slide Attacks on IDEA 320
3.9.2 Slide Attacks on MESH-64 322
3.9.3 Slide Attacks on MESH-96 324
3.9.4 Slide Attacks on MESH-128 325
3.10 Advanced Slide Attacks 326
3.10.1 Advanced Slide Attacks on MESH-64 326
3.10.2 Advanced Slide Attacks on MESH-96 327
3.10.3 Advanced Slide Attacks on MESH-128 327
3.11 Biclique Attacks 327
3.11.1 Biclique Attacks on IDEA 330
3.12 Boomerang Attacks 336
3.12.1 Boomerang Attacks on IDEA 343
3.12.2 Boomerang Attacks on MESH-64 365
3.12.3 Boomerang Attacks on MESH-96 367
3.12.4 Boomerang Attacks on MESH-128 369
3.13 Linear Cryptanalysis 370
3.13.1 LC of PES Under Weak-Key Assumptions 380
3.13.2 LC of IDEA Under Weak-Key Assumptions 384
3.13.3 New Linear Relations for Multiplication 387
3.13.4 LC of MESH-64 Under Weak-Key Assumptions 388
3.13.5 LC of MESH-96 Under Weak-Key Assumptions 391
3.13.6 LC of MESH-128 Under Weak-Key Assumptions 398
3.13.7 LC of MESH-64(8) Under Weak-Key Assumptions 419
3.13.8 LC of MESH-128(8) Under Weak-Key Assumptions 440
3.13.9 LC of WIDEA-n Under Weak-Key Assumptions 449
3.13.10 LC of RIDEA Under Weak-Key Assumptions 451
3.13.11 LC of REESSE3+ Under Weak-Key Assumptions 454

3.13.12 LC of IDEA* Without Weak-Key Assumptions 470
3.13.13 LC of YBC Without Weak-Key Assumptions 472
3.14 Differential-Linear Cryptanalysis 474
3.14.1 DL Analysis of PES 476
3.14.2 DL Analysis of IDEA 476
3.14.3 Higher-Order Differential-Linear Analysis of IDEA 480
3.14.4 DL Analysis of MESH-64 482
3.14.5 DL Analysis of MESH-96 483
3.14.6 DL Analysis of MESH-128 484
3.15 Square/Multiset Attacks 484
3.15.1 Square/Multiset Attacks on IDEA 489
3.15.2 Square/Multiset Attacks on MESH-64 499
3.15.3 Square/Multiset Attacks on MESH-96 502
3.15.4 Square/Multiset Attacks on MESH-128 504
3.15.5 Square/Multiset Attacks on MESH-64(8) 508
3.15.6 Square/Multiset Attacks on MESH-128(8) 512
3.15.7 Square/Multiset Attacks on REESSE3+ 517
3.15.8 Square/Multiset Attacks on FOX/IDEA-NXT 518
3.16 Demirci Attack 524
3.16.1 Demirci Attack on MESH-64 524
3.16.2 Demirci Attack on MESH-96 526
3.16.3 Demirci Attack on MESH-128 527
3.17 Biryukov-Demirci Attack 529
3.17.1 Biryukov-Demirci Attack on IDEA 529
3.17.2 Biryukov-Demirci Attack on MESH-64 540
3.17.3 Biryukov-Demirci Attack on MESH-96 543
3.17.4 Biryukov-Demirci Attack on MESH-128 544
3.17.5 Biryukov-Demirci Attack on MESH-64(8) 546
3.17.6 Biryukov-Demirci Attack on MESH-128(8) 548
3.18 Key-Dependent Distribution Attack 550
3.18.1 Key-Dependent Attacks on IDEA 553
3.19 BDK Attacks 559
3.20 Meet-in-the-Middle Attacks 563
3.20.1 Meet-in-the-Middle Attacks on IDEA 564
3.20.2 More Meet-in-the-Middle Attacks on IDEA 568
3.20.3 Improved Meet-in-the-Middle Attacks on IDEA 572
3.20.4 Meet-in-the-Middle Attack on FOX128 575
3.20.5 Improved ASR Attack on FOX Ciphers 578
3.20.6 Improved ASR Attack on FOX64 579
3.20.7 Improved ASR Attack on FOX128 581
3.21 Related-Key Attacks 583
3.21.1 RK Differential-Linear Attacks on IDEA 584
3.21.2 Related-Key Keyless-BD Attack 586
3.21.3 Related-Key Boomerang Attack on IDEA 587

3.21.4 Further Related-Key Boomerang and Rectangle Attack on IDEA ... 592
References ... 595

4 New Cipher Designs ... 605
4.1 XC_1: Encryption and Decryption Frameworks ... 605
4.1.1 Key Schedule Algorithm ... 608
4.1.2 Differential Analysis ... 608
4.1.3 Linear Analysis ... 612
4.2 XC_2: Encryption and Decryption Frameworks ... 617
4.2.1 Key Schedule Algorithm ... 619
4.2.2 Differential Analysis ... 620
4.2.3 Linear Analysis ... 623
4.3 XC_3: Encryption and Decryption Frameworks ... 625
4.3.1 Key Schedule Algorithm ... 627
4.3.2 Differential Analysis ... 628
4.3.3 Linear Analysis ... 630
4.4 XC_4: Encryption and Decryption Frameworks ... 632
4.4.1 Key Schedule Algorithm ... 634
4.4.2 Differential Analysis ... 634
4.4.3 Linear Analysis ... 636
References ... 638

5 Conclusions ... 639
5.1 The Lai-Massey Design Paradigm ... 639
References ... 656

A Monoids, Groups, Rings and Fields ... 659

B Differential and Linear Branch Number ... 661
B.1 MDS Codes ... 662
B.1.1 How to Construct MDS Matrices? ... 664
B.2 Implementation Costs ... 672

C Substitution Boxes (S-boxes) ... 675
C.1 S-box Definition ... 675
C.2 S-box Representations ... 681
C.2.1 Examples of Real S-boxes ... 703
References ... 718

Index ... 723

Chapter 1
Introduction

The journey of a thousand miles begins with a single step.
– Lao Tzu

Abstract

This chapter provides several background concepts, the context, as well as some terminology and motivations for the material discussed in the rest of the book. In particular, we discuss block ciphers, a fundamental cryptographic primitive which plays a major role in safeguarding sensitive information. Among the several kinds of block ciphers, we focus attention on ciphers that follow the Lai-Massey design paradigm. The first such cipher was the Proposed Encryption Standard (PES) designed by X. Lai and J.L. Massey in 1990, followed by the International Data Encryption Algorithm (IDEA) by the same authors and S. Murphy in 1991. Moreover, other ciphers that appeared afterwards are also described such as the MESH ciphers. In particular, their history, motivation, design criteria, how they differ from other Lai-Massey designs and the attacks known so far are also described.

1.1 Symmetric and Asymmetric Ciphers

In a world increasingly connected by computer networks, where electronic devices became pervasive, there is a growing need to safeguard the confidentiality and authenticity of the increasing volume of sensitive information that flows through these networks. The required protection applies not only to military and governmental networks, but to civilian ones as well, including telephone links, radio broadcasts, pay-TV streaming, banking transactions and electronic commerce data in general.

Cryptography provides several techniques and mechanisms to protect not only the confidentiality of electronic data, but also its authenticity and that of the users involved in the communication process, among other security services [101]. These cryptographic mechanisms are independent of the type

J. Nakahara Jr., *Lai-Massey Cipher Designs*, https://doi.org/10.1007/978-3-319-68273-0_1

of data to be transmitted, be it text, voice, image, sound, telemetry data or their combination.

Meaningful and intelligible information in some appropriate encoding is termed plaintext, while its encrypted form is termed ciphertext.

There are two types of keyed cryptographic algorithms: **symmetric** (or secret-key) and **asymmetric** (or public-key) [101] algorithms.

In symmetric algorithms, the legitimate users share the same key, which must be kept secret and is used both to encrypt and to decrypt information.

In asymmetric algorithms, each user has an (ordered) key pair denoted (e, d). While e is publicly known and used for encryption, d must be kept secret and is used for the decryption procedure. Moreover, the value of e should: (i) be authenticated i.e. strongly associated to its owner, and (ii) not compromise nor provide any kind of information to efficiently derive d. Moreover, the key d can also be used for signature generation, while e is used for signature verification [132, 133, 134].

Therefore, asymmetric algorithms can be used both for encryption and signature purposes, while symmetric algorithms can only provide the encryption functionality.

Of course, cryptographic algorithms by themselves do not solve all of the problems related to information protection. Cryptography is just one (small) piece of the large puzzle of information security. Many other measures (not related to mathematics) are necessary that complement and strengthen the protection provided by cryptographic mechanisms, such as strict (company) policies, human resources screening, strict laws, physical protection, employee training, to mention just a few. Most of the security problems are in fact related to the people who actually use the cryptographic algorithms, and how they (mis)behave [42].

In this book, only symmetric cryptographic algorithms will be discussed. In the symmetric cryptography setting, block ciphers are simply permutation mappings, because of the need to reverse the encryption operation. More specifically, block ciphers are keyed or key-dependent permutations, that is, bijective mappings whose exact behavior depend on a secret key. A permutation mapping performs a rearrangement or re-indexing of the elements in a finite, nonempty set L into a one-to-one correspondence with L itself. Such mappings are also called endomorphic, that is, the domain and range are the same set.

A permutation can be represented by a table representing the full mapping from its domain to itself. As a short example, a permutation π and its inverse π^{-1} on $\mathbb{Z}_{16}$ is depicted in Table 1.1, which lists its *codebook*, that is, the full mapping of its inputs and corresponding outputs. A shortcut notation for π is simply the ordered tuple: $(5, 9, 6, 2, 13, 10, 0, 8, 4, 3, 15, 14, 11, 12, 1, 7)$ in which the domain elements are implicit in the ordering of the image values. For an actual block cipher, describing the whole permutation mapping requires exhausting its codebook. Ideally, each key should determine a different codebook.

Table 1.1 Codebook of $\pi : \mathbb{Z}_{16} \rightarrow \mathbb{Z}_{16}$ mapping each element of its domain to its range.

x	0	1	2	3	4	5	6	7	8	9	10	11	12	13	14	15
$\pi(x)$	5	9	6	2	13	10	0	8	4	3	15	14	11	12	1	7
$\pi^{-1}(x)$	6	14	3	9	8	0	2	15	7	1	5	12	13	4	11	10

Given a permutation mapping π and a starting value x_0, one can construct a chain of values $x_1 = \pi(x_0)$, $x_2 = \pi(x_1) = \pi(\pi(x_0))$, through an iterated process that generates a sequence $x_0, x_1, \ldots, x_t, \ldots$, where x_t is the t-th iterate obtained by applying t times the π mapping to x_0. This sequence is known as an orbit of its starting point x_0. Pick an $x_0 \in \{1, 2, \ldots, n\}$ and consider its orbit x_0, $\pi(x_0)$, $\pi(\pi(x_0))$, ... Since π is bijective and there are only a finite number of elements, for some t, the sequence of iterates will ultimately cycle, that is, $x_t = \pi^t(x_0) = \pi(\pi(...\pi(x_0))) = x_0$. The chain of elements $x_0, \ldots, x_{t-1}$ is called a *cycle* of length t and can be denoted as the ordered tuple $(x_0, \ldots, x_{t-1})$. On one hand, π can be fully described by a single cycle as in Table 1.1, that includes all elements in the domain. However, if there exists y_1 that does not belong to the orbit of x_0, then y_1 has an orbit of its own which is associated with another cycle: $(y_1, \ldots, y_m)$. Eventually all the elements will be associated with one cycle or another, but each element with one cycle only. Then, π can be written as a product of cycles, $\pi = (x_0, \ldots, x_{t-1})(y_1, \ldots, y_m)$ meaning the consecutive scrolling of elements along their corresponding orbits. In general, every permutation can be decomposed into disjoint cycles that is, the cycles have no elements in common. The cycle decomposition of a permutation contains crucial information about it. For example, if we know the cycle lengths of all cycles inside a permutation π, we can compute the least common multiple (LCM) m for which π^m is the identity permutation. This integer m is called the *order* of π in group theory. Some particular permutation mappings are *involutions*, namely, the mapping is its own inverse. It means, its cycle decomposition is composed of pairs of values.

Modern block ciphers, such as IDEA [74], DES [111, 112, 113, 114] and AES (Rijndael) [47] perform permutations over some fixed, finite domains called plaintext space and denoted $\mathcal{P} = \mathbb{Z}_2^n$, for a small integer n.

As any permutation mapping, block ciphers are *length-preserving*, namely, bijective transformations where the domain and range have the same dimensions and are often over the same alphabet. The terminology *block cipher* means that the cipher operates on an internal state consisting of a block of bits of fixed, finite length. For example, a plaintext block (or state) of n bits implies a plaintext space such as $\mathbb{Z}_{2^n}$ or GF(2^n), of size 2^n. Nonetheless, block ciphers can operate on sets whose size is not necessarily a power of 2, such as for credit card number generation. In the latter case, for convenience, the range can be of size 10^t, for some $t > 0$. Examples of these block ciphers include DEAN18 and DEAN27, which stand for Digital Encryption Algorithm for Numbers. They encrypt blocks made of 18 and 27 decimal digits [9].

The block size does not even have to be a power of two. The CTC (Courtois Toy Cipher) is a block cipher operating on 255-bit blocks [29].

Let $E : \mathbb{Z}_2^k \times \mathbb{Z}_2^n \to \mathbb{Z}_2^n$ denote a block cipher's signature, for a k-bit key and n-bit plaintext and ciphertext blocks. For a fixed key, encryption transformations are permutation mappings. That is the reason why block ciphers are called *keyed permutations*.

For a fixed plaintext block, the mapping from key to ciphertext (or for a fixed ciphertext block, the mapping from key to plaintext) is expected to be a one-way function, in the sense that it is efficient to compute $C = E_K(P)$ given P and K (the block cipher E is assumed to be publicly known), while computing P given C (or vice-versa) but not K should be intractable, namely, computationally inefficient.

Block ciphers are often modeled as *random permutations* or black boxes, since this analogy provides a useful abstraction in security arguments such as in modes of operation, and when block ciphers are used as building blocks for the construction of other cryptographic algorithms such as stream ciphers, hash functions and message authentication codes (MAC) [124, 76, 101].

Definition 1.1. (Random Permutation)
A *random permutation*, from a domain $\mathcal{P}$ to itself, is an encryption algorithm selected uniformly at random from the set of all permutations over $\mathcal{P}$.

Definition 1.2. [22]
A secure block cipher should be indistinguishable from a random permutation for every key.

There are $(2^n)! \approx 2^{(n-1)\cdot 2^n}$ possible permutations of the n-bit plaintext space $\mathcal{P} = \mathbb{Z}_2^n$, using Stirling's approximation [101]. This is the size of the space of all permutations over n-bit strings, that is, the space of all permutations of all n-bit strings. This is also the size of the symmetric group on 2^n elements, usually denoted S_{2^n}. Random permutations do not have keys, but, the permutations can be enumerated in lexicographic order, for instance, by assigning an index in increasing lexicographic order. Such an index would have to accommodate $(2^n)!$ values.

To be able to identify each (random) permutation, a separate codebook containing all plaintext-ciphertext pairs is needed to represent each permutation. In contrast, block ciphers with a k-bit key perform up to 2^k *keyed permutations*, namely each permutation of the plaintext space is key-dependent. Thus, there is a large gap between the number of permutations a given block cipher with k-bit key and n-bit block can perform compared to the set of all random permutation over the same plaintext space.

How many keyed permutations are there? Let $T_{k,n}$ denote the set of all block ciphers with k-bit keys and n-bit blocks. If we assume $|T_{k,n}| = (2^n!)^{2^k}$, then many equivalent keys might exist, that is, keys that lead to the same permutation mapping. Alternatively, for each key to represent a distinct permutation mapping, $|T_{k,n}| = \binom{2^n!}{2^k}$, so the choice of the permutations for each key is made taking into account the permutations chosen by other keys.

A design criterion for an n-bit block cipher (with a k-bit key) is to have its permutations selected from $T_{k,n}$ according to the uniform distribution.

If one assumes that different keys generate distinct permutations, then for a block cipher to perform all possible permutations of the plaintext space, the key size should be about $(n-1) \cdot 2^n$ bits long. But, the key size k of typical block ciphers is usually close to the size of the plaintext block, n, or a small multiple of it. Therefore, the number of possible permutations, 2^k, is much smaller than $(2^n!) \approx 2^{(n-1)\cdot 2^n}$. For example, in the IDEA block cipher [74] the key size is $k = 128$ bits, and the block size is $n = 64$ bits. That is, among the $2^{64}! \approx 2^{63 \cdot 2^{64}}$ permutations over $\mathcal{P} = \mathbb{Z}_2^{64}$, IDEA can perform at most a tiny fraction of 2^{128} out of the $2^{63 \cdot 2^{64}}$ possible permutations. Even if all fifty-two 16-bit word subkeys in 8.5-round IDEA were independently generated (this value represents an upper bound on the size of the so called *independent* key space of IDEA), it would still not be enough, because $2^{52 \cdot 16} = 2^{832}$ is insignificant compared to $2^{63 \cdot 2^{64}}$. On the other way around, since $6 \cdot 16 = 96$ subkeys bits are used per round of IDEA, at least $63 \cdot 2^{64}/96 \approx 2^{63.5}$ rounds would be needed, hypothetically, to provide for all possible permutations under independent subkeys. In general, suppose an r-round block cipher, operating on n-bit blocks, and using k-bit subkeys per round. Let us assume that each key generates a different permutation. Thus, this block cipher can perform $2^k \cdot 2^k \cdot \ldots 2^k = 2^{kr}$ different permutations after r rounds. Therefore, the number of rounds until this block cipher could potentially perform the same number of permutations of a random permutation is $(2^n)! = 2^{kr}$, that is, $r = \frac{1}{k} \cdot \log_2(2^n)!$ rounds are needed.

1.2 Iterated (or Product) Ciphers

The majority of modern block ciphers follow an *iterated* design, that is, they consist of the functional composition of one or more key-dependent transformations called a *round* or *iteration*.

Historically, the most prominent block cipher is the Data Encryption Standard (DES) algorithm [111, 112, 113, 114]. DES operates on 64-bit text blocks, under a 56-bit key and iterates sixteen rounds. The successive rounds in a block cipher, for both encryption and decryption, uses subkeys as parameters which are generated in sync with the encryption/decryption operations by a key schedule algorithm (KSA). See Fig. 1.1. Assuming the KSA is independent of the plaintext and ciphertext, the KSA could be executed prior to either operation (if storage space is available for the round subkeys). This approach is particularly interesting if the KSA latency is non-negligible (compared to the encryption/decryption latency).

Usually, there is a small number of different round types in order to minimize the implementation costs in both hardware and software. Moreover, a useful design criterion is to have the same round function used for both encryption and decryption. This way, the difference in both operations is essentially in the order and value of the round subkeys.

The exact number of rounds depends on a trade-off between security and performance. A small number of rounds implies faster operation (better performance) but potentially higher vulnerability to attacks. A larger number of rounds usually implies a block cipher more resilient against cryptanalysis, but also worse performance. Having an arbitrarily large number of rounds (apart from very low performance) may also have security issues [30].

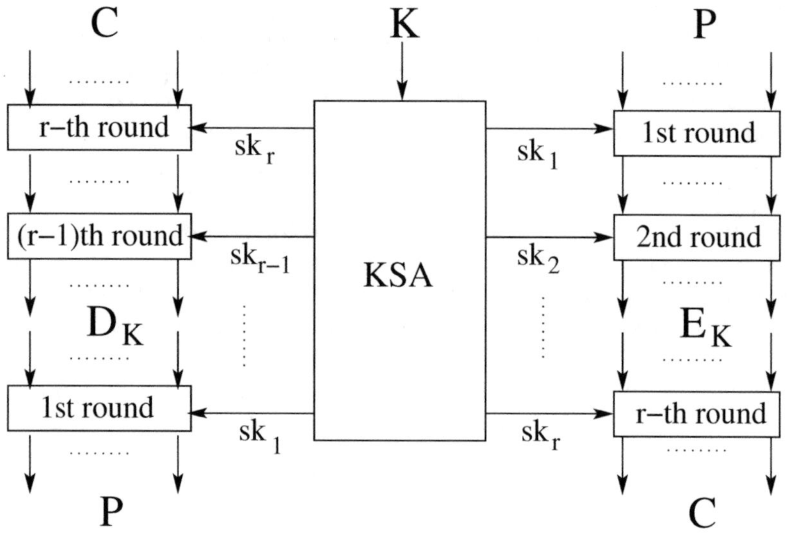

Fig. 1.1 Iterated block ciphers and key schedule algorithms operate in sync for both encryption and decryption.

A natural extension of the principle of composing rounds is the functional composition of block ciphers, also called *multiple encryption* or *cascade constructions*, such as double-DES (two DES instances using two independent 56-bit keys) and triple-DES constructions (three DES instances using two or three independent 56-bit keys). DES permutations do not form an (algebraic) group. But, the set of all DES-like permutations does form a group. This group has about 10^{2499} elements. Lack of a group structure is a strength of DES [64, 63]. If DES were a group, multiple encryption under different keys would not be stronger than single encryption.

Due to meet-in-the-middle attacks, though, multiple encryption has its drawbacks, because the keys used in each single DES instance are independent and may be attacked separately [151].

The design of the DES cipher follows a so called *Feistel Network* paradigm, named after one of the DES designers, Horst Feistel [46]. In a Feistel Network cipher, each round is designed so that part of the block is input into a round function, whose output is combined with the remaining part of the block via a reversible (usually the exclusive-or) operation. See Fig. 1.2. Moreover, the round function does not need to be a bijective mapping. The decryption operation may or may not share the same framework as encryption (for so called generalized Feistel Networks) but is accomplished due to the fact that exclusive-or is an involutory transformation.

In contrast to Feistel ciphers, in the Substitution Permutation Network (SPN) paradigm the whole input text block is transformed by reversible mappings in every round. See Fig. 1.2.

Since 2001, the DES was replaced by Rijndael [47, 37] (in fact, only the 128-bit block cipher member of the Rijndael family) as the new Advanced Encryption Standard (AES) cipher, following a competition organized by the US National Institute of Standards and Technology (NIST) [4].

Basic parameters that are shared by most publicly known block ciphers include: block size (denoted by n), key size (k), number of rounds (r) and the overall cipher design (Feistel Network, SPN, Lai-Massey). There are different criteria to determine each parameter. Moreover, these parameters are not necessarily related. For instance, in DES, $k = 56 < n = 65$, while in IDEA we have $k = 2n$. In the AES, $k \in \{n, 3n/2, 2n\}$. We treat them as independent parameters.

Some initiatives such as the NESSIE project in Europe [123], the Advanced Encryption Standard (AES) competition in USA [4] and the CRYPTREC project in Japan [31], sparked renewed interest in block cipher design and analysis. For an overview, Tables 1.2, 1.3 and 1.4 show an (incomplete) list of known block ciphers and their basic parameters, in alphabetical order. These tables are not meant to be comprehensive.

Fig. 1.2 depicts the three main design frameworks that were adopted by the majority of modern block ciphers listed in Tables 1.2, 1.3 and 1.4: Feistel Networks, SPN and Lai-Massey. The vast majority of the known block ciphers follow either a Feistel Network or an SPN design.

Some recent designs such as KATAN and KTANTAN [40] are hardware oriented, and based on stream ciphers using Nonlinear Feedback Shift Registers (NLFSR). If we consider the transformations performed in the NLFSR as equivalent to those performed by a Feistel Network operating on 1-bit words, then the NLFSR design can be considered as a kind of Feistel Network (and vice-versa [41]).

Each kind of cipher framework has peculiar properties:

- the round function F in Feistel Network ciphers can be an arbitrary mapping, not necessarily a permutation. To allow decryption, the operation used to combined the F output is invertible, for example, bitwise exclusive-or.

Table 1.2 A list of block ciphers and their main parameters.

Cipher	Block Size (bits)	Key Size (bits)	# Rounds	Design	Reference
ABC	256	512	17	SPN	[135]
AKELARRE	128	128	$r \geq 4$	Lai-Massey	[5]
AKE98	128	128	$r \geq 4$	Lai-Massey	[89]
ANUBIS	128	$32t, 4 \leq t \leq 10$	$12-18$	SPN	[10]
ARIA	128	$128; 192; 256$	$12; 14; 16$	SPN	[120]
BEAR	2^{13}–2^{23}	$3 \cdot 2^{13}$–$3 \cdot 2^{23}$	3; 4	Feistel	[6]
Bel-T	128	$128, 192, 256$	8	Feistel/Lai-Massey	[12]
BASEKING	192	192	11	SPN	[33]
BKSQ	96	$96; 144; 192$	$10; 14; 18$	SPN	[38]
BLOWFISH	64	$32-448$	16	Feistel	[136]
CAMELLIA	128	$128; 192; 256$	$18; 24; 24$	Feistel	[8]
CAST-128	64	$40-128$	$12; 16$	Feistel	[2]
CAST-256	128	$128; 192; 256$	48	Feistel	[3]
CDMF	64	40	16	Feistel	[60]
CIKS-1	64	256	8	Feistel	[105]
CIPHERUNICORN-A	128	$128; 192; 256$	16	Feistel	[150]
CIPHERUNICORN-E	64	128	16	Feistel	[115]
CLEFIA	128	$128; 192; 256$	$18; 22; 26$	Feistel	[143]
COCONUT98	64	256	4	Feistel	[152]
CRYPTON v0.5	128	$64 + 32t, 0 \leq t \leq 6$	$r \geq 12$	SPN	[82]
CRYPTON v1.0	128	$8t, 0 \leq t \leq 32$	$r \geq 12$	SPN	[83]
CS	64	128	8	Feistel	[145]
DEAL	128	$128; 192; 256$	$6; 6; 8$	Feistel	[72]
DES	64	56	16	Feistel	[111]
DESX	64	184	16	Feistel	[66]
DFC	128	$t, 0 \leq t \leq 256$	8	Feistel	[52]
E2	128	$128; 192; 256$	12	Feistel	[121]
FEAL-4	64	64	4	Feistel	[140]
FEAL-8	64	64	8	Feistel	[140]
FEAL-N	64	64	$2^x, x > 2$	Feistel	[104]
FEAL-NX	64	128	$2^x, x > 2$	Feistel	[104]
FOX64	64	$0-256$	$12-255$	Lai-Massey	[62]
FOX128	128	$0-256$	$12-255$	Lai-Massey	[62]
FROG	128	$40 + 8t, 0 \leq t \leq 120$	8	SPN	[50]
GOST 28147-89	64	256	32	Feistel	[159]
GRAND CRU	128	128	$10; 12; 14$	SPN	[21]
HIEROCRYPT-3	128	$128; 192; 256$	6; 7; 8	SPN	[148]
HIEROCRYPT-L1	64	128	6	SPN	[149]
ICE-r	64	$64r$	$16r$	Feistel	[73]
IDEA	64	128	8.5	Lai-Massey	[74]
IDEA*	64	128	6.5	Lai-Massey	[80]
KASUMI	64	128	8	Feistel	[1]
KATAN	$32; 48; 64$	80	254	NLFSR	[40]
KHAFRE	64	$64; 128$	$8r, r > 1$	Feistel	[103]
KHAZAD (tweak)	64	128	8	SPN	[11]
KHUFU	64	512	$8r, r \geq 1$	Feistel	[102]

Table 1.3 A list of block ciphers and their main parameters.

Cipher	Block Size (bits)	Key Size (bits)	# Rounds	Design	Reference
KLEIN	64	$64; 80; 96$	$12; 16; 20$	SPN	[53]
KTANTAN	$32; 48; 64$	80	254	NLFSR	[40]
KUZNYECHIK	128	256	10	Feistel	[54]
LION	2^{13}-2^{23}	$3 \cdot 2^{13}$–$3 \cdot 2^{23}$	3	Feistel	[6]
LOKI89	64	64	16	Feistel	[25]
LOKI91	64	64	16	Feistel	[23]
LOKI97	128	$128; 192; 256$	16	Feistel	[24]
LUCIFER	128	128	16	SPN	[142]
MACGUFFIN	64	128	32	Feistel	[18]
MADRYGA	64	64	8	Feistel	[88]
MAGENTA	128	$128; 192; 256$	$6; 6; 8$	Feistel	[59]
MARS	128	$128 - 400$	32	Feistel	[26]
MCRYPTON	64	$64; 96; 128$	12	SPN	[84]
MESH-64	64	128	8.5	Lai-Massey	[110]
MESH-64(8)	64	128	8.5	Lai-Massey	[106]
MESH-96	96	192	10.5	Lai-Massey	[110]
MESH-128	128	256	12.5	Lai-Massey	[110]
MESH-128(8)	128	256	8.5	Lai-Massey	[106]
MISTY1	64	128	$4t \geq 8$	Feistel	[97]
MISTY2	64	128	$4t \geq 12$	Feistel	[98]
MMB (1.0/2.0)	128	128	6	SPN	[126]
NIMBUS	64	128	5	—	[87]
NOEKEON	128	128	16	SPN	[36]
NUSH	$64; 128; 256$	$128; 192; 256$	$9; 17; 33$	Feistel	[78]
PEANUT98	64	576	9	Feistel	[152]
PICCOLO	64	$80; 128$	$25; 31$	Feistel	[139]
Q	128	$128; 256$	$8; 9$	SPN	[100]
RAINBOW	128	128	7	SPN	[79]
RC2	64	$8 - 1024$	18	Feistel	[129]
RC5-$w/r/b$	$2w$	$8b, 0 \leq b \leq 256$	r	Feistel	[130]
RC6-$w/r/b$	$4w \geq 128$	$8b, b \leq 255$	$r \geq 20$	Feistel	[131]
REDOC II	80	160	10	SPN	[32]
RIDEA	64	128	8.5	Lai-Massey	[157]
RIJNDAEL	128	$128; 192; 256$	$10; 12; 14$	SPN	[37]
RIJNDAEL	160	$128; 192; 256$	$11; 12; 14$	SPN	[37]
RIJNDAEL	192	$128; 192; 256$	$12; 12; 14$	SPN	[37]
RIJNDAEL	224	$128; 192; 256$	$13; 14; 14$	SPN	[37]
RIJNDAEL	256	$128; 192; 256$	$14; 14; 14$	SPN	[37]
SAFER K-64	64	64	$8 - 10$	SPN	[91]
SAFER K-128	64	128	$10 - 12$	SPN	[94]
SAFER SK-40	64	40	5	SPN	[92]
SAFER SK-64	64	64	8	SPN	[93]
SAFER SK-128	64	128	$10 - 12$	SPN	[93]
SAFER+	128	$128; 192; 256$	$8; 12; 16$	SPN	[95]
SAFER++	64	128	7	SPN	[96]
SAFER++	128	$128; 256$	$7; 10$	SPN	[96]

Table 1.4 A list of block ciphers and their main parameters.

Cipher	Block Size (bits)	Key Size (bits)	# Rounds	Design	Reference
SC2000	128	128; 192; 256	19; 22; 22	SPN/Feistel	[141]
SEED	128	128	16	Feistel	[19]
SERPENT v.1	128	128; 192; 256	32	SPN	[7]
SHACAL-1	160	512	80	Feistel	[55]
SHACAL-2	256	512	64	Feistel	[55]
SHARK	64	64 − 128	6	SPN	[127]
SIMON32	32	64	32	Feistel	[14]
SIMON48	48	72; 96	36	Feistel	[14]
SIMON64	64	96; 128	42; 44	Feistel	[14]
SIMON96	96	96; 144	52; 54	Feistel	[14]
SIMON128	128	128; 192; 256	68; 69; 72	Feistel	[14]
SPECK32	32	64	22	Feistel	[14]
SPECK48	48	72; 96	22; 23	Feistel	[14]
SPECK64	64	96; 128	26; 27	Feistel	[14]
SPECK96	96	96; 144	28; 29	Feistel	[14]
SPECK128	128	128; 192; 256	32; 33; 34	Feistel	[14]
SKIPJACK	64	80	32	Feistel	[118]
SMS4	128	128	32	Feistel	[144]
SPEED	64; 128; 256	48 − 256	$r \geq 48$	Feistel	[160]
Spectr-H64	64	256	12	Feistel	[44]
SQUARE	128	128	8	SPN	[35]
s^2-DES	64	56	16	Feistel	[68]
s^3-DES	64	56	16	Feistel	[69]
s^5-DES	64	56	16	Feistel	[68]
TEA	64	128	64	Feistel	[155]
TWINE	64	80; 128	36	Feistel	[147]
TWOFISH	128	128; 192; 256	16	Feistel	[137]
2KEY-3DES	64	112	48	Feistel	[114]
3D	512	512	22	SPN	[108]
3KEY-3DES	64	168	48	Feistel	[114]
3-WAY	96	96	11	SPN	[33]
WIDEA-4	256	512	8.5	Lai-Massey	[61]
WIDEA-8	512	1024	8.5	Lai-Massey	[61]
XTEA	64	128	64	Feistel	[116]
YBC	64	128	6.5	Feistel/Lai-Massey	[156]
ZORRO	128	128	24	SPN	[51]

- each internal component of an SPN cipher has to be bijective in order for decryption to be possible.
- in Lai-Massey designs, the internal function F is a permutation mapping and the round function achieves complete (text) diffusion in a single round. Comparatively, text diffusion in Feistel Network and SPN ciphers are usually slower[1] than in Lai-Massey designs. For this reason,

[1] Notable exceptions of SPN ciphers achieving complete (text) diffusion in a single round include Khazad [11] and a modified AES [109].

Lai-Massey ciphers typically adopt a comparatively smaller number of rounds than Feistel Network and SPN ciphers.

But all of these designs also share common features:

- all of them repeat a fixed number of times a computational structure called a *round* or *iteration.*
- they perform keyed permutations over fixed n-bit data blocks, $n \in \{64, 96, 128\}$, but using internal components which are comparatively much smaller than n bits.
- they are all made of internal components placed *in a non-arbitrary sequence* which fulfill the confusion and diffusion properties [138].
- the overall scheme has to be invertible due to the need for decryption.

1.3 Symmetric Cryptosystems

A symmetric (or secret-key) *cryptosystem* or a *cryptographic system* consists of a tuple $(\mathcal{P}, \mathcal{C}, \mathcal{K}, \mathbf{E}, \mathbf{D}, \mathbf{S})$, where

- $\mathcal{P}$ is the plaintext space consisting of a finite, nonempty set of strings over some alphabet Ψ that defines the possible plaintext messages. $|\mathcal{P}|$ denotes its size.
- $\mathcal{C}$ is the ciphertext space consisting of a finite, nonempty set of strings over some alphabet Γ that defines the possible ciphertext messages. $|\mathcal{C}|$ denotes its size.
- $\mathcal{K}$ is the key space comprised of a finite, nonempty set of strings over some alphabet Ω, that defines the possible keys. $|\mathcal{K}|$ denotes its size.
- $\mathbf{E} : \mathcal{K} \times \mathcal{P} \to \mathcal{C}$ denotes the set of encryption algorithms, and $\mathbf{D} : \mathcal{K} \times \mathcal{C} \to \mathcal{P}$ denotes the set of decryption algorithms. Moreover, $\mathbf{D}(K, \mathbf{E}(K, P)) = P$ for all $P \in \mathcal{P}$ and for all $K \in \mathcal{K}$.
- $\mathbf{S} : \mathbb{Z}_2^* \to \mathcal{K}$ denotes the set of key generation algorithms that randomly selects a key from $\mathcal{K}$ based on a random seed of arbitrary size. For each key K originating from $\mathbf{S}(\$)$, where the \$ symbol denotes a random value, the encryption algorithm $\mathbf{E}$ performs a key-dependent transformation from $\mathcal{P}$ to $\mathcal{C}$.

For the particular setting of an n-bit block cipher under a k-bit key, the cryptographic primitive used for encryption has signature

$$E : \mathbb{Z}_2^k \times \mathbb{Z}_2^n \to \mathbb{Z}_2^n,$$

while the primitive for decryption has signature

$$D : \mathbb{Z}_2^k \times \mathbb{Z}_2^n \to \mathbb{Z}_2^n.$$

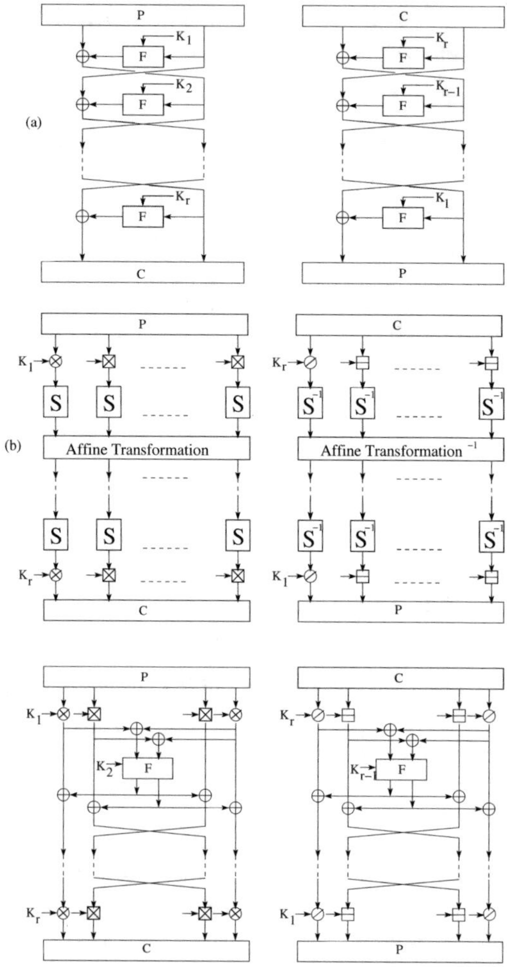

Fig. 1.2 General design frameworks for block ciphers (encryption and decryption). (a) Feistel Networks, (b) SPN, (c) Lai-Massey. P and C are text blocks; $\otimes$ and $\boxtimes$ are invertible operations; $\oslash$ and $\boxminus$ are the inverse operations. S stands for an S-box, and F for a round function.

Therefore, the plaintext and ciphertext spaces are identical, and both mappings are permutations. For a fixed key K, the signatures become simply

$$E_K : \mathbb{Z}_2^n \rightarrow \mathbb{Z}_2^n$$

and

$$D_K : \mathbb{Z}_2^n \rightarrow \mathbb{Z}_2^n.$$

These mappings are each others inverse: $E_K(D_K(C)) = C$ for all $K \in \mathbb{Z}_2^k$ and for all $C \in \mathbb{Z}_2^n$; $D_K(E_K(P)) = P$ for all $K \in \mathbb{Z}_2^k$ and for all $P \in \mathbb{Z}_2^n$.

To avoid unwanted behavior such as equivalent keys [101], the key schedule algorithm should be carefully designed in tandem with both E and D such that $D_K(E_{K'}(P)) \neq P$ for all $P \in \mathbb{Z}_2^n$ and all $K \neq K'$.

In practice, to account for the encryption and decryption of arbitrarily large messages and not only single text blocks, modes of operation (as in Sect. 1.8) are needed to extend the functionality of n-bit blockwise encryption/decryption to m-bit messages, for an arbitrarily large but finite integer m unrelated to n. If there is no ciphertext expansion (for example, due to authentication tags or block extensions), then both the plaintext and ciphertext spaces remain the same: $\mathcal{P} = \mathcal{C} = \cup_{i=1}^{2^m} \mathbb{Z}_2^i$. Otherwise, the ciphertext space gets bigger to accommodate additional data needed for proper decryption. The key space is not affected.

Therefore, while E and D operate strictly of n-bit blocks, **E** and **D** (in a cryptosystem) operate on arbitrarily large and variable (but finite) domains via some mode of operation. Therefore, the boldface and normal type (**E** and E, respectively) will indicate the size of the domain we are operating in.

The key schedule algorithm, **S**, provides secret-key input to both E and D in the correct order and for timely operation. The signature of **S** becomes $S : \mathbb{Z}_2^k \rightarrow \mathbb{Z}_2^*$, where the output size depends on the input size k (and the number of rounds) in E (or D).

1.4 Entropy

Let X be a random variable which takes on a finite number of values $\{x_1, x_2, \ldots, x_m\}$, with probabilities $p_i = \Pr(X = x_i)$ such that $\sum_{i=1}^m p_i = 1$.

Definition 1.3. [101]
The *entropy* or *uncertainty* of X is defined as

$$H(X) = -\sum_{i=1}^{n} p_i \cdot \log_2 p_i, \tag{1.1}$$

where by convention, if $p_i = 0$, the corresponding term is omitted from the sum.

The entropy of X measures the amount of information provided by an observation of X. The entropy is a useful measure for approximating the average number of bits required to encode the elements of X in bits.

Properties of the entropy function include

- $0 \leq H(M) \leq \log n$.
- $H(M) = 0$ if and only if there exists only one i such that $p_i = 1$ and for all $j \neq i$, $p_j = 0$.

- $H(M) = \log_2 n$ if and only if $p_i = 1/n$ for all i.

As an example, let us assume that keys are chosen independently and uniformly at random for use in a given block cipher, that is, $p_i = \Pr(K = K_i) = \frac{1}{K}$. When the size of a secret key is mentioned as k bits, it means that its entropy is k bits.

Therefore, parity check bits like in the DES cipher [111, 112] or any other kind of redundancy added to the key, do not contribute to the effective key size.

Another application of entropy in the context of Lai-Massey ciphers concerns a combination of entropy and differential cryptanalysis in [17].

1.5 Confusion and Diffusion

In [138], Claude Shannon described two design guidelines for the construction of secure ciphers: *confusion* and *diffusion*.

The *confusion* property aims to make the relationship among the plaintext, the ciphertext and the key as complex and involved as possible. In practice, this complex relationship is achieved by using nonlinear operations such as substitution boxes (S-boxes), (modular) multiplication, (modular) addition or subtraction. A brief survey of S-boxes is presented in the appendix.

In contrast, the *diffusion* property means that any change in the plaintext or the key must affect the entire ciphertext. The aim of *diffusion* is to dissipate the redundancy in the plaintext (and key), or its statistical structure, into long range statistics. Therefore, diffusion targets the dependency of ciphertext bits on the plaintext and the key bits. In practice, diffusion is achieved by affine transformations. A modern trend towards efficient diffusion, after the Advanced Encryption Standard (AES) [47], is the use of *Maximum Distance Separable (MDS) codes* [81]. A brief survey of MDS codes is provided in the appendix.

Nonlinear transformations that perform only confusion are not enough to guarantee security. For example, using only s-boxes in a cipher design would require displacing them around the cipher state to effectively reach out every bit of the state, otherwise, there could be a divide-and-conquer attack that exploits the lack of dependency of some input bits (lack of proper diffusion).

Likewise, transformations that perform only diffusion are not enough either. The resulting cipher would be essentially an affine transformation that could be solved efficiently like a system of linear/affine equations (in a known-plaintext setting).

Confusion and diffusion are orthogonal concepts and are not commutative transformations [138]. In fact, the alternate composition of confusion- and diffusion-based transformations that makes ciphers stronger and avoids the drawbacks of each individual transformation alone.

These two concepts still form the main *ad hoc* guidelines for the design of modern block ciphers.

1.6 Security Assumptions and Threat Models

Security assumptions are hypothesis that are set for a particular application environment, to allow assessment of the security of a cryptographic algorithm.

Some typical security assumptions for symmetric ciphers are:

- Kerckhoffs' principles [67]: the adversary knows all the details of the cryptographic algorithm, except the key.
- the ciphertext is publicly known.
- the secret key is fixed during the attack (for an entire message).
- keys are chosen independently and uniformly at random from the key space for each plaintext message.
- the key is statistically independent from the plaintext.
- a key of k bits ideally contains k bits of entropy.

Threat models classify attacks according to the assumed capabilities of the adversary. They include

- **ciphertext-only (CO) attack**: this is a passive attack. The adversary knows the ciphertext and the statistical distribution of the plaintext. The aim is to discover the key or the plaintext. A cipher susceptible to CO attacks is not secure against eavesdropping. CO attacks represent the weakest model since the adversary only observes public information but does not manipulate the text nor try to induce the legitimate parties into adopting a particular behavior.
- **known-plaintext (KP) attack**: this is a passive attack. The adversary knows (fully or partially) pairs of matching plaintext-ciphertext and aims at discovering the key or other unknown plaintexts. This is a realistic scenario assuming the use of common headers or salutations in plaintext messages, or by exploiting the fact that messages in natural languages contain plenty of redundancy (assuming no compression is performed prior to encryption). Data compression (or other invertible transformation) is a useful mechanism to reduce redundancy in the plaintext prior to encryption.
  Sometimes, if the input is ASCII text, KP attacks can be transformed into CO ones, if the attack depends only on the most significant bit of every plaintext byte.
- **chosen-plaintext (CP) attack**: this is an active attack. The adversary can (somehow) choose the plaintext, fully or partially, and obtain the corresponding ciphertext from the legitimate parties. The aim is to recover the key. Resisting CP attacks is a minimum assumption for a modern cipher to be considered secure. The text(s) can be chosen adaptively or

not. In the adaptive case, each plaintext is chosen only after the previous ones have been encrypted and the ciphertext observed. Therefore, each new choice of plaintext is based on previously observed texts. In the non-adaptive case, all plaintexts are chosen at once and the attack is called non-adaptive.

- **chosen-ciphertext (CC) attack**: this is an active attack. The adversary is able to choose the ciphertext, fully or partially, and aims to recover the key or forge ciphertext (hoping that the result is sensible plaintext or advantageous to recover other kinds of information). There are adaptive and non-adaptive variants of CC attack, just like in the CP setting. Potential attack scenarios include the case in which the adversary has temporary access to a tamper-resistant device that decrypts cryptograms of the users' choice (an oracle). Some kinds of attack such as the boomerang technique [154] require the adversary to have both CP and CC capability. There are also ciphers whose decryption operation has weaker diffusion than the encryption scheme, and it is thus advantageous to attack these ciphers from the ciphertext end.
- **related-key (RK) attack**: this is an active attack. The adversary knows or is able to choose a relationship among keys used in different instantiations, but not their actual value. This attack is typically combined with one of the previous ones and is denoted related-key ciphertext-only (RK-CO), related-key known-plaintext (RK-KP), RK-CP, RK-CC, for instance. Related-key attacks test the limits of a cipher's design, in particular the key schedule algorithm (KSA) and its interaction with the encryption/decryption frameworks. Originally, RK attacks were considered only of theoretical interest. Nonetheless, related-key attacks are relevant in practice since block ciphers are widely used as building blocks in many cryptographic constructions. For instance, in some protocols, stream ciphers have resynchronization steps[2] and in some hash functions constructions the key input can be manipulated or be under the control of the adversary, making related-key attacks possible and thus, relevant. In particular, settings in which the key management or key (re)synchronization procedure are not well designed, leaves open the opportunity for related-key attacks, such as meet-in-the-middle attacks, if the user key is not quickly mixed over all round subkeys (key diffusion).
  Related-key attacks may serve as a warning about weak key schedules or weaknesses concerning bad interactions between the encryption framework and the key schedule in certain modes of operation or attack settings. In [70], Knudsen presented an RK attack on the first version of LOKI (called LOKI91), reducing an exhaustive search by almost a factor of four. In [15], Biham introduced RK attacks on DES. In [71], an RK attack was applied to SAFER K-64, and in [65, 66] to other block ciphers. It may be argued that attacks using a chosen or known relation between

---

[2] In stream cipher settings, related-key attacks are known as resynchronization attacks.

keys may be unrealistic. The adversary needs to get encryptions under several keys, in some cases, even in a chosen-plaintext setting. However, there are realistic settings in which an adversary may succeed in obtaining such privilege.

1.7 PRP and SPRP

Following [90], secure n-bit block ciphers are modeled as PseudoRandom Permutations (PRP) [86]. The concept of a PRP means that computationally bounded adversaries, denoted A, allowed a polynomial number q of queries, may distinguish a given block cipher E from a random permutation π, selected uniformly at random from the set RP^n of all $2^n!$ permutations, with negligible advantage given by

$$\mathrm{Adv}_A(q) = |\Pr(k \stackrel{\$}{\leftarrow} \mathcal{K} : A^{E_K} = 1) - \Pr(\pi \stackrel{\$}{\leftarrow} \mathrm{RP}^n : A^{\pi} = 1)|$$

where $y \stackrel{\$}{\leftarrow} \mathcal{Y}$ means y is selected uniformly at random from the set $\mathcal{Y}$, and A^X returns 1 if A believes it is dealing with oracle X; otherwise, A returns 0.

Therefore, 1 means *success* while 0 means *failure*; the term *negligible* (in the parameter n) means that the advantage grows slower than the inverse of any polynomial in n.

If the advantage is negligible even if the adversary is allowed decryption queries ($D_K = E_K^{-1}$) then, the block cipher is called a Strong PseudoRandom Permutation (SPRP) [122].

For negligible advantage, if the encryption queries are only *known* by the adversary, then the block cipher is deemed indistinguishable under known-plaintext attacks (IND-KPA). If the encryption queries are *chosen* by the adversary, then the block cipher is deemed indistinguishable under a chosen-plaintext attack (IND-CPA). If the decryption queries are *chosen non-adaptively* by the adversary, then the block cipher is deemed indistinguishable under a (non-adaptive) chosen-ciphertext attack (IND-CCA1). If the decryption queries are adaptively chosen by the adversary, then the block cipher is deemed indistinguishable under an adaptively chosen-ciphertext attack (IND-CCA2).

If the advantage is non negligible for a single adversary, then if the queries are known to the adversary, the block cipher is not IND-KPA. Analogously, for chosen queries the block cipher is not IND-CPA, and so on.

It is a common assumption to model block ciphers as PRP or SPRP in proofs of security of modes of operation (Sect. 1.8), so that the security of the mode is not affected by eventual weaknesses from the underlying block cipher.

1.8 Modes of Operation

From the block cipher signature $E : \mathbb{Z}_2^k \times \mathbb{Z}_2^n \rightarrow \mathbb{Z}_2^n$, it is clear that E can at most transform one fixed-size (plaintext) block into a fixed-size (ciphertext) block, or vice-versa. In order to operate on arbitrarily large domains in practical applications, *modes of operation* need to be defined to extend the functionality of a mapping E operating on a fixed-size domain to a mapping $\mathbf{E}$ operating on variable-size domains.

Moreover, modes of operation are used to hide patterns in the plaintext, to help protect ciphers against chosen-plaintext (CP) and chosen-ciphertext (CC) attacks and to support faster online encryption with precomputation.

For the DES cipher, only four modes of operation were initially standardized: Electronic Code Book (ECB), Cipher Block Chaining (CBC), Cipher Feedback (CFB) and Output Feedback (OFB). In 2001, the Counter (CTR) mode was standardized by NIST for the AES in a special publication called SP800-38A.

Due to developments in the analysis of DES (and other legacy block ciphers), including the ability to exhaustively search for keys, it was proposed to use multiple (or cascaded) modes of operation, such as triple-DES-ECB mode, triple-DES-CBC mode, CBC-CBC^{-1}-CBC mode, and so on. Expected advantages of using multiple modes include:

- higher security compared to single modes against exhaustive search.
- as fast as single encryption on pipelined hardware.
- it is assumed that adversaries cannot find out the feedback value(s) used during encryption, and thus cannot mount known-plaintext (KP) attacks.

All the triple (cascaded) modes of operation of DES are not much more secure than single modes, as they require up to 2^{56}–2^{66} time complexity and plaintexts. The cryptanalytic techniques to attack multiple modes break one single component mode at a time, and can be based on any technique to attack single modes, including differential, linear, Davies' attack and exhaustive search. The main observation is that, under certain circumstances, the adversary can have control over (or guess) the unknown feedback values, and can thus feed them with the data required to apply cryptanalysis of the single modes. To feed these values, the attacks are usually in the CP or CC settings. The fact that each single instance in multiple modes uses an *independent key* makes the overall scheme susceptible to divide-and-conquer attacks whose complexities are below that of the sum of each component.

The following is a brief summary of conventional modes of operation for n-bit block ciphers [101]. Let $P = (P_1, P_2, \ldots, P_t)$ denote an m-bit plaintext message and $C = (C_1, C_2, \ldots, C_t)$ denote the corresponding ciphertext, where $m = \sum_{i=1}^{t} |P_i|$. Some modes may require random n-bit initial values, denoted IV.

- ECB: the Electronic CodeBook mode was standardized by NIST [117] and ISO [57]. In the ECB mode, each ciphertext block C_i depends only on P_i and vice-versa. In summary:
  - to encrypt P_i, the procedure is
$$C_i = E_K(P_i), \text{ for } i \geq 1.$$
  - to decrypt C_i, the procedure is
$$P_i = D_K(C_i), \text{ for } i \geq 1.$$

  ECB is a memoryless mode. Consequently, (text) diffusion in ECB mode is only blockwise, which means limited error propagation: a ciphertext C_i damaged during transmission only affects the decrypted block P_i. Moreover, independent (parallel) encryption of text blocks is possible since processing P_i does not depend on P_{i-1} nor on P_{i+1}. This mode requires padding the plaintext if $m \not\equiv 0 \bmod n$, since each (block) encryption involves n bits.
  The underlying block cipher is assumed to be a PRP.
- CBC: the Cipher Block Chaining mode was standardized by NIST [117] and ISO [57]. In CBC mode, each ciphertext block is chained to the previous cipher block, and thus depends on all previous plaintext blocks. In summary:
  - the recurrence for encryption is
$$C_i = E_K(P_i \oplus C_{i-1}), \text{ for } i \geq 1.$$
  - the recurrence for decryption is
$$P_i = D_K(C_i) \oplus C_{i-1}, \text{ for } i \geq 1.$$

  and $C_0 = \text{IV}$. In contrast to the ECB mode, the CBC mode preserves full memory from one block encryption to the next. Consequently, (text) diffusion is better compared to the ECB mode due to ciphertext chaining dependence on C_{i-1}. On the other hand, parallel processing is hindered because of this same ciphertext chaining. Error propagation is limited: if C_i is damaged, only P_i and P_{i+1} are affected (not properly decrypted). Note that both E_K and D_K are needed for the encryption and decryption operations (like in the ECB mode).
  The security of the CBC mode under a chosen-plaintext attack was analyzed by Bellare *et al.* in [13] and requires that: (i) the underlying block cipher be a PRP; (ii) the IV be randomly chosen, and be secret for each new message and key. But, the CBC mode is not secure in the chosen-ciphertext setting. Random IV values help avoid replay attacks, that is, replay of previous ciphertexts under the same IV.

Due to the ciphertext feedback, decryption of an arbitrary ciphertext block (random access) requires access only to the previous ciphertext block (or the IV), but updating a single block requires all subsequent blocks to be encrypted again.
Bit-flipping errors in a ciphertext block cause the current and following plaintext blocks to be incorrectly decrypted (finite error propagation). After two non-garbled ciphertext blocks are received, correct decryption resumes. Arbitrary bit insertions/deletions cause loss of block boundary alignment, and therefore, infinite error expansion. Integrity mechanisms, such as CBC-MAC [58] can be used but have to be computed separately from the CBC mode.
The ISO document [57] provides two padding mechanisms to process messages whose total length is not a multiple of n: (i) encrypt the last b-bit segment by combining the plaintext using exclusive-or with the leftmost b bits of $E_K(C_{l-1})$, where C_{l-1} is the previous ciphertext block; (ii) ciphertext stealing: the last two blocks are $\mathrm{msb}_b(C_{l-1})$ and $C_l = E_K(\mathrm{lsb}_{n-b}(C_{l-1})||P_l)$. Decryption is performed in reverse order: first process C_l and then decrypt C_{l-1}.

- b-bit CFB: the b-bit Cipher FeedBack mode was standardized by NIST [117] and ISO [57]. The CFB mode turns any n-bit block cipher into a self-synchronous stream cipher. In summary:
  - the recurrence for encryption is

$$C_i = P_i \oplus \mathrm{lsb}_b(X_{i-1}), \text{ for } i \geq 1.$$

  - the recurrence for decryption is

$$P_i = C_i \oplus \mathrm{lsb}_b(X_{i-1}), \text{ for } i \geq 1.$$

where $X_0 = E_K(\mathrm{IV})$, $X_i = E_K((X_{i-1} \ll b)||\mathrm{msb}_b(C_i))$ for $i \geq 1$, $\mathrm{lsb}_b(x)$ denotes the least significant b bits of x, and $\mathrm{msb}_b(x)$ denotes the b most significant bits of x. Note that at most a b-bit plaintext string P_i is encrypted at a time, by a buffer X_{i-1} that depends on text blocks P_j and C_j for $j \leq i$. Due to the ciphertext chaining (memory) C_i, this mode does not allow parallel processing like in the ECB mode. Note that the CFB mode only needs E_K for both the encryption and decryption operations.
A bit-flipping error in a b-bit cipher segment C_i will affect the decryption of the following $\lceil \frac{n}{b} \rceil$ ciphertext segments, that is, until C_i is shifted out of X_i. Afterwards, correct decryption resumes. The same if a full n-bit block disappears. But, insertion or deletion of an arbitrary s-bit segment, $0 < s < b$, causes loss of segment alignment and therefore, infinite error expansion. Only for the case $b = s = 1$, the CFB mode is able to recover from such errors, a property called self-synchronization (Maurer [99] and Heys [56]).

It is not necessary to use exactly the least significant b bits of X_i to encrypt or decrypt. The choice of which bits to extract from X_i should be arbitrary assuming a secure block cipher.
The efficiency of this mode is b bits per n-bit block processed. For small b, it will require $\lceil \frac{n}{b} \rceil$ times the number of block cipher calls of the ECB mode to process the same amount of data. To contrast, note that full resilience to arbitrary insertion/deletion errors (as discussed previously) implies $b = 1$, which means a considerable decrease in performance: one full n-bit encryption is needed to process one single plaintext bit.
Due to the feedback of ciphertext segments, the CFB processing is inherently sequential. Another consequence is that random access for reading encrypted segments requires availability of the previous $\lceil \frac{n}{b} \rceil$ segments or the IV, and updating a segment requires changing all the following ciphertext segments. A third consequence of text feedback is that it conceals plaintext redundancy.
A non-repeating IV is required for each new message and also for resynchronization. The IV is not required to be secret.

- OFB: the Output FeedBack mode was standardized by NIST [117] and ISO [57]. The OFB mode turns a block cipher into a synchronous stream cipher. In summary:

  – the recurrence for encryption is

$$C_i = P_i \oplus X_i, \text{ for } i \geq 1.$$

  – the recurrence for decryption is

$$P_i = C_i \oplus X_i, \text{ for } i \geq 1.$$

  where $X_1 = E_K(\text{IV})$ and $X_j = E_K(X_{j-1})$, $j > 1$. Note that the key stream X_i exclusive-ored to P_i is text independent, that is, unlike the CFB mode, X_i only depends on the IV and the key K. Therefore, the OFB mode can operate in parallel bitwise since X_i can be precomputed offline (and stored) before encryption or decryption starts. It also means independent blocks P_i can be (re)encrypted without affecting C_j whether $j < i$ or $j > i$.
  The propagation of bit-flipping errors is limited: a bit flipped in C_i only affects P_i. This error propagation is the same as in the One-Time Pad (OTP) [138, 153]. As a consequence, arbitrary plaintext bits can be flipped without detection (especially if the underlying plaintext has high redundancy), unless there is a separate integrity mechanism to detect active tampering.
  Insertion/deletion errors cause infinite error expansion, and require appropriate resynchronization protocols. These errors occur because X_j gets out of sync. Therefore, the OFB mode is not self-synchronizing.

The security of the OFB mode depends on the unpredictability of its keystream. Therefore, a non-repeating IV must be used for each new message and new key, as well as for each resynchronization step. Otherwise, the very same keystream will be generated, and security will be compromised (just like in a OTP). Also, the period of the keystream must be long enough (Gait [48]).

Originally, the OFB mode was parameterized as b-bit OFB, but full n-bit feedback is recommended: $b = n$ following [101]. This recommendation results from the fact that E_K with n-bit feedback, under the assumption of a randomly chosen key, can be modeled as a PRP. Moreover, it was shown by Davies and Parkin in [39] that for a random fixed key and IV, the expected cycle length of X_j is about 2^{n-1}. If the feedback is $b < n$ bits, then the keystream X_j is from a truncated permutation, and the expected cycle length of X_j is much shorter [27]: $2^{n/2}$. Concerns over short periods also involves avoiding weak and semi-weak keys of the underlying block cipher [101]. This last assumption is implied when E_K is assumed to be a PRP.

Due to the output feedback, plaintext redundancy is concealed in OFB mode. Since the keystream is independent of the plaintext, the former can be generated and stored prior to the encryption and decryption processes, therefore, allowing parallel processing of the plaintext and ciphertext, as well as random access capability.

Note also that only E_K is necessary for both the encryption and decryption operations.

- CTR: the counter mode was originally suggested by Diffie and Hellman in [43]. Later, it was reviewed and submitted by Lipmaa *et al.* [85] for the NIST Modes of Operation Workshop [119]. The CTR mode is recommended in a NIST Special Publication [45] and in ISO [57]. The CTR mode consists of processing plaintext blocks with a counter-dependent encrypted output. In summary:

  – the recurrence for encryption is

$$C_i = P_i \oplus E_K(X_i), \text{ for } i \geq 1.$$

  – the recurrence for decryption is

$$P_i = C_i \oplus E_K(X_i), \text{ for } i \geq 1.$$

  where $X_1 = \text{IV}$ and $X_j = E_K(f(X_{j-1}))$ for $j > 1$, with f a simple counter function or a Linear Feedback Shift Register (LSFR). The CTR mode turns a block cipher into a synchronous stream cipher (like the OFB mode). Thus, there is no need for padding, and error propagation is limited (like in a OTP). Just like in OFB mode, only E_K is enough for both the encryption and decryption operations in CTR mode.

Security of the CTR mode, under a chosen-plaintext attack, assumes that the underlying block cipher is a PRP [13], and that the IV counter is non repeating for each new message and for each resynchronization under the same key (to avoid the same keystream to be generated from the IV values). The IV is not required to be secret. The different counter values also help conceal the plaintext redundancy. Notice that for an n-bit block cipher, there are at most 2^n different counter values.

There is no need for padding in the CTR mode, which is essentially a stream mode. In fact, for any plaintext block P_i, not only for the last one, less than n bits can be used for encryption and decryption.

In CTR mode, the keystream is not data-dependent. Consequently, the keystream can be generated and stored beforehand, which allows parallel processing both for encryption and for decryption. Due to the lack of text chaining, ciphertext blocks can be arbitrarily read or modified (random access capability). On the other hand, it also allows malicious change of the order of blocks, which will be undetected unless a separate integrity mechanism is in place. In case of bit-flipping errors, only the bits in the current block is affected: there is no error expansion in this case. In case of arbitrary insertion/deletion, the block alignment is usually lost, and the counter X_i loses synchronization, resulting in infinite error expansion.

Further details about these modes can be found in [101]. Some limitations and drawbacks of these modes can be found in [90].

1.9 Unconditional Security

A cryptosystem is *unconditionally secure* if it is secure against adversaries with unlimited computational resources [74, 138]. Unconditional security, also known as *theoretical security* or *perfect security*, means that breaking such a cipher is impossible.

Following [74, 138], unconditional security is impractical because an unlimited amount of key bits will be needed (that is, generated, stored and transmitted to each legitimate party before starting) to encrypt messages of arbitrary length. An example of an unconditionally secure cipher is the One-Time Pad [153]. However difficult to achieve, this notion of security establishes a sound principle for the use of cryptography in practice, namely: *the key should be changed as often as possible, preferably, for each new message.*

Also, according to [74, 138], unconditional security (in a CO setting) can be achieved through data compression, that is, if the plaintext has no redundancy, the corresponding ciphertext will be secure against ciphertext-only attacks. It requires that plaintext blocks be independent and uniformly random. Whether such compression is possible or not in practice, Shannon's result lead to another useful guideline for cryptographic practice: *plaintext data should be as random as possible, or equivalently, the plaintext entropy should be as high as possible.*

On the other hand, *computational or practical security* refers to cryptosystems secure against adversaries with limited computational resources. This notion of security is linked to practical considerations, for instance, an analysis based on the best known and relevant attacks.

In the practical security setting, attack complexities can be classified in two categories: *pre-processing* and *post-processing*. In the former, processing is done offline and before the actual attack begins. In the latter, the computation is particular to each unknown fixed key and is repeated for each new target key. There may be offline and online costs involved. The offline costs are usually related to pre-processing tasks and which do not depend on the legitimate parties' assistance, while the latter depends on cooperation of the legitimate parties (KP, CP, CC attacks).

Offline/online complexity measurements concern:

- time complexity: there can be pre-computation and post-computation complexities; the former stands for the number of encryptions/decryptions or other units (elementary operations, CPU cycles) made offline by the adversary before the attack actually starts; the latter is the amount of real-time effort (CPU cycles, encryptions/decryptions, memory accesses) required by the adversary during the online phase of an attack.
- data complexity: amount and type of texts obtained from queries to the legitimate parties, under a certain threat model.
- memory complexity: the amount of storage (RAM, cache, hard disk) necessary to accommodate all data needed during an attack, measured according to some appropriate unit.
- other complexity parameters: it may be related to the number of processors (CPU), communication costs, related keys and so on.
- success probability: some attacks are deterministic like the exhaustive key search, which always succeeds. Other attacks are probabilistic and may succeed with variable rate depending on the time/data/memory complexities available. The success probability measures the frequency with which the attack objectives are achieved.

In some contexts, memory may be a cheap resource but time is scarce, or the other way around. Therefore, depending on the attack environment, some complexity parameters may be more relevant than others.

1.10 The Origins of the MESH Ciphers

The largest family of ciphers following the Lai-Massey design framework comprise the MESH ciphers.

The MESH block ciphers [110] were inspired by the International Data Encryption Algorithm (IDEA) [74, 77], which, on its turn, evolved from the Proposed Encryption Standard (PES) block cipher [75].

The IDEA and PES ciphers had a unique design back in 1990/1991 when they were released, since they used no substitution boxes (S-boxes) and did not employ a Feistel Network design such as the Data Encryption Standard (DES) [111, 112, 113], nor a Substitution-Permutation Network (SPN) design, such as the AES cipher [47, 37]. Rather, IDEA and PES's design used an involutory framework sometimes called Lai-Massey scheme due to its designers: Xuejia Lai and James Massey. Their designs were based on a particular combination of three group operations: addition modulo 2^{16}, multiplication modulo $2^{16}+1$ (with $2^{16} \equiv 0$) and bitwise exclusive-or.

The *de facto* block cipher standard in 1990/1991 was the DES and its main components were eight S-boxes whose design criteria were still not disclosed at the time [28]. The fact that IDEA did not have S-boxes made users less suspicious about it, since it looked like there were no trapdoors nor unexplained internal components [128]. In this setting, a trapdoor is a hidden structure embedded in a cipher design whose knowledge would allow one to easily recover privileged information such as the key or to decrypt the ciphertexts. Without knowledge of the trapdoor, the cipher would look secure.

The idea of combining a few group operations (without S-boxes) was also adopted in other block ciphers, such as TEA [155], whose design employed modular addition, bitwise shifts ($\ll$, $\gg$) and exclusive-or. Block ciphers such as RC5 [130] used modular addition, bitwise rotation ($\lll$, $\ggg$) and exclusive-or, from which these designs became know as ARX (Addition-Rotation-Xor). From this point of view, IDEA and MESH could be identified as AMX (Addition-Multiplication-Xor) ciphers.

The absence of S-boxes and the involutory design of the round function in IDEA attracted attention from both designers and practitioners. One of the earliest adopters of IDEA was Phil Zimmermann who used IDEA as a symmetric cipher component in his Pretty Good Privacy (PGP) software for email encryption [49], although the IDEA cipher was a patented algorithm[3].

Concerning S-boxes, it has been noted that the keyed modular multiplication and modular addition in MESH (and IDEA) could be interpreted as keyed S-boxes, and therefore, such S-boxes would change for every new key. It is in fact possible to represent the modular multiplication using two $2^w \times 2^w$-bit S-boxes for $w \in \{2, 4, 8, 16\}$: one S-box based on a discrete logarithm table and another S-box based on a discrete exponentiation table. But, both S-boxes would be keyed by an unknown w-bit subkey. Moreover, due to the memory needed to store such large S-boxes and the possibility of side-channel attacks, it is unlikely that these tables will be used in practice.

---

[3] European patent #0482154, filed 16 May 1991, issued 22 June 1994 and expired 16 May 2011; US patent 5.214.703 issued 25 May 1993, expired 7 January 2012; Japanese patent JP 3225440, expired 16 May 2011.

All Lai-Massey cipher designs are iterated, that is, they repeat a key-dependent round function a fixed number of times. This round function is designed in such a way that it can be reused, with small changes, for both encryption and decryption, that is, essentially by reversing the order and changing the value of round subkeys. Thus, the decryption framework can be derived from the encryption scheme upside-down.

This framework similarity as a design criterion helps save on implementation costs in both hardware and software environments. That means there shall be no need for separate implementations for each scheme. Moreover, it has been noted that this similarity between the encryption and decryption frameworks means that they have the same cryptographic strength, that is, there is no advantage in attacking one scheme instead of the other. As a consequence, analysis under a chosen-plaintext setting shall not be more (or less) powerful than under a chosen-ciphertext setting. Therefore, analysis can be done without loss of generality on the encryption scheme only, and the results would apply for the decryption scheme as well.

If both schemes are (quite) dissimilar, then: (i) analysis would have to be performed on both schemes separately, which would be more expensive; (ii) it will be an unusual design to justify since implementation costs would double for encryption and decryption; (iii) it could raise suspicion of a hidden trapdoor.

An example of a cipher in which the encryption and decryption frameworks are not similar is the Skipjack block cipher [118, 16].

Further, unlike the majority of block cipher designs that iterate many times a weak round function, most of the Lai-Massey designs rather iterate a few times a strong round function.

For instance, the DES cipher always iterates 16 rounds, while MESH ciphers iterate 8, 10 or 12 rounds depending on the block size. For comparison, notice the number of rounds of different ciphers in Tables 1.2, 1.3 and 1.4. The notion of *weak* and *strong* is related to the speed of text (and key) diffusion measured in number of rounds. The faster the diffusion, the stronger the cipher becomes concerning attacks such as the meet-in-the-middle and divide-and-conquer techniques.

While the majority of block cipher designs achieve full text diffusion only after two or more rounds, Lai-Massey designs achieve full text diffusion after a single round due to the construction of the MA-box and the MA half-round. See details in Sects. 2.1, 2.2 and 2.3.

Lai-Massey designs such as IDEA and MESH are *word oriented*, that is, each internal operation is performed on fixed-size w-bit *words*. Typically, $w = 16$ or $w = 8$. An interesting feature of these ciphers is that they are amenable to scale-down variants by successively halving w, while retaining the overall cipher framework. For instance, halving the original 16-bit word of IDEA to 8 and 4 bits, leads to mini-cipher versions with block size 32 and 16 bits, respectively. Not many block ciphers allow such straightforward reduction in size. The availability of mini-versions was used to help test and implement attacks in practice that would otherwise be too costly if applied to the original cipher [77, 74].

On the other hand, to scale up IDEA to larger word sizes is not possible if one wants to keep the same three original group operations, because the multiplication depends on the largest known Fermat prime: $2^{2^4}+1 = 65537$. This means that $\mathbb{Z}^*_{2^{16}+1}$ is a field, but $\mathbb{Z}^*_{2^{2^t}+1}$ is just a ring for $t > 4$ [101]. Only $\phi(2^{2^t}+1)$ elements in $\mathbb{Z}^*_{2^{2^t}+1}$ have a multiplicative inverse, where ϕ is Euler's phi function [101]. Thus, multiplication in $\mathbb{Z}^*_{2^{2^t}+1}$ would not work properly for any key if a ring was used instead of a finite field.

Actually, even if we were interested in word sizes that are not powers of 2, there are no prime numbers of the form 2^t+1, for $17 < t < 207$ (assuming, someone is actually willing to operate on word sizes as large as 200 bits).

Notice that no Lai-Massey cipher was submitted to the AES competition [4], perhaps due to the requirement that the candidates should operate on 128-bit text blocks (or larger).

The MESH ciphers [110, 106] were motivated by the desire to achieve strong Lai-Massey designs operating on text blocks larger than 64 bits. Other approaches that simply tried to increase (for example, double) the original word size $w = 16$ had failed due to problems in the encryption/decryption processes [107].

It was eventually found out that extending the MA-box, while preserving the word size at 16 bits, was a more promising approach. This new strategy allowed us to keep the original word size $w = 16$ and to reuse the three original group operations present in IDEA and PES. Consequently, the MESH ciphers could take advantage of the existing (hardware and software) infrastructure that were already used by IDEA.

Further, the MESH ciphers incorporated new design features aimed at countering attacks against the full-round IDEA and PES, essentially, attacks that exploited the weak key schedule algorithms and the keyed multiplication operation [34]. These features will be described in Chap. 2.

The MESH ciphers follow the Lai-Massey design paradigm initiated with PES and IDEA. Key features include: (i) full diffusion in a single round (without MDS codes); (ii) no S-boxes; (iii) a relatively strong round function iterated a few times; (iv) similarity of the encryption and decryption frameworks.

Therefore, the Lai-Massey cipher design forms an alternative paradigm to the Feistel Network and SPN frameworks.

The IDEA and PES ciphers inspired several other ciphers, such as: Akelarre [5], Yi's cipher [156], the MESH ciphers [110], Ake98 [89], RIDEA [158, 157], IDEA-X [20], IDEA-X/2 [125], IDEA-128 [107], FOX/IDEA-NXT [62], WIDEA-n [61], MESH-64(8) and MESH-128(8) [106], REESSE3+ [146], IDEA* [80] and Bel-T [12].

Ciphers such as IDEA-X, IDEA-X/2 and IDEA-128 were not meant to be used in practice but were prototypes used to demonstrate weaknesses against particular attacks [20, 125, 107]. Also, Yi's cipher [156] and Bel-T [12], do not quite follow the Lai-Massey design paradigm. They are hybrid designs combining features from both Feistel and Lai-Massey designs.

This book surveys, describes and analyzes block ciphers whose design was inspired by the original PES and IDEA ciphers. These ciphers, especially recent developments such as REESSE3+, Bel-T and IDEA*, provide evidence that the Lai-Massey design paradigm is still alive and motivates new ciphers, which are continuously evolving based on novel insights and on third-party cryptanalysis.

References

1. 3GPP: 3GPP TS 35.202 V5.0.0 Third Generation Partnership Project. Technical Specification Group Services and System Aspects; 3G Security; Specification of the 3GPP Confidentiality and Integrity Algorithms. Document 2: KASUMI Specification (Release 5) (2002)
2. Adams, C.M.: The CAST-128 Encryption Algorithm. Request For Comments (RFC) 2144, Network Working Group, Internet Enginnering Task Force (1997)
3. Adams, C.M.: The CAST-256 Encryption Algorithm. First AES Conference, USA (1998)
4. AES: The Advanced Encryption Standard Development Process, https://csrc.nist.gov/ (1997)
5. Alvarez, G., de la Guia, D., Montoya, F., Peinado, A.: Akelarre: a New Block Cipher Algorithm. In: H. Meijer, S. Tavares (eds.) Selected Areas in Cryptography (SAC), pp. 1–14. Springer (1996)
6. Anderson, R., Biham, E.: Two Practical and Provably Secure Block Ciphers: BEAR and LION. In: D. Gollmann (ed.) Fast Software Encryption (FSE), LNCS 1039, pp. 113–120. Springer (1996)
7. Anderson, R.J., Biham, E., Knudsen, L.R.: Serpent and Smartcards. In: J.J. Quisquater, B. Schneier (eds.) Smart Card Research and Application Conference (CARDIS), LNCS 2000, pp. 246–253. Springer (1998)
8. Aoki, K., Ichikawa, T., Kanda, M., Matsui, M., Moriai, S., Nakajima, J., Tokita, T.: Camellia: a 128-bit Block Cipher Suitable for Multiple Platforms. First NESSIE Workshop, Belgium (2000)
9. Baigneres, T.: Quantitative Security of Block Ciphers: Designs and Cryptanalytic Tools. Ph.D. thesis, École Polytechnique Fédérale de Lausanne (EPFL), Switzerland (2008)
10. Barreto, P.S.L.M., Rijmen, V.: The ANUBIS Block Cipher. First NESSIE Workshop, Heverlee, Belgium (2000)
11. Barreto, P.S.L.M., Rijmen, V.: The KHAZAD Legacy-Level Block Cipher. First NESSIE Workshop, Heverlee, Belgium (2000)
12. Belarus: Data Encryption and Integrity Algorithms. Preliminary State Standard of Republic of Belarus (STB P 34.101.31-2007). http://apmi.bsu.by/assets/files/std/belt-spec27.pdf (2007)
13. Bellare, M., Desai, A., Jokipii, E., Rogaway, P.: A Concrete Security Treatment of Symmetric Encryption. In: Proceedings of the 38th Symposium on Foundations of Computer Science. IEEE (1997)
14. Beualieu, R., Shors, D., Smith, J., Treatman-Clark, S., Weeks, B., Wingers, L.: The Simon and Speck Families of Lightweight Block Ciphers. IACR ePrint archive, 2013/404 (2013)

15. Biham, E.: New Types of Cryptanalytic Attacks using Related Keys. In: T. Helleseth (ed.) Advances in Cryptology, Eurocrypt, LNCS 765, pp. 398–409. Springer (1993)
16. Biham, E., Biryukov, A., Shamir, A.: Cryptanalysis of Skipjack Reduced to 31 rounds using Impossible Differentials. In: J. Stern (ed.) Advances in Cryptology, Eurocrypt, LNCS 1592, pp. 12–23. Springer (1999)
17. Biryukov, A., Nakahara Jr, J., Yildirim, H.M.: Differential Entropy Analysis of the IDEA Block Cipher. Journal of Computational and Applied Mathematics **259**, 561–570 (2014)
18. Blaze, M., Schneier, B.: The MacGuffin Block Cipher Algorithm. In: B. Preneel (ed.) Fast Software Encryption (FSE), LNCS 1008, pp. 97–110. Springer (1995)
19. Body, K.N.: Contribution for Korean Candidates of Encryption Algorithm (SEED). ISO/IEC JTC1 SC27 N2563, http://www.kisa.or.kr (2000)
20. Borisov, N., Chew, M., Johnson, R., Wagner, D.: Multiplicative Differentials. In: J. Daemen, V. Rijmen (eds.) Fast Software Encryption (FSE), LNCS 2365, pp. 17–33. Springer (2002)
21. Borst, J.: The Block Cipher: GRAND CRU. First NESSIE Workshop, Heverlee, Belgium (2000)
22. Borst, J.: Block Ciphers: Design, Analysis and Side-Channel Analysis. Ph.D. thesis, Dept. Elektrotechniek, ESAT, Katholieke Universiteit Leuven, Belgium (2001)
23. Brown, L., Kwan, M., Pieprzyk, J., Seberry, J.: Improving Resistance to Differential Cryptanalysis and the Redesign of LOKI. In: H. Imai, R.L. Rivest, T. Matsumoto (eds.) Advances in Cryptology, Asiacrypt, LNCS 739, pp. 36–50. Springer (1991)
24. Brown, L., Pieprzyk, J.: Introducing the New LOKI97 Block Cipher. First AES Conference, California, USA (1998)
25. Brown, L., Pieprzyk, J., Seberry, J.: LOKI - a Cryptographic Primitive for Authentication and Secrecy Applications. In: J. Seberry, J. Pieprzyk (eds.) Advances in Cryptology, Auscrypt, LNCS 453, pp. 229–236. Springer (1990)
26. Burwick, C., Coppersmith, D., D'Avignon, E., Genario, R., Halevi, S., Jutla, C., Matyas Jr, S.M., O'Connor, L., Peyravian, M., Safford, D., Zunic, N.: MARS – a Candidate Cipher for AES. First AES Conference, California, USA (1998)
27. Chambers, W.G.: On Random Mappings and Random Permutations. In: B. Preneel (ed.) Fast Software Encryption (FSE), LNCS 1008, pp. 22–28. Springer (1995)
28. Coppersmith, D.: The Data Encryption Standard (DES) and its Strength against Attacks. IBM Journal of Research and Development **38**(3), 243–250 (1994)
29. Courtois, N.T.: How Fast can be Algebraic Attacks on Block Ciphers? IACR ePrint archive, 2006/168 (2006)
30. Courtois, N.T., Bard, G.V., Ault, S.V.: Statistics of Random Permutations and the Cryptanalysis of Periodic Block Ciphers. IACR ePrint Archive 2009/186 (2009)
31. CRYPTREC: Evaluation of Cryptographic Techniques Project, 2000–2003, http://www.ipa.go.jp/security/enc/cryptrec/index-e.html (2000)
32. Cusick, T.W., Wood, M.C.: The REDOC-II Cryptosystem. In: A.J. Menezes, S.A. Vanstone (eds.) Advances in Cryptology, Crypto, LNCS 537, pp. 545–563. Springer (1990)
33. Daemen, J.: Cipher and Hash Function Design – Strategies based on Linear and Differential Cryptanalysis. Ph.D. thesis, Dept. Elektrotechniek, ESAT, Katholieke Universiteit Leuven, Belgium (1995)
34. Daemen, J., Govaerts, R., Vandewalle, J.: Weak Keys for IDEA. In: D.R. Stinson (ed.) Advances in Cryptology, Crypto, LNCS 773, pp. 224–231. Springer (1993)
35. Daemen, J., Knudsen, L.R., Rijmen, V.: The Block Cipher SQUARE. In: E. Biham (ed.) Fast Software Encryption (FSE), LNCS 1267, pp. 149–165. Springer (1997)
36. Daemen, J., Peeters, M., Van Assche, G., Rijmen, V.: NESSIE Proposal: NOEKEON. First NESSIE Workshop, Heverlee, Belgium (2000)
37. Daemen, J., Rijmen, V.: AES Proposal: Rijndael. First AES Conference, California, USA, http://www.nist.gov/aes (1998)

38. Daemen, J., Rijmen, V.: The Block Cipher BKSQ. In: J.J. Quisquater, B. Schneier (eds.) Smart Card Research and Applications (CARDIS), LNCS 1820, pp. 236–245. Springer (2000)
39. Davies, D.W., Parkin, G.I.P.: The Average Cycle Size of the Key Stream in Output Feedback Encipherment. In: D. Chaum, R.L. Rivest, A.T. Sherman (eds.) Advances in Cryptology, Crypto, pp. 97–98. Plenum Press (1983)
40. De Canniere, C., Dunkelman, O., Knezevic, M.: KATAN and KTANTAN - a Family of Small and Efficient Hardware Oriented Block Ciphers. In: C. Clavier, K. Gaj (eds.) Cryptographic Hardware and Embedded Systems (CHES), LNCS 5747, pp. 272–288. Springer (2009)
41. De Canniere, C., Rechberger, C.: Preimages for Reduced SHA-0 and SHA-1. In: D. Wagner (ed.) Advances in Cryptology, Crypto, LNCS 5157, pp. 179–202. Springer (2008)
42. Diehl, E.: Ten Laws for Security. Springer (2016)
43. Diffie, W., Hellman, M.E.: Privacy and Authentication: an Introduction to Cryptography. Proceedings of the IEEE **67**(3), 397–427 (1979)
44. Doots, N., Moldovyan, A.A., Moldovyan, N.A.: Fast Encryption Algorithm Spectr-H64. In: V.I. Gorodetski, V.A. Skormim, L.J. Popyack (eds.) Information Assurance in Computer Networks: Methods, Models and Architectures for Network Security, LNCS 2052, pp. 275–286. Springer (2001)
45. Dworkin, M.: Recommendations for Block Ciphers Modes of Operation, Methods and Techniques. NIST Special Publication 800-38A, NIST, US Department of Commerce (2001)
46. Feistel, H.: Cryptography and Computer Privacy. Scientific American **228**(5), 15–23 (1973)
47. FIPS197: Advanced Encryption Standard (AES). FIPS PUB 197 Federal Information Processing Standard Publication 197, United States Department of Commerce (2001)
48. Gait, J.: A New Nonlinear Pseudo-Random Number Generator. IEEE Transactions on Software Engineering **SE-3**(5), 359–363 (1977)
49. Garfinkel, S.: PGP: Pretty Good Privacy. O'Reilly and Associates (1994)
50. Georgoudis, D.G., Leroux, D., Chaves, B.S.: FROG. First AES Conference, USA (1998)
51. Gerard, B., Grosso, V., Naya-Plasencia, M., Standaert, F.X.: Block Ciphers that are Easier to Mask: How Far can we go? In: G. Bertoni, J.S. Coron (eds.) Cryptographic Hardware and Embedded Systems (CHES), LNCS 8086, pp. 383–399. Springer (2013)
52. Gilbert, H., Girault, M., Hoogvorst, P., Noilhan, F., Pornin, T., Poupard, G., Stern, J., Vaudenay, S.: Decorrelated Fast Cipher: an AES Candidate. First AES Conference, USA (1998)
53. Gong, Z., Nikova, S., Law, Y.K.: KLEIN: a New Family of Lightweight Block Ciphers. In: A. Juels, C. Paar (eds.) RFID Security and Privacy, LNCS 7055, pp. 1–18. Springer (2011)
54. GOST: Federal Agency on Technical Regulation and Metrology, National Standard of the Russian Federation, GOST R 34.12-2015. http://tc26.ru/en/standard/gost/GOSTR34122015ENG.pdf (2015)
55. Handschuh, H., Naccache, D.: SHACAL. Submission to NESSIE, https://www.cosic.esat.kuleuven.be/nessie/tweaks.html (2001)
56. Heys, H.M.: An Analysis of the Statistical Self-Synchronization of Stream Ciphers. Proceedings of INFOCOM 2001 (2001)
57. ISO: Information Processing - Modes of Operation for an n-bit Block Cipher Algorithm. ISO IEC 10116, International Organization for Standardization - ISO, Geneva, Switzerland (1991)
58. ISO: Information Technology - Security Techniques - Message Authentication Codes - part 1: mechanisms using a block cipher. ISO/IEC 9797-1 (1999)

59. Jacobson Jr, M.J., Huber, K.: The MAGENTA Block Cipher Algorithm. First AES Conference, California, USA (1998)
60. Johnson, D.B., Matyas, S.M., Lee, A.V., Wilkins, J.D.: The Commercial Data Masking Facility (CDMF) Data Privacy Algorithm. IBM Journal of Research and Development **38**(2), 217–225 (1994)
61. Junod, P., Macchetti, M.: Revisiting the IDEA Philosophy. In: O. Dunkelman (ed.) Fast Software Encryption (FSE), LNCS 5665, pp. 277–295. Springer (2009)
62. Junod, P., Vaudenay, S.: FOX: a New Family of Block Ciphers. In: H. Handschuh, M.A. Hasan (eds.) Selected Areas in Cryptography (SAC), LNCS 3357, pp. 114–129. Springer (2004)
63. Kaliski Jr, B.S., Rivest, R.L., Sherman, A.T.: Is the Data Encryption Standard a Group? (preliminary abstract). In: F. Pichler (ed.) Advances in Cryptology, Eurocrypt, LNCS 219, pp. 81–95. Springer (1985)
64. Kaliski Jr, B.S., Rivest, R.L., Sherman, A.T.: Is the Data Encryption Standard a Group? Journal of Cryptology **1**, 3–36 (1988)
65. Kelsey, J., Schneier, B., Wagner, D.: Key-Schedule Cryptanalysis of IDEA, G-DES, GOST, SAFER and triple-DES. In: N. Koblitz (ed.) Advances in Cryptology, Crypto, LNCS 1109, pp. 237–251. Springer (1996)
66. Kelsey, J., Schneier, B., Wagner, D.: Related-Key Cryptanalysis of 3-Way, Biham-DES, CAST, DES-X, NewDES, RC2, and TEA. In: Y. Han, T. Okamoto, S. Qing (eds.) Information and Communication Security (ICICS), LNCS 1334, pp. 233–246. Springer (1997)
67. Kerckhoffs, A.: La Cryptografie Militaire. Journal des Sciences Militaires **9**, 5–83, 161–191 (1883)
68. Kim, K., Lee, S., Park, S., Lee, D.: How to Strengthen DES Against Two Robust Attacks. Proceedings of 1995 Korea–Japan Joint Workshop on Info. Security and Cryptology, JW-ISC'95 (1995)
69. Kim, K., Lee, S., Park, S., Lee, D.: Securing DES S-boxes against Three Robust Cryptanalysis. In: E. Kranakis, P.C. Van Oorschot (eds.) Selected Areas in Cryptography (SAC), pp. 145–157. Springer (1995)
70. Knudsen, L.R.: Cryptanalysis of LOKI'91. In: J. Seberry, Y. Zheng (eds.) Advances in Cryptology, Auscrypt, LNCS 718, pp. 196–208. Springer (1993)
71. Knudsen, L.R.: A Key-Schedule Weakness in SAFER K-64. In: D. Coppersmith (ed.) Advances in Cryptology, Crypto, LNCS 963, pp. 274–286. Springer (1995)
72. Knudsen, L.R.: DEAL – a 128-bit Block Cipher. Technical Report #151, University of Bergen, Dept. of Informatics, Norway (1998)
73. Kwan, M.: The Design of the ICE Encryption Algorithm. In: E. Biham (ed.) Fast Software Encryption (FSE), LNCS 1267, pp. 69–82. Springer (1997)
74. Lai, X.: On the Design and Security of Block Ciphers, *ETH Series in Information Processing*, vol. 1. Hartung-Gorre Verlag, Konstanz (1995)
75. Lai, X., Massey, J.L.: A Proposal for a New Block Encryption Standard. In: I.B. Damgård (ed.) Advances in Cryptology, Eurocrypt, LNCS 473, pp. 389–404. Springer (1990)
76. Lai, X., Massey, J.L.: Hash Functions based on Block Ciphers. In: R.A. Rueppel (ed.) Advances in Cryptology, Eurocrypt, LNCS 658, pp. 55–70 (1993)
77. Lai, X., Massey, J.L., Murphy, S.: Markov Ciphers and Differential Cryptanalysis. In: D.W. Davies (ed.) Advances in Cryptology, Eurocrypt, LNCS 547, pp. 17–38. Springer (1991)
78. LANCrypto: Cryptographic Primitives Based upon the Block Code NUSH. First NESSIE Workshop, Heverlee, Belgium (2000)
79. Lee, C.H., Kim, J.S.: The New Block Cipher Rainbow. Samsung Advanced Institute of Technology (1997)

80. Lerman, L., Nakahara Jr, J., Veshchikov, N.: Improving Block Cipher Design by Rearranging Internal Operations. In: P. Samarati (ed.) 10th International Conference on Security and Cryptography (SECRYPT), pp. 27–38. IEEE (2013)
81. Lidl, R., Niederreiter, H.: Finite Fields. Encyclopedia of Mathematics and its Applications. Cambridge University Press (2008)
82. Lim, C.H.: CRYPTON: a New 128-bit Block Cipher – Specification and Analysis. First AES Conference, USA (1998)
83. Lim, C.H.: A Revised Version of Crypton - CRYPTON version 1.0. In: L.R. Knudsen (ed.) Fast Software Encryption (FSE), LNCS 1636, p. 31. Springer (1999)
84. Lim, C.H., Korkishko, T.: mCrypton - a Lightweight Block Cipher for Security of Low-Cost RFID Tags and Sensors. In: J. Song, T. Kwon, M. Yung (eds.) WISA, LNCS 3786, pp. 243–258. Springer (2005)
85. Lipmaa, H., Rogaway, P., Wagner, D.: CTR-Mode Encryption. Second NIST Modes of Operation Workshop (2001)
86. Luby, M., Rackoff, C.: How to Construct Pseudorandom Permutations from Pseudorandom Functions. SIAM Journal of Computing **17**(2), 373–386 (1988)
87. Machado, A.W.: The Nimbus Cipher. First NESSIE Workshop, Heverlee, Belgium (2000)
88. Madryga, W.E.: A High Performance Encryption Algorithm. In: Computer Security: a Global Challenge. North-Holland: Elsevier Science Publishers (1984)
89. Maranon, G.A.: Contribución al Estudio de la Estructura Interna del Conjunto de Mandelbrot y Aplicaciones en Criptografia. Ph.D. thesis, Facultad de Informatica, Universidad Politecnica de Madrid, Spain (2000)
90. Markowitch, O., Nakahara Jr, J.: Attacks on Single-Pass Confidentiality Modes of Operation. In: Proceedings of the XIV Brazilian Symposium on Information and Computational Systems Security (SBSeg), pp. 84–99. Brazilian Computer Society (SBC) (2014)
91. Massey, J.L.: SAFER K–64: a Byte-Oriented Block-Ciphering Algorithm. In: R. Anderson (ed.) Fast Software Encryption (FSE), LNCS 809, pp. 1–17. Springer (1994)
92. Massey, J.L.: Announcement of a 40-bit Key Schedule for the Cipher SAFER. sci.crypt newsgroup (1995)
93. Massey, J.L.: Announcement of a Strengthened Key Schedule for the Cipher SAFER (Secure And Fast Encryption Routine for both 64 and 128 bit Key Lengths). sci.crypt newsgroup (1995)
94. Massey, J.L.: SAFER K–64: One Year Later. In: B. Preneel (ed.) Fast Software Encryption (FSE), LNCS 1008, pp. 212–241. Springer (1995)
95. Massey, J.L., Khachatrian, G.H., Kuregian, M.K.: Nomination of SAFER+ as Candidate Algorithm for the Advanced Encryption Standard. First AES Conference, California, USA (1998)
96. Massey, J.L., Khachatrian, G.H., Kuregian, M.K.: The SAFER++ Block Encryption Algorithm. First NESSIE Workshop, Heverlee, Belgium (2000)
97. Matsui, M.: Block Encryption Algorithm MISTY. Technical report of IEICE, isec96-11, in Japanese, Mitsubishi Co. (1996)
98. Matsui, M.: New Block Encryption Algorithm MISTY. In: E. Biham (ed.) Fast Software Encryption (FSE), LNCS 1267, pp. 54–68. Springer (1997)
99. Maurer, U.M.: New Approaches to the Design of Self-Synchronizing Stream Ciphers. In: D.W. Davies (ed.) Advances in Cryptology, Eurocrypt, LNCS 547, pp. 458–471. Springer (1991)
100. McBride, L.: Q, a Proposal to NESSIE, v2.00. First NESSIE Workshop, Heverlee, Belgium (2000)
101. Menezes, A.J., Van Oorschot, P., Vanstone, S.A.: Handbook of Applied Cryptography. CRC Press (1997)
102. Merkle, R.C.: A Software Encryption Function. posted to sci.crypt USENET newsgroup, Xerox PARC, Palo Alto (1989)

103. Merkle, R.C.: Fast Software Encryption Functions. In: A.J. Menezes, S.A. Vanstone (eds.) Advances in Cryptology, Crypto, LNCS 537, pp. 476–501. Springer (1991)
104. Miyaguchi, S.: The FEAL Cipher Family. In: A.J. Menezes, S.A. Vanstone (eds.) Advances in Cryptology, Crypto, LNCS 537, pp. 627–638. Springer (1991)
105. Moldovyan, A.A., Moldovyan, N.A.: A Method of the Cryptographical Transformation of Binary Data Blocks. Russian patent number 2141729 Bulletin no. 32 (1999)
106. Nakahara Jr, J.: Faster Variants of the MESH Block Ciphers. In: A. Canteaut, K. Viswanathan (eds.) International Conference on Cryptology in India, Indocrypt, LNCS 3348, pp. 162–174. Springer (2004)
107. Nakahara Jr, J.: On the Design of IDEA-128. Brazilian Symposium on Information and Computational Systems Security (SBSeg) (2005)
108. Nakahara Jr, J.: 3D: a Three-Dimensional Block Cipher. In: M.K. Franklin, L.C.K. Hui, D.S. Wong (eds.) Cryptology and Network Security (CANS), LNCS 5339, pp. 252–267. Springer (2008)
109. Nakahara Jr, J., Abrahao, E.: A New Involutory MDS Matrix for the AES. International Journal of Network Security (IJNS) **9**(2), 109–116 (2009)
110. Nakahara Jr, J., Rijmen, V., Preneel, B., Vandewalle, J.: The MESH Block Ciphers. In: K. Chae, M. Yung (eds.) Information Security Applications (WISA), LNCS 2908, pp. 458–473. Springer (2003)
111. NBS: Data Encryption Standard (DES). FIPS PUB 46, Federal Information Processing Standards Publication 46, U.S. Department of Commerce (1977)
112. NBS: Data Encryption Standard (DES). FIPS PUB 46–1, Federal Information Processing Standards Publication 46–1 (1988)
113. NBS: Data Encryption Standard (DES). FIPS PUB 46–2, Federal Information Processing Standards Publication 46–2 (1993)
114. NBS: Data Encryption Standard (DES). FIPS PUB 46–3, Federal Information Processing Standards Publication 46–3 (1999)
115. NEC: ISO/IEC9979-0019 Register Entry (1998)
116. Needham, R., Wheeler, D.: eXtended Tiny Encryption Algorithm (1997)
117. NIST: DES Modes of Operation. FIPS PUB 81, US Department of Commerce (1980)
118. NIST: Skipjack and KEA Specification, version 2.0. http://csrc.nist.gov/ (1998)
119. NIST: Report on the Symmetric-Key Block Cipher Modes of Operation Workshop. http://csrc.nist.gov/ (2000)
120. NSRI: ARIA Specification, National Security Research Institute. http://210.104.33.10/ARIA/index-e.html (2005)
121. NTT: Specification of E2 – a 128-bit Block Cipher. First AES Conference, California, USA (1998)
122. Patarin, J.: How to Construct Pseudorandom and Super-Pseudorandom Permutations from One Single Pseudorandom Function. In: R.A. Rueppel (ed.) Advances in Cryptology, Eurocrypt, LNCS 658, pp. 256–266. Springer (1992)
123. Preneel, B.: New European Schemes for Signatures, Integrity and Encryption. In: D. Naccache, P. Paillier (eds.) Public-key Cryptography (PKC), LNCS 2274, pp. 297–309. Springer (2002)
124. Preneel, B., Govaerts, R., Vandewalle, J.: Hash Functions based on Block Ciphers: a Synthetic Approach. In: D.R. Stinson (ed.) Advances in Cryptology, Crypto, LNCS 773, pp. 368–378. Springer (1993)
125. Raddum, H.: Cryptanalysis of IDEA-X/2. In: T. Johansson (ed.) Fast Software Encryption (FSE), LNCS 2887, pp. 1–8. Springer (2003)
126. Rijmen, V.: Cryptanalysis and Design of Iterated Block Ciphers. Ph.D. thesis, Dept. Elektrotechniek, ESAT, Katholieke Universiteit Leuven, Belgium (1997)
127. Rijmen, V., Daemen, J., Preneel, B., Bosselaers, A., De Win, E.: The Cipher SHARK. In: D. Gollmann (ed.) Fast Software Encryption (FSE), LNCS 1039, pp. 99–112. Springer (1996)

128. Rijmen, V., Preneel, B.: A Family of Trapdoor Ciphers. In: E. Biham (ed.) Fast Software Encryption (FSE), LNCS 1267, pp. 139–148. Springer (1997)
129. Rivest, R.L.: The RC2 Encryption Algorithm. RSA Data Security, Inc. (1992)
130. Rivest, R.L.: The RC5 Encryption Algorithm. In: B. Preneel (ed.) Fast Software Encryption (FSE), LNCS 1008, pp. 86–96. Springer (1995)
131. Rivest, R.L., Robshaw, M.J.B., Sidney, R., Yin, Y.L.: The RC6 Block Cipher. First AES Conference, USA (1998)
132. Rivest, R.L., Shamir, A., Adleman, L.M.: Cryptographic Communications System and Method, US Patent 4.405.829, Filed Dec. 14, 1977, Issued Sep. 20, 1983, Expired Sep. 20, 2000
133. Rivest, R.L., Shamir, A., Adleman, L.M.: On Digital Signatures and Public-Key Cryptosystems. Tech. rep., MIT Laboratory for Computer Science Technical Memorandum 82 (1977)
134. Rivest, R.L., Shamir, A., Adleman, L.M.: A Method for Obtaining Digital Signatures and Public-Key Cryptosystems. Communication of the ACM **21**(2), 120–126 (1978)
135. Schmidt, D.: ABC – a Block Cipher. IACR ePrint Archive, 2002/062 (2002)
136. Schneier, B.: Description of a New Variable-Length Key, 64-bit Block Cipher (Blowfish). In: R. Anderson (ed.) Fast Software Encryption (FSE), LNCS 809, pp. 191–204. Springer (1994)
137. Schneier, B., Kelsey, J., Whiting, D., Wagner, D., Hall, C., Ferguson, N.: Twofish: a 128-bit Block Cipher. First AES Conference, California, USA (1998)
138. Shannon, C.E.: Communication Theory of Secrecy Systems. Bell System Technical Journal **28**(4), 656–715 (1949)
139. Shibutani, K., Inobe, T., Hiwatari, H., Mitsuda, A., Akishita, T., Shirai, T.: Piccolo: an Ultra-Lightweight Block Cipher. In: B. Preneel, T. Takagi (eds.) Cryptographic Hardware and Embedded Systems (CHES), LNCS 6917, pp. 342–357. Springer (2011)
140. Shimizu, A., Miyaguchi, S.: Fast Data Encipherment Algorithm FEAL. In: D. Chaum, W.L. Price (eds.) Advances in Cryptology, Eurocrypt, LNCS 304, pp. 267–278. Springer (1988)
141. Shimoyama, T., Yanami, H., Yokoyama, K., Takenaka, M., Ito, K., Yajima, J., Torii, N., Tanaka, H.: The SC2000 Block Cipher. First NESSIE Workshop, Heverlee, Belgium (2000)
142. Smith, J.L.: The Design of Lucifer: a Cryptographic Device for Data Communications. Tech report, IBM T. J. Watson Research Center, Yorktown Heights (1971)
143. Sony: CLEFIA. https://www.sony.net/Products/cryptography/clefia (2007)
144. Office of State Commercial Cryptography Administration, P.: The SMS4 Block Cipher (in Chinese). http://www.oscca.gov.cn/ (2006)
145. Stern, J., Vaudenay, S.: CS-Cipher. In: S. Vaudenay (ed.) Fast Software Encryption (FSE), LNCS 1372, pp. 189–205. Springer (1998)
146. Su, S., Lu, S.: A 128-bit Block Cipher based on Three Group Arithmetics. IACR ePrint archive 2014/704 (2014)
147. Suzaki, T., Minematsu, K., Morioka, S., Kobayashi, E.: TWINE: a Lightweight Block Cipher for Multiple Platforms. In: L.R. Knudsen, H. Wu (eds.) Selected Areas in Cryptography (SAC), LNCS 7707, pp. 339–354. Springer (2013)
148. Toshiba: Specification of Hierocrypt-3. First NESSIE Workshop, Heverlee, Belgium (2000)
149. Toshiba: Specification of Hierocrypt-L1. First NESSIE Workshop, Heverlee, Belgium (2000)
150. Tsunoo, Y., Kubo, H., Miyauchi, H., Nakamura, K.: A New 128-bit Block Cipher CIPHERUNICORN-A. The Institute of Electronics, Information and Communication Engineers, Technical Report of IEICE, ISEC2000-5 (2000)
151. Van Oorschot, P.C., Wiener, M.J.: Improving Implementable Meet-In-The-Middle Attacks by Orders of Magnitude. In: N. Koblitz (ed.) Advances in Cryptology, Crypto, LNCS 1109, pp. 229–236. Springer (1996)

152. Vaudenay, S.: Provable Security for Block Cipher by Decorrelation. In: M. Morvan, C. Meinel, D. Krob (eds.) Proceedings of the 15th Annual Symposium on Theoretical Aspects of Computer Science (STACS), LNCS 1373, pp. 249–275. Springer (1998)
153. Vernam, G.: Cipher Printing Telegraph System for Secret Wire and Radio Telegraph Communications. Journal of the IEEE **55**, 109–115 (1926)
154. Wagner, D.: The Boomerang Attack. In: L.R. Knudsen (ed.) Fast Software Encryption (FSE), LNCS 1636, pp. 156–170. Springer (1999)
155. Wheeler, D.J., Needham, R.M.: TEA, a Tiny Encryption Algorithm. In: B. Preneel (ed.) Fast Software Encryption (FSE), LNCS 1008, pp. 363–366. Springer (1995)
156. Yi, X.: On Design and Analysis of a New Block Cipher. In: J. Jaffar, R.H.C. Yap (eds.) Concurrency and Parallelism, Programming, Networking and Security, ASIAN 1996, LNCS 1179, pp. 213–222. Springer (1996)
157. Yıldırım, H.M.: Some Linear Relations for Block Cipher IDEA. Ph.D. thesis, The Middle East Technical University, Turkey (2002)
158. Yıldırım, H.M.: Nonlinearity Properties of the Mixing Operations of the Block Cipher IDEA. In: T. Johansson, S. Maitra (eds.) Progress in Cryptology, Indocrypt, LNCS 2904, pp. 68–81. Springer (2003)
159. Zabotin, I.A., Glazkov, G.P., Isaeva, V.B.: GOST 28147-89, Cryptographic Protection for Data Processing Systems, Cryptographic Transformation Algorithm. Government Standard of the U.S.S.R., Inv. No. 3583, UDC 681.325.6:006.354 (1989)
160. Zheng, Y.: The SPEED Cipher. In: R. Hirschfeld (ed.) Financial Cryptography, LNCS 1997, pp. 71–99. Springer (1997)

Chapter 2
Lai-Massey Block Ciphers

The most important thing in communication is hearing what isn't said.
– Peter Drucker

Abstract

This chapter describes several block ciphers that follow the Lai-Massey design paradigm as exemplified in the PES and IDEA ciphers. Other ciphers that follow this paradigm are MESH-64, MESH-96, MESH-128, MESH-64(8), MESH-128(8), RIDEA, WIDEA-n, FOX/IDEA-NXT, REESSE3+, IDEA* and Bel-T. Therefore, unlike [59], the terminology *Lai-Massey cipher design* in this book applies to a larger variety of ciphers.

The encryption/decryption schemes, including design criteria, origin, basic parameters, as well as the key schedule algorithm of each cipher are described and explained in detail.

2.1 The PES Block Cipher

The Proposed Encryption Standard (PES) block cipher operates on 64-bit text blocks, under a 128-bit key and iterates eight rounds plus an output transformation (OT). PES was designed by X. Lai and J.L. Massey [37] dating back to 1990.

PES uses only three (binary) group operations on 16-bit words: addition modulo 2^{16} (denoted $\boxplus$), multiplication modulo $2^{16}+1$ (denoted $\odot$) with the exception that $2^{16} \equiv 0$, and bitwise exclusive-or (denoted $\oplus$). There are no S-boxes.

Concerning the condition $0 \equiv 2^{16}$ in the finite field $\mathrm{GF}(2^{16}+1)$, there are at least two good reasons for it:

- without it, 0 would have no multiplicative inverse. With this condition, it follows that $0^{-1} \bmod (2^{16}+1) = (2^{16})^{-1} \bmod (2^{16}+1) = 2^{16} = 0$, that is, 0 becomes its own inverse.

J. Nakahara Jr., *Lai-Massey Cipher Designs*, https://doi.org/10.1007/978-3-319-68273-0_2

- without it, 2^{16} would be the only element in $GF(2^{16}+1)$ that could not fit in 16 bits. With this condition, it follows that all elements in $GF(2^{16}+1)$ can be represented in 16 bits.

This condition $0 \equiv 2^{16}$ sets $GF(2^{16}+1)$ apart from several alternative groups (and fields) operating on 16-bit words [46].

The following terminology [37] will be adopted for the three group operations: $(GF(2), \oplus)$, $(\mathbb{Z}_{2^{16}}, \boxplus)$ and $(GF(2^{16}+1), \odot)$. Each group operation and its inverse are used in PES, IDEA, MESH and derived ciphers for the encryption and decryption transformations.

The design criteria for PES (as well as for IDEA and derived ciphers) include

- use of three incompatible group operations on 16-bit words: addition modulo 2^{16}, multiplication modulo $2^{16}+1$ (with the exception that $0 \equiv 2^{16}$) and bitwise exclusive-or.
  The incompatibility property refers to the fact that these operations are, in general, not distributive nor associative when combined. It means that there exist $a, b, c \in Z_2^{16}$ such that:
  - $a \boxplus (b \oplus c) \neq (a \boxplus b) \oplus (a \boxplus c)$
  - $a \boxplus (b \odot c) \neq (a \boxplus b) \odot (a \boxplus c)$
  - $a \oplus (b \boxplus c) \neq (a \oplus b) \boxplus (a \oplus c)$
  - $a \oplus (b \odot c) \neq (a \oplus b) \odot (a \oplus c)$
  - $a \odot (b \oplus c) \neq (a \odot b) \oplus (a \odot c)$
  - $a \odot (b \boxplus c) \neq (a \odot b) \boxplus (a \odot c)$

  hold and also
  - $a \boxplus (b \oplus c) \neq (a \boxplus b) \oplus c$
  - $a \boxplus (b \odot c) \neq (a \boxplus b) \odot c$
  - $a \oplus (b \boxplus c) \neq (a \oplus b) \boxplus c$
  - $a \oplus (b \odot c) \neq (a \oplus b) \odot c$
  - $a \odot (b \oplus c) \neq (a \odot b) \oplus c$
  - $a \odot (b \boxplus c) \neq (a \odot b) \boxplus c$.
- all three arithmetic operations are needed: it has been observed in [12] that using only $\odot$ and $\oplus$ (that is, removing $\boxplus$) is not enough, since the resulting cipher becomes susceptible to a kind of differential attack called multiplicative differential. Following [4]: if $\odot$ is removed, then the resulting cipher can be cryptanalyzed by examining the least significant bit of the words in the whole encryption framework. If $\oplus$ is removed, the resulting cipher becomes affine over $(\mathbb{Z}_{2^{16}}, \boxplus)$.
- no group operation is used twice in succession anywhere along the cipher. For instance, no $\odot$ operation is preceded nor followed by another $\odot$ operation. Likewise for $\oplus$ and $\boxplus$. See the computational graph of PES (for encryption) in Fig. 2.1(a).
- all operators are binary.

- the Multiplication-Addition-box (MA-box) interleaves only $\odot$ and $\boxplus$ in a zigzag pattern, across two layers. Two layers are necessary and sufficient to make all the MA-box outputs depend on all of its inputs. The fact that the MA-box outputs are combined to each of the four input words in a round, means that complete (text) diffusion is achieved in a single round. See Fig. 2.1.

Each operation has distinctive diffusion properties: $\oplus$ is a linear operator and acts bitwise, thus, it has the weakest form of diffusion: the i-th output bit depends only on the i-th bit of the inputs; $\boxplus$ is a nonlinear operator and has unidirectional diffusion from the right-to-left direction following the direction of carry propagation; $\odot$ is a nonlinear operator and has the best diffusion except for weak subkey values. Note that $\oplus$ and $\boxplus$ are T-functions [33, 32].

Bitwise diffusion for $\odot$ is better understood from X. Lai's Low-High algorithm [35] for multiplication in $\mathrm{GF}(2^w+1)$, for $w \in \{2,4,8,16\}$:

Definition 2.1. (Low-High Algorithm for Multiplication in $\mathrm{GF}(2^w+1)$)
Let $a, b \in \mathbb{Z}_{2^w+1}$, $R = ab \bmod 2^w$ and $Q = ab \operatorname{div} 2^w$, $w \in \{2,4,8,16\}$. Then

$$a \odot b = \begin{cases} R-Q & \text{if } R \geq Q \\ R-Q+2^w+1 & \text{if } R < Q \end{cases}$$

where R denotes the remainder (Low-end half) and Q denotes the quotient (High-end half) when ab is divided by 2^w.

The majority of Lai-Massey designs use multiplication for $w = 16$, but the Low-High algorithm also applies to word sizes $w \in \{2,4,8\}$. These three values correspond to the only known Fermat primes, that is, prime numbers of the form $2^{2^w}+1$.

Notice that R and Q are the low and high w-bit words of the $2w$-bit intermediate result of long multiplication in $\mathbb{Z}_{2^w+1}$, respectively. Thus, the multiplication operation in $\mathrm{GF}(2^w+1)$ wraps around all intermediate $2w$ bits into a single w-bit result, instead of truncating it to the low-order w bits, as would happen if multiplication was in $\mathbb{Z}_{2^w}$.

Round subkeys in PES are 16-bit words that are either multiplied or added to the cipher data. In the first case, they are called *multiplicative subkeys*. Otherwise, they are called *additive subkeys*.

Shannon's principles of confusion and diffusion [56] were used in PES. The mixing of nonlinear operators, such as $\odot$ and $\boxplus$, provide for *confusion*, while the MA-box and the construction of the MA half-round, which combines the MA-box outputs to all four words in a block are responsible for *diffusion*.

2.1.1 Encryption and Decryption Frameworks

Fig. 2.1(a) shows a pictorial view of the first full PES encryption round, part of the second round and the output transformation. Fig. 2.1(b) shows the corresponding decryption scheme. The direction of arrows indicates the data flow.

Modular division in GF($2^{16}+1$) is denoted by $\boxdot$, that is, $a \boxdot b = a/b = a \odot b^{-1}$. It is a compact notation to denote the multiplicative inverse operation for decryption.

Let $X^{(i)} = (X_1^{(i)}, X_2^{(i)}, X_3^{(i)}, X_4^{(i)})$ denote the 64-bit text block to the i-th round, where $X_j^{(i)} \in \mathbb{Z}_2^{16}$ for $1 \leq j \leq 4$. So, for instance, $X^{(1)}$ represents a plaintext block.

One full round in PES can be split into a Key-Mixing (KM) and a Multiplication-Addition (MA) half-rounds, in this order. Note that round subkeys are combined with the cipher data only via $\odot$ and $\boxplus$.

The i-th 16-bit subkey of the j-th round is denoted by $Z_i^{(j)}$, with $1 \leq i \leq 6$ and $1 \leq j \leq 9$. Subkey are generated by a key schedule algorithm, which is described in Sect. 2.1.2.

$X^{(i)}$ denotes not only the 64-bit input to the i-th round, but also the input to the i-th KM half-round. The output of the i-th KM half-round is

$$Y^{(i)} = (Y_1^{(i)}, Y_2^{(i)}, Y_3^{(i)}, Y_4^{(i)}),$$

where

$$Y^{(i)} = (X_1^{(i)} \odot Z_1^{(i)}, X_2^{(i)} \odot Z_2^{(i)}, X_3^{(i)} \boxplus Z_3^{(i)}, X_4^{(i)} \boxplus Z_4^{(i)}), \tag{2.1}$$

which becomes the input to the i-th MA half-round. Since PES has 8.5 rounds, $Y^{(9)}$ represents the resulting ciphertext block.

To compute an MA half-round, two temporary values are needed: $(n_i, q_i) = (Y_1^{(i)} \oplus Y_3^{(i)}, Y_2^{(i)} \oplus Y_4^{(i)})$, which form the input to the i-th MA-box. The MA-box is a 32-bit keyed permutation mapping, MA:$(\mathbb{Z}_2^{16})^2 \rightarrow (\mathbb{Z}_2^{16})^2$, that consists of multiplication and addition operations interleaved and in a zigzag order. See Fig. 2.1.

The output of the MA-box is the ordered pair (r_i, s_i), with

$$s_i = ((Z_5^{(i)} \odot n_i) \boxplus q_i) \odot Z_6^{(i)}$$

and

$$r_i = s_i \boxplus (Z_5^{(i)} \odot n_i).$$

Note that in Fig. 2.1 the arrow representing the value $Z_5^{(i)} \odot n_i$ is duplicated in the computational graph of the MA-box, just like s_i.

Note that s_i depends on n_i and q_i, which depend on $(Y_1^{(i)}, Y_2^{(i)}, Y_3^{(i)}, Y_4^{(i)})$, and which ultimately depend on $(X_1^{(i)}, X_2^{(i)}, X_3^{(i)}, X_4^{(i)})$. Since r_i depends on s_i, both MA-box outputs depend on all words in $(X_1^{(i)}, X_2^{(i)}, X_3^{(i)}, X_4^{(i)})$ after only two layers of $\odot$ and $\boxplus$ in the MA-box.

Note that all $\odot$ operations in PES are keyed, that is, one of the operands in all $\odot$ operations is always a round subkey.

The next step is to combine r_i and s_i with the original inputs to the MA half-round, to generate $(Y_1^{(i)} \oplus s_i, Y_2^{(i)} \oplus r_i, Y_3^{(i)} \oplus s_i, Y_4^{(i)} \oplus r_i)$. Thus, r_i and s_i make each word in a block depend nonlinearly on each other word and therefore, complete (text) diffusion is achieved in a single round.

The last step is to swap the two leftmost words with the two rightmost words in a block. The round output becomes

$$X^{(i+1)} = (Y_3^{(i)} \oplus s_i, Y_4^{(i)} \oplus r_i, Y_1^{(i)} \oplus s_i, Y_2^{(i)} \oplus r_i),$$

which also becomes the input to the next round. This procedure is repeated eight times for PES.

The output transformation (OT) consists of a word swap followed by a KM half-round. The word swap is an involution, that is, it is its own inverse.

Therefore, the OT effectively cancels the word swap of the last full round. It is equivalent to simply removing the word swap in the eighth round:

$$X^{(9)} = (Y_1^{(8)} \oplus s_8, Y_2^{(8)} \oplus r_8, Y_3^{(8)} \oplus s_8, Y_4^{(8)} \oplus r_8).$$

The ciphertext becomes

$$C = (X_1^{(9)} \odot Z_1^{(9)}, X_2^{(9)} \odot Z_2^{(9)}, X_3^{(9)} \boxplus Z_3^{(9)}, X_4^{(9)} \boxplus Z_4^{(9)}).$$

The differences between encryption and decryption in PES are the order, value of subkeys and internal operations: (i) in i-th KM half-rounds, for $i \in \{1, 9\}$, (modular) addition is replaced by (modular) subtraction and (modular) multiplication is replaced by (modular) division; (ii) from $2 \leq i \leq 7$, the subkeys $(Z_1^{(i)}, Z_2^{(i)})$ in the KM half-rounds are replaced by $(-Z_3^{(i)}, -Z_4^{(i)})$, and vice-versa, $(Z_3^{(i)}, Z_4^{(i)})$ are replaced by $((Z_1^{(i)})^{-1}, (Z_2^{(i)})^{-1})$.

Consequently, Fig. 2.1(a) no single framework can represent both the encryption and the decryption operations, since they require more than just replacing the order and value of round subkeys (according to Table 2.2). It is also necessary to change the wordwise operations in the internal KM half-rounds $2 \leq i \leq 7$: $\odot$ shall be replace by $\boxminus$, and $\boxplus$ by $\boxdot$.

2.1.2 Key Schedule Algorithm

The key schedule algorithm (KSA) for iterated block ciphers is a deterministic procedure used to *expand* a fixed-length secret key into a longer keystream

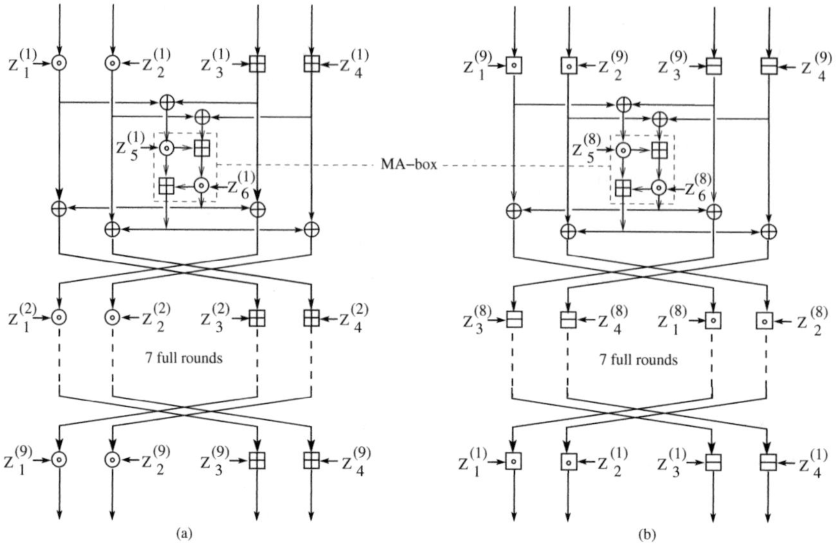

Fig. 2.1 Computational graph of the PES cipher: (a) encryption [37] and (b) decryption. Note the use of modular division (⊡) and subtraction (⊟) in decryption.

that is partitioned into (round) subkeys. The subkeys may be transformed in order to perform either the encryption or the decryption operations.

The key schedule of PES transforms a 128-bit user key $K = (k_0, \ldots, k_{127})$ into fifty-two 16-bit subkeys for either the encryption or the decryption transformations. In total, 832 bits of subkey material are needed across 8.5 rounds. The key schedule of PES operates as follows:

(i) the 128-bit user key K is stored in a register, which is initially partitioned into eight 16-bit words. These words become the first eight subkeys: $Z_1^{(1)}$, $Z_1^{(2)}$, $Z_1^{(3)}$, $Z_1^{(4)}$, $Z_1^{(5)}$, $Z_1^{(6)}$, $Z_2^{(1)}$ and $Z_2^{(2)}$.
(ii) the register is left rotated by 25 bits and is partitioned into eight 16-bit words, which become the next eight round subkeys.
(iii) step (ii) is repeated until 52 subkeys are obtained.

The key schedule of PES is a linear transformation in the sense that there is a linear relationship between key bits and subkey bits, and vice-versa. The user key bits are simply cut and paste into round subkeys. On one hand, it means that subkeys for both encryption and decryption can be very efficiently generated, that is, with the same latency. This subkey generation is fast compared to the time of one encryption. On the other hand, there is no confusion in the key schedule: every subkey bit is a linear function of a single user key bit. This fact also means that there is no diffusion in the key schedule since each subkey bit depends on a single bit of the 128-bit user key. Thus, several bit patterns in the key easily propagate to the round subkeys.

For instance, the key $K = 0^{128}$ (128 bits equal to 0) implies that all subkeys are equal to 0^{16}. Similarly, the key $K = 1^{128}$ implies that all subkeys are equal to 1^{16}.

These facts about the key schedule algorithm are widely exploited in several attacks against the full 8.5-round PES and IDEA ciphers [8, 16, 13, 23, 27, 31].

Curiously, this simple bit permutation in the key schedule of PES is similar to the bit permutation in the key schedule of DES [51], except that there is no key expansion nor parity bits in PES as in the DES.

This simple design of the key schedule of PES in effect helps speed up the subkey generation process, which not only speeds up both the encryption/decryption processes, but also reduces the storage for the subkeys.

It is straightforward to generate the subkeys in sync for encryption, but also to reconstruct the subkeys for decryption: let the bits of the register holding the 128-bit user key be numbered from 0 to 127 in left-to-right order. Table 2.1 shows which user key bits $K = (k_0, k_1, k_2, \ldots, k_{127})$ each subkey $Z_j^{(i)}$ depends on. Consequently, the overlap between bits of K and the subkeys becomes more evident. Note that due to the bitwise rotation in the key schedule, K can be interpreted as a circular buffer. So, the bit range $k_{114} \ldots k_1$, for instance, means the 16 bits from k_{114} up to k_{127} followed by k_0 and k_1, in this order.

Table 2.1 Dependency of subkey bits on the user key bits via the key schedule algorithm of PES.

i	$Z_1^{(i)}$	$Z_2^{(i)}$	$Z_3^{(i)}$	$Z_4^{(i)}$	$Z_5^{(i)}$	$Z_6^{(i)}$
1	$k_0, \ldots, k_{15}$	$k_{16}, \ldots, k_{31}$	$k_{32}, \ldots, k_{47}$	$k_{48}, \ldots, k_{63}$	$k_{64}, \ldots, k_{79}$	$k_{80}, \ldots, k_{95}$
2	$k_{96}, \ldots, k_{111}$	$k_{112}, \ldots, k_{127}$	$k_{25}, \ldots, k_{40}$	$k_{41}, \ldots, k_{56}$	$k_{57}, \ldots, k_{72}$	$k_{73}, \ldots, k_{88}$
3	$k_{89}, \ldots, k_{104}$	$k_{105}, \ldots, k_{120}$	$k_{121}, \ldots, k_8$	$k_9, \ldots, k_{24}$	$k_{50}, \ldots, k_{65}$	$k_{66}, \ldots, k_{81}$
4	$k_{82}, \ldots, k_{97}$	$k_{98}, \ldots, k_{113}$	$k_{114}, \ldots, k_1$	$k_2, \ldots, k_{17}$	$k_{18}, \ldots, k_{33}$	$k_{34}, \ldots, k_{49}$
5	$k_{75}, \ldots, k_{90}$	$k_{91}, \ldots, k_{106}$	$k_{107}, \ldots, k_{122}$	$k_{123}, \ldots, k_{10}$	$k_{11}, \ldots, k_{26}$	$k_{27}, \ldots, k_{42}$
6	$k_{43}, \ldots, k_{58}$	$k_{59}, \ldots, k_{74}$	$k_{100}, \ldots, k_{115}$	$k_{116}, \ldots, k_3$	$k_4, \ldots, k_{19}$	$k_{20}, \ldots, k_{35}$
7	$k_{36}, \ldots, k_{51}$	$k_{52}, \ldots, k_{67}$	$k_{68}, \ldots, k_{83}$	$k_{84}, \ldots, k_{99}$	$k_{125}, \ldots, k_{12}$	$k_{13}, \ldots, k_{28}$
8	$k_{29}, \ldots, k_{44}$	$k_{45}, \ldots, k_{60}$	$k_{61}, \ldots, k_{76}$	$k_{77}, \ldots, k_{92}$	$k_{93}, \ldots, k_{108}$	$k_{109}, \ldots, k_{124}$
OT	$k_{22}, \ldots, k_{37}$	$k_{38}, \ldots, k_{53}$	$k_{54}, \ldots, k_{69}$	$k_{70}, \ldots, k_{85}$	—	—

Following the description of Sect. 2.1.1, the encryption and decryption subkeys of the full 8.5-round PES can be listed as in Table 2.2. Multiplicative subkeys are depicted in boldface type.

Note that either the multiplicative inverse and additive inverse subkeys are used or the inverse operations $\boxdot$ and $\boxminus$ are used for decryption. As an example, subkeys in Table 2.2 for encryption (Fig. 2.1(a)) were simply permuted for decryption (Fig. 2.1(b)) contingent on $\odot$ and $\boxplus$ in Fig. 2.1(a) be replaced for $\boxdot$ and $\boxminus$ in Fig. 2.1(b), respectively.

Note also that the subkeys, in the order given in Fig. 2.1(a), *cannot* be used straight away in the framework of Fig. 2.1(b) because it takes more than just inverting the multiplicative and additive subkeys: the $\odot$ and $\boxplus$ operations in the inner KM half-rounds have to be changed as well.

Table 2.2 Encryption and decryption round subkeys for the PES cipher.

round	Encryption subkeys	Decryption subkeys
1	$\mathbf{Z}_1^{(1)}, \mathbf{Z}_2^{(1)}, Z_3^{(1)}, Z_4^{(1)}, \mathbf{Z}_5^{(1)}, \mathbf{Z}_6^{(1)}$	$\mathbf{Z}_1^{(9)}, \mathbf{Z}_2^{(9)}, Z_3^{(9)}, Z_4^{(9)}, \mathbf{Z}_5^{(8)}, \mathbf{Z}_6^{(8)}$
2	$\mathbf{Z}_1^{(2)}, \mathbf{Z}_2^{(2)}, Z_3^{(2)}, Z_4^{(2)}, \mathbf{Z}_5^{(2)}, \mathbf{Z}_6^{(2)}$	$Z_3^{(8)}, Z_4^{(8)}, \mathbf{Z}_1^{(8)}, \mathbf{Z}_2^{(8)}, \mathbf{Z}_5^{(7)}, \mathbf{Z}_6^{(7)}$
3	$\mathbf{Z}_1^{(3)}, \mathbf{Z}_2^{(3)}, Z_3^{(3)}, Z_4^{(3)}, \mathbf{Z}_5^{(3)}, \mathbf{Z}_6^{(3)}$	$Z_3^{(7)}, Z_4^{(7)}, \mathbf{Z}_1^{(7)}, \mathbf{Z}_2^{(7)}, \mathbf{Z}_5^{(6)}, \mathbf{Z}_6^{(6)}$
4	$\mathbf{Z}_1^{(4)}, \mathbf{Z}_2^{(4)}, Z_3^{(4)}, Z_4^{(4)}, \mathbf{Z}_5^{(4)}, \mathbf{Z}_6^{(4)}$	$Z_3^{(6)}, Z_4^{(6)}, \mathbf{Z}_1^{(6)}, \mathbf{Z}_2^{(6)}, \mathbf{Z}_5^{(5)}, \mathbf{Z}_6^{(5)}$
5	$\mathbf{Z}_1^{(5)}, \mathbf{Z}_2^{(5)}, Z_3^{(5)}, Z_4^{(5)}, \mathbf{Z}_5^{(5)}, \mathbf{Z}_6^{(5)}$	$Z_3^{(5)}, Z_4^{(5)}, \mathbf{Z}_1^{(5)}, \mathbf{Z}_2^{(5)}, \mathbf{Z}_5^{(4)}, \mathbf{Z}_6^{(4)}$
6	$\mathbf{Z}_1^{(6)}, \mathbf{Z}_2^{(6)}, Z_3^{(6)}, Z_4^{(6)}, \mathbf{Z}_5^{(6)}, \mathbf{Z}_6^{(6)}$	$Z_3^{(4)}, Z_4^{(4)}, \mathbf{Z}_1^{(4)}, \mathbf{Z}_2^{(4)}, \mathbf{Z}_5^{(3)}, \mathbf{Z}_6^{(3)}$
7	$\mathbf{Z}_1^{(7)}, \mathbf{Z}_2^{(7)}, Z_3^{(7)}, Z_4^{(7)}, \mathbf{Z}_5^{(7)}, \mathbf{Z}_6^{(7)}$	$Z_3^{(3)}, Z_4^{(3)}, \mathbf{Z}_1^{(3)}, \mathbf{Z}_2^{(3)}, \mathbf{Z}_5^{(2)}, \mathbf{Z}_6^{(2)}$
8	$\mathbf{Z}_1^{(8)}, \mathbf{Z}_2^{(8)}, Z_3^{(8)}, Z_4^{(8)}, \mathbf{Z}_5^{(8)}, \mathbf{Z}_6^{(8)}$	$Z_3^{(2)}, Z_4^{(2)}, \mathbf{Z}_1^{(2)}, \mathbf{Z}_2^{(2)}, \mathbf{Z}_5^{(1)}, \mathbf{Z}_6^{(1)}$
9	$\mathbf{Z}_1^{(9)}, \mathbf{Z}_2^{(9)}, Z_3^{(9)}, Z_4^{(9)}$	$\mathbf{Z}_1^{(1)}, \mathbf{Z}_2^{(1)}, Z_3^{(1)}, Z_4^{(1)}$

2.2 The IDEA Block Cipher

The International Data Encryption Algorithm (IDEA) block cipher operates on 64-bit text blocks, under a 128-bit key and iterates eight rounds plus an output transformation (OT). IDEA was designed by X. Lai, J.L. Massey and S. Murphy [36, 38] and dates back to 1991. IDEA was formerly known as IPES (Improved PES).

The main design differences between PES and IDEA are: (i) the order of $\odot$ and $\boxplus$ in the KM half-rounds, and (ii) the word swap at the end of each MA half-round. Otherwise, the key schedule, the word size, the number of rounds, the key size and the group operations are exactly the same for both ciphers. Still, IDEA was designed to better resist differential cryptanalysis compared to PES [38]. It is interesting to note how these two changes in the PES cipher framework have had such an impact on the security of IDEA concerning the differential cryptanalysis technique [7].

In [12], Borisov *et al.* observed that a weaker version of IDEA called IDEA-X, in which all $\boxplus$ were replaced by $\oplus$ is susceptible to a variant of Differential Cryptanalysis (DC) called multiplicative differentials. They concluded that the use of at least three incompatible group operations is the minimum necessary.

Further analysis by Raddum in [53] showed that replacing only half of all $\boxplus$ by $\oplus$ still leaves IDEA vulnerable multiplicative differentials. Raddum called this variant IDEA-X/2. His research demonstrated that the design of IDEA is tight concerning this variant of differential cryptanalysis.

2.2.1 Encryption and Decryption Frameworks

Fig. 2.2(a) shows a pictorial view of the first full IDEA encryption round, part of the second round and the output transformation. The direction of arrows indicates the data flow. Fig. 2.2(b) shows the corresponding decryption scheme. Note the change of order of subkeys in the KM half-round for decryption, compared to the encryption framework, due to the word swap at the end of each full round.

IDEA is a word-oriented cipher design. Although IDEA operates on fixed-size words, the word size w is not explicit in Fig. 2.2. Valid word sizes are $w \in \{2, 4, 8, 16\}$ bits and Fig. 2.2 applies to all of them. The official word size for IDEA is $w = 16$ bits. The other sizes correspond to mini-cipher versions of IDEA.

Modular division in GF$(2^{16}+1)$ is denoted by $\boxdot$. For instance, $a \boxdot b = a/b = a \odot b^{-1}$. It is a compact notation to denote the multiplicative inverse for decryption.

One full round in IDEA can be split into a Key-Mixing (KM) and a Multiplication-Addition (MA) half-round, in this order. This decomposition of a full round is useful not only to describe the cipher itself, which consists of 8.5 rounds, but also to describe attack details in the next chapter.

The i-th 16-bit subkey of the j-th round is denoted by $Z_j^{(i)}$, with $1 \leq j \leq 6$ and $1 \leq i \leq 9$. Subkey generation by the key schedule algorithm of IDEA is described in Sect. 2.2.2.

Let $X^{(i)} = (X_1^{(i)}, X_2^{(i)}, X_3^{(i)}, X_4^{(i)})$ denote the 64-bit text input block to the i-th round, where $X_j^{(i)} \in \mathbb{Z}_2^{16}$, for $1 \leq j \leq 4$. Therefore, $X^{(1)}$ represents a plaintext block.

The output of the i-th KM half-round is denoted $Y^{(i)} = (Y_1^{(i)}, Y_2^{(i)}, Y_3^{(i)}, Y_4^{(i)})$ where

$$Y^{(i)} = (X_1^{(i)} \odot Z_1^{(i)}, X_2^{(i)} \boxplus Z_2^{(i)}, X_3^{(i)} \boxplus Z_3^{(i)}, X_4^{(i)} \odot Z_4^{(i)}), \qquad (2.2)$$

which becomes the input to the MA half-round. Note how (2.2) differ from (2.1) in PES.

The input to the i-th MA half-round consists of the ordered pair: $(n_i, q_i) = (Y_1^{(i)} \oplus Y_3^{(i)}, Y_2^{(i)} \oplus Y_4^{(i)})$. The MA-box is a 32-bit keyed-permutation that consists of multiplications and additions interleaved and in a sequential order. The output of the MA-box is the ordered pair (r_i, s_i):

$$s_i = ((Z_5^{(i)} \odot n_i) \boxplus q_i) \odot Z_6^{(i)}$$

and

$$r_i = s_i \boxplus (Z_5^{(i)} \odot n_i).$$

Note that s_i depends on both n_i and q_i, and thus, indirectly depends on all words in $(X_1^{(i)}, X_2^{(i)}, X_3^{(i)}, X_4^{(i)})$. Since r_i depends on s_i, all outputs from the MA-box depend on $(X_1^{(i)}, X_2^{(i)}, X_3^{(i)}, X_4^{(i)})$.

The next step is to combine r_i and s_i with the original inputs to the MA half-round, to generate $(Y_1^{(i)} \oplus s_i,\ Y_2^{(i)} \oplus r_i,\ Y_3^{(i)} \oplus s_i,\ Y_4^{(i)} \oplus r_i)$. Due to the dependence of r_i and s_i on all input words to the i-th round, all round output words also depend on each other. Thus, complete diffusion is achieved in a single round.

The last step is to swap the two middle words in a block. The swap function can be denoted $\sigma(A, B, C, D) = (A, C, B, D)$, and it is an involution: $\sigma \circ \sigma(A, B, C, D) = (A, B, C, D)$.

The round output becomes

$$X^{(i+1)} = (Y_1^{(i)} \oplus s_i, Y_3^{(i)} \oplus s_i, Y_2^{(i)} \oplus r_i, Y_4^{(i)} \oplus r_i),$$

which becomes the input to the next round. This procedure is repeated eight times for IDEA.

Note that all $\odot$ operations in IDEA are keyed, that is, one operand in all $\odot$ operations is always a round subkey.

Also, note that while the wordwise operations in a KM half-round can be performed in parallel, the operations in an MA half-round have to be performed sequentially.

The output transformation consists of σ followed by a KM half-round. The presence of σ in the output transformation effectively cancels the σ in the eighth round. It is equivalent to simply removing σ from the last full round.

$$X^{(9)} = (Y_1^{(8)} \oplus s_8, Y_2^{(8)} \oplus r_8, Y_3^{(8)} \oplus s_8, Y_4^{(8)} \oplus r_8).$$

The ciphertext becomes

$$C = (X_1^{(9)} \odot Z_1^{(9)}, X_2^{(9)} \boxplus Z_2^{(9)}, X_3^{(9)} \boxplus Z_3^{(9)}, X_4^{(9)} \odot Z_4^{(9)}).$$

Let (A, B, C, D) denote the input to an MA half-round. In IDEA, σ exchanges only the words B and C. It means that the inputs to the MA-box, $(A \oplus C,\ B \oplus D)$, in consecutive rounds contains words from different positions in a block. It is an interesting feature to improve diffusion across rounds. Comparatively, in PES, the word swap exchanged the pair (A, B) with (C, D), but the input values $(A \oplus C, B \oplus D)$ to the MA-box remain the same after the swap: $(C \oplus A, D \oplus B)$. It means that the word swap in PES did not perform as good diffusion as in IDEA.

Note that there are two invariants in the MA half-round (independent of the subkeys):

$$n_i = Y_1^{(i)} \oplus Y_3^{(i)} = X_1^{(i+1)} \oplus X_2^{(i)}, \tag{2.3}$$

and

$$q_i = Y_2^{(i)} \oplus Y_4^{(i)} = X_3^{(i+1)} \oplus X_4^{(i+1)}. \tag{2.4}$$

The invariant (2.3) is due to the fact that $Y_1^{(i)}$ and $Y_3^{(i)}$ are exclusive-ored to the same value s_i. Likewise, in (2.4), $Y_2^{(i)}$ and $Y_4^{(i)}$ are both exclusive-ored to r_i. These design features allow the MA half-round to be an involution, but also leads to a nonrandom property across the MA half-round. These invariants do not extend to the KM half-round because of the subkey mixing via $\odot$ and $\boxplus$. For instance, $X_1^{(i)} \oplus X_3^{(i)} \neq Y_1^{(i)} \oplus Y_3^{(i)}$. This simple fact helps justifies the presence of the KM half-round.

If X denotes a 64-bit text block, let KM_i denote the i-th KM half-round, MA_i denote the i-th MA half-round and σ denote the word swap. Then, a full round in IDEA can be denoted by the composition

$$\rho_i(X) = \sigma \circ \text{MA}_i \circ \text{KM}_i(X) = \sigma(\text{MA}_i(\text{KM}_i(X))).$$

One full IDEA encryption of a 64-bit plaintext block P into a 64-bit ciphertext block C can be denoted

$$C = \text{KM}_9 \circ \sigma \circ \rho_8 \circ \rho_7 \circ \rho_6 \circ \rho_5 \circ \rho_4 \circ \rho_3 \circ \rho_2 \circ \rho_1(P)$$

where functional composition operates in right-to-left order.

The main differences between encryption and decryption in IDEA are in the KM half-rounds where (modular) addition becomes (modular) subtraction and (modular) multiplication is replaced by (modular) division. Note that both σ and $\sigma \circ \text{MA}_i$ are involutory transformation.

The encryption and decryption frameworks of IDEA are very similar (turning one framework upside-down gives the other), except for the order and value of round subkeys. Consequently, Fig. 2.2(a) can represent both the encryption and the decryption frameworks, by just appropriately changing the order and value of round subkeys according to Table 2.3. That is the reason why only Fig. 2.2(a) is depicted in IDEA descriptions.

2.2.2 Key Schedule Algorithm

The key schedule of IDEA processes a 128-bit user key $K = (k_0, k_1, \ldots, k_{127})$ and generates fifty-two 16-bit subkeys for 8.5 rounds of encryption or decryption transformations. The successive steps of the key schedule algorithm of IDEA are exactly the same as those used by PES in Sect. 2.1.2.

Note that the multiplicative and additive inverse subkeys are needed for decryption. Apart from that fact, the order and value of subkeys for decryp-

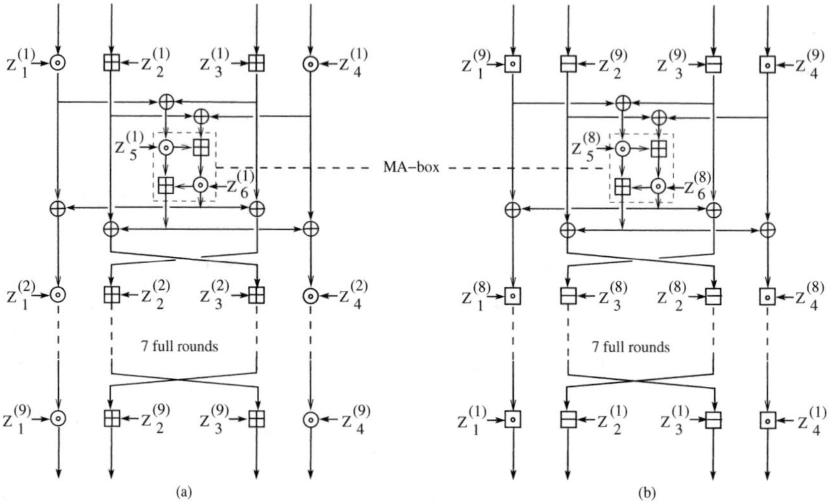

Fig. 2.2 Computational graph of the IDEA cipher: (a) encryption [38] and (b) decryption. Note the use of modular division (⊡) and subtraction (⊟) in decryption.

tion are also different compared to the encryption framework. This means that essentially the same general framework can be used for both schemes.

Table 2.2.1 shows the subkeys for encryption and decryption in IDEA for each round. Multiplicative subkeys are depicted in boldface type.

Note that the subkeys listed in Table 2.3 are permuted from one framework (Fig. 2.2(a)) to the other (Fig. 2.2(b)), contingent on the ⊙ and ⊞ operations be replaced by the ⊡ and ⊟ operations, respectively.

Table 2.3 Encryption and decryption round subkeys for the IDEA cipher.

round	Encryption subkeys	Decryption subkeys
1	$\mathbf{Z_1^{(1)}}, Z_2^{(1)}, Z_3^{(1)}, \mathbf{Z_4^{(1)}}, \mathbf{Z_5^{(1)}}, \mathbf{Z_6^{(1)}}$	$\mathbf{Z_1^{(9)}}, Z_2^{(9)}, Z_3^{(9)}, \mathbf{Z_4^{(9)}}, \mathbf{Z_5^{(8)}}, \mathbf{Z_6^{(8)}}$
2	$\mathbf{Z_1^{(2)}}, Z_2^{(2)}, Z_3^{(2)}, \mathbf{Z_4^{(2)}}, \mathbf{Z_5^{(2)}}, \mathbf{Z_6^{(2)}}$	$\mathbf{Z_1^{(8)}}, Z_3^{(8)}, Z_2^{(8)}, \mathbf{Z_4^{(8)}}, \mathbf{Z_5^{(7)}}, \mathbf{Z_6^{(7)}}$
3	$\mathbf{Z_1^{(3)}}, Z_2^{(3)}, Z_3^{(3)}, \mathbf{Z_4^{(3)}}, \mathbf{Z_5^{(3)}}, \mathbf{Z_6^{(3)}}$	$\mathbf{Z_1^{(7)}}, Z_3^{(7)}, Z_2^{(7)}, \mathbf{Z_4^{(7)}}, \mathbf{Z_5^{(6)}}, \mathbf{Z_6^{(6)}}$
4	$\mathbf{Z_1^{(4)}}, Z_2^{(4)}, Z_3^{(4)}, \mathbf{Z_4^{(4)}}, \mathbf{Z_5^{(4)}}, \mathbf{Z_6^{(4)}}$	$\mathbf{Z_1^{(6)}}, Z_3^{(6)}, Z_2^{(6)}, \mathbf{Z_4^{(6)}}, \mathbf{Z_5^{(5)}}, \mathbf{Z_6^{(5)}}$
5	$\mathbf{Z_1^{(5)}}, Z_2^{(5)}, Z_3^{(5)}, \mathbf{Z_4^{(5)}}, \mathbf{Z_5^{(5)}}, \mathbf{Z_6^{(5)}}$	$\mathbf{Z_1^{(5)}}, Z_3^{(5)}, Z_2^{(5)}, \mathbf{Z_4^{(5)}}, \mathbf{Z_5^{(4)}}, \mathbf{Z_6^{(4)}}$
6	$\mathbf{Z_1^{(6)}}, Z_2^{(6)}, Z_3^{(6)}, \mathbf{Z_4^{(6)}}, \mathbf{Z_5^{(6)}}, \mathbf{Z_6^{(6)}}$	$\mathbf{Z_1^{(4)}}, Z_3^{(4)}, Z_2^{(4)}, \mathbf{Z_4^{(4)}}, \mathbf{Z_5^{(3)}}, \mathbf{Z_6^{(3)}}$
7	$\mathbf{Z_1^{(7)}}, Z_2^{(7)}, Z_3^{(7)}, \mathbf{Z_4^{(7)}}, \mathbf{Z_5^{(7)}}, \mathbf{Z_6^{(7)}}$	$\mathbf{Z_1^{(3)}}, Z_3^{(3)}, Z_2^{(3)}, \mathbf{Z_4^{(3)}}, \mathbf{Z_5^{(2)}}, \mathbf{Z_6^{(2)}}$
8	$\mathbf{Z_1^{(8)}}, Z_2^{(8)}, Z_3^{(8)}, \mathbf{Z_4^{(8)}}, \mathbf{Z_5^{(8)}}, \mathbf{Z_6^{(8)}}$	$\mathbf{Z_1^{(2)}}, Z_3^{(2)}, Z_2^{(2)}, \mathbf{Z_4^{(2)}}, \mathbf{Z_5^{(1)}}, \mathbf{Z_6^{(1)}}$
9	$\mathbf{Z_1^{(9)}}, Z_2^{(9)}, Z_3^{(9)}, \mathbf{Z_4^{(9)}}$	$\mathbf{Z_1^{(1)}}, Z_2^{(1)}, Z_3^{(1)}, \mathbf{Z_4^{(1)}}$

2.3 The MESH Block Ciphers

The main motivation for the design of the MESH ciphers was the desire to enlarge the block size of IDEA beyond 64 bits. Although it is straightforward to extend the bitwise exclusive-or ($\oplus$) and modular addition ($\boxplus$) operations to larger word sizes, the same is not true for modular multiplication ($\odot$).

Note, for instance, that

$$2^{2^5} + 1 = 2^{32} + 1 = 641 * 6700417$$

and

$$2^{2^6} + 1 = 2^{64} + 1 = 274177 * 67280421310721$$

and

$$2^{2^7} + 1 = 2^{128} + 1 = 59649589127497217 * 5704689200685129054721$$

are composite numbers [22], unlike $2^{2^4} + 1 = 65537$ which is prime. These facts mean that $\mathbb{Z}_{2^{2^5}+1}$, $\mathbb{Z}_{2^{2^6}+1}$ and $\mathbb{Z}_{2^{2^7}+1}$ are not finite fields, but only (commutative) rings. See the Appendix, Sect. A for more details on rings.

Thus, since $2^{16} + 1$ is the largest known Fermat prime, there was no way to obtain larger Galois Fields using Fermat primes, and thus larger block sizes by simply increasing the word size. Nonetheless, the design of MESH ciphers actually enlarged the MA-box but preserved the word size to 16 bits, just like in IDEA. This design change only enlarged the MA-box, and that was enough to arrive at larger block sizes as will become clear in Sect. 2.3.1. Moreover, by preserving the same three group operations, the MESH ciphers will profit from the existing infrastructure already in place due to IDEA and PES ciphers.

Note that MESH-64 has the same block size as IDEA but a longer MA-box. This design decision was aimed at making the MA-box stronger against (both known and unknown) attacks.

The MESH cipher family was initially comprised of three members: MESH-64, MESH-96 and MESH-128 [50], where the suffix denotes the block size in bits. In order to preserve the similarity of encryption and decryption in a single framework, the number of rounds had to be an even number. It was decided to set a minimum of 8, 10 and 12 rounds to these ciphers, respectively.

MESH is not an acronym like IDEA or PES, but rather, stands for the interwoven (mesh) pattern of $\odot$, $\boxplus$ and $\oplus$ along the cipher.

A new design criterion in the MESH ciphers was due to the Biryukov-Demirci relation in IDEA: if one tracks the value of the input and output of *only the two middle 16-bit words in a block*, the results are equations (3.273) and (3.274) in Sect. 3.17. See also Fig. 3.27. These relations mean that the two middle words in the computation graph of IDEA only use $\boxplus$ and $\oplus$ operations. The Biryukov-Demirci relation exploits this design feature in IDEA to extract a linear relation involving only the least significant bit value of these two middle words. This relation holds with certainty.

In the MESH ciphers, all 16-bit words in a block in every round (following the computation graphs) involve $\odot$, $\boxplus$ and $\oplus$, which avoids the Biryukov-Demirci attacks (Sect. 3.17). Curiously, the PES cipher had all three group operations in every word along its computational graph.

The key size for the MESH ciphers is always double the block size. It is an attractive feature to

- counter exhaustive key search for the foreseeable future [20].
- counter chosen-key attacks [61, 3], which is an attack exploiting the birthday-paradox effect on the key input.
- counter time-memory trade-off (TMTO) attacks [24, 14].

Concerning the number of rounds:

- the standard approach to determine the minimum number of rounds is to find the largest number of rounds that can be attacked by known techniques, and add a margin of security on top of it. Some of these attack techniques will be described in Chap. 3.
- the concept of pseudorandom permutation (PRP) was formally described by Luby and Rackoff in [40]. See Sect. 1.7. PRPs refer to permutation mappings (over large enough domains) that cannot be distinguished from a uniformly-distributed random permutation in polynomial time. PRPs are often used to describe idealized abstractions of block ciphers which play important roles in symmetric-key cryptography, such as in proofs of security. In these abstractions, though, some components are ignored, such as the key schedule algorithm.
  The following papers give some lower bound on the number of rounds of Lai-Massey designs in general: in [59, 41], it was proved that *three rounds* are not only sufficient but also necessary for the pseudorandomness of Lai-Massey schemes, that is, security against chosen-plaintext attacks. Further, they showed that *four rounds* are not only sufficient but also necessary for the strong pseudorandomness of the Lai-Massey schemes, that is, security against chosen-plaintext-and-chosen-ciphertext attacks within the birthday-paradox bound.
  In [65], Yun *et al.* provided another proofs for Lai-Massey schemes based on the coefficient-H technique in [52] and exploited the relationship between the Lai-Massey scheme and the quasi-Feistel networks. But, this later result is still within the birthday-paradox bound.

Taking these analyses into account, it was decided to set a minimum of 8, 10 and 12 rounds to MESH-64, MESH-96 and MESH-128, respectively.

Another issue is the MA-box design. In MESH-64, MESH-96 and MESH-128 the MA-boxes have at least three interleaved layers of $\odot$ and $\boxplus$ following a zigzag pattern, in comparison with two layers in PES and in IDEA. These longer MA-boxes were designed to better resist differential, linear and future (unknown) attacks. Further, the MA-boxes of MESH-96 and MESH-128 have the novel feature that not all $\odot$ operations involve a subkey directly as an operand, but rather, depend upon the cipher state. This very important operation is called *unkeyed multiplication*.

Nonetheless, all these new MA-boxes are bijective mappings for any subkey value. The bijective property was a criterion meant to avoid nonsurjective (among other) attacks [55].

The new key schedule algorithms of the MESH ciphers were motivated by the following facts:

- in IDEA, all multiplication operations involved a subkey as operand. Since the $\odot$ is the main nonlinear operator, the new key schedule algorithms were designed to avoid weak multiplicative subkeys for any choice of the user key and for as many rounds as possible. Otherwise, all $\odot$ (along a trail) could, in principle, be manipulated by carefully selected weak user keys.
- fast key avalanche: each subkey generated from the user key K quickly depends upon all words of K. More precisely, this dependence is nonlinear and is expressed by the exponents of a primitive polynomial (one polynomial for each MESH cipher). All key schedule algorithms interleave $\boxplus$, $\oplus$ and fixed bitwise rotation. The reasoning for the bit rotation is that both $\boxplus$ and $\oplus$ preserve the relative bit position of their inputs. Without the bit rotation, two related keys, with subkeys differing only in the most significant bit position could propagate this difference across several other subkeys.
- use of fixed constants to avoid patterns in the user key to propagate to the subkeys: for instance, without the constants, the all-zero user key would result in all subkeys to be zero, for any number of rounds, independent of the nonlinear $\boxplus$ operation or the bit rotation.

2.3.1 Encryption and Decryption Frameworks of MESH-64

MESH-64 [50] is an iterated block cipher operating on 64-bit blocks, under a 128-bit key and consisting of eight rounds plus an output transformation (OT). All internal operations are on 16-bit words.

Fig. 2.3 shows a pictorial representation of the encryption and the decryption frameworks of MESH-64.

Let $X^{(i)} = (X_1^{(i)}, X_2^{(i)}, X_3^{(i)}, X_4^{(i)})$ denote the 64-bit input text block to the i-th round, where $X_j^{(i)} \in \mathbb{Z}_2^{16}$, for $1 \leq j \leq 4$ and $1 \leq i \leq 9$.

One full encryption round in MESH-64 can be decomposed into a Key-Mixing (KM) and a Multiplication-Addition (MA) half-round, in this order.

There are two types of KM half-rounds: for even- and odd-numbered rounds. A KM half-round uses four subkeys while an MA half-round uses three subkeys.

The j-th subkey of the i-th round is denoted by $Z_j^{(i)}$, with $1 \leq j \leq 6$ and $1 \leq i \leq 9$. Subkey generation by the key schedule algorithm is discussed in Sect. 2.3.2.

Given the input $X^{(i)}$ to the i-th round, it is also the input to the i-th KM half-round. The output of odd-numbered KM half-rounds is

$$Y^{(i)} = (X_1^{(i)} \odot Z_1^{(i)}, X_2^{(i)} \boxplus Z_2^{(i)}, X_3^{(i)} \boxplus Z_3^{(i)}, X_4^{(i)} \odot Z_4^{(i)}).$$

The output of even-numbered KM half-rounds is

$$Y^{(i)} = (X_1^{(i)} \boxplus Z_1^{(i)}, X_2^{(i)} \odot Z_2^{(i)}, X_3^{(i)} \odot Z_3^{(i)}, X_4^{(i)} \boxplus Z_4^{(i)}).$$

The output of the i-th KM half-round, $Y^{(i)} = (Y_1^{(i)}, Y_2^{(i)}, Y_3^{(i)}, Y_4^{(i)})$, becomes the input to the i-th MA half-round.

The input to the i-th MA half-round is the ordered pair (n_i, q_i), where $n_i = Y_1^{(i)} \oplus Y_3^{(i)}$ and $q_i = Y_2^{(i)} \oplus Y_4^{(i)}$. The MA-box is a 32-bit keyed permutation consisting of three layers of $\odot$ and $\boxplus$ operations interleaved and in a zigzag order. The output of the i-th MA-box of MESH-64 is the ordered pair (r_i, s_i), with

$$r_i = ((((Z_5^{(i)} \odot n_i) \boxplus q_i) \odot Z_6^{(i)}) \boxplus (Z_5^{(i)} \odot n_i)) \odot Z_7^{(i)}$$

and

$$s_i = r_i \boxplus (((Z_5^{(i)} \odot n_i) \boxplus q_i) \odot Z_6^{(i)}).$$

Note that r_i depends on n_i and q_i, both of which depend on ($X_1^{(i)}$, $X_2^{(i)}$, $X_3^{(i)}$, $X_4^{(i)}$). Likewise, s_i depends on r_i, which also depends on all word in ($X_1^{(i)}$, $X_2^{(i)}$, $X_3^{(i)}$, $X_4^{(i)}$). Thus, both output words from the i-th MA-box depend on all input words to the i-th round.

The next step is to combine r_i and s_i with the original inputs to the MA half-round, to generate

$$(Y_1^{(i)} \oplus s_i, Y_2^{(i)} \oplus r_i, Y_3^{(i)} \oplus s_i, Y_4^{(i)} \oplus r_i).$$

Therefore, complete text diffusion is achieved in a single round.

The last step in a full round is to swap the two middle words in a block. This operation is denoted σ and it performs a fixed word permutation: $\sigma(A, B, C, D) = (A, C, B, D)$.

The round output becomes

$$X^{(i+1)} = \sigma(Y_1^{(i)} \oplus s_i, Y_2^{(i)} \oplus r_i, Y_3^{(i)} \oplus s_i, Y_4^{(i)} \oplus r_i) =$$

$$(Y_1^{(i)} \oplus s_i, Y_3^{(i)} \oplus s_i, Y_2^{(i)} \oplus r_i, Y_4^{(i)} \oplus r_i)$$

which becomes the input to the next round. This procedure is repeated eight times for MESH-64.

The output transformation consists of σ followed by an (odd) KM half-round. The word swap in the output transformation cancels the word swap of the eighth round. It is equivalent to simply removing σ from the last full round:

$$X^{(9)} = (Y_1^{(8)} \oplus s_8, Y_2^{(8)} \oplus r_8, Y_3^{(8)} \oplus s_8, Y_4^{(8)} \oplus r_8).$$

The ciphertext becomes

$$C = (X_1^{(9)} \odot Z_1^{(9)}, X_2^{(9)} \boxplus Z_2^{(9)}, X_3^{(9)} \boxplus Z_3^{(9)}, X_4^{(9)} \odot Z_4^{(9)}).$$

If X denotes a 64-bit block, KM_i denotes the i-th KM half-round, MA_i denotes the i-th MA half-round, then, a full round in MESH-64 can be denoted by the composition

$$\tau_i(X) = \sigma \circ \text{MA}_i \circ \text{KM}_i(X) = \sigma(\text{MA}_i(\text{KM}_i(X))).$$

One full MESH-64 encryption of a 64-bit plaintext block P into a 64-bit ciphertext block C can be denoted

$$C = \text{KM}_9 \circ \sigma \circ \tau_8 \circ \tau_7 \circ \tau_6 \circ \tau_5 \circ \tau_4 \circ \tau_3 \circ \tau_2 \circ \tau_1(P)$$

where functional composition operates in right-to-left order.

The main differences between encryption and decryption in MESH-64 are in the KM half-rounds where (modular) addition becomes (modular) subtraction and (modular) multiplication is replaced by (modular) division.

Note that σ and $\sigma \circ \text{MA}_i$ are involutory transformations.

Moreover, note that the order of two middle round subkeys, $Z_2^{(i)}$ and $Z_3^{(i)}$, changes for decryption in Fig. 2.3(b) compared to Fig. 2.3(a), except for $i = 1$ and $i = 9$. This happens because when Fig. 2.3(a) is turned upside-down, σ changes the order of $Z_2^{(i)}$ and $Z_3^{(i)}$, except for the first and the last KM half-rounds.

Consequently, Fig. 2.3(a) alone could represent both the encryption and the decryption frameworks, by just appropriately changing the order and value of subkeys (according to Table 2.5). Therefore, Fig. 2.3(b) is redundant. Nonetheless, we depict both transformations side-by-side to demonstrate their similarity which shows the encryption framework can be adapted to support decryption as well.

The main differences between MESH-64 and IDEA are:

- the MA-box of MESH-64 contains three layers of multiplication and addition interleaved, in a zigzag order, compared to two layers in IDEA.
- odd- and even-numbered rounds in MESH-64 contain a different pattern of $\odot$ and $\boxplus$, to avoid linear trails across consecutive rounds [49]. If one follows any single thread in the computational graph of MESH-64 in Fig. 2.3 across consecutive rounds, one will notice that the operators will

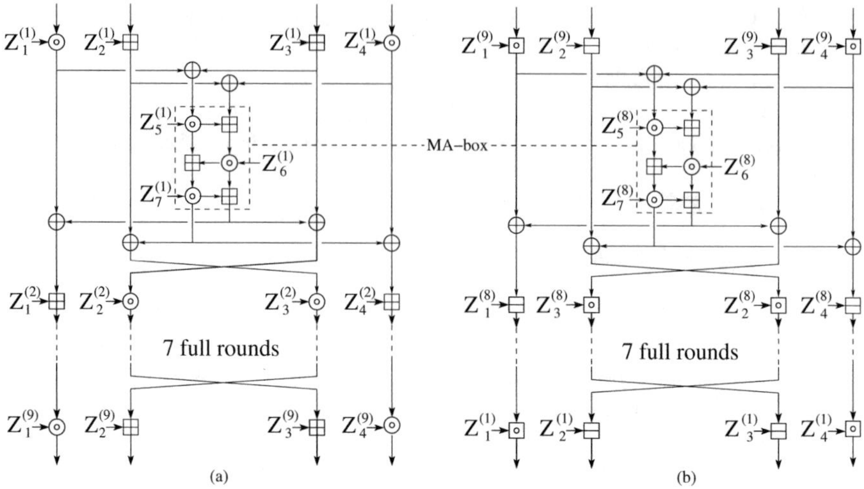

Fig. 2.3 Computational graph of the MESH-64 cipher: (a) encryption [50] and (b) decryption. Note the use of modular division (⊡) and subtraction (⊟) in decryption.

never repeat twice in a row. But, equally important, there is no linear trail between them either. In IDEA, though, the two middle words are only combined via ⊕ and ⊞ across the whole cipher. It means that there is a linear trail involving only the least significant bit of the second plaintext word and the second ciphertext word. These bits are not affected by carry bits, and thus, this linear trail holds with certainty. An attack has been described using such linear trails in [49]. To counter these trails in all MESH ciphers, odd- and even-numbered rounds alternate ⊙ and ⊞. It does not affect the similarity of encryption and decryption frameworks as long as the total number of rounds is even.

- a strengthened key schedule algorithm with fast (key) avalanche and without overlapping bits (as in IDEA) between the user key and the round subkeys.

When MESH-64 was designed, it was believed that a longer MA-box would make it harder to attack each round, since it would require more weak-key conditions to traverse the MA-box either by differential or by linear cryptanalysis. Moreover, the fact that the MA-boxes were longer also implied that there would be more subkeys per round. The later was not a problem: key diffusion was fast enough, further strengthening MESH ciphers against divide-and-conquer techniques such as meet-in-the-middle attacks [11, 19].

In hindsight, a longer MA-box also helped make the MESH ciphers more resilient against unknown attacks at the time the MESH ciphers were designed, such as [6, 64].

2.3.2 Key Schedule Algorithm of MESH-64

In total, MESH-64 requires sixty 16-bit subkeys (960 subkey bits in total) for either the encryption or the decryption of a single 64-bit text block.

The key schedule of MESH-64 uses a nonlinear feedback shift register (NLFSR) design. It works as follows:

- 16-bit constants denoted c_i are iteratively defined as $c_0 = 1$ and $c_i = 3 \cdot c_{i-1}$ for $i \geq 1$. The multiplication operation in $\mathrm{GF}(2)[x]/p(x)$ is denoted simply by $\cdot$ with $p(x)$ a primitive polynomial over $\mathrm{GF}(2)[x]$: $p(x) = x^{16} + x^5 + x^3 + x^2 + 1$. The constant 3 in the construction of c_i represents the polynomial $x + 1$ in $\mathrm{GF}(2)[x]$. See Table 2.4.
- the 128-bit user key $K = (k_0, k_1, k_2, \ldots, k_{127})$ is partitioned into eight 16-bit words K_i for $0 \leq i \leq 7$ and assigned to the subkey $Z_{j+1}^{(1)} = K_j \oplus c_j$ for $0 \leq j \leq 6$ and $Z_1^{(2)} = K_7 \oplus c_7$.
- each subsequent 16-bit subkey is generated iteratively as follows[1]:

$$Z_{l(i)}^{(h(i))} = (((((Z_{l(i-8)}^{(h(i-8))} \boxplus Z_{l(i-7)}^{(h(i-7))}) \oplus Z_{l(i-6)}^{(h(i-6))}) \boxplus Z_{l(i-3)}^{(h(i-3))}) \oplus Z_{l(i-2)}^{(h(i-2))}) \boxplus Z_{l(i-1)}^{(h(i-1))}) \lll 7 \oplus c_i, \quad (2.5)$$

with $8 \leq i \leq 59$; $x \lll 7$ means x is left rotated by 7 bits; $h(i) = \lfloor i/7 \rfloor + 1$ and $l(i) = i \bmod 7 + 1$.

This key schedule algorithm achieves fast key diffusion through the use of the primitive polynomial $q(x) = x^8 + x^7 + x^6 + x^5 + x^2 + x + 1$ in $h(i)$ and $l(i)$, the bitwise rotation and the interleaving of $\oplus$ and $\boxplus$.

To demonstrate the speed of key diffusion, it can be deduced that the subkey $Z_4^{(2)}$ is the first one that depends upon all eight 16-bit words of K. The first eight subkeys depend on a single key word: $Z_1^{(1)} = K_0 \oplus c_0$, $Z_2^{(1)} = K_1 \oplus c_1$, $Z_3^{(1)} = K_2 \oplus c_2$, $Z_4^{(1)} = K_3 \oplus c_3$, $Z_5^{(1)} = K_4 \oplus c_4$, $Z_6^{(1)} = K_5 \oplus c_5$, $Z_7^{(1)} = K_6 \oplus c_6$ and $Z_1^{(2)} = K_7 \oplus c_7$.

Further, for $i = 8$:

$$Z_2^{(2)} = (((((Z_1^{(1)} \boxplus Z_2^{(1)}) \oplus Z_3^{(1)}) \boxplus Z_6^{(1)}) \oplus Z_7^{(1)}) \boxplus Z_1^{(2)}) \lll 7 \oplus c_8,$$

which depends on six words from K.

Next, for $i = 9$:

$$Z_3^{(2)} = (((((Z_2^{(1)} \boxplus Z_3^{(1)}) \oplus Z_4^{(1)}) \boxplus Z_7^{(1)}) \oplus Z_1^{(2)}) \boxplus Z_2^{(2)}) \lll 7 \oplus c_9,$$

which depends on seven words from K, due to the chained dependence on $Z_2^{(2)}$.

Next, for $i = 10$:

[1] Bit rotation has higher precedence than $\oplus$.

Table 2.4 16-bit constants used in the key schedule algorithm of MESH-64 [44].

i	c_{8i}	c_{8i+1}	c_{8i+2}	c_{8i+3}	c_{8i+4}	c_{8i+5}	c_{8i+6}	c_{8i+7}
0	0001_x	0003_x	0005_x	$000f_x$	0011_x	0033_x	0055_x	$00ff_x$
1	0101_x	0303_x	0505_x	$0f0f_x$	1111_x	3333_x	5555_x	$ffff_x$
2	$002c_x$	0074_x	$009c_x$	$01a4_x$	$02ec_x$	0734_x	$095c_x$	$1be4_x$
3	$2c2c_x$	7474_x	$9c9c_x$	$a589_x$	$eeb6_x$	$33f7_x$	5419_x	$fc2b_x$
4	0450_x	$0cf0_x$	1510_x	$3f30_x$	4150_x	$c3f0_x$	$443d_x$	$cc47_x$
5	$54e4_x$	$fd2c_x$	0759_x	$09eb_x$	$1a3d_x$	$2e47_x$	$72c9_x$	$975b_x$
6	$b9c0_x$	$ca6d_x$	$5e9a_x$	$e3ae_x$	$24df_x$	$6d61_x$	$b7a3_x$	$d8c8_x$
7	6975_x	$bb9f_x$	$cc8c_x$	$55b9_x$	$fecb_x$	0370_x	0590_x	$0eb0_x$
8	$13d0_x$	3470_x	$5c90_x$	$e5b0_x$	$2efd_x$	7307_x	9509_x	$bf36_x$
9	$c177_x$	$43b4_x$	$c4dc_x$	$4d49_x$	$d7db_x$	7840_x	$88c0_x$	$996d_x$
10	$ab9a_x$	$fc83_x$	$05a8_x$	$0ef8_x$	1308_x	3518_x	$5f28_x$	$e178_x$
11	$23a5_x$	$64ef_x$	$ad31_x$	$f77e_x$	$19af_x$	$2af1_x$	$7f13_x$	8135_x
12	8372_x	$85bb_x$	$8ee0_x$	$930d_x$	$b53a_x$	$df63_x$	6188_x	$a298_x$
13	$e785_x$	$28a2_x$	$79e6_x$	$8a2a_x$	$9e53_x$	$a2d8_x$	$e745_x$	$29e2_x$
14	$7a26_x$	$8e6a_x$	9293_x	$b798_x$	$d885_x$	$69a2_x$	$bae6_x$	$cf07_x$
15	5124_x	$f36c_x$	1599_x	$3eab_x$	$43fd_x$	$c407_x$	$4c24_x$	$d46c_x$
16	$7c99_x$	$85ab_x$	$8ed0_x$	$935d_x$	$b5ca_x$	$de73_x$	$62b8_x$	$a7c8_x$
17	$e875_x$	$38b2_x$	$49d6_x$	$da7a_x$	$6ea3_x$	$b3e5_x$	$d402_x$	$7c2b_x$
18	$847d_x$	$8caa_x$	$95d3_x$	$be58_x$	$c2c5_x$	4762_x	$c9a6_x$	$5ac7_x$

$$Z_4^{(2)} = (((((Z_3^{(1)} \boxplus Z_4^{(1)}) \oplus Z_5^{(1)}) \boxplus Z_1^{(2)}) \oplus Z_2^{(2)}) \boxplus Z_3^{(2)}) \lll 7 \oplus c_{10},$$

which finally depends on all eight words from K due to the chained dependence on $Z_2^{(2)}$ and $Z_3^{(2)}$. Therefore, in two rounds, key diffusion is complete in MESH-64.

Since both $\oplus$ and $\boxplus$ preserve the relative bit positions of their operands, it was decided to employ a fixed-bit rotation at the end of (2.5). In this way, differences in the most significant bit positions (in a related-key differential setting) that remain intact by $\oplus$ and $\boxplus$, are effectively displaced by the bit rotation.

Finally, the constants c_i aim to avoid bit patterns in the key K to propagate to the subkeys. Otherwise, this key schedule would suffer the same drawbacks as the key schedule of PES and IDEA. For instance, without the c_i constants, the key $K = 0^{128}$ (128 bits equal to zero) would result in all subkeys to be equal to 0^{16}. Likewise, without the c_i, the key $K = 1^{128}$ (128 bits equal to 1) would result in all subkeys equal to 1^{16}.

Table 2.5 lists the encryption and decryption subkeys for each round of MESH-64. Multiplicative subkeys are depicted in boldface type. Note that the order of operations $\odot$ and $\boxplus$ changes in the KM half-rounds for even- and odd-numbered rounds. Moreover, the order of $Z_2^{(i)}$, $Z_3^{(i)}$ change for decryption for $2 \leq i \leq 7$ due to the word swap operation.

Note that subkeys are listed permuted in Table 2.5 for encryption and decryption. But, they cannot be used straight away in both frameworks because the $\odot$ and $\boxplus$ operations have to be replace for $\boxdot$ and $\boxminus$, respectively.

Table 2.5 Encryption and decryption round subkeys of MESH-64 cipher.

round	Encryption subkeys
1	$\mathbf{Z}_1^{(1)}, Z_2^{(1)}, Z_3^{(1)}, \mathbf{Z}_4^{(1)}, \mathbf{Z}_5^{(1)}, \mathbf{Z}_6^{(1)}, \mathbf{Z}_7^{(1)}$
2	$Z_1^{(2)}, \mathbf{Z}_2^{(2)}, \mathbf{Z}_3^{(2)}, Z_4^{(2)}, \mathbf{Z}_5^{(2)}, \mathbf{Z}_6^{(2)}, \mathbf{Z}_7^{(2)}$
3	$\mathbf{Z}_1^{(3)}, Z_2^{(3)}, Z_3^{(3)}, \mathbf{Z}_4^{(3)}, \mathbf{Z}_5^{(3)}, \mathbf{Z}_6^{(3)}, \mathbf{Z}_7^{(3)}$
4	$Z_1^{(4)}, \mathbf{Z}_2^{(4)}, \mathbf{Z}_3^{(4)}, Z_4^{(4)}, \mathbf{Z}_5^{(4)}, \mathbf{Z}_6^{(4)}, \mathbf{Z}_7^{(4)}$
5	$\mathbf{Z}_1^{(5)}, Z_2^{(5)}, Z_3^{(5)}, \mathbf{Z}_4^{(5)}, \mathbf{Z}_5^{(5)}, \mathbf{Z}_6^{(5)}, \mathbf{Z}_7^{(5)}$
6	$Z_1^{(6)}, \mathbf{Z}_2^{(6)}, \mathbf{Z}_3^{(6)}, Z_4^{(6)}, \mathbf{Z}_5^{(6)}, \mathbf{Z}_6^{(6)}, \mathbf{Z}_7^{(6)}$
7	$\mathbf{Z}_1^{(7)}, Z_2^{(7)}, Z_3^{(7)}, \mathbf{Z}_4^{(7)}, \mathbf{Z}_5^{(7)}, \mathbf{Z}_6^{(7)}, \mathbf{Z}_7^{(7)}$
8	$Z_1^{(8)}, \mathbf{Z}_2^{(8)}, \mathbf{Z}_3^{(8)}, Z_4^{(8)}, \mathbf{Z}_5^{(8)}, \mathbf{Z}_6^{(8)}, \mathbf{Z}_7^{(8)}$
9	$\mathbf{Z}_1^{(9)}, Z_2^{(9)}, Z_3^{(9)}, \mathbf{Z}_4^{(9)}$

round	Decryption subkeys
1	$\mathbf{Z}_1^{(9)}, Z_2^{(9)}, Z_3^{(9)}, \mathbf{Z}_4^{(9)}, \mathbf{Z}_5^{(8)}, \mathbf{Z}_6^{(8)}, \mathbf{Z}_7^{(8)}$
2	$Z_1^{(8)}, \mathbf{Z}_3^{(8)}, \mathbf{Z}_2^{(8)}, Z_4^{(8)}, \mathbf{Z}_5^{(7)}, \mathbf{Z}_6^{(7)}, \mathbf{Z}_7^{(7)}$
3	$\mathbf{Z}_1^{(7)}, Z_3^{(7)}, Z_2^{(7)}, \mathbf{Z}_4^{(7)}, \mathbf{Z}_5^{(6)}, \mathbf{Z}_6^{(6)}, \mathbf{Z}_7^{(6)}$
4	$Z_1^{(6)}, \mathbf{Z}_3^{(6)}, \mathbf{Z}_2^{(6)}, Z_4^{(6)}, \mathbf{Z}_5^{(5)}, \mathbf{Z}_6^{(5)}, \mathbf{Z}_7^{(5)}$
5	$\mathbf{Z}_1^{(5)}, Z_3^{(5)}, Z_2^{(5)}, \mathbf{Z}_4^{(5)}, \mathbf{Z}_5^{(4)}, \mathbf{Z}_6^{(4)}, \mathbf{Z}_7^{(4)}$
6	$Z_1^{(4)}, \mathbf{Z}_3^{(4)}, \mathbf{Z}_2^{(4)}, Z_4^{(4)}, \mathbf{Z}_5^{(3)}, \mathbf{Z}_6^{(3)}, \mathbf{Z}_7^{(3)}$
7	$\mathbf{Z}_1^{(3)}, Z_3^{(3)}, Z_2^{(3)}, \mathbf{Z}_4^{(3)}, \mathbf{Z}_5^{(2)}, \mathbf{Z}_6^{(2)}, \mathbf{Z}_7^{(2)}$
8	$Z_1^{(2)}, \mathbf{Z}_3^{(2)}, \mathbf{Z}_2^{(2)}, Z_4^{(2)}, \mathbf{Z}_5^{(1)}, \mathbf{Z}_6^{(1)}, \mathbf{Z}_7^{(1)}$
9	$\mathbf{Z}_1^{(1)}, Z_2^{(1)}, Z_3^{(1)}, \mathbf{Z}_4^{(1)}$

Note that the key schedule algorithm of MESH-64 is invertible: the NLSFR can be clocked backwards to generate subkeys for decryption as easily as for encryption. It is only necessary to store the last eight subkeys (128 bits) instead of K. Let us call $K' = (Z_4^{(8)}, Z_5^{(8)}, Z_6^{(8)}, Z_7^{(8)}, Z_1^{(9)}, Z_2^{(9)}, Z_3^{(9)}, Z_4^{(9)})$.

The last generated subkey for encryption was

$$Z_4^{(9)} = (((((Z_3^{(8)} \boxplus Z_4^{(8)}) \oplus Z_5^{(8)}) \boxplus Z_1^{(9)}) \oplus Z_2^{(9)}) \boxplus Z_3^{(9)}) \lll 7 \oplus c_{59}.$$

We can isolate $Z_3^{(8)}$ from this equation as:

$$Z_3^{(8)} = ((((((Z_4^{(9)} \oplus c_{59}) \ggg 7) \boxminus Z_3^{(9)}) \oplus Z_2^{(9)}) \boxminus Z_1^{(9)}) \oplus Z_5^{(8)}) \boxminus Z_4^{(8)}, \tag{2.6}$$

where $\ggg$ means bitwise rotation to the right.

If we compute and store K' then all terms in the right-hand side of (2.6) are known. This way, the subkeys can be computed in reverse order for decryption. We do not need to always generate the subkeys starting from $Z_1^{(1)}$ and store them, as is usual for most block ciphers.

The last generated subkey (in reverse order) is derived from the equation:

$$Z_2^{(2)} = (((((Z_1^{(1)} \boxplus Z_2^{(1)}) \oplus Z_3^{(1)}) \boxplus Z_6^{(1)}) \oplus Z_7^{(1)}) \boxplus Z_1^{(2)}) \lll 7 \oplus c_8.$$

We just isolate $Z_1^{(1)}$ from it:

$$Z_1^{(1)} = ((((((Z_2^{(2)} \oplus c_8) \ggg 7) \boxminus Z_1^{(2)}) \oplus Z_7^{(1)}) \boxminus Z_6^{(1)}) \oplus Z_3^{(1)}) \boxminus Z_2^{(1)}.$$

In the end, if decryption operations are frequently needed (depending on the mode of operation in use), we only need to compute K' once and store it along with K.

2.3.3 Encryption and Decryption Frameworks of MESH-96

MESH-96 [50] is an iterated block cipher operating on 96-bit blocks, under a 192-bit key and consisting of ten rounds plus an output transformation. All internal operations are on 16-bit words. This is the only MESH variant whose block size is not a power of 2. This is an unusual block size, but not a novelty. There are a few other ciphers operating on 96-bit blocks such as BKSQ [17] and 3-WAY [15].

Fig. 2.4 shows a pictorial representation of the encryption and the decryption frameworks of MESH-96.

Let $X^{(i)} = (X_1^{(i)}, X_2^{(i)}, X_3^{(i)}, X_4^{(i)}, X_5^{(i)}, X_6^{(i)})$ denote the 96-bit input text block to the i-th round, where $X_j^{(i)} \in \mathbb{Z}_2^{16}$, for $1 \leq j \leq 6$ and $1 \leq i \leq 11$.

One full encryption round in MESH-96 can be decomposed into a Key-Mixing (KM) and a Multiplication-Addition (MA) half-round, in this order. There are two types of KM half-rounds: for even- and odd-numbered rounds. A KM half-round uses six subkeys while an MA half-round uses three subkeys.

The j-th 16-bit subkey of the i-th round is denoted by $Z_j^{(i)}$, with $1 \leq j \leq 9$ and $1 \leq i \leq 11$. Subkey generation by the key schedule algorithm is detailed in Sect. 2.3.4.

Given the 96-bit input $X^{(i)}$ to the i-th round, it is also the input to the i-th KM half-round. The output of an odd-numbered KM half-round is

$$Y^{(i)} = (X_1^{(i)} \odot Z_1^{(i)}, X_2^{(i)} \boxplus Z_2^{(i)}, X_3^{(i)} \odot Z_3^{(i)}, X_4^{(i)} \boxplus Z_4^{(i)}, X_5^{(i)} \odot Z_5^{(i)}, X_6^{(i)} \boxplus Z_6^{(i)}).$$

The output of an even-numbered KM half-round is

$$Y^{(i)} = (X_1^{(i)} \boxplus Z_1^{(i)}, X_2^{(i)} \odot Z_2^{(i)}, X_3^{(i)} \boxplus Z_3^{(i)}, X_4^{(i)} \odot Z_4^{(i)}, X_5^{(i)} \boxplus Z_5^{(i)}, X_6^{(i)} \odot Z_6^{(i)}).$$

The output of the i-th KM half-round becomes the input to the i-th MA half-round.

The input to the i-th MA half-round consists of the ordered triple (n_i, q_i, m_i), where

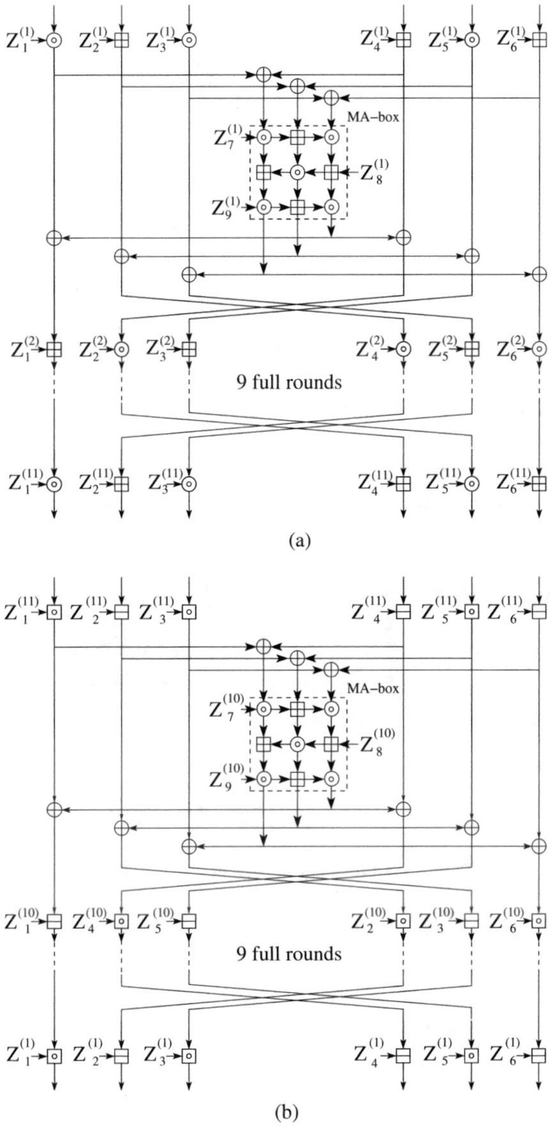

Fig. 2.4 Computational graph of the MESH-96 cipher: (a) encryption [50] and (b) decryption. Note the use of modular division (⊡) and subtraction (⊟) in decryption.

$$n_i = Y_1^{(i)} \oplus Y_4^{(i)},$$

$$q_i = Y_2^{(i)} \oplus Y_5^{(i)}$$

and

$$m_i = Y_3^{(i)} \oplus Y_6^{(i)}.$$

The MA-box of MESH-96 is a 48-bit keyed-permutation that consists of three layers of $\odot$ and $\boxplus$ interleaved and in a zigzag order. The output of the i-th MA-box is the ordered triple (r_i, s_i, t_i), with

$$r_i = ((((((Z_7^{(i)} \odot n_i) \boxplus q_i) \odot m_i) \boxplus Z_8^{(i)}) \odot ((Z_7^{(i)} \odot n_i) \boxplus q_i)) \boxplus (Z_7^{(i)} \odot n_i)) \odot Z_9^{(i)},$$

$$s_i = r_i \boxplus (((((Z_7^{(i)} \odot n_i) \boxplus q_i) \odot m_i) \boxplus Z_8^{(i)}) \odot ((Z_7^{(i)} \odot n_i) \boxplus q_i))$$

and

$$t_i = s_i \odot ((((Z_7^{(i)} \odot n_i) \boxplus q_i) \odot m_i) \boxplus Z_8^{(i)}).$$

An important feature of the MA-box of MESH-96 is that subkeys are not directly input to all $\odot$ operations. These operations are called unkeyed multiplications, and were also used in RIDEA [63, 64]. This is an important design feature of the MESH ciphers that is aimed at strengthening them by making the MA-box less vulnerable to weak subkeys.

The next step is to combine (r_i, s_i, t_i) with the original inputs to the MA half-round:

$$T^{(i)} = (Y_1^{(i)} \oplus t_i, Y_2^{(i)} \oplus s_i, Y_3^{(i)} \oplus r_i, Y_4^{(i)} \oplus t_i, Y_5^{(i)} \oplus s_i, Y_6^{(i)} \oplus r_i).$$

The last step in a full round is to swap the four middle words in a block. For a 96-bit block (A, B, C, D, E, F) the swap operation, denoted σ, performs a fixed word permutation: $\sigma(A, B, C, D, E, F) = (A,\ D,\ E,\ B,\ C,\ F)$.

The i-th full-round output becomes

$$X^{(i+1)} = \sigma(T^{(i)}) = (Y_1^{(i)} \oplus t_i, Y_4^{(i)} \oplus t_i, Y_5^{(i)} \oplus s_i, Y_2^{(i)} \oplus s_i, Y_3^{(i)} \oplus r_i, Y_6^{(i)} \oplus r_i),$$

which becomes the input to the next round. This procedure is repeated ten times for MESH-96.

The output transformation consists of an application of σ followed by an odd-numbered KM half-round. The word swap in the output transformation cancels the word swap of the last full round. It is equivalent to simply removing σ from the tenth round.

$$X^{(11)} = (Y_1^{(10)} \oplus t_{10}, Y_2^{(10)} \oplus s_{10}, Y_3^{(10)} \oplus r_{10}, Y_4^{(10)} \oplus t_{10}, Y_5^{(10)} \oplus s_{10}, Y_6^{(10)} \oplus r_{10}).$$

The ciphertext becomes

$$C = (X_1^{(11)} \odot Z_1^{(11)}, X_2^{(11)} \boxplus Z_2^{(11)}, X_3^{(11)} \odot Z_3^{(11)}, X_4^{(11)} \boxplus Z_4^{(11)}, X_5^{(11)} \odot Z_5^{(11)}, X_6^{(11)} \boxplus Z_6^{(11)}).$$

If X denotes a 96-bit block, KM_i denotes the i-th KM half-round, MA_i denotes the i-th MA half-round, then, a full round can be denoted by

$$\psi_i(X) = \sigma \circ \text{MA}_i \circ \text{KM}_i(X) = \sigma(\text{MA}_i(\text{KM}_i(X))).$$

One full MESH-96 encryption of a 96-bit plaintext block P into a 96-bit ciphertext block C can be denoted

$$C = \mathrm{KM}_{11} \circ \sigma \circ \psi_{10} \circ \psi_9 \circ \psi_8 \circ \psi_7 \circ \psi_6 \circ \psi_5 \circ \psi_4 \circ \psi_3 \circ \psi_2 \circ \psi_1(P)$$

where functional composition operates in right-to-left order.

The main differences between encryption and decryption in MESH-96 are in the KM half-rounds where operations need to be inverted: (modular) addition becomes (modular) subtraction and (modular) multiplication becomes (modular) division.

Note that σ and $\sigma \circ \mathrm{MA}_i$ are involutory transformations, that is, they are their own inverses.

Note that the order of the four middle round subkeys in the KM half-rounds: $Z_2^{(i)}$, $Z_3^{(i)}$, $Z_4^{(i)}$ and $Z_5^{(i)}$, change for decryption in Fig. 2.4(b) (except for $i = 1$ and $i = 11$). This change happens because when Fig. 2.4(a) is turned upside-down, σ changes the order of $(Z_2^{(i)}, Z_3^{(i)})$ and $(Z_4^{(i)}, Z_5^{(i)})$ except for the first and the last KM half-rounds.

Consequently, Fig. 2.4(a) can represent both the encryption and the decryption frameworks, by just appropriately changing the order and value of subkeys (according to Table 2.6).

Although Fig. 2.4(b) is redundant, we depict both operations side-by-side to demonstrate their similarity and to emphasize the different KM half-rounds (which is important to take into account when properly implementing the encryption and decryption algorithms).

2.3.4 Key Schedule Algorithm of MESH-96

In total, MESH-96 requires ninety-six 16-bit subkeys (1536 subkey bits in total) for either encryption or decryption of a single 96-bit text block.

The key schedule of MESH-96 uses a nonlinear feedback shift register (NLFSR) design. An initial set up is the following:

- define iteratively 16-bit constants, denoted c_i, like for MESH-64, as follows: $c_0 = 1$ and $c_i = 3 \cdot c_{i-1}$ for $i \geq 1$. See Table 2.4. Multiplication in $\mathrm{GF}(2)[x]/p(x)$ is denoted simply by $\cdot$ with $p(x)$ a primitive polynomial over $\mathrm{GF}(2)[x]$: $p(x) = x^{16} + x^5 + x^3 + x^2 + 1$. The constant 3 in the construction of c_i represents the polynomial $x + 1$ in $\mathrm{GF}(2)[x]$.
- the 192-bit key $K = (k_0, k_1, k_2, \ldots, k_{191})$ is partitioned into twelve 16-bit words K_i, for $0 \leq i \leq 11$ and assigned to the subkeys $Z_{j+1}^{(1)} = K_j \oplus c_j$ for $0 \leq j \leq 8$, and $Z_{t \bmod 9+1}^{(2)} = K_t \oplus c_t$ for $9 \leq t \leq 11$.
- each subsequent 16-bit subkey is constructed iteratively as follows (note that bit rotation has higher precedence than $\oplus$):

$$Z_{l(i)}^{(h(i))} = (((((Z_{l(i-12)}^{(h(i-12))} \boxplus Z_{l(i-8)}^{(h(i-8))}) \oplus Z_{l(i-6)}^{(h(i-6))}) \boxplus Z_{l(i-4)}^{(h(i-4))}) \oplus Z_{l(i-2)}^{(h(i-2))}) \boxplus Z_{l(i-1)}^{(h(i-1))}) \lll 9 \oplus c_i, \quad (2.7)$$

with $12 \leq i \leq 95$; $x \lll 9$ means x left rotated by 9 bits; $h(i) = \lfloor i/9 \rfloor + 1$ and $l(i) = i \bmod 9 + 1$.

This key schedule was designed to achieve fast key diffusion through the use of the primitive polynomial $r(x) = x^{12} + x^{11} + x^{10} + x^8 + x^6 + x^4 + 1$ in $l(i)$ and $h(i)$, the bitwise rotation and the interleaving of $\oplus$ and $\boxplus$.

Let us deduce that all round subkeys starting from $Z_7^{(2)}$ already depend nonlinearly upon all the twelve key words. Initially, $Z_1^{(1)} = K_0 \oplus c_0$, $Z_2^{(1)} = K_1 \oplus c_1$, $Z_3^{(1)} = K_2 \oplus c_2$, $Z_4^{(1)} = K_3 \oplus c_3$, $Z_5^{(1)} = K_4 \oplus c_4$, $Z_6^{(1)} = K_5 \oplus c_5$, $Z_7^{(1)} = K_6 \oplus c_6$, $Z_8^{(1)} = K_7 \oplus c_7$, $Z_9^{(1)} = K_8 \oplus c_8$, $Z_1^{(2)} = K_9 \oplus c_9$, $Z_2^{(2)} = K_{10} \oplus c_{10}$ and $Z_3^{(2)} = K_{11} \oplus c_{11}$.

Further, for $i = 12$:

$$Z_4^{(2)} = (((((Z_1^{(1)} \boxplus Z_5^{(1)}) \oplus Z_7^{(1)}) \boxplus Z_9^{(1)}) \oplus Z_2^{(2)}) \boxplus Z_3^{(2)}) \lll 9 \oplus c_{12},$$

which depends on six words from K.

For $i = 13$:

$$Z_5^{(2)} = (((((Z_2^{(1)} \boxplus Z_6^{(1)}) \oplus Z_8^{(1)}) \boxplus Z_1^{(2)}) \oplus Z_3^{(2)}) \boxplus Z_4^{(2)}) \lll 9 \oplus c_{13},$$

which depends on eleven words from K, due to the chained dependence on $Z_4^{(2)}$.

For $i = 14$:

$$Z_6^{(2)} = (((((Z_3^{(1)} \boxplus Z_7^{(1)}) \oplus Z_9^{(1)}) \boxplus Z_2^{(2)}) \oplus Z_4^{(2)}) \boxplus Z_5^{(2)}) \lll 9 \oplus c_{14},$$

which still depends on eleven words from K, due to the chained dependence on $Z_5^{(2)}$.

Finally, for $i = 15$:

$$Z_7^{(2)} = (((((Z_4^{(1)} \boxplus Z_8^{(1)}) \oplus Z_1^{(2)}) \boxplus Z_3^{(2)}) \oplus Z_5^{(2)}) \boxplus Z_6^{(2)}) \lll 9 \oplus c_{14},$$

which depends on all twelve words from K. Therefore, in two rounds, key diffusion is complete in MESH-96.

Since both $\oplus$ and $\boxplus$ preserve the relative bit positions of their operands, it was decided to employ a fixed-bit rotation at the end of (2.7). In this way, differences in the most significant bit positions (in a related-key differential setting) that remain intact by $\oplus$ and $\boxplus$, are effectively displaced by the bit rotation.

Finally, the constants c_i aim to avoid bit patterns in K to propagate to the subkeys, otherwise, this key schedule would suffer some of the same drawbacks as the key schedule of PES and IDEA. For instance, without the

c_i constants, the key $K = 0^{192}$ would result in all subkeys equal to zero. Similarly, the user key $K = 1^{192}$ would result in all subkeys equal to 1^{16}.

Table 2.6 lists the encryption and decryption subkeys for each round of MESH-96. Note that the order of operations $\odot$ and $\boxplus$ changes in the KM half-rounds for even- and odd-numbered rounds. Multiplicative subkeys are depicted in boldface type.

Table 2.6 Encryption and decryption round subkeys for MESH-96 cipher.

round	Encryption subkeys
1	$\mathbf{Z_1^{(1)}}, Z_2^{(1)}, \mathbf{Z_3^{(1)}}, Z_4^{(1)}, \mathbf{Z_5^{(1)}}, Z_6^{(1)}, \mathbf{Z_7^{(1)}}, Z_8^{(1)}, \mathbf{Z_9^{(1)}}$
2	$Z_1^{(2)}, \mathbf{Z_2^{(2)}}, Z_3^{(2)}, \mathbf{Z_4^{(2)}}, Z_5^{(2)}, \mathbf{Z_6^{(2)}}, \mathbf{Z_7^{(2)}}, Z_8^{(2)}, \mathbf{Z_9^{(2)}}$
3	$\mathbf{Z_1^{(3)}}, Z_2^{(3)}, \mathbf{Z_3^{(3)}}, Z_4^{(3)}, \mathbf{Z_5^{(3)}}, Z_6^{(3)}, \mathbf{Z_7^{(3)}}, Z_8^{(3)}, \mathbf{Z_9^{(3)}}$
4	$Z_1^{(4)}, \mathbf{Z_2^{(4)}}, Z_3^{(4)}, \mathbf{Z_4^{(4)}}, Z_5^{(4)}, \mathbf{Z_6^{(4)}}, \mathbf{Z_7^{(4)}}, Z_8^{(4)}, \mathbf{Z_9^{(4)}}$
5	$\mathbf{Z_1^{(5)}}, Z_2^{(5)}, \mathbf{Z_3^{(5)}}, Z_4^{(5)}, \mathbf{Z_5^{(5)}}, Z_6^{(5)}, \mathbf{Z_7^{(5)}}, Z_8^{(5)}, \mathbf{Z_9^{(5)}}$
6	$Z_1^{(6)}, \mathbf{Z_2^{(6)}}, Z_3^{(6)}, \mathbf{Z_4^{(6)}}, Z_5^{(6)}, \mathbf{Z_6^{(6)}}, \mathbf{Z_7^{(6)}}, Z_8^{(6)}, \mathbf{Z_9^{(6)}}$
7	$\mathbf{Z_1^{(7)}}, Z_2^{(7)}, \mathbf{Z_3^{(7)}}, Z_4^{(7)}, \mathbf{Z_5^{(7)}}, Z_6^{(7)}, \mathbf{Z_7^{(7)}}, Z_8^{(7)}, \mathbf{Z_9^{(7)}}$
8	$Z_1^{(8)}, \mathbf{Z_2^{(8)}}, Z_3^{(8)}, \mathbf{Z_4^{(8)}}, Z_5^{(8)}, \mathbf{Z_6^{(8)}}, \mathbf{Z_7^{(8)}}, Z_8^{(8)}, \mathbf{Z_9^{(8)}}$
9	$\mathbf{Z_1^{(9)}}, Z_2^{(9)}, \mathbf{Z_3^{(9)}}, Z_4^{(9)}, \mathbf{Z_5^{(9)}}, Z_6^{(9)}, \mathbf{Z_7^{(9)}}, Z_8^{(9)}, \mathbf{Z_9^{(9)}}$
10	$Z_1^{(10)}, \mathbf{Z_2^{(10)}}, Z_3^{(10)}, \mathbf{Z_4^{(10)}}, Z_5^{(10)}, \mathbf{Z_6^{(10)}}, \mathbf{Z_7^{(10)}}, Z_8^{(10)}, \mathbf{Z_9^{(10)}}$
11	$\mathbf{Z_1^{(11)}}, Z_2^{(11)}, \mathbf{Z_3^{(11)}}, Z_4^{(11)}, \mathbf{Z_5^{(11)}}, Z_6^{(11)}$

round	Decryption subkeys
1	$\mathbf{Z_1^{(11)}}, Z_2^{(11)}, \mathbf{Z_3^{(11)}}, Z_4^{(11)}, \mathbf{Z_5^{(11)}}, Z_6^{(11)}, \mathbf{Z_7^{(10)}}, Z_8^{(10)}, \mathbf{Z_9^{(10)}}$
2	$Z_1^{(10)}, \mathbf{Z_4^{(10)}}, Z_5^{(10)}, \mathbf{Z_2^{(10)}}, Z_3^{(10)}, \mathbf{Z_6^{(10)}}, \mathbf{Z_7^{(9)}}, Z_8^{(9)}, \mathbf{Z_9^{(9)}}$
3	$\mathbf{Z_1^{(9)}}, Z_4^{(9)}, \mathbf{Z_5^{(9)}}, Z_2^{(9)}, \mathbf{Z_3^{(9)}}, Z_6^{(9)}, \mathbf{Z_7^{(8)}}, Z_8^{(8)}, \mathbf{Z_9^{(8)}}$
4	$Z_1^{(8)}, \mathbf{Z_4^{(8)}}, Z_5^{(8)}, \mathbf{Z_2^{(8)}}, Z_3^{(8)}, \mathbf{Z_6^{(8)}}, \mathbf{Z_7^{(7)}}, Z_8^{(7)}, \mathbf{Z_9^{(7)}}$
5	$\mathbf{Z_1^{(7)}}, Z_4^{(7)}, \mathbf{Z_5^{(7)}}, Z_2^{(7)}, \mathbf{Z_3^{(7)}}, Z_6^{(7)}, \mathbf{Z_7^{(6)}}, Z_8^{(6)}, \mathbf{Z_9^{(6)}}$
6	$Z_1^{(6)}, \mathbf{Z_4^{(6)}}, Z_5^{(6)}, \mathbf{Z_2^{(6)}}, Z_3^{(6)}, \mathbf{Z_6^{(6)}}, \mathbf{Z_7^{(5)}}, Z_8^{(5)}, \mathbf{Z_9^{(5)}}$
7	$\mathbf{Z_1^{(5)}}, Z_4^{(5)}, \mathbf{Z_5^{(5)}}, Z_2^{(5)}, \mathbf{Z_3^{(5)}}, Z_6^{(5)}, \mathbf{Z_7^{(4)}}, Z_8^{(4)}, \mathbf{Z_9^{(4)}}$
8	$Z_1^{(4)}, \mathbf{Z_4^{(4)}}, Z_5^{(4)}, \mathbf{Z_2^{(4)}}, Z_3^{(4)}, \mathbf{Z_6^{(4)}}, \mathbf{Z_7^{(3)}}, Z_8^{(3)}, \mathbf{Z_9^{(3)}}$
9	$\mathbf{Z_1^{(3)}}, Z_4^{(3)}, \mathbf{Z_5^{(3)}}, Z_2^{(3)}, \mathbf{Z_3^{(3)}}, Z_6^{(3)}, \mathbf{Z_7^{(2)}}, Z_8^{(2)}, \mathbf{Z_9^{(2)}}$
10	$Z_1^{(2)}, \mathbf{Z_4^{(2)}}, Z_5^{(2)}, \mathbf{Z_2^{(2)}}, Z_3^{(2)}, \mathbf{Z_6^{(2)}}, \mathbf{Z_7^{(1)}}, Z_8^{(1)}, \mathbf{Z_9^{(1)}}$
11	$\mathbf{Z_1^{(1)}}, Z_2^{(1)}, \mathbf{Z_3^{(1)}}, Z_4^{(1)}, \mathbf{Z_5^{(1)}}, Z_6^{(1)}$

Note that the key schedule of MESH-96 is invertible: the NLSFR can be clocked backwards to generate subkeys for decryption as easily as for encryption. It is only necessary to store the last twelve subkeys (192 bits) instead of K. Let us denote the last subkeys as $K' = (Z_4^{(10)}, Z_5^{(10)}, Z_6^{(10)}, Z_7^{(10)}, Z_8^{(10)}, Z_9^{(10)}, Z_1^{(1)}, Z_2^{(11)}, Z_3^{(11)}, Z_4^{(11)}, Z_5^{(11)}, Z_6^{(11)})$.

The last generated subkey for encryption was

$$Z_6^{(11)} = (((((Z_3^{(10)} \boxplus Z_7^{(10)}) \oplus Z_9^{(10)}) \boxplus Z_2^{(10)}) \oplus Z_4^{(11)}) \boxplus Z_5^{(11)}) \lll 9 \oplus c_{95}. \tag{2.8}$$

We can isolate $Z_3^{(10)}$ from (2.8). Note that if we know K', then $Z_3^{(10)}$ is the only unknown value:

$$Z_3^{(10)} = ((((((Z_6^{(11)} \oplus c_{95}) \ggg 9) \boxminus Z_5^{(11)}) \oplus Z_4^{(11)}) \boxminus Z_2^{(10)}) \oplus Z_9^{(10)}) \boxminus Z_7^{(10)}. \tag{2.9}$$

If we compute and store K', then, all terms in the right-hand side of (2.9) are known. This way, the subkeys can be computed in reverse order for decryption. It is not necessary to always generate the subkeys starting from $Z_1^{(1)}$ and store them, as is usual for most block ciphers.

The last generated subkey (in reverse order) is derived from the equation:

$$Z_4^{(2)} = (((((Z_1^{(1)} \boxplus Z_5^{(1)}) \oplus Z_7^{(1)}) \boxplus Z_9^{(1)}) \oplus Z_2^{(2)}) \boxplus Z_3^{(2)}) \lll 9 \oplus c_{12}.$$

We just isolate $Z_1^{(1)}$ from it:

$$Z_1^{(1)} = ((((((Z_4^{(2)} \oplus c_{12}) \ggg 9) \boxminus Z_3^{(2)}) \oplus Z_2^{(2)}) \boxminus Z_9^{(1)}) \oplus Z_7^{(1)}) \boxminus Z_5^{(1)}.$$

In the end, if decryption operations are frequently needed (depending on the mode of operation), we only need to compute K' once and store it.

2.3.5 Encryption and Decryption Frameworks of MESH-128

MESH-128 [50] is an iterated block cipher operating on 128-bit blocks, under a 256-bit key and consisting of twelve rounds plus an output transformation. All internal operations are on 16-bit words.

Fig. 2.5 shows a pictorial representation of the encryption and the decryption frameworks of MESH-128.

Let $X^{(i)} = (X_1^{(i)}, X_2^{(i)}, X_3^{(i)}, X_4^{(i)}, X_5^{(i)}, X_6^{(i)}, X_7^{(i)}, X_8^{(i)})$ denote the 128-bit input text block to the i-th round, where $X_j^{(i)} \in \mathbb{Z}_2^{16}$, for $1 \leq j \leq 8$ and $1 \leq i \leq 13$.

One full encryption round in MESH-128 can be decomposed into a Key-Mixing (KM) and a Multiplication-Addition (MA) half-round, in this order. There are two types of KM half-rounds: for even- and odd-numbered rounds. A KM half-round uses eight subkeys while an MA half-round uses four subkeys.

The j-th round subkey of the i-th round is denoted by $Z_j^{(i)}$, with $1 \leq j \leq 12$ and $1 \leq i \leq 13$. Subkey generation by the key schedule algorithm is detailed in Sect. 2.3.6.

Given the input $X^{(i)}$ to the i-th round, it is also the input to the i-th KM half-round.

The output of an odd-numbered KM half-round is

$$Y^{(i)} = (X_1^{(i)} \odot Z_1^{(i)}, X_2^{(i)} \boxplus Z_2^{(i)}, X_3^{(i)} \odot Z_3^{(i)}, X_4^{(i)} \boxplus Z_4^{(i)},\\ X_5^{(i)} \boxplus Z_5^{(i)}, X_6^{(i)} \odot Z_6^{(i)}, X_7^{(i)} \boxplus Z_7^{(i)}, X_8^{(i)} \odot Z_8^{(i)}).$$

The output of an even-numbered KM half-round is

$$Y^{(i)} = (X_1^{(i)} \boxplus Z_1^{(i)}, X_2^{(i)} \odot Z_2^{(i)}, X_3^{(i)} \boxplus Z_3^{(i)}, X_4^{(i)} \odot Z_4^{(i)},\\ X_5^{(i)} \odot Z_5^{(i)}, X_6^{(i)} \boxplus Z_6^{(i)}, X_7^{(i)} \odot Z_7^{(i)}, X_8^{(i)} \boxplus Z_8^{(i)}).$$

The output of the i-th KM half-round, $Y^{(i)}$, becomes the input to the i-th MA half-round.

The input to the i-th MA half-round is the ordered 4-tuple (n_i, q_i, m_i, u_i), where $n_i = Y_1^{(i)} \oplus Y_5^{(i)}$, $q_i = Y_2^{(i)} \oplus Y_6^{(i)}$, $m_i = Y_3^{(i)} \oplus Y_7^{(i)}$ and $u_i = Y_4^{(i)} \oplus Y_8^{(i)}$. The MA-box of MESH-128 is a 64-bit keyed permutation that consists of four layers of $\odot$ and $\boxplus$ interleaved and in a zigzag order.

The output of the i-th MA-box is the 4-tuple (r_i, s_i, t_i, v_i), which is computed as follows:

$$\alpha = ((((n_i \odot Z_9^{(i)}) \boxplus q_i) \odot m_i) \boxplus u_i) \odot Z_{10}^{(i)},$$

$$\beta = \alpha \boxplus (((n_i \odot Z_9^{(i)}) \boxplus q_i) \odot m_i),$$

$$\gamma = ((\beta \odot ((n_i \odot Z_9^{(i)}) \boxplus q_i)) \boxplus (n_i \odot Z_9^{(i)})) \odot Z_{11}^{(i)},$$

$$\delta = \gamma \boxplus (\beta \odot ((n_i \odot Z_9^{(i)}) \boxplus q_i)),$$

and

$$\epsilon = \delta \odot \beta.$$

Further,

$$v_i = (\alpha \boxplus \epsilon) \odot Z_{12}^{(i)},$$

$$t_i = v_i \boxplus \epsilon,$$

$$s_i = t_i \odot \delta,$$

and

$$r_i = s_i \boxplus \gamma.$$

Both the MA-boxes of MESH-128 and that of MESH-96 contain multiplications which are not directly keyed by subkeys but rather depend on mix of subkeys and internal MA-box data. This is a new feature of MESH ciphers that is aimed at strengthening them by making the MA-box less vulnerable to weak subkeys.

The next step is to combine (r_i, s_i, t_i, v_i) with the original inputs to the MA half-round to obtain:

$(Y_1^{(i)} \oplus v_i, Y_2^{(i)} \oplus t_i, Y_3^{(i)} \oplus s_i, Y_4^{(i)} \oplus r_i, Y_5^{(i)} \oplus v_i, Y_6^{(i)} \oplus t_i, Y_7^{(i)} \oplus s_i, Y_8^{(i)} \oplus r_i)$.

The last step is to swap the six middle words in a block. For a 128-bit block (A, B, C, D, E, F, G, H), the word swap operation, denoted σ, performs a fixed word permutation:

$$\sigma(A, B, C, D, E, F, G, H) = (A, E, F, G, B, C, D, H).$$

The round output becomes

$$X^{(i+1)} = (Y_1^{(i)} \oplus v_i, Y_5^{(i)} \oplus v_i, Y_6^{(i)} \oplus t_i, Y_7^{(i)} \oplus s_i,$$

$$Y_2^{(i)} \oplus t_i, Y_3^{(i)} \oplus s_i, Y_4^{(i)} \oplus r_i, Y_8^{(i)} \oplus r_i),$$

which also becomes the input to the next round. This procedure is repeated twelve times for MESH-128.

The output transformation consists of an application of σ followed by a KM half-round. In effect, σ in the output transformation cancels the word swap of the previous full round. It is equivalent to simply removing σ from the last full round:

$$X^{(13)} = (Y_1^{(12)} \oplus v_{12}, Y_5^{(12)} \oplus v_{12}, Y_6^{(12)} \oplus t_{12}, Y_7^{(12)} \oplus s_{12},$$
$$Y_2^{(12)} \oplus t_{12}, Y_3^{(12)} \oplus s_{12}, Y_4^{(12)} \oplus r_{12}, Y_8^{(12)} \oplus r_{12})$$

The ciphertext becomes

$$C = (X_1^{(13)} \odot Z_1^{(13)}, X_2^{(13)} \boxplus Z_2^{(13)}, X_3^{(13)} \odot Z_3^{(13)}, X_4^{(13)} \boxplus Z_4^{(13)},$$
$$X_5^{(13)} \boxplus Z_5^{(13)}, X_6^{(13)} \odot Z_6^{(13)}, X_7^{(13)} \boxplus Z_7^{(13)}, X_8^{(13)} \odot Z_8^{(13)})$$

If X denotes a 128-bit block, KM_i denotes the i-th KM half-round, MA_i denotes the i-th MA half-round, then, a full round in MESH-128 can be denoted by

$$\phi_i(X) = \sigma \circ \text{MA}_i \circ \text{KM}_i(X) = \sigma(\text{MA}_i(\text{KM}_i(X))).$$

One full MESH-128 encryption of a 128-bit plaintext block P into a 128-bit ciphertext block C can be denoted

$$C = \text{KM}_{13} \circ \sigma \circ \bigcirc_{i=1}^{12} \phi_i(P)$$

where functional composition operates in right-to-left order.

The main differences between encryption and decryption in MESH-128 are in the KM half-rounds where operations need to be inverted: (modular) addition becomes (modular) subtraction and (modular) multiplication becomes (modular) division.

Note that σ and $\sigma \circ \text{MA}_i$ are involutory transformations.

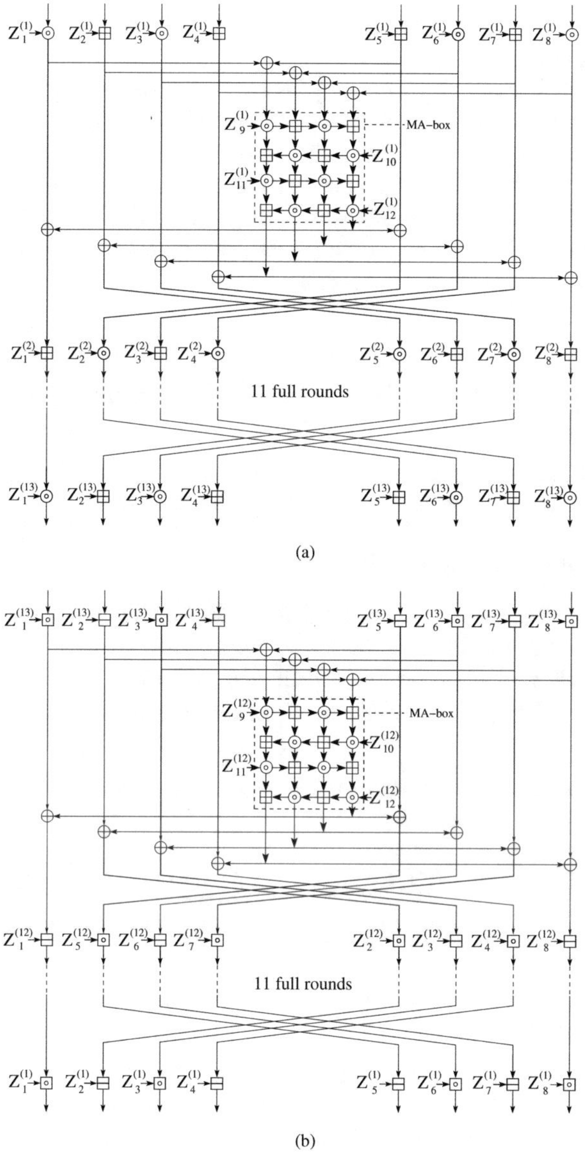

Fig. 2.5 Computational graph of the MESH-128 cipher: (a) encryption [50] and (b) decryption. Note the use of modular division (⊡) and subtraction (⊟) in decryption.

Also, note that the order of six middle round subkeys, $Z_j^{(i)}$ for $j \in \{2,3,4,5,6,7\}$, changes for decryption in Fig. 2.5(b) (except for $i = 1$ and $i = 13$). This change happens because when Fig. 2.5(a) is turned upside-down σ changes the order of $(Z_2^{(i)}, Z_3^{(i)}, Z_4^{(i)})$ and $(Z_5^{(i)}, Z_6^{(i)}, Z_7^{(i)})$ except for the first and the last KM half-rounds.

Consequently, Fig. 2.5(a) can represent both the encryption and the decryption frameworks, by just appropriately changing the order and value of subkeys (according to Table 2.7). Therefore, Fig. 2.5(b) is redundant. Nonetheless, we depict both operations side-by-side to demonstrate their similarity and to point out the different KM half-rounds, which is important to take into account in proper (software and hardware) implementations of the encryption and decryption operations.

2.3.6 Key Schedule Algorithm of MESH-128

MESH-128 requires 152 16-bit subkeys (2432 subkey bits in total) for either encryption or decryption of a single 128-bit text block.

The key schedule of MESH-128 uses a nonlinear feedback shift register (NLFSR) design and consists of:

- the same 16-bit constants c_i as defined for the key schedule of MESH-64 in Sect. 2.3.2. See Table 2.4.
- the 256-bit user key K is partitioned into sixteen 16-bit words K_i, $0 \leq i \leq 15$, and assigned to the first sixteen subkeys $Z_{j+1}^{(1)} = K_j \oplus c_j$ for $0 \leq j \leq 11$ and $Z_{t \bmod 12+1}^{(2)} = K_t \oplus c_t$ for $12 \leq t \leq 15$.
- each subsequent round subkey 16-bit subkey is generated iteratively according to the recurrence[2]

$$Z_{l(i)}^{h(i)} = (((((Z_{l(i-16)}^{h(i-16)} \boxplus Z_{l(i-13)}^{h(i-13)}) \oplus Z_{l(i-12)}^{h(i-12)}) \boxplus Z_{l(i-8)}^{h(i-8)}) \oplus Z_{l(i-2)}^{h(i-2)}) \boxplus Z_{l(i-1)}^{h(i-1)}) \lll 11 \oplus c_i, \qquad (2.10)$$

  for $16 \leq i \leq 151$; $x \lll 11$ means x left rotated by 11 bits; $h(i) = \lfloor i/12 \rfloor + 1$ and $l(i) = i \bmod 12 + 1$.

The key schedule of MESH-128 achieves fast key diffusion due to the use of the primitive polynomial $r(x) = x^{16} + x^{15} + x^{14} + x^8 + x^4 + x^3 + 1$ in $l(i)$ and $h(i)$, the bit rotation and the interleaving of $\oplus$ and $\boxplus$.

Let us deduce that all round subkeys starting from $Z_{10}^{(2)}$ already depend nonlinearly upon all sixteen user key words. Initially, $Z_1^{(1)} = K_0 \oplus c_0$, $Z_2^{(1)} = K_1 \oplus c_1$, $Z_3^{(1)} = K_2 \oplus c_2$, $Z_4^{(1)} = K_3 \oplus c_3$, $Z_5^{(1)} = K_4 \oplus c_4$, $Z_6^{(1)} = K_5 \oplus c_5$, $Z_7^{(1)} = K_6 \oplus c_6$, $Z_8^{(1)} = K_7 \oplus c_7$, $Z_9^{(1)} = K_8 \oplus c_8$, $Z_{10}^{(1)} = K_9 \oplus c_9$, $Z_{11}^{(1)} = K_{10} \oplus c_{10}$, $Z_{12}^{(1)} = K_{11} \oplus c_{11}$, $Z_1^{(2)} = K_{12} \oplus c_{12}$, $Z_2^{(2)} = K_{13} \oplus c_{13}$, $Z_3^{(2)} = K_{14} \oplus c_{14}$ and $Z_4^{(2)} = K_{15} \oplus c_{15}$.

[2] Bit rotation has higher precedence than $\oplus$.

Further, for $i = 16$:

$$Z_5^{(2)} = (((((Z_1^{(1)} \boxplus Z_4^{(1)}) \oplus Z_5^{(1)}) \boxplus Z_9^{(1)}) \oplus Z_3^{(2)}) \boxplus Z_4^{(2)}) \lll 11 \oplus c_{16},$$

which depends on six words from K.

For $i = 17$:

$$Z_6^{(2)} = (((((Z_2^{(1)} \boxplus Z_5^{(1)}) \oplus Z_6^{(1)}) \boxplus Z_{10}^{(1)}) \oplus Z_4^{(2)}) \boxplus Z_5^{(2)}) \lll 11 \oplus c_{17},$$

which depends on nine words from K due to the chained dependence on $Z_5^{(2)}$.

For $i = 18$:

$$Z_7^{(2)} = (((((Z_3^{(1)} \boxplus Z_6^{(1)}) \oplus Z_7^{(1)}) \boxplus Z_{11}^{(1)}) \oplus Z_5^{(2)}) \boxplus Z_6^{(2)}) \lll 11 \oplus c_{18},$$

which depends on twelve words from K due to the chained dependence on $Z_6^{(2)}$.

For $i = 19$:

$$Z_8^{(2)} = (((((Z_4^{(1)} \boxplus Z_7^{(1)}) \oplus Z_8^{(1)}) \boxplus Z_{12}^{(1)}) \oplus Z_6^{(2)}) \boxplus Z_7^{(2)}) \lll 11 \oplus c_{19},$$

which depends on fourteen words from K due to the chained dependence on $Z_7^{(2)}$.

For $i = 20$:

$$Z_9^{(2)} = (((((Z_5^{(1)} \boxplus Z_8^{(1)}) \oplus Z_9^{(1)}) \boxplus Z_1^{(2)}) \oplus Z_7^{(2)}) \boxplus Z_8^{(2)}) \lll 11 \oplus c_{20},$$

which depends on fifteen words from K due to the chained dependence on $Z_8^{(2)}$.

Finally, for $i = 21$:

$$Z_{10}^{(2)} = (((((Z_6^{(1)} \boxplus Z_9^{(1)}) \oplus Z_{10}^{(1)}) \boxplus Z_2^{(2)}) \oplus Z_8^{(2)}) \boxplus Z_9^{(2)}) \lll 11 \oplus c_{21},$$

which depends on all sixteen words from K.

Thus, MESH-128 achieves complete key diffusion in two rounds.

Since both $\oplus$ and $\boxplus$ preserve the relative bit positions of their operands, it was decided to employ a fixed bit rotation at the end of (2.10). In this way, differences in the most significant bit positions (in a related-key differential setting) that remain intact by $\oplus$ and $\boxplus$, are effectively displaced by the bit rotation.

The constants c_i aim to avoid bit patterns in K to propagate across the subkeys, otherwise, this key schedule would suffer some of the same drawbacks as the key schedule of PES and IDEA. For instance, without the c_i, the key $K = 0^{256}$ would result in all subkeys equal to zero.

Table 2.7 lists the encryption and decryption subkeys for each round of MESH-96. Multiplicative subkeys are depicted in boldface type.

Note that the order of operations $\odot$ and $\boxplus$ during encryption changes in the KM half-rounds for even- and odd-numbered rounds.

Table 2.7 Encryption and decryption round subkeys for MESH-128 cipher.

round	Encryption subkeys
1	$\mathbf{Z_1^{(1)}}, Z_2^{(1)}, \mathbf{Z_3^{(1)}}, Z_4^{(1)}, Z_5^{(1)}, \mathbf{Z_6^{(1)}}, Z_7^{(1)}, \mathbf{Z_8^{(1)}}, \mathbf{Z_9^{(1)}}, \mathbf{Z_{10}^{(1)}}, \mathbf{Z_{11}^{(1)}}, \mathbf{Z_{12}^{(1)}}$
2	$Z_1^{(2)}, \mathbf{Z_2^{(2)}}, Z_3^{(2)}, \mathbf{Z_4^{(2)}}, \mathbf{Z_5^{(2)}}, Z_6^{(2)}, \mathbf{Z_7^{(2)}}, Z_8^{(2)}, \mathbf{Z_9^{(2)}}, \mathbf{Z_{10}^{(2)}}, \mathbf{Z_{11}^{(2)}}, \mathbf{Z_{12}^{(2)}}$
3	$\mathbf{Z_1^{(3)}}, Z_2^{(3)}, \mathbf{Z_3^{(3)}}, Z_4^{(3)}, Z_5^{(3)}, \mathbf{Z_6^{(3)}}, Z_7^{(3)}, \mathbf{Z_8^{(3)}}, \mathbf{Z_9^{(3)}}, \mathbf{Z_{10}^{(3)}}, \mathbf{Z_{11}^{(3)}}, \mathbf{Z_{12}^{(3)}}$
4	$Z_1^{(4)}, \mathbf{Z_2^{(4)}}, Z_3^{(4)}, \mathbf{Z_4^{(4)}}, \mathbf{Z_5^{(4)}}, Z_6^{(4)}, \mathbf{Z_7^{(4)}}, Z_8^{(4)}, \mathbf{Z_9^{(4)}}, \mathbf{Z_{10}^{(4)}}, \mathbf{Z_{11}^{(4)}}, \mathbf{Z_{12}^{(4)}}$
5	$\mathbf{Z_1^{(5)}}, Z_2^{(5)}, \mathbf{Z_3^{(5)}}, Z_4^{(5)}, Z_5^{(5)}, \mathbf{Z_6^{(5)}}, Z_7^{(5)}, \mathbf{Z_8^{(5)}}, \mathbf{Z_9^{(5)}}, \mathbf{Z_{10}^{(5)}}, \mathbf{Z_{11}^{(5)}}, \mathbf{Z_{12}^{(5)}}$
6	$Z_1^{(6)}, \mathbf{Z_2^{(6)}}, Z_3^{(6)}, \mathbf{Z_4^{(6)}}, \mathbf{Z_5^{(6)}}, Z_6^{(6)}, \mathbf{Z_7^{(6)}}, Z_8^{(6)}, \mathbf{Z_9^{(6)}}, \mathbf{Z_{10}^{(6)}}, \mathbf{Z_{11}^{(6)}}, \mathbf{Z_{12}^{(6)}}$
7	$\mathbf{Z_1^{(7)}}, Z_2^{(7)}, \mathbf{Z_3^{(7)}}, Z_4^{(7)}, Z_5^{(7)}, \mathbf{Z_6^{(7)}}, Z_7^{(7)}, \mathbf{Z_8^{(7)}}, \mathbf{Z_9^{(7)}}, \mathbf{Z_{10}^{(7)}}, \mathbf{Z_{11}^{(7)}}, \mathbf{Z_{12}^{(7)}}$
8	$Z_1^{(8)}, \mathbf{Z_2^{(8)}}, Z_3^{(8)}, \mathbf{Z_4^{(8)}}, \mathbf{Z_5^{(8)}}, Z_6^{(8)}, \mathbf{Z_7^{(8)}}, Z_8^{(8)}, \mathbf{Z_9^{(8)}}, \mathbf{Z_{10}^{(8)}}, \mathbf{Z_{11}^{(8)}}, \mathbf{Z_{12}^{(8)}}$
9	$\mathbf{Z_1^{(9)}}, Z_2^{(9)}, \mathbf{Z_3^{(9)}}, Z_4^{(9)}, Z_5^{(9)}, \mathbf{Z_6^{(9)}}, Z_7^{(9)}, \mathbf{Z_8^{(9)}}, \mathbf{Z_9^{(9)}}, \mathbf{Z_{10}^{(9)}}, \mathbf{Z_{11}^{(9)}}, \mathbf{Z_{12}^{(9)}}$
10	$Z_1^{(10)}, \mathbf{Z_2^{(10)}}, Z_3^{(10)}, \mathbf{Z_4^{(10)}}, \mathbf{Z_5^{(10)}}, Z_6^{(10)}, \mathbf{Z_7^{(10)}}, Z_8^{(10)}, \mathbf{Z_9^{(10)}}, \mathbf{Z_{10}^{(10)}}, \mathbf{Z_{11}^{(10)}}, \mathbf{Z_{12}^{(10)}}$
11	$\mathbf{Z_1^{(11)}}, Z_2^{(11)}, \mathbf{Z_3^{(11)}}, Z_4^{(11)}, Z_5^{(11)}, \mathbf{Z_6^{(11)}}, Z_7^{(11)}, \mathbf{Z_8^{(11)}}, \mathbf{Z_9^{(11)}}, \mathbf{Z_{10}^{(11)}}, \mathbf{Z_{11}^{(11)}}, \mathbf{Z_{12}^{(11)}}$
12	$Z_1^{(12)}, \mathbf{Z_2^{(12)}}, Z_3^{(12)}, \mathbf{Z_4^{(12)}}, \mathbf{Z_5^{(12)}}, Z_6^{(12)}, \mathbf{Z_7^{(12)}}, Z_8^{(12)}, \mathbf{Z_9^{(12)}}, \mathbf{Z_{10}^{(12)}}, \mathbf{Z_{11}^{(12)}}, \mathbf{Z_{12}^{(12)}}$
13	$\mathbf{Z_1^{(13)}}, Z_2^{(13)}, \mathbf{Z_3^{(13)}}, Z_4^{(13)}, Z_5^{(13)}, \mathbf{Z_6^{(13)}}, Z_7^{(13)}, \mathbf{Z_8^{(13)}}$

round	Decryption subkeys
1	$\mathbf{Z_1^{(13)}}, Z_2^{(13)}, \mathbf{Z_3^{(13)}}, Z_4^{(13)}, Z_5^{(13)}, \mathbf{Z_6^{(13)}}, Z_7^{(13)}, \mathbf{Z_8^{(13)}}, \mathbf{Z_9^{(12)}}, \mathbf{Z_{10}^{(12)}}, \mathbf{Z_{11}^{(12)}}, \mathbf{Z_{12}^{(12)}}$
2	$Z_1^{(12)}, \mathbf{Z_5^{(12)}}, Z_6^{(12)}, \mathbf{Z_7^{(12)}}, \mathbf{Z_2^{(12)}}, Z_3^{(12)}, \mathbf{Z_4^{(12)}}, Z_8^{(12)}, \mathbf{Z_9^{(11)}}, \mathbf{Z_{10}^{(11)}}, \mathbf{Z_{11}^{(11)}}, \mathbf{Z_{12}^{(11)}}$
3	$\mathbf{Z_1^{(11)}}, Z_5^{(11)}, \mathbf{Z_6^{(11)}}, Z_7^{(11)}, Z_2^{(11)}, \mathbf{Z_3^{(11)}}, Z_4^{(11)}, \mathbf{Z_8^{(11)}}, \mathbf{Z_9^{(10)}}, \mathbf{Z_{10}^{(10)}}, \mathbf{Z_{11}^{(10)}}, \mathbf{Z_{12}^{(10)}}$
4	$Z_1^{(10)}, \mathbf{Z_5^{(10)}}, Z_6^{(10)}, \mathbf{Z_7^{(10)}}, \mathbf{Z_2^{(10)}}, Z_3^{(10)}, \mathbf{Z_4^{(10)}}, Z_8^{(10)}, \mathbf{Z_9^{(9)}}, \mathbf{Z_{10}^{(9)}}, \mathbf{Z_{11}^{(9)}}, \mathbf{Z_{12}^{(9)}}$
5	$\mathbf{Z_1^{(9)}}, Z_5^{(9)}, \mathbf{Z_6^{(9)}}, Z_7^{(9)}, Z_2^{(9)}, \mathbf{Z_3^{(9)}}, Z_4^{(9)}, \mathbf{Z_8^{(9)}}, \mathbf{Z_9^{(8)}}, \mathbf{Z_{10}^{(8)}}, \mathbf{Z_{11}^{(8)}}, \mathbf{Z_{12}^{(8)}}$
6	$Z_1^{(8)}, \mathbf{Z_5^{(8)}}, Z_6^{(8)}, \mathbf{Z_7^{(8)}}, \mathbf{Z_2^{(8)}}, Z_3^{(8)}, \mathbf{Z_4^{(8)}}, Z_8^{(8)}, \mathbf{Z_9^{(7)}}, \mathbf{Z_{10}^{(7)}}, \mathbf{Z_{11}^{(7)}}, \mathbf{Z_{12}^{(7)}}$
7	$\mathbf{Z_1^{(7)}}, Z_5^{(7)}, \mathbf{Z_6^{(7)}}, Z_7^{(7)}, Z_2^{(7)}, \mathbf{Z_3^{(7)}}, Z_4^{(7)}, \mathbf{Z_8^{(7)}}, \mathbf{Z_9^{(6)}}, \mathbf{Z_{10}^{(6)}}, \mathbf{Z_{11}^{(6)}}, \mathbf{Z_{12}^{(6)}}$
8	$Z_1^{(6)}, \mathbf{Z_5^{(6)}}, Z_6^{(6)}, \mathbf{Z_7^{(6)}}, \mathbf{Z_2^{(6)}}, Z_3^{(6)}, \mathbf{Z_4^{(6)}}, Z_8^{(6)}, \mathbf{Z_9^{(5)}}, \mathbf{Z_{10}^{(5)}}, \mathbf{Z_{11}^{(5)}}, \mathbf{Z_{12}^{(5)}}$
9	$\mathbf{Z_1^{(5)}}, Z_5^{(5)}, \mathbf{Z_6^{(5)}}, Z_7^{(5)}, Z_2^{(5)}, \mathbf{Z_3^{(5)}}, Z_4^{(5)}, \mathbf{Z_8^{(5)}}, \mathbf{Z_9^{(4)}}, \mathbf{Z_{10}^{(4)}}, \mathbf{Z_{11}^{(4)}}, \mathbf{Z_{12}^{(4)}}$
10	$Z_1^{(4)}, \mathbf{Z_5^{(4)}}, Z_6^{(4)}, \mathbf{Z_7^{(4)}}, \mathbf{Z_2^{(4)}}, Z_3^{(4)}, \mathbf{Z_4^{(4)}}, Z_8^{(4)}, \mathbf{Z_9^{(3)}}, \mathbf{Z_{10}^{(3)}}, \mathbf{Z_{11}^{(3)}}, \mathbf{Z_{12}^{(3)}}$
11	$\mathbf{Z_1^{(3)}}, Z_5^{(3)}, \mathbf{Z_6^{(3)}}, Z_7^{(3)}, Z_2^{(3)}, \mathbf{Z_3^{(3)}}, Z_4^{(3)}, \mathbf{Z_8^{(3)}}, \mathbf{Z_9^{(2)}}, \mathbf{Z_{10}^{(2)}}, \mathbf{Z_{11}^{(2)}}, \mathbf{Z_{12}^{(2)}}$
12	$Z_1^{(2)}, \mathbf{Z_5^{(2)}}, Z_6^{(2)}, \mathbf{Z_7^{(2)}}, \mathbf{Z_2^{(2)}}, Z_3^{(2)}, \mathbf{Z_4^{(2)}}, Z_8^{(2)}, \mathbf{Z_9^{(1)}}, \mathbf{Z_{10}^{(1)}}, \mathbf{Z_{11}^{(1)}}, \mathbf{Z_{12}^{(1)}}$
13	$\mathbf{Z_1^{(1)}}, Z_2^{(1)}, \mathbf{Z_3^{(1)}}, Z_4^{(1)}, Z_5^{(1)}, \mathbf{Z_6^{(1)}}, Z_7^{(1)}, \mathbf{Z_8^{(1)}}$

Note that the key schedule of MESH-128 is invertible: the NLSFR can be clocked backwards to generate subkeys for decryption as easily as for encryption. It is only necessary to store the last sixteen subkeys (256 bits) instead of K. Let us denote it by $K' = (Z_5^{(12)}, Z_6^{(12)}, Z_7^{(12)}, Z_8^{(12)}, Z_9^{(12)}, Z_{10}^{(12)}, Z_{11}^{(12)}, Z_{12}^{(12)}, Z_1^{(13)}, Z_2^{(13)}, Z_3^{(13)}, Z_4^{(13)}, Z_5^{(13)}, Z_6^{(13)}, Z_7^{(13)}, Z_8^{(13)})$.

The last generated subkey for encryption was

$$Z_8^{(13)} = (((((Z_4^{(12)} \boxplus Z_7^{(12)}) \oplus Z_8^{(12)}) \boxplus Z_{12}^{(12)}) \oplus Z_6^{(13)}) \boxplus Z_7^{(13)}) \lll 11 \oplus c_{151}.$$

We can isolate $Z_4^{(12)}$ from this equation which is the only unknown (if we know K'):

$$Z_4^{(12)} = ((((((Z_8^{(13)} \oplus c_{151}) \ggg 11) \boxminus Z_7^{(13)}) \oplus Z_6^{(13)}) \boxminus Z_{12}^{(12)}) \oplus Z_8^{(12)}) \boxminus Z_7^{(12)}. \qquad (2.11)$$

If we compute and store K', then, all terms in the right-hand side of (2.11) are known.

This way, we can compute the subkeys in reverse order for decryption. Therefore, it is not necessary to always generate the subkeys starting from $Z_1^{(1)}$ and store them, as is usual for most block ciphers.

The last generated subkey (in reverse order) is derived from the equation:

$$Z_5^{(2)} = (((((Z_1^{(1)} \boxplus Z_4^{(1)}) \oplus Z_5^{(1)}) \boxplus Z_9^{(1)}) \oplus Z_3^{(2)}) \boxplus Z_4^{(2)}) \lll 11 \oplus c_{16}.$$

We just isolate $Z_1^{(1)}$ from it:

$$Z_1^{(1)} = ((((((Z_5^{(2)} \oplus c_{16}) \ggg 11) \boxminus Z_4^{(2)}) \oplus Z_3^{(2)}) \boxminus Z_9^{(1)}) \oplus Z_5^{(1)}) \boxminus Z_4^{(1)}.$$

In the end, if decryption operations are frequently needed (depending on the mode of operation), we only need to compute K' once and store it (along with K).

2.3.7 Encryption and Decryption Frameworks of MESH-64(8)

MESH-64(8) [45] is an iterated block cipher operating on 64-bit blocks, under a 128-bit key and consisting of eight rounds plus an output transformation. All internal operations are on 8-bit words. Fig. 2.6 shows a pictorial representation of the encryption and decryption frameworks of MESH-64(8).

MESH-64(8) uses the same three group operations of PES and IDEA, but *on 8-bit words* instead of 16-bit words. They are: exclusive-or, denoted $\oplus$, modular addition in $\mathbb{Z}_{2^8}$, denoted $\boxplus$, and modular multiplication over $GF(2^8+1)$, denoted $\odot$, and with $0 \equiv 2^8$.

The design of the MA-box of MESH-64(8) consists of two layers of modular addition and modular multiplication alternated and in a zigzag order.

The number of (full) rounds was set to an even number, so that the encryption and decryption frameworks are similar except for the order and value of round subkeys. See Fig. 2.6.

One full round consists of a Key-Mixing (KM) and a Multiplication-Addition (MA) half-rounds, in this order. A KM half-round uses eight subkeys, while an MA half-round uses two subkeys.

Let $X^{(i)} = (X_1^{(i)}, X_2^{(i)}, X_3^{(i)}, X_4^{(i)}, X_5^{(i)}, X_6^{(i)}, X_7^{(i)}, X_8^{(i)})$ denote the 64-bit input block to the i-th round.

The j-th 16-bit subkey of the i-th round is denoted by $Z_j^{(i)}$, with $1 \leq j \leq 10$ and $1 \leq i \leq 9$. Subkey generation by the key schedule algorithm is detailed in Sect. 2.3.8.

The output to the i-th KM half-round is denoted

$$Y^{(i)} = (Y_1^{(i)}, Y_2^{(i)}, Y_3^{(i)}, Y_4^{(i)}, Y_5^{(i)}, Y_6^{(i)}, Y_7^{(i)}, Y_8^{(i)}).$$

If i is odd, then the output of the i-th KM half-round is computed as

$$Y^{(i)} = (X_1^{(i)} \odot Z_1^{(i)}, X_2^{(i)} \boxplus Z_2^{(i)}, X_3^{(i)} \odot Z_3^{(i)}, X_4^{(i)} \boxplus Z_4^{(i)},$$
$$X_5^{(i)} \boxplus Z_5^{(i)}, X_6^{(i)} \odot Z_6^{(i)}, X_7^{(i)} \boxplus Z_7^{(i)}, X_8^{(i)} \odot Z_8^{(i)}).$$

Otherwise, if i is even, then the i-th KM half-round output is

$$Y^{(i)} = (X_1^{(i)} \boxplus Z_1^{(i)}, X_2^{(i)} \odot Z_2^{(i)}, X_3^{(i)} \boxplus Z_3^{(i)}, X_4^{(i)} \odot Z_4^{(i)},$$
$$X_5^{(i)} \odot Z_5^{(i)}, X_6^{(i)} \boxplus Z_6^{(i)}, X_7^{(i)} \odot Z_7^{(i)}, X_8^{(i)} \boxplus Z_8^{(i)}).$$

The output of the i-th KM half-round is input to the i-th MA half-round. The input to the i-th MA-box is the ordered 4-tuple

$$(n_i, q_i, m_i, u_i) = (Y_1^{(i)} \oplus Y_5^{(i)}, Y_2^{(i)} \oplus Y_6^{(i)}, Y_3^{(i)} \oplus Y_7^{(i)}, Y_4^{(i)} \oplus Y_8^{(i)}).$$

The output of the i-th MA-box is the ordered 4-tuple (r_i, s_i, t_i, v_i), where

$$v_i = ((((n_i \odot Z_9^{(i)}) \boxplus q_i) \odot m_i) \boxplus u_i) \odot Z_{10}^{(i)},$$

$$t_i = (((n_i \odot Z_9^{(i)}) \boxplus q_i) \odot m_i) \boxplus v_i,$$

$$s_i = ((n_i \odot Z_9^{(i)}) \boxplus q_i) \odot t_i$$

and

$$r_i = (n_i \odot Z_9^{(i)}) \boxplus s_i.$$

Finally, the output of the MA half-round is combined with a fixed byte swap operation denoted σ. For a 64-bit block (A, B, C, D, E, F, G, H), σ is defined as follows

$$\sigma(A, B, C, D, E, F, G, H) = (A, E, F, G, B, C, D, H).$$

The i-th MA half-round output is

$$X^{(i+1)} = (Y_1^{(i)} \oplus r_i, Y_2^{(i)} \oplus r_i, Y_3^{(i)} \oplus s_i, Y_4^{(i)} \oplus t_i,$$

$$Y_5^{(i)} \oplus s_i, Y_6^{(i)} \oplus t_i, Y_7^{(i)} \oplus v_i, Y_8^{(i)} \oplus v_i).$$

This procedure is repeated eight times for MESH-64(8). Note that v_i already depends on all words in $Y^{(i)}$ due to its dependence on (n_i, q_i, m_i, u_i). From to the chained dependence of r_i on s_i, which depends on t_i, which depends on v_i, all outputs from the MA-box also depend on all words in $Y^{(i)}$, and consequently, on all words in $X^{(i)}$.

The last step is the output transformation (OT) consisting of an application of σ followed by an odd KM half-round. In effect, σ in the OT cancels the word swap of the previous full round. It is equivalent to removing σ from the last full round:

$$X^{(9)} = (Y_1^{(8)} \oplus r_8, Y_2^{(8)} \oplus s_8, Y_3^{(8)} \oplus t_8, Y_4^{(8)} \oplus v_8,$$
$$Y_5^{(8)} \oplus r_8, Y_6^{(8)} \oplus s_8, Y_7^{(8)} \oplus t_8, Y_8^{(8)} \oplus v_8)$$

The ciphertext becomes

$$C = (X_1^{(9)} \odot Z_1^{(9)}, X_2^{(9)} \boxplus Z_2^{(9)}, X_3^{(9)} \odot Z_3^{(9)}, X_4^{(9)} \boxplus Z_4^{(9)},$$
$$X_5^{(9)} \boxplus Z_5^{(9)}, X_6^{(9)} \odot Z_6^{(9)}, X_7^{(9)} \boxplus Z_7^{(9)}, X_8^{(9)} \odot Z_8^{(9)})$$

If X denotes a 64-bit block, KM_i denotes the i-th KM half-round, MA_i denotes the i-th MA half-round, then, a full round of MESH-64(8) can be denoted by

$$\zeta_i(X) = \sigma \circ \text{MA}_i \circ \text{KM}_i(X) = \sigma(\text{MA}_i(\text{KM}_i(X))).$$

One full MESH-64(8) encryption of a 64-bit plaintext block P into a 64-bit ciphertext block C can be denoted

$$C = \text{KM}_9 \circ \sigma \circ \zeta_8 \circ \zeta_7 \circ \zeta_6 \circ \zeta_5 \circ \zeta_4 \circ \zeta_3 \circ \zeta_2 \circ \zeta_1(P)$$

where functional composition operates in right-to-left order.

2.3.8 Key Schedule Algorithm of MESH-64(8)

In total, the key schedule of MESH-64(8) generates eighty-eight 8-bit round subkeys (704 bits in total) for either the encryption or the decryption of a single 64-bit text block.

The key schedule of MESH-64(8) uses a nonlinear feedback shift register (NLFSR) design and consists of the following steps:

- a preliminary step is to define 8-bit constants c_i as follows: $c_0 = 1$, $c_i = 2 \cdot c_{i-1}$ for $i \geq 1$, where $\cdot$ denotes the multiplication operation in GF(2^8), represented as GF(2)[x]/($p(x)$) where $p(x) = x^8 + x^4 + x^3 + x + 1$ is a primitive polynomial in GF(2)[x]. The constant 2 is represented by the polynomial x.

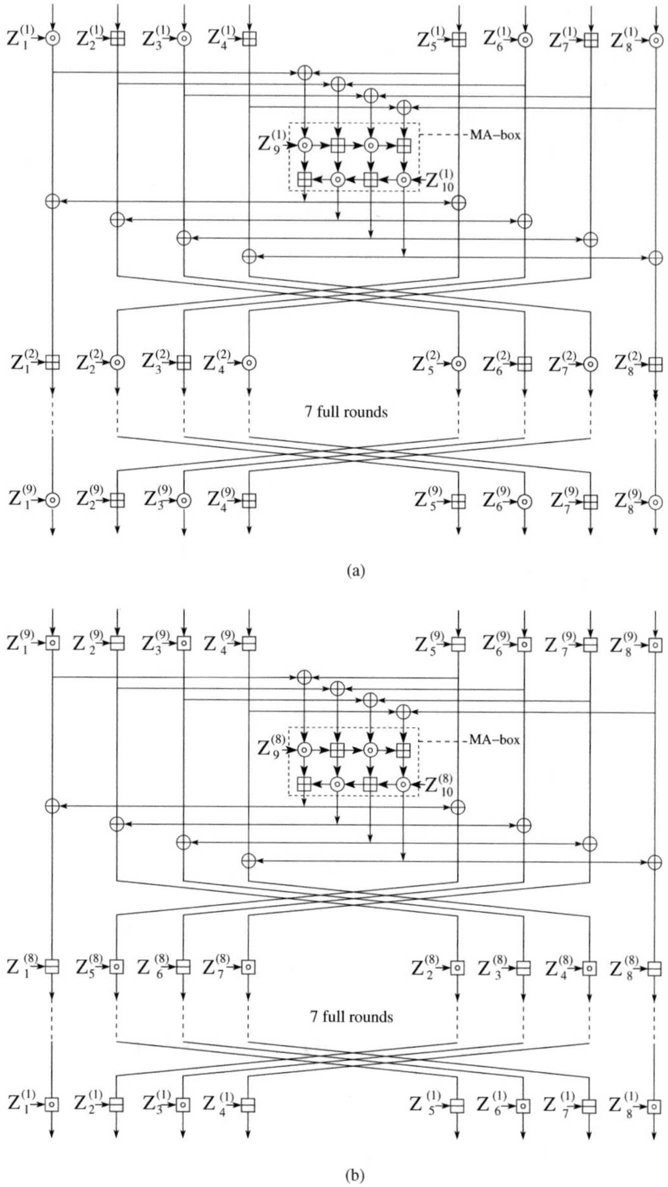

Fig. 2.6 Computational graph of the MESH-64(8) cipher: (a) encryption [45] and (b) decryption. Note the use of modular division (⊡) and subtraction (⊟) in decryption.

- the 128-bit user key K is partitioned into sixteen 8-bit words K_i for $0 \le i \le 15$. The first sixteen subkeys are assigned as $Z_{i+1}^{(1)} = K_i \oplus c_i$ for $0 \le i \le 9$, and $Z_{j \bmod 10+1}^{(2)} = K_j \oplus c_j$ for $10 \le j \le 15$.

- each subsequent round subkey is computed iteratively as follows[3]:

$$Z_{l(i)}^{h(i)} = (((Z_{l(i-16)}^{h(i-16)} \boxplus Z_{l(i-12)}^{h(i-12)}) \oplus Z_{l(i-3)}^{h(i-3)}) \boxplus Z_{l(i-2)}^{h(i-2)}) \lll 1 \oplus c_i, \tag{2.12}$$

  for $16 \leq i \leq 87$, where $x \lll 1$ means x left rotated by 1 bit, $h(i) = \lfloor i/10 \rfloor + 1$ and $l(i) = i \bmod 10 + 1$.

The key schedule of MESH-64(8) achieves complete key diffusion due to the use of the primitive polynomial $r(x) = x^{16} + x^{14} + x^{12} + x^4 + 1$ in $h(i)$ and $l(i)$, the bit rotation and the interleaving of $\oplus$ and $\boxplus$.

Let us deduce that each subkey starting with $Z_7^{(3)}$ already depends non-linearly upon all sixteen words of the key K. Initially, $Z_1^{(1)} = K_0 \oplus c_0$, $Z_2^{(1)} = K_1 \oplus c_1$, $Z_3^{(1)} = K_2 \oplus c_2$, $Z_4^{(1)} = K_3 \oplus c_3$, $Z_5^{(1)} = K_4 \oplus c_4$, $Z_6^{(1)} = K_5 \oplus c_5$, $Z_7^{(1)} = K_6 \oplus c_6$, $Z_8^{(1)} = K_7 \oplus c_7$, $Z_9^{(1)} = K_8 \oplus c_8$, $Z_{10}^{(1)} = K_9 \oplus c_9$, $Z_1^{(2)} = K_{10} \oplus c_{10}$, $Z_2^{(2)} = K_{11} \oplus c_{11}$, $Z_3^{(2)} = K_{12} \oplus c_{12}$, $Z_4^{(2)} = K_{13} \oplus c_{13}$, $Z_5^{(2)} = K_{14} \oplus c_{14}$ and $Z_6^{(2)} = K_{15} \oplus c_{15}$.

Further, for $i = 16$: $Z_7^{(2)} = (((Z_1^{(1)} \boxplus Z_5^{(1)}) \oplus Z_4^{(2)}) \boxplus Z_5^{(2)}) \lll 1 \oplus c_{16}$, which depends on four words from K.

Next, for $i = 17$: $Z_8^{(2)} = (((Z_2^{(1)} \boxplus Z_6^{(1)}) \oplus Z_5^{(2)}) \boxplus Z_6^{(2)}) \lll 1 \oplus c_{17}$, which depends on four words from K.

For $i = 18$: $Z_9^{(2)} = (((Z_3^{(1)} \boxplus Z_7^{(1)}) \oplus Z_6^{(2)}) \boxplus Z_7^{(2)}) \lll 1 \oplus c_{18}$, which depends on seven words from K due to the dependence on $Z_7^{(2)}$.

For $i = 19$: $Z_{10}^{(2)} = (((Z_4^{(1)} \boxplus Z_8^{(1)}) \oplus Z_7^{(2)}) \boxplus Z_8^{(2)}) \lll 1 \oplus c_{19}$, which depends on ten words from K due to the dependence on $Z_7^{(2)}$ and $Z_8^{(2)}$.

For $i = 20$: $Z_1^{(3)} = (((Z_5^{(1)} \boxplus Z_9^{(1)}) \oplus Z_8^{(2)}) \boxplus Z_9^{(2)}) \lll 1 \oplus c_{20}$, which depends on ten words from K due to the dependence on $Z_8^{(2)}$ and $Z_9^{(2)}$.

For $i = 21$: $Z_2^{(3)} = (((Z_6^{(1)} \boxplus Z_{10}^{(1)}) \oplus Z_9^{(2)}) \boxplus Z_{10}^{(2)}) \lll 1 \oplus c_{21}$, which depends on eleven words from K due to the dependence on $Z_9^{(2)}$ and $Z_{10}^{(2)}$.

For $i = 22$: $Z_3^{(3)} = (((Z_7^{(1)} \boxplus Z_1^{(2)}) \oplus Z_{10}^{(2)}) \boxplus Z_1^{(3)}) \lll 1 \oplus c_{22}$, which depends on thirteen words from K due to the dependence on $Z_{10}^{(2)}$ and $Z_1^{(3)}$.

For $i = 23$: $Z_4^{(3)} = (((Z_8^{(1)} \boxplus Z_2^{(2)}) \oplus Z_1^{(3)}) \boxplus Z_2^{(3)}) \lll 1 \oplus c_{23}$, which depends on fourteen words from K due to the dependence on $Z_1^{(3)}$ and $Z_3^{(3)}$.

For $i = 24$: $Z_5^{(3)} = (((Z_9^{(1)} \boxplus Z_3^{(2)}) \oplus Z_2^{(3)}) \boxplus Z_3^{(3)}) \lll 1 \oplus c_{24}$, which depends on fifteen words from K due to the dependence on $Z_2^{(3)}$ and $Z_3^{(3)}$.

For $i = 25$: $Z_6^{(3)} = (((Z_{10}^{(1)} \boxplus Z_4^{(2)}) \oplus Z_3^{(3)}) \boxplus Z_4^{(3)}) \lll 1 \oplus c_{25}$, which depends on fifteen words from K due to the dependence on $Z_3^{(3)}$ and $Z_4^{(3)}$.

Finally, for $i = 26$: $Z_7^{(3)} = (((Z_1^{(2)} \boxplus Z_5^{(2)}) \oplus Z_4^{(3)}) \boxplus Z_5^{(3)}) \lll 1 \oplus c_{26}$, which depends on all sixteen words from K due to the dependence on $Z_4^{(3)}$ and

[3] Bit rotation has higher precedence than $\oplus$.

$Z_5^{(3)}$. Thus, it takes three full rounds for MESH-64(8) to achieve complete key diffusion.

The constants c_i are meant to avoid bit patterns in the user key to propagate to the round subkeys. Without these constants, the user key $K = 0^{128}$ consisting of 128 bits zero would result in all round subkeys to be 0^8.

In [34], evidence was provided to corroborate that the key schedule construction for MESH-64(8) effectively avoids some classes of weak keys. The presence of these weak keys would allow the existence of narrow differential and linear trails across several rounds, like the ones found for IDEA and PES in [16, 48]. These weak keys could further jeopardize the use of MESH-64(8) in hash function constructions by allowing an adversary to control the key input, like in the Miyaguchi-Preneel hash mode [43].

Table 2.8 lists the encryption and decryption subkeys for each round of MESH-64(8). Multiplicative subkeys are depicted in boldface type. Note that the order of operations $\odot$ and $\boxplus$ during encryption changes in the KM half-rounds for even- and odd-numbered rounds.

Table 2.8 Encryption and decryption round subkeys of MESH-64(8) cipher.

round	Encryption subkeys	Decryption subkeys
1	$\mathbf{Z}_{1}^{(1)}, Z_{2}^{(1)}, \mathbf{Z}_{3}^{(1)}, Z_{4}^{(1)}, Z_{5}^{(1)}, \mathbf{Z}_{6}^{(1)}, Z_{7}^{(1)}, \mathbf{Z}_{8}^{(1)}, \mathbf{Z}_{9}^{(1)}, \mathbf{Z}_{10}^{(1)}$	$\mathbf{Z}_{1}^{(9)}, Z_{2}^{(9)}, \mathbf{Z}_{3}^{(9)}, Z_{4}^{(9)}, Z_{5}^{(9)}, \mathbf{Z}_{6}^{(9)}, Z_{7}^{(9)}, \mathbf{Z}_{8}^{(9)}, \mathbf{Z}_{9}^{(8)}, \mathbf{Z}_{10}^{(8)}$
2	$Z_{1}^{(2)}, \mathbf{Z}_{2}^{(2)}, Z_{3}^{(2)}, \mathbf{Z}_{4}^{(2)}, \mathbf{Z}_{5}^{(2)}, Z_{6}^{(2)}, \mathbf{Z}_{7}^{(2)}, Z_{8}^{(2)}, \mathbf{Z}_{9}^{(2)}, \mathbf{Z}_{10}^{(2)}$	$Z_{1}^{(8)}, \mathbf{Z}_{5}^{(8)}, Z_{6}^{(8)}, \mathbf{Z}_{7}^{(8)}, \mathbf{Z}_{2}^{(8)}, Z_{3}^{(8)}, \mathbf{Z}_{4}^{(8)}, Z_{8}^{(8)}, \mathbf{Z}_{9}^{(7)}, \mathbf{Z}_{10}^{(7)}$
3	$\mathbf{Z}_{1}^{(3)}, Z_{2}^{(3)}, \mathbf{Z}_{3}^{(3)}, Z_{4}^{(3)}, Z_{5}^{(3)}, \mathbf{Z}_{6}^{(3)}, Z_{7}^{(3)}, \mathbf{Z}_{8}^{(3)}, \mathbf{Z}_{9}^{(3)}, \mathbf{Z}_{10}^{(3)}$	$\mathbf{Z}_{1}^{(7)}, Z_{5}^{(7)}, \mathbf{Z}_{6}^{(7)}, Z_{7}^{(7)}, Z_{2}^{(7)}, \mathbf{Z}_{3}^{(7)}, Z_{4}^{(7)}, \mathbf{Z}_{8}^{(7)}, \mathbf{Z}_{9}^{(6)}, \mathbf{Z}_{10}^{(6)}$
4	$Z_{1}^{(4)}, \mathbf{Z}_{2}^{(4)}, Z_{3}^{(4)}, \mathbf{Z}_{4}^{(4)}, \mathbf{Z}_{5}^{(4)}, Z_{6}^{(4)}, \mathbf{Z}_{7}^{(4)}, Z_{8}^{(4)}, \mathbf{Z}_{9}^{(4)}, \mathbf{Z}_{10}^{(4)}$	$Z_{1}^{(6)}, \mathbf{Z}_{5}^{(6)}, Z_{6}^{(6)}, \mathbf{Z}_{7}^{(6)}, \mathbf{Z}_{2}^{(6)}, Z_{3}^{(6)}, \mathbf{Z}_{4}^{(6)}, Z_{8}^{(6)}, \mathbf{Z}_{9}^{(5)}, \mathbf{Z}_{10}^{(5)}$
5	$\mathbf{Z}_{1}^{(5)}, Z_{2}^{(5)}, \mathbf{Z}_{3}^{(5)}, Z_{4}^{(5)}, Z_{5}^{(5)}, \mathbf{Z}_{6}^{(5)}, Z_{7}^{(5)}, \mathbf{Z}_{8}^{(5)}, \mathbf{Z}_{9}^{(5)}, \mathbf{Z}_{10}^{(5)}$	$\mathbf{Z}_{1}^{(5)}, Z_{5}^{(5)}, \mathbf{Z}_{6}^{(5)}, Z_{7}^{(5)}, Z_{2}^{(5)}, \mathbf{Z}_{3}^{(5)}, Z_{4}^{(5)}, \mathbf{Z}_{8}^{(5)}, \mathbf{Z}_{9}^{(4)}, \mathbf{Z}_{10}^{(4)}$
6	$Z_{1}^{(6)}, \mathbf{Z}_{2}^{(6)}, Z_{3}^{(6)}, \mathbf{Z}_{4}^{(6)}, \mathbf{Z}_{5}^{(6)}, Z_{6}^{(6)}, \mathbf{Z}_{7}^{(6)}, Z_{8}^{(6)}, \mathbf{Z}_{9}^{(6)}, \mathbf{Z}_{10}^{(6)}$	$Z_{1}^{(4)}, \mathbf{Z}_{5}^{(4)}, Z_{6}^{(4)}, \mathbf{Z}_{7}^{(4)}, \mathbf{Z}_{2}^{(4)}, Z_{3}^{(4)}, \mathbf{Z}_{4}^{(4)}, Z_{8}^{(4)}, \mathbf{Z}_{9}^{(3)}, \mathbf{Z}_{10}^{(3)}$
7	$\mathbf{Z}_{1}^{(7)}, Z_{2}^{(7)}, \mathbf{Z}_{3}^{(7)}, Z_{4}^{(7)}, Z_{5}^{(7)}, \mathbf{Z}_{6}^{(7)}, Z_{7}^{(7)}, \mathbf{Z}_{8}^{(7)}, \mathbf{Z}_{9}^{(7)}, \mathbf{Z}_{10}^{(7)}$	$\mathbf{Z}_{1}^{(3)}, Z_{5}^{(3)}, \mathbf{Z}_{6}^{(3)}, Z_{7}^{(3)}, Z_{2}^{(3)}, \mathbf{Z}_{3}^{(3)}, Z_{4}^{(3)}, \mathbf{Z}_{8}^{(3)}, \mathbf{Z}_{9}^{(2)}, \mathbf{Z}_{10}^{(2)}$
8	$Z_{1}^{(8)}, \mathbf{Z}_{2}^{(8)}, Z_{3}^{(8)}, \mathbf{Z}_{4}^{(8)}, \mathbf{Z}_{5}^{(8)}, Z_{6}^{(8)}, \mathbf{Z}_{7}^{(8)}, Z_{8}^{(8)}, \mathbf{Z}_{9}^{(8)}, \mathbf{Z}_{10}^{(8)}$	$Z_{1}^{(2)}, \mathbf{Z}_{5}^{(2)}, Z_{6}^{(2)}, \mathbf{Z}_{7}^{(2)}, \mathbf{Z}_{2}^{(2)}, Z_{3}^{(2)}, \mathbf{Z}_{4}^{(2)}, Z_{8}^{(2)}, \mathbf{Z}_{9}^{(1)}, \mathbf{Z}_{10}^{(1)}$
9	$\mathbf{Z}_{1}^{(9)}, Z_{2}^{(9)}, \mathbf{Z}_{3}^{(9)}, Z_{4}^{(9)}, Z_{5}^{(9)}, \mathbf{Z}_{6}^{(9)}, Z_{7}^{(9)}, \mathbf{Z}_{8}^{(9)}$	$\mathbf{Z}_{1}^{(1)}, Z_{2}^{(1)}, \mathbf{Z}_{3}^{(1)}, Z_{4}^{(1)}, Z_{5}^{(1)}, \mathbf{Z}_{6}^{(1)}, Z_{7}^{(1)}, \mathbf{Z}_{8}^{(1)}$

2.3.9 Encryption and Decryption Frameworks of MESH-128(8)

MESH-128(8) is an iterated block cipher operating on 128-bit blocks, under a 256-bit key and consisting of eight rounds plus an output transformation (OT) [45]. All internal operations are on 8-bit words. Figure 2.7 depicts the encryption and decryption frameworks for MESH-128(8).

MESH-128(8) uses the same three group operations as MESH-64(8) on 8-bit words.

Each round of MESH-128(8) consists of two halves: a key-mixing (KM) and a multiplication-addition (MA) half-rounds, in this order.

Let $X^{(i)} = (X_1^{(i)}, X_2^{(i)}, X_3^{(i)}, \ldots, X_{15}^{(i)}, X_{16}^{(i)})$, with $X_j^{(i)} \in \mathbb{Z}_2^8$, $1 \leq j \leq 16$, denote the input text block to the i-th round, $1 \leq i \leq 9$.

For i odd, the i-th KM half-round output is

$$Y^{(i)} = (Y_1^{(i)}, Y_2^{(i)}, Y_3^{(i)}, \ldots, Y_{15}^{(i)}, Y_{16}^{(i)}) =$$
$$(X_1^{(i)} \odot Z_1^{(i)}, X_2^{(i)} \boxplus Z_2^{(i)}, X_3^{(i)} \odot Z_3^{(i)}, X_4^{(i)} \boxplus Z_4^{(i)},$$
$$X_5^{(i)} \odot Z_5^{(i)}, X_6^{(i)} \boxplus Z_6^{(i)}, X_7^{(i)} \odot Z_7^{(i)}, X_8^{(i)} \boxplus Z_8^{(i)},$$
$$X_9^{(i)} \boxplus Z_9^{(i)}, X_{10}^{(i)} \odot Z_{10}^{(i)}, X_{11}^{(i)} \boxplus Z_{11}^{(i)}, X_{12}^{(i)} \odot Z_{12}^{(i)},$$
$$X_{13}^{(i)} \boxplus Z_{13}^{(i)}, X_{14}^{(i)} \odot Z_{14}^{(i)}, X_{15}^{(i)} \boxplus Z_{15}^{(i)}, X_{16}^{(i)} \odot Z_{16}^{(i)}).$$

For even i, the i-th KM half-round output is

$$Y^{(i)} = (Y_1^{(i)}, Y_2^{(i)}, Y_3^{(i)}, \ldots, Y_{15}^{(i)}, Y_{16}^{(i)}) =$$
$$(X_1^{(i)} \boxplus Z_1^{(i)}, X_2^{(i)} \odot Z_2^{(i)}, X_3^{(i)} \boxplus Z_3^{(i)}, X_4^{(i)} \odot Z_4^{(i)},$$
$$X_5^{(i)} \boxplus Z_5^{(i)}, X_6^{(i)} \odot Z_6^{(i)}, X_7^{(i)} \boxplus Z_7^{(i)}, X_8^{(i)} \odot Z_8^{(i)},$$
$$X_9^{(i)} \odot Z_9^{(i)}, X_{10}^{(i)} \boxplus Z_{10}^{(i)}, X_{11}^{(i)} \odot Z_{11}^{(i)}, X_{12}^{(i)} \boxplus Z_{12}^{(i)},$$
$$X_{13}^{(i)} \odot Z_{13}^{(i)}, X_{14}^{(i)} \boxplus Z_{14}^{(i)}, X_{15}^{(i)} \odot Z_{15}^{(i)}, X_{16}^{(i)} \boxplus Z_{16}^{(i)}).$$

In both cases, $Y^{(i)}$ becomes the input to the i-th MA half-round.

The input to the i-th MA-box is the ordered 8-tuple:

$$(n_i, q_i, m_i, u_i, a_i, b_i, d_i, e_i),$$

where $n_i = Y_1^{(i)} \oplus Y_9^{(i)}$, $q_i = Y_2^{(i)} \oplus Y_{10}^{(i)}$, $m_i = Y_3^{(i)} \oplus Y_{11}^{(i)}$, $u_i = Y_4^{(i)} \oplus Y_{12}^{(i)}$, $a_i = Y_5^{(i)} \oplus Y_{13}^{(i)}$, $b_i = Y_6^{(i)} \oplus Y_{14}^{(i)}$, $d_i = Y_7^{(i)} \oplus Y_{15}^{(i)}$ and $e_i = Y_8^{(i)} \oplus Y_{16}^{(i)}$.

The MA-box consists of two layers of interleaved $\odot$ and $\boxplus$ operations following a zigzag pattern. The MA-box output is the ordered 8-tuple $(r_i, s_i, t_i, v_i, f_i, g_i, h_i, l_i)$, where

$l_i = (((((((n_i \odot Z_{17}^{(i)}) \boxplus q_i) \odot m_i) \boxplus u_i) \odot a_i) \boxplus b_i) \odot d_i) \boxplus e_i) \odot Z_{18}^{(i)}$,

$h_i = ((((((n_i \odot Z_{17}^{(i)}) \boxplus q_i) \odot m_i) \boxplus u_i) \odot a_i) \boxplus b_i) \odot d_i) \boxplus l_i$,

$g_i = ((((((n_i \odot Z_{17}^{(i)}) \boxplus q_i) \odot m_i) \boxplus u_i) \odot a_i) \boxplus b_i) \odot h_i,$
$f_i = (((((n_i \odot Z_{17}^{(i)}) \boxplus q_i) \odot m_i) \boxplus u_i) \odot a_i) \boxplus g_i,$
$v_i = ((((n_i \odot Z_{17}^{(i)}) \boxplus q_i) \odot m_i) \boxplus u_i) \odot f_i,$
$t_i = (((n_i \odot Z_{17}^{(i)}) \boxplus q_i) \odot m_i) \boxplus v_i,$
$s_i = ((n_i \odot Z_{17}^{(i)}) \boxplus q_i) \odot t_i$ and
$r_i = (n_i \odot Z_{17}^{(i)}) \boxplus s_i.$

For a 128-bit block $(A, B, C, D, E, F, G, H, I, J, K, L, M, N, O, P)$, the byte swap operation performs a fixed permutation and is denoted

$$\sigma(A, B, C, D, E, F, G, H, I, J, K, L, M, N, O, P) =$$

$$(A, I, J, K, L, M, N, O, B, C, D, E, F, G, H, P).$$

The output of the i-th round is

$$X^{(i+1)} = (Y_1^{(i)} \oplus l_i, Y_9^{(i)} \oplus l_i, Y_{10}^{(i)} \oplus h_i, Y_{11}^{(i)} \oplus g_i,$$
$$Y_{12}^{(i)} \oplus f_i, Y_{13}^{(i)} \oplus v_i, Y_{14}^{(i)} \oplus t_i, Y_{15}^{(i)} \oplus s_i,$$
$$Y_2^{(i)} \oplus h_i, Y_3^{(i)} \oplus g_i, Y_4^{(i)} \oplus f_i, Y_5^{(i)} \oplus v_i,$$
$$Y_6^{(i)} \oplus t_i, Y_7^{(i)} \oplus s_i, Y_8^{(i)} \oplus r_i, Y_{16}^{(i)} \oplus r_i),$$

already taking into account σ at the end of the round.

This process is repeated seven more times for MESH-128(8). See Fig. 2.7.

The output transformation is simply an application of σ followed by a (odd) KM half-round. Effectively, the σ in the OT undoes the last σ applied in the previous round.

$$X^{(9)} = (Y_1^{(8)} \oplus l_8, Y_2^{(8)} \oplus h_8, Y_3^{(8)} \oplus g_8, Y_4^{(8)} \oplus f_8,$$
$$Y_5^{(8)} \oplus v_8, Y_6^{(8)} \oplus t_8, Y_7^{(8)} \oplus s_8, Y_8^{(8)} \oplus r_8,$$
$$Y_9^{(8)} \oplus l_8, Y_{10}^{(8)} \oplus h_8, Y_{11}^{(8)} \oplus g_8, Y_{12}^{(8)} \oplus f_8,$$
$$Y_{13}^{(8)} \oplus v_8, Y_{14}^{(8)} \oplus t_8, Y_{15}^{(8)} \oplus s_8, Y_{16}^{(8)} \oplus r_8).$$

The ciphertext becomes

$$C = (X_1^{(9)} \odot Z_1^{(9)}, X_2^{(9)} \boxplus Z_2^{(9)}, X_3^{(9)} \odot Z_3^{(9)}, X_4^{(9)} \boxplus Z_4^{(9)},$$
$$X_5^{(9)} \odot Z_5^{(9)}, X_6^{(9)} \boxplus Z_6^{(9)}, X_7^{(9)} \odot Z_7^{(9)}, X_8^{(9)} \boxplus Z_8^{(9)},$$
$$X_9^{(9)} \boxplus Z_9^{(9)}, X_{10}^{(9)} \odot Z_{10}^{(9)}, X_{11}^{(9)} \boxplus Z_{11}^{(9)}, X_{12}^{(9)} \odot Z_{12}^{(9)},$$
$$X_{13}^{(9)} \boxplus Z_{13}^{(9)}, X_{14}^{(9)} \odot Z_{14}^{(9)}, X_{15}^{(9)} \boxplus Z_{15}^{(9)}, X_{16}^{(9)} \odot Z_{16}^{(9)}).$$

If X denotes a 128-bit block, KM_i denotes the i-th KM half-round, MA_i denotes the i-th MA half-round, then, a full round can be denoted by

$$\chi_i(X) = \sigma \circ \text{MA}_i \circ \text{KM}_i(X) = \sigma(\text{MA}_i(\text{KM}_i(X))).$$

One full MESH-128(8) encryption of a 128-bit plaintext block P into a 128-bit ciphertext block C can be denoted

$$C = \mathrm{KM}_9 \circ \sigma \circ \chi_8 \circ \chi_7 \circ \chi_6 \circ \chi_5 \circ \chi_4 \circ \chi_3 \circ \chi_2 \circ \chi_1(P)$$

where functional composition operates in right-to-left order.

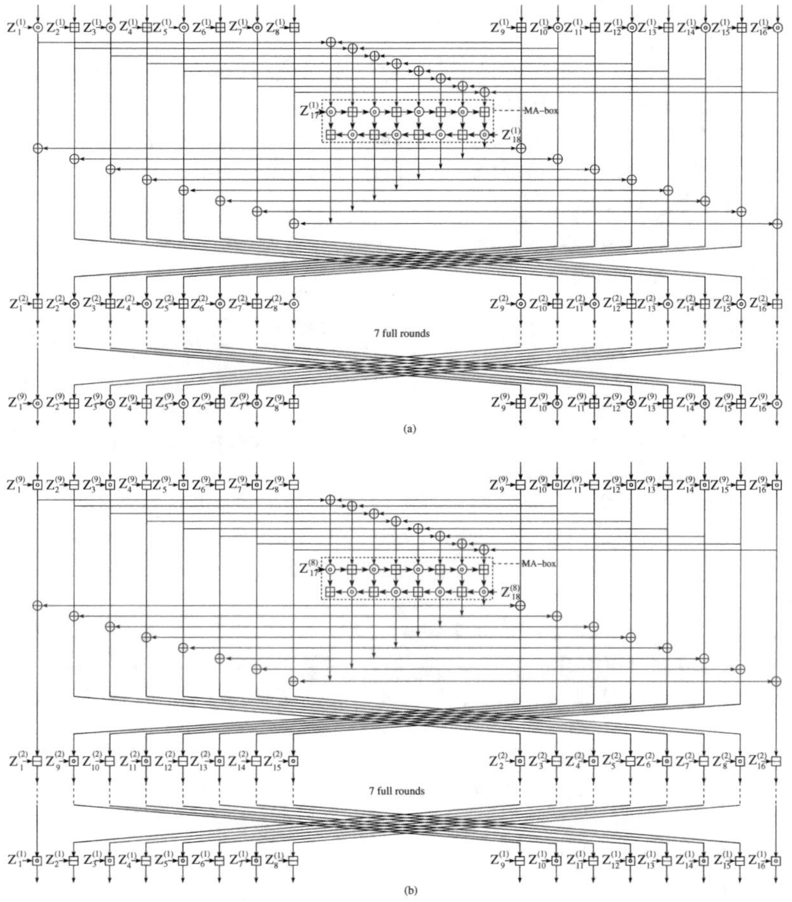

Fig. 2.7 Computational graph of the MESH-128(8) cipher: (a) encryption [45] and (b) decryption. Note the use of modular division (⊡) and subtraction (⊟) in decryption.

2.3.10 Key Schedule Algorithm of MESH-128(8)

In total, 160 8-bit round subkeys (1280 bits in total) are generated by the key schedule of MESH-128(8) for either the encryption or decryption of a single 128-bit text block. Due to a typo in the original publication [45], a wrong polynomial was used, which did not provide fast key diffusion. The correct polynomial should have been $q(x) = x^{32} + x^{31} + x^{30} + x^{10} + 1$.

The key schedule of MESH-128(8) consists of the following steps:

- the same 8-bit constants c_i defined for MESH-64(8) in Sect. 2.3.8 are needed for MESH-128(8).
- the 256-bit user key K is partitioned into thirty-two 8-bit words K_i, for $0 \leq i \leq 31$, and assigned to the first 32 round subkeys: $Z_{i+1}^{(1)} = K_i \oplus c_i$ for $0 \leq i \leq 17$ and $Z_{j \bmod 18+1}^{(2)} = K_j \oplus c_j$ for $18 \leq j \leq 31$.
- each subsequent 8-bit subkey is computed iteratively using a NLFSR design as follows[4]:

$$Z_{l(i)}^{(h(i))} = (((Z_{l(i-32)}^{(h(i-32))} \boxplus Z_{l(i-22)}^{(h(i-22))}) \oplus Z_{l(i-2)}^{(h(i-2))}) \boxplus Z_{l(i-1)}^{(h(i-1))})) \lll 1 \oplus c_i \tag{2.13}$$

for $32 \leq i \leq 159$, where $x \lll 1$ means x left rotated by one bit, $h(i) = \lfloor i/18 \rfloor + 1$ and $l(i) = i \bmod 18 + 1$.

The key schedule of MESH-128(8) achieves complete key diffusion due to the $h(i)$ and $l(i)$ indices, which are based on the primitive polynomial $q(x) = x^{32} + x^{31} + x^{30} + x^{10} + 1$, the bit rotation and the interleaving of $\boxplus$ and $\oplus$.

Let us deduce that each subkey starting with $Z_{16}^{(3)}$ depends upon all 32 user key words. Initially, $Z_1^{(1)} = K_0 \oplus c_0$, $Z_2^{(1)} = K_1 \oplus c_1$, $Z_3^{(1)} = K_2 \oplus c_2$, $Z_4^{(1)} = K_3 \oplus c_3$, $Z_5^{(1)} = K_4 \oplus c_4$, $Z_6^{(1)} = K_5 \oplus c_5$, $Z_7^{(1)} = K_6 \oplus c_6$, $Z_8^{(1)} = K_7 \oplus c_7$, $Z_9^{(1)} = K_8 \oplus c_8$, $Z_{10}^{(1)} = K_9 \oplus c_9$, $Z_{11}^{(1)} = K_{10} \oplus c_{10}$, $Z_{12}^{(1)} = K_{11} \oplus c_{11}$, $Z_{13}^{(1)} = K_{12} \oplus c_{12}$, $Z_{14}^{(1)} = K_{13} \oplus c_{13}$, $Z_{15}^{(1)} = K_{14} \oplus c_{14}$, $Z_{16}^{(1)} = K_{15} \oplus c_{15}$, $Z_{17}^{(1)} = K_{16} \oplus c_{16}$, $Z_{18}^{(1)} = K_{17} \oplus c_{17}$, $Z_1^{(2)} = K_{18} \oplus c_{18}$, $Z_2^{(2)} = K_{19} \oplus c_{19}$, $Z_3^{(2)} = K_{20} \oplus c_{20}$, $Z_4^{(2)} = K_{21} \oplus c_{21}$, $Z_5^{(2)} = K_{22} \oplus c_{22}$, $Z_6^{(2)} = K_{23} \oplus c_{23}$, $Z_7^{(2)} = K_{24} \oplus c_{24}$, $Z_8^{(2)} = K_{25} \oplus c_{25}$, $Z_9^{(2)} = K_{26} \oplus c_{26}$, $Z_{10}^{(2)} = K_{27} \oplus c_{27}$, $Z_{11}^{(2)} = K_{28} \oplus c_{28}$, $Z_{12}^{(2)} = K_{29} \oplus c_{29}$, $Z_{13}^{(2)} = K_{30} \oplus c_{30}$ and $Z_{14}^{(2)} = K_{31} \oplus c_{31}$.

Further, for $i = 32$: $Z_{15}^{(2)} = (((Z_1^{(1)} \boxplus Z_{11}^{(1)}) \oplus Z_{13}^{(2)}) \boxplus Z_{14}^{(2)}) \lll 1 \oplus c_{32}$, which depends on four words from K.

Next, for $i = 33$: $Z_{16}^{(2)} = (((Z_2^{(1)} \boxplus Z_{12}^{(1)}) \oplus Z_{14}^{(2)}) \boxplus Z_{15}^{(2)}) \lll 1 \oplus c_{33}$, which depends on seven words from K.

Next, for $i = 34$: $Z_{17}^{(2)} = (((Z_3^{(1)} \boxplus Z_{13}^{(1)}) \oplus Z_{15}^{(2)}) \boxplus Z_{16}^{(2)}) \lll 1 \oplus c_{34}$, which depends on eight words from K.

---

[4] Bit rotation has higher precedence than $\oplus$.

Next, for $i = 35$: $Z_{18}^{(2)} = (((Z_4^{(1)} \boxplus Z_{14}^{(1)}) \oplus Z_{16}^{(2)}) \boxplus Z_{17}^{(2)}) \lll 1 \oplus c_{35}$, which depends on nine words from K.

Next, for $i = 36$: $Z_1^{(3)} = (((Z_5^{(1)} \boxplus Z_{15}^{(1)}) \oplus Z_{17}^{(2)}) \boxplus Z_{18}^{(2)}) \lll 1 \oplus c_{36}$, which depends on twelve words from K.

Next, for $i = 37$: $Z_2^{(3)} = (((Z_6^{(1)} \boxplus Z_{16}^{(1)}) \oplus Z_{18}^{(2)}) \boxplus Z_1^{(3)}) \lll 1 \oplus c_{37}$, which depends on thirteen words from K.

Next, for $i = 38$: $Z_3^{(3)} = (((Z_7^{(1)} \boxplus Z_{17}^{(1)}) \oplus Z_1^{(3)}) \boxplus Z_2^{(3)}) \lll 1 \oplus c_{38}$, which depends on sixteen words from K.

Next, for $i = 39$: $Z_4^{(3)} = (((Z_8^{(1)} \boxplus Z_{18}^{(1)}) \oplus Z_2^{(3)}) \boxplus Z_3^{(3)}) \lll 1 \oplus c_{39}$, which depends on eighteen words from K.

Next, for $i = 40$: $Z_5^{(3)} = (((Z_9^{(1)} \boxplus Z_1^{(2)}) \oplus Z_3^{(3)}) \boxplus Z_4^{(3)}) \lll 1 \oplus c_{40}$, which depends on twenty words from K.

Next, for $i = 41$: $Z_6^{(3)} = (((Z_{10}^{(1)} \boxplus Z_2^{(2)}) \oplus Z_4^{(3)}) \boxplus Z_5^{(3)}) \lll 1 \oplus c_{41}$, which depends on 22 words from K.

Next, for $i = 42$: $Z_7^{(3)} = (((Z_{11}^{(1)} \boxplus Z_3^{(2)}) \oplus Z_5^{(3)}) \boxplus Z_6^{(3)}) \lll 1 \oplus c_{42}$, which depends on 23 words from K.

Next, for $i = 43$: $Z_8^{(3)} = (((Z_{12}^{(1)} \boxplus Z_4^{(2)}) \oplus Z_6^{(3)}) \boxplus Z_7^{(3)}) \lll 1 \oplus c_{43}$, which depends on 24 words from K.

Next, for $i = 44$: $Z_9^{(3)} = (((Z_{13}^{(1)} \boxplus Z_5^{(2)}) \oplus Z_7^{(3)}) \boxplus Z_8^{(3)}) \lll 1 \oplus c_{44}$, which depends on 25 words from K.

Next, for $i = 45$: $Z_{10}^{(3)} = (((Z_{14}^{(1)} \boxplus Z_6^{(2)}) \oplus Z_8^{(3)}) \boxplus Z_9^{(3)}) \lll 1 \oplus c_{45}$, which depends on 26 words from K.

Next, for $i = 46$: $Z_{11}^{(3)} = (((Z_{15}^{(1)} \boxplus Z_7^{(2)}) \oplus Z_9^{(3)}) \boxplus Z_{10}^{(3)}) \lll 1 \oplus c_{46}$, which depends on 27 words from K.

Next, for $i = 47$: $Z_{12}^{(3)} = (((Z_{16}^{(1)} \boxplus Z_8^{(2)}) \oplus Z_{10}^{(3)}) \boxplus Z_{11}^{(3)}) \lll 1 \oplus c_{47}$, which depends on 28 words from K.

Next, for $i = 48$: $Z_{13}^{(3)} = (((Z_{17}^{(1)} \boxplus Z_9^{(2)}) \oplus Z_{11}^{(3)}) \boxplus Z_{12}^{(3)}) \lll 1 \oplus c_{48}$, which depends on 29 words from K.

Next, for $i = 49$: $Z_{14}^{(3)} = (((Z_{18}^{(1)} \boxplus Z_{10}^{(2)}) \oplus Z_{12}^{(3)}) \boxplus Z_{13}^{(3)}) \lll 1 \oplus c_{49}$, which depends on 30 words from K.

Next, for $i = 50$: $Z_{15}^{(3)} = (((Z_1^{(2)} \boxplus Z_{11}^{(2)}) \oplus Z_{13}^{(3)}) \boxplus Z_{14}^{(3)}) \lll 1 \oplus c_{50}$, which depends on 31 words from K.

Finally, for $i = 51$: $Z_{16}^{(3)} = (((Z_2^{(2)} \boxplus Z_{12}^{(2)}) \oplus Z_{14}^{(3)}) \boxplus Z_{15}^{(3)}) \lll 1 \oplus c_{51}$, which depends on all 32 words from K. Thus, it takes three full rounds for MESH-128(8) to achieve complete key diffusion.

Decryption in MESH-128(8) uses essentially the same framework as for encryption, except for the order and values of some round subkeys.

Table 2.9 lists the encryption and decryption subkeys for each round of MESH-128(8). Note that the order of operations $\odot$ and $\boxplus$ changes for encryption in the KM half-rounds for even- and odd-numbered rounds. Multiplicative subkeys are depicted in boldface type.

Table 2.9 Encryption and decryption round subkeys for MESH-128(8) cipher.

round	Encryption subkeys
1	$\mathbf{Z}_1^{(1)}, Z_2^{(1)}, \mathbf{Z}_3^{(1)}, Z_4^{(1)}, \mathbf{Z}_5^{(1)}, Z_6^{(1)}, \mathbf{Z}_7^{(1)}, Z_8^{(1)}, Z_9^{(1)}, \mathbf{Z}_{10}^{(1)}, Z_{11}^{(1)}, \mathbf{Z}_{12}^{(1)}, Z_{13}^{(1)}, \mathbf{Z}_{14}^{(1)}, Z_{15}^{(1)}, \mathbf{Z}_{16}^{(1)}, \mathbf{Z}_{17}^{(1)}, \mathbf{Z}_{18}^{(1)}$
2	$Z_1^{(2)}, \mathbf{Z}_2^{(2)}, Z_3^{(2)}, \mathbf{Z}_4^{(2)}, Z_5^{(2)}, \mathbf{Z}_6^{(2)}, Z_7^{(2)}, \mathbf{Z}_8^{(2)}, \mathbf{Z}_9^{(2)}, Z_{10}^{(2)}, \mathbf{Z}_{11}^{(2)}, Z_{12}^{(2)}, \mathbf{Z}_{13}^{(2)}, Z_{14}^{(2)}, \mathbf{Z}_{15}^{(2)}, Z_{16}^{(2)}, \mathbf{Z}_{17}^{(2)}, \mathbf{Z}_{18}^{(2)}$
3	$\mathbf{Z}_1^{(3)}, Z_2^{(3)}, \mathbf{Z}_3^{(3)}, Z_4^{(3)}, \mathbf{Z}_5^{(3)}, Z_6^{(3)}, \mathbf{Z}_7^{(3)}, Z_8^{(3)}, Z_9^{(3)}, \mathbf{Z}_{10}^{(3)}, Z_{11}^{(3)}, \mathbf{Z}_{12}^{(3)}, Z_{13}^{(3)}, \mathbf{Z}_{14}^{(3)}, Z_{15}^{(3)}, \mathbf{Z}_{16}^{(3)}, \mathbf{Z}_{17}^{(3)}, \mathbf{Z}_{18}^{(3)}$
4	$Z_1^{(4)}, \mathbf{Z}_2^{(4)}, Z_3^{(4)}, \mathbf{Z}_4^{(4)}, Z_5^{(4)}, \mathbf{Z}_6^{(4)}, Z_7^{(4)}, \mathbf{Z}_8^{(4)}, \mathbf{Z}_9^{(4)}, Z_{10}^{(4)}, \mathbf{Z}_{11}^{(4)}, Z_{12}^{(4)}, \mathbf{Z}_{13}^{(4)}, Z_{14}^{(4)}, \mathbf{Z}_{15}^{(4)}, Z_{16}^{(4)}, \mathbf{Z}_{17}^{(4)}, \mathbf{Z}_{18}^{(4)}$
5	$\mathbf{Z}_1^{(5)}, Z_2^{(5)}, \mathbf{Z}_3^{(5)}, Z_4^{(5)}, \mathbf{Z}_5^{(5)}, Z_6^{(5)}, \mathbf{Z}_7^{(5)}, Z_8^{(5)}, Z_9^{(5)}, \mathbf{Z}_{10}^{(5)}, Z_{11}^{(5)}, \mathbf{Z}_{12}^{(5)}, Z_{13}^{(5)}, \mathbf{Z}_{14}^{(5)}, Z_{15}^{(5)}, \mathbf{Z}_{16}^{(5)}, \mathbf{Z}_{17}^{(5)}, \mathbf{Z}_{18}^{(5)}$
6	$Z_1^{(6)}, \mathbf{Z}_2^{(6)}, Z_3^{(6)}, \mathbf{Z}_4^{(6)}, Z_5^{(6)}, \mathbf{Z}_6^{(6)}, Z_7^{(6)}, \mathbf{Z}_8^{(6)}, \mathbf{Z}_9^{(6)}, Z_{10}^{(6)}, \mathbf{Z}_{11}^{(6)}, Z_{12}^{(6)}, \mathbf{Z}_{13}^{(6)}, Z_{14}^{(6)}, \mathbf{Z}_{15}^{(6)}, Z_{16}^{(6)}, \mathbf{Z}_{17}^{(6)}, \mathbf{Z}_{18}^{(6)}$
7	$\mathbf{Z}_1^{(7)}, Z_2^{(7)}, \mathbf{Z}_3^{(7)}, Z_4^{(7)}, \mathbf{Z}_5^{(7)}, Z_6^{(7)}, \mathbf{Z}_7^{(7)}, Z_8^{(7)}, Z_9^{(7)}, \mathbf{Z}_{10}^{(7)}, Z_{11}^{(7)}, \mathbf{Z}_{12}^{(7)}, Z_{13}^{(7)}, \mathbf{Z}_{14}^{(7)}, Z_{15}^{(7)}, \mathbf{Z}_{16}^{(7)}, \mathbf{Z}_{17}^{(7)}, \mathbf{Z}_{18}^{(7)}$
8	$Z_1^{(8)}, \mathbf{Z}_2^{(8)}, Z_3^{(8)}, \mathbf{Z}_4^{(8)}, Z_5^{(8)}, \mathbf{Z}_6^{(8)}, Z_7^{(8)}, \mathbf{Z}_8^{(8)}, \mathbf{Z}_9^{(8)}, Z_{10}^{(8)}, \mathbf{Z}_{11}^{(8)}, Z_{12}^{(8)}, \mathbf{Z}_{13}^{(8)}, Z_{14}^{(8)}, \mathbf{Z}_{15}^{(8)}, Z_{16}^{(8)}, \mathbf{Z}_{17}^{(8)}, \mathbf{Z}_{18}^{(8)}$
9	$\mathbf{Z}_1^{(9)}, Z_2^{(9)}, \mathbf{Z}_3^{(9)}, Z_4^{(9)}, \mathbf{Z}_5^{(9)}, Z_6^{(9)}, \mathbf{Z}_7^{(9)}, Z_8^{(9)}, Z_9^{(9)}, \mathbf{Z}_{10}^{(9)}, Z_{11}^{(9)}, \mathbf{Z}_{12}^{(9)}, Z_{13}^{(9)}, \mathbf{Z}_{14}^{(9)}, Z_{15}^{(9)}, \mathbf{Z}_{16}^{(9)}$

round	Decryption subkeys
1	$\mathbf{Z}_1^{(9)}, Z_2^{(9)}, \mathbf{Z}_3^{(9)}, Z_4^{(9)}, \mathbf{Z}_5^{(9)}, Z_6^{(9)}, \mathbf{Z}_7^{(9)}, Z_8^{(9)}, Z_9^{(9)}, \mathbf{Z}_{10}^{(9)}, Z_{11}^{(9)}, \mathbf{Z}_{12}^{(9)}, Z_{13}^{(9)}, \mathbf{Z}_{14}^{(9)}, Z_{15}^{(9)}, \mathbf{Z}_{16}^{(9)}, \mathbf{Z}_{17}^{(8)}, \mathbf{Z}_{18}^{(8)}$
2	$Z_1^{(8)}, \mathbf{Z}_9^{(8)}, Z_{10}^{(8)}, \mathbf{Z}_{11}^{(8)}, Z_{12}^{(8)}, \mathbf{Z}_{13}^{(8)}, Z_{14}^{(8)}, \mathbf{Z}_{15}^{(8)}, \mathbf{Z}_2^{(8)}, Z_3^{(8)}, \mathbf{Z}_4^{(8)}, Z_5^{(8)}, \mathbf{Z}_6^{(8)}, Z_7^{(8)}, \mathbf{Z}_8^{(8)}, Z_{16}^{(8)}, \mathbf{Z}_{17}^{(7)}, \mathbf{Z}_{18}^{(7)}$
3	$\mathbf{Z}_1^{(7)}, Z_9^{(7)}, \mathbf{Z}_{10}^{(7)}, Z_{11}^{(7)}, \mathbf{Z}_{12}^{(7)}, Z_{13}^{(7)}, \mathbf{Z}_{14}^{(7)}, Z_{15}^{(7)}, Z_2^{(7)}, \mathbf{Z}_3^{(7)}, Z_4^{(7)}, \mathbf{Z}_5^{(7)}, Z_6^{(7)}, \mathbf{Z}_7^{(7)}, Z_8^{(7)}, \mathbf{Z}_{16}^{(7)}, \mathbf{Z}_{17}^{(6)}, \mathbf{Z}_{18}^{(6)}$
4	$Z_1^{(6)}, \mathbf{Z}_9^{(6)}, Z_{10}^{(6)}, \mathbf{Z}_{11}^{(6)}, Z_{12}^{(6)}, \mathbf{Z}_{13}^{(6)}, Z_{14}^{(6)}, \mathbf{Z}_{15}^{(6)}, \mathbf{Z}_2^{(6)}, Z_3^{(6)}, \mathbf{Z}_4^{(6)}, Z_5^{(6)}, \mathbf{Z}_6^{(6)}, Z_7^{(6)}, \mathbf{Z}_8^{(6)}, Z_{16}^{(6)}, \mathbf{Z}_{17}^{(5)}, \mathbf{Z}_{18}^{(5)}$
5	$\mathbf{Z}_1^{(5)}, Z_9^{(5)}, \mathbf{Z}_{10}^{(5)}, Z_{11}^{(5)}, \mathbf{Z}_{12}^{(5)}, Z_{13}^{(5)}, \mathbf{Z}_{14}^{(5)}, Z_{15}^{(5)}, Z_2^{(5)}, \mathbf{Z}_3^{(5)}, Z_4^{(5)}, \mathbf{Z}_5^{(5)}, Z_6^{(5)}, \mathbf{Z}_7^{(5)}, Z_8^{(5)}, \mathbf{Z}_{16}^{(5)}, \mathbf{Z}_{17}^{(4)}, \mathbf{Z}_{18}^{(4)}$
6	$Z_1^{(4)}, \mathbf{Z}_9^{(4)}, Z_{10}^{(4)}, \mathbf{Z}_{11}^{(4)}, Z_{12}^{(4)}, \mathbf{Z}_{13}^{(4)}, Z_{14}^{(4)}, \mathbf{Z}_{15}^{(4)}, \mathbf{Z}_2^{(4)}, Z_3^{(4)}, \mathbf{Z}_4^{(4)}, Z_5^{(4)}, \mathbf{Z}_6^{(4)}, Z_7^{(4)}, \mathbf{Z}_8^{(4)}, Z_{16}^{(4)}, \mathbf{Z}_{17}^{(3)}, \mathbf{Z}_{18}^{(3)}$
7	$\mathbf{Z}_1^{(3)}, Z_9^{(3)}, \mathbf{Z}_{10}^{(3)}, Z_{11}^{(3)}, \mathbf{Z}_{12}^{(3)}, Z_{13}^{(3)}, \mathbf{Z}_{14}^{(3)}, Z_{15}^{(3)}, Z_2^{(3)}, \mathbf{Z}_3^{(3)}, Z_4^{(3)}, \mathbf{Z}_5^{(3)}, Z_6^{(3)}, \mathbf{Z}_7^{(3)}, Z_7^{(3)}, \mathbf{Z}_{16}^{(3)}, \mathbf{Z}_{17}^{(2)}, \mathbf{Z}_{18}^{(2)}$
8	$Z_1^{(2)}, \mathbf{Z}_9^{(2)}, Z_{10}^{(2)}, \mathbf{Z}_{11}^{(2)}, Z_{12}^{(2)}, \mathbf{Z}_{13}^{(2)}, Z_{14}^{(2)}, \mathbf{Z}_{15}^{(2)}, \mathbf{Z}_2^{(2)}, Z_3^{(2)}, \mathbf{Z}_4^{(2)}, Z_5^{(2)}, \mathbf{Z}_6^{(2)}, Z_7^{(2)}, \mathbf{Z}_8^{(2)}, Z_{16}^{(2)}, \mathbf{Z}_{17}^{(1)}, \mathbf{Z}_{18}^{(1)}$
9	$\mathbf{Z}_1^{(1)}, Z_2^{(1)}, \mathbf{Z}_3^{(1)}, Z_4^{(1)}, \mathbf{Z}_5^{(1)}, Z_6^{(1)}, \mathbf{Z}_7^{(1)}, Z_8^{(1)}, Z_9^{(1)}, \mathbf{Z}_{10}^{(1)}, Z_{11}^{(1)}, \mathbf{Z}_{12}^{(1)}, Z_{13}^{(1)}, \mathbf{Z}_{14}^{(1)}, Z_{15}^{(1)}, \mathbf{Z}_{16}^{(1)}$

2.4 The RIDEA Block Cipher

RIDEA (Reverse IDEA) is an iterated block cipher operating on 64-bit blocks, under a 128-bit key and consisting of eight rounds plus an output transformation. All internal operations are on 16-bit words. RIDEA was designed by Yıldırım [63, 64].

Fig. 2.8 shows a pictorial representation of the encryption and the decryption operations of the full 8.5-round RIDEA. The arrows indicate the direction the data flows.

RIDEA uses the same three group operations on 16-bit words of PES and IDEA: addition modulo 2^{16}, denoted $\boxplus$, multiplication in $\mathrm{GF}(2^{16}+1)$ with $0 \equiv 2^{16}$, denoted $\odot$, and bitwise exclusive-or, denoted $\oplus$.

The main difference between IDEA and RIDEA is the MA-box, which is called RMA-box (Reverse MA-box) in RIDEA. The RMA-box was designed to improve the nonlinearity of the original MA-box of IDEA. In the RMA-box, both input words are fed to $\odot$, but one of the $\odot$ is not keyed directly by a subkey. Rather, the inputs are both key-and-text dependent. This redesign helps protect the $\odot$ against weak subkey values.

Just like IDEA, in RIDEA the two middle 16-bit words only use $\boxplus$ and $\oplus$ operations, which can be exploited by the Biryukov-Demirci relation: if one tracks the value of the input and output of *only the two middle 16-bit words in a block*, the results are equations (3.273) and (3.274) in Sect. 3.17. See also Fig. 3.27. The Biryukov-Demirci relation exploits this design feature in IDEA to extract a linear relation involving only the least significant bit value of these two middle words. This relation holds with certainty.

2.4.1 Encryption and Decryption Frameworks

Let $X^{(i)} = (X_1^{(i)}, X_2^{(i)}, X_3^{(i)}, X_4^{(i)})$ denote the 64-bit input text block to the i-th round, where $X_j^{(i)} \in \mathbb{Z}_2^{16}$, for $1 \leq j \leq 4$ and $1 \leq i \leq 9$.

One full encryption round in RIDEA can be decomposed into a Key-Mixing (KM) and a Reverse-Multiplication-Addition (RMA) half-round, in this order.

The j-th 16-bit subkey of the i-th round is denoted by $Z_j^{(i)}$, with $1 \leq j \leq 6$ and $1 \leq i \leq 9$. Subkey generation by the key schedule is detailed in Sect. 2.4.2.

Given the input $X^{(i)}$ to the i-th round, it is also the input to the i-th KM half-round. The KM half-round output is denoted $Y^{(i)} = (Y_1^{(i)}, Y_2^{(i)}, Y_3^{(i)}, Y_4^{(i)})$ where

$$Y^{(i)} = (X_1^{(i)} \odot Z_1^{(i)}, X_2^{(i)} \boxplus Z_2^{(i)}, X_3^{(i)} \boxplus Z_3^{(i)}, X_4^{(i)} \odot Z_4^{(i)}).$$

The output of the i-th KM half-round is the input to the i-th RMA half-round.

The input to the i-th RMA-box is the ordered pair $(n_i, q_i) = (Y_1^{(i)} \oplus Y_3^{(i)}, Y_2^{(i)} \oplus Y_4^{(i)})$. The RMA-box is a 32-bit keyed permutation consisting of $\odot$ and $\boxplus$ operations. See Fig. 2.8.

The output of the i-th RMA-box is the ordered pair (r_i, s_i), where

$$s_i = ((n_i \odot Z_5^{(i)}) \boxplus Z_6^{(i)}) \odot q_i,$$

$$r_i = (n_i \odot Z_5^{(i)}) \boxplus s_i.$$

Note that there is a multiplication operation in the RMA-box that does not depend directly on a round subkey.

The word swap at the end of a full round in RIDEA is the same as in IDEA: $\sigma(A, B, C, D) = (A, C, B, D)$.

The output of the i-th RMA half-round (including the word swap) becomes the i-th round output:

$$X^{(i+1)} = (Y_1^{(i)} \oplus s_i, Y_3^{(i)} \oplus s_i, Y_2^{(i)} \oplus r_i, Y_4^{(i)} \oplus r_i).$$

The output transformation (OT) in RIDEA consists of an application of σ followed by a KM half-round. In summary, the word swap in the OT cancels the word swap of the previous full round. It is equivalent to simply removing the word swap from the previous round:

$$X^{(9)} = (Y_1^{(8)} \oplus s_8, Y_2^{(8)} \oplus r_8, Y_3^{(8)} \oplus s_8, Y_4^{(8)} \oplus r_8).$$

The ciphertext is

$$C = (X_1^{(9)} \odot Z_1^{(9)}, X_2^{(9)} \boxplus Z_2^{(9)}, X_3^{(9)} \boxplus Z_3^{(9)}, X_4^{(9)} \odot Z_4^{(9)}).$$

In total, there are four $\odot$, four $\boxplus$ and six $\oplus$ operations per full round of RIDEA.

2.4.2 Key Schedule Algorithm

Both ciphers operate on the same key size, the same number of rounds and the same number of subkeys per round.

The key schedule algorithm of RIDEA is the same one used for IDEA. See Sect.2.2.2.

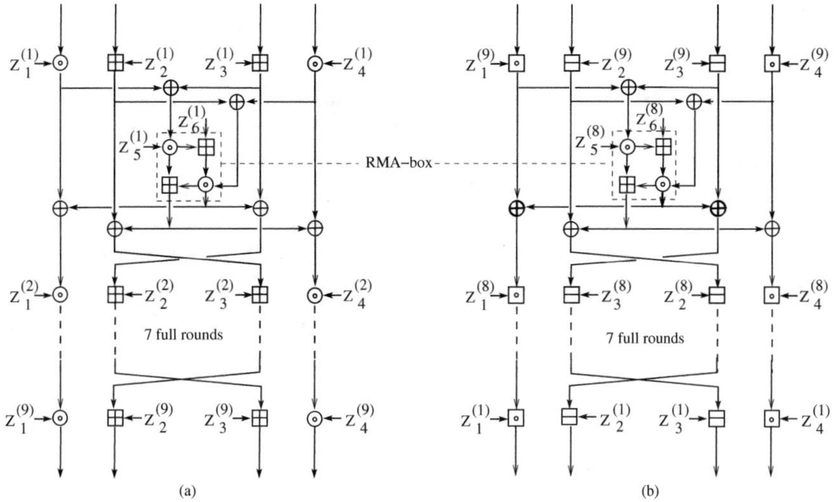

Fig. 2.8 Computational graph of the RIDEA cipher: (a) encryption [64] and (b) decryption. Note the use of modular division (⊡) and subtraction (⊟) in decryption.

2.5 The WIDEA-n Block Ciphers

WIDEA-n, for $n \in \{4, 8\}$, stands for two Wide-block cipher variants of the IDEA cipher [38] operating on $64n$-bit blocks, under a $128n$-bit key and consisting of eight rounds plus an output transformation[5]. The WIDEA-n ciphers were designed by Junod and Macchetti [29].

WIDEA-n is a rare 3-dimensional cipher design, combining n separate instances of the IDEA cipher through an $n \times n$ matrix derived from a Maximum Distance Separable (MDS) code [58, 42, 54, 18] placed inside the MA-box across the n instances of IDEA. Fig. 2.9 shows a pictorial graph of the 3-dimensional state of WIDEA-4. The original MA-box in IDEA becomes a so called MAD-box (Multiply-Add-Diffuse-box) in WIDEA-n [29]. The MAD-box is depicted in blue in Fig. 2.9.

This approach to design large-block cipher variants was aimed at creating a (large) internal permutation component for a hash function construction called HIDEA [28].

[5] © IACR. Published with permission.

2.5.1 Encryption and Decryption Frameworks

For WIDEA-4, there are four copies of IDEA side-by-side, connected by an 4×4 MDS matrix (see Fig. 2.9). Each full round of WIDEA-4 can be partitioned into a KM and a MAD half-round, in this order.

The i-th KM half-round in WIDEA-4 spans all four IDEA instances, and consists of four copies of the KM half-round of IDEA. Likewise, the i-th MAD half-round of WIDEA-4 consists of four copies of the MA half-round of IDEA connected by a 4×4 MDS matrix.

The MDS matrix in WIDEA-4 is the same one used in the AES cipher [21]:

$$\begin{pmatrix} 2\,3\,1\,1 \\ 1\,2\,3\,1 \\ 1\,1\,2\,3 \\ 3\,1\,1\,2 \end{pmatrix} . \tag{2.14}$$

In WIDEA-4, the matrix (2.14) is multiplied on the right by a 4×1 vector $(a, b, c, d)^t$ containing one column of the cipher state of four IDEA instances, where t denotes the vector transposition operation. This matrix product will be denoted simply $\text{MDS}(a, b, c, d)^t$.

For WIDEA-8, there are eight copies of IDEA side-by-side, connected by an 8×8 MDS matrix (depicted in red in Fig. 2.11).

The MDS matrix in WIDEA-8 is (2.15) and comes from the W cipher in the Whirlpool hash function [25]. The semantics for matrix multiplication is the same as for (2.14):

$$\begin{pmatrix} 1\,1\,4\,1\,8\,5\,2\,9 \\ 9\,1\,1\,4\,1\,8\,5\,2 \\ 2\,9\,1\,1\,4\,1\,8\,5 \\ 5\,2\,9\,1\,1\,4\,1\,8 \\ 8\,5\,2\,9\,1\,1\,4\,1 \\ 1\,8\,5\,2\,9\,1\,1\,4 \\ 4\,1\,8\,5\,2\,9\,1\,1 \\ 1\,4\,1\,8\,5\,2\,9\,1 \end{pmatrix} . \tag{2.15}$$

In both (2.14) and (2.15), matrix coefficients and operations are performed over $\text{GF}(2^{16}) = \text{GF}(2)[\text{x}]/(p(x))$, where $p(x) = x^{16} + x^5 + x^3 + x^2 + 1$ is an irreducible polynomial over GF(2). Therefore, WIDEA-n incorporates a fourth algebraic group, $\text{GF}(2^{16})$, to the three original groups in IDEA. The multiplication operator in $\text{GF}(2^{16})$ will be denoted $\otimes$.

The MA-box (Fig. 2.9) in IDEA works as follows: let the input to the i-th MA-box of the i-th IDEA instance, for $1 \leq i \leq n$, be the ordered pair (n_i, q_i), its output be (r_i, s_i) and $(Z_{5,i-1}, Z_{6,i-1})$ denote the round subkeys in the MA-box. We ignore the superscripts since they are not relevant in this setting. Then,

$$s_i = (n_i \odot Z_{5,i-1} \boxplus q_i) \odot Z_{6,i-1}$$

and

$$r_i = p_i \odot Z_{5,i-1} \boxplus s_i,$$

where $\odot$ has higher precedence than $\boxplus$.

For the MAD-box (Fig. 2.11): the MDS matrix in WIDEA-n is positioned inside the original MA-box of IDEA in every round after $n_i \odot Z_{5,i-1} \boxplus q_i$. This way, a single MAD-box of WIDEA-n has output (r'_i, s'_i) such that

- $s'_i = \text{MDS}(n_i \odot Z_{5,i-1} \boxplus q_i) \odot Z_{6,i-1}$ and $r'_i = n_i \odot Z_{5,i-1} \boxplus s'$ for $1 \leq i \leq n$ and $\text{MDS}(n_i \odot Z_{5,i-1} \boxplus q_i)$ stands for the multiplication of an MDS matrix (2.14) or (2.15) by an $n \times 1$ vector containing the n values $n_i \odot Z_{5,i-1} \boxplus q_i$, for $1 \leq i \leq n$.
- every output pair $(r'_i,\ s'_i)$ depends on all elements $(n_i,\ q_i,\ Z_{5,i-1})$ for $1 \leq i \leq n$, but not on $Z_{6,i-1}$.
- the placement of the MDS matrix also implies that its dependence on $(n_i,\ q_i,\ Z_{5,i-1})$ for $1 \leq i \leq n$ is spread to $(r'_i,\ s'_i)$ in all instances of IDEA in every round.
- the MDS matrix is preceded by $\boxplus$ and followed by $\odot$, while inside the matrix computation there is a sequence of $\oplus$ and multiplication operations in GF(2^{16}). Except for the repeated $\oplus$ in the matrix product, no other operation is repeated twice in a row in the MAD-box.
- since the MAD half-round is an involution there is no need to compute the inverse MDS matrix for decryption. See Fig. 2.10 and 2.12.

To allow a compact representation for analysis, and taking into account the 3-dimensional structure of WIDEA-4, the cipher state of WIDEA-4 is denoted by the 4×4 matrix (2.16):

$$\begin{pmatrix} a_{12} & a_{13} & a_{14} & a_{15} \\ a_8 & a_9 & a_{10} & a_{11} \\ a_4 & a_5 & a_6 & a_7 \\ a_0 & a_1 & a_2 & a_3 \end{pmatrix} \tag{2.16}$$

where each a_i, for $0 \leq i \leq 15$, is a 16-bit word. The word numbering in (2.16) follows from Fig. 2.9, where

$$(a_{4(j-1)}, a_{4(j-1)+1}, a_{4(j-1)+2}, a_{4(j-1)+3})$$

represent the state of the j-th IDEA instance for $1 \leq j \leq 4$. The MAD-boxes of the four IDEA instances are connected to each other via the 4×4 MDS matrix (2.14).

Analogously, the cipher state of WIDEA-8 is denoted by the 8×4 matrix:

$$\begin{pmatrix} a_{28} & a_{29} & a_{30} & a_{31} \\ a_{24} & a_{25} & a_{26} & a_{27} \\ a_{20} & a_{21} & a_{22} & a_{23} \\ a_{16} & a_{17} & a_{18} & a_{19} \\ a_{12} & a_{13} & a_{14} & a_{15} \\ a_{8} & a_{9} & a_{10} & a_{11} \\ a_{4} & a_{5} & a_{6} & a_{7} \\ a_{0} & a_{1} & a_{2} & a_{3} \end{pmatrix} \tag{2.17}$$

where each a_i, for $0 \leq i \leq 31$, is a 16-bit word. Word numbering in (2.17) follows from Fig. 2.11 where

$$(a_{4(j-1)}, a_{4(j-1)+1}, a_{4(j-1)+2}, a_{4(j-1)+3})$$

represents the state of the j-th IDEA instance, $1 \leq j \leq 8$. The MAD-boxes of the eight IDEA instances are connected by a single 8×8 MDS matrix (2.15) in every round. It is depicted in blue in Fig. 2.11.

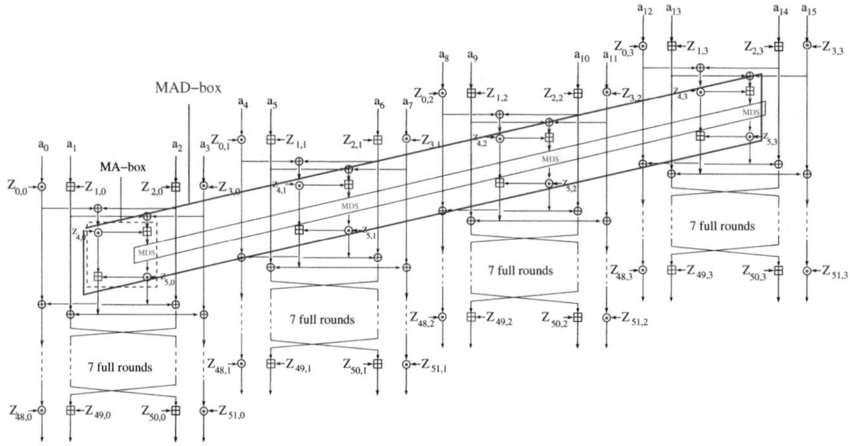

Fig. 2.9 Computational graph of the WIDEA-4 block cipher for encryption [47].

Note that the MAD half-rounds (that contain the MAD-boxes) are involutory mappings, just like in the original IDEA cipher. Therefore, the decryption framework of WIDEA-n follows the same framework as for encryption except for the order and value of round subkeys. See Fig. 2.10 and 2.12.

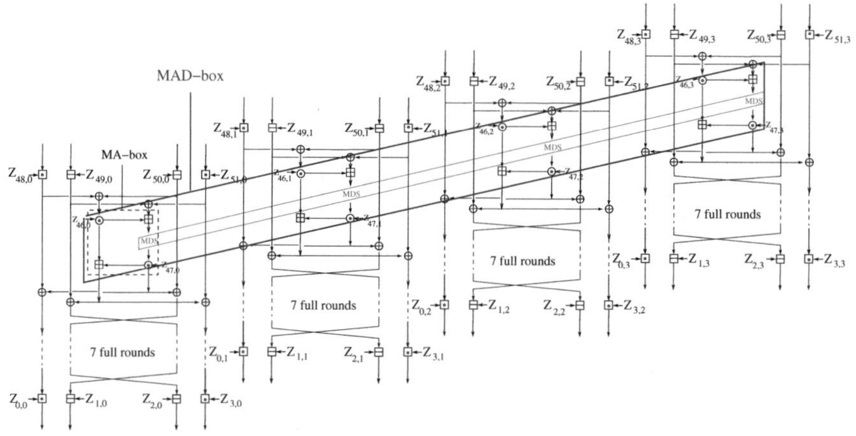

Fig. 2.10 Computational graph of the WIDEA-4 block cipher for decryption.

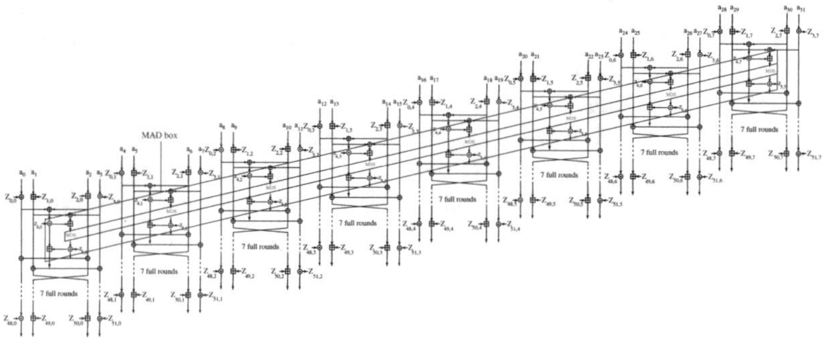

Fig. 2.11 Computational graph of the WIDEA-8 block cipher for encryption [47].

2.5.2 Key Schedule Algorithms

Let Z_i, for $0 \leq i \leq 51$, denote the i-th round subkey used in 8.5-round WIDEA-n, $n \in \{4, 8\}$.

The key schedule algorithm of WIDEA-4 is as follows [29]: due to the 4-way parallelism, each subkey has 64 bits. Thus, each subkey Z_i can be split into four slices $Z_{i,0}$, $Z_{i,1}$, $Z_{i,2}$, $Z_{i,3}$ (see Fig. 2.9). Let K_i, for $0 \leq i \leq 7$, denote the eight 64-bit words representing the 512-bit user key. The round subkeys are computed as follows:

- $Z_i = K_i$, for $0 \leq i \leq 7$.
- $Z_i = ((((Z_{i-1} \oplus Z_{i-8}) \overset{16}{\boxplus} Z_{i-5}) \overset{16}{\lll} 5) \lll 24) \oplus C_{i/8-1}$, for $8 \leq i \leq 51$, $i \equiv 0 \bmod 8$.
- $Z_i = ((((Z_{i-1} \oplus Z_{i-8}) \overset{16}{\boxplus} Z_{i-5}) \overset{16}{\lll} 5) \lll 24)$, for $8 \leq i \leq 51$, $i \not\equiv 0 \bmod 8$.

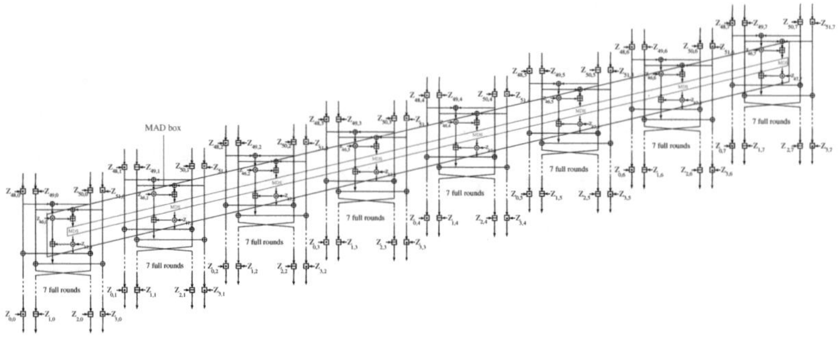

Fig. 2.12 Computational graph of the WIDEA-8 block cipher for decryption.

where operations superscripted with 16 indicate that the operation is carried out over 16-bit slices of Z_i. Otherwise, the operation is carried out across 64-bit words, such as the bitwise left-rotation $\lll 24$. Following [29], $C_0, \ldots, C_5$ are constants inserted every eight rounds. This design uses nonlinear feedback shift registers inspired by the key schedule of MESH ciphers [50].

The key schedule algorithm of WIDEA-8 [29] follows an 8-way parallelism. Each 128-bit subkey Z_i can be partitioned into eight slices $Z_{i,0}, \ldots, Z_{i,7}$ (see Fig. 2.11). Let K_i, for $0 \leq i \leq 7$, denote the eight 128-bit words representing the 1024-bit user key. The round subkeys are computed exactly as for WIDEA-4, except that the subkeys and constants $C_{i/8-1}$ are 128 bits long.

2.6 The FOX/IDEA-NXT Block Ciphers

FOX, also known as IDEA-NXT, is a family of iterated, byte-oriented block ciphers based on the Lai-Massey scheme and designed by Junod and Vaudenay [30]. The main parameters of this cipher family are listed in Table 2.10, where k is a multiple of 8.

Table 2.10 Parameters of the FOX/IDEA-NXT ciphers.

Cipher	Block Size (bits)	Key Size (bits)	#Rounds
FOX64	64	128	16
FOX128	128	256	16
FOX64/k/r	64	k, $0 \leq k \leq 256$, $8 \| k$	r, $12 \leq r \leq 255$
FOX128/k/r	128	k, $0 \leq k \leq 256$, $8 \| k$	r, $12 \leq r \leq 255$

2.6.1 Encryption and Decryption Frameworks

The high-level structure of the FOX ciphers uses the Lai-Massey framework, like the IDEA cipher [36], but the FOX ciphers incorporate an 8-bit bijective substitution box (S-box) and Maximum Distance Separable (MDS) codes, both of which were absent in IDEA. Moreover, the FOX ciphers operate on 8-bit words (instead of 16-bit words in IDEA) and do not use modular addition nor modular multiplication.

All FOX ciphers are byte oriented, and all byte operations are performed in $\mathrm{GF}(2^8) = GF(2)[x]/(p(x))$, where $p(x) = x^8 + x^7 + x^6 + x^5 + x^4 + x^3 + 1$ is an irreducible polynomial over GF(2). FOX64 and FOX64/k/r encryption operations iterate $r - 1$ (full) rounds, denoted *lmor64*, followed by a final round denoted *lmid64*. Formally, *lmor64*, *lmid64*: $\mathbb{Z}_2^{64} \times \mathbb{Z}_2^{64} \rightarrow \mathbb{Z}_2^{64}$, where the inputs are a 64-bit data block and 64-bit subkey, and the output is a 64-bit data block (Fig. 2.13). The encryption of a 64-bit data block P under a key K results in the ciphertext block

$$C = \mathit{lmid64}(\mathit{lmor64}(\ldots \mathit{lmor64}(P, \mathrm{RK}_0), \ldots, \mathrm{RK}_{r-2}), \mathrm{RK}_{r-1}),$$

where the 64-bit subkeys RK_i, $0 \leq i \leq r-1$, are derived from K according to a key schedule algorithm. The decryption transformation uses a round function called *lmio64*: $\mathbb{Z}_2^{64} \times \mathbb{Z}_2^{64} \rightarrow \mathbb{Z}_2^{64}$, and recovers the plaintext block P from a given ciphertext block C and key K as

$$P = \mathit{lmid64}(\mathit{lmio64}(\ldots \mathit{lmio64}(C, \mathrm{RK}_{r-1}), \ldots, \mathrm{RK}_1), \mathrm{RK}_0).$$

The *elmor64* round function is built using a Lai-Massey framework combined with an orthomorphism mapping *or*: $\mathbb{Z}_2^{32} \rightarrow \mathbb{Z}_2^{32}$, defined as $or(a, b) = (b, a \oplus b)$, where $a, b \in \mathbb{Z}_2^{16}$ and $\oplus$ is bitwise exclusive-or. Formally,

$$Y = \mathit{lmor64}(X_L||X_R) =$$

$$or(X_L \oplus \mathrm{f32}(X_L \oplus X_R, \mathrm{RK}_i))||(X_R \oplus \mathrm{f32}(X_L \oplus X_R, \mathrm{RK}_i)),$$

where $||$ denotes bit string concatenation and $Y = Y_L||Y_R$. The *lmid64* function is the same as *lmor64* but without *or*. The inverse of *or* is denoted *io*: $\mathbb{Z}_2^{32} \rightarrow \mathbb{Z}_2^{32}$ and defined as $io(a, b) = (a \oplus b, a)$.

The bijective mapping f32: $\mathbb{Z}_2^{32} \times \mathbb{Z}_2^{64} \rightarrow \mathbb{Z}_2^{32}$ takes a 32-bit data X and a 64-bit round subkey $RK_i = RK_{0i}||RK_{1i}$ as inputs and returns a 32-bit output

$$Y = \mathrm{f32}(X, \mathrm{RK}_i) = \mathrm{sigma4}(\mathrm{mu4}(\mathrm{sigma4}(X \oplus \mathrm{RK}_{0i})) \oplus \mathrm{RK}_{1i}) \oplus \mathrm{RK}_{0i}.$$

Somehow, f32 could be interpreted as the equivalent of the MA-box in IDEA. The transformation sigma4 consists of parallel application of a fixed 8×8-bit S-box denoted S. The S-box inverse is denoted simply S^{-1}. The linear transformation *mu4* consists of a multiplication of the intermediate data with a 4×4 matrix over GF(2^8).

Given a 32-bit input $A = (A_0, A_1, A_2, A_3)$ to mu4, its output is $B = (B_0, B_1, B_2, B_3)$:

$$\begin{pmatrix} B_0 \\ B_1 \\ B_2 \\ B_3 \end{pmatrix} = \begin{pmatrix} 1 & 1 & 1 & x \\ 1 & x^{-1}+1 & x & 1 \\ x^{-1}+1 & x & 1 & 1 \\ x & 1 & x^{-1}+1 & 1 \end{pmatrix} \cdot \begin{pmatrix} A_0 \\ A_1 \\ A_2 \\ A_3 \end{pmatrix}.$$

The 8×8-bit S-box S consists of a 3-round Lai-Massey scheme with three 4×4 S-boxes denoted S_1, S_2 and S_3. See Table 2.11. Let $(A, B) \in \mathbb{Z}_2^4 \times \mathbb{Z}_2^4$ denote the input to this 3-round Lai-Massey scheme. Its 8-bit output is $(C, D) = (\text{or4}(A \oplus S_1[A \oplus B]), B \oplus S_1[A \oplus B])$, where *or4* is a 1-round Feistel Network operating on 4 bits with the identity mapping as round function: *or4*: $\mathbb{Z}_2^2 \times \mathbb{Z}_2^2 \rightarrow \mathbb{Z}_2^2 \times \mathbb{Z}_2^2$ with $or4(a, b) = (b, a \oplus b)$. (C, D) is the output of the second round. The output of the second round is $(E, F) = (or4(C \oplus S_2[C \oplus D]), D \oplus S_2[C \oplus D])$, which is the input to the third round. Finally, the 8-bit output of S is $(E \oplus S_3[E \oplus F], F \oplus S_3[E \oplus F])$.

Table 2.11 4×4-bit S-boxes S_1, S_2 and S_3. The subscript x denotes hexadecimal value.

i	0	1	2	3	4	5	6	7	8	9	10	11	12	13	14	15
$S_1[i]$	2_x	5_x	1_x	9_x	E_x	A_x	C_x	8_x	6_x	4_x	7_x	F_x	D_x	B_x	0_x	3_x
$S_2[i]$	B_x	4_x	1_x	F_x	0_x	3_x	E_x	D_x	A_x	8_x	7_x	5_x	C_x	2_x	9_x	6_x
$S_3[i]$	D_x	A_x	B_x	1_x	4_x	3_x	8_x	9_x	5_x	7_x	2_x	C_x	F_x	0_x	6_x	E_x

FOX128 and FOX128/k/r both iterate $r - 1$ (full) rounds, denoted *elmor128*, followed by a final round denoted *elmid128*. Similarly, a modified round function *elmio128* is used for decryption (Fig.2.14). Formally, *elmor128*, *elmid128* and *elmio128*: $\mathbb{Z}_2^{128} \times \mathbb{Z}_2^{128} \rightarrow \mathbb{Z}_2^{128}$. The encryption of a 128-bit block P under a 128-bit key K results in a 128-bit ciphertext block C as follows:

$$C = \mathit{elmid128}(\mathit{elmor128}(\ldots \mathit{elmor128}(P, \mathrm{RK}_0), \ldots, \mathrm{RK}_{r-2}), \mathrm{RK}_{r-1}),$$

where RK_i for $0 \leq i \leq r-1$ are 128-bit subkeys derived from K using a key schedule algorithm.

Analogously, the decryption of a data block C, given a key K, results in the plaintext block P given by

$$P = \mathit{elmid128}(\mathit{elmio128}(\ldots \mathit{elmio128}(C, \mathrm{RK}_{r-1}), \ldots, \mathrm{RK}_1), \mathrm{RK}_0).$$

The f64 mapping, which is at the core of FOX128/k/r, takes a 64-bit input X, a 128-bit round subkey $RK = RK_0 \| RK_1$ and returns

$$Y = \text{sigma8}(\text{mu8}(\text{sigma8}(X \oplus \text{RK}_0)) \oplus \text{RK}_1) \oplus \text{RK}_0.$$

The function sigma8 consists of eight parallel computations of a nonlinear bijective mapping: an 8-bit S-box.

The linear transformation mu8 consists of a multiplication of the intermediate data with a 8×8 matrix over GF(2^8).

Given a 64-bit input $A = (A_0, A_1, A_2, A_3, A_4, A_5, A_6, A_7)$ to mu8, its output is $B = (B_0, B_1, B_2, B_3, B_4, B_5, B_6, B_7)$:

$$\begin{pmatrix} B_0 \\ B_1 \\ B_2 \\ B_3 \\ B_4 \\ B_5 \\ B_6 \\ B_7 \end{pmatrix} = \begin{pmatrix} 1 & 1 & 1 & 1 & 1 & 1 & 1 & a \\ 1 & a & b & c & d & e & f & 1 \\ a & b & c & d & e & f & 1 & 1 \\ b & c & d & e & f & 1 & a & 1 \\ c & d & e & f & 1 & a & b & 1 \\ d & e & f & 1 & a & b & c & 1 \\ e & f & 1 & a & b & c & d & 1 \\ f & 1 & a & b & c & d & e & 1 \end{pmatrix} \cdot \begin{pmatrix} A_0 \\ A_1 \\ A_2 \\ A_3 \\ A_4 \\ A_5 \\ A_6 \\ A_7 \end{pmatrix}$$

where $a = x+1$, $b = x^7+x$, $c = x$, $d = x^2$, $e = x^7+x^6+x^5+x^4+x^3+x^2$ and $f = x^6+x^5+x^4+x^3+x^2+x$ in $\mathrm{GF}(2)[x]/(x^8+x^7+x^6+x^5+x^4+x^3+1)$.

Some noteworthy properties of the orthomorphism *or* (and its inverse *io*) are:

- *or* has order 3, namely $or(or(or(a,b))) = or^3(a,b) = (a,b)$; the same applies to *io*.
- $(a,b) \oplus or(a,b) \oplus or^2(a,b) = (0,0)$, for any $(a,b) \in \mathbb{Z}_2^{16} \times \mathbb{Z}_2^{16}$.
- $or(a,b) = (a,b) \Leftrightarrow (a,b) = (0,0)$.
- $or^2(a,b) = (a,b) \Leftrightarrow (a,b) = (0,0)$.
- $or(a,b) = io(a,b) \Leftrightarrow (a,b) = (0,0)$.

The orthomorphism *or* is essential for security in the round structure of FOX. Consider a FOX64/k/r version *without the orthomorphism mapping.* Let the plaintext be denoted $P = (P_L, P_R)$, the ciphertext be $C = (C_L, C_R)$, and consider the exclusive-or of the two 32-bit words in a block, at the input and output after r rounds. Since the output of every f32 instance is exclusive-ored to both words in a block at the end of every round, it follows that

$$P_L \oplus P_R = C_L \oplus C_R,$$

that is, the f32 values are canceled in pairs. This invariant relation allows us to distinguish this weak FOX64/k/r version from a random permutation *for any* $r > 0$ by simply comparing the xor of input words with the xor of output words. Note that this invariant is a 32-bit distinguisher that would hold with probability 2^{-32} in a random permutation.

Similarly, in FOX128/k/r *without the orthomorphism*, the exclusive-or of all four 32-bit words of plaintext and ciphertext blocks would also be equal *for any* $r > 0$.

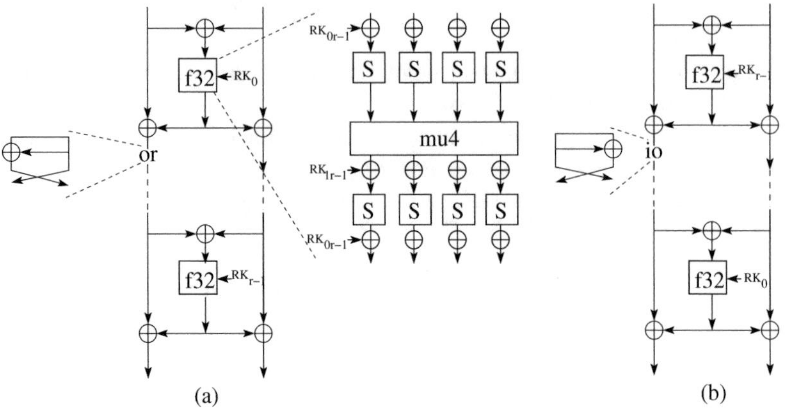

Fig. 2.13 Computational graph of r-round FOX64/k/r: (a) encryption [30], (b) decryption.

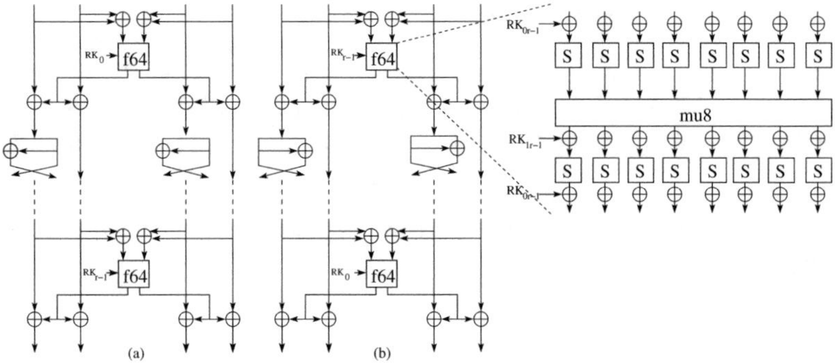

Fig. 2.14 Computational graph of r-round FOX128/k/r: (a) encryption [30], (b) decryption.

2.6.2 Key Schedule Algorithms

The key schedule algorithms of FOX are not lightweight. The authors estimated the key schedules to take about the time for FOX to encrypt six data blocks [30].

There are three key schedule algorithms denoted KS64, KS64h and KS128. Table 2.12 summarizes which key schedule is use in which cipher, and a constant ek. The three key schedule algorithms consist of four parts: a padding part, denoted P, expanding the k-bit user key K into ek bits, a mixing part, denoted M, a diversification part, denoted D, whose core consists mainly of an LFSR, and finally, a nonlinear part denoted NLx, which is actually the only part which differs between the three versions. The later are denoted NL64, NL64h and NL128. When $ek = k$, the P and M parts are omitted.

Table 2.12 Key schedule algorithms used in each FOX cipher.

Cipher	Block size	Key size	Key Schedule	ek
FOX64	64	$0 \leq k \leq 128$	KS64	128
FOX64	64	$136 \leq k \leq 256$	KS64h	256
FOX128	128	$0 \leq k \leq 256$	KS128	256

The P part takes ek and k as input and expands a bit string by $\frac{ek-k}{8}$ bytes. It concatenates the key K with the first $ek-k$ bits of a constant, *pad*, resulting in $PKEY$ as output. The constant *pad* is defined as the first 256 bits of the hexadecimal expansion of $exp(1) - 2 = \sum_{i=0}^{+\infty} 1/i! - 2$, where $exp(1)$ is the base of natural logarithms.

The M part mixes the padded key $PKEY$ with the help of a Fibonacci-like recursion. It takes as input the key $PKEY$ with length ek bits, seen as an array of $\frac{ek}{8}$ bytes $PKEY_{i(8)}$ for $0 \leq i \leq \frac{ek}{8} - 1$, and is processed according to

$$MKEY_{i(8)} = PKEY_{i(8)} \oplus (MKEY_{i-1(8)} + MKEY_{i-2(8)} \bmod 2^8)$$

for $0 \leq i \leq \frac{ek}{8} - 1$, assuming that $MKEY_{-2(8)} = 6A_x$ and $MKEY_{-1(8)} = 76_x$.

The D part takes a key $MKEY$ with bit length equal to ek, the total number of rounds r and the current round number i, with $1 \leq i \leq r$ and modifies $MKEY$ with the help of the output of a 23-bit LFSR. More precisely, $MKEY$ is seen as an array of $\lfloor \frac{ek}{24} \rfloor$ 24-bit values $MKEY_{j(24)}$, with $0 \leq j \leq \lfloor \frac{ek}{24} \rfloor$ concatenated with one residue byte $MKEYRB_{(8)}$ (if $ek = 128$) or two residue bytes $MKEYRB_{(16)}$ (if $ek = 256$) and for $0 \leq j \leq \lfloor \frac{ek}{24} \rfloor - 1$ is modified according to

$$DKEY_{j(24)} = MKEY_{j(24)} \oplus \text{LFSR}((i-1) \cdot \lceil \frac{ek}{24} \rceil + j, r)$$

and the $DKEYRB_{(8)}$ value ($DKEYRB_{(16)}$) is obtained by xoring the 8 (16) most significant bits of $\text{LFSR}((i-1) \cdot \lceil \frac{ek}{24} \rceil + \lfloor \frac{ek}{24} \rfloor, r)$ with $MKEYRB_{(8)}$ ($MKEYRB_{(16)}$), respectively. The stream of pseudorandom values is generated by a 24-bit LFSR. It takes two inputs: the total number r of rounds and the number of preliminary clockings. It is based on the primitive polynomial over GF(2): $x^{24} + x^4 + x^3 + 1$. The register is initially seeded with the value $6a_X \| r_{(8)} \| \overline{r_{(8)}}$, where $r_{(8)}$ is expressed as an 8-bit value and $\overline{x}$ is the bitwise complement of x.

Further details about the key schedule algorithms can be found in [30].

2.7 The REESSE3+ Block Cipher

The REESSE3+ block cipher was designed by Su and Lu and described in [57]. The REESSE3+ cipher follows a Lai-Massey design and operates on 128-bit text blocks, under a 256-bit key and iterating eight rounds plus an output transformation. See Fig. 2.15(a).

REESSE3+ uses the same three group operations of IDEA, over 16-bit words: bitwise exclusive-or, denoted $\oplus$, addition in $\mathbb{Z}_{2^{16}}$, denoted $\boxplus$ and multiplication in $GF(2^{16}+1)$, denoted $\odot$, with the exception that $0 \equiv 2^{16}$.

The authors in [57] did not list explicitly the design criteria for the round function nor for the MA-box of REESSE3+. Nonetheless, both IDEA and REESSE3+ seem to follow similar design principles. For instance, in both ciphers the same group operation is never used twice in succession across the entire cipher. Also, both ciphers contain a so called MA-box (Multiplication-Addition box), a bijective mapping composed of addition and multiplication operations alternated and in a zigzag order. But, IDEA uses a 32-bit *keyed MA-box*, while REESSE3+'s 64-bit MA-box is *not keyed*.

Looking at the 16-bit words along the computational graph of REESSE3+ in Fig. 2.15(a), we note that some of them involve only $\oplus$ and $\odot$, while other words contain only $\boxplus$ and $\oplus$, or all three group operations. Those words that contain only $\boxplus$ and $\oplus$ may be susceptible to the Biryukov-Demirci attacks in Sect. 3.17.

2.7.1 Encryption and Decryption Frameworks

One full encryption round of REESSE3+ can be partitioned into a Key-Mixing (KM) and a Multiplication-Addition (MA) half-round, in this order.

The j-th subkey of the i-th round is denoted by $Z_j^{(i)}$, with $1 \leq j \leq 8$ and $1 \leq i \leq 9$. Subkey generation by the key schedule algorithm is detailed in Sect. 2.7.2.

The 128-bit input $X^{(i)} = (X_1^{(i)}, X_2^{(i)}, X_3^{(i)}, X_4^{(i)}, X_5^{(i)}, X_6^{(i)}, X_7^{(i)}, X_8^{(i)})$ to the i-th round is also the input to the i-th KM half-round. The output of the KM half-round is

$$(X_9^{(i)}, X_{10}^{(i)}, X_{11}^{(i)}, X_{12}^{(i)}, X_{13}^{(i)}, X_{14}^{(i)}, X_{15}^{(i)}, X_{16}^{(i)}) = (X_1^{(i)} \odot Z_1^{(i)}, X_2^{(i)} \boxplus Z_2^{(i)},$$
$$X_3^{(i)} \boxplus Z_3^{(i)}, X_4^{(i)} \odot Z_4^{(i)}, X_5^{(i)} \boxplus Z_5^{(i)}, X_6^{(i)} \odot Z_6^{(i)}, X_7^{(i)} \odot Z_7^{(i)}, X_8^{(i)} \boxplus Z_8^{(i)}),$$

which becomes the input to the MA half-round. The round subkeys which are combined with intermediate cipher data via $\boxplus$ are called additive subkeys. Subkeys which are combined via $\odot$ are called multiplicative subkeys.

To compute an MA half-round, four temporary values are needed:

$$X_{17}^{(i)} = X_9^{(i)} \oplus X_{11}^{(i)},$$

$$X_{18}^{(i)} = X_{10}^{(i)} \oplus X_{12}^{(i)},$$

$$X_{19}^{(i)} = X_{13}^{(i)} \oplus X_{15}^{(i)},$$

and

$$X_{20}^{(i)} = X_{14}^{(i)} \oplus X_{16}^{(i)},$$

which form the input to the MA-box, a 64-bit *unkeyed* permutation mapping that consists of $\odot$ and $\boxplus$ operations only. The output of the i-th MA-box is the 4-tuple $(X_{21}^{(i)}, X_{22}^{(i)}, X_{23}^{(i)}, X_{24}^{(i)})$ where

$$X_{24}^{(i)} = (((X_{17}^{(i)} \odot X_{18}^{(i)}) \boxplus X_{19}^{(i)}) \odot X_{20}^{(i)}),$$

$$X_{21}^{(i)} = X_{24}^{(i)} \boxplus (X_{17}^{(i)} \odot X_{18}^{(i)}),$$

$$X_{22}^{(i)} = X_{24}^{(i)} \boxplus X_{17}^{(i)}$$

and

$$X_{23}^{(i)} = X_{21}^{(i)} \odot X_{20}^{(i)}.$$

Note that $X_{24}^{(i)}$ already depends (nonlinearly) on all four input words $X_{17}^{(i)}$, $X_{18}^{(i)}$, $X_{19}^{(i)}$, $X_{20}^{(i)}$, which depend on all $X_j^{(i)}$, $9 \leq j \leq 16$. Consequently, $X_{23}^{(i)}$, $X_{22}^{(i)}$ and $X_{21}^{(i)}$ also depend on all $X_j^{(i)}$. In summary, complete text diffusion is achieved in a single round.

Note also that all $\odot$ and $\boxplus$ operations inside the MA-box are unkeyed. Thus, the MA-box of REESSE3+ is a fixed 64-bit permutation.

The next step is to combine $(X_{21}^{(i)}, X_{22}^{(i)}, X_{23}^{(i)}, X_{24}^{(i)})$ with the original inputs to the MA half-round, resulting in

$$(X_9^{(i)} \oplus X_{21}^{(i)}, X_{10}^{(i)} \oplus X_{22}^{(i)}, X_{11}^{(i)} \oplus X_{21}^{(i)}, X_{12}^{(i)} \oplus X_{22}^{(i)},$$
$$X_{13}^{(i)} \oplus X_{23}^{(i)}, X_{14}^{(i)} \oplus X_{24}^{(i)}, X_{15}^{(i)} \oplus X_{23}^{(i)}, X_{16}^{(i)} \oplus X_{24}^{(i)}),$$

and full text diffusion is achieved in a single round. This procedure is repeated eight times.

The last step is to swap the six middle words in pairs. The round output becomes

$$(X_9^{(i)} \oplus X_{21}^{(i)}, X_{11}^{(i)} \oplus X_{21}^{(i)}, X_{10}^{(i)} \oplus X_{22}^{(i)}, X_{13}^{(i)} \oplus X_{23}^{(i)},$$
$$X_{12}^{(i)} \oplus X_{22}^{(i)}, X_{15}^{(i)} \oplus X_{23}^{(i)}, X_{14}^{(i)} \oplus X_{24}^{(i)}, X_{16}^{(i)} \oplus X_{24}^{(i)}),$$

which also becomes the input to the next round. See Fig. 2.15(a).

In total, there are seven $\odot$, seven $\boxplus$ and twelve $\oplus$ per full round.

The output transformation (OT) consists of an inverse swap of the middle three pairs of words and a KM half-round with the round subkeys: $Z_1^{(9)}$, $Z_2^{(9)}$, $Z_3^{(9)}$, $Z_4^{(9)}$, $Z_5^{(9)}$, $Z_6^{(9)}$, $Z_7^{(9)}$ and $Z_8^{(9)}$.

The decryption operation consists of the inverse of the output transformation and eight inverse full rounds. Essentially, the additive or the multiplicative inverses of the rounds subkeys are needed, according to Table 2.14. See Fig. 2.15(b).

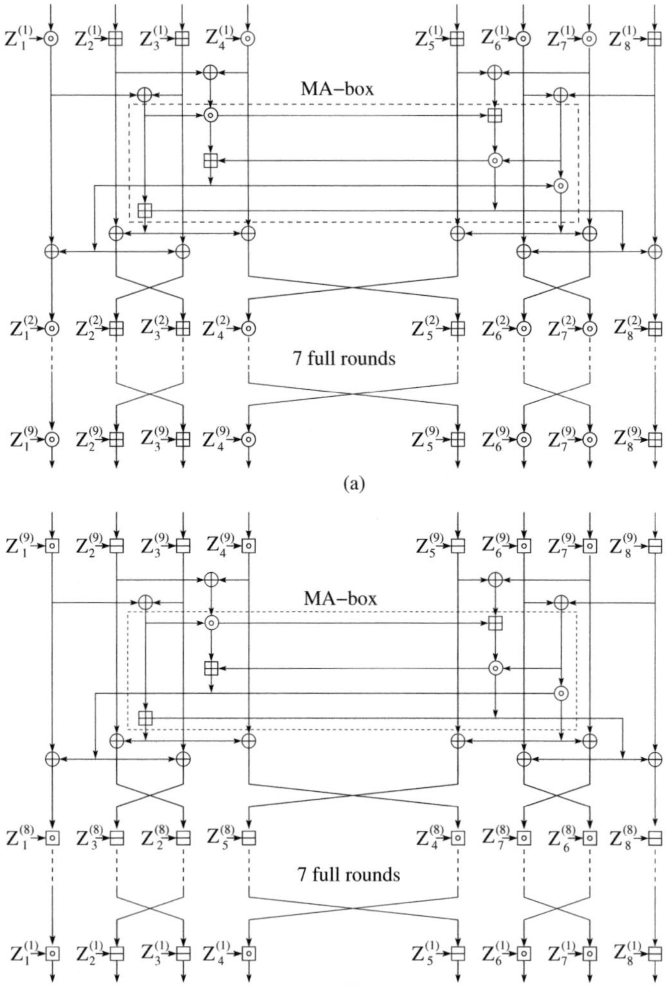

Fig. 2.15 Computational graph of REESSE3+: (a) encryption and (b) decryption. Note the use of modular division (⊡) and subtraction (⊟) in decryption.

2.7.2 Key Schedule Algorithm

There are eight subkeys per round in REESSE3+, and all of them are in the KM half-rounds. In total, 72 16-bit subkeys (1152 bits in total) are needed for either the encryption or decryption of a single 128-bit text block.

The key schedule algorithm of REESSE3+ operates on a 256-bit user key $K = (k_0, k_1, \ldots, k_{254}, k_{255})$. Initially, the leftmost 128 bits $(k_0, k_1, \ldots, k_{127})$ of K are partitioned into eight 16-bit words, and each word is assigned to the first eight round subkeys: $Z_1^{(1)}$, $Z_2^{(1)}$, $Z_3^{(1)}$, $Z_4^{(1)}$, $Z_5^{(1)}$, $Z_6^{(1)}$, $Z_7^{(1)}$, $Z_8^{(1)}$. Next, K is cyclically left shifted by 25 bits, and again its leftmost 128 bits are partitioned into eight 16-bit words, which are assigned to the following eight round subkeys: $Z_1^{(2)}$, $Z_2^{(2)}$, $Z_3^{(2)}$, $Z_4^{(2)}$, $Z_5^{(2)}$, $Z_6^{(2)}$, $Z_7^{(2)}$, $Z_8^{(2)}$. This procedure is repeated until all 72 round subkeys are generated.

Table 2.13 shows the mapping of the 256-bit user key K and the round subkeys. From Table 2.13, it is clear that any encryption and decryption subkeys can be generated independently by just picking the appropriate bits from K and eventually performing an arithmetic operation (division or subtraction, for decryption). Compare Table 2.13 with Table 2.1 that shows the mapping for the key schedule of PES and IDEA.

Table 2.13 Mapping bits from the 256-bit user key $K = (k_0, k_1, \ldots, k_{254}, k_{255})$ of REESSE3+ to the subkeys $Z_j^{(i)}$ for $1 \leq j \leq 8$, $1 \leq i \leq 9$.

i	$Z_1^{(i)}$	$Z_2^{(i)}$	$Z_3^{(i)}$	$Z_4^{(i)}$	$Z_5^{(i)}$	$Z_6^{(i)}$	$Z_7^{(i)}$	$Z_8^{(i)}$
1	$k_0 - k_{15}$	$k_{16} - k_{31}$	$k_{32} - k_{47}$	$k_{48} - k_{63}$	$k_{64} - k_{79}$	$k_{80} - k_{95}$	$k_{96} - k_{111}$	$k_{112} - k_{127}$
2	$k_{25} - k_{40}$	$k_{41} - k_{56}$	$k_{57} - k_{72}$	$k_{73} - k_{88}$	$k_{89} - k_{104}$	$k_{105} - k_{120}$	$k_{121} - k_{136}$	$k_{137} - k_{152}$
3	$k_{50} - k_{65}$	$k_{66} - k_{81}$	$k_{82} - k_{97}$	$k_{98} - k_{113}$	$k_{114} - k_{129}$	$k_{130} - k_{145}$	$k_{146} - k_{161}$	$k_{162} - k_{177}$
4	$k_{75} - k_{90}$	$k_{91} - k_{106}$	$k_{107} - k_{122}$	$k_{123} - k_{138}$	$k_{139} - k_{154}$	$k_{155} - k_{170}$	$k_{171} - k_{186}$	$k_{187} - k_{202}$
5	$k_{100} - k_{115}$	$k_{116} - k_{131}$	$k_{132} - k_{147}$	$k_{148} - k_{163}$	$k_{164} - k_{179}$	$k_{180} - k_{195}$	$k_{196} - k_{211}$	$k_{212} - k_{227}$
6	$k_{125} - k_{140}$	$k_{141} - k_{156}$	$k_{157} - k_{172}$	$k_{173} - k_{188}$	$k_{189} - k_{204}$	$k_{205} - k_{220}$	$k_{221} - k_{236}$	$k_{237} - k_{252}$
7	$k_{150} - k_{165}$	$k_{166} - k_{181}$	$k_{182} - k_{197}$	$k_{198} - k_{213}$	$k_{214} - k_{229}$	$k_{230} - k_{245}$	$k_{246} - k_5$	$k_6 - k_{21}$
8	$k_{175} - k_{190}$	$k_{191} - k_{206}$	$k_{207} - k_{222}$	$k_{223} - k_{238}$	$k_{239} - k_{254}$	$k_{255} - k_{14}$	$k_{15} - k_{30}$	$k_{31} - k_{46}$
9	$k_{200} - k_{215}$	$k_{216} - k_{231}$	$k_{232} - k_{247}$	$k_{248} - k_7$	$k_8 - k_{23}$	$k_{24} - k_{39}$	$k_{40} - k_{55}$	$k_{56} - k_{71}$

Table 2.7.1 shows the subkeys for encryption/decryption in REESSE3+ for each round. Multiplicative subkeys are highlighted in boldface type.

Notice that key diffusion is very slow: only 25 additional bits, out of 256, are added per round, due to the fixed rotation by 25 bits. In total, $\lceil \frac{256}{25} \rceil = 6$ rounds will be required until all 256 key bits are used at least once during the encryption process.

Moreover, there is full overlap between the subkeys and the user key K. Thus, recovery of any subkey (or parts of it) leads to immediate recovery of the same amount of information on the user key K.

Table 2.14 Encryption and decryption round subkeys of REESSE3+ cipher.

round	Encryption subkeys
1	$\mathbf{Z}_1^{(1)}, Z_2^{(1)}, Z_3^{(1)}, \mathbf{Z}_4^{(1)}, Z_5^{(1)}, \mathbf{Z}_6^{(1)}, \mathbf{Z}_7^{(1)}, Z_8^{(1)}$
2	$\mathbf{Z}_1^{(2)}, Z_2^{(2)}, Z_3^{(2)}, \mathbf{Z}_4^{(2)}, Z_5^{(2)}, \mathbf{Z}_6^{(2)}, \mathbf{Z}_7^{(2)}, Z_8^{(2)}$
3	$\mathbf{Z}_1^{(3)}, Z_2^{(3)}, Z_3^{(3)}, \mathbf{Z}_4^{(3)}, Z_5^{(3)}, \mathbf{Z}_6^{(3)}, \mathbf{Z}_7^{(3)}, Z_8^{(3)}$
4	$\mathbf{Z}_1^{(4)}, Z_2^{(4)}, Z_3^{(4)}, \mathbf{Z}_4^{(4)}, Z_5^{(4)}, \mathbf{Z}_6^{(4)}, \mathbf{Z}_7^{(4)}, Z_8^{(4)}$
5	$\mathbf{Z}_1^{(5)}, Z_2^{(5)}, Z_3^{(5)}, \mathbf{Z}_4^{(5)}, Z_5^{(5)}, \mathbf{Z}_6^{(5)}, \mathbf{Z}_7^{(5)}, Z_8^{(5)}$
6	$\mathbf{Z}_1^{(6)}, Z_2^{(6)}, Z_3^{(6)}, \mathbf{Z}_4^{(6)}, Z_5^{(6)}, \mathbf{Z}_6^{(6)}, \mathbf{Z}_7^{(6)}, Z_8^{(6)}$
7	$\mathbf{Z}_1^{(7)}, Z_2^{(7)}, Z_3^{(7)}, \mathbf{Z}_4^{(7)}, Z_5^{(7)}, \mathbf{Z}_6^{(7)}, \mathbf{Z}_7^{(7)}, Z_8^{(7)}$
8	$\mathbf{Z}_1^{(8)}, Z_2^{(8)}, Z_3^{(8)}, \mathbf{Z}_4^{(8)}, Z_5^{(8)}, \mathbf{Z}_6^{(8)}, \mathbf{Z}_7^{(8)}, Z_8^{(8)}$
9	$\mathbf{Z}_1^{(9)}, Z_2^{(9)}, Z_3^{(9)}, \mathbf{Z}_4^{(9)}, Z_5^{(9)}, \mathbf{Z}_6^{(9)}, \mathbf{Z}_7^{(9)}, Z_8^{(9)}$

round	Decryption subkeys
1	$\mathbf{Z}_1^{(9)}, Z_2^{(9)}, Z_3^{(9)}, \mathbf{Z}_4^{(9)}, Z_5^{(9)}, \mathbf{Z}_6^{(9)}, \mathbf{Z}_7^{(9)}, Z_8^{(9)}$
2	$\mathbf{Z}_1^{(8)}, Z_3^{(8)}, Z_2^{(8)}, Z_5^{(8)}, \mathbf{Z}_4^{(8)}, \mathbf{Z}_7^{(8)}, \mathbf{Z}_6^{(8)}, Z_8^{(8)}$
3	$\mathbf{Z}_1^{(7)}, Z_3^{(7)}, Z_2^{(7)}, Z_5^{(7)}, \mathbf{Z}_4^{(7)}, \mathbf{Z}_7^{(7)}, \mathbf{Z}_6^{(7)}, Z_8^{(7)}$
4	$\mathbf{Z}_1^{(6)}, Z_3^{(6)}, Z_2^{(6)}, Z_5^{(6)}, \mathbf{Z}_4^{(6)}, \mathbf{Z}_7^{(6)}, \mathbf{Z}_6^{(6)}, Z_8^{(6)}$
5	$\mathbf{Z}_1^{(5)}, Z_3^{(5)}, Z_2^{(5)}, Z_5^{(5)}, \mathbf{Z}_4^{(5)}, \mathbf{Z}_7^{(5)}, \mathbf{Z}_6^{(5)}, Z_8^{(5)}$
6	$\mathbf{Z}_1^{(4)}, Z_3^{(4)}, Z_2^{(4)}, Z_5^{(4)}, \mathbf{Z}_4^{(4)}, \mathbf{Z}_7^{(4)}, \mathbf{Z}_6^{(4)}, Z_8^{(4)}$
7	$\mathbf{Z}_1^{(3)}, Z_3^{(3)}, Z_2^{(3)}, Z_5^{(3)}, \mathbf{Z}_4^{(3)}, \mathbf{Z}_7^{(3)}, \mathbf{Z}_6^{(3)}, Z_8^{(3)}$
8	$\mathbf{Z}_1^{(2)}, Z_3^{(2)}, Z_2^{(2)}, Z_5^{(2)}, \mathbf{Z}_4^{(2)}, \mathbf{Z}_7^{(2)}, \mathbf{Z}_6^{(2)}, Z_8^{(2)}$
9	$\mathbf{Z}_1^{(1)}, Z_2^{(1)}, Z_3^{(1)}, \mathbf{Z}_4^{(1)}, Z_5^{(1)}, \mathbf{Z}_6^{(1)}, \mathbf{Z}_7^{(1)}, Z_8^{(1)}$

2.8 The IDEA* Block Cipher

IDEA* is block cipher operating on 64-bit text blocks, under a 128-bit user key, and iterating six rounds plus an output transformation. IDEA* was designed by Lerman, Nakahara and Veshchikov [39].

IDEA* employs the same three algebraic operations of IDEA. Consequently, it is expected that IDEA* fits in the same legacy environments used by IDEA, such as PGP/GPG, digital rights management, video scrambling for pay-TV, internet audio/video distribution, government and corporate IT infrastructure protection [39].

2.8.1 Encryption and Decryption Frameworks

IDEA* is the result of a simple strategy: how the rearrangement of the individual cipher components affects the security (and performance) of a block cipher? As an example, what happens in the IDEA block cipher if all the exclusive-or ($\oplus$) and multiplication ($\odot$) operations are swapped? This simple modification has not been reported before in the cryptographic literature, and it implies that in IDEA*, subkeys are no longer a mandatory input to the multiplication operations, meaning that both inputs are variable. IDEA* also uses the unkeyed division operation, denoted $\boxdot$, and defined as $a \boxdot b = a \odot b^{-1} = a/b$, where $a, b \in \mathrm{GF}(2^{16}+1)$. Therefore, if a table of multiplicative inverses is provided, a division costs one multiplication plus a table look-up. Note that unlike exclusive-or, $a \boxdot b \neq b \boxdot a$, so the order of the operands matters in $\boxdot$.

The *unkeyed multiplication* is a primitive operation in MESH-96, MESH-128 and RIDEA. The *unkeyed division* is a new primitive operation. The swap of $\oplus$ by $\odot$ has a considerable impact: there are no more weak multiplicative subkeys in IDEA* regardless of the key schedule algorithm. Moreover, wordwise diffusion is stronger with multiplication because of a wrap-around effect in comparison to the bitwise diffusion in exclusive-or. This fact is corroborated in Lai's Low-High algorithm [35] for multiplication in $\mathrm{GF}(2^{16}+1)$. See Def. 2.1.

Note that swapping $\oplus$ with $\boxplus$ in IDEA would not eliminate weak subkeys, as subkeys would still be a mandatory input to $\odot$. This modified version was called IDEA-X by Borisov *et al.* [12]. They showed multiplicative differential attacks on IDEA-X, which is a weakened version of IDEA. Likewise, swapping $\boxplus$ for $\odot$ in IDEA would not work either because subkeys would still be input to $\odot$ in the KM half-rounds.

IDEA* uses an updated key schedule algorithm with full key diffusion after the third generated subkey, which makes each round subkey quickly depend on all bits of the user key. This design was borrowed from [50] and effectively counters Meet-In-The-Middle (MITM), related-key, slide and advanced slide (among other) attacks. This means that the encryption/decryption frameworks cannot be accidentally or purposefully weakened due to particular bit patterns in the user key. Comparatively, in IDEA, different rounds do not have the same strength because subkey bits overlap with user key bits, and the total key entropy per round can be much lower than 96 bits. In IDEA*, individual key bits cannot be flipped independently without affecting several subkeys at once, thus hindering divide-and-conquer attacks that try to exploit independent subkey bits such as the biclique technique [10].

IDEA* preserves the wordwise structure and the same group operations on 16-bit words in IDEA, as well as the design philosophy of repeating a strong round structure a small number of times, instead of iterating a weak round function a large number of times. IDEA* also adopted the design feature of never repeating the same group operation consecutively along the encryption/decryption frameworks [35]. Moreover, there is full (text) diffusion after every single round.

The original MA-box (with Multiplication and Addition) in IDEA becomes an AX-box in IDEA* (with Addition and eXclusive-or) operations.

One full round in IDEA* consists of two half-rounds: Key-Whitening (KW) and Addition-Xor (AX), in this order. See Fig. 2.16.

Let $X^{(i)} = (X_1^{(i)}, X_2^{(i)}, X_3^{(i)}, X_4^{(i)})$ denote the 64-bit input block to the i-th round, where $X_j^{(i)} \in \mathbb{Z}_2^{16}$ for $1 \leq j \leq 4$ and $1 \leq i \leq 7$. So, for instance, $X^{(1)}$ represents a plaintext block.

The i-th KW half-round transforms $X^{(i)}$ as follows:

$$Y^{(i)} = (X_1^{(i)} \oplus Z_1^{(i)}, X_2^{(i)} \boxplus Z_2^{(i)}, X_3^{(i)} \boxplus Z_3^{(i)}, X_4^{(i)} \oplus Z_4^{(i)}).$$

So, for instance, $Y^{(7)}$ represents a ciphertext block.

The i-th AX half-round contains an AX-box and has a kind of involutory structure. See Fig. 2.16(a) for encryption and Fig. 2.16(b) for decryption.

In more detail, the input to the i-th AX-box is the ordered pair $(n_i, q_i) = (Y_1^{(i)} \boxdot Y_3^{(i)}, Y_2^{(i)} \boxdot Y_4^{(i)})$. Let the ordered pair (r_i, s_i) denote the i-th AX-box output. Then,

$$s_i = ((n_i \oplus Z_5^{(i)}) \boxplus q_i) \oplus Z_6^{(i)},$$

and

$$r_i = (n_i \oplus Z_5^{(i)}) \boxplus s_i.$$

Let $\sigma(A, B, C, D) = (A, C, B, D)$ denote a fixed word permutation.

The i-th round output for encryption becomes

$$X^{(i+1)} = (Y_1^{(i)} \odot s_i, Y_3^{(i)} \odot s_i, Y_2^{(i)} \odot r_i, Y_4^{(i)} \odot r_i),$$

where σ was the last operation performed.

Let $X^{(7)} = (X_1^{(7)}, X_2^{(7)}, X_3^{(7)}, X_4^{(7)})$ denote the 64-bit input to the 7th round. The output transformation consists of $\sigma(X^{(7)})$ followed by a KW half-round.

Thus, the ciphertext is

$$(X_1^{(7)} \oplus Z_1^{(7)}, X_3^{(7)} \boxplus Z_2^{(7)}, X_2^{(7)} \boxplus Z_3^{(7)}, X_4^{(7)} \oplus Z_4^{(7)}).$$

For decryption, the inverse KW-box input becomes

$$Y^{(i)} = (X_1^{(i)} \oplus Z_1^{(i)}, X_2^{(i)} \boxminus Z_2^{(i)}, X_3^{(i)} \boxminus Z_3^{(i)}, X_4^{(i)} \oplus Z_4^{(i)}).$$

Further, the inverse AX half-round contains the same AX-box as for encryption. The input to the AX-box is the ordered pair (n_i, q_i), and its output is the pair (r_i, s_i), the same ones as for encryption.

But, in order to invert the original $\odot$ operations, its inverse operator $\boxdot$ is used. Thus, the encryption and decryption frameworks of IDEA* are like in PES. It is not enough to change the order and recompute the rounds subkeys for each framework. Some $\odot$ operations in the output of the AX half-rounds have to be replaced by $\boxdot$ in the decryption framework, and vice-versa. See Fig. 2.16.

The i-th round output for decryption becomes

$$(Y_1^{(i)} \boxdot s_i, Y_3^{(i)} \boxdot s_i, Y_2^{(i)} \boxdot r_i, Y_4^{(i)} \boxdot r_i),$$

which includes σ as the last transformation.

Therefore, the AX half-round is almost its own inverse, except that $\odot$ are replaced by $\boxdot$ for decryption.

A novelty in IDEA* is the use of unkeyed-$\boxdot$ that is, with both operands variable. IDEA* also uses the unkeyed-$\odot$ operation. Note that in IDEA, one operand in every $\odot$ is always an (unknown) subkey, which may weaken the multiplication depending on the subkey value [16, 8].

Notice that the $\odot$ operation and its inverse operation $\boxdot$ have much better diffusion power than $\oplus$ (which works simply bit-by-bit). This fact is corroborated by Lai's Low-High algorithm [35] for multiplication in GF$(2^{16}+1)$. See Def. 2.1 for $w = 16$. It essentially means that the result of $\odot$ depends on all 32 bits of the extended multiplication.

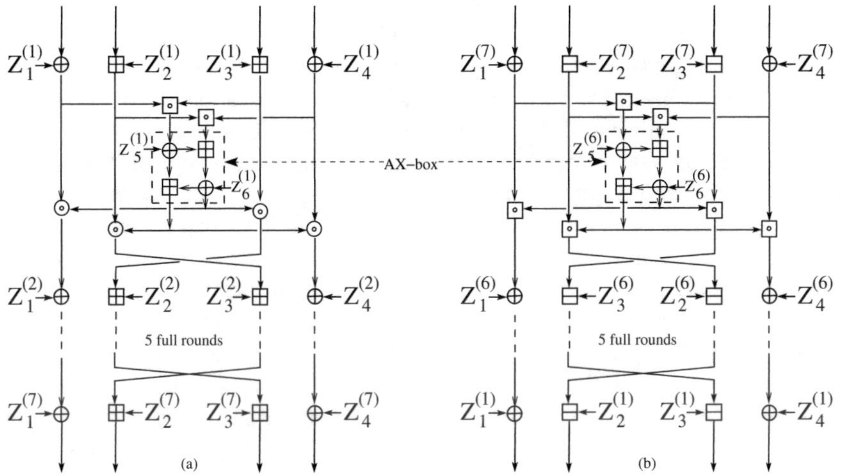

Fig. 2.16 Computational graph of IDEA*: (a) encryption and (b) decryption. Note the use of modular division ($\boxdot$) and subtraction ($\boxminus$) in decryption.

2.8.2 Key Schedule Algorithm

IDEA* iterates 6.5 rounds and uses six 16-bit subkeys per round for a total of 40 subkeys or 640 subkey bits in total. The key schedule of IDEA* adopts the design of the key schedule of MESH-64 [50] in Sect. 2.3.8, using a NLFSR construction.

Let c_i denote 16-bit constants defined as follows: $c_0 = 1$ and $c_i = 3 \cdot c_{i-1}$ for $i \geq 1$, with multiplication in GF(2)$[x]/p(x)$ denoted by $\cdot$ and where $p(x) =$

$x^{16}+x^5+x^3+x^2+1$ is a primitive polynomial in GF(2). The constant 3 is represented by the polynomial $x+1$ in $\mathrm{GF}(2)[x]/p(x)$. See Table 2.4.

Let a 128-bit key K be partitioned into eight 16-bit words K_j, $-7 \le j \le 0$. The elements $K_j \oplus c_{j+7}$ form the eight initial values in the formula (2.18), for $1 \le i \le 40$:

$$K_i = (((((K_{i-8} \boxplus K_{i-7}) \oplus K_{i-6}) \boxplus K_{i-3}) \oplus K_{i-2}) \boxplus K_{i-1}) \lll 7 \oplus c_{i+7} \,. \tag{2.18}$$

The j-th subkey of the i-th round is denoted $Z_j^{(i)}$ for $1 \le j \le 6$ and $1 \le i \le 7$. It is just the element $K_{6(i-1)+j}$. For instance, $Z_1^{(1)} = K_1$ and $Z_2^{(1)} = K_2$.

Low-weight differences in the key schedule (2.18) quickly become unpredictable because of fast key avalanche due to the primitive polynomial $q(x) = x^8+x^7+x^6+x^5+x^2+x+1$ and the interleaving of $\boxplus$, fixed bit rotation ($\lll 7$) and $\oplus$, all of which are efficient and lightweight operations.

Following equation (2.18), one finds out that $Z_3^{(1)}$ is the first subkey that depends on all eight words of K. All following subkeys also fully depend on K due to the chaining of K_{i-1} and K_i in (2.18). Thus, complete key diffusion is achieved even faster than text diffusion in the encryption framework. Moreover, subkey bits in IDEA* depend nonlinearly on each other, unlike the simple bit permutation mapping in the key schedule algorithm of IDEA.

Concerning differentials in the key schedule, analyses have been performed wordwise (with exclusive-or and subtraction) differences with difference value 8000_x in some key words. This difference was chosen because it affects only the most significant bit in a word, and thus propagates across $\boxplus$ and $\oplus$ with certainty. But, this difference does not survive for long in (2.18), quickly becoming heavier Hamming-weight word differences because of the combined use of $\lll$ and the c_i constants. These operations provide fast key diffusion at low cost and destroy algebraic invariants and difference patterns in subkeys, helping thwart related-key (differential) attacks [31, 5] on IDEA*. These operations make the key schedule nonlinear and prevent patterns in the key schedule to propagate or to cancel difference patterns in the encryption framework, further countering MITM [19, 6, 1], slide and advanced slide attacks [9].

The existence of weak keys in IDEA demonstrated: (i) how a strong encryption framework can be compromised by a comparatively weak key schedule algorithm. Although the number of weak keys in differential and linear settings represents a small fraction of the key space [16] it is still more than in any other block cipher, and even larger than the number of weak and semi-weak keys in DES [43] combined; (ii) IDEA is not suitable as a building block in compression function constructions since the key can be chosen or manipulated by an opponent in some hash function modes [48, 60]. Actually,

[48] demonstrated that weak keys are a recurrent problem in IDEA even if the number of rounds was doubled.

An interesting feature of (2.18) is that recovery of round subkeys $Z_j^{(i)}$ do not immediately lead to recovery of the user key K. Note that the eight words K_l, $-7 \leq l \leq 0$ are the initial values in the recurrence (2.18). Suppose, for instance, that the first eight round subkeys $Z_1^{(1)}$, $Z_2^{(1)}$, $Z_3^{(1)}$, $Z_4^{(1)}$, $Z_5^{(1)}$, $Z_6^{(1)}$, $Z_1^{(2)}$, $Z_2^{(2)}$ are recovered. To further derive the user key words K_l for $-7 \leq l \leq 0$, the following system of eight nonlinear equations in eight variables will still have to be solved (the left-hand sides of these equations are known):

$$
\begin{aligned}
Z_1^{(1)} &= (((((K_{-7} \boxplus K_{-6}) \oplus K_{-5}) \boxplus K_{-2}) \oplus K_{-1}) \boxplus K_0) \lll 7 \oplus c_8,\\
Z_2^{(1)} &= (((((K_{-6} \boxplus K_{-5}) \oplus K_{-4}) \boxplus K_{-1}) \oplus K_0) \boxplus Z_1^{(1)}) \lll 7 \oplus c_9,\\
Z_3^{(1)} &= (((((K_{-5} \boxplus K_{-4}) \oplus K_{-3}) \boxplus K_0) \oplus Z_1^{(1)}) \boxplus Z_2^{(1)}) \lll 7 \oplus c_{10},\\
Z_4^{(1)} &= (((((K_{-4} \boxplus K_{-3}) \oplus K_{-2}) \boxplus Z_1^{(1)}) \oplus Z_2^{(1)}) \boxplus Z_3^{(1)}) \lll 7 \oplus c_{11},\\
Z_5^{(1)} &= (((((K_{-3} \boxplus K_{-2}) \oplus K_{-1}) \boxplus Z_2^{(1)}) \oplus Z_3^{(1)}) \boxplus Z_4^{(1)}) \lll 7 \oplus c_{12},\\
Z_6^{(1)} &= (((((K_{-2} \boxplus K_{-1}) \oplus K_0) \boxplus Z_3^{(1)}) \oplus Z_4^{(1)}) \boxplus Z_5^{(1)}) \lll 7 \oplus c_{13},\\
Z_1^{(2)} &= (((((K_{-1} \boxplus K_0) \oplus Z_1^{(1)}) \boxplus Z_4^{(1)}) \oplus Z_5^{(1)}) \boxplus Z_6^{(1)}) \lll 7 \oplus c_{14},\\
Z_2^{(2)} &= (((((K_0 \boxplus Z_1^{(1)}) \oplus Z_2^{(1)}) \boxplus Z_5^{(1)}) \oplus Z_6^{(1)}) \boxplus Z_1^{(2)}) \lll 7 \oplus c_{15},
\end{aligned}
$$

and the same reasoning applies to any sequence of eight consecutive round subkeys.

Also, note that for decryption, (2.18) could be run backwards if the last eight subkeys are stored instead of the original user key K:

$$
\begin{aligned}
K_{33} &= (((((K_{25} \boxplus K_{26}) \oplus K_{27}) \boxplus K_{30}) \oplus K_{31}) \boxplus K_{32}) \lll 7 \oplus c_{40}, && (2.19)\\
K_{34} &= (((((K_{26} \boxplus K_{27}) \oplus K_{28}) \boxplus K_{31}) \oplus K_{32}) \boxplus K_{33}) \lll 7 \oplus c_{41}, && (2.20)\\
K_{35} &= (((((K_{27} \boxplus K_{28}) \oplus K_{29}) \boxplus K_{32}) \oplus K_{33}) \boxplus K_{34}) \lll 7 \oplus c_{42}, && (2.21)\\
K_{36} &= (((((K_{28} \boxplus K_{29}) \oplus K_{30}) \boxplus K_{33}) \oplus K_{34}) \boxplus K_{35}) \lll 7 \oplus c_{43}, && (2.22)\\
K_{37} &= (((((K_{29} \boxplus K_{30}) \oplus K_{31}) \boxplus K_{34}) \oplus K_{35}) \boxplus K_{36}) \lll 7 \oplus c_{44}, && (2.23)\\
K_{38} &= (((((K_{30} \boxplus K_{31}) \oplus K_{32}) \boxplus K_{35}) \oplus K_{36}) \boxplus K_{37}) \lll 7 \oplus c_{45}, && (2.24)\\
K_{39} &= (((((K_{31} \boxplus K_{32}) \oplus K_{33}) \boxplus K_{36}) \oplus K_{37}) \boxplus K_{38}) \lll 7 \oplus c_{46}, && (2.25)\\
K_{40} &= (((((K_{32} \boxplus K_{33}) \oplus K_{34}) \boxplus K_{37}) \oplus K_{38}) \boxplus K_{39}) \lll 7 \oplus c_{47}. && (2.26)
\end{aligned}
$$

From (2.26), the value of K_{32} can be isolated as

$$K_{32} = ((((((K_{40} \oplus c_{47}) \ggg 7) \boxminus K_{39}) \oplus K_{38}) \boxminus K_{37}) \oplus K_{34}) \boxminus K_{33},$$

since all terms on the right-hand side are known. Thus, previous subkeys can be derived analogously using (2.25) to (2.19). Note that each subkey following (2.18) depends nonlinearly on six other subkeys, the farthest of which is eight positions away.

2.9 The Yi's Block Cipher

In [62], X. Yi described a block cipher operating on 64-bit text blocks, under a 128-bit key and iterating six rounds plus an output transformation. The cipher was not named by its author in [62]. To allow easy reference, this cipher will be referred to simply as YBC (Yi's Block Cipher).

YBC uses the same three arithmetic operations of IDEA: bitwise exclusive-or, denoted $\oplus$, addition modulo 2^{16}, denoted $\boxplus$, and multiplication in $GF(2^{16}+1)$, denoted $\odot$, with the exception that $0 \equiv 2^{16}$. Nonetheless, YBC is a hybrid cipher combining features from the Feistel and the Lai-Massey design paradigms.

2.9.1 Encryption and Decryption Frameworks

Let $X^{(i)} = (X_1^{(i)}, X_2^{(i)}, X_3^{(i)}, X_4^{(i)})$ denote the 64-bit input block to the i-th round, where $X_j^{(i)} \in \mathbb{Z}_2^{16}$ for $1 \leq j \leq 4$ and $1 \leq i \leq 7$. So, for instance, $X^{(1)}$ represents a plaintext block.

A full encryption round in YBC consists of a Key-Mixing (KM) and an Addition-Multiplication-eXclusive-or (AMX) half-round, in this order. See Fig. 2.17.

In the i-th KM half-round, $X^{(i)}$ is transformed into $Y^{(i)} = (Y_1^{(i)}, Y_2^{(i)}, Y_3^{(i)}, Y_4^{(i)})$, where $Y_j^{(i)} = X_j^{(i)} \oplus Z_j^{(i)}$ for $1 \leq j \leq 4$. Note that only exclusive-or operations are used in the KM half-round, which avoids the weak-key issues in PES and IDEA.

In the i-th AMX half-round, $Y^{(i)}$ is transformed as follows

$$W_1^{(i)} = Y_1^{(i)} \oplus Y_3^{(i)} \odot (Y_2^{(i)} \boxplus (Y_1^{(i)} \oplus Y_4^{(i)})),$$

$$W_2^{(i)} = Y_2^{(i)} \oplus W_1^{(i)} \odot (Y_3^{(i)} \boxplus (Y_1^{(i)} \oplus Y_4^{(i)})),$$

$$W_3^{(i)} = Y_3^{(i)} \oplus (Y_1^{(i)} \oplus Y_4^{(i)}) \odot (W_1^{(i)} \boxplus W_2^{(i)}),$$

$$W_4^{(i)} = Y_1^{(i)} \oplus Y_4^{(i)} \oplus W_2^{(i)} \odot (W_1^{(i)} \boxplus W_3^{(i)}).$$

Note that there are consecutive $\oplus$ operations in the AMX half-round, which violates one of the original design criteria of Lai-Massey ciphers.

Moreover, the AMX half-round is not keyed, but is rather a fixed permutation mapping, consisting of four repetitions of so called AMX structure where the three arithmetic operations (Addition, Multiplication, eXclusive-or) are combined. This AMX structure is depicted in the computation of $W_1^{(i)}$, $W_2^{(i)}$, $W_3^{(i)}$ and $W_4^{(i)}$.

No reason nor design criteria were provided in [62] for the AMX half-round construction, nor for the particular order of the AMX structures.

Let $T^{(i)} = (W_1^{(i)} \oplus W_2^{(i)}, W_2^{(i)}, W_3^{(i)}, W_4^{(i)})$.

A word swap operator $\sigma_1 : (\mathbb{Z}_2^{16})^4 \rightarrow (\mathbb{Z}_2^{16})^4$ is defined as

$$\sigma_1(A, B, C, D) = (B, A, D, C)$$

and is applied to $T^{(i)}$ that results in the i-th round output $X^{(i+1)} = (W_2^{(i)}, W_1^{(i)} \oplus W_2^{(i)}, W_4^{(i)}, W_3^{(i)})$. This procedure is repeated six times in YBC.

Note that this construction of the AMX half-round guarantees full text diffusion in a single round.

Each full round in YBC uses four $\odot$, four $\boxplus$ and ten $\oplus$. In contrast, IDEA uses four $\odot$, four $\boxplus$ and six $\oplus$ per full round.

The output transformation (OT) undoes σ_1, performs a different word permutation σ_2, and finally performs a KM half-round. Formally, $\sigma_2 : (\mathbb{Z}_2^{16})^4 \rightarrow (\mathbb{Z}_2^{16})^4$ and its operation is defined as

$$\sigma_2(A, B, C, D) = (D, C, B, A).$$

Let $X^{(7)} = (X_1^{(7)}, X_2^{(7)}, X_3^{(7)}, X_4^{(7)})$ denote the input 64-bit block to the 7th round.

Let $Y^{(7)} = \sigma_2(\sigma_1(X^{(7)})) = \sigma_2(X_2^{(7)}, X_1^{(7)}, X_4^{(7)}, X_3^{(7)}) = (X_3^{(7)}, X_4^{(7)}, X_1^{(7)}, X_2^{(7)})$. Then, the ciphertext becomes

$$C = (X_3^{(7)} \oplus Z_1^{(7)}, X_4^{(7)} \oplus Z_2^{(7)}, X_1^{(7)} \oplus Z_3^{(7)}, X_2^{(7)} \oplus Z_4^{(7)}).$$

One full 6.5-round YBC encryption employs 24 $\odot$, 24 $\boxplus$ and 64 $\oplus$. In contrast, one full 8.5-round IDEA encryption uses 34 $\odot$, 34 $\boxplus$ and 48 $\oplus$. If one assumes an $\odot$ costs about three times an $\oplus$ or a $\boxplus$, then one YBC encryption costs about 160 $\odot$ operations, while one IDEA encryption would cost about 184 $\oplus$ operations.

For decryption, each full round is computed backwards. See Fig. 2.17. Note that the AMX structures are computed in reverse order, which makes the encryption and decryption frameworks quite different. Therefore, the similarity of encryption and decryption is not preserved in YBC, unlike the previous Lai-Massey cipher designs.

Note that the AMX half-round is *non-involutory*, unlike the MA half-round of other Lai-Massey designs. Moreover, the AMX structures in Fig. 2.17(a) can be rearranged as depicted in Fig. 2.18(a) and (b). So, the AMX structures actually resemble a 4-round Feistel Network. In this sense, 6.5-round YBC can be interpreted as a 24-round Feistel Network design. Therefore, except for the arithmetic operations, YBC does not actually follow the Lai-Massey design framework.

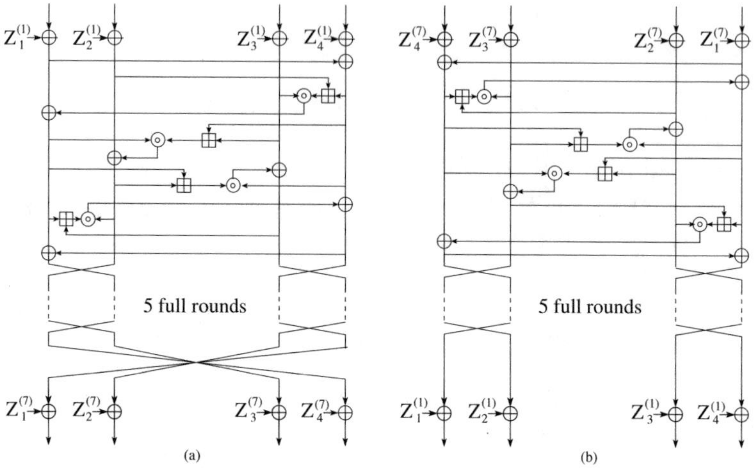

Fig. 2.17 Computational graph of YBC: (a) encryption [62] and (b) decryption.

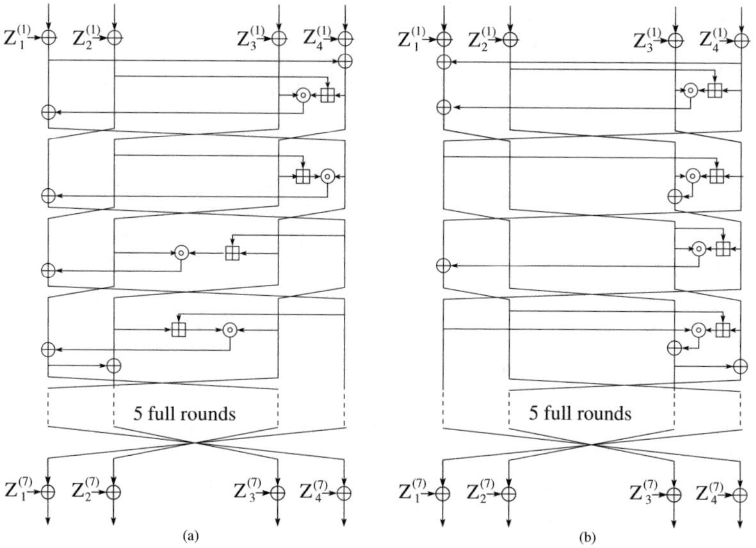

Fig. 2.18 Alternative computational graphs of YBC encryption.

2.9.2 Key Schedule Algorithm

The key schedule algorithm of YBC generates twenty-eight 16-bit round subkeys for the encryption and decryption transformations. The key schedule operates as follows:

(1) let K denote the 128-bit user key. Set a counter $i = 1$.

(2) partition K into eight 16-bit words $A_j^{(i)}$ and $A_j^{(i+1)}$ for $1 \leq j \leq 4$, and set $i = i + 2$.
(3) if $i < 9$, then cyclically shift K to the left by 25 bit positions, and go back to step (2).
(4) let $X_j = A_j^{(1)}$ for $1 \leq j \leq 4$ and let $n = 1$.
(5) let $X_j = A_j^{(n+1)} \oplus X_j$ for $j \in \{1, 3\}$ and $X_j = A_j^{(n+1)} \oplus X_j \oplus ffff_x$ for $j \in \{2, 4\}$.
(6) encrypt (X_1, X_2, X_3, X_4) with YBC under the 128-bit key 0^{128} (which is equivalent to eliminating all subkeys). The resulting 64-bit ciphertext becomes the subkeys $Z_j^{(n)}$, $1 \leq j \leq 4$. Set $n = n + 1$.
(7) if $n < 8$, then return to step (5).

This key schedule algorithm uses the 6.5-round YBC encryption operation seven times to generate all 28 round subkeys or 488 subkey bits in total. This fact serves to limit the number of rounds in YBC. The larger the number of rounds, the more calls to YBC are needed to generate enough round subkeys.

Table 2.15 gives the order of subkeys for encryption and decryption. Note that only the order of subkeys change for each transformation, but not their values since all subkeys are combined with cipher data (state) via bitwise exclusive-or (which is an involutory operation).

Table 2.15 Encryption and decryption round subkeys of YBC cipher.

Round	Encryption subkeys	Decryption subkeys
1	$Z_1^{(1)}, Z_2^{(1)}, Z_3^{(1)}, Z_4^{(1)}$	$Z_4^{(7)}, Z_3^{(7)}, Z_2^{(7)}, Z_1^{(7)}$
2	$Z_1^{(2)}, Z_2^{(2)}, Z_3^{(2)}, Z_4^{(2)}$	$Z_2^{(6)}, Z_1^{(6)}, Z_4^{(6)}, Z_3^{(6)}$
3	$Z_1^{(3)}, Z_2^{(3)}, Z_3^{(3)}, Z_4^{(3)}$	$Z_2^{(5)}, Z_1^{(5)}, Z_4^{(5)}, Z_3^{(5)}$
4	$Z_1^{(4)}, Z_2^{(4)}, Z_3^{(4)}, Z_4^{(4)}$	$Z_2^{(4)}, Z_1^{(4)}, Z_4^{(4)}, Z_3^{(4)}$
5	$Z_1^{(5)}, Z_2^{(5)}, Z_3^{(5)}, Z_4^{(5)}$	$Z_2^{(3)}, Z_1^{(3)}, Z_4^{(3)}, Z_3^{(3)}$
6	$Z_1^{(6)}, Z_2^{(6)}, Z_3^{(6)}, Z_4^{(6)}$	$Z_2^{(2)}, Z_1^{(2)}, Z_4^{(2)}, Z_3^{(2)}$
OT	$Z_1^{(7)}, Z_2^{(7)}, Z_3^{(7)}, Z_4^{(7)}$	$Z_1^{(1)}, Z_2^{(1)}, Z_3^{(1)}, Z_4^{(1)}$

2.10 The Bel-T Block Cipher

Bel-T is the name of the block cipher standard of the Republic of Belarus, according to [2]. It was approved in 2011. The specification of Bel-T, in Russian, is available from the website of the Research Institute for Applied Problems of Mathematics and Informatics of the Belarussian State University.

Bel-T has a hybrid design following the Feistel and the Lai-Massey design paradigm, somehow like the YBC cipher. No design criteria have been disclosed.

2.10.1 Encryption and Decryption Frameworks

Bel-T operates on 128-bit blocks under user keys of 128, 192 or 256 bits, and the encryption/decryption operations iterate eight rounds.

There is an additional 256-bit value called $\theta = \theta_1\|\theta_2\|\ldots\|\theta_8$, with $\theta_i \in \mathbb{Z}_2^{32}$, for $1 \leq i \leq 8$, that depends on the secret key (denoted Z).

The following description is based in [26].

Bel-T's internal operations are on 32-bit words, and include: modular addition and modular subtraction in $\mathbb{Z}_{2^{32}}$, denoted $\boxplus$ and $\boxminus$, respectively; bitwise exclusive-or in $\mathbb{Z}_2^{32}$, denoted $\oplus$.

There are three main mappings denoted G_5, G_{13}, G_{21} : $\mathbb{F}_2^{32} \rightarrow \mathbb{F}_2^{32}$, where G_t is parameterized by a 32-bit word $u = u_1\|u_2\|u_3\|u_4$ with $u_j \in \mathbb{F}_2^8$ for $1 \leq j \leq 4$. The operation of G_t is $G_t = (S[u_1]\|S[u_2]\|S[u_3]\|S[u_4]) \lll t$, where S is an 8×8 S-box specified in [2], and $\lll t$ denotes bitwise cyclic rotation to the left by t bits.

The encryption process consists of eight rounds. The input to the i-th round is denoted by the 4-tuple (a, b, c, d) where $a, b, c, d \in \mathbb{Z}_2^{32}$. The round subkeys Z_j are described in Sect. 2.10.2. The i-th round consists of the following steps (where $e \in \mathbb{Z}_2^{32}$ is an auxiliary variable):

- $b = b \oplus G_5(a \boxplus Z_{7i-6})$
- $c = c \oplus G_{21}(d \boxplus Z_{7i-5})$
- $a = a \boxminus G_{13}(b \boxplus Z_{7i-4})$
- $e = G_{21}(b \boxplus c \boxplus Z_{7i-3}) \oplus i$
- $b = b \boxplus e$
- $c = c \boxminus e$
- $d = d \boxplus G_{13}(c \boxplus Z_{7i-2})$
- $b = b \oplus G_{21}(a \boxplus Z_{7i-1})$
- $c = c \oplus G_5(d \boxplus Z_{7i})$
- $e = a$, $a = b$, $b = e$
- $e = c$, $c = d$, $d = e$
- $e = b$, $b = c$, $c = e$

See Fig. 2.19 for a pictorial representation of the i-th round of Bel-T for encryption.

The decryption operation is similar to the encryption operation except that the subkeys are reversed: Z_{7i} is used instead of Z_{7i-6}, Z_{7i-1} is used instead of Z_{7i-5}, and so on, up to Z_{7i-6} is used instead of Z_{7i}.

- $b = b \oplus G_5(a \boxplus Z_{7i})$
- $c = c \oplus G_{21}(d \boxplus Z_{7i-1})$
- $a = a \boxminus G_{13}(b \boxplus Z_{7i-2})$
- $e = G_{21}(b \boxplus c \boxplus Z_{7i-3}) \oplus i$
- $b = b \boxplus e$
- $c = c \boxminus e$
- $d = d \boxplus G_{13}(c \boxplus Z_{7i-4})$

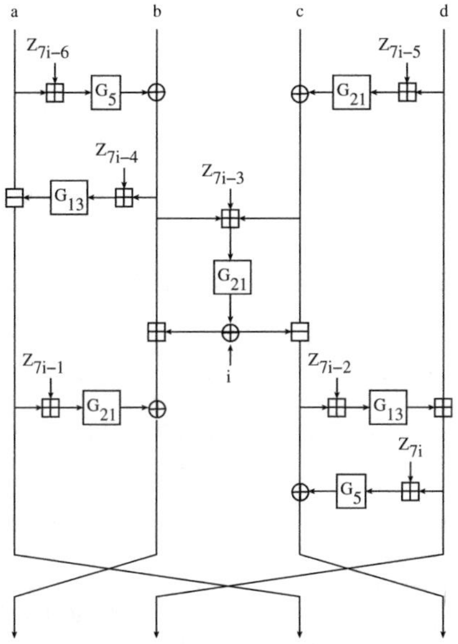

Fig. 2.19 Computational graphs of the i-th round of Bel-T for encryption.

- $b = b \oplus G_{21}(a \boxplus Z_{7i-5})$
- $c = c \oplus G_5(d \boxplus Z_{7i-6})$
- $e = a$, $a = b$, $b = e$
- $e = c$, $c = d$, $d = e$
- $e = b$, $b = c$, $c = e$

2.10.2 Key Schedule Algorithm

The key schedule of Bel-T uses the 256-bit string θ:

- for 128-bit keys: the first four words θ_1, θ_2, θ_3 and θ_4 correspond to the 128-bit key, while $\theta_5 = \theta_1$, $\theta_6 = \theta_2$, $\theta_7 = \theta_3$ and $\theta_8 = \theta_4$.
- for 192-bit keys: the first six words θ_1, ..., θ_6 correspond to the 192-bit key, while $\theta_8 = \theta_1 \oplus \theta_2 \oplus \theta_3$ and $\theta_8 = \theta_4 \oplus \theta_5 \oplus \theta_6$.
- for 256-bit keys: θ is assigned the 256-bit key.

Each round uses seven subkeys. The j-th round subkey of the i-th round is denoted Z_{7i-j}, for $1 \leq i \leq 8$ and $0 \leq j \leq 6$.

The fifty-six round subkeys are 32-bit values denoted $Z_1, \ldots, Z_{56}$. The θ values are assigned repeatedly to the fifty-six round subkeys in a circular fashion, as follows: $Z_1 = \theta_1$, $Z_2 = \theta_2$, ..., $Z_8 = \theta_8$, $Z_9 = \theta_1$, $Z_{10} = \theta_2$, ..., $Z_{56} = \theta_8$.

References

1. Ayaz, E.S., Selçuk, A.A.: Improved DST Cryptanalysis of IDEA. In: E. Biham, A.M. Youssef (eds.) Selected Areas in Cryptography (SAC), LNCS 4356, pp. 1–14. Springer (2006)
2. Belarus: Data Encryption and Integrity Algorithms. Preliminary State Standard of Republic of Belarus (STB P 34.101.31-2007). http://apmi.bsu.by/assets/files/std/belt-spec27.pdf (2007)
3. Biham, E.: How to Decrypt or Even Substitute DES-encrypted Messages in 2^{28} Steps. Information Processing Letters **84**(3), 117–124 (2002)
4. Biham, E., Dunkelman, O., Keller, N.: New Cryptanalytic Results on IDEA. In: X. Lai, K. Chen (eds.) Advances in Cryptology, Asiacrypt, LNCS 4284, pp. 412–427. Springer (2006)
5. Biham, E., Dunkelman, O., Keller, N.: A Unified Approach to Related-key Attacks. In: K. Nyberg (ed.) Fast Software Encryption (FSE), LNCS 5086, pp. 73–96. Springer (2008)
6. Biham, E., Dunkelman, O., Keller, N., Shamir, A.: New Data-Efficient Attacks on Reduced-Round IDEA. IACR ePrint Archive 2011/417 (2011)
7. Biham, E., Shamir, A.: Differential Cryptanalysis of DES-like Cryptosystems (extended abstract). In: A.J. Menezes, S.A. Vanstone (eds.) Advances in Cryptology, Crypto, LNCS 537, pp. 1–19. Springer (1990)
8. Biryukov, A., Nakahara Jr, J., Preneel, B., Vandewalle, J.: New Weak-Key Classes of IDEA. In: R.H. Deng, S. Qing, F. Bao, J. Zhou (eds.) Information and Communications Security (ICICS), LNCS 2513, pp. 315–326. Springer (2002)
9. Biryukov, A., Wagner, D.: Slide Attacks. In: L.R. Knudsen (ed.) Fast Software Encryption (FSE), LNCS 1636, pp. 245–259. Springer (1999)
10. Bogdanov, A., Khovratovich, D., Rechberger, C.: Biclique Cryptanalysis of the Full AES. In: D.H. Lee, X. Wang (eds.) Advances in Cryptology, Asiacrypt, LNCS 7073, pp. 344–371. Springer (2011)
11. Bogdavov, A., Rechberger, C.: A 3-Subset Meet-in-the-Middle Attack: Cryptanalysis of the Lightweight Block Cipher KTANTAN. In: A. Biryukov, G. Gong, D.R. Stinson (eds.) Selected Areas in Cryptography (SAC), LNCS 6544, pp. 229–240. Springer (2010)
12. Borisov, N., Chew, M., Johnson, R., Wagner, D.: Multiplicative Differentials. In: J. Daemen, V. Rijmen (eds.) Fast Software Encryption (FSE), LNCS 2365, pp. 17–33. Springer (2002)
13. Borst, J., Knudsen, L.R., Rijmen, V.: Two Attacks on Reduced IDEA (extended abstract). In: W. Fumy (ed.) Advances in Cryptology, Eurocrypt, LNCS 1233, pp. 1–13. Springer (1997)
14. Borst, J., Preneel, B., Vandewalle, J.: On the Time-Memory Trade-Off Between Exhaustive Search and Table Precomputation. Symposium on Information Theory in the Benelux, Veldhoven, the Netherlands (1998)
15. Daemen, J.: Cipher and Hash Function Design – Strategies based on Linear and Differential Cryptanalysis. Ph.D. thesis, Dept. Elektrotechniek, ESAT, Katholieke Universiteit Leuven, Belgium (1995)
16. Daemen, J., Govaerts, R., Vandewalle, J.: Weak Keys for IDEA. In: D.R. Stinson (ed.) Advances in Cryptology, Crypto, LNCS 773, pp. 224–231. Springer (1993)
17. Daemen, J., Rijmen, V.: The Block Cipher BKSQ. In: J.J. Quisquater, B. Schneier (eds.) Smart Card Research and Applications (CARDIS), LNCS 1820, pp. 236–245. Springer (2000)
18. Dehnavi, S.M., Rishakani, A.M., Shamsabad, M.R.M.: Characterization of MDS Mappings. IACR ePrint archive 2015/002 (2015)

19. Demirci, H., Selçuk, A.A., Türe, E.: A New Meet-in-the-Middle Attack on the IDEA Block Cipher. In: M. Matsui, R. Zuccherato (eds.) Selected Areas in Cryptography (SAC), LNCS 3006, pp. 117–129. Springer (2003)
20. Electronic-Frontier-Foundation: Cracking DES - Secrets of Encryption Research Wiretap Politics and Chip Design. O'Reilly and Associates Inc. (1998)
21. FIPS197: Advanced Encryption Standard (AES). FIPS PUB 197 Federal Information Processing Standard Publication 197, United States Department of Commerce (2001)
22. Guy, R.K.: Unsolved Problems in Number Theory. Springer (2004)
23. Hawkes, P.: Differential-Linear Weak Key Classes of IDEA. In: K. Nyberg (ed.) Advances in Cryptology, Eurocrypt, LNCS 1403, pp. 112–126. Springer (1998)
24. Hellman, M.E.: A Cryptanalytic Time-Memory Trade-Off. IEEE Transactions on Information Theory **IT-26**(4), 401–406 (1980)
25. ISO: Information Technology – Security Techniques – Hash Functions – part 3: Dedicated hash functions. ISO/IEC 10118-3:2004, International Organization for Standardization (2004)
26. Jovanovic, P.: Analysis and Design of Symmetric Cryptographic Algorithms. Ph.D. thesis, University of Passau, Germany (2015)
27. Junod, P.: New Attacks against Reduced-Round Versions of IDEA. In: H. Gilbert, H. Handschuh (eds.) Fast Software Encryption (FSE), LNCS 3557, pp. 384–397. Springer (2005)
28. Junod, P.: IDEA: Past, Present and Future. Early Symmetric Crypto, https://www.cryptolux.org/index.php/Home (2010)
29. Junod, P., Macchetti, M.: Revisiting the IDEA Philosophy. In: O. Dunkelman (ed.) Fast Software Encryption (FSE), LNCS 5665, pp. 277–295. Springer (2009)
30. Junod, P., Vaudenay, S.: FOX: a New Family of Block Ciphers. In: H. Handschuh, M.A. Hasan (eds.) Selected Areas in Cryptography (SAC), LNCS 3357, pp. 114–129. Springer (2004)
31. Kelsey, J., Schneier, B., Wagner, D.: Key-Schedule Cryptanalysis of IDEA, G-DES, GOST, SAFER and triple-DES. In: N. Koblitz (ed.) Advances in Cryptology, Crypto, LNCS 1109, pp. 237–251. Springer (1996)
32. Klimov, A., Shamir, A.: Cryptographic Applications of T-Functions. In: M. Matsui, R.J. Zuccherato (eds.) Selected Areas in Cryptography (SAC), LNCS 3006, pp. 248–261. Springer (2003)
33. Klimov, A., Shamir, A.: A New Class of Invertible Mappings. In: B.S. Kaliski Jr, Çetin K. Koç, C. Paar (eds.) Cryptographic Hardware and Embedded Systems (CHES), LNCS 2523, pp. 470–483. Springer (2003)
34. Lafitte, F., Nakahara Jr, J., Van Heule, D.: Applications of SAT Solvers in Cryptanalysis: Finding Weak Keys and Preimages. Journal of Satisfiability, Boolean Modeling and Computation **9**(1), 1–25 (2014)
35. Lai, X.: On the Design and Security of Block Ciphers. Ph.D. thesis, ETH no. 9752, Swiss Federal Institute of Technology, Zurich (1992)
36. Lai, X.: On the Design and Security of Block Ciphers, *ETH Series in Information Processing*, vol. 1. Hartung-Gorre Verlag, Konstanz (1995)
37. Lai, X., Massey, J.L.: A Proposal for a New Block Encryption Standard. In: I.B. Damgård (ed.) Advances in Cryptology, Eurocrypt, LNCS 473, pp. 389–404. Springer (1990)
38. Lai, X., Massey, J.L., Murphy, S.: Markov Ciphers and Differential Cryptanalysis. In: D.W. Davies (ed.) Advances in Cryptology, Eurocrypt, LNCS 547, pp. 17–38. Springer (1991)
39. Lerman, L., Nakahara Jr, J., Veshchikov, N.: Improving Block Cipher Design by Rearranging Internal Operations. In: P. Samarati (ed.) 10th International Conference on Security and Cryptography (SECRYPT), pp. 27–38. IEEE (2013)
40. Luby, M., Rackoff, C.: How to Construct Pseudorandom Permutations from Pseudorandom Functions. SIAM Journal of Computing **17**(2), 373–386 (1988)

41. Luo, Y., Lai, X., Gong, Z.: Pseudorandomness Analysis of the (Extended) Lai-Massey Scheme. Information Processing Letters **111**(2), 90–96 (2010)
42. MacWilliams, F.J., Sloane, N.J.A.: The Theory of Error-Correcting Codes. North-Holland Mathematical Library. North-Holland Publishing Co. (1977)
43. Menezes, A.J., Van Oorschot, P., Vanstone, S.A.: Handbook of Applied Cryptography. CRC Press (1997)
44. Nakahara Jr, J.: Cryptanalysis and Design of Block Ciphers. Ph.D. thesis, Dept. Elektrotechniek, ESAT, Katholieke Universiteit Leuven, Belgium (2003)
45. Nakahara Jr, J.: Faster Variants of the MESH Block Ciphers. In: A. Canteaut, K. Viswanathan (eds.) International Conference on Cryptology in India, Indocrypt, LNCS 3348, pp. 162–174. Springer (2004)
46. Nakahara Jr, J.: On the Design of IDEA-128. Brazilian Symposium on Information and Computational Systems Security (SBSeg) (2005)
47. Nakahara Jr, J.: Differential and Linear Attacks on the Full WIDEA-n Block Ciphers (under Weak Keys). In: J. Pieprzyk, A.R. Sadeghi, M. Manulis (eds.) International Conference on Cryptology and Network Security (CANS), LNCS 7712, pp. 56–71. Springer (2012)
48. Nakahara Jr, J., Preneel, B., Vandewalle, J.: A Note on Weak-Keys of PES, IDEA and some Extended Variants. In: C. Boyd, W. Mao (eds.) 6th Information Security Conference (ISC), LNCS 2851, pp. 267–279. Springer (2003)
49. Nakahara Jr, J., Preneel, B., Vandewalle, J.: The Biryukov-Demirci Attack on Reduced-Round Version of IDEA and MESH Ciphers. In: H. Wang, J. Pieprzyk, V. Varadharajan (eds.) 9th Australasian Conference on Information Security and Privacy (ACISP), LNCS 3108, pp. 98–109. Springer (2004)
50. Nakahara Jr, J., Rijmen, V., Preneel, B., Vandewalle, J.: The MESH Block Ciphers. In: K. Chae, M. Yung (eds.) Information Security Applications (WISA), LNCS 2908, pp. 458–473. Springer (2003)
51. NBS: Data Encryption Standard (DES). FIPS PUB 46, Federal Information Processing Standards Publication 46, U.S. Department of Commerce (1977)
52. Patarin, J.: The Coefficient H Technique. In: R. Avanzi, F. Sica, L. Keliher (eds.) Selected Areas in Cryptography (SAC), LNCS 5381, pp. 328–345. Springer (2009)
53. Raddum, H.: Cryptanalysis of IDEA-X/2. In: T. Johansson (ed.) Fast Software Encryption (FSE), LNCS 2887, pp. 1–8. Springer (2003)
54. Rijmen, V., Daemen, J., Preneel, B., Bosselaers, A., De Win, E.: The Cipher SHARK. In: D. Gollmann (ed.) Fast Software Encryption (FSE), LNCS 1039, pp. 99–112. Springer (1996)
55. Rijmen, V., Preneel, B., De Win, E.: On Weaknesses of Non-Surjective Round Functions. Designs, Codes and Cryptography **12**(3), 253–266 (1997)
56. Shannon, C.E.: Communication Theory of Secrecy Systems. Bell System Technical Journal **28**(4), 656–715 (1949)
57. Su, S., Lu, S.: A 128-bit Block Cipher based on Three Group Arithmetics. IACR ePrint archive 2014/704 (2014)
58. Vaudenay, S.: On the Need for Multipermutations: Cryptanalysis of MD4 and SAFER. In: B. Preneel (ed.) Fast Software Encryption (FSE), LNCS 1008, pp. 286–297. Springer (1995)
59. Vaudenay, S.: On the Lai-Massey Scheme. In: K.Y. Lam, E. Okamoto, C. Xing (eds.) Advances in Cryptology, Asiacrypt, LNCS 1716, pp. 8–19. Springer (1999)
60. Wei, L., Peyrin, T., Sokolowski, P., Ling, S., Pieprzyk, J., Wang, H.: On the (In)Security of IDEA in Various Hashing Modes. IACR ePrint archive 2012/264 (2012)
61. Winternitz, R., Hellman, M.E.: Chosen-Key Attacks on a Block Cipher. Cryptologia **11**(1), 16–20 (1987)
62. Yi, X.: On Design and Analysis of a New Block Cipher. In: J. Jaffar, R.H.C. Yap (eds.) Concurrency and Parallelism, Programming, Networking and Security, ASIAN 1996, LNCS 1179, pp. 213–222. Springer (1996)

63. Yıldırım, H.M.: Some Linear Relations for Block Cipher IDEA. Ph.D. thesis, The Middle East Technical University, Turkey (2002)
64. Yıldırım, H.M.: Nonlinearity Properties of the Mixing Operations of the Block Cipher IDEA. In: T. Johansson, S. Maitra (eds.) Progress in Cryptology, Indocrypt, LNCS 2904, pp. 68–81. Springer (2003)
65. Yun, A., Park, J.H., Lee, J.: On Lai-Massey and Quasi-Feistel Ciphers. Designs, Codes and Cryptography **58**(1), 45–72 (2011)

Chapter 3
Attacks

The only source of knowledge is experience.
- Albert Einstein

Abstract

This chapter presents an overview of old and new attacks that have (or not) been reported in the literature against PES, IDEA, the MESH ciphers and other Lai-Massey ciphers described in Chap. 2.

We start our discussion with generic black-box attacks such as exhaustive search, dictionary and time-memory trade-off, that apply to any cipher independent of its design, the block size, the key size or the number of rounds.

Further, we describe modern attack techniques such as differential cryptanalysis, linear cryptanalysis, square/multiset and impossible-differential attacks, among others, exemplifying them on specific Lai-Massey ciphers.

3.1 Exhaustive Search (Brute Force) Attack

Exhaustive search (or brute force) is a universal type of attack which tries every possible value of an unknown quantity until a certain condition is satisfied. For instance, given an n-bit block cipher E, under an unknown k-bit key K, and a plaintext-ciphertext pair (P, C), an exhaustive key search (EKS) for K would try every possible k-bit value K' until one value is found that satisfies $E_{K'}(P) = C$. Depending on the relative sizes of n and k, there could be multiple candidates or a unique candidate for K.

The number of matching plaintext-ciphertext pairs for high assurance about the correct key value depends on the *unicity distance* [136], which is $\lceil \frac{k}{n} \rceil$. This value, which represents the data and memory complexities of the attack, is usually small.

This fact motivates k to be as large as possible compared to (a fixed value) n, but even in the extreme case that all round subkeys in a block cipher were independent, $\lceil \frac{k}{n} \rceil$ would still not be exponentially large because: (i) the number of rounds of typical block ciphers is usually small for performance

J. Nakahara Jr., *Lai-Massey Cipher Designs*, https://doi.org/10.1007/978-3-319-68273-0_3

reasons; (ii) the number of subkeys per round is also not large; (iii) large keys would make the key management process more expensive and slower. For instance, assume IDEA with $n = 64$, $k = 128$, $r = 8.5$ rounds and $t = 6 \cdot 16 = 96$ subkey bits per (full) round. Then, for independent subkeys there would be no key schedule and the user key would consist of independent bits. Thus, $k = r \cdot t = 96 \cdot 8 + 16 \cdot 4 = 832$ and $\lceil \frac{k}{n} \rceil = \lceil \frac{832}{64} \rceil = 13$. In real life applications (assuming a known-plaintext setting), it is hardly acceptable that a fixed key is used to encrypt (a message with) less than 13 plaintext-ciphertext blocks before the key has to be changed.

In this example, the search was performed in a known-plaintext setting because a matching pair (P, C) was made available. In a ciphertext-only (CO) setting, an exhaustive search may succeed if enough redundancy can be found in the plaintext (which will affect the unicity distance).

The search procedure can be performed in several ways, and not necessarily following a sequential order of candidate keys across the search space. What does matter is that the entire search space is covered and no element in the search space is covered more than once during the search.

The exhaustive search is a universal kind of attack since it applies equally well to several cryptographic primitives, such as block and stream ciphers and message authentication codes. What changes from one algorithm to another is the unknown value and the condition(s) for stopping the search.

In iterated algorithms, exhaustive search applies to any number of rounds, and is independent of the internal structure of the cryptographic algorithm. In other words, this attack depends only on general parameters of an algorithm such as the block size, the key size, the state size or other values (nonces, IVs, constants). When applied to an unknown key, it is called *exhaustive key search*, when applied to an internal cipher state (of a stream cipher), it is called *state-recovery* attack.

When the attack uses a single CPU, EKS is very time consuming since the cost is *exponential* in the bit size of the search space. But, if computational resources are available, there is no reason to perform the search *sequentially* with a single processor since there is no dependency between different key candidates tested. It is much more effective to perform the search in *parallel*. Therefore, the key size (in bits) is not the correct indicator of an EKS attack complexity.

Consider the EKS attack in a known-plaintext (KP) setting. Assume a k-bit key K and suppose that the best adversary has 2^t CPUs available, for some integer t. Then, the EKS (clock) time would drop from 2^k sequential encryptions to (the time equivalent of) 2^{k-t} sequential encryptions, since the search is performed in parallel.

This strategy has already been used, for instance, by the Electronic Frontier Foundation in [66] in building the DES Cracker which used 1856 dedicated ASIC chips implementing DES, and used to performed an EKS attack on (single) DES. In January 1999, Distributed.net [1] an organization specialized in collecting and managing computer's idle time, used more than 100000 computers to perform an EKS attack on a (single) DES instance in 23 hours. In 2006, a custom hardware design using 120 low-cost FPGAs, called COPACOBANA, was used to perform an EKS attack on a (single) DES instance in 9 days [110].

Consider now the setting of multiple keys: in real-life, it is usual that several parties (in pairs) use the same block cipher, say IDEA, but with different keys. This is the multi-key setting, which translated to an EKS attack means a generalization of the conventional (or single-key) EKS discussed previously.

In a multi-key EKS setting, suppose an adversary is able to have the same plaintext P encrypted under $r > 1$ distinct unknown keys K_1, K_2, $\ldots$, K_r leading to r (distinct) ciphertexts $C_1 = E_{K_1}(P)$, $C_2 = E_{K_2}(P)$, $\ldots$, $C_r = E_{K_r}(P)$. The objective is to find at least one of the r keys. How to do it? Just perform the usual (single-key) EKS attack with random keys K', but instead of checking if $E_{K'}(P)$ matches a single ciphertext, rather compare $E_{K'}(P)$ with all of the C_i, for $1 \leq i \leq r$.

Furthermore, compounding the multi-key EKS with multiple CPUs, it is possible to optimize the search for K_1, K_2, $\ldots$, K_r by splitting the effort into 2^t CPUs. The attack (clock) time drops further to $2^k/(2^t \cdot r)$ sequential encryptions per CPU. Assuming r is a power of 2, say, $r = 2^u$, then the effort becomes 2^{k-t-u} sequential encryptions per CPU. We assume that all ciphertexts are distinct, i.e. $C_i \neq C_j$, for $i \neq j$, and that the cost of comparing bit strings is negligible compared to the cost of a single encryption.

Although expensive on its own, exhaustive search is often used in other kinds of attacks but restricted to a smaller search space, that is, using a divide-and-conquer strategy.

An interesting feature of exhaustive search is that it is an *all-or-nothing* kind of attack, that is, either all unknown bits are recovered at once or none at all. For instance, if all unknown key bits are guessed right, except one bit (in any position), then the adversary is as close to the real solution as if all bits were wrong. An exception is the case of equivalent keys as for instance, in the TEA block cipher [182], that is, keys that perform the same encryption transformation [95]: if K_1 and K_2 are equivalent keys[2], then $E_{K_1}(P) = E_{K_2}(P)$ for all plaintext blocks P. This weakness decreases the search effort since finding a key equivalent to the original one is enough.

Any information about the target element in the search space may reduce the search effort. For instance, if it is known that the target value is odd, then only the odd values need to be searched for. In a space of 2^n values, if it is known that the Hamming weight of the target value is m, then, only $\binom{2^n}{m}$ values need to be searched for, and this is a considerable improvement from the initial search space of size 2^n.

---

[1] http://www.distributed.net

[2] In general, there may be several equivalent key classes. In each class, all keys perform the same encryption transformation.

Exhaustive key search is still used as a *benchmark* attack to determine whether a cryptographic algorithm is considered broken (in practice) or not. For instance, the full 16-round DES cipher had been cryptanalyzed in [131, 30] but only after repeated successful exhaustive key search attacks implemented in hardware [66] were confirmed in practice did NIST withdrew its support of DES, and initiated a competition for an Advanced Encryption Standard (AES) algorithm to replace the DES.

An interesting setting that is related to the exhaustive key search is the quantum or post-quantum setting. Assuming quantum computers become widely available, symmetric ciphers will be affected by Grover's algorithm [75], which can search an unsorted database of size T in time $O(\sqrt{T})$. Under Grover's attack algorithm, the security provided by a block cipher with a k-bit key is equivalent to about $k/2$ bits. The conclusion is that the key length of a block cipher (in a quantum setting) must be at least doubled to keep the same security level as expected in a non-quantum setting.

An interesting feature of most Lai-Massey designs is that the (user) key size is double the block size[3], which helps to mitigate this kind of attack (in a quantum scenario).

On the other hand, doubling the key size would affect the speed of key diffusion in the key schedule algorithm (KSA): the larger the key (for a fixed number of key bits per round) the faster the key diffusion must be in the KSA in order to guarantee that the whole cipher state depends on every key bit (in the least number of rounds). Another drawback is that the key management becomes more expensive for longer keys.

3.2 Dictionary Attack

A dictionary attack, also called Table-Lookup (TLU) attack, assumes a known piece of plaintext that is highly probable to be encrypted. This means a known-plaintext setting.

Such piece of probable text P is encrypted under all possible keys K, and the resulting pair $(C = E_K(P), K)$ is stored in a table T ordered by the C value. When a piece of ciphertext C' is intercepted, a match with an entry (C^*, K^*) in T will indicate a candidate key K^* for the unknown key.

The dictionary attack requires a one-time offline pre-computation step that costs 2^k encryptions (for E using a k-bit key) and storage (memory) of 2^k pairs (C, K). But, the online complexity consists in indexing the T table, which takes constant time.

The attack complexity in a dictionary attack is in the offline pre-computation step and is heavily dependent on a large memory, while the exhaustive search

[3] With the exception of FOX/IDEA-NXT for which the key size is variable.

attack does not rely on memory but is expensive in the time complexity, pushing the effort towards the online phase of the attack. See Table 3.1.

Table 3.1 Comparison of EKS and TLU attacks for an n-bit text block, k-bit key block cipher.

Attack	block size (bits)	key size (bits)	time complexity online	time complexity offline	data complexity	memory complexity
EKS	n	k	2^k	—	$\lceil k/n \rceil$	$\lceil k/n \rceil$
TLU	n	k	—	2^k	$\lceil k/n \rceil$	2^k

3.3 Birthday-Paradox Attacks

The birthday-paradox problem [156, 68] asks for the minimum number n of people such that there is more than 50% chance that at least two people share the same birthday (day and month). Even though years are not of equal length, nor are birthrates uniform all over each year, it is assumed that birthdays are independently and uniformly distributed on the m days of a year. We assume, as an approximation, that taking a random sample of n people is equivalent to a random sample of n birthdays.

The solution to the problem is made easier by looking at the complementary problem: the probability that all n birthdays are different is

$$q = \frac{m \cdot (m-1) \cdot (m-2) \ldots (m-n+1)}{m^n}.$$

As $m \to \infty$, the probability that at least two people share the same birthday [156] tends to

$$1 - q \approx 1 - \exp(-\frac{n(n-1)}{2m} + O(\frac{1}{\sqrt{m}})) \approx 1 - \exp(-\frac{n^2}{2m}). \qquad (3.1)$$

For $n = 23$ people in $m = 365$ days, we have $1 - q \approx 0.507297$.

The apparent paradox comes from the small number $n = 23$ people compared to the length $m = 365$ of days in a year. But, there should be no surprise since there are no restrictions in the choice of the colliding birthdays among n people. This fact is made more evident by the term n^2 in (3.1), which means that the number of potential pairs of people with matching birthdays grows with n^2.

The birthday-paradox problem is a special case of the classical occupancy problem [156]:

Definition 3.1. (Classical Occupancy Problem) [156]
An urn has n balls numbered from 1 to m. Suppose that n balls are drawn from the urn one at a time, with replacement, and their numbers are recorded. The probability that exactly t different balls have been drawn is

$$P(m,n,t) = \left\{ \begin{matrix} \mathrm{n} \\ \mathrm{t} \end{matrix} \right\} \frac{m^{(t)}}{m^n}, \quad 1 \leq t \leq n. \tag{3.2}$$

where

$$\left\{ \begin{matrix} \mathrm{n} \\ \mathrm{t} \end{matrix} \right\} = \frac{1}{t!} \sum_{k=0}^{t} (-1)^{t-k} \binom{t}{k} k^n, \tag{3.3}$$

denotes the Stirling number of the second kind, with the exception that $\left\{ \begin{matrix} 0 \\ 0 \end{matrix} \right\} = 1$. The symbol $\left\{ \begin{matrix} \mathrm{n} \\ \mathrm{t} \end{matrix} \right\}$ counts the number of ways of partitioning a set of n objects into t nonempty subsets. Also, for positive integers m and n such that $m \geq n$, the integer $m^{(n)}$ is defined as

$$m^{(n)} = m(m-1)(m-2)\dots(m-n+1).$$

In [138], the birthday-paradox problem is used in the following setting: let X be a nonempty set and $F : X \to X$ be a random function. Provide the minimum size of two subsets A and B of X in order for a collision to occur between their elements with a probability higher than 50%. Initially, consider the probability p of no collision between A and B: $p = (1 - \frac{|A|}{|X|})^{|B|}$, which, if $|A|$ is much smaller than X, is approximately $e^{-\frac{|A|\cdot|B|}{|X|}}$. Then, the probability of at least one collision is $q = 1 - p \approx 1 - e^{-\frac{|A|\cdot|B|}{|X|}}$.

If $|A| \cdot |B| \approx |X|$, then $q = 1 - e^{-1} \approx 0.63$. In order to achieve a trade-off for the subset sizes, it is usual to set $|A| = |B| = \sqrt{|X|}$.

A countermeasure to attacks based on the birthday-paradox problem is to fix $n = |X|$ to a large enough value, taking into account the square-root reduction in complexity. Due to this reduction scaling, attacks based on the birthday paradox are sometimes known as square-root attacks.

In [189], Yuval discussed an attack on a digital signature scheme by Rabin [164]. Rabin's scheme consisted of a compressed encoding using an encryption function E_K to digitally sign a message M in order to efficiently confirm its content as not been tampered with during transmission. Let $y = E_K(x)$ with $x, y \in \mathbb{Z}_2^n$ and $K \in \mathbb{Z}_2^k$, for n and k nonzero integers.

Yuval's idea was to have two sets A and B. One of them, say A, would contain modified versions of the legitimate message while B would contain modified versions of false, biased messages. The attack would need $2^{n/2}$ random versions in each of A and B, all of which would be digitally signed. Further, a matching between elements of both sets would be searched for. Such a match between digital signatures means that a legitimate message could be replaced by a false one, violating the security assumption of the digital signature scheme. This attack was related to the birthday paradox approach. As pointed out by Nishimura and Sibuya in [156], Yuval's matching procedure uses two sets, instead of a single set (as in the classical birthday-paradox setting).

Another consequence of the birthday paradox for the security of an n-bit block cipher is that repeated occurrences of a ciphertext block can be expected with probability about 0.63 if more than $2^{n/2}+1$ random plaintext blocks are encrypted under the same key, independent of the key size [99]. For the CBC mode of operation, if two ciphertext blocks match, $C_i = C_j$, then the corresponding inputs to the encryption function E_K must also be equal: $C_i = E_K(P_i \oplus C_{i-1}) = E_K(P_j \oplus C_{j-1}) = C_j$, which implies $P_i \oplus C_{i-1} = P_j \oplus C_{j-1}$. This equality means that in a ciphertext-only setting, $C_{i-1} \oplus C_{j-1} = P_i \oplus P_j$, that is, plaintext information is revealed (leaks) from ciphertext data.

Birthday attacks were also considered by Kaliski *et al.* in attacks to determine whether or not the DES cipher had any kind of algebraic structure [93].

In [13], Biham described a known-plaintext attack based on the birthday paradox. His attack can find the k-bit key K of a block cipher in $2^{k/2}$ queries, under a fixed plaintext, with time complexity $2^{k/2}$ encryptions. His attack assumes the adversary can obtain the encryption of some fixed, known plaintext block P under $2^{k/2}$ randomly chosen keys. The data $(E_{K_i}(P), K_i)$, for each candidate K_i, are stored in a table indexed by the first element of the pair. Every ciphertext block $C' = E_{K'}(P)$ is then compared to the first entry (C, K_i) in the table for some K_i. A match indicates a potential candidate K' for K. If K' is the correct key, messages can be forged while K' is still in use. It is expected by the birthday paradox that one of the first $2^{k/2}$ received ciphertexts discloses its encryption key. If the size of the plaintext block is smaller than $|K|$, then additional plaintexts might be requested in order for the attack to find a unique solution (due to the unicity distance). The conclusion is that the theoretical strength of a cipher is upper-bounded by the square root of the key space size.

In [183], Winternitz and Hellman described a generic chosen-key attack. Let E_K be a block cipher under a secret key K. Assume the adversary can choose a message block m and a vector l, which is called a *key flip* vector, and can request the encryption $E_{K \oplus l}(m)$ or the decryption $E^{-1}_{K \oplus l}(m)$. The key-flip vector l is 1 at those positions at which the adversary wishes the key to be changed.

The chosen-key attack operates as follows: let E be a secure block cipher with a k-bit key. Define a new block cipher F as follows: $F_K(m) = E_K(m)$ if the leftmost bit of K is 0, and $F_K(m) = E_{K'}(m) \oplus K$ otherwise, where $K' = K \oplus (1, 0, 0, \ldots, 0)$. Decryption is as follows: $F_K^{-1}(m) = E_K^{-1}(m)$ if the leftmost bit of K is 0; $F_K^{-1}(m) = E_{K'}^{-1}(m \oplus K)$, otherwise. If the adversary has no *a priori* knowledge of E, he can only gain useful information if he encrypts or decrypts under K or K'. In the chosen-key attack on F, the adversary requests two encryptions, $F_K(m)$ and $F_{K'}(m)$, for some convenient message m. If the leftmost bit of K is zero, $F_K(m) \oplus F_{K'}(m) = E_K(m) \oplus (E_{K'\oplus(1,0,\ldots,0)}(m) \oplus K') = K'$. Otherwise, $F_K(m) \oplus F_{K'}(m) = (E_{K'}(m) \oplus K) \oplus E_{K'}(m) = K$. Thus, the adversary can recover all bits of K. For the leftmost bit of K there are only two choices.

A general chosen-key attack uses a meet-in-the-middle (MITM) strategy. It allows exhaustive search to be cut by a factor of the number of key bit flips [183]: suppose the adversary can force 2^p key flips, and has 2^p words of memory. Then, E can be broken in 2^{k-p} additional steps. Using an unrestricted chosen-key attack, any block cipher with an k-bit key can be broken with effort about $2^{k/2}$ in computation and memory [183]: just let $p = k/2$. This fact motivates a design criterion: doubling the key size (if necessary) to account for the effect of a chosen-key attack.

3.3.1 Generalized Birthday Paradox Attack

The setting of the classical birthday paradox consists in finding a collision for a mapping $F : D \to R$, that is, two distinct elements $x, z \in D$ such that $F(x) = F(y)$. This problem can be generalized to a multi-collision setting or an s-collision setting (for any finite s), for which $s = 2$ is the classical case. An s-collision for a mapping F consists of s distinct points $x_1, x_2, \ldots, x_s \in D$ all of which collide $F(x_1) = F(x_2) = \ldots = F(x_s)$.

Previous results on multi-collision attacks include the security of iterated hash functions [88], of identification schemes [74], of signature schemes [49] and of micropayment schemes [166].

In [175], Suzuki *et al.* proved that in order to find an s-collision for a mapping $F : D \to R$, where $|R| = 2^n$, with probability at least 50%, $(s!)^{1/s} \cdot n^{\frac{s-1}{s}}$ distinct values are needed. It is assumed that each image $y \in R$ has the same number of preimages, that is, $|F^{-1}(y)| = |D|/|R|$ for all $y \in R$.

For $s = 2$, the number of values for 50% success rate is about the square root of the range size of F, that is, $\sqrt{2} \cdot n^{1/2}$, which is the classical birthday-paradox setting.

3.4 Time-Memory Trade-Off Attacks

The Time-Memory Trade-Off (TMTO) attack is due to M.E. Hellman [82]. Originally, the TMTO technique was applied on reduced-round versions of the DES cipher. TMTO is a chosen-plaintext attack that trades-off computing time for memory storage, balancing advantages from both the exhaustive search (Sect. 3.1) and the dictionary (Sect. 3.2) attacks.

The TMTO was extensively applied to stream ciphers [8, 35, 63], where it is also referred to as the Babbage-Golic TMTO attack.

Hellman's attack assumes the target cipher is operated in ECB mode, but, otherwise, in other modes such as CBC, CTR, OFB and CFB, the adversary must find a way to choose the appropriate input for encryption. Hong and Sarkar in [83] have demonstrated attacks in those modes and also in a known-plaintext setting, assuming for instance, fixed headings in messages.

Let T denote the time complexity, and M denote the memory needed to attack a given target cipher (operating on n-bit text blocks and using a k-bit key). Two general attacks on such ciphers are:

- exhaustive key search (EKS): $T = O(2^k)$, $M = O(1)$. EKS is a known-plaintext attack. The adversary is assumed to know for some ciphertexts the corresponding plaintexts (the number of text blocks is given by the unicity distance $\lceil \frac{k}{n} \rceil$); the aim is to find the key that is compatible with these texts by searching the key space exhaustively.
- table lookup (TLU) also known as dictionary attack: $T = O(1)$, $M = O(2^k)$. TLU is a chosen-plaintext attack. The adversary pre-computes a table of all possible ciphertexts for a chosen plaintext block (cost: 2^k offline computations) ordered by the ciphertexts (cost: $2^k \cdot \log 2^k = k \cdot 2^k$ for sorting). This pre-computation is a one-time (offline) cost. Further, during the online part of the attack, a ciphertext is given, and the corresponding key is found by table lookup (cost: $\log 2^k = k$, for example, using binary search [55]).

The use of the big-O notation [136] is because there may be lower-order complexity figures such as logarithmic factors, which are ignored.

A TMTO attack is a hybrid technique that optimizes the resources (CPU, memory, storage) required by the exhaustive search and the table lookup attacks. In particular, the TMTO attack requires lower online processing complexity than an exhaustive key search and lower memory complexity than table lookup. Also, unlike the latter, the TMTO is a probabilistic attack. The TMTO is a general approach to invert an arbitrary mapping [70]. This approach balances a one-time *offline* work (the result of the pre-computation is stored in memory) with the time needed to run an *online* search algorithm.

The TMTO approach speeds up the search at the cost of a one-time effort and an increased storage (memory). All of these three techniques, though, share the following features:

- the attacks assume the target mapping as a black box in the sense that no internal detail of the target cipher is exploited in the attack.
- these techniques are independent of the number of rounds of the target cipher. So, increasing the number of rounds does not help protect the cipher against any of the above techniques.
- the attacks can be performed in parallel (once the pre-computation is finished) and on multiple targets.
- they are generic attacks in the sense that they depend on general parameters (key size, block size). Therefore, these attacks set general lower bounds on attack complexities. Consequently, new cipher designs should

have minimum parameter values (key size and block size) set high enough to counter these attacks (from becoming practical).

Consider the mapping from the key input to the ciphertext output when DES is used to encrypt a fixed plaintext block. This mapping is non-invertible, and is often modeled as a one-way function. In general, the mappings used in the TMTO attack are random mappings from finite sets into finite sets. These mappings can be thought of as being implemented by a lookup table in which each entry is chosen uniformly, independently and with replacement from the potential elements in the range. It is convenient to think of the directed graph (Fig. 3.1) associated with each realization of a random mapping f, that is, the graph with vertex set the domain of the mapping, and with a direct edge from vertex x to vertex y if and only if $y = f(x)$. Such graphs consist of a number of components, each of which has exactly one cycle (otherwise, f would not be well defined[4]).

If an arbitrary vertex in such a graph is chosen, and the path beginning there is followed, it will eventually end in a cycle [72]. This is just another way of viewing the fact that the iteration of a mapping from a finite set into itself eventually leads to a periodic repetition of the same values (Pigeon-Hole Principle) [136]. The fact that each component of the graph has only one cycle is equivalent to the fact that the mapping is single-valued.

The number of vertices in a component is referred to as the drainage into that component. In Hellman's method applied to the DES cipher, a mapping from the key space into itself was used, by fixing the plaintext and interpreting the output ciphertext as an element of the key space. In this case, keys are 56 bits long and ciphertexts are 64-bit blocks, and arbitrary reduction functions were used to obtain 56 bits out of 64 bits.

A peculiar graph structure would be one in which the graph consisted only of cycles with no branches draining into them (Fig. 3.2). In this case, the TMTO method could invert this structure in $N^{1/2}$ operations using $N^{1/2}$ words of memory, where $N = 2^k$ is the key space size. Such a cyclic structure only happens for permutation mappings. Under plausible assumptions about the structure of the graphs obtained using the TMTO attack, Hellman showed that inverting it should be effective in $N^{2/3}$ operations and the same amount of memory. If we begin at an arbitrary point in the graph and follow the path leading from it, we will almost certainly enter a cycle before traveling farther than a few multiples of $N^{1/2}$. By choosing different starting points and calculating the lengths of the cycles into which they drain, we can tell when different points drain into the same components. Thus, we can estimate the size of the largest component in a random graph on N points with $N^{1/2}$ operations.

[4] It means that for some preimage x there would be (at least) two distinct images.

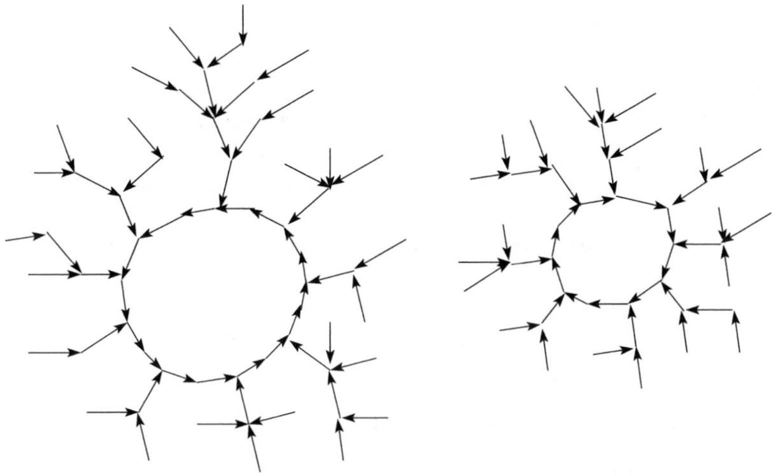

Fig. 3.1 Pictorial representation of a directed graph of powers of a random mapping.

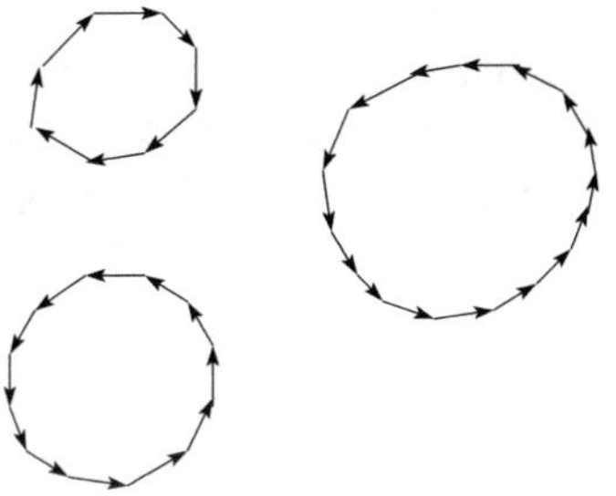

Fig. 3.2 Pictorial representation of a directed graph of powers of a random permutation.

3.4.1 Hellman's Attack

The Hellman's TMTO attack is designed to invert an arbitrary function $E : \mathbb{Z}_2^k \times \mathbb{Z}_2^n \rightarrow \mathbb{Z}_2^n$, keyed by a k-bit key. Let $N = 2^k$ denote the search space size.

Hellman's TMTO attack assumes a fixed plaintext P_0 that can be chosen or known by the adversary. P_0 may represent a typical text block that is highly likely to be encrypted under the unknown key in the online part of the attack. For instance, P_0 may be a standard salutation in the beginning of a message or a common greeting at the end of a typical message.

Parts of the TMTO attack description that follows are based on [62].

In the pre-computation phase, m values $X_{i,0} \in \mathbb{Z}_2^k$, $1 \leq i \leq m$, are chosen randomly from the key space. For $1 \leq i \leq m$ and $1 \leq j \leq t$, define $X_{i,j}$ and a function f by

$$X_{i,j} = f(X_{i,j-1}) = R(E_{X_{i,j-1}}(P_0)). \tag{3.4}$$

The parameters m and t are chosen to trade-off time against memory.

The contrived f function maps n-bit ciphertext blocks into the k-bit key input. This transformation is performed in two steps: (i) $E_{K'}(P_0)$ computation under key $K' = X_{i,j-1}$, (ii) the reduction mapping R.

The mapping $R : \{0,1\}^n \to \{0,1\}^k$ transforms an n-bit plaintext block into a k-bit key, and assumes nothing about the key space except its size. For instance, for DES, R may simply drop the eight parity bits in a 64-bit key.

The $X_{i,0}$ values are denoted SP_i for *starting points*, while the $X_{i,t}$ values are called *end points* and denoted EP_i. A table of $X_{i,j} \in \mathbb{Z}_2^k$ (the images under iterated f) is created

$$\begin{array}{lllll} SP_1 = X_{1,0} & \xrightarrow{f} X_{1,1} = R(E_{X_{1,0}}(P_0)) & \xrightarrow{f} \ldots \xrightarrow{f} & X_{1,t} = EP_1 \\ SP_2 = X_{2,0} & \xrightarrow{f} X_{2,1} = R(E_{X_{2,0}}(P_0)) & \xrightarrow{f} \ldots \xrightarrow{f} & X_{2,t} = EP_2 \\ \ldots & \ldots \quad \ldots & \ldots\ldots\ldots & \ldots \\ SP_m = X_{m,0} & \xrightarrow{f} X_{m,1} = R(E_{X_{m,0}}(P_0)) & \xrightarrow{f} \ldots \xrightarrow{f} & X_{m,t} = EP_m \end{array} \tag{3.5}$$

In each line of (3.5), f is iterated t times.

To reduce memory requirements, all intermediate points between SP_i and EP_i are discarded and only the pairs $(\mathrm{SP}_i,\mathrm{EP}_i)$ are stored, sorted on the EP_i as the result of this pre-computation. The pre-computation cost is $\mathrm{O}(m \cdot t \cdot \log_2 t)$ using, for example, the quicksort algorithm [55] for t elements in a row, for m rows.

Now suppose an unknown key K is used and the ciphertext block $C_0 = E_K(P_0)$ is obtained, for a text block P_0 chosen by the adversary. Let $Y_1 = R(C_0)$ and check if Y_1 is an end point EP_i for some $1 \leq i \leq m$. If $Y_1 = X_{i,t}$, for some i, then either $K = X_{i,t-1}$, since $R(C_0) = X_{i,t} = R(E_{X_{i,t-1}}(P_0))$, or Y_1 had more than one preimage. In the latter case, there is some $X_{l,t-1}$ with $l \neq i$ such that $X_{i,t} = R(E_{X_{l,t-1}}(P_0))$, but the value of l is not known. This latter event, when $X_{i,t-1}$ is a wrong key candidate, is referred to as a *false alarm.* When searching for a key in a table, finding a matching endpoint does not imply that the key is in the table. The key may be part of a chain which has the same endpoint but is not in the table. In that case, generating the chain from the saved starting point does not yield the key (that is why it is called a false alarm). False alarms also occur when a key is in a chain that is part of the table but which merges with other chains. In that case, several starting points correspond to the same end point and several chains may have to be generated until the key is finally found.

If $Y_1 = X_{i,t} = R(C_0) = f(X_{i,t-1}) = R(E_{X_{i,t-1}}(P_0))$ then compute $X_{i,t-1} = f^{t-1}(X_{i,0}) = f^{t-1}(\mathrm{SP}_i)$. Further, check if $X_{i,t-1}$ can decrypt C_0 into P_0 (an n-bit condition for a k-bit key requires $\lceil k/n \rceil$ plaintext-ciphertext pairs to ensure its correctness, according to the unicity distance). If Y_1 is not an end point or a false alarm occurred, then compute $Y_2 = f(Y_1)$, and check if Y_2 is an end point. If it is an end point for some i, then compute $X_{i,t-2} = f^{t-2}(X_{i,0})$

and check if $X_{i,t-2}$ can be the key. If not, compute $Y_3 = f(Y_2)$ and continue in the same way through until $Y_t = f(Y_{t-1})$.

In general, if $Y_j = X_{i,t}$, then $f(Y_{j-1}) = f(X_{i,t-1})$, that is, $f^2(Y_{j-2}) = f^2(X_{i,t-2})$, and so on, $f^{j-1}(Y_1) = f^{j-1}(X_{i,t-j+1})$. Therefore, $f^{j-1}(R(C_0)) = f^{j-1}(X_{i,t-j+1}) = f^j(X_{i,t-j})$, and $R(C_0) = f(X_{i,t-j}) = R(E_{X_{i,t-j}}(P_0))$. Recall that except for the R truncation, C_0 is the encryption of P_0 under the key $X_{i,t-j}$.

Finally, compute $X_{i,t-j} = f^{t-j}(X_{i,0})$ and check if it decrypts C_0 into P_0.

If $Y_j = SP_i$, then the search fails since the SP_is have no known preimage. If we knew C_0 beforehand, then this choice of SP_i could have been avoided, but the pre-computation step is done before knowing C_0.

If all mt elements of the table are distinct, and if k was chosen uniformly at random from all possible values, then the probability of success would be $p_{success} = mt/N$. This value can be compared to t/N for a table lookup with t entries. Actually, the larger mt is, the more probable it is that the elements $X_{i,j}$ will overlap, and it can be shown that $p_{success} \geq 0.63mt/N$, where m and t should be approximately $N^{2/3}$. This restriction will make $p_{success}$ small, and to overcome this $O(N^{2/3})$ complexity, several tables can be created, with different f functions, that is, using different reduction functions. The efficiency of Hellman's scheme is related to the cycle structure of the cipher mapping under the f functions. The estimates for the probability of success and false alarm rate are based on the assumption that the cipher and the f functions behave as random functions. The above reasoning assumed that $|K| = 56 < |C| = 64$, such as in the DES.

In the case of AES and IDEA, in which $|K| \geq |C|$, Kusuda and Matsumoto [111, 112] suggested a modified f function. Let $q = \lceil \frac{k}{n} \rceil$, where $k = |K|$ and $n = |C|$. Prepare q chosen plaintexts P_1, P_2, ..., P_q, and redefine f as $f(k) = R(E_k(P_1)||E_k(P_2)||\ldots||E_k(P_q))$, where $R : \{0,1\}^{qn} \rightarrow \{0,1\}^k$ are linear functions of rank k.

Because of the $N = 2^k$ pre-computation cost, it makes sense to apply the TMTO many times for several pairs (P_0, C_0) representing the encryption under many distinct, random keys, because it amortizes the total cost over several keys. Security concerning many independent keys is an important security goal, and can be proven to be implied by the standard PRP/PRF notion, except that attack costs are reduced by a factor of b, if there are b keys under attack.

Analyzing Hellman's attack, we say that a point p in the key space is *covered* by Table 3.5 in the pre-computing phase if p is equal to one of the $X_{i,j}$, $1 \leq i \leq m$, $1 \leq j \leq t$. Note that a point does not have to appear in the table in order to be covered (since only the SP_is and EP_is are actually stored). When choosing at random a point (key) K from the key space, the TMTO attack is successful if the point chosen is covered by the table. In this case, iterating f at most t times on that point will give an end point. Thus, success means that we can arrive, using Hellman's attack, at a candidate for the unknown key after at most t steps. The probability of success depends on how many distinct points of the key space are covered by the table. If the table covers a subset F of the key space, then, the success probability is $p_{success} = |F|/N$. In the ideal case, when all $X_{i,j}$s are distinct, we have $|F| = mt$ and $p_{success} = mt/N$. Since the amount of memory is m, up to a constant factor, and the time to get to the first candidate is at most t, to maintain a fixed probability of success, only the term mt has to be kept fixed, a clear trade-off between time and memory.

However, when mt is large, we can no longer assume that the points covered by the table are all distinct. When we generate a chain of encryptions, an individual chain can overlap with itself, or two chains can merge into a single chain. Assuming that each new chain element is randomly selected from the set of possible keys, the larger the key space covered, the more likely a new chain will overlap with some existing chain. Any key that does not appear in a chain will not be found. Thus, the TMTO is probabilistic, and the objective is to maximize the probability of success for a given amount of work.

3.4.2 Time/Memory/Data Trade-Off Attacks

In [2], Amirazizi and Hellman argued that there is a further trade-off between time, memory and processor capability allowing further speed up than conventional TMTO attacks by taking parallel processing into account.

In [32, 33], Biryukov *et al.* described a Time/Memory/Data Trade-Off (TMDTO) attack on block ciphers where a new trade-off variable *data*, which is represented by *key*, is added to the attack framework.

This attack on block ciphers is the analogue of the same attack applied to stream ciphers in [35].

It means that a full coverage of the search space N is no longer needed, but rather a coverage of a fraction N/D_k is enough, where D_k is the number of keys searched. Thus, using t/D_k tables instead of t tables means that the memory requirements go down to $M = mt/D_k$, where m is the number of Hellman's tables.

The time requirement becomes $T = t/D_k \cdot t \cdot D_k = t^2$ (less tables to check but for more data points) which is the same figure as in the original Hellman's trade-off attack. The matrix stopping rule is $N = mt^2$, which is the condition to minimize the waste of matrix coverage due to the birthday paradox effect. Using the matrix stopping rule, the trade-off formula becomes

$$N^2 = T \cdot (M \cdot D_K)^2.$$

For comparison, the original trade-off formula was

$$N^2 = T \cdot M^2.$$

The original Hellman's TMTO attack applies to one unknown key: $D_k = 1$, $T = M = 2^{2k/3}$ and pre-computation cost $N = 2^k$. The TMDTO attack searches for several keys at once, that is, the plaintext/ciphertext queries may be encrypted under several keys. The objective is to recover one of these keys. This scenario is called *multiple-key* in contrast to the *single-key scenario* where there is only one unknown key target.

The TMDTO attack for a block cipher using a k-bit key is exemplified in Table 3.2, along with the trade-off attacks [13] and [82].

Table 3.2 Comparison of Time/Memory/Data trade-offs for block ciphers using k-bit keys [32].

Cipher	Key size (bits)	Keys (Data)	Time	Memory	Pre-processing (cost)
DES	56	2^{14}	2^{28}	2^{28}	2^{42}
3key-3DES	168	2^{42}	2^{84}	2^{84}	2^{126}
AES	128	2^{32}	2^{80}	2^{56}	2^{96}
AES	192	2^{48}	2^{96}	2^{96}	2^{144}
AES	256	2^{85}	2^{170}	2^{85}	2^{170}
any cipher	k	$2^{k/4}$	$2^{k/2}$	$2^{k/2}$	$2^{3k/4}$
any cipher	k	$2^{k/3}$	$2^{2k/3}$	$2^{k/3}$	$2^{2k/3}$
any cipher [13]	k	$2^{k/2}$	$2^{k/2}$	$2^{k/2}$	$2^{k/2}$
any cipher [82]	k	1	$2^{2k/3}$	$2^{2k/3}$	2^{k}

3.5 Differential Cryptanalysis

Differential Cryptanalysis (DC) is an analysis technique originally developed by E. Biham and A. Shamir in [28, 29]. DC was initially applied against reduced-round variants of the DES block cipher and later to the full 16-round DES [31, 30].

An earlier work related to DC is the attack of Murphy on the FEAL-4 block cipher [142].

DC is a general analysis technique in the sense that it has been adapted to other cryptographic primitives such as hash functions and stream ciphers [30, 171]. DC became a benchmark technique for any new cryptographic design ever since [50, 98, 145, 117].

The main concept in DC is the notion of *difference*:

Definition 3.2. (Difference of Two Bit Strings)
The **difference** between two bit strings X and X^* is defined as

$$\Delta X = X \otimes (X^*)^{-1},$$

where $\otimes$ is a group operation. $(X^*)^{-1}$ is the inverse of X^* with respect to the $\otimes$ operator.

An implicit bit-string-to-group-element conversion is assumed in the definition of difference. For instance, if the group is Z_{2^n}, then the bit string $X = (x_{n-1}, x_{n-2}, \ldots, x_0)$ is associated with the integer $\sum_{i=0}^{n-1} x_i \cdot 2^i$. Typical bit strings include m-bit words ($m \in \{4, 8, 16\}$) and n-bit text blocks ($n \in \{32, 64, 128\}$).

A motivation for the use of text pairs, instead of a single text block, is that a single random (plaintext) block does not give the adversary much information about what the output block will be (assuming the full details of the encryption algorithm are known, but the key is unknown).

An intuition of the use of *differences of pairs of values* is that the adversary can compare a random text X with another one X^* of the adversaries choice, thus, with a known difference, observing how and why their difference evolve during the encryption process. This ability to compare text values and further to track down and to predict differences makes DC a quite powerful cryptanalytic tool.

Former concepts that implicitly used the idea of comparing pairs of values include: the strict avalanche criterion [73] (without well-defined differences) and the higher-order propagation criteria of degree k [161] (usually with differences of low Hamming Weight).

An initial separation between differences is: the *trivial difference* corresponding to the case $\Delta X = 0$. Otherwise, the difference is termed *nontrivial.*

A typical differential analysis operates in a bottom-up fashion, starting from the smallest cipher components and further extending the analysis to larger ones, taking into account that the output difference from one component should match the input difference of the next component.

Let us show examples to make more concrete the notion of how a difference *propagates* across internal cipher components.

A simple example of difference propagation corresponds to the splitting function $J(X) = (X, X)$ in which the input X is simply duplicated, i.e. the output consists of two copies of the input. In this case, an input difference ΔX is simply split into two identical copies: $\Delta J(X) = J(X) \otimes J(X^*)^{-1} = (X \oplus (X^*)^{-1}, X \oplus (X^*)^{-1}) = (\Delta X, \Delta X)$.

Likewise, across the $\otimes$ operation $Y = X \otimes W$, where X and W are internal cipher data, differences behave linearly: $\Delta Y = (X \otimes W) \otimes (X^* \otimes W^*)^{-1} = X \otimes W \otimes (W^*)^{-1} \otimes (X^*)^{-1} = \Delta X \otimes \Delta W$.

The difference operator $\otimes$ is usually chosen to be the same operator used to combine the (secret) key with internal cipher data. Let $Y = X \otimes K$ be an internal cipher component, where K is a secret key. Let ΔX denote the input difference and $\Delta Y = Y \otimes (Y^*)^{-1}$ denote the output difference. Then,

$$\Delta Y = (X \otimes K) \otimes (X^* \otimes K)^{-1} = X \otimes K \otimes K^{-1} \otimes (X^*)^{-1} = X \otimes (X^*)^{-1} = \Delta X.$$

In this case, the output difference is fully determined from the input difference (with certainty). Moreover, the $\otimes$-based difference is independent of the key, that is, it is *key invariant*. This simple fact makes this technique extremely valuable since the difference propagates independent of the key (which is unknown to the adversary).

In general, as a standard terminology, for a mapping $F(X, K) = X \otimes K$, the input difference ΔX is said to propagate across F, resulting in the output difference ΔY (with a certain probability p), or simply ΔX causes ΔY across F with probability p. This relationship is denoted $\Delta X \rightarrow \Delta Y$, where the probability p and the underlying mapping F are made clear from the context.

For DES, the difference operator used in [28, 29] was bitwise exclusive-or. In this case, $\Delta X = X \oplus X^*$ since X^* is its own inverse under the exclusive-or operation. The exclusive-or as difference operator measures how X and X^* differ bit-by-bit.

Let us consider the case where the difference operator is not the same one used to mix the (unknown) key with internal cipher data. Let $G(X, K) = X \boxplus K$, where $\boxplus$ is addition modulo 2^n in the group $\mathbb{Z}_{2^n}$, and let $\oplus$ be the difference operator. Then, $\Delta X = X \oplus X^*$ and

$$\Delta Y = (X \boxplus K) \oplus (X^* \boxplus K) = X \oplus C \oplus K \oplus X^* \oplus C^* \oplus K = \Delta X \oplus \Delta C,$$

where C is the vector of carry bits of the modular addition of X and K, while C^* is the vector of carry bits of the modular addition of X^* and K. In this case, ΔY depends not only on ΔX, but also on the difference $\Delta C = C \oplus C^*$ between the carry bits and which also depends on K. Unlike in F, the choice of the difference operator in G does not make ΔY key-invariant since the effect of the key still persists but is hidden in the carry bits. Sometimes, this incompatibility between group operator and the difference operator is inevitable, specially if the target cipher uses more than one group operation to combine the (sub)key with internal cipher data. That is the case with IDEA [114] and SAFER [129] block ciphers.

Thus, the output difference ΔY in G changes according to the (unknown) key since the actual ΔY depends on the distribution of carry-bit differences in modular addition. This behavior motivates an exhaustive study of the difference distribution of ΔC.

For a given mapping F, a table that exhaustively and systematically lists all the frequencies of occurrence of each ΔY for each possible ΔX, under a given difference operator $\otimes$, is called the $\otimes$ *Difference Distribution Table (DDT)* of F. If $\otimes$ is clear from the context, it is simply called the DDT of F.

The DDT of F provides an overview of the differential behavior of F concerning a given operator $\otimes$. Therefore, the DDT of F provides a so called *differential profile* of F, since it allows the adversary to identify (for the operator $\otimes$) which input difference(s) ΔX are more advantageous to choose, that is, which one(s) allows the adversary to predict ΔY with the highest probability.

As a small example, Table 3.3 shows the DDT of $G(X, 7)$ for $X \in \mathbb{Z}_{2^4}$ in which all sixteen possible ΔX are listed in the leftmost column. For each ΔX in a row, all sixteen possible ΔY are listed columnwise.

Table 3.3 $\oplus$ DDT of $G(X, 7)$ for $X \in \mathbb{Z}_{2^4}$.

	ΔY															
ΔX	0	1	2	3	4	5	6	7	8	9	10	11	12	13	14	15
0	16	0	0	0	0	0	0	0	0	0	0	0	0	0	0	0
1	0	0	0	8	0	0	0	4	0	0	0	0	0	0	0	4
2	0	0	8	0	0	0	4	0	0	0	0	0	0	0	4	0
3	0	8	0	0	0	4	0	0	0	0	0	0	0	4	0	0
4	0	0	0	0	12	0	0	0	0	0	0	0	4	0	0	0
5	0	0	0	4	0	0	0	8	0	0	0	4	0	0	0	0
6	0	0	4	0	0	0	8	0	0	0	4	0	0	0	0	0
7	0	4	0	0	0	8	0	0	0	4	0	0	0	0	0	0
8	0	0	0	0	0	0	0	0	16	0	0	0	0	0	0	0
9	0	0	0	0	0	0	0	4	0	0	0	8	0	0	0	4
10	0	0	0	0	0	0	4	0	0	0	8	0	0	0	4	0
11	0	0	0	0	0	4	0	0	0	8	0	0	0	4	0	0
12	0	0	0	0	4	0	0	0	0	0	0	0	12	0	0	0
13	0	0	0	4	0	0	0	0	0	0	0	4	0	0	0	8
14	0	0	4	0	0	0	0	0	0	0	4	0	0	0	8	0
15	0	4	0	0	0	0	0	0	0	4	0	0	0	8	0	0

Table 3.4 shows the difference distribution table of $G(X, 4)$ for $X \in \mathbb{Z}_{2^4}$.

Table 3.4 $\oplus$ DDT of $G(X, 4)$ for $X \in \mathbb{Z}_{2^4}$.

	ΔY															
ΔX	0	1	2	3	4	5	6	7	8	9	10	11	12	13	14	15
0	16	0	0	0	0	0	0	0	0	0	0	0	0	0	0	0
1	0	16	0	0	0	0	0	0	0	0	0	0	0	0	0	0
2	0	0	16	0	0	0	0	0	0	0	0	0	0	0	0	0
3	0	0	0	16	0	0	0	0	0	0	0	0	0	0	0	0
4	0	0	0	0	0	0	0	0	0	0	0	0	16	0	0	0
5	0	0	0	0	0	0	0	0	0	0	0	0	0	16	0	0
6	0	0	0	0	0	0	0	0	0	0	0	0	0	0	16	0
7	0	0	0	0	0	0	0	0	0	0	0	0	0	0	0	16
8	0	0	0	0	0	0	0	0	16	0	0	0	0	0	0	0
9	0	0	0	0	0	0	0	0	0	16	0	0	0	0	0	0
10	0	0	0	0	0	0	0	0	0	0	16	0	0	0	0	0
11	0	0	0	0	0	0	0	0	0	0	0	16	0	0	0	0
12	0	0	0	0	16	0	0	0	0	0	0	0	0	0	0	0
13	0	0	0	0	0	16	0	0	0	0	0	0	0	0	0	0
14	0	0	0	0	0	0	16	0	0	0	0	0	0	0	0	0
15	0	0	0	0	0	0	0	16	0	0	0	0	0	0	0	0

In both tables, the distribution of carry-bit differences is sparse and nonuniform distributed: some differences are quite more probable than others. The distribution of ΔY changes from Table 3.3 to Table 3.4 due to the bit patterns in each key K, which affects the distribution of carry bits.

Even though ΔY changes due to the carry-bit distribution, the careful choice of ΔX may still allow the adversary to predict the most probable value(s) of ΔY. For instance, assume $n = 16$ and $\Delta X = 8000_x$. Then,

$$\Delta Y = (X \boxplus K) \oplus (X^* \boxplus K) = X \oplus C \oplus K \oplus X \oplus 8000_x \oplus C^* \oplus K = 8000_x,$$

since the carry-bit difference affects only the most significant bit position and the carry-bit out from that position is discarded (truncated due to reduction modulo 2^{16}). In this very special case, the output difference is fully predictable ($p = 1$) and shows an exceptional situation in which the careful choice of ΔX allows us to propagate the difference ΔX across two algebraically incompatible operators: $\oplus$ and $\boxplus$, independent of the unknown key value. This difference will be used extensively in the differential analysis of Lai-Massey cipher designs.

This difference can be verified in both Table 3.3 and Table 3.4: $\Delta X = 8 \to \Delta Y = 8$ with probability one. This difference is the only nontrivial difference that matches the frequency of the trivial difference.

We can play around with this special value $\Delta X = 8000_x$, by changing the roles of text difference and key. Suppose that the key value is $K = 8000_x$, while ΔX is arbitrary. Then,

$$\Delta Y = (X \boxplus 8000_x) \oplus (X^* \boxplus 8000_x) = X \oplus 8000_x \oplus X^* \oplus 8000_x = \Delta X,$$

that is, the carry-bit difference ΔC vanishes, since adding 8000_x module 2^{16} is equivalent to xoring. This fact holds for any n in $\mathbb{Z}_{2^n}$. This is a first example of a *weak key* in a differential cryptanalysis setting, that is, a key value which causes an unusual (differential) behavior of a given mapping: $\Delta Y = \Delta X$ with a very high probability (in this case, with certainty). Note that this notion of a weak key is particular to a differential cryptanalysis context. There are other definitions of weak keys (and semi-weak keys) for particular ciphers and other contexts, such as for DES[5] [163, 54, 136, 140].

Therefore, the reason why the differential cryptanalysis technique operates in a chosen-plaintext setting is to allow the adversary to choose the input text pairs, and therefore, the input difference ΔX, in order to increase the adversary's chances of obtaining the most probable output difference ΔY (or the difference of his choice). Nonetheless, in [31], Biham and Shamir described how to apply *DC in a known-plaintext (KP) setting*: suppose a DC attack of an n-bit block cipher requires m plaintext pairs and that $2^{n/2}\sqrt{2m}$ random known plaintexts are available. These known plaintexts can generate

[5] A weak key for the DES cipher is a 56-bit user key K such that $\text{DES}_K(P) = \text{DES}_K^{-1}(P)$ for all plaintext blocks P.

up to $2^{n/2}\sqrt{2m}(2^{n/2}\sqrt{2m}-1)/2 \approx 2^n \cdot m$ pairs whose differences can be immediately discovered. It means that there are about $(2^n \cdot m)/2^n = m$ pairs for each possible n-bit difference. In particular, there might be about m pairs with the required difference for attacking the given block cipher.

Continuing with the study of difference propagation, let us consider the mapping $H(X, K) = S[X \oplus K]$, where S is an $n \times m$ nonlinear S-box, and K is a secret key. See the Appendix for an explanation of S-boxes.

S-boxes are typical building blocks of many block ciphers. In the case of H, we have a cipher component which is a combination of an S-box with the mapping F (using the exclusive-or operation). The input difference $\Delta X = X \oplus X^*$ leads to the output difference $\Delta Y = H(X, K) \oplus H(X^*, K) = S[X \oplus K] \oplus S[X^* \oplus K]$.

It is important to observe that propagating differences across consecutive components requires that the output difference from the first component matches the input difference of the next component(s).

Similar to the analysis of $G(X, K)$, there are many possible ΔY that may come out of $H(X, K)$ since there are many possible input pairs (X, X^*) that satisfy ΔX. The usual approach in this case is to study the full distribution of output differences based on S and ΔX.

For an $n \times m$ S-box S, the DDT of S is a table with dimensions $2^n \times 2^m$.

The (i, j) entry, $0 \leq i \leq 2^n - 1$, $0 \leq j \leq 2^m - 1$, of the DDT of S contains the value

$$DDT[i, j] = \#\{(X, X^*)|X \otimes (X^*)^{-1} = i, S[X \otimes K] \otimes (S[X^* \otimes K])^{-1} = j\}, \tag{3.6}$$

which is the frequency with which pairs of values with the input difference i cause pairs with the output difference j across the S-box S.

The i-th row corresponds to $\Delta X = i$ and the j-th column corresponds to $\Delta Y = j$.

The fraction $DDT[i, j]/\#(X, X^*)$ approximates the probability that the input difference i causes the output difference j across the given S-box S.

The largest entry for a nontrivial difference pair in the DDT of an $n \times m$-bit S-box S is denoted $\delta_{\max}$ and the corresponding probability associated with those entries is $\delta_{\max}/\#(X, X^*)$. Such an S-box is called **differentially $\delta_{\max}$-uniform**, which means that all nontrivial entries in its DDT have value less than or equal to $\delta_{\max}$. For instance, for the AES cipher, $\delta_{\max} = 4$. See the Appendix for other examples of S-boxes and their corresponding $\delta_{\max}$.

If $X \in \mathbb{Z}_{2^n}$ and $\Delta X \neq 0$, then $\#(X, X^*) = 2^n$.

Let us assume $\otimes = \oplus$ and the group $\mathbb{Z}_{2^n}$. For a given $\Delta X = X \oplus X^*$, there are several pairs that satisfy it: $(0, \Delta X)$, $(1, 1 \oplus \Delta X)$, $(2, 2 \oplus \Delta X)$, ..., $(2^n - 1, (2^n - 1) \oplus \Delta X)$. In fact, each pair (X, X^*) has a dual pair (Y^*, Y) such that $Y = X^*$ and $X = Y^*$. Since $X \oplus X^* = \Delta X \Rightarrow X^* = X \oplus \Delta X = Y \Rightarrow X = Y \oplus \Delta X = Y^* \Rightarrow \Delta X = Y \oplus Y^*$. Therefore, each entry in the DDT is an even integer.

As an example, let us consider the following 4×4-bit S-box from [14]:

$$S_0 = [3, 8, 15, 1, 10, 6, 5, 11, 14, 13, 4, 2, 7, 0, 9, 12], \tag{3.7}$$

which means $S_0[0] = 3$, $S_0[1] = 8$, ..., $S_0[15] = 12$. Assuming $\oplus$ as difference operator, the DDT of S_0 is depicted in Table 3.5. Note that

Table 3.5 $\oplus$ Difference Distribution Table of the S-box S_0.

ΔX	ΔY															
	0	1	2	3	4	5	6	7	8	9	10	11	12	13	14	15
0	16	0	0	0	0	0	0	0	0	0	0	0	0	0	0	0
1	0	0	0	2	0	2	2	2	0	0	0	2	2	0	4	0
2	0	0	0	0	0	0	0	0	0	2	2	0	4	2	2	4
3	0	2	2	2	0	0	0	2	0	4	0	2	2	0	0	0
4	0	0	0	0	0	0	0	0	0	4	4	0	0	4	4	0
5	0	0	2	0	4	2	0	0	2	0	2	2	0	0	2	0
6	0	2	2	4	0	2	2	4	0	0	0	0	0	0	0	0
7	0	0	2	0	4	2	0	0	2	2	0	2	0	2	0	0
8	0	0	0	2	0	2	2	2	0	0	0	2	2	4	0	0
9	0	2	2	0	0	2	2	0	0	2	2	0	0	2	2	0
10	0	2	2	2	0	0	0	2	0	0	4	2	2	0	0	0
11	0	2	2	0	0	2	2	0	0	0	0	0	4	0	0	4
12	0	2	0	0	4	0	2	0	2	2	0	2	0	2	0	0
13	0	0	0	4	0	0	0	4	4	0	0	0	0	0	0	4
14	0	2	0	0	4	0	2	0	2	0	2	2	0	0	2	0
15	0	2	2	0	0	2	2	0	4	0	0	0	0	0	0	4

- the largest entry in Table 3.5 corresponds to the trivial difference (ΔX, ΔY) = (0, 0). This S-box is invertible (a bijective mapping), which means that the zero input difference can only cause the zero output difference.
- the largest entry for a nontrivial difference is $\delta_{\max} = 4$, which corresponds, for instance, to $\Delta X = 3 \rightarrow \Delta Y = 9$ with an associated probability of $4/16 = 2^{-2}$.
- the sum of the entries in any single row i is 16, which means that all 16 input pairs with difference $\Delta X = i$ are distributed across all possible 4-bit output differences.
- since this S-box is invertible, the sum of the entries in any single column j is also 16, which means that all 16 output pairs with difference $\Delta Y = j$ are distributed across all possible 4-bit input differences.

The overall path of *nontrivial differences* across several cipher components (including across multiple rounds) forms what is called a *differential trail* [59].

Among the nontrivial trails, the most relevant are the narrow trails.

Fig. 3.3(a) shows (in red) an example of *narrow* differential trail across 5-round IDEA, while Fig. 3.3(b) shows (in red) an example of a *wide* differential trail. A narrow trail affects relatively few internal cipher components along

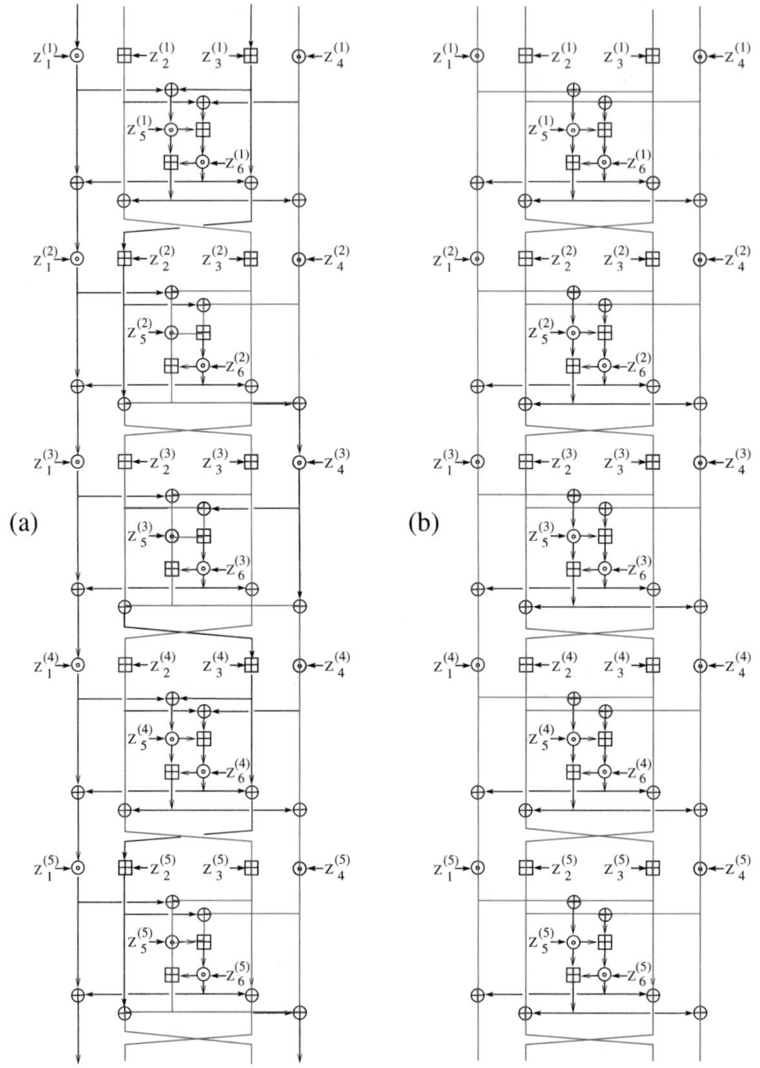

Fig. 3.3 Examples (in red) of a: (a) narrow differential trail; (b) wide differential trail.

its path, while a wide trail encompasses relatively many components along its path. A trail is very useful tool to visualize which components are active in a differential distinguisher. In particular, for Lai-Massey cipher designs, the trails identify which operations are active (have nonzero input/output differences). Relevant operations to account for are the keyed $\odot$, because they lead to weak-subkey conditions in DC.

Intuitively, narrow trails are preferred for cryptanalysis, since they minimize the number of active cipher components. The wide trail shown in Fig. 3.3(b), although affecting all four words in a block, on the other hand succeeds in bypassing the MA-box in every round, and it is by itself 1-round iterative.

An S-box S which has either a nonzero input or nonzero output difference is termed an *active S-box*, in a DC setting. It means, that the S-box effectively participates in a differential trail. Otherwise, S is called a *passive S-box*.

The best choice of difference(s) across an S-box may depend on the type of the S-box, even without knowledge of the full DDT. Advantageous choices exploit the peculiar design of each type of S-box.

Definition 3.3. (Differentially Active Nonsurjective S-box)
A differentially active nonsurjective S-box $S : \mathbb{Z}_2^n \to \mathbb{Z}_2^m$, with $0 < n \leq m$, is a nonsurjective S-box that participates in a differential trail and has only nonzero input difference.

Example of ciphers which use nonsurjective S-boxes include Blowfish [169] and Khufu [137]. Following Def. 3.3, differential analysis of these S-boxes exploit the nonsurjective property, and use only nonzero input difference [178].

Definition 3.4. (Differentially Active Noninjective S-box)
A differentially active noninjective S-box $S : \mathbb{Z}_2^n \to \mathbb{Z}_2^m$, with $n \geq m > 0$, is a noninjective S-box that participates in a differential trail and has only nonzero output difference.

Examples of ciphers that use noninjective S-boxes include DES [155] and s^2-DES [98]. See other examples in Table C.1 in the Appendix.

Following Def. 3.4, differential analysis of these S-boxes exploit the noninjective property by using nonzero output differences only such as [131] on DES. Fig. 3.26(a) shows an example of an (iterative) characteristic for DES that exploits the fact that its S-boxes are noninjective.

To avoid the drawbacks of both noninjective and nonsurjective S-box, a bijective S-box is suggested.

Definition 3.5. (Differentially Active Bijective S-box)
A differentially active bijective S-box $S : \mathbb{Z}_2^n \to \mathbb{Z}_2^n$, is a bijective S-box that participates in a differential trail and has either both zero or both nonzero input/output differences.

An example of bijective S-box is the one used in the AES cipher [71]. Following Def. 3.5, it can be observed from the DDT of the AES S-box[6] that a nonzero entry (difference) corresponds only to nonzero input/output differences.

---

[6] The DDT of the AES S-box is available from the book's webpage http://www.springer.com/978-3-319-68272-3.

See Table C.1 in the Appendix for a listing of bijective S-boxes of several block ciphers.

Propagating differences covering larger and larger components will ultimately lead to trails covering full rounds. A differential trail, for iterated ciphers, covering full rounds leads to the concept of a *differential characteristic* or *differential distinguisher.*

Definition 3.6. [31]
An r-*round differential characteristic* for an n-bit block cipher is a sequence of n-bit text differences defined as an $(r+1)$-tuple $(\delta_0, \delta_1, \ldots, \delta_r)$ where $\Delta P = \delta_0$ is the input plaintext difference and $\Delta C_i = \delta_i$, for $1 \leq i \leq r$, is the i-th round output difference. For each pair (δ_i, δ_{i+1}), with $0 \leq i \leq r-1$, there is an associated transition probability p_i that a text pair with difference δ_{i+1} will be observed at the round output, given an input pair with difference δ_i. This relationship is denoted $\delta_i \stackrel{1r}{\rightarrow} \delta_{i+1}$.

A shortcut notation for a sequence of text differences in a characteristic is $\delta_0 \stackrel{1r}{\rightarrow} \delta_1 \stackrel{1r}{\rightarrow} \ldots \stackrel{1r}{\rightarrow} \delta_r$ indicating a chain of text differences across consecutive rounds.

Definition 3.7. An r-round differential characteristic $\Delta_1 = (\delta_0, \delta_1, \ldots, \delta_r)$ can be concatenated to a t-round characteristic $\Delta_2 = (\delta'_0, \delta'_1, \ldots, \delta'_t)$ if $\delta_r = \delta'_0$. The result is the $(r+t)$-round characteristic $\Delta = (\delta_0, \delta_1, \ldots, \delta_r, \delta'_1, \ldots, \delta'_t)$. The probability of the concatenation of characteristics is approximated by the product of their 1-round characteristics.

Constructing differential trails (and choosing the difference propagation) across internal cipher components requires a trade-off involving several criteria:

- minimize the number of active nonlinear components with nonzero input/output difference: nontrivial differences propagating across nonlinear components typically account for the largest part of the probability in the construction of a trail.
  Looking at the DDT of the S-box S_0, the probability for the trivial difference $(\Delta X, \Delta Y) = (0, 0)$ is the highest among all entries. Using nontrivial difference values, at either the input or the output of a nonlinear component, usually imply lower probability. Therefore, including nonlinear components in the trail is best avoided whenever possible in order to maximize the overall probability. Moreover, since the data complexity of differential attacks depends on the inverse of the probability of the differential trail, one criterion to build effective DC attacks is to minimize the number of active nonlinear components in the differential trail.
- focus on narrow trails: a *narrow trail* contains relatively few components along its path. Otherwise, the trail is called *wide*. See Fig. 3.3. Choosing narrow trails over many consecutive rounds is a greedy approach: the less components with nontrivial differences there are, the higher the overall

probability of the trail (and the lower the attack complexity becomes). This criterion is in agreement with the previous one.
Satisfying this criterion depends on the block cipher design. If the (target) cipher was carefully designed, for instance, including fast and complete text diffusion, then narrow trails (holding with high probability) may not be easy to find and exploit.

- focus on iterative trails: it is essential to propagate differences not only across a single mapping but across multiple mappings, and over many rounds. Building trails in a bottom-up fashion means the focus is on local optimization. Ultimately, when the trail reaches a full round, one will have to restart building the trail from bottom-up, round after round, since the input difference from one round rarely equals the output difference.
A heuristic strategy, to minimize the total effort, is to plan from the very beginning to have trails whose input difference equals the output difference. This way, the trail can be concatenated with itself. These differential characteristics are called *iterative*. Using iterative characteristics, one can construct arbitrarily long characteristics with a fixed decrease in the probability after each concatenation [30].

In a sense, differential trails and characteristics have been used interchangeably. In fact, trails allows us to visualize the propagation of differences in detail, across all internal components of a cipher. Characteristics, on the other hand, are more formal, structured objects, with (ordered) numerical values (differences) and associated probabilities. Therefore, in a sense, they complement each other. For instance, the trail in Fig. 3.3(a) corresponds to the characteristic $\Delta = ((0,\delta,0,\delta), (0,0,\delta,\delta), (0,\delta,\delta,0), (0,\delta,0,\delta), (0,0,\delta,\delta), (0,\delta,\delta,0))$, for 5-round IDEA, where $\delta = 8000_x$. The probability for every round is 1, as long as the active multiplicative subkeys are weak.

If the target cipher is a Markov cipher [114], then the probability of a characteristic is independent of the actual inputs to each round and is computed over all possible round subkeys. This case is summarized in the following definition.

Definition 3.8. [114]
If there is a group operation $\otimes$ for defining differences such that for all choices of α ($\alpha \neq 0$) and β ($\beta \neq 0$) the probability $\Pr(\Delta Y = \beta | \Delta X = \alpha, X = \gamma)$ is independent of γ, when the key K is chosen according to the uniform distribution, then an iterated cipher with round function $Y = F_K(X)$ is a Markov cipher.

If the round subkeys are independent and uniformly distributed, then the individual probabilities p_i for each round are also independent:

$$\Pr(\Delta C_r = \delta_r | \Delta P = \delta_0) = \prod_{i=1}^{r} \Pr(\Delta C_i = \delta_i | \Delta C_{i-1} = \delta_{i-1}).$$

Experimental differential analysis of the DES block cipher in [29] using $\oplus$ as difference operator demonstrated that the concatenation of 1-round characteristics is a good approximation to the real probability for attacks in practice, even when the round subkeys are generated from a deterministic key schedule algorithm.

A concept linked to r-round characteristics is that of text pairs that satisfy the roundwise differences:

Definition 3.9. (Right Pair) [29]
A *right pair* with respect to an r-round characteristic $(\delta_0, \delta_1, \ldots, \delta_r)$ is a text pair (P, P^*) for which $P \otimes (P^*)^{-1} = \delta_0$ and for each round i, with $1 \leq i \leq r$, the encryption of the pair using the unknown, independent subkey k_i leads to the round input difference δ_{i-1} and the round output difference δ_i. Every pair which is not a right pair with respect to the characteristic and independent round subkeys is called a *wrong pair*.

For some characteristics there may be more than one trail leading from a given input difference δ_0 to the output difference δ_r. Moreover, for some attacks, the confirmation of each intermediate round difference δ_i, for $1 \leq i < r$, is not possible or not needed. These facts lead to the concept of a *differential*.

Definition 3.10. (r-round Differential) [117]
An r-round differential is a pair (δ_0, δ_r) of text differences where the input difference is $\Delta P = \delta_0$ and the output difference after r rounds is $\Delta C = \delta_r$.

The probability of a differential is as follows:

Definition 3.11. [101]
The probability of an r-round differential $(\Delta P, \Delta C) = (\delta_0, \delta_r)$ is

$$\Pr(\Delta C = \delta_r | \Delta P = \delta_0) = \sum_{\delta_1} \cdots \sum_{\delta_{r-1}} \prod_{i=1}^{r} \Pr(\Delta C_i = \delta_i | \Delta C_{i-1} = \delta_{i-1}),$$

where $\Delta C_0 = \Delta P$.

A *differential* may generally have a higher probability than a characteristic because the former represents the union of all characteristics with the same (δ_0, δ_r).

The concatenation of characteristics implicitly assumes that:

- the subkeys are independent and uniformly distributed.
- the propagation of differences across a key-dependent round component is statistically independent of the unknown subkeys.
- the final probability is the average over all possible subkey values, as in the following hypothesis.

Hypothesis 1 *(Hypothesis of Stochastic Equivalence) [116]*
For virtually all high probability $(r-1)$-round differentials (α, β), the probability

$$Pr(\Delta C_{r-1} = \beta | \Delta P = \alpha) = Pr(\Delta C_{r-1} = \beta | \Delta P = \alpha, K_1 = k_1, \ldots, K_{r-1} = k_{r-1})$$

holds for a substantial fraction of all round subkeys $(k_1, k_2, \ldots, k_{r-1})$.

The data complexity of a differential attack, given by the number of chosen plaintexts, is inversely proportional to the probability of the characteristic or differential.

Given an n-bit block cipher and a characteristic (or differential) Δ with probability p, $0 < p \leq 1$, about $1/p$ chosen plaintexts are required to observe the output difference of Δ in a real attack, or to obtain one right pair. This result assumes that $p \ggg 2^{-n}$ or that enough plaintext pairs can be generated to satisfy the characteristic, otherwise, even the full codebook will not provide enough text pairs for an attack.

A differential characteristic or differential *distinguisher* forms a cryptanalytic tool with several applications:

- observing the output difference of a characteristic or differential means that the latter can be used to distinguish the given (number of rounds of the given) block cipher E_K from a random permutation π over the same text space. This result follows from the assumption that for E_K, the characteristic (or differential) holds with probability $p \ggg 2^{-n}$, while for π, the output difference of Δ might appear with probability around 2^{-n}.
  In a sense, this distinguisher provides a kind of signature that identifies the target cipher (or a fixed number of consecutive rounds) among all other (keyed) permutations (except for dual ciphers) operating over the same text space.
- it allows us to identify weak-key classes (if the distinguisher depends on weak keys for its construction) efficiently, that is, with few resources: data, memory, time. Examples will be described for Lai-Massey ciphers.
- it is a essential tool for key-recovery attacks which operate on top of the distinguishers, recovering key information in rounds before or after the distinguisher. Examples will be described for Lai-Massey cipher designs.

Key-recovery attacks on $(t+u)$ rounds employ one or more high-probability t-round characteristics, where $u \geq 1$ rounds are located before or after the characteristics. Such attacks are called uR attacks since they recover subkeys from u rounds surrounding the characteristics.

The idea is to guess subkeys from those u rounds, partially encrypt or decrypt the u rounds until a t-round (core) is obtained, with an input/output difference pair (δ_0', δ_t'), which depends on the guessed subkeys. Further, these differences are compared (in full or partially) to the original input/output differences (δ_0, δ_t) of the distinguisher. A (full or partial) match would indicate that the guessed subkeys are a candidate to the correct subkey value(s).

Therefore, a distinguisher allows an adversary to recover a (conveniently small) number of subkey bits over t rounds instead of all subkeys embedded in these t rounds.

An attack that recovers b subkey bits requires in general 2^b counters to keep track of how often each subkey is suggested. Each one of the possible 2^b subkeys is tested exhaustively, and this amount of counters are taken into account in the memory/storage complexity of the attack.

An important concept in a differential key-recovery attack is the notion of signal-to-noise ratio:

Definition 3.12. (Signal-to-Noise Ratio) [30]
The ratio between the number of right pairs and the average count of the incorrect subkeys in a counting scheme is called the *signal-to-noise ratio* of the counting scheme and is denoted S/N.

To find the correct subkeys using a counting scheme, a high-probability characteristic is needed, as well as a sufficiently high number of right pairs. The number of required right pairs depends on the probability p of the characteristic, the number k of subkey bits that are counted altogether, the average count α per analyzed pair (excluding the wrong pairs that can be discarded before the counting), and the fraction β of analyzed pairs among all the pairs. The counters contain an average value of $\frac{m \cdot \alpha \cdot \beta}{2^k}$, where m is the number of created pairs. The right subkey value is counted about $m \cdot p$ times by the right pairs, in addition to the random counts from the wrong pairs. The signal-to-noise ratio of a counting scheme is

$$\text{S/N} = \frac{m \cdot p}{\frac{m \cdot \alpha \cdot \beta}{2^k}} = \frac{2^k \cdot p}{\alpha \cdot \beta}.$$

A consequence of the definition of S/N is that the signal-to-noise ratio of a counting scheme is independent of the number of pairs used in an attack.

In [30], Biham and Shamir related high values of S/N, namely S/N$\ggg 1$, to a small number of right pairs needed for a successful attack, and S/N≤ 1 to an unreasonably high number of right pairs. However, in [48], Borst *et al.* discovered in an attack on reduced-round IDEA that the case S/N> 1 indicates that the right subkey is among the ones most suggested by the characteristic (highest counter values), while the case S/N< 1 indicates that the right subkey value is among the least suggested ones by the characteristic (lowest counter values). The overall consensus is that the case S/N $= 1$ does not allow us to distinguish the correct subkey from the wrong ones.

In the following sections, DC attacks are described against several Lai-Massey cipher designs.

3.5.1 DC of PES Under Weak-Key Assumptions

In [57], Daemen *et al.* described differential attacks on the full 8.5-round IDEA under weak-key assumptions. These assumptions state that some multiplicative subkeys have value 0 or 1, that is, their fifteen most significant bits are zero.

The following differential attacks on PES are based on [149] and use a similar approach as done by Daemen *et al.* on IDEA. Moreover, the analyses in [149] can go beyond the full 8.5-round PES, that is, to more than 8.5 rounds, since the key schedule of PES allows us to arbitrarily extend the number of rounds.

The difference operator is bitwise exclusive-or and the difference is wordwise. The only 16-bit difference values used in the characteristic are $\nu = 8000_x$ and 0000_x, which have particularly how Hamming weight. The difference 8000_x propagates across the modular addition with probability one, since it affects only the most significant bit.

The analysis follows a bottom-up approach. Using wordwise differences, the next step is to cover larger components such as MA-boxes, and further to 1-round differential characteristics. In order to cover full rounds, weak-subkey restrictions will be assumed for the characteristics to hold with the highest possible probability. Under these restrictions, the characteristic will hold with certainty.

Table 3.6 lists all fifteen nontrivial 1-round differential characteristics of PES, under weak-key assumptions, using only wordwise differences 0000_x and 8000_x. They all hold with certainty under the weak-key assumptions. Multiple-round characteristics are obtained by concatenating 1-round characteristics such that the output difference from one characteristic matches the input difference of the next characteristic.

From Table 3.6, the multiple-round (iterative) characteristics are clustered into eleven chains covering 1 or 2 rounds:

- $(0,0,0,\nu) \overset{1r}{\to} (\nu,0,\nu,\nu) \overset{1r}{\to} (0,0,0,\nu)$, which is 2-round iterative; the weak subkeys required across two rounds are: $Z_6^{(i)}$, $Z_1^{(i+1)}$, $Z_6^{(i+1)}$.
- $(0,0,\nu,\nu) \overset{1r}{\to} (\nu,0,0,\nu) \overset{1r}{\to} (0,0,\nu,\nu)$, which is 2-round iterative; the weak subkeys required across two rounds are: $Z_5^{(i)}$, $Z_1^{(i+1)}$, $Z_5^{(i+1)}$.
- $(0,\nu,0,0) \overset{1r}{\to} (\nu,\nu,\nu,0) \overset{1r}{\to} (0,\nu,0,0)$, which is 2-round iterative; the weak subkeys required across two rounds are: $Z_2^{(i)}$, $Z_6^{(i)}$, $Z_1^{(i+1)}$, $Z_2^{(i+1)}$, $Z_6^{(i+1)}$.
- $(0,\nu,\nu,0) \overset{1r}{\to} (\nu,\nu,0,0) \overset{1r}{\to} (0,\nu,\nu,0)$, which is 2-round iterative; the weak subkeys required across two rounds are: $Z_2^{(i)}$, $Z_5^{(i)}$, $Z_1^{(i+1)}$, $Z_2^{(i+1)}$, $Z_5^{(i+1)}$.
- $(0,\nu,\nu,\nu) \overset{1r}{\to} (0,\nu,\nu,\nu)$, which is 1-round iterative; the weak subkeys required across each round are: $Z_2^{(i)}$, $Z_5^{(i)}$, $Z_6^{(i)}$.
- $(0,0,\nu,0) \overset{1r}{\to} (0,0,\nu,0)$, which is 1-round iterative; the weak subkeys required across each round are: $Z_5^{(i)}$, $Z_6^{(i)}$.

Table 3.6 Nontrivial 1-round characteristics of PES using xor differences and $\nu = 8000_x$.

1-round characteristics	Weak Subkeys (i-th round)	Difference across MA-box
$(0,0,0,\nu) \overset{1r}{\to} (\nu,0,\nu,\nu)$	$Z_6^{(i)}$	$(0,\nu) \to (\nu,\nu)$
$(0,0,\nu,0) \overset{1r}{\to} (0,0,\nu,0)$	$Z_5^{(i)}, Z_6^{(i)}$	$(\nu,0) \to (0,\nu)$
$(0,0,\nu,\nu) \overset{1r}{\to} (\nu,0,0,\nu)$	$Z_5^{(i)}$	$(\nu,\nu) \to (\nu,0)$
$(0,\nu,0,0) \overset{1r}{\to} (\nu,\nu,\nu,0)$	$Z_2^{(i)}, Z_6^{(i)}$	$(0,\nu) \to (\nu,\nu)$
$(0,\nu,0,\nu) \overset{1r}{\to} (0,\nu,0,\nu)$	$Z_2^{(i)}$	$(0,0) \to (0,0)$
$(0,\nu,\nu,0) \overset{1r}{\to} (\nu,\nu,0,0)$	$Z_2^{(i)}, Z_5^{(i)}$	$(\nu,\nu) \to (\nu,0)$
$(0,\nu,\nu,\nu) \overset{1r}{\to} (0,\nu,\nu,\nu)$	$Z_2^{(i)}, Z_5^{(i)}, Z_6^{(i)}$	$(\nu,0) \to (0,\nu)$
$(\nu,0,0,0) \overset{1r}{\to} (\nu,0,0,0)$	$Z_1^{(i)}, Z_5^{(i)}, Z_6^{(i)}$	$(\nu,0) \to (0,\nu)$
$(\nu,0,0,\nu) \overset{1r}{\to} (0,0,\nu,\nu)$	$Z_1^{(i)}, Z_5^{(i)}$	$(\nu,\nu) \to (\nu,0)$
$(\nu,0,\nu,0) \overset{1r}{\to} (\nu,0,\nu,0)$	$Z_1^{(i)}$	$(0,0) \to (0,0)$
$(\nu,0,\nu,\nu) \overset{1r}{\to} (0,0,0,\nu)$	$Z_1^{(i)}, Z_6^{(i)}$	$(0,\nu) \to (\nu,\nu)$
$(\nu,\nu,0,0) \overset{1r}{\to} (0,\nu,\nu,0)$	$Z_1^{(i)}, Z_2^{(i)}, Z_5^{(i)}$	$(\nu,\nu) \to (\nu,0)$
$(\nu,\nu,0,\nu) \overset{1r}{\to} (\nu,\nu,0,\nu)$	$Z_1^{(i)}, Z_2^{(i)}, Z_5^{(i)}, Z_6^{(i)}$	$(\nu,0) \to (0,\nu)$
$(\nu,\nu,\nu,0) \overset{1r}{\to} (0,\nu,0,0)$	$Z_1^{(i)}, Z_2^{(i)}, Z_6^{(i)}$	$(0,\nu) \to (\nu,\nu)$
$(\nu,\nu,\nu,\nu) \overset{1r}{\to} (\nu,\nu,\nu,\nu)$	$Z_1^{(i)}, Z_2^{(i)}$	$(0,0) \to (0,0)$

- $(0,\nu,0,\nu) \overset{1r}{\to} (0,\nu,0,\nu)$, which is 1-round iterative; the weak subkey required across each round is: $Z_2^{(i)}$.
- $(\nu,0,0,0) \overset{1r}{\to} (\nu,0,0,0)$, which is 1-round iterative; the weak subkeys required across each round are: $Z_1^{(i)}$, $Z_5^{(i)}$, $Z_6^{(i)}$.
- $(\nu,0,\nu,0) \overset{1r}{\to} (\nu,0,\nu,0)$, which is 1-round iterative; the weak subkey required across each round is: $Z_1^{(i)}$
- $(\nu,\nu,0,\nu) \overset{1r}{\to} (\nu,\nu,0,\nu)$, which is 1-round iterative; the weak subkeys required across each round are: $Z_1^{(i)}$, $Z_2^{(i)}$, $Z_5^{(i)}$, $Z_6^{(i)}$.
- $(\nu,\nu,\nu,\nu) \overset{1r}{\to} (\nu,\nu,\nu,\nu)$, which is 1-round iterative; the weak subkeys required across each round are: $Z_1^{(i)}$, $Z_2^{(i)}$.

Note that there is a particularly large number of 1-round iterative characteristics in PES.

An exhaustive analysis of these chains of characteristics indicates that the 1-round iterative characteristic $(\nu,0,\nu,0) \overset{1r}{\to} (\nu,0,\nu,0)$, iterated across 8.5 rounds, provides the largest weak-key class of size 2^{41}. Note that this characteristic has the minimum number of weak subkeys: only one weak subkey per round. Moreover, it bypasses the MA-box in every round: the input and output differences to the MA-box are both $(0,0)$. For this weak-key class, according to the key schedule algorithm of PES in Sect. 2.1.2, the key bits 0–14, 22–57 and 75–110 have to be zero, while the remaining key bit can hold arbitrary values. The evolution of the weak-key conditions is depicted in Table 3.7.

Table 3.7 8.5-round differential characteristic of PES under weak-key assumptions.

i-th Round	Characteristic	$\text{msb}_{15}(Z_1^{(i)})$	Weak-key Class size
1	$(\nu,0,\nu,0) \stackrel{1r}{\to} (\nu,0,\nu,0)$	0–14	2^{113}
2	$(\nu,0,\nu,0) \stackrel{1r}{\to} (\nu,0,\nu,0)$	96–110	2^{98}
3	$(\nu,0,\nu,0) \stackrel{1r}{\to} (\nu,0,\nu,0)$	89–103	2^{91}
4	$(\nu,0,\nu,0) \stackrel{1r}{\to} (\nu,0,\nu,0)$	82–96	2^{84}
5	$(\nu,0,\nu,0) \stackrel{1r}{\to} (\nu,0,\nu,0)$	75–89	2^{77}
6	$(\nu,0,\nu,0) \stackrel{1r}{\to} (\nu,0,\nu,0)$	43–57	2^{62}
7	$(\nu,0,\nu,0) \stackrel{1r}{\to} (\nu,0,\nu,0)$	36–50	2^{55}
8	$(\nu,0,\nu,0) \stackrel{1r}{\to} (\nu,0,\nu,0)$	29–43	2^{48}
OT	$(\nu,0,\nu,0) \stackrel{0.5r}{\to} (\nu,0,\nu,0)$	22–36	2^{41}

In fact, the characteristic $(\nu,0,\nu,0) \stackrel{1r}{\to} (\nu,0,\nu,0)$ can be iterated beyond 8.5 rounds. For instance, across 17-round PES, $(\nu,0,\nu,0) \stackrel{17r}{\to} (\nu,0,\nu,0)$ holds for a weak-key class of size 2^7. For 17.5 rounds, $(\nu,0,\nu,0) \stackrel{17.5r}{\to} (\nu,0,\nu,0)$ holds only for the all-zero key.

Another interesting characteristic is depicted in Table 3.8, and also holds for a weak-key class of size 2^{41}. The key bits at positions 16–30, 38–73 and 91–126 have to be zero, while key bits in positions 0–15, 31–37, 74–90 and 127 can be arbitrary. This characteristic, under such a weak-key class, holds with certainty and can be used to distinguish the full 8.5-round PES from a random permutation with very low attack complexity.

Table 3.8 8.5-round differential characteristics of PES under weak-key assumptions.

i-th Round	Characteristic	$\text{msb}_{15}(Z_2^{(i)})$	Weak-key Class size
1	$(0,\nu,0,\nu) \stackrel{1r}{\to} (0,\nu,0,\nu)$	16–30	2^{113}
2	$(0,\nu,0,\nu) \stackrel{1r}{\to} (0,\nu,0,\nu)$	112–126	2^{98}
3	$(0,\nu,0,\nu) \stackrel{1r}{\to} (0,\nu,0,\nu)$	105–119	2^{91}
4	$(0,\nu,0,\nu) \stackrel{1r}{\to} (0,\nu,0,\nu)$	98–112	2^{84}
5	$(0,\nu,0,\nu) \stackrel{1r}{\to} (0,\nu,0,\nu)$	91–105	2^{77}
6	$(0,\nu,0,\nu) \stackrel{1r}{\to} (0,\nu,0,\nu)$	59–73	2^{62}
7	$(0,\nu,0,\nu) \stackrel{1r}{\to} (0,\nu,0,\nu)$	52–66	2^{55}
8	$(0,\nu,0,\nu) \stackrel{1r}{\to} (0,\nu,0,\nu)$	45–59	2^{48}
OT	$(0,\nu,0,\nu) \stackrel{0.5r}{\to} (0,\nu,0,\nu)$	38–52	2^{41}

These differential characteristics provide useful information about PES:

- they allow us to study the diffusion power of the MA half-round of PES with respect to modern tools such as the (differential) branch number.

The set of characteristics in Table 3.6 means that the (differential) branch number *per round* of PES is only 2 (under weak-key assumptions). For instance, the 1-round characteristic $(0,0,\nu,0) \stackrel{1r}{\to} (0,0,\nu,0)$ has both the input and output round differences with Hamming Weight one. So, even though the design of PES provides full diffusion in a single round, a detailed analysis of its characteristics (under weak-key assumptions) show that its differential branch number is much lower than ideal. For a 4-word block (such as used by PES), an MDS code using a 4×4 MDS matrix would provide a branch number of $1+4=5$. See Appendix, Sect. B.1 for more details about the branch number.
Since the given 1-round is iterative, the branch number for t-round PES, for $t > 1$, is also 2.

- the characteristics in Table 3.6 hold with certainty, given the corresponding weak-key class. Looking the other way around, these characteristics provide efficient membership tests for keys in a weak-key class. For instance, the characteristic in Table 3.7 allows us to efficiently identify all 2^{41} keys in the class. The test consists of two encryptions with input difference $(\nu,0,\nu,0)$. The sample key belongs to the class (with high probability) if the output difference is $(\nu,0,\nu,0)$ after 8.5-round PES. The difference patterns inside the trail detailed in Table 3.6 are quite sensitive to the given weak-subkey assumptions. So, if even a single multiplicative subkey in the trail is not weak, the difference pattern breaks down and the output difference (after 8.5 rounds) will be unpredictable. For a random key, the chance of the difference $(\nu,0,\nu,0)$ showing up at the output is around 2^{-64}. For a key space of 2^{128} keys, excluding the weak-key class of size 2^{41}, and using two text pairs, the number of keys satisfying $(0,\nu,0,\nu) \stackrel{8.5r}{\to} (0,\nu,0,\nu)$ is $(2^{-64})^2 \cdot (2^{128} - 2^{41}) < 1$.
- the characteristics in Table 3.6, under the given weak-key assumptions, can be used to distinguish the full 8.5-round PES from a random permutation with very low attack complexity, since the characteristics hold with certainty. The cost is essentially two encryptions, two chosen plaintexts and equivalent memory.
The reasoning is the following: given the input difference $(0,\nu,0,\nu)$ to both 8.5-round PES (under a weak-key) and a random 64-bit permutation π, the output difference $(0,\nu,0,\nu)$ will show up with certainty for PES, while for π, it will happen with probability about 2^{-64}. The advantage (see Sect. 1.7) to distinguish one from the other is: $1 - 2^{-64}$, which is non-negligible. Therefore, the full 8.5-round PES cipher is not an ideal primitive, that is, it is not a PRP.
- usually, the larger the number of rounds the smaller the weak-key class becomes. But, the weak-key class size never vanishes for PES, due to its key schedule algorithm. See Sect. 2.1.2. For instance, the 17.5-round characteristic $(\nu,0,\nu,0) \stackrel{17.5r}{\to} (\nu,0,\nu,0)$ holds with certainty only for the all-zero key. Beyond 17.5 rounds, this characteristic still holds, but only

for the key $K = 0^{128}$, which is very special. It is the only weak key of PES for which **any** characteristic can be iterated for **any** number of rounds because $K = 0^{128}$ means that all (multiplicative) subkeys are weak, even those which do not belong to the differential trail.

- a weak-key class of size 2^{41} represents a fraction of only $2^{41-128} = 2^{-87}$ of the key space of PES. Nonetheless, block ciphers are fundamental building blocks for the construction of many other cryptographic primitives, such as hash functions [162]. In some of these hash function constructions, the key input may not be secret, and may even be under the control of the adversary. In these cases, the existence of a single weak key is enough to allow the adversary to choose a weak key in order to manipulate the encryption scheme [120, 39].

3.5.1.1 Key-Recovery Attacks

Weak-key classes can be increased by reducing the number of weak-subkey restrictions. On the one hand, it will cause the distinguisher to reach less rounds, but, at the same time, it will allow subkeys to be recovered from the rounds not covered by the distinguisher.

An example of key-recovery attack on the full 8.5-round PES could use the iterative characteristic $(0, \nu, 0, \nu) \overset{8r}{\to} (0, \nu, 0, \nu)$ across 8 rounds, and recover subkeys from the OT. Removing the OT from the distinguisher means that the weak-key class size increases to 2^{48} keys according to Table 3.7.

This attack essentially guesses subkeys $Z_2^{(9)}$ and $Z_4^{(9)}$, partially decrypt the OT until the output of the eight round, and check if the difference at that point is ν as predicted by the distinguisher.

Since the distinguisher holds with certainty, the attack effort is $2 \cdot 2^{16}$ multiplications per subkey per text pair, in the worst case. It is equivalent to $2^{18}/4$ OT computations, or $2^{16}/17 \approx 2^{12}$ PES computations, assuming 17 half-rounds per full 8.5-round PES computation. The data complexity is two chosen plaintexts, and the memory needed to store the subkeys.

This characteristic does not allow us to recover other subkeys, especially those which follow trivial output (wordwise) differences, such as $Z_1^{(9)}$ and $Z_3^{(9)}$. Use of another characteristic would imply another weak-key class. Therefore, this is only a partial key-recovery attack: instead of the full 128-bit key, only 32 key bits are recovered.

3.5.2 DC of IDEA Under Weak-Key Assumptions

Let $X = (X_1, X_2, X_3, X_4)$ denote a 64-bit text block.

In [114], Lai described a differential analysis of mini-versions[7] of IDEA and PES ciphers under the hybrid difference operator $\otimes$ which, for a text pair (X, X^*), means that

$$X \otimes X^* = (X_1 \odot X_1^*, X_2 \boxplus X_2^*, X_3 \boxplus X_3^*, X_4 \odot X_4^*), \tag{3.8}$$

This difference operator: (i) is compatible with the operations used to mix subkeys bits into the KM half-round; (ii) is used under the assumption that IDEA is a Markov cipher (Def. 3.8) for independent and uniformly distributed random subkeys (not created with the key schedule algorithm of IDEA). Further analysis of mini-versions of IDEA under the difference (3.8) were presented in [52].

In [57], Daemen *et al.* described differential attacks on the full 8.5-round IDEA using differential characteristics under the difference operator $\oplus$. That means that the difference of a text pair (X, X^*) is defined as

$$X \oplus X^* = (X_1 \oplus X_1^*, X_2 \oplus X_2^*, X_3 \oplus X_3^*, X_4 \oplus X_4^*). \tag{3.9}$$

Under the difference operator (3.9), IDEA is not a Markov cipher.

The first step is to obtain 1-round differential characteristics and the corresponding weak-subkey restrictions for the characteristic to hold with certainty. The difference operator is bitwise exclusive-or and the difference is wordwise. The only 16-bit difference values used in the characteristic are $\nu = 8000_x$ and 0000_x. The difference 8000_x propagates across the modular addition with probability one, since it affects only the most significant bit (and the last carry-out bit is truncated due to modular reduction).

Table 3.9 lists all fifteen nontrivial 1-round differential characteristics of IDEA, under weak-key assumptions, using only wordwise differences 0000_x and 8000_x. They hold with certainty under the given weak-subkey assumptions. Multiple-round characteristics are obtained by concatenating 1-round characteristics, such that the output difference of one characteristic matches the input difference of the next characteristic.

From Table 3.9, the multiple-round characteristics are clustered into five chains covering 1, 2, 3 or 6 rounds:

- $(0,0,0,\delta) \stackrel{1r}{\to} (\delta,\delta,\delta,0) \stackrel{1r}{\to} (0,0,0,\delta)$, which is 2-round iterative; the weak subkeys required are: $Z_4^{(i)}$, $Z_6^{(i)}$, $Z_1^{(i+1)}$ and $Z_6^{(i+1)}$.
- $(0,0,\delta,0) \stackrel{1r}{\to} (\delta,0,0,0) \stackrel{1r}{\to} (0,\delta,0,0) \stackrel{1r}{\to} (\delta,\delta,0,\delta) \stackrel{1r}{\to} (0,\delta,\delta,\delta) \stackrel{1r}{\to} (\delta,0,\delta,\delta) \stackrel{1r}{\to} (0,0,\delta,0)$, which is 6-round iterative; the weak subkeys required are: $Z_5^{(i)}$, $Z_6^{(i)}$, $Z_1^{(i+1)}$, $Z_5^{(i+1)}$, $Z_6^{(i+1)}$, $Z_6^{(i+2)}$, $Z_1^{(i+3)}$, $Z_4^{(i+3)}$, $Z_5^{(i+3)}$, $Z_6^{(i+3)}$, $Z_4^{(i+4)}$, $Z_5^{(i+4)}$, $Z_6^{(i+4)}$, $Z_1^{(i+5)}$, $Z_4^{(i+5)}$ and $Z_6^{(i+5)}$.

[7] A mini-version of a block cipher is a scaled down cipher were each internal component is reduced in size, for instance, a 16-bit multiplication operator (in the original cipher) is replace by a 4-bit multiplication operator in the mini-cipher.

Table 3.9 Nontrivial 1-round characteristics for IDEA.

1-round characteristics	Weak subkeys	Difference across the MA-box
$(0,0,0,\delta) \xrightarrow{1r} (\delta,\delta,\delta,0)$	$Z_4^{(i)}, Z_6^{(i)}$	$(0,\delta) \rightarrow (\delta,\delta)$
$(0,0,\delta,0) \xrightarrow{1r} (\delta,0,0,0)$	$Z_5^{(i)}, Z_6^{(i)}$	$(\delta,0) \rightarrow (0,\delta)$
$(0,0,\delta,\delta) \xrightarrow{1r} (0,\delta,\delta,0)$	$Z_4^{(i)}, Z_5^{(i)}$	$(\delta,\delta) \rightarrow (\delta,0)$
$(0,\delta,0,0) \xrightarrow{1r} (\delta,\delta,0,\delta)$	$Z_6^{(i)}$	$(0,\delta) \rightarrow (\delta,\delta)$
$(0,\delta,0,\delta) \xrightarrow{1r} (0,0,\delta,\delta)$	$Z_4^{(i)}$	$(0,0) \rightarrow (0,0)$
$(0,\delta,\delta,0) \xrightarrow{1r} (0,\delta,0,\delta)$	$Z_5^{(i)}$	$(\delta,\delta) \rightarrow (\delta,0)$
$(0,\delta,\delta,\delta) \xrightarrow{1r} (\delta,0,\delta,\delta)$	$Z_4^{(i)}, Z_5^{(i)}, Z_6^{(i)}$	$(\delta,0) \rightarrow (0,\delta)$
$(\delta,0,0,0) \xrightarrow{1r} (0,\delta,0,0)$	$Z_1^{(i)}, Z_5^{(i)}, Z_6^{(i)}$	$(\delta,0) \rightarrow (0,\delta)$
$(\delta,0,0,\delta) \xrightarrow{1r} (\delta,0,\delta,0)$	$Z_1^{(i)}, Z_4^{(i)}, Z_5^{(i)}$	$(\delta,\delta) \rightarrow (\delta,0)$
$(\delta,0,\delta,0) \xrightarrow{1r} (\delta,\delta,0,0)$	$Z_1^{(i)}$	$(0,0) \rightarrow (0,0)$
$(\delta,0,\delta,\delta) \xrightarrow{1r} (0,0,\delta,0)$	$Z_1^{(i)}, Z_4^{(i)}, Z_6^{(i)}$	$(0,\delta) \rightarrow (\delta,\delta)$
$(\delta,\delta,0,0) \xrightarrow{1r} (\delta,0,0,\delta)$	$Z_1^{(i)}, Z_5^{(i)}$	$(\delta,\delta) \rightarrow (\delta,0)$
$(\delta,\delta,0,\delta) \xrightarrow{1r} (0,\delta,\delta,\delta)$	$Z_1^{(i)}, Z_4^{(i)}, Z_5^{(i)}, Z_6^{(i)}$	$(\delta,0) \rightarrow (0,\delta)$
$(\delta,\delta,\delta,0) \xrightarrow{1r} (0,0,0,\delta)$	$Z_1^{(i)}, Z_6^{(i)}$	$(0,\delta) \rightarrow (\delta,\delta)$
$(\delta,\delta,\delta,\delta) \xrightarrow{1r} (\delta,\delta,\delta,\delta)$	$Z_1^{(i)}, Z_4^{(i)}$	$(0,0) \rightarrow (0,0)$

- $(0,0,\delta,\delta) \xrightarrow{1r} (0,\delta,\delta,0) \xrightarrow{1r} (0,\delta,0,\delta) \xrightarrow{1r} (0,0,\delta,\delta)$ which is 3-round iterative; the weak subkeys required are: $Z_4^{(i)}$, $Z_5^{(i)}$, $Z_5^{(i+1)}$ and $Z_4^{(i+2)}$.
- $(\delta,0,0,\delta) \xrightarrow{1r} (\delta,0,\delta,0) \xrightarrow{1r} (\delta,\delta,0,0) \xrightarrow{1r} (\delta,0,0,\delta)$ which is 3-round iterative; the weak subkeys required are: $Z_1^{(i)}$, $Z_4^{(i)}$, $Z_5^{(i)}$, $Z_1^{(i+1)}$, $Z_1^{(i+2)}$ and $Z_5^{(i+2)}$.
- $(\delta,\delta,\delta,\delta) \xrightarrow{1r} (\delta,\delta,\delta,\delta)$ which is 1-round iterative; the weak subkeys required are: $Z_1^{(i)}$ and $Z_4^{(i)}$.

The 3-round iterative characteristic

$$(0,0,\delta,\delta) \xrightarrow{1r} (0,\delta,\delta,0) \xrightarrow{1r} (0,\delta,0,\delta) \xrightarrow{1r} (0,0,\delta,\delta)$$

gives the largest differential weak-key class of IDEA, with 2^{35} keys. Table 3.10 details this differential trail across the full 8.5-round IDEA.

These differential characteristics provide essential information on IDEA:

- this differential analysis using characteristics allows us to study the diffusion power of the MA half-round of IDEA with respect to modern tools such as the (differential) branch number.
  The set of characteristics in Table 3.9 indicates that the (differential) branch number *per round* of IDEA is only 2 (under weak-key assumptions). For instance, the 1-round characteristic $(0,0,\delta,0) \xrightarrow{1r} (\delta,0,0,0)$ has both the input and output round differences with (wordwise) Hamming Weight one. So, even though IDEA provides full diffusion in a single round, a detailed analysis shows that its differential branch number is

Table 3.10 8.5-round differential characteristic of IDEA under weak-key assumptions.

i-th Round	Characteristic	$\mathrm{msb}_{15}(Z_4^{(i)})$	$\mathrm{msb}_{15}(Z_5^{(i)})$	Weak-key Class size
1	$(0,\delta,0,\delta) \stackrel{1r}{\to} (0,0,\delta,\delta)$	48–62	—	2^{113}
2	$(0,0,\delta,\delta) \stackrel{1r}{\to} (0,\delta,\delta,0)$	41–55	57–71	2^{97}
3	$(0,\delta,\delta,0) \stackrel{1r}{\to} (0,\delta,0,\delta)$	—	50–64	2^{97}
4	$(0,\delta,0,\delta) \stackrel{1r}{\to} (0,0,\delta,\delta)$	2–16	—	2^{82}
5	$(0,0,\delta,\delta) \stackrel{1r}{\to} (0,\delta,\delta,0)$	123–9	11–25	2^{66}
6	$(0,\delta,\delta,0) \stackrel{1r}{\to} (0,\delta,0,\delta)$	—	4–18	2^{66}
7	$(0,\delta,0,\delta) \stackrel{1r}{\to} (0,0,\delta,\delta)$	84–98	—	2^{51}
8	$(0,0,\delta,\delta) \stackrel{1r}{\to} (0,\delta,\delta,0)$	77–91	93–107	2^{35}
OT	$(0,\delta,\delta,0) \stackrel{0.5r}{\to} (0,\delta,\delta,0)$	—	—	2^{35}

much lower than ideal. For a 4-word block, such as in IDEA, an MDS code using a 4×4 MDS matrix would provide a branch number of $1+4=5$. See Appendix, Sect. B.1 for more information about the differential branch number.

- given a corresponding weak key, the characteristics in Table 3.9 hold with certainty. Looking the other way around, these characteristics provide efficient membership tests for keys in a weak-key class. For instance, the characteristic in Table 3.10 allows us to efficiently identify all 2^{35} keys in the class. The test consists of two encryptions with input difference $(0,\delta,0,\delta)$. The sample key belongs to the class (with high probability) if the output difference is $(0,\delta,\delta,0)$ after 8.5-round IDEA. The difference patterns inside the trail detailed in Table 3.9 are very sensitive to the given weak-key assumptions. If even a single multiplicative subkey in the trail is not weak, the difference patterns break down and the output difference (after 8.5 rounds) will become unpredictable. For a random key, the chance of the difference $(0,\delta,\delta,0)$ showing up at the output is around 2^{-64}. For a key space of 2^{128} keys, excluding the weak-key class of size 2^{35}, and using two text pairs, the number of keys satisfying $(0,\delta,0,\delta) \stackrel{8.5r}{\to} (0,\delta,\delta,0)$ is $(2^{-64})^2 \cdot (2^{128}-2^{35}) < 1$.
- the characteristics in Table 3.9, under the given weak-key assumptions, can be used to distinguish the full 8.5-round IDEA from a random permutation with very low attack complexity, since the characteristics hold with certainty. The cost is essentially two encryptions, two chosen plaintexts and equivalent memory.
  The reasoning is the following: given the input difference $(0,\delta,0,\delta)$ to both 8.5-round IDEA (under a weak-key) and a random 64-bit permutation π, the output difference $(0,\delta,\delta,0)$ will show up with certainty for IDEA, while for π, it will happen with probability about 2^{-64}. The advantage (see Sect. 1.7 to distinguish one from the other is: $1-2^{-64}$, which

is non-negligible. Therefore, the full 8.5-round IDEA cipher is not an ideal primitive, that is, it is not a PRP.

- usually, the larger the number of rounds the smaller the weak-key class becomes, as can be observed on the rightmost column in Table 3.10. But, the weak-key class size never vanishes for IDEA, due to its key schedule algorithm. See Sect. 2.2.2. For instance, the 12-round characteristic $(\delta,\delta,\delta,\delta) \stackrel{12r}{\rightarrow} (\delta,\delta,\delta,\delta)$ holds with certainty only for two weak keys. Beyond 12 rounds, it still holds but only for $K = 0^{128}$ which is a very special key[8]. It is the only user key of IDEA for which **any** characteristic can be iterated for **any** number of rounds because $K = 0^{128}$ means that all (multiplicative) subkeys are weak, even those which do not belong to the differential trail.

3.5.2.1 Key-Recovery Attacks

Weak-key classes can be increased by reducing the number of weak-subkey restrictions. On the one hand, it will cause the distinguisher to reach less rounds, but, at the same time, it will allow subkeys to be recovered from the rounds not covered by the distinguisher.

An example of key-recovery attack on the full 8.5-round IDEA could use the iterative characteristic $(\delta,\delta,\delta,\delta) \stackrel{1r}{\rightarrow} (\delta,\delta,\delta,\delta)$ across 8 rounds, and recover subkeys from the OT.

This attack essentially guesses subkeys $Z_1^{(9)}$, $Z_2^{(9)}$, $Z_3^{(9)}$ and $Z_4^{(9)}$, partially decrypt the OT until the output of the eight round, and check if the difference at that point is δ as predicted by the distinguisher.

Since the distinguisher holds with certainty, the attack effort is $2 \cdot 2^{16}$ multiplications per subkey per text pair, in the worst case. It is equivalent to $4{\cdot}2^{17}/4$ OT computations, or $2^{17}/17 \approx 2^{13}$ full IDEA computations, assuming 17 half-rounds per full 8.5-round IDEA computation. The data complexity is two chosen plaintexts, and the memory needed to store the subkeys.

But, this characteristic does not allow to recover other subkeys. Use of another characteristic would imply another weak-key class. Therefore, this is only a partial key-recovery attack: instead of the full 128-bit key only 64 key bits are recovered.

[8] It can be argued that a weak-key class of size one does not actually represent a class. Moreover, the effort to identify it is the same as an exhaustive key search (of the class). Nonetheless, the all-zero key has very special properties: it is a valid IDEA key and it actually generates weak subkeys for any trail and for any number of rounds.

3.5.3 DC of MESH-64 Under Weak-Key Assumptions

The following differential analysis of MESH-64 is based on [151] and operates under *weak-key assumptions.* These assumptions state that some multiplicative subkeys have value 0 or 1, that is, their fifteen most significant bits are zero.

The initial approach is to obtain 1-round characteristics and the corresponding weak-key conditions for the characteristics to hold with the highest possible probability. The difference operator is bitwise exclusive-or and the difference is wordwise. The only 16-bit difference values used in the characteristics are $\delta = 8000_x$ and 0000_x. The difference 8000_x propagates across the modular addition with probability one, since it affects only the most significant bit (and the last carry-out bit is truncated due to modular reduction).

Table 3.11 lists all fifteen nontrivial 1-round differential characteristics of MESH-64, under weak-key assumptions for odd- and even-numbered rounds. Under these assumptions, the characteristics hold with certainty. Multiple-round characteristics are obtained by concatenating 1-round characteristics, such that the output difference of one characteristic matches the input difference of the next characteristic.

Table 3.11 Nontrivial 1-round characteristics for MESH-64 under weak-subkey assumptions.

1-round characteristics	Weak subkeys Odd-numbered rounds	Weak subkeys Even-numbered rounds	Difference across the MA-box
$(0,0,0,\delta) \xrightarrow{1r} (0,0,\delta,0)$	$Z_4^{(i)}, Z_6^{(i)}, Z_7^{(i)}$	$Z_6^{(i)}, Z_7^{(i)}$	$(0,\delta) \to (\delta,0)$
$(0,0,\delta,0) \xrightarrow{1r} (\delta,0,0,0)$	$Z_5^{(i)}, Z_6^{(i)}$	$Z_3^{(i)}, Z_5^{(i)}, Z_6^{(i)}$	$(\delta,0) \to (0,\delta)$
$(0,0,\delta,\delta) \xrightarrow{1r} (\delta,0,\delta,0)$	$Z_4^{(i)}, Z_5^{(i)}, Z_7^{(i)}$	$Z_3^{(i)}, Z_5^{(i)}, Z_7^{(i)}$	$(\delta,\delta) \to (\delta,\delta)$
$(0,\delta,0,0) \xrightarrow{1r} (0,0,0,\delta)$	$Z_6^{(i)}, Z_7^{(i)}$	$Z_2^{(i)}, Z_6^{(i)}, Z_7^{(i)}$	$(0,\delta) \to (\delta,0)$
$(0,\delta,0,\delta) \xrightarrow{1r} (0,0,\delta,\delta)$	$Z_4^{(i)}$	$Z_2^{(i)}$	$(0,0) \to (0,0)$
$(0,\delta,\delta,0) \xrightarrow{1r} (\delta,0,0,\delta)$	$Z_5^{(i)}, Z_7^{(i)}$	$Z_2^{(i)}, Z_3^{(i)}, Z_5^{(i)}, Z_7^{(i)}$	$(\delta,\delta) \to (\delta,\delta)$
$(0,\delta,\delta,\delta) \xrightarrow{1r} (\delta,0,\delta,\delta)$	$Z_4^{(i)}, Z_5^{(i)}, Z_6^{(i)}$	$Z_2^{(i)}, Z_3^{(i)}, Z_5^{(i)}, Z_6^{(i)}$	$(\delta,0) \to (0,\delta)$
$(\delta,0,0,0) \xrightarrow{1r} (0,\delta,0,0)$	$Z_1^{(i)}, Z_5^{(i)}, Z_6^{(i)}$	$Z_5^{(i)}, Z_6^{(i)}$	$(\delta,0) \to (0,\delta)$
$(\delta,0,0,\delta) \xrightarrow{1r} (0,\delta,\delta,0)$	$Z_1^{(i)}, Z_4^{(i)}, Z_5^{(i)}, Z_7^{(i)}$	$Z_5^{(i)}, Z_7^{(i)}$	$(\delta,\delta) \to (\delta,\delta)$
$(\delta,0,\delta,0) \xrightarrow{1r} (\delta,\delta,0,0)$	$Z_1^{(i)}$	$Z_3^{(i)}$	$(0,0) \to (0,0)$
$(\delta,0,\delta,\delta) \xrightarrow{1r} (\delta,\delta,\delta,0)$	$Z_1^{(i)}, Z_4^{(i)}, Z_6^{(i)}, Z_7^{(i)}$	$Z_3^{(i)}, Z_6^{(i)}, Z_7^{(i)}$	$(0,\delta) \to (\delta,0)$
$(\delta,\delta,0,0) \xrightarrow{1r} (0,\delta,0,\delta)$	$Z_1^{(i)}, Z_5^{(i)}, Z_7^{(i)}$	$Z_2^{(i)}, Z_5^{(i)}, Z_7^{(i)}$	$(\delta,\delta) \to (\delta,\delta)$
$(\delta,\delta,0,\delta) \xrightarrow{1r} (0,\delta,\delta,\delta)$	$Z_1^{(i)}, Z_4^{(i)}, Z_5^{(i)}, Z_6^{(i)}$	$Z_2^{(i)}, Z_5^{(i)}, Z_6^{(i)}$	$(\delta,0) \to (0,\delta)$
$(\delta,\delta,\delta,0) \xrightarrow{1r} (\delta,\delta,0,\delta)$	$Z_1^{(i)}, Z_6^{(i)}, Z_7^{(i)}$	$Z_2^{(i)}, Z_3^{(i)}, Z_6^{(i)}, Z_7^{(i)}$	$(0,\delta) \to (\delta,0)$
$(\delta,\delta,\delta,\delta) \xrightarrow{1r} (\delta,\delta,\delta,\delta)$	$Z_1^{(i)}, Z_4^{(i)}$	$Z_2^{(i)}, Z_3^{(i)}$	$(0,0) \to (0,0)$

From Table 3.11, the multiple-round characteristics are clustered into five chains covering 1, 2 or 4 rounds:

(1) $(0,0,0,\delta) \stackrel{1r}{\rightarrow} (0,0,\delta,0) \stackrel{1r}{\rightarrow} (\delta,0,0,0) \stackrel{1r}{\rightarrow} (0,\delta,0,0) \stackrel{1r}{\rightarrow} (0,0,0,\delta)$, which is 4-round iterative. Assuming i is odd, the weak subkeys required are: $Z_4^{(i)}$, $Z_6^{(i)}$, $Z_7^{(i)}$, $Z_3^{(i+1)}$, $Z_5^{(i+1)}$, $Z_6^{(i+1)}$, $Z_1^{(i+2)}$, $Z_5^{(i+2)}$, $Z_6^{(i+2)}$, $Z_2^{(i+3)}$, $Z_6^{(i+3)}$ and $Z_7^{(i+3)}$. Assuming i is even, the required weak subkeys are: $Z_6^{(i)}$, $Z_7^{(i)}$, $Z_5^{(i+1)}$, $Z_6^{(i+1)}$, $Z_5^{(i+2)}$, $Z_6^{(i+2)}$, $Z_6^{(i+3)}$ and $Z_7^{(i+3)}$.

(2) $(0,0,\delta,\delta) \stackrel{1r}{\rightarrow} (\delta,0,\delta,0) \stackrel{1r}{\rightarrow} (\delta,\delta,0,0) \stackrel{1r}{\rightarrow} (0,\delta,0,\delta) \stackrel{1r}{\rightarrow} (0,0,\delta,\delta)$, which is 4-round iterative. Assuming i is odd, the weak subkeys required are: $Z_4^{(i)}$, $Z_5^{(i)}$, $Z_7^{(i)}$, $Z_3^{(i+1)}$, $Z_1^{(i+2)}$, $Z_5^{(i+2)}$, $Z_7^{(i+2)}$ and $Z_2^{(i+3)}$. Assuming i is even, the required weak subkeys are: $Z_3^{(i)}$, $Z_5^{(i)}$, $Z_7^{(i)}$, $Z_1^{(i+1)}$, $Z_2^{(i+2)}$, $Z_5^{(i+2)}$, $Z_7^{(i+2)}$ and $Z_4^{(i+3)}$.

(3) $(0,\delta,\delta,0) \stackrel{1r}{\rightarrow} (\delta,0,0,\delta) \stackrel{1r}{\rightarrow} (0,\delta,\delta,0)$, which is 2-round iterative. Assuming i is odd, the weak subkeys required are: $Z_5^{(i)}$, $Z_7^{(i)}$, $Z_5^{(i+1)}$ and $Z_7^{(i+1)}$. Assuming i is even, the required weak subkeys are: $Z_2^{(i)}$, $Z_3^{(i)}$, $Z_5^{(i)}$, $Z_7^{(i)}$, $Z_1^{(i+1)}$, $Z_4^{(i+1)}$, $Z_5^{(i+1)}$ and $Z_7^{(i+1)}$.

(4) $(0,\delta,\delta,\delta) \stackrel{1r}{\rightarrow} (\delta,0,\delta,\delta) \stackrel{1r}{\rightarrow} (\delta,\delta,\delta,0) \stackrel{1r}{\rightarrow} (\delta,\delta,0,\delta) \stackrel{1r}{\rightarrow} (0,\delta,\delta,\delta)$, which is 4-round iterative. Assuming i is odd, the weak subkeys required are: $Z_4^{(i)}$, $Z_5^{(i)}$, $Z_6^{(i)}$, $Z_3^{(i+1)}$, $Z_6^{(i+1)}$, $Z_7^{(i+1)}$, $Z_1^{(i+2)}$, $Z_6^{(i+2)}$, $Z_7^{(i+2)}$, $Z_2^{(i+3)}$, $Z_5^{(i+3)}$ and $Z_6^{(i+3)}$. Assuming i is even, the required weak subkeys are: $Z_2^{(i)}$, $Z_3^{(i)}$, $Z_5^{(i)}$, $Z_6^{(i)}$, $Z_1^{(i+1)}$, $Z_2^{(i+1)}$, $Z_6^{(i+1)}$, $Z_7^{(i+1)}$, $Z_2^{(i+2)}$, $Z_3^{(i+2)}$, $Z_6^{(i+2)}$, $Z_7^{(i+2)}$, $Z_1^{(i+3)}$, $Z_4^{(i+3)}$, $Z_5^{(i+3)}$ and $Z_6^{(i+3)}$.

(5) $(\delta,\delta,\delta,\delta) \stackrel{1r}{\rightarrow} (\delta,\delta,\delta,\delta)$, which is 1-round iterative. Assuming i is odd, the weak subkeys required are: $Z_1^{(i)}$ and $Z_4^{(i)}$. Assuming i even, the required weak subkeys are: $Z_2^{(i)}$ and $Z_3^{(i)}$.

Starting from the first round ($i = 1$), the characteristics that require the least number of weak subkeys per round are (2) and (5). Both require two weak subkeys per round on average. Over eight rounds, the characteristic (2) will be repeated twice, and require 16 weak subkeys. Over 8.5-round, characteristic (5) will be repeated 8.5 times and require 17 weak subkeys.

Note that the use of three layers of $\boxplus$ and $\odot$ (with three multiplicative subkeys) in the MA-box of MESH-64 can be corroborated by these characteristics. The best characteristics bypass the MA-box altogether, since the number of weak subkeys needed to cross an MA half-round can be larger than what is needed to avoid the MA half-round.

Unlike the key schedule algorithms of PES and IDEA, the key schedule of MESH-64 does not consist of a fixed bit permutation, but is more involved. See Sect. 2.3.2. Experiments with mini-versions of the key schedule for a reduced version of MESH-64 with 16-bit blocks and 32-bit keys indicated that each weak subkey restriction (with the three most significant bits equal to zero) is satisfied for a fraction of 2^{-3} of the key space. Assuming this behavior holds for each weak subkey independently, and for larger key schedule as well, such as the one for MESH-64, then t weak subkeys would apply to a fraction of $2^{128-t\cdot 15}$ of the key space, or a weak-key class of this size. Thus, for a nonempty weak-key class, t must satisfy $128 - t \cdot 15 \geq 0$, that is, $t \leq 8$. The estimated weak-key class size is 2^8.

Assuming this bound, both characteristics (2) and (5) can be used to distinguish at most 4-round MESH-64 from a random permutation (under the corresponding weak-key assumptions).

This differential analysis using characteristics allows us to study the diffusion power of the MA half-round of MESH-64 with respect to modern tools such as the (differential) branch number. See Appendix, Sect. B.

The set of characteristics in Table 3.11 indicates that the differential branch number *per round* of MESH-64 is only $1 + 1 = 2$ (under weak-key assumptions). For instance, the 1-round characteristic

$$(\delta, 0, 0, 0) \stackrel{1r}{\rightarrow} (0, \delta, 0, 0)$$

has both the input and output round differences with wordwise Hamming Weight one. So, even though MESH-64 provides full diffusion in a single round, a detailed analysis shows that its differential branch number is much lower than ideal. For a 4-word block, such as in MESH-64, an MDS code using a 4×4 MDS matrix would provide a branch number of $1 + 4 = 5$. See Appendix, Sect. B.1.

3.5.4 DC of MESH-96 Under Weak-Key Assumptions

The following differential analysis of MESH-96 was initiated in [151] and operates under weak-key assumptions.

The analysis follows a bottom-up approach, starting from the smallest components: $\odot$, $\boxplus$ and $\oplus$. MESH-96 is the first MESH cipher to use the unkeyed $\odot$ operator. For MESH-96 (and other Lai-Massey designs that also use unkeyed multiplication), Table 3.12 lists the difference propagation and the corresponding probabilities across $\odot$. Given that $c = a \odot b$, it follows that $\Delta c = (a \odot b) \oplus ((a \oplus \Delta a) \odot (b \oplus \Delta b))$. The difference operator is bitwise exclusive-or and the difference is wordwise (16 bits for MESH-96).

Table 3.13 lists the difference propagation and the corresponding probabilities across $\boxplus$. Given that $c = a \boxplus b$, it follows that $\Delta c = (a \boxplus b) \oplus ((a \oplus \Delta a) \boxplus (b \oplus \Delta b))$.

The next step is to obtain 1-round differential characteristics and the corresponding weak-key restrictions for the characteristics to hold with highest possible probability. The only 16-bit difference values used in the characteristics are $\delta = 8000_x$ and 0000_x.

Table 3.14 and Table 3.15 list all 63 nontrivial 1-round differential characteristics of MESH-96, under weak-key assumptions, using only wordwise

Table 3.12 $\oplus$ difference propagation across a single unkeyed $\odot$ for 8-bit and 16-bit words.

Δa	Δb	Δc	Probability	
			8-bit words	16-bit words
0	0	0	1	1
0	0	δ	0	0
0	δ	0	0	0
0	δ	δ	2^{-8}	2^{-15}
δ	0	0	0	0
δ	0	δ	2^{-8}	2^{-15}
δ	δ	0	2^{-7}	2^{-15}
δ	δ	δ	$2^{-6.89}$	$2^{-14.98}$

Table 3.13 $\oplus$ difference propagation across a single unkeyed $\boxplus$ for 8-bit and 16-bit words.

Δa	Δb	Δc	Probability	
			8-bit words	16-bit words
0	0	0	1	1
0	0	δ	0	0
0	δ	0	0	0
0	δ	δ	1	1
δ	0	0	0	0
δ	0	δ	1	1
δ	δ	0	1	1
δ	δ	δ	0	0

differences 0000_x and 8000_x. These tables provide an exhaustive analysis of how characteristics propagate across a single round of MESH-96, and they are more extensive than the analyses in [151].

Multiple-round characteristics are obtained by concatenating 1-round characteristics, such that the output difference of one characteristic matches the input difference of the next characteristic.

From Tables 3.14 and 3.15, the multiple-round characteristics are clustered into eleven chains covering 1, 2, 3, 4, 6 or 12 rounds:

(1) $(0,0,0,0,0,\delta) \stackrel{1r}{\to} (\delta,\delta,0,0,\delta,0) \stackrel{1r}{\to} (\delta,0,0,0,\delta,\delta) \stackrel{1r}{\to} (\delta,0,0,\delta,0,\delta) \stackrel{1r}{\to} (0,0,0,0,\delta,0) \stackrel{1r}{\to} (\delta,\delta,\delta,0,0,0) \stackrel{1r}{\to} (\delta,0,\delta,0,\delta,0) \stackrel{1r}{\to} (\delta,0,0,\delta,\delta,0) \stackrel{1r}{\to} (0,0,\delta,0,0,0) \stackrel{1r}{\to} (\delta,\delta,0,0,0,\delta) \stackrel{1r}{\to} (\delta,0,\delta,0,0,\delta) \stackrel{1r}{\to} (\delta,0,\delta,\delta,0,0) \stackrel{1r}{\to} (0,0,0,0,0,\delta)$, which is 12-round iterative and holds with probability $2^{-45\cdot 12} = 2^{-540}$. Starting from an odd-numbered round i, the following 29 subkeys have to be weak: $Z_9^{(i)}$, $Z_2^{(i+1)}$, $Z_7^{(i+1)}$, $Z_9^{(i+1)}$, $Z_1^{(i+2)}$, $Z_5^{(i+2)}$, $Z_7^{(i+2)}$, $Z_4^{(i+3)}$, $Z_6^{(i+3)}$, $Z_9^{(i+3)}$, $Z_5^{(i+4)}$, $Z_2^{(i+5)}$, $Z_7^{(i+5)}$, $Z_1^{(i+6)}$, $Z_3^{(i+6)}$, $Z_5^{(i+6)}$, $Z_7^{(i+6)}$, $Z_4^{(i+7)}$, $Z_3^{(i+8)}$, $Z_9^{(i+8)}$, $Z_2^{(i+9)}$, $Z_6^{(i+9)}$, $Z_7^{(i+9)}$, $Z_1^{(i+9)}$, $Z_3^{(i+9)}$, $Z_7^{(i+9)}$, $Z_9^{(i+9)}$, $Z_4^{(i+10)}$ and $Z_9^{(i+10)}$, which means $29/12 \approx 2.41$ weak subkeys/round. Starting from an even-numbered round i, the following 25 subkeys have to be weak: $Z_6^{(i)}$, $Z_9^{(i)}$, $Z_1^{(i+1)}$, $Z_5^{(i+1)}$, $Z_7^{(i+1)}$, $Z_9^{(i+1)}$, $Z_6^{(i+2)}$, $Z_7^{(i+2)}$, $Z_1^{(i+3)}$, $Z_9^{(i+3)}$, $Z_1^{(i+5)}$, $Z_3^{(i+5)}$, $Z_7^{(i+5)}$, $Z_7^{(i+6)}$,

Table 3.14 Nontrivial 1-round characteristics of MESH-96 under weak-subkey restrictions.

1-round characteristics	Weak subkeys $Z_j^{(i)}$ (i-th round)		Difference across the MA-box	Prob.
	j values (odd i)	j values (even i)		
$(0,0,0,0,0,\delta) \stackrel{1r}{\to} (\delta,\delta,0,0,\delta,0)$	9	6; 9	$(0,0,\delta) \to (\delta,0,\delta)$	2^{-45}
$(0,0,0,0,\delta,0) \stackrel{1r}{\to} (\delta,\delta,\delta,0,0,0)$	5		$(0,\delta,0) \to (0,0,\delta)$	2^{-45}
$(0,0,0,0,\delta,\delta) \stackrel{1r}{\to} (0,0,\delta,0,\delta,0)$	5; 9	6; 9	$(0,\delta,\delta) \to (\delta,0,0)$	2^{-30}
$(0,0,0,\delta,0,0) \stackrel{1r}{\to} (0,\delta,\delta,\delta,\delta,\delta)$	7; 9	4; 7; 9	$(\delta,0,0) \to (\delta,\delta,0)$	2^{-45}
$(0,0,0,\delta,0,\delta) \stackrel{1r}{\to} (\delta,0,\delta,\delta,0,\delta)$	7	4; 6; 7	$(\delta,0,\delta) \to (0,\delta,\delta)$	2^{-45}
$(0,0,0,\delta,\delta,0) \stackrel{1r}{\to} (\delta,0,0,\delta,\delta,\delta)$	5; 7; 9	4; 7; 9	$(\delta,\delta,0) \to (\delta,\delta,\delta)$	2^{-15}
$(0,0,0,\delta,\delta,\delta) \stackrel{1r}{\to} (0,\delta,0,\delta,0,\delta)$	5; 7	4; 6; 7	$(\delta,\delta,\delta) \to (0,\delta,0)$	2^{-45}
$(0,0,\delta,0,0,0) \stackrel{1r}{\to} (\delta,\delta,0,0,0,\delta)$	3; 9	9	$(0,0,\delta) \to (\delta,0,\delta)$	2^{-45}
$(0,0,\delta,0,0,\delta) \stackrel{1r}{\to} (0,0,0,0,\delta,\delta)$	3	6	$(0,0,0) \to (0,0,0)$	1
$(0,0,\delta,0,\delta,0) \stackrel{1r}{\to} (0,0,\delta,0,0,\delta)$	3; 5; 9	9	$(0,\delta,\delta) \to (\delta,0,0)$	2^{-30}
$(0,0,\delta,0,\delta,\delta) \stackrel{1r}{\to} (\delta,\delta,\delta,0,\delta,\delta)$	3; 5	6	$(0,\delta,0) \to (0,0,\delta)$	2^{-45}
$(0,0,\delta,\delta,0,0) \stackrel{1r}{\to} (\delta,0,\delta,\delta,\delta,0)$	3; 7	4; 7	$(\delta,0,\delta) \to (0,\delta,\delta)$	2^{-45}
$(0,0,\delta,\delta,0,\delta) \stackrel{1r}{\to} (0,\delta,\delta,\delta,0,0)$	3; 7; 9	4; 6; 7; 9	$(\delta,0,0) \to (\delta,\delta,0)$	2^{-45}
$(0,0,\delta,\delta,\delta,0) \stackrel{1r}{\to} (0,\delta,0,\delta,\delta,0)$	3; 5; 7	4; 7	$(\delta,\delta,\delta) \to (0,\delta,0)$	2^{-45}
$(0,0,\delta,\delta,\delta,\delta) \stackrel{1r}{\to} (\delta,0,0,\delta,0,0)$	3; 5; 7; 9	4; 6; 7; 9	$(\delta,\delta,0) \to (\delta,\delta,\delta)$	2^{-15}
$(0,\delta,0,0,0,0) \stackrel{1r}{\to} (\delta,\delta,0,\delta,0,0)$		2	$(0,\delta,0) \to (0,0,\delta)$	2^{-45}
$(0,\delta,0,0,0,\delta) \stackrel{1r}{\to} (0,0,0,\delta,\delta,0)$	9	2; 6; 9	$(0,\delta,\delta) \to (\delta,0,0)$	2^{-30}
$(0,\delta,0,0,\delta,0) \stackrel{1r}{\to} (0,0,\delta,\delta,0,0)$	5	2	$(0,0,0) \to (0,0,0)$	1
$(0,\delta,0,0,\delta,\delta) \stackrel{1r}{\to} (\delta,\delta,\delta,\delta,\delta,0)$	5; 9	2; 6; 9	$(0,0,\delta) \to (\delta,0,\delta)$	2^{-45}
$(0,\delta,0,\delta,0,0) \stackrel{1r}{\to} (\delta,0,\delta,0,\delta,\delta)$	7; 9	2; 4; 7; 9	$(\delta,\delta,0) \to (\delta,\delta,\delta)$	2^{-15}
$(0,\delta,0,\delta,0,\delta) \stackrel{1r}{\to} (0,\delta,\delta,0,0,\delta)$	7	2; 4; 6; 7	$(\delta,\delta,\delta) \to (0,\delta,0)$	2^{-45}
$(0,\delta,0,\delta,\delta,0) \stackrel{1r}{\to} (0,\delta,0,0,\delta,\delta)$	5; 7; 9	2; 4; 7; 9	$(\delta,0,0) \to (\delta,\delta,0)$	2^{-45}
$(0,\delta,0,\delta,\delta,\delta) \stackrel{1r}{\to} (\delta,0,0,0,0,\delta)$	5; 7	2; 4; 6; 7	$(\delta,0,\delta) \to (0,\delta,\delta)$	2^{-45}
$(0,\delta,\delta,0,0,0) \stackrel{1r}{\to} (0,0,0,\delta,0,\delta)$	3; 9	2; 9	$(0,\delta,\delta) \to (\delta,0,0)$	2^{-30}
$(0,\delta,\delta,0,0,\delta) \stackrel{1r}{\to} (\delta,\delta,0,\delta,\delta,\delta)$	3	2; 6	$(0,\delta,0) \to (0,0,\delta)$	2^{-45}
$(0,\delta,\delta,0,\delta,0) \stackrel{1r}{\to} (\delta,\delta,\delta,\delta,0,\delta)$	3; 5; 9	2; 9	$(0,0,\delta) \to (\delta,0,\delta)$	2^{-45}
$(0,\delta,\delta,0,\delta,\delta) \stackrel{1r}{\to} (0,0,\delta,\delta,\delta,\delta)$	3; 5	2; 6	$(0,0,0) \to (0,0,0)$	1
$(0,\delta,\delta,\delta,0,0) \stackrel{1r}{\to} (0,\delta,\delta,0,\delta,0)$	3; 7	2; 4; 7	$(\delta,\delta,\delta) \to (0,\delta,0)$	2^{-45}
$(0,\delta,\delta,\delta,0,\delta) \stackrel{1r}{\to} (\delta,0,\delta,0,0,0)$	3; 7; 9	2; 4; 6; 7; 9	$(\delta,\delta,0) \to (\delta,\delta,\delta)$	2^{-15}
$(0,\delta,\delta,\delta,\delta,0) \stackrel{1r}{\to} (\delta,0,0,0,\delta,0)$	3; 5; 7	2; 4; 7	$(\delta,0,\delta) \to (0,\delta,\delta)$	2^{-45}
$(0,\delta,\delta,\delta,\delta,\delta) \stackrel{1r}{\to} (0,\delta,0,0,0,0)$	3; 5; 7; 9	2; 4; 6; 7; 9	$(\delta,0,0) \to (\delta,\delta,0)$	2^{-45}
$(\delta,0,0,0,0,0) \stackrel{1r}{\to} (\delta,0,\delta,\delta,\delta,\delta)$	1; 7; 9	7; 9	$(\delta,0,0) \to (\delta,\delta,0)$	2^{-45}

Table 3.15 Nontrivial 1-round characteristics of MESH-96 under weak-subkey conditions (cont.)

1-round characteristics	Weak subkeys $Z_j^{(i)}$ (i-th round) j values (odd i)	j values (even i)	Difference across the MA-box	Prob.
$(\delta,0,0,0,0,\delta) \xrightarrow{1r} (0,\delta,\delta,\delta,0,\delta)$	1; 7	6; 7	$(\delta,0,\delta) \to (0,\delta,\delta)$	2^{-45}
$(\delta,0,0,0,\delta,0) \xrightarrow{1r} (0,\delta,0,\delta,\delta,\delta)$	1; 5; 7; 9	7; 9	$(\delta,\delta,0) \to (\delta,\delta,\delta)$	2^{-15}
$(\delta,0,0,0,\delta,\delta) \xrightarrow{1r} (\delta,0,0,\delta,0,\delta)$	1; 5; 7	6; 7	$(\delta,\delta,\delta) \to (0,\delta,0)$	2^{-45}
$(\delta,0,0,\delta,0,0) \xrightarrow{1r} (\delta,\delta,0,0,0,0)$	1	4	$(0,0,0) \to (0,0,0)$	1
$(\delta,0,0,\delta,0,\delta) \xrightarrow{1r} (0,0,0,0,\delta,0)$	1; 9	4; 6; 9	$(0,0,\delta) \to (\delta,0,\delta)$	2^{-45}
$(\delta,0,0,\delta,\delta,0) \xrightarrow{1r} (0,0,\delta,0,0,0)$	1; 5	4	$(0,\delta,0) \to (0,0,\delta)$	2^{-45}
$(\delta,0,0,\delta,\delta,\delta) \xrightarrow{1r} (\delta,\delta,\delta,0,\delta,0)$	1; 5; 9	4; 6; 9	$(0,\delta,\delta) \to (\delta,0,0)$	2^{-30}
$(\delta,0,\delta,0,0,0) \xrightarrow{1r} (0,\delta,\delta,\delta,\delta,0)$	1; 3; 7	7	$(\delta,0,\delta) \to (0,\delta,\delta)$	2^{-45}
$(\delta,0,\delta,0,0,\delta) \xrightarrow{1r} (\delta,0,\delta,\delta,0,0)$	1; 3; 7; 9	6; 7; 9	$(\delta,0,0) \to (\delta,\delta,0)$	2^{-45}
$(\delta,0,\delta,0,\delta,0) \xrightarrow{1r} (\delta,0,0,\delta,\delta,0)$	1; 3; 5; 7	7	$(\delta,\delta,\delta) \to (0,\delta,0)$	2^{-45}
$(\delta,0,\delta,0,\delta,\delta) \xrightarrow{1r} (0,\delta,0,\delta,0,0)$	1; 3; 5; 7; 9	6; 7; 9	$(\delta,\delta,0) \to (\delta,\delta,\delta)$	2^{-15}
$(\delta,0,\delta,\delta,0,0) \xrightarrow{1r} (0,0,0,0,0,\delta)$	1; 3; 9	4; 9	$(0,0,\delta) \to (\delta,0,\delta)$	2^{-45}
$(\delta,0,\delta,\delta,0,\delta) \xrightarrow{1r} (\delta,\delta,0,0,\delta,\delta)$	1; 3	4; 6	$(0,0,0) \to (0,0,0)$	1
$(\delta,0,\delta,\delta,\delta,0) \xrightarrow{1r} (\delta,\delta,\delta,0,0,\delta)$	1; 3; 5; 9	4; 9	$(0,\delta,\delta) \to (\delta,0,0)$	2^{-30}
$(\delta,0,\delta,\delta,\delta,\delta) \xrightarrow{1r} (0,0,\delta,0,\delta,\delta)$	1; 3; 5	4; 6	$(0,\delta,0) \to (0,0,\delta)$	2^{-45}
$(\delta,\delta,0,0,0,0) \xrightarrow{1r} (0,\delta,\delta,0,\delta,\delta)$	1; 7; 9	2; 7; 9	$(\delta,\delta,0) \to (\delta,\delta,\delta)$	2^{-15}
$(\delta,\delta,0,0,0,\delta) \xrightarrow{1r} (\delta,0,\delta,0,0,\delta)$	1; 7	2; 6; 7	$(\delta,\delta,\delta) \to (0,\delta,0)$	2^{-45}
$(\delta,\delta,0,0,\delta,0) \xrightarrow{1r} (\delta,0,0,0,\delta,\delta)$	1; 5; 7; 9	2; 7; 9	$(\delta,0,0) \to (\delta,\delta,0)$	2^{-45}
$(\delta,\delta,0,0,\delta,\delta) \xrightarrow{1r} (0,\delta,0,0,0,\delta)$	1; 5; 7	2; 6; 7	$(\delta,0,\delta) \to (0,\delta,\delta)$	2^{-45}
$(\delta,\delta,0,\delta,0,0) \xrightarrow{1r} (0,0,0,\delta,0,0)$	1	2; 4	$(0,\delta,0) \to (0,0,\delta)$	2^{-45}
$(\delta,\delta,0,\delta,0,\delta) \xrightarrow{1r} (\delta,\delta,0,\delta,\delta,0)$	1; 9	2; 4; 6; 9	$(0,\delta,\delta) \to (\delta,0,0)$	2^{-30}
$(\delta,\delta,0,\delta,\delta,0) \xrightarrow{1r} (\delta,\delta,\delta,\delta,0,0)$	1; 5	2; 4	$(0,0,0) \to (0,0,0)$	1
$(\delta,\delta,0,\delta,\delta,\delta) \xrightarrow{1r} (0,0,\delta,\delta,\delta,0)$	1; 5; 9	2; 4; 6; 9	$(0,0,\delta) \to (\delta,0,\delta)$	2^{-45}
$(\delta,\delta,\delta,0,0,0) \xrightarrow{1r} (\delta,0,\delta,0,\delta,0)$	1; 3; 7	2; 7	$(\delta,\delta,\delta) \to (0,\delta,0)$	2^{-45}
$(\delta,\delta,\delta,0,0,\delta) \xrightarrow{1r} (0,\delta,\delta,0,0,0)$	1; 3; 7; 9	2; 6; 7; 9	$(\delta,\delta,0) \to (\delta,\delta,\delta)$	2^{-15}
$(\delta,\delta,\delta,0,\delta,0) \xrightarrow{1r} (0,\delta,0,0,\delta,0)$	1; 3; 5; 7	2; 7	$(\delta,0,\delta) \to (0,\delta,\delta)$	2^{-45}
$(\delta,\delta,\delta,0,\delta,\delta) \xrightarrow{1r} (\delta,0,0,0,0,0)$	1; 3; 5; 7; 9	2; 6; 7; 9	$(\delta,0,0) \to (\delta,\delta,0)$	2^{-45}
$(\delta,\delta,\delta,\delta,0,0) \xrightarrow{1r} (\delta,\delta,0,\delta,0,\delta)$	1; 3; 9	2; 4; 9	$(0,\delta,\delta) \to (\delta,0,0)$	2^{-30}
$(\delta,\delta,\delta,\delta,0,\delta) \xrightarrow{1r} (0,0,0,\delta,\delta,\delta)$	1; 3	2; 4; 6	$(0,\delta,0) \to (0,0,\delta)$	2^{-45}
$(\delta,\delta,\delta,\delta,\delta,0) \xrightarrow{1r} (0,0,\delta,\delta,0,\delta)$	1; 3; 5; 9	2; 4; 9	$(0,0,\delta) \to (\delta,0,\delta)$	2^{-45}
$(\delta,\delta,\delta,\delta,\delta,\delta) \xrightarrow{1r} (\delta,\delta,\delta,\delta,\delta,\delta)$	1; 3; 5	2; 4; 6	$(0,0,0) \to (0,0,0)$	1

$Z_1^{(i+7)}$, $Z_5^{(i+7)}$, $Z_9^{(i+8)}$, $Z_1^{(i+9)}$, $Z_7^{(i+9)}$, $Z_6^{(i+10)}$, $Z_7^{(i+10)}$, $Z_9^{(i+10)}$, $Z_1^{(i+11)}$, $Z_3^{(i+11)}$ and $Z_9^{(i+11)}$, which means $25/12 \approx 2$ weak subkeys/round.

(2) $(0,0,0,0,\delta,\delta) \xrightarrow{1r} (0,0,\delta,0,\delta,0) \xrightarrow{1r} (0,0,\delta,0,0,\delta) \xrightarrow{1r} (0,0,0,0,\delta,\delta)$ which is 3-round iterative and holds with probability 2^{-60}. Starting from an odd-numbered round i, the following four subkeys have to be weak: $Z_5^{(i)}$, $Z_9^{(i)}$, $Z_9^{(i+1)}$ and $Z_3^{(i+2)}$, which means $4/3 \approx 1.33$ weak subkeys/round. Starting from an even-numbered round i, the following six subkeys have to be weak: $Z_6^{(i)}$, $Z_9^{(i)}$, $Z_3^{(i+1)}$, $Z_5^{(i+1)}$, $Z_9^{(i+1)}$ and $Z_6^{(i+2)}$, which means $6/3 = 2$ weak subkeys/round.

(3) $(0,0,0,\delta,0,0) \xrightarrow{1r} (0,\delta,\delta,\delta,\delta,\delta) \xrightarrow{1r} (0,\delta,0,0,0,0) \xrightarrow{1r} (\delta,\delta,0,\delta,0,0) \xrightarrow{1r} (0,0,0,\delta,0,0)$, which is 4-round iterative and holds with probability 2^{-180}. Starting from an odd-numbered round i, the following nine subkeys have to be weak: $Z_7^{(i)}$, $Z_9^{(i)}$, $Z_2^{(i+1)}$, $Z_4^{(i+1)}$, $Z_6^{(i+1)}$, $Z_7^{(i+1)}$, $Z_9^{(i+1)}$, $Z_2^{(i+3)}$ and $Z_4^{(i+3)}$, which means $9/4 = 2.25$ weak subkeys/round. Starting from an even-numbered round i, the following nine subkeys have to be weak: $Z_4^{(i)}$, $Z_7^{(i)}$, $Z_9^{(i)}$, $Z_3^{(i+1)}$, $Z_5^{(i+1)}$, $Z_7^{(i+1)}$, $Z_9^{(i+1)}$, $Z_2^{(i+2)}$ and $Z_1^{(i+3)}$, which means 2.25 weak subkeys/round.

(4) $(0,0,0,\delta,0,\delta) \xrightarrow{1r} (\delta,0,\delta,\delta,0,\delta) \xrightarrow{1r} (\delta,\delta,0,0,\delta,\delta) \xrightarrow{1r} (0,\delta,0,0,0,\delta) \xrightarrow{1r} (0,0,0,\delta,\delta,0) \xrightarrow{1r} (\delta,0,0,\delta,\delta,\delta) \xrightarrow{1r} (\delta,\delta,\delta,0,\delta,0) \xrightarrow{1r} (0,\delta,0,0,\delta,0) \xrightarrow{1r} (0,0,\delta,\delta,0,0) \xrightarrow{1r} (\delta,0,\delta,\delta,\delta,0) \xrightarrow{1r} (\delta,\delta,\delta,0,0,\delta) \xrightarrow{1r} (0,\delta,\delta,0,0,0) \xrightarrow{1r} (0,0,0,\delta,0,\delta)$, which is 12-round iterative and holds with probability 2^{-330}. Starting from an odd-numbered round i, the following 30 subkeys have to be weak: $Z_7^{(i)}$, $Z_4^{(i+1)}$, $Z_6^{(i+1)}$, $Z_1^{(i+2)}$, $Z_5^{(i+2)}$, $Z_7^{(i+2)}$, $Z_2^{(i+3)}$, $Z_6^{(i+3)}$, $Z_9^{(i+3)}$, $Z_5^{(i+4)}$, $Z_7^{(i+4)}$, $Z_9^{(i+4)}$, $Z_4^{(i+5)}$, $Z_6^{(i+5)}$, $Z_9^{(i+5)}$, $Z_1^{(i+6)}$, $Z_3^{(i+6)}$, $Z_5^{(i+6)}$, $Z_7^{(i+6)}$, $Z_2^{(i+7)}$, $Z_3^{(i+8)}$, $Z_7^{(i+8)}$, $Z_4^{(i+9)}$, $Z_9^{(i+9)}$, $Z_1^{(i+10)}$, $Z_3^{(i+10)}$, $Z_7^{(i+10)}$, $Z_9^{(i+10)}$, $Z_2^{(i+11)}$ and $Z_9^{(i+11)}$, which means $30/12 = 2.5$ weak subkeys/round. Starting from an even-numbered round i, the following 30 subkeys have to be weak: $Z_4^{(i)}$, $Z_6^{(i)}$, $Z_7^{(i)}$, $Z_1^{(i+1)}$, $Z_3^{(i+1)}$, $Z_2^{(i+2)}$, $Z_6^{(i+2)}$, $Z_7^{(i+2)}$, $Z_9^{(i+3)}$, $Z_4^{(i+4)}$, $Z_7^{(i+4)}$, $Z_9^{(i+4)}$, $Z_1^{(i+5)}$, $Z_5^{(i+5)}$, $Z_9^{(i+5)}$, $Z_2^{(i+6)}$, $Z_7^{(i+6)}$, $Z_5^{(i+7)}$, $Z_4^{(i+8)}$, $Z_7^{(i+8)}$, $Z_1^{(i+9)}$, $Z_3^{(i+9)}$, $Z_5^{(i+9)}$, $Z_9^{(i+9)}$, $Z_2^{(i+10)}$, $Z_6^{(i+10)}$, $Z_7^{(i+10)}$, $Z_9^{(i+10)}$, $Z_3^{(i+11)}$ and $Z_9^{(i+11)}$, which means 2.5 weak subkeys/round.

(5) $(0,0,0,\delta,\delta,\delta) \xrightarrow{1r} (0,\delta,0,\delta,0,\delta) \xrightarrow{1r} (0,\delta,\delta,0,0,\delta) \xrightarrow{1r} (\delta,\delta,0,\delta,\delta,\delta) \xrightarrow{1r} (0,0,\delta,\delta,\delta,0) \xrightarrow{1r} (0,\delta,0,\delta,\delta,0) \xrightarrow{1r} (0,\delta,0,0,\delta,\delta) \xrightarrow{1r} (\delta,\delta,\delta,\delta,\delta,0) \xrightarrow{1r} (0,0,\delta,\delta,0,\delta) \xrightarrow{1r} (0,\delta,\delta,\delta,0,0) \xrightarrow{1r} (0,\delta,\delta,0,\delta,0) \xrightarrow{1r} (\delta,\delta,\delta,\delta,0,\delta) \xrightarrow{1r} (0,0,0,\delta,\delta,\delta)$, which is 12-round iterative and holds with probability 2^{-540}. Starting from an odd-numbered round i, the following 36 subkeys have to be weak: $Z_5^{(i)}$, $Z_7^{(i)}$, $Z_2^{(i+1)}$, $Z_4^{(i+1)}$, $Z_6^{(i+1)}$, $Z_7^{(i+1)}$, $Z_3^{(i+2)}$, $Z_2^{(i+3)}$, $Z_4^{(i+3)}$, $Z_6^{(i+3)}$, $Z_9^{(i+3)}$, $Z_3^{(i+4)}$, $Z_5^{(i+4)}$, $Z_7^{(i+4)}$, $Z_2^{(i+5)}$, $Z_4^{(i+5)}$, $Z_7^{(i+5)}$, $Z_9^{(i+5)}$, $Z_5^{(i+6)}$, $Z_7^{(i+6)}$, $Z_9^{(i+6)}$, $Z_2^{(i+7)}$, $Z_4^{(i+7)}$, $Z_9^{(i+7)}$, $Z_3^{(i+8)}$, $Z_7^{(i+8)}$,

$Z_9^{(i+8)}$, $Z_2^{(i+9)}$, $Z_4^{(i+9)}$, $Z_7^{(i+9)}$, $Z_3^{(i+10)}$, $Z_5^{(i+10)}$, $Z_9^{(i+10)}$, $Z_2^{(i+11)}$, $Z_4^{(i+11)}$ and $Z_6^{(i+11)}$, which means 3 weak subkeys/round. Starting from an even-numbered round i, the following 32 subkeys have to be weak: $Z_4^{(i)}$, $Z_6^{(i)}$, $Z_7^{(i)}$, $Z_7^{(i+1)}$, $Z_2^{(i+2)}$, $Z_6^{(i+2)}$, $Z_1^{(i+3)}$, $Z_5^{(i+3)}$, $Z_9^{(i+3)}$, $Z_4^{(i+4)}$, $Z_7^{(i+4)}$, $Z_5^{(i+5)}$, $Z_7^{(i+5)}$, $Z_9^{(i+5)}$, $Z_2^{(i+6)}$, $Z_4^{(i+6)}$, $Z_7^{(i+6)}$, $Z_9^{(i+6)}$, $Z_1^{(i+7)}$, $Z_3^{(i+7)}$, $Z_5^{(i+7)}$, $Z_9^{(i+7)}$, $Z_4^{(i+8)}$, $Z_6^{(i+8)}$, $Z_7^{(i+8)}$, $Z_9^{(i+8)}$, $Z_3^{(i+9)}$, $Z_7^{(i+9)}$, $Z_2^{(i+10)}$, $Z_9^{(i+10)}$, $Z_1^{(i+11)}$ and $Z_9^{(i+11)}$, which means $32/12 \approx 2.66$ weak subkeys/round.

(6) $(0,0,\delta,0,\delta,\delta) \stackrel{1r}{\to} (\delta,\delta,\delta,0,\delta,\delta) \stackrel{1r}{\to} (\delta,0,0,0,0,0) \stackrel{1r}{\to} (\delta,0,\delta,\delta,\delta,\delta) \stackrel{1r}{\to} (0, 0, \delta, 0, \delta, \delta)$, which is 4-round iterative and holds with probability 2^{-180}. Starting from an odd-numbered round i, the following 11 subkeys have to be weak: $Z_3^{(i)}$, $Z_5^{(i)}$, $Z_2^{(i+1)}$, $Z_6^{(i+1)}$, $Z_7^{(i+1)}$, $Z_9^{(i+1)}$, $Z_1^{(i+2)}$, $Z_7^{(i+2)}$, $Z_9^{(i+2)}$, $Z_4^{(i+3)}$ and $Z_6^{(i+3)}$, which means $11/4 = 2.75$ weak subkeys/round. Starting from an even-numbered round i, the following 11 subkeys have to be weak: $Z_6^{(i)}$, $Z_1^{(i+1)}$, $Z_3^{(i+1)}$, $Z_5^{(i+1)}$, $Z_7^{(i+1)}$, $Z_9^{(i+1)}$, $Z_7^{(i+2)}$, $Z_9^{(i+2)}$, $Z_1^{(i+3)}$, $Z_3^{(i+3)}$ and $Z_5^{(i+3)}$, which means $11/4 = 2.75$ weak subkeys/round.

(7) $(0,0,\delta,\delta,\delta,\delta) \stackrel{1r}{\to} (\delta,0,0,\delta,0,0) \stackrel{1r}{\to} (\delta,\delta,0,0,0,0) \stackrel{1r}{\to} (0,\delta,\delta,0,\delta,\delta) \stackrel{1r}{\to} (0, 0, \delta, \delta, \delta, \delta)$, which is 4-round iterative and holds with probability 2^{-30}. Starting from an odd-numbered round i, the following 10 subkeys have to be weak: $Z_3^{(i)}$, $Z_5^{(i)}$, $Z_7^{(i)}$, $Z_9^{(i)}$, $Z_4^{(i+1)}$, $Z_1^{(i+2)}$, $Z_7^{(i+2)}$, $Z_9^{(i+2)}$, $Z_2^{(i+3)}$ and $Z_6^{(i+2)}$, which means $10/4 = 2.5$ weak subkeys/round. Starting from an even-numbered round i, the following 10 subkeys have to be weak: $Z_4^{(i)}$, $Z_6^{(i)}$, $Z_7^{(i)}$, $Z_9^{(i)}$, $Z_1^{(i+1)}$, $Z_2^{(i+2)}$, $Z_7^{(i+2)}$, $Z_9^{(i+2)}$, $Z_3^{(i+3)}$ and $Z_5^{(i+3)}$, which means 2.5 weak subkeys/round.

(8) $(0,\delta,0,\delta,0,0) \stackrel{1r}{\to} (\delta,0,\delta,0,\delta,\delta) \stackrel{1r}{\to} (0,\delta,0,\delta,0,0)$, which is 2-round iterative and holds with probability 2^{-30}. Starting from an odd-numbered round i, the following five subkeys have to be weak: $Z_7^{(i)}$, $Z_9^{(i)}$, $Z_6^{(i+1)}$, $Z_7^{(i+1)}$ and $Z_9^{(i+1)}$, which means $5/2 = 2.5$ weak subkeys/round. Starting from an even-numbered round i, the following nine subkeys have to be weak: $Z_2^{(i)}$, $Z_4^{(i)}$, $Z_7^{(i)}$, $Z_9^{(i)}$, $Z_1^{(i+1)}$, $Z_3^{(i+1)}$, $Z_5^{(i+1)}$, $Z_7^{(i+1)}$ and $Z_9^{(i+1)}$, which means $9/2 = 4.5$ weak subkeys/round.

(9) $(0,\delta,0,\delta,\delta,\delta) \stackrel{1r}{\to} (\delta,0,0,0,0,\delta) \stackrel{1r}{\to} (0,\delta,\delta,\delta,0,\delta) \stackrel{1r}{\to} (\delta,0,\delta,0,0,0) \stackrel{1r}{\to} (0, \delta, \delta, \delta, \delta, 0) \stackrel{1r}{\to} (\delta,0,0,0,\delta,0) \stackrel{1r}{\to} (0, \delta, 0, \delta, \delta, \delta)$, which is 6-round iterative and holds with probability $2^{-45\cdot 4-15\cdot 2} = 2^{-210}$. Starting from an odd-numbered round i, the following 13 subkeys have to be weak: $Z_5^{(i)}$, $Z_7^{(i)}$, $Z_6^{(i+1)}$, $Z_7^{(i+1)}$, $Z_3^{(i+2)}$, $Z_7^{(i+2)}$, $Z_9^{(i+2)}$, $Z_7^{(i+3)}$, $Z_3^{(i+4)}$, $Z_5^{(i+4)}$, $Z_7^{(i+4)}$, $Z_7^{(i+5)}$ and $Z_9^{(i+5)}$, which means $13/6 \approx 2$ weak subkeys/round. Starting from an even-numbered round i, the following 21 subkeys have to be weak: $Z_2^{(i)}$, $Z_4^{(i)}$, $Z_6^{(i)}$, $Z_7^{(i)}$, $Z_1^{(i+1)}$, $Z_7^{(i+1)}$, $Z_2^{(i+3)}$, $Z_4^{(i+3)}$, $Z_6^{(i+3)}$, $Z_7^{(i+3)}$, $Z_9^{(i+3)}$, $Z_1^{(i+4)}$, $Z_3^{(i+4)}$, $Z_7^{(i+4)}$, $Z_2^{(i+5)}$, $Z_4^{(i+5)}$, $Z_7^{(i+5)}$, $Z_1^{(i+6)}$, $Z_5^{(i+6)}$, $Z_7^{(i+6)}$ and $Z_9^{(i+6)}$, which means $21/6 = 3.5$ weak subkeys/round.

(10) $(\delta,\delta,0,\delta,0,\delta) \stackrel{1r}{\to} (\delta,\delta,0,\delta,\delta,0) \stackrel{1r}{\to} (\delta,\delta,\delta,\delta,0,0) \stackrel{1r}{\to} (\delta,\delta,0,\delta,0,\delta)$, which is 3-round iterative and holds with probability 2^{-60}. Starting from an odd-numbered round i, the following seven subkeys have to be weak: $Z_1^{(i)}$, $Z_9^{(i)}$, $Z_2^{(i+1)}$, $Z_4^{(i+1)}$, $Z_1^{(i+2)}$, $Z_3^{(i+2)}$ and $Z_9^{(i+2)}$, which means $7/3 \approx 2.33$ weak subkeys/round. Starting from an even-numbered round i, the following nine subkeys have to be weak: $Z_2^{(i)}$, $Z_4^{(i)}$, $Z_6^{(i)}$, $Z_9^{(i)}$, $Z_1^{(i+1)}$, $Z_5^{(i+1)}$, $Z_2^{(i+2)}$, $Z_4^{(i+2)}$ and $Z_9^{(i+2)}$, which means 3 weak subkeys/round.

(11) $(\delta,\delta,\delta,\delta,\delta,\delta) \stackrel{1r}{\to} (\delta,\delta,\delta,\delta,\delta,\delta)$, which is 1-round iterative and holds with probability 1. Starting from an odd-numbered round i, the following 3 subkeys have to be weak: $Z_1^{(i)}$, $Z_3^{(i)}$ and $Z_5^{(i)}$, which means 3 weak subkeys/round. Starting from an even-numbered round i, the following 3 subkeys have to be weak: $Z_2^{(i)}$, $Z_4^{(i)}$ and $Z_6^{(i)}$, which means 3 weak subkeys/round.

Starting from the first round, $i = 1$, the characteristic that requires the least amount of weak subkeys is (2): only 1.33 weak subkeys/round on average. Although it is 3-round iterative, it holds with probability 2^{-60}. Characteristic (11), though requiring 3 weak subkeys per round, has the advantage of holding with probability one (if all weak-subkey restrictions are satisfied).

Unlike the key schedule algorithms of PES and IDEA, the key schedule of MESH-96 does not consist of a fixed bit permutation, but is rather more involved. See Sect. 2.3.4. Experiments with mini-versions of the key schedule for a scaled-down version of MESH-96 operating on 24-bit blocks and 48-bit keys indicated that each weak subkey restriction (with the three most significant bits equal to zero) is satisfied for a fraction of 2^{-3} of the key space. Assuming this behavior holds for each weak subkey independently, and for larger key schedules as well, such as the one for MESH-96, then t weak subkeys would apply to a fraction of $2^{192-t\cdot 15}$ of the key space, or a weak-key class of this size. Thus, for a nonempty weak-key class, t must satisfy $192 - t \cdot 15 \geq 0$, that is, $t \leq 12$. The estimated weak-key class size is 2^{12}.

If characteristic (2) is repeated three times across nine rounds, the resulting characteristic will require exactly 12 weak subkeys, which means an estimated weak-key class of 2^{12}. The corresponding probability becomes 2^{-180}. For this characteristic, one could produce up to $2^{4\cdot 16} \cdot 2^{32} \cdot (2^{32} - 1)/2 \approx 2^{125}$ plaintext pairs, taking into account four trivial difference words and two nontrivial difference words. This is not enough to obtain a single right pair.

It means that this characteristic can be repeated at most twice, and be used to distinguish at most 6-round MESH-96 from a random permutation, under the appropriate weak-key assumptions. The weak-key class size is estimated as $2^{192-8\cdot 15} = 2^{72}$. The probability across six rounds is 2^{-120}.

If we use characteristic (11) instead, then, we can reach at most $12/3 = 4$ rounds, given the upper-bound on the estimated number of weak subkeys.

This differential analysis using characteristics also allows us to study the diffusion power of the MA half-round of MESH-96 with respect to modern tools such as the (differential) branch number. See Sect. B in the appendix.

The set of characteristics in Table 3.14 and Table 3.15 means that the differential branch number *per round* of MESH-96 is only 4 (under weak-key assumptions). For instance, the wordwise Hamming Weight of the input and output round differences of the 1-round characteristic

$$(\delta, 0, 0, \delta, 0, 0) \stackrel{1r}{\rightarrow} (\delta, \delta, 0, 0, 0, 0)$$

add up to 4. So, even though MESH-96 provides full diffusion in a single round, the previous analysis shows that its differential branch number is much lower than ideal. For a 6-word block, such as in MESH-96, an MDS code using a 6×6 MDS matrix would provide a branch number of $1 + 6 = 7$. See Appendix, Sect. B.1.

3.5.5 DC of MESH-128 Under Weak-Key Assumptions

The following differential analysis of MESH-128 is more extensive and more detailed than [151]. The strategy is to build the characteristics in a bottom-up fashion, from the smallest components: $\odot$, $\boxplus$ and $\oplus$, up to an MA-box, up to a half-round, further to a full round, and finally to multiple rounds.

The difference operator is bitwise exclusive-or.

Across unkeyed $\odot$ and $\boxplus$, the propagation of 16-bit difference words follows from Table 3.12 and Table 3.13, respectively.

The next step is to obtain all difference patterns across the MA-box of MESH-128, under the corresponding weak-subkey restrictions.

Table 3.16 lists all difference patterns across the MA-box of MESH-128. The difference operator is bitwise exclusive-or and the difference is wordwise. The only 16-bit difference values used are $\delta = 8000_x$ and 0000_x. The difference 8000_x propagates across the modular addition with probability one, since it affects only the most significant bit (and the last carry-out bit is truncated due to modular reduction.

The next step is to obtain from Table 3.16 all nontrivial 1-round differential characteristics and the corresponding weak-key assumptions.

Table 3.17 through Table 3.24 list all 255 nontrivial 1-round differential characteristics of MESH-128, under weak-key assumptions, using only wordwise differences 0000_x and 8000_x. These tables provide an exhaustive analysis of how characteristics propagate across a single round of MESH-128.

Multiple-round characteristics are obtained by concatenating 1-round characteristics, such that the output difference of one characteristic matches the input difference of the next characteristic.

Table 3.16 Wordwise difference patterns across the MA-box of MESH-128 ($\delta = 8000_x$).

Wordwise differences across the MA-box	# active ⊙	Probability	Weak subkey(s)
$(0,0,0,0) \to (0,0,0,0)$	0	1	—
$(0,0,0,\delta) \to (0,\delta,\delta,0)$	3	2^{-45}	$Z_{10}^{(i)}, Z_{11}^{(i)}$
$(0,0,\delta,0) \to (\delta,\delta,\delta,\delta)$	2	2^{-30}	$Z_{10}^{(i)}, Z_{12}^{(i)}$
$(0,0,\delta,\delta) \to (\delta,0,0,\delta)$	3	2^{-45}	$Z_{11}^{(i)}, Z_{12}^{(i)}$
$(0,\delta,0,0) \to (0,\delta,\delta,\delta)$	3	2^{-45}	$Z_{10}^{(i)}, Z_{11}^{(i)}, Z_{12}^{(i)}$
$(0,\delta,0,\delta) \to (0,0,0,\delta)$	3	2^{-45}	$Z_{12}^{(i)}$
$(0,\delta,\delta,0) \to (\delta,0,0,0)$	2	2^{-30}	$Z_{11}^{(i)}$
$(0,\delta,\delta,\delta) \to (\delta,\delta,\delta,0)$	4	2^{-60}	$Z_{10}^{(i)}$
$(\delta,0,0,0) \to (0,0,\delta,0)$	4	2^{-60}	$Z_{9}^{(i)}, Z_{10}^{(i)}$
$(\delta,0,0,\delta) \to (0,\delta,0,0)$	4	2^{-60}	$Z_{9}^{(i)}, Z_{11}^{(i)}$
$(\delta,0,\delta,0) \to (\delta,\delta,0,\delta)$	4	2^{-60}	$Z_{9}^{(i)}, Z_{12}^{(i)}$
$(\delta,0,\delta,\delta) \to (\delta,0,\delta,\delta)$	4	2^{-60}	$Z_{9}^{(i)}, Z_{10}^{(i)}, Z_{11}^{(i)}, Z_{12}^{(i)}$
$(\delta,\delta,0,0) \to (0,\delta,0,\delta)$	2	2^{-30}	$Z_{9}^{(i)}, Z_{11}^{(i)}, Z_{12}^{(i)}$
$(\delta,\delta,0,\delta) \to (0,0,\delta,\delta)$	3	2^{-45}	$Z_{9}^{(i)}, Z_{10}^{(i)}, Z_{12}^{(i)}$
$(\delta,\delta,\delta,0) \to (\delta,0,\delta,0)$	3	2^{-45}	$Z_{9}^{(i)}, Z_{10}^{(i)}, Z_{11}^{(i)}$
$(\delta,\delta,\delta,\delta) \to (\delta,\delta,0,0)$	4	2^{-60}	$Z_{9}^{(i)}$

From Table 3.17 through Table 3.24, the multiple-round characteristics are clustered into 23 characteristics covering 1, 2, 3, 5, 6, 10, 15 or 30 rounds:

(1) $(0,0,0,0,0,0,0,\delta) \stackrel{1r}{\to} (0,0,\delta,\delta,\delta,\delta,0,\delta) \stackrel{1r}{\to} (0,\delta,0,0,\delta,\delta,0,0) \stackrel{1r}{\to} (0,\delta,0,0,0,0,0,0) \stackrel{1r}{\to} (\delta,\delta,\delta,\delta,0,\delta,0,0) \stackrel{1r}{\to} (0,\delta,0,0,0,\delta,0,\delta) \stackrel{1r}{\to} (0,0,0,\delta,0,\delta,0,\delta) \stackrel{1r}{\to} (\delta,\delta,0,\delta,\delta,\delta,\delta,\delta) \stackrel{1r}{\to} (0,0,0,0,0,\delta,0,0) \stackrel{1r}{\to} (\delta,\delta,0,\delta,\delta,\delta,0,0) \stackrel{1r}{\to} (\delta,\delta,0,\delta,0,\delta,\delta,0) \stackrel{1r}{\to} (0,\delta,0,\delta,0,0,0,\delta) \stackrel{1r}{\to} (\delta,\delta,\delta,\delta,0,\delta,\delta,\delta) \stackrel{1r}{\to} (\delta,0,0,\delta,0,\delta,\delta,\delta) \stackrel{1r}{\to} (\delta,0,0,\delta,\delta,0,0,0) \stackrel{1r}{\to} (\delta,\delta,\delta,\delta,\delta,\delta,\delta,0) \stackrel{1r}{\to} (\delta,\delta,0,0,0,0,\delta,0) \stackrel{1r}{\to} (\delta,0,\delta,\delta,0,0,\delta,\delta) \stackrel{1r}{\to} (\delta,0,\delta,\delta,\delta,\delta,\delta,\delta) \stackrel{1r}{\to} (0,0,0,0,\delta,0,\delta,\delta) \stackrel{1r}{\to} (\delta,0,\delta,\delta,\delta,0,\delta,0) \stackrel{1r}{\to} (\delta,\delta,\delta,0,\delta,0,\delta,0) \stackrel{1r}{\to} (0,0,\delta,0,0,0,0,0) \stackrel{1r}{\to} (\delta,\delta,\delta,\delta,\delta,0,\delta,\delta) \stackrel{1r}{\to} (0,0,\delta,0,0,0,\delta,\delta) \stackrel{1r}{\to} (0,0,\delta,0,\delta,0,0,\delta) \stackrel{1r}{\to} (\delta,0,\delta,0,\delta,\delta,\delta,0) \stackrel{1r}{\to} (0,0,0,0,\delta,0,0,0) \stackrel{1r}{\to} (0,\delta,\delta,0,\delta,0,0,0) \stackrel{1r}{\to} (0,\delta,\delta,0,0,\delta,\delta,\delta) \stackrel{1r}{\to} (0,0,0,0,0,0,0,\delta)$, which is 30-round iterative and holds with probability 2^{-1440}.
Starting from an odd-numbered round i, the following 143 subkeys have to be weak: $Z_8^{(i)}$, $Z_{10}^{(i)}$, $Z_{11}^{(i)}$, $Z_4^{(i+1)}$, $Z_5^{(i+1)}$, $Z_9^{(i+1)}$, $Z_{10}^{(i+1)}$, $Z_{11}^{(i+1)}$, $Z_6^{(i+2)}$, $Z_9^{(i+2)}$, $Z_{10}^{(i+2)}$, $Z_2^{(i+3)}$, $Z_{10}^{(i+3)}$, $Z_{11}^{(i+3)}$, $Z_{12}^{(i+3)}$, $Z_1^{(i+4)}$, $Z_3^{(i+4)}$, $Z_6^{(i+4)}$, $Z_9^{(i+4)}$, $Z_{10}^{(i+4)}$, $Z_{11}^{(i+4)}$, $Z_{12}^{(i+4)}$, $Z_2^{(i+5)}$, $Z_{10}^{(i+5)}$, $Z_{11}^{(i+5)}$, $Z_6^{(i+6)}$, $Z_8^{(i+6)}$, $Z_{10}^{(i+6)}$, $Z_{11}^{(i+6)}$, $Z_{12}^{(i+6)}$, $Z_2^{(i+7)}$, $Z_4^{(i+7)}$, $Z_5^{(i+7)}$, $Z_7^{(i+7)}$, $Z_{10}^{(i+7)}$, $Z_{12}^{(i+7)}$, $Z_6^{(i+8)}$, $Z_{10}^{(i+8)}$, $Z_{11}^{(i+8)}$, $Z_{12}^{(i+8)}$, $Z_2^{(i+9)}$, $Z_4^{(i+9)}$, $Z_5^{(i+9)}$, $Z_{10}^{(i+9)}$, $Z_{11}^{(i+9)}$, $Z_1^{(i+10)}$, $Z_6^{(i+10)}$, $Z_9^{(i+10)}$, $Z_{10}^{(i+10)}$, $Z_{11}^{(i+10)}$, $Z_{12}^{(i+10)}$, $Z_2^{(i+11)}$, $Z_4^{(i+11)}$, $Z_{10}^{(i+11)}$, $Z_{11}^{(i+11)}$, $Z_{12}^{(i+11)}$, $Z_1^{(i+12)}$, $Z_3^{(i+12)}$, $Z_6^{(i+12)}$, $Z_8^{(i+12)}$, $Z_9^{(i+12)}$, $Z_{10}^{(i+12)}$, $Z_4^{(i+13)}$,

Table 3.17 Nontrivial 1-round characteristics for MESH-128 (part I).

1-round characteristic	Weak subkeys $Z_j^{(i)}$ (i-th round)		Prob.
	j value (odd i)	j value (even i)	
$(0,0,0,0,0,0,0,\delta) \xrightarrow{1r} (0,0,\delta,\delta,\delta,\delta,0,\delta)$	8; 10; 11	10; 11	2^{-45}
$(0,0,0,0,0,0,\delta,0) \xrightarrow{1r} (\delta,\delta,\delta,0,\delta,\delta,\delta,\delta)$	10; 12	7; 10; 12	2^{-30}
$(0,0,0,0,0,0,\delta,\delta) \xrightarrow{1r} (\delta,\delta,0,\delta,0,0,\delta,0)$	8; 11; 12	7; 11; 12	2^{-45}
$(0,0,0,0,0,\delta,0,0) \xrightarrow{1r} (\delta,\delta,0,\delta,\delta,\delta,0,0)$	6; 10; 11; 12	10; 11; 12	2^{-45}
$(0,0,0,0,0,\delta,0,\delta) \xrightarrow{1r} (\delta,\delta,\delta,0,0,0,0,\delta)$	6; 8; 12	12	2^{-45}
$(0,0,0,0,0,\delta,\delta,0) \xrightarrow{1r} (0,0,\delta,\delta,0,0,\delta,\delta)$	6; 11	7; 11	2^{-30}
$(0,0,0,0,0,\delta,\delta,\delta) \xrightarrow{1r} (0,0,0,0,\delta,\delta,\delta,0)$	6; 8; 10	7; 10	2^{-60}
$(0,0,0,0,\delta,0,0,0) \xrightarrow{1r} (0,\delta,\delta,0,\delta,0,0,0)$	9; 10	5; 9; 10	2^{-60}
$(0,0,0,0,\delta,0,0,\delta) \xrightarrow{1r} (0,\delta,0,\delta,0,\delta,0,\delta)$	8; 9; 11	5; 9; 11	2^{-60}
$(0,0,0,0,\delta,0,\delta,0) \xrightarrow{1r} (\delta,0,0,0,0,\delta,\delta,\delta)$	9; 12	5; 7; 9; 12	2^{-60}
$(0,0,0,0,\delta,0,\delta,\delta) \xrightarrow{1r} (\delta,0,\delta,\delta,\delta,0,\delta,0)$	8; 9; 10; 11; 12	5; 7; 9; 10; 11; 12	2^{-60}
$(0,0,0,0,\delta,\delta,0,0) \xrightarrow{1r} (\delta,0,\delta,\delta,0,\delta,0,0)$	6; 9; 11; 12	5; 9; 11; 12	2^{-30}
$(0,0,0,0,\delta,\delta,0,\delta) \xrightarrow{1r} (\delta,0,0,0,\delta,0,0,\delta)$	6; 8; 9; 10; 12	5; 9; 10; 12	2^{-45}
$(0,0,0,0,\delta,\delta,\delta,0) \xrightarrow{1r} (0,\delta,0,\delta,\delta,0,\delta,\delta)$	6; 9; 10; 11	5; 7; 9; 10; 11	2^{-45}
$(0,0,0,0,\delta,\delta,\delta,\delta) \xrightarrow{1r} (0,\delta,\delta,0,0,\delta,\delta,0)$	6; 8; 9	5; 7; 9	2^{-60}
$(0,0,0,\delta,0,0,0,0) \xrightarrow{1r} (0,0,\delta,\delta,\delta,\delta,\delta,0)$	10; 11	4; 10; 11	2^{-45}
$(0,0,0,\delta,0,0,0,\delta) \xrightarrow{1r} (0,0,0,0,0,0,\delta,\delta)$	8	4	1
$(0,0,0,\delta,0,0,\delta,0) \xrightarrow{1r} (\delta,\delta,0,\delta,0,0,0,\delta)$	11; 12	4; 7; 11; 12	2^{-45}
$(0,0,0,\delta,0,0,\delta,\delta) \xrightarrow{1r} (\delta,\delta,\delta,0,\delta,\delta,0,0)$	8; 10; 12	4; 7; 10; 12	2^{-30}
$(0,0,0,\delta,0,\delta,0,0) \xrightarrow{1r} (\delta,\delta,\delta,0,0,0,\delta,0)$	6; 12	4; 12	2^{-45}
$(0,0,0,\delta,0,\delta,0,\delta) \xrightarrow{1r} (\delta,\delta,0,\delta,\delta,\delta,\delta,\delta)$	6; 8; 10; 11; 12	4; 10; 11; 12	2^{-45}
$(0,0,0,\delta,0,\delta,\delta,0) \xrightarrow{1r} (0,0,0,0,\delta,\delta,0,\delta)$	6; 10	4; 7; 10	2^{-60}
$(0,0,0,\delta,0,\delta,\delta,\delta) \xrightarrow{1r} (0,0,\delta,\delta,0,0,0,0)$	6; 8; 11	4; 7; 11	2^{-30}
$(0,0,0,\delta,\delta,0,0,0) \xrightarrow{1r} (0,\delta,0,\delta,0,\delta,\delta,0)$	9; 11	4; 5; 9; 11	2^{-60}
$(0,0,0,\delta,\delta,0,0,\delta) \xrightarrow{1r} (0,\delta,\delta,0,\delta,0,\delta,\delta)$	8; 9; 10	4; 5; 9; 10	2^{-60}
$(0,0,0,\delta,\delta,0,\delta,0) \xrightarrow{1r} (\delta,0,\delta,\delta,\delta,0,0,\delta)$	9; 10; 11; 12	4; 5; 7; 9; 10; 11; 12	2^{-60}
$(0,0,0,\delta,\delta,0,\delta,\delta) \xrightarrow{1r} (\delta,0,0,0,0,\delta,0,0)$	8; 9; 12	4; 5; 7; 9; 12	2^{-60}
$(0,0,0,\delta,\delta,\delta,0,0) \xrightarrow{1r} (\delta,0,0,0,\delta,0,\delta,0)$	6; 9; 10; 12	4; 5; 9; 10; 12	2^{-45}
$(0,0,0,\delta,\delta,\delta,0,\delta) \xrightarrow{1r} (\delta,0,\delta,\delta,0,\delta,\delta,\delta)$	6; 8; 9; 11; 12	4; 5; 9; 11; 12	2^{-30}
$(0,0,0,\delta,\delta,\delta,\delta,0) \xrightarrow{1r} (0,\delta,\delta,0,0,\delta,0,\delta)$	6; 9	4; 5; 7; 9	2^{-60}
$(0,0,0,\delta,\delta,\delta,\delta,\delta) \xrightarrow{1r} (0,\delta,0,\delta,\delta,0,0,0)$	6; 8; 9; 10; 11	4; 5; 7; 9; 10; 11	2^{-45}
$(0,0,\delta,0,0,0,0,0) \xrightarrow{1r} (\delta,\delta,\delta,\delta,\delta,0,\delta,\delta)$	3; 10; 12	10; 12	2^{-30}

Table 3.18 Nontrivial 1-round characteristics for MESH-128 (part II).

1-round characteristic	Weak subkeys $Z_j^{(i)}$ (i-th round)		Prob.
	j value (odd i)	j value (even i)	
$(0,0,\delta,0,0,0,0,\delta) \xrightarrow{1r} (\delta,\delta,0,0,0,\delta,\delta,0)$	3; 8; 11; 12	11; 12	2^{-45}
$(0,0,\delta,0,0,0,\delta,0) \xrightarrow{1r} (0,0,0,\delta,0,\delta,0,0)$	3	7	1
$(0,0,\delta,0,0,0,\delta,\delta) \xrightarrow{1r} (0,0,\delta,0,\delta,0,0,\delta)$	3; 8; 10; 11	7; 10; 11	2^{-45}
$(0,0,\delta,0,0,\delta,0,0) \xrightarrow{1r} (0,0,\delta,0,0,\delta,\delta,\delta)$	3; 6; 11	11	2^{-30}
$(0,0,\delta,0,0,\delta,0,\delta) \xrightarrow{1r} (0,0,0,\delta,\delta,0,\delta,0)$	3; 6; 8; 10	10	2^{-60}
$(0,0,\delta,0,0,\delta,\delta,0) \xrightarrow{1r} (\delta,\delta,0,0,\delta,0,0,0)$	3; 6; 10; 11; 12	7; 10; 11; 12	2^{-45}
$(0,0,\delta,0,0,\delta,\delta,\delta) \xrightarrow{1r} (\delta,\delta,\delta,\delta,0,\delta,0,\delta)$	3; 6; 8; 12	7; 12	2^{-45}
$(0,0,\delta,0,\delta,0,0,0) \xrightarrow{1r} (\delta,0,0,\delta,0,0,\delta,\delta)$	3; 9; 12	5; 9; 12	2^{-60}
$(0,0,\delta,0,\delta,0,0,\delta) \xrightarrow{1r} (\delta,0,\delta,0,\delta,\delta,\delta,0)$	3; 8; 9; 10; 11; 12	5; 9; 10; 11; 12	2^{-60}
$(0,0,\delta,0,\delta,0,\delta,0) \xrightarrow{1r} (0,\delta,\delta,\delta,\delta,\delta,0,0)$	3; 9; 10	5; 7; 9; 10	2^{-60}
$(0,0,\delta,0,\delta,0,\delta,\delta) \xrightarrow{1r} (0,\delta,0,0,0,0,0,\delta)$	3; 8; 9; 11	5; 7; 9; 11	2^{-60}
$(0,0,\delta,0,\delta,\delta,0,0) \xrightarrow{1r} (0,\delta,0,0,\delta,\delta,\delta,\delta)$	3; 6; 9; 10; 11	5; 9; 10; 11	2^{-45}
$(0,0,\delta,0,\delta,\delta,0,\delta) \xrightarrow{1r} (0,\delta,\delta,\delta,0,0,\delta,0)$	3; 6; 8; 9	5; 9	2^{-60}
$(0,0,\delta,0,\delta,\delta,\delta,0) \xrightarrow{1r} (\delta,0,\delta,0,0,0,0,0)$	3; 6; 9; 11; 12	5; 7; 9; 11; 12	2^{-30}
$(0,0,\delta,0,\delta,\delta,\delta,\delta) \xrightarrow{1r} (\delta,0,0,\delta,\delta,\delta,0,\delta)$	3; 6; 8; 9; 10; 12	5; 7; 9; 10; 12	2^{-45}
$(0,0,\delta,\delta,0,0,0,0) \xrightarrow{1r} (\delta,\delta,0,0,0,\delta,0,\delta)$	3; 11; 12	4; 11; 12	2^{-45}
$(0,0,\delta,\delta,0,0,0,\delta) \xrightarrow{1r} (\delta,\delta,\delta,\delta,\delta,0,0,0)$	3; 8; 10; 12	4; 10; 12	2^{-30}
$(0,0,\delta,\delta,0,0,\delta,0) \xrightarrow{1r} (0,0,\delta,0,\delta,0,\delta,0)$	3; 10; 11	4; 7; 10; 11	2^{-45}
$(0,0,\delta,\delta,0,0,\delta,\delta) \xrightarrow{1r} (0,0,0,\delta,0,\delta,\delta,\delta)$	3; 8	4; 7	1
$(0,0,\delta,\delta,0,\delta,0,0) \xrightarrow{1r} (0,0,0,\delta,\delta,0,0,\delta)$	3; 6; 10	4; 10	2^{-60}
$(0,0,\delta,\delta,0,\delta,0,\delta) \xrightarrow{1r} (0,0,\delta,0,0,\delta,0,0)$	3; 6; 8; 11	4; 11	2^{-30}
$(0,0,\delta,\delta,0,\delta,\delta,0) \xrightarrow{1r} (\delta,\delta,\delta,\delta,0,\delta,\delta,0)$	3; 6; 12	4; 7; 12	2^{-45}
$(0,0,\delta,\delta,0,\delta,\delta,\delta) \xrightarrow{1r} (\delta,\delta,0,0,\delta,0,\delta,\delta)$	3; 6; 8; 10; 11; 12	4; 7; 10; 11; 12	2^{-45}
$(0,0,\delta,\delta,\delta,0,0,0) \xrightarrow{1r} (\delta,0,\delta,0,\delta,\delta,0,\delta)$	3; 9; 10; 11; 12	4; 5; 9; 10; 11; 12	2^{-60}
$(0,0,\delta,\delta,\delta,0,0,\delta) \xrightarrow{1r} (\delta,0,0,\delta,0,0,0,0)$	3; 8; 9; 12	4; 5; 9; 12	2^{-60}
$(0,0,\delta,\delta,\delta,0,\delta,0) \xrightarrow{1r} (0,\delta,0,0,0,0,\delta,0)$	3; 9; 11	4; 5; 7; 9; 11	2^{-60}
$(0,0,\delta,\delta,\delta,0,\delta,\delta) \xrightarrow{1r} (0,\delta,\delta,\delta,\delta,\delta,\delta,\delta)$	3; 8; 9; 10	4; 5; 7; 9; 10	2^{-60}
$(0,0,\delta,\delta,\delta,\delta,0,0) \xrightarrow{1r} (0,\delta,\delta,\delta,0,0,0,\delta)$	3; 6; 9	4; 5; 9	2^{-60}
$(0,0,\delta,\delta,\delta,\delta,0,\delta) \xrightarrow{1r} (0,\delta,0,0,\delta,\delta,0,0)$	3; 6; 8; 9; 10; 11	4; 5; 9; 10; 11	2^{-45}
$(0,0,\delta,\delta,\delta,\delta,\delta,0) \xrightarrow{1r} (\delta,0,0,\delta,\delta,\delta,\delta,0)$	3; 6; 9; 10; 12	4; 5; 7; 9; 10; 12	2^{-45}
$(0,0,\delta,\delta,\delta,\delta,\delta,\delta) \xrightarrow{1r} (\delta,0,\delta,0,0,0,\delta,\delta)$	3; 6; 8; 9; 11; 12	4; 5; 7; 9; 11; 12	2^{-30}
$(0,\delta,0,0,0,0,0,0) \xrightarrow{1r} (\delta,\delta,\delta,\delta,0,\delta,0,0)$	10; 11; 12	2; 10; 11; 12	2^{-45}

Table 3.19 Nontrivial 1-round characteristics for MESH-128 (part III).

1-round characteristic	Weak subkeys $Z_j^{(i)}$ (i-th round)		Prob.
	j value (odd i)	j value (even i)	
$(0,\delta,0,0,0,0,0,\delta) \xrightarrow{1r} (\delta,\delta,0,0,\delta,0,0,\delta)$	8; 12	2; 12	2^{-45}
$(0,\delta,0,0,0,0,\delta,0) \xrightarrow{1r} (0,0,0,\delta,\delta,0,\delta,\delta)$	11	2; 7; 11	2^{-30}
$(0,\delta,0,0,0,0,\delta,\delta) \xrightarrow{1r} (0,0,\delta,0,0,\delta,\delta,0)$	8; 10	2; 7; 10	2^{-60}
$(0,\delta,0,0,0,\delta,0,0) \xrightarrow{1r} (0,0,\delta,0,\delta,0,0,0)$	6	2	1
$(0,\delta,0,0,0,\delta,0,\delta) \xrightarrow{1r} (0,0,0,\delta,0,\delta,0,\delta)$	6; 8; 10; 11	2; 10; 11	2^{-45}
$(0,\delta,0,0,0,\delta,\delta,0) \xrightarrow{1r} (\delta,\delta,0,0,0,\delta,\delta,\delta)$	6; 10; 12	2; 7; 10; 12	2^{-30}
$(0,\delta,0,0,0,\delta,\delta,\delta) \xrightarrow{1r} (\delta,\delta,\delta,\delta,\delta,0,\delta,0)$	6; 8; 11; 12	2; 7; 11; 12	2^{-45}
$(0,\delta,0,0,\delta,0,0,0) \xrightarrow{1r} (\delta,0,0,\delta,\delta,\delta,0,0)$	9; 11; 12	2; 5; 9; 11; 12	2^{-30}
$(0,\delta,0,0,\delta,0,0,\delta) \xrightarrow{1r} (\delta,0,\delta,0,0,0,0,\delta)$	8; 9; 10; 12	2; 5; 9; 10; 12	2^{-45}
$(0,\delta,0,0,\delta,0,\delta,0) \xrightarrow{1r} (0,\delta,\delta,\delta,0,0,\delta,\delta)$	9; 10; 11	2; 5; 7; 9; 10; 11	2^{-45}
$(0,\delta,0,0,\delta,0,\delta,\delta) \xrightarrow{1r} (0,\delta,0,0,\delta,\delta,\delta,0)$	8; 9	2; 5; 7; 9	2^{-60}
$(0,\delta,0,0,\delta,\delta,0,0) \xrightarrow{1r} (0,\delta,0,0,0,0,0,0)$	6; 9; 10	2; 5; 9; 10	2^{-60}
$(0,\delta,0,0,\delta,\delta,0,\delta) \xrightarrow{1r} (0,\delta,\delta,\delta,\delta,\delta,0,\delta)$	6; 8; 9; 11	2; 5; 9; 11	2^{-60}
$(0,\delta,0,0,\delta,\delta,\delta,0) \xrightarrow{1r} (\delta,0,\delta,0,\delta,\delta,\delta,\delta)$	6; 9; 12	2; 5; 7; 9; 12	2^{-60}
$(0,\delta,0,0,\delta,\delta,\delta,\delta) \xrightarrow{1r} (\delta,0,0,\delta,0,0,\delta,0)$	6; 8; 9; 10; 11; 12	2; 5; 7; 9; 10; 11; 12	2^{-60}
$(0,\delta,0,\delta,0,0,0,0) \xrightarrow{1r} (\delta,\delta,0,0,\delta,0,\delta,0)$	12	2; 4; 12	2^{-45}
$(0,\delta,0,\delta,0,0,0,\delta) \xrightarrow{1r} (\delta,\delta,\delta,\delta,0,\delta,\delta,\delta)$	8; 10; 11; 12	2; 4; 10; 11; 12	2^{-45}
$(0,\delta,0,\delta,0,0,\delta,0) \xrightarrow{1r} (0,0,\delta,0,0,\delta,0,\delta)$	10	2; 4; 7; 10	2^{-60}
$(0,\delta,0,\delta,0,0,\delta,\delta) \xrightarrow{1r} (0,0,0,\delta,\delta,0,0,0)$	8; 11	2; 4; 7; 11	2^{-30}
$(0,\delta,0,\delta,0,\delta,0,0) \xrightarrow{1r} (0,0,0,\delta,0,\delta,\delta,0)$	6; 10; 11	2; 4; 10; 11	2^{-45}
$(0,\delta,0,\delta,0,\delta,0,\delta) \xrightarrow{1r} (0,0,\delta,0,\delta,0,\delta,\delta)$	6; 8	2; 4	1
$(0,\delta,0,\delta,0,\delta,\delta,0) \xrightarrow{1r} (\delta,\delta,\delta,\delta,\delta,0,0,\delta)$	6; 11; 12	2; 4; 7; 11; 12	2^{-45}
$(0,\delta,0,\delta,0,\delta,\delta,\delta) \xrightarrow{1r} (\delta,\delta,0,0,0,\delta,0,0)$	6; 8; 10; 12	2; 4; 7; 10; 12	2^{-30}
$(0,\delta,0,\delta,\delta,0,0,0) \xrightarrow{1r} (\delta,0,\delta,0,0,0,\delta,0)$	9; 10; 12	2; 4; 5; 9; 10; 12	2^{-45}
$(0,\delta,0,\delta,\delta,0,0,\delta) \xrightarrow{1r} (\delta,0,0,\delta,\delta,\delta,\delta,\delta)$	8; 9; 11; 12	2; 4; 5; 9; 11; 12	2^{-30}
$(0,\delta,0,\delta,\delta,0,\delta,0) \xrightarrow{1r} (0,\delta,0,0,\delta,\delta,0,\delta)$	9	2; 4; 5; 7; 9	2^{-60}
$(0,\delta,0,\delta,\delta,0,\delta,\delta) \xrightarrow{1r} (0,\delta,\delta,\delta,0,0,0,0)$	8; 9; 10; 11	2; 4; 5; 7; 9; 10; 11	2^{-45}
$(0,\delta,0,\delta,\delta,\delta,0,0) \xrightarrow{1r} (0,\delta,\delta,\delta,\delta,\delta,\delta,0)$	6; 9; 11	2; 4; 5; 9; 11	2^{-60}
$(0,\delta,0,\delta,\delta,\delta,0,\delta) \xrightarrow{1r} (0,\delta,0,0,0,0,\delta,\delta)$	6; 8; 9; 10	2; 4; 5; 9; 10	2^{-60}
$(0,\delta,0,\delta,\delta,\delta,\delta,0) \xrightarrow{1r} (\delta,0,0,\delta,0,0,0,\delta)$	6; 9; 10; 11; 12	2; 4; 5; 7; 9; 10; 11; 12	2^{-60}
$(0,\delta,0,\delta,\delta,\delta,\delta,\delta) \xrightarrow{1r} (\delta,0,\delta,0,\delta,\delta,0,0)$	6; 8; 9; 12	2; 4; 5; 7; 9; 12	2^{-60}
$(0,\delta,\delta,0,0,0,0,0) \xrightarrow{1r} (0,0,0,0,\delta,\delta,\delta,\delta)$	3; 11	2; 11	2^{-30}

Table 3.20 Nontrivial 1-round characteristics for MESH-128 (part IV).

1-round characteristic	Weak subkeys $Z_j^{(i)}$ (i-th round)		Prob.
	j value (odd i)	j value (even i)	
$(0,\delta,\delta,0,0,0,0,\delta) \xrightarrow{1r} (0,0,\delta,\delta,0,0,\delta,0)$	3; 8; 10	2; 10	2^{-60}
$(0,\delta,\delta,0,0,0,\delta,0) \xrightarrow{1r} (\delta,\delta,\delta,0,0,0,0,0)$	3; 10; 11; 12	2; 7; 10; 11; 12	2^{-45}
$(0,\delta,\delta,0,0,0,\delta,\delta) \xrightarrow{1r} (\delta,\delta,0,\delta,\delta,\delta,0,\delta)$	3; 8; 12	2; 7; 12	2^{-45}
$(0,\delta,\delta,0,0,\delta,0,0) \xrightarrow{1r} (\delta,\delta,0,\delta,0,0,\delta,\delta)$	3; 6; 10; 12	2; 10; 12	2^{-30}
$(0,\delta,\delta,0,0,\delta,0,\delta) \xrightarrow{1r} (\delta,\delta,\delta,0,\delta,\delta,\delta,0)$	3; 6; 8; 11; 12	2; 11; 12	2^{-45}
$(0,\delta,\delta,0,0,\delta,\delta,0) \xrightarrow{1r} (0,0,\delta,\delta,\delta,\delta,0,0)$	3; 6	2; 7	1
$(0,\delta,\delta,0,0,\delta,\delta,\delta) \xrightarrow{1r} (0,0,0,0,0,0,0,\delta)$	3; 6; 8; 10; 11	2; 7; 10; 11	2^{-45}
$(0,\delta,\delta,0,\delta,0,0,0) \xrightarrow{1r} (0,\delta,\delta,0,0,\delta,\delta,\delta)$	3; 9; 10; 11	2; 5; 9; 10; 11	2^{-45}
$(0,\delta,\delta,0,\delta,0,0,\delta) \xrightarrow{1r} (0,\delta,0,\delta,\delta,0,\delta,0)$	3; 8; 9	2; 5; 9	2^{-60}
$(0,\delta,\delta,0,\delta,0,\delta,0) \xrightarrow{1r} (\delta,0,0,0,\delta,0,0,0)$	3; 9; 11; 12	2; 5; 7; 9; 11; 12	2^{-30}
$(0,\delta,\delta,0,\delta,0,\delta,\delta) \xrightarrow{1r} (\delta,0,\delta,\delta,0,\delta,0,\delta)$	3; 8; 9; 10; 12	2; 5; 7; 9; 10; 12	2^{-45}
$(0,\delta,\delta,0,\delta,\delta,0,0) \xrightarrow{1r} (\delta,0,\delta,\delta,\delta,0,\delta,\delta)$	3; 6; 9; 12	2; 5; 9; 12	2^{-60}
$(0,\delta,\delta,0,\delta,\delta,0,\delta) \xrightarrow{1r} (\delta,0,0,0,0,\delta,\delta,0)$	3; 6; 8; 9; 10; 11; 12	2; 5; 9; 10; 11; 12	2^{-60}
$(0,\delta,\delta,0,\delta,\delta,\delta,0) \xrightarrow{1r} (0,\delta,0,\delta,0,\delta,0,0)$	3; 6; 9; 10	2; 5; 7; 9; 10	2^{-60}
$(0,\delta,\delta,0,\delta,\delta,\delta,\delta) \xrightarrow{1r} (0,\delta,\delta,0,\delta,0,0,\delta)$	3; 6; 8; 9; 11	2; 5; 7; 9; 11	2^{-60}
$(0,\delta,\delta,\delta,0,0,0,0) \xrightarrow{1r} (0,0,\delta,\delta,0,0,0,\delta)$	3; 10	2; 4; 10	2^{-60}
$(0,\delta,\delta,\delta,0,0,0,\delta) \xrightarrow{1r} (0,0,0,0,\delta,\delta,0,0)$	3; 8; 11	2; 4; 11	2^{-30}
$(0,\delta,\delta,\delta,0,0,\delta,0) \xrightarrow{1r} (\delta,\delta,0,\delta,\delta,\delta,\delta,0)$	3; 12	2; 4; 7; 12	2^{-45}
$(0,\delta,\delta,\delta,0,0,\delta,\delta) \xrightarrow{1r} (\delta,\delta,\delta,0,0,0,\delta,\delta)$	3; 8; 10; 11; 12	2; 4; 7; 10; 11; 12	2^{-45}
$(0,\delta,\delta,\delta,0,\delta,0,0) \xrightarrow{1r} (\delta,\delta,\delta,0,\delta,\delta,0,\delta)$	3; 6; 11; 12	2; 4; 11; 12	2^{-45}
$(0,\delta,\delta,\delta,0,\delta,0,\delta) \xrightarrow{1r} (\delta,\delta,0,\delta,0,0,0,0)$	3; 6; 8; 10; 12	2; 4; 10; 12	2^{-30}
$(0,\delta,\delta,\delta,0,\delta,\delta,0) \xrightarrow{1r} (0,0,0,0,0,0,\delta,0)$	3; 6; 10; 11	2; 4; 7; 10; 11	2^{-45}
$(0,\delta,\delta,\delta,0,\delta,\delta,\delta) \xrightarrow{1r} (0,0,\delta,\delta,\delta,\delta,\delta,\delta)$	3; 6; 8	2; 4; 7	1
$(0,\delta,\delta,\delta,\delta,0,0,0) \xrightarrow{1r} (0,\delta,0,\delta,\delta,0,0,\delta)$	3; 9	2; 4; 5; 9	2^{-60}
$(0,\delta,\delta,\delta,\delta,0,0,\delta) \xrightarrow{1r} (0,\delta,\delta,0,0,\delta,0,0)$	3; 8; 9; 10; 11	2; 4; 5; 9; 10; 11	2^{-45}
$(0,\delta,\delta,\delta,\delta,0,\delta,0) \xrightarrow{1r} (\delta,0,\delta,\delta,0,\delta,\delta,0)$	3; 9; 10; 12	2; 4; 5; 7; 9; 10; 12	2^{-45}
$(0,\delta,\delta,\delta,\delta,0,\delta,\delta) \xrightarrow{1r} (\delta,0,0,0,\delta,0,\delta,\delta)$	3; 8; 9; 11; 12	2; 4; 5; 7; 9; 11; 12	2^{-30}
$(0,\delta,\delta,\delta,\delta,\delta,0,0) \xrightarrow{1r} (\delta,0,0,0,0,\delta,0,\delta)$	3; 6; 9; 10; 11; 12	2; 4; 5; 9; 10; 11; 12	2^{-60}
$(0,\delta,\delta,\delta,\delta,\delta,0,\delta) \xrightarrow{1r} (\delta,0,\delta,\delta,\delta,0,0,0)$	3; 6; 8; 9; 12	2; 4; 5; 9; 12	2^{-60}
$(0,\delta,\delta,\delta,\delta,\delta,\delta,0) \xrightarrow{1r} (0,\delta,\delta,0,\delta,0,\delta,0)$	3; 6; 9; 11	2; 4; 5; 7; 9; 11	2^{-60}
$(0,\delta,\delta,\delta,\delta,\delta,\delta,\delta) \xrightarrow{1r} (0,\delta,0,\delta,0,\delta,\delta,\delta)$	3; 6; 8; 9; 10	2; 4; 5; 7; 9; 10	2^{-60}
$(\delta,0,0,0,0,0,0,0) \xrightarrow{1r} (\delta,0,\delta,0,\delta,0,0,0)$	1; 9; 10	9; 10	2^{-60}

Table 3.21 Nontrivial 1-round characteristics for MESH-128 (part V).

1-round characteristic	Weak subkeys $Z_j^{(i)}$ (i-th round)		Prob.
	j value (odd i)	j value (even i)	
$(\delta,0,0,0,0,0,0,\delta) \xrightarrow{1r} (\delta,0,0,\delta,0,\delta,0,\delta)$	1; 8; 9; 11	9; 11	2^{-60}
$(\delta,0,0,0,0,0,\delta,0) \xrightarrow{1r} (0,\delta,0,0,0,\delta,\delta,\delta)$	1; 9; 12	7; 9; 12	2^{-60}
$(\delta,0,0,0,0,0,\delta,\delta) \xrightarrow{1r} (0,\delta,\delta,\delta,\delta,0,\delta,0)$	1; 8; 9; 10; 11; 12	7; 9; 10; 11; 12	2^{-60}
$(\delta,0,0,0,0,\delta,0,0) \xrightarrow{1r} (0,\delta,\delta,\delta,0,\delta,0,0)$	1; 6; 9; 11; 12	9; 11; 12	2^{-30}
$(\delta,0,0,0,0,\delta,0,\delta) \xrightarrow{1r} (0,\delta,0,0,\delta,0,0,\delta)$	1; 6; 8; 9; 10; 12	9; 10; 12	2^{-45}
$(\delta,0,0,0,0,\delta,\delta,0) \xrightarrow{1r} (\delta,0,0,\delta,\delta,0,\delta,\delta)$	1; 6; 9; 10; 11	7; 9; 10; 11	2^{-45}
$(\delta,0,0,0,0,\delta,\delta,\delta) \xrightarrow{1r} (\delta,0,\delta,0,0,\delta,\delta,0)$	1; 6; 8; 9	7; 9	2^{-60}
$(\delta,0,0,0,\delta,0,0,0) \xrightarrow{1r} (\delta,\delta,0,0,0,0,0,0)$	1	5	1
$(\delta,0,0,0,\delta,0,0,\delta) \xrightarrow{1r} (\delta,\delta,\delta,\delta,\delta,\delta,0,\delta)$	1; 8; 10; 11	5; 10; 11	2^{-45}
$(\delta,0,0,0,\delta,0,\delta,0) \xrightarrow{1r} (0,0,\delta,0,\delta,\delta,\delta,\delta)$	1; 10; 12	5; 7; 10; 12	2^{-30}
$(\delta,0,0,0,\delta,0,\delta,\delta) \xrightarrow{1r} (0,0,0,\delta,0,0,\delta,0)$	1; 8; 11; 12	5; 7; 11; 12	2^{-45}
$(\delta,0,0,0,\delta,\delta,0,0) \xrightarrow{1r} (0,0,0,\delta,\delta,\delta,0,0)$	1; 6; 10; 11; 12	5; 10; 11; 12	2^{-45}
$(\delta,0,0,0,\delta,\delta,0,\delta) \xrightarrow{1r} (0,0,\delta,0,0,0,0,\delta)$	1; 6; 8; 12	5; 12	2^{-45}
$(\delta,0,0,0,\delta,\delta,\delta,0) \xrightarrow{1r} (\delta,\delta,\delta,\delta,0,0,\delta,\delta)$	1; 6; 11	5; 7; 11	2^{-30}
$(\delta,0,0,0,\delta,\delta,\delta,\delta) \xrightarrow{1r} (\delta,\delta,0,0,\delta,\delta,\delta,0)$	1; 6; 8; 10	5; 7; 10	2^{-60}
$(\delta,0,0,\delta,0,0,0,0) \xrightarrow{1r} (\delta,0,0,\delta,0,\delta,\delta,0)$	1; 9; 11	4; 9; 11	2^{-60}
$(\delta,0,0,\delta,0,0,0,\delta) \xrightarrow{1r} (\delta,0,\delta,0,\delta,0,\delta,\delta)$	1; 8; 9; 10	4; 9; 10	2^{-60}
$(\delta,0,0,\delta,0,0,\delta,0) \xrightarrow{1r} (0,\delta,\delta,\delta,\delta,0,0,\delta)$	1; 9; 10; 11; 12	4; 7; 9; 10; 11; 12	2^{-60}
$(\delta,0,0,\delta,0,0,\delta,\delta) \xrightarrow{1r} (0,\delta,0,0,0,\delta,0,0)$	1; 8; 9; 12	4; 7; 9; 12	2^{-60}
$(\delta,0,0,\delta,0,\delta,0,0) \xrightarrow{1r} (0,\delta,0,0,\delta,0,\delta,0)$	1; 6; 9; 10; 12	4; 9; 10; 12	2^{-45}
$(\delta,0,0,\delta,0,\delta,0,\delta) \xrightarrow{1r} (0,\delta,\delta,\delta,0,\delta,\delta,\delta)$	1; 6; 8; 9; 11; 12	4; 9; 11; 12	2^{-30}
$(\delta,0,0,\delta,0,\delta,\delta,0) \xrightarrow{1r} (\delta,0,\delta,0,0,\delta,0,\delta)$	1; 6; 9	4; 7; 9	2^{-60}
$(\delta,0,0,\delta,0,\delta,\delta,\delta) \xrightarrow{1r} (\delta,0,0,\delta,\delta,0,0,0)$	1; 6; 8; 9; 10; 11	4; 7; 9; 10; 11	2^{-45}
$(\delta,0,0,\delta,\delta,0,0,0) \xrightarrow{1r} (\delta,\delta,\delta,\delta,\delta,\delta,\delta,0)$	1; 10; 11	4; 5; 10; 11	2^{-45}
$(\delta,0,0,\delta,\delta,0,0,\delta) \xrightarrow{1r} (\delta,\delta,0,0,0,0,\delta,\delta)$	1; 8	4; 5	1
$(\delta,0,0,\delta,\delta,0,\delta,0) \xrightarrow{1r} (0,0,0,\delta,0,0,0,\delta)$	1; 11; 12	4; 5; 7; 11; 12	2^{-45}
$(\delta,0,0,\delta,\delta,0,\delta,\delta) \xrightarrow{1r} (0,0,\delta,0,\delta,\delta,0,0)$	1; 8; 10; 12	4; 5; 7; 10; 12	2^{-30}
$(\delta,0,0,\delta,\delta,\delta,0,0) \xrightarrow{1r} (0,0,\delta,0,0,0,\delta,0)$	1; 6; 12	4; 5; 12	2^{-45}
$(\delta,0,0,\delta,\delta,\delta,0,\delta) \xrightarrow{1r} (0,0,0,\delta,\delta,\delta,\delta,\delta)$	1; 6; 8; 10; 11; 12	4; 5; 10; 11; 12	2^{-45}
$(\delta,0,0,\delta,\delta,\delta,\delta,0) \xrightarrow{1r} (\delta,\delta,0,0,\delta,\delta,0,\delta)$	1; 6; 10	4; 5; 7; 10	2^{-60}
$(\delta,0,0,\delta,\delta,\delta,\delta,\delta) \xrightarrow{1r} (\delta,\delta,\delta,\delta,0,0,0,0)$	1; 6; 8; 11	4; 5; 7; 11	2^{-30}
$(\delta,0,\delta,0,0,0,0,0) \xrightarrow{1r} (0,\delta,0,\delta,0,0,\delta,\delta)$	1; 3; 9; 12	9; 12	2^{-60}

Table 3.22 Nontrivial 1-round characteristics for MESH-128 (part VI).

1-round characteristic	Weak subkeys $Z_j^{(i)}$ (i-th round)		Prob.
	j value (odd i)	j value (even i)	
$(\delta,0,\delta,0,0,0,0,\delta) \xrightarrow{1r} (0,\delta,\delta,0,\delta,\delta,\delta,0)$	1; 3; 8; 9; 10; 11; 12	9; 10; 11; 12	2^{-60}
$(\delta,0,\delta,0,0,0,\delta,0) \xrightarrow{1r} (\delta,0,\delta,\delta,\delta,\delta,0,0)$	1; 3; 9; 10	7; 9; 10	2^{-60}
$(\delta,0,\delta,0,0,0,\delta,\delta) \xrightarrow{1r} (\delta,0,0,0,0,0,0,\delta)$	1; 3; 8; 9; 11	7; 9; 11	2^{-60}
$(\delta,0,\delta,0,0,\delta,0,0) \xrightarrow{1r} (\delta,0,0,0,\delta,\delta,\delta,\delta)$	1; 3; 6; 9; 10; 11	9; 10; 11	2^{-45}
$(\delta,0,\delta,0,0,\delta,0,\delta) \xrightarrow{1r} (\delta,0,\delta,\delta,0,0,\delta,0)$	1; 3; 6; 8; 9	9	2^{-60}
$(\delta,0,\delta,0,0,\delta,\delta,0) \xrightarrow{1r} (0,\delta,\delta,0,0,0,0,0)$	1; 3; 6; 9; 11; 12	7; 9; 11; 12	2^{-30}
$(\delta,0,\delta,0,0,\delta,\delta,\delta) \xrightarrow{1r} (0,\delta,0,\delta,\delta,\delta,0,\delta)$	1; 3; 6; 8; 9; 10; 12	7; 9; 10; 12	2^{-45}
$(\delta,0,\delta,0,\delta,0,0,0) \xrightarrow{1r} (0,0,\delta,\delta,\delta,0,\delta,\delta)$	1; 3; 10; 12	5; 10; 12	2^{-30}
$(\delta,0,\delta,0,\delta,0,0,\delta) \xrightarrow{1r} (0,0,0,0,0,\delta,\delta,0)$	1; 3; 8; 11; 12	5; 11; 12	2^{-45}
$(\delta,0,\delta,0,\delta,0,\delta,0) \xrightarrow{1r} (\delta,\delta,0,\delta,0,\delta,0,0)$	1; 3	5; 7	1
$(\delta,0,\delta,0,\delta,0,\delta,\delta) \xrightarrow{1r} (\delta,\delta,\delta,0,\delta,0,0,\delta)$	1; 3; 8; 10; 11	5; 7; 10; 11	2^{-45}
$(\delta,0,\delta,0,\delta,\delta,0,0) \xrightarrow{1r} (\delta,\delta,\delta,0,0,\delta,\delta,\delta)$	1; 3; 6; 11	5; 11	2^{-30}
$(\delta,0,\delta,0,\delta,\delta,0,\delta) \xrightarrow{1r} (\delta,\delta,0,\delta,\delta,0,\delta,0)$	1; 3; 6; 8; 10	5; 10	2^{-60}
$(\delta,0,\delta,0,\delta,\delta,\delta,0) \xrightarrow{1r} (0,0,0,0,\delta,0,0,0)$	1; 3; 6; 10; 11; 12	5; 7; 10; 11; 12	2^{-45}
$(\delta,0,\delta,0,\delta,\delta,\delta,\delta) \xrightarrow{1r} (0,0,\delta,\delta,0,\delta,0,\delta)$	1; 3; 6; 8; 12	5; 7; 12	2^{-45}
$(\delta,0,\delta,\delta,0,0,0,0) \xrightarrow{1r} (0,\delta,\delta,0,\delta,\delta,0,\delta)$	1; 3; 9; 10; 11; 12	4; 9; 10; 11; 12	2^{-60}
$(\delta,0,\delta,\delta,0,0,0,\delta) \xrightarrow{1r} (0,\delta,0,\delta,0,0,0,0)$	1; 3; 8; 9; 12	4; 9; 12	2^{-60}
$(\delta,0,\delta,\delta,0,0,\delta,0) \xrightarrow{1r} (\delta,0,0,0,0,0,\delta,0)$	1; 3; 9; 11	4; 7; 9; 11	2^{-60}
$(\delta,0,\delta,\delta,0,0,\delta,\delta) \xrightarrow{1r} (\delta,0,\delta,\delta,\delta,\delta,\delta,\delta)$	1; 3; 8; 9; 10	4; 7; 9; 10	2^{-60}
$(\delta,0,\delta,\delta,0,\delta,0,0) \xrightarrow{1r} (\delta,0,\delta,\delta,0,0,0,\delta)$	1; 3; 6; 9	4; 9	2^{-60}
$(\delta,0,\delta,\delta,0,\delta,0,\delta) \xrightarrow{1r} (\delta,0,0,0,\delta,\delta,0,0)$	1; 3; 6; 8; 9; 10; 11	4; 9; 10; 11	2^{-45}
$(\delta,0,\delta,\delta,0,\delta,\delta,0) \xrightarrow{1r} (0,0,\delta,0,\delta,\delta,\delta,0)$	1; 3; 6; 9; 10; 12	4; 7; 9; 10; 12	2^{-45}
$(\delta,0,\delta,\delta,0,\delta,\delta,\delta) \xrightarrow{1r} (0,\delta,\delta,0,0,0,\delta,\delta)$	1; 3; 6; 8; 9; 11; 12	4; 7; 9; 11; 12	2^{-30}
$(\delta,0,\delta,\delta,\delta,0,0,0) \xrightarrow{1r} (0,0,0,0,0,\delta,0,\delta)$	1; 3; 11; 12	4; 5; 11; 12	2^{-45}
$(\delta,0,\delta,\delta,\delta,0,0,\delta) \xrightarrow{1r} (0,0,\delta,\delta,\delta,0,0,0)$	1; 3; 8; 10; 12	4; 5; 10; 12	2^{-30}
$(\delta,0,\delta,\delta,\delta,0,\delta,0) \xrightarrow{1r} (\delta,\delta,\delta,0,\delta,0,\delta,0)$	1; 3; 10; 11	4; 5; 7; 10; 11	2^{-45}
$(\delta,0,\delta,\delta,\delta,0,\delta,\delta) \xrightarrow{1r} (\delta,\delta,0,\delta,0,\delta,\delta,\delta)$	1; 3; 8	4; 5; 7	1
$(\delta,0,\delta,\delta,\delta,\delta,0,0) \xrightarrow{1r} (\delta,\delta,0,\delta,\delta,0,0,\delta)$	1; 3; 6; 10	4; 5; 10	2^{-60}
$(\delta,0,\delta,\delta,\delta,\delta,0,\delta) \xrightarrow{1r} (\delta,\delta,\delta,0,0,\delta,0,0)$	1; 3; 6; 8; 11	4; 5; 11	2^{-30}
$(\delta,0,\delta,\delta,\delta,\delta,\delta,0) \xrightarrow{1r} (0,0,\delta,\delta,0,\delta,\delta,0)$	1; 3; 6; 12	4; 5; 7; 12	2^{-45}
$(\delta,0,\delta,\delta,\delta,\delta,\delta,\delta) \xrightarrow{1r} (0,0,0,0,\delta,0,\delta,\delta)$	1; 3; 6; 8; 10; 11; 12	4; 5; 7; 10; 11; 12	2^{-45}
$(\delta,\delta,0,0,0,0,0,0) \xrightarrow{1r} (0,\delta,0,\delta,\delta,\delta,0,0)$	1; 9; 11; 12	2; 9; 11; 12	2^{-30}

Table 3.23 Nontrivial 1-round characteristics for MESH-128 (part VII).

1-round characteristic	Weak subkeys $Z_j^{(i)}$ (i-th round)		Prob.
	j value (odd i)	j value (even i)	
$(\delta,\delta,0,0,0,0,0,\delta) \xrightarrow{1r} (0,\delta,\delta,0,0,0,0,\delta)$	1; 8; 9; 10; 12	2; 9; 10; 12	2^{-45}
$(\delta,\delta,0,0,0,0,\delta,0) \xrightarrow{1r} (\delta,0,\delta,\delta,0,0,\delta,\delta)$	1; 9; 10; 11	2; 7; 9; 10; 11	2^{-45}
$(\delta,\delta,0,0,0,0,\delta,\delta) \xrightarrow{1r} (\delta,0,0,0,\delta,\delta,\delta,0)$	1; 8; 9	2; 7; 9	2^{-60}
$(\delta,\delta,0,0,0,\delta,0,0) \xrightarrow{1r} (\delta,0,0,0,0,0,0,0)$	1; 6; 9; 10	2; 9; 10	2^{-60}
$(\delta,\delta,0,0,0,\delta,0,\delta) \xrightarrow{1r} (\delta,0,\delta,\delta,\delta,\delta,0,\delta)$	1; 6; 8; 9; 11	2; 9; 11	2^{-60}
$(\delta,\delta,0,0,0,\delta,\delta,0) \xrightarrow{1r} (0,\delta,\delta,0,\delta,\delta,\delta,\delta)$	1; 6; 9; 12	2; 7; 9; 12	2^{-60}
$(\delta,\delta,0,0,0,\delta,\delta,\delta) \xrightarrow{1r} (0,\delta,0,\delta,0,0,\delta,0)$	1; 6; 8; 9; 10; 11; 12	2; 7; 9; 10; 11; 12	2^{-60}
$(\delta,\delta,0,0,\delta,0,0,0) \xrightarrow{1r} (0,0,\delta,\delta,0,\delta,0,0)$	1; 10; 11; 12	2; 5; 10; 11; 12	2^{-45}
$(\delta,\delta,0,0,\delta,0,0,\delta) \xrightarrow{1r} (0,0,0,0,\delta,0,0,\delta)$	1; 8; 12	2; 5; 12	2^{-45}
$(\delta,\delta,0,0,\delta,0,\delta,0) \xrightarrow{1r} (\delta,\delta,0,\delta,\delta,0,\delta,\delta)$	1; 11	2; 5; 7; 11	2^{-30}
$(\delta,\delta,0,0,\delta,0,\delta,\delta) \xrightarrow{1r} (\delta,\delta,\delta,0,0,\delta,\delta,0)$	1; 8; 10	2; 5; 7; 10	2^{-60}
$(\delta,\delta,0,0,\delta,\delta,0,0) \xrightarrow{1r} (\delta,\delta,\delta,0,\delta,0,0,0)$	1; 6	2; 5	1
$(\delta,\delta,0,0,\delta,\delta,0,\delta) \xrightarrow{1r} (\delta,\delta,0,\delta,0,\delta,0,\delta)$	1; 6; 8; 10; 11	2; 5; 10; 11	2^{-45}
$(\delta,\delta,0,0,\delta,\delta,\delta,0) \xrightarrow{1r} (0,0,0,0,0,\delta,\delta,\delta)$	1; 6; 10; 12	2; 5; 7; 10; 12	2^{-30}
$(\delta,\delta,0,0,\delta,\delta,\delta,\delta) \xrightarrow{1r} (0,0,\delta,\delta,\delta,0,\delta,0)$	1; 6; 8; 11; 12	2; 5; 7; 11; 12	2^{-45}
$(\delta,\delta,0,\delta,0,0,0,0) \xrightarrow{1r} (0,\delta,\delta,0,0,0,\delta,0)$	1; 9; 10; 12	2; 4; 9; 10; 12	2^{-45}
$(\delta,\delta,0,\delta,0,0,0,\delta) \xrightarrow{1r} (0,\delta,0,\delta,\delta,\delta,\delta,\delta)$	1; 8; 9; 11; 12	2; 4; 9; 11; 12	2^{-30}
$(\delta,\delta,0,\delta,0,0,\delta,0) \xrightarrow{1r} (\delta,0,0,0,\delta,\delta,0,\delta)$	1; 9	2; 4; 7; 9	2^{-60}
$(\delta,\delta,0,\delta,0,0,\delta,\delta) \xrightarrow{1r} (\delta,0,\delta,\delta,0,0,0,0)$	1; 8; 9; 10; 11	2; 4; 7; 9; 10; 11	2^{-45}
$(\delta,\delta,0,\delta,0,\delta,0,0) \xrightarrow{1r} (\delta,0,\delta,\delta,\delta,\delta,\delta,0)$	1; 6; 9; 11	2; 4; 9; 11	2^{-60}
$(\delta,\delta,0,\delta,0,\delta,0,\delta) \xrightarrow{1r} (\delta,0,0,0,0,0,\delta,\delta)$	1; 6; 8; 9; 10	2; 4; 9; 10	2^{-60}
$(\delta,\delta,0,\delta,0,\delta,\delta,0) \xrightarrow{1r} (0,\delta,0,\delta,0,0,0,\delta)$	1; 6; 9; 10; 11; 12	2; 4; 7; 9; 10; 11; 12	2^{-60}
$(\delta,\delta,0,\delta,0,\delta,\delta,\delta) \xrightarrow{1r} (0,\delta,\delta,0,\delta,\delta,0,0)$	1; 6; 8; 9; 12	2; 4; 7; 9; 12	2^{-60}
$(\delta,\delta,0,\delta,\delta,0,0,0) \xrightarrow{1r} (0,0,0,0,\delta,0,\delta,0)$	1; 12	2; 4; 5; 12	2^{-45}
$(\delta,\delta,0,\delta,\delta,0,0,\delta) \xrightarrow{1r} (0,0,\delta,\delta,0,\delta,\delta,\delta)$	1; 8; 10; 11; 12	2; 4; 5; 10; 11; 12	2^{-45}
$(\delta,\delta,0,\delta,\delta,0,\delta,0) \xrightarrow{1r} (\delta,\delta,\delta,0,0,\delta,0,\delta)$	1; 10	2; 4; 5; 7; 10	2^{-60}
$(\delta,\delta,0,\delta,\delta,0,\delta,\delta) \xrightarrow{1r} (\delta,\delta,0,\delta,\delta,0,0,0)$	1; 8; 11	2; 4; 5; 7; 11	2^{-30}
$(\delta,\delta,0,\delta,\delta,\delta,0,0) \xrightarrow{1r} (\delta,\delta,0,\delta,0,\delta,\delta,0)$	1; 6; 10; 11	2; 4; 5; 10; 11	2^{-45}
$(\delta,\delta,0,\delta,\delta,\delta,0,\delta) \xrightarrow{1r} (\delta,\delta,\delta,0,\delta,0,\delta,\delta)$	1; 6; 8	2; 4; 5	1
$(\delta,\delta,0,\delta,\delta,\delta,\delta,0) \xrightarrow{1r} (0,0,\delta,\delta,\delta,0,0,\delta)$	1; 6; 11; 12	2; 4; 5; 7; 11; 12	2^{-45}
$(\delta,\delta,0,\delta,\delta,\delta,\delta,\delta) \xrightarrow{1r} (0,0,0,0,0,\delta,0,0)$	1; 6; 8; 10; 12	2; 4; 5; 7; 10; 12	2^{-30}
$(\delta,\delta,\delta,0,0,0,0,0) \xrightarrow{1r} (\delta,0,\delta,0,0,\delta,\delta,\delta)$	1; 3; 9; 10; 11	2; 9; 10; 11	2^{-45}

Table 3.24 Nontrivial 1-round characteristics for MESH-128 (part VIII).

1-round characteristic	Weak subkeys $Z_j^{(i)}$ (i-th round)		Prob.
	j value (odd i)	j value (even i)	
$(\delta,\delta,\delta,0,0,0,0,\delta) \xrightarrow{1r} (\delta,0,0,\delta,\delta,0,\delta,0)$	1; 3; 8; 9	2; 9	2^{-60}
$(\delta,\delta,\delta,0,0,0,\delta,0) \xrightarrow{1r} (0,\delta,0,0,\delta,0,0,0)$	1; 3; 9; 11; 12	2; 7; 9; 11; 12	2^{-30}
$(\delta,\delta,\delta,0,0,0,\delta,\delta) \xrightarrow{1r} (0,\delta,\delta,\delta,0,\delta,0,\delta)$	1; 3; 8; 9; 10; 12	2; 7; 9; 10; 12	2^{-45}
$(\delta,\delta,\delta,0,0,\delta,0,0) \xrightarrow{1r} (0,\delta,\delta,\delta,\delta,0,\delta,\delta)$	1; 3; 6; 9; 12	2; 9; 12	2^{-60}
$(\delta,\delta,\delta,0,0,\delta,0,\delta) \xrightarrow{1r} (0,\delta,0,0,0,\delta,\delta,0)$	1; 3; 6; 8; 9; 10; 11; 12	2; 9; 10; 11; 12	2^{-60}
$(\delta,\delta,\delta,0,0,\delta,\delta,0) \xrightarrow{1r} (\delta,0,0,\delta,0,\delta,0,0)$	1; 3; 6; 9; 10	2; 7; 9; 10	2^{-60}
$(\delta,\delta,\delta,0,0,\delta,\delta,\delta) \xrightarrow{1r} (\delta,0,\delta,0,\delta,0,0,\delta)$	1; 3; 6; 8; 9; 11	2; 7; 9; 11	2^{-60}
$(\delta,\delta,\delta,0,\delta,0,0,0) \xrightarrow{1r} (\delta,\delta,0,0,\delta,\delta,\delta,\delta)$	1; 3; 11	2; 5; 11	2^{-30}
$(\delta,\delta,\delta,0,\delta,0,0,\delta) \xrightarrow{1r} (\delta,\delta,\delta,\delta,0,0,\delta,0)$	1; 3; 8; 10	2; 5; 10	2^{-60}
$(\delta,\delta,\delta,0,\delta,0,\delta,0) \xrightarrow{1r} (0,0,\delta,0,0,0,0,0)$	1; 3; 10; 11; 12	2; 5; 7; 10; 11; 12	2^{-45}
$(\delta,\delta,\delta,0,\delta,0,\delta,\delta) \xrightarrow{1r} (0,0,0,\delta,\delta,\delta,0,\delta)$	1; 3; 8; 12	2; 5; 7; 12	2^{-45}
$(\delta,\delta,\delta,0,\delta,\delta,0,0) \xrightarrow{1r} (0,0,0,\delta,0,0,\delta,\delta)$	1; 3; 6; 10; 12	2; 5; 10; 12	2^{-30}
$(\delta,\delta,\delta,0,\delta,\delta,0,\delta) \xrightarrow{1r} (0,0,\delta,0,\delta,\delta,\delta,0)$	1; 3; 6; 8; 11; 12	2; 5; 11; 12	2^{-45}
$(\delta,\delta,\delta,0,\delta,\delta,\delta,0) \xrightarrow{1r} (\delta,\delta,\delta,\delta,\delta,\delta,0,0)$	1; 3; 6	2; 5; 7	1
$(\delta,\delta,\delta,0,\delta,\delta,\delta,\delta) \xrightarrow{1r} (\delta,\delta,0,0,0,0,0,\delta)$	1; 3; 6; 8; 10; 11	2; 5; 7; 10; 11	2^{-45}
$(\delta,\delta,\delta,\delta,0,0,0,0) \xrightarrow{1r} (\delta,0,0,\delta,\delta,0,0,\delta)$	1; 3; 9	2; 4; 9	2^{-60}
$(\delta,\delta,\delta,\delta,0,0,0,\delta) \xrightarrow{1r} (\delta,0,\delta,0,0,\delta,0,0)$	1; 3; 8; 9; 10; 11	2; 4; 9; 10; 11	2^{-45}
$(\delta,\delta,\delta,\delta,0,0,\delta,0) \xrightarrow{1r} (0,\delta,\delta,\delta,0,\delta,\delta,0)$	1; 3; 9; 10; 12	2; 4; 7; 9; 10; 12	2^{-45}
$(\delta,\delta,\delta,\delta,0,0,\delta,\delta) \xrightarrow{1r} (0,\delta,0,0,\delta,0,\delta,\delta)$	1; 3; 8; 9; 11; 12	2; 4; 7; 9; 11; 12	2^{-30}
$(\delta,\delta,\delta,\delta,0,\delta,0,0) \xrightarrow{1r} (0,\delta,0,0,0,\delta,0,\delta)$	1; 3; 6; 9; 10; 11; 12	2; 4; 9; 10; 11; 12	2^{-60}
$(\delta,\delta,\delta,\delta,0,\delta,0,\delta) \xrightarrow{1r} (0,\delta,\delta,\delta,\delta,0,0,0)$	1; 3; 6; 8; 9; 12	2; 4; 9; 12	2^{-60}
$(\delta,\delta,\delta,\delta,0,\delta,\delta,0) \xrightarrow{1r} (\delta,0,\delta,0,\delta,0,\delta,0)$	1; 3; 6; 9; 11	2; 4; 7; 9; 11	2^{-60}
$(\delta,\delta,\delta,\delta,0,\delta,\delta,\delta) \xrightarrow{1r} (\delta,0,0,\delta,0,\delta,\delta,\delta)$	1; 3; 6; 8; 9; 10	2; 4; 7; 9; 10	2^{-60}
$(\delta,\delta,\delta,\delta,\delta,0,0,0) \xrightarrow{1r} (\delta,\delta,\delta,\delta,0,0,0,\delta)$	1; 3; 10	2; 4; 5; 10	2^{-60}
$(\delta,\delta,\delta,\delta,\delta,0,0,\delta) \xrightarrow{1r} (\delta,\delta,0,0,\delta,\delta,0,0)$	1; 3; 8; 11	2; 4; 5; 11	2^{-30}
$(\delta,\delta,\delta,\delta,\delta,0,\delta,0) \xrightarrow{1r} (0,0,0,\delta,\delta,\delta,\delta,0)$	1; 3; 12	2; 4; 5; 7; 12	2^{-45}
$(\delta,\delta,\delta,\delta,\delta,0,\delta,\delta) \xrightarrow{1r} (0,0,\delta,0,0,0,\delta,\delta)$	1; 3; 8; 10; 11; 12	2; 4; 5; 7; 10; 11; 12	2^{-45}
$(\delta,\delta,\delta,\delta,\delta,\delta,0,0) \xrightarrow{1r} (0,0,\delta,0,\delta,\delta,0,\delta)$	1; 3; 6; 11; 12	2; 4; 5; 11; 12	2^{-45}
$(\delta,\delta,\delta,\delta,\delta,\delta,0,\delta) \xrightarrow{1r} (0,0,0,\delta,0,0,0,0)$	1; 3; 6; 8; 10; 12	2; 4; 5; 10; 12	2^{-30}
$(\delta,\delta,\delta,\delta,\delta,\delta,\delta,0) \xrightarrow{1r} (\delta,\delta,0,0,0,0,\delta,0)$	1; 3; 6; 10; 11	2; 4; 5; 7; 10; 11	2^{-45}
$(\delta,\delta,\delta,\delta,\delta,\delta,\delta,\delta) \xrightarrow{1r} (\delta,\delta,\delta,\delta,\delta,\delta,\delta,\delta)$	1; 3; 6; 8	2; 4; 5; 7	1

$$Z_7^{(i+13)}, Z_9^{(i+13)}, Z_{10}^{(i+13)}, Z_{11}^{(i+13)}, Z_1^{(i+14)}, Z_{10}^{(i+14)}, Z_{11}^{(i+14)}, Z_2^{(i+15)}, Z_4^{(i+15)},$$
$$Z_5^{(i+15)}, Z_7^{(i+15)}, Z_{10}^{(i+15)}, Z_{11}^{(i+15)}, Z_1^{(i+16)}, Z_9^{(i+16)}, Z_{10}^{(i+16)}, Z_{11}^{(i+16)}, Z_4^{(i+17)},$$
$$Z_7^{(i+17)}, Z_9^{(i+17)}, Z_{10}^{(i+17)}, Z_1^{(i+18)}, Z_3^{(i+18)}, Z_6^{(i+18)}, Z_8^{(i+18)}, Z_{10}^{(i+18)}, Z_{11}^{(i+18)},$$
$$Z_{12}^{(i+18)}, Z_5^{(i+19)}, Z_7^{(i+19)}, Z_9^{(i+19)}, Z_{10}^{(i+19)}, Z_{11}^{(i+19)}, Z_{12}^{(i+19)}, Z_1^{(i+20)}, Z_3^{(i+20)},$$
$$Z_{10}^{(i+20)}, Z_{11}^{(i+20)}, Z_2^{(i+21)}, Z_5^{(i+21)}, Z_7^{(i+21)}, Z_{10}^{(i+21)}, Z_{11}^{(i+21)}, Z_{12}^{(i+21)}, Z_3^{(i+22)},$$
$$Z_{10}^{(i+22)}, Z_{12}^{(i+22)}, Z_2^{(i+23)}, Z_4^{(i+23)}, Z_5^{(i+23)}, Z_7^{(i+23)}, Z_{10}^{(i+23)}, Z_{11}^{(i+23)}, Z_{12}^{(i+23)},$$

$Z_3^{(i+24)}$, $Z_8^{(i+24)}$, $Z_{10}^{(i+24)}$, $Z_{11}^{(i+24)}$, $Z_5^{(i+25)}$, $Z_9^{(i+25)}$, $Z_{10}^{(i+25)}$, $Z_{11}^{(i+25)}$, $Z_{12}^{(i+25)}$, $Z_1^{(i+26)}$, $Z_3^{(i+26)}$, $Z_6^{(i+26)}$, $Z_{10}^{(i+26)}$, $Z_{11}^{(i+26)}$, $Z_{12}^{(i+26)}$, $Z_5^{(i+27)}$, $Z_9^{(i+27)}$, $Z_{10}^{(i+27)}$, $Z_3^{(i+28)}$, $Z_9^{(i+28)}$, $Z_{10}^{(i+28)}$, $Z_{11}^{(i+28)}$, $Z_2^{(i+29)}$, $Z_7^{(i+29)}$, $Z_{10}^{(i+29)}$ and $Z_{11}^{(i+29)}$.

Starting from an even-numbered round i, the following 137 subkeys have to be weak: $Z_{10}^{(i)}$, $Z_{11}^{(i)}$, $Z_3^{(i+1)}$, $Z_6^{(i+1)}$, $Z_8^{(i+1)}$, $Z_9^{(i+1)}$, $Z_{10}^{(i+1)}$, $Z_{11}^{(i+1)}$, $Z_2^{(i+2)}$, $Z_5^{(i+2)}$, $Z_9^{(i+2)}$, $Z_{10}^{(i+2)}$, $Z_{10}^{(i+3)}$, $Z_{11}^{(i+3)}$, $Z_{12}^{(i+3)}$, $Z_2^{(i+4)}$, $Z_4^{(i+4)}$, $Z_9^{(i+4)}$, $Z_{10}^{(i+4)}$, $Z_{11}^{(i+4)}$, $Z_{12}^{(i+4)}$, $Z_6^{(i+5)}$, $Z_8^{(i+5)}$, $Z_{10}^{(i+5)}$, $Z_{11}^{(i+5)}$, $Z_4^{(i+6)}$, $Z_{10}^{(i+6)}$, $Z_{11}^{(i+6)}$, $Z_{12}^{(i+6)}$, $Z_1^{(i+7)}$, $Z_6^{(i+7)}$, $Z_8^{(i+7)}$, $Z_{10}^{(i+7)}$, $Z_{12}^{(i+7)}$, $Z_{10}^{(i+8)}$, $Z_{11}^{(i+8)}$, $Z_{12}^{(i+8)}$, $Z_1^{(i+9)}$, $Z_6^{(i+9)}$, $Z_{10}^{(i+9)}$, $Z_{11}^{(i+9)}$, $Z_2^{(i+10)}$, $Z_4^{(i+10)}$, $Z_7^{(i+10)}$, $Z_9^{(i+10)}$, $Z_{10}^{(i+10)}$, $Z_{11}^{(i+10)}$, $Z_{12}^{(i+10)}$, $Z_8^{(i+11)}$, $Z_{10}^{(i+11)}$, $Z_{11}^{(i+11)}$, $Z_{12}^{(i+11)}$, $Z_2^{(i+12)}$, $Z_4^{(i+12)}$, $Z_7^{(i+12)}$, $Z_9^{(i+12)}$, $Z_{10}^{(i+12)}$, $Z_1^{(i+13)}$, $Z_6^{(i+13)}$, $Z_8^{(i+13)}$, $Z_9^{(i+13)}$, $Z_{10}^{(i+13)}$, $Z_{11}^{(i+13)}$, $Z_4^{(i+14)}$, $Z_5^{(i+14)}$, $Z_{10}^{(i+14)}$, $Z_{11}^{(i+14)}$, $Z_1^{(i+15)}$, $Z_3^{(i+15)}$, $Z_6^{(i+15)}$, $Z_{10}^{(i+15)}$, $Z_{11}^{(i+15)}$, $Z_2^{(i+16)}$, $Z_7^{(i+16)}$, $Z_9^{(i+16)}$, $Z_{10}^{(i+16)}$, $Z_{11}^{(i+16)}$, $Z_1^{(i+17)}$, $Z_3^{(i+17)}$, $Z_8^{(i+17)}$, $Z_9^{(i+17)}$, $Z_{10}^{(i+17)}$, $Z_4^{(i+18)}$, $Z_5^{(i+18)}$, $Z_7^{(i+18)}$, $Z_{10}^{(i+18)}$, $Z_{11}^{(i+18)}$, $Z_{12}^{(i+18)}$, $Z_8^{(i+19)}$, $Z_9^{(i+19)}$, $Z_{10}^{(i+19)}$, $Z_{11}^{(i+19)}$, $Z_{12}^{(i+19)}$, $Z_4^{(i+20)}$, $Z_5^{(i+20)}$, $Z_7^{(i+20)}$, $Z_{10}^{(i+20)}$, $Z_{11}^{(i+20)}$, $Z_1^{(i+21)}$, $Z_3^{(i+21)}$, $Z_{10}^{(i+21)}$, $Z_{11}^{(i+21)}$, $Z_{12}^{(i+21)}$, $Z_{10}^{(i+22)}$, $Z_{12}^{(i+22)}$, $Z_1^{(i+23)}$, $Z_3^{(i+23)}$, $Z_8^{(i+23)}$, $Z_{10}^{(i+23)}$, $Z_{11}^{(i+23)}$, $Z_{12}^{(i+23)}$, $Z_7^{(i+24)}$, $Z_{10}^{(i+24)}$, $Z_{11}^{(i+24)}$, $Z_3^{(i+25)}$, $Z_8^{(i+25)}$, $Z_9^{(i+25)}$, $Z_{10}^{(i+25)}$, $Z_{11}^{(i+25)}$, $Z_{12}^{(i+25)}$, $Z_5^{(i+26)}$, $Z_7^{(i+26)}$, $Z_{10}^{(i+26)}$, $Z_{11}^{(i+26)}$, $Z_{12}^{(i+26)}$, $Z_9^{(i+27)}$, $Z_{10}^{(i+27)}$, $Z_2^{(i+28)}$, $Z_5^{(i+28)}$, $Z_9^{(i+28)}$, $Z_{10}^{(i+28)}$, $Z_{11}^{(i+28)}$, $Z_3^{(i+29)}$, $Z_6^{(i+29)}$, $Z_8^{(i+29)}$, $Z_{10}^{(i+29)}$ and $Z_{11}^{(i+29)}$.

(2) $(0,0,0,0,0,0,\delta,0) \xrightarrow{1r} (\delta,\delta,\delta,0,\delta,\delta,\delta,\delta) \xrightarrow{1r} (\delta,\delta,0,0,0,0,0,\delta) \xrightarrow{1r} (0,\delta,\delta,0,0,0,0,\delta) \xrightarrow{1r} (0,0,\delta,\delta,0,0,\delta,0) \xrightarrow{1r} (0,0,\delta,0,\delta,0,\delta,0) \xrightarrow{1r} (0,\delta,\delta,\delta,\delta,\delta,0,0) \xrightarrow{1r} (\delta,0,0,0,0,\delta,0,\delta) \xrightarrow{1r} (0,\delta,0,0,\delta,0,0,\delta) \xrightarrow{1r} (\delta,0,\delta,0,0,0,0,\delta) \xrightarrow{1r} (0,\delta,\delta,0,\delta,\delta,\delta,0) \xrightarrow{1r} (0,\delta,0,\delta,0,\delta,0,0) \xrightarrow{1r} (0,0,0,\delta,0,\delta,\delta,0) \xrightarrow{1r} (0,0,0,0,\delta,\delta,0,\delta) \xrightarrow{1r} (\delta,0,0,0,\delta,0,0,\delta) \xrightarrow{1r} (\delta,\delta,\delta,\delta,\delta,\delta,0,\delta) \xrightarrow{1r} (0,0,0,\delta,0,0,0,0) \xrightarrow{1r} (0,0,\delta,\delta,\delta,\delta,\delta,0) \xrightarrow{1r} (\delta,0,0,\delta,\delta,\delta,\delta,0) \xrightarrow{1r} (\delta,\delta,0,0,\delta,\delta,0,\delta) \xrightarrow{1r} (\delta,\delta,0,\delta,0,\delta,0,\delta) \xrightarrow{1r} (\delta,0,0,0,0,0,\delta,\delta) \xrightarrow{1r} (0,\delta,\delta,\delta,\delta,0,\delta,0) \xrightarrow{1r} (\delta,0,\delta,\delta,0,\delta,\delta,0) \xrightarrow{1r} (0,\delta,0,\delta,\delta,\delta,\delta,0) \xrightarrow{1r} (\delta,0,0,\delta,0,0,0,\delta) \xrightarrow{1r} (\delta,0,\delta,0,\delta,0,\delta,\delta) \xrightarrow{1r} (\delta,\delta,\delta,0,\delta,0,0,\delta) \xrightarrow{1r} (\delta,\delta,\delta,\delta,0,0,\delta,0) \xrightarrow{1r} (0,\delta,\delta,\delta,0,\delta,\delta,0) \xrightarrow{1r} (0,0,0,0,0,0,\delta,0)$, which is 30-round iterative and holds with probability 2^{-1500}.

Starting from an odd-numbered round i, the following 121 subkeys have to be weak: $Z_{10}^{(i)}$, $Z_{12}^{(i)}$, $Z_2^{(i+1)}$, $Z_5^{(i+1)}$, $Z_7^{(i+1)}$, $Z_{10}^{(i+1)}$, $Z_{11}^{(i+1)}$, $Z_1^{(i+2)}$, $Z_8^{(i+2)}$, $Z_9^{(i+2)}$, $Z_{10}^{(i+2)}$, $Z_{12}^{(i+2)}$, $Z_2^{(i+3)}$, $Z_{10}^{(i+3)}$, $Z_3^{(i+4)}$, $Z_{10}^{(i+4)}$, $Z_{11}^{(i+4)}$, $Z_5^{(i+5)}$, $Z_7^{(i+5)}$, $Z_9^{(i+5)}$, $Z_{10}^{(i+5)}$, $Z_3^{(i+6)}$, $Z_6^{(i+6)}$, $Z_9^{(i+6)}$, $Z_{10}^{(i+6)}$, $Z_{11}^{(i+6)}$, $Z_{12}^{(i+6)}$, $Z_9^{(i+7)}$, $Z_{10}^{(i+7)}$, $Z_{12}^{(i+7)}$, $Z_8^{(i+8)}$, $Z_9^{(i+8)}$, $Z_{10}^{(i+8)}$, $Z_{12}^{(i+8)}$, $Z_9^{(i+9)}$, $Z_{10}^{(i+9)}$, $Z_{11}^{(i+9)}$, $Z_{12}^{(i+9)}$, $Z_3^{(i+10)}$, $Z_6^{(i+10)}$, $Z_9^{(i+10)}$, $Z_{10}^{(i+10)}$, $Z_2^{(i+11)}$, $Z_4^{(i+11)}$, $Z_{10}^{(i+11)}$, $Z_{11}^{(i+11)}$, $Z_6^{(i+12)}$, $Z_{10}^{(i+12)}$, $Z_5^{(i+13)}$, $Z_9^{(i+13)}$, $Z_{10}^{(i+13)}$, $Z_{12}^{(i+13)}$, $Z_1^{(i+14)}$, $Z_8^{(i+14)}$, $Z_{10}^{(i+14)}$, $Z_{11}^{(i+14)}$, $Z_2^{(i+15)}$, $Z_4^{(i+15)}$, $Z_5^{(i+15)}$, $Z_{10}^{(i+15)}$, $Z_{12}^{(i+15)}$, $Z_{10}^{(i+16)}$, $Z_{11}^{(i+16)}$,

$Z_4^{(i+17)}, Z_5^{(i+17)}, Z_7^{(i+17)}, Z_9^{(i+17)}, Z_{10}^{(i+17)}, Z_{12}^{(i+17)}, Z_1^{(i+18)}, Z_6^{(i+18)}, Z_{10}^{(i+18)}, Z_2^{(i+19)}, Z_5^{(i+19)}, Z_{10}^{(i+19)}, Z_{11}^{(i+19)}, Z_1^{(i+20)}, Z_6^{(i+20)}, Z_8^{(i+20)}, Z_9^{(i+20)}, Z_{10}^{(i+20)}, Z_7^{(i+21)}, Z_9^{(i+21)}, Z_{10}^{(i+21)}, Z_{11}^{(i+21)}, Z_{12}^{(i+21)}, Z_3^{(i+22)}, Z_9^{(i+22)}, Z_{10}^{(i+22)}, Z_{12}^{(i+22)}, Z_4^{(i+23)}, Z_7^{(i+23)}, Z_9^{(i+23)}, Z_{10}^{(i+23)}, Z_{12}^{(i+23)}, Z_6^{(i+24)}, Z_9^{(i+24)}, Z_{10}^{(i+24)}, Z_{11}^{(i+24)}, Z_{12}^{(i+24)}, Z_4^{(i+25)}, Z_9^{(i+25)}, Z_{10}^{(i+25)}, Z_1^{(i+26)}, Z_3^{(i+26)}, Z_8^{(i+26)}, Z_{10}^{(i+26)}, Z_{11}^{(i+26)}, Z_2^{(i+27)}, Z_5^{(i+27)}, Z_{10}^{(i+27)}, Z_1^{(i+28)}, Z_3^{(i+28)}, Z_9^{(i+28)}, Z_{10}^{(i+28)}, Z_{12}^{(i+28)}, Z_2^{(i+29)}, Z_4^{(i+29)}, Z_7^{(i+29)}, Z_{10}^{(i+29)}$ and $Z_{11}^{(i+29)}$.

Starting from an even-numbered round i, the following 143 subkeys have to be weak: $Z_7^{(i)}, Z_{10}^{(i)}, Z_{12}^{(i)}, Z_1^{(i+1)}, Z_3^{(i+1)}, Z_6^{(i+1)}, Z_8^{(i+1)}, Z_{10}^{(i+1)}, Z_{11}^{(i+1)}, Z_2^{(i+2)}, Z_9^{(i+2)}, Z_{10}^{(i+2)}, Z_{12}^{(i+2)}, Z_3^{(i+3)}, Z_8^{(i+3)}, Z_{10}^{(i+3)}, Z_4^{(i+4)}, Z_7^{(i+4)}, Z_{10}^{(i+4)}, Z_{11}^{(i+4)}, Z_3^{(i+5)}, Z_9^{(i+5)}, Z_{10}^{(i+5)}, Z_2^{(i+6)}, Z_4^{(i+6)}, Z_5^{(i+6)}, Z_9^{(i+6)}, Z_{10}^{(i+6)}, Z_{11}^{(i+6)}, Z_{12}^{(i+6)}, Z_1^{(i+7)}, Z_6^{(i+7)}, Z_8^{(i+7)}, Z_9^{(i+7)}, Z_{10}^{(i+7)}, Z_{12}^{(i+7)}, Z_2^{(i+8)}, Z_5^{(i+8)}, Z_9^{(i+8)}, Z_{10}^{(i+8)}, Z_{12}^{(i+8)}, Z_1^{(i+9)}, Z_3^{(i+9)}, Z_8^{(i+9)}, Z_9^{(i+9)}, Z_{10}^{(i+9)}, Z_{11}^{(i+9)}, Z_{12}^{(i+9)}, Z_2^{(i+10)}, Z_5^{(i+10)}, Z_7^{(i+10)}, Z_9^{(i+10)}, Z_{10}^{(i+10)}, Z_6^{(i+11)}, Z_{10}^{(i+11)}, Z_{11}^{(i+11)}, Z_4^{(i+12)}, Z_7^{(i+12)}, Z_{10}^{(i+12)}, Z_6^{(i+13)}, Z_8^{(i+13)}, Z_9^{(i+13)}, Z_{10}^{(i+13)}, Z_{12}^{(i+13)}, Z_5^{(i+14)}, Z_{10}^{(i+14)}, Z_{11}^{(i+14)}, Z_1^{(i+15)}, Z_3^{(i+15)}, Z_6^{(i+15)}, Z_8^{(i+15)}, Z_{10}^{(i+15)}, Z_{12}^{(i+15)}, Z_4^{(i+16)}, Z_{10}^{(i+16)}, Z_{11}^{(i+16)}, Z_3^{(i+17)}, Z_6^{(i+17)}, Z_9^{(i+17)}, Z_{10}^{(i+17)}, Z_{12}^{(i+17)}, Z_4^{(i+18)}, Z_5^{(i+18)}, Z_7^{(i+18)}, Z_{10}^{(i+18)}, Z_1^{(i+19)}, Z_6^{(i+19)}, Z_8^{(i+19)}, Z_{10}^{(i+19)}, Z_{11}^{(i+19)}, Z_2^{(i+20)}, Z_4^{(i+20)}, Z_9^{(i+20)}, Z_{10}^{(i+20)}, Z_1^{(i+21)}, Z_8^{(i+21)}, Z_9^{(i+21)}, Z_{10}^{(i+21)}, Z_{11}^{(i+21)}, Z_{12}^{(i+21)}, Z_2^{(i+22)}, Z_4^{(i+22)}, Z_5^{(i+22)}, Z_7^{(i+22)}, Z_9^{(i+22)}, Z_{10}^{(i+22)}, Z_{12}^{(i+22)}, Z_1^{(i+23)}, Z_3^{(i+23)}, Z_6^{(i+23)}, Z_9^{(i+23)}, Z_{10}^{(i+23)}, Z_{12}^{(i+23)}, Z_2^{(i+24)}, Z_4^{(i+24)}, Z_5^{(i+24)}, Z_7^{(i+24)}, Z_9^{(i+24)}, Z_{10}^{(i+24)}, Z_{11}^{(i+24)}, Z_{12}^{(i+24)}, Z_1^{(i+25)}, Z_8^{(i+25)}, Z_9^{(i+25)}, Z_{10}^{(i+25)}, Z_5^{(i+26)}, Z_7^{(i+26)}, Z_{10}^{(i+26)}, Z_{11}^{(i+26)}, Z_1^{(i+27)}, Z_3^{(i+27)}, Z_8^{(i+27)}, Z_{10}^{(i+27)}, Z_2^{(i+28)}, Z_4^{(i+28)}, Z_7^{(i+28)}, Z_9^{(i+28)}, Z_{10}^{(i+28)}, Z_{12}^{(i+28)}, Z_3^{(i+29)}, Z_6^{(i+29)}, Z_{10}^{(i+29)}$ and $Z_{11}^{(i+29)}$.

(3) $(0,0,0,0,0,0,\delta,\delta) \stackrel{1r}{\to} (\delta,\delta,0,\delta,0,0,\delta,0) \stackrel{1r}{\to} (\delta,0,0,0,\delta,\delta,0,\delta) \stackrel{1r}{\to} (0,0,\delta,0,0,0,0,\delta) \stackrel{1r}{\to} (\delta,\delta,0,0,0,\delta,\delta,0) \stackrel{1r}{\to} (0,\delta,\delta,0,\delta,\delta,\delta,\delta) \stackrel{1r}{\to} (0,\delta,\delta,0,\delta,0,0,\delta) \stackrel{1r}{\to} (0,\delta,0,\delta,\delta,0,\delta,0) \stackrel{1r}{\to} (0,\delta,0,0,\delta,\delta,0,\delta) \stackrel{1r}{\to} (0,\delta,\delta,\delta,\delta,\delta,0,\delta) \stackrel{1r}{\to} (\delta,0,\delta,\delta,\delta,0,0,0) \stackrel{1r}{\to} (0,0,0,0,0,\delta,0,\delta) \stackrel{1r}{\to} (\delta,\delta,\delta,0,0,0,0,\delta) \stackrel{1r}{\to} (\delta,0,0,\delta,\delta,0,\delta,0) \stackrel{1r}{\to} (0,0,0,\delta,0,0,0,\delta) \stackrel{1r}{\to} (0,0,0,0,0,0,\delta,\delta)$, which is 15-round iterative and holds with probability 2^{-750}.

Starting from an odd-numbered round i, the following 54 subkeys have to be weak: $Z_8^{(i)}, Z_{11}^{(i)}, Z_{12}^{(i)}, Z_2^{(i+1)}, Z_4^{(i+1)}, Z_7^{(i+1)}, Z_9^{(i+1)}, Z_1^{(i+2)}, Z_6^{(i+2)}, Z_8^{(i+2)}, Z_{12}^{(i+2)}, Z_{11}^{(i+3)}, Z_{12}^{(i+3)}, Z_1^{(i+4)}, Z_6^{(i+4)}, Z_9^{(i+4)}, Z_{12}^{(i+4)}, Z_2^{(i+5)}, Z_5^{(i+5)}, Z_7^{(i+5)}, Z_9^{(i+5)}, Z_{11}^{(i+5)}, Z_3^{(i+6)}, Z_8^{(i+6)}, Z_9^{(i+6)}, Z_2^{(i+7)}, Z_4^{(i+7)}, Z_5^{(i+7)}, Z_7^{(i+7)}, Z_9^{(i+7)}, Z_6^{(i+8)}, Z_8^{(i+8)}, Z_9^{(i+8)}, Z_{11}^{(i+8)}, Z_2^{(i+9)}, Z_4^{(i+9)}, Z_5^{(i+9)}, Z_9^{(i+9)}, Z_{12}^{(i+9)}, Z_1^{(i+10)}, Z_3^{(i+10)}, Z_{11}^{(i+10)}, Z_{12}^{(i+10)}, Z_{12}^{(i+11)}, Z_1^{(i+12)}, Z_3^{(i+12)}, Z_8^{(i+12)}, Z_9^{(i+12)}, Z_4^{(i+13)}, Z_5^{(i+13)}, Z_7^{(i+13)}, Z_{11}^{(i+13)}, Z_{12}^{(i+13)}$ and $Z_8^{(i+14)}$.

Starting from an even-numbered round i, the following 46 subkeys have to be weak: $Z_7^{(i)}$, $Z_{11}^{(i)}$, $Z_{12}^{(i)}$, $Z_1^{(i+1)}$, $Z_9^{(i+1)}$, $Z_5^{(i+2)}$, $Z_{12}^{(i+2)}$, $Z_3^{(i+3)}$, $Z_8^{(i+3)}$, $Z_{11}^{(i+3)}$, $Z_{12}^{(i+3)}$, $Z_2^{(i+4)}$, $Z_7^{(i+4)}$, $Z_9^{(i+4)}$, $Z_{12}^{(i+4)}$, $Z_3^{(i+5)}$, $Z_6^{(i+5)}$, $Z_8^{(i+5)}$, $Z_9^{(i+5)}$, $Z_{11}^{(i+5)}$, $Z_2^{(i+6)}$, $Z_5^{(i+6)}$, $Z_9^{(i+6)}$, $Z_9^{(i+7)}$, $Z_2^{(i+8)}$, $Z_5^{(i+8)}$, $Z_9^{(i+8)}$, $Z_{11}^{(i+8)}$, $Z_3^{(i+9)}$, $Z_6^{(i+9)}$, $Z_8^{(i+9)}$, $Z_9^{(i+9)}$, $Z_{12}^{(i+9)}$, $Z_4^{(i+10)}$, $Z_5^{(i+10)}$, $Z_{11}^{(i+10)}$, $Z_{12}^{(i+10)}$, $Z_6^{(i+11)}$, $Z_8^{(i+11)}$, $Z_{12}^{(i+11)}$, $Z_2^{(i+12)}$, $Z_9^{(i+12)}$, $Z_1^{(i+13)}$, $Z_{11}^{(i+13)}$, $Z_{12}^{(i+13)}$ and $Z_4^{(i+14)}$.

(4) $(0,0,0,0,0,\delta,\delta,0) \xrightarrow{1r} (0,0,\delta,\delta,0,0,\delta,\delta) \xrightarrow{1r} (0,0,0,\delta,0,\delta,\delta,\delta) \xrightarrow{1r} (0,0,\delta,\delta,0,0,0,0) \xrightarrow{1r} (\delta,\delta,0,0,0,\delta,0,\delta) \xrightarrow{1r} (\delta,0,\delta,\delta,\delta,\delta,0,\delta) \xrightarrow{1r} (\delta,\delta,\delta,0,0,\delta,0,0) \xrightarrow{1r} (0,\delta,\delta,\delta,\delta,0,\delta,\delta) \xrightarrow{1r} (\delta,0,0,0,\delta,0,\delta,\delta) \xrightarrow{1r} (0,0,0,\delta,0,0,\delta,0) \xrightarrow{1r} (\delta,\delta,0,\delta,0,0,0,\delta) \xrightarrow{1r} (0,\delta,0,\delta,\delta,\delta,\delta,\delta) \xrightarrow{1r} (\delta,0,\delta,0,\delta,\delta,0,0) \xrightarrow{1r} (\delta,\delta,\delta,0,0,\delta,\delta,\delta) \xrightarrow{1r} (\delta,0,\delta,0,\delta,0,0,\delta) \xrightarrow{1r} (0,0,0,0,0,\delta,\delta,0)$, which is 15-round iterative and holds with probability 2^{-600}.

Starting from an odd-numbered round i, the following 62 subkeys have to be weak: $Z_6^{(i)}$, $Z_{11}^{(i)}$, $Z_4^{(i+1)}$, $Z_7^{(i+1)}$, $Z_6^{(i+2)}$, $Z_8^{(i+2)}$, $Z_{11}^{(i+2)}$, $Z_4^{(i+3)}$, $Z_{11}^{(i+3)}$, $Z_{12}^{(i+3)}$, $Z_1^{(i+4)}$, $Z_6^{(i+4)}$, $Z_8^{(i+4)}$, $Z_9^{(i+4)}$, $Z_{11}^{(i+4)}$, $Z_4^{(i+5)}$, $Z_5^{(i+5)}$, $Z_{11}^{(i+5)}$, $Z_1^{(i+6)}$, $Z_3^{(i+6)}$, $Z_6^{(i+6)}$, $Z_9^{(i+6)}$, $Z_{12}^{(i+6)}$, $Z_2^{(i+7)}$, $Z_4^{(i+7)}$, $Z_5^{(i+7)}$, $Z_7^{(i+7)}$, $Z_9^{(i+7)}$, $Z_{11}^{(i+7)}$, $Z_{12}^{(i+7)}$, $Z_1^{(i+8)}$, $Z_8^{(i+8)}$, $Z_{11}^{(i+8)}$, $Z_{12}^{(i+8)}$, $Z_4^{(i+9)}$, $Z_7^{(i+9)}$, $Z_{11}^{(i+9)}$, $Z_{12}^{(i+9)}$, $Z_1^{(i+10)}$, $Z_8^{(i+10)}$, $Z_9^{(i+10)}$, $Z_{11}^{(i+10)}$, $Z_{12}^{(i+10)}$, $Z_2^{(i+11)}$, $Z_4^{(i+11)}$, $Z_5^{(i+11)}$, $Z_7^{(i+11)}$, $Z_9^{(i+11)}$, $Z_{12}^{(i+11)}$, $Z_1^{(i+12)}$, $Z_3^{(i+12)}$, $Z_6^{(i+12)}$, $Z_{11}^{(i+12)}$, $Z_2^{(i+13)}$, $Z_7^{(i+13)}$, $Z_9^{(i+13)}$, $Z_{11}^{(i13)}$, $Z_1^{(i+14)}$, $Z_3^{(i+14)}$, $Z_8^{(i+14)}$, $Z_{11}^{(i+14)}$ and $Z_{12}^{(i+14)}$.

Starting from an even-numbered round i, the following 52 subkeys have to be weak: $Z_7^{(i)}$, $Z_{11}^{(i)}$, $Z_3^{(i+1)}$, $Z_8^{(i+1)}$, $Z_4^{(i+2)}$, $Z_7^{(i+2)}$, $Z_{11}^{(i+2)}$, $Z_3^{(i+3)}$, $Z_{11}^{(i+3)}$, $Z_{12}^{(i+3)}$, $Z_2^{(i+4)}$, $Z_9^{(i+4)}$, $Z_{11}^{(i+4)}$, $Z_1^{(i+5)}$, $Z_3^{(i+5)}$, $Z_6^{(i+5)}$, $Z_8^{(i+5)}$, $Z_{11}^{(i+5)}$, $Z_2^{(i+6)}$, $Z_9^{(i+6)}$, $Z_{12}^{(i+6)}$, $Z_3^{(i+7)}$, $Z_8^{(i+7)}$, $Z_9^{(i+7)}$, $Z_{11}^{(i+7)}$, $Z_{12}^{(i+7)}$, $Z_5^{(i+8)}$, $Z_7^{(i+8)}$, $Z_{11}^{(i+8)}$, $Z_{12}^{(i+8)}$, $Z_{11}^{(i+9)}$, $Z_{12}^{(i+9)}$, $Z_2^{(i+10)}$, $Z_4^{(i+10)}$, $Z_9^{(i+10)}$, $Z_{11}^{(i+10)}$, $Z_{12}^{(i+10)}$, $Z_6^{(i+11)}$, $Z_8^{(i+11)}$, $Z_9^{(i+11)}$, $Z_{12}^{(i+11)}$, $Z_5^{(i+12)}$, $Z_{11}^{(i+12)}$, $Z_1^{(i+13)}$, $Z_3^{(i+13)}$, $Z_6^{(i+13)}$, $Z_8^{(i+13)}$, $Z_9^{(i+13)}$, $Z_{11}^{(i+13)}$, $Z_5^{(i+14)}$, $Z_{11}^{(i+14)}$ and $Z_{12}^{(i+14)}$.

(5) $(0,0,0,0,0,\delta,\delta,\delta) \xrightarrow{1r} (0,0,0,0,\delta,\delta,\delta,0) \xrightarrow{1r} (0,\delta,0,\delta,\delta,0,\delta,\delta) \xrightarrow{1r} (0,\delta,\delta,\delta,0,0,0,0) \xrightarrow{1r} (0,0,\delta,\delta,0,0,0,\delta) \xrightarrow{1r} (\delta,\delta,\delta,\delta,\delta,0,0,0) \xrightarrow{1r} (\delta,\delta,\delta,\delta,0,0,0,\delta) \xrightarrow{1r} (\delta,0,\delta,0,0,\delta,0,0) \xrightarrow{1r} (\delta,0,0,0,\delta,\delta,\delta,\delta) \xrightarrow{1r} (\delta,\delta,0,0,\delta,\delta,\delta,0) \xrightarrow{1r} (0,0,0,0,0,\delta,\delta,\delta)$, which is 10-round iterative and holds with probability 2^{-480}.

Starting from an odd-numbered round i, the following 41 subkeys have to be weak: $Z_6^{(i)}$, $Z_8^{(i)}$, $Z_{10}^{(i)}$, $Z_5^{(i+1)}$, $Z_7^{(i+1)}$, $Z_9^{(i+1)}$, $Z_{10}^{(i+1)}$, $Z_{11}^{(i+1)}$, $Z_8^{(i+2)}$, $Z_9^{(i+2)}$, $Z_{10}^{(i+2)}$, $Z_{11}^{(i+2)}$, $Z_2^{(i+3)}$, $Z_4^{(i+3)}$, $Z_{10}^{(i+3)}$, $Z_3^{(i+4)}$, $Z_8^{(i+4)}$, $Z_{10}^{(i+4)}$, $Z_{12}^{(i+4)}$, $Z_2^{(i+5)}$, $Z_4^{(i+5)}$, $Z_5^{(i+5)}$, $Z_{10}^{(i+5)}$, $Z_1^{(i+6)}$, $Z_3^{(i+6)}$, $Z_8^{(i+6)}$, $Z_9^{(i+6)}$, $Z_{10}^{(i+6)}$, $Z_{11}^{(i+6)}$, $Z_9^{(i+7)}$, $Z_{10}^{(i+7)}$, $Z_{11}^{(i+7)}$, $Z_1^{(i+8)}$, $Z_6^{(i+8)}$, $Z_8^{(i+8)}$, $Z_{10}^{(i+8)}$, $Z_2^{(i+9)}$, $Z_5^{(i+9)}$, $Z_7^{(i+9)}$, $Z_{10}^{(i+9)}$ and $Z_{12}^{(i+9)}$.

Starting from an even-numbered round i, the following 39 subkeys have to be weak: $Z_7^{(i)}$, $Z_{10}^{(i)}$, $Z_6^{(i+1)}$, $Z_9^{(i+1)}$, $Z_{10}^{(i+1)}$, $Z_{11}^{(i+1)}$, $Z_2^{(i+2)}$, $Z_4^{(i+2)}$, $Z_5^{(i+2)}$, $Z_7^{(i+2)}$, $Z_9^{(i+2)}$, $Z_{10}^{(i+2)}$, $Z_{11}^{(i+2)}$, $Z_3^{(i+3)}$, $Z_{10}^{(i+3)}$, $Z_4^{(i+4)}$, $Z_{10}^{(i+4)}$, $Z_{12}^{(i+4)}$, $Z_1^{(i+5)}$, $Z_3^{(i+5)}$, $Z_{10}^{(i+5)}$, $Z_2^{(i+6)}$, $Z_4^{(i+6)}$, $Z_9^{(i+6)}$, $Z_{10}^{(i+6)}$, $Z_{11}^{(i+6)}$, $Z_1^{(i+7)}$, $Z_3^{(i+7)}$, $Z_6^{(i+7)}$, $Z_9^{(i+7)}$, $Z_{10}^{(i+7)}$, $Z_{11}^{(i+7)}$, $Z_5^{(i+8)}$, $Z_7^{(i+8)}$, $Z_{10}^{(i+8)}$, $Z_1^{(i+9)}$, $Z_6^{(i+9)}$, $Z_{10}^{(i+9)}$ and $Z_{12}^{(i+9)}$.

(6) $(0,0,0,0,\delta,0,0,\delta) \xrightarrow{1r} (0,\delta,0,\delta,0,\delta,0,\delta) \xrightarrow{1r} (0,0,\delta,0,\delta,0,\delta,\delta) \xrightarrow{1r} (0,\delta,0,0,0,0,0,\delta) \xrightarrow{1r} (\delta,\delta,0,0,\delta,0,0,\delta) \xrightarrow{1r} (0,0,0,0,\delta,0,0,\delta)$, which is 5-round iterative and holds with probability 2^{-210}.
Starting from an odd-numbered round i, the following 14 subkeys have to be weak: $Z_8^{(i)}$, $Z_9^{(i)}$, $Z_{11}^{(i)}$, $Z_2^{(i+1)}$, $Z_4^{(i+1)}$, $Z_3^{(i+2)}$, $Z_8^{(i+2)}$, $Z_9^{(i+2)}$, $Z_{11}^{(i+2)}$, $Z_2^{(i+3)}$, $Z_{12}^{(i+3)}$, $Z_1^{(i+4)}$, $Z_8^{(i+4)}$ and $Z_{12}^{(i+4)}$.
Starting from an even-numbered round i, the following 14 subkeys have to be weak: $Z_5^{(i)}$, $Z_9^{(i)}$, $Z_{11}^{(i)}$, $Z_6^{(i+1)}$, $Z_8^{(i+1)}$, $Z_5^{(i+2)}$, $Z_7^{(i+2)}$, $Z_9^{(i+2)}$, $Z_{11}^{(i+2)}$, $Z_8^{(i+3)}$, $Z_{12}^{(i+3)}$, $Z_2^{(i+4)}$, $Z_5^{(i+4)}$ and $Z_{12}^{(i+4)}$.

(7) $(0,0,0,0,\delta,0,\delta,0) \xrightarrow{1r} (\delta,0,0,0,0,\delta,\delta,\delta) \xrightarrow{1r} (\delta,0,\delta,0,0,\delta,\delta,0) \xrightarrow{1r} (0,\delta,\delta,0,0,0,0,0) \xrightarrow{1r} (0,0,0,0,\delta,\delta,\delta,\delta) \xrightarrow{1r} (0,\delta,\delta,0,0,\delta,\delta,0) \xrightarrow{1r} (0,0,\delta,\delta,\delta,\delta,0,0) \xrightarrow{1r} (0,\delta,\delta,\delta,0,0,0,\delta) \xrightarrow{1r} (0,0,0,0,\delta,\delta,0,0) \xrightarrow{1r} (\delta,0,\delta,\delta,0,\delta,0,0) \xrightarrow{1r} (\delta,0,\delta,\delta,0,0,0,\delta) \xrightarrow{1r} (0,\delta,0,\delta,0,0,0,0) \xrightarrow{1r} (\delta,\delta,0,0,\delta,0,\delta,0) \xrightarrow{1r} (\delta,\delta,0,\delta,\delta,0,\delta,\delta) \xrightarrow{1r} (\delta,\delta,0,\delta,\delta,0,0,0) \xrightarrow{1r} (0,0,0,0,\delta,0,\delta,0)$, which is 15-round iterative and holds with probability 2^{-630}.
Starting from an odd-numbered round i, the following 46 subkeys have to be weak: $Z_9^{(i)}$, $Z_{12}^{(i)}$, $Z_7^{(i+1)}$, $Z_9^{(i+1)}$, $Z_1^{(i+2)}$, $Z_3^{(i+2)}$, $Z_6^{(i+2)}$, $Z_9^{(i+2)}$, $Z_{11}^{(i+2)}$, $Z_{12}^{(i+2)}$, $Z_2^{(i+3)}$, $Z_{11}^{(i+3)}$, $Z_6^{(i+4)}$, $Z_8^{(i+4)}$, $Z_9^{(i+4)}$, $Z_2^{(i+5)}$, $Z_7^{(i+5)}$, $Z_3^{(i+6)}$, $Z_6^{(i+6)}$, $Z_9^{(i+6)}$, $Z_2^{(i+7)}$, $Z_4^{(i+7)}$, $Z_{11}^{(i+7)}$, $Z_6^{(i+8)}$, $Z_9^{(i+8)}$, $Z_{11}^{(i+8)}$, $Z_{12}^{(i+8)}$, $Z_4^{(i+9)}$, $Z_9^{(i+9)}$, $Z_1^{(i+10)}$, $Z_3^{(i+10)}$, $Z_8^{(i+10)}$, $Z_9^{(i+10)}$, $Z_{12}^{(i+10)}$, $Z_2^{(i+11)}$, $Z_4^{(i+11)}$, $Z_{12}^{(i+11)}$, $Z_1^{(i+12)}$, $Z_{11}^{(i+12)}$, $Z_2^{(i+13)}$, $Z_4^{(i+13)}$, $Z_5^{(i+13)}$, $Z_7^{(i+13)}$, $Z_{11}^{(i+13)}$, $Z_1^{(i+14)}$ and $Z_{12}^{(i+14)}$.
Starting from an even-numbered round i, the following 48 subkeys have to be weak: $Z_5^{(i)}$, $Z_7^{(i)}$, $Z_9^{(i)}$, $Z_{12}^{(i)}$, $Z_1^{(i+1)}$, $Z_6^{(i+1)}$, $Z_8^{(i+1)}$, $Z_9^{(i+1)}$, $Z_7^{(i+2)}$, $Z_9^{(i+2)}$, $Z_{11}^{(i+2)}$, $Z_{12}^{(i+2)}$, $Z_3^{(i+3)}$, $Z_{11}^{(i+3)}$, $Z_5^{(i+4)}$, $Z_7^{(i+4)}$, $Z_9^{(i+4)}$, $Z_3^{(i+5)}$, $Z_6^{(i+5)}$, $Z_4^{(i+6)}$, $Z_5^{(i+6)}$, $Z_9^{(i+6)}$, $Z_3^{(i+7)}$, $Z_8^{(i+7)}$, $Z_{11}^{(i+7)}$, $Z_5^{(i+8)}$, $Z_9^{(i+8)}$, $Z_{11}^{(i+8)}$, $Z_{12}^{(i+8)}$, $Z_1^{(i+9)}$, $Z_3^{(i+9)}$, $Z_6^{(i+9)}$, $Z_9^{(i+9)}$, $Z_4^{(i+10)}$, $Z_9^{(i+10)}$, $Z_{12}^{(i+10)}$, $Z_{12}^{(i+11)}$, $Z_2^{(i+12)}$, $Z_5^{(i+12)}$, $Z_7^{(i+12)}$, $Z_{11}^{(i+12)}$, $Z_1^{(i+13)}$, $Z_8^{(i+13)}$, $Z_{11}^{(i+13)}$, $Z_2^{(i+14)}$, $Z_4^{(i+14)}$, $Z_5^{(i+14)}$ and $Z_{12}^{(i+14)}$.

(8) $(0,0,0,\delta,0,0,\delta,\delta) \xrightarrow{1r} (\delta,\delta,\delta,0,\delta,\delta,0,0) \xrightarrow{1r} (0,0,0,\delta,0,0,\delta,\delta)$, which is 2-round iterative and holds with probability 2^{-60}.
Starting from an odd-numbered round i, the following seven subkeys have to be weak: $Z_8^{(i)}$, $Z_{10}^{(i)}$, $Z_{12}^{(i)}$, $Z_2^{(i+1)}$, $Z_5^{(i+1)}$, $Z_{10}^{(i+1)}$ and $Z_{12}^{(i+1)}$.

Starting from an even-numbered round i, the following nine subkeys have to be weak: $Z_4^{(i)}$, $Z_7^{(i)}$, $Z_{10}^{(i)}$, $Z_{12}^{(i)}$, $Z_1^{(i+1)}$, $Z_3^{(i+1)}$, $Z_6^{(i+1)}$, $Z_{10}^{(i+1)}$ and $Z_{12}^{(i+1)}$.

(9) $(0,0,0,\delta,0,\delta,0,0) \xrightarrow{1r} (\delta,\delta,\delta,0,0,0,\delta,0) \xrightarrow{1r} (0,\delta,0,0,\delta,0,0,0) \xrightarrow{1r} (\delta,0,0,\delta,\delta,\delta,0,0) \xrightarrow{1r} (0,0,\delta,0,0,0,\delta,0) \xrightarrow{1r} (0,0,0,\delta,0,\delta,0,0)$, which is 5-round iterative and holds with probability 2^{-150}.

Starting from an odd-numbered round i, the following 14 subkeys have to be weak: $Z_6^{(i)}$, $Z_{12}^{(i)}$, $Z_2^{(i+1)}$, $Z_7^{(i+1)}$, $Z_9^{(i+1)}$, $Z_{11}^{(i+2)}$, $Z_{12}^{(i+2)}$, $Z_9^{(i+3)}$, $Z_{11}^{(i+3)}$, $Z_{12}^{(i+3)}$, $Z_4^{(i+4)}$, $Z_5^{(i+4)}$, $Z_{12}^{(i+4)}$ and $Z_3^{(i+5)}$.

Starting from an even-numbered round i, the following 16 subkeys have to be weak: $Z_4^{(i)}$, $Z_{12}^{(i)}$, $Z_1^{(i+1)}$, $Z_3^{(i+1)}$, $Z_9^{(i+1)}$, $Z_{11}^{(i+1)}$, $Z_{12}^{(i+1)}$, $Z_2^{(i+2)}$, $Z_5^{(i+2)}$, $Z_9^{(i+2)}$, $Z_{11}^{(i+2)}$, $Z_{12}^{(i+2)}$, $Z_1^{(i+3)}$, $Z_6^{(i+3)}$, $Z_{12}^{(i+3)}$ and $Z_7^{(i+4)}$.

(10) $(0,0,0,\delta,\delta,0,0,0) \xrightarrow{1r} (0,\delta,0,\delta,0,\delta,\delta,0) \xrightarrow{1r} (\delta,\delta,\delta,\delta,\delta,0,0,\delta) \xrightarrow{1r} (\delta,\delta,0,0,\delta,\delta,0,0) \xrightarrow{1r} (\delta,\delta,\delta,0,\delta,0,0,0) \xrightarrow{1r} (\delta,\delta,0,0,\delta,\delta,\delta,\delta) \xrightarrow{1r} (0,0,\delta,\delta,\delta,0,\delta,0) \xrightarrow{1r} (0,\delta,0,0,0,0,\delta,0) \xrightarrow{1r} (0,0,0,\delta,\delta,0,\delta,\delta) \xrightarrow{1r} (\delta,0,0,0,0,\delta,0,0) \xrightarrow{1r} (0,\delta,\delta,\delta,0,\delta,0,0) \xrightarrow{1r} (\delta,\delta,\delta,0,\delta,\delta,0,\delta) \xrightarrow{1r} (0,0,\delta,0,\delta,\delta,\delta,0) \xrightarrow{1r} (\delta,0,\delta,0,0,0,0,0) \xrightarrow{1r} (0,\delta,0,\delta,0,0,\delta,\delta) \xrightarrow{1r} (0,0,0,\delta,\delta,0,0,0)$, which is 15-round iterative and holds with probability 2^{-600}.

Starting from an odd-numbered round i, the following 50 subkeys have to be weak: $Z_9^{(i)}$, $Z_{11}^{(i)}$, $Z_2^{(i+1)}$, $Z_4^{(i+1)}$, $Z_7^{(i+1)}$, $Z_{11}^{(i+1)}$, $Z_{12}^{(i+1)}$, $Z_1^{(i+2)}$, $Z_3^{(i+2)}$, $Z_8^{(i+2)}$, $Z_{11}^{(i+2)}$, $Z_2^{(i+3)}$, $Z_5^{(i+3)}$, $Z_1^{(i+4)}$, $Z_3^{(i+4)}$, $Z_{11}^{(i+4)}$, $Z_2^{(i+5)}$, $Z_5^{(i+5)}$, $Z_7^{(i+5)}$, $Z_{11}^{(i+5)}$, $Z_{12}^{(i+5)}$, $Z_3^{(i+6)}$, $Z_9^{(i+6)}$, $Z_{11}^{(i+6)}$, $Z_2^{(i+7)}$, $Z_7^{(i+7)}$, $Z_{11}^{(i+7)}$, $Z_8^{(i+8)}$, $Z_9^{(i+8)}$, $Z_{12}^{(i+8)}$, $Z_9^{(i+9)}$, $Z_{11}^{(i+9)}$, $Z_{12}^{(i+9)}$, $Z_3^{(i+10)}$, $Z_6^{(i+10)}$, $Z_{11}^{(i+10)}$, $Z_{12}^{(i+10)}$, $Z_2^{(i+11)}$, $Z_5^{(i+11)}$, $Z_{11}^{(i+11)}$, $Z_{12}^{(i+11)}$, $Z_3^{(i+12)}$, $Z_6^{(i+12)}$, $Z_9^{(i+12)}$, $Z_{11}^{(i+12)}$, $Z_{12}^{(i+12)}$, $Z_9^{(i+13)}$, $Z_{12}^{(i+13)}$, $Z_8^{(i+14)}$ and $Z_{11}^{(i+14)}$.

Starting from an even-numbered round i, the following 60 subkeys have to be weak: $Z_4^{(i)}$, $Z_5^{(i)}$, $Z_9^{(i)}$, $Z_{11}^{(i)}$, $Z_6^{(i+1)}$, $Z_{11}^{(i+1)}$, $Z_{12}^{(i+1)}$, $Z_2^{(i+2)}$, $Z_4^{(i+2)}$, $Z_5^{(i+2)}$, $Z_{11}^{(i+2)}$, $Z_1^{(i+3)}$, $Z_6^{(i+3)}$, $Z_2^{(i+4)}$, $Z_5^{(i+4)}$, $Z_{11}^{(i+4)}$, $Z_1^{(i+5)}$, $Z_6^{(i+5)}$, $Z_8^{(i+5)}$, $Z_{11}^{(i+5)}$, $Z_{12}^{(i+5)}$, $Z_4^{(i+6)}$, $Z_5^{(i+6)}$, $Z_7^{(i+6)}$, $Z_9^{(i+6)}$, $Z_{11}^{(i+6)}$, $Z_{11}^{(i+7)}$, $Z_4^{(i+8)}$, $Z_5^{(i+8)}$, $Z_7^{(i+8)}$, $Z_9^{(i+8)}$, $Z_{12}^{(i+8)}$, $Z_1^{(i+9)}$, $Z_6^{(i+9)}$, $Z_9^{(i+9)}$, $Z_{11}^{(i+9)}$, $Z_{12}^{(i+9)}$, $Z_2^{(i+10)}$, $Z_4^{(i+10)}$, $Z_{11}^{(i+10)}$, $Z_{12}^{(i+10)}$, $Z_1^{(i+11)}$, $Z_3^{(i+11)}$, $Z_6^{(i+11)}$, $Z_8^{(i+11)}$, $Z_{11}^{(i+11)}$, $Z_{12}^{(i+11)}$, $Z_5^{(i+12)}$, $Z_7^{(i+12)}$, $Z_9^{(i+12)}$, $Z_{11}^{(i+12)}$, $Z_{12}^{(i+12)}$, $Z_1^{(i+13)}$, $Z_3^{(i+13)}$, $Z_9^{(i+13)}$, $Z_{12}^{(i+13)}$, $Z_2^{(i+14)}$, $Z_4^{(i+14)}$, $Z_7^{(i+14)}$ and $Z_{11}^{(i+14)}$.

(11) $(0,0,0,\delta,\delta,0,0,\delta) \xrightarrow{1r} (0,\delta,\delta,0,\delta,0,\delta,\delta) \xrightarrow{1r} (\delta,0,\delta,\delta,0,\delta,0,\delta) \xrightarrow{1r} (\delta,0,0,0,\delta,\delta,0,0) \xrightarrow{1r} (0,0,0,\delta,\delta,\delta,0,0) \xrightarrow{1r} (\delta,0,0,0,\delta,0,\delta,0) \xrightarrow{1r} (0,0,\delta,0,\delta,\delta,\delta,\delta) \xrightarrow{1r} (\delta,0,0,\delta,\delta,\delta,0,\delta) \xrightarrow{1r} (0,0,0,\delta,\delta,\delta,\delta,\delta) \xrightarrow{1r} (0,\delta,0,\delta,\delta,0,0,0) \xrightarrow{1r} (\delta,0,\delta,0,0,0,\delta,0) \xrightarrow{1r} (\delta,0,\delta,\delta,\delta,\delta,0,0) \xrightarrow{1r} (\delta,\delta,0,\delta,\delta,0,0,\delta) \xrightarrow{1r} (0,0,\delta,\delta,0,\delta,\delta,\delta) \xrightarrow{1r} (\delta,\delta,0,0,\delta,0,\delta,\delta) \xrightarrow{1r} (\delta,\delta,\delta,0,0,\delta,\delta,0) \xrightarrow{1r} (\delta,0,0,\delta,0,\delta,0,0) \xrightarrow{1r} (0,\delta,0,0,\delta,0,\delta,0) \xrightarrow{1r} (0,\delta,\delta,\delta,0,0,\delta,\delta) \xrightarrow{1r} (\delta,\delta,\delta,0,0,0,\delta,\delta) \xrightarrow{1r} (0,\delta,\delta,\delta,0,\delta,0,\delta) \xrightarrow{1r} (\delta,\delta,0,\delta,0,0,0,$

0) $\xrightarrow{1r}$ $(0, \delta, \delta, 0, 0, 0, \delta, 0)$ $\xrightarrow{1r}$ $(\delta, \delta, \delta, 0, 0, 0, 0, 0)$ $\xrightarrow{1r}$ $(\delta, 0, \delta, 0, 0, \delta, \delta, \delta)$ $\xrightarrow{1r}$ $(0, \delta, 0, \delta, \delta, \delta, 0, \delta)$ $\xrightarrow{1r}$ $(0, \delta, 0, 0, 0, 0, \delta, \delta)$ $\xrightarrow{1r}$ $(0, 0, \delta, 0, 0, \delta, \delta, 0)$ $\xrightarrow{1r}$ $(\delta, \delta, 0, 0, \delta, 0, 0, 0)$ $\xrightarrow{1r}$ $(0, 0, \delta, \delta, 0, \delta, 0, 0)$ $\xrightarrow{1r}$ $(0, 0, 0, \delta, \delta, 0, 0, \delta)$, which is 30-round iterative and holds with probability 2^{-1440}.

Starting from an odd-numbered round i, the following 137 subkeys have to be weak: $Z_8^{(i)}$, $Z_9^{(i)}$, $Z_{10}^{(i)}$, $Z_2^{(i+1)}$, $Z_5^{(i+1)}$, $Z_7^{(i+1)}$, $Z_9^{(i+1)}$, $Z_{10}^{(i+1)}$, $Z_{12}^{(i+1)}$, $Z_1^{(i+2)}$, $Z_3^{(i+2)}$, $Z_6^{(i+2)}$, $Z_8^{(i+2)}$, $Z_9^{(i+2)}$, $Z_{10}^{(i+2)}$, $Z_{11}^{(i+2)}$, $Z_5^{(i+3)}$, $Z_{10}^{(i+3)}$, $Z_{11}^{(i+3)}$, $Z_{12}^{(i+3)}$, $Z_6^{(i+4)}$, $Z_9^{(i+4)}$, $Z_{10}^{(i+4)}$, $Z_{12}^{(i+4)}$, $Z_5^{(i+5)}$, $Z_7^{(i+5)}$, $Z_{10}^{(i+5)}$, $Z_{12}^{(i+5)}$, $Z_3^{(i+6)}$, $Z_6^{(i+6)}$, $Z_8^{(i+6)}$, $Z_9^{(i+6)}$, $Z_{10}^{(i+6)}$, $Z_{12}^{(i+6)}$, $Z_4^{(i+7)}$, $Z_5^{(i+7)}$, $Z_{10}^{(i+7)}$, $Z_{11}^{(i+7)}$, $Z_{12}^{(i+7)}$, $Z_6^{(i+8)}$, $Z_8^{(i+8)}$, $Z_9^{(i+8)}$, $Z_{10}^{(i+8)}$, $Z_{11}^{(i+8)}$, $Z_2^{(i+9)}$, $Z_4^{(i+9)}$, $Z_5^{(i+9)}$, $Z_9^{(i+9)}$, $Z_{10}^{(i+9)}$, $Z_{12}^{(i+9)}$, $Z_1^{(i+10)}$, $Z_3^{(i+10)}$, $Z_9^{(i+10)}$, $Z_{10}^{(i+10)}$, $Z_4^{(i+11)}$, $Z_5^{(i+11)}$, $Z_{10}^{(i+11)}$, $Z_1^{(i+12)}$, $Z_8^{(i+12)}$, $Z_{10}^{(i+12)}$, $Z_{11}^{(i+12)}$, $Z_{12}^{(i+12)}$, $Z_4^{(i+13)}$, $Z_7^{(i+13)}$, $Z_{10}^{(i+13)}$, $Z_{11}^{(i+13)}$, $Z_{12}^{(i+13)}$, $Z_1^{(i+14)}$, $Z_8^{(i+14)}$, $Z_{10}^{(i+14)}$, $Z_2^{(i+15)}$, $Z_7^{(i+15)}$, $Z_9^{(i+15)}$, $Z_{10}^{(i+15)}$, $Z_1^{(i+16)}$, $Z_6^{(i+16)}$, $Z_9^{(i+16)}$, $Z_{10}^{(i+16)}$, $Z_{12}^{(i+16)}$, $Z_2^{(i+17)}$, $Z_5^{(i+17)}$, $Z_7^{(i+17)}$, $Z_9^{(i+17)}$, $Z_{10}^{(i+17)}$, $Z_{11}^{(i+17)}$, $Z_3^{(i+18)}$, $Z_8^{(i+18)}$, $Z_{10}^{(i+18)}$, $Z_{11}^{(i+18)}$, $Z_{12}^{(i+18)}$, $Z_2^{(i+19)}$, $Z_7^{(i+19)}$, $Z_9^{(i+19)}$, $Z_{10}^{(i+19)}$, $Z_{12}^{(i+19)}$, $Z_3^{(i+20)}$, $Z_6^{(i+20)}$, $Z_8^{(i+20)}$, $Z_{10}^{(i+20)}$, $Z_{12}^{(i+20)}$, $Z_2^{(i+21)}$, $Z_4^{(i+21)}$, $Z_9^{(i+21)}$, $Z_{10}^{(i+21)}$, $Z_{12}^{(i+21)}$, $Z_3^{(i+22)}$, $Z_{10}^{(i+22)}$, $Z_{11}^{(i+22)}$, $Z_{12}^{(i+22)}$, $Z_2^{(i+23)}$, $Z_9^{(i+23)}$, $Z_{10}^{(i+23)}$, $Z_{11}^{(i+23)}$, $Z_1^{(i+24)}$, $Z_3^{(i+24)}$, $Z_6^{(i+24)}$, $Z_8^{(i+24)}$, $Z_9^{(i+24)}$, $Z_{10}^{(i+24)}$, $Z_{12}^{(i+24)}$, $Z_2^{(i+25)}$, $Z_4^{(i+25)}$, $Z_5^{(i+25)}$, $Z_9^{(i+25)}$, $Z_{10}^{(i+25)}$, $Z_8^{(i+26)}$, $Z_{10}^{(i+26)}$, $Z_7^{(i+27)}$, $Z_{10}^{(i+27)}$, $Z_{11}^{(i+27)}$, $Z_{12}^{(i+27)}$, $Z_1^{(i+28)}$, $Z_{10}^{(i+28)}$, $Z_{11}^{(i+28)}$, $Z_{12}^{(i+28)}$, $Z_4^{(i+29)}$ and $Z_{10}^{(i+29)}$.

Starting from an even-numbered round i, the following 134 subkeys have to be weak: $Z_4^{(i)}$, $Z_5^{(i)}$, $Z_9^{(i)}$, $Z_{10}^{(i)}$, $Z_3^{(i+1)}$, $Z_8^{(i+1)}$, $Z_9^{(i+1)}$, $Z_{10}^{(i+1)}$, $Z_{12}^{(i+1)}$, $Z_4^{(i+2)}$, $Z_9^{(i+2)}$, $Z_{10}^{(i+2)}$, $Z_{11}^{(i+2)}$, $Z_1^{(i+3)}$, $Z_6^{(i+3)}$, $Z_{10}^{(i+3)}$, $Z_{11}^{(i+3)}$, $Z_{12}^{(i+3)}$, $Z_4^{(i+4)}$, $Z_5^{(i+4)}$, $Z_9^{(i+4)}$, $Z_{10}^{(i+4)}$, $Z_{12}^{(i+4)}$, $Z_1^{(i+5)}$, $Z_{10}^{(i+5)}$, $Z_{12}^{(i+5)}$, $Z_5^{(i+6)}$, $Z_7^{(i+6)}$, $Z_9^{(i+6)}$, $Z_{10}^{(i+6)}$, $Z_{12}^{(i+6)}$, $Z_1^{(i+7)}$, $Z_6^{(i+7)}$, $Z_8^{(i+7)}$, $Z_{10}^{(i+7)}$, $Z_{11}^{(i+7)}$, $Z_{12}^{(i+7)}$, $Z_4^{(i+8)}$, $Z_5^{(i+8)}$, $Z_7^{(i+8)}$, $Z_9^{(i+8)}$, $Z_{10}^{(i+8)}$, $Z_{11}^{(i+8)}$, $Z_9^{(i+9)}$, $Z_{10}^{(i+9)}$, $Z_{12}^{(i+9)}$, $Z_7^{(i+10)}$, $Z_9^{(i+10)}$, $Z_{10}^{(i+10)}$, $Z_1^{(i+11)}$, $Z_3^{(i+11)}$, $Z_6^{(i+11)}$, $Z_{10}^{(i+11)}$, $Z_2^{(i+12)}$, $Z_4^{(i+12)}$, $Z_5^{(i+12)}$, $Z_{10}^{(i+12)}$, $Z_{11}^{(i+12)}$, $Z_{12}^{(i+12)}$, $Z_3^{(i+13)}$, $Z_6^{(i+13)}$, $Z_8^{(i+13)}$, $Z_{10}^{(i+13)}$, $Z_{11}^{(i+13)}$, $Z_{12}^{(i+13)}$, $Z_2^{(i+14)}$, $Z_5^{(i+14)}$, $Z_7^{(i+14)}$, $Z_{10}^{(i+14)}$, $Z_1^{(i+15)}$, $Z_3^{(i+15)}$, $Z_6^{(i+15)}$, $Z_9^{(i+15)}$, $Z_{10}^{(i+15)}$, $Z_4^{(i+16)}$, $Z_9^{(i+16)}$, $Z_{10}^{(i+16)}$, $Z_{12}^{(i+16)}$, $Z_9^{(i+17)}$, $Z_{10}^{(i+17)}$, $Z_{11}^{(i+17)}$, $Z_2^{(i+18)}$, $Z_4^{(i+18)}$, $Z_7^{(i+18)}$, $Z_{10}^{(i+18)}$, $Z_{11}^{(i+18)}$, $Z_{12}^{(i+18)}$, $Z_1^{(i+19)}$, $Z_3^{(i+19)}$, $Z_8^{(i+19)}$, $Z_9^{(i+19)}$, $Z_{10}^{(i+19)}$, $Z_{12}^{(i+19)}$, $Z_2^{(i+20)}$, $Z_4^{(i+20)}$, $Z_{10}^{(i+20)}$, $Z_{12}^{(i+20)}$, $Z_1^{(i+21)}$, $Z_9^{(i+21)}$, $Z_{10}^{(i+21)}$, $Z_{12}^{(i+21)}$, $Z_2^{(i+22)}$, $Z_7^{(i+22)}$, $Z_{10}^{(i+22)}$, $Z_{11}^{(i+22)}$, $Z_{12}^{(i+22)}$, $Z_1^{(i+23)}$, $Z_3^{(i+23)}$, $Z_9^{(i+23)}$, $Z_{10}^{(i+23)}$, $Z_{11}^{(i+23)}$, $Z_7^{(i+24)}$, $Z_9^{(i+24)}$, $Z_{10}^{(i+24)}$, $Z_{12}^{(i+24)}$, $Z_6^{(i+25)}$, $Z_8^{(i+25)}$, $Z_9^{(i+25)}$, $Z_{10}^{(i+25)}$, $Z_2^{(i+26)}$, $Z_7^{(i+26)}$, $Z_{10}^{(i+26)}$, $Z_3^{(i+27)}$, $Z_6^{(i+27)}$, $Z_{10}^{(i+27)}$, $Z_{11}^{(i+27)}$, $Z_{12}^{(i+27)}$, $Z_2^{(i+28)}$, $Z_5^{(i+28)}$, $Z_{10}^{(i+28)}$, $Z_{11}^{(i+28)}$, $Z_{12}^{(i+28)}$, $Z_3^{(i+29)}$, $Z_6^{(i+29)}$ and $Z_{10}^{(i+29)}$.

(12) $(0,0,0,\delta,\delta,0,\delta,0) \stackrel{1r}{\rightarrow} (\delta,0,\delta,\delta,\delta,0,0,\delta) \stackrel{1r}{\rightarrow} (0,0,\delta,\delta,\delta,0,0,0) \stackrel{1r}{\rightarrow} (\delta, 0, \delta, 0, \delta, \delta, 0, \delta) \stackrel{1r}{\rightarrow} (\delta,\delta,0,\delta,\delta,0,\delta,0) \stackrel{1r}{\rightarrow} (\delta,\delta,\delta,0,0,\delta,0,\delta) \stackrel{1r}{\rightarrow} (0, \delta, 0, 0, 0, \delta, \delta, 0) \stackrel{1r}{\rightarrow} (\delta,\delta,0,0,0,\delta,\delta,\delta) \stackrel{1r}{\rightarrow} (0,\delta,0,\delta,0,0,\delta,0) \stackrel{1r}{\rightarrow} (0, 0, \delta, 0, 0, \delta, 0, \delta) \stackrel{1r}{\rightarrow} (0,0,0,\delta,\delta,0,\delta,0)$, which is 10-round iterative and holds with probability 2^{-540}.
Starting from an odd-numbered round i, the following 33 subkeys have to be weak: $Z_9^{(i)}$, $Z_{10}^{(i)}$, $Z_{11}^{(i)}$, $Z_{12}^{(i)}$, $Z_4^{(i+1)}$, $Z_5^{(i+1)}$, $Z_{10}^{(i+1)}$, $Z_{12}^{(i+1)}$, $Z_3^{(i+2)}$, $Z_9^{(i+2)}$, $Z_{10}^{(i+2)}$, $Z_{11}^{(i+2)}$, $Z_{12}^{(i+2)}$, $Z_5^{(i+3)}$, $Z_{10}^{(i+3)}$, $Z_1^{(i+4)}$, $Z_{10}^{(i+4)}$, $Z_2^{(i+5)}$, $Z_9^{(i+5)}$, $Z_{10}^{(i+5)}$, $Z_{11}^{(i+5)}$, $Z_{12}^{(i+5)}$, $Z_6^{(i+6)}$, $Z_{10}^{(i+6)}$, $Z_{12}^{(i+6)}$, $Z_2^{(i+7)}$, $Z_7^{(i+7)}$, $Z_9^{(i+7)}$, $Z_{10}^{(i+7)}$, $Z_{11}^{(i+7)}$, $Z_{12}^{(i+7)}$, $Z_{10}^{(i+8)}$ and $Z_{10}^{(i+9)}$.
Starting from an even-numbered round i, the following 55 subkeys have to be weak: $Z_4^{(i)}$, $Z_5^{(i)}$, $Z_7^{(i)}$, $Z_9^{(i)}$, $Z_{10}^{(i)}$, $Z_{11}^{(i)}$, $Z_{12}^{(i)}$, $Z_1^{(i+1)}$, $Z_3^{(i+1)}$, $Z_8^{(i+1)}$, $Z_{10}^{(i+1)}$, $Z_{12}^{(i+1)}$, $Z_4^{(i+2)}$, $Z_5^{(i+2)}$, $Z_9^{(i+2)}$, $Z_{10}^{(i+2)}$, $Z_{11}^{(i+2)}$, $Z_{12}^{(i+2)}$, $Z_1^{(i+3)}$, $Z_3^{(i+3)}$, $Z_6^{(i+3)}$, $Z_8^{(i+3)}$, $Z_{10}^{(i+3)}$, $Z_2^{(i+4)}$, $Z_4^{(i+4)}$, $Z_5^{(i+4)}$, $Z_7^{(i+4)}$, $Z_{10}^{(i+4)}$, $Z_1^{(i+5)}$, $Z_3^{(i+5)}$, $Z_6^{(i+5)}$, $Z_8^{(i+5)}$, $Z_9^{(i+5)}$, $Z_{10}^{(i+5)}$, $Z_{11}^{(i+5)}$, $Z_{12}^{(i+5)}$, $Z_2^{(i+6)}$, $Z_7^{(i+6)}$, $Z_{10}^{(i+6)}$, $Z_{12}^{(i+6)}$, $Z_1^{(i+7)}$, $Z_6^{(i+7)}$, $Z_8^{(i+7)}$, $Z_9^{(i+7)}$, $Z_{10}^{(i+7)}$, $Z_{11}^{(i+7)}$, $Z_{12}^{(i+7)}$, $Z_2^{(i+8)}$, $Z_4^{(i+8)}$, $Z_7^{(i+8)}$, $Z_{10}^{(i+8)}$, $Z_3^{(i+9)}$, $Z_6^{(i+9)}$, $Z_8^{(i+9)}$ and $Z_{10}^{(i+9)}$.

(13) $(0,0,0,\delta,\delta,\delta,0,\delta) \stackrel{1r}{\rightarrow} (\delta,0,\delta,\delta,0,\delta,\delta,\delta) \stackrel{1r}{\rightarrow} (0,\delta,\delta,0,0,0,\delta,\delta) \stackrel{1r}{\rightarrow} (\delta, \delta, 0, \delta, \delta, \delta, 0, \delta) \stackrel{1r}{\rightarrow} (\delta,\delta,\delta,0,\delta,0,\delta,\delta) \stackrel{1r}{\rightarrow} (0,0,0,\delta,\delta,\delta,0,\delta)$, which is 5-round iterative and holds with probability 2^{-150}.
Starting from an odd-numbered round i, the following 20 subkeys have to be weak: $Z_6^{(i)}$, $Z_8^{(i)}$, $Z_9^{(i)}$, $Z_{11}^{(i)}$, $Z_{12}^{(i)}$, $Z_4^{(i+1)}$, $Z_7^{(i+1)}$, $Z_9^{(i+1)}$, $Z_{11}^{(i+1)}$, $Z_{12}^{(i+1)}$, $Z_3^{(i+2)}$, $Z_8^{(i+2)}$, $Z_{12}^{(i+2)}$, $Z_2^{(i+3)}$, $Z_4^{(i+3)}$, $Z_5^{(i+3)}$, $Z_1^{(i+4)}$, $Z_3^{(i+4)}$, $Z_8^{(i+4)}$ and $Z_{12}^{(i+4)}$.
Starting from an even-numbered round i, the following 22 subkeys have to be weak: $Z_4^{(i)}$, $Z_5^{(i)}$, $Z_9^{(i)}$, $Z_{11}^{(i)}$, $Z_{12}^{(i)}$, $Z_1^{(i+1)}$, $Z_3^{(i+1)}$, $Z_6^{(i+1)}$, $Z_8^{(i+1)}$, $Z_9^{(i+1)}$, $Z_{11}^{(i+1)}$, $Z_{12}^{(i+1)}$, $Z_2^{(i+2)}$, $Z_7^{(i+2)}$, $Z_{12}^{(i+2)}$, $Z_1^{(i+3)}$, $Z_6^{(i+3)}$, $Z_8^{(i+3)}$, $Z_2^{(i+4)}$, $Z_5^{(i+4)}$, $Z_7^{(i+4)}$ and $Z_{12}^{(i+4)}$.

(14) $(0,0,0,\delta,\delta,\delta,\delta,0) \stackrel{1r}{\rightarrow} (0,\delta,\delta,0,0,\delta,0,\delta) \stackrel{1r}{\rightarrow} (\delta,\delta,\delta,0,\delta,\delta,\delta,0) \stackrel{1r}{\rightarrow} (\delta, \delta, \delta, \delta, \delta, \delta, 0, 0) \stackrel{1r}{\rightarrow} (0,0,\delta,0,\delta,\delta,0,\delta) \stackrel{1r}{\rightarrow} (0,\delta,\delta,\delta,0,0,\delta,0) \stackrel{1r}{\rightarrow} (\delta, \delta, 0, \delta, \delta, \delta, \delta, 0) \stackrel{1r}{\rightarrow} (0,0,\delta,\delta,\delta,0,0,\delta) \stackrel{1r}{\rightarrow} (\delta,0,0,\delta,0,0,0,0) \stackrel{1r}{\rightarrow} (\delta, 0, 0, \delta, 0, \delta, \delta, 0) \stackrel{1r}{\rightarrow} (\delta,0,\delta,0,0,\delta,0,\delta) \stackrel{1r}{\rightarrow} (\delta,0,\delta,\delta,0,0,\delta,0) \stackrel{1r}{\rightarrow} (\delta, 0, 0, 0, 0, 0, \delta, 0) \stackrel{1r}{\rightarrow} (0,\delta,0,0,0,\delta,\delta,\delta) \stackrel{1r}{\rightarrow} (\delta,\delta,\delta,\delta,\delta,0,\delta,0) \stackrel{1r}{\rightarrow} (0, 0, 0, \delta, \delta, \delta, \delta, 0)$, which is 15-round iterative and holds with probability 2^{-750}.
Starting from an odd-numbered round i, the following 54 subkeys have to be weak: $Z_6^{(i)}$, $Z_9^{(i)}$, $Z_2^{(i+1)}$, $Z_{11}^{(i+1)}$, $Z_{12}^{(i+1)}$, $Z_1^{(i+2)}$, $Z_3^{(i+2)}$, $Z_6^{(i+2)}$, $Z_2^{(i+3)}$, $Z_4^{(i+3)}$, $Z_5^{(i+3)}$, $Z_{11}^{(i+3)}$, $Z_{12}^{(i+3)}$, $Z_3^{(i+4)}$, $Z_6^{(i+4)}$, $Z_8^{(i+4)}$, $Z_9^{(i+4)}$, $Z_2^{(i+5)}$, $Z_4^{(i+5)}$, $Z_7^{(i+5)}$, $Z_{12}^{(i+5)}$, $Z_1^{(i+6)}$, $Z_6^{(i+6)}$, $Z_{11}^{(i+6)}$, $Z_{12}^{(i+6)}$, $Z_4^{(i+7)}$, $Z_5^{(i+7)}$, $Z_9^{(i+7)}$, $Z_{12}^{(i+7)}$, $Z_1^{(i+8)}$, $Z_9^{(i+8)}$, $Z_{11}^{(i+8)}$, $Z_4^{(i+9)}$, $Z_7^{(i+9)}$, $Z_9^{(i+9)}$, $Z_1^{(i+10)}$,

$Z_3^{(i+10)}, Z_6^{(i+10)}, Z_8^{(i+10)}, Z_9^{(i+10)}, Z_4^{(i+11)}, Z_7^{(i+11)}, Z_9^{(i+11)}, Z_{11}^{(i+11)}, Z_1^{(i+12)}, Z_9^{(i+12)}, Z_{12}^{(i+12)}, Z_2^{(i+13)}, Z_7^{(i+13)}, Z_{11}^{(i+13)}, Z_{12}^{(i+13)}, Z_1^{(i+14)}, Z_3^{(i+14)}$ and $Z_{12}^{(i+14)}$.

Starting from an even-numbered round i, the following 53 subkeys have to be weak: $Z_4^{(i)}, Z_5^{(i)}, Z_7^{(i)}, Z_9^{(i)}, Z_3^{(i+1)}, Z_6^{(i+1)}, Z_8^{(i+1)}, Z_{11}^{(i+1)}, Z_{12}^{(i+1)}, Z_2^{(i+2)}, Z_5^{(i+2)}, Z_7^{(i+2)}, Z_1^{(i+3)}, Z_3^{(i+3)}, Z_6^{(i+3)}, Z_{11}^{(i+3)}, Z_{12}^{(i+3)}, Z_5^{(i+4)}, Z_9^{(i+4)}, Z_3^{(i+5)}, Z_{12}^{(i+5)}, Z_2^{(i+6)}, Z_4^{(i+6)}, Z_5^{(i+6)}, Z_7^{(i+6)}, Z_{11}^{(i+6)}, Z_{12}^{(i+6)}, Z_3^{(i+7)}, Z_8^{(i+7)}, Z_9^{(i+7)}, Z_{12}^{(i+7)}, Z_4^{(i+8)}, Z_9^{(i+8)}, Z_{11}^{(i+8)}, Z_1^{(i+9)}, Z_6^{(i+9)}, Z_9^{(i+9)}, Z_9^{(i+10)}, Z_1^{(i+11)}, Z_3^{(i+11)}, Z_9^{(i+11)}, Z_{11}^{(i+11)}, Z_7^{(i+12)}, Z_9^{(i+12)}, Z_{12}^{(i+12)}, Z_6^{(i+13)}, Z_8^{(i+13)}, Z_{11}^{(i+13)}, Z_{12}^{(i+13)}, Z_2^{(i+14)}, Z_4^{(i+14)}, Z_5^{(i+14)}, Z_7^{(i+14)}$ and $Z_{12}^{(i+14)}$.

(15) $(0,0,\delta,0,0,\delta,0,0) \xrightarrow{1r} (0,0,\delta,0,0,\delta,\delta,\delta) \xrightarrow{1r} (\delta,\delta,\delta,\delta,0,\delta,0,\delta) \xrightarrow{1r} (0,\delta,\delta,\delta,\delta,0,0,0) \xrightarrow{1r} (0,\delta,0,\delta,\delta,0,0,\delta) \xrightarrow{1r} (\delta,0,0,\delta,\delta,\delta,\delta,\delta) \xrightarrow{1r} (\delta,\delta,\delta,\delta,0,0,0,0) \xrightarrow{1r} (\delta,0,0,\delta,\delta,0,0,\delta) \xrightarrow{1r} (\delta,\delta,0,0,0,0,\delta,\delta) \xrightarrow{1r} (\delta,0,0,0,\delta,\delta,\delta,0) \xrightarrow{1r} (\delta,\delta,\delta,\delta,0,0,\delta,\delta) \xrightarrow{1r} (0,\delta,0,0,\delta,0,\delta,\delta) \xrightarrow{1r} (0,\delta,0,0,\delta,\delta,\delta,0) \xrightarrow{1r} (\delta,0,\delta,0,\delta,\delta,\delta,\delta) \xrightarrow{1r} (0,0,\delta,\delta,0,\delta,0,\delta) \xrightarrow{1r} (0,0,\delta,0,0,\delta,0,0)$, which is 15-round iterative and holds with probability 2^{-570}.

Starting from an odd-numbered round i, the following 54 subkeys have to be weak: $Z_3^{(i)}, Z_6^{(i)}, Z_{11}^{(i)}, Z_7^{(i+1)}, Z_{12}^{(i+1)}, Z_1^{(i+2)}, Z_3^{(i+2)}, Z_6^{(i+2)}, Z_8^{(i+2)}, Z_9^{(i+2)}, Z_{12}^{(i+2)}, Z_2^{(i+3)}, Z_4^{(i+3)}, Z_5^{(i+3)}, Z_9^{(i+3)}, Z_8^{(i+4)}, Z_9^{(i+4)}, Z_{11}^{(i+4)}, Z_{12}^{(i+4)}, Z_4^{(i+5)}, Z_5^{(i+5)}, Z_7^{(i+5)}, Z_{11}^{(i+5)}, Z_1^{(i+6)}, Z_3^{(i+6)}, Z_9^{(i+6)}, Z_4^{(i+7)}, Z_5^{(i+7)}, Z_1^{(i+8)}, Z_8^{(i+8)}, Z_9^{(i+8)}, Z_5^{(i+9)}, Z_7^{(i+9)}, Z_{11}^{(i+9)}, Z_1^{(i+10)}, Z_3^{(i+10)}, Z_8^{(i+10)}, Z_9^{(i+10)}, Z_{11}^{(i+10)}, Z_{12}^{(i+10)}, Z_2^{(i+11)}, Z_5^{(i+11)}, Z_7^{(i+11)}, Z_9^{(i+11)}, Z_6^{(i+12)}, Z_9^{(i+12)}, Z_{12}^{(i+12)}, Z_5^{(i+13)}, Z_7^{(i+13)}, Z_{12}^{(i+13)}, Z_3^{(i+14)}, Z_6^{(i+14)}, Z_8^{(i+14)}$ and $Z_{11}^{(i+14)}$.

Starting from an even-numbered round i, the following 52 subkeys have to be weak: $Z_{11}^{(i)}, Z_3^{(i+1)}, Z_6^{(i+1)}, Z_8^{(i+1)}, Z_{12}^{(i+1)}, Z_2^{(i+2)}, Z_4^{(i+2)}, Z_9^{(i+2)}, Z_{12}^{(i+2)}, Z_3^{(i+3)}, Z_9^{(i+3)}, Z_2^{(i+4)}, Z_4^{(i+4)}, Z_5^{(i+4)}, Z_9^{(i+4)}, Z_{11}^{(i+4)}, Z_{12}^{(i+4)}, Z_1^{(i+5)}, Z_6^{(i+5)}, Z_8^{(i+5)}, Z_{11}^{(i+5)}, Z_2^{(i+6)}, Z_4^{(i+6)}, Z_9^{(i+6)}, Z_1^{(i+7)}, Z_8^{(i+7)}, Z_2^{(i+8)}, Z_7^{(i+8)}, Z_9^{(i+8)}, Z_1^{(i+9)}, Z_6^{(i+9)}, Z_{11}^{(i+9)}, Z_2^{(i+10)}, Z_4^{(i+10)}, Z_7^{(i+10)}, Z_9^{(i+10)}, Z_{11}^{(i+10)}, Z_{12}^{(i+10)}, Z_8^{(i+11)}, Z_9^{(i+11)}, Z_2^{(i+12)}, Z_5^{(i+12)}, Z_7^{(i+12)}, Z_9^{(i+12)}, Z_{12}^{(i+12)}, Z_1^{(i+13)}, Z_3^{(i+13)}, Z_6^{(i+13)}, Z_8^{(i+13)}, Z_{12}^{(i+13)}, Z_4^{(i+14)}$ and $Z_{11}^{(i+14)}$.

(16) $(0,0,\delta,0,\delta,0,0,0) \xrightarrow{1r} (\delta,0,0,\delta,0,0,\delta,\delta) \xrightarrow{1r} (0,\delta,0,0,0,\delta,0,0) \xrightarrow{1r} (0,0,\delta,0,\delta,0,0,0)$, which is 3-round iterative and holds with probability 2^{-120}. Starting from an odd-numbered round i, the following eight subkeys have to be weak: $Z_3^{(i)}, Z_9^{(i)}, Z_{12}^{(i)}, Z_4^{(i+1)}, Z_7^{(i+1)}, Z_9^{(i+1)}, Z_{12}^{(i+1)}$ and $Z_6^{(i+2)}$. Starting from an even-numbered round i, the following eight subkeys have to be weak: $Z_5^{(i)}, Z_9^{(i)}, Z_{12}^{(i)}, Z_1^{(i+1)}, Z_8^{(i+1)}, Z_9^{(i+1)}, Z_{12}^{(i+1)}$ and $Z_2^{(i+2)}$.

(17) $(0,0,\delta,0,\delta,\delta,0,0) \xrightarrow{1r} (0,\delta,0,0,\delta,\delta,\delta,\delta) \xrightarrow{1r} (\delta,0,0,\delta,0,0,\delta,0) \xrightarrow{1r} (0, \delta, \delta, \delta, \delta, 0, 0, \delta) \xrightarrow{1r} (0,\delta,\delta,0,0,\delta,0,0) \xrightarrow{1r} (\delta,\delta,0,\delta,0,0,\delta,\delta) \xrightarrow{1r} (\delta, 0, \delta, \delta, 0, 0, 0, 0) \xrightarrow{1r} (0,\delta,\delta,0,\delta,\delta,0,\delta) \xrightarrow{1r} (\delta,0,0,0,0,\delta,\delta,0) \xrightarrow{1r} (\delta, 0, 0, \delta, \delta, 0, \delta, \delta) \xrightarrow{1r} (0,0,\delta,0,\delta,\delta,0,0)$, which is 10-round iterative and holds with probability 2^{-480}.
Starting from an odd-numbered round i, the following 55 subkeys have to be weak: $Z_3^{(i)}$, $Z_6^{(i)}$, $Z_9^{(i)}$, $Z_{10}^{(i)}$, $Z_{11}^{(i)}$, $Z_2^{(i+1)}$, $Z_5^{(i+1)}$, $Z_7^{(i+1)}$, $Z_9^{(i+1)}$, $Z_{10}^{(i+1)}$, $Z_{11}^{(i+1)}$, $Z_{12}^{(i+1)}$, $Z_1^{(i+2)}$, $Z_9^{(i+2)}$, $Z_{10}^{(i+2)}$, $Z_{11}^{(i+2)}$, $Z_{12}^{(i+2)}$, $Z_2^{(i+3)}$, $Z_4^{(i+3)}$, $Z_5^{(i+3)}$, $Z_9^{(i+3)}$, $Z_{10}^{(i+3)}$, $Z_{11}^{(i+3)}$, $Z_3^{(i+4)}$, $Z_6^{(i+4)}$, $Z_{10}^{(i+4)}$, $Z_{12}^{(i+4)}$, $Z_2^{(i+5)}$, $Z_4^{(i+5)}$, $Z_7^{(i+5)}$, $Z_9^{(i+5)}$, $Z_{10}^{(i+5)}$, $Z_{11}^{(i+5)}$, $Z_1^{(i+6)}$, $Z_3^{(i+6)}$, $Z_9^{(i+6)}$, $Z_{10}^{(i+6)}$, $Z_{11}^{(i+6)}$, $Z_{12}^{(i+6)}$, $Z_2^{(i+7)}$, $Z_5^{(i+7)}$, $Z_9^{(i+7)}$, $Z_{10}^{(i+7)}$, $Z_{11}^{(i+7)}$, $Z_{12}^{(i+7)}$, $Z_1^{(i+8)}$, $Z_6^{(i+8)}$, $Z_9^{(i+8)}$, $Z_{10}^{(i+8)}$, $Z_{11}^{(i+8)}$, $Z_4^{(i+9)}$, $Z_5^{(i+9)}$, $Z_7^{(i+9)}$, $Z_{10}^{(i+9)}$ and $Z_{12}^{(i+9)}$.
Starting from an even-numbered round i, the following 49 subkeys have to be weak: $Z_5^{(i)}$, $Z_9^{(i)}$, $Z_{10}^{(i)}$, $Z_{11}^{(i)}$, $Z_6^{(i+1)}$, $Z_8^{(i+1)}$, $Z_9^{(i+1)}$, $Z_{10}^{(i+1)}$, $Z_{11}^{(i+1)}$, $Z_{12}^{(i+1)}$, $Z_4^{(i+2)}$, $Z_7^{(i+2)}$, $Z_9^{(i+2)}$, $Z_{10}^{(i+2)}$, $Z_{11}^{(i+2)}$, $Z_{12}^{(i+2)}$, $Z_3^{(i+3)}$, $Z_8^{(i+3)}$, $Z_9^{(i+3)}$, $Z_{10}^{(i+3)}$, $Z_{11}^{(i+3)}$, $Z_2^{(i+4)}$, $Z_{10}^{(i+4)}$, $Z_{12}^{(i+4)}$, $Z_1^{(i+5)}$, $Z_8^{(i+5)}$, $Z_9^{(i+5)}$, $Z_{10}^{(i+5)}$, $Z_{11}^{(i+5)}$, $Z_4^{(i+6)}$, $Z_9^{(i+6)}$, $Z_{10}^{(i+6)}$, $Z_{11}^{(i+6)}$, $Z_{12}^{(i+6)}$, $Z_3^{(i+7)}$, $Z_6^{(i+7)}$, $Z_8^{(i+7)}$, $Z_9^{(i+7)}$, $Z_{10}^{(i+7)}$, $Z_{11}^{(i+7)}$, $Z_{12}^{(i+7)}$, $Z_7^{(i+8)}$, $Z_9^{(i+8)}$, $Z_{10}^{(i+8)}$, $Z_{11}^{(i+8)}$, $Z_1^{(i+9)}$, $Z_8^{(i+9)}$, $Z_{10}^{(i+9)}$ and $Z_{12}^{(i+9)}$.

(18) $(0,0,\delta,\delta,0,\delta,\delta,0) \xrightarrow{1r} (\delta,\delta,\delta,\delta,0,\delta,\delta,0) \xrightarrow{1r} (\delta,0,\delta,0,\delta,0,\delta,0) \xrightarrow{1r} (\delta, \delta, 0, \delta, 0, \delta, 0, 0) \xrightarrow{1r} (\delta,0,\delta,\delta,\delta,\delta,\delta,0) \xrightarrow{1r} (0,0,\delta,\delta,0,\delta,\delta,0)$, which is 5-round iterative and holds with probability 2^{-210}.
Starting from an odd-numbered round i, the following 18 subkeys have to be weak: $Z_3^{(i)}$, $Z_6^{(i)}$, $Z_{12}^{(i)}$, $Z_2^{(i+1)}$, $Z_4^{(i+1)}$, $Z_7^{(i+1)}$, $Z_9^{(i+1)}$, $Z_{11}^{(i+1)}$, $Z_1^{(i+2)}$, $Z_3^{(i+2)}$, $Z_2^{(i+3)}$, $Z_4^{(i+3)}$, $Z_9^{(i+3)}$, $Z_{11}^{(i+3)}$, $Z_1^{(i+4)}$, $Z_3^{(i+4)}$, $Z_6^{(i+4)}$ and $Z_{12}^{(i+4)}$.
Starting from an even-numbered round i, the following 18 subkeys have to be weak: $Z_4^{(i)}$, $Z_7^{(i)}$, $Z_{12}^{(i)}$, $Z_1^{(i+1)}$, $Z_3^{(i+1)}$, $Z_6^{(i+1)}$, $Z_9^{(i+1)}$, $Z_{11}^{(i+1)}$, $Z_5^{(i+2)}$, $Z_7^{(i+2)}$, $Z_1^{(i+3)}$, $Z_6^{(i+3)}$, $Z_9^{(i+3)}$, $Z_{11}^{(i+3)}$, $Z_4^{(i+4)}$, $Z_5^{(i+4)}$, $Z_7^{(i+4)}$ and $Z_{12}^{(i+4)}$

(19) $(0,0,\delta,\delta,\delta,0,\delta,\delta) \xrightarrow{1r} (0,\delta,\delta,\delta,\delta,\delta,\delta,\delta) \xrightarrow{1r} (0,\delta,0,\delta,0,\delta,\delta,\delta) \xrightarrow{1r} (\delta, \delta, 0, 0, 0, \delta, 0, 0) \xrightarrow{1r} (\delta,0,0,0,0,0,0,0) \xrightarrow{1r} (\delta,0,\delta,0,\delta,0,0,0) \xrightarrow{1r} (0, 0, \delta, \delta, \delta, 0, \delta, \delta)$, which is 6-round iterative and holds with probability 2^{-300}.
Starting from an odd-numbered round i, the following 23 subkeys have to be weak: $Z_3^{(i)}$, $Z_8^{(i)}$, $Z_9^{(i)}$, $Z_{10}^{(i)}$, $Z_2^{(i+1)}$, $Z_4^{(i+1)}$, $Z_5^{(i+1)}$, $Z_7^{(i+1)}$, $Z_9^{(i+1)}$, $Z_{10}^{(i+1)}$, $Z_6^{(i+2)}$, $Z_8^{(i+2)}$, $Z_{10}^{(i+2)}$, $Z_{12}^{(i+2)}$, $Z_2^{(i+3)}$, $Z_9^{(i+3)}$, $Z_{10}^{(i+3)}$, $Z_1^{(i+4)}$, $Z_9^{(i+4)}$, $Z_{10}^{(i+4)}$, $Z_5^{(i+5)}$, $Z_{10}^{(i+5)}$ and $Z_{12}^{(i+5)}$.
Starting from an even-numbered round i, the following 25 subkeys have to be weak: $Z_4^{(i)}$, $Z_5^{(i)}$, $Z_7^{(i)}$, $Z_9^{(i)}$, $Z_{10}^{(i)}$, $Z_3^{(i+1)}$, $Z_6^{(i+1)}$, $Z_8^{(i+1)}$, $Z_9^{(i+1)}$, $Z_{10}^{(i+1)}$, $Z_2^{(i+2)}$, $Z_4^{(i+2)}$, $Z_7^{(i+2)}$, $Z_{10}^{(i+2)}$, $Z_{12}^{(i+2)}$, $Z_1^{(i+3)}$, $Z_6^{(i+3)}$, $Z_9^{(i+3)}$, $Z_{10}^{(i+3)}$, $Z_9^{(i+4)}$, $Z_{10}^{(i+4)}$, $Z_1^{(i+5)}$, $Z_3^{(i+5)}$, $Z_{10}^{(i+5)}$ and $Z_{12}^{(i+5)}$.

(20) $(0,0,\delta,\delta,\delta,\delta,\delta,\delta) \xrightarrow{1r} (\delta,0,\delta,0,0,0,\delta,\delta) \xrightarrow{1r} (\delta,0,0,0,0,0,0,\delta) \xrightarrow{1r} (\delta,0,0,\delta,0,\delta,0,\delta) \xrightarrow{1r} (0,\delta,\delta,\delta,0,\delta,\delta,\delta) \xrightarrow{1r} (0,0,\delta,\delta,\delta,\delta,\delta,\delta)$, which is 5-round iterative and holds with probability 2^{-180}.
Starting from an odd-numbered round i, the following 20 subkeys have to be weak: $Z_3^{(i)}$, $Z_6^{(i)}$, $Z_8^{(i)}$, $Z_9^{(i)}$, $Z_{11}^{(i)}$, $Z_{12}^{(i)}$, $Z_7^{(i+1)}$, $Z_9^{(i+1)}$, $Z_{11}^{(i+1)}$, $Z_1^{(i+2)}$, $Z_8^{(i+2)}$, $Z_9^{(i+2)}$, $Z_{11}^{(i+2)}$, $Z_4^{(i+3)}$, $Z_9^{(i+3)}$, $Z_{11}^{(i+3)}$, $Z_{12}^{(i+3)}$, $Z_3^{(i+4)}$, $Z_6^{(i+4)}$ and $Z_8^{(i+4)}$.
Starting from an even-numbered round i, the following 22 subkeys have to be weak: $Z_4^{(i)}$, $Z_5^{(i)}$, $Z_7^{(i)}$, $Z_9^{(i)}$, $Z_{11}^{(i)}$, $Z_{12}^{(i)}$, $Z_1^{(i+1)}$, $Z_3^{(i+1)}$, $Z_8^{(i+1)}$, $Z_9^{(i+1)}$, $Z_{11}^{(i+1)}$, $Z_9^{(i+2)}$, $Z_{11}^{(i+2)}$, $Z_1^{(i+3)}$, $Z_6^{(i+3)}$, $Z_8^{(i+3)}$, $Z_9^{(i+3)}$, $Z_{11}^{(i+3)}$, $Z_{12}^{(i+3)}$, $Z_2^{(i+4)}$, $Z_4^{(i+4)}$ and $Z_7^{(i+4)}$.

(21) $(0,\delta,0,\delta,\delta,\delta,0,0) \xrightarrow{1r} (0,\delta,\delta,\delta,\delta,\delta,\delta,0) \xrightarrow{1r} (0,\delta,\delta,0,\delta,0,\delta,0) \xrightarrow{1r} (\delta,0,0,0,\delta,0,0,0) \xrightarrow{1r} (\delta,\delta,0,0,0,0,0,0) \xrightarrow{1r} (0,\delta,0,\delta,\delta,\delta,0,0)$, which is 5-round iterative and holds with probability 2^{-180}.
Starting from an odd-numbered round i, the following 18 subkeys have to be weak: $Z_6^{(i)}$, $Z_9^{(i)}$, $Z_{11}^{(i)}$, $Z_2^{(i+1)}$, $Z_4^{(i+1)}$, $Z_5^{(i+1)}$, $Z_7^{(i+1)}$, $Z_9^{(i+1)}$, $Z_{11}^{(i+1)}$, $Z_3^{(i+2)}$, $Z_9^{(i+2)}$, $Z_{11}^{(i+2)}$, $Z_{12}^{(i+2)}$, $Z_5^{(i+3)}$, $Z_1^{(i+4)}$, $Z_9^{(i+4)}$, $Z_{11}^{(i+4)}$ and $Z_{12}^{(i+4)}$.
Starting from an even-numbered round i, the following 20 subkeys have to be weak: $Z_2^{(i)}$, $Z_4^{(i)}$, $Z_5^{(i)}$, $Z_9^{(i)}$, $Z_{11}^{(i)}$, $Z_3^{(i+1)}$, $Z_6^{(i+1)}$, $Z_9^{(i+1)}$, $Z_{11}^{(i+1)}$, $Z_2^{(i+2)}$, $Z_5^{(i+2)}$, $Z_7^{(i+2)}$, $Z_9^{(i+2)}$, $Z_{11}^{(i+2)}$, $Z_{12}^{(i+2)}$, $Z_1^{(i+3)}$, $Z_2^{(i+4)}$, $Z_9^{(i+4)}$, $Z_{11}^{(i+4)}$ and $Z_{12}^{(i+4)}$

(22) $(0,\delta,\delta,0,\delta,\delta,0,0) \xrightarrow{1r} (\delta,0,\delta,\delta,\delta,0,\delta,\delta) \xrightarrow{1r} (\delta,\delta,0,\delta,0,\delta,\delta,\delta) \xrightarrow{1r} (0,\delta,\delta,0,\delta,\delta,0,0)$, which is 3-round iterative and holds with probability 2^{-120}.
Starting from an odd-numbered round i, the following 12 subkeys have to be weak: $Z_3^{(i)}$, $Z_6^{(i)}$, $Z_9^{(i)}$, $Z_{12}^{(i)}$, $Z_4^{(i+1)}$, $Z_5^{(i+1)}$, $Z_7^{(i+1)}$, $Z_1^{(i+2)}$, $Z_6^{(i+2)}$, $Z_8^{(i+2)}$, $Z_9^{(i+2)}$ and $Z_{12}^{(i+2)}$.
Starting from an even-numbered round i, the following 12 subkeys have to be weak: $Z_2^{(i)}$, $Z_5^{(i)}$, $Z_9^{(i)}$, $Z_{12}^{(i)}$, $Z_1^{(i+1)}$, $Z_3^{(i+1)}$, $Z_8^{(i+1)}$, $Z_2^{(i+2)}$, $Z_4^{(i+2)}$, $Z_7^{(i+2)}$, $Z_9^{(i+2)}$ and $Z_{12}^{(i+2)}$.

(23) $(\delta,\delta,\delta,\delta,\delta,\delta,\delta,\delta) \xrightarrow{1r} (\delta,\delta,\delta,\delta,\delta,\delta,\delta,\delta)$, which is 1-round iterative and holds with probability 1.
Starting from an odd-numbered round i, the following four subkeys have to be weak: $Z_1^{(i)}$, $Z_3^{(i)}$, $Z_6^{(i)}$ and $Z_8^{(i)}$.
Starting from an even-numbered round i, the following four subkeys have to be weak: $Z_2^{(i)}$, $Z_4^{(i)}$, $Z_5^{(i)}$ and $Z_7^{(i)}$.

Starting from the first round, $i = 1$, the characteristic that requires the least amount of weak subkeys is (16): only 2.66 weak subkeys/round on average. Although it is 3-round iterative, it holds with probability 2^{-120}. Characteristic (8) comes next, with only 3.5 weak subkeys/round on average and with probability 2^{-60}. Finally, we have characteristic (23), with 4 weak subkeys/round and with probability 1.

Unlike the key schedule algorithms of PES and IDEA, the key schedule of MESH-128 does not consist of a fixed bit permutation, but is rather more involved. See Sect. 2.3.6. Experiments with mini-versions of the key schedule for a scaled-down version of MESH-128 operating on 32-bit blocks and 64-bit keys indicated that each weak subkey restriction (with the three most significant bits equal to zero) is satisfied for a fraction of 2^{-3} of the key space. Assuming this behavior holds for each weak subkey independently, and for larger key schedules as well, such as the one for MESH-128, then t weak subkeys would apply to a fraction of $2^{256-t\cdot 15}$ of the key space, or a weak-key class of this size. Thus, for a nonempty weak-key class, t must satisfy $256 - t \cdot 15 \geq 0$, that is, $t \leq 17$.

If characteristic (16) is repeated twice, it will require 16 weak subkeys, and cover six rounds with probability 2^{-240}. If characteristic (8) is repeated twice, it will require 14 weak subkeys and cover four rounds with probability 2^{-120}. If characteristic (23) is repeated four times, it will require 16 weak subkeys and cover four rounds with probability 1.

For a 128-bit block cipher, one could produce up to $2^{6\cdot 16} \cdot 2^{32}(2^{32}-1)/2 \approx 2^{159}$ plaintext pairs, taking into account the six trivial difference words and two nontrivial difference words in characteristic (16). This amount is still insufficient to satisfy (16).

In summary, the best trade-off is characteristic (23), which allows us to distinguish 4-round MESH-128 from a random permutation, under sixteen weak subkeys.

This differential analysis using characteristics also allows us to study the diffusion power of the MA half-round of MESH-128 with respect to modern tools such as the (differential) branch number (Appendix, Sect. B).

The set of characteristics in Table 3.17 and Table 3.24 indicates that the differential branch number *per round* of MESH-128 is only 4 (under weak-key assumptions). For instance, the wordwise Hamming Weight of the input and output differences of the 1-round characteristic

$$(\delta, 0, 0, 0, 0, 0, 0, 0) \stackrel{1r}{\to} (\delta, 0, \delta, 0, \delta, 0, 0, 0)$$

add up to 4. So, even though MESH-128 provides full diffusion in a single round, a detailed analysis shows that its differential branch number is much lower than ideal. For an 8-word block, such as in MESH-128, an MDS code using a 8×8 MDS matrix would rather provide a branch number of $1+8=9$. See Appendix, Sect. B.1, for an explanation of the differential branch number.

3.5.6 DC of MESH-64(8) Under Weak-Key Assumptions

The following differential analysis of MESH-64(8) is novel and operates under weak-key assumptions.

The analysis follows a bottom-up approach, starting from the smallest components: $\odot$, $\boxplus$ and $\oplus$, up to an MA-box, up to a half-round, further up to a full round, and finally to multiple rounds.

Across unkeyed $\odot$ and $\boxplus$, the propagation of 8-bit difference words follows from Table 3.12 and Table 3.13, respectively.

The next step is to obtain all difference patterns across the MA-box of MESH-64(8), under the corresponding weak-subkey restrictions. The only 8-bit difference values used are $\delta = 80_x$ and 00_x. Table 3.25 lists all such difference patterns across the MA-box of MESH-64(8).

Table 3.25 Wordwise differences across the MA-box of MESH-64(8) ($\delta = 80_x$).

Wordwise differences across the MA-box	# active $\odot$	Probability	Weak subkey(s)
$(0,0,0,0) \rightarrow (0,0,0,0)$	0	1	—
$(0,0,0,\delta) \rightarrow (\delta,\delta,\delta,\delta)$	1	2^{-7}	$Z_{10}^{(i)}$
$(0,0,\delta,0) \rightarrow (0,0,0,\delta)$	1	2^{-7}	$Z_{10}^{(i)}$
$(0,0,\delta,\delta) \rightarrow (\delta,\delta,\delta,0)$	2	2^{-14}	—
$(0,\delta,0,0) \rightarrow (\delta,\delta,0,\delta)$	2	2^{-14}	$Z_{10}^{(i)}$
$(0,\delta,0,\delta) \rightarrow (0,0,\delta,0)$	2	2^{-14}	—
$(0,\delta,\delta,0) \rightarrow (\delta,\delta,0,0)$	2	2^{-14}	—
$(0,\delta,\delta,\delta) \rightarrow (0,0,\delta,\delta)$	2	2^{-14}	$Z_{10}^{(i)}$
$(\delta,0,0,0) \rightarrow (0,\delta,0,\delta)$	2	2^{-14}	$Z_9^{(i)}, Z_{10}^{(i)}$
$(\delta,0,0,\delta) \rightarrow (\delta,0,\delta,0)$	2	2^{-14}	$Z_9^{(i)}$
$(\delta,0,\delta,0) \rightarrow (0,\delta,0,0)$	2	2^{-14}	$Z_9^{(i)}$
$(\delta,0,\delta,\delta) \rightarrow (\delta,0,\delta,\delta)$	2	2^{-14}	$Z_9^{(i)}, Z_{10}^{(i)}$
$(\delta,\delta,0,0) \rightarrow (\delta,0,0,0)$	0	1	$Z_9^{(i)}$
$(\delta,\delta,0,\delta) \rightarrow (0,\delta,\delta,\delta)$	1	2^{-7}	$Z_9^{(i)}, Z_{10}^{(i)}$
$(\delta,\delta,\delta,0) \rightarrow (\delta,0,0,\delta)$	1	2^{-7}	$Z_9^{(i)}, Z_{10}^{(i)}$
$(\delta,\delta,\delta,\delta) \rightarrow (0,\delta,\delta,0)$	2	2^{-14}	$Z_9^{(i)}$

The next step is to obtain from Table 3.25 all nontrivial 1-round differential characteristics and the corresponding weak subkeys restrictions.

Table 3.26 through Table 3.33 list all 255 nontrivial 1-round differential characteristics of MESH-64(8), under weak-key assumptions, using only wordwise differences 00_x and 80_x. These tables provide an exhaustive analysis of how characteristics propagate across a single round of MESH-64(8).

Multiple-round characteristics for MESH-64(8) are obtained by concatenating 1-round characteristics such that the output difference of one characteristic matches the input difference of the next characteristic.

Table 3.26 1-round characteristics for MESH-64(8) (part I).

1-round characteristic	Weak subkeys $Z_j^{(i)}$ (i-th round) j value (odd i)	j value (even i)	Prob.
$(0,0,0,0,0,0,0,\delta) \xrightarrow{1r} (\delta,\delta,\delta,\delta,\delta,\delta,\delta,0)$	8;10	10	2^{-7}
$(0,0,0,0,0,0,\delta,0) \xrightarrow{1r} (0,0,0,\delta,0,0,\delta,\delta)$	10	7;10	2^{-7}
$(0,0,0,0,0,0,\delta,\delta) \xrightarrow{1r} (\delta,\delta,\delta,0,\delta,\delta,0,\delta)$	8	7	2^{-14}
$(0,0,0,0,0,\delta,0,0) \xrightarrow{1r} (\delta,\delta,0,0,\delta,0,\delta,\delta)$	6;10	10	2^{-14}
$(0,0,0,0,0,\delta,0,\delta) \xrightarrow{1r} (0,0,\delta,\delta,0,\delta,0,\delta)$	6;8	—	2^{-14}
$(0,0,0,0,0,\delta,\delta,0) \xrightarrow{1r} (\delta,\delta,0,\delta,\delta,0,0,0)$	6	7	2^{-14}
$(0,0,0,0,0,\delta,\delta,\delta) \xrightarrow{1r} (0,0,\delta,0,0,\delta,\delta,0)$	6;8;10	7;10	2^{-14}
$(0,0,0,0,\delta,0,0,0) \xrightarrow{1r} (0,\delta,\delta,0,\delta,0,\delta,\delta)$	9; 10	5;9;10	2^{-14}
$(0,0,0,0,\delta,0,0,\delta) \xrightarrow{1r} (\delta,0,0,\delta,0,\delta,0,\delta)$	8;9	5;9	2^{-14}
$(0,0,0,0,\delta,0,\delta,0) \xrightarrow{1r} (0,\delta,\delta,\delta,\delta,0,0,0)$	9	5;7;9	2^{-14}
$(0,0,0,0,\delta,0,\delta,\delta) \xrightarrow{1r} (\delta,0,0,0,0,\delta,\delta,0)$	8;9;10	5;7;9;10	2^{-14}
$(0,0,0,0,\delta,\delta,0,0) \xrightarrow{1r} (\delta,0,\delta,0,0,0,0,0)$	6;9	5;9	1
$(0,0,0,0,\delta,\delta,0,\delta) \xrightarrow{1r} (0,\delta,0,\delta,\delta,\delta,\delta,0)$	6;8;9;10	5;9;10	2^{-7}
$(0,0,0,0,\delta,\delta,\delta,0) \xrightarrow{1r} (\delta,0,\delta,\delta,0,0,\delta,\delta)$	6;9;10	5;7;9;10	2^{-7}
$(0,0,0,0,\delta,\delta,\delta,\delta) \xrightarrow{1r} (0,\delta,0,0,\delta,\delta,0,\delta)$	6;8;9	5;7;9	2^{-14}
$(0,0,0,\delta,0,0,0,0) \xrightarrow{1r} (\delta,\delta,\delta,\delta,\delta,\delta,0,\delta)$	10	4;10	2^{-7}
$(0,0,0,\delta,0,0,0,\delta) \xrightarrow{1r} (0,0,0,0,0,0,\delta,\delta)$	8	4	1
$(0,0,0,\delta,0,0,\delta,0) \xrightarrow{1r} (\delta,\delta,\delta,0,\delta,\delta,\delta,0)$	—	4;7	2^{-14}
$(0,0,0,\delta,0,0,\delta,\delta) \xrightarrow{1r} (0,0,0,\delta,0,0,0,0)$	8;10	4;7;10	2^{-7}
$(0,0,0,\delta,0,\delta,0,0) \xrightarrow{1r} (0,0,\delta,\delta,0,\delta,\delta,0)$	6	4	2^{-14}
$(0,0,0,\delta,0,\delta,0,\delta) \xrightarrow{1r} (\delta,\delta,0,0,\delta,0,0,0)$	6;8;10	4;10	2^{-14}
$(0,0,0,\delta,0,\delta,\delta,0) \xrightarrow{1r} (0,0,\delta,0,0,\delta,0,\delta)$	6;10	4;7;10	2^{-14}
$(0,0,0,\delta,0,\delta,\delta,\delta) \xrightarrow{1r} (\delta,\delta,0,\delta,\delta,0,\delta,\delta)$	6;8	4;7	2^{-14}
$(0,0,0,\delta,\delta,0,0,0) \xrightarrow{1r} (\delta,0,0,\delta,0,\delta,\delta,0)$	9	4;5;9	2^{-14}
$(0,0,0,\delta,\delta,0,0,\delta) \xrightarrow{1r} (0,\delta,\delta,0,\delta,0,0,0)$	8;9;10	4;5;9;10	2^{-14}
$(0,0,0,\delta,\delta,0,\delta,0) \xrightarrow{1r} (\delta,0,0,0,0,\delta,0,\delta)$	9;10	4;5;7;9;10	2^{-14}
$(0,0,0,\delta,\delta,0,\delta,\delta) \xrightarrow{1r} (0,\delta,\delta,\delta,\delta,0,\delta,\delta)$	8;9	4;5;7;9	2^{-14}
$(0,0,0,\delta,\delta,\delta,0,0) \xrightarrow{1r} (0,\delta,0,\delta,\delta,\delta,0,\delta)$	6;9;10	4;5;9;10	2^{-7}
$(0,0,0,\delta,\delta,\delta,0,\delta) \xrightarrow{1r} (\delta,0,\delta,0,0,0,\delta,\delta)$	6;8;9	4;5;9	1
$(0,0,0,\delta,\delta,\delta,\delta,0) \xrightarrow{1r} (0,\delta,0,0,\delta,\delta,\delta,0)$	6;9	4;5;7;9	2^{-14}
$(0,0,0,\delta,\delta,\delta,\delta,\delta) \xrightarrow{1r} (\delta,0,\delta,\delta,0,0,0,0)$	6;8;9;10	4;5;7;9; 10	2^{-7}
$(0,0,\delta,0,0,0,0,0) \xrightarrow{1r} (0,0,0,0,0,\delta,\delta,\delta)$	3;10	10	2^{-7}

Table 3.27 1-round characteristics for MESH-64(8) (part II).

1-round characteristic	Weak subkeys $Z_j^{(i)}$ (i-th round)		Prob.
	j value (odd i)	j value (even i)	
$(0,0,\delta,0,0,0,0,\delta) \xrightarrow{1r} (\delta,\delta,\delta,\delta,\delta,0,0,\delta)$	3;8	—	2^{-14}
$(0,0,\delta,0,0,0,\delta,0) \xrightarrow{1r} (0,0,0,\delta,0,\delta,0,0)$	3	7	1
$(0,0,\delta,0,0,0,\delta,\delta) \xrightarrow{1r} (\delta,\delta,\delta,0,\delta,0,\delta,0)$	3;8;10	7;10	2^{-7}
$(0,0,\delta,0,0,\delta,0,0) \xrightarrow{1r} (\delta,\delta,0,0,\delta,\delta,0,0)$	3;6	—	2^{-14}
$(0,0,\delta,0,0,\delta,0,\delta) \xrightarrow{1r} (0,0,\delta,\delta,0,0,\delta,0)$	3;6;8;10	10	2^{-14}
$(0,0,\delta,0,0,\delta,\delta,0) \xrightarrow{1r} (\delta,\delta,0,\delta,\delta,\delta,\delta,\delta)$	3;6;10	7;10	2^{-14}
$(0,0,\delta,0,0,\delta,\delta,\delta) \xrightarrow{1r} (0,0,\delta,0,0,0,0,\delta)$	3;6;8;	7	2^{-14}
$(0,0,\delta,0,\delta,0,0,0) \xrightarrow{1r} (0,\delta,\delta,0,\delta,\delta,0,0)$	3;9	5;9	2^{-14}
$(0,0,\delta,0,\delta,0,0,\delta) \xrightarrow{1r} (\delta,0,0,\delta,0,0,\delta,0)$	3;8;9;10	5;9;10	2^{-14}
$(0,0,\delta,0,\delta,0,\delta,0) \xrightarrow{1r} (0,\delta,\delta,\delta,\delta,\delta,\delta,\delta)$	3;9;10	5;7;9;10	2^{-14}
$(0,0,\delta,0,\delta,0,\delta,\delta) \xrightarrow{1r} (\delta,0,0,0,0,0,0,\delta)$	3;8;9	5;7;9	2^{-14}
$(0,0,\delta,0,\delta,\delta,0,0) \xrightarrow{1r} (\delta,0,\delta,0,0,\delta,\delta,\delta)$	3;6;9;10	5;9;10	2^{-7}
$(0,0,\delta,0,\delta,\delta,0,\delta) \xrightarrow{1r} (0,\delta,0,\delta,\delta,0,0,\delta)$	3;6;8;9	5;9	2^{-14}
$(0,0,\delta,0,\delta,\delta,\delta,0) \xrightarrow{1r} (\delta,0,\delta,\delta,0,\delta,0,0)$	3;6;9	5;7;9	1
$(0,0,\delta,0,\delta,\delta,\delta,\delta) \xrightarrow{1r} (0,\delta,0,0,\delta,0,\delta,0)$	3;6;8;9;10	5;7;9;10	2^{-7}
$(0,0,\delta,\delta,0,0,0,0) \xrightarrow{1r} (\delta,\delta,\delta,\delta,\delta,0,\delta,0)$	3	4	2^{-14}
$(0,0,\delta,\delta,0,0,0,\delta) \xrightarrow{1r} (0,0,0,0,0,\delta,0,0)$	3;8;10	4;10	2^{-7}
$(0,0,\delta,\delta,0,0,\delta,0) \xrightarrow{1r} (\delta,\delta,0,\delta,0,0,\delta)$	3;10	4;7;10	2^{-7}
$(0,0,\delta,\delta,0,0,\delta,\delta) \xrightarrow{1r} (0,0,0,\delta,0,\delta,\delta,\delta)$	3;8	4;7	1
$(0,0,\delta,\delta,0,\delta,0,0) \xrightarrow{1r} (0,0,\delta,\delta,0,0,0,\delta)$	3;6;10	4;10	2^{-14}
$(0,0,\delta,\delta,0,\delta,0,\delta) \xrightarrow{1r} (\delta,\delta,0,0,\delta,\delta,\delta,\delta)$	3;6;8	4	2^{-14}
$(0,0,\delta,\delta,0,\delta,\delta,0) \xrightarrow{1r} (0,0,\delta,0,0,0,\delta,0)$	3;6	4;7	2^{-14}
$(0,0,\delta,\delta,0,\delta,\delta,\delta) \xrightarrow{1r} (\delta,\delta,0,\delta,\delta,\delta,0,0)$	3;6;8;10	4;7;10	2^{-14}
$(0,0,\delta,\delta,\delta,0,0,0) \xrightarrow{1r} (\delta,0,0,\delta,0,0,0,\delta)$	3;9;10	4;5;9;10	2^{-14}
$(0,0,\delta,\delta,\delta,0,0,\delta) \xrightarrow{1r} (0,\delta,\delta,0,\delta,\delta,\delta,\delta)$	3;8;9	4;5;9	2^{-14}
$(0,0,\delta,\delta,\delta,0,\delta,0) \xrightarrow{1r} (\delta,0,0,0,0,0,\delta,0)$	3;9	4;5;7;9	2^{-14}
$(0,0,\delta,\delta,\delta,0,\delta,\delta) \xrightarrow{1r} (0,\delta,\delta,\delta,\delta,\delta,0,0)$	3;8;9;10	4;5;7;9;10	2^{-14}
$(0,0,\delta,\delta,\delta,\delta,0,0) \xrightarrow{1r} (0,\delta,0,\delta,\delta,0,\delta,0)$	3;6;9	4;5;9	2^{-14}
$(0,0,\delta,\delta,\delta,\delta,0,\delta) \xrightarrow{1r} (\delta,0,\delta,0,0,\delta,0,0)$	3;6;8;9;10	4;5;9;10	2^{-7}
$(0,0,\delta,\delta,\delta,\delta,\delta,0) \xrightarrow{1r} (0,\delta,0,0,\delta,0,0,\delta)$	3;6;9;10	4;5;7;9; 10	2^{-7}
$(0,0,\delta,\delta,\delta,\delta,\delta,\delta) \xrightarrow{1r} (\delta,0,\delta,\delta,0,\delta,\delta,\delta)$	3;6;8;9	4;5;7;9	1
$(0,\delta,0,0,0,0,0,0) \xrightarrow{1r} (\delta,\delta,\delta,0,0,0,\delta,\delta)$	10	2;10	2^{-14}

Table 3.28 1-round characteristics for MESH-64(8) (part III).

1-round characteristic	Weak subkeys $Z_j^{(i)}$ (i-th round) j value (odd i)	j value (even i)	Prob.
$(0,\delta,0,0,0,0,0,\delta) \xrightarrow{1r} (0,0,0,\delta,\delta,\delta,0,\delta)$	8	2	2^{-14}
$(0,\delta,0,0,0,0,\delta,0) \xrightarrow{1r} (\delta,\delta,\delta,\delta,0,0,0,0)$	—	2;7	2^{-14}
$(0,\delta,0,0,0,0,\delta,\delta) \xrightarrow{1r} (0,0,0,0,\delta,\delta,\delta,0)$	8;10	2;7;10	2^{-14}
$(0,\delta,0,0,0,\delta,0,0) \xrightarrow{1r} (0,0,\delta,0,\delta,0,0,0)$	6	2	1
$(0,\delta,0,0,0,\delta,0,\delta) \xrightarrow{1r} (\delta,\delta,0,\delta,0,\delta,\delta,0)$	6;8;10	2;10	2^{-7}
$(0,\delta,0,0,0,\delta,\delta,0) \xrightarrow{1r} (0,0,\delta,\delta,\delta,0,\delta,\delta)$	6;10	2;7;10	2^{-7}
$(0,\delta,0,0,0,\delta,\delta,\delta) \xrightarrow{1r} (\delta,\delta,0,0,0,\delta,0,\delta)$	6;8	2;7	2^{-14}
$(0,\delta,0,0,\delta,0,0,0) \xrightarrow{1r} (\delta,0,0,0,\delta,0,0,0)$	9	2;5;9	1
$(0,\delta,0,0,\delta,0,0,\delta) \xrightarrow{1r} (0,\delta,\delta,\delta,0,\delta,\delta,0)$	8;9;10	2;5;9;10	2^{-7}
$(0,\delta,0,0,\delta,0,\delta,0) \xrightarrow{1r} (\delta,0,0,\delta,\delta,0,\delta,\delta)$	9;10	2;5;7;9;10	2^{-7}
$(0,\delta,0,0,\delta,0,\delta,\delta) \xrightarrow{1r} (0,\delta,\delta,0,0,\delta,0,\delta)$	8;9	2;5;7;9	2^{-14}
$(0,\delta,0,0,\delta,\delta,0,0) \xrightarrow{1r} (0,\delta,0,0,0,0,\delta,\delta)$	6;9;10	2;5;9;10	2^{-14}
$(0,\delta,0,0,\delta,\delta,0,\delta) \xrightarrow{1r} (\delta,0,\delta,\delta,\delta,\delta,0,\delta)$	6;8;9	2;5;9	2^{-14}
$(0,\delta,0,0,\delta,\delta,\delta,0) \xrightarrow{1r} (0,\delta,0,\delta,0,0,0,0)$	6;9	2;5;7;9	2^{-14}
$(0,\delta,0,0,\delta,\delta,\delta,\delta) \xrightarrow{1r} (\delta,0,\delta,0,\delta,\delta,\delta,0)$	6;8;9;10	2;5;7;9; 10	2^{-14}
$(0,\delta,0,\delta,0,0,0,0) \xrightarrow{1r} (0,0,0,\delta,\delta,\delta,\delta,0)$	—	2;4	2^{-14}
$(0,\delta,0,\delta,0,0,0,\delta) \xrightarrow{1r} (\delta,\delta,\delta,0,0,0,0,0)$	8;10	2;4;10	2^{-14}
$(0,\delta,0,\delta,0,0,\delta,0) \xrightarrow{1r} (0,0,0,0,\delta,\delta,0,\delta)$	10	2;4;7;10	2^{-14}
$(0,\delta,0,\delta,0,0,\delta,\delta) \xrightarrow{1r} (\delta,\delta,\delta,\delta,0,0,\delta,\delta)$	8	2;4;7	2^{-14}
$(0,\delta,0,\delta,0,\delta,0,0) \xrightarrow{1r} (\delta,\delta,0,\delta,0,\delta,0,\delta)$	6;10	2;4;10	2^{-7}
$(0,\delta,0,\delta,0,\delta,0,\delta) \xrightarrow{1r} (0,0,\delta,0,\delta,0,\delta,\delta)$	6;8	2;4	1
$(0,\delta,0,\delta,0,\delta,\delta,0) \xrightarrow{1r} (\delta,\delta,0,0,0,\delta,\delta,0)$	6	2;4;7	2^{-14}
$(0,\delta,0,\delta,0,\delta,\delta,\delta) \xrightarrow{1r} (0,0,\delta,\delta,\delta,0,0,0)$	6;8;10	2;4;7;10	2^{-7}
$(0,\delta,0,\delta,\delta,0,0,0) \xrightarrow{1r} (0,\delta,\delta,\delta,0,\delta,0,\delta)$	9;10	2;4;5;9;10	2^{-7}
$(0,\delta,0,\delta,\delta,0,0,\delta) \xrightarrow{1r} (\delta,0,0,0,\delta,0,\delta,\delta)$	8;9	2;4;5;9	1
$(0,\delta,0,\delta,\delta,0,\delta,0) \xrightarrow{1r} (0,\delta,\delta,0,0,\delta,\delta,0)$	9	2;4;5;7;9	2^{-14}
$(0,\delta,0,\delta,\delta,0,\delta,\delta) \xrightarrow{1r} (\delta,0,0,\delta,\delta,0,0,0)$	8;9;10	2;4;5;7;9;10	2^{-7}
$(0,\delta,0,\delta,\delta,\delta,0,0) \xrightarrow{1r} (\delta,0,\delta,\delta,\delta,\delta,\delta,0)$	6;9	2;4;5;9	2^{-14}
$(0,\delta,0,\delta,\delta,\delta,0,\delta) \xrightarrow{1r} (0,\delta,0,0,0,0,0,0)$	6;8;9;10	2;4;5;9; 10	2^{-14}
$(0,\delta,0,\delta,\delta,\delta,\delta,0) \xrightarrow{1r} (\delta,0,\delta,0,\delta,\delta,0,\delta)$	6;9;10	2;4;5;7;9; 10	2^{-14}
$(0,\delta,0,\delta,\delta,\delta,\delta,\delta) \xrightarrow{1r} (0,\delta,0,\delta,0,0,\delta,\delta)$	6;8;9	2;4;5;7;9	2^{-14}
$(0,\delta,\delta,0,0,0,0,0) \xrightarrow{1r} (\delta,\delta,\delta,0,0,\delta,0,0)$	3	2	2^{-14}

Table 3.29 1-round characteristics for MESH-64(8) (part IV).

1-round characteristic	Weak subkeys $Z_j^{(i)}$ (i-th round)		Prob.
	j value (odd i)	j value (even i)	
$(0,\delta,\delta,0,0,0,0,\delta) \xrightarrow{1r} (0,0,0,\delta,\delta,0,\delta,0)$	3;8;10	2;10	2^{-14}
$(0,\delta,\delta,0,0,0,\delta,0) \xrightarrow{1r} (\delta,\delta,\delta,\delta,0,\delta,\delta,\delta)$	3;10	2;7;10	2^{-14}
$(0,\delta,\delta,0,0,0,\delta,\delta) \xrightarrow{1r} (0,0,0,0,\delta,0,0,\delta)$	3;8	2;7	2^{-14}
$(0,\delta,\delta,0,0,\delta,0,0) \xrightarrow{1r} (0,0,\delta,0,\delta,\delta,\delta,\delta)$	3;6;10	2;10	2^{-7}
$(0,\delta,\delta,0,0,\delta,0,\delta) \xrightarrow{1r} (\delta,\delta,0,\delta,0,0,0,\delta)$	3;6;8	2	2^{-14}
$(0,\delta,\delta,0,0,\delta,\delta,0) \xrightarrow{1r} (0,0,\delta,\delta,\delta,\delta,0,0)$	3;6	2;7	1
$(0,\delta,\delta,0,0,\delta,\delta,\delta) \xrightarrow{1r} (\delta,\delta,0,0,0,0,\delta,0)$	3;6;8;10	2;7;10	2^{-7}
$(0,\delta,\delta,0,\delta,0,0,0) \xrightarrow{1r} (\delta,0,0,0,\delta,\delta,\delta,\delta)$	3;9;10	2;5;9;10	2^{-7}
$(0,\delta,\delta,0,\delta,0,0,\delta) \xrightarrow{1r} (0,\delta,\delta,\delta,0,0,0,\delta)$	3;8;9	2;5;9	2^{-14}
$(0,\delta,\delta,0,\delta,0,\delta,0) \xrightarrow{1r} (\delta,0,0,\delta,\delta,\delta,0,0)$	3;9	2;5;7;9	1
$(0,\delta,\delta,0,\delta,0,\delta,\delta) \xrightarrow{1r} (0,\delta,\delta,0,0,0,\delta,0)$	3;8;9;10	2;5;7;9;10	2^{-7}
$(0,\delta,\delta,0,\delta,\delta,0,0) \xrightarrow{1r} (0,\delta,0,0,0,\delta,0,0)$	3;6;9	2;5;9	2^{-14}
$(0,\delta,\delta,0,\delta,\delta,0,\delta) \xrightarrow{1r} (\delta,0,\delta,\delta,\delta,0,\delta,0)$	3;6;8;9;10	2;5;9;10	2^{-14}
$(0,\delta,\delta,0,\delta,\delta,\delta,0) \xrightarrow{1r} (0,\delta,0,\delta,0,\delta,\delta,\delta)$	3;6;9;10	2;5;7;9;10	2^{-14}
$(0,\delta,\delta,0,\delta,\delta,\delta,\delta) \xrightarrow{1r} (\delta,0,\delta,0,\delta,0,0,\delta)$	3;6;8;9	2;5;7;9	2^{-14}
$(0,\delta,\delta,\delta,0,0,0,0) \xrightarrow{1r} (0,0,0,\delta,\delta,0,0,\delta)$	3;10	2;4;10	2^{-14}
$(0,\delta,\delta,\delta,0,0,0,\delta) \xrightarrow{1r} (\delta,\delta,\delta,0,0,\delta,\delta,\delta)$	3;8	2;4	2^{-14}
$(0,\delta,\delta,\delta,0,0,\delta,0) \xrightarrow{1r} (0,0,0,0,\delta,0,\delta,0)$	3	2;4;7	2^{-14}
$(0,\delta,\delta,\delta,0,0,\delta,\delta) \xrightarrow{1r} (\delta,\delta,\delta,\delta,0,\delta,0,0)$	3;8;10	2;4;7;10	2^{-14}
$(0,\delta,\delta,\delta,0,\delta,0,0) \xrightarrow{1r} (\delta,\delta,0,\delta,0,0,\delta,0)$	3;6	2;4	2^{-14}
$(0,\delta,\delta,\delta,0,\delta,0,\delta) \xrightarrow{1r} (0,0,\delta,0,\delta,\delta,0,0)$	3;6;8;10	2;4;10	2^{-7}
$(0,\delta,\delta,\delta,0,\delta,\delta,0) \xrightarrow{1r} (\delta,\delta,0,0,0,0,0,\delta)$	3;6;10	2;4;7;10	2^{-7}
$(0,\delta,\delta,\delta,0,\delta,\delta,\delta) \xrightarrow{1r} (0,0,\delta,\delta,\delta,\delta,\delta,\delta)$	3;6;8	2;4;7	1
$(0,\delta,\delta,\delta,\delta,0,0,0) \xrightarrow{1r} (0,\delta,\delta,\delta,0,0,\delta,0)$	3;9	2;4;5;9	2^{-14}
$(0,\delta,\delta,\delta,\delta,0,0,\delta) \xrightarrow{1r} (\delta,0,0,0,\delta,\delta,0,0)$	3;8;9;10	2;4;5;9; 10	2^{-7}
$(0,\delta,\delta,\delta,\delta,0,\delta,0) \xrightarrow{1r} (0,\delta,\delta,0,0,0,0,\delta)$	3;9;10	2;4;5;7;9; 10	2^{-7}
$(0,\delta,\delta,\delta,\delta,0,\delta,\delta) \xrightarrow{1r} (\delta,0,0,\delta,\delta,\delta,\delta,\delta)$	3;8;9	2;4;5;7;9	1
$(0,\delta,\delta,\delta,\delta,\delta,0,0) \xrightarrow{1r} (\delta,0,\delta,\delta,\delta,0,0,\delta)$	3;6;9;10	2;4;5;9; 10	2^{-14}
$(0,\delta,\delta,\delta,\delta,\delta,0,\delta) \xrightarrow{1r} (0,\delta,0,0,0,\delta,\delta,\delta)$	3;6;8;9	2;4;5;9	2^{-14}
$(0,\delta,\delta,\delta,\delta,\delta,\delta,0) \xrightarrow{1r} (\delta,0,\delta,0,\delta,0,\delta,0)$	3;6;9	2;4;5;7;9	2^{-14}
$(0,\delta,\delta,\delta,\delta,\delta,\delta,\delta) \xrightarrow{1r} (0,\delta,0,\delta,0,\delta,0,0)$	3;6;8;9;10	2;4;5;7;9;10	2^{-14}
$(\delta,0,0,0,0,0,0,0) \xrightarrow{1r} (\delta,0,\delta,0,\delta,0,\delta,\delta)$	1;9;10	9;10	2^{-14}

Table 3.30 1-round characteristics for MESH-64(8) (part V).

1-round characteristic	Weak subkeys $Z_j^{(i)}$ (i-th round) j value (odd i)	j value (even i)	Prob.
$(\delta,0,0,0,0,0,0,\delta) \xrightarrow{1r} (0,\delta,0,\delta,0,\delta,0,\delta)$	1;8;9	9	2^{-14}
$(\delta,0,0,0,0,0,\delta,0) \xrightarrow{1r} (\delta,0,\delta,\delta,\delta,0,0,0)$	1;9	7;9	2^{-14}
$(\delta,0,0,0,0,0,\delta,\delta) \xrightarrow{1r} (0,\delta,0,0,0,\delta,\delta,0)$	1;8;9;10	7;9;10	2^{-14}
$(\delta,0,0,0,0,\delta,0,0) \xrightarrow{1r} (0,\delta,\delta,0,0,0,0,0)$	1;6;9	9	1
$(\delta,0,0,0,0,\delta,0,\delta) \xrightarrow{1r} (\delta,0,0,\delta,\delta,\delta,\delta,0)$	1;6;8;9;10	9;10	2^{-7}
$(\delta,0,0,0,0,\delta,\delta,0) \xrightarrow{1r} (0,\delta,\delta,\delta,0,0,\delta,\delta)$	1;6;9;10	7;9;10	2^{-7}
$(\delta,0,0,0,0,\delta,\delta,\delta) \xrightarrow{1r} (\delta,0,0,0,\delta,\delta,0,\delta)$	1;6;8;9	7;9	2^{-14}
$(\delta,0,0,0,\delta,0,0,0) \xrightarrow{1r} (\delta,\delta,0,0,0,0,0,0)$	1	5	1
$(\delta,0,0,0,\delta,0,0,\delta) \xrightarrow{1r} (0,0,\delta,\delta,\delta,\delta,\delta,0)$	1;8;10	5;10	2^{-7}
$(\delta,0,0,0,\delta,0,\delta,0) \xrightarrow{1r} (\delta,\delta,0,\delta,0,0,\delta,\delta)$	1;10	5;7;10	2^{-7}
$(\delta,0,0,0,\delta,0,\delta,\delta) \xrightarrow{1r} (0,0,\delta,0,\delta,\delta,0,\delta)$	1;8	5;7	2^{-14}
$(\delta,0,0,0,\delta,\delta,0,0) \xrightarrow{1r} (0,0,0,0,\delta,0,\delta,\delta)$	1;6;10	5;10	2^{-14}
$(\delta,0,0,0,\delta,\delta,0,\delta) \xrightarrow{1r} (\delta,\delta,\delta,\delta,0,\delta,0,\delta)$	1;6;8	5	2^{-14}
$(\delta,0,0,0,\delta,\delta,\delta,0) \xrightarrow{1r} (0,0,0,\delta,\delta,0,0,0)$	1;6	5;7	2^{-14}
$(\delta,0,0,0,\delta,\delta,\delta,\delta) \xrightarrow{1r} (\delta,\delta,\delta,0,0,\delta,\delta,0)$	1;6;8;10	5;7;10	2^{-14}
$(\delta,0,0,\delta,0,0,0,0) \xrightarrow{1r} (0,\delta,0,\delta,0,\delta,\delta,0)$	1;9	4;9	2^{-14}
$(\delta,0,0,\delta,0,0,0,\delta) \xrightarrow{1r} (\delta,0,\delta,0,\delta,0,0,0)$	1;8;9;10	4;9;10	2^{-14}
$(\delta,0,0,\delta,0,0,\delta,0) \xrightarrow{1r} (0,\delta,0,0,0,\delta,0,\delta)$	1;9;10	4;7;9;10	2^{-14}
$(\delta,0,0,\delta,0,0,\delta,\delta) \xrightarrow{1r} (\delta,0,\delta,\delta,\delta,0,\delta,\delta)$	1;8;9	4;7;9	2^{-14}
$(\delta,0,0,\delta,0,\delta,0,0) \xrightarrow{1r} (\delta,0,0,\delta,\delta,\delta,0,\delta)$	1;6;9;10	4;9;10	2^{-7}
$(\delta,0,0,\delta,0,\delta,0,\delta) \xrightarrow{1r} (0,\delta,\delta,0,0,0,\delta,\delta)$	1;6;8;9	4;9	1
$(\delta,0,0,\delta,0,\delta,\delta,0) \xrightarrow{1r} (\delta,0,0,0,\delta,\delta,\delta,0)$	1;6;9	4;7;9	2^{-14}
$(\delta,0,0,\delta,0,\delta,\delta,\delta) \xrightarrow{1r} (0,\delta,\delta,\delta,0,0,0,0)$	1;6;8;9;10	4;7;9; 10	2^{-7}
$(\delta,0,0,\delta,\delta,0,0,0) \xrightarrow{1r} (0,0,\delta,\delta,\delta,\delta,0,\delta)$	1;10	4;5;10	2^{-7}
$(\delta,0,0,\delta,\delta,0,0,\delta) \xrightarrow{1r} (\delta,\delta,0,0,0,0,\delta,\delta)$	1;8	4;5	1
$(\delta,0,0,\delta,\delta,0,\delta,0) \xrightarrow{1r} (0,0,\delta,0,\delta,\delta,\delta,0)$	1	4;5;7	2^{-14}
$(\delta,0,0,\delta,\delta,0,\delta,\delta) \xrightarrow{1r} (\delta,\delta,0,\delta,0,0,0,0)$	1;8;10	4;5;7;10	2^{-7}
$(\delta,0,0,\delta,\delta,\delta,0,0) \xrightarrow{1r} (\delta,\delta,\delta,\delta,0,\delta,\delta,0)$	1;6	4;5	2^{-14}
$(\delta,0,0,\delta,\delta,\delta,0,\delta) \xrightarrow{1r} (0,0,0,0,\delta,0,0,0)$	1;6;8;10	4;5;10	2^{-14}
$(\delta,0,0,\delta,\delta,\delta,\delta,0) \xrightarrow{1r} (\delta,\delta,\delta,0,0,\delta,0,\delta)$	1;6;10	4;5;7;10	2^{-14}
$(\delta,0,0,\delta,\delta,\delta,\delta,\delta) \xrightarrow{1r} (0,0,0,\delta,\delta,0,\delta,\delta)$	1;6;8	4;5;7	2^{-14}
$(\delta,0,\delta,0,0,0,0,0) \xrightarrow{1r} (\delta,0,\delta,0,\delta,\delta,0,0)$	1;3;9	9	2^{-14}

Table 3.31 1-round characteristics for MESH-64(8) (part VI).

1-round characteristic	Weak subkeys $Z_j^{(i)}$ (i-th round)		Prob.
	j value (odd i)	j value (even i)	
$(\delta,0,\delta,0,0,0,0,\delta) \xrightarrow{1r} (0,\delta,0,\delta,0,0,\delta,0)$	1;3;8;9;10	9;10	2^{-14}
$(\delta,0,\delta,0,0,0,\delta,0) \xrightarrow{1r} (\delta,0,\delta,\delta,\delta,\delta,\delta,\delta)$	1;3;9;10	7;9;10	2^{-14}
$(\delta,0,\delta,0,0,0,\delta,\delta) \xrightarrow{1r} (0,\delta,0,0,0,0,0,\delta)$	1;3;8;9	7;9	2^{-14}
$(\delta,0,\delta,0,0,\delta,0,0) \xrightarrow{1r} (0,\delta,\delta,0,0,\delta,\delta,\delta)$	1;3;6;9;10	9; 10	2^{-7}
$(\delta,0,\delta,0,0,\delta,0,\delta) \xrightarrow{1r} (\delta,0,0,\delta,\delta,0,0,\delta)$	1;3;6;8;9	9	2^{-14}
$(\delta,0,\delta,0,0,\delta,\delta,0) \xrightarrow{1r} (0,\delta,\delta,\delta,0,\delta,0,0)$	1;3;6;9	7;9	1
$(\delta,0,\delta,0,0,\delta,\delta,\delta) \xrightarrow{1r} (\delta,0,0,0,\delta,0,\delta,0)$	1;3;6;8;9;10	7;9;10	2^{-7}
$(\delta,0,\delta,0,\delta,0,0,0) \xrightarrow{1r} (\delta,\delta,0,0,0,\delta,\delta,\delta)$	1;3;10	5;10	2^{-7}
$(\delta,0,\delta,0,\delta,0,0,\delta) \xrightarrow{1r} (0,0,\delta,\delta,\delta,0,0,\delta)$	1;3;8	5	2^{-14}
$(\delta,0,\delta,0,\delta,0,\delta,0) \xrightarrow{1r} (\delta,\delta,0,\delta,0,\delta,0,0)$	1;3	5;7	1
$(\delta,0,\delta,0,\delta,0,\delta,\delta) \xrightarrow{1r} (0,0,\delta,0,\delta,0,\delta,0)$	1;3;8;10	5;7;10	2^{-7}
$(\delta,0,\delta,0,\delta,\delta,0,0) \xrightarrow{1r} (0,0,0,0,\delta,\delta,0,0)$	1;3;6	5	2^{-14}
$(\delta,0,\delta,0,\delta,\delta,0,\delta) \xrightarrow{1r} (\delta,\delta,\delta,\delta,0,0,\delta,0)$	1;3;6;8;10	5;10	2^{-14}
$(\delta,0,\delta,0,\delta,\delta,\delta,0) \xrightarrow{1r} (0,0,0,\delta,\delta,\delta,\delta,\delta)$	1;3;6;10	5;7;10	2^{-14}
$(\delta,0,\delta,0,\delta,\delta,\delta,\delta) \xrightarrow{1r} (\delta,\delta,\delta,0,0,0,0,\delta)$	1;3;6;8	5;7	2^{-14}
$(\delta,0,\delta,\delta,0,0,0,0) \xrightarrow{1r} (0,\delta,0,\delta,0,0,0,\delta)$	1;3;9;10	4;9;10	2^{-14}
$(\delta,0,\delta,\delta,0,0,0,\delta) \xrightarrow{1r} (\delta,0,\delta,0,\delta,\delta,\delta,\delta)$	1;3;8;9	4;9	2^{-14}
$(\delta,0,\delta,\delta,0,0,\delta,0) \xrightarrow{1r} (0,\delta,0,0,0,0,\delta,0)$	1;3;9	4;7;9	2^{-14}
$(\delta,0,\delta,\delta,0,0,\delta,\delta) \xrightarrow{1r} (\delta,0,\delta,\delta,\delta,\delta,0,0)$	1;3;8;9;10	4;7;9; 10	2^{-14}
$(\delta,0,\delta,\delta,0,\delta,0,0) \xrightarrow{1r} (\delta,0,0,\delta,\delta,0,\delta,0)$	1;3;6;9	4;9	2^{-14}
$(\delta,0,\delta,\delta,0,\delta,0,\delta) \xrightarrow{1r} (0,\delta,\delta,0,0,\delta,0,0)$	1;3;6;8;9;10	4;9; 10	2^{-7}
$(\delta,0,\delta,\delta,0,\delta,\delta,0) \xrightarrow{1r} (\delta,0,0,0,\delta,0,0,\delta)$	1;3;6;9;10	4;7;9; 10	2^{-7}
$(\delta,0,\delta,\delta,0,\delta,\delta,\delta) \xrightarrow{1r} (0,\delta,\delta,\delta,0,\delta,\delta,\delta)$	1;3;6;8;9	4;7;9	1
$(\delta,0,\delta,\delta,\delta,0,0,0) \xrightarrow{1r} (0,0,\delta,\delta,\delta,0,\delta,0)$	1;3	4;5	2^{-14}
$(\delta,0,\delta,\delta,\delta,0,0,\delta) \xrightarrow{1r} (\delta,\delta,0,0,0,\delta,0,0)$	1;3;8;10	4;5;10	2^{-7}
$(\delta,0,\delta,\delta,\delta,0,\delta,0) \xrightarrow{1r} (0,0,\delta,0,\delta,0,0,\delta)$	1;3;10	4;5;7;10	2^{-7}
$(\delta,0,\delta,\delta,\delta,0,\delta,\delta) \xrightarrow{1r} (\delta,\delta,0,\delta,0,\delta,\delta,\delta)$	1;3;8	4;5;7	1
$(\delta,0,\delta,\delta,\delta,\delta,0,0) \xrightarrow{1r} (\delta,\delta,\delta,\delta,0,0,0,\delta)$	1;3;6;10	4;5;10	2^{-14}
$(\delta,0,\delta,\delta,\delta,\delta,0,\delta) \xrightarrow{1r} (0,0,0,0,\delta,\delta,\delta,\delta)$	1;3;6;8	4;5	2^{-14}
$(\delta,0,\delta,\delta,\delta,\delta,\delta,0) \xrightarrow{1r} (\delta,\delta,\delta,0,0,0,\delta,0)$	1;3;6	4;5;7	2^{-14}
$(\delta,0,\delta,\delta,\delta,\delta,\delta,\delta) \xrightarrow{1r} (0,0,0,\delta,\delta,\delta,0,0)$	1;3;6;8;10	4;5;7;10	2^{-14}
$(\delta,\delta,0,0,0,0,0,0) \xrightarrow{1r} (0,\delta,0,0,\delta,0,0,0)$	1;9	2;9	1

Table 3.32 1-round characteristics for MESH-64(8) (part VII).

1-round characteristic	Weak subkeys $Z_j^{(i)}$ (i-th round) j value (odd i)	j value (even i)	Prob.
$(\delta,\delta,0,0,0,0,0,\delta) \xrightarrow{1r} (\delta,0,\delta,\delta,0,\delta,\delta,0)$	1;8;9;10	2;9;10	2^{-7}
$(\delta,\delta,0,0,0,0,\delta,0) \xrightarrow{1r} (0,\delta,0,\delta,\delta,0,\delta,\delta)$	1;9;10	2;7;9;10	2^{-7}
$(\delta,\delta,0,0,0,0,\delta,\delta) \xrightarrow{1r} (\delta,0,\delta,0,0,\delta,0,\delta)$	1;8;9	2;7;9	2^{-14}
$(\delta,\delta,0,0,0,\delta,0,0) \xrightarrow{1r} (\delta,0,0,0,0,0,\delta,\delta)$	1;6;9;10	2;9; 10	2^{-14}
$(\delta,\delta,0,0,0,\delta,0,\delta) \xrightarrow{1r} (0,\delta,\delta,\delta,\delta,\delta,0,\delta)$	1;6;8;9	2;9	2^{-14}
$(\delta,\delta,0,0,0,\delta,\delta,0) \xrightarrow{1r} (\delta,0,0,\delta,0,0,0,0)$	1;6;9	2;7;9	2^{-14}
$(\delta,\delta,0,0,0,\delta,\delta,\delta) \xrightarrow{1r} (0,\delta,\delta,0,\delta,\delta,\delta,0)$	1;6;8;9;10	2;7;9; 10	2^{-14}
$(\delta,\delta,0,0,\delta,0,0,0) \xrightarrow{1r} (0,0,\delta,0,0,0,\delta,\delta)$	1;10	2;5;10	2^{-14}
$(\delta,\delta,0,0,\delta,0,0,\delta) \xrightarrow{1r} (\delta,\delta,0,\delta,\delta,\delta,0,\delta)$	1;8	2;5	2^{-14}
$(\delta,\delta,0,0,\delta,0,\delta,0) \xrightarrow{1r} (0,0,\delta,\delta,0,0,0,0)$	1	2;5;7	2^{-14}
$(\delta,\delta,0,0,\delta,0,\delta,\delta) \xrightarrow{1r} (\delta,\delta,0,0,\delta,\delta,\delta,0)$	1;8;10	2;5;7;10	2^{-14}
$(\delta,\delta,0,0,\delta,\delta,0,0) \xrightarrow{1r} (\delta,\delta,\delta,0,\delta,0,0,0)$	1;6	2;5	1
$(\delta,\delta,0,0,\delta,\delta,0,\delta) \xrightarrow{1r} (0,0,0,\delta,0,\delta,\delta,0)$	1;6;8;10	2;5;10	2^{-7}
$(\delta,\delta,0,0,\delta,\delta,\delta,0) \xrightarrow{1r} (\delta,\delta,\delta,\delta,\delta,0,\delta,\delta)$	1;6;10	2;5;7;10	2^{-7}
$(\delta,\delta,0,0,\delta,\delta,\delta,\delta) \xrightarrow{1r} (0,0,0,0,0,\delta,0,\delta)$	1;6;8	2;5;7	2^{-14}
$(\delta,\delta,0,\delta,0,0,0,0) \xrightarrow{1r} (\delta,0,\delta,\delta,0,\delta,0,\delta)$	1;9;10	2;4;9;10	2^{-7}
$(\delta,\delta,0,\delta,0,0,0,\delta) \xrightarrow{1r} (0,\delta,0,0,\delta,0,\delta,\delta)$	1;8;9	2;4;9	1
$(\delta,\delta,0,\delta,0,0,\delta,0) \xrightarrow{1r} (\delta,0,\delta,0,0,\delta,\delta,0)$	1;9	2;4;7;9	2^{-14}
$(\delta,\delta,0,\delta,0,0,\delta,\delta) \xrightarrow{1r} (0,\delta,0,\delta,\delta,0,0,0)$	1;8;9;10	2;4;7;9; 10	2^{-7}
$(\delta,\delta,0,\delta,0,\delta,0,0) \xrightarrow{1r} (0,\delta,\delta,\delta,\delta,\delta,\delta,0)$	1;6;9	2;4;9	2^{-14}
$(\delta,\delta,0,\delta,0,\delta,0,\delta) \xrightarrow{1r} (\delta,0,0,0,0,0,0,0)$	1;6;8;9;10	2;4;9; 10	2^{-14}
$(\delta,\delta,0,\delta,0,\delta,\delta,0) \xrightarrow{1r} (0,\delta,\delta,0,\delta,\delta,0,\delta)$	1;6;9;10	2;4;7;9; 10	2^{-14}
$(\delta,\delta,0,\delta,0,\delta,\delta,\delta) \xrightarrow{1r} (\delta,0,0,\delta,0,0,\delta,\delta)$	1;6;8;9	2;4;7;9	2^{-14}
$(\delta,\delta,0,\delta,\delta,0,0,0) \xrightarrow{1r} (\delta,\delta,0,\delta,\delta,\delta,\delta,0)$	1	2;4;5	2^{-14}
$(\delta,\delta,0,\delta,\delta,0,0,\delta) \xrightarrow{1r} (0,0,\delta,0,0,0,0,0)$	1;8;10	2;4;5;10	2^{-14}
$(\delta,\delta,0,\delta,\delta,0,\delta,0) \xrightarrow{1r} (\delta,\delta,0,0,\delta,\delta,0,\delta)$	1;10	2;4;5;7;10	2^{-14}
$(\delta,\delta,0,\delta,\delta,0,\delta,\delta) \xrightarrow{1r} (0,0,\delta,\delta,0,0,\delta,\delta)$	1;8	2;4;5;7	2^{-14}
$(\delta,\delta,0,\delta,\delta,\delta,0,0) \xrightarrow{1r} (0,0,0,\delta,0,\delta,0,\delta)$	1;6;10	2;4;5;10	2^{-7}
$(\delta,\delta,0,\delta,\delta,\delta,0,\delta) \xrightarrow{1r} (\delta,\delta,\delta,0,\delta,0,\delta,\delta)$	1;6;8	2;4;5	1
$(\delta,\delta,0,\delta,\delta,\delta,\delta,0) \xrightarrow{1r} (0,0,0,0,0,\delta,\delta,0)$	1;6	2;4;5;7	2^{-14}
$(\delta,\delta,0,\delta,\delta,\delta,\delta,\delta) \xrightarrow{1r} (\delta,\delta,\delta,\delta,\delta,0,0,0)$	1;6;8;10	2;4;5;7;10	2^{-7}
$(\delta,\delta,\delta,0,0,0,0,0) \xrightarrow{1r} (0,\delta,0,0,\delta,\delta,\delta,\delta)$	1;3;9;10	2;9; 10	2^{-7}

Table 3.33 1-round characteristics for MESH-64(8) (part VIII).

1-round characteristic	Weak subkeys $Z_j^{(i)}$ (i-th round) j value (odd i)	j value (even i)	Prob.
$(\delta,\delta,\delta,0,0,0,0,\delta) \xrightarrow{1r} (\delta,0,\delta,\delta,0,0,0,\delta)$	1;3;8;9	2;9	2^{-14}
$(\delta,\delta,\delta,0,0,0,\delta,0) \xrightarrow{1r} (0,\delta,0,\delta,\delta,\delta,0,0)$	1;3;9	2;7;9	1
$(\delta,\delta,\delta,0,0,0,\delta,\delta) \xrightarrow{1r} (\delta,0,\delta,0,0,0,\delta,0)$	1;3;8;9;10	2;7;9; 10	2^{-7}
$(\delta,\delta,\delta,0,0,\delta,0,0) \xrightarrow{1r} (\delta,0,0,0,0,\delta,0,0)$	1;3;6;9	2;9	2^{-14}
$(\delta,\delta,\delta,0,0,\delta,0,\delta) \xrightarrow{1r} (0,\delta,\delta,\delta,\delta,0,\delta,0)$	1;3;6;8;9;10	2;9; 10	2^{-14}
$(\delta,\delta,\delta,0,0,\delta,\delta,0) \xrightarrow{1r} (\delta,0,0,\delta,0,\delta,\delta,\delta)$	1;3;6;9;10	2;7;9; 10	2^{-14}
$(\delta,\delta,\delta,0,0,\delta,\delta,\delta) \xrightarrow{1r} (0,\delta,\delta,0,\delta,0,0,\delta)$	1;3;6;8;9	2;7;9	2^{-14}
$(\delta,\delta,\delta,0,\delta,0,0,0) \xrightarrow{1r} (0,0,\delta,0,0,\delta,0,0)$	1;3	2;5	2^{-14}
$(\delta,\delta,\delta,0,\delta,0,0,\delta) \xrightarrow{1r} (\delta,\delta,0,\delta,\delta,0,\delta,0)$	1;3;8;10	2;5;10	2^{-14}
$(\delta,\delta,\delta,0,\delta,0,\delta,0) \xrightarrow{1r} (0,0,\delta,\delta,0,\delta,\delta,\delta)$	1;3;10	2;5;7;10	2^{-14}
$(\delta,\delta,\delta,0,\delta,0,\delta,\delta) \xrightarrow{1r} (\delta,\delta,0,0,\delta,0,0,\delta)$	1;3;8	2;5;7	2^{-14}
$(\delta,\delta,\delta,0,\delta,\delta,0,0) \xrightarrow{1r} (\delta,\delta,\delta,0,\delta,\delta,\delta,\delta)$	1;3;6;10	2;5;10	2^{-7}
$(\delta,\delta,\delta,0,\delta,\delta,0,\delta) \xrightarrow{1r} (0,0,0,\delta,0,0,0,\delta)$	1;3;6;8	2;5	2^{-14}
$(\delta,\delta,\delta,0,\delta,\delta,\delta,0) \xrightarrow{1r} (\delta,\delta,\delta,\delta,\delta,\delta,0,0)$	1;3;6	2;5;7	1
$(\delta,\delta,\delta,0,\delta,\delta,\delta,\delta) \xrightarrow{1r} (0,0,0,0,0,0,\delta,0)$	1;3;6;8;10	2;5;7;10	2^{-7}
$(\delta,\delta,\delta,\delta,0,0,0,0) \xrightarrow{1r} (\delta,0,\delta,\delta,0,0,\delta,0)$	1;3;9	2;4;9	2^{-14}
$(\delta,\delta,\delta,\delta,0,0,0,\delta) \xrightarrow{1r} (0,\delta,0,0,\delta,\delta,0,0)$	1;3;8;9;10	2;4;9;10	2^{-7}
$(\delta,\delta,\delta,\delta,0,0,\delta,0) \xrightarrow{1r} (\delta,0,\delta,0,0,0,0,\delta)$	1;3;9;10	2;4;7;9;10	2^{-7}
$(\delta,\delta,\delta,\delta,0,0,\delta,\delta) \xrightarrow{1r} (0,\delta,0,\delta,\delta,\delta,\delta,\delta)$	1;3;8;9	2;4;7;9	1
$(\delta,\delta,\delta,\delta,0,\delta,0,0) \xrightarrow{1r} (0,\delta,\delta,\delta,\delta,0,0,\delta)$	1;3;6;9;10	2;4;9; 10	2^{-14}
$(\delta,\delta,\delta,\delta,0,\delta,0,\delta) \xrightarrow{1r} (\delta,0,0,0,0,\delta,\delta,\delta)$	1;3;6;8;9	2;4;9	2^{-14}
$(\delta,\delta,\delta,\delta,0,\delta,\delta,0) \xrightarrow{1r} (0,\delta,\delta,0,\delta,0,\delta,0)$	1;3;6;9	2;4;7;9	2^{-14}
$(\delta,\delta,\delta,\delta,0,\delta,\delta,\delta) \xrightarrow{1r} (\delta,0,0,\delta,0,\delta,0,0)$	1;3;6;8;9;10	2;4;7;9; 10	2^{-14}
$(\delta,\delta,\delta,\delta,\delta,0,0,0) \xrightarrow{1r} (\delta,\delta,0,\delta,\delta,0,0,\delta)$	1;3;10	2;4;5;10	2^{-14}
$(\delta,\delta,\delta,\delta,\delta,0,0,\delta) \xrightarrow{1r} (0,0,\delta,0,0,\delta,\delta,\delta)$	1;3;8	2;4;5	2^{-14}
$(\delta,\delta,\delta,\delta,\delta,0,\delta,0) \xrightarrow{1r} (\delta,\delta,0,0,\delta,0,\delta,0)$	1;3	2;4;5;7	2^{-14}
$(\delta,\delta,\delta,\delta,\delta,0,\delta,\delta) \xrightarrow{1r} (0,0,\delta,\delta,0,\delta,0,0)$	1;3;8;10	2;4;5;7;10	2^{-14}
$(\delta,\delta,\delta,\delta,\delta,\delta,0,0) \xrightarrow{1r} (0,0,0,\delta,0,0,\delta,0)$	1;3;6	2;4;5	2^{-14}
$(\delta,\delta,\delta,\delta,\delta,\delta,0,\delta) \xrightarrow{1r} (\delta,\delta,\delta,0,\delta,\delta,0,0)$	1;3;6;8;10	2;4;5;10	2^{-7}
$(\delta,\delta,\delta,\delta,\delta,\delta,\delta,0) \xrightarrow{1r} (0,0,0,0,0,0,0,\delta)$	1;3;6;10	2;4;5;7;10	2^{-7}
$(\delta,\delta,\delta,\delta,\delta,\delta,\delta,\delta) \xrightarrow{1r} (\delta,\delta,\delta,\delta,\delta,\delta,\delta,\delta)$	1;3;6;8	2;4;5;7	1

From Table 3.26 through Table 3.33, the multiple-round characteristics are clustered into 65 chains covering 1, 2, 3 or 6 rounds:

(1) $(0,0,0,0,0,0,0,\delta) \xrightarrow{1r} (\delta,\delta,\delta,\delta,\delta,\delta,\delta,0) \xrightarrow{1r} (0,0,0,0,0,0,0,\delta)$, which is 2-round iterative and holds with probability 2^{-14}. Starting from an odd-numbered round i, the following seven subkeys have to be weak: $Z_8^{(i)}$, $Z_{10}^{(i)}$, $Z_2^{(i+1)}$, $Z_4^{(i+1)}$, $Z_5^{(i+1)}$, $Z_7^{(i+1)}$ and $Z_{10}^{(i+1)}$. Starting from an even-

numbered round i, the following five subkeys have to be weak: $Z_{10}^{(i)}$, $Z_1^{(i)}$, $Z_3^{(i)}$, $Z_6^{(i)}$ and $Z_{10}^{(i)}$.

(2) $(0,0,0,0,0,0,\delta,0) \stackrel{1r}{\rightarrow} (0,0,0,\delta,0,0,\delta,\delta) \stackrel{1r}{\rightarrow} (0,0,0,\delta,0,0,0,0) \stackrel{1r}{\rightarrow} (\delta, \delta, \delta, \delta, \delta, \delta, 0, \delta) \stackrel{1r}{\rightarrow} (\delta,\delta,\delta,0,\delta,\delta,0,0) \stackrel{1r}{\rightarrow} (\delta,\delta,\delta,0,\delta,\delta,\delta,\delta) \stackrel{1r}{\rightarrow} (0, 0, 0, 0, 0, 0, \delta, 0)$, which is 6-round iterative and holds with probability 2^{-42}. Starting from an odd-numbered round i, the following 17 subkeys have to be weak: $Z_{10}^{(i)}$, $Z_4^{(i+1)}$, $Z_7^{(i+1)}$, $Z_{10}^{(i+1)}$, $Z_{10}^{(i+2)}$, $Z_2^{(i+3)}$, $Z_4^{(i+3)}$, $Z_5^{(i+3)}$, $Z_{10}^{(i+3)}$, $Z_1^{(i+4)}$, $Z_3^{(i+4)}$, $Z_6^{(i+4)}$, $Z_{10}^{(i+4)}$, $Z_2^{(i+5)}$, $Z_5^{(i+5)}$, $Z_7^{(i+5)}$ and $Z_{10}^{(i+5)}$. Starting from an even-numbered round i, the following 19 subkeys have to be weak: $Z_7^{(i)}$, $Z_{10}^{(i)}$, $Z_8^{(i+1)}$, $Z_{10}^{(i+1)}$, $Z_4^{(i+2)}$, $Z_{10}^{(i+2)}$, $Z_1^{(i+3)}$, $Z_3^{(i+3)}$, $Z_6^{(i+3)}$, $Z_8^{(i+3)}$, $Z_{10}^{(i+3)}$, $Z_2^{(i+4)}$, $Z_5^{(i+4)}$, $Z_{10}^{(i+4)}$, $Z_1^{(i+5)}$, $Z_3^{(i+5)}$, $Z_6^{(i+5)}$, $Z_8^{(i+5)}$ and $Z_{10}^{(i+5)}$.

(3) $(0,0,0,0,0,0,\delta,\delta) \stackrel{1r}{\rightarrow} (\delta,\delta,\delta,0,\delta,\delta,0,\delta) \stackrel{1r}{\rightarrow} (0,0,0,\delta,0,0,0,\delta) \stackrel{1r}{\rightarrow} (0, 0, 0, 0, 0, 0, \delta, \delta)$, which is 3-round iterative and holds with probability 2^{-28}. Starting from an odd-numbered round i, the following four subkeys have to be weak: $Z_8^{(i)}$, $Z_2^{(i+1)}$, $Z_5^{(i+1)}$ and $Z_8^{(i+2)}$. Starting from an even-numbered round i, the following six subkeys have to be weak: $Z_7^{(i)}$, $Z_1^{(i+1)}$, $Z_3^{(i+1)}$, $Z_6^{(i+1)}$, $Z_8^{(i+1)}$ and $Z_4^{(i+2)}$.

(4) $(0,0,0,0,0,\delta,0,0) \stackrel{1r}{\rightarrow} (\delta,\delta,0,0,\delta,0,\delta,\delta) \stackrel{1r}{\rightarrow} (\delta,\delta,0,0,\delta,\delta,\delta,0) \stackrel{1r}{\rightarrow} (\delta, \delta, \delta, \delta, \delta, 0, \delta, \delta) \stackrel{1r}{\rightarrow} (0,0,\delta,\delta,0,\delta,0,0) \stackrel{1r}{\rightarrow} (0,0,\delta,\delta,0,0,0,\delta) \stackrel{1r}{\rightarrow} (0, 0, 0, 0, 0, \delta, 0, 0)$, which is 6-round iterative and holds with probability 2^{-70}. Starting from an odd-numbered round i, the following 19 subkeys have to be weak: $Z_6^{(i)}$, $Z_{10}^{(i)}$, $Z_2^{(i+1)}$, $Z_5^{(i+1)}$, $Z_7^{(i+1)}$, $Z_{10}^{(i+1)}$, $Z_1^{(i+2)}$, $Z_6^{(i+2)}$, $Z_{10}^{(i+2)}$, $Z_2^{(i+3)}$, $Z_4^{(i+3)}$, $Z_5^{(i+3)}$, $Z_7^{(i+3)}$, $Z_{10}^{(i+3)}$, $Z_3^{(i+4)}$, $Z_6^{(i+4)}$, $Z_{10}^{(i+4)}$, $Z_4^{(i+5)}$ and $Z_{10}^{(i+5)}$. Starting from an even-numbered round i, the following 17 subkeys have to be weak: $Z_{10}^{(i)}$, $Z_1^{(i+1)}$, $Z_8^{(i+1)}$, $Z_{10}^{(i+1)}$, $Z_2^{(i+2)}$, $Z_5^{(i+2)}$, $Z_7^{(i+2)}$, $Z_{10}^{(i+2)}$, $Z_1^{(i+3)}$, $Z_3^{(i+3)}$, $Z_8^{(i+3)}$, $Z_{10}^{(i+3)}$, $Z_4^{(i+4)}$, $Z_{10}^{(i+4)}$, $Z_3^{(i+5)}$, $Z_8^{(i+5)}$ and $Z_{10}^{(i+5)}$.

(5) $(0,0,0,0,0,\delta,0,\delta) \stackrel{1r}{\rightarrow} (0,0,\delta,\delta,0,\delta,0,\delta) \stackrel{1r}{\rightarrow} (\delta,\delta,0,0,\delta,\delta,\delta,\delta) \stackrel{1r}{\rightarrow} (0, 0, 0, 0, 0, \delta, 0, \delta)$, which is 3-round iterative and holds with probability 2^{-42}. Starting from an odd-numbered round i, the following six subkeys have to be weak: $Z_6^{(i)}$, $Z_8^{(i)}$, $Z_4^{(i+1)}$, $Z_1^{(i+2)}$, $Z_6^{(i+2)}$ and $Z_8^{(i+2)}$. Starting from an even-numbered round i, the following six subkeys have to be weak: $Z_3^{(i+1)}$, $Z_6^{(i+1)}$, $Z_8^{(i+1)}$, $Z_2^{(i+2)}$, $Z_5^{(i+2)}$ and $Z_7^{(i+2)}$.

(6) $(0,0,0,0,0,\delta,\delta,0) \stackrel{1r}{\rightarrow} (\delta,\delta,0,\delta,\delta,0,0,0) \stackrel{1r}{\rightarrow} (\delta,\delta,0,\delta,\delta,\delta,\delta,0) \stackrel{1r}{\rightarrow} (0, 0, 0, 0, 0, \delta, \delta, 0)$, which is 3-round iterative and holds with probability 2^{-42}. Starting from an odd-numbered round i, the following six subkeys have to be weak: $Z_6^{(i)}$, $Z_2^{(i+1)}$, $Z_4^{(i+1)}$, $Z_5^{(i+1)}$, $Z_1^{(i+2)}$ and $Z_6^{(i+2)}$. Starting from an even-numbered round i, the following six subkeys have to be weak: $Z_7^{(i)}$, $Z_1^{(i+1)}$, $Z_2^{(i+2)}$, $Z_4^{(i+2)}$, $Z_5^{(i+2)}$ and $Z_7^{(i+2)}$.

(7) $(0,0,0,0,0,\delta,\delta,\delta) \stackrel{1r}{\rightarrow} (0,0,\delta,0,0,\delta,\delta,0) \stackrel{1r}{\rightarrow} (\delta,\delta,0,\delta,\delta,\delta,\delta,\delta) \stackrel{1r}{\rightarrow} (\delta, \delta, \delta, \delta, \delta, 0, 0, 0) \stackrel{1r}{\rightarrow} (\delta,\delta,0,\delta,\delta,0,0,\delta) \stackrel{1r}{\rightarrow} (0,0,\delta,0,0,0,0,0) \stackrel{1r}{\rightarrow} (0, 0, 0, 0, 0, \delta, \delta, \delta)$, which is 6-round iterative and holds with probability 2^{-70}. Starting from an odd-numbered round i, the following 17 subkeys have to be weak: $Z_6^{(i)}$, $Z_8^{(i)}$, $Z_{10}^{(i)}$, $Z_7^{(i+1)}$, $Z_{10}^{(i+1)}$, $Z_1^{(i+2)}$, $Z_6^{(i+2)}$, $Z_8^{(i+2)}$, $Z_{10}^{(i+2)}$, $Z_2^{(i+3)}$, $Z_4^{(i+3)}$, $Z_5^{(i+3)}$, $Z_{10}^{(i+3)}$, $Z_1^{(i+4)}$, $Z_8^{(i+4)}$, $Z_{10}^{(i+4)}$ and $Z_{10}^{(i+5)}$. Starting from an even-numbered round i, the following 19 subkeys have to be weak: $Z_7^{(i)}$, $Z_{10}^{(i)}$, $Z_3^{(i+1)}$, $Z_6^{(i+1)}$, $Z_{10}^{(i+1)}$, $Z_2^{(i+2)}$, $Z_4^{(i+2)}$, $Z_5^{(i+2)}$, $Z_7^{(i+2)}$, $Z_{10}^{(i+2)}$, $Z_1^{(i+3)}$, $Z_3^{(i+3)}$, $Z_{10}^{(i+3)}$, $Z_2^{(i+4)}$, $Z_4^{(i+4)}$, $Z_5^{(i+4)}$, $Z_{10}^{(i+4)}$, $Z_3^{(i+5)}$ and $Z_{10}^{(i+5)}$.

(8) $(0,0,0,0,\delta,0,0,0) \stackrel{1r}{\rightarrow} (0,\delta,\delta,0,\delta,0,\delta,\delta) \stackrel{1r}{\rightarrow} (0,\delta,\delta,0,0,0,\delta,0) \stackrel{1r}{\rightarrow} (\delta, \delta, \delta, \delta, 0, \delta, \delta, \delta) \stackrel{1r}{\rightarrow} (\delta,0,0,\delta,0,\delta,0,0) \stackrel{1r}{\rightarrow} (\delta,0,0,\delta,\delta,\delta,0,\delta) \stackrel{1r}{\rightarrow} (0, 0, 0, 0, \delta, 0, 0, 0)$, which is 6-round iterative and holds with probability 2^{-70}. Starting from an odd-numbered round i, the following 21 subkeys have to be weak: $Z_9^{(i)}$, $Z_{10}^{(i)}$, $Z_2^{(i+1)}$, $Z_5^{(i+1)}$, $Z_7^{(i+1)}$, $Z_9^{(i+1)}$, $Z_{10}^{(i+1)}$, $Z_3^{(i+2)}$, $Z_{10}^{(i+2)}$, $Z_2^{(i+3)}$, $Z_4^{(i+3)}$, $Z_7^{(i+3)}$, $Z_9^{(i+3)}$, $Z_{10}^{(i+3)}$, $Z_1^{(i+4)}$, $Z_6^{(i+4)}$, $Z_9^{(i+4)}$, $Z_{10}^{(i+4)}$, $Z_4^{(i+5)}$, $Z_5^{(i+5)}$ and $Z_{10}^{(i+5)}$. Starting from an even-numbered round i, the following 23 subkeys have to be weak: $Z_5^{(i)}$, $Z_9^{(i)}$, $Z_{10}^{(i)}$, $Z_3^{(i+1)}$, $Z_8^{(i+1)}$, $Z_9^{(i+1)}$, $Z_{10}^{(i+1)}$, $Z_2^{(i+2)}$, $Z_7^{(i+2)}$, $Z_{10}^{(i+2)}$, $Z_1^{(i+3)}$, $Z_3^{(i+3)}$, $Z_6^{(i+3)}$, $Z_8^{(i+3)}$, $Z_9^{(i+3)}$, $Z_{10}^{(i+3)}$, $Z_4^{(i+4)}$, $Z_9^{(i+4)}$, $Z_{10}^{(i+4)}$, $Z_1^{(i+5)}$, $Z_6^{(i+5)}$, $Z_8^{(i+5)}$ and $Z_{10}^{(i+5)}$.

(9) $(0,0,0,0,\delta,0,0,\delta) \stackrel{1r}{\rightarrow} (\delta,0,0,\delta,0,\delta,0,\delta) \stackrel{1r}{\rightarrow} (0,\delta,\delta,0,0,0,\delta,\delta) \stackrel{1r}{\rightarrow} (0, 0, 0, 0, \delta, 0, 0, \delta)$, which is 3-round iterative and holds with probability 2^{-28}. Starting from an odd-numbered round i, the following six subkeys have to be weak: $Z_8^{(i)}$, $Z_9^{(i)}$, $Z_4^{(i+1)}$, $Z_9^{(i+1)}$, $Z_3^{(i+2)}$ and $Z_8^{(i+2)}$. Starting from an even-numbered round i, the following eight subkeys have to be weak: $Z_5^{(i)}$, $Z_9^{(i)}$, $Z_1^{(i+1)}$, $Z_6^{(i+1)}$, $Z_8^{(i+1)}$, $Z_9^{(i+1)}$, $Z_2^{(i+2)}$ and $Z_7^{(i+2)}$.

(10) $(0,0,0,0,\delta,0,\delta,0) \stackrel{1r}{\rightarrow} (0,\delta,\delta,\delta,\delta,0,0,0) \stackrel{1r}{\rightarrow} (0,\delta,\delta,\delta,0,0,\delta,0) \stackrel{1r}{\rightarrow} (0, 0, 0, 0, \delta, 0, \delta, 0)$, which is 3-round iterative and holds with probability 2^{-42}. Starting from an odd-numbered round i, the following six subkeys have to be weak: $Z_9^{(i)}$, $Z_2^{(i+1)}$, $Z_4^{(i+1)}$, $Z_5^{(i+1)}$, $Z_9^{(i+1)}$ and $Z_3^{(i+2)}$. Starting from an even-numbered round i, the following eight subkeys have to be weak: $Z_5^{(i)}$, $Z_7^{(i)}$, $Z_9^{(i)}$, $Z_3^{(i+1)}$, $Z_9^{(i+1)}$, $Z_2^{(i+2)}$, $Z_4^{(i+2)}$ and $Z_7^{(i+2)}$.

(11) $(0,0,0,0,\delta,0,\delta,\delta) \stackrel{1r}{\rightarrow} (\delta,0,0,0,0,\delta,\delta,0) \stackrel{1r}{\rightarrow} (0,\delta,\delta,\delta,0,0,\delta,\delta) \stackrel{1r}{\rightarrow} (\delta, \delta, \delta, \delta, 0, \delta, 0, 0) \stackrel{1r}{\rightarrow} (0,\delta,\delta,\delta,\delta,0,0,\delta) \stackrel{1r}{\rightarrow} (\delta,0,0,0,\delta,\delta,0,0) \stackrel{1r}{\rightarrow} (0, 0, 0, 0, \delta, 0, \delta, \delta)$, which is 6-round iterative and holds with probability 2^{-70}. Starting from an odd-numbered round i, the following 19 subkeys have to be weak: $Z_8^{(i)}$, $Z_9^{(i)}$, $Z_{10}^{(i)}$, $Z_7^{(i+1)}$, $Z_9^{(i+1)}$, $Z_{10}^{(i+1)}$, $Z_3^{(i+2)}$, $Z_8^{(i+2)}$, $Z_{10}^{(i+2)}$, $Z_2^{(i+3)}$, $Z_4^{(i+3)}$, $Z_9^{(i+3)}$, $Z_{10}^{(i+3)}$, $Z_3^{(i+4)}$, $Z_8^{(i+4)}$, $Z_9^{(i+4)}$, $Z_{10}^{(i+4)}$, $Z_5^{(i+5)}$ and $Z_{10}^{(i+5)}$. Starting from an even-numbered round i, the following 25 subkeys have to be weak: $Z_5^{(i)}$, $Z_7^{(i)}$, $Z_9^{(i)}$, $Z_{10}^{(i)}$, $Z_1^{(i+1)}$, $Z_6^{(i+1)}$, $Z_9^{(i+1)}$, $Z_{10}^{(i+1)}$,

$Z_2^{(i+2)}$, $Z_4^{(i+2)}$, $Z_7^{(i+2)}$, $Z_{10}^{(i+2)}$, $Z_1^{(i+3)}$, $Z_3^{(i+3)}$, $Z_6^{(i+3)}$, $Z_9^{(i+3)}$, $Z_{10}^{(i+3)}$, $Z_2^{(i+4)}$, $Z_4^{(i+4)}$, $Z_5^{(i+4)}$, $Z_9^{(i+4)}$, $Z_{10}^{(i+4)}$, $Z_1^{(i+5)}$, $Z_6^{(i+5)}$ and $Z_{10}^{(i+5)}$.

(12) $(0,0,0,0,\delta,\delta,0,0) \xrightarrow{1r} (\delta,0,\delta,0,0,0,0,0) \xrightarrow{1r} (\delta,0,\delta,0,\delta,\delta,0,0) \xrightarrow{1r} (0, 0, 0, 0, \delta, \delta, 0, 0)$, which is 3-round iterative and holds with probability 2^{-28}. Starting from an odd-numbered round i, the following six subkeys have to be weak: $Z_6^{(i)}$, $Z_9^{(i)}$, $Z_9^{(i+1)}$, $Z_1^{(i+2)}$, $Z_3^{(i+2)}$ and $Z_6^{(i+2)}$. Starting from an even-numbered round i, the following six subkeys have to be weak: $Z_5^{(i)}$, $Z_9^{(i)}$, $Z_1^{(i+1)}$, $Z_3^{(i+1)}$, $Z_9^{(i+1)}$ and $Z_5^{(i+2)}$.

(13) $(0,0,0,0,\delta,\delta,0,\delta) \xrightarrow{1r} (0,\delta,0,\delta,\delta,\delta,\delta,0) \xrightarrow{1r} (\delta,0,\delta,0,\delta,\delta,0,\delta) \xrightarrow{1r} (\delta, \delta, \delta, 0, 0, \delta, 0) \xrightarrow{1r} (\delta,0,\delta,0,0,0,0,\delta) \xrightarrow{1r} (0,\delta,0,\delta,0,0,\delta,0) \xrightarrow{1r} (0, 0, 0, 0, \delta, \delta, 0, \delta)$, which is 6-round iterative and holds with probability 2^{-70}. Starting from an odd-numbered round i, the following 29 subkeys have to be weak: $Z_6^{(i)}$, $Z_8^{(i)}$, $Z_9^{(i)}$, $Z_{10}^{(i)}$, $Z_2^{(i+1)}$, $Z_4^{(i+1)}$, $Z_5^{(i+1)}$, $Z_7^{(i+1)}$, $Z_9^{(i+1)}$, $Z_{10}^{(i+1)}$, $Z_1^{(i+2)}$, $Z_3^{(i+2)}$, $Z_6^{(i+2)}$, $Z_8^{(i+2)}$, $Z_{10}^{(i+2)}$, $Z_2^{(i+3)}$, $Z_4^{(i+3)}$, $Z_7^{(i+3)}$, $Z_9^{(i+3)}$, $Z_{10}^{(i+3)}$, $Z_1^{(i+4)}$, $Z_3^{(i+4)}$, $Z_8^{(i+4)}$, $Z_9^{(i+4)}$, $Z_{10}^{(i+4)}$, $Z_2^{(i+5)}$, $Z_4^{(i+5)}$, $Z_7^{(i+5)}$ and $Z_{10}^{(i+5)}$. Starting from an even-numbered round i, the following 15 subkeys have to be weak: $Z_5^{(i)}$, $Z_9^{(i)}$, $Z_{10}^{(i)}$, $Z_6^{(i+1)}$, $Z_9^{(i+1)}$, $Z_{10}^{(i+1)}$, $Z_5^{(i+2)}$, $Z_{10}^{(i+2)}$, $Z_1^{(i+3)}$, $Z_3^{(i+3)}$, $Z_9^{(i+3)}$, $Z_{10}^{(i+3)}$, $Z_9^{(i+4)}$, $Z_{10}^{(i+4)}$ and $Z_{10}^{(i+5)}$.

(14) $(0,0,0,0,\delta,\delta,\delta,0) \xrightarrow{1r} (\delta,0,\delta,\delta,0,0,\delta,\delta) \xrightarrow{1r} (\delta,0,\delta,\delta,\delta,\delta,0,0) \xrightarrow{1r} (\delta, \delta, \delta, \delta, 0, 0, 0, \delta) \xrightarrow{1r} (0,\delta,0,0,\delta,\delta,0,0) \xrightarrow{1r} (0,\delta,0,0,0,0,\delta,\delta) \xrightarrow{1r} (0, 0, 0, 0, \delta, \delta, \delta, 0)$, which is 6-round iterative and holds with probability 2^{-70}. Starting from an odd-numbered round i, the following 21 subkeys have to be weak: $Z_6^{(i)}$, $Z_9^{(i)}$, $Z_{10}^{(i)}$, $Z_4^{(i+1)}$, $Z_7^{(i+1)}$, $Z_9^{(i+1)}$, $Z_{10}^{(i+1)}$, $Z_1^{(i+2)}$, $Z_3^{(i+2)}$, $Z_6^{(i+2)}$, $Z_{10}^{(i+2)}$, $Z_2^{(i+3)}$, $Z_4^{(i+3)}$, $Z_9^{(i+3)}$, $Z_{10}^{(i+3)}$, $Z_6^{(i+4)}$, $Z_9^{(i+4)}$, $Z_{10}^{(i+4)}$, $Z_2^{(i+5)}$, $Z_7^{(i+5)}$ and $Z_{10}^{(i+5)}$. Starting from an even-numbered round i, the following 23 subkeys have to be weak: $Z_5^{(i)}$, $Z_7^{(i)}$, $Z_9^{(i)}$, $Z_{10}^{(i)}$, $Z_1^{(i+1)}$, $Z_3^{(i+1)}$, $Z_8^{(i+1)}$, $Z_9^{(i+1)}$, $Z_{10}^{(i+1)}$, $Z_4^{(i+2)}$, $Z_5^{(i+2)}$, $Z_{10}^{(i+2)}$, $Z_1^{(i+3)}$, $Z_3^{(i+3)}$, $Z_8^{(i+3)}$, $Z_9^{(i+3)}$, $Z_{10}^{(i+3)}$, $Z_2^{(i+4)}$, $Z_5^{(i+4)}$, $Z_9^{(i+4)}$, $Z_{10}^{(i+4)}$, $Z_8^{(i+5)}$ and $Z_{10}^{(i+5)}$.

(15) $(0,0,0,0,\delta,\delta,\delta,\delta) \xrightarrow{1r} (0,\delta,0,0,\delta,\delta,0,\delta) \xrightarrow{1r} (\delta,0,\delta,\delta,\delta,\delta,0,\delta) \xrightarrow{1r} (0, 0, 0, 0, \delta, \delta, \delta, \delta)$, which is 3-round iterative and holds with probability 2^{-42}. Starting from an odd-numbered round i, the following ten subkeys have to be weak: $Z_6^{(i)}$, $Z_8^{(i)}$, $Z_9^{(i)}$, $Z_2^{(i+1)}$, $Z_5^{(i+1)}$, $Z_9^{(i+1)}$, $Z_1^{(i+2)}$, $Z_3^{(i+2)}$, $Z_6^{(i+2)}$ and $Z_8^{(i+2)}$. Starting from an even-numbered round i, the following eight subkeys have to be weak: $Z_5^{(i)}$, $Z_7^{(i)}$, $Z_9^{(i)}$, $Z_6^{(i+1)}$, $Z_8^{(i+1)}$, $Z_9^{(i+1)}$, $Z_4^{(i+2)}$ and $Z_5^{(i+2)}$.

(16) $(0,0,0,\delta,0,0,\delta,0) \xrightarrow{1r} (\delta,\delta,\delta,0,\delta,\delta,\delta,0) \xrightarrow{1r} (\delta,\delta,\delta,\delta,\delta,\delta,0,0) \xrightarrow{1r} (0, 0, 0, \delta, 0, 0, \delta, 0)$, which is 3-round iterative and holds with probability 2^{-28}. Starting from an odd-numbered round i, the following six subkeys have to be weak: $Z_2^{(i+1)}$, $Z_5^{(i+1)}$, $Z_7^{(i+1)}$, $Z_1^{(i+2)}$, $Z_3^{(i+2)}$ and $Z_6^{(i+2)}$. Starting

from an even-numbered round i, the following eight subkeys have to be weak: $Z_4^{(i)}$, $Z_7^{(i)}$, $Z_1^{(i+1)}$, $Z_3^{(i+1)}$, $Z_6^{(i+1)}$, $Z_2^{(i+2)}$, $Z_4^{(i+2)}$ and $Z_5^{(i+2)}$.

(17) $(0,0,0,\delta,0,\delta,0,0) \stackrel{1r}{\rightarrow} (0,0,\delta,\delta,0,\delta,\delta,0) \stackrel{1r}{\rightarrow} (0,0,\delta,0,0,0,\delta,0) \stackrel{1r}{\rightarrow} (0, 0, 0, \delta, 0, \delta, 0, 0)$, which is 3-round iterative and holds with probability 2^{-28}. Starting from an odd-numbered round i, the following four subkeys have to be weak: $Z_6^{(i)}$, $Z_4^{(i+1)}$, $Z_7^{(i+1)}$ and $Z_3^{(i+2)}$. Starting from an even-numbered round i, the following four subkeys have to be weak: $Z_4^{(i)}$, $Z_3^{(i+1)}$, $Z_6^{(i+1)}$ and $Z_7^{(i+2)}$.

(18) $(0,0,0,\delta,0,\delta,0,\delta) \stackrel{1r}{\rightarrow} (\delta,\delta,0,0,\delta,0,0,0) \stackrel{1r}{\rightarrow} (0,0,\delta,0,0,0,\delta,\delta) \stackrel{1r}{\rightarrow} (\delta, \delta, \delta, 0, \delta, 0, \delta, 0) \stackrel{1r}{\rightarrow} (0,0,\delta,\delta,0,\delta,\delta,\delta) \stackrel{1r}{\rightarrow} (\delta,\delta,0,\delta,\delta,\delta,0,0) \stackrel{1r}{\rightarrow} (0, 0, 0, \delta, 0, \delta, 0, \delta)$, which is 6-round iterative and holds with probability 2^{-70}. Starting from an odd-numbered round i, the following 21 subkeys have to be weak: $Z_6^{(i)}$, $Z_8^{(i)}$, $Z_{10}^{(i)}$, $Z_2^{(i+1)}$, $Z_5^{(i+1)}$, $Z_{10}^{(i+1)}$, $Z_3^{(i+2)}$, $Z_8^{(i+2)}$, $Z_{10}^{(i+2)}$, $Z_2^{(i+3)}$, $Z_5^{(i+3)}$, $Z_7^{(i+3)}$, $Z_{10}^{(i+3)}$, $Z_3^{(i+4)}$, $Z_6^{(i+4)}$, $Z_8^{(i+4)}$, $Z_{10}^{(i+4)}$, $Z_2^{(i+5)}$, $Z_4^{(i+5)}$, $Z_5^{(i+5)}$ and $Z_{10}^{(i+5)}$. Starting from an even-numbered round i, the following 15 subkeys have to be weak: $Z_4^{(i)}$, $Z_{10}^{(i)}$, $Z_1^{(i+1)}$, $Z_{10}^{(i+1)}$, $Z_7^{(i+2)}$, $Z_{10}^{(i+2)}$, $Z_1^{(i+3)}$, $Z_3^{(i+3)}$, $Z_{10}^{(i+3)}$, $Z_4^{(i+4)}$, $Z_7^{(i+4)}$, $Z_{10}^{(i+4)}$, $Z_1^{(i+5)}$, $Z_6^{(i+5)}$ and $Z_{10}^{(i+5)}$.

(19) $(0,0,0,\delta,0,\delta,\delta,0) \stackrel{1r}{\rightarrow} (0,0,\delta,0,0,\delta,0,\delta) \stackrel{1r}{\rightarrow} (0,0,\delta,\delta,0,0,\delta,0) \stackrel{1r}{\rightarrow} (\delta, \delta, \delta, 0, \delta, 0, 0, \delta) \stackrel{1r}{\rightarrow} (\delta,\delta,0,\delta,\delta,0,\delta,0) \stackrel{1r}{\rightarrow} (\delta,\delta,0,0,\delta,\delta,0,\delta) \stackrel{1r}{\rightarrow} (0, 0, 0, \delta, 0, \delta, \delta, 0)$, which is 6-round iterative and holds with probability 2^{-70}. Starting from an odd-numbered round i, the following 13 subkeys have to be weak: $Z_6^{(i)}$, $Z_{10}^{(i)}$, $Z_{10}^{(i+)}$, $Z_3^{(i+2)}$, $Z_{10}^{(i+2)}$, $Z_2^{(i+3)}$, $Z_5^{(i+3)}$, $Z_{10}^{(i+3)}$, $Z_1^{(i+4)}$, $Z_{10}^{(i+4)}$, $Z_2^{(i+5)}$, $Z_5^{(i+5)}$ and $Z_{10}^{(i+5)}$. Starting from an even-numbered round i, the following 23 subkeys have to be weak: $Z_4^{(i)}$, $Z_7^{(i)}$, $Z_{10}^{(i)}$, $Z_3^{(i+1)}$, $Z_6^{(i+1)}$, $Z_8^{(i+1)}$, $Z_{10}^{(i+1)}$, $Z_4^{(i+2)}$, $Z_7^{(i+2)}$, $Z_{10}^{(i+2)}$, $Z_1^{(i+3)}$, $Z_3^{(i+3)}$, $Z_8^{(i+3)}$, $Z_{10}^{(i+3)}$, $Z_2^{(i+4)}$, $Z_4^{(i+4)}$, $Z_5^{(i+4)}$, $Z_7^{(i+4)}$, $Z_{10}^{(i+4)}$, $Z_1^{(i+5)}$, $Z_6^{(i+5)}$, $Z_8^{(i+5)}$ and $Z_{10}^{(i+5)}$.

(20) $(0,0,0,\delta,0,\delta,\delta,\delta) \stackrel{1r}{\rightarrow} (\delta,\delta,0,\delta,\delta,0,\delta,\delta) \stackrel{1r}{\rightarrow} (0,0,\delta,\delta,0,0,\delta,\delta) \stackrel{1r}{\rightarrow} (0, 0, 0, \delta, 0, \delta, \delta, \delta)$, which is 3-round iterative and holds with probability 2^{-28}. Starting from an odd-numbered round i, the following eight subkeys have to be weak: $Z_6^{(i)}$, $Z_8^{(i)}$, $Z_2^{(i+1)}$, $Z_4^{(i+1)}$, $Z_5^{(i+1)}$, $Z_7^{(i+1)}$, $Z_3^{(i+2)}$ and $Z_8^{(i+2)}$. Starting from an even-numbered round i, the following six subkeys have to be weak: $Z_4^{(i)}$, $Z_7^{(i)}$, $Z_1^{(i+1)}$, $Z_8^{(i+1)}$, $Z_4^{(i+2)}$ and $Z_7^{(i+2)}$.

(21) $(0,0,0,\delta,\delta,0,0,0) \stackrel{1r}{\rightarrow} (\delta,0,0,\delta,0,\delta,\delta,0) \stackrel{1r}{\rightarrow} (\delta,0,0,0,\delta,\delta,\delta,0) \stackrel{1r}{\rightarrow} (0, 0, 0, \delta, \delta, 0, 0, 0)$, which is 3-round iterative and holds with probability 2^{-42}. Starting from an odd-numbered round i, the following six subkeys have to be weak: $Z_9^{(i)}$, $Z_4^{(i+1)}$, $Z_7^{(i+1)}$, $Z_9^{(i+1)}$, $Z_1^{(i+2)}$ and $Z_6^{(i+2)}$. Starting from an even-numbered round i, the following eight subkeys have to be weak: $Z_4^{(i)}$, $Z_5^{(i)}$, $Z_9^{(i)}$, $Z_1^{(i+1)}$, $Z_6^{(i+1)}$, $Z_9^{(i+1)}$, $Z_5^{(i+2)}$ and $Z_7^{(i+2)}$.

(22) $(0,0,0,\delta,\delta,0,0,\delta) \stackrel{1r}{\rightarrow} (0,\delta,\delta,0,\delta,0,0,0) \stackrel{1r}{\rightarrow} (\delta,0,0,0,\delta,\delta,\delta,\delta) \stackrel{1r}{\rightarrow} (\delta, \delta, \delta, 0, 0, \delta, \delta, 0) \stackrel{1r}{\rightarrow} (\delta,0,0,\delta,0,\delta,\delta,\delta) \stackrel{1r}{\rightarrow} (0,\delta,\delta,\delta,0,0,0,0) \stackrel{1r}{\rightarrow} (0, 0, 0, \delta, \delta, 0, 0, \delta)$, which is 6-round iterative and holds with probability 2^{-70}. Starting from an odd-numbered round i, the following 23 subkeys have to be weak: $Z_8^{(i)}$, $Z_9^{(i)}$, $Z_{10}^{(i)}$, $Z_2^{(i+1)}$, $Z_5^{(i+1)}$, $Z_9^{(i+1)}$, $Z_{10}^{(i+1)}$, $Z_1^{(i+2)}$, $Z_6^{(i+2)}$, $Z_8^{(i+2)}$, $Z_{10}^{(i+2)}$, $Z_2^{(i+3)}$, $Z_7^{(i+3)}$, $Z_9^{(i+3)}$, $Z_{10}^{(i+3)}$, $Z_1^{(i+4)}$, $Z_6^{(i+4)}$, $Z_8^{(i+4)}$, $Z_9^{(i+4)}$, $Z_{10}^{(i+4)}$, $Z_2^{(i+5)}$, $Z_4^{(i+5)}$ and $Z_{10}^{(i+5)}$. Starting from an even-numbered round i, the following 21 subkeys have to be weak: $Z_4^{(i)}$, $Z_5^{(i)}$, $Z_9^{(i)}$, $Z_{10}^{(i)}$, $Z_3^{(i+1)}$, $Z_9^{(i+1)}$, $Z_{10}^{(i+1)}$, $Z_5^{(i+2)}$, $Z_7^{(i+2)}$, $Z_{10}^{(i+2)}$, $Z_1^{(i+3)}$, $Z_3^{(i+3)}$, $Z_6^{(i+3)}$, $Z_9^{(i+3)}$, $Z_{10}^{(i+3)}$, $Z_4^{(i+4)}$, $Z_7^{(i+4)}$, $Z_9^{(i+4)}$, $Z_{10}^{(i+4)}$, $Z_3^{(i+5)}$ and $Z_{10}^{(i+5)}$.

(23) $(0,0,0,\delta,\delta,0,\delta,0) \stackrel{1r}{\rightarrow} (\delta,0,0,0,0,\delta,0,\delta) \stackrel{1r}{\rightarrow} (\delta,0,0,\delta,\delta,\delta,\delta,0) \stackrel{1r}{\rightarrow} (\delta, \delta, \delta, 0, 0, \delta, 0, \delta) \stackrel{1r}{\rightarrow} (0,\delta,\delta,\delta,\delta,0,\delta,0) \stackrel{1r}{\rightarrow} (0,\delta,\delta,0,0,0,0,\delta) \stackrel{1r}{\rightarrow} (0, 0, 0, \delta, \delta, 0, \delta, 0)$, which is 6-round iterative and holds with probability 2^{-70}. Starting from an odd-numbered round i, the following 15 subkeys have to be weak: $Z_9^{(i)}$, $Z_{10}^{(i)}$, $Z_9^{(i+1)}$, $Z_{10}^{(i+1)}$, $Z_1^{(i+2)}$, $Z_6^{(i+2)}$, $Z_{10}^{(i+2)}$, $Z_2^{(i+3)}$, $Z_9^{(i+3)}$, $Z_{10}^{(i+3)}$, $Z_3^{(i+4)}$, $Z_9^{(i+4)}$, $Z_{10}^{(i+4)}$, $Z_2^{(i+5)}$ and $Z_{10}^{(i+5)}$. Starting from an even-numbered round i, the following 29 subkeys have to be weak: $Z_4^{(i)}$, $Z_5^{(i)}$, $Z_7^{(i)}$, $Z_9^{(i)}$, $Z_{10}^{(i)}$, $Z_1^{(i+1)}$, $Z_6^{(i+1)}$, $Z_8^{(i+1)}$, $Z_9^{(i+1)}$, $Z_{10}^{(i+1)}$, $Z_4^{(i+2)}$, $Z_5^{(i+2)}$, $Z_7^{(i+2)}$, $Z_{10}^{(i+2)}$, $Z_1^{(i+3)}$, $Z_3^{(i+3)}$, $Z_6^{(i+3)}$, $Z_8^{(i+3)}$, $Z_9^{(i+3)}$, $Z_{10}^{(i+3)}$, $Z_2^{(i+4)}$, $Z_4^{(i+4)}$, $Z_5^{(i+4)}$, $Z_7^{(i+4)}$, $Z_9^{(i+4)}$, $Z_{10}^{(i+4)}$, $Z_3^{(i+5)}$, $Z_8^{(i+5)}$ and $Z_{10}^{(i+5)}$.

(24) $(0,0,0,\delta,\delta,0,\delta,\delta) \stackrel{1r}{\rightarrow} (0,\delta,\delta,\delta,\delta,0,\delta,\delta) \stackrel{1r}{\rightarrow} (\delta,0,0,\delta,\delta,\delta,\delta,\delta) \stackrel{1r}{\rightarrow} (0, 0, 0, \delta, \delta, 0, \delta, \delta)$, which is 3-round iterative and holds with probability 2^{-28}. Starting from an odd-numbered round i, the following ten subkeys have to be weak: $Z_8^{(i)}$, $Z_9^{(i)}$, $Z_2^{(i+1)}$, $Z_4^{(i+1)}$, $Z_5^{(i+1)}$, $Z_7^{(i+1)}$, $Z_9^{(i+1)}$, $Z_1^{(i+2)}$, $Z_6^{(i+2)}$ and $Z_8^{(i+2)}$. Starting from an even-numbered round i, the following ten subkeys have to be weak: $Z_4^{(i)}$, $Z_5^{(i)}$, $Z_7^{(i)}$, $Z_9^{(i)}$, $Z_3^{(i+1)}$, $Z_8^{(i+1)}$, $Z_9^{(i+1)}$, $Z_4^{(i+2)}$, $Z_5^{(i+2)}$ and $Z_7^{(i+2)}$.

(25) $(0,0,0,\delta,\delta,\delta,0,0) \stackrel{1r}{\rightarrow} (0,\delta,0,\delta,\delta,\delta,0,\delta) \stackrel{1r}{\rightarrow} (0,\delta,0,0,0,0,0,0) \stackrel{1r}{\rightarrow} (\delta, \delta, \delta, 0, 0, 0, \delta, \delta) \stackrel{1r}{\rightarrow} (\delta,0,\delta,0,0,0,\delta,0) \stackrel{1r}{\rightarrow} (\delta,0,\delta,\delta,\delta,\delta,\delta,\delta) \stackrel{1r}{\rightarrow} (0, 0, 0, \delta, \delta, \delta, 0, 0)$, which is 6-round iterative and holds with probability 2^{-70}. Starting from an odd-numbered round i, the following 21 subkeys have to be weak: $Z_6^{(i)}$, $Z_9^{(i)}$, $Z_{10}^{(i)}$, $Z_2^{(i+1)}$, $Z_4^{(i+1)}$, $Z_5^{(i+1)}$, $Z_9^{(i+1)}$, $Z_{10}^{(i+1)}$, $Z_{10}^{(i+2)}$, $Z_2^{(i+3)}$, $Z_7^{(i+3)}$, $Z_9^{(i+3)}$, $Z_{10}^{(i+3)}$, $Z_1^{(i+4)}$, $Z_3^{(i+4)}$, $Z_9^{(i+4)}$, $Z_{10}^{(i+4)}$, $Z_4^{(i+5)}$, $Z_5^{(i+5)}$, $Z_7^{(i+5)}$ and $Z_{10}^{(i+5)}$. Starting from an even-numbered round i, the following 23 subkeys have to be weak: $Z_4^{(i)}$, $Z_5^{(i)}$, $Z_9^{(i)}$, $Z_{10}^{(i)}$, $Z_6^{(i+1)}$, $Z_8^{(i+1)}$, $Z_9^{(i+1)}$, $Z_{10}^{(i+1)}$, $Z_2^{(i+2)}$, $Z_{10}^{(i+2)}$, $Z_1^{(i+3)}$, $Z_3^{(i+3)}$, $Z_8^{(i+3)}$, $Z_9^{(i+3)}$, $Z_{10}^{(i+3)}$, $Z_7^{(i+4)}$, $Z_9^{(i+4)}$, $Z_{10}^{(i+4)}$, $Z_1^{(i+5)}$, $Z_3^{(i+5)}$, $Z_6^{(i+5)}$, $Z_8^{(i+5)}$ and $Z_{10}^{(i+5)}$.

(26) $(0,0,0,\delta,\delta,\delta,0,\delta) \stackrel{1r}{\rightarrow} (\delta,0,\delta,0,0,0,\delta,\delta) \stackrel{1r}{\rightarrow} (0,\delta,0,0,0,0,0,\delta) \stackrel{1r}{\rightarrow} (0, 0, 0, \delta, \delta, \delta, 0, \delta)$, which is 3-round iterative and holds with probability 2^{-28}. Starting from an odd-numbered round i, the following six subkeys

have to be weak: $Z_6^{(i)}$, $Z_8^{(i)}$, $Z_9^{(i)}$, $Z_7^{(i+1)}$, $Z_9^{(i+1)}$ and $Z_8^{(i+2)}$. Starting from an even-numbered round i, the following eight subkeys have to be weak: $Z_4^{(i)}$, $Z_5^{(i)}$, $Z_9^{(i)}$, $Z_1^{(i+1)}$, $Z_3^{(i+1)}$, $Z_8^{(i+1)}$, $Z_9^{(i+1)}$ and $Z_2^{(i+2)}$.

(27) $(0,0,0,\delta,\delta,\delta,\delta,0) \stackrel{1r}{\to} (0,\delta,0,0,\delta,\delta,\delta,0) \stackrel{1r}{\to} (0,\delta,0,\delta,0,0,0,0) \stackrel{1r}{\to} (0, 0, 0, \delta, \delta, \delta, \delta, 0)$, which is 3-round iterative and holds with probability 2^{-42}. Starting from an odd-numbered round i, the following six subkeys have to be weak: $Z_6^{(i)}$, $Z_9^{(i)}$, $Z_2^{(i+1)}$, $Z_5^{(i+1)}$, $Z_7^{(i+1)}$ and $Z_9^{(i+1)}$. Starting from an even-numbered round i, the following eight subkeys have to be weak: $Z_4^{(i)}$, $Z_5^{(i)}$, $Z_7^{(i)}$, $Z_9^{(i)}$, $Z_6^{(i+1)}$, $Z_9^{(i+1)}$, $Z_2^{(i+2)}$ and $Z_4^{(i+2)}$.

(28) $(0,0,0,\delta,\delta,\delta,\delta,\delta) \stackrel{1r}{\to} (\delta,0,\delta,\delta,0,0,0,0) \stackrel{1r}{\to} (0,\delta,0,\delta,0,0,0,\delta) \stackrel{1r}{\to} (\delta, \delta, \delta, 0, 0, 0, 0, 0) \stackrel{1r}{\to} (0,\delta,0,0,\delta,\delta,\delta,\delta) \stackrel{1r}{\to} (\delta,0,\delta,0,\delta,\delta,\delta,0) \stackrel{1r}{\to} (0, 0, 0, \delta, \delta, \delta, \delta, \delta)$, which is 6-round iterative and holds with probability 2^{-70}. Starting from an odd-numbered round i, the following 19 subkeys have to be weak: $Z_6^{(i)}$, $Z_8^{(i)}$, $Z_9^{(i)}$, $Z_{10}^{(i)}$, $Z_4^{(i+1)}$, $Z_9^{(i+1)}$, $Z_{10}^{(i+1)}$, $Z_8^{(i+2)}$, $Z_{10}^{(i+2)}$, $Z_2^{(i+3)}$, $Z_9^{(i+3)}$, $Z_{10}^{(i+3)}$, $Z_6^{(i+4)}$, $Z_8^{(i+4)}$, $Z_9^{(i+4)}$, $Z_{10}^{(i+4)}$, $Z_5^{(i+5)}$, $Z_7^{(i+5)}$ and $Z_{10}^{(i+5)}$. Starting from an even-numbered round i, the following 25 subkeys have to be weak: $Z_4^{(i)}$, $Z_5^{(i)}$, $Z_7^{(i)}$, $Z_9^{(i)}$, $Z_{10}^{(i)}$, $Z_1^{(i+1)}$, $Z_3^{(i+1)}$, $Z_9^{(i+1)}$, $Z_{10}^{(i+1)}$, $Z_2^{(i+2)}$, $Z_4^{(i+2)}$, $Z_{10}^{(i+2)}$, $Z_1^{(i+3)}$, $Z_3^{(i+3)}$, $Z_9^{(i+3)}$, $Z_{10}^{(i+3)}$, $Z_2^{(i+4)}$, $Z_5^{(i+4)}$, $Z_7^{(i+4)}$, $Z_9^{(i+4)}$, $Z_{10}^{(i+4)}$, $Z_1^{(i+5)}$, $Z_3^{(i+5)}$, $Z_6^{(i+5)}$ and $Z_{10}^{(i+5)}$.

(29) $(0,0,\delta,0,0,0,0,\delta) \stackrel{1r}{\to} (\delta,\delta,\delta,\delta,\delta,0,0,\delta) \stackrel{1r}{\to} (0,0,\delta,0,0,\delta,\delta,\delta) \stackrel{1r}{\to} (0, 0, \delta, 0, 0, 0, 0, \delta)$, which is 3-round iterative and holds with probability 2^{-42}. Starting from an odd-numbered round i, the following eight subkeys have to be weak: $Z_3^{(i)}$, $Z_8^{(i)}$, $Z_2^{(i+1)}$, $Z_4^{(i+1)}$, $Z_5^{(i+1)}$, $Z_3^{(i+2)}$, $Z_6^{(i+2)}$ and $Z_8^{(i+2)}$. Starting from an even-numbered round i, the following four subkeys have to be weak: $Z_1^{(i+1)}$, $Z_3^{(i+1)}$, $Z_8^{(i+1)}$ and $Z_7^{(i+2)}$.

(30) $(0,0,\delta,0,0,\delta,0,0) \stackrel{1r}{\to} (\delta,\delta,0,0,\delta,\delta,0,0) \stackrel{1r}{\to} (\delta,\delta,\delta,0,\delta,0,0,0) \stackrel{1r}{\to} (0, 0, \delta, 0, 0, \delta, 0, 0)$, which is 3-round iterative and holds with probability 2^{-28}. Starting from an odd-numbered round i, the following six subkeys have to be weak: $Z_3^{(i)}$, $Z_6^{(i)}$, $Z_2^{(i+1)}$, $Z_5^{(i+1)}$, $Z_1^{(i+2)}$ and $Z_3^{(i+2)}$. Starting from an even-numbered round i, the following four subkeys have to be weak: $Z_1^{(i+1)}$, $Z_6^{(i+1)}$, $Z_2^{(i+2)}$ and $Z_5^{(i+2)}$.

(31) $(0,0,\delta,0,\delta,0,0,0) \stackrel{1r}{\to} (0,\delta,\delta,0,\delta,\delta,0,0) \stackrel{1r}{\to} (0,\delta,0,0,0,\delta,0,0) \stackrel{1r}{\to} (0, 0, \delta, 0, \delta, 0, 0, 0)$, which is 3-round iterative and holds with probability 2^{-28}. Starting from an odd-numbered round i, the following six subkeys have to be weak: $Z_3^{(i)}$, $Z_9^{(i)}$, $Z_2^{(i+1)}$, $Z_5^{(i+1)}$, $Z_9^{(i+1)}$ and $Z_6^{(i+2)}$. Starting from an even-numbered round i, the following six subkeys have to be weak: $Z_5^{(i)}$, $Z_9^{(i)}$, $Z_3^{(i+1)}$, $Z_6^{(i+1)}$, $Z_9^{(i+1)}$ and $Z_2^{(i+2)}$.

(32) $(0,0,\delta,0,\delta,0,0,\delta) \stackrel{1r}{\to} (\delta,0,0,\delta,0,0,\delta,0) \stackrel{1r}{\to} (0,\delta,0,0,0,\delta,0,\delta) \stackrel{1r}{\to} (\delta, \delta, 0, \delta, 0, \delta, \delta, 0) \stackrel{1r}{\to} (0,\delta,\delta,0,\delta,\delta,0,\delta) \stackrel{1r}{\to} (\delta,0,\delta,\delta,\delta,0,\delta,0) \stackrel{1r}{\to} (0, 0, \delta, 0, \delta, 0, 0, \delta)$, which is 6-round iterative and holds with probability 2^{-70}. Starting from an odd-numbered round i, the following 25 subkeys have

to be weak: $Z_3^{(i)}$, $Z_8^{(i)}$, $Z_9^{(i)}$, $Z_{10}^{(i)}$, $Z_4^{(i+1)}$, $Z_7^{(i+1)}$, $Z_9^{(i+1)}$, $Z_{10}^{(i+1)}$, $Z_6^{(i+2)}$, $Z_8^{(i+2)}$, $Z_{10}^{(i+2)}$, $Z_2^{(i+3)}$, $Z_4^{(i+3)}$, $Z_7^{(i+3)}$, $Z_9^{(i+3)}$, $Z_{10}^{(i+3)}$, $Z_3^{(i+4)}$, $Z_6^{(i+4)}$, $Z_8^{(i+4)}$, $Z_9^{(i+4)}$, $Z_{10}^{(i+4)}$, $Z_4^{(i+5)}$, $Z_5^{(i+5)}$, $Z_7^{(i+5)}$ and $Z_{10}^{(i+5)}$. Starting from an even-numbered round i, the following 19 subkeys have to be weak: $Z_5^{(i)}$, $Z_9^{(i)}$, $Z_{10}^{(i)}$, $Z_1^{(i+1)}$, $Z_9^{(i+1)}$, $Z_{10}^{(i+1)}$, $Z_2^{(i+2)}$, $Z_{10}^{(i+2)}$, $Z_1^{(i+3)}$, $Z_6^{(i+3)}$, $Z_9^{(i+3)}$, $Z_{10}^{(i+3)}$, $Z_2^{(i+4)}$, $Z_5^{(i+4)}$, $Z_9^{(i+4)}$, $Z_{10}^{(i+4)}$, $Z_1^{(i+5)}$, $Z_3^{(i+5)}$ and $Z_{10}^{(i+5)}$.

(33) $(0,0,\delta,0,\delta,0,\delta,0) \stackrel{1r}{\to} (0,\delta,\delta,\delta,\delta,\delta,\delta,\delta) \stackrel{1r}{\to} (0,\delta,0,\delta,0,\delta,0,0) \stackrel{1r}{\to} (\delta,\delta,0,\delta,0,\delta,0,\delta) \stackrel{1r}{\to} (\delta,0,0,0,0,0,0,0) \stackrel{1r}{\to} (\delta,0,\delta,0,\delta,0,\delta,\delta) \stackrel{1r}{\to} (0,0,\delta,0,\delta,0,\delta,0)$, which is 6-round iterative and holds with probability 2^{-70}. Starting from an odd-numbered round i, the following 21 subkeys have to be weak: $Z_3^{(i)}$, $Z_9^{(i)}$, $Z_{10}^{(i)}$, $Z_2^{(i+1)}$, $Z_4^{(i+1)}$, $Z_5^{(i+1)}$, $Z_7^{(i+1)}$, $Z_9^{(i+1)}$, $Z_{10}^{(i+1)}$, $Z_6^{(i+2)}$, $Z_{10}^{(i+2)}$, $Z_2^{(i+3)}$, $Z_4^{(i+3)}$, $Z_9^{(i+3)}$, $Z_{10}^{(i+3)}$, $Z_1^{(i+4)}$, $Z_9^{(i+4)}$, $Z_{10}^{(i+4)}$, $Z_5^{(i+5)}$, $Z_7^{(i+5)}$ and $Z_{10}^{(i+5)}$. Starting from an even-numbered round i, the following 23 subkeys have to be weak: $Z_5^{(i)}$, $Z_7^{(i)}$, $Z_9^{(i)}$, $Z_{10}^{(i)}$, $Z_3^{(i+1)}$, $Z_6^{(i+1)}$, $Z_8^{(i+1)}$, $Z_9^{(i+1)}$, $Z_{10}^{(i+1)}$, $Z_2^{(i+2)}$, $Z_4^{(i+2)}$, $Z_{10}^{(i+2)}$, $Z_1^{(i+3)}$, $Z_6^{(i+3)}$, $Z_8^{(i+3)}$, $Z_9^{(i+3)}$, $Z_{10}^{(i+3)}$, $Z_9^{(i+4)}$, $Z_{10}^{(i+4)}$, $Z_1^{(i+5)}$, $Z_3^{(i+5)}$, $Z_8^{(i+5)}$ and $Z_{10}^{(i+5)}$.

(34) $(0,0,\delta,0,\delta,0,\delta,\delta) \stackrel{1r}{\to} (\delta,0,0,0,0,0,0,\delta) \stackrel{1r}{\to} (0,\delta,0,\delta,0,\delta,0,\delta) \stackrel{1r}{\to} (0,0,\delta,0,\delta,0,\delta,\delta)$, which is 3-round iterative and holds with probability 2^{-28}. Starting from an odd-numbered round i, the following six subkeys have to be weak: $Z_3^{(i)}$, $Z_8^{(i)}$, $Z_9^{(i)}$, $Z_9^{(i+1)}$, $Z_6^{(i+2)}$ and $Z_8^{(i+2)}$. Starting from an even-numbered round i, the following eight subkeys have to be weak: $Z_5^{(i)}$, $Z_7^{(i)}$, $Z_9^{(i)}$, $Z_1^{(i+1)}$, $Z_8^{(i+1)}$, $Z_9^{(i+1)}$, $Z_2^{(i+2)}$ and $Z_4^{(i+2)}$.

(35) $(0,0,\delta,0,\delta,\delta,0,0) \stackrel{1r}{\to} (\delta,0,\delta,0,0,\delta,\delta,\delta) \stackrel{1r}{\to} (\delta,0,0,0,\delta,0,\delta,0) \stackrel{1r}{\to} (\delta,\delta,0,\delta,0,0,\delta,\delta) \stackrel{1r}{\to} (0,\delta,0,\delta,\delta,0,0,0) \stackrel{1r}{\to} (0,\delta,\delta,\delta,0,\delta,0,\delta) \stackrel{1r}{\to} (0,0,\delta,0,\delta,\delta,0,0)$, which is 6-round iterative and holds with probability 2^{-42}. Starting from an odd-numbered round i, the following 19 subkeys have to be weak: $Z_3^{(i)}$, $Z_6^{(i)}$, $Z_9^{(i)}$, $Z_{10}^{(i)}$, $Z_7^{(i+1)}$, $Z_9^{(i+1)}$, $Z_{10}^{(i+1)}$, $Z_1^{(i+2)}$, $Z_{10}^{(i+2)}$, $Z_2^{(i+3)}$, $Z_4^{(i+3)}$, $Z_7^{(i+3)}$, $Z_9^{(i+3)}$, $Z_{10}^{(i+3)}$, $Z_9^{(i+4)}$, $Z_{10}^{(i+4)}$, $Z_2^{(i+5)}$, $Z_4^{(i+5)}$ and $Z_{10}^{(i+5)}$. Starting from an even-numbered round i, the following 25 subkeys have to be weak: $Z_5^{(i)}$, $Z_9^{(i)}$, $Z_{10}^{(i)}$, $Z_1^{(i+1)}$, $Z_3^{(i+1)}$, $Z_6^{(i+1)}$, $Z_8^{(i+1)}$, $Z_9^{(i+1)}$, $Z_{10}^{(i+1)}$, $Z_5^{(i+2)}$, $Z_7^{(i+2)}$, $Z_{10}^{(i+2)}$, $Z_1^{(i+3)}$, $Z_8^{(i+3)}$, $Z_9^{(i+3)}$, $Z_{10}^{(i+3)}$, $Z_2^{(i+4)}$, $Z_4^{(i+4)}$, $Z_5^{(i+4)}$, $Z_9^{(i+4)}$, $Z_{10}^{(i+4)}$, $Z_3^{(i+5)}$, $Z_6^{(i+5)}$, $Z_8^{(i+5)}$ and $Z_{10}^{(i+5)}$.

(36) $(0,0,\delta,0,\delta,\delta,0,\delta) \stackrel{1r}{\to} (0,\delta,0,\delta,\delta,0,0,\delta) \stackrel{1r}{\to} (\delta,0,0,0,\delta,0,\delta,\delta) \stackrel{1r}{\to} (0,0,\delta,0,\delta,\delta,0,\delta)$, which is 3-round iterative and holds with probability 2^{-28}. Starting from an odd-numbered round i, the following ten subkeys have to be weak: $Z_3^{(i)}$, $Z_6^{(i)}$, $Z_8^{(i)}$, $Z_9^{(i)}$, $Z_2^{(i+1)}$, $Z_4^{(i+1)}$, $Z_5^{(i+1)}$, $Z_9^{(i+1)}$, $Z_1^{(i+2)}$ and $Z_8^{(i+2)}$. Starting from an even-numbered round i, the following six subkeys have to be weak: $Z_5^{(i)}$, $Z_9^{(i)}$, $Z_8^{(i+1)}$, $Z_9^{(i+1)}$, $Z_5^{(i+2)}$ and $Z_7^{(i+2)}$.

(37) $(0,0,\delta,0,\delta,\delta,\delta,0) \xrightarrow{1r} (\delta,0,\delta,\delta,0,\delta,0,0) \xrightarrow{1r} (\delta,0,0,\delta,\delta,0,\delta,0) \xrightarrow{1r} (0,0,\delta,0,\delta,\delta,\delta,0)$, which is 3-round iterative and holds with probability 2^{-28}. Starting from an odd-numbered round i, the following six subkeys have to be weak: $Z_3^{(i)}$, $Z_6^{(i)}$, $Z_9^{(i)}$, $Z_4^{(i+1)}$, $Z_9^{(i+1)}$ and $Z_1^{(i+2)}$. Starting from an even-numbered round i, the following ten subkeys have to be weak: $Z_5^{(i)}$, $Z_7^{(i)}$, $Z_9^{(i)}$, $Z_1^{(i+1)}$, $Z_3^{(i+1)}$, $Z_6^{(i+1)}$, $Z_9^{(i+1)}$, $Z_4^{(i+2)}$, $Z_5^{(i+2)}$ and $Z_7^{(i+2)}$.

(38) $(0,0,\delta,0,\delta,\delta,\delta,\delta) \xrightarrow{1r} (0,\delta,0,0,\delta,0,\delta,0) \xrightarrow{1r} (\delta,0,0,\delta,\delta,0,\delta,\delta) \xrightarrow{1r} (\delta,\delta,0,\delta,0,0,0,0) \xrightarrow{1r} (\delta,0,\delta,\delta,0,\delta,0,\delta) \xrightarrow{1r} (0,\delta,\delta,0,0,\delta,0,0) \xrightarrow{1r} (0,0,\delta,0,\delta,\delta,\delta,\delta)$, which is 6-round iterative and holds with probability 2^{-42}. Starting from an odd-numbered round i, the following 25 subkeys have to be weak: $Z_3^{(i)}$, $Z_6^{(i)}$, $Z_8^{(i)}$, $Z_9^{(i)}$, $Z_{10}^{(i)}$, $Z_2^{(i+1)}$, $Z_5^{(i+1)}$, $Z_7^{(i+1)}$, $Z_9^{(i+1)}$, $Z_{10}^{(i+1)}$, $Z_1^{(i+2)}$, $Z_8^{(i+2)}$, $Z_{10}^{(i+2)}$, $Z_2^{(i+3)}$, $Z_4^{(i+3)}$, $Z_9^{(i+3)}$, $Z_{10}^{(i+3)}$, $Z_1^{(i+4)}$, $Z_3^{(i+4)}$, $Z_6^{(i+4)}$, $Z_8^{(i+4)}$, $Z_9^{(i+4)}$, $Z_{10}^{(i+4)}$, $Z_2^{(i+5)}$ and $Z_{10}^{(i+5)}$. Starting from an even-numbered round i, the following 19 subkeys have to be weak: $Z_5^{(i)}$, $Z_7^{(i)}$, $Z_9^{(i)}$, $Z_{10}^{(i)}$, $Z_9^{(i+1)}$, $Z_{10}^{(i+1)}$, $Z_4^{(i+2)}$, $Z_5^{(i+2)}$, $Z_7^{(i+2)}$, $Z_{10}^{(i+2)}$, $Z_1^{(i+3)}$, $Z_9^{(i+3)}$, $Z_{10}^{(i+3)}$, $Z_4^{(i+4)}$, $Z_9^{(i+4)}$, $Z_{10}^{(i+4)}$, $Z_3^{(i+5)}$, $Z_6^{(i+5)}$ and $Z_{10}^{(i+5)}$.

(39) $(0,0,\delta,\delta,0,0,0,0) \xrightarrow{1r} (\delta,\delta,\delta,\delta,\delta,0,\delta,0) \xrightarrow{1r} (\delta,\delta,0,0,\delta,0,\delta,0) \xrightarrow{1r} (0,0,\delta,\delta,0,0,0,0)$, which is 3-round iterative and holds with probability 2^{-42}. Starting from an odd-numbered round i, the following six subkeys have to be weak: $Z_3^{(i)}$, $Z_2^{(i+1)}$, $Z_4^{(i+1)}$, $Z_5^{(i+1)}$, $Z_7^{(i+1)}$ and $Z_1^{(i+2)}$. Starting from an even-numbered round i, the following six subkeys have to be weak: $Z_4^{(i)}$, $Z_1^{(i+1)}$, $Z_3^{(i+1)}$, $Z_2^{(i+2)}$, $Z_5^{(i+2)}$ and $Z_7^{(i+2)}$.

(40) $(0,0,\delta,\delta,\delta,0,0,0) \xrightarrow{1r} (\delta,0,0,\delta,0,0,0,\delta) \xrightarrow{1r} (\delta,0,\delta,0,\delta,0,0,0) \xrightarrow{1r} (\delta,\delta,0,0,0,\delta,\delta,\delta) \xrightarrow{1r} (0,\delta,\delta,0,\delta,\delta,\delta,0) \xrightarrow{1r} (0,\delta,0,\delta,0,\delta,\delta,\delta) \xrightarrow{1r} (0,0,\delta,\delta,\delta,0,0,0)$, which is 6-round iterative and holds with probability 2^{-70}. Starting from an odd-numbered round i, the following 21 subkeys have to be weak: $Z_3^{(i)}$, $Z_9^{(i)}$, $Z_{10}^{(i)}$, $Z_4^{(i+1)}$, $Z_9^{(i+1)}$, $Z_{10}^{(i+1)}$, $Z_1^{(i+2)}$, $Z_3^{(i+2)}$, $Z_{10}^{(i+2)}$, $Z_2^{(i+3)}$, $Z_7^{(i+3)}$, $Z_9^{(i+3)}$, $Z_{10}^{(i+3)}$, $Z_3^{(i+4)}$, $Z_6^{(i+4)}$, $Z_9^{(i+4)}$, $Z_{10}^{(i+4)}$, $Z_2^{(i+5)}$, $Z_4^{(i+5)}$, $Z_7^{(i+5)}$ and $Z_{10}^{(i+5)}$. Starting from an even-numbered round i, the following 23 subkeys have to be weak: $Z_4^{(i)}$, $Z_5^{(i)}$, $Z_9^{(i)}$, $Z_{10}^{(i)}$, $Z_1^{(i+1)}$, $Z_8^{(i+1)}$, $Z_9^{(i+1)}$, $Z_{10}^{(i+1)}$, $Z_5^{(i+2)}$, $Z_{10}^{(i+2)}$, $Z_1^{(i+3)}$, $Z_6^{(i+3)}$, $Z_8^{(i+3)}$, $Z_9^{(i+3)}$, $Z_{10}^{(i+3)}$, $Z_2^{(i+4)}$, $Z_5^{(i+4)}$, $Z_7^{(i+4)}$, $Z_9^{(i+4)}$, $Z_{10}^{(i+4)}$, $Z_6^{(i+5)}$, $Z_8^{(i+5)}$ and $Z_{10}^{(i+5)}$.

(41) $(0,0,\delta,\delta,\delta,0,0,\delta) \xrightarrow{1r} (0,\delta,\delta,0,\delta,\delta,\delta,\delta) \xrightarrow{1r} (\delta,0,\delta,0,\delta,0,0,\delta) \xrightarrow{1r} (0,0,\delta,\delta,\delta,0,0,\delta)$, which is 3-round iterative and holds with probability 2^{-42}. Starting from an odd-numbered round i, the following ten subkeys have to be weak: $Z_3^{(i)}$, $Z_8^{(i)}$, $Z_9^{(i)}$, $Z_2^{(i+1)}$, $Z_5^{(i+1)}$, $Z_7^{(i+1)}$, $Z_9^{(i+1)}$, $Z_1^{(i+2)}$, $Z_3^{(i+2)}$ and $Z_8^{(i+2)}$. Starting from an even-numbered round i, the following eight subkeys have to be weak: $Z_4^{(i)}$, $Z_5^{(i)}$, $Z_9^{(i)}$, $Z_3^{(i+1)}$, $Z_6^{(i+1)}$, $Z_8^{(i+1)}$, $Z_9^{(i+1)}$ and $Z_5^{(i+2)}$.

(42) $(0,0,\delta,\delta,\delta,0,\delta,0) \xrightarrow{1r} (\delta,0,0,0,0,0,\delta,0) \xrightarrow{1r} (\delta,0,\delta,\delta,\delta,0,0,0) \xrightarrow{1r} (0,0,\delta,\delta,\delta,0,\delta,0)$, which is 3-round iterative and holds with probability 2^{-42}. Starting from an odd-numbered round i, the following six subkeys have to be weak: $Z_3^{(i)}$, $Z_9^{(i)}$, $Z_7^{(i+1)}$, $Z_9^{(i+1)}$, $Z_1^{(i+2)}$ and $Z_3^{(i+2)}$. Starting from an even-numbered round i, the following eight subkeys have to be weak: $Z_4^{(i)}$, $Z_5^{(i)}$, $Z_7^{(i)}$, $Z_9^{(i)}$, $Z_1^{(i+1)}$, $Z_9^{(i+1)}$, $Z_4^{(i+2)}$ and $Z_5^{(i+2)}$.

(43) $(0,0,\delta,\delta,\delta,0,\delta,\delta) \xrightarrow{1r} (0,\delta,\delta,\delta,\delta,\delta,0,0) \xrightarrow{1r} (\delta,0,\delta,\delta,\delta,0,0,\delta) \xrightarrow{1r} (\delta,\delta,0,0,0,\delta,0,0) \xrightarrow{1r} (\delta,0,0,0,0,0,\delta,\delta) \xrightarrow{1r} (0,\delta,0,0,0,\delta,\delta,0) \xrightarrow{1r} (0,0,\delta,\delta,\delta,0,\delta,\delta)$, which is 6-round iterative and holds with probability 2^{-70}. Starting from an odd-numbered round i, the following 23 subkeys have to be weak: $Z_3^{(i)}$, $Z_8^{(i)}$, $Z_9^{(i)}$, $Z_{10}^{(i)}$, $Z_2^{(i+1)}$, $Z_4^{(i+1)}$, $Z_5^{(i+1)}$, $Z_9^{(i+1)}$, $Z_{10}^{(i+1)}$, $Z_1^{(i+2)}$, $Z_3^{(i+2)}$, $Z_8^{(i+2)}$, $Z_{10}^{(i+2)}$, $Z_2^{(i+3)}$, $Z_9^{(i+3)}$, $Z_{10}^{(i+3)}$, $Z_1^{(i+4)}$, $Z_8^{(i+4)}$, $Z_9^{(i+4)}$, $Z_{10}^{(i+4)}$, $Z_2^{(i+5)}$, $Z_7^{(i+5)}$ and $Z_{10}^{(i+5)}$. Starting from an even-numbered round i, the following 21 subkeys have to be weak: $Z_4^{(i)}$, $Z_5^{(i)}$, $Z_7^{(i)}$, $Z_9^{(i)}$, $Z_{10}^{(i)}$, $Z_3^{(i+1)}$, $Z_6^{(i+1)}$, $Z_9^{(i+1)}$, $Z_{10}^{(i+1)}$, $Z_4^{(i+2)}$, $Z_5^{(i+2)}$, $Z_{10}^{(i+2)}$, $Z_1^{(i+3)}$, $Z_6^{(i+3)}$, $Z_9^{(i+3)}$, $Z_{10}^{(i+3)}$, $Z_7^{(i+4)}$, $Z_9^{(i+4)}$, $Z_{10}^{(i+4)}$, $Z_6^{(i+5)}$ and $Z_{10}^{(i+5)}$.

(44) $(0,0,\delta,\delta,\delta,\delta,0,0) \xrightarrow{1r} (0,\delta,0,\delta,\delta,0,\delta,0) \xrightarrow{1r} (0,\delta,\delta,0,0,\delta,\delta,0) \xrightarrow{1r} (0,0,\delta,\delta,\delta,\delta,0,0)$, which is 3-round iterative and holds with probability 2^{-28}. Starting from an odd-numbered round i, the following ten subkeys have to be weak: $Z_3^{(i)}$, $Z_6^{(i)}$, $Z_9^{(i)}$, $Z_2^{(i+1)}$, $Z_4^{(i+1)}$, $Z_5^{(i+1)}$, $Z_7^{(i+1)}$, $Z_9^{(i+1)}$, $Z_3^{(i+2)}$ and $Z_6^{(i+2)}$. Starting from an even-numbered round i, the following six subkeys have to be weak: $Z_4^{(i)}$, $Z_5^{(i)}$, $Z_9^{(i)}$, $Z_9^{(i+1)}$, $Z_2^{(i+2)}$ and $Z_7^{(i+2)}$.

(45) $(0,0,\delta,\delta,\delta,\delta,0,\delta) \xrightarrow{1r} (\delta,0,\delta,0,0,\delta,0,0) \xrightarrow{1r} (0,\delta,\delta,0,0,\delta,\delta,\delta) \xrightarrow{1r} (\delta,\delta,0,0,0,0,\delta,0) \xrightarrow{1r} (0,\delta,0,\delta,\delta,0,\delta,\delta) \xrightarrow{1r} (\delta,0,0,\delta,\delta,0,0,0) \xrightarrow{1r} (0,0,\delta,\delta,\delta,\delta,0,\delta)$, which is 6-round iterative and holds with probability 2^{-42}. Starting from an odd-numbered round i, the following 21 subkeys have to be weak: $Z_3^{(i)}$, $Z_6^{(i)}$, $Z_8^{(i)}$, $Z_9^{(i)}$, $Z_{10}^{(i)}$, $Z_9^{(i+1)}$, $Z_{10}^{(i+1)}$, $Z_3^{(i+2)}$, $Z_6^{(i+2)}$, $Z_8^{(i+2)}$, $Z_{10}^{(i+2)}$, $Z_2^{(i+3)}$, $Z_7^{(i+3)}$, $Z_9^{(i+3)}$, $Z_{10}^{(i+3)}$, $Z_8^{(i+4)}$, $Z_9^{(i+4)}$, $Z_{10}^{(i+4)}$, $Z_4^{(i+5)}$, $Z_5^{(i+5)}$ and $Z_{10}^{(i+5)}$. Starting from an even-numbered round i, the following 23 subkeys have to be weak: $Z_4^{(i)}$, $Z_5^{(i)}$, $Z_9^{(i)}$, $Z_{10}^{(i)}$, $Z_1^{(i+1)}$, $Z_3^{(i+1)}$, $Z_6^{(i+1)}$, $Z_9^{(i+1)}$, $Z_{10}^{(i+1)}$, $Z_2^{(i+2)}$, $Z_7^{(i+2)}$, $Z_{10}^{(i+2)}$, $Z_1^{(i+3)}$, $Z_9^{(i+3)}$, $Z_{10}^{(i+3)}$, $Z_2^{(i+4)}$, $Z_4^{(i+4)}$, $Z_5^{(i+4)}$, $Z_7^{(i+4)}$, $Z_9^{(i+4)}$, $Z_{10}^{(i+4)}$, $Z_1^{(i+5)}$ and $Z_{10}^{(i+5)}$.

(46) $(0,0,\delta,\delta,\delta,\delta,\delta,0) \xrightarrow{1r} (0,\delta,0,0,\delta,0,0,\delta) \xrightarrow{1r} (0,\delta,\delta,\delta,0,\delta,\delta,0) \xrightarrow{1r} (\delta,\delta,0,0,0,0,0,\delta) \xrightarrow{1r} (\delta,0,\delta,\delta,0,\delta,\delta,0) \xrightarrow{1r} (\delta,0,0,0,\delta,0,0,\delta) \xrightarrow{1r} (0,0,\delta,\delta,\delta,\delta,\delta,0)$, which is 6-round iterative and holds with probability 2^{-42}. Starting from an odd-numbered round i, the following 21 subkeys have to be weak: $Z_3^{(i)}$, $Z_6^{(i)}$, $Z_9^{(i)}$, $Z_{10}^{(i)}$, $Z_2^{(i+1)}$, $Z_5^{(i+1)}$, $Z_9^{(i+1)}$, $Z_{10}^{(i+1)}$, $Z_3^{(i+2)}$, $Z_6^{(i+2)}$, $Z_{10}^{(i+2)}$, $Z_2^{(i+3)}$, $Z_9^{(i+3)}$, $Z_{10}^{(i+3)}$, $Z_1^{(i+4)}$, $Z_3^{(i+4)}$, $Z_6^{(i+4)}$, $Z_9^{(i+4)}$, $Z_{10}^{(i+4)}$, $Z_5^{(i+5)}$ and $Z_{10}^{(i+5)}$. Starting from an even-numbered round i, the following 23 subkeys have to be weak: $Z_4^{(i)}$, $Z_5^{(i)}$, $Z_7^{(i)}$, $Z_9^{(i)}$, $Z_{10}^{(i)}$, $Z_8^{(i+1)}$, $Z_9^{(i+1)}$, $Z_{10}^{(i+1)}$, $Z_2^{(i+2)}$,

$Z_4^{(i+2)}$, $Z_7^{(i+2)}$, $Z_{10}^{(i+2)}$, $Z_1^{(i+3)}$, $Z_8^{(i+3)}$, $Z_9^{(i+3)}$, $Z_{10}^{(i+3)}$, $Z_4^{(i+4)}$, $Z_7^{(i+4)}$, $Z_9^{(i+4)}$, $Z_{10}^{(i+4)}$, $Z_1^{(i+5)}$, $Z_8^{(i+5)}$ and $Z_{10}^{(i+5)}$.

(47) $(0,0,\delta,\delta,\delta,\delta,\delta,\delta) \overset{1r}{\to} (\delta,0,\delta,\delta,0,\delta,\delta,\delta) \overset{1r}{\to} (0,\delta,\delta,\delta,0,\delta,\delta,\delta) \overset{1r}{\to} (0, 0, \delta, \delta, \delta, \delta, \delta, \delta)$, which is 3-round iterative and holds with probability 1. Starting from an odd-numbered round i, the following ten subkeys have to be weak: $Z_3^{(i)}$, $Z_6^{(i)}$, $Z_8^{(i)}$, $Z_9^{(i)}$, $Z_4^{(i+1)}$, $Z_7^{(i+1)}$, $Z_9^{(i+1)}$, $Z_3^{(i+2)}$, $Z_6^{(i+2)}$ and $Z_8^{(i+2)}$. Starting from an even-numbered round i, the following 12 subkeys have to be weak: $Z_4^{(i)}$, $Z_5^{(i)}$, $Z_7^{(i)}$, $Z_9^{(i)}$, $Z_1^{(i+1)}$, $Z_3^{(i+1)}$, $Z_6^{(i+1)}$, $Z_8^{(i+1)}$, $Z_9^{(i+1)}$, $Z_2^{(i+2)}$, $Z_4^{(i+2)}$ and $Z_7^{(i+2)}$.

(48) $(0,\delta,0,0,0,0,\delta,0) \overset{1r}{\to} (\delta,\delta,\delta,\delta,0,0,0,0) \overset{1r}{\to} (\delta,0,\delta,\delta,0,0,\delta,0) \overset{1r}{\to} (0, \delta, 0, 0, 0, 0, \delta, 0)$, which is 3-round iterative and holds with probability 2^{-42}. Starting from an odd-numbered round i, the following six subkeys have to be weak: $Z_2^{(i+1)}$, $Z_4^{(i+1)}$, $Z_9^{(i+1)}$, $Z_1^{(i+2)}$, $Z_3^{(i+2)}$ and $Z_9^{(i+2)}$. Starting from an even-numbered round i, the following eight subkeys have to be weak: $Z_2^{(i)}$, $Z_7^{(i)}$, $Z_1^{(i+1)}$, $Z_3^{(i+1)}$, $Z_9^{(i+1)}$, $Z_4^{(i+2)}$, $Z_7^{(i+2)}$ and $Z_9^{(i+2)}$.

(49) $(0,\delta,0,0,0,\delta,\delta,\delta) \overset{1r}{\to} (\delta,\delta,0,0,0,\delta,0,\delta) \overset{1r}{\to} (0,\delta,\delta,\delta,\delta,\delta,0,\delta) \overset{1r}{\to} (0, \delta, 0, 0, 0, \delta, \delta, \delta)$, which is 3-round iterative and holds with probability 2^{-42}. Starting from an odd-numbered round i, the following eight subkeys have to be weak: $Z_6^{(i)}$, $Z_8^{(i)}$, $Z_2^{(i+1)}$, $Z_9^{(i+1)}$, $Z_3^{(i+2)}$, $Z_6^{(i+2)}$, $Z_8^{(i+2)}$ and $Z_9^{(i+2)}$. Starting from an even-numbered round i, the following ten subkeys have to be weak: $Z_2^{(i)}$, $Z_7^{(i)}$, $Z_1^{(i+1)}$, $Z_6^{(i+1)}$, $Z_8^{(i+1)}$, $Z_9^{(i+1)}$, $Z_2^{(i+2)}$, $Z_4^{(i+2)}$, $Z_5^{(i+2)}$ and $Z_9^{(i+2)}$.

(50) $(0,\delta,0,0,\delta,0,0,0) \overset{1r}{\to} (\delta,0,0,0,\delta,0,0,0) \overset{1r}{\to} (\delta,\delta,0,0,0,0,0,0) \overset{1r}{\to} (0, \delta, 0, 0, \delta, 0, 0, 0)$, which is 3-round iterative and holds with probability 1. Starting from an odd-numbered round i, the following four subkeys have to be weak: $Z_9^{(i)}$, $Z_5^{(i+1)}$, $Z_1^{(i+2)}$ and $Z_9^{(i+2)}$. Starting from an even-numbered round i, the following six subkeys have to be weak: $Z_2^{(i)}$, $Z_5^{(i)}$, $Z_9^{(i)}$, $Z_1^{(i+1)}$, $Z_2^{(i+2)}$ and $Z_9^{(i+2)}$.

(51) $(0,\delta,0,0,\delta,0,\delta,\delta) \overset{1r}{\to} (0,\delta,\delta,0,0,\delta,0,\delta) \overset{1r}{\to} (\delta,\delta,0,\delta,0,0,0,\delta) \overset{1r}{\to} (0, \delta, 0, 0, \delta, 0, \delta, \delta)$, which is 3-round iterative and holds with probability 2^{-28}. Starting from an odd-numbered round i, the following six subkeys have to be weak: $Z_8^{(i)}$, $Z_9^{(i)}$, $Z_2^{(i+1)}$, $Z_1^{(i+2)}$, $Z_8^{(i+2)}$ and $Z_9^{(i+2)}$. Starting from an even-numbered round i, the following ten subkeys have to be weak: $Z_2^{(i)}$, $Z_5^{(i)}$, $Z_7^{(i)}$, $Z_9^{(i)}$, $Z_3^{(i+1)}$, $Z_6^{(i+1)}$, $Z_8^{(i+1)}$, $Z_2^{(i+2)}$, $Z_4^{(i+2)}$ and $Z_9^{(i+2)}$.

(52) $(0,\delta,0,\delta,0,0,\delta,\delta) \overset{1r}{\to} (\delta,\delta,\delta,\delta,0,0,\delta,\delta) \overset{1r}{\to} (0,\delta,0,\delta,\delta,\delta,\delta,\delta) \overset{1r}{\to} (0, \delta, 0, \delta, 0, 0, \delta, \delta)$, which is 3-round iterative and holds with probability 2^{-28}. Starting from an odd-numbered round i, the following eight subkeys have to be weak: $Z_8^{(i)}$, $Z_2^{(i+1)}$, $Z_4^{(i+1)}$, $Z_7^{(i+1)}$, $Z_9^{(i+1)}$, $Z_6^{(i+2)}$, $Z_8^{(i+2)}$ and $Z_9^{(i+2)}$. Starting from an even-numbered round i, the following 12 subkeys have to be weak: $Z_2^{(i)}$, $Z_4^{(i)}$, $Z_7^{(i)}$, $Z_1^{(i+1)}$, $Z_3^{(i+1)}$, $Z_8^{(i+1)}$, $Z_9^{(i+1)}$, $Z_2^{(i+2)}$, $Z_4^{(i+2)}$, $Z_5^{(i+2)}$, $Z_7^{(i+2)}$ and $Z_9^{(i+2)}$.

(53) $(0,\delta,0,\delta,0,\delta,\delta,0) \stackrel{1r}{\rightarrow} (\delta,\delta,0,0,0,\delta,\delta,0) \stackrel{1r}{\rightarrow} (\delta,0,0,\delta,0,0,0,0) \stackrel{1r}{\rightarrow} (0, \delta, 0, \delta, 0, \delta, \delta, 0)$, which is 3-round iterative and holds with probability 2^{-42}. Starting from an odd-numbered round i, the following six subkeys have to be weak: $Z_6^{(i)}$, $Z_2^{(i+1)}$, $Z_7^{(i+1)}$, $Z_9^{(i+1)}$, $Z_1^{(i+2)}$ and $Z_9^{(i+2)}$. Starting from an even-numbered round i, the following eight subkeys have to be weak: $Z_2^{(i)}$, $Z_4^{(i)}$, $Z_7^{(i)}$, $Z_1^{(i+1)}$, $Z_6^{(i+1)}$, $Z_9^{(i+1)}$, $Z_4^{(i+2)}$ and $Z_9^{(i+2)}$.

(54) $(0,\delta,0,\delta,\delta,\delta,0,0) \stackrel{1r}{\rightarrow} (\delta,0,\delta,\delta,\delta,\delta,\delta,0) \stackrel{1r}{\rightarrow} (\delta,\delta,\delta,0,0,0,\delta,0) \stackrel{1r}{\rightarrow} (0, \delta, 0, \delta, \delta, \delta, 0, 0)$, which is 3-round iterative and holds with probability 2^{-28}. Starting from an odd-numbered round i, the following eight subkeys have to be weak: $Z_6^{(i)}$, $Z_9^{(i)}$, $Z_4^{(i+1)}$, $Z_5^{(i+1)}$, $Z_7^{(i+1)}$, $Z_1^{(i+2)}$, $Z_3^{(i+2)}$ and $Z_9^{(i+2)}$. Starting from an even-numbered round i, the following ten subkeys have to be weak: $Z_2^{(i)}$, $Z_4^{(i)}$, $Z_5^{(i)}$, $Z_9^{(i)}$, $Z_1^{(i+1)}$, $Z_3^{(i+1)}$, $Z_6^{(i+1)}$, $Z_2^{(i+2)}$, $Z_7^{(i+2)}$ and $Z_9^{(i+2)}$.

(55) $(0,\delta,\delta,0,0,0,0,0) \stackrel{1r}{\rightarrow} (\delta,\delta,\delta,0,0,\delta,0,0) \stackrel{1r}{\rightarrow} (\delta,0,0,0,0,\delta,0,0) \stackrel{1r}{\rightarrow} (0, \delta, \delta, 0, 0, 0, 0, 0)$, which is 3-round iterative and holds with probability 2^{-28}. Starting from an odd-numbered round i, the following six subkeys have to be weak: $Z_3^{(i)}$, $Z_2^{(i+1)}$, $Z_9^{(i+1)}$, $Z_1^{(i+2)}$, $Z_6^{(i+2)}$ and $Z_9^{(i+2)}$. Starting from an even-numbered round i, the following six subkeys have to be weak: $Z_2^{(i)}$, $Z_1^{(i+1)}$, $Z_3^{(i+1)}$, $Z_6^{(i+1)}$, $Z_9^{(i+1)}$ and $Z_9^{(i+2)}$.

(56) $(0,\delta,\delta,0,\delta,0,0,\delta) \stackrel{1r}{\rightarrow} (0,\delta,\delta,\delta,0,0,0,\delta) \stackrel{1r}{\rightarrow} (\delta,\delta,\delta,0,0,\delta,\delta,\delta) \stackrel{1r}{\rightarrow} (0, \delta, \delta, 0, \delta, 0, 0, \delta)$, which is 3-round iterative and holds with probability 2^{-42}. Starting from an odd-numbered round i, the following ten subkeys have to be weak: $Z_3^{(i)}$, $Z_8^{(i)}$, $Z_9^{(i)}$, $Z_2^{(i+1)}$, $Z_4^{(i+1)}$, $Z_1^{(i+2)}$, $Z_3^{(i+2)}$, $Z_6^{(i+2)}$, $Z_8^{(i+2)}$ and $Z_9^{(i+2)}$. Starting from an even-numbered round i, the following eight subkeys have to be weak: $Z_2^{(i)}$, $Z_5^{(i)}$, $Z_9^{(i)}$, $Z_3^{(i+1)}$, $Z_8^{(i+1)}$, $Z_2^{(i+2)}$, $Z_7^{(i+2)}$ and $Z_9^{(i+2)}$.

(57) $(0,\delta,\delta,0,\delta,0,\delta,0) \stackrel{1r}{\rightarrow} (\delta,0,0,\delta,\delta,\delta,0,0) \stackrel{1r}{\rightarrow} (\delta,\delta,\delta,\delta,0,\delta,\delta,0) \stackrel{1r}{\rightarrow} (0, \delta, \delta, 0, \delta, 0, \delta, 0)$, which is 3-round iterative and holds with probability 2^{-28}. Starting from an odd-numbered round i, the following eight subkeys have to be weak: $Z_3^{(i)}$, $Z_9^{(i)}$, $Z_4^{(i+1)}$, $Z_5^{(i+1)}$, $Z_1^{(i+2)}$, $Z_3^{(i+2)}$, $Z_6^{(i+2)}$ and $Z_9^{(i+2)}$. Starting from an even-numbered round i, the following ten subkeys have to be weak: $Z_2^{(i)}$, $Z_5^{(i)}$, $Z_7^{(i)}$, $Z_9^{(i)}$, $Z_1^{(i+1)}$, $Z_6^{(i+1)}$, $Z_2^{(i+2)}$, $Z_4^{(i+2)}$, $Z_7^{(i+2)}$ and $Z_9^{(i+2)}$.

(58) $(0,\delta,\delta,\delta,0,\delta,0,0) \stackrel{1r}{\rightarrow} (\delta,\delta,0,\delta,0,0,\delta,0) \stackrel{1r}{\rightarrow} (\delta,0,\delta,0,0,\delta,\delta,0) \stackrel{1r}{\rightarrow} (0, \delta, \delta, \delta, 0, \delta, 0, 0)$, which is 3-round iterative and holds with probability 2^{-28}. Starting from an odd-numbered round i, the following ten subkeys have to be weak: $Z_3^{(i)}$, $Z_6^{(i)}$, $Z_2^{(i+1)}$, $Z_4^{(i+1)}$, $Z_7^{(i+1)}$, $Z_9^{(i+1)}$, $Z_1^{(i+2)}$, $Z_3^{(i+2)}$, $Z_6^{(i+2)}$ and $Z_9^{(i+2)}$. Starting from an even-numbered round i, the following six subkeys have to be weak: $Z_2^{(i)}$, $Z_4^{(i)}$, $Z_1^{(i+1)}$, $Z_9^{(i+1)}$, $Z_7^{(i+2)}$ and $Z_9^{(i+2)}$.

(59) $(0,\delta,\delta,\delta,\delta,\delta,\delta,0) \stackrel{1r}{\rightarrow} (\delta,0,\delta,0,\delta,0,\delta,0) \stackrel{1r}{\rightarrow} (\delta,\delta,0,\delta,0,\delta,0,0) \stackrel{1r}{\rightarrow} (0, \delta, \delta, \delta, \delta, \delta, \delta, 0)$, which is 3-round iterative and holds with probability 2^{-28}.

Starting from an odd-numbered round i, the following eight subkeys have to be weak: $Z_3^{(i)}$, $Z_6^{(i)}$, $Z_9^{(i)}$, $Z_5^{(i+1)}$, $Z_7^{(i+1)}$, $Z_1^{(i+2)}$, $Z_6^{(i+2)}$ and $Z_9^{(i+2)}$. Starting from an even-numbered round i, the following ten subkeys have to be weak: $Z_2^{(i)}$, $Z_4^{(i)}$, $Z_5^{(i)}$, $Z_7^{(i)}$, $Z_9^{(i)}$, $Z_1^{(i+1)}$, $Z_3^{(i+1)}$, $Z_2^{(i+2)}$, $Z_4^{(i+2)}$ and $Z_9^{(i+2)}$.

(60) $(\delta,0,0,0,0,\delta,\delta,\delta) \stackrel{1r}{\to} (\delta,0,0,0,\delta,\delta,0,\delta) \stackrel{1r}{\to} (\delta,\delta,\delta,\delta,0,\delta,0,\delta) \stackrel{1r}{\to} (\delta, 0, 0, 0, 0, \delta, \delta, \delta)$, which is 3-round iterative and holds with probability 2^{-42}. Starting from an odd-numbered round i, the following ten subkeys have to be weak: $Z_1^{(i)}$, $Z_6^{(i)}$, $Z_8^{(i)}$, $Z_9^{(i)}$, $Z_5^{(i+1)}$, $Z_1^{(i+2)}$, $Z_3^{(i+2)}$, $Z_6^{(i+2)}$, $Z_8^{(i+2)}$ and $Z_9^{(i+2)}$. Starting from an even-numbered round i, the following eight subkeys have to be weak: $Z_7^{(i)}$, $Z_9^{(i)}$, $Z_1^{(i+1)}$, $Z_6^{(i+1)}$, $Z_8^{(i+1)}$, $Z_2^{(i+2)}$, $Z_4^{(i+2)}$ and $Z_9^{(i+2)}$.

(61) $(\delta,0,0,\delta,0,0,\delta,\delta) \stackrel{1r}{\to} (\delta,0,\delta,\delta,\delta,0,\delta,\delta) \stackrel{1r}{\to} (\delta,\delta,0,\delta,0,\delta,\delta,\delta) \stackrel{1r}{\to} (\delta, 0, 0, \delta, 0, 0, \delta, \delta)$, which is 3-round iterative and holds with probability 2^{-28}. Starting from an odd-numbered round i, the following ten subkeys have to be weak: $Z_1^{(i)}$, $Z_8^{(i)}$, $Z_9^{(i)}$, $Z_4^{(i+1)}$, $Z_5^{(i+1)}$, $Z_7^{(i+1)}$, $Z_1^{(i+2)}$, $Z_6^{(i+2)}$, $Z_8^{(i+2)}$ and $Z_9^{(i+2)}$. Starting from an even-numbered round i, the following ten subkeys have to be weak: $Z_4^{(i)}$, $Z_7^{(i)}$, $Z_9^{(i)}$, $Z_1^{(i+1)}$, $Z_3^{(i+1)}$, $Z_8^{(i+1)}$, $Z_2^{(i+2)}$, $Z_4^{(i+2)}$, $Z_7^{(i+2)}$ and $Z_9^{(i+2)}$.

(62) $(\delta,0,0,\delta,\delta,0,0,\delta) \stackrel{1r}{\to} (\delta,\delta,0,0,0,0,\delta,\delta) \stackrel{1r}{\to} (\delta,0,\delta,0,0,\delta,0,\delta) \stackrel{1r}{\to} (\delta, 0, 0, \delta, \delta, 0, 0, \delta)$, which is 3-round iterative and holds with probability 2^{-28}. Starting from an odd-numbered round i, the following ten subkeys have to be weak: $Z_1^{(i)}$, $Z_8^{(i)}$, $Z_2^{(i+1)}$, $Z_7^{(i+1)}$, $Z_9^{(i+1)}$, $Z_1^{(i+2)}$, $Z_3^{(i+2)}$, $Z_6^{(i+2)}$, $Z_8^{(i+2)}$ and $Z_9^{(i+2)}$. Starting from an even-numbered round i, the following six subkeys have to be weak: $Z_4^{(i)}$, $Z_5^{(i)}$, $Z_1^{(i+1)}$, $Z_8^{(i+1)}$, $Z_9^{(i+1)}$ and $Z_9^{(i+2)}$.

(63) $(\delta,0,\delta,0,\delta,\delta,\delta,\delta) \stackrel{1r}{\to} (\delta,\delta,\delta,0,0,0,0,\delta) \stackrel{1r}{\to} (\delta,0,\delta,\delta,0,0,0,\delta) \stackrel{1r}{\to} (\delta, 0, \delta, 0, \delta, \delta, \delta, \delta)$, which is 3-round iterative and holds with probability 2^{-42}. Starting from an odd-numbered round i, the following ten subkeys have to be weak: $Z_1^{(i)}$, $Z_3^{(i)}$, $Z_6^{(i)}$, $Z_8^{(i)}$, $Z_2^{(i+1)}$, $Z_9^{(i+1)}$, $Z_1^{(i+2)}$, $Z_3^{(i+2)}$, $Z_8^{(i+2)}$ and $Z_9^{(i+2)}$. Starting from an even-numbered round i, the following eight subkeys have to be weak: $Z_5^{(i)}$, $Z_7^{(i)}$, $Z_1^{(i+1)}$, $Z_3^{(i+1)}$, $Z_8^{(i+1)}$, $Z_9^{(i+1)}$, $Z_4^{(i+2)}$ and $Z_9^{(i+2)}$.

(64) $(\delta,\delta,0,0,\delta,0,0,\delta) \stackrel{1r}{\to} (\delta,\delta,0,\delta,\delta,\delta,0,\delta) \stackrel{1r}{\to} (\delta,\delta,\delta,0,\delta,0,\delta,\delta) \stackrel{1r}{\to} (\delta, \delta, 0, 0, \delta, 0, 0, \delta)$, which is 3-round iterative and holds with probability 2^{-28}. Starting from an odd-numbered round i, the following eight subkeys have to be weak: $Z_1^{(i)}$, $Z_8^{(i)}$, $Z_2^{(i+1)}$, $Z_4^{(i+1)}$, $Z_5^{(i+1)}$, $Z_1^{(i+2)}$, $Z_3^{(i+2)}$ and $Z_8^{(i+2)}$. Starting from an even-numbered round i, the following eight subkeys have to be weak: $Z_2^{(i)}$, $Z_5^{(i)}$, $Z_1^{(i+1)}$, $Z_6^{(i+1)}$, $Z_8^{(i+1)}$, $Z_2^{(i+2)}$, $Z_5^{(i+2)}$ and $Z_7^{(i+2)}$.

(65) $(\delta,\delta,\delta,\delta,\delta,\delta,\delta,\delta) \stackrel{1r}{\to} (\delta,\delta,\delta,\delta,\delta,\delta,\delta,\delta)$, which is 1-round iterative and holds with probability 1. Starting from an odd-numbered round i, the

following four subkeys have to be weak: $Z_1^{(i)}$, $Z_3^{(i)}$, $Z_6^{(i)}$ and $Z_8^{(i)}$. Starting from an even-numbered round i, the following four subkeys have to be weak: $Z_2^{(i)}$, $Z_4^{(i)}$, $Z_5^{(i)}$ and $Z_7^{(i)}$.

Starting from the first round, $i = 1$, the characteristics that require the least amount of weak subkeys are (3), (17) and (50): 1.33 weak subkeys per round on average. On the other hand, the characteristics that hold with the highest probability, under weak-key assumptions, are (47), (50) and (65). A trade-off is characteristic (50) which is 3-round iterative and hold with certainty, under four weak subkeys.

Unlike the key schedule algorithms of PES and IDEA, the key schedule of MESH-128 does not consist of a fixed bit permutation, but is rather more involved. See Sect. 2.3.8. Experiments with mini-versions of the key schedule for a scaled-down version of MESH-64(8) operating on 16-bit blocks and 32-bit keys indicated that each weak subkey restriction (with the three most significant bits equal to zero) is satisfied for a fraction of 2^{-3} of the key space. Assuming this behavior holds for each weak subkey independently, and for larger key schedules as well, such as the one for MESH-64(8), then t weak subkeys would apply to a fraction of $2^{128-t\cdot 7}$ of the key space, or a weak-key class of this size. Thus, for a nonempty weak-key class, t must satisfy $128 - t \cdot 7 \geq 0$, that is, $t \leq 18$.

If characteristic (50) is repeated three times, then it will require 12 weak subkeys and cover nine rounds with probability 1. Therefore, it is possible to distinguish the full 8.5-round MESH-64(8) from a random permutation, under the appropriate weak-key assumptions, with very low attack complexity, since this characteristic holds with certainty. The cost is essentially two encryptions, two chosen plaintexts and equivalent memory. The weak-key class size is estimated as $2^{128-12\cdot 7} = 2^{44}$.

The reasoning is the following: given the input difference of characteristic (50) to both 8.5-round MESH-64(8) (under a weak-key) and a random 64-bit permutation π, the output difference of (50) will show up with certainty for MESH-64(8), while for π, it will happen with probability about 2^{-64}. The advantage (see Sect. 1.7 to distinguish one from the other is: $1 - 2^{-64}$, which is non-negligible. Therefore, the full 8.5-round MESH-64(8) cipher is not an ideal primitive, that is, it is not a PRP.

This differential analysis using characteristics also allows us to study the diffusion power of the MA half-round of MESH-64(8) with respect to modern tools such as the differential branch number (Appendix, Sect. B).

The set of characteristics in Table 3.26 through Table 3.33 indicates that the differential branch number *per round* of MESH-64(8) is only 4 (under weak-key assumptions). For instance, the wordwise Hamming Weight of the input and output differences of the 1-round characteristic

$$(\delta, \delta, 0, 0, 0, 0, 0, 0) \stackrel{1r}{\rightarrow} (0, \delta, 0, 0, \delta, 0, 0, 0)$$

add up to 4. So, even though MESH-64(8) provides full diffusion in a single round, the previous analysis shows that its differential branch number is much lower than ideal. For an 8-word block, such as in MESH-64(8), an MDS code using a 8×8 MDS matrix would provide a branch number of $1 + 8 = 9$. See Appendix, Sect. B.1, for an explanation of the differential branch number.

3.5.7 DC of MESH-128(8) Under Weak-Key Assumptions

The following differential analysis of MESH-64 is novel and more extensive than [151] and operates under weak-key assumptions.

The analysis follows a bottom-up approach, starting from the smallest components: $\odot$, $\boxplus$ and $\oplus$, up to an MA-box, up to a half-round, further up to a full round, and finally to multiple rounds.

Across unkeyed $\odot$ and $\boxplus$, the propagation of 16-bit difference words follows from Table 3.12 and Table 3.13, respectively.

The next step is to obtain all difference patterns across the MA-box of MESH-128(8), under the corresponding weak-subkey restrictions.

Tables 3.34 through 3.41 lists all 255 nontrivial difference patterns across the MA-box of MESH-128(8). The difference operator is bitwise exclusive-or.

The next step is to obtain 1-round differential characteristics and the corresponding weak-subkey restrictions for the characteristics to hold with highest possible probability. The only 8-bit difference values used are $\delta = 80_x$ and 00_x.

In total, there are 65535 nontrivial 1-round differential characteristics[9] for MESH-128(8).

A sample of these 65535 1-round characteristics is in Table 3.42.

Multiple-round characteristics are obtained by concatenating 1-round characteristics, such that the output difference of the first one matches the input difference of the next characteristic.

From the 65535 1-round characteristics, multiple-round characteristics are clustered into 11009 chains. They are too many characteristics to list here. A small sample is detailed next.

- $(0,0,0,0,0,0,0,0,0,0,0,0,0,0,0,\delta) \stackrel{1r}{\rightarrow} (\delta, \delta, \delta, \delta, \delta, \delta, \delta, \delta, \delta, \delta, \delta, \delta, \delta, \delta, \delta, 0) \stackrel{1r}{\rightarrow} (0,0,0,0,0,0,0,0,0,0,0,0,0,0,0,\delta)$, which is 2-round iterative and holds with probability 2^{-48}. Starting from an odd-numbered round i, the following 11 subkeys have to be weak: $Z_{16}^{(i)}$, $Z_{18}^{(i)}$, $Z_2^{(i+1)}$, $Z_4^{(i+1)}$, $Z_6^{(i+1)}$, $Z_8^{(i+1)}$, $Z_9^{(i+1)}$, $Z_{11}^{(i+1)}$, $Z_{13}^{(i+1)}$, $Z_{15}^{(i+1)}$ and $Z_{18}^{(i+1)}$. Starting from

[9] There is no room to list them all here. Rather, these 1-round characteristics are available from the book's website http://www.springer.com/978-3-319-68272-3.

Table 3.34 Nontrivial difference propagation across the MA-box of MESH-128(8) (part I).

input-output difference	Weak subkeys (i-th round)	Probability
$(0,0,0,0,0,0,0,\delta) \xrightarrow{1r} (\delta,\delta,\delta,\delta,\delta,\delta,\delta,\delta)$	$Z_{18}^{(i)}$	2^{-24}
$(0,0,0,0,0,0,\delta,0) \xrightarrow{1r} (0,0,0,0,0,0,0,\delta)$	$Z_{18}^{(i)}$	2^{-8}
$(0,0,0,0,0,0,\delta,\delta) \xrightarrow{1r} (\delta,\delta,\delta,\delta,\delta,\delta,\delta,0)$	—	2^{-32}
$(0,0,0,0,0,\delta,0,0) \xrightarrow{1r} (\delta,\delta,\delta,\delta,\delta,\delta,0,\delta)$	$Z_{18}^{(i)}$	2^{-32}
$(0,0,0,0,0,\delta,0,\delta) \xrightarrow{1r} (0,0,0,0,0,0,\delta,0)$	—	2^{-15}
$(0,0,0,0,0,\delta,\delta,0) \xrightarrow{1r} (\delta,\delta,\delta,\delta,\delta,\delta,0,0)$	—	2^{-31}
$(0,0,0,0,0,\delta,\delta,\delta) \xrightarrow{1r} (0,0,0,0,0,0,\delta,\delta)$	$Z_{18}^{(i)}$	2^{-14}
$(0,0,0,0,\delta,0,0,0) \xrightarrow{1r} (0,0,0,0,0,\delta,0,\delta)$	$Z_{18}^{(i)}$	2^{-24}
$(0,0,0,0,\delta,0,0,\delta) \xrightarrow{1r} (\delta,\delta,\delta,\delta,\delta,0,\delta,0)$	—	2^{-39}
$(0,0,0,0,\delta,0,\delta,0) \xrightarrow{1r} (0,0,0,0,0,\delta,0,0)$	—	2^{-23}
$(0,0,0,0,\delta,0,\delta,\delta) \xrightarrow{1r} (\delta,\delta,\delta,\delta,\delta,0,\delta,\delta)$	$Z_{18}^{(i)}$	2^{-38}
$(0,0,0,0,\delta,\delta,0,0) \xrightarrow{1r} (\delta,\delta,\delta,\delta,\delta,0,0,0)$	—	2^{-24}
$(0,0,0,0,\delta,\delta,0,\delta) \xrightarrow{1r} (0,0,0,0,0,\delta,\delta,\delta)$	$Z_{18}^{(i)}$	2^{-16}
$(0,0,0,0,\delta,\delta,\delta,0) \xrightarrow{1r} (\delta,\delta,\delta,\delta,\delta,0,0,\delta)$	$Z_{18}^{(i)}$	2^{-32}
$(0,0,0,0,\delta,\delta,\delta,\delta) \xrightarrow{1r} (0,0,0,0,0,\delta,\delta,0)$	—	2^{-24}
$(0,0,0,\delta,0,0,0,0) \xrightarrow{1r} (\delta,\delta,\delta,\delta,0,\delta,0,\delta)$	$Z_{18}^{(i)}$	2^{-40}
$(0,0,0,\delta,0,0,0,\delta) \xrightarrow{1r} (0,0,0,0,\delta,0,\delta,0)$	—	2^{-30}
$(0,0,0,\delta,0,0,\delta,0) \xrightarrow{1r} (\delta,\delta,\delta,\delta,0,\delta,0,0)$	—	2^{-39}
$(0,0,0,\delta,0,0,\delta,\delta) \xrightarrow{1r} (0,0,0,0,\delta,0,\delta,\delta)$	$Z_{18}^{(i)}$	2^{-29}
$(0,0,0,\delta,0,\delta,0,0) \xrightarrow{1r} (0,0,0,0,\delta,0,0,0)$	—	2^{-15}
$(0,0,0,\delta,0,\delta,0,\delta) \xrightarrow{1r} (\delta,\delta,\delta,\delta,0,\delta,\delta,\delta)$	$Z_{18}^{(i)}$	2^{-32}
$(0,0,0,\delta,0,\delta,\delta,0) \xrightarrow{1r} (0,0,0,0,\delta,0,0,\delta)$	$Z_{18}^{(i)}$	2^{-23}
$(0,0,0,\delta,0,\delta,\delta,\delta) \xrightarrow{1r} (\delta,\delta,\delta,\delta,0,\delta,\delta,0)$	—	2^{-40}
$(0,0,0,\delta,\delta,0,0,0) \xrightarrow{1r} (\delta,\delta,\delta,\delta,0,0,0,0)$	—	2^{-23}
$(0,0,0,\delta,\delta,0,0,\delta) \xrightarrow{1r} (0,0,0,0,\delta,\delta,\delta,\delta)$	$Z_{18}^{(i)}$	2^{-22}
$(0,0,0,\delta,\delta,0,\delta,0) \xrightarrow{1r} (\delta,\delta,\delta,\delta,0,0,0,\delta)$	$Z_{18}^{(i)}$	2^{-31}
$(0,0,0,\delta,\delta,0,\delta,\delta) \xrightarrow{1r} (0,0,0,0,\delta,\delta,\delta,0)$	—	2^{-30}
$(0,0,0,\delta,\delta,\delta,0,0) \xrightarrow{1r} (0,0,0,0,\delta,\delta,0,\delta)$	$Z_{18}^{(i)}$	2^{-30}
$(0,0,0,\delta,\delta,\delta,0,\delta) \xrightarrow{1r} (\delta,\delta,\delta,\delta,0,0,\delta,0)$	—	2^{-38}
$(0,0,0,\delta,\delta,\delta,\delta,0) \xrightarrow{1r} (0,0,0,0,\delta,\delta,0,0)$	—	2^{-29}
$(0,0,0,\delta,\delta,\delta,\delta,\delta) \xrightarrow{1r} (\delta,\delta,\delta,\delta,0,0,\delta,\delta)$	$Z_{18}^{(i)}$	2^{-37}
$(0,0,\delta,0,0,0,0,0) \xrightarrow{1r} (0,0,0,\delta,0,\delta,0,\delta)$	$Z_{18}^{(i)}$	2^{-40}

Table 3.35 Nontrivial difference propagation across the MA-box of MESH-128(8) (part II).

input-output difference	Weak subkeys (i-th round)	Probability
$(0,0,\delta,0,0,0,0,\delta) \xrightarrow{1r} (\delta,\delta,\delta,0,\delta,0,\delta,0)$	—	2^{-46}
$(0,0,\delta,0,0,0,\delta,0) \xrightarrow{1r} (0,0,0,\delta,0,\delta,0,0)$	—	2^{-39}
$(0,0,\delta,0,0,0,\delta,\delta) \xrightarrow{1r} (\delta,\delta,\delta,0,\delta,0,\delta,\delta)$	$Z_{18}^{(i)}$	2^{-45}
$(0,0,\delta,0,0,\delta,0,0) \xrightarrow{1r} (\delta,\delta,\delta,0,\delta,0,0,0)$	—	2^{-31}
$(0,0,\delta,0,0,\delta,0,\delta) \xrightarrow{1r} (0,0,0,\delta,0,\delta,\delta,\delta)$	$Z_{18}^{(i)}$	2^{-32}
$(0,0,\delta,0,0,\delta,\delta,0) \xrightarrow{1r} (\delta,\delta,\delta,0,\delta,0,0,\delta)$	$Z_{18}^{(i)}$	2^{-39}
$(0,0,\delta,0,0,\delta,\delta,\delta) \xrightarrow{1r} (0,0,0,\delta,0,\delta,\delta,0)$	—	2^{-40}
$(0,0,\delta,0,\delta,0,0,0) \xrightarrow{1r} (0,0,0,\delta,0,0,0,0)$	—	2^{-23}
$(0,0,\delta,0,\delta,0,0,\delta) \xrightarrow{1r} (\delta,\delta,\delta,0,\delta,\delta,\delta,\delta)$	$Z_{18}^{(i)}$	2^{-38}
$(0,0,\delta,0,\delta,0,\delta,0) \xrightarrow{1r} (0,0,0,\delta,0,0,0,\delta)$	$Z_{18}^{(i)}$	2^{-31}
$(0,0,\delta,0,\delta,0,\delta,\delta) \xrightarrow{1r} (\delta,\delta,\delta,0,\delta,\delta,\delta,0)$	—	2^{-46}
$(0,0,\delta,0,\delta,\delta,0,0) \xrightarrow{1r} (\delta,\delta,\delta,0,\delta,\delta,0,\delta)$	$Z_{18}^{(i)}$	2^{-46}
$(0,0,\delta,0,\delta,\delta,0,\delta) \xrightarrow{1r} (0,0,0,\delta,0,0,\delta,0)$	—	2^{-38}
$(0,0,\delta,0,\delta,\delta,\delta,0) \xrightarrow{1r} (\delta,\delta,\delta,0,\delta,\delta,0,0)$	—	2^{-45}
$(0,0,\delta,0,\delta,\delta,\delta,\delta) \xrightarrow{1r} (0,0,0,\delta,0,0,\delta,\delta)$	$Z_{18}^{(i)}$	2^{-37}
$(0,0,\delta,\delta,0,0,0,0) \xrightarrow{1r} (\delta,\delta,\delta,0,0,0,0,0)$	—	2^{-16}
$(0,0,\delta,\delta,0,0,0,\delta) \xrightarrow{1r} (0,0,0,\delta,\delta,\delta,\delta,\delta)$	$Z_{18}^{(i)}$	2^{-24}
$(0,0,\delta,\delta,0,0,\delta,0) \xrightarrow{1r} (\delta,\delta,\delta,0,0,0,0,\delta)$	$Z_{18}^{(i)}$	2^{-24}
$(0,0,\delta,\delta,0,0,\delta,\delta) \xrightarrow{1r} (0,0,0,\delta,\delta,\delta,\delta,0)$	—	2^{-32}
$(0,0,\delta,\delta,0,\delta,0,0) \xrightarrow{1r} (0,0,0,\delta,\delta,\delta,0,\delta)$	$Z_{18}^{(i)}$	2^{-32}
$(0,0,\delta,\delta,0,\delta,0,\delta) \xrightarrow{1r} (\delta,\delta,\delta,0,0,0,\delta,0)$	—	2^{-31}
$(0,0,\delta,\delta,0,\delta,\delta,0) \xrightarrow{1r} (0,0,0,\delta,\delta,\delta,0,0)$	—	2^{-31}
$(0,0,\delta,\delta,0,\delta,\delta,\delta) \xrightarrow{1r} (\delta,\delta,\delta,0,0,0,\delta,\delta)$	$Z_{18}^{(i)}$	2^{-30}
$(0,0,\delta,\delta,\delta,0,0,0) \xrightarrow{1r} (\delta,\delta,\delta,0,0,\delta,0,\delta)$	$Z_{18}^{(i)}$	2^{-40}
$(0,0,\delta,\delta,\delta,0,0,\delta) \xrightarrow{1r} (0,0,0,\delta,\delta,0,\delta,0)$	—	2^{-39}
$(0,0,\delta,\delta,\delta,0,\delta,0) \xrightarrow{1r} (\delta,\delta,\delta,0,0,\delta,0,0)$	—	2^{-39}
$(0,0,\delta,\delta,\delta,0,\delta,\delta) \xrightarrow{1r} (0,0,0,\delta,\delta,0,\delta,\delta)$	$Z_{18}^{(i)}$	2^{-38}
$(0,0,\delta,\delta,\delta,\delta,0,0) \xrightarrow{1r} (0,0,0,\delta,\delta,0,0,0)$	—	2^{-24}
$(0,0,\delta,\delta,\delta,\delta,0,\delta) \xrightarrow{1r} (\delta,\delta,\delta,0,0,\delta,\delta,\delta)$	$Z_{18}^{(i)}$	2^{-32}
$(0,0,\delta,\delta,\delta,\delta,\delta,0) \xrightarrow{1r} (0,0,0,\delta,\delta,0,0,\delta)$	$Z_{18}^{(i)}$	2^{-32}
$(0,0,\delta,\delta,\delta,\delta,\delta,\delta) \xrightarrow{1r} (\delta,\delta,\delta,0,0,\delta,\delta,0)$	—	2^{-40}
$(0,\delta,0,0,0,0,0,0) \xrightarrow{1r} (\delta,\delta,0,\delta,0,\delta,0,\delta)$	$Z_{18}^{(i)}$	2^{-48}

Table 3.36 Nontrivial difference propagation across the MA-box of MESH-128(8) (part III).

input-output difference	Weak subkeys (i-th round)	Probability
$(0,\delta,0,0,0,0,0,\delta) \xrightarrow{1r} (0,0,\delta,0,\delta,0,\delta,0)$	—	2^{-45}
$(0,\delta,0,0,0,0,\delta,0) \xrightarrow{1r} (\delta,\delta,0,\delta,0,\delta,0,0)$	—	2^{-47}
$(0,\delta,0,0,0,0,\delta,\delta) \xrightarrow{1r} (0,0,\delta,0,\delta,0,\delta,\delta)$	$Z_{18}^{(i)}$	2^{-44}
$(0,\delta,0,0,0,\delta,0,0) \xrightarrow{1r} (0,0,\delta,0,\delta,0,0,0)$	—	2^{-30}
$(0,\delta,0,0,0,\delta,0,\delta) \xrightarrow{1r} (\delta,\delta,0,\delta,0,\delta,\delta,\delta)$	$Z_{18}^{(i)}$	2^{-40}
$(0,\delta,0,0,0,\delta,\delta,0) \xrightarrow{1r} (0,0,\delta,0,\delta,0,0,\delta)$	$Z_{18}^{(i)}$	2^{-38}
$(0,\delta,0,0,0,\delta,\delta,\delta) \xrightarrow{1r} (\delta,\delta,0,\delta,0,\delta,\delta,0)$	—	2^{-48}
$(0,\delta,0,0,\delta,0,0,0) \xrightarrow{1r} (\delta,\delta,0,\delta,0,0,0,0)$	—	2^{-31}
$(0,\delta,0,0,\delta,0,0,\delta) \xrightarrow{1r} (0,0,\delta,0,\delta,\delta,\delta,\delta)$	$Z_{18}^{(i)}$	2^{-37}
$(0,\delta,0,0,\delta,0,\delta,0) \xrightarrow{1r} (\delta,\delta,0,\delta,0,0,0,\delta)$	$Z_{18}^{(i)}$	2^{-39}
$(0,\delta,0,0,\delta,0,\delta,\delta) \xrightarrow{1r} (0,0,\delta,0,\delta,\delta,\delta,0)$	—	2^{-45}
$(0,\delta,0,0,\delta,\delta,0,0) \xrightarrow{1r} (0,0,\delta,0,\delta,\delta,0,\delta)$	$Z_{18}^{(i)}$	2^{-45}
$(0,\delta,0,0,\delta,\delta,0,\delta) \xrightarrow{1r} (\delta,\delta,0,\delta,0,0,\delta,0)$	—	2^{-46}
$(0,\delta,0,0,\delta,\delta,\delta,0) \xrightarrow{1r} (0,0,\delta,0,\delta,\delta,0,0)$	—	2^{-44}
$(0,\delta,0,0,\delta,\delta,\delta,\delta) \xrightarrow{1r} (\delta,\delta,0,\delta,0,0,\delta,\delta)$	$Z_{18}^{(i)}$	2^{-45}
$(0,\delta,0,\delta,0,0,0,0) \xrightarrow{1r} (0,0,\delta,0,0,0,0,0)$	—	2^{-15}
$(0,\delta,0,\delta,0,0,0,\delta) \xrightarrow{1r} (\delta,\delta,0,\delta,\delta,\delta,\delta,\delta)$	$Z_{18}^{(i)}$	2^{-32}
$(0,\delta,0,\delta,0,0,\delta,0) \xrightarrow{1r} (0,0,\delta,0,0,0,0,\delta)$	$Z_{18}^{(i)}$	2^{-23}
$(0,\delta,0,\delta,0,0,\delta,\delta) \xrightarrow{1r} (\delta,\delta,0,\delta,\delta,\delta,\delta,0)$	—	2^{-40}
$(0,\delta,0,\delta,0,\delta,0,0) \xrightarrow{1r} (\delta,\delta,0,\delta,\delta,\delta,0,\delta)$	$Z_{18}^{(i)}$	2^{-40}
$(0,\delta,0,\delta,0,\delta,0,\delta) \xrightarrow{1r} (0,0,\delta,0,0,0,\delta,0)$	—	2^{-30}
$(0,\delta,0,\delta,0,\delta,\delta,0) \xrightarrow{1r} (\delta,\delta,0,\delta,\delta,\delta,0,0)$	—	2^{-39}
$(0,\delta,0,\delta,0,\delta,\delta,\delta) \xrightarrow{1r} (0,0,\delta,0,0,0,\delta,\delta)$	$Z_{18}^{(i)}$	2^{-29}
$(0,\delta,0,\delta,\delta,0,0,0) \xrightarrow{1r} (0,0,\delta,0,0,\delta,0,\delta)$	$Z_{18}^{(i)}$	2^{-39}
$(0,\delta,0,\delta,\delta,0,0,\delta) \xrightarrow{1r} (\delta,\delta,0,\delta,\delta,0,\delta,0)$	—	2^{-47}
$(0,\delta,0,\delta,\delta,0,\delta,0) \xrightarrow{1r} (0,0,\delta,0,0,\delta,0,0)$	—	2^{-38}
$(0,\delta,0,\delta,\delta,0,\delta,\delta) \xrightarrow{1r} (\delta,\delta,0,\delta,\delta,0,\delta,\delta)$	$Z_{18}^{(i)}$	2^{-46}
$(0,\delta,0,\delta,\delta,\delta,0,0) \xrightarrow{1r} (\delta,\delta,0,\delta,\delta,0,0,0)$	—	2^{-32}
$(0,\delta,0,\delta,\delta,\delta,0,\delta) \xrightarrow{1r} (0,0,\delta,0,0,\delta,\delta,\delta)$	$Z_{18}^{(i)}$	2^{-31}
$(0,\delta,0,\delta,\delta,\delta,\delta,0) \xrightarrow{1r} (\delta,\delta,0,\delta,\delta,0,0,\delta)$	$Z_{18}^{(i)}$	2^{-40}
$(0,\delta,0,\delta,\delta,\delta,\delta,\delta) \xrightarrow{1r} (0,0,\delta,0,0,\delta,\delta,0)$	—	2^{-39}
$(0,\delta,\delta,0,0,0,0,0) \xrightarrow{1r} (\delta,\delta,0,0,0,0,0,0)$	—	2^{-15}

Table 3.37 Nontrivial difference propagation across the MA-box of MESH-128(8) (part IV).

input-output difference	Weak subkeys (i-th round)	Probability
$(0,\delta,\delta,0,0,0,0,\delta) \stackrel{1r}{\to} (0,0,\delta,\delta,\delta,\delta,\delta,\delta)$	$Z_{18}^{(i)}$	2^{-30}
$(0,\delta,\delta,0,0,0,\delta,0) \stackrel{1r}{\to} (\delta,\delta,0,0,0,0,0,\delta)$	$Z_{18}^{(i)}$	2^{-23}
$(0,\delta,\delta,0,0,0,\delta,\delta) \stackrel{1r}{\to} (0,0,\delta,\delta,\delta,\delta,\delta,0)$	—	2^{-38}
$(0,\delta,\delta,0,0,\delta,0,0) \stackrel{1r}{\to} (0,0,\delta,\delta,\delta,\delta,0,\delta)$	$Z_{18}^{(i)}$	2^{-38}
$(0,\delta,\delta,0,0,\delta,0,\delta) \stackrel{1r}{\to} (\delta,\delta,0,0,0,0,\delta,0)$	—	2^{-30}
$(0,\delta,\delta,0,0,\delta,\delta,0) \stackrel{1r}{\to} (0,0,\delta,\delta,\delta,\delta,0,0)$	—	2^{-37}
$(0,\delta,\delta,0,0,\delta,\delta,\delta) \stackrel{1r}{\to} (\delta,\delta,0,0,0,0,\delta,\delta)$	$Z_{18}^{(i)}$	2^{-29}
$(0,\delta,\delta,0,\delta,0,0,0) \stackrel{1r}{\to} (\delta,\delta,0,0,0,\delta,0,\delta)$	$Z_{18}^{(i)}$	2^{-39}
$(0,\delta,\delta,0,\delta,0,0,\delta) \stackrel{1r}{\to} (0,0,\delta,\delta,\delta,0,\delta,0)$	—	2^{-45}
$(0,\delta,\delta,0,\delta,0,\delta,0) \stackrel{1r}{\to} (\delta,\delta,0,0,0,\delta,0,0)$	—	2^{-38}
$(0,\delta,\delta,0,\delta,0,\delta,\delta) \stackrel{1r}{\to} (0,0,\delta,\delta,\delta,0,\delta,\delta)$	$Z_{18}^{(i)}$	2^{-44}
$(0,\delta,\delta,0,\delta,\delta,0,0) \stackrel{1r}{\to} (0,0,\delta,\delta,\delta,0,0,0)$	—	2^{-30}
$(0,\delta,\delta,0,\delta,\delta,0,\delta) \stackrel{1r}{\to} (\delta,\delta,0,0,0,\delta,\delta,\delta)$	$Z_{18}^{(i)}$	2^{-31}
$(0,\delta,\delta,0,\delta,\delta,\delta,0) \stackrel{1r}{\to} (0,0,\delta,\delta,\delta,0,0,\delta)$	$Z_{18}^{(i)}$	2^{-38}
$(0,\delta,\delta,0,\delta,\delta,\delta,\delta) \stackrel{1r}{\to} (\delta,\delta,0,0,0,\delta,\delta,0)$	—	2^{-39}
$(0,\delta,\delta,\delta,0,0,0,0) \stackrel{1r}{\to} (0,0,\delta,\delta,0,\delta,0,\delta)$	$Z_{18}^{(i)}$	2^{-46}
$(0,\delta,\delta,\delta,0,0,0,\delta) \stackrel{1r}{\to} (\delta,\delta,0,0,\delta,0,\delta,0)$	—	2^{-45}
$(0,\delta,\delta,\delta,0,0,\delta,0) \stackrel{1r}{\to} (0,0,\delta,\delta,0,\delta,0,0)$	—	2^{-45}
$(0,\delta,\delta,\delta,0,0,\delta,\delta) \stackrel{1r}{\to} (\delta,\delta,0,0,\delta,0,\delta,\delta)$	$Z_{18}^{(i)}$	2^{-44}
$(0,\delta,\delta,\delta,0,\delta,0,0) \stackrel{1r}{\to} (\delta,\delta,0,0,\delta,0,0,0)$	—	2^{-30}
$(0,\delta,\delta,\delta,0,\delta,0,\delta) \stackrel{1r}{\to} (0,0,\delta,\delta,0,\delta,\delta,\delta)$	$Z_{18}^{(i)}$	2^{-38}
$(0,\delta,\delta,\delta,0,\delta,\delta,0) \stackrel{1r}{\to} (\delta,\delta,0,0,\delta,0,0,\delta)$	$Z_{18}^{(i)}$	2^{-38}
$(0,\delta,\delta,\delta,0,\delta,\delta,\delta) \stackrel{1r}{\to} (0,0,\delta,\delta,0,\delta,\delta,0)$	—	2^{-46}
$(0,\delta,\delta,\delta,\delta,0,0,0) \stackrel{1r}{\to} (0,0,\delta,\delta,0,0,0,0)$	—	2^{-29}
$(0,\delta,\delta,\delta,\delta,0,0,\delta) \stackrel{1r}{\to} (\delta,\delta,0,0,\delta,\delta,\delta,\delta)$	$Z_{18}^{(i)}$	2^{-37}
$(0,\delta,\delta,\delta,\delta,0,\delta,0) \stackrel{1r}{\to} (0,0,\delta,\delta,0,0,0,\delta)$	$Z_{18}^{(i)}$	2^{-37}
$(0,\delta,\delta,\delta,\delta,0,\delta,\delta) \stackrel{1r}{\to} (\delta,\delta,0,0,\delta,\delta,\delta,0)$		2^{-45}
$(0,\delta,\delta,\delta,\delta,\delta,0,0) \stackrel{1r}{\to} (\delta,\delta,0,0,\delta,\delta,0,\delta)$	$Z_{18}^{(i)}$	2^{-45}
$(0,\delta,\delta,\delta,\delta,\delta,0,\delta) \stackrel{1r}{\to} (0,0,\delta,\delta,0,0,\delta,0)$	—	2^{-44}
$(0,\delta,\delta,\delta,\delta,\delta,\delta,0) \stackrel{1r}{\to} (\delta,\delta,0,0,\delta,\delta,0,0)$	—	2^{-44}
$(0,\delta,\delta,\delta,\delta,\delta,\delta,\delta) \stackrel{1r}{\to} (0,0,\delta,\delta,0,0,\delta,\delta)$	$Z_{18}^{(i)}$	2^{-43}
$(\delta,0,0,0,0,0,0,0) \stackrel{1r}{\to} (0,\delta,0,\delta,0,\delta,0,\delta)$	$Z_{17}^{(i)}, Z_{18}^{(i)}$	2^{-48}

Table 3.38 Nontrivial difference propagation across the MA-box of MESH-128(8) (part V).

input-output difference	Weak subkeys (i-th round)	Probability
$(\delta,0,0,0,0,0,0,\delta) \xrightarrow{1r} (\delta,0,\delta,0,\delta,0,\delta,0)$	$Z_{17}^{(i)}$	2^{-45}
$(\delta,0,0,0,0,0,\delta,0) \xrightarrow{1r} (0,\delta,0,\delta,0,\delta,0,0)$	$Z_{17}^{(i)}$	2^{-47}
$(\delta,0,0,0,0,0,\delta,\delta) \xrightarrow{1r} (\delta,0,\delta,0,\delta,0,\delta,\delta)$	$Z_{17}^{(i)}, Z_{18}^{(i)}$	2^{-44}
$(\delta,0,0,0,0,\delta,0,0) \xrightarrow{1r} (\delta,0,\delta,0,\delta,0,0,0)$	$Z_{17}^{(i)}$	2^{-30}
$(\delta,0,0,0,0,\delta,0,\delta) \xrightarrow{1r} (0,\delta,0,\delta,0,\delta,\delta,\delta)$	$Z_{17}^{(i)}, Z_{18}^{(i)}$	2^{-40}
$(\delta,0,0,0,0,\delta,\delta,0) \xrightarrow{1r} (\delta,0,\delta,0,\delta,0,0,\delta)$	$Z_{17}^{(i)}, Z_{18}^{(i)}$	2^{-38}
$(\delta,0,0,0,0,\delta,\delta,\delta) \xrightarrow{1r} (0,\delta,0,\delta,0,\delta,\delta,0)$	$Z_{17}^{(i)}$	2^{-48}
$(\delta,0,0,0,\delta,0,0,0) \xrightarrow{1r} (0,\delta,0,\delta,0,0,0,0)$	$Z_{17}^{(i)}$	2^{-31}
$(\delta,0,0,0,\delta,0,0,\delta) \xrightarrow{1r} (\delta,0,\delta,0,\delta,\delta,\delta,\delta)$	$Z_{17}^{(i)}, Z_{18}^{(i)}$	2^{-37}
$(\delta,0,0,0,\delta,0,\delta,0) \xrightarrow{1r} (0,\delta,0,\delta,0,0,0,\delta)$	$Z_{17}^{(i)}, Z_{18}^{(i)}$	2^{-39}
$(\delta,0,0,0,\delta,0,\delta,\delta) \xrightarrow{1r} (\delta,0,\delta,0,\delta,\delta,\delta,0)$	$Z_{17}^{(i)}$	2^{-45}
$(\delta,0,0,0,\delta,\delta,0,0) \xrightarrow{1r} (\delta,0,\delta,0,\delta,\delta,0,\delta)$	$Z_{17}^{(i)}, Z_{18}^{(i)}$	2^{-45}
$(\delta,0,0,0,\delta,\delta,0,\delta) \xrightarrow{1r} (0,\delta,0,\delta,0,0,\delta,0)$	$Z_{17}^{(i)}$	2^{-46}
$(\delta,0,0,0,\delta,\delta,\delta,0) \xrightarrow{1r} (\delta,0,\delta,0,\delta,\delta,0,0)$	$Z_{17}^{(i)}$	2^{-44}
$(\delta,0,0,0,\delta,\delta,\delta,\delta) \xrightarrow{1r} (0,\delta,0,\delta,0,0,\delta,\delta)$	$Z_{17}^{(i)}, Z_{18}^{(i)}$	2^{-45}
$(\delta,0,0,\delta,0,0,0,0) \xrightarrow{1r} (\delta,0,\delta,0,0,0,0,0)$	$Z_{17}^{(i)}$	2^{-15}
$(\delta,0,0,\delta,0,0,0,\delta) \xrightarrow{1r} (0,\delta,0,\delta,\delta,\delta,\delta,\delta)$	$Z_{17}^{(i)}, Z_{18}^{(i)}$	2^{-32}
$(\delta,0,0,\delta,0,0,\delta,0) \xrightarrow{1r} (\delta,0,\delta,0,0,0,0,\delta)$	$Z_{17}^{(i)}, Z_{18}^{(i)}$	2^{-23}
$(\delta,0,0,\delta,0,0,\delta,\delta) \xrightarrow{1r} (0,\delta,0,\delta,\delta,\delta,\delta,0)$	$Z_{17}^{(i)}$	2^{-40}
$(\delta,0,0,\delta,0,\delta,0,0) \xrightarrow{1r} (0,\delta,0,\delta,\delta,\delta,0,\delta)$	$Z_{17}^{(i)}, Z_{18}^{(i)}$	2^{-40}
$(\delta,0,0,\delta,0,\delta,0,\delta) \xrightarrow{1r} (\delta,0,\delta,0,0,0,\delta,0)$	$Z_{17}^{(i)}$	2^{-30}
$(\delta,0,0,\delta,0,\delta,\delta,0) \xrightarrow{1r} (0,\delta,0,\delta,\delta,\delta,0,0)$	$Z_{17}^{(i)}$	2^{-39}
$(\delta,0,0,\delta,0,\delta,\delta,\delta) \xrightarrow{1r} (\delta,0,\delta,0,0,0,\delta,\delta)$	$Z_{17}^{(i)}, Z_{18}^{(i)}$	2^{-29}
$(\delta,0,0,\delta,\delta,0,0,0) \xrightarrow{1r} (\delta,0,\delta,0,0,\delta,0,\delta)$	$Z_{17}^{(i)}, Z_{18}^{(i)}$	2^{-39}
$(\delta,0,0,\delta,\delta,0,0,\delta) \xrightarrow{1r} (0,\delta,0,\delta,\delta,0,\delta,0)$	$Z_{17}^{(i)}$	2^{-47}
$(\delta,0,0,\delta,\delta,0,\delta,0) \xrightarrow{1r} (\delta,0,\delta,0,0,\delta,0,0)$	$Z_{17}^{(i)}$	2^{-38}
$(\delta,0,0,\delta,\delta,0,\delta,\delta) \xrightarrow{1r} (0,\delta,0,\delta,\delta,0,\delta,\delta)$	$Z_{17}^{(i)}, Z_{18}^{(i)}$	2^{-46}
$(\delta,0,0,\delta,\delta,\delta,0,0) \xrightarrow{1r} (0,\delta,0,\delta,\delta,0,0,0)$	$Z_{17}^{(i)}$	2^{-32}
$(\delta,0,0,\delta,\delta,\delta,0,\delta) \xrightarrow{1r} (\delta,0,\delta,0,0,\delta,\delta,\delta)$	$Z_{17}^{(i)}, Z_{18}^{(i)}$	2^{-31}
$(\delta,0,0,\delta,\delta,\delta,\delta,0) \xrightarrow{1r} (0,\delta,0,\delta,\delta,0,0,\delta)$	$Z_{17}^{(i)}, Z_{18}^{(i)}$	2^{-40}
$(\delta,0,0,\delta,\delta,\delta,\delta,\delta) \xrightarrow{1r} (\delta,0,\delta,0,0,\delta,\delta,0)$	$Z_{17}^{(i)}$	2^{-39}
$(\delta,0,\delta,0,0,0,0,0) \xrightarrow{1r} (0,\delta,0,0,0,0,0,0)$	$Z_{17}^{(i)}$	2^{-15}

Table 3.39 Nontrivial difference propagation across the MA-box of MESH-128(8) (part VI).

input-output difference	Weak subkeys (i-th round)	Probability
$(\delta,0,\delta,0,0,0,0,\delta) \xrightarrow{1r} (\delta,0,\delta,\delta,\delta,\delta,\delta,\delta)$	$Z_{17}^{(i)}, Z_{18}^{(i)}$	2^{-30}
$(\delta,0,\delta,0,0,0,\delta,0) \xrightarrow{1r} (0,\delta,0,0,0,0,0,\delta)$	$Z_{17}^{(i)}, Z_{18}^{(i)}$	2^{-23}
$(\delta,0,\delta,0,0,0,\delta,\delta) \xrightarrow{1r} (\delta,0,\delta,\delta,\delta,\delta,\delta,0)$	$Z_{17}^{(i)}$	2^{-38}
$(\delta,0,\delta,0,0,\delta,0,0) \xrightarrow{1r} (\delta,0,\delta,\delta,\delta,\delta,0,\delta)$	$Z_{17}^{(i)}, Z_{18}^{(i)}$	2^{-38}
$(\delta,0,\delta,0,0,\delta,0,\delta) \xrightarrow{1r} (0,\delta,0,0,0,0,\delta,0)$	$Z_{17}^{(i)}$	2^{-30}
$(\delta,0,\delta,0,0,\delta,\delta,0) \xrightarrow{1r} (\delta,0,\delta,\delta,\delta,\delta,0,0)$	$Z_{17}^{(i)}$	2^{-37}
$(\delta,0,\delta,0,0,\delta,\delta,\delta) \xrightarrow{1r} (0,\delta,0,0,0,0,\delta,\delta)$	$Z_{17}^{(i)}, Z_{18}^{(i)}$	2^{-29}
$(\delta,0,\delta,0,\delta,0,0,0) \xrightarrow{1r} (0,\delta,0,0,0,\delta,0,\delta)$	$Z_{17}^{(i)}, Z_{18}^{(i)}$	2^{-39}
$(\delta,0,\delta,0,\delta,0,0,\delta) \xrightarrow{1r} (\delta,0,\delta,\delta,\delta,0,\delta,0)$	$Z_{17}^{(i)}$	2^{-45}
$(\delta,0,\delta,0,\delta,0,\delta,0) \xrightarrow{1r} (0,\delta,0,0,0,\delta,0,0)$	$Z_{17}^{(i)}$	2^{-38}
$(\delta,0,\delta,0,\delta,0,\delta,\delta) \xrightarrow{1r} (\delta,0,\delta,\delta,\delta,0,\delta,\delta)$	$Z_{17}^{(i)}, Z_{18}^{(i)}$	2^{-44}
$(\delta,0,\delta,0,\delta,\delta,0,0) \xrightarrow{1r} (\delta,0,\delta,\delta,\delta,0,0,0)$	$Z_{17}^{(i)}$	2^{-30}
$(\delta,0,\delta,0,\delta,\delta,0,\delta) \xrightarrow{1r} (0,\delta,0,0,0,\delta,\delta,\delta)$	$Z_{17}^{(i)}, Z_{18}^{(i)}$	2^{-31}
$(\delta,0,\delta,0,\delta,\delta,\delta,0) \xrightarrow{1r} (\delta,0,\delta,\delta,\delta,0,0,\delta)$	$Z_{17}^{(i)}, Z_{18}^{(i)}$	2^{-38}
$(\delta,0,\delta,0,\delta,\delta,\delta,\delta) \xrightarrow{1r} (0,\delta,0,0,0,\delta,\delta,0)$	$Z_{17}^{(i)}$	2^{-39}
$(\delta,0,\delta,\delta,0,0,0,0) \xrightarrow{1r} (\delta,0,\delta,\delta,0,\delta,0,\delta)$	$Z_{17}^{(i)}, Z_{18}^{(i)}$	2^{-46}
$(\delta,0,\delta,\delta,0,0,0,\delta) \xrightarrow{1r} (0,\delta,0,0,\delta,0,\delta,0)$	$Z_{17}^{(i)}$	2^{-45}
$(\delta,0,\delta,\delta,0,0,\delta,0) \xrightarrow{1r} (\delta,0,\delta,\delta,0,\delta,0,0)$	$Z_{17}^{(i)}$	2^{-45}
$(\delta,0,\delta,\delta,0,0,\delta,\delta) \xrightarrow{1r} (0,\delta,0,0,\delta,0,\delta,\delta)$	$Z_{17}^{(i)}, Z_{18}^{(i)}$	2^{-44}
$(\delta,0,\delta,\delta,0,\delta,0,0) \xrightarrow{1r} (0,\delta,0,0,\delta,0,0,0)$	$Z_{17}^{(i)}$	2^{-30}
$(\delta,0,\delta,\delta,0,\delta,0,\delta) \xrightarrow{1r} (\delta,0,\delta,\delta,0,\delta,\delta,\delta)$	$Z_{17}^{(i)}, Z_{18}^{(i)}$	2^{-38}
$(\delta,0,\delta,\delta,0,\delta,\delta,0) \xrightarrow{1r} (0,\delta,0,0,\delta,0,0,\delta)$	$Z_{17}^{(i)}, Z_{18}^{(i)}$	2^{-38}
$(\delta,0,\delta,\delta,0,\delta,\delta,\delta) \xrightarrow{1r} (\delta,0,\delta,\delta,0,\delta,\delta,0)$	$Z_{17}^{(i)}$	2^{-46}
$(\delta,0,\delta,\delta,\delta,0,0,0) \xrightarrow{1r} (\delta,0,\delta,\delta,0,0,0,0)$	$Z_{17}^{(i)}$	2^{-29}
$(\delta,0,\delta,\delta,\delta,0,0,\delta) \xrightarrow{1r} (0,\delta,0,0,\delta,\delta,\delta,\delta)$	$Z_{17}^{(i)}, Z_{18}^{(i)}$	2^{-37}
$(\delta,0,\delta,\delta,\delta,0,\delta,0) \xrightarrow{1r} (\delta,0,\delta,\delta,0,0,0,\delta)$	$Z_{17}^{(i)}, Z_{18}^{(i)}$	2^{-37}
$(\delta,0,\delta,\delta,\delta,0,\delta,\delta) \xrightarrow{1r} (0,\delta,0,0,\delta,\delta,\delta,0)$	$Z_{17}^{(i)}$	2^{-45}
$(\delta,0,\delta,\delta,\delta,\delta,0,0) \xrightarrow{1r} (0,\delta,0,0,\delta,\delta,0,\delta)$	$Z_{17}^{(i)}, Z_{18}^{(i)}$	2^{-45}
$(\delta,0,\delta,\delta,\delta,\delta,0,\delta) \xrightarrow{1r} (\delta,0,\delta,\delta,0,0,\delta,0)$	$Z_{17}^{(i)}$	2^{-44}
$(\delta,0,\delta,\delta,\delta,\delta,\delta,0) \xrightarrow{1r} (0,\delta,0,0,\delta,\delta,0,0)$	$Z_{17}^{(i)}$	2^{-44}
$(\delta,0,\delta,\delta,\delta,\delta,\delta,\delta) \xrightarrow{1r} (\delta,0,\delta,\delta,0,0,\delta,\delta)$	$Z_{17}^{(i)}, Z_{18}^{(i)}$	2^{-43}
$(\delta,\delta,0,0,0,0,0,0) \xrightarrow{1r} (\delta,0,0,0,0,0,0,0)$	$Z_{17}^{(i)}$,	2^{-0}

Table 3.40 Nontrivial difference propagation across the MA-box of MESH-128(8) (part VII).

input-output difference	Weak subkeys (i-th round)	Probability
$(\delta,\delta,0,0,0,0,0,\delta) \stackrel{1r}{\rightarrow} (0,\delta,\delta,\delta,\delta,\delta,\delta,\delta)$	$Z_{17}^{(i)}, Z_{18}^{(i)}$	2^{-24}
$(\delta,\delta,0,0,0,0,\delta,0) \stackrel{1r}{\rightarrow} (\delta,0,0,0,0,0,0,\delta)$	$Z_{17}^{(i)}, Z_{18}^{(i)}$	2^{-8}
$(\delta,\delta,0,0,0,0,\delta,\delta) \stackrel{1r}{\rightarrow} (0,\delta,\delta,\delta,\delta,\delta,\delta,0)$	$Z_{17}^{(i)}$	2^{-32}
$(\delta,\delta,0,0,0,\delta,0,0) \stackrel{1r}{\rightarrow} (0,\delta,\delta,\delta,\delta,\delta,0,\delta)$	$Z_{17}^{(i)}, Z_{18}^{(i)}$	2^{-32}
$(\delta,\delta,0,0,0,\delta,0,\delta) \stackrel{1r}{\rightarrow} (\delta,0,0,0,0,0,\delta,0)$	$Z_{17}^{(i)}$	2^{-15}
$(\delta,\delta,0,0,0,\delta,\delta,0) \stackrel{1r}{\rightarrow} (0,\delta,\delta,\delta,\delta,\delta,0,0)$	$Z_{17}^{(i)}$	2^{-31}
$(\delta,\delta,0,0,0,\delta,\delta,\delta) \stackrel{1r}{\rightarrow} (\delta,0,0,0,0,0,\delta,\delta)$	$Z_{17}^{(i)}, Z_{18}^{(i)}$	2^{-14}
$(\delta,\delta,0,0,\delta,0,0,0) \stackrel{1r}{\rightarrow} (\delta,0,0,0,0,\delta,0,\delta)$	$Z_{17}^{(i)}, Z_{18}^{(i)}$	2^{-24}
$(\delta,\delta,0,0,\delta,0,0,\delta) \stackrel{1r}{\rightarrow} (0,\delta,\delta,\delta,\delta,0,\delta,0)$	$Z_{17}^{(i)}$	2^{-39}
$(\delta,\delta,0,0,\delta,0,\delta,0) \stackrel{1r}{\rightarrow} (\delta,0,0,0,0,\delta,0,0)$	$Z_{17}^{(i)}$	2^{-23}
$(\delta,\delta,0,0,\delta,0,\delta,\delta) \stackrel{1r}{\rightarrow} (0,\delta,\delta,\delta,\delta,0,\delta,\delta)$	$Z_{17}^{(i)}, Z_{18}^{(i)}$	2^{-38}
$(\delta,\delta,0,0,\delta,\delta,0,0) \stackrel{1r}{\rightarrow} (0,\delta,\delta,\delta,\delta,0,0,0)$	$Z_{17}^{(i)}$	2^{-24}
$(\delta,\delta,0,0,\delta,\delta,0,\delta) \stackrel{1r}{\rightarrow} (\delta,0,0,0,0,\delta,\delta,\delta)$	$Z_{17}^{(i)}, Z_{18}^{(i)}$	2^{-16}
$(\delta,\delta,0,0,\delta,\delta,\delta,0) \stackrel{1r}{\rightarrow} (0,\delta,\delta,\delta,\delta,0,0,\delta)$	$Z_{17}^{(i)}, Z_{18}^{(i)}$	2^{-32}
$(\delta,\delta,0,0,\delta,\delta,\delta,\delta) \stackrel{1r}{\rightarrow} (\delta,0,0,0,0,\delta,\delta,0)$	$Z_{17}^{(i)}$	2^{-24}
$(\delta,\delta,0,\delta,0,0,0,0) \stackrel{1r}{\rightarrow} (0,\delta,\delta,\delta,0,\delta,0,\delta)$	$Z_{17}^{(i)}, Z_{18}^{(i)}$	2^{-40}
$(\delta,\delta,0,\delta,0,0,0,\delta) \stackrel{1r}{\rightarrow} (\delta,0,0,0,\delta,0,\delta,0)$	$Z_{17}^{(i)}$	2^{-30}
$(\delta,\delta,0,\delta,0,0,\delta,0) \stackrel{1r}{\rightarrow} (0,\delta,\delta,\delta,0,\delta,0,0)$	$Z_{17}^{(i)}$	2^{-39}
$(\delta,\delta,0,\delta,0,0,\delta,\delta) \stackrel{1r}{\rightarrow} (\delta,0,0,0,\delta,0,\delta,\delta)$	$Z_{17}^{(i)}, Z_{18}^{(i)}$	2^{-29}
$(\delta,\delta,0,\delta,0,\delta,0,0) \stackrel{1r}{\rightarrow} (\delta,0,0,0,\delta,0,0,0)$	$Z_{17}^{(i)}$	2^{-15}
$(\delta,\delta,0,\delta,0,\delta,0,\delta) \stackrel{1r}{\rightarrow} (0,\delta,\delta,\delta,0,\delta,\delta,\delta)$	$Z_{17}^{(i)}, Z_{18}^{(i)}$	2^{-32}
$(\delta,\delta,0,\delta,0,\delta,\delta,0) \stackrel{1r}{\rightarrow} (\delta,0,0,0,\delta,0,0,\delta)$	$Z_{17}^{(i)}, Z_{18}^{(i)}$	2^{-23}
$(\delta,\delta,0,\delta,0,\delta,\delta,\delta) \stackrel{1r}{\rightarrow} (0,\delta,\delta,\delta,0,\delta,\delta,0)$	$Z_{17}^{(i)}$	2^{-40}
$(\delta,\delta,0,\delta,\delta,0,0,0) \stackrel{1r}{\rightarrow} (0,\delta,\delta,\delta,0,0,0,0)$	$Z_{17}^{(i)}$	2^{-23}
$(\delta,\delta,0,\delta,\delta,0,0,\delta) \stackrel{1r}{\rightarrow} (\delta,0,0,0,\delta,\delta,\delta,\delta)$	$Z_{17}^{(i)}, Z_{18}^{(i)}$	2^{-22}
$(\delta,\delta,0,\delta,\delta,0,\delta,0) \stackrel{1r}{\rightarrow} (0,\delta,\delta,\delta,0,0,0,\delta)$	$Z_{17}^{(i)}, Z_{18}^{(i)}$	2^{-31}
$(\delta,\delta,0,\delta,\delta,0,\delta,\delta) \stackrel{1r}{\rightarrow} (\delta,0,0,0,\delta,\delta,\delta,0)$	$Z_{17}^{(i)}$	2^{-30}
$(\delta,\delta,0,\delta,\delta,\delta,0,0) \stackrel{1r}{\rightarrow} (\delta,0,0,0,\delta,\delta,0,\delta)$	$Z_{17}^{(i)}, Z_{18}^{(i)}$	2^{-30}
$(\delta,\delta,0,\delta,\delta,\delta,0,\delta) \stackrel{1r}{\rightarrow} (0,\delta,\delta,\delta,0,0,\delta,0)$	$Z_{17}^{(i)}$	2^{-38}
$(\delta,\delta,0,\delta,\delta,\delta,\delta,0) \stackrel{1r}{\rightarrow} (\delta,0,0,0,\delta,\delta,0,0)$	$Z_{17}^{(i)}$	2^{-29}
$(\delta,\delta,0,\delta,\delta,\delta,\delta,\delta) \stackrel{1r}{\rightarrow} (0,\delta,\delta,\delta,0,0,\delta,\delta)$	$Z_{17}^{(i)}, Z_{18}^{(i)}$	2^{-37}
$(\delta,\delta,\delta,0,0,0,0,0) \stackrel{1r}{\rightarrow} (\delta,0,0,\delta,0,\delta,0,\delta)$	$Z_{17}^{(i)}, Z_{18}^{(i)}$	2^{-40}

Table 3.41 Nontrivial difference propagation across the MA-box of MESH-128(8) (part VIII).

input-output difference	Weak subkeys (i-th round)	Probability
$(\delta,\delta,\delta,0,0,0,0,\delta) \stackrel{1r}{\rightarrow} (0,\delta,\delta,0,\delta,0,\delta,0)$	$Z_{17}^{(i)}$	2^{-46}
$(\delta,\delta,\delta,0,0,0,\delta,0) \stackrel{1r}{\rightarrow} (\delta,0,0,\delta,0,\delta,0,0)$	$Z_{17}^{(i)}$	2^{-39}
$(\delta,\delta,\delta,0,0,0,\delta,\delta) \stackrel{1r}{\rightarrow} (0,\delta,\delta,0,\delta,0,\delta,\delta)$	$Z_{17}^{(i)}, Z_{18}^{(i)}$	2^{-45}
$(\delta,\delta,\delta,0,0,\delta,0,0) \stackrel{1r}{\rightarrow} (0,\delta,\delta,0,\delta,0,0,0)$	$Z_{17}^{(i)}$	2^{-31}
$(\delta,\delta,\delta,0,0,\delta,0,\delta) \stackrel{1r}{\rightarrow} (\delta,0,0,\delta,0,\delta,\delta,\delta)$	$Z_{17}^{(i)}, Z_{18}^{(i)}$	2^{-32}
$(\delta,\delta,\delta,0,0,\delta,\delta,0) \stackrel{1r}{\rightarrow} (0,\delta,\delta,0,\delta,0,0,\delta)$	$Z_{17}^{(i)}, Z_{18}^{(i)}$	2^{-39}
$(\delta,\delta,\delta,0,0,\delta,\delta,\delta) \stackrel{1r}{\rightarrow} (\delta,0,0,\delta,0,\delta,\delta,0)$	$Z_{17}^{(i)}$	2^{-40}
$(\delta,\delta,\delta,0,\delta,0,0,0) \stackrel{1r}{\rightarrow} (\delta,0,0,\delta,0,0,0,0)$	$Z_{17}^{(i)}$	2^{-23}
$(\delta,\delta,\delta,0,\delta,0,0,\delta) \stackrel{1r}{\rightarrow} (0,\delta,\delta,0,\delta,\delta,\delta,\delta)$	$Z_{17}^{(i)}, Z_{18}^{(i)}$	2^{-38}
$(\delta,\delta,\delta,0,\delta,0,\delta,0) \stackrel{1r}{\rightarrow} (\delta,0,0,\delta,0,0,0,\delta)$	$Z_{17}^{(i)}, Z_{18}^{(i)}$	2^{-31}
$(\delta,\delta,\delta,0,\delta,0,\delta,\delta) \stackrel{1r}{\rightarrow} (0,\delta,\delta,0,\delta,\delta,\delta,0)$	$Z_{17}^{(i)}$	2^{-46}
$(\delta,\delta,\delta,0,\delta,\delta,0,0) \stackrel{1r}{\rightarrow} (0,\delta,\delta,0,\delta,\delta,0,\delta)$	$Z_{17}^{(i)}, Z_{18}^{(i)}$	2^{-46}
$(\delta,\delta,\delta,0,\delta,\delta,0,\delta) \stackrel{1r}{\rightarrow} (\delta,0,0,\delta,0,0,\delta,0)$	$Z_{17}^{(i)}$	2^{-38}
$(\delta,\delta,\delta,0,\delta,\delta,\delta,0) \stackrel{1r}{\rightarrow} (0,\delta,\delta,0,\delta,\delta,0,0)$	$Z_{17}^{(i)}$	2^{-45}
$(\delta,\delta,\delta,0,\delta,\delta,\delta,\delta) \stackrel{1r}{\rightarrow} (\delta,0,0,\delta,0,0,\delta,\delta)$	$Z_{17}^{(i)}, Z_{18}^{(i)}$	2^{-37}
$(\delta,\delta,\delta,\delta,0,0,0,0) \stackrel{1r}{\rightarrow} (0,\delta,\delta,0,0,0,0,0)$	$Z_{17}^{(i)}$	2^{-16}
$(\delta,\delta,\delta,\delta,0,0,0,\delta) \stackrel{1r}{\rightarrow} (\delta,0,0,\delta,\delta,\delta,\delta,\delta)$	$Z_{17}^{(i)}, Z_{18}^{(i)}$	2^{-24}
$(\delta,\delta,\delta,\delta,0,0,\delta,0) \stackrel{1r}{\rightarrow} (0,\delta,\delta,0,0,0,0,\delta)$	$Z_{17}^{(i)}, Z_{18}^{(i)}$	2^{-24}
$(\delta,\delta,\delta,\delta,0,0,\delta,\delta) \stackrel{1r}{\rightarrow} (\delta,0,0,\delta,\delta,\delta,\delta,0)$	$Z_{17}^{(i)}$	2^{-32}
$(\delta,\delta,\delta,\delta,0,\delta,0,0) \stackrel{1r}{\rightarrow} (\delta,0,0,\delta,\delta,\delta,0,\delta)$	$Z_{17}^{(i)}, Z_{18}^{(i)}$	2^{-32}
$(\delta,\delta,\delta,\delta,0,\delta,0,\delta) \stackrel{1r}{\rightarrow} (0,\delta,\delta,0,0,0,\delta,0)$	$Z_{17}^{(i)}$	2^{-31}
$(\delta,\delta,\delta,\delta,0,\delta,\delta,0) \stackrel{1r}{\rightarrow} (\delta,0,0,\delta,\delta,\delta,0,0)$	$Z_{17}^{(i)}$	2^{-31}
$(\delta,\delta,\delta,\delta,0,\delta,\delta,\delta) \stackrel{1r}{\rightarrow} (0,\delta,\delta,0,0,0,\delta,\delta)$	$Z_{17}^{(i)}, Z_{18}^{(i)}$	2^{-30}
$(\delta,\delta,\delta,\delta,\delta,0,0,0) \stackrel{1r}{\rightarrow} (0,\delta,\delta,0,0,\delta,0,\delta)$	$Z_{17}^{(i)}, Z_{18}^{(i)}$	2^{-40}
$(\delta,\delta,\delta,\delta,\delta,0,0,\delta) \stackrel{1r}{\rightarrow} (\delta,0,0,\delta,\delta,0,\delta,0)$	$Z_{17}^{(i)}$	2^{-39}
$(\delta,\delta,\delta,\delta,\delta,0,\delta,0) \stackrel{1r}{\rightarrow} (0,\delta,\delta,0,0,\delta,0,0)$	$Z_{17}^{(i)}$	2^{-39}
$(\delta,\delta,\delta,\delta,\delta,0,\delta,\delta) \stackrel{1r}{\rightarrow} (\delta,0,0,\delta,\delta,0,\delta,\delta)$	$Z_{17}^{(i)}, Z_{18}^{(i)}$	2^{-38}
$(\delta,\delta,\delta,\delta,\delta,\delta,0,0) \stackrel{1r}{\rightarrow} (\delta,0,0,\delta,\delta,0,0,0)$	$Z_{17}^{(i)}$	2^{-24}
$(\delta,\delta,\delta,\delta,\delta,\delta,0,\delta) \stackrel{1r}{\rightarrow} (0,\delta,\delta,0,0,\delta,\delta,\delta)$	$Z_{17}^{(i)}, Z_{18}^{(i)}$	2^{-32}
$(\delta,\delta,\delta,\delta,\delta,\delta,\delta,0) \stackrel{1r}{\rightarrow} (\delta,0,0,\delta,\delta,0,0,\delta)$	$Z_{17}^{(i)}, Z_{18}^{(i)}$	2^{-32}
$(\delta,\delta,\delta,\delta,\delta,\delta,\delta,\delta) \stackrel{1r}{\rightarrow} (0,\delta,\delta,0,0,\delta,\delta,0)$	$Z_{17}^{(i)}$	2^{-40}

an even-numbered round i, the following 9 subkeys have to be weak: $Z_{18}^{(i)}$, $Z_1^{(i+1)}$, $Z_3^{(i+1)}$, $Z_5^{(i+1)}$, $Z_7^{(i+1)}$, $Z_{10}^{(i+1)}$, $Z_{12}^{(i+1)}$, $Z_{14}^{(i+1)}$ and $Z_{18}^{(i+1)}$.

- $(0,0,0,0,0,0,0,0,0,0,0,0,0,0,\delta,0) \stackrel{1r}{\rightarrow} (\delta,\ \delta,\ 0,\ 0,\ 0,\ 0,\ 0,\ \delta,\ 0,\ 0,\ 0,\ 0,\ 0,\ 0,\ 0,\ 0) \stackrel{1r}{\rightarrow} (0,\delta,\delta,\delta,\delta,\delta,\delta,\delta,0,\delta,\delta,\delta,\delta,\delta,\delta,0) \stackrel{1r}{\rightarrow} (\delta,\ \delta,\ 0,\ 0,\ 0,\ 0,\ 0,\ 0,\ 0,\ 0,\ 0,\ 0,\ 0,\ 0,\ 0,\ \delta) \stackrel{1r}{\rightarrow} (0,\delta,\delta,\delta,\delta,\delta,\delta,\delta,0,\delta,\delta,\delta,\delta,\delta,0,\delta) \stackrel{1r}{\rightarrow} (\delta,\ \delta,\ \delta,\ \delta,\ \delta,\ \delta,$

Table 3.42 First 32 nontrivial 1-round characteristics for MESH-128(8).

1-round characteristic	Weak subkeys $Z_j^{(i)}$)		Prob.
	j (odd i)	j (even i)	
$(0,0,0,0,0,0,0,0,0,0,0,0,0,0,0,\delta) \xrightarrow{1r} (\delta,\delta,\delta,\delta,\delta,\delta,\delta,\delta,\delta,\delta,\delta,\delta,\delta,\delta,\delta,0)$	16, 18	18	2^{-24}
$(0,0,0,0,0,0,0,0,0,0,0,0,0,0,\delta,0) \xrightarrow{1r} (\delta,\delta,0,0,0,0,0,\delta,0,0,0,0,0,0,0,0)$	18	15, 18	2^{-8}
$(0,0,0,0,0,0,0,0,0,0,0,0,0,0,\delta,\delta) \xrightarrow{1r} (0,0,\delta,\delta,\delta,\delta,\delta,0,\delta,\delta,\delta,\delta,\delta,\delta,\delta,0)$	16	15	2^{-32}
$(0,0,0,0,0,0,0,0,0,0,0,0,0,\delta,0,0) \xrightarrow{1r} (\delta,\delta,0,\delta,\delta,\delta,0,\delta,0,\delta,\delta,\delta,\delta,\delta,\delta,\delta)$	14, 18	18	2^{-32}
$(0,0,0,0,0,0,0,0,0,0,0,0,0,\delta,0,\delta) \xrightarrow{1r} (0,0,\delta,0,0,0,\delta,0,\delta,0,0,0,0,0,0,\delta)$	14, 16	—	2^{-15}
$(0,0,0,0,0,0,0,0,0,0,0,0,0,\delta,\delta,0) \xrightarrow{1r} (0,0,0,\delta,\delta,\delta,0,0,0,\delta,\delta,\delta,\delta,\delta,\delta,\delta)$	14	15	2^{-31}
$(0,0,0,0,0,0,0,0,0,0,0,0,0,\delta,\delta,\delta) \xrightarrow{1r} (\delta,\delta,\delta,0,0,0,\delta,\delta,\delta,0,0,0,0,0,0,\delta)$	14, 16, 18	15, 18	2^{-14}
$(0,0,0,0,0,0,0,0,0,0,0,0,\delta,0,0,0) \xrightarrow{1r} (\delta,\delta,0,\delta,0,\delta,0,0,0,\delta,0,0,0,0,0,0)$	18	13, 18	2^{-24}
$(0,0,0,0,0,0,0,0,0,0,0,0,\delta,0,0,\delta) \xrightarrow{1r} (0,0,\delta,0,\delta,0,\delta,\delta,\delta,0,\delta,\delta,\delta,\delta,\delta,0)$	16	13	2^{-39}
$(0,0,0,0,0,0,0,0,0,0,0,0,\delta,0,\delta,0) \xrightarrow{1r} (0,0,0,\delta,0,\delta,0,\delta,0,\delta,0,0,0,0,0,0)$	—	13, 15	2^{-23}
$(0,0,0,0,0,0,0,0,0,0,0,0,\delta,0,\delta,\delta) \xrightarrow{1r} (\delta,\delta,\delta,0,\delta,0,\delta,0,\delta,0,\delta,\delta,\delta,\delta,\delta,0)$	16, 18	13, 15, 18	2^{-38}
$(0,0,0,0,0,0,0,0,0,0,0,0,\delta,\delta,0,0) \xrightarrow{1r} (0,0,0,0,\delta,0,0,\delta,0,0,\delta,\delta,\delta,\delta,\delta,\delta)$	14	13	2^{-24}
$(0,0,0,0,0,0,0,0,0,0,0,0,\delta,\delta,0,\delta) \xrightarrow{1r} (\delta,\delta,\delta,\delta,0,\delta,\delta,0,\delta,\delta,0,0,0,0,0,\delta)$	14, 16, 18	13, 18	2^{-16}
$(0,0,0,0,0,0,0,0,0,0,0,0,\delta,\delta,\delta,0) \xrightarrow{1r} (\delta,\delta,0,0,\delta,0,0,0,0,0,\delta,\delta,\delta,\delta,\delta,\delta)$	14, 18	13, 15, 18	2^{-32}
$(0,0,0,0,0,0,0,0,0,0,0,0,\delta,\delta,\delta,\delta) \xrightarrow{1r} (0,0,\delta,\delta,0,\delta,\delta,\delta,\delta,\delta,0,0,0,0,0,\delta)$	14, 16	13, 15	2^{-24}
$(0,0,0,0,0,0,0,0,0,0,0,\delta,0,0,0,0) \xrightarrow{1r} (\delta,\delta,0,\delta,\delta,\delta,\delta,\delta,0,\delta,0,\delta,\delta,\delta,\delta,\delta)$	12, 18	18	2^{-40}
$(0,0,0,0,0,0,0,0,0,0,0,\delta,0,0,0,\delta) \xrightarrow{1r} (0,0,\delta,0,0,0,0,0,\delta,0,\delta,0,0,0,0,\delta)$	12, 16	—	2^{-30}
$(0,0,0,0,0,0,0,0,0,0,0,\delta,0,0,\delta,0) \xrightarrow{1r} (0,0,0,\delta,\delta,\delta,\delta,0,0,\delta,0,\delta,\delta,\delta,\delta,\delta)$	12	15	2^{-39}
$(0,0,0,0,0,0,0,0,0,0,0,\delta,0,0,\delta,\delta) \xrightarrow{1r} (\delta,\delta,\delta,0,0,0,0,\delta,\delta,0,\delta,0,0,0,0,\delta)$	12, 16, 18	15, 18	2^{-29}
$(0,0,0,0,0,0,0,0,0,0,0,\delta,0,\delta,0,0) \xrightarrow{1r} (0,0,0,0,0,0,\delta,0,0,0,\delta,0,0,0,0,0)$	12, 14	—	2^{-15}
$(0,0,0,0,0,0,0,0,0,0,0,\delta,0,\delta,0,\delta) \xrightarrow{1r} (\delta,\delta,\delta,\delta,\delta,\delta,0,\delta,\delta,\delta,0,\delta,\delta,\delta,\delta,0)$	12, 14, 16, 18	18	2^{-32}
$(0,0,0,0,0,0,0,0,0,0,0,\delta,0,\delta,\delta,0) \xrightarrow{1r} (\delta,\delta,0,0,0,0,\delta,\delta,0,0,\delta,0,0,0,0,0)$	12, 14, 18	15, 18	2^{-23}
$(0,0,0,0,0,0,0,0,0,0,0,\delta,0,\delta,\delta,\delta) \xrightarrow{1r} (0,0,\delta,\delta,\delta,\delta,0,0,\delta,\delta,0,\delta,\delta,\delta,\delta,0)$	12, 14, 16	15	2^{-40}
$(0,0,0,0,0,0,0,0,0,0,0,\delta,\delta,0,0,0) \xrightarrow{1r} (0,0,0,0,\delta,0,\delta,\delta,0,0,0,\delta,\delta,\delta,\delta,\delta)$	12	13	2^{-23}
$(0,0,0,0,0,0,0,0,0,0,0,\delta,\delta,0,0,\delta) \xrightarrow{1r} (\delta,\delta,\delta,\delta,0,\delta,0,0,\delta,\delta,\delta,0,0,0,0,\delta)$	12, 16, 18	13, 18	2^{-22}
$(0,0,0,0,0,0,0,0,0,0,0,\delta,\delta,0,\delta,0) \xrightarrow{1r} (\delta,\delta,0,0,\delta,0,\delta,0,0,0,0,\delta,\delta,\delta,\delta,\delta)$	12, 18	13, 15, 18	2^{-31}
$(0,0,0,0,0,0,0,0,0,0,0,\delta,\delta,0,\delta,\delta) \xrightarrow{1r} (0,0,\delta,\delta,0,\delta,0,\delta,\delta,\delta,\delta,0,0,0,0,\delta)$	12, 16	13, 15	2^{-30}
$(0,0,0,0,0,0,0,0,0,0,0,\delta,\delta,\delta,0,0) \xrightarrow{1r} (\delta,\delta,0,\delta,0,\delta,\delta,0,0,\delta,\delta,0,0,0,0,0)$	12, 14, 18	13, 18	2^{-30}
$(0,0,0,0,0,0,0,0,0,0,0,\delta,\delta,\delta,0,\delta) \xrightarrow{1r} (0,0,\delta,0,\delta,0,0,\delta,\delta,0,0,\delta,\delta,\delta,\delta,0)$	12, 14, 16	13	2^{-38}
$(0,0,0,0,0,0,0,0,0,0,0,\delta,\delta,\delta,\delta,0) \xrightarrow{1r} (0,0,0,\delta,0,\delta,\delta,\delta,0,\delta,\delta,0,0,0,0,0)$	12, 14	13, 15	2^{-29}
$(0,0,0,0,0,0,0,0,0,0,0,\delta,\delta,\delta,\delta,\delta) \xrightarrow{1r} (\delta,\delta,\delta,0,\delta,0,0,0,\delta,0,0,\delta,\delta,\delta,\delta,0)$	12, 14, 16, 18	13, 15, 18	2^{-37}
$(0,0,0,0,0,0,0,0,0,0,\delta,0,0,0,0,0) \xrightarrow{1r} (\delta,\delta,0,0,0,\delta,0,0,0,\delta,0,\delta,0,0,0,0)$	18	11, 18	2^{-40}

$\delta, 0, \delta, \delta, \delta, \delta, \delta, \delta, \delta, \delta) \stackrel{1r}{\rightarrow} (0,0,0,0,0,0,0,0,0,0,0,0,0,0,\delta,0)$, which is 6-round iterative and holds with probability 2^{-112}.

Starting from an odd-numbered round, the following 31 subkeys have to be weak: $Z_{18}^{(i)}$, $Z_{2}^{(i+1)}$, $Z_{8}^{(i+1)}$, $Z_{17}^{(i+1)}$, $Z_{18}^{(i+1)}$, $Z_{3}^{(i+2)}$, $Z_{5}^{(i+2)}$, $Z_{7}^{(i+2)}$, $Z_{10}^{(i+2)}$, $Z_{12}^{(i+2)}$, $Z_{14}^{(i+2)}$, $Z_{18}^{(i+2)}$, $Z_{2}^{(i+3)}$, $Z_{17}^{(i+3)}$, $Z_{18}^{(i+3)}$, $Z_{3}^{(i+4)}$, $Z_{5}^{(i+4)}$, $Z_{7}^{(i+4)}$, $Z_{10}^{(i+4)}$, $Z_{12}^{(i+4)}$, $Z_{14}^{(i+4)}$, $Z_{16}^{(i+4)}$, $Z_{18}^{(i+4)}$, $Z_{2}^{(i+5)}$, $Z_{4}^{(i+5)}$, $Z_{6}^{(i+5)}$, $Z_{9}^{(i+5)}$, $Z_{11}^{(i+5)}$, $Z_{13}^{(i+5)}$, $Z_{15}^{(i+5)}$ and $Z_{18}^{(i+5)}$.

Starting from an even-numbered round, the following 33 subkeys have to be weak: $Z_{15}^{(i)}$, $Z_{18}^{(i)}$, $Z_{1}^{(i+1)}$, $Z_{17}^{(i+1)}$, $Z_{18}^{(i+1)}$, $Z_{2}^{(i+2)}$, $Z_{4}^{(i+2)}$, $Z_{6}^{(i+2)}$, $Z_{8}^{(i+2)}$, $Z_{11}^{(i+2)}$, $Z_{13}^{(i+2)}$, $Z_{15}^{(i+2)}$, $Z_{18}^{(i+2)}$, $Z_{1}^{(i+3)}$, $Z_{16}^{(i+3)}$, $Z_{17}^{(i+3)}$, $Z_{18}^{(i+3)}$, $Z_{2}^{(i+4)}$, $Z_{4}^{(i+4)}$, $Z_{6}^{(i+4)}$, $Z_{8}^{(i+4)}$, $Z_{11}^{(i+4)}$, $Z_{13}^{(i+4)}$, $Z_{18}^{(i+4)}$, $Z_{1}^{(i+5)}$, $Z_{3}^{(i+5)}$, $Z_{5}^{(i+5)}$, $Z_{7}^{(i+5)}$, $Z_{10}^{(i+5)}$, $Z_{12}^{(i+5)}$, $Z_{14}^{(i+5)}$, $Z_{16}^{(i+5)}$ and $Z_{18}^{(i+5)}$.

- $(0,0,0,0,0,0,0,0,0,0,0,0,0,0,\delta,\delta) \stackrel{1r}{\rightarrow} (0, 0, \delta, \delta, \delta, \delta, \delta, 0, \delta, \delta, \delta, \delta, \delta, \delta, \delta, 0) \stackrel{1r}{\rightarrow} (0,\delta,\delta,\delta,\delta,\delta,\delta,\delta,0,\delta,\delta,\delta,\delta,\delta,\delta,\delta) \stackrel{1r}{\rightarrow} (0, 0, \delta, \delta, \delta, \delta, \delta, \delta, \delta, \delta, \delta, \delta, \delta, \delta, \delta, \delta) \stackrel{1r}{\rightarrow} (0,\delta,\delta,\delta,\delta,\delta,\delta,\delta,0,\delta,\delta,\delta,\delta,\delta,0,0) \stackrel{1r}{\rightarrow} (0, 0, 0, 0, 0, 0, 0, \delta, 0, 0, 0, 0, 0, 0, 0, \delta) \stackrel{1r}{\rightarrow} (0,0,0,0,0,0,0,0,0,0,0,0,0,0,\delta,\delta)$, which is 6-round iterative and holds with probability 2^{-64}.

  Starting from an odd-numbered round, the following 30 subkeys have to be weak: $Z_{16}^{(i)}$, $Z_{4}^{(i+1)}$, $Z_{6}^{(i+1)}$, $Z_{9}^{(i+1)}$, $Z_{11}^{(i+1)}$, $Z_{13}^{(i+1)}$, $Z_{15}^{(i+1)}$, $Z_{17}^{(i+1)}$, $Z_{3}^{(i+2)}$, $Z_{5}^{(i+2)}$, $Z_{7}^{(i+2)}$, $Z_{10}^{(i+2)}$, $Z_{12}^{(i+2)}$, $Z_{14}^{(i+2)}$, $Z_{16}^{(i+2)}$, $Z_{4}^{(i+3)}$, $Z_{6}^{(i+3)}$, $Z_{8}^{(i+3)}$, $Z_{9}^{(i+3)}$, $Z_{11}^{(i+3)}$, $Z_{13}^{(i+3)}$, $Z_{15}^{(i+3)}$, $Z_{17}^{(i+3)}$, $Z_{3}^{(i+4)}$, $Z_{5}^{(i+4)}$, $Z_{7}^{(i+4)}$, $Z_{10}^{(i+4)}$, $Z_{12}^{(i+4)}$, $Z_{14}^{(i+4)}$, and $Z_{8}^{(i+5)}$.

  Starting from an even-numbered round, the following 30 subkeys have to be weak: $Z_{15}^{(i)}$, $Z_{3}^{(i+1)}$, $Z_{5}^{(i+1)}$, $Z_{7}^{(i+1)}$, $Z_{10}^{(i+1)}$, $Z_{12}^{(i+1)}$, $Z_{14}^{(i+1)}$, $Z_{17}^{(i+1)}$, $Z_{2}^{(i+2)}$, $Z_{4}^{(i+2)}$, $Z_{6}^{(i+2)}$, $Z_{8}^{(i+2)}$, $Z_{11}^{(i+2)}$, $Z_{13}^{(i+2)}$, $Z_{15}^{(i+2)}$, $Z_{3}^{(i+3)}$, $Z_{5}^{(i+3)}$, $Z_{7}^{(i+3)}$, $Z_{10}^{(i+3)}$, $Z_{12}^{(i+3)}$, $Z_{14}^{(i+3)}$, $Z_{16}^{(i+3)}$, $Z_{17}^{(i+3)}$, $Z_{2}^{(i+4)}$, $Z_{4}^{(i+4)}$, $Z_{6}^{(i+4)}$, $Z_{8}^{(i+4)}$, $Z_{11}^{(i+4)}$, $Z_{13}^{(i+4)}$ and $Z_{16}^{(i+5)}$.

Starting from the first round, $i = 1$, the characteristics that require the least number of weak subkeys are:

(i) $(0,0,0,0,0,0,0,0,0,0,\delta,0,\delta,0,0,0) \stackrel{1r}{\rightarrow} (0, 0, 0, \delta, 0, 0, 0, 0, 0, 0, 0, \delta, 0, 0, 0, 0) \stackrel{1r}{\rightarrow} (0,0,0,0,\delta,0,0,0,0,0,\delta,0,0,0,0,0) \stackrel{1r}{\rightarrow} (0, 0, 0, \delta, 0, \delta, 0, 0, 0, 0, 0, 0, 0, 0, 0, 0) \stackrel{1r}{\rightarrow} (0,0,0,0,\delta,0,0,0,0,0,0,0,\delta,0,0,0) \stackrel{1r}{\rightarrow} (0, 0, 0, 0, 0, \delta, 0, 0, 0, 0, 0, \delta, 0, 0, 0, 0) \stackrel{1r}{\rightarrow} (0,0,0,0,0,0,0,0,0,0,\delta,0,\delta,0,0,0)$, which is 6-round iterative and hold with probability 2^{-76}.

For odd-numbered rounds, the following six subkeys have to be weak: $Z_{4}^{(i+1)}$, $Z_{5}^{(i+2)}$, $Z_{4}^{(i+3)}$, $Z_{6}^{(i+3)}$, $Z_{5}^{(i+4)}$ and $Z_{6}^{(i+5)}$. For even-numbered rounds, the following six subkeys have to be weak: $Z_{11}^{(i)}$, $Z_{13}^{(i)}$, $Z_{12}^{(i+1)}$, $Z_{11}^{(i+2)}$, $Z_{13}^{(i+4)}$ and $Z_{12}^{(i+5)}$.

(ii) $(0,0,0,0,0,0,0,0,0,0,0,\delta,0,\delta,0,0) \stackrel{1r}{\rightarrow} (0, 0, 0, 0, 0, 0, \delta, 0, 0, 0, \delta, 0, 0, 0, 0, 0) \stackrel{1r}{\rightarrow} (0,0,0,0,0,\delta,0,0,0,\delta,0,\delta,0,\delta,0,0) \stackrel{1r}{\rightarrow} (0, 0, \delta, 0, \delta, 0, 0, 0, 0, 0, 0, 0, 0, 0, 0, 0) \stackrel{1r}{\rightarrow} (0,0,0,0,0,\delta,0,0,0,\delta,0,0,0,0,0,0) \stackrel{1r}{\rightarrow} (0, 0, \delta, 0, \delta, 0, \delta, 0, 0, 0, \delta, 0, 0, 0, 0, 0) \stackrel{1r}{\rightarrow} (0,0,0,0,0,0,0,0,0,0,0,\delta,0,\delta,0,0)$, which is 6-round iterative and hold with probability 2^{-145}.
For odd-numbered rounds, the following eight subkeys have to be weak: $Z_{12}^{(i)}$, $Z_{14}^{(i)}$, $Z_{11}^{(i+1)}$, $Z_{10}^{(i+2)}$, $Z_{12}^{(i+2)}$, $Z_{14}^{(i+2)}$, $Z_{10}^{(i+4)}$ and $Z_{11}^{(i+5)}$.
For even-numbered rounds, the following eight subkeys have to be weak: $Z_7^{(i+1)}$, $Z_6^{(i+2)}$, $Z_3^{(i+3)}$, $Z_5^{(i+3)}$, $Z_6^{(i+4)}$, $Z_3^{(i+5)}$, $Z_5^{(i+5)}$ and $Z_7^{(i+5)}$.

Unlike the key schedule algorithms of PES and IDEA, the key schedule of MESH-128(8) does not consist of a fixed bit permutation, but is rather more involved. See Sect. 2.3.10.

Experiments with mini-versions of the key schedule for a scaled-down version of MESH-64(8) operating on 16-bit blocks and 32-bit keys indicated that each weak-subkey restriction (with the three most significant bits equal to zero) is satisfied for a fraction of 2^{-3} of the key space. This key schedule is very similar to the one for MESH-128(8). Assuming this behavior holds for each weak subkey independently, and for larger key schedules as well, such as the one for MESH-128(8), then t weak subkeys would apply to a fraction of $2^{256-t\cdot 7}$ of the key space, or a weak-key class of this size. Thus, for a nonempty weak-key class, t must satisfy $256 - t \cdot 7 \geq 0$, that is, $t \leq 36$.

On the other hand, the characteristic that holds with the highest probability, under weak-key assumptions, follows a wide trail: $(\delta, \delta, \delta, \delta, \delta, \delta, \delta, \delta, \delta, \delta, \delta, \delta, \delta, \delta, \delta, \delta) \stackrel{1r}{\rightarrow} (\delta, \delta, \delta, \delta, \delta, \delta, \delta, \delta, \delta, \delta, \delta, \delta, \delta, \delta, \delta, \delta)$, but it requires eight weak subkeys per round.

So, the best trade-off is the characteristic (i).

Extending the characteristic in (i) across 8.5 rounds leads to a probability of 2^{-99} and nine weak subkeys: $Z_4^{(1)}$, $Z_5^{(2)}$, $Z_4^{(3)}$, $Z_6^{(3)}$, $Z_5^{(4)}$, $Z_6^{(5)}$, $Z_4^{(7)}$, $Z_5^{(8)}$ and $Z_4^{(9)}$. The weak-key class size is estimated as $2^{256-9\cdot 7} = 2^{193}$.

For a 128-bit block cipher, one could produce up to $2^{14\cdot 8} \cdot 2^{16}(2^{16}-1)/2 \approx 2^{143}$ plaintext pairs, taking into account the 14 trivial difference bytes and two nontrivial difference bytes in characteristic (i). Thus, one can distinguish the full 8.5-round MESH-128(8) from a random permutation (under nine weak subkeys) and using about 2^{99} plaintext pairs.

Although the full 8.5-round MESH-128(8) cipher can be covered by characteristic (i), the attack complexity is not polynomial in the cipher parameters. See Sect. 1.7.

This differential analysis using characteristics also allows us to study the diffusion power of MESH-128(8) with respect to the differential branch number (Appendix, Sect. B).

Analyzing the set of 65535 1-round characteristics for MESH-128(8), we found out that the differential branch number *per round* of MESH-128(8) is only 4 (under weak-key assumptions). For instance, the sum of the wordwise Hamming Weight of the input and output differences of the 1-round characteristic

$$(\delta,0,0,0,0,0,\delta,0,\delta,0,0,0,0,0,0,0) \stackrel{1r}{\rightarrow} (0,0,0,0,0,0,0,0,0,0,0,0,0,\delta,0,0)$$

is 4. So, even though MESH-128(8) provides full diffusion in a single round, the previous analysis shows that its differential branch number is much lower than ideal. For a 16-word block, such as in MESH-128(8), an MDS code using a 16×16 MDS matrix would rather provide a branch number of $1+16=17$. See Appendix, Sect. B.1, for an explanation of the differential branch number.

3.5.8 DC of WIDEA-n Under Weak-Key Assumptions

The analyses in this section were adapted from [145, 113].

Information about the WIDEA-n ciphers is provided in Sect. 2.5.

A differential analysis of WIDEA-n uses a bottom-up strategy with Table 3.43 that exhaustively lists all 1-round differential characteristics of (one IDEA instance of) WIDEA-4, in the encryption direction, using wordwise difference $\delta = 8000_x$, exclusive-or as difference operator and under weak-key assumptions. The subscript x indicates hexadecimal value.

In general, a weak key of a block cipher is a user key which leads to a nonrandom behavior of the cipher. Ideally, a block cipher should have no weak keys. In particular, for IDEA and WIDEA-n, a weak key causes some multiplicative subkeys to have value 0 or 1. In both IDEA and WIDEA-n, these multiplicative subkeys are mandatory inputs to the $\odot$ operation over GF$(2^{16}+1)$, where $0 \equiv 2^{16}$ by construction [117].

The difference 8000_x propagates across $\oplus$ and $\boxplus$ with certainty for any subkey value because the only active bit difference is in the most significant bit position.

All these characteristics hold with probability 1 under weak-key assumptions. Thus, the main purpose of weak keys is that they lead to weak subkeys in specific positions across WIDEA-n, allowing differential trails to hold with certainty. The third column in Table 3.43 shows the difference propagation across the MA-box (of one IDEA instance) and consequently, shows if the MA-box is differentially active or not. An MA/MAD-box is *differentially active* if its input difference is nonzero. Otherwise, it is called passive.

The best characteristics are distinguished based on some criteria: (i) minimize the number of weak subkeys per round; (ii) reach the largest number of rounds; (iii) lead to iterative difference patterns. Under these conditions, the best choices include the 3-round characteristic

$$(0,0,\delta,\delta) \stackrel{1r}{\rightarrow} (0,\delta,\delta,0) \stackrel{1r}{\rightarrow} (0,\delta,0,\delta) \stackrel{1r}{\rightarrow} (0,0,\delta,\delta), \qquad (3.10)$$

Table 3.43 1-round characteristics across one IDEA instance of WIDEA-4 ($\delta = 8000_x$).

1-round characteristics	Weak Subkeys (j-th round)	diff. across MA-box
$(0,0,0,\delta) \stackrel{1r}{\to} (\delta,\delta,\delta,0)$	$Z_{6(j-1)+3}, Z_{6(j-1)+5}$	$(0,\delta) \to (\delta,\delta)$
$(0,0,\delta,0) \stackrel{1r}{\to} (\delta,0,0,0)$	$Z_{6(j-1)+4}, Z_{6(j-1)+5}$	$(\delta,0) \to (0,\delta)$
$(0,0,\delta,\delta) \stackrel{1r}{\to} (0,\delta,\delta,0)$	$Z_{6(j-1)+3}, Z_{6(j-1)+4}$	$(\delta,\delta) \to (\delta,0)$
$(0,\delta,0,0) \stackrel{1r}{\to} (\delta,\delta,0,\delta)$	$Z_{6(j-1)+5}$	$(0,\delta) \to (\delta,\delta)$
$(0,\delta,0,\delta) \stackrel{1r}{\to} (0,0,\delta,\delta)$	$Z_{6(j-1)+3}$	$\mathbf{(0,0) \to (0,0)}$
$(0,\delta,\delta,0) \stackrel{1r}{\to} (0,\delta,0,\delta)$	$Z_{6(j-1)+4}$	$(\delta,\delta) \to (\delta,0)$
$(0,\delta,\delta,\delta) \stackrel{1r}{\to} (\delta,0,\delta,\delta)$	$Z_{6(j-1)+3}, Z_{6(j-1)+4}, Z_{6(j-1)+5}$	$(\delta,0) \to (0,\delta)$
$(\delta,0,0,0) \stackrel{1r}{\to} (0,\delta,0,0)$	$Z_{6(j-1)}, Z_{6(j-1)+4}, Z_{6(j-1)+5}$	$(\delta,0) \to (0,\delta)$
$(\delta,0,0,\delta) \stackrel{1r}{\to} (\delta,0,\delta,0)$	$Z_{6(j-1)}, Z_{6(j-1)+3}, Z_{6(j-1)+4}$	$(\delta,\delta) \to (\delta,0)$
$(\delta,0,\delta,0) \stackrel{1r}{\to} (\delta,\delta,0,0)$	$Z_{6(j-1)}$	$\mathbf{(0,0) \to (0,0)}$
$(\delta,0,\delta,\delta) \stackrel{1r}{\to} (0,0,\delta,0)$	$Z_{6(j-1)}, Z_{6(j-1)+3}, Z_{6(j-1)+5}$	$(0,\delta) \to (\delta,\delta)$
$(\delta,\delta,0,0) \stackrel{1r}{\to} (\delta,0,0,\delta)$	$Z_{6(j-1)}, Z_{6(j-1)+4}$	$(\delta,\delta) \to (\delta,0)$
$(\delta,\delta,0,\delta) \stackrel{1r}{\to} (0,\delta,\delta,\delta)$	$Z_{6(j-1)}, Z_{6(j-1)+3}, Z_{6(j-1)+4}, Z_{6(j-1)+5}$	$(\delta,0) \to (0,\delta)$
$(\delta,\delta,\delta,0) \stackrel{1r}{\to} (0,0,0,\delta)$	$Z_{6(j-1)}, Z_{6(j-1)+5}$	$(0,\delta) \to (\delta,\delta)$
$(\delta,\delta,\delta,\delta) \stackrel{1r}{\to} (\delta,\delta,\delta,\delta)$	$Z_{6(j-1)}, Z_{6(j-1)+3}$	$\mathbf{(0,0) \to (0,0)}$

with four weak subkeys: $Z_{6(j-1)+3}$, $Z_{6(j-1)+4}$, Z_{6j+4}, $Z_{6(j+1)+3}$ starting from round j, for $j \geq 1$. The terminology $Z_i^{(j)}$ denotes the i-subkey of the j-th round in IDEA. The notation Z_l applies to WIDEA-n, where $l = 6(j-1) + i - 1$, since there are six subkeys per round in IDEA.

All rotations of (3.10), for instance, starting from $(0,\delta,\delta,0)$ instead of $(0,0,\delta,\delta)$, result in equivalent characteristics.

Another relevant choice is the 1-round iterative characteristic

$$(\delta,\delta,\delta,\delta) \stackrel{1r}{\to} (\delta,\delta,\delta,\delta), \tag{3.11}$$

with two weak-subkey assumptions: $Z_{6(j-1)}$, $Z_{6(j-1)+3}$ starting from round j.

Next, we describe attacks on WIDEA-n that bypass all the MDS matrices across the full 8.5-round WIDEA-n.

3.5.8.1 Differential Attack Using One IDEA Instance Only

We use (3.11), a 1-round iterative characteristic whose differential trail bypasses *all* MA-boxes, that is, all MA-boxes are *passive* (the input and output differences are zero). In Table 3.43, this fact is depicted with differences in boldface. Extending it to WIDEA-4, one IDEA instance will follow the differential pattern (3.11), while the other three IDEA instances will have only the trivial input difference. See the trail in red in Fig. 3.4. This means that all MAD-boxes will be passive. In other words, the (trivial) fixed point is exploited

$$MDS(0,0,0,0)^T = (0,0,0,0)^T,$$

where the superscript T denotes the transposition operation. In other words, if the weak-subkey assumptions are satisfied, then the differential trail (3.11) concatenated with itself will propagate across a single 8.5-round IDEA instance (out of n instances) in WIDEA-n, effectively bypassing the MDS diffusion layer in every round. This attack holds independent of which MDS matrix is used. Note that our attack does not contradict the branch number of the MDS matrix [71], but rather exploit the trivial fixed point.

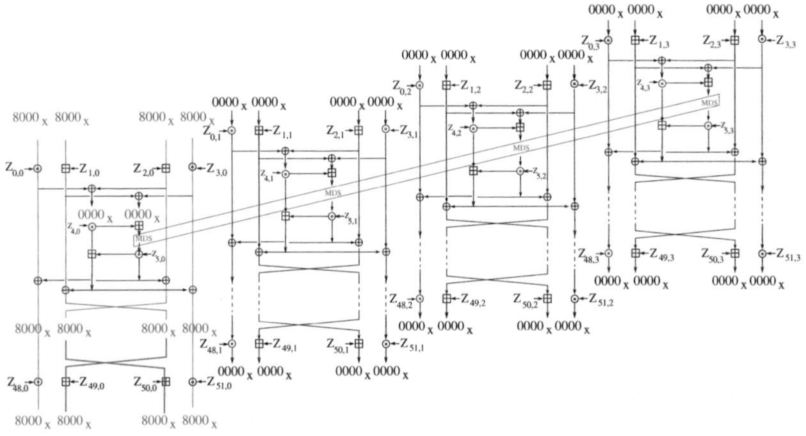

Fig. 3.4 Examples of a narrow differential trail (depicted in red) in WIDEA-4 using a fix-point of the MAD-box.

Thus, the following 1-round iterative characteristic results, using (3.11), activates a single IDEA instance of WIDEA-4:

$$\begin{pmatrix} 0\,0\,0\,0 \\ 0\,0\,0\,0 \\ 0\,0\,0\,0 \\ \delta\,\delta\,\delta\,\delta \end{pmatrix} \overset{1r}{\rightarrow} \begin{pmatrix} 0\,0\,0\,0 \\ 0\,0\,0\,0 \\ 0\,0\,0\,0 \\ \delta\,\delta\,\delta\,\delta \end{pmatrix}. \tag{3.12}$$

Without loss of generality, only the last row of the state was activated. The same reasoning applies if any of the other IDEA instances were chosen (although the weak-key class would not be the same). Thus, the fifteen most significant bits (MSB) of the following eighteen subkeys should be zero across 8.5 rounds of one IDEA instance: $Z_{6(j-1)}$, $Z_{6(j-1)+3}$, for $1 \leq j \leq 9$. Under these conditions, the output difference pattern in (3.12) will appear as ciphertext difference with certainty. In any case, (3.12) demonstrates a narrow differential trail in WIDEA-4 and thus, serves as concrete evidence that text diffusion across a single round of WIDEA-4 is very weak, even with the use of MDS codes.

Assuming the key schedule of WIDEA-4 behaves as a random mapping and the weak-subkey conditions hold independently, then in WIDEA-4's key space of size 2^{512}, the weak subkeys would represent a class of about $2^{512-15\cdot 18} = 2^{242}$ keys. The reasoning is that each 16-bit weak subkey can be either 0 or 1, and eighteen of them need to be weak. For WIDEA-8, the same reasoning applies, but the weak-key class size is estimated at $2^{1024-15\cdot 18} = 2^{754}$.

In either case, it is possible to distinguish the full 8.5-round WIDEA-4 from a random permutation π, under weak-key assumptions, with low attack complexity, due to the characteristics exploiting narrow trails and covering the full-round cipher with certainty.

The reasoning is the following: given the input difference of characteristic (3.12) to both the full 8.5-round WIDEA-4 (under a weak key) and a random 64-bit permutation π, the output difference of (3.12) will show up with certainty for WIDEA-4, while for π, it will happen with probability about $2^{-4\cdot 64} = 2^{-256}$. The advantage (see Sect. 1.7) to distinguish one from the other is: $1 - 2^{-256}$, which is non-negligible. Therefore, the full 8.5-round WIDEA-4 cipher is not an ideal primitive, that is, it is not a PRP. A similar reasoning applies to WIDEA-8.

In order to avoid this distinguishing attack using (3.12), more than $512/15 = 34$ weak-subkey restrictions would be required, since each weak subkey implies that its fifteen most significant bits be zero. This means that WIDEA-4 would need more than $34/2 = 17$ rounds, which means more than double the original number of rounds, since each round requires two weak subkeys. But, in this case, performance would decrease by 50%.

A (partial) key-recovery attack on the full 8.5-round WIDEA-4, using (3.12), can obtain the subkeys $Z_{48,0}$ and $Z_{51,0}$ of the last MA half-round. In this case, only sixteen subkeys need be weak, which implies a weak-key class of about $2^{512-15\cdot 16} = 2^{272}$ keys. The output difference after eight rounds, restricted to one IDEA instance, is $(\delta, \delta, \delta, \delta)$. For the additive subkeys, the δ difference propagates across to the ciphertext, but not for $Z_{48,0}$ and $Z_{51,0}$. This means a 16-bit condition for each 16-bit subkey. Thus, one chosen pair of texts is enough. Decrypting two multiplications in a half-round in one IDEA instance is equivalent to $\frac{1}{17\cdot 2\cdot 4}$ of a full WIDEA-4, that is, the cost becomes $\frac{2^{32}}{17\cdot 8} \approx 2^{25}$ WIDEA-4 encryptions. There are 17 half-rounds in 8.5-round IDEA. Memory cost is negligible. Time complexity for WIDEA-8 becomes 2^{24} encryptions since WIDEA-8 contains eight copies of IDEA. Moreover, the weak-key class size becomes $2^{1024-15\cdot 16} = 2^{784}$. Recovering subkeys from the other IDEA instances is not possible because of the zero differences in (3.12). If we shift the $(\delta, \delta, \delta, \delta)$ pattern to another row of the state in (3.12), then *another* weak key would be required that causes weak subkeys in that same part of the cipher state.

3.5.8.2 Differential Attack Using All IDEA Instances

The following analyses apply to WIDEA-4. Using the 3-round iterative linear relation (3.10) means that there are active MAD-boxes along the differential trail. See Table 3.43. This means we need to exploit another fixed point of the AES MDS matrix:

$$\mathrm{MDS}(\delta,\delta,\delta,\delta)^T = (\delta,\delta,\delta,\delta)^T.$$

When a MAD-box is active, all four copies in WIDEA-4 have to be attacked simultaneously so that the same value appears inside each MAD-box. This means that the input to the active MDS matrices is $(\delta,\delta,\delta,\delta)$. Applying it to the MDS matrix results in $2\cdot\delta \oplus 3\cdot\delta \oplus \delta \oplus \delta = (2\oplus 3\oplus 1\oplus 1)\cdot\delta = \delta$ in all four rows since the MDS matrix in AES is circulant.

In other words, this fixed point exploits the fact that the exclusive-or of the coefficients in a line (or column) of the MDS matrix exclusive-ors to 1. This is an important interaction between the differential pattern $(\delta,\delta,\delta,\delta)$ and the coefficients of the AES MDS matrix.

This attack does not contradict the branch number of the MDS matrix [71], but rather exploit a nontrivial fixed point.

Note that this property does not hold for the 8×8 MDS matrix in WIDEA-8 since in the latter, the exclusive-or sum of the coefficients in a line or column is 3. A consequence of this finding is an additional criterion for block cipher designs that employ matrices from MDS codes in the way they are used in WIDEA-4: *carefully select the coefficients in these matrices in order to avoid fixed-point differences.*

Thus, we arrive at the following 3-round iterative characteristic:

$$\begin{pmatrix} 0\,\delta\,0\,\delta \\ 0\,\delta\,0\,\delta \\ 0\,\delta\,0\,\delta \\ 0\,\delta\,0\,\delta \end{pmatrix} \stackrel{1r}{\rightarrow} \begin{pmatrix} 0\,0\,\delta\,\delta \\ 0\,0\,\delta\,\delta \\ 0\,0\,\delta\,\delta \\ 0\,0\,\delta\,\delta \end{pmatrix} \stackrel{1r}{\rightarrow} \begin{pmatrix} 0\,\delta\,\delta\,0 \\ 0\,\delta\,\delta\,0 \\ 0\,\delta\,\delta\,0 \\ 0\,\delta\,\delta\,0 \end{pmatrix} \stackrel{1r}{\rightarrow} \begin{pmatrix} 0\,\delta\,0\,\delta \\ 0\,\delta\,0\,\delta \\ 0\,\delta\,0\,\delta \\ 0\,\delta\,0\,\delta \end{pmatrix}. \quad (3.13)$$

In summary, we exploited a combined symmetry in the AES MDS matrix, the differential pattern and the four identical copies of IDEA in WIDEA-4. Across 8.5 rounds, the following eleven 64-bit subkeys need to be weak: Z_3, Z_9, Z_{10}, Z_{16}, Z_{21}, Z_{27}, Z_{28}, Z_{34}, Z_{39}, Z_{45} and Z_{46} across all four IDEA instances. These weak subkeys imply conditions on $11\cdot 4\cdot 15 = 660 > 512$ bits. In a key space of 2^{512} keys, we do not expect any weak-key class to satisfy all these conditions simultaneously. It is a negative result. Analogous conclusions hold for WIDEA-8.

3.5.8.3 WIDEA-n in Davies-Meyer Mode

In [91], the authors suggested to use WIDEA-n as a compression function in Davies-Meyer (DM) mode, because the key size is double the block size. The number of rounds, though, was increased from 8.5 to 10.5 to provide a larger security margin. The hash digest could range from 224 bits up to 512 bits, as in the SHA-2 hash function family [158] by truncation of the last chaining variable. The DM mode for a compression function construction is as follows [136]: the i-th chaining value is

$$H_i = H_{i-1} \oplus E_{m_i}(H_{i-1}), \tag{3.14}$$

where $H_0 = IV$ is the initial value, m_i is the i-th message block and $E_x(y)$ is a block cipher with key x and plaintext y. In particular, E is WIDEA-n, $|m_i|$ is $128n$ bits, $|H_i|$ is $64n$ bits.

The issue of weak subkeys in WIDEA-n is quite relevant in a hash function setting. In (3.14), the message to be hashed becomes the key input and can be chosen by the adversary (to be a weak key).

We point to the following consequences from the results in the previous sections when WIDEA-n is used in DM mode:

- *semi free-start collision*: suppose we can set H_{i-1} with difference (3.12) for WIDEA-4. If m_i is a weak key that leads to weak subkeys as required in Sect. 3.5.8.1, then $H_{i-1} = E_{m_i}(H_{i-1})$, that is, H_i contains only zero word differences according to (3.14). It is a semi free-start collision because only the chaining variable has nonzero difference [136]. The same reasoning applies to WIDEA-8, using the same difference (3.12), but extended to a 1024-bit state. Note that this attack is independent of the MDS matrix used.
- *truncation*: suppose the output transformation in a (hypothetical) hash function using WIDEA-n in the compression function simply truncates the output to the least significant 192 bits for WIDEA-4, or to the least significant 448 bits for WIDEA-8 (2.17). In both cases, we assume that at least 64 bits are cut off from the last chaining variable. More bits can be dropped, but 64 bits is enough for our attack. Suppose we have a differential trail like (3.12) in H_{i-1} but with weak subkeys up to the 8th round. These trails reduce the number of required weak subkeys to sixteen instead of eighteen (increasing the weak-key class size), but the input and output difference patterns are not the same. This fact implies that in DM mode, the exclusive-or between H_{i-1} and $E_{m_i}(H_{i-1})$ will not vanish. But, on the other hand, the nonzero difference words are isolated in a single 64-bit piece of the state. If that 64-bit piece is in the most significant part of the state, it will be truncated and we have a collision since the rest of the state has only zero word difference. In this way, we use the output transformation of the hash function to an attacker's advantage, if we can control the difference to remain in the part of the

state that is going to be truncated. This collision has to happen in the last message block hashed. If the message length is, say, at most $2^{128}-1$ bits, then the last 128 bits are reserved for the message length.
Assume the first 64 bits are variable, so we can control the difference in it. For WIDEA-4, the remaining $256-64-128=64$ bits are reserved for padding. For WIDEA-8, the remaining $1024-64-128=832$ bits are reserved for padding. So, this is feasible, since we only need nonzero difference in the most significant 64 bits, while the rest of the state has zero difference.

- a 2-round iterative linear relation such as

$$\begin{pmatrix} 0\,0\,0\,0 \\ 0\,0\,0\,0 \\ 0\,0\,0\,0 \\ \gamma\,\gamma\,\gamma\,\gamma \end{pmatrix} \stackrel{1r}{\rightarrow} \begin{pmatrix} 0\,0\,0\,0 \\ 0\,0\,0\,0 \\ 0\,0\,0\,0 \\ \gamma\,\gamma\,\gamma\,\gamma \end{pmatrix}, \tag{3.15}$$

for WIDEA-4 $= E$, using relation (3.173), implies the linear relation $H_{i-1}\cdot\varGamma_1 = E_{m_i}(H_{i-1})\cdot\varGamma_1$, where $\varGamma_1$ is any of the masks in a state in (3.15). Applying this iterative relation to a single compression function in DM mode, leads to

$$H_i \cdot \varGamma_1 = 0, \tag{3.16}$$

that is, the linear relation does not depend on H_{i-1}, due to the feed-forward in the DM mode, and the splitting of the $\varGamma_1$ mask[10]. Relation (3.16) could be used to distinguish the compression function using WIDEA-4 in DM mode from a random function. Note that (3.16) depends only on the output H_i, and could be applied to the hash digest only if the masked bits are not truncated. This linear relation have implications for WIDEA-n in applications such as pseudorandom number generation, since the masked bit would leak information in the output bitstream.
Similar reasoning applies for WIDEA-8 as well since the reasoning concerns one single IDEA instance (out of n).

3.5.8.4 Weak Keys

Typical design criteria for key schedule algorithms include: high performance, compact code, low latency, resistance to side-channel analysis, and avoidance of weak keys and equivalent keys.

The presence of *weak keys* means a failure of both the key schedule and the encryption/decryption algorithm's designs, since the existence of bit patterns in the key should not be a threat if there is no shortcut attack in the encryption/decryption frameworks that can exploit those keys. Likewise, nonrandom cipher behavior that is unrelated to bit patterns in the key or

[10] Actually, the right-hand-side is $m_i.\varGamma$, with m_i the key input, but this value is fixed since m_i and $\varGamma$ are fixed.

in the subkeys does not mean the latter are weak. Keys are considered weak only when the apparent weakness propagates from the key schedule to the encryption algorithm, and leads to nonrandom cipher behavior and eventually to (efficient) attacks.

An important question to address is whether weak keys actually exist in WIDEA-n that can generate the weak subkeys required in the attacks in Sect. 3.5.8, (3.13.9) and (3.5.8.3). According to the key schedule algorithm of WIDEA-4 in Sect. 2, the pattern (3.12) repeated across six rounds, requires that the most significant fifteen bits (corresponding to the first IDEA instance in WIDEA-4) of the following subkeys to be zero: Z_0, Z_3, Z_6, Z_9, Z_{12}, Z_{15}, Z_{18}, Z_{21}, Z_{24}, Z_{27}, Z_{30} and Z_{33}. Satisfying Z_0, Z_3 and Z_6 means imposing restrictions on the user key. There are nine subkey conditions left, where the variables in boldface are either 0 or 1:

$$\mathbf{Z_9} = (((Z_8 \oplus Z_1) \overset{16}{\boxplus} Z_4) \overset{16}{\lll} 5) \lll 24, \tag{3.17}$$

$$\mathbf{Z_{12}} = (((Z_{11} \oplus Z_4) \overset{16}{\boxplus} Z_7) \overset{16}{\lll} 5) \lll 24, \tag{3.18}$$

$$\mathbf{Z_{15}} = (((Z_{14} \oplus Z_7) \overset{16}{\boxplus} Z_{10}) \overset{16}{\lll} 5) \lll 24, \tag{3.19}$$

$$\mathbf{Z_{18}} = (((Z_{17} \oplus Z_{10}) \overset{16}{\boxplus} Z_{13}) \overset{16}{\lll} 5) \lll 24, \tag{3.20}$$

$$\mathbf{Z_{21}} = (((Z_{20} \oplus Z_{13}) \overset{16}{\boxplus} Z_{16}) \overset{16}{\lll} 5) \lll 24, \tag{3.21}$$

$$\mathbf{Z_{24}} = ((((Z_{23} \oplus Z_{16}) \overset{16}{\boxplus} Z_{19}) \overset{16}{\lll} 5) \lll 24) \oplus C_2, \tag{3.22}$$

$$\mathbf{Z_{27}} = (((Z_{26} \oplus Z_{19}) \overset{16}{\boxplus} Z_{22}) \overset{16}{\lll} 5) \lll 24, \tag{3.23}$$

$$\mathbf{Z_{30}} = (((Z_{29} \oplus Z_{22}) \overset{16}{\boxplus} Z_{25}) \overset{16}{\lll} 5) \lll 24, \tag{3.24}$$

$$\mathbf{Z_{33}} = (((Z_{32} \oplus Z_{25}) \overset{16}{\boxplus} Z_{28}) \overset{16}{\lll} 5) \lll 24. \tag{3.25}$$

Assuming that the key schedule algorithm behaves as a random mapping, generating (approximately) uniformly distributed subkeys, we expect each subkey to have equal chance to assume a value in the range $[0, \ldots, 2^{64} - 1]$ for WIDEA-4, or $[0, \ldots, 2^{128} - 1]$ for WIDEA-8. Therefore, we assume each 16-bit subkey for each IDEA instance to have approximately the same chance to have values in the range $[0, \ldots, 2^{16} - 1]$. For the particular values 0 and 1 the chance is 2^{-16} for each. Under these assumptions (3.17)-(3.25), the weak-key class sizes were estimated in Sect. 3.5.8, (3.13.9) and (3.5.8.3).

In order to have some experimental evidence on the existence or absence of weak keys, experiments were carried out using Boolean satisfiability tools such as SAT solvers. Boolean satisfiability (SAT) refers to the problem of determining the existence of values for Boolean variables such that a given propositional formula evaluates to *true*. When trying to solve instances of this problem, different outcomes are possible: either the instance has no solution (and the solver terminates), or a solution is found (and possibly output by the solver), or the solver is unable to answer (the solver is incomplete) [11].

SAT solvers were introduced for the first time as a tool for cryptanalysis in [128]. For our purposes, a SAT solver is a deterministic algorithm that is given a formula *in Conjunctive Normal Form (CNF)* and that always outputs a satisfying valuation of its variables, in case such a valuation exists. Otherwise, the SAT solver outputs that the SAT instance is unsatisfiable (the solver is complete). A CNF is a conjunction of clauses, where a clause is a disjunction of literals, and a literal is a propositional variable or its negation.

SAT-based cryptanalysis is reminiscent of algebraic attacks, since both approaches aim at solving an encoding of the computations underlying a cryptographic algorithm in the secret variables, such as key bits. In the case of algebraic attacks, the encoding is a system of multivariate equations where variables, over the finite field GF(2) (resp. $GF(2^8)$ or $GF(2^{32})$), correspond to bits (resp. bytes or words). Then, the system of equations is solved in the secret variables using various algebraic techniques [9]. In the case of SAT-based cryptanalysis, the encoding is a CNF, where the propositional variables correspond to the bits used in the computations. This encoding is then handed to an off-the-shelf SAT solver, instead of an algebraic solver, in order to recover the value of secret variables.

Algebraic attacks are mounted in two phases: (i) generating a system of equations, (ii) solving the system in the secret variables. It is straightforward to convert a SAT instance into a system of equations over GF(2). For instance, the CNF formula $(x_1 \vee \neg x_2) \wedge (x_2 \vee \neg x_4)$ can be written as the system of equations

$$\begin{aligned} 1 &= \mathsf{ANF}(x_1 \vee \neg x_2) \\ 1 &= \mathsf{ANF}(x_2 \wedge \neg x_4) \end{aligned} \tag{3.26}$$

where $\mathsf{ANF}(\cdot)$ denotes a function that returns the Algebraic Normal Form (ANF) of the input Boolean function.

3.5.8.5 Weak Keys in WIDEA-4

Let us define an order for each IDEA instance in WIDEA-n: the one whose subkeys are in the most significant 64-bit position will be referred to as the first instance, and so forth until the n-th IDEA instance.

The state of WIDEA-4 can represented by a 4×4 matrix of 16-bit words. Each row of the state corresponds to an IDEA instance.

The best differential and linear distinguishers found for the full 8.5-round WIDEA-4, under weak-key assumptions, depicted in (3.12) and (3.174) are 1-round iterative and have the form

$$\begin{pmatrix} 0 & 0 & 0 & 0 \\ 0 & 0 & 0 & 0 \\ 0 & 0 & 0 & 0 \\ \Delta X & \Delta X & \Delta X & \Delta X \end{pmatrix} \stackrel{1r}{\rightarrow} \begin{pmatrix} 0 & 0 & 0 & 0 \\ 0 & 0 & 0 & 0 \\ 0 & 0 & 0 & 0 \\ \Delta X & \Delta X & \Delta X & \Delta X \end{pmatrix}, \quad (3.27)$$

where ΔX represents nonzero difference words. In summary, the attack focuses on a single IDEA instance (out of four) in order to minimize the number of weak-subkey assumptions. In a differential cryptanalysis setting, $\Delta X = 8000_x$ is a 16-bit xor difference.

Suppose a differential setting. For a weak key, the 1-round iterative distinguisher (3.27) holds with certainty. But, for a random, not necessarily weak key, (3.27) holds with probability $(2^{-16})^2$ per round (there are two $\odot$ operations per round). Therefore, weak keys allows us to distinguish the full 8.5-round WIDEA-4 with certainty, which means a complexity of two chosen plaintexts and equivalent encryption effort. For a random key, (3.27) repeated 8.5 times would hold with (negligible) probability $(2^{-32})^9 = 2^{-288}$. This is a concrete example of a nonrandom behavior caused by weak keys.

The corresponding weak subkeys consist of 64-bit values whose fifteen most significant bits (MSBs) are zero. In WIDEA-4, this means that the following eighteen subkeys should be weak in their most significant 16 bits: Z_0, Z_3, Z_6, Z_9, Z_{12}, Z_{15}, Z_{18}, Z_{21}, Z_{24}, Z_{27}, Z_{30}, Z_{33}, Z_{36}, Z_{39}, Z_{42}, Z_{45}, Z_{48} and Z_{51}.

For attacking different IDEA instances in WIDEA-4, the pattern of fifteen zero bits must move to other non-overlapping 16-bit pieces of the 64-bit subkey Z_i.

Table 3.44 to Table 3.47 show examples of weak 512-bit keys for WIDEA-4 involving only one of the four IDEA instances. These weak keys were found by a SAT solver called CryptoMiniSAT [172]. Values in bold indicate the weak subkey bits. The top part of Table 3.44 shows a weak user key for the first IDEA instance in WIDEA-4, while the bottom half shows the weak subkeys. The non-weak subkeys are not depicted. The 15-bit weak subkey piece is highlighted in boldface type. Similar reasoning applies to the other tables.

Table 3.44 Weak 512-bit user key and weak subkeys of WIDEA-4 for the first IDEA instance according to (3.27).

$Z_0 : \mathbf{0000}c316e08c890b_x$,	$Z_1 : 6948c5b2bf00b2ac_x$,	$Z_2 : 23c3a1b6abcc2cd4_x$,
$Z_3 : \mathbf{0000}1c314c914bdc_x$,	$Z_4 : b040247dd96b60c7_x$,	$Z_5 : 17742274e751c5e6_x$,
$Z_6 : \mathbf{0000}c420e5e0ea07_x$,	$Z_7 : 5cb031399c53f5e7_x$.	
$Z_9 : \mathbf{0000}3044e3b98425_x$,	$Z_{12} : \mathbf{0000}f9fdad0d44be_x$,	$Z_{15} : \mathbf{0000}50d144558401_x$,
$Z_{18} : \mathbf{0000}fed459e1becc_x$,	$Z_{21} : \mathbf{0000}d19360a3715d_x$,	$Z_{24} : \mathbf{0000}b1896d117091_x$,
$Z_{27} : \mathbf{0000}a9dfc4026dda_x$,	$Z_{30} : \mathbf{0000}5602a4d5e716_x$,	$Z_{33} : \mathbf{0000}1f321f43ebb1_x$,
$Z_{36} : \mathbf{0000}653cf1c8f17b_x$,	$Z_{39} : \mathbf{0000}55f818f450b2_x$,	$Z_{42} : \mathbf{0000}2a7b6bb35033_x$,
$Z_{45} : \mathbf{0000}4fa80b12afb0_x$,	$Z_{48} : \mathbf{0000}3a03feaa8739_x$,	$Z_{51} : \mathbf{0000}f3c0f427d54b_x$.

Table 3.45 Weak 512-bit user key and weak subkeys of WIDEA-4 for the second IDEA instance [113].

$Z_0 : 395d\mathbf{0000}0b1972f2_x$,	$Z_1 : 14687352eebfa3a3_x$,	$Z_2 : eda06123101bbcae_x$,
$Z_3 : 86a0\mathbf{0000}5b003dd8_x$,	$Z_4 : a386b84284567a12_x$,	$Z_5 : a469c03e47fbb15b_x$,
$Z_6 : 7324\mathbf{0000}8b1de93a_x$,	$Z_7 : d7560a9fbb483f76_x$.	
$Z_9 : 6170\mathbf{0000}498dd14f_x$,	$Z_{12} : c705\mathbf{0000}9fa883f0_x$,	$Z_{15} : 7e10\mathbf{0000}2655ab9b_x$,
$Z_{18} : f189\mathbf{0000}562e5f20_x$,	$Z_{21} : 55cd\mathbf{0000}d2ce3064_x$,	$Z_{24} : b4d6\mathbf{0000}c56fd176_x$,
$Z_{27} : 00e2\mathbf{0000}a8980eee_x$,	$Z_{30} : 1a1e\mathbf{0000}13d13a3e_x$,	$Z_{33} : 414c\mathbf{0000}fc266ff6_x$,
$Z_{36} : 4531\mathbf{0000}65bbfe76_x$,	$Z_{39} : d005\mathbf{0000}fef80032_x$,	$Z_{42} : 30b3\mathbf{0000}55671aaa_x$,
$Z_{45} : 0125\mathbf{0000}3968554f_x$,	$Z_{48} : 9574\mathbf{0000}5ddfbe99_x$,	$Z_{51} : d744\mathbf{0000}27705035_x$.

Table 3.46 Weak 512-bit user key and weak subkeys of WIDEA-4 for the third IDEA instance [113].

$Z_0 : 59c63711\mathbf{0000}05ec_x$,	$Z_1 : 7352e14d3f6f1e79_x$,	$Z_2 : 04602ecc119a73b9_x$,
$Z_3 : 63fde770\mathbf{0001}8fc5_x$,	$Z_4 : 6ed237962c1a1f3a_x$,	$Z_5 : ff9dc3e17cd9f303_x$,
$Z_6 : 210a0fbc\mathbf{0000}57a5_x$,	$Z_7 : 15e4659e944c63a5_x$.	
$Z_9 : 51b0c1f0\mathbf{0000}7f64_x$,	$Z_{12} : a3053bb3\mathbf{0001}723b_x$,	$Z_{15} : e167203a\mathbf{0001}a146_x$,
$Z_{18} : fca94ab0\mathbf{0001}29a9_x$,	$Z_{21} : 128e7dec\mathbf{0001}4652_x$,	$Z_{24} : bf57a8da\mathbf{0000}f389_x$,
$Z_{27} : 2e6357bf\mathbf{0001}b385_x$,	$Z_{30} : f2f34081\mathbf{0001}b4f9_x$,	$Z_{33} : fb2ac214\mathbf{0000}b785_x$,
$Z_{36} : 36dd251f\mathbf{0001}29f5_x$,	$Z_{39} : 5ca022ff\mathbf{0000}6dd7_x$,	$Z_{42} : f264e49e\mathbf{0001}8923_x$,
$Z_{45} : 76525390\mathbf{0000}8174_x$,	$Z_{48} : 6ea0d6fd\mathbf{0000}afb2_x$,	$Z_{51} : 5f5940bb\mathbf{0000}7c57_x$.

Table 3.47 Weak 512-bit user key and weak subkeys of WIDEA-4 for the fourth IDEA instance [113].

$Z_0 : 1d4c31917771\mathbf{0001}_x$,	$Z_1 : d94189cbd37575e4_x$,	$Z_2 : d591a3515b7bee44_x$,
$Z_3 : 5f83fa2d40b7\mathbf{0001}_x$,	$Z_4 : d283c7edd6b1a0b2_x$,	$Z_5 : dc7670a5e493c19a_x$,
$Z_6 : b8efbcfe3d5c\mathbf{0001}_x$,	$Z_7 : aae76a1b39f1fe8e_x$.	
$Z_9 : c86834008b1a\mathbf{0000}_x$,	$Z_{12} : 11142242c55a\mathbf{0001}_x$,	$Z_{15} : 9604d7afd613\mathbf{0000}_x$,
$Z_{18} : 65e14192bdda\mathbf{0000}_x$,	$Z_{21} : d3631317e7c2\mathbf{0001}_x$,	$Z_{24} : 2a2603d8ea9e\mathbf{0000}_x$,
$Z_{27} : d8b77b8e38cb\mathbf{0001}_x$,	$Z_{30} : c158b5526da8\mathbf{0001}_x$,	$Z_{33} : 647a08e80c87\mathbf{0000}_x$,
$Z_{36} : 475dd2004fb0\mathbf{0001}_x$,	$Z_{39} : 6818491dc5af\mathbf{0001}_x$,	$Z_{42} : 7eeb1fcfe969\mathbf{0001}_x$,
$Z_{45} : e3539a838b60\mathbf{0000}_x$,	$Z_{48} : bdc17d1aed49\mathbf{0000}_x$,	$Z_{51} : 278f011942f7\mathbf{0001}_x$.

The given weak keys are just examples, but they already contradict the claims of [91] that no weak keys existed for WIDEA-4. Finding these weak keys took only a few seconds of a SAT solver. Thousands of other weak keys were found just as fast. Therefore, although the key schedule of WIDEA-4 (Sect. 2.5.2) is more involved than IDEA's, weak keys still exist.

Note that the weak key in Table 3.44 allows us to attack only the first IDEA instance in WIDEA-4. Therefore, this weak key cannot be used to attack any of the other three IDEA instances, and vice-versa. Consequently, each weak key belongs to a different, non-overlapping weak-key class. The non-overlapping nature of these classes guarantees that the size of their union is maximal (compared to the overlapping case).

We have also searched for weak keys in all four IDEA instances simultaneously in WIDEA-4. The corresponding SAT instance was unsatisfiable, therefore, proving that no such weak keys exist.

The estimate of the weak-key class size in [145] is based on the assumption that each weak subkey requires fifteen bits to be zero. Across 8.5 rounds, with two weak subkeys per round, and attacking two IDEA instances at once, means $9 \cdot 2 \cdot 2 \cdot 15 = 540$ bits restricted to zero (if each subkey condition holds independently). But, this means a probability of 2^{-540}, which for a 512-bit key, means no weak key is expected to exist.

3.5.8.6 Weak keys in WIDEA-8

SAT solvers also found weak 1024-bit keys for WIDEA-8, for each of the eight IDEA instances separately. Therefore, each such weak key belongs to a different weak-key class.

The state of WIDEA-8 can be represented by an 8×4 matrix of 16-bit words. Each row corresponds to an IDEA instance. The best differential and linear distinguishers for the full 8.5-round WIDEA-8, under weak-key assumptions, are 1-round iterative and have the form

$$\begin{pmatrix} 0 & 0 & 0 & 0 \\ 0 & 0 & 0 & 0 \\ 0 & 0 & 0 & 0 \\ 0 & 0 & 0 & 0 \\ 0 & 0 & 0 & 0 \\ 0 & 0 & 0 & 0 \\ 0 & 0 & 0 & 0 \\ \Delta X & \Delta X & \Delta X & \Delta X \end{pmatrix} \overset{1r}{\rightarrow} \begin{pmatrix} 0 & 0 & 0 & 0 \\ 0 & 0 & 0 & 0 \\ 0 & 0 & 0 & 0 \\ 0 & 0 & 0 & 0 \\ 0 & 0 & 0 & 0 \\ 0 & 0 & 0 & 0 \\ 0 & 0 & 0 & 0 \\ \Delta X & \Delta X & \Delta X & \Delta X \end{pmatrix}, \qquad (3.28)$$

where ΔX stands for nonzero difference words. in other words, the attack focuses on a single IDEA instance at a time (out of eight). In a differential cryptanalysis setting, $\Delta X = 8000_x$ is a 16-bit xor difference.

The corresponding weak subkeys consist of 128-bit values whose fifteen most significant bits (MSBs) are zero. In WIDEA-8, this means that the same eighteen subkeys as in WIDEA-4 should be weak.

The top part of Table 3.48 shows a weak key for the first IDEA instance in WIDEA-8, while the bottom half shows the round subkeys. The 15-bit weak subkey piece is in boldface type. Table 3.49 through to Table 3.55 show weak keys for the other IDEA instances in WIDEA-8.

Table 3.48 Weak 1024-bit user key and weak subkeys of WIDEA-8 for the first IDEA instance according to (3.28) [113].

$Z_0 : \mathbf{0001}c9c63160bceeac68c62c4d16be85_x$,	$Z_1 : aea162405f37f9d07babc9d3f32a531d_x$,
$Z_2 : ba122e73a3da189d4879b0e05af6d08b_x$,	$Z_3 : \mathbf{0000}aee346b6a872f3b9a7f1f9b3f5d4_x$,
$Z_4 : bdefc6e9fe7c328113798a7b84bd93a3_x$,	$Z_5 : 09569d8d3b26ffd6fbfbbf5186794b24_x$,
$Z_6 : \mathbf{0001}41b5ef0c87fdca638e8c32d7231b_x$,	$Z_7 : fc2f60e69dc434a23666d040470117ac_x$.
$Z_9 : \mathbf{0001}735387097c70e2ad98f803c1f268_x$,	$Z_{12} : \mathbf{0001}c0cf121ebe551e37cc5e343df4cc_x$,
$Z_{15} : \mathbf{0001}bf1718de290b722dc61fc9c73852_x$,	$Z_{18} : \mathbf{0001}4b8c62faab56adcba849880bdc5a_x$,
$Z_{21} : \mathbf{0001}3dd5e3acdb23bf6c553bddf98c2f_x$,	$Z_{24} : \mathbf{0000}4291bed30f6867003c59dba003ff_x$,
$Z_{27} : \mathbf{0001}f986b4abffe9016fc6a91bbf8603_x$,	$Z_{30} : \mathbf{0001}9e310c7a27fdc67d3854325ffc99_x$,
$Z_{33} : \mathbf{0000}81bb2ea6ee8fdcc8bf4d2057fef7_x$,	$Z_{36} : \mathbf{0001}9c4b08cc2785b5f14767b0ae7cf1_x$,
$Z_{39} : \mathbf{0000}e0e35c2505baff6510fefb155e0a_x$,	$Z_{42} : \mathbf{0000}a7b12bbfafdef197ca0f03acb055_x$,
$Z_{45} : \mathbf{0000}7e7af512af2be5fe6093f34142e9_x$,	$Z_{48} : \mathbf{0000}15b14f4fbbff5542104f9046818e_x$,
$Z_{51} : \mathbf{0001}ca4b0125f706f0b65baaca691528_x$.	

Table 3.49 Weak 1024-bit user key and weak subkeys of WIDEA-8 for the second IDEA instance [113].

$Z_0 : 542d\mathbf{0000}bbd83bc8ac9707b155f63bd3_x$,	$Z_1 : 735266ee520753f77df9fab137eff93f_x$,
$Z_2 : be979d90686e51da5932877b69d63846_x$,	$Z_3 : f1d6\mathbf{0000}9cacff2b71662da75594117b_x$,
$Z_4 : 2a36af25e7789f0958478b75c263393e_x$,	$Z_5 : 5bac73e09b4c0b45de379e4d0cc563e5_x$,
$Z_6 : 6221\mathbf{0001}cf0172f9dd181aef3ff10779_x$,	$Z_7 : df41e88f0dd657efcc4345a3020f738b_x$.
$Z_9 : 3bd7\mathbf{0001}3511f5601bd543ed9dc4b775_x$,	$Z_{12} : 5200\mathbf{0001}cb2d7e11e45d8259362ca32f_x$,
$Z_{15} : 0299\mathbf{0001}6fdda9694747fbc316e383a5_x$,	$Z_{18} : 08f5\mathbf{0000}ea90399d4b193435d640b587_x$,
$Z_{21} : 9b6f\mathbf{0001}e90b0f0abd63878134da0a99_x$,	$Z_{24} : faf9\mathbf{0001}db18ac2bbfe85bcddc83b3b9_x$,
$Z_{27} : e248\mathbf{0000}60bab3782c406652450e7042_x$,	$Z_{30} : 7c20\mathbf{0001}e0057b9fe5c87565875c1e2c_x$,
$Z_{33} : c6e1\mathbf{0001}0e5683458326f1a7466e1175_x$,	$Z_{36} : cf8e\mathbf{0001}5101ce2d00d51ace5d1ad73d_x$,
$Z_{39} : 0b18\mathbf{0001}45d30475c4b55d0b5c2b1f45_x$,	$Z_{42} : c49d\mathbf{0001}50155ddf831d13ac3abd1f3c_x$,
$Z_{45} : 386c\mathbf{0001}2ac4819818a3bf5bf7c81f45_x$,	$Z_{48} : 1a51\mathbf{0001}ec3ce8fad9924a09fd3a4bf8_x$,
$Z_{51} : 4860\mathbf{0000}e1db34e24029667275c79ed1_x$.	

3.5.9 DC of RIDEA Under Weak-Key Assumptions

The first step is to obtain 1-round differential characteristics and the corresponding weak-subkey restrictions for the characteristic to hold with the highest possible probability.

The difference operator is bitwise exclusive-or and the difference is wordwise (16 bits). The only 16-bit difference values used in the characteristic are $\nu = 8000_x$ and 0000_x.

Across unkeyed $\odot$ and $\boxplus$, the propagation of 16-bit word differences follows from Table 3.12 and Table 3.13, respectively.

Table 3.56 lists all fifteen nontrivial 1-round differential characteristics of RIDEA, under weak-key assumptions.

Table 3.50 Weak 1024-bit user key and weak subkeys of WIDEA-8 for the third IDEA instance [113].

$Z_0 : ce0369cc\mathbf{0001}1fa0112b402378607d8c_x$,	$Z_1 : 52d228190db8b2fb36fe7cb6a4923ff3_x$,
$Z_2 : 7f60ec624ce8c9fda66e8e6025db6591_x$,	$Z_3 : e9052c24\mathbf{0001}b497ea4bde45c769b4eb_x$,
$Z_4 : 5a78636ecb0bbf3fa4ce55b9e0a8f7f7_x$,	$Z_5 : 1449c25a95e64dc878d7522b28a2c65b_x$,
$Z_6 : 526b94b7\mathbf{0001}2ce2af34521b35ec9470_x$,	$Z_7 : f3ffb172b7787e30bec74c2816a52b56_x$.
$Z_9 : 7c61371e\mathbf{0001}3a10e2c35c6cc249ed49_x$,	$Z_{12} : 5d466be3\mathbf{0001}46c122fabc38864dad00_x$,
$Z_{15} : 1396cd24\mathbf{0001}f0eee7acae17e12b0ec3_x$,	$Z_{18} : 349fe13c\mathbf{0000}82a0d2bb621ffe4bdf29_x$,
$Z_{21} : 5a23fe80\mathbf{0001}ae1e500849dbbcb11637_x$,	$Z_{24} : 6218c3c2\mathbf{0001}9cfcb889b17d3edbce3d_x$,
$Z_{27} : 7a8da3c3\mathbf{0001}de83074f6c3e91ffd6de_x$,	$Z_{30} : 9fedb129\mathbf{0001}ae017b413bf6907eccda_x$,
$Z_{33} : 56e66105\mathbf{0001}e731739e5e987ecd0559_x$,	$Z_{36} : 9e7ab6ab\mathbf{0001}d98c633231d7f731dd53_x$,
$Z_{39} : 3b795af3\mathbf{0001}ebb2a1ef8f8fb18de7a3_x$,	$Z_{42} : 26063962\mathbf{0000}7e5eb98ace84a35eb394_x$,
$Z_{45} : c0dc6ceb\mathbf{0001}c60a21b2ff48f7adffa6_x$,	$Z_{48} : ee13bbf2\mathbf{0001}5f5b5d2893bda45e0879_x$,
$Z_{51} : 8a4cbdb3\mathbf{0000}3d37a407ac5dd7763ef0_x$.	

Table 3.51 Weak 1024-bit user key and weak subkeys of WIDEA-8 for the fourth IDEA instance [113].

$Z_0 : 271684859e1a\mathbf{0000}545c90f7861619c8_x$,	$Z_1 : 77c21d2c72e99573add29cdd403172cb_x$,
$Z_2 : c583ce79a6b3be73c2c9ec870eb33d8c_x$,	$Z_3 : 509a5df54e27\mathbf{0001}de2a1ba34aa04294_x$,
$Z_4 : feb294de96090aac23518277da3a01ed_x$,	$Z_5 : 41d5bf9ab09075d409f7ea767db1ff8b_x$,
$Z_6 : 40e086b9c8a8\mathbf{0001}4d617ed4298a2d05_x$,	$Z_7 : a42694d8640b5fffd1dbeb2fb899de21_x$.
$Z_9 : 559502891869\mathbf{0001}3e70573dc5bc2421_x$,	$Z_{12} : ea60df10d962\mathbf{0000}abe129e59814fa73_x$,
$Z_{15} : 7be16da082b3\mathbf{0001}97bfc19ba1f44882_x$,	$Z_{18} : ab4ed68bb908\mathbf{0000}5acfb041e22feb75_x$,
$Z_{21} : b5ee1beb5541\mathbf{0000}701589ede87cde39_x$,	$Z_{24} : a08b4e2020e5\mathbf{0001}bd5c2e2971c72430_x$,
$Z_{27} : 377a08bda7bd\mathbf{0000}e5ab34125c386a83_x$,	$Z_{30} : fd22dc3a43d7\mathbf{0000}c0f11760e5d0dda2_x$,
$Z_{33} : c8f6cad75eb7\mathbf{0000}dfabf9995c47c6eb_x$,	$Z_{36} : caa5fae11445\mathbf{0001}cda5832804c6099a_x$,
$Z_{39} : 27a3fbc49936\mathbf{0000}a8662231bf511031_x$,	$Z_{42} : bf0c15c6ebfe\mathbf{0001}3505a5e90a740adc_x$,
$Z_{45} : 98c684c2facc\mathbf{0000}b5e6db1acc9b9af9_x$,	$Z_{48} : 17604504ae20\mathbf{0001}a2fdc63e7d1288ab_x$,
$Z_{51} : 1049c36b2c65\mathbf{0000}b7d683c4b52aed56_x$.	

Multiple-round characteristics for RIDEA are obtained by concatenating 1-round characteristics such that the output difference of one characteristic matches the input difference of the next characteristic.

From Table 3.56, the multiple-round characteristics are clustered into five chains covering 1, 2, 3 or 6 rounds:

(i) $(0,0,0,\delta) \overset{1r}{\to} (\delta,\delta,\delta,0) \overset{1r}{\to} (0,0,0,\delta)$, which is 2-round iterative and holds with probability 2^{-30}. The weak subkeys required are: $Z_4^{(i)}$ and $Z_1^{(i+1)}$.

(ii) $(0,0,\delta,0) \overset{1r}{\to} (\delta,0,0,0) \overset{1r}{\to} (0,\delta,0,0) \overset{1r}{\to} (\delta,\delta,0,\delta) \overset{1r}{\to} (0,\delta,\delta,\delta) \overset{1r}{\to} (\delta,0,\delta,\delta) \overset{1r}{\to} (0,0,\delta,0)$, which is 6-round iterative and holds with probability 2^{-90}. The weak subkeys required are: $Z_5^{(i)}$, $Z_1^{(i+1)}$, $Z_5^{(i+1)}$, $Z_1^{(i+3)}$, $Z_4^{(i+3)}$, $Z_5^{(i+3)}$, $Z_4^{(i+4)}$, $Z_5^{(i+4)}$, $Z_1^{(i+5)}$ and $Z_4^{(i+5)}$.

Table 3.52 Weak 1024-bit user key and weak subkeys of WIDEA-8 for the fifth IDEA instance [113].

$Z_0 : dc46b3c130f2c632\mathbf{0000}f934498aa261_x$,	$Z_1 : 23c47bd870045f66226b5403b7818766_x$,
$Z_2 : 4bc944e3f9d5d7a801a7a3b11d54c07a_x$,	$Z_3 : 977fe5a7a2f23dea\mathbf{0001}3a585c5c4b6c_x$,
$Z_4 : dbbc32be55fd2b512ec881770ce640c2_x$,	$Z_5 : f7c60b939bf735e3db7b6cf6c403eea7_x$,
$Z_6 : 34d8c582b201fe40\mathbf{0001}4ce16e5feb64_x$,	$Z_7 : c9d2b3dfb6577915984e7cb86638ba4b_x$.
$Z_9 : 1aa154fd7a965cdd\mathbf{0001}317406bb0b27_x$,	$Z_{12} : 928a95a07688367\mathbf{2000}005f1c2011c62_x$,
$Z_{15} : ac0c29221f1fcb16\mathbf{0001}4c6b76de7dcd_x$,	$Z_{18} : d03332e9e85707bd\mathbf{0001}6796aa0cfb09_x$,
$Z_{21} : be07dd8c80c3eed0\mathbf{0001}150ce428a399_x$,	$Z_{24} : 787a08faedae6be0\mathbf{0000}b96a04780beb_x$,
$Z_{27} : bb31de5de85fc4a1\mathbf{0001}4a6a7311c464_x$,	$Z_{30} : b8470b10c6cc8b09\mathbf{0001}6cc7cb4d44c2_x$,
$Z_{33} : df038adacf67869d\mathbf{0000}8509b3186327_x$,	$Z_{36} : 4936f8193fd70b0e\mathbf{0001}067ba3122b51_x$,
$Z_{39} : 247e02f2014fed88\mathbf{0001}de12fca9cf79_x$,	$Z_{42} : 5de90b8ac77d7d93\mathbf{0001}046c08fab387_x$,
$Z_{45} : 4a75c5e98d071b8b\mathbf{0001}94c27f19283e_x$,	$Z_{48} : 6886fb8ece25f7e8\mathbf{0001}3be02e34d781_x$,
$Z_{51} : 49168c6f5bfd0874\mathbf{0000}6a2f183c90d4_x$.	

Table 3.53 Weak 1024-bit user key and weak subkeys of WIDEA-8 for the sixth IDEA instance [113].

$Z_0 : 48175bed72710162acf2\mathbf{0001}c4ce4e1b_x$,	$Z_1 : 8697b91966d475e00d7cac9450f11e78_x$,
$Z_2 : 7a448cb7f2f7144078185192ad45b2c3_x$,	$Z_3 : 10994b4b70f6a90d2bd7\mathbf{0001}8f07f7e0_x$,
$Z_4 : 06eb0f2048330f57917d60215a5ff33a_x$,	$Z_5 : ed2c50d87fdf9a4de109750039e950bc_x$,
$Z_6 : 0155b95592ad5a8181c4\mathbf{0001}2d9888ab_x$,	$Z_7 : cbfd21bde970ba6e83bb27fe3b39ab4c_x$.
$Z_9 : 4c5c72c6f14413aa7bb8\mathbf{0000}ac3dddf7_x$,	$Z_{12} : 25018878b50f030ab80e\mathbf{0000}5400c5dd_x$,
$Z_{15} : 1b4b22f41fe38790163\mathbf{8000}03b2fbc6d_x$,	$Z_{18} : 08f50f63e7ab756f977f\mathbf{0001}95386318_x$,
$Z_{21} : 4f37cf5d570ea163051\mathbf{6000}12818004f_x$,	$Z_{24} : 47a158ff1e7522d7e0bd\mathbf{0001}36429cfc_x$,
$Z_{27} : a7814b3e3f3fb9166728\mathbf{0001}c8cb1ba2_x$,	$Z_{30} : 1a91068663f99e4b0d3f\mathbf{0000}30171c51_x$,
$Z_{33} : a4dfb85fecfb1561910e\mathbf{0001}6be92cbf_x$,	$Z_{36} : a30c3bcc05fd7e3fecfe\mathbf{0001}92ea1450_x$,
$Z_{39} : d739858fa3059d7ec999\mathbf{0001}a01c355b_x$,	$Z_{42} : 51d1515fa7873cbff8f9\mathbf{0000}49ce9d16_x$,
$Z_{45} : 175bb8f5a1c45e289f84\mathbf{0001}d649e88f_x$,	$Z_{48} : 38f96070f5cea1b03eb0\mathbf{0000}0fbc9f4f_x$,
$Z_{51} : eb9cee03b4b94f18c3ff\mathbf{0000}c8f01ed8_x$.	

(iii) $(0, 0, \delta, \delta) \xrightarrow{1r} (0, \delta, \delta, 0) \xrightarrow{1r} (0, \delta, 0, \delta) \xrightarrow{1r} (0, 0, \delta, \delta)$ which is 3-round iterative and holds with probability 2^{-30}. The weak subkeys required are: $Z_4^{(i)}$, $Z_5^{(i)}$, $Z_5^{(i+1)}$ and $Z_4^{(i+2)}$.

(iv) $(\delta, 0, 0, \delta) \xrightarrow{1r} (\delta, 0, \delta, 0) \xrightarrow{1r} (\delta, \delta, 0, 0) \xrightarrow{1r} (\delta, 0, 0, \delta)$ which is 3-round iterative and holds with probability 2^{-30}. The weak subkeys required are: $Z_1^{(i)}$, $Z_5^{(i)}$, $Z_1^{(i+1)}$, $Z_1^{(i+2)}$, $Z_4^{(i+2)}$ and $Z_5^{(i+2)}$.

(v) $(\delta, \delta, \delta, \delta) \xrightarrow{1r} (\delta, \delta, \delta, \delta)$ which is 1-round iterative and holds with probability 1. The weak subkeys required are: $Z_1^{(i)}$ and $Z_4^{(i)}$.

The characteristics (i) and (iii) require the least number of weak subkeys per round, while characteristic (v) holds with the largest probability.

Considering the key schedule of RIDEA described in Sect. 2.4.2, Table 3.57 lists the evolution of weak-key class size for characteristic (i) per round, for RIDEA. Overall, characteristic (i) holds for a weak-key class of size 2^{24} for

Table 3.54 Weak 1024-bit user key and weak subkeys of WIDEA-8 for the seventh IDEA instance [113].

$Z_0 : ef54735c2c732a7fb4817009\mathbf{0000}d4d8_x$,	$Z_1 : c9c697819a33561fbe9b3fc9ebbb425c_x$,
$Z_2 : 81a4a2f1a1222eb8892b8af1bab30416_x$,	$Z_3 : 1d5869ad16ac69728f909e63\mathbf{0000}023a_x$,
$Z_4 : 34e6b1d745b0302f6011507ca8a2a0e4_x$,	$Z_5 : aa2fd64fc7ca837cad105a832d70536c_x$,
$Z_6 : 1ad19efc2df74cc3692fe9f1\mathbf{0001}edd3_x$,	$Z_7 : ec25fd09bd08537d589ff69871be4833_x$.
$Z_9 : c27ec9361957b335282880e8\mathbf{0001}d894_x$,	$Z_{12} : 8326112376c91dedbfec47e5\mathbf{0001}e44d_x$,
$Z_{15} : fa08cd48a05e71f0090046d0\mathbf{0001}f9c8_x$,	$Z_{18} : f4c4c6d0be42bf3f76dfc0b6\mathbf{0001}85f8_x$,
$Z_{21} : b3b7908b7b529cfae017c2f9\mathbf{0000}55cf_x$,	$Z_{24} : e8e974cf3ee47d8c9597bc92\mathbf{0001}ed72_x$,
$Z_{27} : b5b60aedfe1fbd40c29de103\mathbf{0001}4dd1_x$,	$Z_{30} : 745de8bc200782a32c854025\mathbf{0001}2cce_x$,
$Z_{33} : 321c03444bd3f3277b9e33db\mathbf{0001}9fc0_x$,	$Z_{36} : d9d015c92f25f4571d5ce55b\mathbf{0000}85a6_x$,
$Z_{39} : 4b51e32877de1f8fea8d10fb\mathbf{0001}415d_x$,	$Z_{42} : b0f9e40984e0d1ffe04b8399\mathbf{0001}3a8d_x$,
$Z_{45} : bf96dc4eb91e5fec224bcdc4\mathbf{0000}faa9_x$,	$Z_{48} : 3a9366ef0e681c3cb5fecbf9\mathbf{0000}0229_x$,
$Z_{51} : cebc2946a713845558a2b9b7\mathbf{0000}b355_x$.	

Table 3.55 Weak 1024-bit user key and weak subkeys of WIDEA-8 for the eighth IDEA instance [113].

$Z_0 : dc7853502880fdf51dc9e795e226\mathbf{0001}_x$,	$Z_1 : 8089ea6c425c1be032fcf0123c779012_x$,
$Z_2 : 8c888bcc4f27c6d35338a6013b6a1ca6_x$,	$Z_3 : 3e619a3dcc1878be7a02a4767961\mathbf{0001}_x$,
$Z_4 : b4e89acbe72198f7a752385ab3961373_x$,	$Z_5 : 8bfdaa848980f7071bbaff068eaa6fce_x$,
$Z_6 : c1e72b1cac7ac68fb604ad52ba5e\mathbf{0001}_x$,	$Z_7 : 062046a11ac16410bd69f29368c0bffe_x$.
$Z_9 : 6ee52038de98c90d89fedad15812\mathbf{0000}_x$,	$Z_{12} : 44a13341c7f395eecb5d78bd4ddc\mathbf{0000}_x$,
$Z_{15} : b3a3e8d0a4562ab0be193a017085\mathbf{0001}_x$,	$Z_{18} : 19fbf668afcf206f9d7a7541a90f\mathbf{0001}_x$,
$Z_{21} : d1c397c903c6ae05e7afef1606aa\mathbf{0001}_x$,	$Z_{24} : 38ad902cf66d2e181cd2eff87195\mathbf{0001}_x$,
$Z_{27} : 95d1c81cd5100769f2e4b58c01a1\mathbf{0001}_x$,	$Z_{30} : 7db276c7e391b6d1c4737480d5e7\mathbf{0001}_x$,
$Z_{33} : 5476b3c749117c4528ab70ffd739\mathbf{0000}_x$,	$Z_{36} : 31d2c9f6074f88c60c492b072efb\mathbf{0000}_x$,
$Z_{39} : abf726da107d99ff449119fff7ce\mathbf{0000}_x$,	$Z_{42} : 88a4e430332cb6ff0a84a394a14a\mathbf{0000}_x$,
$Z_{45} : 038f0541c19465814e1630857bdc\mathbf{0000}_x$,	$Z_{48} : e79c3b83500fb127f1dde6ebfe62\mathbf{0001}_x$,
$Z_{51} : 68ca63c026e38b4c3d087c457b04\mathbf{0000}_x$.	

the full 8.5-round RIDEA and with probability 2^{-135}.

Likewise, Table 3.58 lists the evolution of weak-key class size for characteristic (iii) per round, for RIDEA. Overall, characteristic (iii) holds for a weak-key class of size 2^{23} for the full 8.5-round RIDEA and with probability 2^{-90}.

A better trade-off may be characteristic (v), which holds with certainty under weak-key assumptions. Table 3.59 lists the evolution of weak-key class size for characteristic (v) per round, for RIDEA. The weak-key class for characteristic (v) has only 2^{12} keys, but since it holds with certainty (under these weak-key restrictions) it can be used to distinguish the full 8-round RIDEA from a random permutation with the same low attack complexity as in PES and IDEA: the cost is essentially two encryptions, two chosen plaintexts and equivalent memory. Therefore, the full 8.5-round RIDEA cipher is not an ideal primitive, that is, it is not a PRP. See Sect. 1.7.

Table 3.56 Nontrivial 1-round characteristics for RIDEA.

1-round characteristics	Weak subkeys	Prob.	Difference across the MA-box
$(0,0,0,\delta) \xrightarrow{1r} (\delta,\delta,\delta,0)$	$Z_4^{(i)}$	2^{-15}	$(0,\delta) \to (\delta,\delta)$
$(0,0,\delta,0) \xrightarrow{1r} (\delta,0,0,0)$	$Z_5^{(i)}$	2^{-15}	$(\delta,0) \to (0,\delta)$
$(0,0,\delta,\delta) \xrightarrow{1r} (0,\delta,\delta,0)$	$Z_4^{(i)}, Z_5^{(i)}$	2^{-15}	$(\delta,\delta) \to (\delta,0)$
$(0,\delta,0,0) \xrightarrow{1r} (\delta,\delta,0,\delta)$	—	2^{-15}	$(0,\delta) \to (\delta,\delta)$
$(0,\delta,0,\delta) \xrightarrow{1r} (0,0,\delta,\delta)$	$Z_4^{(i)}$	1	$(0,0) \to (0,0)$
$(0,\delta,\delta,0) \xrightarrow{1r} (0,\delta,0,\delta)$	$Z_5^{(i)}$	2^{-15}	$(\delta,\delta) \to (\delta,0)$
$(0,\delta,\delta,\delta) \xrightarrow{1r} (\delta,0,\delta,\delta)$	$Z_4^{(i)}, Z_5^{(i)}$	2^{-15}	$(\delta,0) \to (0,\delta)$
$(\delta,0,0,0) \xrightarrow{1r} (0,\delta,0,0)$	$Z_1^{(i)}, Z_5^{(i)}$	2^{-15}	$(\delta,0) \to (0,\delta)$
$(\delta,0,0,\delta) \xrightarrow{1r} (\delta,0,\delta,0)$	$Z_1^{(i)}, Z_5^{(i)}$	2^{-15}	$(\delta,\delta) \to (\delta,0)$
$(\delta,0,\delta,0) \xrightarrow{1r} (\delta,\delta,0,0)$	$Z_1^{(i)}$	1	$(0,0) \to (0,0)$
$(\delta,0,\delta,\delta) \xrightarrow{1r} (0,0,\delta,0)$	$Z_1^{(i)}, Z_4^{(i)}$	2^{-15}	$(0,\delta) \to (\delta,\delta)$
$(\delta,\delta,0,0) \xrightarrow{1r} (\delta,0,0,\delta)$	$Z_1^{(i)}, Z_4^{(i)}, Z_5^{(i)}$	2^{-15}	$(\delta,\delta) \to (\delta,0)$
$(\delta,\delta,0,\delta) \xrightarrow{1r} (0,\delta,\delta,\delta)$	$Z_1^{(i)}, Z_4^{(i)}, Z_5^{(i)}$	2^{-15}	$(\delta,0) \to (0,\delta)$
$(\delta,\delta,\delta,0) \xrightarrow{1r} (0,0,0,\delta)$	$Z_1^{(i)}$	2^{-15}	$(0,\delta) \to (\delta,\delta)$
$(\delta,\delta,\delta,\delta) \xrightarrow{1r} (\delta,\delta,\delta,\delta)$	$Z_1^{(i)}, Z_4^{(i)}$	1	$(0,0) \to (0,0)$

Table 3.57 Evolution of weak-key class size of characteristic (i) for RIDEA.

Weak subkeys	Key bits (in key schedule)	Weak-key Class Size (bits)
$Z_4^{(1)}$	48–62	2^{113}
$Z_1^{(2)}$	96–110	2^{98}
$Z_4^{(3)}$	112–126	2^{83}
$Z_1^{(4)}$	32–46	2^{69}
$Z_4^{(5)}$	48–62	2^{55}
$Z_1^{(6)}$	96–110	2^{50}
$Z_4^{(7)}$	112–126	2^{50}
$Z_1^{(8)}$	32–46	2^{36}
$Z_4^{(9)}$	48–62	2^{24}

Similarly to the case of PES and IDEA, this differential characteristic allows us to study the diffusion power of RIDEA with respect to the differential branch number. See Appendix, Sect. B.

The set of characteristics in Table 3.56 indicates that the differential branch number *per round* of RIDEA is only 2 (under weak-key assumptions). For instance, the 1-round characteristic

$$(0,0,\delta,0) \xrightarrow{1r} (\delta,0,0,0)$$

has both the input and output differences with Hamming Weight equal to one. So, even though the design of RIDEA provides full diffusion in a single

Table 3.58 Evolution of weak-key class size of characteristic (iii) for RIDEA.

Weak subkeys	Key bits (in key schedule)	Weak-key Class Size (bits)
$Z_4^{(1)}, Z_5^{(1)}$	48–78	2^{98}
$Z_5^{(2)}$	32–46	2^{97}
$Z_4^{(3)}$	112–126	2^{82}
$Z_4^{(4)}, Z_5^{(4)}$	80–110	2^{66}
$Z_5^{(5)}$	64–78	2^{66}
$Z_4^{(6)}$	16–30	2^{52}
$Z_4^{(7)}$	112–126	2^{37}
$Z_5^{(7)}, Z_5^{(8)}$	0–14, 96–110	2^{28}
$Z_4^{(9)}$	48–62	2^{23}

Table 3.59 Evolution of weak-key class size of characteristic (v) for RIDEA.

Weak subkeys	Key bits (in key schedule)	Weak-key Class Size (bits)
$Z_1^{(1)}, Z_4^{(1)}$	0–14, 48–62	2^{96}
$Z_1^{(2)}$	96–110	2^{76}
$Z_4^{(2)}$	16-30	2^{76}
$Z_1^{(3)}, Z_4^{(3)}$	64–78, 112–126	2^{60}
$Z_1^{(4)}, Z_4^{(4)}$	32–46, 80–94	2^{53}
$Z_1^{(5)}, Z_4^{(5)}$	0–14, 48–62	2^{41}
$Z_1^{(6)}$	96–110	2^{34}
$Z_4^{(6)}$	16–30	2^{34}
$Z_1^{(7)}, Z_4^{(7)}$	64–78, 112–126	2^{29}
$Z_1^{(8)}, Z_4^{(8)}$	32–46, 80–94	2^{22}
$Z_1^{(9)}, Z_4^{(9)}$	0–14, 48–62	2^{12}

round, the previous analysis of its characteristics shows that its differential branch number is much lower than ideal. For a 4-word block, such as in RIDEA, an MDS code using a 4×4 MDS matrix would rather provide a branch number of $1 + 4 = 5$. See Appendix, Sect. B.1, for an explanation of the differential branch number.

The fact that RIDEA uses the same key schedule algorithm as PES and IDEA has far reaching consequences. Usually, the larger the number of rounds the smaller the weak-key class becomes. But, the weak-key class size never vanishes for RIDEA for the same reason as for PES and IDEA. For instance, the 1-round characteristic $(\delta, \delta, \delta, \delta) \stackrel{1r}{\rightarrow} (\delta, \delta, \delta, \delta)$ holds with certainty for any number of rounds for the all-zero key $K = 0^{128}$, which is a very special user key. It is the only weak key of RIDEA for which **any** characteristic can be iterated for **any** number of rounds because $K = 0^{128}$ means that all multiplicative subkeys are weak, even those which do not belong to the differential trail (of any characteristic).

Another observation: although a weak-key class of size 2^{24} represents a fraction of only $2^{24-128} = 2^{-104}$ of the key space of RIDEA, it is important to remember that block ciphers are fundamental building blocks for the construction of many other cryptographic primitives, such as hash functions [162]. In some of these hash function constructions, the key input is not secret and may even be under the control of the adversary. In these cases, the existence of a single weak key is enough to allow the adversary to choose a weak key in order to manipulate the encryption scheme [120, 39]. See Sect. 3.5.8.3.

3.5.10 DC of REESSE3+ Under Weak-Key Assumptions

For differential cryptanalysis (DC) [30] of REESSE3+ [173], the difference operator used is bitwise exclusive-or, $\oplus$.

The initial approach is to obtain all difference patterns across the MA-box of REESSE3+, under the corresponding weak-subkey restrictions. The strategy is to build the characteristics in a bottom-up fashion, from the smallest components: $\odot$, $\boxplus$ and $\oplus$, up to an MA-box, up to a half-round, further up to a full round, and finally to multiple rounds.

The only 16-bit difference values used in the characteristics are 8000_x and 0000_x.

Let $\delta = X \oplus X*$ denote a wordwise difference. Under multiplication with a subkey Z, the input difference $\delta = 8000_x$ causes the output difference: $\nabla = (X \odot Z) \oplus ((X \oplus \delta) \odot Z) = (X \odot Z) \oplus ((X \oplus 8000_x) \odot Z)$.

Note that $-X = X \oplus 8000_x$, and $0 \equiv 2^{16} \equiv -1 \bmod (2^{16}+1)$.

Under a weak subkey $Z \in \{0, 1\}$, ∇ becomes either:

- $(X \odot 1) \oplus ((X \oplus 8000_x) \odot 1) = X \oplus X \oplus 8000_x = 8000_x$,
- $(X \odot 0) \oplus ((X \oplus 8000_x) \odot 0) = (X \odot (-1)) \oplus (-X \odot (-1)) = (-X) \oplus X = 8000_x$.

So, in both cases, $\delta = 8000_x$ is a fix-point difference for the multiplication operation under a weak subkey.

Across unkeyed $\odot$ and $\boxplus$, the propagation of 16-bit difference words follows from Table 3.12 and Table 3.13, respectively.

Table 3.60 lists all wordwise difference patterns across the MA-box of REESSE3+ as well as the corresponding probability.

REESSE3+ contains unkeyed $\odot$ operations in its MA-box, just like in RIDEA, MESH-96 and MESH-128. But, unlike the latter, the MA-box of REESSE3+ does not contain subkeys. It means that no weak-subkey assumption can be used to facilitate difference propagation across its MA-box.

The next step is to obtain from Table 3.60 all nontrivial 1-round differential characteristics and the corresponding weak-keys restrictions.

Table 3.60 All wordwise difference patterns across the MA-box of REESSE3+.

Wordwise differences across the MA-box	# active $\odot$	Probability
$(0,0,0,0) \to (0,0,0,0)$	0	1
$(0,0,0,\delta) \to (\delta,\delta,0,\delta)$	2	2^{-30}
$(0,0,\delta,0) \to (\delta,\delta,\delta,\delta)$	2	2^{-30}
$(0,0,\delta,\delta) \to (0,0,\delta,0)$	2	2^{-30}
$(0,\delta,0,0) \to (0,\delta,0,\delta)$	2	2^{-30}
$(0,\delta,0,\delta) \to (\delta,0,0,0)$	3	2^{-45}
$(0,\delta,\delta,0) \to (\delta,0,\delta,0)$	2	2^{-30}
$(0,\delta,\delta,\delta) \to (0,\delta,\delta,\delta)$	3	2^{-45}
$(\delta,0,0,0) \to (0,0,0,\delta)$	2	2^{-30}
$(\delta,0,0,\delta) \to (\delta,\delta,0,0)$	3	2^{-45}
$(\delta,0,\delta,0) \to (\delta,\delta,\delta,0)$	2	2^{-30}
$(\delta,0,\delta,\delta) \to (0,0,\delta,\delta)$	3	2^{-45}
$(\delta,\delta,0,0) \to (0,\delta,0,0)$	1	2^{-15}
$(\delta,\delta,0,\delta) \to (\delta,0,0,\delta)$	3	2^{-45}
$(\delta,\delta,\delta,0) \to (\delta,0,\delta,\delta)$	3	2^{-45}
$(\delta,\delta,\delta,\delta) \to (0,\delta,\delta,0)$	3	2^{-45}

Tables 3.61 through 3.68 list all 255 nontrivial 1-round differential characteristics of REESSE3+, under weak-key assumptions. These tables provide an exhaustive analysis of how characteristics propagate across a single round of REESSE3+, and they are quite more extensive than the analyses in [146].

Multi-round characteristics are obtained by concatenating 1-round characteristics such that the output difference of the top one matches the input difference of the next one.

From the 255 nontrivial 1-round characteristics in Tables 3.61 through 3.68, multiple-round characteristics can be obtained. They are clustered into 29 disjoint chains.

(1) $(0,0,0,0,0,0,0,\delta) \xrightarrow{1r} (\delta,\delta,\delta,0,\delta,0,\delta,0) \xrightarrow{1r} (\delta,\delta,0,\delta,\delta,\delta,\delta,\delta) \xrightarrow{1r} (\delta, 0, \delta, \delta, \delta, \delta, 0, 0) \xrightarrow{1r} (\delta,\delta,\delta,0,0,\delta,0,\delta) \xrightarrow{1r} (\delta,\delta,0,0,\delta,0,0,0) \xrightarrow{1r} (0, \delta, \delta, 0, 0, \delta, \delta, \delta) \xrightarrow{1r} (\delta,0,\delta,\delta,0,0,0,0) \xrightarrow{1r} (\delta,\delta,\delta,0,0,0,\delta,\delta) \xrightarrow{1r} (\delta, \delta, 0, \delta, \delta, 0, \delta, 0) \xrightarrow{1r} (\delta,0,\delta,\delta,\delta,\delta,\delta,\delta) \xrightarrow{1r} (\delta,\delta,\delta,\delta,0,\delta,0,0) \xrightarrow{1r} (0, 0, 0, 0, 0, 0, 0, \delta)$, which is a 12-round iterative characteristic holding with probability 2^{-420}. Starting from the i-th round, the following 28 subkeys have to be weak: $Z_1^{(i+1)}$, $Z_7^{(i+1)}$, $Z_1^{(i+2)}$, $Z_4^{(i+2)}$, $Z_6^{(i+2)}$, $Z_7^{(i+2)}$, $Z_1^{(i+3)}$, $Z_4^{(i+3)}$, $Z_6^{(i+3)}$, $Z_1^{(i+4)}$, $Z_6^{(i+4)}$, $Z_1^{(i+5)}$, $Z_6^{(i+6)}$, $Z_7^{(i+6)}$, $Z_1^{(i+7)}$, $Z_4^{(i+7)}$, $Z_1^{(i+8)}$, $Z_7^{(i+8)}$, $Z_1^{(i+9)}$, $Z_4^{(i+9)}$, $Z_7^{(i+9)}$, $Z_1^{(i+10)}$, $Z_4^{(i+10)}$, $Z_6^{(i+10)}$, $Z_7^{(i+10)}$, $Z_1^{(i+11)}$, $Z_4^{(i+11)}$ and $Z_6^{(i+11)}$.

(2) $(0,0,0,0,0,0,\delta,0) \xrightarrow{1r} (\delta,\delta,\delta,\delta,\delta,0,\delta,\delta) \xrightarrow{1r} (0,0,0,\delta,0,\delta,\delta,0) \xrightarrow{1r} (0, 0, \delta, \delta, 0, 0, 0, \delta) \xrightarrow{1r} (\delta,0,0,0,\delta,0,\delta,0) \xrightarrow{1r} (\delta,0,0,\delta,0,\delta,\delta,\delta) \xrightarrow{1r} (0, \delta, 0, \delta, \delta, 0, 0, 0) \xrightarrow{1r} (\delta,\delta,0,0,0,\delta,\delta,\delta) \xrightarrow{1r} (0,\delta,\delta,\delta,0,0,0,0) \xrightarrow{1r} (0, \delta, \delta, 0, \delta, 0, \delta, \delta) \xrightarrow{1r} (\delta,0,\delta,\delta,0,\delta,\delta,0) \xrightarrow{1r} (\delta,\delta,\delta,\delta,0,0,0,\delta) \xrightarrow{1r} (0, 0, 0, 0, 0, 0, \delta, 0)$, which is a 12-round iterative characteristic holding with probability

Table 3.61 Nontrivial 1-round characteristics for REESSE3+ (part I).

1-round characteristic	Weak subkeys $Z_j^{(i)}$ j values	Prob.
$(0,0,0,0,0,0,0,\delta) \xrightarrow{1r} (\delta,\delta,\delta,0,\delta,0,\delta,0)$	—	2^{-30}
$(0,0,0,0,0,0,\delta,0) \xrightarrow{1r} (\delta,\delta,\delta,\delta,\delta,0,\delta,\delta)$	7	2^{-30}
$(0,0,0,0,0,0,\delta,\delta) \xrightarrow{1r} (0,0,0,\delta,0,0,0,\delta)$	7	2^{-30}
$(0,0,0,0,0,\delta,0,0) \xrightarrow{1r} (\delta,\delta,\delta,0,\delta,0,0,\delta)$	6	2^{-30}
$(0,0,0,0,0,\delta,0,\delta) \xrightarrow{1r} (0,0,0,0,0,0,\delta,\delta)$	6	1
$(0,0,0,0,0,\delta,\delta,0) \xrightarrow{1r} (0,0,0,\delta,0,0,\delta,0)$	6; 7	2^{-30}
$(0,0,0,0,0,\delta,\delta,\delta) \xrightarrow{1r} (\delta,\delta,\delta,\delta,\delta,0,0,0)$	6; 7	2^{-30}
$(0,0,0,0,\delta,0,0,0) \xrightarrow{1r} (\delta,\delta,\delta,0,\delta,\delta,\delta,\delta)$	—	2^{-30}
$(0,0,0,0,\delta,0,0,\delta) \xrightarrow{1r} (0,0,0,0,0,\delta,0,\delta)$	—	2^{-30}
$(0,0,0,0,\delta,0,\delta,0) \xrightarrow{1r} (0,0,0,\delta,0,\delta,0,0)$	7	1
$(0,0,0,0,\delta,0,\delta,\delta) \xrightarrow{1r} (\delta,\delta,\delta,\delta,\delta,\delta,\delta,0)$	7	2^{-30}
$(0,0,0,0,\delta,\delta,0,0) \xrightarrow{1r} (0,0,0,0,0,\delta,\delta,0)$	6	2^{-30}
$(0,0,0,0,\delta,\delta,0,\delta) \xrightarrow{1r} (\delta,\delta,\delta,0,\delta,\delta,0,0)$	6	2^{-30}
$(0,0,0,0,\delta,\delta,\delta,0) \xrightarrow{1r} (\delta,\delta,\delta,\delta,\delta,\delta,0,\delta)$	6; 7	2^{-30}
$(0,0,0,0,\delta,\delta,\delta,\delta) \xrightarrow{1r} (0,0,0,\delta,0,\delta,\delta,\delta)$	6; 7	1
$(0,0,0,\delta,0,0,0,0) \xrightarrow{1r} (0,0,\delta,0,0,0,\delta,\delta)$	4	2^{-30}
$(0,0,0,\delta,0,0,0,\delta) \xrightarrow{1r} (\delta,\delta,0,0,\delta,0,0,\delta)$	4	2^{-45}
$(0,0,0,\delta,0,0,\delta,0) \xrightarrow{1r} (\delta,\delta,0,\delta,\delta,0,0,0)$	4; 7	2^{-30}
$(0,0,0,\delta,0,0,\delta,\delta) \xrightarrow{1r} (0,0,\delta,\delta,0,0,\delta,0)$	4; 7	2^{-45}
$(0,0,0,\delta,0,\delta,0,0) \xrightarrow{1r} (\delta,\delta,0,0,\delta,0,\delta,0)$	4; 6	2^{-45}
$(0,0,0,\delta,0,\delta,0,\delta) \xrightarrow{1r} (0,0,\delta,0,0,0,0,0)$	4; 6	2^{-30}
$(0,0,0,\delta,0,\delta,\delta,0) \xrightarrow{1r} (0,0,\delta,\delta,0,0,0,\delta)$	4; 6; 7	2^{-45}
$(0,0,0,\delta,0,\delta,\delta,\delta) \xrightarrow{1r} (\delta,\delta,0,\delta,\delta,0,\delta,\delta)$	4; 6; 7	2^{-30}
$(0,0,0,\delta,\delta,0,0,0) \xrightarrow{1r} (\delta,\delta,0,0,\delta,\delta,0,0)$	4	2^{-30}
$(0,0,0,\delta,\delta,0,0,\delta) \xrightarrow{1r} (0,0,\delta,0,0,\delta,\delta,0)$	4	2^{-45}
$(0,0,0,\delta,\delta,0,\delta,0) \xrightarrow{1r} (0,0,\delta,\delta,0,\delta,\delta,\delta)$	4; 7	2^{-30}
$(0,0,0,\delta,\delta,0,\delta,\delta) \xrightarrow{1r} (\delta,\delta,0,\delta,\delta,\delta,0,\delta)$	4; 7	2^{-45}
$(0,0,0,\delta,\delta,\delta,0,0) \xrightarrow{1r} (0,0,\delta,0,0,\delta,0,\delta)$	4; 6	2^{-45}
$(0,0,0,\delta,\delta,\delta,0,\delta) \xrightarrow{1r} (\delta,\delta,0,0,\delta,\delta,\delta,\delta)$	4; 6	2^{-30}
$(0,0,0,\delta,\delta,\delta,\delta,0) \xrightarrow{1r} (\delta,\delta,0,\delta,\delta,\delta,\delta,0)$	4; 6; 7	2^{-45}
$(0,0,0,\delta,\delta,\delta,\delta,\delta) \xrightarrow{1r} (0,0,\delta,\delta,0,\delta,0,0)$	4; 6; 7	2^{-30}
$(0,0,\delta,0,0,0,0,0) \xrightarrow{1r} (0,\delta,0,0,0,0,\delta,\delta)$	—	2^{-30}

Table 3.62 Nontrivial 1-round characteristics for REESSE3+ (part II).

1-round characteristic	Weak subkeys $Z_j^{(i)}$ j values	Prob.
$(0,0,\delta,0,0,0,0,\delta) \xrightarrow{1r} (\delta,0,\delta,0,\delta,0,0,\delta)$	—	2^{-45}
$(0,0,\delta,0,0,0,\delta,0) \xrightarrow{1r} (\delta,0,\delta,\delta,\delta,0,0,0)$	7	2^{-30}
$(0,0,\delta,0,0,0,\delta,\delta) \xrightarrow{1r} (0,\delta,0,\delta,0,0,\delta,0)$	7	2^{-45}
$(0,0,\delta,0,0,\delta,0,0) \xrightarrow{1r} (\delta,0,\delta,0,\delta,0,\delta,0)$	6	2^{-45}
$(0,0,\delta,0,0,\delta,0,\delta) \xrightarrow{1r} (0,\delta,0,0,0,0,0,0)$	6	2^{-30}
$(0,0,\delta,0,0,\delta,\delta,0) \xrightarrow{1r} (0,\delta,0,\delta,0,0,0,\delta)$	6; 7	2^{-45}
$(0,0,\delta,0,0,\delta,\delta,\delta) \xrightarrow{1r} (\delta,0,\delta,\delta,\delta,0,\delta,\delta)$	6; 7	2^{-30}
$(0,0,\delta,0,\delta,0,0,0) \xrightarrow{1r} (\delta,0,\delta,0,\delta,\delta,0,0)$	—	2^{-30}
$(0,0,\delta,0,\delta,0,0,\delta) \xrightarrow{1r} (0,\delta,0,0,0,\delta,\delta,0)$	—	2^{-45}
$(0,0,\delta,0,\delta,0,\delta,0) \xrightarrow{1r} (0,\delta,0,\delta,0,\delta,\delta,\delta)$	7	2^{-30}
$(0,0,\delta,0,\delta,0,\delta,\delta) \xrightarrow{1r} (\delta,0,\delta,\delta,\delta,\delta,0,\delta)$	7	2^{-45}
$(0,0,\delta,0,\delta,\delta,0,0) \xrightarrow{1r} (0,\delta,0,0,0,\delta,0,\delta)$	6	2^{-45}
$(0,0,\delta,0,\delta,\delta,0,\delta) \xrightarrow{1r} (\delta,0,\delta,0,\delta,\delta,\delta,\delta)$	6	2^{-30}
$(0,0,\delta,0,\delta,\delta,\delta,0) \xrightarrow{1r} (\delta,0,\delta,\delta,\delta,\delta,\delta,0)$	6; 7	2^{-45}
$(0,0,\delta,0,\delta,\delta,\delta,\delta) \xrightarrow{1r} (0,\delta,0,\delta,0,\delta,0,0)$	6; 7	2^{-30}
$(0,0,\delta,\delta,0,0,0,0) \xrightarrow{1r} (0,\delta,\delta,0,0,0,0,0)$	4	2^{-15}
$(0,0,\delta,\delta,0,0,0,\delta) \xrightarrow{1r} (\delta,0,0,0,\delta,0,\delta,0)$	4	2^{-45}
$(0,0,\delta,\delta,0,0,\delta,0) \xrightarrow{1r} (\delta,0,0,\delta,\delta,0,\delta,\delta)$	4; 7	2^{-45}
$(0,0,\delta,\delta,0,0,\delta,\delta) \xrightarrow{1r} (0,\delta,\delta,\delta,0,0,0,\delta)$	4; 7	2^{-45}
$(0,0,\delta,\delta,0,\delta,0,0) \xrightarrow{1r} (\delta,0,0,0,\delta,0,0,\delta)$	4; 6	2^{-45}
$(0,0,\delta,\delta,0,\delta,0,\delta) \xrightarrow{1r} (0,\delta,\delta,0,0,0,\delta,\delta)$	4; 6	2^{-15}
$(0,0,\delta,\delta,0,\delta,\delta,0) \xrightarrow{1r} (0,\delta,\delta,\delta,0,0,\delta,0)$	4; 6; 7	2^{-45}
$(0,0,\delta,\delta,0,\delta,\delta,\delta) \xrightarrow{1r} (\delta,0,0,\delta,\delta,0,0,0)$	4; 6; 7	2^{-45}
$(0,0,\delta,\delta,\delta,0,0,0) \xrightarrow{1r} (\delta,0,0,0,\delta,\delta,\delta,\delta)$	4	2^{-45}
$(0,0,\delta,\delta,\delta,0,0,\delta) \xrightarrow{1r} (0,\delta,\delta,0,0,\delta,0,\delta)$	4	2^{-45}
$(0,0,\delta,\delta,\delta,0,\delta,0) \xrightarrow{1r} (0,\delta,\delta,\delta,0,\delta,0,0)$	4; 7	2^{-15}
$(0,0,\delta,\delta,\delta,0,\delta,\delta) \xrightarrow{1r} (\delta,0,0,\delta,\delta,\delta,\delta,0)$	4; 7	2^{-45}
$(0,0,\delta,\delta,\delta,\delta,0,0) \xrightarrow{1r} (0,\delta,\delta,0,0,\delta,\delta,0)$	4; 6	2^{-45}
$(0,0,\delta,\delta,\delta,\delta,0,\delta) \xrightarrow{1r} (\delta,0,0,0,\delta,\delta,0,0)$	4; 6	2^{-45}
$(0,0,\delta,\delta,\delta,\delta,\delta,0) \xrightarrow{1r} (\delta,0,0,\delta,\delta,\delta,0,\delta)$	4; 6; 7	2^{-45}
$(0,0,\delta,\delta,\delta,\delta,\delta,\delta) \xrightarrow{1r} (0,\delta,\delta,\delta,0,\delta,\delta,\delta)$	4; 6; 7	2^{-15}
$(0,\delta,0,0,0,0,0,0) \xrightarrow{1r} (0,0,0,0,\delta,0,\delta,\delta)$	—	2^{-30}

Table 3.63 Nontrivial 1-round characteristics for REESSE3+ (part III).

1-round characteristic	Weak subkeys $Z_j^{(i)}$ j values	Prob.
$(0,\delta,0,0,0,0,0,\delta) \stackrel{1r}{\to} (\delta,\delta,\delta,0,0,0,0,\delta)$	—	2^{-45}
$(0,\delta,0,0,0,0,\delta,0) \stackrel{1r}{\to} (\delta,\delta,\delta,\delta,0,0,0,0)$	7	2^{-30}
$(0,\delta,0,0,0,0,\delta,\delta) \stackrel{1r}{\to} (0,0,0,\delta,\delta,0,\delta,0)$	7	2^{-45}
$(0,\delta,0,0,0,\delta,0,0) \stackrel{1r}{\to} (\delta,\delta,\delta,0,0,0,\delta,0)$	6	2^{-45}
$(0,\delta,0,0,0,\delta,0,\delta) \stackrel{1r}{\to} (0,0,0,0,\delta,0,0,0)$	6	2^{-30}
$(0,\delta,0,0,0,\delta,\delta,0) \stackrel{1r}{\to} (0,0,0,\delta,\delta,0,0,\delta)$	6; 7	2^{-45}
$(0,\delta,0,0,0,\delta,\delta,\delta) \stackrel{1r}{\to} (\delta,\delta,\delta,\delta,0,0,\delta,\delta)$	6; 7	2^{-30}
$(0,\delta,0,0,\delta,0,0,0) \stackrel{1r}{\to} (\delta,\delta,\delta,0,0,\delta,0,0)$	—	2^{-30}
$(0,\delta,0,0,\delta,0,0,\delta) \stackrel{1r}{\to} (0,0,0,0,\delta,\delta,\delta,0)$	—	2^{-45}
$(0,\delta,0,0,\delta,0,\delta,0) \stackrel{1r}{\to} (0,0,0,\delta,\delta,\delta,\delta,\delta)$	7	2^{-30}
$(0,\delta,0,0,\delta,0,\delta,\delta) \stackrel{1r}{\to} (\delta,\delta,\delta,\delta,0,\delta,0,\delta)$	7	2^{-45}
$(0,\delta,0,0,\delta,\delta,0,0) \stackrel{1r}{\to} (0,0,0,0,\delta,\delta,0,\delta)$	6	2^{-45}
$(0,\delta,0,0,\delta,\delta,0,\delta) \stackrel{1r}{\to} (\delta,\delta,\delta,0,0,\delta,\delta,\delta)$	6	2^{-30}
$(0,\delta,0,0,\delta,\delta,\delta,0) \stackrel{1r}{\to} (\delta,\delta,\delta,\delta,0,\delta,\delta,0)$	6; 7	2^{-45}
$(0,\delta,0,0,\delta,\delta,\delta,\delta) \stackrel{1r}{\to} (0,0,0,\delta,\delta,\delta,0,0)$	6; 7	2^{-30}
$(0,\delta,0,\delta,0,0,0,0) \stackrel{1r}{\to} (0,0,\delta,0,\delta,0,0,0)$	4	1
$(0,\delta,0,\delta,0,0,0,\delta) \stackrel{1r}{\to} (\delta,\delta,0,0,0,0,\delta,0)$	4	2^{-30}
$(0,\delta,0,\delta,0,0,\delta,0) \stackrel{1r}{\to} (\delta,\delta,0,\delta,0,0,\delta,\delta)$	4; 7	2^{-30}
$(0,\delta,0,\delta,0,0,\delta,\delta) \stackrel{1r}{\to} (0,0,\delta,\delta,\delta,0,0,\delta)$	4; 7	2^{-30}
$(0,\delta,0,\delta,0,\delta,0,0) \stackrel{1r}{\to} (\delta,\delta,0,0,0,0,0,\delta)$	4; 6	2^{-30}
$(0,\delta,0,\delta,0,\delta,0,\delta) \stackrel{1r}{\to} (0,0,\delta,0,\delta,0,\delta,\delta)$	4; 6	1
$(0,\delta,0,\delta,0,\delta,\delta,0) \stackrel{1r}{\to} (0,0,\delta,\delta,\delta,0,\delta,0)$	4; 6; 7	2^{-30}
$(0,\delta,0,\delta,0,\delta,\delta,\delta) \stackrel{1r}{\to} (\delta,\delta,0,\delta,0,0,0,0)$	4; 6; 7	2^{-30}
$(0,\delta,0,\delta,\delta,0,0,0) \stackrel{1r}{\to} (\delta,\delta,0,0,0,\delta,\delta,\delta)$	4	2^{-30}
$(0,\delta,0,\delta,\delta,0,0,\delta) \stackrel{1r}{\to} (0,0,\delta,0,\delta,\delta,0,\delta)$	4	2^{-30}
$(0,\delta,0,\delta,\delta,0,\delta,0) \stackrel{1r}{\to} (0,0,\delta,\delta,\delta,\delta,0,0)$	4; 7	1
$(0,\delta,0,\delta,\delta,0,\delta,\delta) \stackrel{1r}{\to} (\delta,\delta,0,\delta,0,\delta,\delta,0)$	4; 7	2^{-30}
$(0,\delta,0,\delta,\delta,\delta,0,0) \stackrel{1r}{\to} (0,0,\delta,0,\delta,\delta,\delta,0)$	4; 6	2^{-30}
$(0,\delta,0,\delta,\delta,\delta,0,\delta) \stackrel{1r}{\to} (\delta,\delta,0,0,0,\delta,0,0)$	4; 6	2^{-30}
$(0,\delta,0,\delta,\delta,\delta,\delta,0) \stackrel{1r}{\to} (\delta,\delta,0,\delta,0,\delta,0,\delta)$	4; 6; 7	2^{-30}
$(0,\delta,0,\delta,\delta,\delta,\delta,\delta) \stackrel{1r}{\to} (0,0,\delta,\delta,\delta,\delta,\delta,\delta)$	4; 6; 7	1
$(0,\delta,\delta,0,0,0,0,0) \stackrel{1r}{\to} (0,\delta,0,0,\delta,0,0,0)$	—	2^{-15}

Table 3.64 Nontrivial 1-round characteristics for REESSE3+ (part IV).

1-round characteristic	Weak subkeys $Z_j^{(i)}$ j values	Prob.
$(0,\delta,\delta,0,0,0,0,\delta) \xrightarrow{1r} (\delta,0,\delta,0,0,0,\delta,0)$	—	2^{-45}
$(0,\delta,\delta,0,0,0,\delta,0) \xrightarrow{1r} (\delta,0,\delta,\delta,0,0,\delta,\delta)$	7	2^{-45}
$(0,\delta,\delta,0,0,0,\delta,\delta) \xrightarrow{1r} (0,\delta,0,\delta,\delta,0,0,\delta)$	7	2^{-45}
$(0,\delta,\delta,0,0,\delta,0,0) \xrightarrow{1r} (\delta,0,\delta,0,0,0,0,\delta)$	6	2^{-45}
$(0,\delta,\delta,0,0,\delta,0,\delta) \xrightarrow{1r} (0,\delta,0,0,\delta,0,\delta,\delta)$	6	2^{-15}
$(0,\delta,\delta,0,0,\delta,\delta,0) \xrightarrow{1r} (0,\delta,0,\delta,\delta,0,\delta,0)$	6; 7	2^{-45}
$(0,\delta,\delta,0,0,\delta,\delta,\delta) \xrightarrow{1r} (\delta,0,\delta,\delta,0,0,0,0)$	6; 7	2^{-45}
$(0,\delta,\delta,0,\delta,0,0,0) \xrightarrow{1r} (\delta,0,\delta,0,0,\delta,\delta,\delta)$	—	2^{-45}
$(0,\delta,\delta,0,\delta,0,0,\delta) \xrightarrow{1r} (0,\delta,0,0,\delta,\delta,0,\delta)$	—	2^{-45}
$(0,\delta,\delta,0,\delta,0,\delta,0) \xrightarrow{1r} (0,\delta,0,\delta,\delta,\delta,0,0)$	7	2^{-15}
$(0,\delta,\delta,0,\delta,0,\delta,\delta) \xrightarrow{1r} (\delta,0,\delta,\delta,0,\delta,\delta,0)$	7	2^{-45}
$(0,\delta,\delta,0,\delta,\delta,0,0) \xrightarrow{1r} (0,\delta,0,0,\delta,\delta,\delta,0)$	6	2^{-45}
$(0,\delta,\delta,0,\delta,\delta,0,\delta) \xrightarrow{1r} (\delta,0,\delta,0,0,\delta,0,0)$	6	2^{-45}
$(0,\delta,\delta,0,\delta,\delta,\delta,0) \xrightarrow{1r} (\delta,0,\delta,\delta,0,\delta,0,\delta)$	6; 7	2^{-45}
$(0,\delta,\delta,0,\delta,\delta,\delta,\delta) \xrightarrow{1r} (0,\delta,0,\delta,\delta,\delta,\delta,\delta)$	6; 7	2^{-15}
$(0,\delta,\delta,\delta,0,0,0,0) \xrightarrow{1r} (0,\delta,\delta,0,\delta,0,\delta,\delta)$	4	2^{-30}
$(0,\delta,\delta,\delta,0,0,0,\delta) \xrightarrow{1r} (\delta,0,0,0,0,0,0,\delta)$	4	2^{-45}
$(0,\delta,\delta,\delta,0,0,\delta,0) \xrightarrow{1r} (\delta,0,0,\delta,0,0,0,0)$	4; 7	2^{-30}
$(0,\delta,\delta,\delta,0,0,\delta,\delta) \xrightarrow{1r} (0,\delta,\delta,\delta,\delta,0,\delta,0)$	4; 7	2^{-45}
$(0,\delta,\delta,\delta,0,\delta,0,0) \xrightarrow{1r} (\delta,0,0,0,0,0,\delta,0)$	4; 6	2^{-45}
$(0,\delta,\delta,\delta,0,\delta,0,\delta) \xrightarrow{1r} (0,\delta,\delta,0,\delta,0,0,0)$	4; 6	2^{-30}
$(0,\delta,\delta,\delta,0,\delta,\delta,0) \xrightarrow{1r} (0,\delta,\delta,\delta,\delta,0,0,\delta)$	4; 6; 7	2^{-45}
$(0,\delta,\delta,\delta,0,\delta,\delta,\delta) \xrightarrow{1r} (\delta,0,0,\delta,0,0,\delta,\delta)$	4; 6; 7	2^{-30}
$(0,\delta,\delta,\delta,\delta,0,0,0) \xrightarrow{1r} (\delta,0,0,0,0,\delta,0,0)$	4	2^{-30}
$(0,\delta,\delta,\delta,\delta,0,0,\delta) \xrightarrow{1r} (0,\delta,\delta,0,\delta,\delta,\delta,0)$	4	2^{-45}
$(0,\delta,\delta,\delta,\delta,0,\delta,0) \xrightarrow{1r} (0,\delta,\delta,\delta,\delta,\delta,\delta,\delta)$	4; 7	2^{-30}
$(0,\delta,\delta,\delta,\delta,0,\delta,\delta) \xrightarrow{1r} (\delta,0,0,\delta,0,\delta,0,\delta)$	4; 7	2^{-45}
$(0,\delta,\delta,\delta,\delta,\delta,0,0) \xrightarrow{1r} (0,\delta,\delta,0,\delta,\delta,0,\delta)$	4; 6	2^{-45}
$(0,\delta,\delta,\delta,\delta,\delta,0,\delta) \xrightarrow{1r} (\delta,0,0,0,0,\delta,\delta,\delta)$	4; 6	2^{-30}
$(0,\delta,\delta,\delta,\delta,\delta,\delta,0) \xrightarrow{1r} (\delta,0,0,\delta,0,\delta,\delta,0)$	4; 6; 7	2^{-45}
$(0,\delta,\delta,\delta,\delta,\delta,\delta,\delta) \xrightarrow{1r} (0,\delta,\delta,\delta,\delta,\delta,0,0)$	4; 6; 7	2^{-30}
$(\delta,0,0,0,0,0,0,0) \xrightarrow{1r} (\delta,0,0,0,0,0,\delta,\delta)$	1	2^{-30}

Table 3.65 Nontrivial 1-round characteristics for REESSE3+ (part V).

1-round characteristic	Weak subkeys $Z_j^{(i)}$ j values	Prob.
$(\delta,0,0,0,0,0,0,\delta) \overset{1r}{\rightarrow} (0,\delta,\delta,0,\delta,0,0,\delta)$	1	2^{-45}
$(\delta,0,0,0,0,0,\delta,0) \overset{1r}{\rightarrow} (0,\delta,\delta,\delta,\delta,0,0,0)$	1; 7	2^{-30}
$(\delta,0,0,0,0,0,\delta,\delta) \overset{1r}{\rightarrow} (\delta,0,0,\delta,0,0,\delta,0)$	1; 7	2^{-45}
$(\delta,0,0,0,0,\delta,0,0) \overset{1r}{\rightarrow} (0,\delta,\delta,0,\delta,0,\delta,0)$	1; 6	2^{-45}
$(\delta,0,0,0,0,\delta,0,\delta) \overset{1r}{\rightarrow} (\delta,0,0,0,0,0,0,0)$	1; 6	2^{-30}
$(\delta,0,0,0,0,\delta,\delta,0) \overset{1r}{\rightarrow} (\delta,0,0,\delta,0,0,0,\delta)$	1; 6; 7	2^{-45}
$(\delta,0,0,0,0,\delta,\delta,\delta) \overset{1r}{\rightarrow} (0,\delta,\delta,\delta,\delta,0,\delta,\delta)$	1; 6; 7	2^{-30}
$(\delta,0,0,0,\delta,0,0,0) \overset{1r}{\rightarrow} (0,\delta,\delta,0,\delta,\delta,0,0)$	1	2^{-30}
$(\delta,0,0,0,\delta,0,0,\delta) \overset{1r}{\rightarrow} (\delta,0,0,0,0,\delta,\delta,0)$	1	2^{-45}
$(\delta,0,0,0,\delta,0,\delta,0) \overset{1r}{\rightarrow} (\delta,0,0,\delta,0,\delta,\delta,\delta)$	1; 7	2^{-30}
$(\delta,0,0,0,\delta,0,\delta,\delta) \overset{1r}{\rightarrow} (0,\delta,\delta,\delta,\delta,\delta,0,\delta)$	1; 7	2^{-45}
$(\delta,0,0,0,\delta,\delta,0,0) \overset{1r}{\rightarrow} (\delta,0,0,0,0,\delta,0,\delta)$	1; 6	2^{-45}
$(\delta,0,0,0,\delta,\delta,0,\delta) \overset{1r}{\rightarrow} (0,\delta,\delta,0,\delta,\delta,\delta,\delta)$	1; 6	2^{-30}
$(\delta,0,0,0,\delta,\delta,\delta,0) \overset{1r}{\rightarrow} (0,\delta,\delta,\delta,\delta,\delta,\delta,0)$	1; 6; 7	2^{-45}
$(\delta,0,0,0,\delta,\delta,\delta,\delta) \overset{1r}{\rightarrow} (\delta,0,0,\delta,0,\delta,0,0)$	1; 6; 7	2^{-30}
$(\delta,0,0,\delta,0,0,0,0) \overset{1r}{\rightarrow} (\delta,0,\delta,0,0,0,0,0)$	1; 4	2^{-15}
$(\delta,0,0,\delta,0,0,0,\delta) \overset{1r}{\rightarrow} (0,\delta,0,0,\delta,0,\delta,0)$	1; 4	2^{-45}
$(\delta,0,0,\delta,0,0,\delta,0) \overset{1r}{\rightarrow} (0,\delta,0,\delta,\delta,0,\delta,\delta)$	1; 4; 7	2^{-45}
$(\delta,0,0,\delta,0,0,\delta,\delta) \overset{1r}{\rightarrow} (\delta,0,\delta,\delta,0,0,0,\delta)$	1; 4; 7	2^{-45}
$(\delta,0,0,\delta,0,\delta,0,0) \overset{1r}{\rightarrow} (0,\delta,0,0,\delta,0,0,\delta)$	1; 4; 6	2^{-45}
$(\delta,0,0,\delta,0,\delta,0,\delta) \overset{1r}{\rightarrow} (\delta,0,\delta,0,0,0,\delta,\delta)$	1; 4; 6	2^{-15}
$(\delta,0,0,\delta,0,\delta,\delta,0) \overset{1r}{\rightarrow} (\delta,0,\delta,\delta,0,0,\delta,0)$	1; 4; 6; 7	2^{-45}
$(\delta,0,0,\delta,0,\delta,\delta,\delta) \overset{1r}{\rightarrow} (0,\delta,0,\delta,\delta,0,0,0)$	1; 4; 6; 7	2^{-45}
$(\delta,0,0,\delta,\delta,0,0,0) \overset{1r}{\rightarrow} (0,\delta,0,0,\delta,\delta,\delta,\delta)$	1; 4	2^{-45}
$(\delta,0,0,\delta,\delta,0,0,\delta) \overset{1r}{\rightarrow} (\delta,0,\delta,0,0,\delta,0,\delta)$	1; 4	2^{-45}
$(\delta,0,0,\delta,\delta,0,\delta,0) \overset{1r}{\rightarrow} (\delta,0,\delta,\delta,0,\delta,0,0)$	1; 4; 7	2^{-15}
$(\delta,0,0,\delta,\delta,0,\delta,\delta) \overset{1r}{\rightarrow} (0,\delta,0,\delta,\delta,\delta,\delta,0)$	1; 4; 7	2^{-45}
$(\delta,0,0,\delta,\delta,\delta,0,0) \overset{1r}{\rightarrow} (\delta,0,\delta,0,0,\delta,\delta,0)$	1; 4; 6	2^{-45}
$(\delta,0,0,\delta,\delta,\delta,0,\delta) \overset{1r}{\rightarrow} (0,\delta,0,0,\delta,\delta,0,0)$	1; 4; 6	2^{-45}
$(\delta,0,0,\delta,\delta,\delta,\delta,0) \overset{1r}{\rightarrow} (0,\delta,0,\delta,\delta,\delta,0,\delta)$	1; 4; 6; 7	2^{-45}
$(\delta,0,0,\delta,\delta,\delta,\delta,\delta) \overset{1r}{\rightarrow} (\delta,0,\delta,\delta,0,\delta,\delta,\delta)$	1; 4; 6; 7	2^{-15}
$(\delta,0,\delta,0,0,0,0,0) \overset{1r}{\rightarrow} (\delta,\delta,0,0,0,0,0,0)$	1	1

Table 3.66 Nontrivial 1-round characteristics for REESSE3+ (part VI).

1-round characteristic	Weak subkeys $Z_j^{(i)}$ j values	Prob.
$(\delta,0,\delta,0,0,0,0,\delta) \xrightarrow{1r} (0,0,\delta,0,\delta,0,\delta,0)$	1	2^{-30}
$(\delta,0,\delta,0,0,0,\delta,0) \xrightarrow{1r} (0,0,\delta,\delta,\delta,0,\delta,\delta)$	1; 7	2^{-30}
$(\delta,0,\delta,0,0,0,\delta,\delta) \xrightarrow{1r} (\delta,\delta,0,\delta,0,0,0,\delta)$	1; 7	2^{-30}
$(\delta,0,\delta,0,0,\delta,0,0) \xrightarrow{1r} (0,0,\delta,0,\delta,0,0,\delta)$	1; 6	2^{-30}
$(\delta,0,\delta,0,0,\delta,0,\delta) \xrightarrow{1r} (\delta,\delta,0,0,0,0,\delta,\delta)$	1; 6	1
$(\delta,0,\delta,0,0,\delta,\delta,0) \xrightarrow{1r} (\delta,\delta,0,\delta,0,0,\delta,0)$	1; 6; 7	2^{-30}
$(\delta,0,\delta,0,0,\delta,\delta,\delta) \xrightarrow{1r} (0,0,\delta,\delta,\delta,0,0,0)$	1; 6; 7	2^{-30}
$(\delta,0,\delta,0,\delta,0,0,0) \xrightarrow{1r} (0,0,\delta,0,\delta,\delta,\delta,\delta)$	1	2^{-30}
$(\delta,0,\delta,0,\delta,0,0,\delta) \xrightarrow{1r} (\delta,\delta,0,0,0,\delta,0,\delta)$	1	2^{-30}
$(\delta,0,\delta,0,\delta,0,\delta,0) \xrightarrow{1r} (\delta,\delta,0,\delta,0,\delta,0,0)$	1; 7	1
$(\delta,0,\delta,0,\delta,0,\delta,\delta) \xrightarrow{1r} (0,0,\delta,\delta,\delta,\delta,\delta,0)$	1; 7	2^{-30}
$(\delta,0,\delta,0,\delta,\delta,0,0) \xrightarrow{1r} (\delta,\delta,0,0,0,\delta,\delta,0)$	1; 6	2^{-30}
$(\delta,0,\delta,0,\delta,\delta,0,\delta) \xrightarrow{1r} (0,0,\delta,0,\delta,\delta,0,0)$	1; 6	2^{-30}
$(\delta,0,\delta,0,\delta,\delta,\delta,0) \xrightarrow{1r} (0,0,\delta,\delta,\delta,\delta,0,\delta)$	1; 6; 7	2^{-30}
$(\delta,0,\delta,0,\delta,\delta,\delta,\delta) \xrightarrow{1r} (\delta,\delta,0,\delta,0,\delta,\delta,\delta)$	1; 6; 7	1
$(\delta,0,\delta,\delta,0,0,0,0) \xrightarrow{1r} (\delta,\delta,\delta,0,0,0,\delta,\delta)$	1; 4	2^{-30}
$(\delta,0,\delta,\delta,0,0,0,\delta) \xrightarrow{1r} (0,0,0,0,\delta,0,0,\delta)$	1; 4	2^{-45}
$(\delta,0,\delta,\delta,0,0,\delta,0) \xrightarrow{1r} (0,0,0,\delta,\delta,0,0,0)$	1; 4; 7	2^{-30}
$(\delta,0,\delta,\delta,0,0,\delta,\delta) \xrightarrow{1r} (\delta,\delta,\delta,\delta,0,0,\delta,0)$	1; 4; 7	2^{-45}
$(\delta,0,\delta,\delta,0,\delta,0,0) \xrightarrow{1r} (0,0,0,0,\delta,0,\delta,0)$	1; 4; 6	2^{-45}
$(\delta,0,\delta,\delta,0,\delta,0,\delta) \xrightarrow{1r} (\delta,\delta,\delta,0,0,0,0,0)$	1; 4; 6	2^{-30}
$(\delta,0,\delta,\delta,0,\delta,\delta,0) \xrightarrow{1r} (\delta,\delta,\delta,\delta,0,0,0,\delta)$	1; 4; 6; 7	2^{-45}
$(\delta,0,\delta,\delta,0,\delta,\delta,\delta) \xrightarrow{1r} (0,0,0,\delta,\delta,0,\delta,\delta)$	1; 4; 6; 7	2^{-30}
$(\delta,0,\delta,\delta,\delta,0,0,0) \xrightarrow{1r} (0,0,0,0,\delta,\delta,0,0)$	1; 4	2^{-30}
$(\delta,0,\delta,\delta,\delta,0,0,\delta) \xrightarrow{1r} (\delta,\delta,\delta,0,0,\delta,\delta,0)$	1; 4	2^{-45}
$(\delta,0,\delta,\delta,\delta,0,\delta,0) \xrightarrow{1r} (\delta,\delta,\delta,\delta,0,\delta,\delta,\delta)$	1; 4; 7	2^{-30}
$(\delta,0,\delta,\delta,\delta,0,\delta,\delta) \xrightarrow{1r} (0,0,0,\delta,\delta,\delta,0,\delta)$	1; 4; 7	2^{-45}
$(\delta,0,\delta,\delta,\delta,\delta,0,0) \xrightarrow{1r} (\delta,\delta,\delta,0,0,\delta,0,\delta)$	1; 4; 6	2^{-45}
$(\delta,0,\delta,\delta,\delta,\delta,0,\delta) \xrightarrow{1r} (0,0,0,0,\delta,\delta,\delta,\delta)$	1; 4; 6	2^{-30}
$(\delta,0,\delta,\delta,\delta,\delta,\delta,0) \xrightarrow{1r} (0,0,0,\delta,\delta,\delta,\delta,0)$	1; 4; 6; 7	2^{-45}
$(\delta,0,\delta,\delta,\delta,\delta,\delta,\delta) \xrightarrow{1r} (\delta,\delta,\delta,\delta,0,\delta,0,0)$	1; 4; 6; 7	2^{-30}
$(\delta,\delta,0,0,0,0,0,0) \xrightarrow{1r} (\delta,0,0,0,\delta,0,0,0)$	1	2^{-15}

Table 3.67 Nontrivial 1-round characteristics for REESSE3+ (part VII).

1-round characteristic	Weak subkeys $Z_j^{(i)}$ j values	Prob.
$(\delta,\delta,0,0,0,0,0,\delta) \stackrel{1r}{\to} (0,\delta,\delta,0,0,0,\delta,0)$	1	2^{-45}
$(\delta,\delta,0,0,0,0,\delta,0) \stackrel{1r}{\to} (0,\delta,\delta,\delta,0,0,\delta,\delta)$	1; 7	2^{-45}
$(\delta,\delta,0,0,0,0,\delta,\delta) \stackrel{1r}{\to} (\delta,0,0,\delta,\delta,0,0,\delta)$	1; 7	2^{-45}
$(\delta,\delta,0,0,0,\delta,0,0) \stackrel{1r}{\to} (0,\delta,\delta,0,0,0,0,\delta)$	1; 6	2^{-45}
$(\delta,\delta,0,0,0,\delta,0,\delta) \stackrel{1r}{\to} (\delta,0,0,0,\delta,0,\delta,\delta)$	1; 6	2^{-15}
$(\delta,\delta,0,0,0,\delta,\delta,0) \stackrel{1r}{\to} (\delta,0,0,\delta,\delta,0,\delta,0)$	1; 6; 7	2^{-45}
$(\delta,\delta,0,0,0,\delta,\delta,\delta) \stackrel{1r}{\to} (0,\delta,\delta,\delta,0,0,0,0)$	1; 6; 7	2^{-45}
$(\delta,\delta,0,0,\delta,0,0,0) \stackrel{1r}{\to} (0,\delta,\delta,0,0,\delta,\delta,\delta)$	1	2^{-45}
$(\delta,\delta,0,0,\delta,0,0,\delta) \stackrel{1r}{\to} (\delta,0,0,0,\delta,\delta,0,\delta)$	1	2^{-45}
$(\delta,\delta,0,0,\delta,0,\delta,0) \stackrel{1r}{\to} (\delta,0,0,\delta,\delta,\delta,0,0)$	1; 7	2^{-15}
$(\delta,\delta,0,0,\delta,0,\delta,\delta) \stackrel{1r}{\to} (0,\delta,\delta,\delta,0,\delta,\delta,0)$	1; 7	2^{-45}
$(\delta,\delta,0,0,\delta,\delta,0,0) \stackrel{1r}{\to} (\delta,0,0,0,\delta,\delta,\delta,0)$	1; 6	2^{-45}
$(\delta,\delta,0,0,\delta,\delta,0,\delta) \stackrel{1r}{\to} (0,\delta,\delta,0,0,\delta,0,0)$	1; 6	2^{-45}
$(\delta,\delta,0,0,\delta,\delta,\delta,0) \stackrel{1r}{\to} (0,\delta,\delta,\delta,0,\delta,0,\delta)$	1; 6; 7	2^{-45}
$(\delta,\delta,0,0,\delta,\delta,\delta,\delta) \stackrel{1r}{\to} (\delta,0,0,\delta,\delta,\delta,\delta,\delta)$	1; 6; 7	2^{-15}
$(\delta,\delta,0,\delta,0,0,0,0) \stackrel{1r}{\to} (\delta,0,\delta,0,\delta,0,\delta,\delta)$	1; 4	2^{-30}
$(\delta,\delta,0,\delta,0,0,0,\delta) \stackrel{1r}{\to} (0,\delta,0,0,0,0,0,\delta)$	1; 4	2^{-45}
$(\delta,\delta,0,\delta,0,0,\delta,0) \stackrel{1r}{\to} (0,\delta,0,\delta,0,0,0,0)$	1; 4; 7	2^{-30}
$(\delta,\delta,0,\delta,0,0,\delta,\delta) \stackrel{1r}{\to} (\delta,0,\delta,\delta,\delta,0,\delta,0)$	1; 4; 7	2^{-45}
$(\delta,\delta,0,\delta,0,\delta,0,0) \stackrel{1r}{\to} (0,\delta,0,0,0,0,\delta,0)$	1; 4; 6	2^{-45}
$(\delta,\delta,0,\delta,0,\delta,0,\delta) \stackrel{1r}{\to} (\delta,0,\delta,0,\delta,0,0,0)$	1; 4; 6	2^{-30}
$(\delta,\delta,0,\delta,0,\delta,\delta,0) \stackrel{1r}{\to} (\delta,0,\delta,\delta,\delta,0,0,\delta)$	1; 4; 6; 7	2^{-45}
$(\delta,\delta,0,\delta,0,\delta,\delta,\delta) \stackrel{1r}{\to} (0,\delta,0,\delta,0,0,\delta,\delta)$	1; 4; 6; 7	2^{-30}
$(\delta,\delta,0,\delta,\delta,0,0,0) \stackrel{1r}{\to} (0,\delta,0,0,0,\delta,0,0)$	1; 4	2^{-30}
$(\delta,\delta,0,\delta,\delta,0,0,\delta) \stackrel{1r}{\to} (\delta,0,\delta,0,\delta,\delta,\delta,0)$	1; 4	2^{-45}
$(\delta,\delta,0,\delta,\delta,0,\delta,0) \stackrel{1r}{\to} (\delta,0,\delta,\delta,\delta,\delta,\delta,\delta)$	1; 4; 7	2^{-30}
$(\delta,\delta,0,\delta,\delta,0,\delta,\delta) \stackrel{1r}{\to} (0,\delta,0,\delta,0,\delta,0,\delta)$	1; 4; 7	2^{-45}
$(\delta,\delta,0,\delta,\delta,\delta,0,0) \stackrel{1r}{\to} (\delta,0,\delta,0,\delta,\delta,0,\delta)$	1; 4; 6	2^{-45}
$(\delta,\delta,0,\delta,\delta,\delta,0,\delta) \stackrel{1r}{\to} (0,\delta,0,0,0,\delta,\delta,\delta)$	1; 4; 6	2^{-30}
$(\delta,\delta,0,\delta,\delta,\delta,\delta,0) \stackrel{1r}{\to} (0,\delta,0,\delta,0,\delta,\delta,0)$	1; 4; 6; 7	2^{-45}
$(\delta,\delta,0,\delta,\delta,\delta,\delta,\delta) \stackrel{1r}{\to} (\delta,0,\delta,\delta,\delta,\delta,0,0)$	1; 4; 6; 7	2^{-30}
$(\delta,\delta,\delta,0,0,0,0,0) \stackrel{1r}{\to} (\delta,\delta,0,0,\delta,0,\delta,\delta)$	1	2^{-30}

Table 3.68 Nontrivial 1-round characteristics for REESSE3+ (part VIII).

1-round characteristic	Weak subkeys $Z_j^{(i)}$ j values	Prob.
$(\delta,\delta,\delta,0,0,0,0,\delta) \xrightarrow{1r} (0,0,\delta,0,0,0,0,\delta)$	1	2^{-45}
$(\delta,\delta,\delta,0,0,0,\delta,0) \xrightarrow{1r} (0,0,\delta,\delta,0,0,0,0)$	1; 7	2^{-30}
$(\delta,\delta,\delta,0,0,0,\delta,\delta) \xrightarrow{1r} (\delta,\delta,0,\delta,\delta,0,\delta,0)$	1; 7	2^{-45}
$(\delta,\delta,\delta,0,0,\delta,0,0) \xrightarrow{1r} (0,0,\delta,0,0,0,\delta,0)$	1; 6	2^{-45}
$(\delta,\delta,\delta,0,0,\delta,0,\delta) \xrightarrow{1r} (\delta,\delta,0,0,\delta,0,0,0)$	1; 6	2^{-30}
$(\delta,\delta,\delta,0,0,\delta,\delta,0) \xrightarrow{1r} (\delta,\delta,0,\delta,\delta,0,0,\delta)$	1; 6; 7	2^{-45}
$(\delta,\delta,\delta,0,0,\delta,\delta,\delta) \xrightarrow{1r} (0,0,\delta,\delta,0,0,\delta,\delta)$	1; 6; 7	2^{-30}
$(\delta,\delta,\delta,0,\delta,0,0,0) \xrightarrow{1r} (0,0,\delta,0,0,\delta,0,0)$	1	2^{-30}
$(\delta,\delta,\delta,0,\delta,0,0,\delta) \xrightarrow{1r} (\delta,\delta,0,0,\delta,\delta,\delta,0)$	1	2^{-45}
$(\delta,\delta,\delta,0,\delta,0,\delta,0) \xrightarrow{1r} (\delta,\delta,0,\delta,\delta,\delta,\delta,\delta)$	1; 7	2^{-30}
$(\delta,\delta,\delta,0,\delta,0,\delta,\delta) \xrightarrow{1r} (0,0,\delta,\delta,0,\delta,0,\delta)$	1; 7	2^{-45}
$(\delta,\delta,\delta,0,\delta,\delta,0,0) \xrightarrow{1r} (\delta,\delta,0,0,\delta,\delta,0,\delta)$	1; 6	2^{-45}
$(\delta,\delta,\delta,0,\delta,\delta,0,\delta) \xrightarrow{1r} (0,0,\delta,0,0,\delta,\delta,\delta)$	1; 6	2^{-30}
$(\delta,\delta,\delta,0,\delta,\delta,\delta,0) \xrightarrow{1r} (0,0,\delta,\delta,0,\delta,\delta,0)$	1; 6; 7	2^{-45}
$(\delta,\delta,\delta,0,\delta,\delta,\delta,\delta) \xrightarrow{1r} (\delta,\delta,0,\delta,\delta,\delta,0,0)$	1; 6; 7	2^{-30}
$(\delta,\delta,\delta,\delta,0,0,0,0) \xrightarrow{1r} (\delta,\delta,\delta,0,\delta,0,0,0)$	1; 4	1
$(\delta,\delta,\delta,\delta,0,0,0,\delta) \xrightarrow{1r} (0,0,0,0,0,0,\delta,0)$	1; 4	2^{-30}
$(\delta,\delta,\delta,\delta,0,0,\delta,0) \xrightarrow{1r} (0,0,0,\delta,0,0,\delta,\delta)$	1; 4; 7	2^{-30}
$(\delta,\delta,\delta,\delta,0,0,\delta,\delta) \xrightarrow{1r} (\delta,\delta,\delta,\delta,\delta,0,0,\delta)$	1; 4; 7	2^{-30}
$(\delta,\delta,\delta,\delta,0,\delta,0,0) \xrightarrow{1r} (0,0,0,0,0,0,0,\delta)$	1; 4; 6	2^{-30}
$(\delta,\delta,\delta,\delta,0,\delta,0,\delta) \xrightarrow{1r} (\delta,\delta,\delta,0,\delta,0,\delta,\delta)$	1; 4; 6	1
$(\delta,\delta,\delta,\delta,0,\delta,\delta,0) \xrightarrow{1r} (\delta,\delta,\delta,\delta,\delta,0,\delta,0)$	1; 4; 6; 7	2^{-30}
$(\delta,\delta,\delta,\delta,0,\delta,\delta,\delta) \xrightarrow{1r} (0,0,0,\delta,0,0,0,0)$	1; 4; 6; 7	2^{-30}
$(\delta,\delta,\delta,\delta,\delta,0,0,0) \xrightarrow{1r} (0,0,0,0,0,\delta,\delta,\delta)$	1; 4	2^{-30}
$(\delta,\delta,\delta,\delta,\delta,0,0,\delta) \xrightarrow{1r} (\delta,\delta,\delta,0,\delta,\delta,0,\delta)$	1; 4	2^{-30}
$(\delta,\delta,\delta,\delta,\delta,0,\delta,0) \xrightarrow{1r} (\delta,\delta,\delta,\delta,\delta,\delta,0,0)$	1; 4; 7	1
$(\delta,\delta,\delta,\delta,\delta,0,\delta,\delta) \xrightarrow{1r} (0,0,0,\delta,0,\delta,\delta,0)$	1; 4; 7	2^{-30}
$(\delta,\delta,\delta,\delta,\delta,\delta,0,0) \xrightarrow{1r} (\delta,\delta,\delta,0,\delta,\delta,\delta,0)$	1; 4; 6	2^{-30}
$(\delta,\delta,\delta,\delta,\delta,\delta,0,\delta) \xrightarrow{1r} (0,0,0,0,0,\delta,0,0)$	1; 4; 6	2^{-30}
$(\delta,\delta,\delta,\delta,\delta,\delta,\delta,0) \xrightarrow{1r} (0,0,0,\delta,0,\delta,0,\delta)$	1; 4; 6; 7	2^{-30}
$(\delta,\delta,\delta,\delta,\delta,\delta,\delta,\delta) \xrightarrow{1r} (\delta,\delta,\delta,\delta,\delta,\delta,\delta,\delta)$	1; 4; 6; 7	1

2^{-450}. Starting from the i-th round, the following 26 subkeys have to be weak: $Z_7^{(i)}$, $Z_1^{(i+1)}$, $Z_4^{(i+1)}$, $Z_7^{(i+1)}$, $Z_4^{(i+2)}$, $Z_6^{(i+2)}$, $Z_7^{(i+2)}$, $Z_4^{(i+3)}$, $Z_1^{(i+4)}$, $Z_7^{(i+4)}$, $Z_1^{(i+5)}$, $Z_4^{(i+5)}$, $Z_6^{(i+5)}$, $Z_7^{(i+5)}$, $Z_4^{(i+6)}$, $Z_1^{(i+7)}$, $Z_6^{(i+7)}$, $Z_7^{(i+7)}$, $Z_4^{(i+8)}$, $Z_7^{(i+9)}$, $Z_1^{(i+10)}$, $Z_4^{(i+10)}$, $Z_6^{(i+10)}$, $Z_7^{(i+10)}$, $Z_1^{(i+11)}$ and $Z_4^{(i+11)}$.

(3) $(0,0,0,0,0,0,\delta,\delta) \xrightarrow{1r} (0,0,0,\delta,0,0,0,\delta) \xrightarrow{1r} (\delta,\delta,0,0,\delta,0,0,\delta) \xrightarrow{1r} (\delta,0,0,0,\delta,\delta,0,\delta) \xrightarrow{1r} (0,\delta,\delta,0,\delta,\delta,\delta,\delta) \xrightarrow{1r} (0,\delta,0,\delta,\delta,\delta,\delta,\delta) \xrightarrow{1r} (0,0,\delta,\delta,$

δ, δ, δ, δ) $\stackrel{1r}{\rightarrow}$ $(0,\delta,\delta,\delta,0,\delta,\delta,\delta)$ $\stackrel{1r}{\rightarrow}$ $(\delta,0,0,\delta,0,0,\delta,\delta)$ $\stackrel{1r}{\rightarrow}$ $(\delta, 0, \delta, \delta, 0, 0, 0, \delta)$ $\stackrel{1r}{\rightarrow}$ $(0,0,0,0,\delta,0,0,\delta)$ $\stackrel{1r}{\rightarrow}$ $(0,0,0,0,0,\delta,0,\delta)$ $\stackrel{1r}{\rightarrow}$ $(0, 0, 0, 0, 0, 0, \delta, \delta)$, which is a 12-round iterative characteristic holding with probability 2^{-330}. Starting from the i-th round, the following 22 subkeys have to be weak: $Z_7^{(i)}$, $Z_4^{(i+1)}$, $Z_1^{(i+2)}$, $Z_1^{(i+3)}$, $Z_6^{(i+3)}$, $Z_6^{(i+4)}$, $Z_7^{(i+4)}$, $Z_4^{(i+5)}$, $Z_6^{(i+5)}$, $Z_7^{(i+5)}$, $Z_4^{(i+6)}$, $Z_6^{(i+6)}$, $Z_7^{(i+6)}$, $Z_4^{(i+7)}$, $Z_6^{(i+7)}$, $Z_7^{(i+7)}$, $Z_1^{(i+8)}$, $Z_4^{(i+8)}$, $Z_7^{(i+8)}$, $Z_1^{(i+9)}$, $Z_4^{(i+9)}$ and $Z_6^{(i+11)}$.

(4) $(0,0,0,0,0,\delta,0,0)$ $\stackrel{1r}{\rightarrow}$ $(\delta,\delta,\delta,0,\delta,0,0,\delta)$ $\stackrel{1r}{\rightarrow}$ $(\delta,\delta,0,0,\delta,\delta,\delta,0)$ $\stackrel{1r}{\rightarrow}$ $(0, \delta, \delta, \delta, 0, \delta, 0, \delta)$ $\stackrel{1r}{\rightarrow}$ $(0,\delta,\delta,0,\delta,0,0,0)$ $\stackrel{1r}{\rightarrow}$ $(\delta,0,\delta,0,0,\delta,\delta,\delta)$ $\stackrel{1r}{\rightarrow}$ $(0, 0, \delta, \delta, \delta, 0, 0, 0)$ $\stackrel{1r}{\rightarrow}$ $(\delta,0,0,0,\delta,\delta,\delta,\delta)$ $\stackrel{1r}{\rightarrow}$ $(\delta,0,0,\delta,0,\delta,0,0)$ $\stackrel{1r}{\rightarrow}$ $(0, \delta, 0, 0, \delta, 0, 0, \delta)$ $\stackrel{1r}{\rightarrow}$ $(0,0,0,0,\delta,\delta,\delta,0)$ $\stackrel{1r}{\rightarrow}$ $(\delta,\delta,\delta,\delta,\delta,\delta,0,\delta)$ $\stackrel{1r}{\rightarrow}$ $(0, 0, 0, 0, 0, \delta, 0, 0)$, which is a 12-round iterative characteristic holding with probability 2^{-450}. Starting from the i-th round, the following 22 subkeys have to be weak: $Z_6^{(i)}$, $Z_1^{(i+1)}$, $Z_1^{(i+2)}$, $Z_6^{(i+2)}$, $Z_7^{(i+2)}$, $Z_4^{(i+3)}$, $Z_6^{(i+3)}$, $Z_1^{(i+5)}$, $Z_6^{(i+5)}$, $Z_7^{(i+5)}$, $Z_4^{(i+6)}$, $Z_1^{(i+7)}$, $Z_6^{(i+7)}$, $Z_7^{(i+7)}$, $Z_1^{(i+8)}$, $Z_4^{(i+8)}$, $Z_6^{(i+8)}$, $Z_6^{(i+10)}$, $Z_7^{(i+10)}$, $Z_1^{(i+11)}$, $Z_4^{(i+11)}$ and $Z_6^{(i+11)}$.

(5) $(0,0,0,0,0,\delta,\delta,0)$ $\stackrel{1r}{\rightarrow}$ $(0,0,0,\delta,0,0,\delta,0)$ $\stackrel{1r}{\rightarrow}$ $(\delta,\delta,0,\delta,\delta,0,0,0)$ $\stackrel{1r}{\rightarrow}$ $(0, \delta, 0, 0, 0, \delta, 0, 0)$ $\stackrel{1r}{\rightarrow}$ $(\delta,\delta,\delta,0,0,0,\delta,0)$ $\stackrel{1r}{\rightarrow}$ $(0,0,\delta,\delta,0,0,0,0)$ $\stackrel{1r}{\rightarrow}$ $(0, \delta, \delta, 0, 0, 0, 0, 0)$ $\stackrel{1r}{\rightarrow}$ $(0,\delta,0,0,\delta,0,0,0)$ $\stackrel{1r}{\rightarrow}$ $(\delta,\delta,\delta,0,0,\delta,0,0)$ $\stackrel{1r}{\rightarrow}$ $(0, 0, \delta, 0, 0, 0, \delta, 0)$ $\stackrel{1r}{\rightarrow}$ $(\delta,0,\delta,\delta,\delta,0,0,0)$ $\stackrel{1r}{\rightarrow}$ $(0,0,0,0,\delta,\delta,0,0)$ $\stackrel{1r}{\rightarrow}$ $(0, 0, 0, 0, 0, \delta, \delta, 0)$, which is a 12-round iterative characteristic holding with probability 2^{-360}. Starting from the i-th round, the following 16 subkeys have to be weak: $Z_6^{(i)}$, $Z_7^{(i)}$, $Z_4^{(i+1)}$, $Z_7^{(i+1)}$, $Z_1^{(i+2)}$, $Z_4^{(i+2)}$, $Z_6^{(i+2)}$, $Z_1^{(i+4)}$, $Z_7^{(i+4)}$, $Z_4^{(i+5)}$, $Z_1^{(i+8)}$, $Z_6^{(i+8)}$, $Z_7^{(i+9)}$, $Z_1^{(i+10)}$, $Z_4^{(i+10)}$ and $Z_6^{(i+11)}$.

(6) $(0,0,0,0,0,\delta,\delta,\delta)$ $\stackrel{1r}{\rightarrow}$ $(\delta,\delta,\delta,\delta,\delta,0,0,0)$ $\stackrel{1r}{\rightarrow}$ $(0,0,0,0,0,\delta,\delta,\delta)$, which is a 2-round iterative characteristic holding with probability 2^{-60}. Starting from the i-th round, the following four subkeys have to be weak: $Z_6^{(i)}$, $Z_7^{(i)}$, $Z_1^{(i+1)}$ and $Z_4^{(i+1)}$.

(7) $(0,0,0,0,\delta,0,0,0)$ $\stackrel{1r}{\rightarrow}$ $(\delta,\delta,\delta,0,\delta,\delta,\delta,\delta)$ $\stackrel{1r}{\rightarrow}$ $(\delta,\delta,0,\delta,\delta,\delta,0,0)$ $\stackrel{1r}{\rightarrow}$ $(\delta, 0, \delta, 0, \delta, \delta, 0, \delta)$ $\stackrel{1r}{\rightarrow}$ $(0,0,\delta,0,\delta,\delta,0,0)$ $\stackrel{1r}{\rightarrow}$ $(0,\delta,0,0,0,\delta,0,\delta)$ $\stackrel{1r}{\rightarrow}$ $(0, 0, 0, 0, \delta, 0, 0, 0)$, which is a 6-round iterative characteristic holding with probability 2^{-210}. Starting from the i-th round, the following ten subkeys have to be weak: $Z_1^{(i+1)}$, $Z_6^{(i+1)}$, $Z_7^{(i+1)}$, $Z_1^{(i+2)}$, $Z_4^{(i+2)}$, $Z_6^{(i+2)}$, $Z_1^{(i+3)}$, $Z_6^{(i+3)}$, $Z_6^{(i+4)}$ and $Z_6^{(i+4)}$.

(8) $(0,0,0,0,\delta,0,\delta,0)$ $\stackrel{1r}{\rightarrow}$ $(0,0,0,\delta,0,\delta,0,0)$ $\stackrel{1r}{\rightarrow}$ $(\delta,\delta,0,0,\delta,0,\delta,0)$ $\stackrel{1r}{\rightarrow}$ $(\delta, 0, 0, \delta, \delta, \delta, 0, 0)$ $\stackrel{1r}{\rightarrow}$ $(\delta,0,\delta,0,0,\delta,\delta,0)$ $\stackrel{1r}{\rightarrow}$ $(\delta,\delta,0,\delta,0,0,\delta,0)$ $\stackrel{1r}{\rightarrow}$ $(0, \delta, 0, \delta, 0, 0, 0, 0)$ $\stackrel{1r}{\rightarrow}$ $(0,0,\delta,0,\delta,0,0,0)$ $\stackrel{1r}{\rightarrow}$ $(\delta,0,\delta,0,\delta,\delta,0,0)$ $\stackrel{1r}{\rightarrow}$ $(\delta, \delta, 0, 0, 0, \delta, \delta, 0)$ $\stackrel{1r}{\rightarrow}$ $(\delta,0,0,\delta,\delta,0,\delta,0)$ $\stackrel{1r}{\rightarrow}$ $(\delta,0,\delta,\delta,0,\delta,0,0)$ $\stackrel{1r}{\rightarrow}$ $(0, 0, 0, 0, \delta, 0, \delta, 0)$, which is a 12-round iterative characteristic holding with probability

2^{-330}. Starting from the i-th round, the following 26 subkeys have to be weak: $Z_7^{(i)}$, $Z_4^{(i+1)}$, $Z_6^{(i+1)}$, $Z_1^{(i+2)}$, $Z_7^{(i+2)}$, $Z_1^{(i+3)}$, $Z_4^{(i+3)}$, $Z_6^{(i+3)}$, $Z_1^{(i+4)}$, $Z_6^{(i+4)}$, $Z_7^{(i+4)}$, $Z_1^{(i+5)}$, $Z_4^{(i+5)}$, $Z_7^{(i+5)}$, $Z_4^{(i+6)}$, $Z_1^{(i+8)}$, $Z_6^{(i+8)}$, $Z_1^{(i+9)}$, $Z_6^{(i+9)}$, $Z_7^{(i+9)}$, $Z_1^{(i+10)}$, $Z_4^{(i+10)}$, $Z_7^{(i+10)}$, $Z_1^{(i+11)}$, $Z_4^{(i+11)}$ and $Z_6^{(i+11)}$.

(9) $(0,0,0,0,\delta,0,\delta,\delta) \stackrel{1r}{\to} (\delta,\delta,\delta,\delta,\delta,\delta,\delta,0) \stackrel{1r}{\to} (0,0,0,\delta,0,\delta,0,\delta) \stackrel{1r}{\to} (0,0,\delta,0,0,0,0,0) \stackrel{1r}{\to} (0,\delta,0,0,0,0,\delta,\delta) \stackrel{1r}{\to} (0,0,0,\delta,\delta,0,\delta,0) \stackrel{1r}{\to} (0,0,\delta,\delta,0,\delta,\delta,\delta) \stackrel{1r}{\to} (\delta,0,0,\delta,\delta,0,0,0) \stackrel{1r}{\to} (0,\delta,0,0,\delta,\delta,\delta,\delta) \stackrel{1r}{\to} (0,0,0,\delta,\delta,\delta,0,0) \stackrel{1r}{\to} (0,0,\delta,0,0,\delta,0,\delta) \stackrel{1r}{\to} (0,\delta,0,0,0,0,0,0) \stackrel{1r}{\to} (0,0,0,0,\delta,0,\delta,\delta)$, which is a 12-round iterative characteristic holding with probability 2^{-420}. Starting from the i-th round, the following 20 subkeys have to be weak: $Z_7^{(i)}$, $Z_1^{(i+1)}$, $Z_4^{(i+1)}$, $Z_6^{(i+1)}$, $Z_7^{(i+1)}$, $Z_4^{(i+2)}$, $Z_6^{(i+2)}$, $Z_7^{(i+4)}$, $Z_4^{(i+5)}$, $Z_7^{(i+5)}$, $Z_4^{(i+6)}$, $Z_6^{(i+6)}$, $Z_7^{(i+6)}$, $Z_1^{(i+7)}$, $Z_4^{(i+7)}$, $Z_6^{(i+8)}$, $Z_7^{(i+8)}$, $Z_4^{(i+9)}$, $Z_6^{(i+9)}$ and $Z_6^{(i+10)}$.

(10) $(0,0,0,0,\delta,\delta,0,\delta) \stackrel{1r}{\to} (\delta,\delta,\delta,0,\delta,\delta,0,0) \stackrel{1r}{\to} (\delta,\delta,0,0,\delta,\delta,0,\delta) \stackrel{1r}{\to} (0,\delta,\delta,0,0,\delta,0,0) \stackrel{1r}{\to} (\delta,0,\delta,0,0,0,0,\delta) \stackrel{1r}{\to} (0,0,\delta,0,\delta,0,\delta,0) \stackrel{1r}{\to} (0,\delta,0,\delta,0,\delta,\delta,\delta) \stackrel{1r}{\to} (\delta,\delta,0,\delta,0,0,0,0) \stackrel{1r}{\to} (\delta,0,\delta,0,\delta,0,\delta,\delta) \stackrel{1r}{\to} (0,0,\delta,\delta,\delta,\delta,\delta,0) \stackrel{1r}{\to} (\delta,0,0,\delta,\delta,\delta,0,\delta) \stackrel{1r}{\to} (0,\delta,0,0,\delta,\delta,0,0) \stackrel{1r}{\to} (0,0,0,0,\delta,\delta,0,\delta)$, which is a 12-round iterative characteristic holding with probability 2^{-450}. Starting from the i-th round, the following 22 subkeys have to be weak: $Z_6^{(i)}$, $Z_1^{(i+1)}$, $Z_6^{(i+1)}$, $Z_1^{(i+2)}$, $Z_6^{(i+2)}$, $Z_6^{(i+3)}$, $Z_1^{(i+4)}$, $Z_7^{(i+5)}$, $Z_4^{(i+6)}$, $Z_6^{(i+6)}$, $Z_7^{(i+6)}$, $Z_1^{(i+7)}$, $Z_4^{(i+7)}$, $Z_1^{(i+8)}$, $Z_7^{(i+8)}$, $Z_4^{(i+9)}$, $Z_6^{(i+9)}$, $Z_7^{(i+9)}$, $Z_1^{(i+10)}$, $Z_4^{(i+10)}$, $Z_6^{(i+10)}$ and $Z_6^{(i+11)}$.

(11) $(0,0,0,0,\delta,\delta,\delta,\delta) \stackrel{1r}{\to} (0,0,0,\delta,0,\delta,\delta,\delta) \stackrel{1r}{\to} (\delta,\delta,0,\delta,\delta,0,\delta,\delta) \stackrel{1r}{\to} (0,\delta,0,\delta,0,\delta,0,\delta) \stackrel{1r}{\to} (0,0,\delta,0,\delta,0,\delta,\delta) \stackrel{1r}{\to} (\delta,0,\delta,\delta,\delta,\delta,0,\delta) \stackrel{1r}{\to} (0,0,0,0,\delta,\delta,\delta,\delta)$, which is a 6-round iterative characteristic holding with probability 2^{-150}. Starting from the i-th round, the following 14 subkeys have to be weak: $Z_6^{(i)}$, $Z_7^{(i)}$, $Z_4^{(i+1)}$, $Z_6^{(i+1)}$, $Z_7^{(i+1)}$, $Z_1^{(i+2)}$, $Z_4^{(i+2)}$, $Z_7^{(i+2)}$, $Z_4^{(i+3)}$, $Z_6^{(i+3)}$, $Z_7^{(i+4)}$, $Z_1^{(i+5)}$, $Z_4^{(i+5)}$ and $Z_6^{(i+5)}$.

(12) $(0,0,0,\delta,0,0,0,0) \stackrel{1r}{\to} (0,0,\delta,0,0,0,\delta,\delta) \stackrel{1r}{\to} (0,\delta,0,\delta,0,0,\delta,0) \stackrel{1r}{\to} (\delta,\delta,0,\delta,0,0,\delta,\delta) \stackrel{1r}{\to} (\delta,0,\delta,\delta,\delta,0,\delta,0) \stackrel{1r}{\to} (\delta,\delta,\delta,\delta,0,\delta,\delta,\delta) \stackrel{1r}{\to} (0,0,0,\delta,0,0,0,0)$, which is a 6-round iterative characteristic holding with probability 2^{-210}. Starting from the i-th round, the following 14 subkeys have to be weak: $Z_4^{(i)}$, $Z_7^{(i+1)}$, $Z_4^{(i+2)}$, $Z_7^{(i+2)}$, $Z_1^{(i+3)}$, $Z_4^{(i+3)}$, $Z_7^{(i+3)}$, $Z_1^{(i+4)}$, $Z_4^{(i+4)}$, $Z_7^{(i+4)}$, $Z_1^{(i+5)}$, $Z_4^{(i+5)}$, $Z_6^{(i+5)}$ and $Z_7^{(i+5)}$.

(13) $(0,0,0,\delta,0,0,\delta,\delta) \stackrel{1r}{\to} (0,0,\delta,\delta,0,0,\delta,0) \stackrel{1r}{\to} (\delta,0,0,\delta,\delta,0,\delta,\delta) \stackrel{1r}{\to} (0,\delta,0,\delta,\delta,\delta,\delta,0) \stackrel{1r}{\to} (\delta,\delta,0,\delta,0,\delta,0,\delta) \stackrel{1r}{\to} (\delta,0,\delta,0,\delta,0,0,0) \stackrel{1r}{\to} (0,0,\delta,0,\delta,\delta,\delta,\delta) \stackrel{1r}{\to} (0,\delta,0,\delta,0,\delta,0,0) \stackrel{1r}{\to} (\delta,\delta,0,0,0,0,0,\delta) \stackrel{1r}{\to} (0,\delta,\delta,0,0,0,\delta,0) \stackrel{1r}{\to} (\delta,0,\delta,\delta,0,0,\delta,\delta) \stackrel{1r}{\to} (\delta,\delta,\delta,\delta,0,0,\delta,0) \stackrel{1r}{\to} (0,0,0,\delta,0,0,\delta,\delta)$, which is a 12-round iterative characteristic holding with probability

2^{-450}. Starting from the i-th round, the following 26 subkeys have to be weak: $Z_4^{(i)}$, $Z_7^{(i)}$, $Z_4^{(i+1)}$, $Z_7^{(i+1)}$, $Z_1^{(i+2)}$, $Z_4^{(i+2)}$, $Z_7^{(i+2)}$, $Z_4^{(i+3)}$, $Z_6^{(i+3)}$, $Z_7^{(i+3)}$, $Z_1^{(i+4)}$, $Z_4^{(i+4)}$, $Z_6^{(i+4)}$, $Z_1^{(i+5)}$, $Z_6^{(i+6)}$, $Z_7^{(i+6)}$, $Z_4^{(i+7)}$, $Z_6^{(i+7)}$, $Z_1^{(i+8)}$, $Z_7^{(i+9)}$, $Z_1^{(i+10)}$, $Z_4^{(i+10)}$, $Z_7^{(i+10)}$, $Z_1^{(i+11)}$, $Z_4^{(i+11)}$ and $Z_7^{(i+11)}$.

(14) $(0,0,0,\delta,\delta,0,0,0) \stackrel{1r}{\rightarrow} (\delta,\delta,0,0,\delta,\delta,0,0) \stackrel{1r}{\rightarrow} (\delta,0,0,0,\delta,\delta,\delta,0) \stackrel{1r}{\rightarrow} (0,\delta,\delta,\delta,\delta,\delta,\delta,0) \stackrel{1r}{\rightarrow} (\delta,0,0,\delta,0,\delta,\delta,0) \stackrel{1r}{\rightarrow} (\delta,0,\delta,\delta,0,0,\delta,0) \stackrel{1r}{\rightarrow} (0,0,0,\delta,\delta,0,0,0)$, which is a 6-round iterative characteristic holding with probability 2^{-240}. Starting from the i-th round, the following 16 subkeys have to be weak: $Z_4^{(i)}$, $Z_1^{(i+1)}$, $Z_6^{(i+1)}$, $Z_1^{(i+2)}$, $Z_6^{(i+2)}$, $Z_7^{(i+2)}$, $Z_4^{(i+3)}$, $Z_6^{(i+3)}$, $Z_7^{(i+3)}$, $Z_1^{(i+4)}$, $Z_4^{(i+4)}$, $Z_6^{(i+4)}$, $Z_7^{(i+4)}$, $Z_1^{(i+5)}$, $Z_4^{(i+5)}$ and $Z_7^{(i+5)}$.

(15) $(0,0,0,\delta,\delta,0,0,\delta) \stackrel{1r}{\rightarrow} (0,0,\delta,0,0,\delta,\delta,0) \stackrel{1r}{\rightarrow} (0,\delta,0,\delta,0,0,0,\delta) \stackrel{1r}{\rightarrow} (\delta,\delta,0,0,0,0,\delta,0) \stackrel{1r}{\rightarrow} (0,\delta,\delta,\delta,0,0,\delta,\delta) \stackrel{1r}{\rightarrow} (0,\delta,\delta,\delta,\delta,0,\delta,0) \stackrel{1r}{\rightarrow} (0,\delta,\delta,\delta,\delta,\delta,\delta,\delta) \stackrel{1r}{\rightarrow} (0,\delta,\delta,\delta,\delta,\delta,0,0) \stackrel{1r}{\rightarrow} (0,\delta,\delta,0,\delta,\delta,0,\delta) \stackrel{1r}{\rightarrow} (\delta,0,\delta,0,0,\delta,0,0) \stackrel{1r}{\rightarrow} (0,0,\delta,0,\delta,0,0,\delta) \stackrel{1r}{\rightarrow} (0,\delta,0,0,0,\delta,\delta,0) \stackrel{1r}{\rightarrow} (0,0,0,\delta,\delta,0,0,\delta)$, which is a 12-round iterative characteristic holding with probability 2^{-480}. Starting from the i-th round, the following 20 subkeys have to be weak: $Z_4^{(i)}$, $Z_6^{(i+1)}$, $Z_7^{(i+1)}$, $Z_4^{(i+2)}$, $Z_1^{(i+3)}$, $Z_7^{(i+3)}$, $Z_4^{(i+4)}$, $Z_7^{(i+4)}$, $Z_4^{(i+5)}$, $Z_7^{(i+5)}$, $Z_4^{(i+6)}$, $Z_6^{(i+6)}$, $Z_7^{(i+6)}$, $Z_4^{(i+7)}$, $Z_6^{(i+7)}$, $Z_6^{(i+8)}$, $Z_1^{(i+9)}$, $Z_6^{(i+9)}$, $Z_6^{(i+11)}$ and $Z_7^{(i+11)}$.

(16) $(0,0,0,\delta,\delta,0,\delta,\delta) \stackrel{1r}{\rightarrow} (\delta,\delta,0,\delta,\delta,\delta,0,\delta) \stackrel{1r}{\rightarrow} (0,\delta,0,0,0,\delta,\delta,\delta) \stackrel{1r}{\rightarrow} (\delta,\delta,\delta,\delta,0,0,\delta,\delta) \stackrel{1r}{\rightarrow} (\delta,\delta,\delta,\delta,\delta,0,0,\delta) \stackrel{1r}{\rightarrow} (\delta,\delta,\delta,0,\delta,\delta,0,\delta) \stackrel{1r}{\rightarrow} (0,0,\delta,0,0,\delta,\delta,\delta) \stackrel{1r}{\rightarrow} (\delta,0,\delta,\delta,\delta,0,\delta,\delta) \stackrel{1r}{\rightarrow} (0,0,0,\delta,\delta,\delta,0,\delta) \stackrel{1r}{\rightarrow} (\delta,\delta,0,0,\delta,\delta,\delta,\delta) \stackrel{1r}{\rightarrow} (\delta,0,0,\delta,\delta,\delta,\delta,\delta) \stackrel{1r}{\rightarrow} (\delta,0,\delta,\delta,0,\delta,\delta,\delta) \stackrel{1r}{\rightarrow} (0,0,0,\delta,\delta,0,\delta,\delta)$, which is a 12-round iterative characteristic holding with probability 2^{-360}. Starting from the i-th round, the following 32 subkeys have to be weak: $Z_4^{(i)}$, $Z_7^{(i)}$, $Z_1^{(i+1)}$, $Z_4^{(i+1)}$, $Z_6^{(i+1)}$, $Z_6^{(i+2)}$, $Z_7^{(i+2)}$, $Z_1^{(i+3)}$, $Z_4^{(i+3)}$, $Z_7^{(i+3)}$, $Z_1^{(i+4)}$, $Z_4^{(i+4)}$, $Z_1^{(i+5)}$, $Z_6^{(i+5)}$, $Z_6^{(i+6)}$, $Z_7^{(i+6)}$, $Z_1^{(i+7)}$, $Z_4^{(i+7)}$, $Z_7^{(i+7)}$, $Z_4^{(i+8)}$, $Z_6^{(i+8)}$, $Z_1^{(i+9)}$, $Z_6^{(i+9)}$, $Z_7^{(i+9)}$, $Z_1^{(i+10)}$, $Z_4^{(i+10)}$, $Z_6^{(i+10)}$, $Z_7^{(i+10)}$, $Z_1^{(i+11)}$, $Z_4^{(i+11)}$, $Z_6^{(i+11)}$ and $Z_7^{(i+11)}$.

(17) $(0,0,0,\delta,\delta,\delta,\delta,0) \stackrel{1r}{\rightarrow} (\delta,\delta,0,\delta,\delta,\delta,\delta,0) \stackrel{1r}{\rightarrow} (0,\delta,0,\delta,0,\delta,\delta,0) \stackrel{1r}{\rightarrow} (0,0,\delta,\delta,\delta,0,\delta,0) \stackrel{1r}{\rightarrow} (0,\delta,\delta,\delta,0,\delta,0,0) \stackrel{1r}{\rightarrow} (\delta,0,0,0,0,0,\delta,0) \stackrel{1r}{\rightarrow} (0,\delta,\delta,\delta,\delta,0,0,0) \stackrel{1r}{\rightarrow} (\delta,0,0,0,0,\delta,0,0) \stackrel{1r}{\rightarrow} (0,\delta,\delta,0,\delta,0,\delta,0) \stackrel{1r}{\rightarrow} (0,\delta,0,\delta,\delta,\delta,0,0) \stackrel{1r}{\rightarrow} (0,0,\delta,0,\delta,\delta,\delta,0) \stackrel{1r}{\rightarrow} (\delta,0,\delta,\delta,\delta,\delta,\delta,0) \stackrel{1r}{\rightarrow} (0,0,0,\delta,\delta,\delta,\delta,0)$, which is a 12-round iterative characteristic holding with probability 2^{-420}. Starting from the i-th round, the following 28 subkeys have to be weak: $Z_4^{(i)}$, $Z_6^{(i)}$, $Z_7^{(i)}$, $Z_1^{(i+1)}$, $Z_4^{(i+1)}$, $Z_6^{(i+1)}$, $Z_7^{(i+1)}$, $Z_4^{(i+2)}$, $Z_6^{(i+2)}$, $Z_7^{(i+2)}$, $Z_4^{(i+3)}$, $Z_7^{(i+3)}$, $Z_4^{(i+4)}$, $Z_6^{(i+4)}$, $Z_1^{(i+5)}$, $Z_7^{(i+5)}$, $Z_4^{(i+6)}$, $Z_1^{(i+7)}$, $Z_6^{(i+7)}$, $Z_7^{(i+8)}$, $Z_4^{(i+9)}$, $Z_6^{(i+9)}$, $Z_6^{(i+10)}$, $Z_7^{(i+10)}$, $Z_1^{(i+11)}$, $Z_4^{(i+11)}$, $Z_6^{(i+11)}$ and $Z_7^{(i+11)}$.

(18) $(0,0,0,\delta,\delta,\delta,\delta,\delta) \xrightarrow{1r} (0,0,\delta,\delta,0,\delta,0,0) \xrightarrow{1r} (\delta,0,0,0,\delta,0,0,\delta) \xrightarrow{1r} (\delta,0,0,0,0,\delta,\delta,0) \xrightarrow{1r} (\delta,0,0,\delta,0,0,0,\delta) \xrightarrow{1r} (0,\delta,0,0,\delta,0,\delta,0) \xrightarrow{1r} (0,0,0,\delta,\delta,\delta,\delta,\delta)$, which is a 6-round iterative characteristic holding with probability 2^{-240}. Starting from the i-th round, the following 12 subkeys have to be weak: $Z_4^{(i)}$, $Z_6^{(i)}$, $Z_7^{(i)}$, $Z_4^{(i+1)}$, $Z_6^{(i+1)}$, $Z_1^{(i+2)}$, $Z_1^{(i+3)}$, $Z_6^{(i+3)}$, $Z_7^{(i+3)}$, $Z_1^{(i+4)}$, $Z_4^{(i+4)}$ and $Z_7^{(i+5)}$.

(19) $(0,0,\delta,0,0,0,0,\delta) \xrightarrow{1r} (\delta,0,\delta,0,\delta,0,0,\delta) \xrightarrow{1r} (\delta,\delta,0,0,0,\delta,0,\delta) \xrightarrow{1r} (\delta,0,0,0,\delta,0,\delta,\delta) \xrightarrow{1r} (0,\delta,\delta,\delta,\delta,\delta,0,\delta) \xrightarrow{1r} (\delta,0,0,0,0,\delta,\delta,\delta) \xrightarrow{1r} (0,\delta,\delta,\delta,0,\delta,\delta) \xrightarrow{1r} (\delta,0,0,\delta,0,\delta,0,\delta) \xrightarrow{1r} (\delta,0,\delta,0,0,0,\delta,\delta) \xrightarrow{1r} (\delta,\delta,0,\delta,0,0,0,\delta) \xrightarrow{1r} (0,\delta,0,0,0,0,0,\delta) \xrightarrow{1r} (\delta,\delta,\delta,0,0,0,0,\delta) \xrightarrow{1r} (0,0,\delta,0,0,0,0,\delta)$, which is a 12-round iterative characteristic holding with probability 2^{-420}. Starting from the i-th round, the following 20 subkeys have to be weak: $Z_1^{(i+1)}$, $Z_1^{(i+2)}$, $Z_6^{(i+2)}$, $Z_1^{(i+3)}$, $Z_7^{(i+3)}$, $Z_4^{(i+4)}$, $Z_6^{(i+4)}$, $Z_1^{(i+5)}$, $Z_6^{(i+5)}$, $Z_7^{(i+5)}$, $Z_4^{(i+6)}$, $Z_7^{(i+6)}$, $Z_1^{(i+7)}$, $Z_4^{(i+7)}$, $Z_6^{(i+7)}$, $Z_1^{(i+8)}$, $Z_7^{(i+8)}$, $Z_1^{(i+9)}$, $Z_4^{(i+9)}$ and $Z_1^{(i+11)}$.

(20) $(0,0,\delta,0,0,\delta,0,0) \xrightarrow{1r} (\delta,0,\delta,0,\delta,0,\delta,0) \xrightarrow{1r} (\delta,\delta,0,\delta,0,\delta,0,0) \xrightarrow{1r} (0,\delta,0,0,0,0,\delta,0) \xrightarrow{1r} (\delta,\delta,\delta,\delta,0,0,0,0) \xrightarrow{1r} (\delta,\delta,\delta,0,\delta,0,0,0) \xrightarrow{1r} (0,0,\delta,0,0,\delta,0,0)$, which is a 6-round iterative characteristic holding with probability 2^{-150}. Starting from the i-th round, the following ten subkeys have to be weak: $Z_6^{(i)}$, $Z_1^{(i+1)}$, $Z_7^{(i+1)}$, $Z_1^{(i+2)}$, $Z_4^{(i+2)}$, $Z_6^{(i+2)}$, $Z_7^{(i+3)}$, $Z_1^{(i+4)}$, $Z_4^{(i+4)}$ and $Z_1^{(i+5)}$.

(21) $(0,0,\delta,0,\delta,\delta,0,\delta) \xrightarrow{1r} (\delta,0,\delta,0,\delta,\delta,\delta,\delta) \xrightarrow{1r} (\delta,\delta,0,\delta,0,\delta,\delta,\delta) \xrightarrow{1r} (0,\delta,0,\delta,0,0,\delta,\delta) \xrightarrow{1r} (0,0,\delta,\delta,\delta,0,0,\delta) \xrightarrow{1r} (0,\delta,\delta,0,0,\delta,0,\delta) \xrightarrow{1r} (0,\delta,0,0,\delta,0,\delta,\delta) \xrightarrow{1r} (\delta,\delta,\delta,\delta,0,\delta,0,\delta) \xrightarrow{1r} (\delta,\delta,\delta,0,\delta,0,\delta,\delta) \xrightarrow{1r} (0,0,\delta,\delta,0,\delta,0,\delta) \xrightarrow{1r} (0,\delta,\delta,0,0,0,\delta,\delta) \xrightarrow{1r} (0,\delta,0,\delta,\delta,0,0,\delta) \xrightarrow{1r} (0,0,\delta,0,\delta,\delta,0,\delta)$, which is a 12-round iterative characteristic holding with probability 2^{-330}. Starting from the i-th round, the following 22 subkeys have to be weak: $Z_6^{(i)}$, $Z_1^{(i+1)}$, $Z_6^{(i+1)}$, $Z_7^{(i+1)}$, $Z_1^{(i+2)}$, $Z_4^{(i+2)}$, $Z_6^{(i+2)}$, $Z_7^{(i+2)}$, $Z_4^{(i+3)}$, $Z_7^{(i+3)}$, $Z_4^{(i+4)}$, $Z_6^{(i+5)}$, $Z_7^{(i+6)}$, $Z_1^{(i+7)}$, $Z_4^{(i+7)}$, $Z_6^{(i+7)}$, $Z_1^{(i+8)}$, $Z_7^{(i+8)}$, $Z_4^{(i+9)}$, $Z_6^{(i+9)}$, $Z_7^{(i+10)}$ and $Z_4^{(i+11)}$.

(22) $(0,0,\delta,\delta,0,0,\delta,\delta) \xrightarrow{1r} (0,\delta,\delta,\delta,0,0,0,\delta) \xrightarrow{1r} (\delta,0,0,0,0,0,0,\delta) \xrightarrow{1r} (0,\delta,\delta,0,\delta,0,0,\delta) \xrightarrow{1r} (0,\delta,0,0,\delta,\delta,0,\delta) \xrightarrow{1r} (\delta,\delta,\delta,0,0,\delta,\delta,\delta) \xrightarrow{1r} (0,0,\delta,\delta,0,0,\delta,\delta)$, which is a 6-round iterative characteristic holding with probability 2^{-240}. Starting from the i-th round, the following eight subkeys have to be weak: $Z_4^{(i)}$, $Z_7^{(i)}$, $Z_4^{(i+1)}$, $Z_1^{(i+2)}$, $Z_6^{(i+4)}$, $Z_1^{(i+5)}$, $Z_6^{(i+5)}$ and $Z_7^{(i+5)}$.

(23) $(0,0,\delta,\delta,0,\delta,\delta,0) \xrightarrow{1r} (0,\delta,\delta,\delta,0,0,\delta,0) \xrightarrow{1r} (\delta,0,0,\delta,0,0,0,0) \xrightarrow{1r} (\delta,0,\delta,0,0,0,0,0) \xrightarrow{1r} (\delta,\delta,0,0,0,0,0,0) \xrightarrow{1r} (\delta,0,0,0,\delta,0,0,0) \xrightarrow{1r} (0,\delta,\delta,0,\delta,\delta,0,0) \xrightarrow{1r} (0,\delta,0,0,\delta,\delta,\delta,0) \xrightarrow{1r} (\delta,\delta,\delta,\delta,0,\delta,\delta,0) \xrightarrow{1r} (\delta,\delta,\delta,\delta,\delta,0,\delta,0) \xrightarrow{1r} (\delta,\delta,\delta,\delta,\delta,\delta,0,0) \xrightarrow{1r} (\delta,\delta,\delta,0,\delta,\delta,\delta,0) \xrightarrow{1r} (0,0,\delta,\delta,0,\delta,\delta,0)$, which is a

12-round iterative characteristic holding with probability 2^{-330}. Starting from the i-th round, the following 26 subkeys have to be weak: $Z_4^{(i)}$, $Z_6^{(i)}$, $Z_7^{(i)}$, $Z_4^{(i+1)}$, $Z_7^{(i+1)}$, $Z_1^{(i+2)}$, $Z_4^{(i+2)}$, $Z_1^{(i+3)}$, $Z_1^{(i+4)}$, $Z_1^{(i+5)}$, $Z_6^{(i+6)}$, $Z_6^{(i+7)}$, $Z_7^{(i+7)}$, $Z_1^{(i+8)}$, $Z_4^{(i+8)}$, $Z_6^{(i+8)}$, $Z_7^{(i+8)}$, $Z_1^{(i+9)}$, $Z_4^{(i+9)}$, $Z_7^{(i+9)}$, $Z_1^{(i+10)}$, $Z_4^{(i+10)}$, $Z_6^{(i+10)}$, $Z_1^{(i+11)}$, $Z_6^{(i+11)}$ and $Z_7^{(i+11)}$.

(24) $(0,0,\delta,\delta,\delta,0,\delta,\delta) \xrightarrow{1r} (\delta,0,0,\delta,\delta,\delta,\delta,0) \xrightarrow{1r} (0,\delta,0,\delta,\delta,\delta,0,\delta) \xrightarrow{1r} (\delta, \delta, 0, 0, 0, \delta, 0, 0) \xrightarrow{1r} (0,\delta,\delta,0,0,0,0,\delta) \xrightarrow{1r} (\delta,0,\delta,0,0,0,\delta,0) \xrightarrow{1r} (0, 0, \delta, \delta, \delta, 0, \delta, \delta)$, which is a 6-round iterative characteristic holding with probability 2^{-150}. Starting from the i-th round, the following 12 subkeys have to be weak: $Z_4^{(i)}$, $Z_7^{(i)}$, $Z_1^{(i+1)}$, $Z_4^{(i+1)}$, $Z_6^{(i+1)}$, $Z_7^{(i+1)}$, $Z_4^{(i+2)}$, $Z_6^{(i+2)}$, $Z_1^{(i+3)}$, $Z_6^{(i+3)}$, $Z_1^{(i+5)}$ and $Z_7^{(i+5)}$.

(25) $(0,0,\delta,\delta,\delta,\delta,0,0) \xrightarrow{1r} (0,\delta,\delta,0,0,\delta,\delta,0) \xrightarrow{1r} (0,\delta,0,\delta,\delta,0,\delta,0) \xrightarrow{1r} (0, 0, \delta, \delta, \delta, \delta, 0, 0)$, which is a 3-round iterative characteristic holding with probability 2^{-90}. Starting from the i-th round, the following six subkeys have to be weak: $Z_4^{(i)}$, $Z_6^{(i)}$, $Z_6^{(i+1)}$, $Z_7^{(i+1)}$, $Z_4^{(i+2)}$ and $Z_7^{(i+2)}$.

(26) $(0,0,\delta,\delta,\delta,\delta,0,\delta) \xrightarrow{1r} (\delta,0,0,0,\delta,\delta,0,0) \xrightarrow{1r} (\delta,0,0,0,0,\delta,0,\delta) \xrightarrow{1r} (\delta, 0, 0, 0, 0, 0, 0, 0) \xrightarrow{1r} (\delta,0,0,0,0,0,\delta,\delta) \xrightarrow{1r} (\delta,0,0,\delta,0,0,\delta,0) \xrightarrow{1r} (0, \delta, 0, \delta, \delta, 0, \delta, \delta) \xrightarrow{1r} (\delta,\delta,0,\delta,0,\delta,\delta,0) \xrightarrow{1r} (\delta,0,\delta,\delta,\delta,0,0,\delta) \xrightarrow{1r} (\delta, \delta, \delta, 0, 0, \delta, \delta, 0) \xrightarrow{1r} (\delta,\delta,0,\delta,\delta,0,0,\delta) \xrightarrow{1r} (\delta,0,\delta,0,\delta,\delta,\delta,0) \xrightarrow{1r} (0, 0, \delta, \delta, \delta, \delta, 0, \delta)$, which is a 12-round iterative characteristic holding with probability 2^{-480}. Starting from the i-th round, the following 28 subkeys have to be weak: $Z_4^{(i)}$, $Z_6^{(i)}$, $Z_1^{(i+1)}$, $Z_6^{(i+1)}$, $Z_1^{(i+2)}$, $Z_6^{(i+2)}$, $Z_1^{(i+3)}$, $Z_1^{(i+4)}$, $Z_7^{(i+4)}$, $Z_1^{(i+5)}$, $Z_4^{(i+5)}$, $Z_7^{(i+5)}$, $Z_4^{(i+6)}$, $Z_7^{(i+6)}$, $Z_1^{(i+7)}$, $Z_4^{(i+7)}$, $Z_6^{(i+7)}$, $Z_7^{(i+7)}$, $Z_1^{(i+8)}$, $Z_4^{(i+8)}$, $Z_1^{(i+9)}$, $Z_6^{(i+9)}$, $Z_7^{(i+9)}$, $Z_1^{(i+10)}$, $Z_4^{(i+10)}$, $Z_1^{(i+11)}$, $Z_6^{(i+11)}$ and $Z_7^{(i+11)}$.

(27) $(0,\delta,\delta,0,\delta,\delta,\delta,0) \xrightarrow{1r} (\delta,0,\delta,\delta,0,\delta,0,\delta) \xrightarrow{1r} (\delta,\delta,\delta,0,0,0,0,0) \xrightarrow{1r} (\delta, \delta, 0, 0, \delta, 0, \delta, \delta) \xrightarrow{1r} (0,\delta,\delta,\delta,0,\delta,\delta,0) \xrightarrow{1r} (0,\delta,\delta,\delta,\delta,0,0,\delta) \xrightarrow{1r} (0, \delta, \delta, 0, \delta, \delta, \delta, 0)$, which is a 6-round iterative characteristic holding with probability 2^{-240}. Starting from the i-th round, the following 12 subkeys have to be weak: $Z_6^{(i)}$, $Z_7^{(i)}$, $Z_1^{(i+1)}$, $Z_4^{(i+1)}$, $Z_6^{(i+1)}$, $Z_1^{(i+2)}$, $Z_1^{(i+3)}$, $Z_7^{(i+3)}$, $Z_4^{(i+4)}$, $Z_6^{(i+4)}$, $Z_7^{(i+4)}$ and $Z_4^{(i+5)}$.

(28) $(\delta,0,0,\delta,\delta,0,0,\delta) \xrightarrow{1r} (\delta,0,\delta,0,0,\delta,0,\delta) \xrightarrow{1r} (\delta,\delta,0,0,0,0,\delta,\delta) \xrightarrow{1r} (\delta, 0, 0, \delta, \delta, 0, 0, \delta)$, which is a 3-round iterative characteristic holding with probability 2^{-90}. Starting from the i-th round, the following six subkeys have to be weak: $Z_1^{(i)}$, $Z_4^{(i)}$, $Z_1^{(i+1)}$, $Z_6^{(i+1)}$, $Z_1^{(i+2)}$ and $Z_7^{(i+2)}$.

(29) $(\delta,\delta,\delta,\delta,\delta,\delta,\delta,\delta) \xrightarrow{1r} (\delta,\delta,\delta,\delta,\delta,\delta,\delta,\delta)$, which is a 1-round iterative characteristic holding with probability 1. Starting from the i-th round, the following four subkeys have to be weak: $Z_1^{(i)}$, $Z_4^{(i)}$, $Z_6^{(i)}$ and $Z_7^{(i)}$.

Constructing multiple-round differential characteristics involves balancing several criteria:

(i) covering as many rounds as possible.
(ii) choose preferably *narrow trails*, which means trails with low wordwise Hamming Weight differences. The objective is to keep the overall probability as high as possible by activating as few nonlinear components as possible per round.
(iii) minimize the number of weak-subkey assumptions as well as the number of active unkeyed $\odot$.

According to criterion (iii), the characteristics (5) and (22) require the least number of weak subkeys per round: about 1.33, but (5) holds with probability 2^{-360}, while (22) holds with probability 2^{-240}.

Since (5) is 12-round iterative, it means an average probability of 2^{-30} per round, while for (22), which is 6-round iterative, it means an average probability of 2^{-40} per round.

Considering only the probability, characteristic (29) holds with certainty but requires four weak subkeys per round. Moreover, it forms a wide trail. On the other hand, (29) bypasses the MA-box altogether in every round, thus, minimizing the number of unkeyed $\odot$, which is not the case for (5) not (22). That is why (29) holds with much higher probability than the others. Bypassing the MA-box means that the input and output differences to the MA are both $(0, 0, 0, 0)$.

Concerning criterion (i), all three characteristics are iterative and can cover an arbitrary number of rounds. Concerning criterion (ii), characteristics (5) and (22) both contain narrower trails than (29).

Balancing competing criteria typically requires some trade-off.

Taking into account that REESSE3+ iterates 8.5 rounds and has a block size of 128 bits, there are not enough text pairs that can be formed to satisfy characteristics (5) or (22) across 8.5 rounds, due to their low probability.

In summary, the best trade-off is achieved with the 1-round iterative characteristic (29).

It is straightforward to implement a search algorithm to count how many user keys exist that lead to weak subkeys at the positions $Z_1^{(i)}$, $Z_4^{(i)}$, $Z_6^{(i)}$ and $Z_7^{(i)}$, for as many rounds i as needed.

Finding the number of weak keys is a simple counting procedure. First consider that REESSE3+ uses a 256-bit key and it key schedule algorithm consists of simple bit shifting operation (see Sect. 2.7.2.

A weak subkey means that its most significant fifteen bits have to be zero. By applying this simple reasoning to each of the four multiplicative subkeys listed above, and checking how many of the 256 key bits are not affected, we find how many bits are free to hold any value. For instance, for the first round, $4 \cdot 15 = 60$ bits of the user key K will have to be zero, while the remaining $256 - 60 = 196$ bits are free to hold any value. This means that for a subset of 2^{196} keys K, characteristic (29) will hold with certainty.

Further, concatenating (29) with itself, that is, covering two rounds, will require zeroing 60 more bits, but due to overlapping bits which are already zero (from the first round), only 46 bits are set to zero. This leaves 150 free bits. It means a weak-key class of 2^{150} keys K satisfying the 2-round characteristic using (29).

Table 3.69 describes the evolution of the weak-key class sizes for characteristic (29) for an increasing number of rounds.

Table 3.69 Evolution of weak-key class size in characteristic (29) for the full 8.5-round REESSE3+.

Weak subkeys	Weak-key class size
$Z_1^{(1)}, Z_4^{(1)}, Z_6^{(1)}, Z_7^{(1)}$	2^{196}
$Z_1^{(2)}, Z_4^{(2)}, Z_6^{(2)}, Z_7^{(2)}$	2^{150}
$Z_1^{(3)}, Z_4^{(3)}, Z_6^{(3)}, Z_7^{(3)}$	2^{124}
$Z_1^{(4)}, Z_4^{(4)}, Z_6^{(4)}, Z_7^{(4)}$	2^{100}
$Z_1^{(5)}, Z_4^{(5)}, Z_6^{(5)}, Z_7^{(5)}$	2^{76}
$Z_1^{(6)}, Z_4^{(6)}, Z_6^{(6)}, Z_7^{(6)}$	2^{52}
$Z_1^{(7)}, Z_4^{(7)}, Z_6^{(7)}, Z_7^{(7)}$	2^{33}
$Z_1^{(8)}, Z_4^{(8)}, Z_6^{(8)}, Z_7^{(8)}$	2^{23}
$Z_1^{(9)}, Z_4^{(9)}, Z_6^{(9)}, Z_7^{(9)}$	2^{15}

For 8.5 rounds, which covers all KM half-rounds, there are 2^{15} weak keys.

Therefore, there are 2^{15} keys, in a key space of 2^{256} keys, for which (29) concatenated with itself nine times becomes a differential distinguisher for the full 8.5-round REESSE3+. Such keys have free bits in the following fifteen positions: 20, 45, 121, 122, 123, 124, 125, 126, 127, 128, 151, 176, 201, 226, 251 (recall that K is numbered in left-to-right order from 0 to 255).

Even though REESSE3+ was defined with 8.5 rounds, one can extend it to more rounds, since the key schedule algorithm can generate subkeys for as many rounds as needed. Let us assume, hypothetically, that REESSE3+ had double the number of rounds (16.5) instead. We can continue with the previous analysis, up to 16 rounds (not including the final output transformation). The result is a weak-key class containing 2 weak keys. Only the bit position 126 in K is free to hold any value.

This means that 16-round REESSE3+ still has weak keys. The existence of even a single weak-key value is enough to jeopardize the use of REESSE3+ in the construction of other cryptographic primitives, such as hash function, via some modes of operation [162] in which the key entry is part of the message to be hashed. For instance, an adversary could choose a weak key as message, and further manipulate the other entries to lead to a collision. This analysis has already been performed for IDEA and PES ciphers in [149]. See Sect. 3.5.8.3.

Another consequence of the key schedule design of REESSE3+ is that the full 8.5-round REESSE3+ can be distinguished from a random permutation with very low attack complexity, since the characteristic holds with certainty. Apart from the weak-key assumptions, the cost is the same as that of the attacks on PES and IDEA: essentially two encryptions, two chosen plaintexts and equivalent memory.

The reasoning is the following: given the input difference $(\delta, \delta, \delta, \delta, \delta, \delta, \delta, \delta)$ to both 8.5-round REESSE3+ (under a weak-key) and a random 128-bit permutation π, the output difference $(\delta, \delta, \delta, \delta, \delta, \delta, \delta, \delta)$ will show up with certainty for REESSE3+, while for π, it will happen with probability about 2^{-128}. The advantage (see Sect. 1.7) to distinguish one from the other is: $1 - 2^{-128}$, which is non-negligible. Therefore, the full 8.5-round REESSE3+ cipher is not an ideal primitive, that is, it is not a PRP.

3.5.11 DC of IDEA Without Weak-Key Assumptions*

The following differential analysis of IDEA* is based on [121].

The initial approach is to obtain 1-round differential characteristics and further derive multi-round characteristics.

The difference operator is bitwise exclusive-or and it operates wordwise. The only 16-bit difference values used are $\delta = 8000_x$ and 0000_x.

Across unkeyed $\odot$ and $\boxplus$, the propagation of 16-bit difference words follows from Table 3.12 and Table 3.13, respectively.

Table 3.70 lists all fifteen nontrivial 1-round differential characteristics of IDEA* using only wordwise differences 0000_x and 8000_x. There are *no weak keys nor weak-subkey assumptions* for these characteristics.

Table 3.70 Nontrivial 1-round characteristics for IDEA* ($\delta = 8000_x$).

1-round characteristics	Differences across the MA-box	Prob.	# Active (unkeyed) $\odot$	# Active (unkeyed) $\boxdot$
$(0,0,0,\delta) \xrightarrow{1r} (\delta,\delta,\delta,0)$	$(0,\delta) \rightarrow (\delta,\delta)$	$2^{-74.98}$	4	1
$(0,0,\delta,0) \xrightarrow{1r} (\delta,0,0,0)$	$(\delta,0) \rightarrow (0,\delta)$	2^{-45}	2	1
$(0,0,\delta,\delta) \xrightarrow{1r} (0,\delta,\delta,0)$	$(\delta,\delta) \rightarrow (\delta,0)$	2^{-75}	3	2
$(0,\delta,0,0) \xrightarrow{1r} (\delta,\delta,0,\delta)$	$(0,\delta) \rightarrow (\delta,\delta)$	$2^{-74.98}$	4	1
$(0,\delta,0,\delta) \xrightarrow{1r} (0,0,\delta,\delta)$	$(0,0) \rightarrow (0,0)$	2^{-45}	2	1
$(0,\delta,\delta,0) \xrightarrow{1r} (0,\delta,0,\delta)$	$(\delta,\delta) \rightarrow (\delta,0)$	2^{-75}	3	2
$(0,\delta,\delta,\delta) \xrightarrow{1r} (\delta,0,\delta,\delta)$	$(\delta,0) \rightarrow (0,\delta)$	2^{-90}	4	2
$(\delta,0,0,0) \xrightarrow{1r} (0,\delta,0,0)$	$(\delta,0) \rightarrow (0,\delta)$	2^{-45}	2	1
$(\delta,0,0,\delta) \xrightarrow{1r} (\delta,0,\delta,0)$	$(\delta,\delta) \rightarrow (\delta,0)$	2^{-75}	3	2
$(\delta,0,\delta,0) \xrightarrow{1r} (\delta,\delta,0,0)$	$(0,0) \rightarrow (0,0)$	2^{-45}	2	1
$(\delta,0,\delta,\delta) \xrightarrow{1r} (0,0,\delta,0)$	$(0,\delta) \rightarrow (\delta,\delta)$	2^{-90}	4	2
$(\delta,\delta,0,0) \xrightarrow{1r} (\delta,0,0,\delta)$	$(\delta,\delta) \rightarrow (\delta,0)$	2^{-75}	3	2
$(\delta,\delta,0,\delta) \xrightarrow{1r} (0,\delta,\delta,\delta)$	$(\delta,0) \rightarrow (0,\delta)$	2^{-90}	4	2
$(\delta,\delta,\delta,0) \xrightarrow{1r} (0,0,0,\delta)$	$(0,\delta) \rightarrow (\delta,\delta)$	2^{-90}	4	2
$(\delta,\delta,\delta,\delta) \xrightarrow{1r} (\delta,\delta,\delta,\delta)$	$(0,0) \rightarrow (0,0)$	2^{-90}	4	2

Multiple-round characteristics are obtained by concatenating 1-round characteristics, such that the output difference of the top one matches the input difference of the characteristic in the next round.

From Table 3.70, there are five multiple-round characteristics:

(1) $(0,0,0,\delta) \stackrel{1r}{\rightarrow} (\delta,\delta,\delta,0) \stackrel{1r}{\rightarrow} (0,0,0,\delta)$, which is 2-round iterative and holds with probability $2^{-164.98}$.

(2) $(0,0,\delta,0) \stackrel{1r}{\rightarrow} (\delta,0,0,0) \stackrel{1r}{\rightarrow} (0,\delta,0,0) \stackrel{1r}{\rightarrow} (\delta,\delta,0,\delta) \stackrel{1r}{\rightarrow} (0,\delta,\delta,\delta) \stackrel{1r}{\rightarrow} (\delta,0,\delta,\delta) \stackrel{1r}{\rightarrow} (0,0,\delta,0)$, which is 6-round iterative and holds with probability $2^{-434.98}$.

(3) $(0,0,\delta,\delta) \stackrel{1r}{\rightarrow} (0,\delta,\delta,0) \stackrel{1r}{\rightarrow} (0,\delta,0,\delta) \stackrel{1r}{\rightarrow} (0,0,\delta,\delta)$, which is 3-round iterative and holds with probability 2^{-195}.

(4) $(\delta,0,0,\delta) \stackrel{1r}{\rightarrow} (\delta,0,\delta,0) \stackrel{1r}{\rightarrow} (\delta,\delta,0,0) \stackrel{1r}{\rightarrow} (\delta,0,0,\delta)$, which is 3-round iterative and holds with probability 2^{-195}.

(5) $(\delta,\delta,\delta,\delta) \stackrel{1r}{\rightarrow} (\delta,\delta,\delta,\delta)$, which is 1-round iterative and holds with probability 2^{-90}.

Studying these multi-round characteristics and taking into account the block size of IDEA*, at most one round can be covered. For instance, $(0,0,\delta,0) \stackrel{1r}{\rightarrow} (\delta,0,0,0)$ which holds with probability 2^{-45}.

Up to $2^{15} \cdot (2^{16})^3 = 2^{63}$ text pairs can be created with an input difference of the form $(0,0,\delta,0)$. Extending this characteristic or any of the others leads to a probability lower than 2^{-90}, and there will be not enough text pairs to obtain a right pair.

This differential analysis using characteristics also allows us to study the diffusion power of IDEA* with respect to modern tools such as the differential branch number (Appendix, Sect. B).

Analyzing the set of nontrivial 1-round characteristics in Table 3.70, we observe that the differential branch number *per round* of IDEA* is only 2 (without any weak-key assumption). For instance, the sum of the (wordwise) Hamming Weight of the input and output differences of the 1-round characteristic

$$(0,0,\delta,0) \stackrel{1r}{\rightarrow} (\delta,0,0,0)$$

is 2. So, even though IDEA* provides full (text) diffusion in a single round, the previous analysis shows that its differential branch number is much lower than ideal. For a 4-word block, such as in IDEA*, an MDS code using a 4×4 MDS matrix would rather provide a branch number of 5. See Appendix, Sect. B.1, for an explanation of the differential branch number.

3.5.12 DC of YBC Without Weak-Key Assumptions

The following differential analysis of YBC is unpublished and follows a top-down approach.

The strategy is to obtain 1-round differential characteristics (without any need for weak-subkey restrictions) and to further combined them into multi-round characteristics.

The difference operator is bitwise exclusive-or and it operates wordwise. The only 16-bit difference values used are $\delta = 8000_x$ and 0000_x.

Across unkeyed $\odot$ and $\boxplus$, the propagation of 16-bit difference words follows from Table 3.12 and Table 3.13, respectively.

Table 3.71 lists all fifteen nontrivial 1-round differential characteristics of YBC using only wordwise differences 0000_x and 8000_x. There are *no weak keys nor weak-subkey assumptions* for these characteristics.

Table 3.71 Nontrivial 1-round characteristics for YBC ($\delta = 8000_x$).

1-round characteristics	Probability	# Active (unkeyed) $\odot$
$(0,0,0,\delta) \xrightarrow{1r} (0,\delta,0,0)$	2^{-60}	4
$(0,0,\delta,0) \xrightarrow{1r} (0,0,\delta,0)$	2^{-60}	4
$(0,0,\delta,\delta) \xrightarrow{1r} (0,\delta,\delta,0)$	2^{-30}	2
$(0,\delta,0,0) \xrightarrow{1r} (0,\delta,0,\delta)$	2^{-45}	3
$(0,\delta,0,\delta) \xrightarrow{1r} (0,0,0,\delta)$	2^{-45}	3
$(0,\delta,\delta,0) \xrightarrow{1r} (0,\delta,\delta,\delta)$	2^{-45}	3
$(0,\delta,\delta,\delta) \xrightarrow{1r} (0,0,\delta,\delta)$	2^{-45}	3
$(\delta,0,0,0) \xrightarrow{1r} (\delta,0,0,0)$	2^{-60}	4
$(\delta,0,0,\delta) \xrightarrow{1r} (\delta,\delta,0,0)$	2^{-30}	2
$(\delta,0,\delta,0) \xrightarrow{1r} (\delta,0,\delta,0)$	2^{-60}	4
$(\delta,0,\delta,\delta) \xrightarrow{1r} (\delta,\delta,\delta,0)$	2^{-60}	4
$(\delta,\delta,0,0) \xrightarrow{1r} (\delta,\delta,0,\delta)$	2^{-45}	3
$(\delta,\delta,0,\delta) \xrightarrow{1r} (\delta,0,0,\delta)$	2^{-45}	3
$(\delta,\delta,\delta,0) \xrightarrow{1r} (\delta,\delta,\delta,\delta)$	2^{-45}	3
$(\delta,\delta,\delta,\delta) \xrightarrow{1r} (\delta,0,\delta,\delta)$	2^{-45}	3

Multiple-round characteristics are obtained by concatenating 1-round characteristics, such that the output difference of the top one matches the input difference of the characteristic in the next round.

From Table 3.71, there are seven multiple-round characteristics:

(1) $(0,0,0,\delta) \xrightarrow{1r} (0,\delta,0,0) \xrightarrow{1r} (0,\delta,0,\delta) \xrightarrow{1r} (0,0,0,\delta)$, which is 3-round iterative and holds with probability 2^{-150}.

(2) $(0,0,\delta,\delta) \xrightarrow{1r} (0,\delta,\delta,0) \xrightarrow{1r} (0,\delta,\delta,\delta) \xrightarrow{1r} (0,0,\delta,\delta)$, which is 3-round iterative and holds with probability 2^{-120}.

(3) $(\delta,0,0,\delta) \xrightarrow{1r} (\delta,\delta,0,0) \xrightarrow{1r} (\delta,\delta,0,\delta) \xrightarrow{1r} (\delta,0,0,\delta)$, which is 3-round iterative and holds with probability 2^{-120}.

(4) $(\delta,\delta,\delta,0) \xrightarrow{1r} (\delta,\delta,\delta,\delta) \xrightarrow{1r} (\delta,0,\delta,\delta) \xrightarrow{1r} (\delta,\delta,\delta,0)$, which is 3-round iterative and holds with probability 2^{-150}.

(5) $(0,0,\delta,0) \stackrel{1r}{\to} (0,0,\delta,0)$, which is 1-round iterative and holds with probability 2^{-60}.
(6) $(\delta,0,0,0) \stackrel{1r}{\to} (\delta,0,0,0)$, which is 1-round iterative and holds with probability 2^{-60}.
(7) $(\delta,0,\delta,0) \stackrel{1r}{\to} (\delta,0,\delta,0)$, which is 1-round iterative and holds with probability 2^{-60}.

Studying these multi-round characteristics and taking into account the block size of YBC, at most one round can be covered. For instance, $(0,0,\delta,\delta) \stackrel{1r}{\to} (0,\delta,\delta,0)$ which holds with probability 2^{-30}.

Up to $(2^{15})^2 \cdot (2^{16})^2 = 2^{62}$ text pairs can be created with an input difference of the form $(0,0,\delta,\delta)$. Extending this characteristic or any of the others leads to a probability lower or equal to 2^{-75}, and there will be not enough text pairs to obtain a right pair.

This differential analysis using characteristics also allows us to study the diffusion power of YBC with respect to modern tools such as the differential branch number (Appendix, Sect. B).

Analyzing the set of nontrivial 1-round characteristics in Table 3.71, we observe that the differential branch number *per round* of YBC is only 2 (without any weak-key assumption). For instance, the sum of the (wordwise) Hamming Weight of the input and output differences of the 1-round characteristic

$$(0,0,0,\delta) \stackrel{1r}{\to} (0,\delta,0,0)$$

is 2. So, even though YBC provides full (text) diffusion in a single round, the previous analysis shows that its differential branch number is much lower than ideal. For a 4-word block, such as in YBC, an MDS code using a 4×4 MDS matrix would rather provide a branch number of 5. See Appendix, Sect. B.1, for an explanation of the differential branch number.

3.6 Truncated Differential Cryptanalysis

The truncated differential technique was originally described by Knudsen and Berson and applied to the SAFER block cipher in [103]. Other analyses followed on SAFER [184], Skipjack [106], IDEA [48] and 3D [108] block ciphers.

Truncated differential cryptanalysis (TDC) exploits the word-oriented structure of modern ciphers such as IDEA, Skipjack, MESH and AES. In contrast, bit-oriented ciphers such as DES and Serpent are less interesting targets for truncated differentials as far as there is no well-defined set of consecutive bits (a word) to keep track of during the analysis. The concept of truncated differential is as follows:

Definition 3.13. (Truncated Differential) [99]
A differential that predicts only parts of an n-bit value is called a *truncated differential*. More formally, let (a, b) denote an i-round differential. If a' is a sub-sequence of a and b' is a sub-sequence of b, then (a', b') is called an i-round truncated differential.

In a truncated differential there are parts of a (block) difference that are not relevant. Therefore, a truncated differential can be seen as a collection of differentials.

In TDC attacks, the exact value of each bit of the blockwise difference is not relevant. That is the rationale behind the name of this attack: the difference is truncated. An example of truncated differential distinguisher only considers two types of (wordwise) differences: zero and nonzero (word difference), where the latter can contain an arbitrary nonzero value.

Let us discuss three kinds of difference operators for a truncated differential analysis of Lai-Massey designs such as the IDEA and MESH ciphers.

There is a certain degree of compatibility between difference operators based on $\oplus$, $\odot$ and $\boxplus$ concerning truncated differences: let $\Delta_{\boxminus} = x \boxminus x^*$ denote a subtraction difference, $\Delta_{\oplus} = x \oplus x^*$ denote an exclusive-or difference and $\Delta_{\boxdot} = x \boxdot x^*$ denote a multiplicative difference.

Theorem 3.1. $\Delta_{\boxminus} = 0 \iff \Delta_{\oplus} = 0 \iff \Delta_{\boxdot} = 1$, *and likewise*, $\Delta_{\boxminus} \neq 0 \iff \Delta_{\oplus} \neq 0 \iff \Delta_{\boxdot} \neq 1$.

Proof. For instance, $\Delta_{\boxminus} = 0$ means $x \boxminus x^* = 0$ which means $x = x^*$. Thus, $x \oplus x^* = 0$ and $\Delta_{\oplus} = 0$, and vice-versa. Similarly, $\Delta_{\oplus} = 0$ means $x \oplus x^* = 0$, that is $x = x^*$, or $x/x^* = 1$, that is, $x \boxdot x^* = 1$ and thus, $\Delta_{\boxdot} = 1$, and vice-versa. The same reasoning backward leads to the reverse chain of implications.

Likewise, $\Delta_{\boxminus} \neq 0$ means $x \boxminus x^* \neq 0$, i.e. $x \neq x^*$. Thus, $x \oplus x^* \neq 0$ and $\Delta_{\oplus} \neq 0$. Also, since $x \neq x^k*$, then $x/x^* \neq 1$, that is, $x \boxdot x^* \neq 1$ and thus, $\Delta_{\boxdot} \neq 1$. Similarly, the same reasoning backwards leads to the reverse chain of implications, which completes the proof.

These relations are useful for truncated differentials which deal exactly with the *zero versus nonzero* nature of each kind of (wordwise) difference, and this aspect is respected across the three group operations, independent of the exact value of Δ.

3.6.1 TDC of IDEA

The truncated differential analysis in this section is adapted from [48]. The difference operator is bitwise exclusive-or, that is, the difference between two bit strings X and X^* (of the same length) will be denoted $\Delta X = X \oplus X^*$. Under this difference operator IDEA is not a Markov cipher [117] (see also Def. 3.8).

So, the probability of the differential used in the attack will actually depend on the (unknown) key. Therefore, the Hypothesis of Stochastic Equivalence [117] (see (1)) will not hold for this difference operator.

The terminology used to describe a 1-round truncated differential for IDEA is similar to that used for characteristics, but with some additional details:

$$(a,b,c,d) \stackrel{p_1}{\to} (e,f,g,h) \stackrel{(e\oplus g,f\oplus h)\stackrel{p_2}{\to}(k,l)}{\longrightarrow} (e\oplus l, g\oplus l, f\oplus k, h\oplus k),$$

where (a,b,c,d) denotes the 4-word input difference to a KM half-round; (e,f,g,h) denotes the 4-word output difference from the KM half-round and thus, the input difference to the MA half-round. The leftmost part: $(a,b,c,d) \stackrel{p_1}{\to} (e,f,g,h)$ means that the difference across the KM half-round holds with probability p_1. The input difference to the MA-box is $(e\oplus g, f\oplus h)$, while the output difference is (k,l), and the MA-box transition probability is p_2. The round output is the 4-word tuple $(e\oplus l, g\oplus l, f\oplus k, h\oplus k)$.

A 3.5-round truncated differential distinguisher for IDEA is depicted in (3.29):

$$\begin{aligned}
(A,0,B,0) &\stackrel{2^{-16}}{\to} (C,0,C,0) \stackrel{(0,0)\stackrel{1}{\to}(0,0)}{\longrightarrow} (C,C,0,0)\\
(C,C,0,0) &\stackrel{1}{\to} (D,E,0,0) \stackrel{(D,E)\stackrel{2^{-32}}{\to}(E,D)}{\longrightarrow} (0,D,0,E)\\
(0,D,0,E) &\stackrel{2^{-16}}{\to} (0,F,0,F) \stackrel{(0,0)\stackrel{1}{\to}(0,0)}{\longrightarrow} (0,0,F,F)\\
&(0,0,F,F) \stackrel{1}{\to} (0,0,G,H),
\end{aligned} \tag{3.29}$$

where $A,B,C,D,E,F,G,H \in Z_2^{16}$ are arbitrary nonzero 16-bit differences. The probability of (3.29) is $2^{-16}\cdot 2^{-32}\cdot 2^{-16} = 2^{-64}$, and this probability is computed over all choices of the inputs to a round and to the MA half-rounds, and over all choices of the round subkeys. It is a truncated differential because only the zero difference words are predicted along 3.5 rounds.

The transition $(A,0,B,0) \stackrel{2^{-16}}{\to} (C,0,C,0)$ has probability 2^{-32} under the requirement that both A and B wordwise differences become the same difference C across a KM half-round. The same reasoning applies to the transition $(0,D,0,E) \stackrel{2^{-16}}{\to} (0,F,0,F)$. The transition $(D,E) \stackrel{2^{-32}}{\to} (E,D)$ across an MA-box has probability 2^{-32} under the requirement that the full 32-bit output difference is a predicted value: the reverse of the input difference.

There is a dual 3.5-round truncated distinguisher with the same probability, depicted in (3.30):

$$(0,A,0,B) \stackrel{2^{-16}}{\to} (0,C,0,C) \stackrel{(0,0)\stackrel{1}{\to}(0,0)}{\longrightarrow} (0,0,C,C)$$
$$(0,0,C,C) \stackrel{1}{\to} (0,0,D,E) \stackrel{(D,E)\stackrel{2^{-32}}{\to}(E,D)}{\longrightarrow} (D,0,E,0)$$
$$(D,0,E,0) \stackrel{2^{-16}}{\to} (F,0,F,0) \stackrel{(0,0)\stackrel{1}{\to}(0,0)}{\longrightarrow} (F,F,0,0)$$
$$(F,F,0,0) \stackrel{1}{\to} (G,H,0,0). \qquad (3.30)$$

Lets $P = (p_1, p_2, p_3, p_4)$ and $P^* = (p_1^*, p_2^*, p_3^*, p_4^*)$ denote a plaintext pair with difference $\Delta = P \oplus P^* = (\Delta p_1, \Delta p_2, \Delta p_3, \Delta p_4)$. The corresponding ciphertexts are denoted $C = (c_1, c_2, c_3, c_4)$ and $C^* = (c_1^*, c_2^*, c_3^*, c_4^*)$, respectively, and their difference is $\Delta C = C \oplus C^*$.

Consider a structure of 2^{32} texts where p_2 and p_4 are fixed to arbitrary values, while p_1 and p_3 take on all possible values. From a single structure, up to $2^{32}(2^{32}-1)/2 \approx 2^{63}$ text pairs can be created with differences of the form $(0, \Delta p_2, 0, \Delta p_4)$. Using (3.29), every structure provides about $2^{-64} \cdot 2^{63} = 0.5$ right pairs.

This distinguisher requires that $\Delta c_1 = \Delta c_3 = 0$. On average, only $2^{63} \cdot 2^{-32} = 2^{31}$ pairs will satisfy this filtering condition. For each surviving pair, and for all possible values of $Z_1^{(1)}$ and $Z_3^{(1)}$, check if

$$(p_1 \odot Z_1^{(1)}) \oplus (p_1^* \odot Z_1^{(1)}) = (p_3 \boxplus Z_3^{(1)}) \oplus (p_3^* \boxplus Z_3^{(1)}), \qquad (3.31)$$

is satisfied. On average, (3.31) holds for $2^{32} \cdot 2^{-16} = 2^{16}$ candidate values of $(Z_1^{(1)}, Z_3^{(1)})$ because it is an equality of 16-bit values. Similarly, check for which subkeys $(Z_1^{(4)}, Z_3^{(4)})$ the following equality holds:

$$(c_3 \boxminus Z_3^{(4)}) \oplus (c_3^* \boxminus Z_3^{(4)}) = (c_4 \odot (Z_1^{(4)})^{-1}) \oplus (c_4^* \odot (Z_1^{(4)})^{-1}). \quad (3.32)$$

In total, each surviving pair suggests $2^{16} \cdot 2^{16} = 2^{32}$ 64-bit subkey values, and therefore, each structure suggests $2^{31} \cdot 2^{32} = 2^{63}$ subkeys.

For a right pair, the filtering conditions (3.31) and (3.32) will be successful for the correct subkey value. Since there are $2^{63} \cdot 2^{-64} = 0.5$ right pairs, every structure will suggest the correct subkey value 0.5 times. Among all the subkey values suggested, wrong pairs may also suggest the correct subkey value.

For the correct subkey value, the input difference to the second round has the form $(C, C, 0, 0)$, the output difference of the third round will have the form $(0, 0, F, F)$, and the input difference of the third round will be $(0, \tilde{D}, 0, \tilde{E})$. The difference after the second KM half-round will have the form $(D, E, 0, 0)$, and the output difference of this round will be $(0, \tilde{D}, 0, \tilde{E})$. This means that $D = \tilde{D}$ and $E = \tilde{E}$ because of the construction of the distinguisher (3.29).

Thus, if the correct value of the subkey is suggested by a pair of plaintexts, it must be a right pair. In summary, for every structure used, there will be 0.5

right pairs which suggest the correct subkey value and 2^{31} wrong pairs which on average suggest a wrong subkey value 0.5 times. Thus, the traditional attack by Biham and Shamir in [30] will not work because the signal-to-noise ratio S/N=$\frac{0.5}{0.5} = 1$, meaning that the correct subkey value cannot be distinguished from any other value.

Nonetheless, the probability of (3.29) depends very much on the secret key. For some keys, the probability is less than the average 2^{-64}, while for other keys, the probability is larger than the average. The key search method of this truncated differential attack is extended to cover the cases where the probability of the distinguisher for the correct subkey value is not the average probability over all keys. The larger this difference in probability, the faster the attack becomes.

If this difference in probability is large enough, and if we assume that wrong subkeys values are suggested randomly and uniformly by the attack, the correct subkey value will be found using sufficiently many text pairs.

According to the key schedule of IDEA (Sect. 2.2.2), 16 bits of ($Z_1^{(1)}$, $Z_3^{(1)}$, $Z_1^{(4)}$, $Z_3^{(4)}$) overlap, that is, are redundant: $Z_1^{(1)}$ corresponds to bits numbered 0–15 of the user key; $Z_3^{(1)}$ corresponds to bits 32–47; $Z_1^{(4)}$ to bit 98–113 and $Z_3^{(4)}$ to bits 2–17.

Moreover, because of the absence of carry bit out of the highest-order bit in the modular addition operation, for $Z_3^{(1)}$ and $Z_3^{(4)}$, subkeys that differ only in those bits cannot be distinguished. Thus, these bits will not be considered.

Consequently, we recover $64 - 16 = 48$ bits instead of 64, and the memory requirements of the attack decreases accordingly.

In order to have experimental data and evidence to corroborate the effectiveness of the attack, two mini-versions of IDEA were constructed: (i) IDEA(32) operates on 32-bit blocks under a 64-bit key, with operations on 8-bit words: $\boxplus$ in $\mathbb{Z}_{2^8}$, $\oplus$ in $\mathbb{Z}_2^8$ and $\odot$ in GF(2^8+1) with $0 \equiv 2^8$; (ii) IDEA(16) operates on 16-bit blocks under a 32-bit key, with operations on 4-bit words: $\boxplus$ in $\mathbb{Z}_{2^4}$, $\oplus$ in $\mathbb{Z}_2^4$ and $\odot$ in GF(2^4+1) with $0 \equiv 2^4$. The key schedule algorithms of IDEA(32) and IDEA(16) were modified to accommodate smaller key sizes, but kept the bit permutation design of the key schedule of IDEA. The objective is to have cipher prototypes small enough to simulate the attacks with manageable complexity, and at the same time, obtain estimates on the complexity of the attack on the original IDEA cipher.

For IDEA(32) and IDEA(16), under the respective key schedule algorithms, 7 and 3 bits overlap, respectively. Consequently, we search for 23 subkey bits and 11 subkeys bits, respectively. The second truncated differential (3.30) can be used to recover additional subkey bits.

3.6.1.1 Experimental Results

An implementation of the attack on IDEA(16) first computed the probability of the distinguisher (3.29) for all keys by exhaustive search. Table 3.72 lists the probabilities for several classes of keys. The average probability over all keys was estimated to $2^{-16.5}$. The dependency of the probability on the key

Table 3.72 Probability of (3.29) for 3.5-round IDEA(16) over all keys.

# Keys/All keys	Probability
13%	0
12%	$0 < p \leq 2^{-18}$
21%	$2^{-18} < p \leq 2^{-17}$
30%	$2^{-17} < p \leq 2^{-16}$
14%	$2^{-16} < p \leq 2^{-15}$
10%	$2^{-15} < p \leq 1.$

value stems mostly from the second round of (3.29) where the difference (D, E) in the input to the MA-box must become (E, D) at its output.

The most interesting keys are those that deviate the most from the average probability. As an extreme case, for 13% of the key space, the probability of the distinguisher is zero. The figures in Table 3.72 also indicate that the attack will not work for key classes whose probabilities are too close to the average probability for all choices of the key: 2^{-16} for IDEA(16).

Table 3.73 lists the results of 1000 attacks on IDEA(16), for randomly-chosen keys and for increasing number of chosen plaintexts. The candidate

Table 3.73 Average number of chosen plaintexts needed to attack 3.5-round IDEA(16) in 1000 trials.

# Keys/All keys	# Structures	# Chosen plaintexts
25%	16	2^{12}
40%	32	2^{13}
51%	64	2^{14}
59%	128	2^{15}
67%	256	2^{16}

subkeys were ranked as in [132], and it was verified whether the correct subkey value was among the eight most suggested or the eight least suggested candidates. Thus, the attack returns 16 suggestions for 11 subkey bits.

Using all plaintexts (the whole codebook), the correct subkey value is among those 16 suggested candidates in about 6% of all cases.

An implementation of the attack on IDEA(32) also estimated the probability of (3.29) for different key classes. The results are listed in Table 3.74. Based on 160 experiments with randomly-chosen keys, the average probability of (3.29) over all keys was estimated to $2^{-32.7}$. This value is slightly less than the initial estimate of 2^{-32}. The difference is due to the fact that the MA-box is not a random permutation mapping.

Table 3.74 Probability of (3.29) for 3.5-round IDEA(32) for several key classes.

# Keys/All keys	Probability
14%	$0 < p \leq 2^{-35}$
10%	$2^{-35} < p \leq 2^{-33}$
31%	$2^{-33} < p \leq 2^{-32.5}$
45%	$2^{-32.5} < p$

The attack on IDEA(32) was implemented for 100 different randomly chosen keys, using up to 2048 structures. The results are described in Table 3.75.

The results on IDEA(16) and IDEA(32) will be used to estimate the number of chosen plaintexts needed to attack IDEA using the same truncated differential (3.29).

Table 3.75 Average number of CP needed in a truncated differential attack on 3.5-round IDEA(32) in 100 trials.

# Keys/All keys	# Structures	# Chosen Plaintexts (CP)
1%	16	2^{20}
7%	64	2^{22}
15%	128	2^{23}
31%	256	2^{24}
54%	512	2^{25}
65%	1024	2^{26}
83%	2048	2^{27}

From Table 3.73, one finds 25% (resp. 51%) of the keys using $2^{3n/4}$ (resp. $2^{7n/8}$) chosen plaintexts, using $n = 16$, for IDEA(16). From Table 3.75, one finds 1% (resp. 83%) of the keys using $2^{5n/8}$ (resp. $2^{7n/8}$) chosen plaintexts, using $n = 32$, for IDEA(32). Thus, the number of recovered keys increases for large block sizes with relatively the same amount of data. It is predicted that a similar increase will occur for the same attack on IDEA.

Once a structure has been analyzed, it is discarded from memory and a new structure is stored in memory and analyzed. Thus, the memory requirements for the attack on IDEA is 2^{32} 64-bit words. The workload is the estimated number of operations needed to perform the attack, measured as the number of encryptions of the given number of rounds of IDEA.

The 2^{32} ciphertexts in a structure are hashed in a table on the values of c_1 and c_2, since for a right pair, the pairs of values (c_1, c_2) have to be equal. The workload of the hashing and storage of the ciphertexts are small compared to the time of the rest of the attack. For each pair that survives the filtering, we try all 2^{16} values of the affected keys of each side of equation (3.31). These tests can be sped up by precomputing a table of all possible multiplications $\odot$. This table would have size 2^{32} 16-bit words and be computed offline (in a preprocessing stage). Let us estimate that a single multiplication costs the equivalent of 3.5 modular additions, and that an addition, an exclusive-or and a table lookup cost the same effort.

So, equation (3.31) costs two $\odot$, two $\boxplus$ and two $\oplus$, which means eleven $\oplus$. In comparison, 3.5-round IDEA costs fourteen $\odot$, fourteen $\boxplus$ and eighteen $\oplus$, which means 81 $\oplus$.

Thus, the workload is about $2^{15} \cdot \frac{11}{81} \approx 2^{12}$ 3.5-round IDEA encryptions for every analyzed pair. In total, the workload is about (# encryptions/# analyzed pairs)(#analyzed pairs/# all pairs)(#all pairs/structure) $= 2^{12} \cdot 2^{-32} \cdot 2^{63} = 2^{43}$ encryptions/structure. Due to overlapping bits between the subkeys in the first KM half-round and in the output transformation, the second part of the key recovery using equation (3.32) is faster than the first part using equation (3.31), and is thus ignored in the workload estimation.

The estimated number of chosen plaintexts (CP) and workload for the attack on IDEA are depicted in Table 3.76. The attack so far recovered 48 bits

Table 3.76 Estimated number of CP for a truncated differential attack on 3.5-round IDEA using 2^{32} words of memory.

# Keys/All keys	# Structures	# CP	Workload
$> 1\%$	2^8	2^{40}	2^{51}
$> 31\%$	2^{16}	2^{48}	2^{59}
$> 83\%$	2^{24}	2^{56}	2^{67}

of the 128-bit user key. One can do a similar attack using the second truncated differential distinguisher (3.30). As noted earlier, the key dependency of the probability of the first differential (3.29) comes mostly from the second round of the differential. Since the second round is the same for both distinguishers, one can expect that for a fixed key, the probabilities of the two differentials are very close. After the attack with the second distinguisher, one has all 64 bits of the first KM half-round, and all 64-bits of the output transformation.

Concerning truncated differential distinguishers beyond 3.5 rounds, note that there are longer ones than (3.29), for instance:

$$\begin{aligned}
(A,0,B,0) &\overset{2^{-16}}{\to} (C,0,C,0) \overset{(0,0)\overset{1}{\to}(0,0)}{\longrightarrow} (C,C,0,0)\\
(C,C,0,0) &\overset{1}{\to} (D,E,0,0) \overset{(D,E)\overset{2^{-32}}{\to}(E,D)}{\longrightarrow} (0,D,0,E)\\
(0,D,0,E) &\overset{2^{-16}}{\to} (0,F,0,F) \overset{(0,0)\overset{1}{\to}(0,0)}{\longrightarrow} (0,0,F,F)\\
(0,0,F,F) &\overset{1}{\to} (0,0,G,H) \overset{(G,H)\overset{2^{-32}}{\to}(H,G)}{\longrightarrow} (G,0,H,0), \qquad (3.33)
\end{aligned}$$

which is a 4-round iterative truncated differential for IDEA with probability 2^{-96}, and where A, B, C, D, E, F, G, H are 16-bit differences. Each line describes the propagation of differences across a full round: the leftmost arrow indicates the propagation across a KM half-round; the rightmost arrow indicates the difference propagation across an MA half-round. A structure following the input difference of (3.33) can generate up to $2^{32}(2^{32}-1)/2 \approx 2^{63}$ text pairs. Only pairs for which $\Delta c_2 = \Delta c_4 = 0$ are considered.

For a single structure, $2^{63} \cdot 2^{-32} = 2^{31}$ pairs satisfy the output difference of (3.33). Each such pair suggests 2^{32} 64-bit subkeys. The probability that a wrong subkey is suggested is $\frac{2^{32}}{2^{64}} = 2^{-32}$.

Thus, the signal-to-noise ratio is $\mathrm{S/N} = \frac{2^{-96}}{2^{-32} \cdot 2^{31}} = 2^{-95}$, which means an unreasonably large number of right pairs will be needed. But, from the probability of (3.33) and the number of pairs that can be generated per structure: $2^{-96} \cdot 2^{63} = 2^{-33} < 1$ right pair can be generated per structure. Thus, (3.33) is not useful for an attack on IDEA. This is a negative result.

Similarly, consider the following 4-round iterative truncated differential for IDEA:

$$\begin{aligned}
(A,0,0,0) &\overset{1}{\rightarrow} (B,0,0,0) \overset{(B,0)\overset{2^{-32}}{\rightarrow}(0,B)}{\longrightarrow} (0,B,0,0)\\
(0,B,0,0) &\overset{1}{\rightarrow} (0,C,0,0) \overset{(0,C)\overset{2^{-32}}{\rightarrow}(C,0)}{\longrightarrow} (0,0,0,C)\\
(0,0,0,C) &\overset{1}{\rightarrow} (0,0,0,D) \overset{(0,D)\overset{2^{-32}}{\rightarrow}(D,0)}{\longrightarrow} (0,0,D,0)\\
(0,0,D,0) &\overset{1}{\rightarrow} (0,0,E,0) \overset{(E,0)\overset{2^{-32}}{\rightarrow}(0,E)}{\longrightarrow} (E,0,0,0), \qquad (3.34)
\end{aligned}$$

that holds with probability 2^{-128}, and where A, B, C, D, E are 16-bit differences.

A structure following the input difference of (3.34) can generate up to $2^{16}(2^{16}-1)/2 \approx 2^{31}$ text pairs. Only pairs for which $\Delta c_2 = \Delta c_3 = \Delta c_4 = 0$ are considered. For a single structure, $2^{31} \cdot 2^{-48} = 2^{-17} < 1$ pairs satisfy the output difference of (3.34). Each pair suggests 2^{16} 32-bit subkeys. The probability that a wrong subkey is suggested is $\frac{2^{16}}{2^{32}} = 2^{-16}$. Thus, the signal-to-noise ratio is $\mathrm{S/N} = \frac{2^{-128}}{2^{-16} \cdot 2^{-17}} = 2^{-95}$, which again means an unreasonably large number of right pairs would be needed for an attack. Note that $2^{-128} \cdot 2^{31} = 2^{-97} < 1$ right pairs are suggested per structure. Since no pairs satisfy the output difference, no pairs survive this filtering, and thus, no subkey is ever suggested. Thus, although (3.34) cover more rounds than (3.29), the former is not useful for an attack on IDEA.

3.6.2 TDC of MESH-64

The truncated differential analysis of the MESH ciphers follows a similar approach to that of Borst *et al.* in [48]. The difference operator is bitwise exclusive-or. Under this difference operator, the MESH ciphers are not Markov ciphers, and consequently, the Hypothesis of Stochastic Equivalence (hypothesis 1 and [116]) does not hold.

For the truncated differential of MESH-64 [143, 151], we use the same operator and terminology as the analysis on IDEA in Sect. 3.6.1. Let $P = (p_1, p_2, p_3, p_4)$ and $P^* = (p_1^*, p_2^*, p_3^*, p_4^*)$ denote two chosen plaintext blocks. A text structure consists of 2^{32} plaintexts, where p_1 and p_3 take all possible values, while p_2 and p_4 are fixed. One such text structure can generate up to $2^{32} \cdot (2^{32}-1)/2 \approx 2^{63}$ text pairs with difference $P \oplus P^* = (A, 0, B, 0)$, where $A, B \in \mathbb{Z}_2^{16}$.

A 3.5-round truncated differential distinguisher for MESH-64 is depicted in (3.35):

$$\begin{aligned}
(A,0,B,0) &\overset{2^{-16}}{\to} (C,0,C,0) \overset{(0,0)\overset{1}{\to}(0,0)}{\longrightarrow} (C,C,0,0) \\
(C,C,0,0) &\overset{1}{\to} (D,E,0,0) \overset{(D,E)\overset{2^{-32}}{\to}(E,D)}{\longrightarrow} (0,D,0,E) \\
(0,D,0,E) &\overset{2^{-16}}{\to} (0,F,0,F) \overset{(0,0)\overset{1}{\to}(0,0)}{\longrightarrow} (0,0,F,F) \\
(0,0,F,F) &\overset{1}{\to} (0,0,G,H),
\end{aligned} \tag{3.35}$$

where $A, B, C, D, E, F, G, H \in Z_2^{16}$ are 16-bit differences. The difference propagation is described one round per line. The arrows indicate the propagation (and transition probability) of 4-word differences across each half-round. In particular, the 4-word difference on the left-hand side of an arrow, e.g. $(A, 0, B, 0)$, indicates that it leads to the difference on the right-hand side, $(C, 0, C, 0)$, across the KM half-round with the probability depicted on top of the arrow: 2^{-16}. The middle 4-word difference, $(C, 0, C, 0)$, further leads to the round output difference, $(C, C, 0, 0)$, across the MA half-round with the probability depicted on top of the arrow: 1. The difference (and corresponding probability) across the MA-box is indicated on top of the inner arrow, as for instance, $(0,0) \overset{1}{\to} (0,0)$.

The 3.5-round truncated differential (3.35) has average probability 2^{-64} over all keys. By symmetry, another 3.5-round truncated differential is depicted in (3.36):

$$
\begin{aligned}
(0, A, 0, B) &\overset{2^{-16}}{\to} (0, C, 0, C) \overset{(0,0)\overset{1}{\to}(0,0)}{\longrightarrow} (0, 0, C, C) \\
(0, 0, C, C) &\overset{1}{\to} (0, 0, D, E) \overset{(D,E)\overset{2^{-32}}{\to}(E,D)}{\longrightarrow} (D, 0, E, 0) \\
(D, 0, E, 0) &\overset{2^{-16}}{\to} (F, 0, F, 0) \overset{(0,0)\overset{1}{\to}(0,0)}{\longrightarrow} (F, F, 0, 0) \\
(F, F, 0, 0) &\overset{1}{\to} (G, H, 0, 0).
\end{aligned} \tag{3.36}
$$

Using (3.35) to attack 4-round MESH-64 allows us to recover subkeys $Z_1^{(1)}$, $Z_3^{(1)}$, $Z_3^{(4)}$ and $Z_4^{(4)}$ by filtering those candidates that satisfy:

$$(p_1 \odot Z_1^{(1)}) \oplus (p_1^* \odot Z_1^{(1)}) = (p_3 \boxplus Z_3^{(1)}) \oplus (p_3^* \boxplus Z_3^{(1)}), \tag{3.37}$$

and

$$(c_3 \odot (Z_3^{(4)})^{-1}) \oplus (c_3^* \odot (Z_3^{(4)})^{-1}) = (c_4 \boxminus Z_4^{(4)}) \oplus (c_4^* \boxminus Z_4^{(4)}). \tag{3.38}$$

In this last equation, recall that the operations in the KM half-round vary for odd- and even-numbered rounds.

On average, about 2^{16} pairs $(Z_1^{(1)}, Z_3^{(1)})$ satisfy (3.37), because this equality implies a 16-bit condition. Likewise, about 2^{16} pairs $(Z_3^{(4)}, Z_4^{(4)})$ shall satisfy (3.38).

According to the key schedule algorithm of MESH-64, the latter two subkeys are computed as follows:

$$\mathbf{Z}_3^{(4)} = (((((Z_2^{(3)} \boxplus Z_3^{(3)}) \oplus Z_4^{(3)}) \boxplus Z_7^{(3)}) \oplus Z_1^{(4)}) \boxplus Z_2^{(4)}) \lll 7 \oplus c_{23} \tag{3.39}$$
$$\mathbf{Z}_4^{(4)} = (((((Z_3^{(3)} \boxplus Z_4^{(3)}) \oplus Z_5^{(3)}) \boxplus Z_1^{(4)}) \oplus Z_2^{(4)}) \boxplus Z_3^{(4)}) \lll 7 \oplus c_{24} \tag{3.40}$$

This means that $(Z_3^{(4)}, Z_4^{(4)})$ cannot be uniquely determined from knowledge of $(Z_1^{(1)}, Z_3^{(1)})$ only. Thus, this part of the attack recovers 64 subkey bits. The most significant bits (MSB) of $Z_3^{(1)}$ and $Z_4^{(4)}$) cannot be uniquely determined due to the absence of the carry bit from the most significant bit position.

Each filtered pair suggests 2^{32} 64-bit subkeys. Therefore, the probability of a wrong subkey being suggested is $2^{32} \cdot 2^{-64} = 2^{-32}$. The signal-to-noise ratio is

$$S/N = \frac{2^{-64}}{2^{-32} \cdot 2^{-32}} = 1.$$

From the viewpoint of the original differential attack in [29], it would appear that the truncated differential (3.35) cannot distinguish the correct subkey from the wrong ones. However, the probability of (3.35) is very much key dependent, implying that for some keys the probability is higher than the average (S/N> 1), while for other keys, it is lower (S/N< 1). The farther the actual probability of (3.35) is from the average, the faster the correct subkeys can be identified. Similar to [48], a ranking of the eight most suggested and the eight least suggested subkey candidates were obtained during attack simulations. If the correct subkey is in this ranking, then the attack is considered successful.

Attacks were implemented on 3.5-round mini-version of MESH-64 with 8-bit blocks, denoted MESH-64[8]. The results are depicted in Table 3.77. Similarly, the results of attack simulations on a 16-bit block mini-version of MESH-64 denoted MESH-64(16), for a sample of 2^{10} randomly chosen keys, are depicted in Table 3.78. The average probability computed from the 2^{10} keys was about $2^{-15.62}$, compared to the expected probability 2^{-16}.

Table 3.77 Success probability and data complexity of differential attack on 3.5-round MESH-64[8].

# Keys/2^{10}	# Structures	# CP
10.54%	1	2^4
10.74%	2	2^5
11.23%	4	2^6
12.40%	8	2^7
13.47%	16	2^8

Table 3.78 Estimated probability of the truncated differential (3.35) for 3.5-round MESH-64(16).

# Keys/2^{10}	Probability
3.12%	$p = 0$
5.56%	$0 < p \leq 2^{-18}$
16.89%	$2^{-18} < p \leq 2^{-17}$
31.83%	$2^{-17} < p \leq 2^{-16}$
30.85%	$2^{-16} < p \leq 2^{-15}$
11.71%	$2^{-15} < p \leq 1$

The results of the attack implementations on MESH-64(16) are summarized in Table 3.79, and indicate that a smaller fraction of the subkeys could be identified by the truncated differential (3.35) compared to the same attack on a mini-version of IDEA [48]. This may be a consequence of the longer MA-box in MESH-64.

According to Table 3.77, the success rates for attacks on 3.5-round MESH-64[8] are about 11.23%, 12.40% and 13.47% using $2^{3n/4}$, $2^{7n/8}$ and 2^n chosen plaintexts, respectively, where $n = 8$ is the block size. Similarly, Table 3.79 shows that the success rates for MESH-64(16) are about 12.01%, 25.19% and 40.33%, respectively, with $2^{3n/4}$, $2^{7n/8}$ and 2^n chosen plaintexts, respectively, where $n = 16$ is the block size.

Table 3.79 Success probability and data complexity of the truncated differential attack on 3.5-round MESH-64(16) and IDEA(16).

# Keys/2^{10} in IDEA(16)	#Keys/2^{10} in MESH-64(16)	#Structures	# CP
25%	11.81%	16	2^{12}
40%	16.79%	32	2^{13}
51%	25.09%	64	2^{14}
59%	33.30%	128	2^{15}
67%	39.06%	256	2^{16}

It is estimated that for MESH-64(32), namely, a mini-version with 32-bit blocks, and for MESH-64, the success rates of about 12.84%, 51.7% and 80% can be achieved with about $2^{3n/4}$, $2^{7n/8}$ and 2^n chosen plaintexts, respectively, where n is the block size.

Estimates for the attack complexity on MESH-64, using (3.35), to recover $Z_1^{(1)}$, $Z_3^{(1)}$, $Z_3^{(4)}$ and $Z_4^{(4)}$ are as follows: for each structure used, the input plaintext pairs are filtered according to the zero output differences of the truncated differential. For the surviving pairs, some subkeys at both ends of the differential are guessed exhaustively, and are selected depending on equation (3.37). The eight most suggested and the eight least suggested subkey candidates are collected and ranked. If the correct subkey is in this ranking, the attack is considered successful. The final attack complexity is measured in terms of the number of 3.5-round encryptions (or equivalent effort) in solving all equations.

The complexity calculations use the following formula: (#structures)(# surviving pairs/structure) (# equations)(# key pairs to find/equation)(# operations per 3.5 rounds). The number of structures, for 80% success rate (the same as in [48] for IDEA), is estimated as $2^{64-32} = 2^{32}$, since each structure contains 2^{32} plaintexts. The number of surviving pairs per structure is $2^{32}(2^{32}-1) \cdot 2^{-1} \cdot 2^{-32} \approx 2^{31}$. The number of equations to be satisfied is two: (3.37) and (3.38). The number of subkeys satisfying each equation is about 2^{16}. The number of operations per equation compared to the number of operations in 3.5 rounds is about $\frac{6}{46} \approx 2^{-3}$. A table of 2^{32} 16-bit words can be precomputed to speed up the multiplications in this part of the attack.

Therefore, the time complexity is $2^{32} \cdot 2^{31} \cdot 2 \cdot 2^{16} \cdot 2^{-3} = 2^{77}$ 3.5-round MESH-64 encryptions. The data complexity is 2^{64} chosen plaintexts (the whole codebook). The memory complexity is 2^{32} 64-bit blocks.

Similarly, the truncated differential (3.36) allows us to recover $Z_2^{(1)}$, $Z_4^{(1)}$, $Z_1^{(4)}$ and $Z_2^{(4)}$ with the same attack complexities, by using the equations

$$(p_2 \boxplus Z_2^{(1)}) \oplus (p_2^* \boxplus Z_2^{(1)}) = (p_4 \odot Z_4^{(1)}) \oplus (p_4^* \odot Z_4^{(1)}), \qquad (3.41)$$

and

$$(c_1 \boxminus Z_1^{(4)}) \oplus (c_1^* \odot Z_1^{(4)}) = (c_2 \odot (Z_2^{(4)})-1) \oplus (c_2^* \odot (Z_2^{(4)})^{-1}). \quad (3.42)$$

The remaining 64 key bits can be recovered by exhaustive search. This step does not affect the overall time complexity.

An attack on 4-round MESH-64 can guess the subkeys $(Z_5^{(4)}, Z_6^{(4)}, Z_7^{(4)})$ and apply the previous attack on 3.5 rounds. But, the time complexity increases to $2^{78+48} = 2^{128}$ 4-round computations (the same as exhaustive key search). The data complexity remains 2^{64} CP.

3.6.3 TDC of MESH-96

For the truncated differential of MESH-96 [143, 151], the difference operator is exclusive-or and the same terminology as in the analysis of IDEA in Sect. 3.6.1 is used. Let $P = (p_1, p_2, p_3, p_4, p_5, p_6)$ and $P^* = (p_1^*, p_2^*, p_3^*, p_4^*, p_5^*, p_6^*)$ denote two chosen plaintext blocks. Let us define a text structure consisting of 2^{64} plaintexts, where p_1, p_2, p_4 and p_5 take all possible values, while p_3 and p_6 are fixed. One such text structure can generate up to $2^{64} \cdot (2^{64}-1)/2 \approx 2^{127}$ text pairs with difference $P \oplus P^* = (A, B, 0, C, D, 0)$, where $A, B, C, D \in \mathbb{Z}_2^{16}$.

A 3.5-round truncated differential for MESH-96 is depicted in (3.43) across each half-round.

$$\begin{gathered}
(A,B,0,C,D,0) \stackrel{2^{-32}}{\rightarrow} (E,F,0,E,F,0) \stackrel{(0,0,0)\stackrel{1}{\rightarrow}(0,0,0)}{\longrightarrow} (E,E,F,F,0,0) \\
(E,E,F,F,0,0) \stackrel{2^{-16}}{\rightarrow} (G,H,I,G,0,0) \stackrel{(0,H,I)\stackrel{2^{-48}}{\rightarrow}(I,H,G)}{\longrightarrow} (0,0,H,0,0,I) \\
(0,0,H,0,0,I) \stackrel{2^{-16}}{\rightarrow} (0,0,J,0,0,J) \stackrel{(0,0,0)\stackrel{1}{\rightarrow}(0,0,0)}{\longrightarrow} (0,0,0,0,J,J) \\
(0,0,0,0,J,J) \stackrel{1}{\rightarrow} (0,0,0,0,K,L), \qquad (3.43)
\end{gathered}$$

where $A, B, C, D, E, F, G, H, I, J, K, L \in Z_2^{16}$ are 16-bit differences. The arrows indicate the direction of propagation of 6-word differences across each half-round (in the encryption direction). In particular, the leftmost arrow in a line indicates that the 6-word difference on the left-hand side, for instance, $(A, B, 0, C, D, 0)$, leads to the difference on the right-hand side, for instance, $(E, F, 0, E, F, 0)$, across the KM half-round with the probability depicted on top of the arrow. The middle 6-word difference, e.g. $(E, F, 0, E, F, 0)$, further leads to the round output difference, for instance, $(E, E, F, F, 0, 0)$, across the MA half-round with the probability indicated on top of the arrow. The difference (and probability) across the MA-box is indicated on top of the inner arrow, as for instance, $(0, 0, 0) \stackrel{1}{\rightarrow} (0, 0, 0)$.

The 3.5-round truncated differential (3.43) has average probability 2^{-112} over all keys. By symmetry, there are two other 3.5-round truncated differentials depicted in (3.44) and (3.45):

$$(A,0,B,C,0,D) \overset{2^{-32}}{\rightarrow} (E,0,F,E,0,F) \overset{(0,0,0)\overset{1}{\rightarrow}(0,0,0)}{\longrightarrow} (E,E,0,0,F,F)$$
$$(E,E,0,0,F,F) \overset{2^{-16}}{\rightarrow} (G,H,0,0,H,I) \overset{(G,0,I)\overset{2^{-48}}{\rightarrow}(I,H,G)}{\longrightarrow} (0,G,0,0,I,0)$$
$$(0,G,0,0,I,0) \overset{2^{-16}}{\rightarrow} (0,J,0,0,J,0) \overset{(0,0,0)\overset{1}{\rightarrow}(0,0,0)}{\longrightarrow} (0,0,J,J,0,0)$$
$$(0,0,J,J,0,0) \overset{1}{\rightarrow} (0,0,K,L,0,0), \tag{3.44}$$

and

$$(0,A,B,0,C,D) \overset{2^{-32}}{\rightarrow} (0,E,F,0,E,F) \overset{(0,0,0)\overset{1}{\rightarrow}(0,0,0)}{\longrightarrow} (0,0,E,E,F,F)$$
$$(0,0,E,E,F,F) \overset{2^{-16}}{\rightarrow} (0,0,G,H,I,G) \overset{(H,I,0)\overset{2^{-48}}{\rightarrow}(G,I,H)}{\longrightarrow} (H,0,0,I,0,0)$$
$$(H,0,0,I,0,0) \overset{2^{-16}}{\rightarrow} (J,0,0,J,0,0) \overset{(0,0,0)\overset{1}{\rightarrow}(0,0,0)}{\longrightarrow} (J,J,0,0,0,0)$$
$$(J,J,0,0,0,0) \overset{1}{\rightarrow} (K,L,0,0,0,0). \tag{3.45}$$

A truncated differential attack on 3.5-round MESH-96 using (3.43) will recover subkeys $Z_1^{(1)}$, $Z_2^{(1)}$, $Z_4^{(1)}$, $Z_5^{(1)}$, $Z_5^{(4)}$ and $Z_6^{(4)}$, by filtering subkey candidates satisfying the following equations:

$$(p_1 \odot Z_1^{(1)}) \oplus (p_1^* \odot Z_1^{(1)}) = (p_4 \boxplus Z_4^{(1)}) \oplus (p_4^* \boxplus Z_4^{(1)}), \tag{3.46}$$
$$(p_2 \boxplus Z_2^{(1)}) \oplus (p_2^* \boxplus Z_2^{(1)}) = (p_5 \odot Z_5^{(1)}) \oplus (p_5^* \odot Z_5^{(1)}), \tag{3.47}$$

and

$$(c_5 \boxminus Z_5^{(4)}) \oplus (c_5^* \boxminus Z_5^{(4)}) = (c_6 \odot (Z_6^{(4)})^{-1}) \oplus (c_6^* \odot (Z_6^{(4)})^{-1}). \tag{3.48}$$

In the last equation, recall that the operations in the KM half-round vary for odd- and even-numbered rounds.

The estimate on the attack complexity on 3.5-round MESH-96 has a similar reasoning as for MESH-64: (# structures) (#surviving pairs/structure) (# equations) (#key pairs to find/equation) (#operations/3.5 rounds). The number of structures, for about 80% success rate, is estimated as $2^{96-64} = 2^{32}$, since each structure contains 2^{64} plaintexts. The number of surviving pairs per structure is $2^{64} \cdot (2^{64}-1) \cdot 2^{-1} \cdot 2^{-64} \approx 2^{63}$. The number of equations to be satisfied is three: (3.46), (3.47) and (3.48). The number of subkeys satisfying each equation is about 2^{16} (there are 32 subkey bits per equation, and each equation means a 16-bit condition). The number of operations per equation, compared to the number of operations in 3.5 rounds, is about $\frac{6}{78} = \frac{1}{13}$.

The time complexity is estimated at $2^{32} \cdot 2^{63} \cdot 3 \cdot 2^{16} \cdot \frac{1}{13} \approx 2^{109}$ 3.5-round MESH-96 encryptions. The data complexity is 2^{96} chosen plaintexts (the whole codebook). The memory requirements corresponds to 2^{64} 96-bit blocks.

The differential (3.44) can be used to further recover $Z_3^{(1)}$, $Z_5^{(1)}$, $Z_3^{(4)}$ and $Z_4^{(4)}$. Finally, (3.45) can be used to recover $Z_1^{(4)}$ and $Z_2^{(4)}$. In total, the first 96 user key bits are directly recovered (from the first KM half-round). The remaining 96 user key bits can be found by exhaustive search (without affecting the time complexity).

An attack on 4-round MESH-96 can guess the subkeys $Z_7^{(4)}$, $Z_8^{(4)}$ and $Z_9^{(4)}$ of an additional KM half-round, and apply the previous attack on 3.5 rounds. The time complexity increases to $2^{109+48} = 2^{157}$ 4-round encryptions. The data complexity remains 2^{96} CP, the whole codebook. The memory complexity remains 2^{64} 96-bit blocks since the same storage can be reused for each stage of the attack.

Concerning truncated differential distinguishers beyond 3.5 rounds, there are longer ones than (3.43), for instance:

$$(A,0,0,B,0,0) \stackrel{2^{-16}}{\rightarrow} (C,0,0,C,0,0) \stackrel{(0,0,0)\stackrel{1}{\rightarrow}(0,0,0)}{\longrightarrow} (C,C,0,0,0,0)$$

$$(C,C,0,0,0,0) \stackrel{1}{\rightarrow} (D,E,0,0,0,0) \stackrel{(D,E,0)\stackrel{2^{-32}}{\rightarrow}(F,E,D)}{\longrightarrow} (0,D,E,0,F,F)$$

$$(0,D,E,0,F,F) \stackrel{2^{-32}}{\rightarrow} (0,G,H,0,G,H) \stackrel{(0,0,0)\stackrel{1}{\rightarrow}(0,0,0)}{\longrightarrow} (0,0,G,G,H,H)$$

$$(0,0,G,G,H,H) \stackrel{2^{-16}}{\rightarrow} (0,0,I,J,K,I) \stackrel{(J,K,0)\stackrel{2^{-48}}{\rightarrow}(I,K,J)}{\longrightarrow} (J,0,0,K,0,0), \tag{3.49}$$

which is a 4-round iterative truncated differential for MESH-96 with probability 2^{-144}, and where A, B, C, D, E, F, G, H, I, J and K are 16-bit differences. A structure following the input difference of (3.49) can generate up to $2^{32}(2^{32}-1)/2 \approx 2^{63}$ text pairs. Only pairs for which $\Delta c_2 = \Delta c_3 = \Delta c_5 = \Delta c_6 = 0$ are considered. This last criterion means a 64-bit condition. For a single structure, $2^{63} \cdot 2^{-64} = 1/2$ pairs satisfy the output difference of (3.49). Each pair suggests 2^{48} 96-bit subkeys. The probability that a wrong subkey is suggested is $\frac{2^{48}}{2^{96}} = 2^{-48}$.

The signal-to-noise ratio is $\mathrm{S/N} = \frac{2^{-144}}{2^{-64}.2^{-1}} = 2^{-79}$, which means an unreasonably large number of right pairs would be needed for an attack. Note that $2^{-144} \cdot 2^{63} = 2^{-81} < 1$ right pairs are suggested per structure. Since no pairs satisfy the output difference, no pairs survive this filtering, and thus, no subkey is ever suggested. Therefore, although (3.49) covers more rounds than (3.43), the former is not useful for an attack on MESH-96.

3.6.4 TDC of MESH-128

For the truncated differential of MESH-128 [151, 143], the exclusive-or is used as the difference operator and also the same terminology as in the analysis of IDEA in Sect. 3.6.1 is used. Let $P = (p_1, p_2, p_3, p_4, p_5, p_6, p_7, p_8)$ and $P^* = (p_1^*, p_2^*, p_3^*, p_4^*, p_5^*, p_6^*, p_7^*, p_8^*)$ denote two chosen plaintext blocks. Let us define a text structure consisting of 2^{64} plaintexts, where p_1, p_4, p_5 and p_8 take all possible values, while p_2, p_3, p_6 and p_7 are fixed constants. One such text structure can generate up to $2^{64} \cdot (2^{64} - 1)/2 \approx 2^{127}$ text pairs with difference $P \oplus P^* = (A, 0, 0, B, C, 0, 0, D)$, where $A, B, C, D \in \mathbb{Z}_2^{16}$.

A 3.5-round truncated differential for MESH-128 is depicted in (3.50) across each half-round.

$$\begin{array}{c}
(A,0,0,B,C,0,0,D) \stackrel{2^{-32}}{\rightarrow} (E,0,0,F,E,0,0,F) \stackrel{(0,0,0,0) \stackrel{1}{\rightarrow} (0,0,0,0)}{\rightarrow} (E,E,0,0,0,0,F,F) \\
(E,E,0,0,0,0,F,F) \stackrel{1}{\rightarrow} (G,H,0,0,0,0,I,J) \stackrel{(G,H,I,J) \stackrel{2^{-64}}{\rightarrow} (J,I,H,G)}{\rightarrow} (0,G,H,0,0,I,J,0) \\
(0,G,H,0,0,I,J,0) \stackrel{2^{-32}}{\rightarrow} (0,K,L,0,0,K,L,0) \stackrel{(0,0,0,0) \stackrel{1}{\rightarrow} (0,0,0,0)}{\rightarrow} (0,0,K,L,K,L,0,0) \\
(0,0,K,L,K,L,0,0) \stackrel{1}{\rightarrow} (0,0,M,N,O,P,0,0),
\end{array} \tag{3.50}$$

where $A, B, C, D, E, F, G, H, I, J, K, L, M, N, O, P \in Z_2^{16}$. The arrows indicate the propagation of 8-word differences across each half-round (in the encryption direction). In particular, the leftmost arrow in a line indicates that the 8-word difference on the left-hand side, for instance, $(A, 0, 0, B, C, 0, 0, D)$, leads to the difference on the right-hand side, for instance, $(E,\ 0,\ 0,\ F,\ E,\ 0,\ 0,\ F)$, across the KM half-round, with the probability depicted on top of the arrow. The middle 8-word difference, for example, $(E,\ 0,\ 0,\ F,\ E,\ 0,\ 0,\ F)$, further leads to the round output difference, for instance, $(E,\ E,\ 0,\ 0,\ 0,\ 0,\ F,\ F)$, across the MA half-round with the probability indicated on top of inner the arrow. The difference (and probability) across the MA-box is indicated on top of the inner arrow, as for instance, $(0, 0, 0, 0) \stackrel{1}{\rightarrow} (0, 0, 0, 0)$.

By symmetry, there is another 3.5-round truncated differential for MESH-96 depicted in (3.51):

$$\begin{array}{c}
(0,A,B,0,0,C,D,0) \stackrel{2^{-32}}{\rightarrow} (0,E,F,0,0,E,F,0) \stackrel{(0,0,0,0) \stackrel{1}{\rightarrow} (0,0,0,0)}{\longrightarrow} (0,0,E,F,E,F,0,0) \\
(0,0,E,F,E,F,0,0) \stackrel{1}{\rightarrow} (0,0,G,H,I,J,0,0) \stackrel{(I,J,G,H) \stackrel{2^{-64}}{\rightarrow} (H,G,J,I)}{\longrightarrow} (I,0,0,G,J,0,0,H) \\
(I,0,0,G,J,0,0,H) \stackrel{2^{-32}}{\rightarrow} (K,0,0,L,K,0,0,L) \stackrel{(0,0,0,0) \stackrel{1}{\rightarrow} (0,0,0,0)}{\longrightarrow} (K,K,0,0,0,0,L,L) \\
(K,K,0,0,0,0,L,L) \stackrel{1}{\rightarrow} (M,N,0,0,0,0,O,P).
\end{array} \tag{3.51}$$

A truncated differential attack on 3.5-round MESH-128 using (3.50) follows a similar procedure as the attack on MESH-96. The attack recovers subkeys $Z_1^{(1)}$, $Z_4^{(1)}$, $Z_5^{(1)}$, $Z_6^{(1)}$, $Z_3^{(4)}$, $Z_4^{(4)}$, $Z_5^{(4)}$ and $Z_6^{(4)}$, by filtering subkey candidates that satisfy the equations:

$$(p_1 \odot Z_1^{(1)}) \oplus (p_1^* \odot Z_1^{(1)}) = (p_5 \boxplus Z_5^{(1)}) \oplus (p_5^* \boxplus Z_5^{(1)}), \quad (3.52)$$

$$(p_4 \boxplus Z_4^{(1)}) \oplus (p_4^* \boxplus Z_4^{(1)}) = (p_8 \odot Z_8^{(1)}) \oplus (p_8^* \odot Z_8^{(1)}), \quad (3.53)$$

$$(c_3 \boxminus Z_3^{(4)}) \oplus (c_3^* \boxminus Z_3^{(4)}) = (c_5 \odot (Z_5^{(4)})^{-1}) \oplus (c_5^* \odot (Z_5^{(4)})^{-1}), \quad (3.54)$$

and

$$(c_4 \odot (Z_4^{(4)})^{-1}) \oplus (c_4^* \odot (Z_4^{(4)})^{-1}) = (c_6 \boxminus Z_6^{(4)}) \oplus (c_6^* \boxminus Z_6^{(4)}). \quad (3.55)$$

In the last equation, recall that the operations in the KM half-round vary for odd- and even-numbered rounds.

An estimate on the attack complexity is similarly based on the same approach used for attacking MESH-64, and use the following formula: (# structures) * (#surviving pairs/structure) * (#equations) * (# key pairs to find/equation) * (# operations per 3.5 rounds).

The number of structures is estimated as $2^{128]64} = 2^{64}$, since each structure contains 2^{64} plaintexts. The number of surviving pairs per structure is $2^{64} \cdot (2^{64} - 1) \cdot 2^{-1} \cdot 2^{-64} \approx 2^{63}$. The number of equations to be satisfied is four: (3.52), (3.53), (3.54) and (3.55). The number of subkeys satisfying each equation is about 2^{16} (there are 32 subkey bits per equation, and each equation means a 16-bit condition). The number of operations per equation compared to the number of operations in 3.5-round MESH-128 is about $\frac{6}{116}$.

The time complexity is estimated as $2^{64} \cdot 2^{63} \cdot 4 \cdot 2^{16} \cdot \frac{6}{116} \approx 2^{141}$ 3.5-round MESH-128 encryptions. The data complexity is 2^{128} chosen plaintexts (the whole codebook). The memory requirements correspond to 2^{64} 128-bit blocks.

In a following stage, the truncated differential (3.51) can be used to recover $Z_2^{(1)}$, $Z_3^{(1)}$, $Z_6^{(1)}$, $Z_7^{(1)}$, $Z_1^{(4)}$, $Z_2^{(4)}$, $Z_7^{(4)}$ and $Z_8^{(4)}$, with the same attack complexities as with (3.50). The final time complexity becomes 2^{142} 3.5-round MESH-128 encryptions. The data complexity remains 2^{128} chosen plaintexts, the whole codebook. The memory complexity remains 2^{64} 128-bit blocks.

The remaining 128 user key bits can be recovered by exhaustive search (without affecting the time complexity).

An attack on 4-round MESH-128 can guess the subkeys $Z_9^{(4)}$, $Z_{10}^{(4)}$, $Z_{11}^{(4)}$ and $Z_{12}^{(4)}$ of an additional KM half-round, and apply the previous attack on 3.5 rounds, resulting in a time complexity of $2^{142+64} = 2^{206}$ 4-round encryptions.

Concerning truncated differential distinguishers beyond 3.5 rounds, note that there are longer ones than (3.50), for instance:

$$(A, B, C, 0, D, E, F, 0) \overset{2^{-48}}{\rightarrow} (G, H, I, 0, G, H, I, 0) \overset{(0,0,0,0) \overset{1}{\rightarrow} (0,0,0,0)}{\longrightarrow} (G, G, H, I, H, I, 0, 0)$$

$$(G, G, H, I, H, I, 0, 0) \overset{2^{-32}}{\rightarrow} (J, K, L, M, J, K, 0, 0) \overset{(0,0,L,M) \overset{2^{-64}}{\rightarrow} (M,L,K,J)}{\longrightarrow} (0, 0, 0, L, 0, 0, 0, M)$$

$$(0, 0, 0, L, 0, 0, 0, M) \overset{2^{-16}}{\rightarrow} (0, 0, 0, N, 0, 0, 0, N) \overset{(0,0,0,0) \overset{1}{\rightarrow} (0,0,0,0)}{\longrightarrow} (0, 0, 0, 0, 0, 0, N, N)$$

$$(0, 0, 0, 0, 0, 0, N, N) \overset{1}{\rightarrow} (0, 0, 0, 0, 0, 0, O, P) \overset{(0,0,O,P) \overset{2^{-32}}{\rightarrow} (P,O,Q,R)}{\longrightarrow} (R, R, Q, 0, Q, O, P, 0), \quad (3.56)$$

which is a 4-round iterative truncated differential with probability 2^{-192}, and where A, B, C, D, E, F, G, H, I, J, K, L, M, N, O, P, Q and R are 16-bit differences. A structure following the input difference of (3.56) can generate up to $2^{96}(2^{96} - 1)/2 \approx 2^{191}$ text pairs. Only pairs for which $\Delta c_4 = \Delta c_8 = 0$ are considered. This last criterion means a 32-bit condition. For a single structure, $2^{191} \cdot 2^{-32} = 2^{159}$ pairs satisfy the output difference of (3.56). Each pair suggests 2^{96} 192-bit subkeys. The probability that a wrong subkey is suggested is $\frac{2^{96}}{2^{192}} = 2^{-96}$.

The signal-to-noise ratio is S/N $= \frac{2^{-192}}{2^{-32}.2^{159}} = 2^{-319}$, which means an unreasonably large number of right pairs would be needed for an attack. Note that $2^{-192} \cdot 2^{191} < 1$ right pairs are suggested per structure. Since no pairs satisfy the output difference, no pairs survive this filtering, and thus, no subkey is ever suggested. Therefore, (3.56) is not useful for an attack on MESH-128.

Another example of iterative truncated differential for MESH-128, with many more than 3.5 rounds is:

$$(A, B, 0, 0, C, D, 0, 0) \overset{2^{-32}}{\rightarrow} (E, F, 0, 0, E, F, 0, 0) \overset{(0,0,0,0) \overset{1}{\rightarrow} (0,0,0,0)}{\longrightarrow} (E, E, F, 0, F, 0, 0, 0)$$

$$(E, E, F, 0, F, 0, 0, 0) \overset{2^{-16}}{\rightarrow} (G, H, I, 0, G, 0, 0, 0) \overset{(0,H,I,0) \overset{2^{-64}}{\rightarrow} (0,I,H,0)}{\longrightarrow} (G, G, H, I, 0, 0, 0, 0)$$

$$(G, G, H, I, 0, 0, 0, 0) \overset{1}{\rightarrow} (J, K, L, M, 0, 0, 0, 0) \overset{(J,K,L,M) \overset{2^{-64}}{\rightarrow} (M,L,K,J)}{\longrightarrow} (0, J, K, L, 0, 0, 0, M)$$

$$(0, J, K, L, 0, 0, 0, M) \overset{2^{-16}}{\rightarrow} (0, N, O, P, 0, 0, 0, P) \overset{(0,N,O,0) \overset{2^{-64}}{\rightarrow} (0,O,N,0)}{\longrightarrow} (0, 0, N, O, 0, 0, P, P),$$

$$(0, 0, N, O, 0, 0, P, P) \overset{2^{-32}}{\rightarrow} (0, 0, Q, R, 0, 0, Q, R) \overset{(0,0,0,0) \overset{1}{\rightarrow} (0,0,0,0)}{\longrightarrow} (0, 0, 0, Q, 0, Q, R, R),$$

$$(0, 0, 0, Q, 0, Q, R, R) \overset{2^{-16}}{\rightarrow} (0, 0, 0, S, 0, T, U, S) \overset{(0,T,U,0) \overset{2^{-64}}{\rightarrow} (0,U,T,0)}{\longrightarrow} (0, 0, 0, 0, T, U, S, S),$$

$$(0, 0, 0, 0, T, U, S, S) \overset{1}{\rightarrow} (0, 0, 0, 0, V, W, X, Y) \overset{(V,W,X,Y) \overset{2^{-64}}{\rightarrow} (Y,X,W,V)}{\longrightarrow} (V, 0, 0, 0, W, X, Y, 0),$$

$$(V, 0, 0, 0, W, X, Y, 0) \overset{2^{-16}}{\rightarrow} (Z, 0, 0, 0, Z, a, b, 0) \overset{(0,a,b,0) \overset{2^{-64}}{\rightarrow} (0,b,a,0)}{\longrightarrow} (z, z, 0, 0, a, b, 0, 0), \quad (3.57)$$

which is an 8-round iterative truncated differential with probability 2^{-512}, and where A, B, C, ..., Z, a and b are 16-bit differences. A structure following the input difference of (3.57) can generate up to $2^{64}(2^{64} - 1)/2 \approx 2^{127}$ text pairs. Only pairs for which $\Delta c_3 = \Delta c_4 = \Delta c_7 = \Delta c_8 = 0$ are considered. This last criterion means a 64-bit condition. For a single structure, $2^{127} \cdot 2^{-64} = 2^{63}$ pairs satisfy the output difference of (3.57). Each pair suggests 2^{64} 128-bit subkeys. The probability that a wrong subkey is suggested is $\frac{2^{64}}{2^{128}} = 2^{-64}$.

The signal-to-noise ratio is $\mathrm{S/N} = \frac{2^{-512}}{2^{-64} \cdot 2^{63}} = 2^{-511}$, which means an unreasonably large number of right pairs would be needed for an attack. Note that $2^{-512} \cdot 2^{127} < 1$ right pairs are suggested per structure. Since no pairs satisfy the output difference, no pairs survive this filtering, and thus, no subkey is ever suggested. Therefore, although (3.57) covers many more rounds than (3.50)), the former is not useful for an attack on MESH-128.

3.6.5 TDC of WIDEA-n

The truncated differential analysis in this section[11] is based on [122].

The difference operator used in bitwise exclusive-or.

The truncated differential distinguisher used for WIDEA-n only distinguishes between zero and nonzero differences. The exact value of the nonzero difference is not relevant.

In WIDEA-n, $n \in \{4, 8\}$, the n copies of IDEA are connected by an $n \times n$ MDS matrix. The truncated differential attack in [122] exploits a narrow differential trail in the input to the MDS matrix.

The attack strategy is to have active differences across a single IDEA instance, thus, bypassing the diffusion layer (the MDS matrix) in every round. This step is accomplished by causing zero input difference to every MDS matrix in every round.

Recalling the notation in Sect. 2.5.1, let (2.16) denote the internal state of WIDEA-4 and (2.17) denote the internal state of WIDEA-8.

Consider a text structure where the first IDEA instance in WIDEA-4, (a_0, a_1, a_2, a_3), assumes all possible 64-bit values, while the remaining words always take on an arbitrary constant value. From a single structure, one can obtain up to $2^{64}(2^{64}-1)/2 \approx 2^{127}$ text pairs with differences of the form (3.58):

$$\Delta A = \begin{pmatrix} 0 & 0 & 0 & 0 \\ 0 & 0 & 0 & 0 \\ 0 & 0 & 0 & 0 \\ \Delta a_0 & \Delta a_1 & \Delta a_2 & \Delta a_3 \end{pmatrix}. \tag{3.58}$$

Similarly, for WIDEA-8, the input plaintext difference has the form (3.59):

[11] © IACR. Published with permission.

$$\Delta A = \begin{pmatrix} 0 & 0 & 0 & 0 \\ 0 & 0 & 0 & 0 \\ 0 & 0 & 0 & 0 \\ 0 & 0 & 0 & 0 \\ 0 & 0 & 0 & 0 \\ 0 & 0 & 0 & 0 \\ 0 & 0 & 0 & 0 \\ \Delta a_0 & \Delta a_1 & \Delta a_2 & \Delta a_3 \end{pmatrix}. \tag{3.59}$$

Across the KM half-round in the first IDEA instance of WIDEA-4, the difference (3.12) keep the same format (although the nonzero word differences may contain other nonzero values).

The input difference to the first MAD-box of the first IDEA instance has the form $(\Delta p_1, \Delta q_1)$, where $\Delta p_1 = p_1 \oplus p_1^*$, with $p_1 = (a_0 \odot Z_{0,0}) \oplus (a_2 \boxplus Z_{2,0})$, and $p_1^* = (a_0^* \odot Z_{0,0}) \oplus (a_2^* \boxplus Z_{2,0})$. Similarly, $\Delta q_1 = q_1 \oplus q_1^*$, with $q_1 = (a_1 \boxplus Z_{1,0}) \oplus (a_3 \odot Z_{3,0})$ and $q_1^* = (a_1^* \boxplus Z_{1,0}) \oplus (a_3^* \odot Z_{3,0})$.

The input to the MDS matrix from the first MAD-box is $t_1 = p_1 \odot Z_{4,0} \boxplus q_1$, which is a 16-bit value. Similarly, the corresponding difference, Δt_1, is a 16-bit value. Thus, with probability 2^{-16}, $\Delta t_1 = 0$. The remaining input differences to the MDS matrix from the other IDEA instances are zero since the plaintext difference is zero. Therefore, with probability 2^{-16}, the MDS matrix is not activated, and consequently, the nonzero differences remains only in the first IDEA instance. In summary, the trivial n-tuple fixed-point difference $(0, \ldots, 0)$ is used:

$$\text{MDS}(0, \ldots, 0)^T = (0, \ldots, 0)^T,$$

where the superscript T denotes the transpose operator. A similar reasoning applies to WIDEA-8.

The round output difference is an n-tuple with the form (3.58), that is, we obtain a 1-round iterative truncated differential with probability 2^{-16}. Repeated across 8.5 rounds, and bypassing the MDS matrix in every round of WIDEA-n, it gives a truncated differential distinguisher with probability $(2^{-16})^8 = 2^{-128}$. The last KM half-round does not affect the probability nor the format of the (3.58).

Note that this distinguisher applies to any key. It does not require weak-key nor weak-subkey assumptions.

The probability that a difference of the form (3.58) will not show up after 8.5-round WIDEA-4 after a single text pair is $1 - 2^{-128}$. After two structures it becomes $(1 - 2^{-128})^{2^{127} \cdot 2} \approx e^{-1}$. The probability that one right pair with difference of the form (3.58) will show up is therefore: $1 - e^{-1} \approx 0.63$. This same reasoning applies to WIDEA-8.

Comparatively, for a random permutation operating on 256-bit blocks, the probability that an input difference of the form (3.58) causes an output difference with this same pattern is $2^{-64 \cdot 3} = 2^{-192}$, because of the zero block differences. Note that $2^{-128} \ggg 2^{-192}$. Moreover, using a similar reasoning, the probability that one right pair with difference of the form (3.58) will show up at the output of a random permutation is: $1-(1-2^{-192})^{2^{128}} \approx 1-e^{2^{-64}} \to 0$.

Similarly, for a random permutation operating on $64 \cdot 8$-bit blocks, the probability that one right pair with difference of the form (3.59) causes an output difference with this same shape is $2^{-64\cdot 7} = 2^{-448}$, because of the zero block differences. The probability that one right pair with difference of the form (3.59) will show up at the output is: $1-(1-2^{-448})^{2^{128}} \approx 1-e^{2^{-320}} \to 0$.

In summary, it means that WIDEA-n, for $n \in \{4, 8\}$, can be efficiently distinguished from a random permutation (over the same plaintext space).

A distinguishing attack on WIDEA-n, using two structures, can be efficiently constructed using a hash table indexed by the ciphertexts in the IDEA instances which are not active. Without loss of generality, let us assume that only the input to the first IDEA instance is active.

The attack consists in encrypting texts from the structures and storing only the partial ciphertext $(a_4, a_5, \ldots, a_{4n})$ in a hash table. As soon as two ciphertexts with the same value of $(a_4, a_5, \ldots, a_{4n})$ appear, a right pair is found, because it corresponds to a ciphertext pair with difference (3.58) or (3.59).

If this ciphertext pair is found, then WIDEA-n is identified. Otherwise, a random permutation is identified.

The time complexity is 2^{65} encryptions. The data complexity is 2^{65} chosen plaintexts. The memory complexity is 2^{64} text blocks (to store the hash table).

3.6.5.1 Key-Recovery Attack on WIDEA-n

The attack starts by recovering the subkeys of the first round of WIDEA-n using a right pair (X, X^*) obtained from the distinguishing attack. Consider the 16-bit values (t_1, t_1^*) in the MAD-box that is input to the MDS matrix in the first rounds:

$$t_1 = (((x_0 \odot Z_{0,0}) \oplus (x_2 \boxplus Z_{2,0})) \odot Z_{4,0}) \boxplus ((x_1 \boxplus Z_{1,0}) \oplus (x_3 \odot Z_{3,0})),$$

$$t_1^* = (((x_0^* \odot Z_{0,0}) \oplus (x_2^* \boxplus Z_{2,0})) \odot Z_{4,0}) \boxplus ((x_1^* \boxplus Z_{1,0}) \oplus (x_3^* \odot Z_{3,0})).$$

Since (X, X^*) is a right pair, it means that $\Delta t_1 = t_1 \oplus t_1^* = 0$, that is $t_1 = t_1^*$:

$$\begin{aligned}(((x_0 \odot Z_{0,0}) \oplus (x_2 \boxplus Z_{2,0})) \odot Z_{4,0}) \boxminus (((x_0^* \odot Z_{0,0}) \oplus (x_2^* \boxplus Z_{2,0})) \odot Z_{4,0}) \\ = ((x_1^* \boxplus Z_{1,0}) \oplus (x_3^* \odot Z_{3,0})) \boxminus ((x_1 \boxplus Z_{1,0}) \oplus (x_3 \odot Z_{3,0})). \quad (3.60)\end{aligned}$$

In (3.60), the left-hand side depends only on $Z_{0,0}$, $Z_{2,0}$ and $Z_{4,0}$, while the right-hand side depends only on $Z_{1,0}$ and $Z_{3,0}$.

Let

$$F_i(X, X^*, Z_{0,i}, Z_{2,i}, Z_{4,i}) = (((x_{0,i} \odot Z_{0,i}) \oplus (x_{2,i} \boxplus Z_{2,i})) \odot Z_{4,i}) \boxminus (((x^*_{0,i} \odot Z_{0,i}) \oplus (x^*_{2,i} \boxplus Z_{2,i})) \odot Z_{4,i}), \quad (3.61)$$

and

$$G_i(X, X^*, Z_{1,i}, Z_{3,i}) = ((x^*_{1,i} \boxplus Z_{1,i}) \oplus (x^*_{3,i} \odot Z_{3,i})) \boxminus ((x_{1,i} \boxplus Z_{1,i}) \oplus (x_{3,i} \odot Z_{3,i})), \quad (3.62)$$

which denote the left-hand side and right-hand side of (3.60), respectively, for the i-th round, where $0 \leq i < 8$.

The five subkeys $Z_{0,i}$, $Z_{1,i}$, $Z_{2,i}$, $Z_{3,i}$ and $Z_{4,i}$ can be recovered using a meet-in-the-middle approach: compute $F_0(X, X^*, k_0, k_2, k_4)$ for all (k_0, k_2, k_4), and make a list; compute $G_i(X, X^*, k_1, k_3)$ for all (k_1, k_3) and look for matches in the list. Since the output of F_0 has 16 bits and there are 2^{48} possibilities for k_0, k_2, k_4, each output is assigned to an average of 2^{32} subkey triplets. Similarly, for G_0, each output is assigned to an average of 2^{16} subkey candidates.

Several right pairs will be needed to uniquely find the correct subkey value. To achieve a strong filtering, simultaneous matchings will be performed for F_0 and G_0 over all right pairs t, that is, for the concatenation of all t F_0 values and all t G_0 values: $\|\|_{j=1}^{t} F_0(X^{(j)}, X^{*(j)}, k_0, k_2, k_4) = \|\|_{j=1}^{t} G_i(X^{(j)}, X^{*(j)}, k_1, k_3)$.

This filtering cannot recover the most significant bit of $Z_{1,0}$ because the effect of this bit on G_i is linear.

Each right pair allows a matching of 16-bit values, and 79 subkey bits are recovered at once.

We expect $t = 5$ right pairs to be enough, but implementations of the attack showed that the filtering given by each pair is not independent. Further experiments showed that $t = 8$ right pairs are enough to isolate a single subkey candidate.

Thus, the 79 subkey bits can be recovered with complexity $8 \cdot (2^{48} + 2^{32}) \approx 2^{51}$ computations of F_0 and G_0. Looking closely, a single computation of F_0 and G_0 corresponds to two KM half-rounds and an MA-box. So, eight computations of F_0 and G_0 corresponds roughly to a single IDEA encryption, which is a fraction of $\frac{1}{n}$ of WIDEA-n. So, the cost is $2^{48}/n$ WIDEA-n encryptions.

The same attack can be repeated to the other IDEA instances to recover the rest of the subkeys Z_0, Z_1, Z_2, Z_3 and Z_4. But, the right pairs in each case will have to follow other differential trails in other IDEA instances. Since there are n IDEA instances, the time complexity becomes $n \cdot 2^{48}/n = 2^{48}$ WIDEA-n encryptions. The data complexity becomes $n \cdot 8 \cdot 2^{65} = n \cdot 2^{68}$ chosen plaintexts. The memory complexity remains 2^{64} text blocks, since memory can be reused for attacking each IDEA instance separately,

Once the subkeys Z_0, Z_1, Z_2, Z_3 and Z_4 are recovered, the inputs to the MDS matrix in the first round can be computed, as well as its outputs. Further, the parallel IDEA instances can be analyzed independently again.

The next step is to recover the subkey Z_5, i, $0 \leq i < n$, and compute the state after the first round.

The meet-in-the-middle approach can then be applied to the second round to recover $Z_{6,i}$, $Z_{7,i}$, $Z_{8,i}$, $Z_{9,i}$ and $Z_{10,i}$. This means the time complexity becomes $2^{16} \cdot 2^{48} \cdot n$ WIDEA-n encryptions.

According to the key schedule of WIDEA-n, Sect. 2.5.2, the user key can be reconstructed from knowledge of Z_j for $0 \leq j \leq 7$.

Concerning the most significant bit of $Z_{1,i}$, $0 \leq i < n$, there are 2^n possible values. Instead of an exhaustive search for these n subkey bits, the authors of [122] used the fact that the unknown bits have a linear effect on the MDS matrix operation. The coefficients of the MDS matrix of WIDEA-8 have value between 1 and 9 (see (2.15) in Sect. 2.5.1). Therefore, any linear combination of these coefficients have value between 0 and 15. For WIDEA-4, the MDS matrix coefficients have value between 1 and 3 (see (2.14) in Sect. 2.5.1), and any linear combination of these coefficients have value between 0 and 3.

Instead of guessing the n subkey bits, the linear combination effect on the MDS matrix output can be recovered, which is of the form $t \otimes 8000_x$, with $0 \leq t < 4$ for WIDEA-4, or $0 \leq t < 16$ for WIDEA-8. Multiplication in $GF(2^{16})$ is denoted by $\otimes$. This means a 2-bit and a 4-bit unknown value, respectively. This linear combination is the value actually needed to recover $Z_{5,i}$ later.

The attack complexity becomes $n \cdot 4 \cdot 2^2 \cdot 2^{64} = 2^{68} \cdot n$ encryptions for WIDEA-4, and $n \cdot 8 \cdot 2^4 \cdot 2^{64} = n \cdot 2^{71}$ encryptions for WIDEA-8.

Let $K = (k_0, k_1, k_2, k_3, k_4, k_5, k_6, k_7)$ denote the user key, where $|k_j|$=64 bits for WIDEA-4 and $|k_j|$= 128 bits for WIDEA-8, $0 \leq j \leq 7$.

From the key schedule algorithms of WIDEA-n, recovery of Z_0, Z_1, Z_2, Z_3, Z_4 and Z_5 leads to immediate reconstruction of the user key, because the first eight subkeys are assigned as $Z_j = k_j$, $0 \leq j \leq 7$.

But, according to the key schedule,

$$Z_8 = ((((Z_7 \oplus Z_0) \overset{16}{\boxplus} Z_3) \overset{16}{\lll} 5) \lll 24) \oplus C_0,$$

$$Z_9 = (((Z_8 \oplus Z_1) \overset{16}{\boxplus} Z_4) \overset{16}{\lll} 5) \lll 24,$$

$$Z_{10} = (((Z_9 \oplus Z_2) \overset{16}{\boxplus} Z_5) \overset{16}{\lll} 5) \lll 24,$$

and

$$Z_{11} = (((Z_{10} \oplus Z_3) \overset{16}{\boxplus} Z_6) \overset{16}{\lll} 5) \lll 24.$$

These equations show that the attack on the second round just need to guess Z_6 and Z_7, since the other subkeys can be immediately derived from Z_j, $0 \leq j \leq 7$. Consequently, the attack complexity is dominated by the cost of the key-recovery in the first round.

3.7 Multiplicative Differential Analysis

In [46], Borisov *et al.* described a differential analysis technique using multiplication as the difference operator. This variant of DC became knows as *multiplicative differential cryptanalysis* and was applied to the Nimbus, xmx, MultiSwap and to a modified variants of the IDEA cipher called IDEA-X where all $\boxplus$ operators were replaced by $\oplus$. Consequently, there are consecutive $\oplus$ operators along the encryption framework of IDEA-X.

The analysis of IDEA-X is more relevant to our context. For the analysis of the other ciphers, we refer to [46].

For a pair of bit-strings (X, X^*), their *multiplicative difference* in $GF(2^{16}+1)$ is defined as

$$\varDelta X = X \odot (X^*)^{-1} = X \boxdot X^*,$$

where $\boxdot$ denotes modular division in $GF(2^{16}+1)$.

So, for a keyed multiplication with a subkey Z_i, the output difference is independent of Z_i: $(X \odot Z_i) \odot (X^* \odot Z_i)^{-1} = X \odot Z_i \odot Z_i^{-1} \odot (X^*)^{-1} = X \boxdot X^*$.

In IDEA-X, there are only two arithmetic operations: $\oplus$ and $\odot$. Borison *et al.* used the following relation to attack IDEA-X:

$$-X \bmod (2^{16}+1) = X \oplus FFFD_x \iff \mathrm{lsb}_2(X) = 1, \tag{3.63}$$

which implies a weak-subkey condition for $\oplus$ operations keyed by subkeys Z_i: $\mathrm{lsb}_2(Z_i)$=1. Under this condition, (3.63) holds with certainty.

The attack of [46] on the whole 8.5-round IDEA-X used a wide-trail differential of the form

$$(-1,-1,-1,-1) \stackrel{1r}{\to} (-1,-1,-1,-1). \tag{3.64}$$

It is a 1-round iterative differential that holds with probability 2^{-4}, under the weak-subkey conditions $\mathrm{lsb}_2(Z_1^{(i)}) = \mathrm{lsb}_2(Z_4^{(i)}) = 1$.

The difference (3.63) allows us to propagate differences across both $\oplus$ and $\odot$ operators for any number of rounds of IDEA-X. But, most interestingly, using (3.64) the difference to the MA-boxes has the form $(1, 1)$, which is the trivial difference for multiplicative differentials: $X \odot (X^*)^{-1} = 1$ means that $X = X^*$, and therefore, $X \oplus X^* = 0$.

Therefore, there is actually no need to replace the $\boxplus$ operators in the MA half-rounds. In [165], Raddum reevaluated the attack by Borisov *et al.*, and replaced only the $\boxplus$ operators in the KM half-rounds by $\oplus$. The resulting cipher was named IDEA-X/2. His attack still covered the whole 8.5-round IDEA-X/2, which is closer to IDEA than IDEA-X. But, still, none of them explained why they could not adapt the multiplicative differentials to the original IDEA cipher.

Looking at (3.63), we need to find out how to extend it to the $\boxplus$ operator. In order to extend (3.63), we study the probability that the 16-bit difference

$FFFD_x$ propagates across $\boxplus$. There are two cases to analyze: (i) keyed-$\boxplus$ such as $Y = X \boxplus Z_i$ with a subkey Z_i, and (ii) unkeyed-$\boxplus$ such as $Y = X \boxplus W$. Case (i) occurs only in the KM half-round, while case (ii) happens only in the MA-boxes.

For case (i), an exhaustive analysis for a few subkeys indicated that $FFFD_x \stackrel{\boxplus}{\rightarrow} FFFD_X$ (propagation across a single keyed-$\boxplus$) for even-valued subkeys Z_i holds with probability 2^{-2}, while for odd-values subkeys the probability varies. Nonetheless, the latter probability was high for the first few sampled subkeys: $\frac{10922}{2^{16}}$ for $Z_i = 3$, $\frac{19660}{2^{16}}$ for $Z_i = 5$ and $\frac{14044}{2^{16}}$ for $Z_i = 7$. A more extensive analysis is required.

For case (ii), an exhaustive analysis was performed over all 32-bit inputs. The results are summarized in Table 3.80. In this case, the structure under analysis has the form $Y = X \boxplus W$ where Y, X, W are 16-bit words; the input differences are ΔX and ΔW and the output difference is ΔY.

Table 3.80 Probability of propagation of exclusive-or difference $FFFD_x$ across unkeyed-$\boxplus$.

ΔX	ΔW	ΔY	probability
$FFFD_x$	$FFFD_x$	$FFFD_x$	$2^{-15.415}$
$FFFD_x$	0000_x	0000_x	$2^{-16.321}$
0000_x	$FFFD_x$	0000_x	$2^{-16.321}$
$FFFD_x$	0000_x	$FFFD_x$	$2^{-15.717}$
0000_x	$FFFD_x$	$FFFD_x$	$2^{-15.717}$

Case (ii) and Table 3.80 are quite helpful for searching for narrow trails across multiple rounds of IDEA. But, given the low probabilities in Table 3.80 for each active $\boxplus$ operator, it is unlikely that these differential trails could reach the whole 8.5-round IDEA.

If one would use the 1-round differential (3.64) in IDEA, then only case (i) is needed, since the input difference to the MA-boxes would be trivial. The issue in this case is that the probability to cross the KM half-round depends on the value of additive subkeys. Weak-subkey conditions such as (3.63) can be imposed on the multiplicative subkeys, but for the additive subkeys the probability is not uniform. If an adversary assumes a probability of 2^{-2} for all additive subkeys (all such subkeys should be even-valued), then the original multiplicative differential probability of $(2^{-4})^8 = 2^{-32}$ would be compounded with the probability $(2^{-2} \cdot 2^{-2})^8 = 2^{-32}$ due to the additive subkeys in the KM half-rounds.

An effective multiplicative differential attack on the full 8.5-round IDEA cipher seems improbable given the conditions imposed on the plaintext difference in [165, 46], the combined probability of 2^{-64} (which is probably lower depending of the unknown additive subkeys), the weak-subkey conditions and their effect on the additive subkeys and whether they would all be compatible under the key schedule algorithm of IDEA. The full analysis is left as an open problem.

3.8 Impossible-Differential Cryptanalysis

The origins of the Impossible-Differential (ID) technique dates back to the works of Knudsen [102], Borst *et al.* [48] and Biham *et al.* [15, 16].

The ID analyses use differences of text pairs like in DC, but while the latter looks for characteristics or differentials that hold with the highest possible probability, the former looks for differentials that never hold. Therefore, a differential with probability zero is called an *impossible differential.*

One technique to construct ID distinguishers is known as *miss-in-the-middle* [15]. The main idea of this technique is to find two events holding with certainty (probability one) which cannot be satisfied simultaneously. This means that their combination is an impossible event. In a differential context, the events consist of two differentials, each of which holds with certainty. The distinguisher consists of their concatenation into a single, longer differential, which cannot be satisfied because their combination leads to a contradiction (at the point where both differentials meet). That is the reason for its name: miss-in-the-middle. This construction will be exemplified later on by several attacks on Lai-Massey cipher designs.

Once the existence of the impossible event is proved for a cipher, it can be used to distinguish the given cipher from a random permutation.

Key-recovery attacks in the ID setting are constructed by guessing subkey bits in rounds surrounding a given ID distinguisher. All subkey candidates that lead to the impossible event are wrong because the distinguisher never holds. Therefore, the ID distinguisher plays the role of a sieve, systematically filtering the wrong subkey candidates. The subkey(s) not suggested by the distinguisher include the correct subkey value.

From [15], consider an n-bit block cipher $E(.)$, a set of input differences $\mathcal{P}$ of cardinality 2^p and a corresponding set of output differences $\mathcal{Q}$ of cardinality 2^q. Suppose that no difference from $\mathcal{P}$ can cause an output difference from $\mathcal{Q}$ across $E(.)$. How many chosen texts should be requested in order to distinguish $E(.)$ from a random permutation?

For $E(.)$, any difference selected from $\mathcal{P}$ will never lead to a difference from $\mathcal{Q}$, so, $E(.)$ can be identified with certainty. But, for a random permutation π, the probability that a single difference from $\mathcal{P}$ will not lead to a difference in $\mathcal{Q}$ is $1-\frac{2^q}{2^n}$. After t chosen differences from $\mathcal{P}$, this probability becomes $(1-\frac{2^q}{2^n})^t$, assuming independence between text differences in $\mathcal{P}$. This probability is approximately

$$(1-1/2^{n-q})^{2^{n-q}\cdot\frac{t}{2^{n-q}}} \approx e^{-t/2^{n-q}}.$$

From the definition of PRP in Sect. 1.7, the advantage to distinguish $E(.)$ from a random permutation π is: $1 - e^{-t/2^{n-q}}$. In general, with $t = 2^{n-q}$ chosen pairs with differences from $\mathcal{P}$, the advantage becomes $1 - e^{-1} \approx 0.63$. For larger t, this advantage increases. This quantity t can be obtained with less chosen plaintexts by using *structures*, which is a standard technique for saving chosen plaintexts in differential attacks [30].

In the optimal case, one can use structures of 2^p texts which contains about $2^p \cdot (2^p - 1)/2 \approx 2^{2p-1}$ pairs with differences from $\mathcal{P}$. In this case, $2^{n-q}/2^{2p-1}$ structures are required, and the number of chosen texts used by the distinguishing attack is about $2^{n-p-q+1}$, assuming $2p < n - q + 1$. Thus, the higher is $p + q$, the better is the distinguisher based on the impossible event, because less chosen texts are needed.

In terms of the signal-to-noise ratio, we have that $S/N = 0$ for the counting scheme since the right subkey is never suggested by the ID distinguisher.

Another approach to search for ID distinguishers, suggested by Biham *et al.* in [15] and is based on the overall structure of the target cipher. The procedure to find ID distinguishers involves encrypting many text pairs under all possible keys, and discard all output differences, since they are not impossible. Therefore, by eliminating (or sieving), only differentials that never occur will be left. This search technique, although generic, requires too many differentials and keys to try exhaustively.

Biham *et al.* [15] also suggested the use of truncated differentials in which differences are defined wordwise and only the zero and nonzero difference status matter. As a consequence, the number of text differences is reduced to the number of combinations of zero/nonzero wordwise differences. But, one may not come up with differentials in which an input pair leads to output differences having zero difference within word boundaries (the cipher may not even be word-oriented, such as DES). To solve both problems, it was suggested to analyze scaled-down cipher variants in which the nature of the cipher components and its overall structure are preserved, but the size is reduced in order to allow an exhaustive search for impossible differentials. To preserve the nature of a cipher component means for example, that (large) permutations are replaced by (smaller) permutations, or (large) noninjective mappings are substituted by (smaller) noninjective mappings. This technique is called *shrinking*.

ID attacks have been applied to several block ciphers, either following a Feistel Network, an SPN or a Lai-Massey design, such as Skipjack [15], IDEA and Khufu [16], Twofish [26], Misty [109] and Rijndael [27].

3.8.1 ID Analysis of IDEA

This section is based on [16] and describes ID attacks on reduced-round variants of the IDEA cipher.

First, a construction of an ID distinguisher will be provided. In [16], a 2.5-round ID distinguisher was described for IDEA using the miss-in-the-middle construction technique and using exclusive-or as difference operator. The distinguisher starts from an MA half-round, that is, it covers the following sequence of consecutive half-rounds: MA$\circ$ KM $\circ$ MA $\circ$ KM $\circ$ MA.

The input differences in $\mathcal{P}$ have the form $(a, 0, a, 0)$ where 0 and $a \neq 0$ are 16-bit differences. The output differences in $\mathcal{Q}$ have the form $(b, b, 0, 0)$, where $b \neq 0$ is a 16-bit difference.

The terminology

$$(a, 0, a, 0) \overset{2.5r}{\not\rightarrow} (b, b, 0, 0), \tag{3.65}$$

will be used to denote that the leftmost 64-bit difference can never cause the rightmost 64-bit difference across 2.5-round IDEA, under the restrictions for a and b. The justification for why (3.65) holds with probability zero is as follows:

- consider a text pair with input difference $(a, 0, a, 0)$ for $a \neq 0$. For such a pair, the inputs to the first MA half-round are $(0, 0)$, as well as its outputs. Thus, the difference after the first MA half-round becomes $(a, a, 0, 0)$, including the swap of the two middle words. After the following KM half-round, the nonzero difference a is transformed by $Z_1^{(i)}$ and $Z_3^{(i)}$ into nonzero differences c and d, respectively, and the output difference has the form $(c, d, 0, 0)$.
- similarly, from the bottom-up direction, the output difference $(b, b, 0, 0)$ becomes $(b, 0, b, 0)$ at the input to the last MA half-round, because of the word swap σ (recall that the inverse of MA half-round without σ is an involution). The difference before the last KM half-round has the form $(e, 0, f, 0)$ for $e \neq 0$ and $f \neq 0$.
- thus, the input difference to the middle MA half-round has the form $(c, d, 0, 0)$ while its output difference has the form $(e, 0, f, 0)$. The output difference before the word swap is $(e, f, 0, 0)$. From these differences and the MA half-round construction, we conclude that the inputs to the MA-box are (c, d) from the top-down and (e, f) from the bottom-up, and are nonzero in both coordinates. The output difference is $(c \oplus e, d \oplus f)$ which equals $(0, 0)$ from the exclusive-or after the MA-box. This is a contradiction since the MA-box is a permutation mapping. Therefore, there can be no text pair satisfying both $(a, 0, a, 0)$ and $(b, b, 0, 0)$ simultaneously across the 2.5-round IDEA variant MA$\circ$ KM $\circ$ MA $\circ$ KM $\circ$ MA.

Likewise, another 2.5-round ID distinguisher for IDEA can be demonstrated:

$$(0, a, 0, a) \overset{2.5r}{\not\rightarrow} (0, 0, b, b), \tag{3.66}$$

The ID distinguishers (3.65) and (3.66) can be used to distinguish 2.5-round IDEA from a random permutation. Moreover, key-recovery attacks can be mounted on top of the distinguishers.

3.8.1.1 ID Attack on 3.5-round IDEA

We attack the first 3.5 rounds of IDEA using (3.65) placed after the first KM half-round. Plaintexts are denoted by $X = (X_1, X_2, X_3, X_4)$ and ciphertext after 3.5-rounds are denoted by $Y = (Y_1, Y_2, Y_3, Y_4)$. The differences are denoted $\Delta X = (\Delta X_1, \Delta X_2, \Delta X_3, \Delta X_4)$ and $\Delta Y = (\Delta Y_1, \Delta Y_2, \Delta Y_3, \Delta Y_4)$, respectively.

The attack proceeds as follows:

- choose a structure of 2^{32} plaintexts with $\Delta X_2 = \Delta X_4 = 0$, and with all possibilities for ΔX_1 and ΔX_3. That means a single structure can generate $2^{32}(2^{32}-1)/2 \approx 2^{63}$ pairs with input difference from (3.65).
- collect about 2^{31} pairs from the structure whose ciphertext difference satisfy $\Delta Y_3 = 0 = \Delta Y_4$. The reasoning is that each of these 16-bit zero differences impose a 16-bit condition, and thus, $2^{63}/(2^{16})^2 = 2^{31}$ pairs survive this filtering.
- for each surviving pair:
  - try all the 2^{32} possible subkeys $(Z_1^{(1)}, Z_3^{(1)})$ that affect ΔX_1 and ΔX_3, and partially encrypt the pair of values with these differences. For instance, let $X_1 \oplus X_1' = \Delta X_1$ and $X_3 \oplus X_3' = \Delta X_3$. Then, $T_1 = X_1 \odot Z_1^{(1)}$, $T_1' = X_1' \odot Z_1^{(1)}$, $T_3 = X_3 \boxplus Z_3^{(1)}$ and $T_3' = X_3' \boxplus Z_3^{(1)}$. Collect about 2^{16} possible 32-bit subkey candidates satisfying $\Delta T_1 = \Delta T_3$. This is a 16-bit equality imposing a 16-bit condition. Thus, we expect $2^{32}/2^{16} = 2^{16}$ subkey candidates will survive.

    For the given (X_1, X_1'), compute for each candidate z of $Z_1^{(1)}$ the difference $(X_1 \odot z) \oplus (X_1' \odot z)$, and store it in a table D as $D[(X_1 \odot z) \oplus (X_1' \odot z)] = z$. This step costs 2^{16} difference computations and equivalent storage.

    For each candidate w of $Z_3^{(1)}$, compute $(X_3 \boxplus w) \oplus (X_3' \boxplus w)$ and get a matching z from D as $D[(X_3 \boxplus w) \oplus (X_3' \boxplus w)]$. Thus, 2^{16} pairs (z, w) of candidates for $(Z_1^{(1)}, Z_3^{(1)})$ can be obtained. Assuming $\odot$s cost much more than $\boxplus$s, the cost of computing $(X_1 \odot z) \oplus (X_1' \odot z)$ dominates over $(X_3 \boxplus w) \oplus (X_3' \boxplus w)$. As stated in [16], this step can be efficiently performed in 2^{16} time and storage.
  - try all the 2^{32} possible subkeys $(Z_1^{(4)}, Z_2^{(4)})$ that affect ΔY_1 and ΔY_2, and partially decrypt the pair of values with these differences. For instance, let $Y_1 \oplus Y_1' = \Delta Y_1$ and $Y_2 \oplus Y_2' = \Delta Y_2$. Then, $U_1 = Y_1 \odot (Z_1^{(4)})^{-1}$, $U_1' = Y_1' \odot (Z_1^{(4)})^{-1}$, $U_2 = Y_2 \boxminus Z_2^{(4)}$ and $U_2' = Y_2' \boxminus Z_2^{(4)}$. Collect about 2^{16} possible 32-bit subkey candidates satisfying $\Delta U_1 =$

ΔU_2. This is a 16-bit equality imposing a 16-bit condition. Thus, we expect $2^{32}/2^{16} = 2^{16}$ subkey candidates will survive. Similarly to the previous item, as as stated in [16], this step can be efficiently performed in 2^{16} time and storage.
  - make a list of all the 2^{32} 64-bit subkey candidates combining the two previous items. These subkeys cannot be the correct ones because otherwise, they lead to a text pair satisfying the impossible differential.

- repeat this analysis for each of the 2^{31} pairs obtained from each structure, and use a total of 90 structures. Each pair defines a list of about 2^{32} wrong subkeys. Compute the union of the lists of wrong 64-bit subkeys they suggest. It is expected that after 90 structures, the number of wrong subkey candidates remaining is

$$2^{64} \cdot (1 - 2^{-32})^{2^{31} \cdot 90} \approx 2^{64} \cdot e^{-45} \approx 0.5 < 1.$$

  Therefore, only the correct subkey value is expected to survive.
- the symmetric differential (3.66) can be used to recover 46 additional subkey bits. According to the key schedule of IDEA in Sect. 2.2.2, 16 bits overlap with the 64 bits already found, and two bits, numbered 17 and 18, are shared by the first and fourth rounds. Finally, 18 other bits can be found by exhaustive search to complete the full 128-bit user key.

This attack requires $90 \cdot 2^{32} \approx 2^{38.5}$ chosen plaintexts and $90 \cdot 2^{31} \cdot (2^{16} + 2^{16}) \approx 2^{54.5}$ steps, where a *step* involves a difference computation that is essentially two $\odot$ and one $\oplus$. Let us assume that 2.5-round IDEA contains ten $\odot$, ten $\boxplus$ and twelve $\oplus$, and that each $\odot$ costs about three time a single $\oplus$, and that a single $\boxplus$ costs the same as a single $\oplus$.

Then, the time complexity becomes $90 \cdot 2^{48} \cdot \frac{7}{52} \approx 2^{52.59}$ 2.5 IDEA computations. Memory complexity is about $90 \cdot 2^{31+17} \approx 2^{54.49}$ 16-bit word, or $2^{52.49}$ text blocks.

3.8.1.2 ID Attack on 4-round IDEA

Further, the attack can be extended to 4-round IDEA, from the (beginning of the) second to the (end of the) fifth round.

From the key schedule of IDEA, not all sets of rounds are equally strong. Overlapping bits in the subkeys allows us to reduce the attack complexity by guessing less subkey bits. Although the ID attacks do not use weak keys as in DC and LC, the overlapping of bits in the key schedule is yet a recurrent weakness that is widely exploited in key-recovery attacks.

This attack will use the differential (3.65). Placing this differential in the middle of the second round means recovering the following subkeys: $Z_1^{(2)}$, $Z_3^{(2)}$, $Z_1^{(5)}$, $Z_2^{(5)}$, $Z_5^{(5)}$ and $Z_6^{(5)}$. These subkeys amount to 96 bits, but according to the key schedule of IDEA, 17 of these bits overlap, leaving 69 independent key bits.

The attack initially recovers $Z_5^{(5)}$ and $Z_6^{(5)}$ of the last MA half-round, and for each guess apply the previous attack on 3.5 rounds as follows:

- for each of 2^{32} possibilities of $(Z_5^{(5)}, Z_6^{(5)})$:
  - decrypt the last MA half-round for all structures using the guessed subkeys
  - for each structure, find all pairs with zero difference in the third and fourth words, leaving about 2^{31} pairs per structure
  - for each surviving pair:
    - at this point $Z_3^{(2)}$ is known because $(Z_5^{(5)}, Z_6^{(5)})$ contain all 16 bits of $Z_3^{(2)}$. Let $T = (T_1, T_2, T_3, T_4)$ denote the input to the third round and $T' = (T_1', T_2', T_3', T_4')$ denote the corresponding text in the pair. We calculate the difference $(Z_2^{(3)} \boxplus T_3) \oplus (Z_3^{(2)} \boxplus T_3')$ and find $Z_1^{(2)}$ which leads to the same difference in the first word: $(Z_1^{(2)} \odot T_1) \oplus (Z_1^{(2)} \odot T_1')$. On average, only one $Z_1^{(2)}$ is suggested per pair.
    - similarly, find $(Z_1^{(5)}, Z_2^{(5)})$ which cause equal difference at the fifth round. Since $Z_1^{(5)}$ and $Z_2^{(5)}$) share 11 key bits, according to the key schedule of IDEA, there are 2^5 choices of subkey pairs and thus about 2^5 choices of newly found 37 subkeys bits. These choices are impossible that is, wrong values.
  - we need 52 structures to filter out all the wrong subkeys because we fixed many key bits at the outermost loop: $2^{37} \cdot (1 - \frac{2^5}{2^{37}})^{2^{31} \cdot 52} \approx 2^{37} \cdot e^{-26} \approx 2^{-0.51} < 1$.
- after analyzing all the structures, only a few subkey values remain. These values can be verified by exhaustive search.

This attack requires $52 \cdot 2^{32} \approx 2^{38}$ chosen plaintexts packed into structures. The total complexity consists of $2^{32} \cdot 2^{38}$ MA half-rounds, which are equivalent to about 2^{67} 4-round encryptions plus $2^{32} \cdot 2^{37} \cdot 2^5 = 2^{74}$ steps. When these steps are performed efficiently, they are equivalent to about 2^{70} 4-round encryptions.

The remaining $128 - 69 = 59$ key bits can be recovered by exhaustive search.

3.8.1.3 ID Attack on 4.5-round IDEA

An attack on 4.5-round IDEA starts after the first KM half-round, that is, covering the first MA half-round until the end of the fifth round.

In addition to the 64 key bits considered in the previous attack on 4 rounds, we need to find the four subkeys of two additional MA half-rounds. However, according to the key schedule algorithm of IDEA, only 16 of the 64 key bits are new, and the other 48 bits are either shared with the set found on the previous attack or shared between the first and the last MA half-rounds. Therefore, it suffices to guess 80 key bits in order to verify whether the ID distinguisher occurs. These key bits are numbered 13–43, 65–112 and cover the following subkeys: $Z_5^{(1)}$, $Z_6^{(1)}$, $Z_1^{(2)}$, $Z_3^{(2)}$, $Z_1^{(5)}$, $Z_2^{(5)}$, $Z_5^{(5)}$ and $Z_6^{(5)}$. The attack is as follows:

- get the ciphertexts of all 2^{64} possible plaintexts (under the unknown key), that is, the whole codebook.
- let $T = (T_1, T_2, T_3, T_4)$ denote the input to the second round, and ΔT $=(\Delta T_1, \Delta T_2, \Delta T_3, \Delta T_4)$ the corresponding difference. Define a structure to be a set of 2^{32} texts whose encryption have T_2 and T_4 fixed to some arbitrary constant values, while T_1 and T_3 range over all possible values. Unlike the previous attack on 4 rounds, these structures are based on the intermediate values rather than on the plaintexts.
- try all the 2^{80} possible values of the 80-bit subkey. For each such guess:
  - prepare a structure and use the trial subkey bits to partially decrypt it by one MA half-round using $Z_5^{(1)}$ and $Z_6^{(1)}$, to get the 2^{32} plaintexts.
  - for each plaintext, find the corresponding ciphertext and partially decrypt the topmost and bottommost half-rounds using the subkeys $Z_5^{(5)}$, $Z_6^{(5)}$, $Z_1^{(5)}$ and $Z_2^{(5)}$. Partially encrypt all pairs in the structure with $Z_1^{(2)}$ and $Z_3^{(2)}$.
  - let $U = (U_1, U_2, U_3, U_4)$ denote the input to the fifth round, and $Y = (Y_1, Y_2, Y_3, Y_4)$ denote the output from the fifth round. Check whether there is some pair in the structure that satisfies the combined 64-bit condition: $\Delta T_1 = \Delta T_3$, $\Delta U_1 = \Delta U_2$, $\Delta Y_3 = 0$ and $\Delta Y_4 = 0$.
  - if there is such an impossible pair in the structure, then the trial 80-bit subkey is wrong.
  - if there is no such pair in the structure, try again with another structure.
  - if no pair is found after trying 100 structures, consider the 80-bit subkey value as the correct value.
- assuming that a unique 80-bit value survives the previous step, the remaining 48 key bits can be found by exhaustive search.

This attack requires the full codebook (2^{64} plaintexts), and finds the 128-bit user key within 2^{112} steps using about 2^{32} blocks of memory.

Following the conversion in the attack on 4 rounds, the time complexity is equivalent to about 2^{108} 4.5-round IDEA encryptions.

3.8.2 ID Analysis of MESH-64

ID analysis of MESH-64 follows from [151, 143] and is based on a similar analysis of Biham *et al.* on IDEA in [16]. But, since the key schedule algorithm of MESH-64 does not consist of just a bit permutation, it is not possible to reduce the attack complexity by exploiting overlapping bits in the subkeys (as in IDEA).

The longest ID distinguishers found for MESH-64 cover 2.5 rounds and are the same ones used for IDEA: $(a, 0, a, 0) \overset{2.5r}{\not\rightarrow} (b, b, 0, 0)$ and $(0, a, 0, a) \overset{2.5r}{\not\rightarrow} (0, 0, b, b)$, where $a \neq 0$, $b \neq 0$ and the crossed arrow indicates that the leftmost 64-bit difference can never cause the rightmost 64-bit difference across 2.5-round MESH-64. These distinguishers start from an MA half-round, and only depend on the fact that the MA-box is a permutation mapping.

The difference operator used is exclusive-or.

Let $X^{(i)} = (X_1^{(i)}, X_2^{(i)}, X_3^{(i)}, X_4^{(i)})$ denote the input to the i-th round, and $Y^{(i)} = (Y_1^{(i)}, Y_2^{(i)}, Y_3^{(i)}, Y_4^{(i)})$ denote the input to the i-th MA half-round. So, $X^{(1)}$ denotes the plaintext and $Y^{(9)}$ denotes the ciphertext after 8.5-round MESH-64. Let $X^{(i)*} = (X_1^{(i)*}, X_2^{(i)*}, X_3^{(i)*}, X_4^{(i)*})$ denote the other element in the difference pair with $X^{(i)}$. So, their difference is denoted $\Delta X^{(i)} = (\Delta X_1^{(i)}, \Delta X_2^{(i)}, \Delta X_3^{(i)}, \Delta X_4^{(i)})$, with $\Delta X_j^{(i)} = X_j^{(i)} \oplus X_J^{(i)*}$ for $1 \leq j \leq 4$.

An attack on the first 3.5 rounds of MESH-64 proceeds as follows:

- choose a text structure consisting of 2^{32} plaintexts with fixed $X_2^{(1)}$ and $X_4^{(1)}$, while $X_1^{(1)}$ and $X_3^{(1)}$ assume all possible 16-bit values. There are $2^{32} \cdot (2^{32} - 1)/2 \approx 2^{63}$ plaintext pairs with difference $(\Delta X_1^{(1)}, 0, \Delta X_3^{(1)}, 0)$ in a single structure. Collect about $2^{63}/(2^{16})^2 = 2^{31}$ pairs from the structure whose ciphertext difference after 3.5 rounds satisfies $\Delta Y_3^{(4)} = 0 = \Delta Y_4^{(4)}$. For each such pair, try all the 2^{32} possible values of $(Z_1^{(1)}, Z_3^{(1)})$ and partially encrypt the two elements in the pair: $(X_1^{(1)}, X_3^{(1)})$ and $(X_1^{(1)*}, X_3^{(1)*})$.
- collect about 2^{16} subkeys $(Z_1^{(1)}, Z_3^{(1)})$ satisfying the difference

$$(X_1^{(1)} \odot Z_1^{(1)}) \oplus (X_1^{(1)*} \odot Z_1^{(1)}) = (X_3^{(1)} \boxplus Z_3^{(1)}) \oplus (X_3^{(1)*} \boxplus Z_3^{(1)}).$$

  For each candidate z of $Z_1^{(1)}$, construct a table D indexed by the 16-bit difference as $D[(X_1^{(1)} \odot z) \oplus (X_1^{(1)*} \odot z)] = z$, that is, z is stored in D in the position indicated by the difference. Next, for each candidate w of $Z_3^{(1)}$, find the corresponding z by $D[(X_3^{(1)} \boxplus w) \oplus (X_3^{(1)*} \boxplus w)]$. Thus, for each z there is a matching w. If one considers that each $(X_1^{(1)} \odot z) \oplus (X_1^{(1)*} \odot z)$ costs significantly more than each $(X_3^{(1)} \boxplus w) \oplus (X_3^{(1)*} \boxplus w)$, then cost of this matching is essentially 2^{16} difference computations and 2^{16} memory (for the table D).

- next, try all the 2^{32} possible values for $(Z_1^{(4)}, Z_2^{(4)})$ and partially decrypt $(Y_1^{(4)}, Y_2^{(4)})$ in each of the two elements of the pair. Collect about 2^{16} subkeys $(Z_1^{(4)}, Z_2^{(4)})$ such that $\Delta X_1^{(4)} = \Delta X_2^{(4)}$, according to the ID distinguisher, that is:

$$(Y_1^{(4)} \boxminus Z_1^{(4)}) \oplus (Y_1^{(4)*} \boxminus Z_1^{(4)}) = (Y_2^{(4)} \odot (Z_2^{(4)})^{-1}) \oplus (Y_2^{(4)*} \odot (Z_2^{(4)})^{-1}).$$

  Similar to the previous item, this step costs about 2^{16} difference computations and 2^{16} memory.
- make a list of all 2^{32} 64-bit subkeys $(Z_1^{(1)}, Z_3^{(1)}, Z_1^{(4)}, Z_2^{(4)})$. Those subkeys cannot be the correct values because they encrypt a pair of the impossible differential. Each pair defines a list of about 2^{32} 64-bit wrong subkey candidates. It is expected that after 90 structures, the number of remaining wrong subkeys is

$$2^{64}(1 - \frac{2^{32}}{2^{64}})^{2^{31} \cdot 90} \approx 2^{64}/e^{45} \approx 2^{-0.92} \approx 0.52 < 1$$

  and the correct subkey value can be identified (uniquely).

This attack requires $90 \cdot 2^{32} \approx 2^{38.5}$ chosen plaintexts. The memory complexity is dominated by $2^{64}/64 = 2^{58}$ blocks for sieving 64-bit subkey candidates. The time complexity is $2^{31} \cdot 90 \cdot (2^{16} + 2^{16}) \approx 2^{54.5}$ difference computations.

The 2.5-round ID distinguisher $(0, a, 0, a) \overset{2.5r}{\not\to} (0, 0, b, b)$, with $a, b \neq 0$, can be used to recover the additional 64-bit subkeys $(Z_2^{(1)}, Z_4^{(1)}, Z_3^{(4)}, Z_4^{(4)})$. According to the key schedule of MESH-64 there is no overlap between the bits of the two sets of 64-bit subkeys.

Each difference computation amounts to three operations: either two $\odot$s and one $\oplus$, or two $\boxplus$s and one $\oplus$. The most expensive operation is $\odot$.

Considering that there are 17 $\odot$s across 3.5 rounds, then the time complexity could be approximated as $\frac{1}{17} \cdot 2 \cdot 2^{54.5} \approx 2^{52.5}$ 3.5-round encryptions. The data complexity is $2 \cdot 2^{38.5} = 2^{39.5}$ chosen plaintexts.

In summary, the first 64 user key bits are recovered as $(Z_1^{(1)}, Z_2^{(1)}, Z_3^{(1)}, Z_4^{(1)})$. The other 64-bit subkeys $(Z_1^{(4)}, Z_2^{(4)}, Z_3^{(4)}, Z_4^{(4)})$ do not allow to reconstruct the missing 64-bit part of the user key. These key bits can be obtained by exhaustive search. The final time complexity is 2^{64} 3.5-round encryptions.

An attack on 4-round MESH-64 can recover the 48 bits $(Z_5^{(4)}, Z_6^{(4)}, Z_7^{(4)})$ from an additional MA half-round and apply the previous attack on 3.5 rounds. The time complexity would increase to $2^{64+48} = 2^{112}$ 4-round encryptions. The memory and data complexity remain the same as for the 3.5-round attack.

Extending the attack further to 4.5 rounds means guessing $Z_j^{(5)}$ for $1 \leq j \leq 4$ and applying the previous attack on 4 rounds. According to the key schedule of MESH-64, the dependency among these subkeys are as follows:

$$Z_4^{(5)} = (((((Z_3^{(4)} \boxplus Z_4^{(4)}) \oplus Z_5^{(4)}) \boxplus Z_1^{(5)}) \oplus Z_2^{(5)}) \boxplus Z_3^{(5)}) \lll 7 \oplus c_{31}, \tag{3.67}$$

$$Z_3^{(5)} = (((((Z_2^{(4)} \boxplus Z_3^{(4)}) \oplus Z_4^{(4)}) \boxplus Z_7^{(4)}) \oplus Z_1^{(5)}) \boxplus Z_2^{(5)}) \lll 7 \oplus c_{30}, \tag{3.68}$$

$$Z_2^{(5)} = (((((Z_1^{(4)} \boxplus Z_2^{(4)}) \oplus Z_3^{(4)}) \boxplus Z_6^{(4)}) \oplus Z_7^{(4)}) \boxplus Z_1^{(5)}) \lll 7 \oplus c_{29}, \tag{3.69}$$

$$Z_1^{(5)} = (((((Z_7^{(3)} \boxplus Z_1^{(4)}) \oplus Z_2^{(4)}) \boxplus Z_5^{(4)}) \oplus Z_6^{(4)}) \boxplus Z_7^{(4)}) \lll 7 \oplus c_{28}. \tag{3.70}$$

Equations (3.67) to (3.70) show that it is not possible to reduce the number of guessed subkey bits: each subkey Z_j depends on six other subkeys, the farthest been eight position behind Z_j. Overall, the complexity rises to $2^{112+64} = 2^{176}$ 4.5-round computations, which is more than the effort of an exhaustive key search.

If the initial attack on 3.5 rounds had started from the second round instead of the first round, then the attack on 4.5 rounds could have guessed the three subkeys of the first MA half-round, but: (i) only 48 user key bits would be recovered in this case: $Z_5^{(1)}$, $Z_6^{(1)}$ and $Z_7^{(1)}$, instead of the 64 bits $Z_1^{(1)}$, $Z_2^{(1)}$, $Z_3^{(1)}$ and $Z_4^{(1)}$ in the original attack; (ii) the time complexity would be $2^{112+48} = 2^{160}$ still larger than an exhaustive key search.

3.8.3 ID Analysis of MESH-96

ID analysis of MESH-96 is based on [151, 143] and on a similar analysis of Biham *et al.* on IDEA in [16]. But, since the key schedule of MESH-96 does not consist of just a bit permutation, it is not possible to reduce the attack complexity by exploiting overlapping bits in the subkeys of MESH-96 (as in IDEA).

The largest ID distinguishers found for MESH-96 cover 2.5 rounds starting from an MA half-round. These are similar ID distinguishers as found for IDEA and MESH-64. The absence of longer ID distinguishers helps corroborate the design choices of MESH-96 (at least against ID attacks) especially concerning the fact that full text diffusion in achieved in every single round.

The difference operator is bitwise exclusive-or.

One of the ID distinguishers found for 2.5-round MESH-96 is

$$(a,0,0,a,0,0) \overset{2.5r}{\not\to} (b,b,c,c,0,0), \qquad (3.71)$$

where $a, b, c \neq 0$ and the crossed arrow means that the leftmost 96-bit difference can never cause the rightmost 96-bit difference across 2.5-round MESH-96.

The proof is as follows: consider a text pair with difference $(a, 0, 0, a, 0, 0)$, for $a \neq 0$, at the input to an MA half-round. For such a pair, the input and output differences of the MA-box are $(0, 0, 0)$. Therefore, the output difference of the MA half-round has the form $(a, a, 0, 0, 0, 0)$ after the word swap.

After the following KM half-round, the nonzero differences are transformed by $Z_1^{(i)}$ and $Z_2^{(i)}$ and become $(d, e, 0, 0, 0, 0)$, for 16-bit differences $d, e \neq 0$.

Similarly, from the bottom-up, consider a text pair with difference (b, b, c, c, 0, 0), for wordwise differences $b, c \neq 0$ after 2.5 rounds. For such a pair, the difference before the last MA half-round has the form $(b, c, 0, b, c, 0)$ due to the word swap. Notice that the inverse of the MA half-round without the word swap is an involutory transformation.

Further, the difference before the last KM half-round is transformed by $Z_1^{(i+2)}$, $Z_2^{(i+2)}$, $Z_4^{(i+2)}$ and $Z_5^{(i+2)}$ and has the form $(f, g, 0, h, i, 0)$ for 16-bit differences $f, g, h, i \neq 0$.

Therefore, the input to the middle MA half-round has the form (d, e, 0, 0, 0, 0) while the corresponding output difference has the form $(f, g, 0, h, i, 0)$. The difference before the swap of the four middle words is $(f, h, i, g, 0, 0)$. In this same middle MA half-round, the input difference to the MA-box is nonzero: $(d, e, 0)$, while the corresponding output difference is also nonzero: $(i, e \oplus h, d \oplus f) = (0, 0, g)$, which can be deduced from the exclusive-or of the input and output differences of the MA half-round.

But, the last equality involving output difference values from the MA-box indicates that $i = 0$, which is a contradiction. Therefore, there can be no text pair simultaneously satisfying both the input and output of (3.71) across 2.5-round MESH-96.

This kind of miss-in-the-middle construction is not the same one used for MESH-64 because the fact that the MA-box of MESH-96 is a permutation mapping was not necessary to demonstrate a contradiction. This fact means that (3.71) would apply even if the MA-box of MESH-96 was not bijective.

Table 3.81 lists several other 2.5-round distinguishers for MESH-96. The proof of each distinguisher follows a similar reasoning to that provided for (3.71) or for (3.65). All these distinguishers start from an MA half-round and the 16-bit differences $a, b, c \neq 0$.

The ID distinguishers in Table 3.81 marked with the ‡ symbol are not useful for an attack because there are too many zero word differences in the

Table 3.81 2.5-round ID distinguishers for MESH-96.

$(a,0,0,a,0,0) \overset{2.5r}{\not\rightarrow} (b,b,0,0,0,0)$ ‡	$(a,0,0,a,0,0) \overset{2.5r}{\not\rightarrow} (0,0,b,b,0,0)$ ‡
$(a,0,0,a,0,0) \overset{2.5r}{\not\rightarrow} (0,0,0,0,b,b)$ ‡	$(a,0,0,a,0,0) \overset{2.5r}{\not\rightarrow} (b,b,c,c,0,0)$
$(a,0,0,a,0,0) \overset{2.5r}{\not\rightarrow} (b,b,0,0,c,c)$	$(0,a,0,0,a,0) \overset{2.5r}{\not\rightarrow} (b,b,0,0,0,0)$ ‡
$(0,a,0,0,a,0) \overset{2.5r}{\not\rightarrow} (0,0,b,b,0,0)$ ‡	$(0,a,0,0,a,0) \overset{2.5r}{\not\rightarrow} (0,0,0,0,b,b)$ ‡
$(0,a,0,0,a,0) \overset{2.5r}{\not\rightarrow} (b,b,c,c,0,0)$	$(0,a,0,0,a,0) \overset{2.5r}{\not\rightarrow} (0,0,b,b,c,c)$
$(0,0,a,0,0,a) \overset{2.5r}{\not\rightarrow} (b,b,0,0,0,0)$ ‡	$(0,0,a,0,0,a) \overset{2.5r}{\not\rightarrow} (0,0,b,b,0,0)$ ‡
$(0,0,a,0,0,a) \overset{2.5r}{\not\rightarrow} (0,0,0,0,b,b)$ ‡	$(0,0,a,0,0,a) \overset{2.5r}{\not\rightarrow} (b,b,0,0,c,c)$
$(0,0,a,0,0,a) \overset{2.5r}{\not\rightarrow} (0,0,b,b,c,c)$	$(a,b,0,a,b,0) \overset{2.5r}{\not\rightarrow} (c,c,0,0,0,0)$
$(a,b,0,a,b,0) \overset{2.5r}{\not\rightarrow} (0,0,c,c,0,0)$	$(a,0,b,a,0,b) \overset{2.5r}{\not\rightarrow} (c,c,0,0,0,0)$
$(a,0,b,a,0,b) \overset{2.5r}{\not\rightarrow} (0,0,0,0,c,c)$	$(0,a,b,0,a,b) \overset{2.5r}{\not\rightarrow} (0,0,c,c,0,0)$
$(0,a,b,0,a,b) \overset{2.5r}{\not\rightarrow} (0,0,0,0,c,c)$	

output. For instance, consider $(a,0,0,a,0,0) \overset{2.5r}{\not\rightarrow} (b,b,0,0,0,0)$. The number of plaintexts that can be generated depends on the number of nonzero input word differences. In this case, there are two nonzero words, each of which is 16 bits long. Thus, the number of plaintexts pairs with the input difference $(a,0,0,a,0,0)$ is $(2^{16})^2 \cdot ((2^{16})^2-1)/2 \approx 2^{63}$ (which forms a *structure*). On the other hand, the number of pairs that satisfy the output difference depends on the number of zero word differences, which in this case, is four. Therefore, the number of input pairs that survives this filtering is $2^{63}/(2^{16})^4 < 1$. That means that no text pair produced from the nonzero input word differences will survive the filtering due to the zero output word differences. In other words, this ID distinguisher cannot be identified.

On the other hand, consider $(a,0,0,a,0,0) \overset{2.5r}{\not\rightarrow} (b,b,0,0,c,c)$, which also contains two nonzero input word differences, and thus, can generate up to 2^{63} text pairs. In this case, the output difference contains only two zero word differences. It means that $2^{63}/(2^{16})^2 = 2^{31}$ text pairs survive the filtering, which allows this ID distinguisher to be identified.

3.8.3.1 ID Attack on 3.5-round MESH-96

Consider the 2.5-round ID distinguisher (3.71) covering the first KM half-round until the end of the third round.

Let $X^{(i)} = (X_1^{(i)}, X_2^{(i)}, X_3^{(i)}, X_4^{(i)}, X_5^{(i)}, X_6^{(i)})$ denote the i-th round input, $Y^{(i)} = (Y_1^{(i)}, Y_2^{(i)}, Y_3^{(i)}, Y_4^{(i)}, Y_5^{(i)}, Y_6^{(i)})$ denote the input to the i-th MA half-round, $X^{(i)*}$ and $Y^{(i)*}$ denote the corresponding elements in the pair for the differences $\Delta X^{(i)}$ and $\Delta Y^{(i)}$, respectively.

The attack proceeds as follows:

- choose a structure of 2^{32} plaintexts with fixed values for $X_2^{(1)}$, $X_3^{(1)}$, $X_5^{(1)}$ and $X_6^{(1)}$, while $X_1^{(1)}$ and $X_4^{(1)}$ assume all possible 32-bit values. There are $2^{32}(2^{32}-1)/2 \approx 2^{63}$ text pairs in such a structure, each with a difference of the form $(\Delta X_1^{(1)}, 0, 0, \Delta X_4^{(1)}, 0, 0)$. Collect about $2^{63}/(2^{16})^2 = 2^{31}$ pairs from this structure whose ciphertext difference satisfies $\Delta Y_5^{(4)} = 0 = \Delta Y_6^{(4)}$. For each such pair, try all 2^{32} values of $(Z_1^{(1)}, Z_4^{(1)})$ and partially encrypt $(X_1^{(1)}, X_4^{(1)})$ and $(X_1^{(1)*}, X_4^{(1)*})$. Collect about 2^{16} 32-bit subkey candidates that satisfy $\Delta Y_1^{(1)} = \Delta Y_4^{(1)}$.
  For instance, create a table D such that $D[(X_1^{(1)} \odot z) \oplus (X_1^{(1)*} \odot z)] = z$ for each candidate z for $Z_1^{(1)}$. Further, for each 16-bit candidate w for $Z_4^{(1)}$, find a matching z in D by indexing it as $D[(X_4^{(1)} \boxplus w) \oplus (X_4^{(1)*} \boxplus w)]$. Assuming $\odot$s cost more than $\boxplus$s, the cost of this matching is about 2^{16} difference computations, and 2^{16} memory for storing the D table.
- next, guess 2^{64} subkeys $(Z_1^{(4)}, Z_2^{(4)}, Z_3^{(4)}, Z_4^{(4)})$ and partially decrypt $(Y_1^{(4)}, Y_2^{(4)}, Y_3^{(4)}, Y_4^{(4)})$ across the KM half-round.
  Collect about 2^{32} 64-bit subkey candidates satisfying the equalities $\Delta X_1^{(4)} = \Delta X_2^{(4)}$ and $\Delta X_3^{(4)} = \Delta X_4^{(4)}$. This procedure uses a single table D as done previously, but this time, it is repeated for $(Z_1^{(4)}, Z_2^{(4)})$ and $(Z_3^{(4)}, Z_4^{(4)})$. The cost is about 2^{18} difference computations and 2^{16} memory (for the D table).
- next, we make a list of all $2^{16} \cdot 2^{32} = 2^{48}$ 96-bit subkeys, combining the previous two steps. These subkeys cannot be the correct value because they encrypt a text pair that leads to the 2.5-round impossible differential. According to the key schedule of MESH-96:

$$Z_1^{(4)} = (((((Z_7^{(2)} \boxplus Z_2^{(3)}) \oplus Z_4^{(3)}) \boxplus Z_6^{(3)}) \oplus Z_8^{(3)}) \boxplus Z_9^{(3)}) \lll 9 \oplus c_{27}. \tag{3.72}$$

$$Z_2^{(4)} = (((((Z_8^{(2)} \boxplus Z_3^{(3)}) \oplus Z_5^{(3)}) \boxplus Z_7^{(3)}) \oplus Z_9^{(3)}) \boxplus Z_1^{(4)}) \lll 9 \oplus c_{28}. \tag{3.73}$$

$$Z_3^{(4)} = (((((Z_9^{(2)} \boxplus Z_4^{(3)}) \oplus Z_6^{(3)}) \boxplus Z_8^{(3)}) \oplus Z_1^{(4)}) \boxplus Z_2^{(4)}) \lll 9 \oplus c_{29}. \tag{3.74}$$

$$Z_4^{(4)} = (((((Z_1^{(3)} \boxplus Z_5^{(3)}) \oplus Z_7^{(3)}) \boxplus Z_9^{(3)}) \oplus Z_2^{(4)}) \boxplus Z_3^{(4)}) \lll 9 \oplus c_{30}. \tag{3.75}$$

Equations (3.72) to (3.75) show that knowledge of $(Z_1^{(1)}, Z_4^{(1)})$ is not enough to uniquely determine $(Z_1^{(4)}, Z_2^{(4)}, Z_3^{(4)}, Z_4^{(4)})$. Therefore, no reduction is the effort for the key sieving phase is possible due to redundancy in the key schedule of MESH-96.

- each text pair defines a list of about 2^{48} wrong 96-bit subkeys. Compute the union of the list of wrong subkeys suggested. It is expected that after 2^{24} structures, the number of remaining wrong subkeys is

$$2^{96} \cdot (1 - \frac{2^{48}}{2^{96}})^{2^{31} \cdot 2^{24}} = 2^{96}(1 - 2^{-48})^{2^{48} \cdot 2^7} \approx 2^{96}/e^{128} \approx 2^{-88.66} < 1.$$

Thus, the correct subkey can be uniquely identified.

This attack requires $2^{32} \cdot 2^{24} = 2^{56}$ chosen plaintexts. The memory complexity is $2^{56} \cdot (2^{16} + 2^{16}) + 2^{96}/96 \approx 2^{89.41}$ blocks for sieving the correct 96-bit subkey. The time complexity is $2^{31} \cdot 2^{24} \cdot (2^{16} + 2^{16} + 2^{16}) \approx 2^{72.5}$ difference computations. A single difference computation is dominated by the cost of $\odot$, and there are 27 $\odot$s across 3.5 rounds of MESH-96. So, the time complexity is estimated as $2^{72.5}/27 \approx 2^{67.74}$ 3.5-round MESH-96 computations.

Another 2.5-round ID distinguisher, $(0, a, 0, 0, a, 0) \overset{2.5r}{\not\to} (0, 0, b, b, c, c)$, allows us to recover $(Z_2^{(1)}, Z_5^{(1)}, Z_5^{(4)}, Z_6^{(4)})$. The subkeys $Z_3^{(4)}$ and $Z_4^{(4)}$ were already recovered. The data complexity is $2^{32} \cdot 2^7 = 2^{39}$ CP, since there are less subkey bits to recover this time. The memory complexity is only 2^{61} bytes, and the same $2^{89.41}$-block storage can be used again. Since there are less subkey bits to recover, 2^7 are needed and the time complexity is $2^{31} \cdot 2^7(2^{16} + 2^{16}) = 2^{55}$ difference computations or $2^{55}/27 \approx 2^{50.25}$ 3.5-round computations.

Finally, the 2.5-round ID distinguisher $(0, 0, a, 0, 0, a) \overset{2.5r}{\not\to} (b, b, 0, 0, c, c)$, can be used to recover $Z_3^{(1)}$ and $Z_6^{(1)}$. The subkeys $Z_1^{(4)}$, $Z_2^{(4)}$, $Z_5^{(4)}$, $Z_6^{(4)}$ were already recovered previously. The data complexity of this phase is 2^{32} CP (only a single structure is enough). The memory complexity is unchanged. The time complexity is equivalent to $2^{31} \cdot 2^{16} = 2^{47}$ difference computations or $2^{47}/27 \approx 2^{42.24}$ 3.5-round encryptions.

In summary, the first 96 user key bits were recovered. The remaining 96 key bits can be obtained by exhaustive search, which leads to a final time complexity of 2^{96} 3.5-round encryptions. The overall data complexity is dominated by the first phase of the key-recovery attack: 2^{56} CP. The memory complexity is also dominated by the first phase (which recovered more key bits): $2^{89.41}$ blocks.

To extend the ID attack to 4-round MESH-96, one can guess the subkeys $(Z_7^{(4)}, Z_8^{(4)}, Z_9^{(4)})$ of an additional MA half-round and apply the previous attack on 3.5 rounds.

According to the key schedule of MESH-96:

$$Z_7^{(4)} = (((((Z_4^{(3)} \boxplus Z_8^{(3)}) \oplus Z_1^{(4)}) \boxplus Z_3^{(4)}) \oplus Z_5^{(4)}) \boxplus Z_6^{(4)}) \lll 9 \oplus c_{33}. \tag{3.76}$$

$$Z_8^{(4)} = (((((Z_5^{(3)} \boxplus Z_9^{(3)}) \oplus Z_2^{(4)}) \boxplus Z_4^{(4)}) \oplus Z_6^{(4)}) \boxplus Z_7^{(4)}) \lll 9 \oplus c_{34}. \tag{3.77}$$

$$Z_9^{(4)} = (((((Z_6^{(3)} \boxplus Z_1^{(4)}) \oplus Z_3^{(4)}) \boxplus Z_5^{(4)}) \oplus Z_7^{(4)}) \boxplus Z_8^{(4)}) \lll 9 \oplus c_{35}. \tag{3.78}$$

Equations (3.76) to (3.78) show that knowledge of ($Z_1^{(4)}$, $Z_2^{(4)}$, $Z_3^{(4)}$, $Z_4^{(4)}$, $Z_5^{(4)}$ and $Z_6^{(4)}$) is not enough to uniquely determine ($Z_7^{(4)}$, $Z_8^{(4)}$, $Z_9^{(4)}$).

Therefore, the time complexity increases to $2^{96} \cdot 2^{48} = 2^{144}$ 4-round encryptions. The data complexity remains at 2^{56} CP. The memory complexity is $2^{89.41}$ blocks.

Extending the ID attack further to 4.5-round MESH-96 would require guessing the subkeys of an additional KM half-round: $Z_1^{(5)}$, $Z_2^{(5)}$, $Z_3^{(5)}$, $Z_4^{(5)}$, $Z_5^{(5)}$ and $Z_6^{(5)}$.

Taking into account the key schedule algorithm of MESH-96, the dependency among the subkeys does not allow to reduce the number of subkey bits guessed. The time complexity would become $2^{144+96} = 2^{240}$ 4.5-round encryptions, which is larger than the exhaustive key search effort.

3.8.4 ID Analysis of MESH-128

The following ID analyses of MESH-128 are based on [151, 143] and on a similar analysis of Biham *et al.* on IDEA [16]. But, since the key schedule of MESH-128 does not consist of just a bit permutation, it is not possible to reduce the attack complexity by exploiting overlapping bits in the subkeys of MESH-128 (as in IDEA).

The longest ID distinguishers found for MESH-128 cover 2.5 rounds starting from an MA half-round.

Table 3.82 and Table 3.83 list the 2.5-round ID distinguishers found for MESH-128. The ones marked with $*$ are not useful for an attack because there are too many zero wordwise differences in the output. That means that no text pair will survive filtering. For instance, consider the 2.5-round ID distinguisher

$$(a, 0, 0, 0, a, 0, 0, 0) \overset{2.5r}{\not\rightarrow} (b, b, 0, 0, 0, 0, 0, 0).$$

The number of plaintexts that can be generated depends on the number of nonzero input word differences. In this case, there are two nonzero words, each of which is 16 bits long. Thus, the number of plaintexts pairs with the input difference of the form $(a, 0, 0, 0, a, 0, 0, 0)$ is $(2^{16})^2 \cdot ((2^{16})^2 - 1)/2 \approx 2^{63}$. On the other hand, the number of pairs that satisfy the output difference depends on the number of zero word differences, which in this case, is six. Therefore, the number of input pairs that survives this filtering is $2^{63}/(2^{16})^6 < 1$. That means that no text pair produced from the nonzero input word differences will survive the filtering due to the zero output word differences. In other words, this ID distinguisher cannot be identified.

Now, consider the ID distinguisher

$$(a, 0, 0, 0, a, 0, 0, 0) \overset{2.5r}{\not\to} (b, b, c, d, c, d, 0, 0),$$

which also contains two nonzero input word differences, and thus, can generate up to 2^{63} pairs. But, the output difference contains only two zero word differences. It means that $2^{63}/(2^{16})^2 = 2^{31}$ text pairs survive the filtering, and this ID distinguisher can be identified.

The following 2.5-round ID distinguisher will be used in an attack on MESH-128:

$$(a, b, 0, 0, a, b, 0, 0) \overset{2.5r}{\not\to} (c, c, d, e, d, e, 0, 0). \tag{3.79}$$

The proof that (3.79) is an impossible differential is as follows: consider a text pair with input difference $(a, b, 0, 0, a, b, 0, 0)$ for 16-bit differences $a, b \neq 0$, at the input to an MA half-round. In such a pair, the difference after the MA half-round has the form $(a, a, b, 0, b, 0, 0, 0)$. After the KM half-round, this difference is transformed by $Z_1^{(i+1)}$, $Z_2^{(i+1)}$, $Z_3^{(i+1)}$, $Z_5^{(i+1)}$, and becomes $(f, g, h, 0, i, 0, 0, 0)$ for 16-bit differences $f, g, h, i \neq 0$.

Similarly, consider a text pair with difference $(c, c, d, e, d, e, 0, 0)$ after 2.5 rounds, for 16-bit differences $c, d, e \neq 0$. For such a pair, the difference before the last MA half-round has the form $(c, d, e, 0, c, d, e, 0)$ because the MA half-round without the word swap is its own inverse. Further, the difference before the last KM half-round this difference is transformed by $Z_1^{(i+2)}$, $Z_2^{(i+2)}$, $Z_3^{(i+2)}$, $Z_5^{(i+2)}$, $Z_6^{(i+2)}$, $Z_7^{(i+2)}$, and becomes $(j, k, l, 0, m, n, o, 0)$ for 16-bit differences $j, k, l, m, n, o \neq 0$.

Therefore, the input and output differences of the middle MA half-round are $(f, g, h, 0, i, 0, 0, 0)$ and $(j, k, l, 0, m, n, o, 0)$, respectively. The difference before the swap of the six middle words is $(j, m, n, o, k, l, 0, 0)$.

In this middle MA half-round the input difference to the MA-box is nonzero: $(f \oplus i, g, h, 0)$, while the output difference is $(o, h \oplus n, g \oplus m, f \oplus j) = (0, 0, l, i \oplus k)$. The last equality can be deduced from the exclusive-or of word differences from both the input and output of the middle MA half-round. This last equality indicates that $o = 0$, which is a contradiction. Conse-

Table 3.82 2.5-round ID distinguishers for MESH-128.

$(a,0,0,0,a,0,0,0) \overset{2.5r}{\not\to} (b,b,0,0,0,0,0,0)$ *	$(a,0,0,0,a,0,0,0) \overset{2.5r}{\not\to} (0,0,b,0,b,0,0,0)$ *
$(a,0,0,0,a,0,0,0) \overset{2.5r}{\not\to} (0,0,0,b,0,b,0,0)$ *	$(a,0,0,0,a,0,0,0) \overset{2.5r}{\not\to} (0,0,0,0,0,0,b,b)$ *
$(a,0,0,0,a,0,0,0) \overset{2.5r}{\not\to} (b,b,c,0,c,0,0,0)$ *	$(a,0,0,0,a,0,0,0) \overset{2.5r}{\not\to} (b,b,0,c,0,c,0,0)$ *
$(a,0,0,0,a,0,0,0) \overset{2.5r}{\not\to} (b,b,0,0,0,0,c,c)$ *	$(a,0,0,0,a,0,0,0) \overset{2.5r}{\not\to} (0,0,b,c,b,c,0,0)$ *
$(a,0,0,0,a,0,0,0) \overset{2.5r}{\not\to} (0,0,b,0,b,0,c,c)$ *	$(a,0,0,0,a,0,0,0) \overset{2.5r}{\not\to} (0,0,0,b,0,b,c,c)$ *
$(a,0,0,0,a,0,0,0) \overset{2.5r}{\not\to} (b,b,c,d,c,d,0,0)$	$(a,0,0,0,a,0,0,0) \overset{2.5r}{\not\to} (b,b,c,0,c,0,d,d)$
$(a,0,0,0,a,0,0,0) \overset{2.5r}{\not\to} (b,b,0,c,0,c,d,d)$	$(a,b,0,0,a,b,0,0) \overset{2.5r}{\not\to} (c,c,0,0,0,0,0,0)$
$(a,b,0,0,a,b,0,0) \overset{2.5r}{\not\to} (0,0,c,0,c,0,0,0)$	$(a,b,0,0,a,b,0,0) \overset{2.5r}{\not\to} (0,0,0,c,0,c,0,0)$
$(a,b,0,0,a,b,0,0) \overset{2.5r}{\not\to} (0,0,0,0,0,0,c,c)$	$(a,b,0,0,a,b,0,0) \overset{2.5r}{\not\to} (c,c,0,d,0,d,0,0)$
$(a,b,0,0,a,b,0,0) \overset{2.5r}{\not\to} (c,c,0,0,0,0,d,d)$	$(a,b,0,0,a,b,0,0) \overset{2.5r}{\not\to} (0,0,c,d,c,d,0,0)$
$(a,b,0,0,a,b,0,0) \overset{2.5r}{\not\to} (0,0,c,0,c,0,d,d)$	$(a,b,0,0,a,b,0,0) \overset{2.5r}{\not\to} (c,c,d,e,d,e,0,0)$
$(a,b,0,0,a,b,0,0) \overset{2.5r}{\not\to} (c,c,d,0,d,0,e,e)$	$(0,a,0,0,0,a,0,0) \overset{2.5r}{\not\to} (b,b,0,0,0,0,0,0)$ *
$(0,a,0,0,0,a,0,0) \overset{2.5r}{\not\to} (0,0,b,0,b,0,0,0)$ *	$(0,a,0,0,0,a,0,0) \overset{2.5r}{\not\to} (0,0,0,b,0,b,0,0)$ *
$(0,a,0,0,0,a,0,0) \overset{2.5r}{\not\to} (0,0,0,0,0,0,b,b)$ *	$(0,a,0,0,0,a,0,0) \overset{2.5r}{\not\to} (b,b,c,0,c,0,0,0)$ *
$(0,a,0,0,0,a,0,0) \overset{2.5r}{\not\to} (b,b,0,c,0,c,0,0)$ *	$(0,a,0,0,0,a,0,0) \overset{2.5r}{\not\to} (b,b,0,0,0,0,c,c)$ *
$(0,a,0,0,0,a,0,0) \overset{2.5r}{\not\to} (0,0,b,c,b,c,0,0)$ *	$(0,a,0,0,0,a,0,0) \overset{2.5r}{\not\to} (0,0,b,0,b,0,c,c)$ *
$(0,a,0,0,0,a,0,0) \overset{2.5r}{\not\to} (0,0,0,b,0,b,c,c)$ *	$(0,a,0,0,0,a,0,0) \overset{2.5r}{\not\to} (b,b,c,d,c,d,0,0)$
$(0,a,0,0,0,a,0,0) \overset{2.5r}{\not\to} (b,b,c,0,c,0,d,d)$	$(0,a,0,0,0,a,0,0) \overset{2.5r}{\not\to} (0,0,b,c,b,c,d,d)$
$(0,0,a,0,0,0,a,0) \overset{2.5r}{\not\to} (b,b,0,0,0,0,0,0)$ *	$(0,0,a,0,0,0,a,0) \overset{2.5r}{\not\to} (0,0,b,0,b,0,0,0)$ *
$(0,0,a,0,0,0,a,0) \overset{2.5r}{\not\to} (0,0,0,b,0,b,0,0)$ *	$(0,0,a,0,0,0,a,0) \overset{2.5r}{\not\to} (0,0,0,0,0,0,b,b)$ *
$(0,0,a,0,0,0,a,0) \overset{2.5r}{\not\to} (b,b,c,0,c,0,0,0)$ *	$(0,0,a,0,0,0,a,0) \overset{2.5r}{\not\to} (b,b,0,c,0,c,0,0)$ *
$(0,0,a,0,0,0,a,0) \overset{2.5r}{\not\to} (b,b,0,0,0,0,c,c)$ *	$(0,0,a,0,0,0,a,0) \overset{2.5r}{\not\to} (0,0,b,c,b,c,0,0)$ *
$(0,0,a,0,0,0,a,0) \overset{2.5r}{\not\to} (0,0,b,0,b,0,c,c)$ *	$(0,0,a,0,0,0,a,0) \overset{2.5r}{\not\to} (0,0,0,b,0,b,c,c)$ *
$(0,0,a,0,0,0,a,0) \overset{2.5r}{\not\to} (b,b,c,d,c,d,0,0)$	$(0,0,a,0,0,0,a,0) \overset{2.5r}{\not\to} (b,b,0,c,0,c,d,d)$
$(0,0,a,0,0,0,a,0) \overset{2.5r}{\not\to} (0,0,b,c,b,c,d,d)$	$(0,0,a,0,0,0,a,0) \overset{2.5r}{\not\to} (b,b,c,d,c,d,e,e)$ *
$(0,0,0,a,0,0,0,a) \overset{2.5r}{\not\to} (b,b,0,0,0,0,0,0)$ *	$(0,0,0,a,0,0,0,a) \overset{2.5r}{\not\to} (0,0,b,0,b,0,0,0)$ *
$(0,0,0,a,0,0,0,a) \overset{2.5r}{\not\to} (0,0,0,b,0,b,0,0)$ *	$(0,0,0,a,0,0,0,a) \overset{2.5r}{\not\to} (0,0,0,0,0,0,b,b)$ *
$(0,0,0,a,0,0,0,a) \overset{2.5r}{\not\to} (b,b,c,0,c,0,0,0)$ *	$(0,0,0,a,0,0,0,a) \overset{2.5r}{\not\to} (b,b,0,c,0,c,0,0)$ *
$(0,0,0,a,0,0,0,a) \overset{2.5r}{\not\to} (b,b,0,0,0,0,c,c)$ *	$(0,0,0,a,0,0,0,a) \overset{2.5r}{\not\to} (0,0,b,c,b,c,0,0)$ *
$(0,0,0,a,0,0,0,a) \overset{2.5r}{\not\to} (0,0,b,0,b,0,c,c)$ *	$(0,0,0,a,0,0,0,a) \overset{2.5r}{\not\to} (0,0,0,b,0,b,c,c)$ *
$(0,0,0,a,0,0,0,a) \overset{2.5r}{\not\to} (b,b,c,0,c,0,d,d)$	$(0,0,0,a,0,0,0,a) \overset{2.5r}{\not\to} (b,b,0,c,0,c,d,d)$
$(0,0,0,a,0,0,0,a) \overset{2.5r}{\not\to} (0,0,b,c,b,c,d,d)$	$(a,0,b,0,a,0,b,0) \overset{2.5r}{\not\to} (c,c,0,0,0,0,0,0)$
$(a,0,b,0,a,0,b,0) \overset{2.5r}{\not\to} (0,0,c,0,c,0,0,0)$	$(a,0,b,0,a,0,b,0) \overset{2.5r}{\not\to} (0,0,0,c,0,c,0,0)$

Table 3.83 2.5-round ID distinguishers for MESH-128.

$(a,0,b,0,a,0,b,0) \overset{2.5r}{\not\rightarrow} (0,0,0,0,0,0,c,c)$	$(a,0,b,0,a,0,b,0) \overset{2.5r}{\not\rightarrow} (c,c,d,0,d,0,0,0)$
$(a,0,b,0,a,0,b,0) \overset{2.5r}{\not\rightarrow} (c,c,0,d,0,d,0,0)$	$(a,0,b,0,a,0,b,0) \overset{2.5r}{\not\rightarrow} (c,c,0,0,0,0,d,d)$
$(a,0,b,0,a,0,b,0) \overset{2.5r}{\not\rightarrow} (0,0,c,d,c,d,0,0)$	$(a,0,b,0,a,0,b,0) \overset{2.5r}{\not\rightarrow} (0,0,0,c,0,c,d,d)$
$(a,0,b,0,a,0,b,0) \overset{2.5r}{\not\rightarrow} (c,c,d,e,d,e,0,0)$	$(a,0,b,0,a,0,b,0) \overset{2.5r}{\not\rightarrow} (c,c,0,d,0,d,e,e)$
$(a,0,0,b,a,0,0,b) \overset{2.5r}{\not\rightarrow} (c,c,0,0,0,0,0,0)$	$(a,0,0,b,a,0,0,b) \overset{2.5r}{\not\rightarrow} (0,0,c,0,c,0,0,0)$
$(a,0,0,b,a,0,0,b) \overset{2.5r}{\not\rightarrow} (0,0,0,c,0,c,0,0)$	$(a,0,0,b,a,0,0,b) \overset{2.5r}{\not\rightarrow} (0,0,0,0,0,0,c,c)$
$(a,0,0,b,a,0,0,b) \overset{2.5r}{\not\rightarrow} (c,c,d,0,d,0,0,0)$	$(a,0,0,b,a,0,0,b) \overset{2.5r}{\not\rightarrow} (c,c,0,d,0,d,0,0)$
$(a,0,0,b,a,0,0,b) \overset{2.5r}{\not\rightarrow} (c,c,0,0,0,0,d,d)$	$(a,0,0,b,a,0,0,b) \overset{2.5r}{\not\rightarrow} (0,0,c,0,c,0,d,d)$
$(a,0,0,b,a,0,0,b) \overset{2.5r}{\not\rightarrow} (0,0,0,c,0,c,d,d)$	$(0,a,b,0,0,a,b,0) \overset{2.5r}{\not\rightarrow} (c,c,0,0,0,0,0,0)$
$(0,a,b,0,0,a,b,0) \overset{2.5r}{\not\rightarrow} (0,0,c,0,c,0,0,0)$	$(0,a,b,0,0,a,b,0) \overset{2.5r}{\not\rightarrow} (0,0,0,c,0,c,0,0)$
$(0,a,b,0,0,a,b,0) \overset{2.5r}{\not\rightarrow} (0,0,0,0,0,0,c,c)$	$(0,a,b,0,0,a,b,0) \overset{2.5r}{\not\rightarrow} (c,c,d,0,d,0,0,0)$
$(0,a,b,0,0,a,b,0) \overset{2.5r}{\not\rightarrow} (c,c,0,d,0,d,0,0)$	$(0,a,b,0,0,a,b,0) \overset{2.5r}{\not\rightarrow} (0,0,c,d,c,d,0,0)$
$(0,a,b,0,0,a,b,0) \overset{2.5r}{\not\rightarrow} (0,0,c,0,c,0,d,d)$	$(0,a,b,0,0,a,b,0) \overset{2.5r}{\not\rightarrow} (0,0,0,c,0,c,d,d)$
$(0,a,0,b,0,a,0,b) \overset{2.5r}{\not\rightarrow} (c,c,0,0,0,0,0,0)$	$(0,a,0,b,0,a,0,b) \overset{2.5r}{\not\rightarrow} (0,0,c,0,c,0,0,0)$
$(0,a,0,b,0,a,0,b) \overset{2.5r}{\not\rightarrow} (0,0,0,c,0,c,0,0)$	$(0,a,0,b,0,a,0,b) \overset{2.5r}{\not\rightarrow} (0,0,0,0,0,0,c,c)$
$(0,a,0,b,0,a,0,b) \overset{2.5r}{\not\rightarrow} (c,c,d,0,d,0,0,0)$	$(0,a,0,b,0,a,0,b) \overset{2.5r}{\not\rightarrow} (c,c,0,0,0,0,d,d)$
$(0,a,0,b,0,a,0,b) \overset{2.5r}{\not\rightarrow} (0,0,c,d,c,d,0,0)$	$(0,a,0,b,0,a,0,b) \overset{2.5r}{\not\rightarrow} (0,0,0,c,0,c,d,d)$
$(0,a,0,b,0,a,0,b) \overset{2.5r}{\not\rightarrow} (c,c,d,0,d,0,e,e)$	$(0,a,0,b,0,a,0,b) \overset{2.5r}{\not\rightarrow} (0,0,c,d,c,d,e,e)$
$(0,0,a,b,a,b,0,0) \overset{2.5r}{\not\rightarrow} (c,c,0,0,0,0,0,0)$	$(0,0,a,b,0,0,a,b) \overset{2.5r}{\not\rightarrow} (0,0,c,0,c,0,0,0)$
$(0,0,a,b,a,b,0,0) \overset{2.5r}{\not\rightarrow} (0,0,0,c,0,c,0,0)$	$(0,0,a,b,0,0,a,b) \overset{2.5r}{\not\rightarrow} (0,0,0,0,0,0,c,c)$
$(0,0,a,b,a,b,0,0) \overset{2.5r}{\not\rightarrow} (c,c,0,d,0,d,0,0)$	$(0,0,a,b,0,0,a,b) \overset{2.5r}{\not\rightarrow} (c,c,0,0,0,0,d,d)$
$(0,0,a,b,a,b,0,0) \overset{2.5r}{\not\rightarrow} (0,0,c,d,c,d,0,0)$	$(0,0,a,b,0,0,a,b) \overset{2.5r}{\not\rightarrow} (0,0,c,0,c,0,d,d)$
$(0,0,a,b,a,b,0,0) \overset{2.5r}{\not\rightarrow} (c,c,0,d,0,d,e,e)$	$(0,0,a,b,0,0,a,b) \overset{2.5r}{\not\rightarrow} (0,0,c,d,c,d,e,e)$
$(a,b,c,0,a,b,c,0) \overset{2.5r}{\not\rightarrow} (d,d,0,0,0,0,0,0)$	$(a,b,c,0,a,b,c,0) \overset{2.5r}{\not\rightarrow} (0,0,d,0,d,0,0,0)$
$(a,b,c,0,a,b,c,0) \overset{2.5r}{\not\rightarrow} (0,0,0,d,0,d,0,0)$	$(a,b,c,0,a,b,c,0) \overset{2.5r}{\not\rightarrow} (d,d,e,0,e,0,0,0)$
$(a,b,c,0,a,b,c,0) \overset{2.5r}{\not\rightarrow} (d,d,0,e,0,e,0,0)$	$(a,b,0,c,a,b,0,c) \overset{2.5r}{\not\rightarrow} (d,d,0,0,0,0,0,0)$
$(a,b,0,c,a,b,0,c) \overset{2.5r}{\not\rightarrow} (0,0,d,0,d,0,0,0)$	$(a,b,0,c,a,b,0,c) \overset{2.5r}{\not\rightarrow} (0,0,0,0,0,0,d,d)$
$(a,b,0,c,a,b,0,c) \overset{2.5r}{\not\rightarrow} (d,d,e,0,e,0,0,0)$	$(a,b,0,c,a,b,0,c) \overset{2.5r}{\not\rightarrow} (0,0,d,0,d,0,e,e)$
$(a,0,b,c,a,0,b,c) \overset{2.5r}{\not\rightarrow} (d,d,0,0,0,0,0,0)$	$(a,0,b,c,a,0,b,c) \overset{2.5r}{\not\rightarrow} (0,0,0,d,0,d,0,0)$
$(a,0,b,c,a,0,b,c) \overset{2.5r}{\not\rightarrow} (0,0,0,0,0,0,d,d)$	$(a,0,b,c,a,0,b,c) \overset{2.5r}{\not\rightarrow} (d,d,0,e,0,e,0,0)$
$(a,0,b,c,a,0,b,c) \overset{2.5r}{\not\rightarrow} (0,0,0,d,0,d,e,e)$	$(0,a,b,c,0,a,b,c) \overset{2.5r}{\not\rightarrow} (0,0,d,0,d,0,0,0)$
$(0,a,b,c,0,a,b,c) \overset{2.5r}{\not\rightarrow} (0,0,0,d,0,d,0,0)$	$(0,a,b,c,0,a,b,c) \overset{2.5r}{\not\rightarrow} (0,0,0,0,0,0,d,d)$
$(0,a,b,c,0,a,b,c) \overset{2.5r}{\not\rightarrow} (0,0,d,0,d,0,e,e)$	$(0,a,b,c,0,a,b,c) \overset{2.5r}{\not\rightarrow} (0,0,0,d,0,d,e,e)$

quently, there can be no text pair satisfying simultaneously the input differences $(a, b, 0, 0, a, b, 0, 0)$ and the output difference $(c, c, d, e, d, e, 0, 0)$ across 2.5-round MESH-128.

This construction of an ID distinguisher using the miss-in-the-middle technique does not depend on the fact that the MA-box of MESH-128 is a permutation mapping (as in the proof of the ID distinguishers of IDEA). Therefore, this ID distinguisher applies even if the MA-box of MESH-128 was not a bijective mapping.

The fact that all ID distinguishers found so far cover at most 2.5 rounds of MESH-64, MESH-96 and MESH-128 helps corroborate the design decision of the MESH ciphers. In particular, the fact that full diffusion is achieved in a single round is an effective measure to curtail the length of ID distinguishers. Thus, even though the block size of the MESH ciphers have been increasing from 64 up to 128 bits, the ID distinguishers (so far) have stayed at 2.5 rounds. Finding longer (and useful) ID distinguishers, or proving if they exist or not, are left as open problems.

3.8.4.1 Attack on 3.5-round MESH-128

A key-recovery attack on 3.5-round MESH-128 will use the ID distinguisher (3.79).

Let $X_j^{(i)}$, for $1 \leq j \leq 8$, denote the i-th round input text; let $Y_j^{(i)}$, for $1 \leq j \leq 8$, denote the text input to the i-th MA half-round. The corresponding elements in the pairs are denoted $X_j^{(i)*}$ and $Y_j^{(i)*}$, respectively. The differences will be denoted $\Delta X_j^{(i)}$ and $\Delta Y_j^{(i)}$, respectively.

The attack proceeds as follows:

- choose a structure of 2^{64} plaintexts, with fixed values for $X_3^{(1)}$, $X_4^{(1)}$, $X_7^{(1)}$ and $X_8^{(1)}$, while $X_1^{(1)}$, $X_2^{(1)}$, $X_5^{(1)}$ and $X_6^{(1)}$ assume all possible values. There are about $2^{64}(2^{64}-1)/2 \approx 2^{127}$ text pairs in such a structure, with differences of the form $(\Delta X_1^{(1)}, \Delta X_2^{(1)}, 0, 0, \Delta X_5^{(1)}, \Delta X_6^{(1)}, 0, 0)$.
- collect about $2^{127}/(2^{16})^2 = 2^{95}$ pairs from the structure whose ciphertext difference satisfies $\Delta Y_7^{(4)} = 0 = \Delta Y_8^{(4)}$. For each such a pair, try all 2^{64} subkey candidates for $(Z_1^{(1)}, Z_2^{(1)}, Z_5^{(1)}, Z_6^{(1)})$ and partially encrypt $(X_i^{(1)}, X_i^{(1)*})$ for $i \in \{1, 2, 5, 6\}$ across the KM half-round. Collect about 2^{32} 64-bit subkeys satisfying $\Delta Y_1^{(1)} = \Delta Y_5^{(1)}$ and $\Delta Y_2^{(1)} = \Delta Y_6^{(1)}$.
  This procedure uses a table D of 2^{16} entries. For each candidate z for $Z_1^{(1)}$, compute and store $D[(X_1^{(1)} \odot z) \oplus (X_1^{(1)*} \odot z)] = z$. For each candidate w for $Z_5^{(1)}$, find a matching z by computing $D[(X_5^{(1)} \boxplus w) \oplus (X_5^{(1)*} \boxplus w)]$. Assuming a single $\odot$ costs (much) more than $\boxplus$, the cost for this matching procedure is about 2^{16} difference computations, and 2^{16} memory (for the D table). A similar procedure applies to $Z_2^{(1)}$ and $Z_6^{(1)}$.

- next, try all 2^{96} subkey candidates for $Z_i^{(4)}$, for $1 \leq i \leq 6$, and partially decrypt $Y_i^{(4)}$ across the KM half-round. Collect about 2^{48} 96-bit subkeys satisfying $\Delta X_1^{(4)} = \Delta X_2^{(4)}$, $\Delta X_3^{(4)} = \Delta X_5^{(4)}$ and $\Delta X_4^{(4)} = \Delta X_6^{(4)}$. Similar to the previous matching procedure, these matchings cost about $3 \cdot 2^{16}$ difference computations and $3 \cdot 2^{16}$ memory.
- make a list of all $2^{32} \cdot 2^{48} = 2^{80}$ 160-bit subkey candidates by combining the previous two steps. None of these subkeys can be the correct subkey value because they lead to a pair of the impossible differential.
From the key schedule algorithm of MESH-128:

$$Z_1^{(4)} = (((((Z_9^{(2)} \boxplus Z_{12}^{(2)}) \oplus Z_1^{(3)}) \boxplus Z_5^{(3)}) \oplus Z_{11}^{(3)}) \boxplus Z_{12}^{(3)}) \lll 11 \oplus c_{36}. \tag{3.80}$$

$$Z_2^{(4)} = (((((Z_{10}^{(2)} \boxplus Z_1^{(3)}) \oplus Z_2^{(3)}) \boxplus Z_6^{(3)}) \oplus Z_{12}^{(3)}) \boxplus Z_1^{(4)}) \lll 11 \oplus c_{37}. \tag{3.81}$$

$$Z_3^{(4)} = (((((Z_{11}^{(2)} \boxplus Z_2^{(3)}) \oplus Z_3^{(3)}) \boxplus Z_7^{(3)}) \oplus Z_1^{(4)}) \boxplus Z_2^{(4)}) \lll 11 \oplus c_{38}. \tag{3.82}$$

$$Z_4^{(4)} = (((((Z_{12}^{(2)} \boxplus Z_3^{(3)}) \oplus Z_4^{(3)}) \boxplus Z_8^{(3)}) \oplus Z_2^{(4)}) \boxplus Z_3^{(4)}) \lll 11 \oplus c_{39}. \tag{3.83}$$

$$Z_5^{(4)} = (((((Z_1^{(3)} \boxplus Z_4^{(3)}) \oplus Z_5^{(3)}) \boxplus Z_9^{(3)}) \oplus Z_3^{(4)}) \boxplus Z_4^{(4)}) \lll 11 \oplus c_{40}. \tag{3.84}$$

$$Z_6^{(4)} = (((((Z_2^{(3)} \boxplus Z_5^{(3)}) \oplus Z_6^{(3)}) \boxplus Z_{10}^{(3)}) \oplus Z_4^{(4)}) \boxplus Z_5^{(4)}) \lll 11 \oplus c_{41}. \tag{3.85}$$

Thus, knowledge of $(Z_1^{(1)}, Z_2^{(1)}, Z_5^{(1)}, Z_6^{(1)})$ is not enough to uniquely determine $Z_i^{(4)}$, for $1 \leq i \leq 6$. Therefore, no reduction in the effort for the key sieving phase is possible for MESH-128, in comparison with the ID attacks on IDEA.
- each text pair defines a list of about 2^{80} wrong 160-bit subkeys. Compute the union of these suggested subkeys. It is expected that after 2^{88} pairs in a single structure, the number of remaining wrong subkeys is 2^{160} .

$(1 - \frac{2^{80}}{2^{160}})^{2^{88}} \approx 2^{-24.66} < 1$. Thus, the correct subkey can be uniquely identified.

The attack takes 2^{64} chosen plaintexts. The memory complexity is dominated by $(2^{88} \cdot 2^{16} + 2^{160})/128 \approx 2^{153}$ blocks for sieving the correct subkey. The memory for the D tables is $2^{64}(2^{16} + 3 \cdot 2^{16})/128 = 2^{75}$ blocks.

The time complexity is $2^{88} \cdot (2 \cdot 2^{16} + 3 \cdot 2^{16}) \approx 2^{106}$ difference computations. If we consider that $\odot$ dominate the difference computations, and that there are 40 $\odot$s across 3.5-round MESH-128, then, the time complexity is estimated at $2^{106}/40 \approx 2^{100.67}$ 3.5-round MESH-128 encryptions.

The 2.5-round ID distinguisher $(0, 0, a, b, 0, 0, a, b) \overset{2.5r}{\not\rightarrow} (c, c, 0, d, 0, d, e, e)$ can be used to further recover $(Z_3^{(1)}, Z_4^{(1)}, Z_7^{(1)}, Z_8^{(1)}, Z_1^{(4)}, Z_2^{(4)}, Z_4^{(4)}, Z_6^{(4)}, Z_7^{(4)}, Z_8^{(4)})$. Since $Z_i^{4)}$ for $1 \leq i \leq 6$, were already recovered, we focus on the other six subkeys (which amount to 96 subkey bits). The same memory as in the previous step can be reused.

The data complexity is still 2^{64} CP (a single structure). In fact, only 2^{55} plaintexts that satisfy $\Delta Y_3^{(4)} = 0 = \Delta Y_5^{(4)}$ are needed. The expected number of wrong subkeys remaining is $2^{96}(1 - 2^{-48})^{2^{48} \cdot 2^7} \approx \frac{2^{96}}{e^{128}} < 1$.

The time complexity of this phase is $2^{55}(2 \cdot 2^{16} + 2^{16}) \approx 2^{72.5}$ difference computations or about $2^{72.5}/40 \approx 2^{67.17}$ 3.5-round MESH-128 encryptions.

In total, the first 128 user key bits are recovered. From the key schedule of MESH-128, knowledge of $Z_i^{(4)}$, for $1 \leq i \leq 8$, does not provide enough information to deduce the remaining 128 user key bits, which can be recovered by exhaustive search. Therefore, the final time complexity is 2^{128} 3.5-round MESH-128 encryptions.

To extend the attack to 4-round MESH-128, we can guess the subkeys $(Z_9^{(4)}, Z_{10}^{(4)}, Z_{11}^{(4)}, Z_{12}^{(4)})$ of an additional MA half-round and apply the previous attack on 3.5 rounds.

According to the key schedule algorithm of MESH-128:

$$Z_9^{(4)} = (((((Z_5^{(3)} \boxplus Z_8^{(3)}) \oplus Z_9^{(3)}) \boxplus Z_1^{(4)}) \oplus Z_7^{(4)}) \boxplus Z_8^{(4)}) \lll 11 \oplus c_{44}. \tag{3.86}$$

$$Z_{10}^{(4)} = (((((Z_6^{(3)} \boxplus Z_9^{(3)}) \oplus Z_{10}^{(3)}) \boxplus Z_2^{(4)}) \oplus Z_8^{(4)}) \boxplus Z_9^{(4)}) \lll 11 \oplus c_{45}. \tag{3.87}$$

$$Z_{11}^{(4)} = (((((Z_7^{(3)} \boxplus Z_{10}^{(3)}) \oplus Z_{11}^{(3)}) \boxplus Z_3^{(4)}) \oplus Z_9^{(4)}) \boxplus Z_{10}^{(4)}) \lll 11 \oplus c_{46}. \tag{3.88}$$

$$Z_{12}^{(4)} = (((((Z_8^{(3)} \boxplus Z_{11}^{(3)}) \oplus Z_{12}^{(3)}) \boxplus Z_4^{(4)}) \oplus Z_{10}^{(4)}) \boxplus Z_{11}^{(4)}) \lll 11 \oplus c_{47}. \quad (3.89)$$

According to equation (3.86) to (3.89), knowledge of ($Z_1^{(4)}$, $Z_2^{(4)}$, $Z_3^{(4)}$, $Z_4^{(4)}$, $Z_5^{(4)}$, $Z_6^{(4)}$, $Z_7^{(4)}$ and $Z_8^{(4)}$ is not enough to uniquely determine ($Z_9^{(4)}$, $Z_{10}^{(4)}$, $Z_{11}^{(4)}$, $Z_{12}^{(4)}$).

Therefore, the time complexity increases to $2^{128} \cdot 2^{64} = 2^{192}$ 4-round encryptions.

3.8.5 ID Analysis of MESH-64(8)

The following ID analyses of MESH-64(8) have not been published before, and are based on similar analyses of Lai-Massey cipher designs in previous sections.

The longest ID distinguishers found for MESH-64(8) cover 2.5 rounds starting from an MA half-round.

The following 2.5-round ID distinguisher will be used in an attack on 3.5-round MESH-64(8):

$$(a, b, 0, 0, a, b, 0, 0) \overset{2.5r}{\not\rightarrow} (c, c, 0, d, 0, d, 0, 0). \quad (3.90)$$

The proof that it is an impossible differential is as follows: consider a text pair with input difference $(a, b, 0, 0, a, b, 0, 0)$ for 8-bit differences $a, b \neq 0$, at the input to the r-th MA half-round. In such a pair, the difference after the MA half-round has the form $(a, a, b, 0, b, 0, 0, 0)$. After the KM half-round, this difference is transformed by $Z_1^{(r+1)}$, $Z_2^{(r+1)}$, $Z_3^{(r+1)}$, $Z_5^{(r+1)}$, and becomes $(e, f, g, 0, h, 0, 0, 0)$ for 8-bit differences $e, f, g, h \neq 0$.

Similarly, consider a text pair with difference $(c, c, 0, d, 0, d, 0, 0)$ after 2.5 rounds, for 8-bit differences $c, d \neq 0$. For such a pair, the difference before the last MA half-round has the form $(c, 0, d, 0, c, 0, d, 0)$ because the MA half-round without the word swap is its own inverse. Further, the difference before the last KM half-round is transformed by $Z_1^{(r+2)}$, $Z_3^{(r+2)}$, $Z_5^{(r+2)}$, $Z_7^{(r+2)}$ and becomes $(i, 0, j, 0, k, 0, l, 0)$ for 8-bit differences $i, j, k, l \neq 0$.

Therefore, the input and output differences of the middle MA half-round are $(e, f, g, 0, h, 0, 0, 0)$ and $(i, 0, j, 0, k, 0, l, 0)$, respectively. The difference before the swap of the six middle words is $(i, k, 0, l, 0, j, 0, 0)$.

In this middle MA half-round the input difference to the MA-box is nonzero: $(e \oplus h, f, g, 0)$, while the output difference is $(e \oplus i, f \oplus k, g, l) = (h, j, 0, 0)$. The last equality can be deduced from the exclusive-or of word differences from both the input and output of the middle MA half-round. This last equality indicates that $g = 0$ and $l = 0$, both of which are contradictions. Consequently, there are no text pairs satisfying simultaneously both the input difference $(a, b, 0, 0, a, b, 0, 0)$ and the output difference $(c, c, 0, d, 0, d, 0, 0)$ across 2.5-round MESH-64(8).

Note that this proof that (3.90) did not use the fact that the MA-box of MESH-64(8) is a permutation mapping. Therefore, this ID distinguisher holds even if the MA-box was not bijective.

3.8.5.1 Attack on 3.5-round MESH-64(8)

A key-recovery attack on 3.5-round MESH-64(8) will use the ID distinguisher (3.90).

Let $X_j^{(i)}$, for $1 \leq j \leq 8$, denote the i-th round input text; let $Y_j^{(i)}$, for $1 \leq j \leq 8$, denote the text input to the i-th MA half-round. The corresponding elements in the pairs are denoted $X_j^{(i)*}$ and $Y_j^{(i)*}$, respectively. The differences will be denoted $\Delta X_j^{(i)}$ and $\Delta Y_j^{(i)}$, respectively.

The attack proceeds as follows:

- choose a structure of 2^{32} plaintexts, with fixed values for $X_3^{(1)}$, $X_4^{(1)}$, $X_7^{(1)}$ and $X_8^{(1)}$, while $X_1^{(1)}$, $X_2^{(1)}$, $X_5^{(1)}$ and $X_6^{(1)}$ assume all possible values. There are about $2^{32}(2^{32}-1)/2 \approx 2^{63}$ text pairs in such a structure, with differences of the form $(\Delta X_1^{(1)}, \Delta X_2^{(1)}, 0, 0, \Delta X_5^{(1)}, \Delta X_6^{(1)}, 0, 0)$.
- collect about $2^{63}/(2^8)^4 = 2^{31}$ pairs from the structure whose ciphertext difference satisfies $\Delta Y_3^{(4)} = \Delta Y_5^{(4)} = \Delta Y_7^{(4)} = \Delta Y_8^{(4)} = 0$. For each such a pair, try al 2^{32} subkey candidates for $(Z_1^{(1)}, Z_2^{(1)}, Z_5^{(1)}, Z_6^{(1)})$ and partially encrypt $(X_j^{(1)}, X_j^{(1)*})$ for $j \in \{1, 2, 5, 6\}$ across the KM half-round. Collect about 2^{16} 32-bit subkeys satisfying $\Delta Y_1^{(1)} = \Delta Y_5^{(1)}$ and $\Delta Y_2^{(1)} = \Delta Y_6^{(1)}$.
  This procedure uses a table D of 2^8 entries. For each candidate z for $Z_1^{(1)}$, compute $D[(X_1^{(1)} \odot z) \oplus (X_1^{(1)*} \odot z)] = z$. For each candidate w for $Z_5^{(1)}$, find a matching z by computing $D[(X_5^{(1)} \boxplus w) \oplus (X_5^{(1)*} \boxplus w)]$.
  Assuming $\odot$ costs (much) more than $\boxplus$, the cost for this matching procedure is about 2^8 difference computations, and 2^8 memory (for the D table). A similar procedure applies to $Z_2^{(1)}$ and $Z_6^{(1)}$.
- next, try all 2^{32} subkey candidates for $Z_1^{(4)}$, $Z_2^{(4)}$, $Z_4^{(4)}$, $Z_6^{(4)}$ and partially decrypt $(Y_1^{(4)}, Y_2^{(4)}, Y_4^{(4)}, Y_6^{(4)})$ across the KM half-round. Collect about 2^{16} 32-bit subkeys satisfying $\Delta X_1^{(4)} = \Delta X_2^{(4)}$ and $\Delta X_4^{(4)} = \Delta X_6^{(4)}$. Similar to the previous matching procedure, these matchings cost about $2 \cdot 2^8$ difference computations and $2 \cdot 2^8$ memory.
- make a list of all $2^{16} \cdot 2^{16} = 2^{32}$ 64-bit subkey candidates by combining the previous two steps. None of these subkeys can be the correct subkey value because they lead to a pair of the impossible differential.
  From the key schedule algorithm of MESH-64(8):

$$Z_1^{(4)} = (((Z_5^{(2)} \boxplus Z_9^{(2)}) \oplus Z_8^{(3)}) \boxplus Z_9^{(3)}) \lll 1 \oplus c_{30}. \quad (3.91)$$

$$Z_2^{(4)} = (((Z_6^{(2)} \boxplus Z_{10}^{(2)}) \oplus Z_9^{(3)}) \boxplus Z_{10}^{(3)}) \lll 1 \oplus c_{31}. \quad (3.92)$$

$$Z_4^{(4)} = (((Z_8^{(2)} \boxplus Z_2^{(3)}) \oplus Z_1^{(4)}) \boxplus Z_2^{(4)}) \lll 1 \oplus c_{33}. \quad (3.93)$$

$$Z_6^{(4)} = (((Z_{10}^{(2)} \boxplus Z_4^{(3)}) \oplus Z_3^{(4)}) \boxplus Z_4^{(4)}) \lll 1 \oplus c_{34}. \quad (3.94)$$

Thus, knowledge of $(Z_1^{(1)}, Z_2^{(1)}, Z_5^{(1)}, Z_6^{(1)})$ is not enough to uniquely determine $Z_1^{(4)}$, $Z_2^{(4)}$, $Z_4^{(4)}$, $Z_6^{(4)}$. Therefore, no reduction in the effort for the key sieving step is possible for MESH-64(8), in comparison to the ID attacks on IDEA.

- each text pair defines a list of about 2^{32} wrong 64-bit subkeys. Compute the union of these suggested subkeys. It is expected that after 2^7 structures, the number of remaining wrong subkeys is $2^{64} \cdot (1 - \frac{2^{32}}{2^{64}})^{2^{31} \cdot 2^7} \approx \frac{2^{64}}{e^{64}} < 1$. Thus, the correct subkey can be uniquely identified.

The attack takes $2^{32} \cdot 2^7 = 2^{39}$ chosen plaintexts. The memory complexity is dominated by $2^{64}/64 \approx 2^{58}$ blocks for sieving the correct subkey. The time complexity is $2^{31} \cdot 2^7(3 \cdot 2^8 + 2 \cdot 2^8) \approx 2^{47.5}$ difference computations. If we consider that $\odot$s dominate the difference computations, and that there are 28 $\odot$s across 3.5-round MESH-64(8), then, the time complexity is estimated at $2^{47.5}/28 \approx 2^{42.69}$ 3.5-round encryptions.

The 2.5-round ID distinguisher $(0, 0, a, b, 0, 0, a, b) \overset{2.5r}{\nrightarrow} (c, c, 0, d, 0, d, 0, 0)$ can be used to further recover $(Z_3^{(1)}, Z_4^{(1)}, Z_7^{(1)}, Z_8^{(1)})$. Note that the output difference is the same as in (3.90) So, $(Z_1^{(4)}, Z_2^{(4)}, Z_4^{(4)}, Z_6^{(4)})$ is already known and we only recover subkeys from the first round. Since there are less subkeys to recover in this step, the time complexity is dominated by the first step: $2^{42.69}$ 3.5-round encryptions. The same memory as in the previous step can be reused. The data complexity is also reduced $2^{32} \cdot 2^6 = 2^{38}$ CP, and the overall data complexity becomes $2^{38} + 2^{39} \approx 2^{39.5}$ CP.

In total, the first 64 user key bits are recovered. From the key schedule of MESH-64(8), knowledge of $(Z_1^{(4)}, Z_2^{(4)}, Z_4^{(4)}, Z_6^{(4)})$ does not provide enough information to deduce the remaining 64 user key bits, which can be recovered by exhaustive search. Therefore, the final time complexity is 2^{64} 3.5-round MESH-64(8) encryptions.

To extend the attack to 4-round MESH-64(8), we can guess the subkeys $(Z_9^{(4)}, Z_{10}^{(4)})$ of an additional MA half-round and apply the previous attack on 3.5 rounds.

According to the key schedule of MESH-64(8):

$$Z_9^{(4)} = (((Z_3^{(3)} \boxplus Z_7^{(3)}) \oplus Z_6^{(4)}) \boxplus Z_7^{(4)}) \lll 1 \oplus c_{38}. \quad (3.95)$$

$$Z_{10}^{(4)} = (((Z_4^{(3)} \boxplus Z_8^{(3)}) \oplus Z_7^{(4)}) \boxplus Z_8^{(4)}) \lll 1 \oplus c_{39}. \quad (3.96)$$

According to equations (3.95) and (3.96), knowledge of ($Z_1^{(4)}$, $Z_2^{(4)}$, $Z_4^{(4)}$ and $Z_6^{(4)}$ is not enough to uniquely determine ($Z_9^{(4)}$, $Z_{10}^{(4)}$), or vice-versa.

Therefore, the time complexity increases to $2^{64} \cdot 2^{16} = 2^{80}$ 4-round encryptions.

3.8.6 ID Analysis of MESH-128(8)

The following ID analyses of MESH-128(8) have not been published before, and are based on similar analyses of Lai-Massey cipher designs in previous sections.

The longest ID distinguishers found for MESH-128(8) cover 2.5 rounds starting from an MA half-round.

The following 2.5-round ID distinguisher will be used in an attack on MESH-128(8):

$$(a,b,c,d,0,0,0,0,a,b,c,d,0,0,0,0) \overset{2.5r}{\not\to} (0,0,0,e,f,g,h,0,0,e,f,g,h,0,0,0). \quad (3.97)$$

The proof that 3.97 is an impossible differential is as follows: consider a text pair with input difference $(a,b,c,d,0,0,0,0,a,b,c,d,0,0,0,0)$ for 8-bit differences $a,b,c,d \neq 0$, at the input to the r-th MA half-round. In such a pair, the difference after the MA half-round has the form (a, a, b, c, d, 0, 0, 0, b, c, d, 0, 0, 0, 0, 0). After the KM half-round, this difference is transformed by $Z_y^{(r+1)}$, for $1 \leq y \leq 16$, and becomes (i, j, k, l, m, 0, 0, 0, n, o, p, 0, 0, 0, 0, 0) for 8-bit differences i, j, k, l, m, n, o, $p \neq 0$.

Similarly, consider a text pair with difference (0, 0, 0, e, f, g, h, 0, 0, e, f, g, h, 0, 0, 0) after 2.5 rounds, for 8-bit differences $e,f,g,h \neq 0$. For such a pair, the difference before the last MA half-round has the form $(0,0,e,f,g,h,0,0,0,0,e,f,g,h,0,0)$ because the MA half-round without the word swap is its own inverse. Further, the difference before the last KM half-round is transformed by $Z_y^{(r+2)}$, for $1 \leq y \leq 16$ and becomes $(0,0,q,r,s,t,0,0,0,0,u,v,w,x,0,0)$ for 8-bit differences q, r, s, t, u, v, w, $x \neq 0$.

Therefore, the input and output differences of the middle MA half-round are $(i, j, k, l, m, 0, 0, 0, n, o, p, 0, 0, 0, 0, 0)$ and $(0, 0, q, r, s, t, 0, 0, 0, 0, u, v, w, x, 0, 0)$, respectively. The difference before the swap of the fourteen middle words is $(0, 0, 0, u, v, w, x, 0, 0, q, r, s, t, 0, 0, 0)$.

In this middle MA half-round, the input difference to the MA-box is nonzero: $(i \oplus n, j \oplus o, k \oplus p, l, m, 0, 0, 0)$, while the output difference is $(i, j, k, l \oplus u, m \oplus v, w, x, 0) = (n, o \oplus q, p \oplus r, s, t, 0, 0, 0)$. The last equality can be deduced from the exclusive-or of word differences from both the input and output of the middle MA half-round. This last equality indicates that $w = 0$ and $x = 0$, both of which are contradictions. Consequently, there are no text pairs satisfying simultaneously both the input and output differences of (3.97) across 2.5-round MESH-128(8).

3.8.6.1 Attack on 3.5-round MESH-128(8)

A key-recovery attack on 3.5-round MESH-128(8) will use the ID distinguisher (3.97).

Let $X_j^{(i)}$, for $1 \leq j \leq 16$, denote the i-th round input text. Let $Y_j^{(i)}$, for $1 \leq j \leq 16$, denote the text input to the i-th MA half-round. The corresponding elements in the pairs are denoted $X_j^{(i)*}$ and $Y_j^{(i)*}$, respectively. The differences will be denoted $\Delta X_j^{(i)}$ and $\Delta Y_j^{(i)}$, respectively.

The attack proceeds as follows:

- choose a structure of 2^{64} plaintexts, with fixed values for $X_j^{(1)}$, for $j \in \{5, 6, 7, 8, 13, 14, 15, 16\}$, while $X_l^{(1)}$, for $l \in \{1, 2, 3, 4, 9, 10, 11, 12\}$, assume all possible values. There are about $2^{64}(2^{64}-1)/2 \approx 2^{127}$ text pairs in such a structure, with differences of the form $(\Delta X_1^{(1)}, \Delta X_2^{(1)}, \Delta X_3^{(1)}, \Delta X_4^{(1)}, 0, 0, 0, 0, \Delta X_9^{(1)}, \Delta X_{10}^{(1)}, \Delta X_{11}^{(1)}, \Delta X_{12}^{(1)}, 0, 0, 0, 0)$.
- collect about $2^{127}/(2^8)^8 = 2^{63}$ pairs from the structure whose ciphertext difference satisfies $\Delta Y_j^{(4)} = 0$ for $j \in \{1, 2, 3, 8, 9, 14, 15, 16\}$. For each such a pair, try all 2^{64} subkey candidates for $Z_j^{(1)}$, for $j \in \{1, 2, 3, 4, 9, 10, 11, 12\}$ and partially encrypt $(X_j^{(1)}, X_j^{(1)*})$ across the KM half-round. Collect about 2^{32} 64-bit subkeys satisfying $\Delta Y_1^{(1)} = \Delta Y_9^{(1)}$, $\Delta Y_2^{(1)} = \Delta Y_{10}^{(1)}$, $\Delta Y_3^{(1)} = \Delta Y_{11}^{(1)}$, and $\Delta Y_4^{(1)} = \Delta Y_{12}^{(1)}$.
  This procedure uses a table D of 2^8 entries. For instance, for each candidate z for $Z_1^{(1)}$, compute $D[(X_1^{(1)} \odot z) \oplus (X_1^{(1)*} \odot z)] = z$. For each candidate w for $Z_9^{(1)}$, find a matching z by computing the entry of $D[(X_9^{(1)} \boxplus w) \oplus (X_9^{(1)*} \boxplus w)]$.
  Assuming $\odot$ costs (much) more than $\boxplus$, the cost for this matching procedure is about 2^8 difference computations, and 2^8 memory (for the D

table). A similar procedure applies to $Z_j^{(1)}$ for $j \in \{2, 3, 4, 9, 10, 11, 12\}$. The total cost is $4 \cdot 2^8 = 2^{10}$ difference computations and $4 \cdot 2^8$ memory.

- next, try all 2^{64} subkey candidates for $Z_l^{(4)}$, for $l \in \{4, 5, 6, 7, 10, 11, 12, 13\}$ and partially decrypt $Y_l^{(4)}$ across the KM half-round. Collect about 2^{32} 64-bit subkeys satisfying $\Delta X_4^{(4)} = \Delta X_{10}^{(4)}$, $\Delta X_5^{(4)} = \Delta X_{11}^{(4)}$, $\Delta X_6^{(4)} = \Delta X_{12}^{(4)}$ and $\Delta X_7^{(4)} = \Delta X_{13}^{(4)}$. Similar to the previous matching procedure, these matchings cost about $4 \cdot 2^8$ difference computations and $4 \cdot 2^8$ memory.
- make a list of all $2^{32} \cdot 2^{32} = 2^{64}$ 128-bit subkey candidates by combining the previous two steps. None of these subkeys can be the correct subkey value because they lead to a pair of the impossible differential. From the key schedule algorithm of MESH-128(8):

$$Z_4^{(4)} = (((Z_8^{(2)} \boxplus Z_{18}^{(2)}) \oplus Z_2^{(4)}) \boxplus Z_3^{(4)}) \lll 1 \oplus c_{57}.$$

$$Z_5^{(4)} = (((Z_9^{(2)} \boxplus Z_1^{(3)}) \oplus Z_3^{(4)}) \boxplus Z_4^{(4)}) \lll 1 \oplus c_{58}.$$

$$Z_6^{(4)} = (((Z_{10}^{(2)} \boxplus Z_2^{(3)}) \oplus Z_4^{(4)}) \boxplus Z_5^{(4)}) \lll 1 \oplus c_{59}.$$

$$Z_7^{(4)} = (((Z_{11}^{(2)} \boxplus Z_3^{(3)}) \oplus Z_5^{(4)}) \boxplus Z_6^{(4)}) \lll 1 \oplus c_{60}.$$

$$Z_{10}^{(4)} = (((Z_{14}^{(2)} \boxplus Z_6^{(3)}) \oplus Z_8^{(4)}) \boxplus Z_9^{(4)}) \lll 1 \oplus c_{63}.$$

$$Z_{11}^{(4)} = (((Z_{15}^{(2)} \boxplus Z_7^{(3)}) \oplus Z_9^{(4)}) \boxplus Z_{10}^{(4)}) \lll 1 \oplus c_{64}.$$

$$Z_{12}^{(4)} = (((Z_{16}^{(2)} \boxplus Z_8^{(3)}) \oplus Z_{10}^{(4)}) \boxplus Z_{11}^{(4)}) \lll 1 \oplus c_{65}.$$

$$Z_{13}^{(4)} = (((Z_{17}^{(2)} \boxplus Z_9^{(3)}) \oplus Z_{11}^{(4)}) \boxplus Z_{12}^{(4)}) \lll 1 \oplus c_{66}.$$

Thus, knowledge of $Z_j^{(1)}$, for $j \in \{1, 2, 3, 4, 9, 10, 11, 12\}$ is not enough to uniquely determine $Z_l^{(4)}$, for $l \in \{4, 5, 6, 7, 10, 11, 12, 13\}$. Therefore, no reduction in the effort for the key sieving phase is possible for MESH-128(8), in comparison with the ID attacks on IDEA.

- each text pair defines a list of about 2^{64} wrong 128-bit subkeys. Compute the union of these suggested subkeys. It is expected that after 2^8 structures, the number of remaining wrong subkeys is

$$2^{128} \cdot (1 - \frac{2^{64}}{2^{128}})^{2^{63} \cdot 2^8} \approx \frac{2^{128}}{e^{128}} < 1.$$

Thus, the correct subkey can be uniquely identified.

The attack takes $2^{64} \cdot 2^8 = 2^{72}$ chosen plaintexts. The memory complexity is dominated by $2^{128}/128 = 2^{121}$ blocks for sieving the correct subkey. The time complexity is $2^{63} \cdot 2^8(2^{10} + 2^{10}) = 2^{82}$ difference computations. If we consider that $\odot$ dominate the difference computations, and that there are 56 $\odot$s across 3.5-round MESH-128(8), then, the time complexity is estimated at $2^{82}/56 \approx 2^{76.19}$ 3.5-round encryptions.

The 2.5-round ID distinguisher

$$(0,0,0,0,a,b,c,d,0,0,0,0,a,b,c,d) \overset{2.5r}{\not\rightarrow} (0,0,0,e,f,g,h,0,0,e,f,g,h,0,0,0), \tag{3.98}$$

(whose proof follows a similar reasoning to that for (3.97)) can be used to further recover $Z_j^{(1)}$ for $j \in \{5,6,7,8,13,14,15,16\}$. Note that the output difference of (3.98) is the same as in (3.97) So, the subkeys of the fourth round were already recovered and we only guess subkeys from the first round. Since there are less subkeys to recover in this step, the time complexity is dominated by the first step: $2^{76.19}$ 3.5-round encryptions.

The same memory as in the previous step can be reused. The data complexity is also reduced $2^{64} \cdot 2^7 = 2^{71}$ CP, and the overall data complexity becomes $2^{71} + 2^{72} \approx 2^{72.5}$ CP.

In total, the first 128 user key bits are recovered. From the key schedule of MESH-128(8), knowledge of the subkeys $Z_j^{(4)}$, for $j \in \{4,5,6,7,10,11,12,13\}$ does not provide enough information to deduce the remaining 128 user key bits, which can be recovered by exhaustive search. Therefore, the final time complexity is 2^{128} 3.5-round encryptions.

To extend the attack to 4 rounds, we can guess the subkeys $(Z_{17}^{(4)}, Z_{18}^{(4)})$ of an additional MA half-round and apply the previous attack on 3.5 rounds.

According to the key schedule algorithm of MESH-128(8):

$$Z_{17}^{(4)} = (((Z_3^{(3)} \boxplus Z_{13}^{(3)}) \oplus Z_{15}^{(4)}) \boxplus Z_{16}^{(4)}) \lll 1 \oplus c_{70}. \tag{3.99}$$

$$Z_{18}^{(4)} = (((Z_4^{(3)} \boxplus Z_{14}^{(3)}) \oplus Z_{16}^{(4)}) \boxplus Z_{17}^{(4)}) \lll 1 \oplus c_{71}. \tag{3.100}$$

From equations (3.99) and (3.100), knowledge of $Z_j^{(4)}$, for $j \in \{4, 5, 6, 7, 10, 11, 12, 13\}$ is not enough to uniquely determine $(Z_{17}^{(4)}, Z_{18}^{(4)})$, or vice-versa.

Therefore, the time complexity increases to $2^{128} \cdot 2^{16} = 2^{144}$ 4-round MESH-128(8) encryptions.

3.8.7 ID Analysis of FOX/IDEA-NXT

The analyses in this section are based on [76, 186]. The longest ID distinguisher found for FOX/IDEA-NXT cover up to 4 rounds and are constructed using the miss-in-the-middle technique [15, 16].

In [186], Wu *et al.* found an ID distinguisher for 4-round FOX64 (such that the last round does not contain the **or** function):

$$(0, x, 0, x, 0, x, 0, x) \overset{4r}{\not\rightarrow} (y_1, y_2, y_1, y_3, y_1, y_2, y_1, y_3), \tag{3.101}$$

where $x, y_1, y_2, y_3 \neq 0$ are byte differences. The difference operator is bitwise exclusive-or.

Consider pairs (P, P^*) with difference $(0, x, 0, x, 0, x, 0, x)$ for $x \neq 0$. The input difference to the first f32 function has the form $(0, 0, 0, 0)$ and since it is a bijective mapping, the corresponding output difference is also $(0, 0, 0, 0)$. After the **or** function the output difference of the first round becomes $(0, x, 0, 0, 0, x, 0, x)$.

The input difference to f32 in the second round has the form $(0, 0, 0, x)$. Across the first S-box layer, the difference remains $(0, 0, 0, x)$, but after mu4 the difference becomes (x_1, x_2, x_3, x_4) where all $x_j \neq 0$, $1 \leq j \leq 4$ since mu4 contains an MDS matrix.

The output difference of the second round (after the **or** function) has the form $(a_3, a_4, a_1 \oplus a_3, a_2 \oplus a_4 \oplus x, a_1, a_2 \oplus x, a_3, a_4 \oplus x)$.

At the output of the fourth round consider a ciphertext pair (C, C^*) with difference $(c_1, c_2, c_1, c_3, c_1, c_2, c_1, c_3)$, with $c_j \neq 0$, $1 \leq j \leq 4$.

This fourth round does not include the **or** function (or it could easily be removed, since it is key independent). Thus, the input to this fourth round's f32 function has the form $(0, 0, 0, 0)$. Consequently, the output difference of the third round is the same as in the fourth round.

Applying the inverse **or** function, the difference becomes $(0$, $c_2 \oplus c_3$, c_1, c_2, c_1, c_2, c_1, $c_3)$. Now, matching the input and output of the third round (before the **or** function), we have that the input of f32 is $(a_3$, a_4, $a_1 \oplus a_3$, $a_2 \oplus a_4 \oplus x) \oplus (a_1$, $a_2 \oplus x$, a_3, $a_4 \oplus x)$ from the top-down direction, and $(0$, $c_2 \oplus c_3$, c_1, $c_2) \oplus (c_1$, c_2, c_1, $c_3)$ from the bottom-up direction.

Therefore, we have the following equalities: $a_3 \oplus a_1 = c_1 \oplus c_2$, $a_4 \oplus a_2 \oplus x = c_2 \oplus c_3 \oplus c_2$, $a_1 \oplus a_3 \oplus a_3 = c_1 \oplus c_1$ and $a_2 \oplus a_4 \oplus x = c_2 \oplus c_3$. The third equality gives $a_1 = 0$ which is a contradiction.

Thus, (3.101) is an ID distinguisher for 4-round FOX64.

Another 4-round impossible-differential for FOX64 was described in [76], and it has the form $(\alpha, \alpha) \overset{4r}{\not\to} (\text{or}(\alpha), \alpha)$, where $\alpha \in \mathbb{Z}_2^{32}$ is a nonzero difference word. The difference operator is exclusive-or, and further details about this distinguisher are provided in [76]. Moreover, ID distinguishers for 5-, 6- and 7-round modified Lai-Massey cipher designs were also described in [76]. But, these distinguishers depend on particular types of round functions and orthomorphism transformations, and they do not apply to FOX64.

For FOX128, Wu *et al.* in [186] found a 4-round ID distinguisher (the last round does not contain the **or** function):

$$(0,x,0,x,0,x,0,x,0,0,0,0,0,0,0,0) \overset{4r}{\not\to} (y_1,y_2,y_1,y_3,y_1,y_2,y_1,y_3,t_1,t_2,t_1,t_3,t_1,t_2,t_1,t_3), \tag{3.102}$$

where $x, y_1, y_2, y_3 \neq 0$ are byte differences. The difference operator is bitwise exclusive-or.

Consider a plaintext pair (P, P^*) with difference (0, x, 0, x, 0, x, 0, x, 0, 0, 0, 0, 0, 0, 0, 0). The input difference to the first f64 function has the form $(0,0,0,0,0,0,0,0)$, and so does its output difference.

The first round output difference becomes (0, x, 0, 0, 0, x, 0, x, 0, 0, 0, 0, 0, 0, 0, 0). The input difference to the second f64 function has the form $(0,0,0,x,0,0,0,0)$. This same difference remains after the subkey layer and the S-box layer. But after mu8, the difference becomes (a_1, a_2, a_3, a_4, a_5, a_6, a_7, a_8) where $a_j \neq 0$ for $1 \leq j \leq 8$ because mu8 contains an MDS matrix.

Before the **or** functions, the difference is (a_1, $a_2 \oplus x$, a_3, a_4, a_1, $a_2 \oplus x$, a_3, $a_4 \oplus x$, a_5, a_6, a_7, a_8, a_5, a_6, a_7, a_8). After the **or** layer, the second round output difference becomes (a_3, a_4, $a_1 \oplus a_3$, $a_2 \oplus a_4 \oplus x$, a_1, $a_2 \oplus x$, a_3, $a_4 \oplus x$, a_7, a_8, $a_5 \oplus a_7$, $a_6 \oplus a_8$, a_5, a_6, a_7, a_8).

From the fourth round, consider a ciphertext pair (C, C^*) with difference (c_1, c_2, c_1, c_3, c_1, c_2, c_1, c_3, d_1, d_2, d_1, d_3, d_1, d_2, d_1, d_3). Since there are no **or** functions and the input to the f64 function is $(0,0,0,0,0,0,0,0,)$, the input difference to the fourth round also has this same format. Before the **or** functions of the third round, the difference becomes (0, $c_2 \oplus c_3$, c_1, c_2, c_1, c_2, c_1, c_3, 0, $d_2 \oplus d_3$, d_1, d_2, d_1, d_2, d_1, d_3).

Thus, from the top-down direction, the input difference to the third round has the form (a_3, a_4, $a_1 \oplus a_3$, $a_2 \oplus a_4 \oplus x$, a_1, $a_2 \oplus x$, a_3, $a_4 \oplus x$, a_7, a_8, $a_5 \oplus a_7$, $a_6 \oplus a_8$, a_5, a_6, a_7, a_8). From the bottom-up direction, the output difference from the third round has the form (0, $c_2 \oplus c_3$, c_1, c_2, c_1, c_2, c_1, c_3, 0, $d_2 \oplus d_3$, d_1, d_2, d_1, d_2, d_1, d_3).

Thus, the input to the f64 function of the third round is ($a_3 \oplus a_1$, $a_4 \oplus a_2 \oplus x$, a_1, a_2, $a_7 \oplus a_5$, $a_8 \oplus a_6$, a_5, a_6), and also (c_1, c_3, 0, $c_2 \oplus c_3$, d_1, d_3, 0, $d_2 \oplus d_3$). Consequently, $a_1 = 0$ and $a_5 = 0$, both of which are contradictions.

In summary, (3.102) is an ID distinguisher for 4-round FOX128.

Another 4-round impossible-differential for FOX128 was described in [76], and it has the form $(\alpha, \alpha, \beta, \beta) \overset{4r}{\nrightarrow} (\mathrm{or}(\alpha), \alpha, \mathrm{or}(\beta), \beta)$, where $\alpha \in \mathbb{Z}_2^{32}$ is a nonzero difference word and $\beta \in \mathbb{Z}_2^{32}$. The difference operator is exclusive-or, and further details about this distinguisher are provided in [76].

3.8.7.1 ID Attack on 5-round FOX64

A key-recovery attack on 5-round FOX64 can be applied to its first five rounds. Round subkeys are guessed in the first round, while the ID distinguisher (3.101) covers the last four rounds.

The adversary selects plaintexts of the form $(t_1, t_2, t_3, t_4, t_1 \oplus c_1, t_5, t_3 \oplus c_2, t_4 \oplus c_3)$, for constants c_1, c_2 and c_3, while t_1, t_2, t_3, t_4, t_5 take on all possible byte values. This means 2^{40} chosen plaintexts whose pairwise differences have the form $(x_1, x_2, x_3, x_4, x_1, x_5, x_3, x_4)$.

Select those pairs whose difference satisfies $x_2 \neq x_5$. There are $2^{40}(2^{40} - 1)/2 \approx 2^{79}$ pairs in total, and a fraction of 2^{-8} of them satisfies $x_2 = x_5$. So, the pairs that we need account for about $2^{79} - 2^{79}/2^8 \approx 2^{79}$.

Next, encrypt those pairs across 5-round FOX64 and keep only those pairs whose output difference is as in (3.101). The output difference of (3.101) holds with a probability $(2^{-8})^5$ for a random permutation. Therefore, we expect about $2^{79} \cdot 2^{-40} = 2^{39}$ pairs survive this filtering.

So far, the plaintext selection step required 2^{40} encryptions and 2^{79} text blocks comparisons (exclusive-ors).

For the 2^{39} remaining pairs, the input difference to the first f32 function is $(0, x_2 \oplus x_5, 0, 0)$. We are interested in the output difference $(x_1, x_2, x_3, x_2 \oplus x_4 \oplus x_5)$ from f32. When this output difference appears, the output difference from the second round will have the form $(0,\ x_2 \oplus x_5,\ 0,\ x_2 \oplus x_5,\ 0,\ x_2 \oplus x_5,\ 0,\ x_2 \oplus x_5)$, which matches the input difference of the ID distinguisher (3.101).

Every subkey $(\mathrm{RK}_0, \mathrm{RK}_1)$ in the first f32 for which the difference in (3.101) holds is a wrong subkey value.

Therefore, for each remaining pair, we discard the wrong subkeys for which the output difference of f32 in the first round has the form $\Delta v = (x_1, x_2, x_3, x_2 \oplus x_4 \oplus x_5)$. The procedure is as follows: for each remaining pair (P, P^*):

(1) select a 32-bit candidate RK_0,
(2) compute $u = mu4(sigma4(P \oplus \mathrm{RK}_0))$, $u^* = mu4(sigma4(P^* \oplus \mathrm{RK}_0))$ and $\Delta u = u \oplus u^*$,
(3) find a pair (t, t^*) that satisfies $t \oplus t^* = \Delta u$, and $sigma4(t) \oplus sigma4(t^*) = \Delta v$,
(4) recover RK_1 from (t, t^*) and (u, u^*).

Step (1) requires 2^{32} guesses. For each RK_0, step (4) will give one RK_1 on average, and their combination is a wrong 64-bit subkey value. Thus, each text pair will indicate 2^{32} wrong 64-bit subkey candidates.

After analyzing 2^{39} pairs, the number of wrong subkey candidates remaining is estimated as

$$2^{64}(1-2^{-32})^{2^{39}} \approx 2^{64}/e^{128} \approx 2^{-118} < 1,$$

and only the correct subkey value is expected to survive.

The time complexity of this step is $2^{40} + 2^{39} \cdot 2^{32}/4 = 2^{69}$ 5-round FOX64 encryptions. Concerning the 2^{79} text block comparisons, 1-round FOX64 contains 8 S-boxes, 25 exclusive-ors and 7 multiplications in GF(2^{16}). Let us estimate that those operations are roughly equivalent to 50 exclusive-ors. So, the 2^{79} text block comparisons are roughly equivalent to $2 \cdot 2^{79}/(5 \cdot 50) \approx 2^{80-7} = 2^{73}$ 5-round FOX64 encryptions.

The data complexity is 2^{40} chosen plaintexts.

The memory complexity is 2^{64} bits, or $2^{64}/64 = 2^{58}$ text blocks, to filter the wrong 64-bit subkey candidates. The remaining $128 - 64 = 64$ user key bits can be recovered by exhaustive key search. The time complexity is still dominated by the subkey filtering step.

This attack can be extended to 6-round FOX64 by guessing the subkeys of the first round and applying the previous attack on the last five rounds. The time complexity increases to $2^{64} \cdot 2^{73} = 2^{137}$ 6-round FOX64 encryptions. The data and memory complexities remain the same as in the attack on five rounds. Since the key size of FOX64 is 128 bits, this attack only applies to FOX64/k/r where $|k| > 133$ bits.

3.8.7.2 ID Attack on 5-round FOX128

A key-recovery attack on 5-round FOX128 can be applied to its first five rounds. Round subkeys are guessed in the first round, while the ID distinguisher (3.102) covers the last four rounds.

The adversary selects plaintexts of the form $(t_1,\ t_2,\ t_3,\ t_4,\ t_1 \oplus c_1,\ t_5,\ t_3 \oplus c_2,\ t_4 \oplus c_3,\ t_6,\ t_7,\ t_8,\ t_9,\ t_6,\ t_7,\ t_8,\ t_9)$, for constants c_1, c_2, c_3, while $t_1, t_2, t_3, t_4, t_5, t_6, t_7, t_8, t_9$ take on all possible byte values. This means $(2^8)^9 = 2^{72}$ chosen plaintexts whose pairwise differences have the form $(x_1,\ x_2,\ x_3,\ x_4,\ x_1,\ x_5,\ x_3,\ x_4,\ x_6,\ x_7,\ x_8,\ x_9,\ x_6,\ x_7,\ x_8,\ x_9)$.

Select those pairs whose difference satisfies $x_2 \neq x_5$. There are $2^{72}(2^{72}-1)/2 \approx 2^{143}$ pairs in total, and a fraction of 2^{-8} of them satisfies $x_2 = x_5$. So, the pairs that we need account for about $2^{143} - 2^{143}/2^8 \approx 2^{143}$.

Next, encrypt those pairs across 5-round FOX128 and keep only those pairs whose output difference is as in (3.102). The output difference of (3.102) holds with a probability $(2^{-8})^9 = 2^{-72}$ for a random permutation. Therefore, we expect about $2^{143} \cdot 2^{-72} = 2^{71}$ pairs to survive this filtering. So far, the plaintext selection step required 2^{72} encryptions and 2^{143} text blocks comparisons (exclusive-ors).

For the 2^{63} remaining pairs, the input difference to the first f64 function is (0, $x_2 \oplus x_5$, 0, 0, 0, 0, 0, 0). We are interested in the output difference (x_1, x_2, x_3, $x_2 \oplus x_4 \oplus x_5$, x_1, x_2, x_3, x_4) from f64. When this output difference appears, the output difference from the second round will have the form (0, $x_2 \oplus x_5$, 0, $x_2 \oplus x_5$, 0, $x_2 \oplus x_5$, 0, $x_2 \oplus x_5$, 0, 0, 0, 0, 0, 0, 0, 0), which matches the input difference of the ID distinguisher (3.102).

Every subkey ($\mathrm{RK}_0, \mathrm{RK}_1$) in the first f64 function for which the difference in (3.102) holds is a wrong subkey value.

Therefore, for each remaining pair, we discard the wrong subkeys for which the output difference of f64 in the first round has the form $\Delta v = (x_1, x_2, x_3, x_2 \oplus x_4 \oplus x_5, x_1, x_2, x_3, x_4)$. The procedure is as follows: for each remaining pair (P, P^*):

(1) select a 64-bit candidate RK_0,
(2) compute $u = mu8(sigma8(P \oplus \mathrm{RK}_0))$, $u^* = mu8(sigma8(P^* \oplus \mathrm{RK}_0))$ and $\Delta u = u \oplus u^*$,
(3) find a pair (t, t^*) that satisfies $t \oplus t^* = \Delta u$, and $sigma8(t) \oplus sigma8(t^*) = \Delta v$,
(4) recover RK_1 from (t, t^*) and (u, u^*).

Step (1) requires 2^{64} guesses. For each RK_0, step (4) will give one RK_1 on average, and their combination is a wrong 128-bit subkey value. Thus, each text pair will indicate 2^{64} wrong 128-bit subkey candidates.

After analyzing 2^{71} pairs, the number of wrong subkey candidates remaining is estimated as

$$2^{128}(1 - 2^{-64})^{2^{71}} \approx 2^{128}/e^{128} \approx 2^{-118} < 1,$$

and only the correct subkey value is expected to survive.

The time complexity of this step is $2^{72} + 2^{71} \cdot 2^{64}/4 = 2^{133}$ 5-round FOX128 encryptions. Concerning the 2^{143} text block comparisons, 1-round FOX128 contains 16 S-boxes, 82 exclusive-ors and 43 multiplications in $\mathrm{GF}(2^{16})$. Let us estimate that those operations are roughly equivalent to 184 exclusive-ors. So, the 2^{133} text block comparisons are roughly equivalent to $2 \cdot 2^{133}/(5 \cdot 184) \approx 2^{134-10} = 2^{123}$ 5-round FOX128 encryptions, which is not relevant compared to the subkey filtering step.

The data complexity is 2^{72} chosen plaintexts (CP).

The memory complexity is 2^{128} bits, or $2^{128}/128 = 2^{121}$ text blocks, to filter the wrong 128-bit subkey candidates. The remaining $256 - 128 = 128$ user key bits can be recovered by exhaustive key search. The time complexity is still dominated by the subkey filtering step 2^{133} 5-round FOX128 encryptions..

This attack can be extended to 6-round FOX128 by guessing the subkeys of the first round and applying the previous attack on the last five rounds. The time complexity increases to $2^{128} \cdot 2^{133} = 2^{261}$ 6-round encryptions. The data and memory complexities remain the same as in the attack on five rounds. Since the key size of FOX128 is 256 bits, this attack only applies to FOX128/k/r where $|k| > 261$ bits.

3.8.8 ID Analysis of REESSE3+

The analyses in this section are based on [146].

The longest ID distinguishers found for REESSE3+ cover up to 2.5 rounds and are constructed using the miss-in-the-middle technique [15, 16]. The distinguishers always start from an MA half-round, for instance,

$$(a, 0, a, 0, 0, b, 0, b) \overset{2.5r}{\not\rightarrow} (c, c, 0, 0, 0, 0, d, d), \tag{3.103}$$

where $a, b, c, d \neq 0$ are nonzero 16-bit differences. The reasoning is as follows:

- from the input of an MA half-round, the input difference to the MA-box becomes $(0, 0, 0, 0)$ since all word pairs xored prior to the MA-box result in zero difference. The output of the MA-box is also $(0, 0, 0, 0)$, since the MA-box is a (fixed) permutation.
- the MA half-round output is $(a, a, 0, 0, 0, 0, b, b)$, already taking into account the word swap.
- after the KM half-round, the difference becomes $(e, f, 0, 0, 0, 0, g, h)$, where $e, f, g, h \neq 0$ are 16-bit differences.
- from the bottom-up direction after 2.5 rounds, the difference $(c,\ c,\ 0,\ 0,\ 0,\ 0,\ d,\ d)$ leads to $(c,\ 0,\ c,\ 0,\ 0,\ d,\ 0,\ d)$ at the input to the last MA half-round because of the word swap and since an MA half-round (without the word swap) is its own inverse.
- the input to the last KM half-round has the form $(i, 0, j, 0, 0, k, 0, l)$, where $i, j, k, l \neq 0$ are 16-bit differences.
- the output from the middle MA half-round, before the word swap, becomes $(i, j, 0, 0, 0, 0, k, l)$.
- so, the input to the MA-box in the middle MA half-round is nonzero: (e, f, g, h), while its output is $(e \oplus i, f \oplus j, g \oplus k, h \oplus l) = (0, 0, 0, 0)$. The left-hand side of the last equality can be deduced from the exclusive-or of the nonzero input and output difference words to the MA half-round. The right-hand side of the equality can also be deduced from the zero input and output difference words to the MA half-round.
- so, on the one hand, the input difference to the MA-box is nonzero, but its output difference is zero. This is a contradiction, since the MA-box is a permutation mapping.
- therefore, (3.103) is an ID distinguisher for 2.5-round REESSE3+, where the crossed arrow indicates the difference on the left-hand side can never cause the difference on the right-hand side of the arrow (after 2.5 rounds).

Analogously, by pairing other word differences both before and after 2.5-round REESSE3+, the following ID distinguisher can be demonstrated:

$$(0, a, 0, a, b, 0, b, 0) \overset{2.5r}{\not\to} (0, 0, c, c, d, d, 0, 0), \tag{3.104}$$

where $a, b, c, d \neq 0$ are 16-bit differences.

Following the key schedule algorithm of REESSE3+, assume we recover the whole 128-bit round subkey from the i-th and $(i+3)$-th KM half-rounds surrounding a 2.5-round ID distinguisher. Let us denote them by K_i and K_{i+3}. They are related as $K_{i+3} = K_i \lll 75$. Since both K_i and K_{i+3} are derived from the same 256-bit user key, they share $128 - 75 = 53$ bits independent of i. Thus, without loss of generality, we attack the first 3.5 rounds.

The attack, using (3.103), works as follows: choose a structure of 2^{64} plaintexts of the form $(X, 0, Y, 0, 0, U, 0, V)$ where $X, Y, U, V \in \mathbb{Z}_2^{16}$ assume all possible 16-bit values, while the remaining four words are fixed constants. One such structure leads to $2^{64}(2^{64} - 1)/2 \approx 2^{127}$ text pairs with differences following the pattern $(\delta_1, 0, \delta_2, 0, 0, \delta_3, 0, \delta_4)$.

Let a plaintext block in the structure be denoted $(p_1, p_2, p_3, p_4, p_5, p_6, p_7, p_8)$, and a ciphertext block be denoted $(c_1, c_2, c_3, c_4, c_5, c_6, c_7, c_8)$.

At the ciphertext end, collect about $2^{127}/2^{64} = 2^{63}$ pairs which have zero difference in the four middle word positions, according to the difference pattern $(\delta_5, \delta_6, 0, 0, 0, 0, \delta_7, \delta_8)$.

For each remaining pair

- for all the 2^{32} possible values of $(Z_1^{(1)}, Z_3^{(1)})$ partially encrypt the topmost KM half-round

$$q_1 = (p_1 \odot Z_1^{(1)}) \oplus ((p_1 \oplus \delta_1) \odot Z_1^{(1)})$$

  and

$$r_1 = (p_3 \boxplus Z_3^{(1)}) \oplus ((p_3 \oplus \delta_2) \boxplus Z_3^{(1)})$$

  before the 2.5-round ID distinguisher and keep the subkey pairs whose differences are equal: $q_1 = r_1$. For all 2^{16} values of $Z_1^{(1)}$, compute and store q_1 in a table of 2^{16} words (in memory). Compute the other difference r_1 for all 2^{16} values of $Z_3^{(1)}$ and look for a match in the table. This step costs 2^{16} words of memory and 2^{17} computations of q_1 and r_1. About $2^{32}/2^{16} = 2^{16}$ of them shall match.
- similarly, for all the 2^{32} possible values of $(Z_6^{(1)}, Z_8^{(1)})$ partially encrypt the topmost KM half-round

$$q_2 = (p_6 \odot Z_6^{(1)}) \oplus ((p_6 \oplus \delta_3) \odot Z_6^{(1)})$$

  and

$$r_2 = (p_8 \boxplus Z_8^{(1)}) \oplus ((p_8 \oplus \delta_4) \boxplus Z_8^{(1)})$$

before the 2.5-round ID distinguisher and keep the subkey pairs whose differences are equal: $q_2 = r_2$. This step costs the same as the previous step. About $2^{32}/2^{16} = 2^{16}$ of them shall match.

- try all the 2^{32} possible values of $(Z_1^{(4)}, Z_2^{(4)})$ and partially decrypt the bottommost KM half-round

$$q_3 = (c_1 \odot (Z_1^{(4)})^{-1}) \oplus ((c_1 \oplus \delta_5) \odot (Z_1^{(4)})^{-1})$$

  and

$$r_3 = (c_2 \boxminus Z_2^{(4)}) \oplus ((c_2 \oplus \delta_6) \boxminus Z_2^{(4)})$$

  and keep the subkey pairs whose differences are equal: $q_3 = r_3$. About $2^{32}/2^{16} = 2^{16}$ of them shall match.

- likewise, try all the 2^{32} possible values of $(Z_7^{(4)}, Z_8^{(4)})$ and partially decrypt the bottommost KM half-round

$$q_4 = (c_7 \odot (Z_7^{(4)})^{-1}) \oplus ((c_7 \oplus \delta_7) \odot (Z_7^{(4)})^{-1})$$

  and

$$r_4 = (c_8 \boxminus Z_8^{(4)}) \oplus ((c_8 \oplus \delta_8) \boxminus Z_8^{(4)})$$

  and keep the subkey pairs whose differences are equal: $q_4 = r_4$. About $2^{32}/2^{16} = 2^{16}$ of them shall match.

- make a list of all the $(2^{16})^4 = 2^{64}$ 128-bit subkeys combining the previous steps. These subkeys cannot be the correct subkey value because they satisfy the differences of an impossible differential.

Each pair defines a list of about 2^{64} wrong subkeys. Compute the union of these lists of wrong 128-bit subkeys. It is expected that after 178 structures, the number of remaining wrong subkeys is

$$2^{128} \cdot (1 - 2^{-64})^{178 \cdot 2^{63}} \approx 2^{128} \cdot e^{-89} < 1,$$

and only the correct subkey value is expected to remain.

To mark the wrong subkey values, a bit vector of 2^{64} bits (2^{57} blocks) can be used. Initially, it is empty. Once a wrong subkey is found, it is marked in this vector. The non-marked positions indicate subkey values that are not impossible.

This attack requires $178 \cdot 2^{64} \approx 2^{71}$ chosen plaintexts (CP) and about $4 \cdot 2^{63} \cdot 2^{17} \cdot 1/7 \approx 2^{79}$ 3.5-round computations. The estimate is that a full KM half-round accounts for 1/7 of 3.5 rounds.

This phase recovers subkeys $(Z_1^{(1)}, Z_3^{(1)}, Z_6^{(1)}, Z_8^{(1)}, Z_1^{(4)}, Z_2^{(4)}, Z_7^{(4)}, Z_8^{(4)})$.

Repeat the same procedure above using the ID distinguisher (3.104) to recover another set of 128-bit subkeys: $(Z_2^{(1)}, Z_4^{(1)}, Z_5^{(1)}, Z_7^{(1)}, Z_3^{(4)}, Z_4^{(4)}, Z_5^{(4)}, Z_6^{(4)})$.

Recalling the key schedule algorithm of REESSE3+ (Sect. 2.7.2), the subkeys $(Z_6^{(1)}, Z_8^{(1)})$ share sixteen key bits $(k_{112}, \ldots, k_{127})$, with the subkeys $(Z_3^{(4)}, Z_4^{(4)})$. Likewise, the subkeys $(Z_5^{(1)}, Z_7^{(1)})$ share 17 key bits, $(k_{25}, \ldots, k_{79}, k_{96}, \ldots, k_{106})$, with the subkeys $(Z_1^{(4)}, Z_2^{(4)})$.

Taking into account this bit overlap, there are 17 less key bits to recover with the ID distinguisher (3.104). Thus, the time and memory used is dominated by the attack using (3.103). The data complexity doubles. The same memory for the attack using (3.103) can be reused for (3.104).

These distinguishers allow a key-recovery attack on 3.5-round REESSE3+. In fact, since the MA half-round is key-independent, the attack also reaches four rounds.

*3.8.9 ID Analysis of IDEA**

The differential analysis of IDEA* in this section is based on [121].

Unlike the ID analyses in previous Lai-Massey cipher designs, the ID distinguisher (3.65):

$$(a, 0, a, 0) \overset{2.5r}{\nrightarrow} (b, b, 0, 0)$$

for 16-bit differences $a, b \neq 0$, beginning at an AX half-round, does not hold for IDEA*, because differences across $\oplus$ behave differently across $\odot$ and $\boxdot$. This distinguisher breaks down at the first AX half-round, at the input to the first AX-box, where the two nonzero inputs with difference a meet at a $\boxdot$ operation. In previous Lai-Massey designs, such as IDEA and MESH ciphers, the inputs to the MA-box were combined via $\oplus$ so that $a \oplus a = 0$, for all a.

Let $W = X \odot Y$. It is possible to have nonzero differences $\Delta X = X \oplus X^*$ and $\Delta Y = Y \oplus Y^*$ at the input but still $\Delta W = X \boxdot Y \oplus (X \oplus \Delta X) \boxdot (Y \oplus \Delta Y)$ could be either zero or nonzero, and in both cases with nonzero probability. In particular, we are interested in $\Delta X = 8000_x$ and $\Delta Y = 8000_x$, but it has been verified exhaustively that the probability is close to 2^{-16}, independent of the particular values for ΔX or ΔY. See Table 3.84. Note that $\boxdot$ stands for modular division, and thus, it is a non-commutative operator: $a \boxdot b = a/b \neq b/a = b \boxdot a$.

Table 3.84 Examples of xor differences crossing unkeyed $\boxdot$: $W = X \boxdot Y$.

ΔX	ΔY	ΔW	Probability
8000_x	8000_x	0000_x	2^{-15}
8000_x	8000_x	8000_x	$2^{-14.98}$
8000_x	0000_x	0000_x	0
8000_x	0000_x	8000_x	2^{-16}
0000_x	8000_x	0000_x	0
0000_x	8000_x	8000_x	2^{-15}

In any case, the fact that these probabilities are far from 1 means that (3.65) cannot be used for IDEA*. It does not mean that (alternative) ID distinguisher exist, but so far, they have not been found. It is left as an open problem.

For comparison purposes, note in Table 3.85 how xor differences propagate (or not) across $\odot$.

Table 3.85 Examples of exclusive-or differences crossing unkeyed $\odot$: $W = X \odot Y$.

ΔX	ΔY	ΔW	Probability
8000_x	0000_x	0000_x	0
8000_x	0000_x	8000_x	2^{-16}
8000_x	8000_x	0000_x	2^{-15}
8000_x	8000_x	8000_x	$2^{-14.98}$

3.9 Slide Attacks

Most of modern block ciphers adopt a single type of round transformation that is repeated a certain number of times. Examples include DES, AES and IDEA. This design philosophy aims to reduce the implementation cost and the overall performance (since there will be one single type of round transformation to implement for either encryption or decryption).

Since all round transformations are the same, the only parameter that differs between rounds is the (round) subkey. Thus, if the round subkeys are identical (for a particular user key), then all round transformations will be identical. Depending on the key schedule algorithm, this phenomenon may happen for a number of user keys. Ideally, this should never happen. But, if it happens, then a number of techniques, such as the slide and advanced slide attacks, were developed to exploit this phenomenon.

These issues imply a design criterion to the key schedule algorithm: the round subkeys should not be identical for different rounds (either consecutive rounds or regularly separated rounds).

The slide attack was presented by Biryukov and Wagner in [37] as a known-plaintext (sometimes chosen-plaintext) attack on iterated block ciphers. This attack exploits the (potential) self-similarity of round transformations (denoted F_k).

The simplest self-similarity condition for an r-round iterated cipher means that $F_{k_i} = F_{k_j}$ for all $1 \leq i, j \leq r$, where F_{k_i} is the i-th round function parameterized by the subkey k_i. This condition implies that the subkeys are periodic (per round).

If the self-similarity condition is satisfied, this attack is independent of the number of rounds of the target cipher.

More formally, let E_K denote an n-bit block cipher with r rounds, that can be modeled as r iterations of a key-dependent permutation F:

$$E_K(.) = F_{k_r}(F_{k_{r-1}}(\ldots(F_{k_1}(.))\ldots)), \tag{3.105}$$

where k_i is the i-th round subkey. The self-similarity condition for one round means that

$$E_K(.) = F_{k'}(F_{k'}(\ldots(F_{k'}(.))\ldots)), \tag{3.106}$$

that is, E_K degenerates into r chained iterations of $F_{k'}$.

The attack consists in sliding one instance of E_K against another copy of E_K so that the two schemes are, for instance, one round out of phase. If (P_1, C_1) and (P_2, C_2) are two plaintext/ciphertext pairs, and assuming the self-similarity condition holds, then one obtains $F_{k'}(P_1) = P_2$ and $F_{k'}(C_1) = C_2$, which is called a *slid pair*. Further, the adversary obtains $2^{n/2}$ known (P_i, C_i) pairs and looks for a slid pair. Due to the birthday paradox effect, it is expected that about one pair of indices (i, j) exist among the $2^{n/2}$ pairs such that $F_{k'}(P_i) = P_j$ and $F_{k'}(C_i) = C_j$, which gives a slid pair. It is assumed that the round function is weak, in the sense that it is computationally efficient to recover the round subkey k' given (P_i, C_i), (P_j, C_j), $F_{k'}(P_i) = P_j$ and $F_{k'}(C_i) = C_j$, that is, from a single round.

Thus, the attack is independent of the number of rounds due to the periodicity of round subkeys. Slide attacks have been reported on Treyfer [169], on variants of the DES cipher [154], on Blowfish [169], and on the stream cipher WAKE [53].

3.9.1 Slide Attacks on IDEA

For the IDEA cipher, the simplest self-similarity criterion is to assume that all subkeys are identical per round, that is, for the KM half-round:

$$Z_i^{(l)} = Z_i^{(j)}, 1 \leq i \leq 4, 1 \leq l, j \leq 9, \tag{3.107}$$

and for the MA-half-round:

$$Z_i^{(l)} = Z_i^{(j)}, 5 \leq i \leq 7, 1 \leq l, j \leq 8. \tag{3.108}$$

For $i = 1$, according to the key schedule algorithm of IDEA, it means that the key bits numbered 0–15 are equal to those in the range 96–111, and also in the range 85–104, and 82–97 and 75–90 and 43–58 and 36–51 and 29–44 and 22–37.

For $i = 2$, it means that the key bits numbered 16–31 are equal to those in the range 112–127, and to those in the range 105–120, and 98–113 and 91–106 and 59–74 and 52–67 and 45–60 and 38–53.

Bitwise, it means that the key bits in $I_1 = [0, 96, 89, 82, 75, 43, 36, 29, 22]$ must all have the same value. So, for instance, $k_0 = k_{96} = k_{89} = \ldots = k_{22}$. Likewise, for the key bits in the sets $I_2 = [1, 97, 90, 83, 76, 44, 37, 30, 23]$, I_3 = [2, 98, 91, 84, 77, 45, 38, 31, 24], I_4 = [3, 99, 92, 85, 78, 46, 39, 32, 25], I_5 = [4, 100, 93, 86, 79, 47, 40, 33, 26], I_6 = [5, 101, 94, 87, 80, 48, 41, 34, 27], I_7 = [6, 102, 95, 88, 81, 49, 42, 35, 28], I_8 = [7, 103, 96, 89, 82, 50, 43, 36, 29], I_9 = [8, 104, 97, 90, 83, 51, 44, 37, 30], I_{10} = [9, 105, 98, 91, 84, 52, 45, 38, 31], I_{11} = [10, 106, 99, 92, 85, 53, 46, 39, 32], I_{12} = [11, 107, 100, 93, 86, 54, 47, 40, 33], I_{13} = [12, 108, 101, 94, 87, 55, 48, 41, 34], I_{14} = [13, 109, 102, 95, 88, 56, 49, 42, 35], I_{15} = [14, 110, 103, 96, 89, 57, 50, 43, 36], I_{16} = [15, 111, 104, 97, 90, 58, 51, 44, 37], I_{17} = [16, 112, 105, 98, 91, 59, 52, 45, 38], I_{18} = [17, 113, 106, 99, 92, 60, 53, 46, 39], I_{19} = [18, 114, 107, 100, 93, 61, 54, 47, 40], I_{20} = [19, 115, 108, 101, 94, 62, 55, 48, 41], I_{21} = [20, 116, 109, 102, 95, 63, 56, 49, 42], I_{22} = [21, 117, 110, 103, 96, 64, 57, 50, 43], I_{23} = [22, 118, 111, 104, 97, 65, 58, 51, 44], I_{24} = [23, 119, 112, 105, 98, 66, 59, 52, 45], I_{25} = [24, 120, 113, 106, 99, 67, 60, 53, 46], I_{26} = [25, 121, 114, 107, 100, 68, 61, 54, 47], I_{27} = [26, 122, 115, 108, 101, 69, 62, 55, 48], I_{28} = [27, 123, 116, 109, 102, 70, 63, 56, 49], I_{29} = [28, 124, 117, 110, 103, 71, 64, 57, 50], I_{30} = [29, 125, 118, 111, 104, 72, 65, 58, 51], I_{31} = [30, 126, 119, 112, 105, 73, 66, 59, 52], I_{32} = [31, 127, 120, 113, 106, 74, 67, 60, 53]. Note that all 128 user key bits are included in at least one of these sets. It means that these sets cover all bits of the user key.

Note that I_1 and I_8 share bit k_{96}. It means the all key bits in both sets are equal. The union of I_1 and I_8 is denoted $I_{1,8} = I_1 \cup I_8$. Likewise, I_2 and I_9 share bit k_{97}, so they can be merged into $I_{2,9} = I_2 \cup I_9$. Likewise, I_3 and I_{10} share bit k_{98}, so they can be merged into $I_{3,10}$. Also, I_4 and I_{11} share key bit k_{99}. Thus, they are merged into $I_{4,11}$. Also, I_5 and I_{12} share bit k_{100}. Thus, they are merged into $I_{5,10}$. Also, I_6 and I_{13} share bit k_{101}. Thus, they are merged into $I_{6,13}$. Also, I_7 and I_{14} share bit k_{102}. Thus, they are merged into $I_{7,14}$.

Further, $I_{1,8}$ and I_{15} share bit k_{103} and are merged into $I_{1,8,15}$. $I_{2,9}$ and I_{16} share bit k_{104}, and are merged into $I_{2,9,16}$. $I_{3,10}$ and I_{17} share bit k_{105}, and are merged into $I_{3,10,17}$. $I_{4,11}$ and I_{18} share bit k_{106}, and are merged into $I_{4,11,18}$. $I_{5,12}$ and I_{19} share bit k_{107}, and are merged into $I_{5,12,19}$. $I_{6,13}$ and I_{20} share bit k_{108}, and are merged into $I_{6,13,20}$. $I_{7,14}$ and I_{21} share bit k_{109}, and are merged into $I_{7,14,21}$. $I_{1,8,15}$ and I_{22} share bit k_{96}, and are merged into $I_{1,8,15,22}$. $I_{2,9,16}$ and I_{23} share bit k_{97}, and are merged into $I_{2,9,16,23}$. $I_{3,10,17}$ and I_{24} share bit k_{98}, and are merged into $I_{3,10,17,24}$. $I_{4,11,18}$ and I_{25} share bit k_{99}, and are merged into $I_{4,11,18,25}$. $I_{5,12,19}$ and I_{26} share bit k_{100}, and are merged into $I_{5,12,19,26}$. $I_{6,13,20}$ and I_{27} share bit k_{101}, and are merged into $I_{6,13,20,27}$. $I_{7,14,21}$ and I_{28} share bit k_{102}, and are merged into $I_{7,14,21,28}$.

$I_{1,8,15,22}$ and I_{29} share bit k_{117}, and are merged into $I_{1,8,15,22,29}$. $I_{2,9,16,23}$ and I_{30} share bit k_{118}, and are merged into $I_{2,9,16,23,30}$. $I_{3,10,17,24}$ and I_{31} share bit k_{119}, and are merged into $I_{3,10,17,24,31}$. $I_{4,11,18,25}$ and I_{32} share bit k_{120}, and are merged into $I_{4,11,18,25,32}$.

$I_{2,9,16,23,30}$ and $I_{1,8,15,22,29}$ share bit k_{22}, and are merged into the combined set $I_{1,2,8,9,15,16,22,23,29,30}$.

Likewise, $I_{1,2,8,9,15,16,22,23,29,30}$ and $I_{3,10,17,24,31}$ share bit k_{23}, and are merged into $I_{1,2,3,8,9,10,15,16,17,22,23,24,29,30,31}$.

$I_{1,2,3,8,9,10,15,16,17,22,23,24,29,30,31}$ and $I_{4,11,18,25,32}$ share bit k_{24}, and are merged into $I_{1,2,3,4,8,9,10,11,15,16,17,18,22,23,24,25,29,30,31,32}$.

$I_{1,2,3,4,8,9,10,11,15,16,17,18,22,23,24,25,29,30,31,32}$ and $I_{5,12,19,26}$ share bit k_{25}, and are merged into $I_{1,2,3,4,5,8,9,10,11,12,15,16,17,18,19,22,23,24,25,26,29,30,31,32}$.

$I_{1,2,3,4,5,8,9,10,11,12,15,16,17,18,19,22,23,24,25,26,29,30,31,32}$ and $I_{6,13,20,27}$ share bit k_{26}, and become $I_{1,2,3,4,5,6,8,9,10,11,12,13,15,16,17,18,19,20,22,23,24,25,26,27,29,30,31,32}$.

$I_{1,2,3,4,5,6,8,9,10,11,12,13,15,16,17,18,19,20,22,23,24,25,26,27,29,30,31,32}$ and $I_{7,14,21,28}$ share k_{27} and become $I_{1,2,3,4,5,6,7,8,9,10,11,12,13,14,15,16,17,18,19,20,21,22,\ldots,32}$.

Therefore, in order for the self-similarity condition to be satisfied for IDEA, all user key bits must be identical. There are only two possibilities: either $K = 0^{128}$ or $K = 1^{128}$. Since either key can be tested using only two encryptions with two plaintext/ciphertext pairs, it is much cheaper than using advanced slide techniques.

3.9.2 Slide Attacks on MESH-64

For the MESH ciphers, due to the different operations in the even- and odd-numbered KM half-rounds, the encryption schemes must be slid by at least two rounds (in general, by an even number of rounds).

Concerning MESH-64, the analysis results are negative. For example, the slide attack using encryption schemes slid by two rounds does not work.

A necessary condition to be satisfied for a slide attack on (the full 8.5-round) MESH-64 is the round self-similarity. Note that all even-numbered rounds are equal except for the subkeys, and likewise for the odd-numbered rounds.

Suppose for a moment that all such conditions could be satisfied. This means that all odd-round subkeys might be equal:

$$Z_j^{(1)} = Z_j^{(3)} = Z_j^{(5)} = Z_j^{(7)} = Z_j^{(9)}, 1 \leq j \leq 4, \tag{3.109}$$

$$Z_j^{(1)} = Z_j^{(3)} = Z_j^{(5)} = Z_j^{(7)}, 5 \leq j \leq 7, \tag{3.110}$$

as well as all the even-round subkeys:

$$Z_j^{(2)} = Z_j^{(4)} = Z_j^{(6)} = Z_j^{(8)}, 1 \leq j \leq 7. \tag{3.111}$$

Using the key schedule algorithm of MESH-64, these conditions imply, for instance, that

$$Z_1^{(3)} = (((((Z_7^{(1)} \boxplus Z_1^{(2)}) \oplus Z_2^{(2)}) \boxplus Z_5^{(2)}) \oplus Z_6^{(2)}) \boxplus Z_7^{(2)}) \lll 7 \oplus c_{14}$$

and

$$Z_1^{(5)} = (((((Z_7^{(3)} \boxplus Z_1^{(4)}) \oplus Z_2^{(4)}) \boxplus Z_5^{(4)}) \oplus Z_6^{(4)}) \boxplus Z_7^{(4)}) \lll 7 \oplus c_{28}$$

must be equal. Therefore, from (3.109), (3.110) and (3.111), it follows that $c_{14} = c_{28}$ must hold. But according to Table 2.4 in the key schedule algorithm of MESH-64, $c_{14} = 5555_x$ and $c_{28} = eeb6_x$, which is a contradiction.

The conclusion is that the round subkeys of MESH-64 cannot satisfy the self-similarity conditions for a 2-round slide attack. This attack helps corroborate the design decisions of the key schedule algorithm of MESH-64.

Similarly, consider now a slide attack on MESH-64 using encryption schemes slid by four rounds. Again, a necessary condition to be satisfied on the full 8.5-round MESH-64 is the round self-similarity which means the following equalities must hold:

$$Z_j^{(1)} = Z_j^{(5)} = Z_j^{(9)}, 1 \leq j \leq 7, \tag{3.112}$$

$$Z_j^{(3)} = Z_j^{(7)}, 1 \leq j \leq 7, \tag{3.113}$$

$$Z_j^{(2)} = Z_j^{(6)}, 1 \leq j \leq 7, \tag{3.114}$$

$$Z_j^{(4)} = Z_j^{(8)}, 1 \leq j \leq 7. \tag{3.115}$$

Using the key schedule of MESH-64, these conditions imply, for instance, that

$$Z_2^{(2)} = (((((Z_1^{(1)} \boxplus Z_2^{(1)}) \oplus Z_3^{(1)}) \boxplus Z_6^{(1)}) \oplus Z_7^{(1)}) \boxplus Z_1^{(2)}) \lll 7 \oplus c_8$$

and

$$Z_2^{(6)} = (((((Z_1^{(5)} \boxplus Z_2^{(5)}) \oplus Z_3^{(5)}) \boxplus Z_6^{(5)}) \oplus Z_7^{(5)}) \boxplus Z_1^{(6)}) \lll 7 \oplus c_{36}$$

must be equal. Therefore, from (3.114) and (3.112), it follows that $c_8 = c_{36}$ must hold. But according to the key schedule algorithm of MESH-64, $c_8 = 0101_x$ and $c_{36} = 4150_x$, which is a contradiction. Similar contradictions apply for the remaining pairs of round subkeys such as $(Z_3^{(2)}, Z_3^{(6)})$, $(Z_4^{(2)}, Z_4^{(6)})$, $(Z_5^{(2)}, Z_5^{(6)})$ and so on.

Therefore, the round subkeys of MESH-64 cannot satisfy the self-similarity conditions for a 4-round slide attack.

Similar reasoning applies to larger gaps between the slid pairs, but MESH-64 has only 8.5 rounds.

3.9.3 Slide Attacks on MESH-96

For the MESH ciphers, due to the different operations in the even- and odd-numbered KM half-rounds, the encryption schemes must be slid by at least two rounds (in general, by an even number of rounds).

Concerning MESH-96, the analysis results are negative. For example, the slide attack using encryption schemes slid by two rounds does not work.

A necessary condition to be satisfied for a slide attack on (the full 8.5-round) MESH-96 is the round self-similarity. Note that all even-numbered rounds are equal except for the subkeys, and likewise for the odd-numbered rounds.

Suppose for a moment that all such condition could be satisfied. This means that all odd-round subkeys might be equal:

$$Z_j^{(1)} = Z_j^{(3)} = Z_j^{(5)} = Z_j^{(7)} = Z_j^{(9)} = Z_j^{(11)}, 1 \leq j \leq 6, \tag{3.116}$$

$$Z_j^{(1)} = Z_j^{(3)} = Z_j^{(5)} = Z_j^{(7)}, 7 \leq j \leq 9, \tag{3.117}$$

as well as all the even-round subkeys:

$$Z_j^{(2)} = Z_j^{(4)} = Z_j^{(6)} = Z_j^{(8)} = Z_j^{(10)}, 1 \leq j \leq 9. \tag{3.118}$$

Using the key schedule of MESH-96, these conditions imply, for instance, that

$$Z_4^{(2)} = (((((Z_1^{(1)} \boxplus Z_5^{(1)}) \oplus Z_7^{(1)}) \boxplus Z_9^{(1)}) \oplus Z_2^{(2)}) \boxplus Z_3^{(2)}) \lll 9 \oplus c_{12}$$

and

$$Z_4^{(4)} = (((((Z_1^{(3)} \boxplus Z_5^{(3)}) \oplus Z_7^{(3)}) \boxplus Z_9^{(3)}) \oplus Z_2^{(4)}) \boxplus Z_3^{(4)}) \lll 9 \oplus c_{30}$$

must be equal. Therefore, from (3.116), (3.117) and (3.118), it follows that $c_{12} = c_{30}$ must hold. But according to Table 2.4, $c_{12} = 1111_x$ and $c_{30} = 5419_x$, which is a contradiction.

The conclusion is that the round subkeys of MESH-96 cannot satisfy the self-similarity conditions for a 2-round slide attack.

This attack helps corroborate the design decisions of the key schedule algorithm of MESH-96.

3.9.4 Slide Attacks on MESH-128

For the MESH ciphers, due to the different operations in the even- and odd-numbered KM half-rounds, the encryption schemes must be slid by at least two rounds (in general, by an even number of rounds).

Concerning MESH-128, the analysis results are negative. For example, the slide attack using encryption schemes slid by two rounds does not work for MESH-128.

A necessary condition to be satisfied for a slide attack on (the full 8.5-round) MESH-128 is the round self-similarity. Note that all even-numbered rounds are equal except for the subkeys, and likewise for the odd-numbered rounds.

Suppose for a moment that all such condition could be satisfied. This means that all odd-round subkeys might be equal:

$$Z_j^{(1)} = Z_j^{(3)} = Z_j^{(5)} = Z_j^{(7)} = Z_j^{(9)} = Z_j^{(11)} = Z_j^{(13)}, 1 \leq j \leq 8, \quad (3.119)$$

$$Z_j^{(1)} = Z_j^{(3)} = Z_j^{(5)} = Z_j^{(7)} = Z_j^{(9)} = Z_j^{(11)}, 9 \leq j \leq 12, \quad (3.120)$$

as well as all the even-round subkeys:

$$Z_j^{(2)} = Z_j^{(4)} = Z_j^{(6)} = Z_j^{(8)} = Z_j^{(10)} = Z_j^{(12)}, 1 \leq j \leq 12. \quad (3.121)$$

Using the key schedule of MESH-128, these conditions imply, for instance, that

$$Z_5^{(2)} = (((((Z_1^{(1)} \boxplus Z_4^{(1)}) \oplus Z_5^{(1)}) \boxplus Z_9^{(1)}) \oplus Z_3^{(2)}) \boxplus Z_4^{(2)}) \lll 11 \oplus c_{16}$$

and

$$Z_5^{(4)} = (((((Z_1^{(3)} \boxplus Z_4^{(3)}) \oplus Z_5^{(3)}) \boxplus Z_9^{(3)}) \oplus Z_3^{(4)}) \boxplus Z_4^{(4)}) \lll 11 \oplus c_{40}$$

must be equal. Therefore, from (3.119), (3.120) and (3.121), it follows that $c_{16} = c_{40}$ must hold. But according to Table 2.4, $c_{16} = 002c_x$ and $c_{40} = 54e4_x$, which is a contradiction.

The conclusion is that the round subkeys of MESH-128 cannot satisfy the self-similarity conditions for a 2-round slide attack.

This attack helps corroborate the design decisions of the key schedule algorithm of MESH-128.

3.10 Advanced Slide Attacks

The advanced slide attack (also called Slide with a Twist) was described by Biryukov and Wagner in [38]. It is an extension of the Slide attack, because the self-similarity conditions are generalized in the sense that slid versions of an encryption and a decryption scheme are compared.

A t-round slided self-similarity condition for an r-round iterated cipher means that $F_{k_{i+t}} = F^{-1}_{k_{r-i+t}}$, for all $1 \leq i \leq r$, where F_{k_i} is the i-th round function parameterized by the subkey k_i. Notice that in this case, the two schemes are out of sync by t rounds.

3.10.1 Advanced Slide Attacks on MESH-64

For the MESH ciphers, due to the different operations in the even- and odd-numbered KM half-rounds, the encryption and decryption schemes must be slided by at least two rounds (in general, by a nonzero even number of rounds).

Concerning MESH-64, the analysis results are negative. For example, the advanced slide attack using an encryption scheme and a decryption scheme (not slided) does not work.

Suppose for a moment that an advanced slide attack could hold for MESH-64. The self-similarity conditions include (3.109), (3.110), (3.111) and additionally:

$$Z_j^{(3)} = (Z_j^{(9)})^{-1}, \text{ for } j \in \{1, 4\}, \tag{3.122}$$

$$Z_j^{(3)} = -Z_j^{(9)}, \text{ for } j \in \{2, 3\}, \tag{3.123}$$

$$Z_j^{(3)} = Z_j^{(8)}, \text{ for } j \in \{5, 6, 7\}. \tag{3.124}$$

And similarly, for the subkeys of the other rounds.

Note that due to the fact that one scheme is in the encryption direction while the other scheme is in the decryption direction, some subkeys have to match the multiplicative inverse or the additive inverse of the slided scheme. Moreover, note that the matching involves the KM and the MA half-rounds of different rounds.

The condition (3.122) on multiplicative subkeys imply that $Z_i^{(3)}, Z_i^{(9)} \in \{0, 1\}$ because 0 and 1 are the only values which are their own multiplicative inverses.

Likewise, the condition (3.123) on additive subkeys imply that $Z_j^{(3)}, Z_j^{(9)} \in \{0, 8000_x\}$ because these are the only values that are their own additive inverses.

These particular values turn these keyed-additions and keyed-multiplications into involutions.

Also, according to the key schedule of MESH-64, in particular:

$$Z_1^{(3)} = (((((Z_7^{(1)} \boxplus Z_1^{(2)}) \oplus Z_2^{(2)}) \boxplus Z_5^{(2)}) \oplus Z_6^{(2)}) \boxplus Z_7^{(2)}) \lll 7 \oplus c_{14},$$

while

$$Z_1^{(9)} = (((((Z_7^{(7)} \boxplus Z_1^{(8)}) \oplus Z_2^{(8)}) \boxplus Z_5^{(8)}) \oplus Z_6^{(8)}) \boxplus Z_7^{(8)}) \lll 7 \oplus c_{56}.$$

The conditions (3.109), (3.110) and (3.111) imply that $c_{14} = c_{56}$, but according to the key schedule of MESH-64, $c_{14} = 5555_x$ and $c_{56} = 6975_x$, which is a contradiction. Therefore, the round subkeys of MESH-64 cannot satisfy the self-similarity conditions for a 2-round advanced slide attack.

This attack helps corroborate the design decisions of the key schedule algorithm of MESH-64.

3.10.2 Advanced Slide Attacks on MESH-96

For the MESH ciphers, due to the different operations in the even- and odd-numbered KM half-rounds, the encryption schemes must be slided by at least two rounds (in general, by an even number of rounds).

Concerning MESH-96, the analysis results are negative. For example, the advanced slide attack using encryption schemes slid by two rounds does not work for MESH-96. The reasoning is analogous to the proofs shown for MESH-64 in Sect. 3.10.1.

3.10.3 Advanced Slide Attacks on MESH-128

For the MESH ciphers, due to the different operations in the even- and odd-numbered KM half-rounds, the encryption schemes must be slided by at least two rounds (in general, by an even number of rounds).

Concerning MESH-128, the analysis results are negative. For example, the advanced slide attack using encryption schemes slid by two rounds does not work for MESH-128. The reasoning is analogous to the proofs shown for MESH-64 in Sect. 3.10.1.

3.11 Biclique Attacks

The biclique technique was first described by Bogdanov *et al.* in [43] against the full AES cipher. Biclique key-recovery attacks are based on the Meet-in-the-Middle (MITM) approach, but uses an important tool from the domain of hash function cryptanalysis: initial structures [168]. Bicliques (which comes from the concept of *complete bipartite graphs*) combined ideas from a number of techniques including MITM attacks on block ciphers [45, 65, 42] and in hash function cryptanalysis [5, 6, 168].

In the biclique attack on block ciphers, the key space is partitioned into key classes so that the keys in a class can be efficiently tested using the MITM framework.

Bicliques do not impose conditions on the key schedule algorithm in the sense that biclique attacks are not a subset of related-key attacks.

The key-space partitions can be described in several ways. For linear key schedule algorithms such as in IDEA, three sets of key bits are used: K^b, K^f and K^g. In a key class, the value K^g is fixed, and hence enumerates the classes, while K^b and K^f take on all possible values.

A biclique is essentially a set of internal cipher states which are constructed either in the first or in the last rounds of a cipher and mapped to each other by carefully chosen keys. Let f be the mapping describing the first rounds of a target cipher. Then, a biclique for a set K^g is a set of states $\{P_i\}$, $\{S_j\}$ such that

$$P_i \underset{f}{\overset{K^b=i,K^f=j}{\longrightarrow}} S_j.$$

Keys in a set are tested as follows: the adversary asks for the encryption of plaintexts P_i and gets ciphertexts C_i. Further, the adversary checks if

$$\exists\, i,j : S_j \underset{g}{\overset{K^b=i,K^f=j}{\longrightarrow}} C_i,$$

where g maps the states S_j to the ciphertexts. A biclique is said to have dimension d if both K^b and K^F have d bits.

Each key set is tested separately. There are two approaches to test keys in a set. In the first approach, an adversary uses an intermediate variable v that can be computed in both the encryption and decryption directions:

$$S_j \underset{g_1}{\overset{K^f=j}{\longrightarrow}} v \overset{?}{=} v \underset{g_2}{\overset{K^b=i}{\longleftarrow}} C_i.$$

The functions g_1 and g_2 are called *chunks*. See Fig. 3.5. The computational complexity of testing a single set is $C_{\text{biclique}} + 2^{|K^f|} \cdot C_{g_1} + 2^{|K^b|} \cdot C_{g_2} + C_{\text{recheck}}$, where C_{g_1} and C_{g_2} are the costs for computing v, C_{biclique} is the biclique construction cost, and C_{recheck} is the cost of rechecking key candidates in other state bits or other (P_i, C_i) pairs. The full attack complexity is the product of the total number of sets.

In the second approach, an adversary is unable to find a variable with these properties. Then, the adversary has to test each key individually:

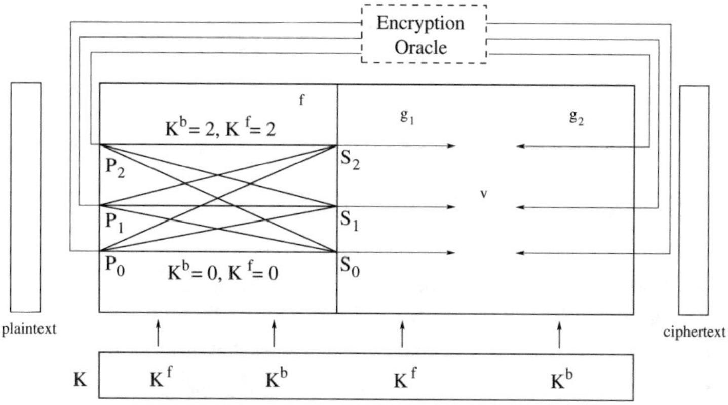

Fig. 3.5 Key testing with a biclique of three plaintexts and three internal cipher states [97].

$$S_j \overset{K^b=i,K^f=j}{\underset{g_1}{\longrightarrow}} v \overset{?}{=} v \overset{K^b=i,K^f=j}{\underset{g_2}{\longleftarrow}} C_i,$$

but reuses the computations of chunks, which are defined as parts of g_1 and g_2 that are independent of K^b and K^f, respectively. This technique was called the independent-biclique approach in [43] due to the use of independent differential trails in the biclique construction.

A straightforward way to construct bicliques is to use related-key differentials that do not share active nonlinear components. Let (P, S, K) denote a 3-tuple of plaintext, internal cipher state and key. Let also K^f and K^b be tuples of key bits.

Proposition 3.1. *[43]*
Suppose that the tuple (P_0, S_0, K_0) conform to the two sets of related-key differential trails:

$$0 \overset{\Delta K^f=\Delta_j^K}{\underset{f}{\longrightarrow}} \Delta_j;\ \nabla_i \overset{\Delta K^b=\nabla_i^K}{\underset{f}{\longrightarrow}} 0, \tag{3.125}$$

that share no active nonlinear transformations. Then, the states $P_i = P_0 \oplus \nabla_i$ and $S_j = S_0 \oplus \nabla_j$ form a biclique for a set of keys defined by K_0.

Narrow Bicliques: a direct application of the independent differential technique limits the length of a biclique to the number of rounds needed to achieve full diffusion.

One of the appealing properties of biclique attacks is the fact that memory complexity is low, that is, only exponential in the dimension of the biclique which is usually a small constant.

The basic idea is that all those computations that do not depend on plaintext or ciphertext queries to oracles can be stored and reused for multiple key recoveries. With this idea, the computational complexity to recover the first key remains the same, but for subsequent key recoveries, less computations are needed.

3.11.1 Biclique Attacks on IDEA

The biclique attack on IDEA was first described in [97]. In this attack, the Biryukov-Demirci relation in Sect. 3.17 will be used.

Let (n_i, q_i) and (r_i, s_i) denote the inputs and outputs of the i-th MA-box of IDEA, respectively. Let $X^{(i)} = (X_1^{(i)}, X_2^{(i)}, X_3^{(i)}, X_4^{(i)})$ denote the input text block to the i-th round; it also denotes the output from the $(i-1)$-th round for $i > 1$. Let $Y^{(i)} = (Y_1^{(i)}, Y_2^{(i)}, Y_3^{(i)}, Y_4^{(i)})$ denote the output text block from the i-th KM half-round.

Let $t_i = n_i \odot Z_5^{(i)}$, so that $r_i = t_i \boxplus s_i$.

The Biryukov-Demirci relation across 2-round IDEA involves the two middle words $(X_2^{(i)}, X_3^{(i)})$:

$$((X_2^{(i)} \boxplus Z_2^{(i)}) \oplus (s_i \boxplus t_i)) \boxplus Z_3^{(i+1)} = X_2^{(i+2)} \oplus s_{i+1}, \qquad (3.126)$$

$$((X_3^{(i)} \boxplus Z_3^{(i)}) \oplus s_i) \boxplus Z_2^{(i+1)} = X_3^{(i+2)} \oplus (s_{i+1} \boxplus t_{i+1}). \qquad (3.127)$$

Combining (3.126), (3.127), restricting the equality to the least significant bit and redistributing the terms results in:

$$\begin{aligned} &\text{lsb}_1(X_2^{(i)} \oplus X_3^{(i)} \oplus Z_2^{(i)} \oplus Z_3^{(i)} \oplus t_i) = \\ &\text{lsb}_1(X_2^{(i+2)} \oplus X_3^{(i+2)} \oplus t_{i+1} \oplus Z_2^{(i+1)} \oplus Z_3^{(i+1)}). \end{aligned} \qquad (3.128)$$

Therefore, for the matching step in the meet-in-the-middle attack, it is enough to compute $X_2^{(i)}$, $X_3^{(i)}$ and t_i in the forward direction, and $X_2^{(i+2)}$, $X_3^{(i+2)}$ and t_{i+1} in the backward direction. To compute t_i, it is enough to compute $X_1^{(i+2)}$, $X_2^{(i+2)}$ and to know $Z_5^{(i+1)}$. If some bits of $Z_2^{(i+1)}$ and $Z_3^{(i+1)}$ belong to K^b or K^f, they are distributed to the corresponding sides of the equation.

The Biryukov-Demirci relation essentially excludes six multiplications or about 1.5 rounds from the matching part.

Improved filtering: it is possible to improve the filtering provided by the Biryukov-Demirci (BD) relation by considering more than one bit in (3.126) and (3.127). Consider $X_2^{(i)}$, $X_3^{(i)}$, t_i, $X_2^{(i+2)}$, $X_3^{(i+2)}$, t_{i+1}, $Z_2^{(i)}$, $Z_3^{(i)}$, $Z_2^{(i+1)}$ and $Z_3^{(i+1)}$ as known parameters, and denote the left-hand side and the right-hand sides of (3.126) and (3.127) as: $L_1(s_i)$, $R_1(s_{i+1})$, $L_2(s_i)$ and $R_2(s_{i+1})$, respectively.

If k bits of s_i are known, we can compute k bits of $L_1(s_i)$ and $L_2(s_i)$, and $k+1$ bits of $L_1(s_i) \oplus L_2(s_i)$. Similarly, if k bits of s_{i+1} are known, we can compute k bits of $R_1(s_{i+1})$ and $R_2(s_{i+1})$, and also $k+1$ bits of $R_1(s_{i+1}) \oplus R_2(s_{i+1})$. Some values of the parameters are incompatible with any choice of s_i and s_{i+1}. In order to improve the filtering, some bits of s_i and s_{i+1} will be guessed and exclude more parameter choices.

For instance, if we guess one bit of s_i and s_{i+1}, we can compute one bit of L_1 and L_2 and two bits of $L_1 \oplus L_2$ in the forward (encryption) direction, and one bit of R_1 and R_2 and two bits of $R_1 \oplus R_2$ in the backwards (decryption) direction. These values are input to a hash table for every value of s_i and s_{i+1}, and we look for a match between the forward and the backward values, for some s_i and s_{i+1} values. There is a match with probability $\frac{3}{8}$, which means that we have a filtering of 1.41 bits. This shows that the parameters will be compatible with probability $\frac{1}{2} \cdot \frac{3}{4} = \frac{3}{8}$ which has been verified experimentally. It can be shown that we have a filtering of roughly two bits when guessing three bits of s_i and s_{i+1}. In this case, L_1 and L_2, and R_1 and R_2 have to be evaluated for eight values of s_i and s_{i+1}, which still costs less than one IDEA evaluation (evaluate four 4-bit values of L_1, L_2, R_1 and R_2 in parallel using 16-bit operations).

3.11.1.1 Biclique Attack on the Full 8.5-round IDEA

The approach towards IDEA is to construct a short biclique of high dimension and cover the remaining rounds with the independent-biclique technique. To find an optimal configuration of K^f and K^b bits, and also of the matching position, a short search program was written. First, it is necessary to figure out the longest biclique that efficiently covers 1.5 rounds. Then, compute the maximum chunk length and the minimum matching cost. Further, select for K^f the bits that form chunks after the first round, and for K^b the bits that form long chunks ending with the ciphertext.

The following key partition was chosen according to the search program. The result is a biclique of dimension 3:

- K^g (guess): bits $K_{0,\ldots,40,42,\ldots,47,50,\ldots,124}$ of the user key
- K^f (forwards): bits $K_{125,\ldots,127}$ of the user key
- K^b (backwards): bits $K_{41,48,49}$ of the user key

The key partition of the full 8.5-round IDEA into a biclique, chunks and the matching part are described in Table 3.86. See also Fig. 3.6. By the attack algorithm, each chunk is computed 2^3 times per key set, and the operations in the matching part are computed for each key. The Biryukov-Demirci relation (3.128) serves as internal variable for the matching in rounds 4-6:

$$\underbrace{\mathrm{lsb}_1(X_2^{(4)} \oplus X_3^{(4)} \oplus Z_2^{(4)} \oplus Z_3^{(4)} \oplus t_4)}_{\text{encryption direction}} \stackrel{?}{=} \underbrace{\mathrm{lsb}_1(X_2^{(6)} \oplus X_3^{(6)} \oplus t_5 \oplus Z_2^{(5)} \oplus Z_3^{(5)})}_{\text{decryption direction}}.$$

Table 3.86 Round partition for the attack on 8.5-round IDEA. It is split into 4 parts: biclique, two chunks (either K^b or K^f are not used) and matching (where both K^b and K^f are used) [97].

Round i	$Z_1^{(i)}$	$Z_2^{(i)}$	$Z_3^{(i)}$	$Z_4^{(i)}$	$Z_5^{(i)}$	$Z_6^{(i)}$
	$\odot$	$\boxplus$	$\boxplus$	$\odot$	$\odot$	$\odot$
			Biclique			
1	0–15	16–31	32–47	48–63	64–79	80–95
2	96–111	112–127↓	25–40	41–56↑		
			Chunk 1 (K^b not used			
2					57–72	73–88
3	89–104	105–120	121–8↓	9–24	50–65	66–81
4	82–97	98–113	114–1	2-17	18–33	
			Matching			
4						34–49↑
5	75–90	91–106	107–122	123–10	11–26	27–42
6	43–58	59–74	100–115	116–3	4–19	20–35
7	36–51	52–67	68–83	84–99	125–12↓	13–28
			Chunk 2 (K^f not used)			
8	29–44	45–60	61–76	77–92	93–108	109–124
9	22-37	38–53↑	54–69	70-85		

A straightforward way to construct a biclique with the given key partition would be: fix K^g and choose arbitrarily a plaintext P_0. For $K^b = 0$ and each value of K^f compute the internal states S_0, S_1, …, S_7 that are tuples of variables $(Y_1^{(2)}, Y_2^{(2)}, Y_3^{(2)}, Y_4^{(2)})$. Consider S_0 and for $K^f = 0$ and each value of K^b compute plaintexts P_0, P_1, …, P_7. Since differentials resulting from the key differences in K^b and K^f do not interleave, these plaintexts and states form a biclique:

$$P_i \overset{K^b=i, K^f=j}{\underset{f}{\longrightarrow}} S_j.$$

However, we do not control the plaintexts P_1, …, P_7. Since 2^{122} bicliques are constructed, we are likely to cover the full codebook. To reduce the data complexity, a more complicated biclique construction algorithm was implemented, which enforces particular plaintext bits to zero in every biclique.

The improved algorithm is the following:

- Fix K^g, $K^f = K^b = 00$.
- Choose arbitrary $(Y_3^{(2)}, Y_4^{(2)}, n_1)$.
  - for each K^b (eight options), compute the output of the MA function

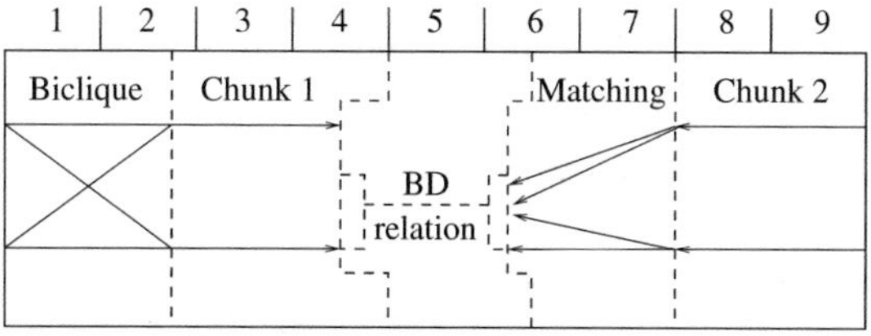

Fig. 3.6 Attack on the full 8.5-round IDEA. Given the biclique, partially encrypt the output states (Chunk 1), ask for ciphertexts, partially decrypt them (Chunk 2) and match with the help of the Biryukov-Demirci relation (it allows us to ignore about 1.5 rounds of computation) [97].

  - for each K^b (eight options), compute $X_2^{(1)}$, the second word of the plaintext.
  - check if the $\text{lsb}_5(X_2^{(1)})$ are zero. Other sets of bits would work as well. If not, choose other values for $(Y_3^{(2)}, Y_4^{(2)}, n_1)$.
  - note that this implies that the $\text{lsb}_5(t_1)$ and $\text{lsb}_5(s_1)$ are the same for all K^b. Therefore, the $\text{lsb}_5(X_3^{(1)})$ will also be the same for all K^b.

- choose a value w with the five least significant bits set to zero.
  - use w as $X_3^{(1)}$ with $K^b = 000$ and compute $X_1^{(2)}$ and $X_2^{(2)}$.
  - for each K^b (eight options), compute $X_1^{(1)}$ from $X_1^{(2)}$ and $X_2^{(2)}$.
  - check if the $\text{lsb}_1(X_1^{(1)})$ is zero. If not, choose another w. If all the w have been tried, choose another value for $(Y_3^{(2)}, Y_4^{(2)}, n_1)$.
- compute other plaintext words for each K^b. Derive plaintexts P_0, P_1, ..., P_7.
- vary K^f and derive internal states S_0, S_1, ..., S_7.

Therefore, a single biclique can be constructed in 2^{40} time, and 11 plaintext bits are set to zero: 15, 27–31, and 43–47. Notice that the key bits numbered 16–25, 32–40, 57–63, 96–124 are neutral for the biclique and can be flipped without violating its plaintext property. Therefore, the biclique can be reused for 2^{55} key sets and hence make the amortized cost negligible.

The first part of the construction requires that there exists a choice of $(Y_3^{(2)}, Y_4^{(2)}, n_1)$ so that the $\text{lsb}_5(X_2^{(1)}) = 0$. This is expected to be the case for a proportion $1 - e^{-8} > 99.9\%$ of the keys. For the full construction, we have a 58-bit condition, which will be satisfied by a proportion $1 - e^{-11} \approx 99.99\%$ of the keys.

We can also control some more bits of $X_1^{(1)}$, but this leads to a 61-bit condition, which is satisfied by 95% of the keys. Alternatively, we can look for a 6-bit match in $X_2^{(1)}$ and $X_3^{(1)}$, and a 1-bit match in $X_1^{(1)}$. The latter gives control over 13 bits of plaintext, but it only works for less than $1 - 1/e \approx 63\%$

of the keys. Various trade-offs can be achieved by changing the dimension of the biclique. For instance:

- 23 bits of a dimension-2 biclique and success rate $(1-e^{-4})(1-e^{-5}) > 97.5\%$
- 24 bits of a dimension-2 biclique and success rate $(1-e^{-4})(1-e^{-2}) > 84\%$
- 11 bits of a dimension-3 biclique and success rate $(1-e^{-8})(1-e^{-11}) > 99.9\%$
- 12 bits of a dimension-3 biclique and success rate $(1-e^{-8})(1-e^{-3}) > 94\%$
- 5 bits of a dimension-4 biclique and success rate $(1-e^{-16})(1-e^{-14}) > 99.9999\%$

Each biclique tests 2^6 keys. The first chunk employs 9 multiplications, the second chunk 6 multiplications, the matching part 13 multiplications (and hence 7 multiplications when using the Biryukov-Demirci relation). We recheck $Y_1^{(5)}$, for which we need 2 multiplications: for $Z_6^{(4)}$ and $Z_1^{(5)}$. A negligible proportion of keys is rechecked on the full state and on another plaintext/ciphertext pair. Therefore, 2^6 keys are tested with $9 \cdot 8 + 6 \cdot 8 + (5 + \frac{1}{2} + \frac{1}{2}) \cdot 64 + 2 \cdot 32 = 632$ multiplications $\approx 2^{4.06}$ IDEA computations. The total time complexity is $2^{126.06}$ IDEA computations.

We can expand K^b to four bits for the cost of increased data complexity. As we would be able to spend only four degrees of freedom per plaintext, the data complexity becomes 2^{59} chosen plaintexts and the number of multiplications for 2^7 keys is $9 \cdot 8 + 6 \cdot 16 + (5 + \frac{1}{2} + \frac{1}{2}) \cdot 128 + 2 \cdot 64 = 1064$ which yields a total complexity of about $2^{125.97}$ IDEA computations.

3.11.1.2 Biclique Attack on 7.5-round IDEA

To attack 7.5-round IDEA, a biclique is constructed in the first 1.5 rounds. The key partitioning is as follows:

- K^g (guess): bits $K_{0,\ldots,24,25,\ldots,40,42,\ldots,99,125,\ldots,127}$
- K^f (forwards): bits $K_{100,\ldots,124}$
- K^b (backwards): bits $K_{25,41}$

The differentials based on K^f and K^b do not interleave in the first 1.5 rounds. Therefore, we can construct a biclique in a similar way to the attack on full IDEA. However, the differential generated by K^b affects the full plaintext. To reduce the data complexity, we construct two bicliques for a key set so that the differential generated by K^b vanishes at the input to the MA half-round.

For the first biclique, we fix $K_{25} \oplus K_{41} = 0$ and for the second biclique, $K_{25} \oplus K_{41} = 1$. Consequently, a difference in K^b generates simultaneous differences in $Z_3^{(2)}$ and $Z_2^{(4)}$. Denote the difference in $Z_3^{(2)}$ by ∇ (generated by K_{25}) and in $Z_2^{(4)}$ by ∇' (generated by K_{41}). We want the difference in $X_2^{(4)}$ to be equal to ∇ so that the MA half-round has zero input difference.

See Fig. 3.7. Differential trail is depicted in red. We fulfill this condition by random trials. Bicliques are therefore, constructed as follows:

- fix $X_1^{(1)} = X_2^{(1)} = X_3^{(1)} = 0$.
- choose arbitrary values for $X_4^{(1)}$:
  - generate internal cipher states for the biclique.
  - check whether the MA half-round has zero input difference. If not, try another value for $X_4^{(1)}$.

A single pair of bicliques is generated in less than 2^{16} IDEA computations. This value is amortized since we can derive 2^{19} more bicliques by changing key bits 96–104 (and recomputing $Y_1^{(2)}$), 125–127 (recompute $Y_2^{(2)}$) and 57–63 of $Z_4^{(1)}$ (and recomputing the plaintext). Thus, an amortized cost to construct a biclique is negligible.

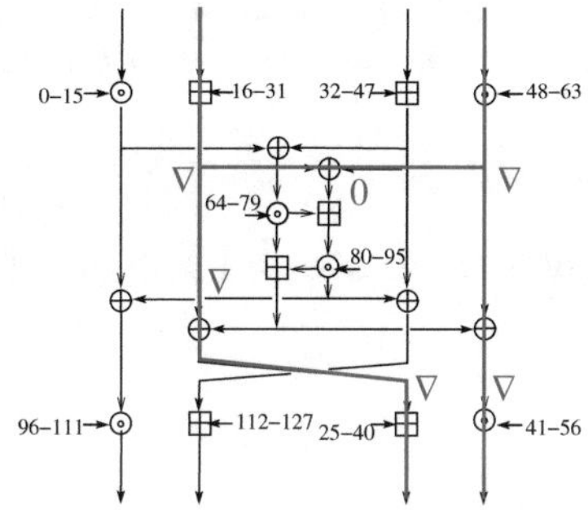

Fig. 3.7 Biclique ∇ differential in the attack on 7.5-round IDEA. Key bit indexes are shown instead of the subkeys [97].

Since the ∇ differential affects the most significant bit of $X_2^{(1)}$ only, the plaintexts generated in bicliques have 47 bits fixed to zero. Therefore, the data complexity does not exceed 2^{17}. The full computation complexity of the attack is as follows:

$$2^{107}(C_{\text{biclique}} + 2^{13}C_{\text{Chunk 1}} + 2C_{\text{Chunk 2}} + 2^{13}C_{\text{recheck}}),$$

where 2^{106} key sets are tested with two bicliques each. The amortized biclique construction cost is negligible. Note that the multiplication by $Z_5^{(2)}$ in round 2 is also amortized as the change in key bits 57–63 does not affect it. Therefore, the total number of multiplications in the first chunk is 1+4+2 = 7, in the second chunk: 1+3+4+2= = 10, and to recheck: 3 (to compute the full n_5 in both directions). The total complexity is therefore: $2^{127} \cdot \frac{10}{30} = 2^{126.5}$.

The time complexity can decrease at the cost of increased data complexity. Let us assign one more bit of K^b so that there are 8 values of K^b. We spend

64 bits of freedom in the internal cipher state to fix 13 bits of each biclique plaintext, as in the attack on the full IDEA. Then, the complexity is estimated as:

$$2^{100}(2^{25} \cdot \frac{8}{30} + 2^3 \cdot \frac{10}{30} + 2^{28}\frac{1}{2} \cdot \frac{2}{30}) = 2^{124.1}.$$

The complexity can be reduced further by using the improved BD filtering on two bits, which filters out $\frac{5}{8}$ of the candidates. First, we consider bits 112-113, 105–106, 121–122, 114–115 as part of K^g instead of K^f, so that all the keys involved in the BD relation are part of K^g. We also use some precomputations. For each K^b, compute t_5 and evaluate $R_1(s_5)$ and $R_2(s_5)$ for two guesses of s_5. Then, consider the potential candidates from the forward chunk. For each possible 1-bit value of $X_2^{(4)}$, $X_3^{(4)}$ and t_4, plus the second bit of $X_2^{(4)} \oplus X_3^{(4)} \oplus t_4$, guess one bit of s_4 in order to compute $L_1(s_4)$ and $L_2(s_4)$. Then, we can filter the corresponding candidates for K^b: we expect three candidates on average. For each K^f, just use this table to recover the candidates. For the complexity evaluation, we assume that finding a match in the hash table costs the same as one multiplication. This yields a complexity of

$$2^{108}(2^{17} \cdot \frac{8}{30} + 2^3 \cdot \frac{10 + 2 + 2^5}{30} + 2^{20} \cdot \frac{3}{8}\frac{2}{30}) = 2^{123.9}.$$

Other attacks on IDEA variants with less than 7.5 rounds are described in [97].

3.12 Boomerang Attacks

In [179], Wagner described the framework of a new cryptanalysis technique called the *boomerang attack*. It is a chosen-text attack that uses advanced differential-style techniques.

Let E_K denote the encryption operation of a target block cipher E under an unknown key K. Let us assume that E_K can be decomposed into two parts as $E_K = E_1 \circ E_0 = E_1(E_0)$, where each of E_1 and E_0 does not necessarily correspond to half of E_K. From the decryption direction, we have $D_K = E_K^{-1}$, and therefore, $D_K = E_0^{-1} \circ E_1^{-1}$.

In a conventional differential attack, plaintext pairs are carefully chosen with a particular difference, and the propagation of differential patterns are tracked across as many rounds of E_K as possible while keeping the overall probability as high as possible.

If these goals can be accomplished for, say, t rounds, then (i) if E_K has t rounds, and the corresponding data complexity is less than the full codebook, then the t-round cipher can be distinguished from a random permutation; (ii) otherwise, a t-round reduced cipher variant can be distinguished from a random permutation, and eventually, a key-recovery attack can be performed around the t-round distinguisher.

The boomerang technique, on the other hand, does not require the whole cipher E_K to be covered by a single high-probability differential. Therefore, instead of a comparatively *long* differential covering most (if not all of) E_K, a boomerang distinguisher exploits *short* high-probability differentials across both E_0 and E_1, and these differentials are not necessarily correlated to each other, but jointly the differentials can cover most (if not all of) E_K.

A boomerang attack requires the adversary to have the capability to perform both chosen-plaintext and chosen-ciphertext queries, some of which will be adaptive.

In general, there are two types of boomerangs, called *top-down* and *bottom-up*. See Fig. 3.8. In a top-down boomerang, the chosen-ciphertext queries are adaptive in the sense that the adversary first obtains ciphertexts which are the result of two chosen-plaintext queries to the encryption oracle. Afterwards, the adversary appropriately chooses another ciphertext pair, related to the given ones, and feeds the former to the decryption oracle, obtaining the corresponding plaintext pair.

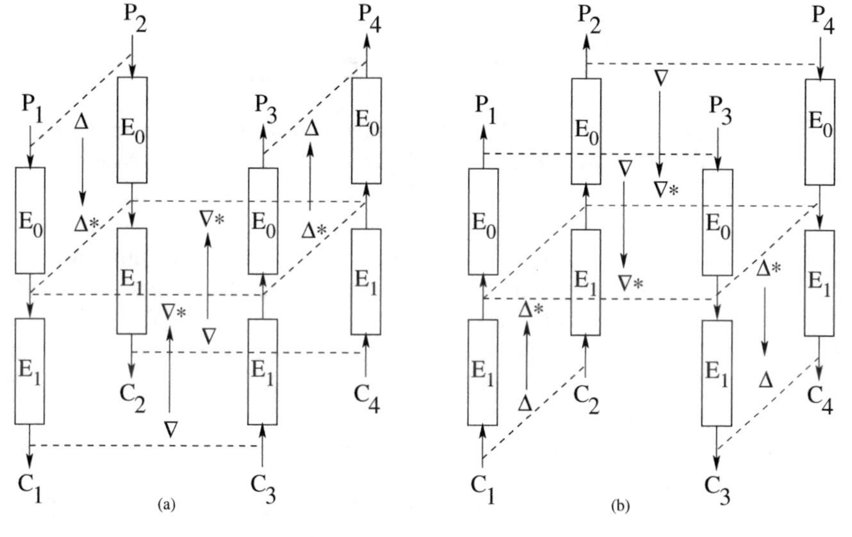

Fig. 3.8 (a) Top-down boomerang and (b) bottom-up boomerang distinguishers. Note the direction of the arrows which indicates the direction differences propagate.

A bottom-up boomerang starts with two chosen-ciphertext queries, and from the resulting plaintexts, a new plaintext pair is chosen. The latter plaintexts are fed to the encryption oracle, which provides the corresponding ciphertexts.

Let us describe in more detail the construction of a top-down boomerang distinguisher. The construction of a bottom-up boomerang distinguisher is analogous.

Let us assume that the difference operator is bitwise exclusive-or.

The attack starts by choosing a plaintext pair (P_1, P_2) such that their difference is a fixed value $\Delta = P_1 \oplus P_2$, and such that $\Delta \to \Delta^*$ holds across E_0 with a non-negligible probability denoted by p. Therefore, $\Delta^* = E_0(P_1) \oplus E_0(P_2)$.

Let $(C_1, C_2) = (E_K(P_1), E_K(P_2))$ and let $\nabla = C_1 \oplus C_3 = C_2 \oplus C_4$ be an adaptively-chosen ciphertext difference such that $\nabla \to \nabla^*$ holds across E_1^{-1} with a non-negligible probability denoted by q. Therefore,

$$\nabla^* = E_1^{-1}(C_1) \oplus E_1^{-1}(C_3) = E_1^{-1}(C_2) \oplus E_1^{-1}(C_4).$$

The corresponding plaintexts are $(P_3, P_4) = (D_K(C_3), D_K(C_4))$ as requested to the decryption oracle. If the difference patterns $\Delta \to \Delta^*$ and $\nabla \to \nabla^*$ occur as predicted in the boomerang (see Fig. 3.8(a)) then the difference in between E_0 and E_1 in the (C_3, C_4) pair will be

$$E_1^{-1}(C_3) \oplus E_1^{-1}(C_4) = \nabla^* \oplus E_1^{-1}(C_1) \oplus \nabla^* \oplus E_1^{-1}(C_2) = \Delta^*.$$

Further, if $\Delta^* \to \Delta$ holds across E_0^{-1}, then,

$$E_0^{-1}(E_1^{-1}(C_3)) \oplus E_0^{-1}(E_1^{-1}(C_4)) = D_K(C_3) \oplus D_K(C_4) = \Delta.$$

Paraphrasing Wagner in [179]: .. *this is why we call it the boomerang attack: when you send (the differences) properly, it always comes back to you.*

Thus, the boomerang is detected by checking whether $P_3 \oplus P_4 = \Delta$.

Note that the boomerang distinguisher uses two independent differentials: $\Delta \to \Delta^*$ and $\nabla \to \nabla^*$, each of which covers only part of E_K, but, together, they cover the whole E_K. These differentials are unrelated but a necessary condition is that $\Delta^* \neq \nabla^*$. Note that $\Delta^* = E_0(P_1) \oplus E_0(P_2)$ and $\nabla^* = E_1^{-1}(C_1) \oplus E_1^{-1}(C_3) = E_0(P_1) \oplus E_0(P_3)$. Thus, $E_0(P_1) = E_0(P_2) \oplus \Delta^*$ and $E_0(P_1) = E_0(P_3) \oplus \nabla^*$. Therefore, $E_0(P_2) \oplus \Delta^* = E_0(P_3) \oplus \nabla^*$. If $\Delta^* = \nabla^*$, then $E_0(P_2) = E_0(P_3)$, which means $P_2 = P_3$. Consequently, it is important to avoid colliding differences in between E_0 and E_1 in order to have a quartet of four distinct plaintexts.

A *right quartet* is defined as a 4-tuple (P_1, P_2, C_3, C_4) for which all four differential patterns $\Delta \to \Delta^*$, $\nabla \to \nabla*$ (twice) and $\Delta^* \to \Delta$ hold simultaneously. Note that (C_3, C_4) are chosen adaptively (in a top-down boomerang) depending on (C_1, C_2), while $(P_3, P_4) = (D_K(C_3), D_K(C_4))$.

A *quartet* for a top-down boomerang consists of two chosen plaintexts (P_1, P_2), and two adaptively-chosen ciphertexts (C_3, C_4). Analogously, a quartet for a bottom-up boomerang consists of two chosen ciphertexts (C_1, C_2) and two adaptively-chosen plaintexts (P_3, P_4).

The probability of observing the difference Δ in the plaintext pairs (P_3, P_4) is estimated as

$$\Pr(\Delta = P_3 \oplus P_4) = \Pr(\Delta \to \Delta^*)^2 \cdot \Pr(\nabla \to \nabla^*)^2 = p^2 \cdot q^2.$$

For a random permutation, the probability that $P_3 \oplus P_4 = \Delta$ is 2^{-n}, where n is the block size of the target cipher E. In order to distinguish E from a random permutation using a boomerang distinguisher, $(p \cdot q)^2 \ggg 2^{-n}$ must hold, that is,

$$p \cdot q \ggg 2^{-n/2}.$$

In a (top-down) boomerang distinguisher, the following property was assumed:

$$\Pr(\Delta \rightarrow \Delta^* \text{ across } E_0) = \Pr(\Delta^* \rightarrow \Delta \text{ across } E_0^{-1}), \tag{3.129}$$

which is true for differential characteristics but not necessarily true for truncated differentials.

It was observed in [19, 17] that the differences Δ^* and ∇^* in between E_0 and E_1 do not need to be fixed. Rather, all possible Δ^* and ∇^* can be counted on simultaneously as long as $\Delta^* \neq \nabla^*$ and the 2nd-order differential relation (collision) are satisfied [115, 100]. Therefore, a right quartet for E has probability $\hat{p}^2 \cdot \hat{q}^2$ where

$$\hat{p} = \sqrt{\sum_{\forall \Delta^*} \Pr^2(\Delta \rightarrow \Delta^*)}, \tag{3.130}$$

and

$$\hat{q} = \sqrt{\sum_{\forall \nabla^*} \Pr^2(\nabla \rightarrow \nabla^*)}, \tag{3.131}$$

where the sums run over all nontrivial differences Δ^* and ∇^*, respectively. Under (3.130) and (3.131), the condition for a boomerang distinguisher to effectively distinguish a target n-bit block cipher from a random permutation becomes

$$\hat{p} \cdot \hat{q} \ggg 2^{-n/2}.$$

In [94], Kelsey *et al.* described a variant of the boomerang technique called *amplified boomerang attack*. While the original boomerang attack used two unrelated 1st-order differentials $\Delta \rightarrow \Delta^*$ and $\nabla \rightarrow \nabla^*$, they are nonetheless connected by a 2nd-order differential relation in between E_0 and E_1. This relation stands for a collision:

$$\Delta^* \oplus \nabla^* \oplus \nabla^* \oplus \Delta^* = 0.$$

The idea of amplified boomerangs is to use larger amounts of chosen-plaintext pairs to obtain 2nd-order differential relations in between E_0 and E_1 using the birthday-paradox effect. This attack uses so called *text structures*. A text structure is a tool aimed at reducing the total number of chosen text blocks needed to form (text) pairs with a give difference. A (text) structure consists of a set of text blocks that provide several pairs for one or more given differences. For instance, a structure composed of four 4-word blocks in which the first word contains random nonzero values and the other words contain fixed constants, can generate up to $\binom{4}{2} = 6$ pairs with differences of the form $(\delta, 0, 0, 0)$, while a conventional approach (without structures) would require 12 blocks to get the same six pairs.

The amplified boomerang attack uses structures of structures in order to obtain 2nd-order differential relations for many text pairs at once.

The following paragraphs summarizing the amplified boomerang attack are based on [17].

Suppose we are dealing with a n-bit block cipher E that can be decomposed as $E_K = E_1 \circ E_0 = E_1(E_0)$ for a user key K. Assume there is a differential $\Delta \to \Delta^*$ holding with probability p across E_0 and another differential $\nabla \to \nabla^*$ holding with probability q across E_1, with $pq \ggg 2^{-n/2}$.

Assume we have N random plaintext pairs with difference Δ. Let (P_i, P_j, P_l, P_t) denote a quartet of plaintexts such that $P_i \oplus P_j = P_l \oplus P_t = \Delta$. Each pair has probability p to satisfy the differential $\Delta \to \Delta^*$ across E_0.

Let (X_i, X_j, X_l, X_t) denote the intermediate values after E_0. So, for instance, $X_i = E_0(P_i)$.

The relevant 2nd-order differential relation is when $X_i \oplus X_j = \Delta^*$, $X_l \oplus X_t = \Delta^*$ and $X_i \oplus X_l = \nabla^*$ because in this case $X_j \oplus X_t = (X_i \oplus \Delta^*) \oplus (X_l \oplus \Delta^*) = \nabla^*$. When the difference ∇^* propagates across E_1 we look for the ciphertexts (C_i, C_j, C_l, C_t) that satisfy

$$C_i \oplus C_l = C_j \oplus C_t = \nabla. \tag{3.132}$$

A quartet satisfying all of these differential requirements is called a right quartet for the amplified boomerang. Equation 3.132 indicates how an amplified boomerang distinguisher is detected.

Note that unlike the original boomerang distinguisher, the differences in an amplified boomerang propagate in only one direction, so the boomerang does not return to its point of origin.

If we have N plaintext pairs with difference Δ a fraction of p satisfies $\Delta \to \Delta^*$ across E_0. So, Np pairs will have difference Δ^* at the input to E_1, and about $(Np)^2/2$ quartets consisting of two such pars.

Assuming that the intermediate encrypted values are distributed uniformly over all possible values, then with probability 2^{-n}, we obtain X_i and X_l such that $X_i \oplus X_l = \nabla^*$. When (X_i, X_l) satisfy this difference, also (X_j, X_t) satisfy it. Likewise, (X_i, X_t) also satisfy $X_i \oplus X_t = \nabla^*$ with probability 2^{-n}. Therefore, given two pairs, (X_i, X_j) and (X_l, X_t), there are two ways to use them as a quartet, with probability 2^{1-n}. In summary, we have $\binom{Np}{2} \cdot 2^{1-n}$ quartets which satisfy the differential requirements. Each pair satisfies $\nabla \to \nabla^*$ with probability q. Starting with N pairs $\{(P_i, P_j), (P_l, P_t)\}$, the expected number of right quartets is

$$\binom{Np}{2} \cdot 2^{1-n} \cdot q^2 \cdot 2^{-n} = (Npq)^2 \cdot 2^{-n}.$$

The amplified boomerang distinguisher counts plaintext quartets (P_i, P_j, P_l, P_t) that satisfy $C_i \oplus C_l = C_j \oplus C_t = \nabla$. For a random permutation, the expected number of right quartets is $N^2 \cdot 2^{-2n}$, since there are N^2 possible quartets: there are $N^2/2$ pairs of pairs, and each pair of pairs can create two quartets, for instance, $\{(P_i, P_j), (P_l, P_t)\}$ and $\{(P_i, P_j), (P_t, P_l)\}$.

For each pair (P_i, P_t) or (P_j, P_l), the probability of obtaining a specific difference is 2^{-n}. If $(Npq)^2 \cdot 2^{-n} > N^2 \cdot 2^{-2n}$, that is, if $pq > 2^{-n/2}$, then the amplified boomerang counts more quartets than in the random case. When N is sufficiently large, the amplified boomerang can distinguish between E and a random permutation (based on the number of right pairs).

In relation to the original boomerang technique, amplified boomerangs use only chosen-plaintext queries (for top-down difference propagation) or only chosen-ciphertext queries (for bottom-up difference propagation). See Fig. 3.9.

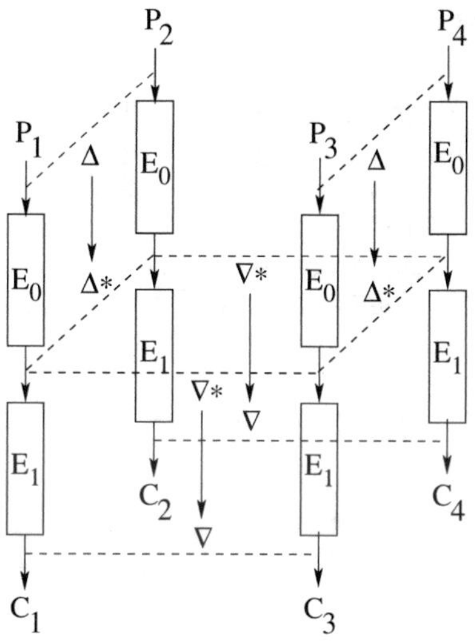

Fig. 3.9 Amplified boomerang distinguisher. The arrows indicate the direction in which differences propagate.

Note that the original boomerang distinguisher requires fewer queries than amplified boomerangs because in the latter enough right pairs are needed in order to have a high chance of obtaining an internal collision (a 2nd-order differential relation): $\Delta^* \oplus \nabla^* \oplus \nabla^* \oplus \Delta^* = 0$. This is the trade-off for switching from a CPACC to a CP (or from a CCACP to a CC) setting.

Note also that in the boomerang attack, the pair (P_3, P_4) that satisfy $P_3 \oplus P_4 = \Delta$ is known from a right quartet. But, for the amplified boomerang, differences propagate in one direction only, so the boomerang does not return.

Moreover, an efficient strategy is needed to pair texts in order to find (j,l) such that $E_0(P_j) \oplus E_0(P_l) = \nabla^*$.

In [17], Biham *et al.* described the *rectangle attack* which like the amplified boomerang technique only requires chosen-plaintext (or only chosen-ciphertext) queries, but not both. In a rectangle attack, all possible intermediate differences Δ^* and ∇^* are considered such that $\Delta \to \Delta^*$ and $\nabla^* \to \nabla$. This new approach improves over the amplified boomerang technique since the probability of obtaining a right quartet in a rectangle attacks becomes $2^{-n/2} \cdot \hat{p} \cdot \hat{q}$, where $\hat{p}$ is as defined in (3.130) and $\hat{q} = \sqrt{\sum_{\forall\nabla^*} \Pr^2(\nabla^* \to \nabla)}$ (note the direction of difference propagation is different from that of a boomerang).

In [125, 124], Lu described a combination of the impossible differential and the boomerang techniques called the Impossible-Boomerang Attack (IBA), and applied it to reduced-round versions of the AES cipher. IBA requires both chosen-plaintext and chosen-ciphertext queries, like in conventional boomerang attacks. Moreover, the differentials used to cover the E_0 and E_1 parts of the target cipher E have to hold with certainty (probability one) in order to satisfy the miss-in-the-middle criterion for the impossible-differential part of the distinguisher.

In [125], a target cipher E_K is decomposed as $E_K = E_1 \circ E_0 = E_1(E_0)$ as in the boomerang attack. Moreover, it is assumed that there exist differentials $\Delta_1 \to \Delta_1^*$ and $\Delta_2 \to \Delta_2^*$ propagating across E_0, as well as differentials $\nabla_1 \to \nabla_1^*$ and $\nabla_2 \to \nabla_2^*$ holding across E_1^{-1}. See Fig. 3.10. The miss-in-the-middle criterion for all of these differentials is

$$\Delta_1^* \oplus \Delta_2^* \oplus \nabla_1^* \oplus \nabla_2^* \neq 0. \tag{3.133}$$

Let (P_1, P_2) be a plaintext pair satisfying $\Delta_1 = P_1 \oplus P_2$. Let (P_3, P_4) be a plaintext pair satisfying $\Delta_2 = P_3 \oplus P_4$. Let (C_1, C_3) denote the ciphertext pair satisfying $C_1 \oplus C_3 = \nabla_1$. Let (C_2, C_4) denote the ciphertext pair satisfying $C_2 \oplus C_4 = \nabla_2$. Then, for an impossible-boomerang distinguisher, the following equations cannot be satisfied simultaneously:

$$E_K(P_1) \oplus E_K(P_3) = \nabla_1,$$

$$E_K(P_1 \oplus \Delta_1) \oplus EK(P_3 \oplus \Delta_2) = E_K(P_2) \oplus E_K(P_4) = \nabla_2.$$

This impossible-boomerang distinguisher can be denoted $(\Delta_1, \Delta_2) \not\to (\nabla_1, \nabla_2)$. See Fig. 3.10.

Like in the case of the rectangle distinguisher, it is possible to generalize (3.133). Consider all possible differences across E_0 starting with Δ_1 and Δ_2. Then, $\Delta_1 \to \{\delta_1, \delta_2, \ldots, \delta_m\}$, $\Delta_2 \to \{\delta_1^*, \delta_2^*, \ldots, \delta_{m'}^*\}$, where $\Pr(\Delta_1 \to \delta_i) > 0$ for $1 \leq i \leq m$, $\Pr(\Delta_2 \to \delta_j^*) > 0$ for $1 \leq j \leq m'$, and $\sum_{\forall i} \Pr(\Delta_1 \to \delta_i) = 1$, $\sum_{\forall j} \Pr(\Delta_2 \to \delta_j^*) = 1$.

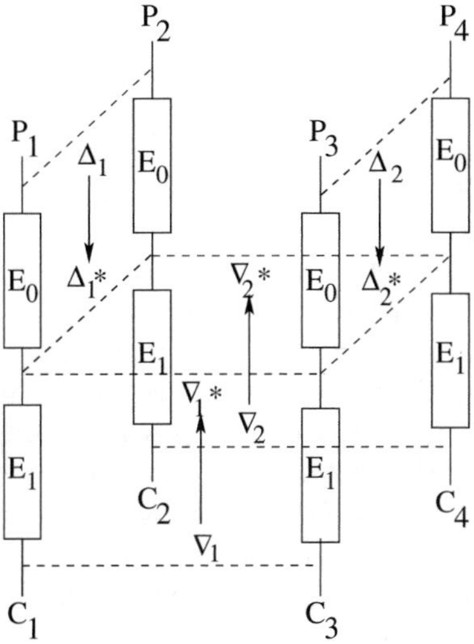

Fig. 3.10 Pictorial representation of an impossible-boomerang distinguisher. The arrows indicate the direction in which differences propagate.

Likewise, consider all possible differences across E_1^{-1} starting from ∇_1 and ∇_2. Then, $\nabla_1 \to \{\eta_1, \eta_2, \ldots, \eta_t\}$, $\nabla_2 \to \{\eta_1^*, \eta_2^*, \ldots, \eta_{t'}^*\}$, where $\Pr(\nabla_1 \to \eta_i) > 0$, $\Pr(\nabla_2 \to \eta_j^*) > 0$ for $1 \leq i \leq t$, $1 \leq j \leq t'$, and $\sum_{\forall i} \Pr(\nabla_1 \to \eta_i) = 1$, $\sum_{\forall j} \Pr(\nabla_2 \to \eta_j^*) = 1$.

This way, it is not necessary to have a single differential that accounts for all differences across E_0 and E_1^{-1}. On the other hand, the miss-in-the-middle criterion becomes

$$\delta_i \oplus \delta_j^* \oplus \eta_l \oplus \eta_v^* \neq 0, \qquad (3.134)$$

which must hold for all $1 \leq i \leq m$, $1 \leq j \leq m'$, $1 \leq l \leq t$ and $1 \leq v \leq t'$.

3.12.1 Boomerang Attacks on IDEA

This section is based[12] on [34, 143].

Definition 3.14. (Boomerang Distinguisher)
A sequence of differential patterns $\Delta \to \Delta^*$, $\nabla \to \nabla^*$ and $\Delta^* \to \Delta$, that propagate according to the differential construction of a boomerang, across a given number of rounds of a target cipher, with non-negligible probability, forms a *boomerang distinguisher*.

[12] © Katholieke Universiteit Leuven (KUL). Published with permission.

In principle, boomerang distinguishers can be used to test if the unknown user key belongs a particular key class. This membership test is successful if the boomerangs can propagate across the given target cipher instance. In the case of IDEA, the key is said to belong to a weak-key class.

Key-recovery attacks can be used on top of the distinguisher to recover subkeys in rounds before or after the distinguisher.

Moreover, boomerang can be sent either top-down or bottom-up, depending on the location of the guessed subkey bits.

Some choices of the differences Δ and ∇ will leave some half-rounds (in IDEA) not covered by either differential pattern. These MA and KM half-rounds are called *gaps* and can be crossed with truncated differentials. This technique was used by Wagner in [179] to attack a 16-round variant of the Khufu block cipher. Fig. 3.11 shows a pictorial representation of a *gap* in a top-down boomerang. In this case, $E_K = E_2 \circ E_1 \circ E_0$, where E_1 stands for a few half-rounds in between E_0 and E_2, and which are not covered by either $\Delta \to \Delta^*$ or $\nabla \to \nabla^*$.

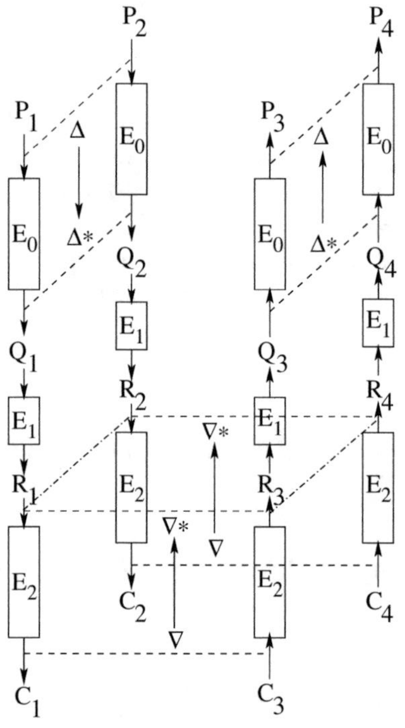

Fig. 3.11 Pictorial representation of a *gap* (E_1) in a top-down boomerang. E_1 represents a few half-rounds.

Truncated differentials are used to cause some differential patterns $\Delta \to \Delta^*$ (or $\nabla \to \nabla^*$) to propagate across two parallel facets of a boomerang. In these cases, the differences propagate across one or more half-rounds with certainty (probability one) independent of the particular value of the unknown difference.

The attacks to be described further are based on [34, 143] and identify new weak-key classes of IDEA, larger than the ones discovered by Daemen *et al.* in [56] and not covered by the key classes discovered by Hawkes in [81] using differential-linear membership tests.

The benefits of using boomerang distinguishers in IDEA are two-fold: (i) they pose different constraints on the key schedule than differential and differential-linear distinguishers; (ii) the freedom to choose unrelated differences Δ and ∇ to cover E_0 and E_1 helps minimize the key-bit constraints. Thus, new weak-key classes are likely to be found.

The difference operator is exclusive-or and text differences consist of 4-tuples of 16-bit words.

Nonzero wordwise differences have the form $\nu = 8000_x$ because this difference propagates across keyed-$\boxplus$ and $\oplus$ with certainty (for the former, whatever the subkey value). Concerning modular multiplication, we assume weak subkeys in order to allow propagation of ν across $\odot$. The weak-subkey condition means that the fifteen most significant bit of the subkey have to be zero.

There are too many possibilities for the differences Δ, Δ^*, ∇ and ∇^*, apart from the variable number of rounds for each differential in IDEA, both from the top-down and from the bottom-up directions. Therefore, a computer program was created to search exhaustively for boomerangs in IDEA taking into account several trade-offs: (i) cover as many rounds of IDEA as possible; (ii) minimize the number of weak-key conditions in order to maximize the weak-key class size.

One additional constraint in this search was the size of the gaps, that is, consecutive half-rounds not covered by any of the difference patterns of the boomerang. Moreover, the subkeys in the gaps are not necessarily restricted. The objective is to increase the weak-key classes (without imposing further restrictions on the key) at the cost of higher data and time complexities for the membership tests.

Another contribution of the boomerang attack framework consists in not covering either the topmost or the bottommost KM half-round, assuming the adversary can guess the required multiplicative subkeys or use special structures to construct appropriate differences to cross the KM half-rounds.

The first boomerang to be described is labeled (11) in Table 3.87. This boomerang consists of a differential pattern $\Delta \to \Delta^*$ covering the first 6 rounds of IDEA, and $\nabla \to \nabla^*$ covering the last 2.5 rounds. There are no gaps. This boomerang covers the full 8.5-round IDEA and can detect members of a weak-key class of size $2^{128-82} = 2^{46}$ using a single quartet (two chosen plaintexts and two chosen ciphertexts) and four IDEA computations. This attack works either with a top-down ($\uparrow$) or a bottom-up ($\downarrow$) boomerang.

Another boomerang, designated as (12) in Table 3.87, consists of the differentials $\Delta \rightarrow \Delta^*$ covering the first six rounds of IDEA, and $\nabla \rightarrow \nabla^*$ covering the last two rounds. There is a gap consisting of the 7th KM half-round that is not covered by either differential pattern. See Fig. 3.12. However, $Z_1^{(7)} = 0$ and the $\text{msb}_{15}(Z_4^{(7)})$ are not restricted. It means that for a boomerang to cross this gap only a collision in the fourth word difference at ∇^+ and ∇^{++} is needed. That is, if the fourth words of ∇^+ and ∇^{++} are the same then the 2nd-order relation $\Delta^* \oplus \nabla^+ \oplus \nabla^{++} = \Delta^*$ will be satisfied and the last leg of the boomerang will be complete. Experiments with 1024 keys demonstrated that 2048 quartets are enough for boomerangs to cross this gap for more than 99% of the keys. The boomerang direction can be either top-down ($\uparrow$) or bottom-up ($\downarrow$). The attack complexity is $4 \cdot 2^{12} = 2^{14}$ chosen texts and 2^{14} IDEA computations.

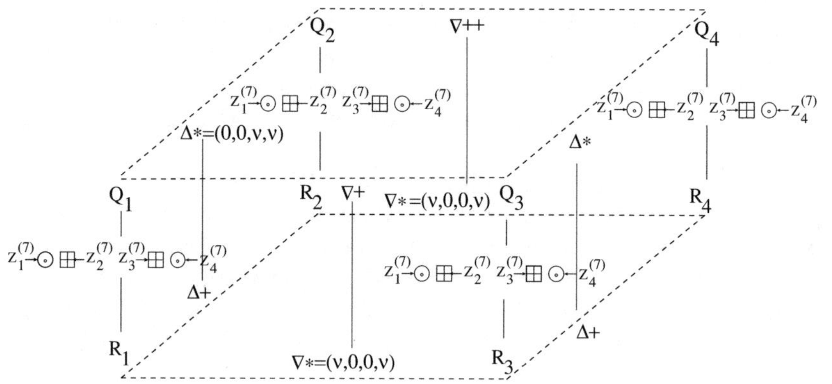

Fig. 3.12 Gap for the boomerang distinguisher labeled (12) in Table 3.87 consisting of the 7th KM half-round in IDEA.

Another boomerang distinguisher, designated as (13) in Table 3.87 consists of the differentials $\Delta \rightarrow \Delta^*$ covering the first 2.5 rounds and $\nabla \rightarrow \nabla^*$ covering the last 5.5 rounds. There is a gap consisting of the third MA half-round in which $\text{lsb}_2(Z_5^{(3)}) = 0$ and $Z_6^{(3)} = 0$. See Fig. 3.13.

The careful choice of Δ^* and ∇^* allows the gap to be crossed with a single quartet. The explanation is as follows: let the inputs/outputs of the gap for each text in a quartet be denoted $(Q_1, R_1) = ((a, b, c, d), (A, B, C, D))$, $(Q_2, R_2) = ((a \oplus 8000_x, b \oplus 8000_x, c, d \oplus 8000_x), (A', B', C', D')$, $(Q_3, R_3) = ((X, Y, U, W))$, $(A, B \oplus 8000_x, C, D \oplus 8000_x)$ because $R_3 = R_1 \oplus \nabla^*$, and $(Q_4, R_4) = ((X', Y', U', W')$, $(A', B' \oplus 8000_x, C', D' \oplus 8000_x))$ because $R_4 = R_2 \oplus \nabla^*$.

For (Q_1, R_1) let the inputs/outputs of the MA-box be denoted $((a \oplus c, b \oplus d), (o_1, o_2)$. For (Q_2, R_2) similarly, $((a \oplus c \oplus 8000_x, b \oplus d)$, $(u_1, u_2))$. For (Q_3, R_3), similarly, $((a \oplus c \oplus 8000_x, b \oplus d \oplus 8000_x), (o_1', o_2'))$, because of the output values R_1. For (Q_4, R_4), similarly, $((a \oplus c, b \oplus d \oplus 8000_x), (u'_1, u'_2))$ because of the output values R_4.

For (Q_3, R_3):

$$X \oplus U = a \oplus c \oplus 8000_x = A \oplus B \oplus 8000_x, \tag{3.135}$$

$$Y \oplus W = b \oplus d \oplus 8000_x = C \oplus D \oplus 8000_x. \tag{3.136}$$

From (Q_4, R_4):

$$X' \oplus U' = a \oplus c = A' \oplus B' \oplus 8000_x, \tag{3.137}$$

$$Y' \oplus W' = b \oplus d \oplus 8000_x = C' \oplus D' \oplus 8000_x. \tag{3.138}$$

From (3.135) and (3.137):

$$X \oplus X' = U \oplus U' \oplus 8000_x. \tag{3.139}$$

From (3.136) and (3.138):

$$Y \oplus Y' = W \oplus W'. \tag{3.140}$$

Notice that $Q_1 \oplus Q_2 = \Delta^*$, and that $Z_6^{(3)} = 0$. These facts imply that

$$o_1 = ((a \oplus c) \odot Z_5^{(3)} \boxplus (b \oplus d)) \odot Z_6^{(3)} \boxplus (a \oplus c) \odot Z_5^{(3)} = -(b \oplus d),$$

and

$$u_1 = ((a \oplus c \oplus 8000_x) \odot Z_5^{(3)} \boxplus (b \oplus d)) \odot Z_6^{(3)} \boxplus (a \oplus c \oplus 8000_x) \odot Z_5^{(3)} = -(b \oplus d),$$

that is, $o_1 \oplus u_1 = 0$. Similarly, $o'_1 = u'_1$. Since $o_1 = b \oplus C$ and $u_1 = b \oplus C' \oplus 8000_x$, it follows that $C \oplus C' = 8000_x$. Also, since $o'_1 = Y \oplus C$ and $u'_1 = Y' \oplus C'$, it follows that

$$Y \oplus Y' = C \oplus C' = 8000_x. \tag{3.141}$$

Moreover, from $R_1 \oplus R_3 = \nabla^*$, and the structure of the MA half-round, $o'_2 = ((a \oplus c \oplus 8000_x) \odot Z_5^{(3)} \boxplus (b \oplus d \oplus 8000_x)) \odot Z_6^{(3)} = -(a \oplus c \oplus 8000_x) \odot Z_5^{(3)} - (b \oplus d \oplus 8000_x)$, $u'_2 = ((a \oplus c) \odot Z_5^{(3)} \boxplus (b \oplus d \oplus 8000_x)) \odot Z_6^{(3)} = -(a \oplus c) \odot Z_5^{(3)} - (b \oplus d \oplus 8000_x)$, $o_2 = ((a \oplus c) \odot Z_5^{(3)} \boxplus (b \oplus d)) \odot Z_6^{(3)} = -(a \oplus c) \odot Z_5^{(3)} - (b \oplus d)$, and $u_2 = ((a \oplus c \oplus 8000_x) \odot Z_5^{(3)} \boxplus (b \oplus d)) \odot Z_6^{(3)} = -(a \oplus c \oplus 8000_x) \odot Z_5^{(3)} - (b \oplus d)$. Therefore, $u_2 \oplus o_2 = u'_2 \oplus o'_2$, which implies that

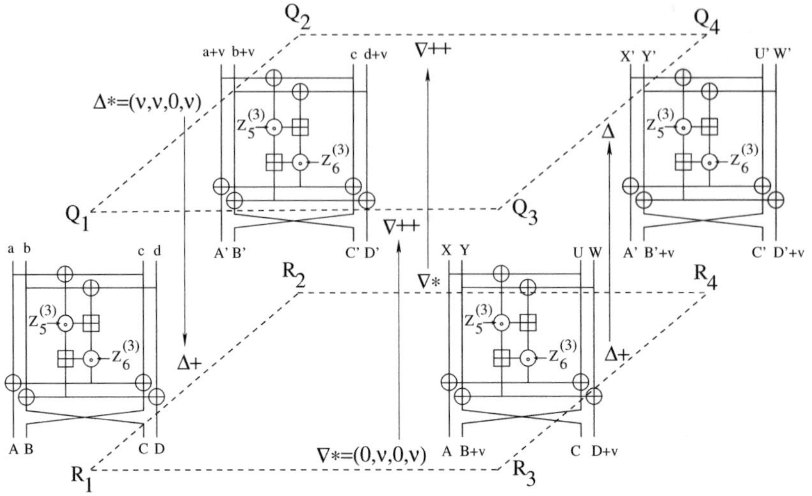

Fig. 3.13 Gap for the boomerang distinguisher labeled (13) in Table 3.87 consisting of the third MA half-round in IDEA.

$$(a \oplus 8000_x \oplus A') \oplus (a \oplus A) = (X \oplus A) \oplus (X' \oplus A') \Rightarrow X \oplus X' = 8000_X. \tag{3.142}$$

From (3.139), (3.140), (3.141) and (3.142) it follows that

$$\Delta = (X \oplus X', Y \oplus Y', U \oplus U', W \oplus W') = (8000_x, 8000_x, 0000_x, 8000_x) = \Delta^*.$$

This means that the boomerang crosses the gap with one quartet only, independent of the value of $Z_5^{(3)}$. The attack complexity corresponds to one quartet (two chosen plaintexts and two chosen ciphertexts) and four IDEA computations.

Another boomerang distinguisher, from the bottom-up direction, designated as (14) in Table 3.87 consists of the differentials $\Delta \rightarrow \Delta^*$ covering the last six rounds of IDEA, and $\nabla \rightarrow \nabla^*$ covering the first 1.5-rounds. There is a gap consisting of the second MA half-round in which $\text{lsb}_5(Z_6^{(2)}) = 0$. This gap can be crossed with a single quartet due to the careful choice of Δ^* and ∇^*. The explanation is as follows: let the inputs/outputs of the gap for each text in a quartet (see Fig. 3.14) be denoted $(Q_1, R_1) = ((a', b', c', d'), (A, B, C, D))$, $(Q_2, R_2) = ((a, b, c, d,), (A, B \oplus 8000_x, C, D \oplus 8000_x))$, $(Q_3, R_3) = ((a' \oplus 8000_x, b' \oplus 8000_x, c', d'), (X, Y, U, W))$ because $Q_3 = Q_1 \oplus \nabla^*$, and $(Q_4, R_4) = ((a \oplus 8000_x, b \oplus 8000_x, c, d), (X', Y', U', W'))$ because $Q_4 = Q_2 \oplus \nabla^*$. For (Q_1, R_1) and (Q_4, R_4), the input/outputs of the MA-box are $(A \oplus B, C \oplus D)$ because $(a \oplus 8000_x) \oplus c = A \oplus B$ and $(b \oplus 8000_x) \oplus d = C \oplus D$ from (Q_2, R_2). For (Q_2, R_2) and (Q_3, R_3), the inputs/outputs of the MA-box are

$(A \oplus B \oplus 8000_x, C \oplus D \oplus 8000_x)$ because $(a' \oplus 8000_x) \oplus c' = A \oplus B \oplus 8000_x$ and $(b' \oplus 8000_x) \oplus d' = C \oplus D \oplus 8000_x$ from (Q_1, R_1).

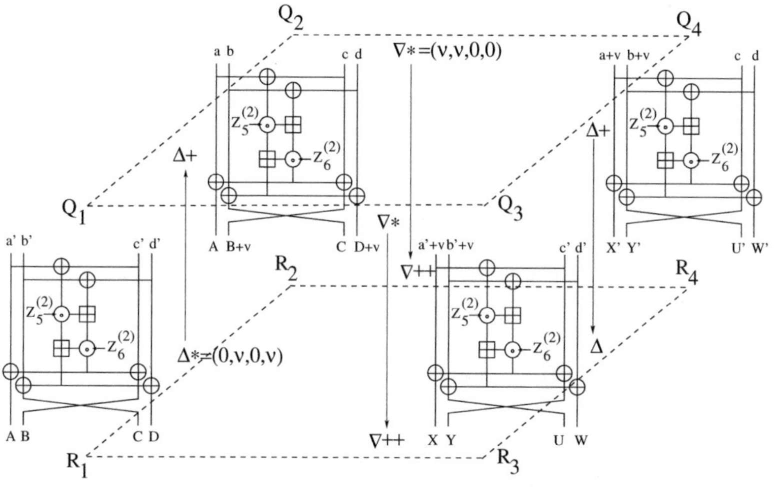

Fig. 3.14 Gap for the boomerang distinguisher labeled (14) in Table 3.87 consisting of the second MA half-round in IDEA.

The following relationships can be derived from (Q_3, R_3):

$$X \oplus Y = A \oplus B \oplus 8000_X, \tag{3.143}$$

$$U \oplus W = C \oplus D \oplus 8000_X, \tag{3.144}$$

and from (Q_4, R_4):

$$X' \oplus Y' = A \oplus B, \tag{3.145}$$

$$U' \oplus W' = C \oplus D. \tag{3.146}$$

From (3.143) and (3.145), it follows that

$$X \oplus X' = Y \oplus Y' \oplus 8000_x. \tag{3.147}$$

From (3.144) and (3.146), it follows that

$$U \oplus U' = W \oplus W' \oplus 8000_x. \tag{3.148}$$

Notice that $Z_5^{(2)}$ and $Z_6^{(2)}$ are fixed and that the MA-box inputs to Q_2 and Q_3 are equal, as well as for Q_1 and Q_4. Therefore, the corresponding outputs, denoted (u_1, u_2), for Q_2 and Q_3 are the same, and also the outputs, denoted (o_1, o_2) for Q_1 and Q_4 are equal. It follows that

$$u_1 = b \oplus C = b' \oplus 8000_x \oplus U, \tag{3.149}$$

$$u_2 = c \oplus B \oplus 8000_x = c' \oplus Y, \tag{3.150}$$

$$o_1 = b' \oplus C = b \oplus 8000_x \oplus U', \tag{3.151}$$

$$o_2 = c' \oplus B = c \oplus Y'. \tag{3.152}$$

From (3.149) and (3.151), it follows that

$$U \oplus U' = 0. \tag{3.153}$$

From (3.150) and (3.152), it follows that

$$Y \oplus Y' = 8000_X. \tag{3.154}$$

From (3.147), (3.148), (3.153) and (3.154), it follows that

$$\Delta = (X \oplus X', Y \oplus Y', U \oplus U', W \oplus W') = (0000_x, 8000_x, 0000_x, 8000_x) = \Delta^*,$$

that is, the gap can be crossed with a single quartet, independent of the values of $Z_5^{(2)}$ and $Z_6^{(2)}$.

The last KM half-round in this boomerang is not covered by either differential pattern, which means the boomerang direction is bottom-up. Moreover, $\text{lsb}_2(Z_4^{(9)}) = 0$.

The attack proceeds as follows: prepare two sets of ciphertexts in which the first and second words are equal, the third words differ by 8000_x and the fourth words have 2^9 random values each. This text structure contains 2^{18} text pairs with difference of the form $(0000_x, 0000_x, 8000_x, \delta)$. After decryption by the last KM half-round, each possible 16-bit difference is suggested by δ and $Z_4^{(9)}$ about $2^{18-16} = 4$ times, on average, and that is enough to cross the gap. The boomerang can be recognized by checking the 48-bit difference $(0000_x, 0000_x, 8000_x)$ at the first three words of $C_3 \oplus C_4$. The attack complexity is $2 \cdot (2^9 + 2^9) = 2^{11}$ chosen texts and 2^{11} IDEA computations.

Comparatively, a key-recovery attack for the $\text{msb}_{14}(Z_4^{9)})$ would require 2^{14} guesses and one text quartet per guess to cross the gap, resulting in $2^{14} \cdot 4 = 2^{16}$ chosen texts and 2^{16} IDEA computations.

Another boomerang distinguisher, designated as (15) in Table 3.87, consists of the differentials $\Delta \to \Delta^*$ covering the first three rounds of IDEA, and

$\nabla \rightarrow \nabla^*$ covering the last 4.5 rounds. There is a gap consisting of the 4th MA half-round in which $\text{msb}_7(Z_5^{(4)}) = 0$. See Fig. 3.15. The careful choice of Δ^* and ∇^* allows this gap to be crossed with a single quartet. The explanation is the following: let the inputs/outputs of the gap for each text in a quartet be denoted $(Q_1, R_1) = ((a, b, c, d), (A, B, C, D))$, $(Q_2, R_2) = ((a, b, c \oplus 8000_x, d \oplus 8000_x), (A', B', C', D'))$, $(Q_3, R_3) = ((X, Y, U, W), (A, B \oplus 8000_x, C.D \oplus 8000_x))$ because $R_3 = R_1 \oplus \nabla^*$, and $(Q_4, R_4) = ((X', Y', U', W'), (A', B' \oplus 8000_x, C', D' \oplus 8000_x))$ because $R_4 = R_2 \oplus \nabla^*$. For (Q_1, R_1), the inputs/outputs of the MA-box are $(a \oplus c, b \oplus d)$. For (Q_2, R_2), the inputs/outputs of the MA box are $(a \oplus c \oplus 8000_x, b \oplus d \oplus 8000_x)$. For (Q_3, R_3), similarly, $(a \oplus c \oplus 8000_x, b \oplus d \oplus 8000_x)$ because $A \oplus B = a \oplus c$ and $C \oplus D = b \oplus d$, from (Q_1, R_1). For (Q_4, R_4), similarly, $(a \oplus c, b \oplus d)$ because $A' \oplus B' = a \oplus c \oplus 8000_x$ and $C' \oplus D' = b \oplus d \oplus 8000_x$, from (Q_2, R_2). The following relations can be derived from (Q_3, R_3):

$$X \oplus U = a \oplus c \oplus 8000_X, \tag{3.155}$$

$$Y \oplus W = b \oplus d \oplus 8000_X, \tag{3.156}$$

and from (Q_4, R_4):

$$X' \oplus U' = a \oplus c, \tag{3.157}$$

$$Y' \oplus W' = b \oplus d. \tag{3.158}$$

From (3.155) and (3.157), it follows that

$$X \oplus X' = U \oplus U' \oplus 8000_x. \tag{3.159}$$

From (3.156) and (3.158), it follows that

$$Y \oplus Y' = W \oplus W' \oplus 8000_x. \tag{3.160}$$

Note that $Z_5^{(4)}$ and $Z_6^{(4)}$ are fixed, and the MA-box inputs for Q_1 and Q_4 are equal, as well as the MA-box inputs for Q_2 and Q_3. Therefore, the corresponding MA-box output for Q_1 and Q_4, denoted (o_1, o_2) are the same and also the MA-box outputs for Q_2 and Q_3, denoted (u_1, u_2). It follows that

$$o_1 = b \oplus C' = Y \oplus C, \tag{3.161}$$

$$o_2 = c \oplus 8000_x \oplus B' = U \oplus B \oplus 8000_x, \tag{3.162}$$

$$u_1 = Y' \oplus C' = b \oplus C, \tag{3.163}$$

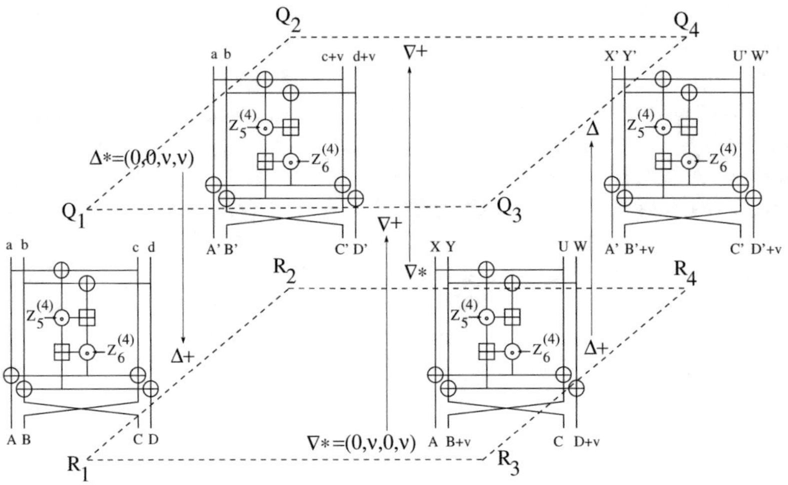

Fig. 3.15 Gap for the boomerang distinguisher labeled (15) in Table 3.87 consisting of the 4th MA half-round in IDEA.

$$u_2 = U' \oplus B' \oplus 8000_x = c \oplus B. \tag{3.164}$$

From (3.161) and (3.163), it follows that

$$Y \oplus Y' = 0000_x. \tag{3.165}$$

From (3.162) and (3.164), it follows that

$$U \oplus U' = 8000_x. \tag{3.166}$$

From (3.159), (3.160), (3.165) and (3.166), it follows that

$$\Delta = (X \oplus X', Y \oplus Y', U \oplus U', W \oplus W') = (0000_x, 0000_x, 8000_x, 8000_x) = \Delta^*,$$

that is, the gap can be crossed with a single quartet, independent of the values of $Z_5^{(4)}$ and $Z_6^{(4)}$.

The first KM half-round is not covered by either differential pattern, which implies that the boomerang direction is top-down. Moreover, $\mathrm{lsb}_7(Z_4^{(1)}) = 0$. The attack proceeds as follows: prepare two sets of plaintexts in which the first and second words are equal, the third words differ by 8000_x and the fourth words take 2^9 random values each. This text structure contains 2^{18} pairs with difference $(0000_x, 0000_x, 8000_x, \delta)$. After encryption by the first KM half-round, each possible 16-bit difference is suggested by δ and $Z_4^{(1)}$ about $2^{18-16} = 4$ times, on average, which is enough to cross the gap. The attack complexity is $2 \cdot (2^9 + 2^9) = 2^{11}$ chosen texts and 2^{11} IDEA computations.

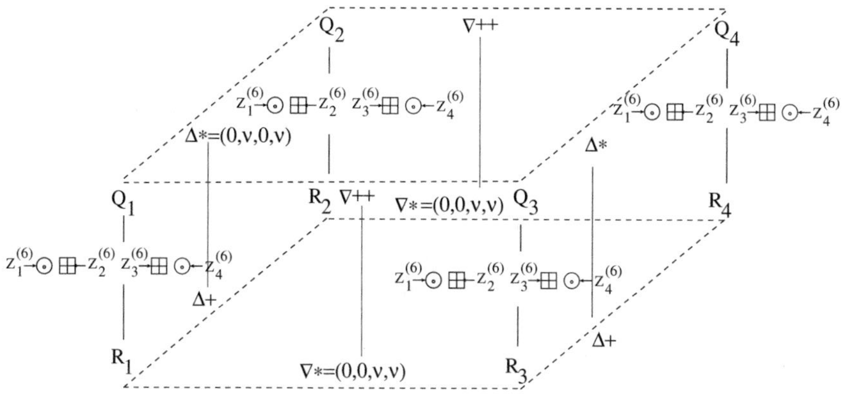

Fig. 3.16 Gap for the boomerang distinguisher labeled (16) in Table 3.87 consisting of the 6th KM half-round in IDEA.

Another boomerang distinguisher is designated as (16) in Table 3.87 and consists of the differentials $\Delta \rightarrow \Delta^*$ covering 4.5 rounds, and $\nabla \rightarrow \nabla^*$ covering the last 3 rounds. There is a gap consisting of the sixth KM half-round in which $\text{lsb}_7(Z_4^{(6)}) = 0$. See Fig. 3.16. The approach to cross this gap is to use enough quartets in order to cause a collision in the fourth difference word in ∇^+ and ∇^{++}, so that the quartet structure will cause $\Delta = \Delta^*$, similar to the boomerang labeled (12). Experiments with 1024 randomly chosen keys from the weak-key class demonstrated that 2^9 quartets are enough for boomerangs to cross the gap for more than 98% of the keys. The first KM half-round is not covered by either differential pattern, which implies that the boomerang direction is top-down. Moreover, $\text{lsb}_7(Z_4^{(1)}) = 0$.

The attack proceeds as follows: prepare two sets of plaintexts in which the first and second words are equal, the third words differ by 8000_x and the fourth words contain 2^{13} random values each. This text structure contains 2^{26} pairs with difference $(0000_x, 0000_x, 8000_x, \delta)$. After encryption by the first KM half-round, each possible 16-bit difference is suggested by δ and $Z_4^{(1)}$ about $2^{26-16} = 2^{10}$ times, on average, which is enough to cross the gap. The attack complexity is $2 \cdot (2^{13} + 2^{13}) = 2^{15}$ chosen texts and 2^{15} IDEA computations.

Comparatively, a key-recovery attack for the $\text{msb}_9(Z_4^{(1)})$ requires 2^9 guesses and 2^9 quartets to cross the gap per guess, resulting in $2^{9+9} \cdot 4 = 2^{20}$ chosen texts and 2^{20} IDEA computations.

Another boomerang distinguisher, designated as (17) in Table 3.87, consists of the differentials $\Delta \rightarrow \Delta^*$ covering 2.5 rounds, and $\nabla \rightarrow \nabla^*$ covering the first 4.5 rounds. There is a gap consisting of the 5th MA half-round and the 6th KM half-round in which $\text{msb}_8(Z_5^{(5)}) = 0$, $\text{lsb}_2(Z_6^{(5)}) = 0$ and $\text{lsb}_9(Z_4^{(6)}) = 0$. See Fig. 3.17. The approach to cross this gap is to use enough

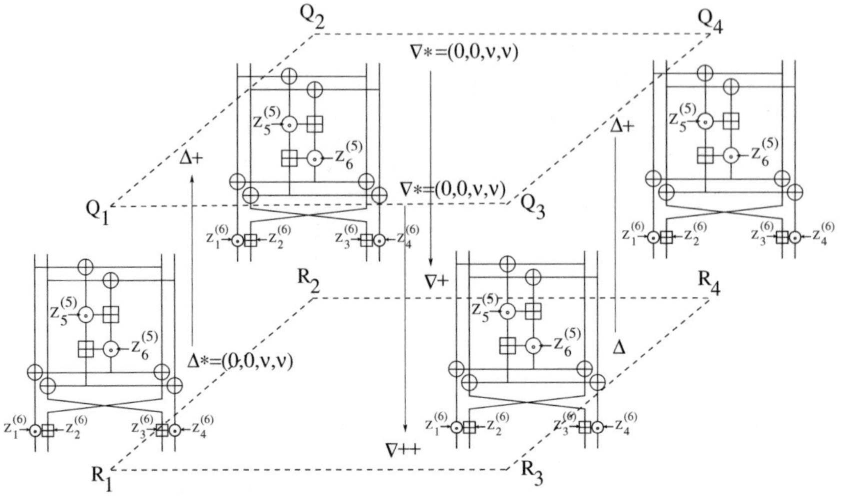

Fig. 3.17 Gap for the boomerang distinguisher labeled (17) in Table 3.87 consisting of the 5th MA and the 6th KM half-rounds in IDEA.

quartets in order to cause a collision between ∇^+ and ∇^{++} in opposite facets of the gap, so that the quartet structure will result in $\Delta = \Delta^*$. Experiments with 1024 randomly chosen keys from the weak-key class indicated that 2^{24} quartets are enough for boomerangs to cross this gap for more than 80% of the keys. The last KM half-round is not covered by either differential pattern, which implies that the boomerang direction is bottom-up. Moreover, $\text{lsb}_9(Z_4^{(9)}) = 0$ and $\text{msb}_2(Z_4^{(9)}) = 0$.

The attack proceeds as follows: prepare two sets of ciphertexts in which the first and third words are equal, the second words differ by 8000_x and the fourth words contain 2^{16} values each. This text structure contains 2^{32} pairs with difference $(0000_x, 8000_x, 0000_x, \delta)$. After decryption by the last KM half-round, each possible 16-bit difference is suggested by δ and $Z_4^{(9)}$ about $2^{32-16} = 2^{16}$ times, on average. Using 256 structures provides the number of right quartets needed to cross this gap. The attack complexity is $2^8 \cdot 2 \cdot (2^{16} + 2^{16}) = 2^{26}$ chosen texts and 2^{26} IDEA computations. Comparatively, a key-recovery attack on the $\text{lsb}_5(Z_4^{(9)} \gg 9)$ would require 2^5 guesses and 2^{24} quartets to cross the gap per guess, resulting in $2^{5+24} \cdot 4 = 2^{31}$ chosen texts and 2^{31} IDEA computations.

Another boomerang distinguisher, designated as (18) in Table 3.87, consists of the differentials $\Delta \to \Delta^*$ covering the last six rounds, and $\nabla \to \nabla^*$ covering the first round. There is a gap consisting of the full second round in which $\text{msb}_{12}(Z_1^{(2)}) = 0$ and $\text{lsb}_5(Z_6^{(2)}) = 0$. See Fig. 3.18. The approach to cross this gap is to use enough quartets in order to obtain a collision between ∇^+ and ∇^{++} in opposite facets of the gap, so that the quartet

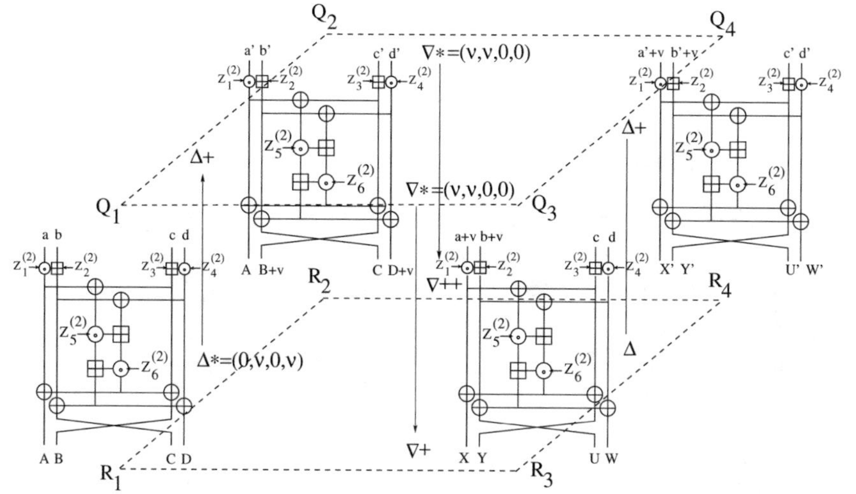

Fig. 3.18 Gap for the boomerang distinguisher labeled (18) in Table 3.87 consisting of the second full round in IDEA.

structure will result in $\Delta = \Delta^*$. Experiments with 1024 randomly chosen keys from the weak-key class demonstrated that 2^{24} quartets are enough for the boomerang to cross this gap for about 90% of the keys. The last KM half-round is not covered by either differential pattern, which implies that the boomerang direction is bottom-up. Moreover, $\text{lsb}_2(Z_4^{(9)}) = 0$.

The attack proceeds as follows: prepare two sets of ciphertexts in which the first and second words are equal, the third words differ by 8000_x and the fourth words take 2^{16} values each. This text structure contains 2^{32} pairs with difference $(0000_x, 0000_x, 8000_x, \delta)$. After decryption by the last KM half-round, each possible 16-bit difference is suggested by δ and $Z_4^{(9)}$ about $2^{32-16} = 2^{16}$ times, on average. Using 256 structures provides the necessary number of right quartets to cross the gap.

The attack complexity is $2^8 \cdot 2 \cdot (2^{16} + 2^{16}) = 2^{26}$ chosen texts and 2^{26} IDEA computations. Comparatively, a key-recovery attack for the $\text{msb}_{14}(Z_4^{(9)})$ requires 2^{14} guesses and 2^{24} quartets per guess, resulting in $2^{14+24} \cdot 4 = 2^{40}$ chosen texts and 2^{40} IDEA computations.

Another boomerang distinguisher, designated as (19) in Table 3.87, consists of the differentials $\Delta \rightarrow \Delta^*$ covering 1.5 rounds, and $\nabla \rightarrow \nabla^*$ covering 5.5 rounds. There is a gap consisting of the third MA half-round in which $\text{lsb}_9(Z_6^{(3)}) = 0$. See Fig. 3.19. Compare to the boomerang in Fig. 3.13, this gap cannot be crossed with a single quartet, because both $Z_5^{(3)}$ and $Z_6^{(3)}$ are not fully restricted. The approach to cross this gap is to use enough quartets in order to obtain a collision between ∇^+ and ∇^{++}, so that the quartet structure will result in $\Delta = \Delta^*$. Experiments with 1024 randomly chosen

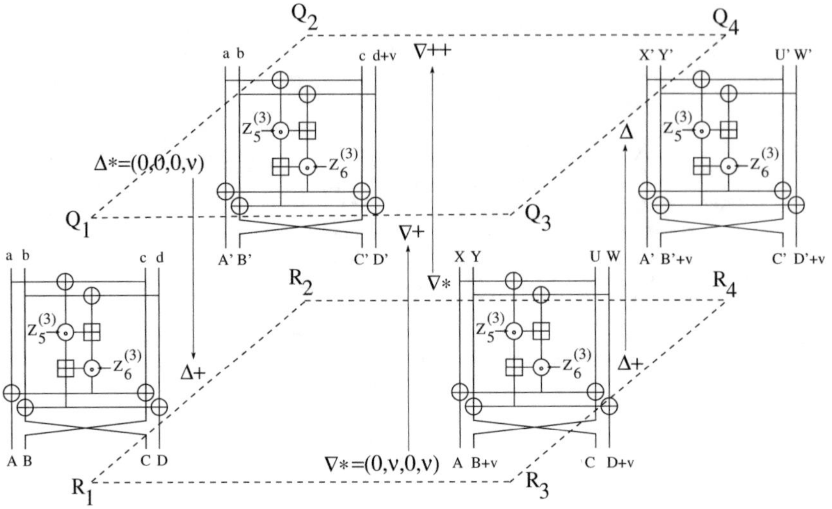

Fig. 3.19 Gap for the boomerang distinguisher labeled (19) in Table 3.87 consisting of the third MA half-round in IDEA.

keys from the weak-key class demonstrated that 2^{16} quartets are enough for boomerangs to cross this gap for more than 99% of the keys. The first KM half-round is not covered by either differential pattern, which implies that the boomerang direction is top-down. Moreover, $Z_4^{(1)}$ is not restricted.

The attack proceeds as follows: prepare two sets of plaintexts in which the first, second and third words are equal, while the fourth words contain 2^{16} random values each. This text structure contains 2^{32} pairs with difference $(0000_x, 0000_x, 0000_x, \delta)$. After decryption by the first KM half-round, each possible 16-bit difference is suggested by δ and $Z_4^{(1)}$ about $2^{32-16} = 2^{16}$ times, on average, which is enough to cross the gap. The attack complexity is $2 \cdot (2^{16} + 2^{16}) = 2^{18}$ chosen texts and 2^{18} IDEA computations. Comparatively, a key-recovery attack on $Z_4^{(1)}$ would require 2^{16} guesses and 2^{16} quartets per guess to cross the gap, resulting in $2^{16+16} \cdot 4 = 2^{34}$ chosen texts and 2^{34} IDEA computations.

Another boomerang distinguisher, designated as (21) in Table 3.87, consists of the differentials $\Delta \rightarrow \Delta^*$ covering 3 rounds, and $\nabla \rightarrow \nabla^*$ covering 4 rounds. There is a gap consisting of the 4th MA half-round and the 5th KM half-round, for which $\text{msb}_8(Z_5^{(4)}) = 0$ and $\text{lsb}_7(Z_4^{(5)}) = 0$.

The approach to cross the gap is to use enough quartets in order to obtain a collision between the differences ∇^+ and ∇^{++}, so that the quartet structure will result in $\Delta = \Delta^*$. Experiments with 1024 randomly chosen keys from the weak-key class demonstrated that 2^{24} quartets are enough for boomerangs to cross the gap for more than 90% of the keys. The first KM half-round is

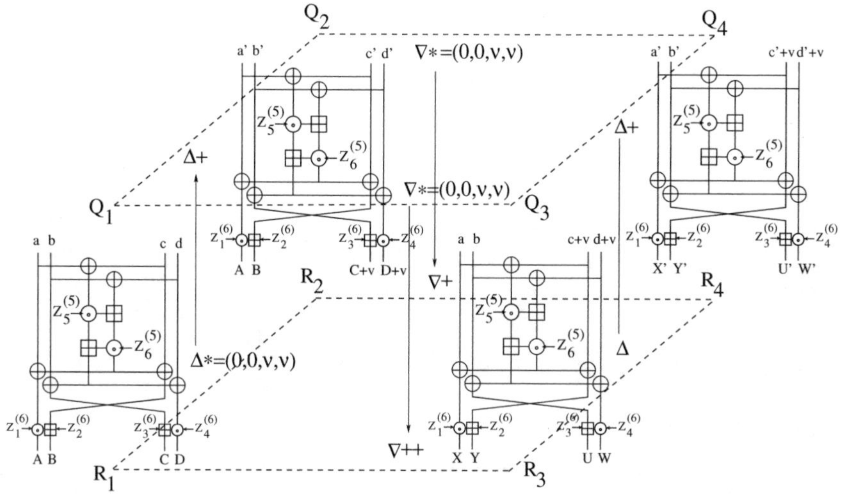

Fig. 3.20 Gap for the boomerang distinguisher labeled (21) in Table 3.87 consisting of the 4th MA and 5th KM half-rounds in IDEA.

not covered by either differential pattern, which implies that the boomerang direction is top-down. Moreover, $Z_4^{(1)}$ is not restricted.

The attack proceeds as follows: prepare two sets of plaintexts in which the first, second and third words are equal, while the fourth words contain 2^{16} random values each. This text structure contains 2^{32} pairs with difference $(0000_x, 0000_x, 0000_x, \delta)$. After encryption by the first KM half-round, each possible 16-bit difference is suggested by δ and $Z_4^{(1)}$ about $2^{32-16} = 2^{16}$ times, on average. Using 256 structures provides the necessary number of right quartets to cross the gap.

The attack complexity is $2^8 \cdot 2 \cdot (2^{16} + 2^{16}) = 2^{26}$ chosen texts and 2^{26} IDEA computations.

Comparatively, a key-recovery attack on $Z_4^{(1)}$ would require 2^{16} guesses and 2^{24} quartets per guess to cross the gap, resulting in $2^{16+24} \cdot 4 = 2^{42}$ chosen texts and 2^{42} IDEA computations.

The next boomerang distinguisher is the largest one found for IDEA. It is similar to Daemen's differential weak-key class and is designated (23) in Table 3.87. It consists of the differential $\Delta \to \Delta^*$ covering 6 rounds and $\nabla \to \nabla^*$ covering the 8th MA half-round. There is a gap consisting of the full 7th round, and the 8th KM half-round, for which $Z_5^{(7)} = 0$ and $\text{msb}_{13}(Z_6^{(7)}) = 0$. See Fig. 3.21. The approach to cross the gap is to use enough quartets in order to obtain a collision between ∇^+ and ∇^{++}, so that the quartet structure will result in $\Delta = \Delta^*$. Experiments with 1024 randomly chosen keys from the weak-key class demonstrated that 2^{14} quartets are enough for the boomerangs to cross this gap for 25% of the keys. The use of more quartets, such as 2^{19}

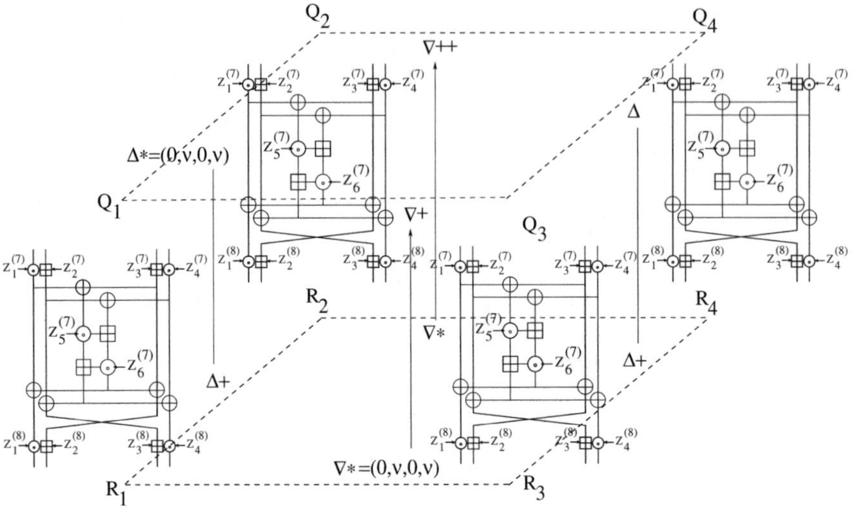

Fig. 3.21 Gap for the boomerang distinguisher labeled (23) in Table 3.87 consisting of the full 7th round and the 8th KM half-round in IDEA.

allows us to identify about 32% of the weak-key class. The last KM half-round is not covered by either differential pattern, which implies that the boomerang direction is bottom-up. Moreover, $\text{msb}_2(Z_4^{(9)}) = 0$.

The attack proceeds as follows: prepare two sets of ciphertexts in which the first and third words are equal, the second words differ by 8000_x and the fourth words contain 2^{14} random values each. This text structure contains 2^{28} pairs with difference $(0000_x, 8000_x, 0000_x, \delta)$. After decryption by the last KM half-round, each possible 16-bit difference is suggested by δ and $Z_4^{(9)}$ about $2^{28-16} = 2^{14}$ times, on average, which is enough to cross the gap and identify a weak-key class of size $\frac{1}{4} \cdot 2^{66} = 2^{64}$.

The attack complexity is $2 \cdot (2^{14} + 2^{14}) = 2^{16}$ chosen texts and 2^{16} IDEA computations. Comparatively, a key-recovery attack for the $\text{msb}_{14}(Z_4^{(9)})$ would require 2^{14} guesses and 2^{14} quartets to cross the gap to identify $\frac{1}{4}$ of the keys in the weak-key class, resulting in $2^{14+14} \cdot 4 = 2^{30}$ chosen texts and 2^{30} IDEA computations.

As an example of boomerang distinguishers with larger gaps, consider the boomerang labeled (22) in Table 3.87. It consists of differentials $\Delta \to \Delta^*$ and $\nabla \to \nabla^*$, both covering 3.5 rounds of IDEA. There is a gap consisting of the 4th MA half-round and the 5th full round, for which $Z_1^{(5)} = 0$, $\text{lsb}_7(Z_4^{(5)}) = 0$ and $\text{msb}_{13}(Z_5^{(5)}) = 0$. See Fig. 3.22. In order to cross the gap, enough quartets are used to cause a collision between ∇^+ and ∇^{++}, so that the quartet structure will result in $\Delta = \Delta^*$. Experiments with 1024 randomly chosen keys from the weak-key class demonstrated that 2^{24} quartets are enough for

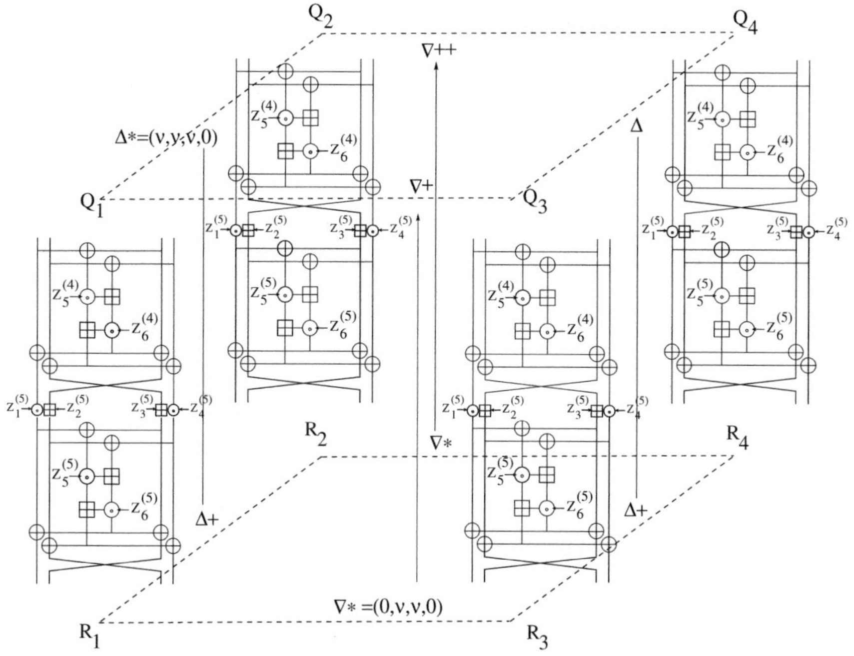

Fig. 3.22 Gap for the boomerang distinguisher labeled (22) in Table 3.87 consisting of the 4th MA half-round and the full 5th round in IDEA.

boomerangs to cross this gap for 25% of the keys. The first KM half-round is not covered by either differential pattern, which implies that the boomerang direction is top-down. Moreover, $Z_4^{(1)}$ is not restricted.

The attack proceeds as follows: prepare two sets of plaintexts in which the first, second and third words are equal, while the fourth word contains all 2^{16} values each. This text structure contains 2^{32} pairs with difference $(0000_x, 0000_x, 0000_x, \delta)$. After encryption by the first KM half-round, each possible 16-bit difference is suggested by δ and $Z_4^{(1)}$ about $2^{32-16} = 2^{16}$ times. Using 256 structures provides the necessary number of right quartets to cross the gap and identify a weak-key class of size $\frac{1}{4} \cdot 2^{63} = 2^{61}$. The attack complexities are $2^8 \cdot 2 \cdot (2^{16} + 2^{16}) = 2^{26}$ chosen texts and 2^{26} IDEA computations.

The last boomerang for the full 8.5-round IDEA is designated as (20) in Table 3.87. It consists of the differentials $\Delta \to \Delta^*$ covering 1.5 rounds, and $\nabla \to \nabla^*$ covering 5.5 rounds. There is a gap consisting of the second MA half-round and the full third round for which $\text{lsb}_{12}(Z_6^{(2)}) = 0$, $Z_1^{(3)} = 0$, $Z_4^{(3)} = 0$ and $\text{lsb}_5(Z_6^{(3)}) = 0$. See Fig. 3.23. In order to cross the gap, enough quartets are required to cause a collision between ∇^+ and ∇^{++}, so that the quartet structure will result in $\Delta = \Delta^*$. Experiments with 1024 randomly chosen

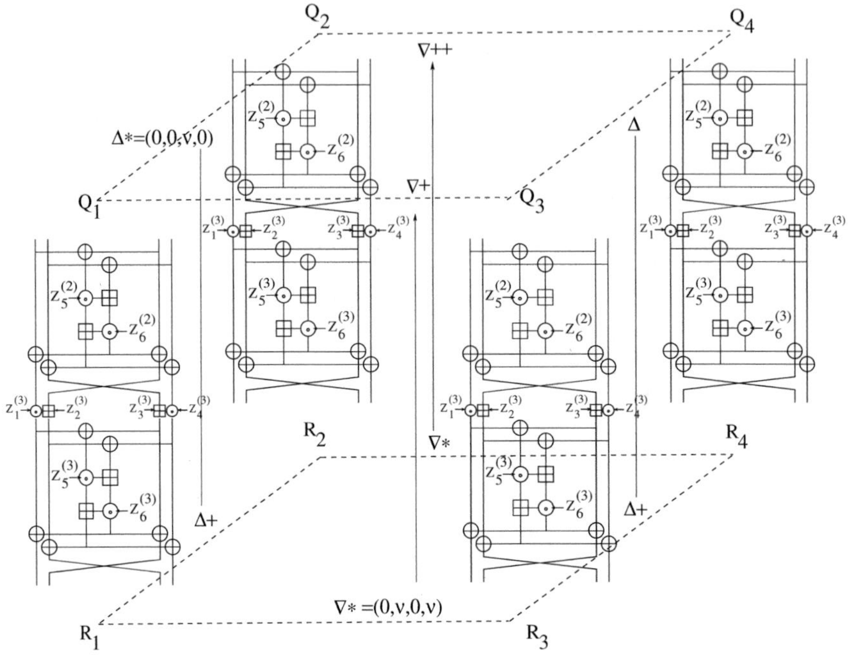

Fig. 3.23 Gap for the boomerang distinguisher labeled (20) in Table 3.87 consisting of the second MA half-round and the full third round in IDEA.

keys from the weak-key class demonstrated that 2^{23} quartets are enough for boomerangs to cross the gap for about 2^{-6} of the keys. The first KM half-round is not covered by either differential pattern, which implies that the boomerang direction is top-down. Moreover, $Z_1^{(1)} = 0$, but $Z_4^{(1)}$ is not restricted.

The attack proceeds as follows: prepare two plaintexts sets in which the second words are equal, the first and third words differ by 8000_x, and the fourth words contain all 2^{16} values each. This structure contains 2^{32} pairs with difference $(8000_x, 0000_x, 8000_x, \delta)$. After encryption by the first KM half-round, each possible 16-bit difference is suggested by δ and $Z_4^{(1)}$ about $2^{32-16} = 2^{16}$ times. Using 128 structures provides the necessary number of right quartets to cross the gap and identify a weak-key class of size $2^{-6} \cdot 2^{66} = 2^{60}$. The attack complexities are $2^7 \cdot 2 \cdot (2^{16} + 2^{16}) = 2^{25}$ chosen texts and 2^{25} IDEA computations.

3.12.1.1 Boomerang Attacks on Reduced-Round IDEA

The computer program constructed to find boomerangs for the full 8.5-round IDEA was also adapted to search for boomerangs in reduced-round versions of the IDEA cipher.

An example of such boomerangs covers 4-round IDEA from the 4th to the 7th round (inclusive) and is designated as (1) in Table 3.87. This boomerang consists of the differential $\Delta \rightarrow \Delta^*$ covering 1.5 rounds and $\nabla \rightarrow \nabla^*$ covering 1 round. There is a gap consisting of the full 6th round, in which $\text{lsb}_9(Z_5^{(6)}) = 0$ and $\text{msb}_{13}(Z_6^{(6)}) = 0$. This gap is similar to the one in the boomerang labeled (22) in Table 3.87, and the procedure to cross it is also similar. The 4th KM half-round is not covered by either differential pattern, which implies that the boomerang direction is top-down. Notice that $\text{lsb}_7(Z_4^{(4)}) = 0$.

The attack proceeds as follows: consider two text sets in which the first and second words are equal, the third words differ by 8000_x and the fourth words assume 2^{15} random values each. This text structure can form 2^{30} text pairs of the form $(0000_x, 0000_x, 8000_x, \delta)$. It is expected that each 16-bit difference after the 4th KM half-round is suggested by δ and $Z_4^{(4)}$ about $2^{30-16} = 2^{14}$ times, on average. Experiments with 1024 randomly chosen keys from the weak-key class demonstrated that 2^{14} quartets are enough for boomerangs to cross the gap for 50% of the keys.

The attack complexity is $2 \cdot (2^{14} + 2^{14}) = 2^{16}$ chosen texts and 2^{16} 4-round IDEA computations.

The next boomerang distinguisher is designated as (2) in Table 3.87, and covers 4.5-round IDEA, from the 4th until the 8th KM half-round. It consists of the differentials $\Delta \rightarrow \Delta^*$ and $\nabla \rightarrow \nabla^*$ both covering 1.5 rounds. There is a gap consisting of the 5th MA half-round and the 6th KM half-round, in which $\text{msb}_8(Z_5^{(5)}) = 0$ and $\text{lsb}_9(Z_4^{(6)}) = 0$. This gap is similar to the one in the boomerang labeled (17) in Table 3.87.

The 8th KM half-round is not covered by either differential pattern, which implies that the boomerang direction is bottom-up. Notice that $Z_4^{(8)}$ is not restricted.

The attack proceeds as follows: consider two text sets in which the first and third words are equal, the second words differ by 8000_x and the fourth words assume all 2^{16} possible values each. This text structure can form 2^{32} text pairs of the form $(0000_x, 8000_x, 0000_x, \delta)$. It is expected that each 16-bit difference after the 8th KM half-round is suggested by δ and $Z_4^{(8)}$ about $2^{32-16} = 2^{16}$ times. Experiments with 1024 randomly chosen keys from the weak-key class demonstrated that 2^{20} quartets allows boomerangs to cross the gap for about 50% of the keys. It means that the boomerang can identify a weak-key class of size 2^{103}. The attack complexities are $2^4 \cdot 2 \cdot (2^{16} + 2^{16}) = 2^{22}$ chosen texts and 2^{22} 4.5-round IDEA computations.

The next boomerang distinguisher is designated as (3) in Table 3.87, and covers 4.5-round IDEA from the third round until the 7th KM half-round. It consists of the differentials $\varDelta \to \varDelta^*$ and $\nabla \to \nabla^*$ both covering 1.5 rounds. There is a gap consisting of the 4th MA half-round in which $\text{msb}_6(Z_5^{(4)}) = 0$, and 5th KM half-round in which $\text{lsb}_9(Z_4^{(5)}) = 0$. This gap is similar to the one in the boomerang labeled (17) in Table 3.87.

The 7th KM half-round is not covered by either differential pattern, which implies that the boomerang direction is bottom-up. Notice that $Z_4^{(7)}$ is not restricted.

The attack proceeds as follows: consider two text sets in which the first and third words are equal, the second words differ by 8000_x and the fourth words assume al 2^{16} possible values each. This text structure can form 2^{32} text pairs of the form $(0000_x, 8000_x, 0000_x, \delta)$. It is expected that each 16-bit difference after decrypting the 7th KM half-round is suggested by δ and $Z_4^{(7)}$ about $2^{32-16} = 2^{16}$ times. Experiments with 1024 randomly chosen keys from the weak-key class demonstrated that 2^{20} quartets are enough for about 50% of the keys to cross the gap. This trade-off allows identification of a weak-key class of size 2^{103}.

The attack complexities are $2^4 \cdot 2 \cdot (2^{16} + 2^{16}) = 2^{22}$ chosen texts and 2^{22} 4.5-round IDEA computations.

The next boomerang distinguisher is designated as (4) in Table 3.87, and covers 5-round IDEA, from the second until the 6th round. It consists of the differentials $\varDelta \to \varDelta^*$ covering 1.5 rounds, and $\nabla \to \nabla^*$ covering 3 rounds. There is a gap consisting of the third MA half-round, in which both $Z_5^{(3)}$ and $Z_6^{(3)}$ are not restricted. This gap is similar to the one in the boomerang labeled (14) in Table 3.87. This means that one quartet is enough to cross this gap.

The second KM half-round is not covered by either differential pattern, which implies that the boomerang direction is top-down. Notice that $Z_4^{(2)}$ is not restricted.

The attack proceeds as follows: consider two text sets in which the first and third words are equal, the second words differ by 8000_X and the fourth words assume 2^8 random values each. This text structure can form 2^{16} text pairs of the form $(0000_x, 8000_x, 0000_x, \delta)$. It is expected that each 16-bit difference after encrypting the second KM half-round is suggested by δ and $Z_4^{(2)}$ about once, on average.

The attack complexities are $2 \cdot (2^8 + 2^8) = 2^{10}$ chosen texts and 2^{10} 5-round IDEA computations.

The next boomerang distinguisher is designated as (5) in Table 3.87, and covers 5-round IDEA, from the third until the seventh round. It consists of the differentials $\varDelta \to \varDelta^*$ covering 1.5 rounds, and $\nabla \to \nabla^*$ covering 3 rounds. There is a gap consisting of the 4th MA half-round, in which $\text{msb}_1(Z_5^{(4)}) = 0$. Note that $Z_6^{(4)}$ is not restricted. This gap is similar to the one in the boomerang labeled (14) in Table 3.87. This means that one quartet is enough to cross this gap. The third KM half-round is not covered by either differential pattern, which implies that the boomerang direction is top-down. Notice that $\mathrm{msb}_{10}(Z_4^{(3)}) = 0$.

The attack proceeds as follows: consider two text sets in which the first and third words are equal, the second words differ by 8000_x and the fourth words assume 2^8 random values each. This text structure can form 2^{16} text pairs of the form $(0000_x, 8000_x, 0000_x, \delta)$. It is expected that each 16-bit difference after decrypting the third KM half-round is suggested by δ and $Z_4^{(3)}$ about once, on average.

The attack complexities are $2 \cdot (2^8 + 2^8) = 2^{10}$ chosen texts and 2^{10} 5-round IDEA computations.

The next boomerang distinguisher is designated as (6) in Table 3.87, and covers 5-round IDEA, from the third until the seventh round. It consists of the differentials $\Delta \rightarrow \Delta^*$ covering 3 rounds, and $\nabla \rightarrow \nabla^*$ covering 1.5 rounds. There is a gap consisting of the 6th KM half-round, in which $\mathrm{lsb}_2(Z_4^{(6)}) = 0$. The careful choice of differences $\Delta^* = (0000_x, 8000_x, 0000_x, 8000_x)$ and $\nabla^* = (8000_x, 0000_x, 0000_x, 0000_x)$ allows boomerangs to cross the gap with a single quartet, independent of $Z_1^{(6)}$ and $Z_4^{(6)}$. This gap is similar to the one in boomerang labeled (14) in Table 3.87. The boomerang direction can be either top-down or bottom-up.

The attack complexities are 4 chosen texts and four 5-round IDEA computations.

The next boomerang is designated as (7) in Table 3.87, and covers 5-round IDEA, from the third until the seventh round. It consists of the differentials $\Delta \rightarrow \Delta^*$ covering 3 rounds, and $\nabla \rightarrow \nabla^*$ covering 1 round. There is a gap consisting of the full 6th round, where $\mathrm{lsb}_2(Z_4^{(6)}) = 0$, $Z_5^{(6)} = 0$ and $\mathrm{msb}_{13}(Z_6^{(6)}) = 0$. This gap is similar to the one in the boomerang labeled (18) in Table 3.87.

The boomerang direction can be either top-down or bottom-up. Experiments with 1024 randomly chosen keys from the weak-key class demonstrated that 2^{11} quartets are enough for boomerangs to cross this gap for 90% of the keys. The attack complexity is $4 \cdot 2^{11} = 2^{13}$ chosen texts and 2^{13} 5-round IDEA computations.

The next boomerang is designated as (8) in Table 3.87, and covers 5.5-round IDEA, from the third round until the eighth KM half-round. It consists of the differentials $\Delta \rightarrow \Delta^*$ covering 3 rounds, and $\nabla \rightarrow \nabla^*$ covering 1.5 rounds. There is a gap consisting of the 6th KM half-round in which $\mathrm{lsb}_2(Z_4^{(6)}) = 0$. This gap is similar to the one in the boomerang labeled (7) in Table 3.87. Only one quartet is enough to cross this gap. The 8th KM half-round is not covered by either differential pattern, which implies that the boomerang direction is bottom-up. Notice that $\mathrm{msb}_6(Z_1^{(8)}) = 0$, while $Z_4^{(8)}$ is not restricted.

The attack proceeds as follows: consider two text sets in which the first words assume 2^8 random values each, the second words differ by 8000_x, the third words are equal and the fourth words assume 2^8 random values each. This text structure can form $(2^8 \cdot 2^8)^2 = 2^{32}$ pairs of the form $(\delta_1, 8000_x, 0000_x, \delta_2)$. It is expected that each pair of 16-bit differences after decrypting the 8th KM half-round is suggested by $(\delta_1, Z_1^{(8)})$ and $(\delta_2, Z_4^{(8)})$ about once, on average.

The attack complexities are $4 \cdot 2^8 \cdot 2^8 = 2^{18}$ chosen texts and 2^{18} 5.5-round IDEA computations.

The next boomerang distinguisher is designated as (9) in Table 3.87 and covers 5.5-round IDEA, from the third round until the eighth KM half-round. It consists of the differentials $\Delta \to \Delta^*$ covering 3 rounds, and $\nabla \to \nabla^*$ covering 1 round. There is a gap consisting of the full 6th round in which $\text{lsb}_2(Z_4^{(6)}) = 0$, $Z_5^{(6)} = 0$ and $\text{msb}_{13}(Z_6^{(6)}) = 0$. This gap is similar to the one in the boomerang labeled (18) in Table 3.87. The 8th KM half-round is not covered by either differential pattern, which implies that the boomerang direction is bottom-up. Notice that $\text{msb}_4(Z_1^{(8)}) = 0$, while $Z_4^{(8)}$ is not restricted.

The attack proceeds as follows: consider two text sets in which the first words assume 2^{11} random values each, the second words differ by 8000_x, the third words are equal and the fourth words assume 2^{11} random values each. This text structure can form $(2^{11} \cdot 2^{11})^2 = 2^{44}$ text pairs of the form $(\delta_1, 8000_x, 0000_x, \delta_2)$. It is expected that each pair of 16-bit differences after decrypting the 8th KM half-round is suggested by $(\delta_1, Z_1^{(8)})$ and $(\delta_2, Z_4^{(8)})$ about $2^{44-32} = 2^{12}$ times, on average. Experiments with 1024 randomly chosen keys from the weak-key class demonstrated that 2^{12} quartets are enough for boomerangs to cross this gap for more than 90% of the keys.

The attack complexities are $4 \cdot 2^{11} \cdot 2^{11} = 2^{24}$ chosen texts and 2^{24} 5.5-round IDEA computations.

The last boomerang distinguisher is labeled (10) in Table 3.87 and covers 6-round IDEA, from the second until the seventh round. It consists of the differentials $\Delta \to \Delta^*$ covering 1 round, and $\nabla \to \nabla^*$ covering 4 rounds. There is a gap consisting of the third MA half-round in which both $Z_5^{(3)}$ and $Z_6^{(3)}$ are not restricted. This gap is similar to the one in the boomerang labeled (15) in Table 3.87 that is, only one quartet is enough to cross this gap.

The second KM half-round is not covered by either differential pattern, which implies that the boomerang direction is top-down. Notice that $Z_4^{(2)}$ is not restricted.

The attack proceeds as follows: consider two text sets in which the first and third words are equal, the second words differ by 8000_x and the fourth words assume 2^8 random values each. This text structure can form 2^{16} text pairs of the form $(0000_x, 8000_x, 0000_x, \delta)$. It is expected that each 16-bit difference after decrypting the second KM half-round is suggested by δ and $Z_4^{(2)}$ about once, on average.

The attack complexities are $2 \cdot (2^8 + 2^8) = 2^{10}$ chosen texts and 2^{10} 6-round IDEA computations.

Table 3.87 summarizes the boomerang attack complexities on both reduced-round and the full 8.5-round IDEA cipher. In the upper-half of Table 3.87, the attacks are on reduced-round versions of IDEA. Thus, there is no inverse swap of the middle words in the last round.

In the 5th column, top-down boomerangs are denoted by $\downarrow$, while bottom-up boomerangs are denoted by $\uparrow$. If the direction of the boomerang can be either up or down, then it is denoted $\updownarrow$.

In the 4th column, the half-round range from 0 until 16. The terminology $|WKC|$ denotes the weak-key class size. In the 7th and 8th columns, the symbol 0 stands for the difference 0000_x, $\nu = 8000_x$ and $*$ denotes an arbitrary difference used to cause difference ν after multiplication by an unrestricted key.

3.12.2 Boomerang Attacks on MESH-64

Boomerang attacks on the MESH ciphers operate under weak-key assumptions, just like in IDEA. Nonetheless, the key schedule algorithms of the MESH ciphers are not linear like in IDEA, that is, the KSAs of the MESH ciphers do not provide a simple, linear mapping between the user key bits and the round subkey bits. Therefore, the weak-key class sizes will be estimated. The approach to determine this size is similar to that used in linear attacks in Sect. 3.13.2: each multiplicative subkey that is required to be weak will imply a fraction of 2^{-15} of the key space.

The difference operator used in bitwise exclusive-or, and nonzero wordwise difference will have the value $\nu = 8000_x$.

Table 3.11 in Sect. 3.5.3 summarizes 1-round characteristics for MESH-64 under weak-subkey assumptions for both odd-numbered and even-numbered rounds.

Using this table of 1-round characteristics, the longest boomerangs found cover 5.5-round MESH-64, and uses the differential patterns $\Delta \rightarrow \Delta^*$, where

$$\Delta = (0,\nu,0,\nu) \stackrel{1r}{\rightarrow} (0,0,\nu,\nu) \stackrel{1r}{\rightarrow} (\nu,0,\nu,0) \stackrel{1r}{\rightarrow} (\nu,\nu,0,0) \stackrel{0.5r}{\rightarrow} (\nu,\nu,0,0) = \Delta^*$$

across the first 3.5 rounds, and

$$\nabla = (\nu,\nu,0,0) \stackrel{1r}{\rightarrow} (\nu,0,\nu,0) = \nabla^*$$

across one round. Both characteristics hold with certainty provided the following (multiplicative) subkeys are weak: $Z_4^{(1)}$, $Z_3^{(2)}$, $Z_5^{(2)}$, $Z_7^{(2)}$, $Z_1^{(3)}$, $Z_2^{(4)}$, $Z_1^{(5)} \in \{0,1\}$. These characteristics were chosen in order to minimize the number of weak-subkey assumptions and therefore, to maximize the weak-key class size, which is estimated as $2^{128-15\cdot 7} = 2^{23}$, if all restrictions hold independently.

Table 3.87 Boomerang distinguishers for IDEA cipher.

Label	#Rounds	Hawkes $\|WKC\|$	New $\|WKC\|$	Half-Rounds	Flow	Weak-Key Bit positions	$\Delta \rightarrow \Delta^*$	$\nabla \rightarrow \nabla^*$	Complexity Data	Complexity Time
(1)	4	2^{99}	2^{105}	6-13	$\downarrow$	11-32	$(0,0,\nu,*) \rightarrow (0,\nu,0,\nu)$	$(\nu,\nu,0,\nu) \rightarrow (0,\nu,0,0)$	2^{16}	2^{16}
(2)	4.5	2^{97}	2^{103}	6-14	$\uparrow$	0-18, 123-127	$(0,\nu,0,\nu) \rightarrow (0,0,\nu,\nu)$	$(0,\nu,0,*) \rightarrow (0,0,\nu,\nu)$	2^{22}	2^{22}
(3)	4.5	2^{97}	2^{103}	4-12	$\uparrow$	2-25	$(0,\nu,0,\nu) \rightarrow (0,0,\nu,\nu)$	$(0,\nu,0,*) \rightarrow (0,0,\nu,\nu)$	2^{22}	2^{22}
(4)	5	2^{84}	2^{97}	2-11	$\downarrow$	0-25, 123-127	$(0,\nu,0,*) \rightarrow (0,0,\nu,\nu)$	$(0,\nu,0,\nu) \rightarrow (0,\nu,0,\nu)$	2^{10}	2^{10}
(5)	5	2^{84}	2^{97}	4-13	$\downarrow$	0-18, 116-127	$(0,\nu,0,*) \rightarrow (0,0,\nu,\nu)$	$(0,\nu,0,\nu) \rightarrow (0,\nu,0,\nu)$	2^{10}	2^{10}
(6)	5	2^{84}	2^{95}	4-13	$\updownarrow$	2-34	$(0,\nu,0,\nu) \rightarrow (0,\nu,0,\nu)$	$(\nu,\nu,0,\nu) \rightarrow (\nu,0,0,0)$	4	4
(7)	5	2^{84}	2^{97}	4-13	$\updownarrow$	2-32	$(0,\nu,0,\nu) \rightarrow (0,\nu,0,\nu)$	$(\nu,\nu,0,\nu) \rightarrow (0,\nu,0,0)$	2^{13}	2^{13}
(8)	5.5	2^{82}	2^{95}	4-14	$\uparrow$	2-34	$(0,\nu,0,\nu) \rightarrow (0,\nu,0,\nu)$	$(*,\nu,0,*) \rightarrow (\nu,0,0,0)$	2^{18}	2^{18}
(9)	5.5	2^{82}	2^{97}	4-14	$\uparrow$	2-32	$(0,\nu,0,\nu) \rightarrow (0,\nu,0,\nu)$	$(*,\nu,0,*) \rightarrow (0,\nu,0,0)$	2^{24}	2^{24}
(10)	6	2^{82}	2^{83}	2-13	$\downarrow$	0-32, 116-127	$(0,\nu,0,*) \rightarrow (0,0,\nu,\nu)$	$(0,\nu,0,\nu) \rightarrow (0,\nu,\nu,0)$	2^{10}	2^{10}
(11)	8.5	2^{63}	2^{46}	0-16	$\updownarrow$	0-25, 41-91, 123-127	$(0,\nu,0,\nu) \rightarrow (0,\nu,0,\nu)$	$(0,\nu,0,\nu) \rightarrow (0,\nu,\nu,0)$	4	4
(12)	8.5	2^{63}	2^{51}	0-16	$\updownarrow$	0-71, 123-127	$(0,\nu,0,\nu) \rightarrow (0,0,\nu,\nu)$	$(\nu,0,\nu,0) \rightarrow (\nu,0,0,\nu)$	2^{14}	2^{14}
(13)	8.5	2^{63}	2^{53}	0-16	$\updownarrow$	0-25, 64-107, 123-127	$(\nu,0,0,0) \rightarrow (\nu,\nu,0,\nu)$	$(0,\nu,\nu,0) \rightarrow (0,\nu,0,\nu)$	4	4
(14)	8.5	2^{63}	2^{56}	0-16	$\uparrow$	0-32, 84-110, 116-127	$(0,0,\nu,\nu) \rightarrow (0,\nu,0,\nu)$	$(\nu,0,\nu,0) \rightarrow (\nu,\nu,0,0)$	2^{11}	2^{11}
(15)	8.5	2^{63}	2^{57}	0-16	$\downarrow$	0-23, 57-91, 116-127	$(0,0,\nu,*) \rightarrow (0,0,\nu,\nu)$	$(0,\nu,0,\nu) \rightarrow (0,\nu,0,\nu)$	2^{11}	2^{11}
(16)	8.5	2^{63}	2^{57}	0-16	$\downarrow$	0-32, 57-91, 125-127	$(0,0,\nu,*) \rightarrow (0,\nu,0,\nu)$	$(0,\nu,0,\nu) \rightarrow (0,0,\nu,\nu)$	2^{15}	2^{15}
(17)	8.5	2^{63}	2^{58}	0-16	$\uparrow$	0-18, 41-71, 77-91, 123-127	$(0,\nu,0,*) \rightarrow (0,0,\nu,\nu)$	$(0,\nu,0,\nu) \rightarrow (0,0,\nu,\nu)$	2^{26}	2^{26}
(18)	8.5	2^{63}	2^{59}	0-16	$\uparrow$	0-32, 84-107, 116-127	$(0,0,\nu,*) \rightarrow (0,\nu,0,\nu)$	$(\nu,0,\nu,0) \rightarrow (\nu,\nu,0,0)$	2^{26}	2^{26}
(19)	8.5	2^{63}	2^{50}	0-16	$\downarrow$	0-25, 73-110, 123-127	$(0,0,0,*) \rightarrow (0,0,0,\nu)$	$(0,\nu,\nu,0) \rightarrow (0,\nu,0,\nu)$	2^{18}	2^{18}
(20)	8.5	2^{63}	2^{60}	0-16	$\downarrow$	0-25, 77-107, 123-127	$(\nu,0,\nu,*) \rightarrow (0,0,\nu,0)$	$(0,\nu,\nu,0) \rightarrow (0,\nu,0,\nu)$	2^{25}	2^{25}
(21)	8.5	2^{63}	2^{61}	0-16	$\downarrow$	4-25, 66-110	$(0,0,0,*) \rightarrow (\nu,\nu,\nu,0)$	$(0,\nu,\nu,0) \rightarrow (0,\nu,0,\nu)$	2^{26}	2^{26}
(22)	8.5	2^{63}	2^{62}	0-16	$\downarrow$	4-23, 66-110	$(0,0,0,*) \rightarrow (\nu,\nu,\nu,0)$	$(0,\nu,\nu,0) \rightarrow (0,\nu,\nu,0)$	2^{26}	2^{26}
(23)	8.5	2^{63}	2^{64}	0-16	$\uparrow$	0-25, 41-71, 123-127	$(0,\nu,0,\nu) \rightarrow (0,\nu,0,\nu)$	$(0,\nu,0,*) \rightarrow (0,\nu,0,\nu)$	2^{16}	2^{16}

There is a gap consisting of the 4th MA half-round. This gap can be crossed with a single quartet. This gap is similar to the one in the boomerangs (13) and (14) in Table 3.87.

The 6th KM half-round is not covered by either differential pattern, which implies that the boomerang direction is bottom-up. This means that text structures will be created to bypass the last KM half-round, determining the direction of the boomerang. This is the same procedure as in the boomerang (14) in Table 3.87.

The attack proceeds as follows: prepare two sets of ciphertexts in which the first words differ by ν, the third and fourth words are equal and the second words assume 2^8 different random values each. This text structure can form 2^{16} pairs with difference $(\nu, \delta, 0000_x, 0000_x)$. After decryption by the last KM half-round, each possible 16-bit difference is suggested by δ and $Z_2^{(6)}$ about $2^{2 \cdot 8-16} = 1$, on average. That is enough to cross the gap.

The attack complexity is $2 \cdot (2^8 + 2^8) = 2^{10}$ chosen texts and 2^{10} 5.5-round MESH-64 computations.

3.12.3 Boomerang Attacks on MESH-96

Boomerang attacks on the MESH ciphers operate under weak-key assumptions, just like in IDEA. Nonetheless, the key schedule algorithms of the MESH ciphers are not linear like in IDEA, that is, they do not provide a simple, linear mapping between the user key bits and the round subkey bits. Therefore, the weak-key class sizes will be estimated. The approach to determine this size is similar to that used in linear attacks in Sect. 3.13.2: each multiplicative subkey that is required to be weak will imply a fraction of 2^{-15} of the key space.

The difference operator used in bitwise exclusive-or, and nonzero wordwise difference will have the value $\nu = 8000_x$.

We start with a listing of 1-round characteristics for MESH-96. Tables 3.14 and 3.15 in Sect. 3.5.4 summarize these characteristics, under weak-subkey conditions for both odd-numbered and even-numbered rounds.

Unlike MESH-64, there are unkeyed $\odot$ operations in the MA-box of MESH-96. In order to cover the largest number of rounds and to maximize the probability of the differentials, and consequently of the whole boomerang, we exploit only 1-round characteristics that bypass these unkeyed $\odot$ operations.

The longest boomerangs found for MESH-96 can cover 5.5 rounds. For instance,

$$\Delta = (\nu, 0, 0, \nu, 0, 0) \stackrel{1r}{\rightarrow} (\nu, \nu, 0, 0, 0, 0) \stackrel{0.5r}{\rightarrow} (\nu, \nu, 0, 0, 0, 0) = \Delta^*$$

covering the first 1.5 rounds, and

$$\nabla = \nabla^* = (\nu, \nu, \nu, \nu, \nu, \nu)$$

covering 3 rounds. It is an iterative characteristic. The differential patterns were chosen in order to minimize the number of weak-subkey restrictions.

Both characteristics hold with certainty, provided the following subkeys are weak: $Z_1^{(1)}$, $Z_2^{(2)}$, $Z_1^{(3)}$, $Z_3^{(3)}$, $Z_5^{(3)}$, $Z_2^{(4)}$, $Z_4^{(4)}$, $Z_6^{(4)}$, $Z_1^{(5)}$, $Z_3^{(5)}$, $Z_5^{(5)}$. Assuming each subkey restriction holds independently, the weak-key class size is estimated as $2^{192-15\cdot 11} = 2^{27}$.

There is a gap consisting of the second MA half-round, but only one quartet is enough to cross it. This gap is analogous to that of the boomerangs (13) and (14) in Table 3.87. See Fig. 3.24. The 6th KM half-round is not covered by either differential pattern, which implies that the boomerang direction is bottom-up.

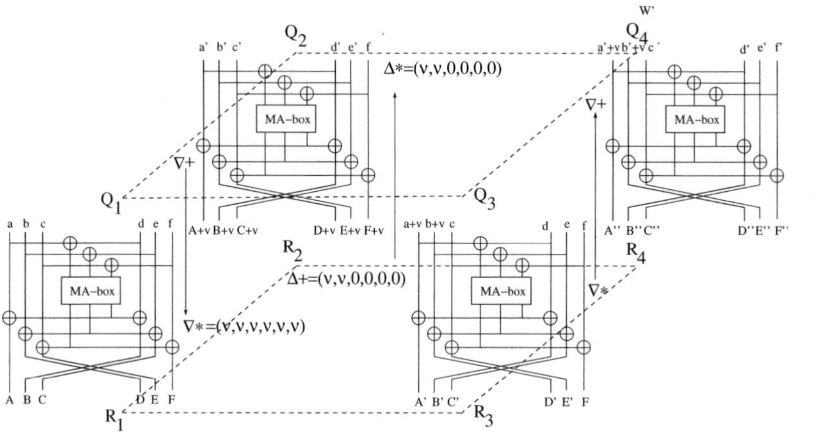

Fig. 3.24 Gap for the boomerang distinguisher for MESH-96 consisting of the second MA half-round.

The attack proceeds as follows: prepare two sets of ciphertexts in which the first, third and fifth words differ by ν, while the second, fourth and sixth words assume 2^8 random values each. This text structure can form $(2^8 \cdot 2^8 \cdot 2^8)^2 = 2^{48}$ pairs of the form $(\nu, \delta_1, \nu, \delta_2, \nu, \delta_3)$. After decryption of the last KM half-round, it is expected that each possible 16-bit difference is suggested by $(\delta_1, Z_2^{(6)})$, by $(\delta_2, Z_4^{(6)})$ and by $(\delta_3, Z_6^{(6)})$ about $2^{48-3\cdot 16} = 1$, on average, and that is enough for the boomerang to cross the gap.

The attack complexity is $4 \cdot (2^8)^3 = 2^{26}$ chosen texts and 2^{26} 5.5-round MESH-96 computations.

3.12.4 Boomerang Attacks on MESH-128

Boomerang attacks on the MESH ciphers operate under weak-key assumptions, just like in IDEA. Nonetheless, the key schedule algorithms of the MESH ciphers are not linear like in IDEA, that is, they do not provide a simple, linear mapping between the user key bits and the round subkey bits. Therefore, the weak-key class sizes will be estimated. The approach to determine this size is similar to that used in linear attacks in Sect. 3.13.2: each multiplicative subkey that is required to be weak will imply a fraction of 2^{-15} of the key space.

The difference operator used in bitwise exclusive-or, and nonzero wordwise difference will have the value $\nu = 8000_x$.

We start with a listing of 1-round characteristics for MESH-128. Tables 3.17 through 3.24 in Sect. 3.5.5 summarize these characteristics, under weak-subkey conditions for both odd-numbered and even-numbered rounds.

Unlike MESH-64, there are unkeyed $\odot$ operations in the MA-box of MESH-128. In order to cover the largest possible number of rounds and to maximize the probability of the differentials, and consequently of the whole boomerang, we exploit only 1-round characteristics that bypass these unkeyed $\odot$ operations.

Based on the given 1-round characteristics and the unkeyed-$\odot$ operations, the longest boomerangs found can cover 5.5 rounds of MESH-128. One such boomerang consists of the differentials

$$\Delta = (\nu, 0, 0, 0, \nu, 0, 0, 0) \stackrel{1r}{\rightarrow} (\nu, \nu, 0, 0, 0, 0, 0, 0) \stackrel{0.5r}{\rightarrow} (\nu, \nu, 0, 0, 0, 0, 0, 0) = \Delta^*,$$

and

$$\nabla = \nabla^* = (\nu, \nu, \nu, \nu, \nu, \nu, \nu, \nu)$$

covering 3 rounds. It is an iterative characteristic. Both characteristics hold with certainty, provided the following subkeys are weak: $Z_1^{(1)}$, $Z_2^{(2)}$, $Z_1^{(3)}$, $Z_3^{(3)}$, $Z_6^{(3)}$, $Z_8^{(3)}$, $Z_2^{(4)}$, $Z_4^{(4)}$, $Z_5^{(4)}$, $Z_7^{(4)}$, $Z_1^{(5)}$, $Z_3^{(5)}$, $Z_6^{(5)}$, $Z_8^{(5)}$. Assuming each weak-subkey restriction holds independently, the weak-key class size is estimated as $2^{256-15\cdot 14} = 2^{46}$.

There is a gap consisting of the second MA half-round, but only on quartet is enough to cross it. This situation is analogous to that of the boomerangs (13) and (14) in Table 3.87. The 6th KM half-round is not covered by either differential pattern, which implies that the boomerang direction is bottom-up.

The attack proceeds as follows: prepare two sets of ciphertexts in which the first, third, sixth and eighth words differ by ν, while the second, fourth, fifth and seventh words assume 2^8 random values each. This text structure can form $(2^8 \cdot 2^8 \cdot 2^8 \cdot 2^8)^2 = 2^{64}$ pairs of the form $(\nu, \delta_1, \nu, \delta_2, \delta_3, \nu, \delta_4, \nu)$. After decryption by the last KM half-round, it is expected that each possible

16-bit difference is suggested by $(\delta_1, Z_2^{(6)})$, by $(\delta_2, Z_4^{(6)})$, by $(\delta_3, Z_5^{(6)})$ and by $(\delta_4, Z_7^{(6)})$ about $2^{64-4\cdot 16} = 1$, on average, and that is enough to cross the gap.

The attack complexities are $4 \cdot (2^8)^4 = 2^{34}$ chosen texts and 2^{34} 5.5-round MESH-128 computations.

3.13 Linear Cryptanalysis

Linear cryptanalysis (LC) is a known-plaintext attack extensively developed by M. Matsui and originally applied to the FEAL block cipher [135].

Previous works and ideas related to LC include patterns in the S-boxes of the DES and attacks on FEAL-4 and FEAL-6 block ciphers [170, 176, 167, 131, 132, 130].

Similar to DC, LC became a benchmark analysis technique for any new block cipher design, as well as for other cryptographic primitives such as stream ciphers [64, 141].

Basic tools in LC involve the concepts of a *dot product* and a *linear relation*:

Definition 3.15. [143]
For n-bit strings $X = x_{n-1}x_{n-2}\ldots x_0$ and bitmask $\Gamma X = [i_1, i_2, \ldots, i_t]$, where i_j denotes the bit positions in X such that $x_{i_j} = 1$, the dot product (or inner product) of X and ΓX is denoted

$$X[i_1, i_2, \ldots, i_t] = X \cdot \Gamma X = x_{i_1} \oplus x_{i_2} \oplus \ldots \oplus x_{i_t}.$$

ΓX is called a bitmask (or simply mask), and the resulting bitwise AND-product corresponds to the 1-bit result (or *parity*) of X under ΓX.

In general, we distinguish between trivial bitmasks ($\Gamma X = 0$) and non-trivial ones ($\Gamma X \neq 0$).

Definition 3.16. A *linear relation* for a mapping F is a dot-product involving its input X and output $F(X)$ under fixed bitmasks ΓX and ΓY, interpreted as bit strings. The linear relation models a (nonzero) correlation among these bits. This relationship is denoted

$$X \cdot \Gamma X \oplus F(X) \cdot \Gamma Y = 0. \tag{3.167}$$

The linear relation (3.167) is depicted as an equality, but in fact, most of the linear relations are approximations that hold with a certain probability and bias.

Let, p_F denote the probability that (3.167) holds:

$$p_F = \Pr(X \cdot \Gamma X = F(X) \cdot \Gamma Y), \text{ with } 0 \leq p_F \leq 1. \tag{3.168}$$

The *deviation* associated with (3.168) is simply $p'_F = p_F - \frac{1}{2}$ and denotes how the parity of (3.167) deviates from that of the uniform distribution. The bias of (3.167) is denoted simply

$$\epsilon = |p_F - \tfrac{1}{2}|. \tag{3.169}$$

Typical linear analyses of block ciphers are performed using a bottom-up approach: from the smallest components up to larger ones.

The path formed by nontrivial linear masks via the concatenation of (nontrivial) bitmasks into linear relations across consecutive components of a cipher forms the so called *linear trail(s)*. This means that a linear trail connects internal components which effectively participate in a linear relation.

For instance, Fig. 3.25(a) shows (in red) an example of *narrow* linear trail across 5-round IDEA, while Fig. 3.25(b) shows (in red) an example of *wide* linear trail. A narrow trail affects relatively few internal cipher components along its path, while a wide trail encompasses relatively many components along its path. A trail is very useful to visualize which components are active in a linear distinguisher. In particular, for Lai-Massey designs, the trails identify which keyed operations are active. The most relevant keyed operations is the keyed-$\odot$, because they lead to weak-subkey conditions in LC. Intuitively, narrow trails are preferred for cryptanalysis, since they minimize the number of linearly active cipher components and therefore help keep the overall bias as high as possible. The wide trail shown in Fig. 3.25(b), although affecting all four words in a block, on the other hand succeeds in bypassing the MA-box in every round. Moreover, this trail represents a 1-round iterative linear relation.

Not coincidentally, linear trails and linear relations have been used almost interchangeably. On the one hand, linear trails allows us to visualize the propagation of bitmasks in details, across all internal cipher components. Linear relations, on the other hand, are more formal, structured objects, with ordered numerical values (bitmasks) and associated biases. In a sense, they complement each other. For instance, the trail depicted in red in Fig. 3.25(a) corresponds to the 5-round linear relation $\Gamma = ((\gamma, 0, \gamma, 0), (\gamma, \gamma, 0, 0), (0, \gamma, \gamma, 0), (\gamma, 0, \gamma, 0), (\gamma, \gamma, 0, 0), (0, \gamma, \gamma, 0))$, where $\gamma = 1$. The bias for every round is 2^{-1}, as long as the active multiplicative subkeys along the trail are weak.

Consider the linear mapping: $Y = X \oplus W$ with $X, Y, W \in \mathbb{Z}_2^n$. Let the bitmasks be denoted $\Gamma X = [i_1, \ldots, i_t]$, $\Gamma Y = [j_1, \ldots, j_r]$ and $\Gamma W = [l_1, \ldots, l_g] \in \mathbb{Z}_2^n$, with $0 \leq t, r, g < n$. The corresponding linear relation is

$$Y \cdot \Gamma Y = X \cdot \Gamma X \oplus W \cdot \Gamma W. \tag{3.170}$$

If there is an index d, with $0 \leq d \leq n-1$, such that i_d, j_d and l_d are not simultaneously zero or not simultaneously nonzero, then (3.170) is violated, that is, at the d-th bit position, we end up with $y_{i_d} = x_{j_d}$ or $y_{i_d} = w_{l_d}$ or $x_{i_d} \oplus w_{l_d} = 0$. In this case, the bias becomes $\epsilon = 0$. Therefore, the bitmasks must be equal: $\Gamma Y = \Gamma X = \Gamma W$ and $t = r = g$, which means (3.170) is satisfied bitwise and therefore $\epsilon = |1 - 1/2| = 2^{-1}$.

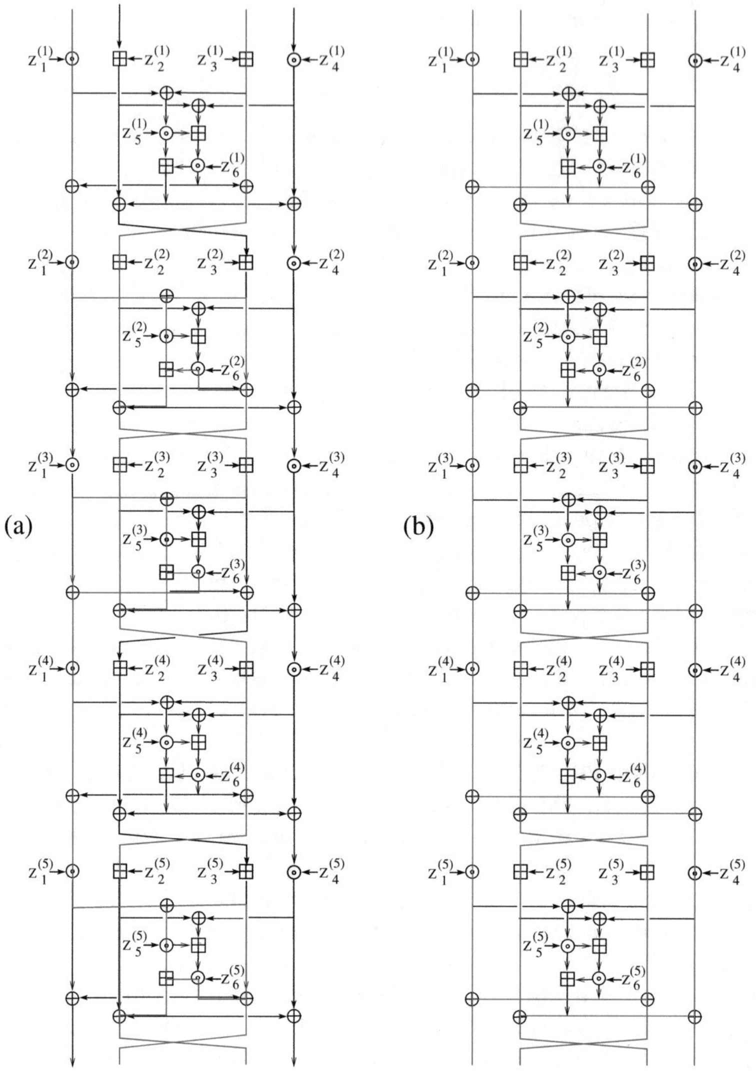

Fig. 3.25 Examples (in red) of a: (a) narrow linear trail; (b) wide linear trail.

As another example, let $F(X) = (X, X)$ denote the splitting function that simply outputs two copies of its input. This is a very usual internal component in several Feistel Network ciphers [67]. Let ΓI denote the input bit mask and $(\Gamma O_1, \Gamma O_2)$ denote the output bit masks. The linear relation is $X \cdot \Gamma I = F(X) \cdot (\Gamma O_1, \Gamma O_2) = X \cdot \Gamma O_1 \oplus X \cdot \Gamma O_2$. Therefore, $X \cdot \Gamma I = X \cdot (\Gamma O_1 \oplus \Gamma O_2)$ must hold for a nonzero bias.

This relationship between the input and output masks for F is known as the *branching rule for bitmasks*: the input mask is the exclusive-or of the output masks.

The same reasoning applies to mappings that split the input into more than two copies. If $G(X) = (X, X, X)$, let ΓI and $(\Gamma O_1, \Gamma O_2, \Gamma O_3)$ denote the input and output masks, respectively. Then, $\Gamma I = \Gamma O_1 \oplus \Gamma O_2 \oplus \Gamma O_3$ might hold to guarantee a nonzero bias ϵ.

Next, consider the nonlinear mapping $Y = M(X, W) = X \boxplus W$, where $X, Y, W \in \mathbb{Z}_{2^n}$, X, Y are internal cipher data and $\boxplus$ denote addition modulo 2^n. Again, let the bit masks be denoted $\Gamma X = [i_1, \ldots, i_t]$, $\Gamma Y = [j_1, \ldots, j_r]$ and $\Gamma W = [l_1, \ldots, l_g] \in \mathbb{Z}_2^n$.

The most promising linear relation for M involves the least significant bit (LSB) position because it is free from carry bits: $\Gamma Y = \Gamma X = \Gamma W = 1$. This means that the linear relation becomes $y_0 = x_0 \oplus w_0$, which always holds. In this case, $\epsilon = |1 - 1/2| = 2^{-1}$.

For the second LSB, if we want to keep all bitmasks equal, then $\Gamma Y = \Gamma X = \Gamma W = 2$ and the carry bit into that position will decrease the bias since $y_1 = x_1 \oplus w_1 \oplus x_1 \cdot w_1$, which is a nonlinear relation. In this case, the approximation $y_1 = x_1 \oplus w_1$ will hold as long as $x_1 \cdot w_1 = 0$, which happens with probability 3/4, assuming randomly and uniformly distributed x_1 and w_1. Thus, $\epsilon = |3/4 - 1/2| = 2^{-2}$.

Likewise, approximating bits in more significant positions and keeping all bitmasks equal will decrease the bias exponentially fast because the carry bits avalanche in the right-to-left direction during the modular addition computation. Consider the third LSB position: $y_2 = x_2 \oplus w_2 \oplus (x_1 \oplus w_1) \cdot x_0 \cdot w_0 \oplus x_1 \cdot w_1$. In order to satisfy $y_2 = x_2 \oplus w_2$, we require that $(x_1 \oplus w_1) \cdot x_0 \cdot w_0 \oplus x_1 \cdot w_1 = 0$, which happens with probability $10/16 = 5/8$ and the bias is $\epsilon = |5/8 - 1/2| = 2^{-3}$.

If W were a secret (key), say K, then $Y = A(X, K) = X \boxplus K$ and the linear relation would become $Y \cdot \Gamma Y = (X \boxplus K) \cdot \Gamma X$. Since K is secret, thus unknown, we cannot approximate it individually. Similar to the discussion of the mapping $G(X, K) = X \boxplus K$ in Sect. 3.5, the linear analysis of $A(X, K)$ becomes key dependent, and a more comprehensive study is needed.

An exhaustive linear analysis including all possible combinations of input/output bitmasks across a give mapping F is called the *Linear Approximation Table (LAT)* of F. If $F : \mathbb{Z}_2^n \to \mathbb{Z}_2^m$, then its LAT is a $2^n \times 2^m$ table. The (i, j)-th entry $d_{i,j}$ of this LAT contains the deviation of the parity of the linear relation involving the input (i) and the output (j) bitmasks, over all possible input values:

$$d_{i,j} = \#\{X \in \mathbb{Z}_2^n | X \cdot \Gamma X = F(X) \cdot \Gamma Y\} - 2^{n-1}.$$

As an example, Table 3.88 shows the LAT of $A(X, 10)$ in $\mathbb{Z}_{2^4}$.

If $K = 15$ instead, then Table 3.89 shows the LAT of $A(X, K)$. The bias of the linear relations corresponding to the (i, j)-th entry in the LAT is

Table 3.88 LAT of $A(X, 10)$ for $X \in \mathbb{Z}_{2^4}$.

ΓX	ΓY															
	0	1	2	3	4	5	6	7	8	9	10	11	12	13	14	15
0	8	0	0	0	0	0	0	0	0	0	0	0	0	0	0	0
1	0	8	0	0	0	0	0	0	0	0	0	0	0	0	0	0
2	0	0	-8	0	0	0	0	0	0	0	0	0	0	0	0	0
3	0	0	0	-8	0	0	0	0	0	0	0	0	0	0	0	0
4	0	0	0	0	0	0	-8	0	0	0	0	0	0	0	0	0
5	0	0	0	0	0	0	0	-8	0	0	0	0	0	0	0	0
6	0	0	0	0	8	0	0	0	0	0	0	0	0	0	0	0
7	0	0	0	0	0	8	0	0	0	0	0	0	0	0	0	0
8	0	0	0	0	0	0	0	0	-4	0	4	0	4	0	4	0
9	0	0	0	0	0	0	0	0	0	-4	0	4	0	4	0	4
10	0	0	0	0	0	0	0	0	-4	0	4	0	-4	0	-4	0
11	0	0	0	0	0	0	0	0	0	-4	0	4	0	-4	0	-4
12	0	0	0	0	0	0	0	0	-4	0	-4	0	-4	0	4	0
13	0	0	0	0	0	0	0	0	0	-4	0	-4	0	-4	0	4
14	0	0	0	0	0	0	0	0	4	0	4	0	-4	0	4	0
15	0	0	0	0	0	0	0	0	0	4	0	4	0	-4	0	4

Table 3.89 LAT of $A(X, 15)$ for $X \in \mathbb{Z}_{2^4}$.

ΓX	ΓY															
	0	1	2	3	4	5	6	7	8	9	10	11	12	13	14	15
0	8	0	0	0	0	0	0	0	0	0	0	0	0	0	0	0
1	0	-8	0	0	0	0	0	0	0	0	0	0	0	0	0	0
2	0	0	0	8	0	0	0	0	0	0	0	0	0	0	0	0
3	0	0	-8	0	0	0	0	0	0	0	0	0	0	0	0	0
4	0	0	0	0	4	4	4	-4	0	0	0	0	0	0	0	0
5	0	0	0	0	-4	-4	4	-4	0	0	0	0	0	0	0	0
6	0	0	0	0	-4	4	4	4	0	0	0	0	0	0	0	0
7	0	0	0	0	-4	4	-4	-4	0	0	0	0	0	0	0	0
8	0	0	0	0	0	0	0	0	6	2	2	-2	2	-2	-2	2
9	0	0	0	0	0	0	0	0	-2	-6	2	-2	2	-2	-2	2
10	0	0	0	0	0	0	0	0	-2	2	2	6	2	-2	-2	2
11	0	0	0	0	0	0	0	0	-2	2	-6	-2	2	-2	-2	2
12	0	0	0	0	0	0	0	0	-2	2	2	-2	6	2	2	-2
13	0	0	0	0	0	0	0	0	-2	2	2	-2	-2	-6	2	-2
14	0	0	0	0	0	0	0	0	-2	2	2	-2	-2	2	2	6
15	0	0	0	0	0	0	0	0	-2	2	2	-2	-2	2	-6	-2

$$|d_{i,j}/2^n - 1/2|.$$

Note that the distribution of values in each LAT changes since the carry-bit distribution also changes (due to different K values).

Another usual nonlinear component in many modern block ciphers is a substitution box (S-box). See the appendix for definitions and properties of S-boxes.

Like modular addition, S-boxes provide for the *confusion* property in block ciphers. In ciphers such as DES [155] and Serpent [14], S-boxes are the only nonlinear components.

S-boxes used in practice typically have small dimensions, for instance, 3×3, 4×4 or 8×8, which makes them very efficient to store and compute.

Let $H(X) = S[X \oplus K]$ denote an internal cipher component combining an $n \times m$ S-box, with a secret key $K \in \mathbb{Z}_{2^n}$.

The input mask ΓX propagates intact across the $\oplus$ operation since it is a linear operation. So, the input mask to S is also ΓX, and the overall bias will be the bias of the linear approximation of S only. It means that a linear analysis of $S[X \oplus K]$ is the same as that of $S[X]$. It can be interpreted as if the fixed, unknown K just performed a permutation of the input data X, which does not affect the bitmask nor the bias of S.

Similar to the reasoning in a DC setting, for varying output masks ΓY the corresponding bias changes as well. For a fixed S, an exhaustive analysis that includes all possible output masks for all possible input masks ΓX is the Linear Approximation Table (LAT) of S.

For an $n \times m$-bit S-box S, the LAT has dimension $2^n \times 2^m$, accounting for all possible (trivial and nontrivial) 2^n input and 2^m output bitmasks.

The (i, j)-th entry in the LAT of an S-box S accounts for the value $d_{i,j} - 2^{n-1}$, where

$$d_{i,j} = \#\{X \in \mathbb{Z}_2^n | X \cdot i = S[X] \cdot j\}$$

and represents the deviation of the frequency with which the linear relation with bitmasks $(\Gamma X, \Gamma Y) = (i, j)$ holds for S. The bias due to the (i, j) entry of the LAT of S is

$$\epsilon_{i,j} = |d_{i,j}/2^n - 1/2| = |d_{i,j} - 2^{n-1}|/2^n.$$

The farther away $d_{i,j}$ is from 2^{n-1}, the closer it is to a linear or affine mapping.

The largest nontrivial entries in the LAT of S indicate the most biased linear relations across S.

Following the terminology in the Appendix, Sect. C, $\gamma_{\max}$ denotes the largest entry, corresponding to nontrivial bitmasks, in the LAT of an S-box S. Then, the S-box is termed linearly $\gamma_{\max}$-uniform [159].

As an example, consider the 4×4 S-box S_0 in [14]:

$$S_0 = [3, 8, 15, 1, 10, 6, 5, 11, 14, 13, 4, 2, 7, 0, 9, 12], \qquad (3.171)$$

which means $S_0[0] = 3$, $S_0[1] = 8$, ..., $S_0[15] = 12$. The LAT of S_0 is depicted in Table 3.90. Note that

- the largest entry in Table 3.90 corresponds to the trivial bitmask pair $(\Gamma X, \Gamma Y) = (0, 0)$. This S-box is invertible (a bijective mapping), which means that the zero input mask can only cause the zero output mask.

Table 3.90 Linear Approximation Table of the S-box S_0.

ΓX	ΓY															
	0	1	2	3	4	5	6	7	8	9	10	11	12	13	14	15
0	8	0	0	0	0	0	0	0	0	0	0	0	0	0	0	0
1	0	-2	-2	0	-2	0	0	-2	0	-2	2	4	-2	0	-4	2
2	0	2	-2	0	0	2	2	4	0	-2	-2	4	0	-2	2	0
3	0	0	-4	0	2	-2	-2	-2	0	-4	0	0	2	2	2	-2
4	0	0	0	0	0	0	0	0	0	0	0	0	4	-4	-4	-4
5	0	2	-2	4	-2	-4	0	2	0	2	2	0	2	0	0	2
6	0	-2	2	0	0	-2	-2	4	-4	-2	-2	0	0	2	-2	0
7	0	0	0	-4	2	-2	2	2	4	0	0	0	2	2	-2	2
8	0	-2	-2	0	2	0	0	2	0	-2	2	-4	-2	-4	0	2
9	0	0	0	0	0	4	-4	0	0	0	0	0	4	0	0	4
10	0	4	0	0	2	-2	-2	-2	0	0	-4	0	-2	-2	-2	2
11	0	-2	2	0	4	-2	-2	0	0	2	2	4	0	-2	2	0
12	0	-2	2	4	2	0	4	-2	0	-2	-2	0	2	0	0	2
13	0	-4	-4	0	0	0	0	0	0	4	-4	0	0	0	0	0
14	0	0	0	4	2	2	-2	2	4	0	0	0	-2	2	-2	-2
15	0	-2	2	0	-4	-2	-2	0	4	-2	-2	0	0	-2	2	0

- the largest entry for a nontrivial bitmask is $\gamma_{\max} = 4$, which corresponds, for instance, to $\Gamma X = 1 \rightarrow \Gamma Y = 11$ with an associated bias of $4/16 = 2^{-2}$.
- the magnitude of the sum of the entries in any single row i is 8.
- since this S-box is invertible, the magnitude of the sum of the entries in any single column j is also 8.

An S-box which has either a nonzero input or nonzero output bitmask is termed *active* in a LC setting. It means, that the S-box effectively participates in a linear trail. Otherwise, the S-box is called *passive*.

The best linear approximations to an S-box may depend on the nature of the S-box.

Definition 3.17. (Linearly active nonsurjective S-box)
A *linearly active* (or simply *active*) nonsurjective S-box $S : \mathbb{Z}_2^n \rightarrow \mathbb{Z}_2^m$, with $0 < n \leq m$, is a nonsurjective S-box that participates in a linear approximation and has only nonzero output bitmask.

Examples of ciphers which use nonsurjective S-boxes include Blowfish [169] and Khufu [137]. See Table C.1 in the Appendix.

Linear analysis of nonsurjective S-boxes has shown that linear approximations using nonzero masks only at the output [144] are more effective than those using nonzero masks at both the input and output.

Definition 3.18. (Linearly active noninjective S-box)
A *linearly active* (or simply *active*) noninjective S-box $S : \mathbb{Z}_2^n \rightarrow \mathbb{Z}_2^m$, with $n \geq m > 0$, is a noninjective S-box that participates in a linear approximation and has only nonzero input bitmask.

Examples of block ciphers that use noninjective S-boxes include DES [154] and s^2-DES [98].

Linear approximations of noninjective S-boxes may use both nonzero input and output masks, but experiments have shown that exploiting linear relationships involving only nonzero input masks lead to more effective attacks [130]. Fig. 3.26(b) shows an example of an (iterative) linear relation for DES that exploits the fact that its S-boxes are noninjective.

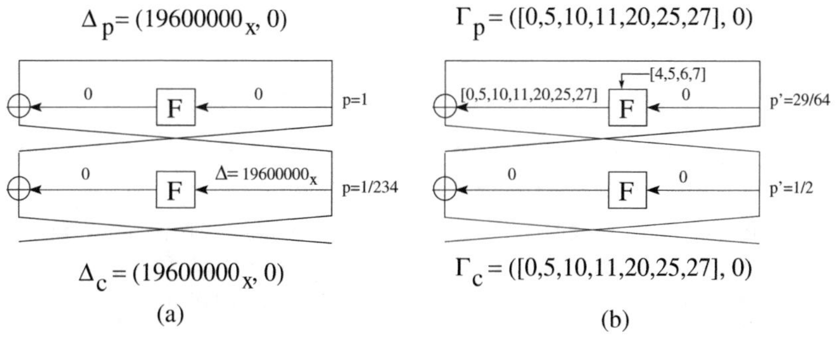

Fig. 3.26 (a) 2-round iterative characteristic for DES; (b) 2-round iterative linear relation for DES.

To avoid the drawbacks of both noninjective and nonsurjective S-boxes, a bijective S-box is suggested.

Definition 3.19. (Linearly active bijective S-box)
A *linearly active* (or simply *active*) bijective S-box $S : \mathbb{Z}_2^n \rightarrow \mathbb{Z}_2^n$, is a bijective S-box that participates in a linear approximation whose input/output bitmasks are either both zero or both nonzero.

Many modern block ciphers use bijective S-boxes, such as the AES cipher[13] [71]. Table C.1 in the Appendix provides more examples of bijective S-boxes.

A general approach to maximize the bias in linear relations covering larger and larger cipher components is to minimize the number of nonlinear components included in the linear trails. This technique is summarized in the following:

Hypothesis 2 *(Optimization Hypothesis) [81]*
The optimal linear approximation(s) to a cipher corresponds to linear relations that minimize the number of approximations to the nonlinear operations.

---

[13] The LAT of the AES S-box is available from the book's webpage http://www.springer.com/978-3-319-68272-3, along with its DDT. This LAT has nonzero bias only if both the input and output bitmasks are nonzero.

The concatenation of linear approximations of individual cipher components into a single approximation of a larger component operates under the assumption that the inputs to each component are independent and uniformly distributed [160]. Under this assumption, the overall bias of the concatenation of two or more linear relations is given by the Piling-Up lemma.

Lemma 3.1. *(Piling-Up Lemma) [131]*
For m independent, binary random variables $\theta_1, \theta_2, \ldots, \theta_m$, with nonzero biases $\epsilon_i = |p_i - \frac{1}{2}|$, for $1 \leq i \leq m$, the bias of $\theta = \theta_1 \oplus \theta_2 \oplus \ldots \theta_m$ can be approximated by

$$\epsilon = 2^{m-1} \cdot \prod_{i=1}^{m} \epsilon_i.$$

Studying the propagation of linear bitmasks across internal cipher components requires a trade-off between several criteria:

- minimizing the number of nonlinear components with nonzero input or output bitmasks: nontrivial masks propagating across nonlinear components typically account for the bulk of the bias in the construction of a linear trail.
  Looking at the LAT of the S-box S_0, the bias for the trivial difference $(\Gamma X, \Gamma Y) = (0, 0)$ is the highest among all entries. Using nontrivial bitmasks at either the input or the output of a nonlinear component, usually implies lower bias. Therefore, including nonlinear components in the trail is best avoided whenever possible in order to maximize the overall bias. Moreover, since the data complexity of linear attacks depends on the square of the inverse of the bias of the linear relation, one criterion to build effective LC attacks is to minimize the number of nonlinear components in any linear trail.
- focus on narrow trails: a *narrow trail* contains relatively few components along its route. Otherwise, the trail is called *wide*. The preference for narrow trials is a greedy approach: the less components in the trail, the higher the overall bias.
- focus on iterative trails: it is essential to propagate bitmasks not only across a single mapping but across multiple, consecutive mappings. Building trails in a bottom-up fashion means the focus is on local optimization. Ultimately, when the trail reaches a full round, one will have to restart building the trail from the bottom up, round after round, since the input mask from one round rarely equals the output mask.
  A strategy to minimize the total effort is to plan from the very beginning to have trails whose input bitmask equals the output bitmask. This way, the trail can be concatenated to itself. Those linear relations are called *iterative*. Using iterative linear relations, one can construct arbitrarily long linear relations with a fixed decrease in the bias after each concatenation [30].

Propagating bitmasks across larger and larger components will ultimately lead to trails covering full rounds. A linear relation covering full rounds leads to the concept of a multi-round linear relation or a (linear) distinguisher. This concept is the linear analogue of a differential characteristic in DC.

Definition 3.20. (r-round Linear Relation)
An *r-round linear relation* for an n-bit block cipher is an ordered sequence of n-bit masks defined as an $(r+1)$-tuple $(\gamma_0, \gamma_1, \ldots, \gamma_r)$ where $\Gamma P = \gamma_0$ is the input plaintext mask and $\Gamma C_i = \gamma_i$, for $1 \leq i \leq r$, is the i-th round output mask. For each pair (γ_i, γ_{i+1}), with $0 \leq i \leq r-1$, there is an associated transition bias $\epsilon_i > 0$ that the input text block with mask γ_i is correlated with the output text block with mask γ_{i+1}. This relationship is denoted $\gamma_i \stackrel{1r}{\rightarrow} \gamma_{i+1}$.

Similar to the concept of differentials in DC, the analogous concept in LC is the *linear hull* [159].

Definition 3.21. (r-round Linear Hull)
An *r-round linear hull* is a pair of bitmasks (γ_0, γ_r) where the input bitmask is $\Gamma P = \gamma_0$ and the output bitmask after r rounds is $\Gamma C = \gamma_r$.

Sometimes, there may be more than one single linear trail connecting the given input and output bitmasks $(\Gamma P, \Gamma C)$. This phenomenon is called a *linear hull*: an ensemble of disjoint linear trails with the same initial and final bitmasks [152].

A pragmatic approach to the linear analysis of iterated ciphers is to search for *iterative* linear relations, since these ciphers typically repeat the same round structure.

In [130], Matsui reported on iterative linear relations for DES in which two neighboring S-boxes were simultaneously active in the same round. Subsequently, Blocher and Dichtl in [40] described experiments on these iterative linear relations of DES and showed that the application of the Piling-Up Lemma was not justified because neighboring S-boxes in DES share some of its input bits and therefore the inputs are not independent nor uniformly distributed.

Nonetheless, Matsui's linear relations using at most one active S-box per round [132] are well approximated by the Piling-Up Lemma.

Another hypothesis that is often implicitly assumed in the construction of multiple-round linear relations is:

Hypothesis 3 *[77, 160]*
The bias of an effective linear relation is virtually independent of the key.

The term *effective* means that the linear relation has nonzero bias.

In the case of Feistel ciphers, such as the DES, Matsui presented in [133] a branch-and-bound algorithm to search for r-round linear relations, $3 \leq r \leq 20$, with the highest possible bias. Matsui's algorithm consists of an heuristic search in a graph where nodes represent 1-round linear relations with nonzero bias, and the edges connect linear relations that can be concatenated to cover multiple rounds. This algorithm requires an underestimation for the bias of the best i-round linear relation, for $1 \leq i \leq r$, in order to find the best r-round linear relation, that is, with the highest bias. The more precise the underestimation is, the faster the processing of the algorithm.

Many analogies (or dualities) can be observed between DC and LC. Some of them have been reported previously in [51].

Table 3.91 lists some concepts in DC and the corresponding concept in LC.

Table 3.91 Dual concepts in DC and LC.

Differential Cryptanalysis	Linear Cryptanalysis
difference	bitmask
differential characteristic	linear relation
differential trail	linear trail
probability $p > 0$	bias $\epsilon > 0$
Data Complexity*: $\approx c/p$	Data Complexity*: $\approx d/\epsilon^2$
DDT (of a mapping F)	LAT (of a mapping F)
CP setting	KP setting
$\Delta X \neq 0$ for nonsurjective S-boxes	$\Gamma Y \neq 0$ for nonsurjective S-boxes
$\Delta Y \neq 0$ for noninjetive S-boxes	$\Gamma X \neq 0$ for noninjetive S-boxes
differential	linear hull
2-round iter.charact. for DES: Fig. 3.26(a)	2-round iter.lin. rel. for DES: Fig. 3.26(b)
Impossible differentials [102]	Zero-correlation distinguishers [44]

$\star$: c and d are constants.

3.13.1 LC of PES Under Weak-Key Assumptions

In [57], Daemen *et al.* described differential attacks on the full 8.5-round IDEA under weak-key assumptions.

The linear analysis in this section is based on [149] and consists of a similar analysis as done by Daemen *et al.* on IDEA. Moreover, this analysis go beyond the full PES, that is, beyond 8.5 rounds assuming the key schedule of PES can be used to generate subkeys for more than 8.5 rounds.

The initial approach is to obtain 1-round linear relations and the corresponding weak-subkeys conditions for the linear relations to hold with maximum bias 2^{-1}. Only two 16-bit masks are used to construct the linear distinguishers: 0 and $\gamma = 1$. The bitmask 1 propagates across the modular addition with maximum bias, since it approximates only the least significant bit position which is not affected by carry bits.

The use of linear approximations exploiting the least significant bit position has been studied by Hawkes and O'Connor in [80, 81]. Using such approximations in LC minimizes the number of approximations to $\odot$ operations. Furthermore, the bias of this linear approximation using only the least significant bit (of a word) can be determined using the Piling-Up Lemma [131] (assuming independent round subkeys).

In [78], Harpes *et al.* studied a generalization of LC of IDEA using the quadratic residue function[14] [136] to approximate the $\odot$ operation with high bias. For the $\oplus$ and $\boxplus$ operators, the bitmasks used approximated only the least significant bit. Nonetheless, their approximation did not cover more than a single round.

In [57], Daemen *et al.* noted that the following property holds for all $x \in \mathbb{Z}_2^{16}$:

$$0 \odot x = \overline{x} \boxplus 2,$$

where $\overline{x}$ is the bitwise complement of x. Also, following [79], the following properties also hold with certainty and were used in [57] to derive linear relations for PES and IDEA:

$$\begin{array}{rlr}
x \oplus x^* = 2^{15} \Leftrightarrow (x \boxplus Z) \oplus (x^* \boxplus Z) = 2^{15} & & \forall\, Z \in \mathbb{Z}_2^{16} \\
x \oplus x^* = 2^{15} \Leftrightarrow (x \odot Z) \oplus (x^* \odot Z) = 2^{15}, & & \text{if and only if } Z \in \{0,1\} \\
1 \cdot x \oplus 1 \cdot Z = 1 \cdot (x \boxplus Z), & & \forall\, Z \in \mathbb{Z}_2^{16} \\
1 \cdot x \oplus 1 \cdot Z \oplus 1 = 1 \cdot (x \odot Z), & & \text{if } Z \in \{0,1\}
\end{array}$$

Table 3.92 lists all fifteen nontrivial 1-round linear relations of PES, under weak-key assumptions and wordwise linear bitmasks 0 and $\gamma = 1$.

Multiple-round linear relations are obtained by concatenating 1-round linear relations such that the output mask of one relation matches the input mask of the next relation. This way and under the corresponding weak-key assumptions, the final linear relation will hold with maximum bias 2^{-1}.

From Table 3.92, the multi-round linear relations are clustered into eleven relations covering 1 or 2 rounds:

- $(0,0,\gamma,0) \overset{1r}{\rightarrow} (0,\gamma,\gamma,\gamma) \overset{1r}{\rightarrow} (0,0,\gamma,0)$, which is 2-round iterative; the weak subkeys required are: $Z_5^{(i)}$, $Z_6^{(i)}$, $Z_2^{(i+1)}$, $Z_5^{(i+1)}$ and $Z_6^{(i+1)}$.
- $(0,0,\gamma,\gamma) \overset{1r}{\rightarrow} (0,\gamma,\gamma,0) \overset{1r}{\rightarrow} (0,0,\gamma,\gamma)$, which is 2-round iterative; the weak subkeys required are: $Z_5^{(i)}$, $Z_2^{(i+1)}$ and $Z_5^{(i+1)}$.
- $(\gamma,0,0,0) \overset{1r}{\rightarrow} (\gamma,\gamma,0,\gamma) \overset{1r}{\rightarrow} (\gamma,0,0,0)$, which is 2-round iterative; the weak subkeys required are: $Z_1^{(i)}$, $Z_5^{(i)}$, $Z_6^{(i)}$, $Z_1^{(i+1)}$, $Z_2^{(i+1)}$, $Z_5^{(i+1)}$ and $Z_6^{(i+1)}$.
- $(\gamma,0,0,\gamma) \overset{1r}{\rightarrow} (\gamma,\gamma,0,0) \overset{1r}{\rightarrow} (\gamma,0,0,\gamma)$, which is 2-round iterative; the weak subkeys required are: $Z_1^{(i)}$, $Z_5^{(i)}$, $Z_1^{(i+1)}$, $Z_2^{(i+1)}$ and $Z_5^{(i+1)}$.

[14] An element $q \in \mathbb{Z}$ is a quadratic residue modulo m if there exists $x \in \mathbb{Z}$ such that $x^2 \equiv q \bmod m$. For prime m, there are $(m+1)/2$ quadratic residues and $(m-1)/2$ quadratic non-residues.

Table 3.92 Nontrivial 1-round linear relations of PES.

1-round linear relations	Weak Subkeys (i-th round)	Linear Rel. across MA-box
$(0,0,0,\gamma) \stackrel{1r}{\rightarrow} (0,0,0,\gamma)$	$Z_6^{(i)}$	$(0,\gamma) \rightarrow (\gamma,0)$
$(0,0,\gamma,0) \stackrel{1r}{\rightarrow} (0,\gamma,\gamma,\gamma)$	$Z_5^{(i)}, Z_6^{(i)}$	$(\gamma,\gamma) \rightarrow (0,\gamma)$
$(0,0,\gamma,\gamma) \stackrel{1r}{\rightarrow} (0,\gamma,\gamma,0)$	$Z_5^{(i)}$	$(\gamma,0) \rightarrow (\gamma,\gamma)$
$(0,\gamma,0,0) \stackrel{1r}{\rightarrow} (0,\gamma,0,0)$	$Z_2^{(i)}, Z_6^{(i)}$	$(0,\gamma) \rightarrow (\gamma,0)$
$(0,\gamma,0,\gamma) \stackrel{1r}{\rightarrow} (0,\gamma,0,\gamma)$	$Z_2^{(i)}$	$(0,0) \rightarrow (0,0)$
$(0,\gamma,\gamma,0) \stackrel{1r}{\rightarrow} (0,0,\gamma,\gamma)$	$Z_2^{(i)}, Z_5^{(i)}$	$(\gamma,0) \rightarrow (\gamma,\gamma)$
$(0,\gamma,\gamma,\gamma) \stackrel{1r}{\rightarrow} (0,0,\gamma,0)$	$Z_2^{(i)}, Z_5^{(i)}, Z_6^{(i)}$	$(\gamma,\gamma) \rightarrow (0,\gamma)$
$(\gamma,0,0,0) \stackrel{1r}{\rightarrow} (\gamma,\gamma,0,\gamma)$	$Z_1^{(i)}, Z_5^{(i)}, Z_6^{(i)}$	$(\gamma,\gamma) \rightarrow (0,\gamma)$
$(\gamma,0,0,\gamma) \stackrel{1r}{\rightarrow} (\gamma,\gamma,0,0)$	$Z_1^{(i)}, Z_5^{(i)}$	$(\gamma,0) \rightarrow (\gamma,\gamma)$
$(\gamma,0,\gamma,0) \stackrel{1r}{\rightarrow} (\gamma,0,\gamma,0)$	$Z_1^{(i)}$	$(0,0) \rightarrow (0,0)$
$(\gamma,0,\gamma,\gamma) \stackrel{1r}{\rightarrow} (\gamma,0,\gamma,\gamma)$	$Z_1^{(i)}, Z_6^{(i)}$	$(0,\gamma) \rightarrow (\gamma,0)$
$(\gamma,\gamma,0,0) \stackrel{1r}{\rightarrow} (\gamma,0,0,\gamma)$	$Z_1^{(i)}, Z_2^{(i)}, Z_5^{(i)}$	$(\gamma,0) \rightarrow (\gamma,\gamma)$
$(\gamma,\gamma,0,\gamma) \stackrel{1r}{\rightarrow} (\gamma,0,0,0)$	$Z_1^{(i)}, Z_2^{(i)}, Z_5^{(i)}, Z_6^{(i)}$	$(\gamma,\gamma) \rightarrow (0,\gamma)$
$(\gamma,\gamma,\gamma,0) \stackrel{1r}{\rightarrow} (\gamma,\gamma,\gamma,0)$	$Z_1^{(i)}, Z_2^{(i)}, Z_6^{(i)}$	$(0,\gamma) \rightarrow (\gamma,0)$
$(\gamma,\gamma,\gamma,\gamma) \stackrel{1r}{\rightarrow} (\gamma,\gamma,\gamma,\gamma)$	$Z_1^{(i)}, Z_2^{(i)}$	$(0,0) \rightarrow (0,0)$

- $(0,0,0,\gamma) \stackrel{1r}{\rightarrow} (0,0,0,\gamma)$, which is 1-round iterative; the weak subkey required is $Z_6^{(i)}$.
- $(0,\gamma,0,0) \stackrel{1r}{\rightarrow} (0,\gamma,0,0)$, which is 1-round iterative; the weak subkeys required are $Z_2^{(i)}$ and $Z_6^{(i)}$.
- $(0,\gamma,0,\gamma) \stackrel{1r}{\rightarrow} (0,\gamma,0,\gamma)$, which is 1-round iterative; the weak subkey required is $Z_2^{(i)}$.
- $(\gamma,0,\gamma,0) \stackrel{1r}{\rightarrow} (\gamma,0,\gamma,0)$, which is 1-round iterative; the weak subkey required is $Z_1^{(i)}$.
- $(\gamma,0,\gamma,\gamma) \stackrel{1r}{\rightarrow} (\gamma,0,\gamma,\gamma)$, which is 1-round iterative; the weak subkeys required are: $Z_1^{(i)}$ and $Z_6^{(i)}$.
- $(\gamma,\gamma,\gamma,0) \stackrel{1r}{\rightarrow} (\gamma,\gamma,\gamma,0)$, which is 1-round iterative; the weak subkeys required are: $Z_1^{(i)}$, $Z_2^{(i)}$ and $Z_6^{(i)}$.
- $(\gamma,\gamma,\gamma,\gamma) \stackrel{1r}{\rightarrow} (\gamma,\gamma,\gamma,\gamma)$, which is 1-round iterative; the weak subkeys required are: $Z_1^{(i)}$ and $Z_2^{(i)}$.

There is a particularly large number of 1-round iterative linear relations in PES.

Based on Table 3.92 and the key schedule algorithm of PES (Sect. 2.1.2), the largest weak-key class for 8.5-round PES uses the iterative linear relation $(0,0,0,\gamma) \stackrel{1r}{\rightarrow} (0,0,0,\gamma)$, which holds with bias 2^{-1}, under the conditions listed in Table 3.93. Note that this linear relation minimizes the number of weak subkeys: there is only one weak subkey per round. Essentially, the key bits in positions 13–48, 66–94 and 109–123 have to be zero, while the remaining 48 key bits can have any value. It results in a weak-key class of size 2^{48}.

Table 3.93 8.5-round linear relation of PES under weak-key assumptions.

i-th Round	Linear Relation	$\text{msb}_{15}(Z_6^{(i)})$	Weak-key Class size
1	$(0,0,0,\gamma) \stackrel{1r}{\rightarrow} (0,0,0,\gamma)$	80–94	2^{113}
2	$(0,0,0,\gamma) \stackrel{1r}{\rightarrow} (0,0,0,\gamma)$	73–87	2^{106}
3	$(0,0,0,\gamma) \stackrel{1r}{\rightarrow} (0,0,0,\gamma)$	66–80	2^{99}
4	$(0,0,0,\gamma) \stackrel{1r}{\rightarrow} (0,0,0,\gamma)$	34–48	2^{84}
5	$(0,0,0,\gamma) \stackrel{1r}{\rightarrow} (0,0,0,\gamma)$	27–41	2^{77}
6	$(0,0,0,\gamma) \stackrel{1r}{\rightarrow} (0,0,0,\gamma)$	20–34	2^{70}
7	$(0,0,0,\gamma) \stackrel{1r}{\rightarrow} (0,0,0,\gamma)$	13–27	2^{63}
8	$(0,0,0,\gamma) \stackrel{1r}{\rightarrow} (0,0,0,\gamma)$	109–123	2^{48}
OT	$(0,0,0,\gamma) \stackrel{0.5r}{\rightarrow} (0,0,0,\gamma)$	—	2^{48}

In fact, the linear relation $(0,0,0,\gamma) \stackrel{1r}{\rightarrow} (0,0,0,\gamma)$ can be iterated beyond 8.5 rounds. For instance, across 16-round PES, $(0,0,0,\gamma) \stackrel{16r}{\rightarrow} (0,0,0,\gamma)$ holds for a weak-key class of size 2^{10}. For 17-round PES, $(0,0,0,\gamma) \stackrel{17r}{\rightarrow} (0,0,0,\gamma)$ holds only for the all-zero key $K = 0^{128}$. Actually, the all-zero key is the only user key for which *any* linear relation can be iterated for any number of rounds.

Another interesting linear relation is $(0,\gamma,0,\gamma) \stackrel{1r}{\rightarrow} (0,\gamma,0,\gamma)$, which for 17-round PES, holds for a weak-key class of size 2^7. For 17.5 rounds, it holds only for the all-zero key. On the other hand, shorter linear relations allows us to obtain larger weak-key classes, while also recovering subkeys at one of both ends of the linear relation.

Apart from key-recovery attacks, these linear relations provide useful information on PES:

- they help to study the diffusion power of the MA half-round of PES with respect to modern tools such as the linear branch number. See the Appendix, Sect. B.
  The set of linear relations in Table 3.92 indicates that the linear branch number *per round* of PES is only 2 (under weak-key assumptions). For instance, the 1-round relation $(0,0,0,\gamma) \stackrel{1r}{\rightarrow} (0,0,0,\gamma)$ has both the input and output masks with Hamming Weight one. So, even though PES provides full diffusion in a single round, a detailed analysis show that its linear branch number is much lower than ideal. For a 4-word block, such as in PES, an MDS code using a 4×4 MDS matrix would rather provide a branch number of $1 + 4 = 5$. See Appendix, Sect. B.1.
  In fact, since the given 1-round linear relation is iterative, the linear branch number for t-round PES, for $t > 1$, is also 2.

- given the appropriate weak key, the linear relations in Table 3.93 hold with maximum bias 2^{-1}. Looking the other way around, these linear relations can also be used as efficient membership tests for keys in a weak-key class. For instance, the relation $(0,0,0,\gamma) \overset{8.5r}{\rightarrow} (0,0,0,\gamma)$ allows us to efficiently identify all 2^{48} keys in the class. The test consists of $c \cdot (2^{-1})^{-2} = 4c$ encryptions (under known plaintexts) with a fixed constant c. For the DES cipher, Matsui showed experimentally that $c = 8$, for high success linear attacks.
  The sample key belongs to the weak-key class if the parity of the linear relation is constant after 8.5-round PES. The linear patterns inside any trail detailed in Table 3.92 are very sensitive to the given weak-subkey assumptions. So, if even a single multiplicative subkey in the trail is not weak, the linear patterns break down and the output parity (after 8.5 rounds) will not be constant. For a random key, the chance for the parity to be constant for the relation $(0,0,0,\gamma) \overset{8.5r}{\rightarrow} (0,0,0,\gamma)$ (as well as any other fixed bitmask), using d known plaintexts, is 2^{-d}.
  The parity computation involves 64 bits in a block. For a key space of 2^{128} keys, excluding the weak-key class of size 2^{48}, and using $4c$ known plaintexts, the number of keys satisfying a given linear relation is $(2^{-64})^{4c} \cdot (2^{128} - 2^{48}) < 1$, assuming $c \geq 1$.
- the linear relations in Table 3.92, under the corresponding weak-key assumptions, hold with maximum bias and can be used to distinguish the full 8.5-round PES from a random permutation with very low attack complexity. The cost is essentially $4c$ encryptions and the same number of known plaintexts and memory. Consequently, PES is not a PRP. See Sect. 1.7.
- usually, the larger the number of rounds of a given cipher the smaller the weak-key class becomes. But, the weak-key class size of PES never vanishes, due to the design of the key schedule algorithm. See Sect. 2.1.2. For instance, the 17-round linear relation $(0,0,0,\gamma) \overset{17r}{\rightarrow} (0,0,0,\gamma)$ holds with maximum bias only for the all-zero key. Beyond 17 rounds, it still holds but only for the all-zero key. In fact, the key $K = 0^{128}$ is very special. It is the only user key of PES for which **any** linear relation can be iterated for **any** number of rounds because $K = 0^{128}$ means that all (multiplicative) subkeys are weak, even those which do not belong to the linear trail.

3.13.2 LC of IDEA Under Weak-Key Assumptions

In [57], Daemen *et al.* described linear cryptanalytic attacks on the full 8.5-round IDEA using linear relations with low Hamming Weight (exploiting only the least significant bit of some 16-bit words in a block).

The first step is to obtain 1-round linear relations and the corresponding weak-subkeys restrictions for the relations to hold with maximum bias 2^{-1}. Only two 16-bit masks are used in the linear relations: 0 and $\gamma = 1$. The bitmask 1 propagates across the modular addition with maximum bias, since it approximates only the least significant bit position which is not affected by carry bits.

Table 3.94 lists all fifteen nontrivial 1-round linear relations of IDEA, under weak-key assumptions, and wordwise linear bitmasks 0 and $\gamma = 1$. They hold with maximum bias under weak-subkey assumptions. Multiple-round linear

Table 3.94 Nontrivial 1-round linear relations for IDEA.

1-round linear relation	Weak subkeys	Linear relation across MA-box
$(0,0,0,\gamma) \stackrel{1r}{\to} (0,0,\gamma,0)$	$Z_4^{(i)}, Z_6^{(i)}$	$(0,\gamma) \to (\gamma,0)$
$(0,0,\gamma,0) \stackrel{1r}{\to} (\gamma,0,\gamma,\gamma)$	$Z_5^{(i)}, Z_6^{(i)}$	$(\gamma,\gamma) \to (0,\gamma)$
$(0,0,\gamma,\gamma) \stackrel{1r}{\to} (\gamma,0,0,\gamma)$	$Z_4^{(i)}, Z_5^{(i)}$	$(\gamma,0) \to (\gamma,\gamma)$
$(0,\gamma,0,0) \stackrel{1r}{\to} (0,0,0,\gamma)$	$Z_6^{(i)}$	$(0,\gamma) \to (\gamma,0)$
$(0,\gamma,0,\gamma) \stackrel{1r}{\to} (0,0,\gamma,\gamma)$	$Z_4^{(i)}$	$(0,0) \to (0,0)$
$(0,\gamma,\gamma,0) \stackrel{1r}{\to} (\gamma,0,\gamma,0)$	$Z_5^{(i)}$	$(\gamma,0) \to (\gamma,\gamma)$
$(0,\gamma,\gamma,\gamma) \stackrel{1r}{\to} (\gamma,0,0,0)$	$Z_4^{(i)}, Z_5^{(i)}, Z_6^{(i)}$	$(\gamma,\gamma) \to (0,\gamma)$
$(\gamma,0,0,0) \stackrel{1r}{\to} (0,\gamma,\gamma,\gamma)$	$Z_1^{(i)}, Z_5^{(i)}, Z_6^{(i)}$	$(\gamma,\gamma) \to (0,\gamma)$
$(\gamma,0,0,\gamma) \stackrel{1r}{\to} (0,\gamma,0,\gamma)$	$Z_1^{(i)}, Z_4^{(i)}, Z_5^{(i)}$	$(\gamma,0) \to (\gamma,\gamma)$
$(\gamma,0,\gamma,0) \stackrel{1r}{\to} (\gamma,\gamma,0,0)$	$Z_1^{(i)}$	$(0,0) \to (0,0)$
$(\gamma,0,\gamma,\gamma) \stackrel{1r}{\to} (\gamma,\gamma,\gamma,0)$	$Z_1^{(i)}, Z_4^{(i)}, Z_6^{(i)}$	$(0,\gamma) \to (\gamma,0)$
$(\gamma,\gamma,0,0) \stackrel{1r}{\to} (0,\gamma,\gamma,0)$	$Z_1^{(i)}, Z_5^{(i)}$	$(\gamma,0) \to (\gamma,\gamma)$
$(\gamma,\gamma,0,\gamma) \stackrel{1r}{\to} (0,\gamma,0,0)$	$Z_1^{(i)}, Z_4^{(i)}, Z_5^{(i)}, Z_6^{(i)}$	$(\gamma,\gamma) \to (0,\gamma)$
$(\gamma,\gamma,\gamma,0) \stackrel{1r}{\to} (\gamma,\gamma,0,\gamma)$	$Z_1^{(i)}, Z_6^{(i)}$	$(0,\gamma) \to (\gamma,0)$
$(\gamma,\gamma,\gamma,\gamma) \stackrel{1r}{\to} (\gamma,\gamma,\gamma,\gamma)$	$Z_1^{(i)}, Z_4^{(i)}$	$(0,0) \to (0,0)$

relations are obtained by concatenating 1-round linear relations, such that the output mask from one relation matches the input mask to the next relation. This way and under weak-key assumptions, the final linear relation will hold with maximum bias 2^{-1}.

From Table 3.94, the multiple-round linear relations are clustered into five relations covering 1, 2, 3 or 6 rounds:

(1) $(0,0,0,\gamma) \stackrel{1r}{\to} (0,0,\gamma,0) \stackrel{1r}{\to} (\gamma,0,\gamma,\gamma) \stackrel{1r}{\to} (\gamma,\gamma,\gamma,0) \stackrel{1r}{\to} (\gamma,\gamma,0,\gamma) \stackrel{1r}{\to} (0,\gamma,0,0) \stackrel{1r}{\to} (0,0,0,\gamma)$, which is 6-round iterative; the weak subkeys required are: $Z_4^{(i)}$, $Z_6^{(i)}$, $Z_5^{(i+1)}$, $Z_6^{(i+1)}$, $Z_1^{(i+2)}$, $Z_4^{(i+2)}$, $Z_6^{(i+2)}$, $Z_1^{(i+3)}$, $Z_6^{(i+3)}$, $Z_1^{(i+4)}$, $Z_4^{(i+4)}$, $Z_5^{(i+4)}$, $Z_6^{(i+4)}$ and $Z_6^{(i+5)}$.

(2) $(0,0,\gamma,\gamma) \stackrel{1r}{\to} (\gamma,0,0,\gamma) \stackrel{1r}{\to} (0,\gamma,0,\gamma) \stackrel{1r}{\to} (0,0,\gamma,\gamma)$, which is 3-round iterative; the weak subkeys required are: $Z_4^{(i)}$, $Z_5^{(i)}$, $Z_1^{(i+1)}$, $Z_4^{(i+1)}$, $Z_5^{(i+1)}$ and $Z_4^{(i+2)}$.

(3) $(0,\gamma,\gamma,0) \stackrel{1r}{\rightarrow} (\gamma,0,\gamma,0) \stackrel{1r}{\rightarrow} (\gamma,\gamma,0,0) \stackrel{1r}{\rightarrow} (0,\gamma,\gamma,0)$ which is 3-round iterative; the weak subkeys required are: $Z_5^{(i)}$, $Z_1^{(i+1)}$, $Z_1^{(i+2)}$ and $Z_5^{(i+2)}$.

(4) $(0,\gamma,\gamma,\gamma) \stackrel{1r}{\rightarrow} (\gamma,0,0,0) \stackrel{1r}{\rightarrow} (0,\gamma,\gamma,\gamma)$, which is 2-round iterative; the weak subkeys required are: $Z_4^{(i)}$, $Z_5^{(i)}$, $Z_6^{(i)}$, $Z_1^{(i+1)}$, $Z_5^{(i+1)}$ and $Z_6^{(i+1)}$.

(5) $(\gamma,\gamma,\gamma,\gamma) \stackrel{1r}{\rightarrow} (\gamma,\gamma,\gamma,\gamma)$ which is 1-round iterative; the weak subkeys required are: $Z_1^{(i)}$ and $Z_4^{(i)}$.

Based on Table 3.94, the largest weak-key class for 8.5-round IDEA uses the 3-round iterative linear relation (3), which holds with bias 2^{-1} under the appropriate weak-key conditions. Note that this linear relation minimizes the number of weak subkeys among the five multiple-round linear relations: only 12 weak subkeys across the full 8.5-round IDEA.

According to the key schedule algorithm of IDEA, the key bits in positions 0–25, 29–71 and 75–110 have to be zero, while the remaining 23 key bits can have any value. It results in a weak-key class of size 2^{23}. Table 3.95 details the linear relation and the weak subkeys round-by-round.

Table 3.95 8.5-round linear relation of IDEA under weak-key assumptions.

i-th Round	Linear Relation	$\text{msb}_{15}(Z_1^{(i)})$	$\text{msb}_{15}(Z_5^{(i)})$	Weak-key Class size
1	$(\gamma,0,\gamma,0) \stackrel{1r}{\rightarrow} (\gamma,\gamma,0,0)$	0–14	—	2^{113}
2	$(\gamma,\gamma,0,0) \stackrel{1r}{\rightarrow} (0,\gamma,\gamma,0)$	96–110	57–71	2^{83}
3	$(0,\gamma,\gamma,0) \stackrel{1r}{\rightarrow} (\gamma,0,\gamma,0)$	—	50–64	2^{76}
4	$(\gamma,0,\gamma,0) \stackrel{1r}{\rightarrow} (\gamma,\gamma,0,0)$	82–96	—	2^{62}
5	$(\gamma,\gamma,0,0) \stackrel{1r}{\rightarrow} (0,\gamma,\gamma,0)$	75–89	11–25	2^{44}
6	$(0,\gamma,\gamma,0) \stackrel{1r}{\rightarrow} (\gamma,0,\gamma,0)$	—	4–18	2^{44}
7	$(\gamma,0,\gamma,0) \stackrel{1r}{\rightarrow} (\gamma,\gamma,0,0)$	36–50	—	2^{30}
8	$(\gamma,\gamma,0,0) \stackrel{1r}{\rightarrow} (0,\gamma,\gamma,0)$	29–44	93–107	2^{23}
OT	$(0,\gamma,\gamma,0) \stackrel{0.5r}{\rightarrow} (0,\gamma,\gamma,0)$	—	—	2^{23}

Shorter linear relations allows us to obtain larger weak-key classes, while also recovering subkeys at one or both ends of the linear relation.

Apart from key-recovery attacks, these linear relations provide useful information on IDEA:

- they help to study the diffusion power of IDEA with respect to modern tools such as the linear branch number. See the Appendix, Sect. B. The set of linear relations in Table 3.94 indicates that the linear branch number *per round* of IDEA is only 2 (under weak-key assumptions). For instance, the 1-round relation $(0,0,0,\gamma) \stackrel{1r}{\rightarrow} (0,0,\gamma,0)$ has both the input and output masks with Hamming Weight one. Thus, even though IDEA provides full diffusion in a single round, a detailed analysis shows that its linear branch number is much lower than ideal. For a 4-word block, such as in IDEA, an MDS code using a 4×4 MDS matrix would rather provide a linear branch number of $1 + 4 = 5$. See Appendix, Sect. B.1.

- given an appropriate weak key, the linear relations in Table 3.94 hold with maximum bias 2^{-1}. Looking the other way around, these linear relations can also be used as efficient membership tests for keys in a weak-key class. For instance, the relation $(\gamma, 0, \gamma, 0) \stackrel{8.5r}{\rightarrow} (\gamma, 0, \gamma, 0)$ allows us to efficiently identify all 2^{23} keys in the class. The test consists of $c \cdot (2^{-1})^{-2} = 4c$ encryptions (under known plaintexts) with a fixed constant c. For DES, Matsui showed experimentally that $c = 8$, for high success linear attacks. The sample key belongs to the weak-key class (with high probability) if the parity of the linear relation is constant after 8.5-round IDEA. The linear patterns inside any trail detailed in Table 3.94 are very sensitive to the given weak-subkey assumptions. So, if even a single multiplicative subkey in the trail is not weak, the linear patterns break down and the output bitmask (after 8.5 rounds) will be unpredictable. For a random key, the chance for the parity to hold constant for the relation $(\gamma, 0, \gamma, 0) \stackrel{8.5r}{\rightarrow} (\gamma, 0, \gamma, 0)$ (as well as any other fixed bitmask), using d known plaintexts, is 2^{-d}.
  The parity computation involves 64 bits in a block. For a key space of 2^{128} keys, excluding the weak-key class of size 2^{23}, and using $4c$ known plaintexts, the number of keys satisfying a given linear relation is $(2^{-64})^{4c} \cdot (2^{128} - 2^{23}) < 1$, assuming $c \geq 1$.
- the linear relations in Table 3.94, under the corresponding weak-key assumptions, hold with maximum bias and can be used to distinguish the full 8.5-round IDEA from a random permutation with very low attack complexity. The cost is essentially $4c$ encryptions and the same number of known plaintexts and memory. Consequently, IDEA is not a PRP. See Sect. 1.7.
- usually, the larger the number of rounds of a cipher the smaller the weak-key class becomes. But, the weak-key class size never vanishes for IDEA, due to the key schedule algorithm. See Sect. 2.2.2. For instance, the 14-round linear relation $(0, 0, \gamma, \gamma) \stackrel{14r}{\rightarrow} (0, \gamma, 0, \gamma)$ holds with maximum bias only for a weak-key class of size 2^4. Beyond 14 rounds, it still holds but only for the all-zero key. In fact, the key $K = 0^{128}$ is the only user key of IDEA for which **any** linear relation can be iterated for **any** number of rounds because $K = 0^{128}$ means that all (multiplicative) subkeys are weak, even those subkeys that do not belong to the linear trail.

3.13.3 New Linear Relations for Multiplication

In [187, 188], Yıldırım described new linear relations with high bias involving weak multiplicative subkeys for IDEA. These new linear relations use bit masks covering the two least significant bits of 16-bit words across the $\odot$ operation, and hold with certainty, that is, with maximum bias.

Let $X, Y, Z \in \mathrm{GF}(2^{16}+1)$ be 16-bit words such that $Y = X \odot Z$, where $X = (x_{15}, \ldots, x_0)$, $Y = (y_{15}, \ldots, y_0)$ and $Z = (z_{15}, \ldots, z_0)$. Table 3.96 lists linear relations for a single keyed $\odot$ operation. The first two rows in Table 3.96 represent linear relations described by Daemen in [57], that exploit only the least significant bits of the operands. The last four rows show the new linear relations from Yıldırım that extend Daemen's linear relations up to the two least significant bits of a word (except that bitmask 0002_x is missing).

Table 3.96 Weak subkey values and bitmasks for linear relation $X \cdot \Gamma X \oplus Z \cdot \Gamma Z = Y \cdot \Gamma Y$, with $Y = X \odot Z$.

Subkey Z	ΓX	ΓY	ΓZ
0	0001_x	0001_x	0001_x
1	0001_x	0001_x	0001_x
2	0001_x	0003_x	0001_x
$2^{16}-1$	0001_x	0003_x	0001_x
2^{15}	0003_x	0001_x	0001_x
$2^{15}+1$	0003_x	0001_x	0001_x

One small issue with Yıldırım weak subkeys is for instance that 2 and $2^{16}-1$ contain opposite bit patterns. Note that the $\mathrm{msb}_{14}(2) = 00000000000000_2$, while the $\mathrm{msb}_{14}(2^{16}-1) = 11111111111111_2$. These patterns contrast with Daemen's weak subkeys 0 and 1, both of whose fifteen most significant bits are zero. The latter weak subkeys allows us to identify weak-key classes in IDEA by imposing that the fifteen most significant bits be zero. The former weak subkeys, though, impose conflicting bit patterns: while 2 requires the 14 most significant bits to be zero, $2^{16}-1$ requires the same bits to be 1.

3.13.4 LC of MESH-64 Under Weak-Key Assumptions

The following linear analysis of MESH-64 is based on [151].

The initial approach is to obtain 1-round linear relations (under wordwise bitmasks 0 and 1) and the corresponding weak-subkeys conditions for the linear relations to hold with highest possible bias. Only two 16-bit masks are used in the linear distinguishers: 0 and $\gamma = 1$. The bitmask 1 propagates across the modular addition with maximum bias, since it approximates only the least significant bit position, which is not affected by carry bits.

Table 3.97 lists all fifteen 1-round linear relations of MESH-64, under weak-key assumptions, and wordwise bitmasks 0 and $\gamma = 1$. Bias is always 2^{-1} under the given weak-subkey conditions.

Table 3.97 Nontrivial 1-round linear relations of MESH-64.

1-round linear relation	Weak Subkeys (i-th round)		Linear Relation
	Odd-numbered rounds	Even-numbered rounds	across MA-box
$(0,0,0,\gamma) \overset{1r}{\to} (0,0,\gamma,0)$	$Z_4^{(i)}, Z_6^{(i)}, Z_7^{(i)}$	$Z_6^{(i)}, Z_7^{(i)}$	$(0,\gamma) \to (\gamma,0)$
$(0,0,\gamma,0) \overset{1r}{\to} (\gamma,0,0,0)$	$Z_5^{(i)}, Z_7^{(i)}$	$Z_3^{(i)}, Z_5^{(i)}, Z_7^{(i)}$	$(\gamma,0) \to (0,\gamma)$
$(0,0,\gamma,\gamma) \overset{1r}{\to} (\gamma,0,\gamma,0)$	$Z_4^{(i)}, Z_5^{(i)}, Z_6^{(i)}$	$Z_3^{(i)}, Z_5^{(i)}, Z_6^{(i)}$	$(\gamma,\gamma) \to (\gamma,\gamma)$
$(0,\gamma,0,0) \overset{1r}{\to} (0,0,0,\gamma)$	$Z_6^{(i)}, Z_7^{(i)}$	$Z_2^{(i)}, Z_6^{(i)}, Z_7^{(i)}$	$(0,\gamma) \to (\gamma,0)$
$(0,\gamma,0,\gamma) \overset{1r}{\to} (0,0,\gamma,\gamma)$	$Z_4^{(i)}$	$Z_2^{(i)}$	$(0,0) \to (0,0)$
$(0,\gamma,\gamma,0) \overset{1r}{\to} (\gamma,0,0,\gamma)$	$Z_5^{(i)}, Z_6^{(i)}$	$Z_2^{(i)}, Z_3^{(i)}, Z_5^{(i)}, Z_6^{(i)}$	$(\gamma,\gamma) \to (\gamma,\gamma)$
$(0,\gamma,\gamma,\gamma) \overset{1r}{\to} (\gamma,0,\gamma,\gamma)$	$Z_4^{(i)}, Z_5^{(i)}, Z_7^{(i)}$	$Z_2^{(i)}, Z_3^{(i)}, Z_5^{(i)}, Z_7^{(i)}$	$(\gamma,0) \to (0,\gamma)$
$(\gamma,0,0,0) \overset{1r}{\to} (0,\gamma,0,0)$	$Z_1^{(i)}, Z_5^{(i)}, Z_7^{(i)}$	$Z_5^{(i)}, Z_7^{(i)}$	$(\gamma,0) \to (0,\gamma)$
$(\gamma,0,0,\gamma) \overset{1r}{\to} (0,\gamma,\gamma,0)$	$Z_1^{(i)}, Z_4^{(i)}, Z_5^{(i)}, Z_6^{(i)}$	$Z_5^{(i)}, Z_6^{(i)}$	$(\gamma,\gamma) \to (\gamma,\gamma)$
$(\gamma,0,\gamma,0) \overset{1r}{\to} (\gamma,\gamma,0,0)$	$Z_1^{(i)}$	$Z_3^{(i)}$	$(0,0) \to (0,0)$
$(\gamma,0,\gamma,\gamma) \overset{1r}{\to} (\gamma,\gamma,\gamma,0)$	$Z_1^{(i)}, Z_4^{(i)}, Z_6^{(i)}, Z_7^{(i)}$	$Z_3^{(i)}, Z_6^{(i)}, Z_7^{(i)}$	$(0,\gamma) \to (\gamma,0)$
$(\gamma,\gamma,0,0) \overset{1r}{\to} (0,\gamma,0,\gamma)$	$Z_1^{(i)}, Z_5^{(i)}, Z_6^{(i)}$	$Z_2^{(i)}, Z_5^{(i)}, Z_6^{(i)}$	$(\gamma,\gamma) \to (\gamma,\gamma)$
$(\gamma,\gamma,0,\gamma) \overset{1r}{\to} (0,\gamma,\gamma,\gamma)$	$Z_1^{(i)}, Z_4^{(i)}, Z_5^{(i)}, Z_7^{(i)}$	$Z_2^{(i)}, Z_5^{(i)}, Z_7^{(i)}$	$(\gamma,0) \to (0,\gamma)$
$(\gamma,\gamma,\gamma,0) \overset{1r}{\to} (\gamma,\gamma,0,\gamma)$	$Z_1^{(i)}, Z_6^{(i)}, Z_7^{(i)}$	$Z_2^{(i)}, Z_3^{(i)}, Z_6^{(i)}, Z_7^{(i)}$	$(0,\gamma) \to (\gamma,0)$
$(\gamma,\gamma,\gamma,\gamma) \overset{1r}{\to} (\gamma,\gamma,\gamma,\gamma)$	$Z_1^{(i)}, Z_4^{(i)}$	$Z_2^{(i)}, Z_3^{(i)}$	$(0,0) \to (0,0)$

Multiple-round linear relations are obtained by concatenating 1-round linear relations and deriving the corresponding fraction of keys from the key space for which the relation holds. This fraction of keys are derived from the restrictions on subkeys in the 1-round linear relations. Nonetheless, the key schedule of MESH-64 does not have a linear mapping of subkey bits to user key bits such as in IDEA. The fraction of keys for the linear relations in MESH-64 was estimated from the weak-key class sizes obtained by exhaustive search in a mini-version of MESH-64 with 16-bit blocks, denoted MESH-64(16), where the key size was set to 32 bits.

Analysis of MESH-64(16) indicates that each subkey restriction (most significant 3 bits set to zero) is satisfied for a fraction of 2^{-3} or less of the key space. This consistent behavior was used to estimate the fraction of subkeys (and the weak-key class size) that satisfy a linear relation for MESH-64 as 2^{-15} per subkey.

Based on Table 3.97, the largest linear weak-key classes of linear relations starting from an odd-numbered round (for instance, the first round) are:

- $(0,\gamma,0,\gamma) \overset{1r}{\to} (0,0,\gamma,\gamma) \overset{1r}{\to} (\gamma,0,\gamma,0) \overset{1r}{\to} (\gamma,\gamma,0,0) \overset{1r}{\to} (0,\gamma,0,\gamma)$, which is 4-round iterative; the weak subkeys required are: $Z_4^{(1)}$, $Z_3^{(2)}$, $Z_5^{(2)}$, $Z_6^{(2)}$, $Z_1^{(3)}$, $Z_1^{(4)}$, $Z_5^{(4)}$ and $Z_6^{(4)}$.
  For MESH-64(16), this relation holds for a weak-key class of size 4, which corresponds to a fraction of $4 \cdot 2^{-32} = 2^{-28}$ of its key space. This fraction is less than $2^{-3 \cdot 4} = 2^{-24}$, that is to be expected if each subkey restriction

held independently. For MESH-64, the weak-key class size is estimated to be $2^{128-15\cdot 8} = 2^8$ for 4 rounds at most.

- $(\gamma, 0, \gamma, 0) \stackrel{1r}{\rightarrow} (\gamma, \gamma, 0, 0) \stackrel{1r}{\rightarrow} (0, \gamma, 0, \gamma) \stackrel{1r}{\rightarrow} (0, 0, \gamma, \gamma) \stackrel{1r}{\rightarrow} (\gamma, 0, \gamma, 0)$, which is 4-round iterative; the weak subkeys required are: $Z_1^{(1)}$, $Z_1^{(2)}$, $Z_5^{(2)}$, $Z_6^{(2)}$, $Z_4^{(3)}$, $Z_4^{(4)}$, $Z_5^{(4)}$ and $Z_6^{(4)}$.
  For MESH-64(16), this relation holds for a weak-key class of size 5, which is a fraction of $5 \cdot 2^{-3} \approx 2^{-30}$ of its key space. This fraction is less than $2^{-3\cdot 4} = 2^{-24}$, that is to be expected if each subkey restriction held independently. For MESH-64, the weak-key class size is estimated to be $2^{128-15\cdot 8} = 2^8$ for 4 rounds at most.

Attacks can also start from an even-numbered round. In this setting, the longest linear relations (for instance, starting from the second round) are:

- $(0, 0, 0, \gamma) \stackrel{1r}{\rightarrow} (0, 0, \gamma, 0) \stackrel{1r}{\rightarrow} (\gamma, 0, 0, 0) \stackrel{1r}{\rightarrow} (0, \gamma, 0, 0) \stackrel{1r}{\rightarrow} (0, 0, 0, \gamma)$, which is 4-round iterative; the weak subkeys required are: $Z_6^{(2)}$, $Z_7^{(2)}$, $Z_5^{(3)}$, $Z_7^{(3)}$, $Z_5^{(4)}$, $Z_7^{(4)}$, $Z_6^{(5)}$ and $Z_7^{(5)}$.
  For MESH-64(16), the weak-key class size is 31, which is a fraction of $31 \cdot 2^{-32} \approx 2^{-27}$ of its key space. This fraction is smaller than the $2^{-3\cdot 8} = 2^{-24}$ that would be expected if each subkey restriction held independently. For MESH-64, the weak-key class size is estimated to be $2^{128-15\cdot 8} = 2^8$, and the distinguisher would hold for 4 rounds at most.
- $(0, 0, \gamma, \gamma) \stackrel{1r}{\rightarrow} (\gamma, 0, \gamma, 0) \stackrel{1r}{\rightarrow} (\gamma, \gamma, 0, 0) \stackrel{1r}{\rightarrow} (0, \gamma, 0, \gamma) \stackrel{1r}{\rightarrow} (0, 0, \gamma, \gamma)$, which is 4-round iterative; the weak subkeys required are: $Z_3^{(2)}$, $Z_5^{(2)}$, $Z_6^{(2)}$, $Z_1^{(3)}$, $Z_2^{(4)}$, $Z_5^{(4)}$, $Z_6^{(4)}$ and $Z_4^{(5)}$. For MESH-64(16), the weak-key class size is 94, which is a fraction of $94 \cdot 2^{-32} \approx 2^{-25.44}$ of its key space. For MESH-64, the weak-key class size is estimated to be $2^{128-15\cdot 8} = 2^8$, and the distinguisher would hold for 4 rounds at most.
- $(\gamma, 0, 0, 0) \stackrel{1r}{\rightarrow} (0, \gamma, 0, 0) \stackrel{1r}{\rightarrow} (0, 0, 0, \gamma) \stackrel{1r}{\rightarrow} (0, 0, \gamma, 0) \stackrel{1r}{\rightarrow} (\gamma, 0, 0, 0)$, which is 4-round iterative; the weak subkeys required are: $Z_5^{(2)}$, $Z_7^{(2)}$, $Z_6^{(3)}$, $Z_7^{(3)}$, $Z_6^{(4)}$, $Z_7^{(4)}$, $Z_5^{(5)}$ and $Z_7^{(5)}$. For MESH-64(16), the weak-key class size is 115, which is a fraction of $115 \cdot 2^{-32} \approx 2^{-25.15}$ of its key space. For MESH-64, the weak-key class size is estimated to be $2^{128-15\cdot 8} = 2^8$, and the distinguisher would hold for 4 rounds at most.
- $(\gamma, 0, 0, \gamma) \stackrel{1r}{\rightarrow} (0, \gamma, \gamma, 0) \stackrel{1r}{\rightarrow} (\gamma, 0, 0, \gamma) \stackrel{1r}{\rightarrow} (0, \gamma, \gamma, 0) \stackrel{1r}{\rightarrow} (\gamma, 0, 0, \gamma)$, which is 2-round iterative; the weak subkeys required (for 4 rounds) are: $Z_5^{(2)}$, $Z_6^{(2)}$, $Z_5^{(3)}$, $Z_6^{(3)}$, $Z_5^{(4)}$, $Z_6^{(4)}$, $Z_5^{(5)}$ and $Z_6^{(5)}$. For MESH-64(16), the weak-key class size is 29, which is a fraction of $29 \cdot 2^{-32} \approx 2^{-27}$ of its key space. For MESH-64, the weak-key class size is estimated to be $2^{128-15\cdot 8} = 2^8$, and the distinguisher would hold for 4 rounds at most.
- $(\gamma, 0, \gamma, 0) \stackrel{1r}{\rightarrow} (\gamma, \gamma, 0, 0) \stackrel{1r}{\rightarrow} (0, \gamma, 0, \gamma) \stackrel{1r}{\rightarrow} (0, 0, \gamma, \gamma) \stackrel{1r}{\rightarrow} (\gamma, 0, \gamma, 0)$, which is 4-round iterative; the weak subkeys required are: $Z_3^{(2)}$, $Z_1^{(3)}$, $Z_5^{(3)}$, $Z_6^{(3)}$, $Z_2^{(4)}$, $Z_4^{(5)}$, $Z_5^{(5)}$ and $Z_6^{(5)}$. For MESH-64(16), the weak-key class size is

2, which is a fraction of $2 \cdot 2^{-32} = 2^{-31}$ of its key space. For MESH-64, the weak-key class size is estimated to be $2^{128-15\cdot 8} = 2^8$, and the distinguisher would hold for 4 rounds at most.

- $(\gamma,\gamma,0,0) \stackrel{1r}{\to} (0,\gamma,0,\gamma) \stackrel{1r}{\to} (0,0,\gamma,\gamma) \stackrel{1r}{\to} (\gamma,0,\gamma,0) \stackrel{1r}{\to} (\gamma,\gamma,0,0)$, which is 4-round iterative; the weak subkeys required are: $Z_2^{(2)}$, $Z_5^{(2)}$, $Z_6^{(2)}$, $Z_4^{(3)}$, $Z_3^{(4)}$, $Z_5^{(4)}$, $Z_6^{(4)}$ and $Z_1^{(5)}$. For MESH-64(16), the weak-key class size is 5, which is a fraction of $5 \cdot 2^{-32} \approx 2^{-30}$ of its key space. For MESH-64, the weak-key class size is estimated to be $2^{128-15\cdot 8} = 2^8$, and the distinguisher would hold for 4 rounds at most.
- $(\gamma,\gamma,\gamma,\gamma) \stackrel{1r}{\to} (\gamma,\gamma,\gamma,\gamma)$, which is 1-round iterative; the weak subkeys required (for 4 rounds) are: $Z_2^{(2)}$, $Z_3^{(2)}$, $Z_1^{(3)}$, $Z_4^{(3)}$, $Z_2^{(4)}$, $Z_3^{(4)}$, $Z_1^{(5)}$ and $Z_4^{(5)}$. For MESH-64(16), the weak-key class size is 11, which is a fraction of $11 \cdot 2^{-32} \approx 2^{-28.54}$ of its key space. For MESH-64, the weak-key class size is estimated to be less than $2^{128-15\cdot 8} = 2^8$, and the distinguisher would hold for 4 rounds at most.

This linear analysis also allows us to study the diffusion power of MESH-64 with respect to modern tools such as the linear branch number. See Appendix, Sect. B.

The set of linear relations in Table 3.97 means that the linear branch number *per round* of MESH-64 is only 2 (under weak-key assumptions). For instance, the 1-round relation $(0,0,0,\gamma) \stackrel{1r}{\to} (0,0,\gamma,0)$ has both the input and output masks with Hamming Weight one. So, even though MESH-64 provides full diffusion in a single round, a detailed analysis shows that its linear branch number is much lower than ideal. For a 4-word block, such as in MESH-64, an MDS code using a 4×4 MDS matrix would rather provide a linear branch number of $1+4=5$. See Appendix, Sect. B.1.

3.13.5 LC of MESH-96 Under Weak-Key Assumptions

The following linear analysis of MESH-96 is based on [151].

The analysis follows a bottom-up approach, starting from the smallest components: $\odot$, $\boxplus$ and $\oplus$. MESH-96 is the first MESH cipher to use the unkeyed $\odot$ operator.

For MESH-96 (and further Lai-Massey designs), Table 3.98 lists the bitmask propagation and the corresponding biases across the unkeyed-$\odot$ operation (under wordwise bitmasks 0 and $\gamma = 1$). Given that $c = a \odot b$, the corresponding linear relation is $c \cdot \Gamma c = (a \odot b) \cdot \Gamma = a \cdot \Gamma a \oplus b \cdot \Gamma b$. A similar linear approximation holds for $\boxdot$.

Table 3.99 lists the bitmask propagation and the corresponding biases across $\boxplus$ (under wordwise bitmasks 0 and 1). Given that $c = a \boxplus b$, the corresponding linear relation is: $c \cdot \Gamma c = (a \boxplus b) \cdot \Gamma = a \cdot \Gamma a \oplus b \cdot \Gamma b$. The

Table 3.98 Bitmask propagation across a single unkeyed $\odot$.

Γa	Γb	Γc	Bias	
			8-bit words	16-bit words
0	0	0	2^{-1}	2^{-1}
0	0	γ	0	0
0	γ	0	0	0
0	γ	γ	0	0
γ	0	0	0	0
γ	0	γ	0	0
γ	γ	0	0	0
γ	γ	γ	$2^{-6.50}$	$2^{-13.47}$

bitmask 1 propagates across the modular addition with maximum bias, since it approximates only the least significant bit position, which is not affected by carry bits.

Table 3.99 Bitmask propagation across a single unkeyed $\boxplus$; $\gamma = 1$.

Γa	Γb	Γc	Bias	
			8-bit words	16-bit words
0	0	0	2^{-1}	2^{-1}
0	0	γ	0	0
0	γ	0	0	0
0	γ	γ	0	0
γ	0	0	0	0
γ	0	γ	0	0
γ	γ	0	0	0
γ	γ	γ	2^{-1}	2^{-1}

The next step is to obtain 1-round linear relations and the corresponding weak-subkey conditions for the linear relations to hold with highest possible bias. Only two 16-bit masks are used in the linear distinguishers: 0 and $\gamma = 1$.

Tables 3.100 and 3.101 list all 63 nontrivial 1-round linear relations of MESH-96. These tables provide an exhaustive analysis of how linear relations propagate across a single round of MESH-96. These tables provide a more extensive coverage than the analyses in [151]. The bias was calculated using the Piling-up lemma [131].

Multiple-round linear relations are obtained by concatenating 1-round linear relations such that the output bitmask from one relation matches the input bitmask of the next relation.

From Tables 3.100 and 3.101, the multiple-round linear relations are clustered into eleven relations covering 1, 2, 3, 4, 6 or 12 rounds:

(1) $(0,0,0,0,0,\gamma) \stackrel{1r}{\rightarrow} (\gamma,\gamma,0,0,\gamma,0) \stackrel{1r}{\rightarrow} (\gamma,0,0,0,\gamma,\gamma) \stackrel{1r}{\rightarrow} (\gamma,0,0,\gamma,0,\gamma) \stackrel{1r}{\rightarrow}$
$(0, 0, 0, 0, \gamma, 0) \stackrel{1r}{\rightarrow} (\gamma,\gamma,\gamma,0,0,0) \stackrel{1r}{\rightarrow} (\gamma,0,\gamma,0,\gamma,0) \stackrel{1r}{\rightarrow} (\gamma,0,0,\gamma,\gamma,0) \stackrel{1r}{\rightarrow}$

Table 3.100 Nontrivial 1-round linear relations of MESH-96 under weak-subkey conditions.

1-round linear relations	Weak Subkeys $Z_j^{(i)}$ (i-th round) j values (odd i)	j values (even i)	Linear Relation across MA-box	Bias
$(0,0,0,0,0,\gamma) \stackrel{1r}{\rightarrow} (\gamma,\gamma,0,0,\gamma,0)$	7; 9	6; 7; 9	$(\gamma,0,\gamma) \rightarrow (\gamma,0,0)$	$2^{-25.94}$
$(0,0,0,0,\gamma,0) \stackrel{1r}{\rightarrow} (\gamma,\gamma,\gamma,0,0,0)$	5; 7; 9	7; 9	$(\gamma,0,0) \rightarrow (0,\gamma,0)$	2^{-1}
$(0,0,0,0,\gamma,\gamma) \stackrel{1r}{\rightarrow} (0,0,\gamma,0,\gamma,0)$	5	6	$(0,0,\gamma) \rightarrow (\gamma,\gamma,0)$	$2^{-25.94}$
$(0,0,0,\gamma,0,0) \stackrel{1r}{\rightarrow} (0,\gamma,\gamma,\gamma,\gamma,\gamma)$	9	4; 9	$(0,\gamma,\gamma) \rightarrow (0,0,\gamma)$	$2^{-25.94}$
$(0,0,0,\gamma,0,\gamma) \stackrel{1r}{\rightarrow} (\gamma,0,\gamma,\gamma,0,\gamma)$	7	4; 6; 7	$(\gamma,\gamma,0) \rightarrow (\gamma,0,\gamma)$	$2^{-25.94}$
$(0,0,0,\gamma,\gamma,0) \stackrel{1r}{\rightarrow} (\gamma,0,0,\gamma,\gamma,\gamma)$	5; 7	4; 7	$(\gamma,\gamma,\gamma) \rightarrow (0,\gamma,\gamma)$	$2^{-25.94}$
$(0,0,0,\gamma,\gamma,\gamma) \stackrel{1r}{\rightarrow} (0,\gamma,0,\gamma,0,\gamma)$	5; 9	4; 6; 9	$(0,\gamma,0) \rightarrow (\gamma,\gamma,\gamma)$	$2^{-25.94}$
$(0,0,\gamma,0,0,0) \stackrel{1r}{\rightarrow} (\gamma,\gamma,0,0,0,\gamma)$	3; 7; 9	7; 9	$(\gamma,0,\gamma) \rightarrow (\gamma,0,0)$	$2^{-25.94}$
$(0,0,\gamma,0,0,\gamma) \stackrel{1r}{\rightarrow} (0,0,0,0,\gamma,\gamma)$	3	6	$(0,0,0) \rightarrow (0,0,0)$	2^{-1}
$(0,0,\gamma,0,\gamma,0) \stackrel{1r}{\rightarrow} (0,0,\gamma,0,0,\gamma)$	3; 5	—	$(0,0,\gamma) \rightarrow (\gamma,\gamma,0)$	$2^{-25.94}$
$(0,0,\gamma,0,\gamma,\gamma) \stackrel{1r}{\rightarrow} (\gamma,\gamma,\gamma,0,\gamma,\gamma)$	3; 5; 7; 9	6; 7; 9	$(\gamma,0,0) \rightarrow (0,\gamma,0)$	2^{-1}
$(0,0,\gamma,\gamma,0,0) \stackrel{1r}{\rightarrow} (\gamma,0,\gamma,\gamma,\gamma,0)$	3; 7	4; 7	$(\gamma,\gamma,0) \rightarrow (\gamma,0,\gamma)$	$2^{-25.94}$
$(0,0,\gamma,\gamma,0,\gamma) \stackrel{1r}{\rightarrow} (0,\gamma,\gamma,\gamma,0,0)$	3; 9	4; 6; 9	$(0,\gamma,\gamma) \rightarrow (0,0,\gamma)$	$2^{-25.94}$
$(0,0,\gamma,\gamma,\gamma,0) \stackrel{1r}{\rightarrow} (0,\gamma,0,\gamma,\gamma,0)$	3; 5; 9	4; 9	$(0,\gamma,0) \rightarrow (\gamma,\gamma,\gamma)$	$2^{-25.94}$
$(0,0,\gamma,\gamma,\gamma,\gamma) \stackrel{1r}{\rightarrow} (\gamma,0,0,\gamma,0,0)$	3; 5; 7	4; 6; 7	$(\gamma,\gamma,\gamma) \rightarrow (0,\gamma,\gamma)$	$2^{-25.94}$
$(0,\gamma,0,0,0,0) \stackrel{1r}{\rightarrow} (\gamma,\gamma,0,\gamma,0,0)$	7; 9	2; 7; 9	$(\gamma,0,0) \rightarrow (0,\gamma,0)$	2^{-1}
$(0,\gamma,0,0,0,\gamma) \stackrel{1r}{\rightarrow} (0,0,0,\gamma,\gamma,0)$	—	2; 6	$(0,0,\gamma) \rightarrow (\gamma,\gamma,0)$	$2^{-25.94}$
$(0,\gamma,0,0,\gamma,0) \stackrel{1r}{\rightarrow} (0,0,\gamma,\gamma,0,0)$	5	2	$(0,0,0) \rightarrow (0,0,0)$	2^{-1}
$(0,\gamma,0,0,\gamma,\gamma) \stackrel{1r}{\rightarrow} (\gamma,\gamma,\gamma,\gamma,\gamma,0)$	5; 7; 9	2; 6; 7; 9	$(\gamma,0,\gamma) \rightarrow (\gamma,0,0)$	$2^{-25.94}$
$(0,\gamma,0,\gamma,0,0) \stackrel{1r}{\rightarrow} (\gamma,0,\gamma,0,\gamma,\gamma)$	7	2; 4; 7	$(\gamma,\gamma,\gamma) \rightarrow (0,\gamma,\gamma)$	$2^{-25.94}$
$(0,\gamma,0,\gamma,0,\gamma) \stackrel{1r}{\rightarrow} (0,\gamma,\gamma,0,0,\gamma)$	9	2; 4; 6; 9	$(0,\gamma,0) \rightarrow (\gamma,\gamma,\gamma)$	$2^{-25.94}$
$(0,\gamma,0,\gamma,\gamma,0) \stackrel{1r}{\rightarrow} (0,\gamma,0,0,\gamma,\gamma)$	5; 9	2; 4; 9	$(0,\gamma,\gamma) \rightarrow (0,0,\gamma)$	$2^{-25.94}$
$(0,\gamma,0,\gamma,\gamma,\gamma) \stackrel{1r}{\rightarrow} (\gamma,0,0,0,0,\gamma)$	5; 7	2; 4; 6; 7	$(\gamma,\gamma,0) \rightarrow (\gamma,0,\gamma)$	$2^{-25.94}$
$(0,\gamma,\gamma,0,0,0) \stackrel{1r}{\rightarrow} (0,0,0,\gamma,0,\gamma)$	3	2	$(0,0,\gamma) \rightarrow (\gamma,\gamma,0)$	$2^{-25.94}$
$(0,\gamma,\gamma,0,0,\gamma) \stackrel{1r}{\rightarrow} (\gamma,\gamma,0,\gamma,\gamma,\gamma)$	3; 7; 9	2; 6; 7; 9	$(\gamma,0,0) \rightarrow (0,\gamma,0)$	2^{-1}
$(0,\gamma,\gamma,0,\gamma,0) \stackrel{1r}{\rightarrow} (\gamma,\gamma,\gamma,\gamma,0,\gamma)$	3; 5; 7; 9	2; 7; 9	$(\gamma,0,\gamma) \rightarrow (\gamma,0,0)$	$2^{-25.94}$
$(0,\gamma,\gamma,0,\gamma,\gamma) \stackrel{1r}{\rightarrow} (0,0,\gamma,\gamma,\gamma,\gamma)$	3; 5	2; 6	$(0,0,0) \rightarrow (0,0,0)$	2^{-1}
$(0,\gamma,\gamma,\gamma,0,0) \stackrel{1r}{\rightarrow} (0,\gamma,\gamma,0,\gamma,0)$	3; 9	2; 4; 9	$(0,\gamma,0) \rightarrow (\gamma,\gamma,\gamma)$	$2^{-25.94}$
$(0,\gamma,\gamma,\gamma,0,\gamma) \stackrel{1r}{\rightarrow} (\gamma,0,\gamma,0,0,0)$	3; 7	2; 4; 6; 7	$(\gamma,\gamma,\gamma) \rightarrow (0,\gamma,\gamma)$	$2^{-25.94}$
$(0,\gamma,\gamma,\gamma,\gamma,0) \stackrel{1r}{\rightarrow} (\gamma,0,0,0,\gamma,0)$	3; 5; 7	2; 4; 7	$(\gamma,\gamma,0) \rightarrow (\gamma,0,\gamma)$	$2^{-25.94}$
$(0,\gamma,\gamma,\gamma,\gamma,\gamma) \stackrel{1r}{\rightarrow} (0,\gamma,0,0,0,0)$	3; 5; 9	2; 4; 6; 9	$(0,\gamma,\gamma) \rightarrow (0,0,\gamma)$	$2^{-25.94}$
$(\gamma,0,0,0,0,0) \stackrel{1r}{\rightarrow} (\gamma,0,\gamma,\gamma,\gamma,\gamma)$	1; 9	9	$(0,\gamma,\gamma) \rightarrow (0,0,\gamma)$	$2^{-25.94}$

Table 3.101 Nontrivial 1-round linear relations of MESH-96 under weak-subkey conditions (cont.)

1-round linear relations	Weak Subkeys $Z_j^{(i)}$ (i-th round)		Linear Relation	Bias
	j value (odd i)	j values (even i)	across MA-box	
$(\gamma,0,0,0,0,\gamma) \xrightarrow{1r} (0,\gamma,\gamma,\gamma,0,\gamma)$	1; 7	6; 7	$(\gamma,\gamma,0) \rightarrow (\gamma,0,\gamma)$	$2^{-25.94}$
$(\gamma,0,0,0,\gamma,0) \xrightarrow{1r} (0,\gamma,0,\gamma,\gamma,\gamma)$	1; 5; 7	7	$(\gamma,\gamma,\gamma) \rightarrow (0,\gamma,\gamma)$	$2^{-25.94}$
$(\gamma,0,0,0,\gamma,\gamma) \xrightarrow{1r} (\gamma,0,0,\gamma,0,\gamma)$	1; 5; 9	6; 9	$(0,\gamma,0) \rightarrow (\gamma,\gamma,\gamma)$	$2^{-25.94}$
$(\gamma,0,0,\gamma,0,0) \xrightarrow{1r} (\gamma,\gamma,0,0,0,0)$	1	4	$(0,0,0) \rightarrow (0,0,0)$	2^{-1}
$(\gamma,0,0,\gamma,0,\gamma) \xrightarrow{1r} (0,0,0,0,\gamma,0)$	1; 7; 9	4; 6; 7; 9	$(\gamma,0,\gamma) \rightarrow (\gamma,0,0)$	$2^{-25.94}$
$(\gamma,0,0,\gamma,\gamma,0) \xrightarrow{1r} (0,0,\gamma,0,0,0)$	1; 5; 7; 9	4; 7; 9	$(\gamma,0,0) \rightarrow (0,\gamma,0)$	2^{-1}
$(\gamma,0,0,\gamma,\gamma,\gamma) \xrightarrow{1r} (\gamma,\gamma,\gamma,0,\gamma,0)$	1; 5	4; 6	$(0,0,\gamma) \rightarrow (\gamma,\gamma,0)$	$2^{-25.94}$
$(\gamma,0,\gamma,0,0,0) \xrightarrow{1r} (0,\gamma,\gamma,\gamma,\gamma,0)$	1; 3; 7	7	$(\gamma,\gamma,0) \rightarrow (\gamma,0,\gamma)$	$2^{-25.94}$
$(\gamma,0,\gamma,0,0,\gamma) \xrightarrow{1r} (\gamma,0,\gamma,\gamma,0,0)$	1; 3; 9	6; 9	$(0,\gamma,\gamma) \rightarrow (0,0,\gamma)$	$2^{-25.94}$
$(\gamma,0,\gamma,0,\gamma,0) \xrightarrow{1r} (\gamma,0,0,\gamma,\gamma,0)$	1; 3; 5; 9	9	$(0,\gamma,0) \rightarrow (\gamma,\gamma,\gamma)$	$2^{-25.94}$
$(\gamma,0,\gamma,0,\gamma,\gamma) \xrightarrow{1r} (0,\gamma,0,\gamma,0,0)$	1; 3; 5; 7	6; 7	$(\gamma,\gamma,\gamma) \rightarrow (0,\gamma,\gamma)$	$2^{-25.94}$
$(\gamma,0,\gamma,\gamma,0,0) \xrightarrow{1r} (0,0,0,0,0,\gamma)$	1; 3; 7; 9	4; 7; 9	$(\gamma,0,\gamma) \rightarrow (\gamma,0,0)$	$2^{-25.94}$
$(\gamma,0,\gamma,\gamma,0,\gamma) \xrightarrow{1r} (\gamma,\gamma,0,0,\gamma,\gamma)$	1; 3	4; 6	$(0,0,0) \rightarrow (0,0,0)$	2^{-1}
$(\gamma,0,\gamma,\gamma,\gamma,0) \xrightarrow{1r} (\gamma,\gamma,\gamma,0,0,\gamma)$	1; 3; 5	4	$(0,0,\gamma) \rightarrow (\gamma,\gamma,0)$	$2^{-25.94}$
$(\gamma,0,\gamma,\gamma,\gamma,\gamma) \xrightarrow{1r} (0,0,\gamma,0,\gamma,\gamma)$	1; 3; 5; 7; 9	4; 6; 7; 9	$(\gamma,0,0) \rightarrow (0,\gamma,0)$	2^{-1}
$(\gamma,\gamma,0,0,0,0) \xrightarrow{1r} (0,\gamma,\gamma,0,\gamma,\gamma)$	1; 7	2; 7	$(\gamma,\gamma,\gamma) \rightarrow (0,\gamma,\gamma)$	$2^{-25.94}$
$(\gamma,\gamma,0,0,0,\gamma) \xrightarrow{1r} (\gamma,0,\gamma,0,0,\gamma)$	1; 9	2; 6; 9	$(0,\gamma,0) \rightarrow (\gamma,\gamma,\gamma)$	$2^{-25.94}$
$(\gamma,\gamma,0,0,\gamma,0) \xrightarrow{1r} (\gamma,0,0,0,\gamma,\gamma)$	1; 5; 9	2; 9	$(0,\gamma,\gamma) \rightarrow (0,0,\gamma)$	$2^{-25.94}$
$(\gamma,\gamma,0,0,\gamma,\gamma) \xrightarrow{1r} (0,\gamma,0,0,0,\gamma)$	1; 5; 7	2; 6; 7	$(\gamma,\gamma,0) \rightarrow (\gamma,0,\gamma)$	$2^{-25.94}$
$(\gamma,\gamma,0,\gamma,0,0) \xrightarrow{1r} (0,0,0,\gamma,0,0)$	1; 7; 9	2; 4; 7; 9	$(\gamma,0,0) \rightarrow (0,\gamma,0)$	2^{-1}
$(\gamma,\gamma,0,\gamma,0,\gamma) \xrightarrow{1r} (\gamma,\gamma,0,\gamma,\gamma,0)$	1	2; 4; 6	$(0,0,\gamma) \rightarrow (\gamma,\gamma,0)$	$2^{-25.94}$
$(\gamma,\gamma,0,\gamma,\gamma,0) \xrightarrow{1r} (\gamma,\gamma,\gamma,\gamma,0,0)$	1; 5	2; 4	$(0,0,0) \rightarrow (0,0,0)$	2^{-1}
$(\gamma,\gamma,0,\gamma,\gamma,\gamma) \xrightarrow{1r} (0,0,\gamma,\gamma,\gamma,0)$	1; 5; 7; 9	2; 4; 6; 7; 9	$(\gamma,0,\gamma) \rightarrow (\gamma,0,0)$	$2^{-25.94}$
$(\gamma,\gamma,\gamma,0,0,0) \xrightarrow{1r} (\gamma,0,\gamma,0,\gamma,0)$	1; 3; 9	2; 9	$(0,\gamma,0) \rightarrow (\gamma,\gamma,\gamma)$	$2^{-25.94}$
$(\gamma,\gamma,\gamma,0,0,\gamma) \xrightarrow{1r} (0,\gamma,\gamma,0,0,0)$	1; 3; 7	2; 6; 7	$(\gamma,\gamma,\gamma) \rightarrow (0,\gamma,\gamma)$	$2^{-25.94}$
$(\gamma,\gamma,\gamma,0,\gamma,0) \xrightarrow{1r} (0,\gamma,0,0,\gamma,0)$	1; 3; 5; 7	2; 7	$(\gamma,\gamma,0) \rightarrow (\gamma,0,\gamma)$	$2^{-25.94}$
$(\gamma,\gamma,\gamma,0,\gamma,\gamma) \xrightarrow{1r} (\gamma,0,0,0,0,0)$	1; 3; 5; 9	2; 6; 9	$(0,\gamma,\gamma) \rightarrow (0,0,\gamma)$	$2^{-25.94}$
$(\gamma,\gamma,\gamma,\gamma,0,0) \xrightarrow{1r} (\gamma,\gamma,0,\gamma,0,\gamma)$	1; 3	2; 4	$(0,0,1) \rightarrow (\gamma,\gamma,0)$	$2^{-25.94}$
$(\gamma,\gamma,\gamma,\gamma,0,\gamma) \xrightarrow{1r} (0,0,0,\gamma,\gamma,\gamma)$	1; 3; 7; 9	2; 4; 6; 7; 9	$(\gamma,0,0) \rightarrow (0,\gamma,0)$	2^{-1}
$(\gamma,\gamma,\gamma,\gamma,\gamma,0) \xrightarrow{1r} (0,0,\gamma,\gamma,0,\gamma)$	1; 3; 5; 7; 9	2; 4; 7; 9	$(\gamma,0,\gamma) \rightarrow (\gamma,0,0)$	$2^{-25.94}$
$(\gamma,\gamma,\gamma,\gamma,\gamma,\gamma) \xrightarrow{1r} (\gamma,\gamma,\gamma,\gamma,\gamma,\gamma)$	1; 3; 5	2; 4; 6	$(0,0,0) \rightarrow (0,0,0)$	2^{-1}

$(0,0,\gamma,0,0,0) \stackrel{1r}{\rightarrow} (\gamma,\gamma,0,0,0,\gamma) \stackrel{1r}{\rightarrow} (\gamma,0,\gamma,0,0,\gamma) \stackrel{1r}{\rightarrow} (\gamma,0,\gamma,\gamma,0,0) \stackrel{1r}{\rightarrow} (0,0,0,0,0,\gamma)$, which is 12-round iterative and holds with bias $2^{-250.4}$.
Starting from an odd-numbered round i, the following 35 subkeys have to be weak: $Z_7^{(i)}$, $Z_9^{(i)}$, $Z_2^{(i+1)}$, $Z_9^{(i+1)}$, $Z_1^{(i+2)}$, $Z_5^{(i+2)}$, $Z_9^{(i+2)}$, $Z_4^{(i+3)}$, $Z_6^{(i+3)}$, $Z_7^{(i+3)}$, $Z_9^{(i+3)}$, $Z_5^{(i+4)}$, $Z_7^{(i+4)}$, $Z_9^{(i+4)}$, $Z_2^{(i+5)}$, $Z_9^{(i+5)}$, $Z_1^{(i+6)}$, $Z_3^{(i+6)}$, $Z_5^{(i+6)}$, $Z_9^{(i+6)}$, $Z_4^{(i+7)}$, $Z_7^{(i+7)}$, $Z_9^{(i+7)}$, $Z_3^{(i+8)}$, $Z_7^{(i+8)}$, $Z_9^{(i+8)}$, $Z_2^{(i+9)}$, $Z_6^{(i+9)}$, $Z_9^{(i+9)}$, $Z_1^{(i+10)}$, $Z_3^{(i+10)}$, $Z_9^{(i+10)}$, $Z_4^{(i+11)}$, $Z_7^{(i+11)}$ and $Z_9^{(i+11)}$, which means $35/12 \approx 3$ weak subkeys/round.
Starting from an even-numbered round i, the following 31 subkeys have to be weak: $Z_6^{(i)}$, $Z_7^{(i)}$, $Z_9^{(i)}$, $Z_1^{(i+1)}$, $Z_5^{(i+1)}$, $Z_9^{(i+1)}$, $Z_6^{(i+2)}$, $Z_9^{(i+2)}$, $Z_1^{(i+3)}$, $Z_7^{(i+3)}$, $Z_9^{(i+3)}$, $Z_7^{(i+4)}$, $Z_9^{(i+4)}$, $Z_1^{(i+5)}$, $Z_3^{(i+5)}$, $Z_9^{(i+5)}$, $Z_9^{(i+6)}$, $Z_1^{(i+7)}$, $Z_5^{(i+7)}$, $Z_7^{(i+7)}$, $Z_9^{(i+7)}$, $Z_7^{(i+8)}$, $Z_9^{(i+8)}$, $Z_1^{(i+9)}$, $Z_9^{(i+9)}$, $Z_6^{(i+10)}$, $Z_9^{(i+10)}$, $Z_1^{(i+11)}$, $Z_3^{(i+11)}$, $Z_7^{(i+11)}$ and $Z_9^{(i+11)}$, which means $31/12 \approx 2.58$ weak subkeys/round.

(2) $(0,0,0,0,\gamma,\gamma) \stackrel{1r}{\rightarrow} (0,0,\gamma,0,\gamma,0) \stackrel{1r}{\rightarrow} (0,0,\gamma,0,0,\gamma) \stackrel{1r}{\rightarrow} (0,0,0,0,\gamma,\gamma)$ which is 3-round iterative and holds with bias $2^{-50.88}$.
Starting from an odd-numbered round i, the following two subkeys have to be weak: $Z_5^{(i)}$ and $Z_3^{(i+2)}$, which means $2/3 \approx 0.66$ weak subkeys/round.
Starting from an even-numbered round i, the following four subkeys have to be weak: $Z_6^{(i)}$, $Z_3^{(i+1)}$, $Z_5^{(i+1)}$ and $Z_6^{(i+2)}$, which means $4/3 \approx 1.33$ weak subkeys/round.

(3) $(0,0,0,\gamma,0,0) \stackrel{1r}{\rightarrow} (0,\gamma,\gamma,\gamma,\gamma,\gamma) \stackrel{1r}{\rightarrow} (0,\gamma,0,0,0,0) \stackrel{1r}{\rightarrow} (\gamma,\gamma,0,\gamma,0,0) \stackrel{1r}{\rightarrow} (0,0,0,\gamma,0,0)$, which is 4-round iterative and holds with bias $2^{-50.88}$.
Starting from an odd-numbered round i, the following 11 subkeys have to be weak: $Z_9^{(i)}$, $Z_2^{(i+1)}$, $Z_4^{(i+1)}$, $Z_6^{(i+1)}$, $Z_9^{(i+1)}$, $Z_7^{(i+2)}$, $Z_9^{(i+2)}$, $Z_2^{(i+3)}$, $Z_4^{(i+3)}$, $Z_7^{(i+3)}$ and $Z_9^{(i+3)}$, which means $11/4 = 2.75$ weak subkeys/round.
Starting from an even-numbered round i, the following 11 subkeys have to be weak: $Z_4^{(i)}$, $Z_9^{(i)}$, $Z_3^{(i+1)}$, $Z_5^{(i+1)}$, $Z_9^{(i+1)}$, $Z_2^{(i+2)}$, $Z_7^{(i+2)}$, $Z_9^{(i+2)}$, $Z_1^{(i+3)}$, $Z_7^{(i+3)}$ and $Z_9^{(i+3)}$, which means $11/4 = 2.75$ weak subkeys/round.

(4) $(0,0,0,\gamma,0,\gamma) \stackrel{1r}{\rightarrow} (\gamma,0,\gamma,\gamma,0,\gamma) \stackrel{1r}{\rightarrow} (\gamma,\gamma,0,0,\gamma,\gamma) \stackrel{1r}{\rightarrow} (0,\gamma,0,0,0,\gamma) \stackrel{1r}{\rightarrow} (0,0,0,\gamma,\gamma,0) \stackrel{1r}{\rightarrow} (\gamma,0,0,\gamma,\gamma,\gamma) \stackrel{1r}{\rightarrow} (\gamma,\gamma,\gamma,0,\gamma,0) \stackrel{1r}{\rightarrow} (0,\gamma,0,0,\gamma,0) \stackrel{1r}{\rightarrow} (0,0,\gamma,\gamma,0,0) \stackrel{1r}{\rightarrow} (\gamma,0,\gamma,\gamma,\gamma,0) \stackrel{1r}{\rightarrow} (\gamma,\gamma,\gamma,0,0,\gamma) \stackrel{1r}{\rightarrow} (0,\gamma,\gamma,0,0,0) \stackrel{1r}{\rightarrow} (0,0,0,\gamma,0,\gamma)$, which is 12-round iterative and holds with bias $2^{-250.40}$.
Starting from an odd-numbered round i, the following 24 subkeys have to be weak: $Z_7^{(i)}$, $Z_4^{(i+1)}$, $Z_6^{(i+1)}$, $Z_1^{(i+2)}$, $Z_5^{(i+2)}$, $Z_7^{(i+2)}$, $Z_2^{(i+3)}$, $Z_6^{(i+3)}$, $Z_5^{(i+4)}$, $Z_7^{(i+4)}$, $Z_4^{(i+5)}$, $Z_6^{(i+5)}$, $Z_1^{(i+6)}$, $Z_3^{(i+6)}$, $Z_5^{(i+6)}$, $Z_7^{(i+6)}$, $Z_2^{(i+7)}$, $Z_3^{(i+8)}$, $Z_7^{(i+8)}$, $Z_4^{(i+9)}$, $Z_1^{(i+10)}$, $Z_3^{(i+10)}$, $Z_7^{(i+10)}$ and $Z_2^{(i+11)}$, which means $24/12 = 2$ weak subkeys/round.
Starting from an even-numbered round i, the following 24 subkeys have to be weak: $Z_4^{(i)}$, $Z_6^{(i)}$, $Z_7^{(i)}$, $Z_1^{(i+1)}$, $Z_3^{(i+1)}$, $Z_2^{(i+2)}$, $Z_6^{(i+2)}$, $Z_7^{(i+2)}$,

$Z_4^{(i+4)}$, $Z_7^{(i+4)}$, $Z_1^{(i+5)}$, $Z_5^{(i+5)}$, $Z_2^{(i+6)}$, $Z_7^{(i+6)}$, $Z_5^{(i+7)}$, $Z_4^{(i+8)}$, $Z_7^{(i+8)}$, $Z_1^{(i+9)}$, $Z_3^{(i+9)}$, $Z_5^{(i+9)}$, $Z_2^{(i+10)}$, $Z_6^{(i+10)}$, $Z_7^{(i+10)}$ and $Z_3^{(i+11)}$, which means $24/12 = 2$ weak subkeys/round.

(5) $(0,0,0,\gamma,\gamma,\gamma) \xrightarrow{1r} (0,\gamma,0,\gamma,0,\gamma) \xrightarrow{1r} (0,\gamma,\gamma,0,0,\gamma) \xrightarrow{1r} (\gamma,\gamma,0,\gamma,\gamma,\gamma) \xrightarrow{1r} (0,0,\gamma,\gamma,\gamma,0) \xrightarrow{1r} (0,\gamma,0,\gamma,\gamma,0) \xrightarrow{1r} (0,\gamma,0,0,\gamma,\gamma) \xrightarrow{1r} (\gamma,\gamma,\gamma,\gamma,\gamma,0) \xrightarrow{1r} (0,0,\gamma,\gamma,0,\gamma) \xrightarrow{1r} (0,\gamma,\gamma,\gamma,0,0) \xrightarrow{1r} (0,\gamma,\gamma,0,\gamma,0) \xrightarrow{1r} (\gamma,\gamma,\gamma,\gamma,0,\gamma) \xrightarrow{1r} (0,0,0,\gamma,\gamma,\gamma)$, which is 12-round iterative and holds with bias $2^{-250.40}$.
Starting from an odd-numbered round i, the following 40 subkeys have to be weak: $Z_5^{(i)}$, $Z_9^{(i)}$, $Z_2^{(i+1)}$, $Z_4^{(i+1)}$, $Z_6^{(i+1)}$, $Z_9^{(i+1)}$, $Z_3^{(i+2)}$, $Z_7^{(i+2)}$, $Z_9^{(i+2)}$, $Z_2^{(i+3)}$, $Z_4^{(i+3)}$, $Z_6^{(i+3)}$, $Z_7^{(i+3)}$, $Z_9^{(i+3)}$, $Z_3^{(i+4)}$, $Z_5^{(i+4)}$, $Z_9^{(i+4)}$, $Z_4^{(i+5)}$, $Z_9^{(i+5)}$, $Z_5^{(i+6)}$, $Z_7^{(i+6)}$, $Z_9^{(i+6)}$, $Z_2^{(i+7)}$, $Z_4^{(i+7)}$, $Z_7^{(i+7)}$, $Z_9^{(i+7)}$, $Z_3^{(i+8)}$, $Z_9^{(i+8)}$, $Z_2^{(i+9)}$, $Z_4^{(i+9)}$, $Z_9^{(i+9)}$, $Z_3^{(i+10)}$, $Z_5^{(i+10)}$, $Z_7^{(i+10)}$, $Z_9^{(i+10)}$, $Z_2^{(i+11)}$, $Z_4^{(i+11)}$, $Z_6^{(i+11)}$, $Z_7^{(i+11)}$ and $Z_9^{(i+11)}$, which means $40/12 \approx 3.33$ weak subkeys/round.
Starting from an even-numbered round i, the following 37 subkeys have to be weak: $Z_4^{(i)}$, $Z_6^{(i)}$, $Z_9^{(i)}$, $Z_9^{(i+1)}$, $Z_2^{(i+2)}$, $Z_6^{(i+2)}$, $Z_7^{(i+2)}$, $Z_9^{(i+2)}$, $Z_1^{(i+3)}$, $Z_5^{(i+3)}$, $Z_7^{(i+3)}$, $Z_9^{(i+3)}$, $Z_4^{(i+4)}$, $Z_9^{(i+4)}$, $Z_5^{(i+5)}$, $Z_9^{(i+5)}$, $Z_2^{(i+6)}$, $Z_6^{(i+6)}$, $Z_7^{(i+6)}$, $Z_9^{(i+6)}$, $Z_1^{(i+7)}$, $Z_3^{(i+7)}$, $Z_5^{(i+7)}$, $Z_7^{(i+7)}$, $Z_9^{(i+7)}$, $Z_4^{(i+8)}$, $Z_6^{(i+8)}$, $Z_9^{(i+8)}$, $Z_3^{(i+9)}$, $Z_9^{(i+9)}$, $Z_2^{(i+10)}$, $Z_7^{(i+10)}$, $Z_9^{(i+10)}$, $Z_1^{(i+11)}$, $Z_3^{(i+11)}$, $Z_7^{(i+11)}$ and $Z_9^{(i+11)}$, which means $37/12 \approx 3$ weak subkeys/round.

(6) $(0,0,\gamma,0,\gamma,\gamma) \xrightarrow{1r} (\gamma,\gamma,\gamma,0,\gamma,\gamma) \xrightarrow{1r} (\gamma,0,0,0,0,0) \xrightarrow{1r} (\gamma,0,\gamma,\gamma,\gamma,\gamma) \xrightarrow{1r} (0,0,\gamma,0,\gamma,\gamma)$, which is 4-round iterative and holds with bias $2^{-50.88}$.
Starting from an odd-numbered round i, the following 13 subkeys have to be weak: $Z_3^{(i)}$, $Z_5^{(i)}$, $Z_7^{(i)}$, $Z_9^{(i)}$, $Z_2^{(i+1)}$, $Z_6^{(i+1)}$, $Z_9^{(i+1)}$, $Z_1^{(i+2)}$, $Z_9^{(i+2)}$, $Z_4^{(i+3)}$, $Z_6^{(i+3)}$, $Z_7^{(i+3)}$ and $Z_9^{(i+3)}$, which means $13/4 \approx 3$ weak subkeys/round.
Starting from an even-numbered round i, the following 13 subkeys have to be weak: $Z_6^{(i)}$, $Z_7^{(i)}$, $Z_9^{(i)}$, $Z_1^{(i+1)}$, $Z_3^{(i+1)}$, $Z_5^{(i+1)}$, $Z_9^{(i+1)}$, $Z_9^{(i+2)}$, $Z_1^{(i+3)}$, $Z_3^{(i+3)}$, $Z_5^{(i+3)}$, $Z_7^{(i+3)}$ and $Z_9^{(i+3)}$, which means $13/4 \approx 3$ weak subkeys/round.

(7) $(0,0,\gamma,\gamma,\gamma,\gamma) \xrightarrow{1r} (\gamma,0,0,\gamma,0,0) \xrightarrow{1r} (\gamma,\gamma,0,0,0,0) \xrightarrow{1r} (0,\gamma,\gamma,0,\gamma,\gamma) \xrightarrow{1r} (0,0,\gamma,\gamma,\gamma,\gamma)$, which is 4-round iterative and holds with bias $2^{-50.88}$.
Starting from an odd-numbered round i, the following 7 subkeys have to be weak: $Z_3^{(i)}$, $Z_7^{(i)}$, $Z_4^{(i+1)}$, $Z_1^{(i+2)}$, $Z_7^{(i+2)}$, $Z_2^{(i+3)}$ and $Z_6^{(i+3)}$, which means $7/4 \approx 2$ weak subkeys/round.
Starting from an even-numbered round i, the following 8 subkeys have to be weak: $Z_4^{(i)}$, $Z_6^{(i)}$, $Z_7^{(i)}$, $Z_1^{(i+1)}$, $Z_2^{(i+1)}$, $Z_7^{(i+1)}$, $Z_3^{(i+2)}$ and $Z_5^{(i+2)}$, which means $8/4 = 2$ weak subkeys/round.

(8) $(0,\gamma,0,\gamma,0,0) \xrightarrow{1r} (\gamma,0,\gamma,0,\gamma,\gamma) \xrightarrow{1r} (0,\gamma,0,\gamma,0,0)$, which is 2-round iterative and holds with bias $2^{-50.88}$.

Starting from an odd-numbered round i, the following 3 subkeys have to be weak: $Z_7^{(i)}$, $Z_6^{(i+1)}$ and $Z_7^{(i+1)}$, which means $3/2 = 1.5$ weak subkeys/round.
Starting from an even-numbered round i, the following 7 subkeys have to be weak: $Z_2^{(i)}$, $Z_4^{(i)}$, $Z_7^{(i)}$, $Z_1^{(i+1)}$, $Z_3^{(i+1)}$, $Z_5^{(i+1)}$ and $Z_7^{(i+1)}$, which means $7/2 = 3.5$ weak subkeys/round.

(9) $(0,\gamma,0,\gamma,\gamma,\gamma) \stackrel{1r}{\rightarrow} (\gamma,0,0,0,0,\gamma) \stackrel{1r}{\rightarrow} (0,\gamma,\gamma,\gamma,0,\gamma) \stackrel{1r}{\rightarrow} (\gamma,0,\gamma,0,0,0) \stackrel{1r}{\rightarrow} (0,\ \gamma,\ \gamma,\ \gamma,\ \gamma,\ 0) \stackrel{1r}{\rightarrow} (\gamma,0,0,0,\gamma,0) \stackrel{1r}{\rightarrow} (0,\gamma,0,\gamma,\gamma,\gamma)$, which is 6-round iterative and holds with bias $2^{-150.64}$.
Starting from an odd-numbered round i, the following 11 subkeys have to be weak: $Z_5^{(i)}$, $Z_7^{(i)}$, $Z_6^{(i+1)}$, $Z_7^{(i+1)}$, $Z_3^{(i+2)}$, $Z_7^{(i+2)}$, $Z_7^{(i+3)}$, $Z_3^{(i+4)}$, $Z_5^{(i+4)}$, $Z_7^{(i+4)}$ and $Z_7^{(i+5)}$, which means $11/6 \approx 2$ weak subkeys/round.
Starting from an even-numbered round i, the following 19 subkeys have to be weak: $Z_2^{(i)}$, $Z_4^{(i)}$, $Z_6^{(i)}$, $Z_7^{(i)}$, $Z_1^{(i+1)}$, $Z_7^{(i+1)}$, $Z_2^{(i+2)}$, $Z_4^{(i+2)}$, $Z_6^{(i+2)}$, $Z_7^{(i+2)}$, $Z_1^{(i+3)}$, $Z_3^{(i+3)}$, $Z_7^{(i+3)}$, $Z_2^{(i+4)}$, $Z_4^{(i+4)}$, $Z_7^{(i+4)}$, $Z_1^{(i+5)}$, $Z_5^{(i+5)}$ and $Z_7^{(i+5)}$, which means $19/6 \approx 3$ weak subkeys/round.

(10) $(\gamma,\gamma,0,\gamma,0,\gamma) \stackrel{1r}{\rightarrow} (\gamma,\gamma,0,\gamma,\gamma,0) \stackrel{1r}{\rightarrow} (\gamma,\gamma,\gamma,\gamma,0,0) \stackrel{1r}{\rightarrow} (\gamma,\gamma,0,\gamma,0,\gamma)$ which is 3-round iterative and holds with bias $2^{-50.88}$.
Starting from an odd-numbered round i, the following 5 subkeys have to be weak: $Z_1^{(i)}$, $Z_2^{(i+1)}$, $Z_4^{(i+1)}$, $Z_1^{(i+2)}$ and $Z_3^{(i+2)}$, which means $5/3 \approx 1.66$ weak subkeys/round.
Starting from an even-numbered round i, the following 7 subkeys have to be weak: $Z_2^{(i)}$, $Z_4^{(i)}$, $Z_6^{(i)}$, $Z_1^{(i+1)}$, $Z_5^{(i+1)}$, $Z_2^{(i+2)}$ and $Z_4^{(i+2)}$, which means $7/3 \approx 2.33$ weak subkeys/round.

(11) $(\gamma,\gamma,\gamma,\gamma,\gamma,\gamma) \stackrel{1r}{\rightarrow} (\gamma,\gamma,\gamma,\gamma,\gamma,\gamma)$, which is 1-round iterative and holds with bias 2^{-1}.
Starting from an odd-numbered round i, the following 3 subkeys have to be weak: $Z_1^{(i)}$, $Z_3^{(i)}$ and $Z_5^{(i)}$, which means 3 weak subkeys/round.
Starting from an even-numbered round i, the following 3 subkeys have to be weak: $Z_2^{(i)}$, $Z_4^{(i)}$ and $Z_6^{(i)}$, which means 3 weak subkeys/round.

Similar to the linear analysis of MESH-64, the key schedule of MESH-96 does not consist of a simple bit permutation like in IDEA. The fraction of keys that satisfies a weak-subkey restriction in MESH-96 was estimated (like in the case of MESH-64) to be about 2^{-15} per subkey. This estimate is in line with the assumption that a weak subkey is either 0 or 1, since the fifteen most significant bits have to be zero. Under this assumption, and since MESH-96 operates under a 192-bit key, at most $\lfloor 192/15 \rfloor = 12$ weak subkeys are allowed (for a nonempty weak-key class).

Under this estimate, linear relation (2) requires the minimum number of weak subkeys: only two every three rounds. But, it holds with a bias of $2^{-50.88}$. To attack six rounds, the data complexity is $c/(2\cdot(2^{-50.88})^2)^2 = c\cdot 2^{201.52}$

known-plaintexts are needed (for a fixed constant $c \geq 1$). But, the codebook size of MESH-96 is only 2^{192} text blocks.

Since linear relation (2) is 3-round iterative with biases $2^{-25.94}$, $2^{-25.94}$ and 2^{-1} for the first, second and third rounds, respectively, one can cover four rounds at most. Thus, at most 4-round MESH-96 can be distinguished from a random permutation. This attack would require three weak subkeys and $c/(2 \cdot 2^{-25.94} \cdot 2^{-50.88})^2 = c \cdot 2^{150.64}$ known plaintexts and equivalent 4-round MESH-96 computations.

Alternatively, linear relation (11) would also allow to distinguish 4-round MESH-96 from a random permutation, but with only $c \cdot (2^{-1})^2 = 4c$ known plaintexts. Nonetheless, it would require 12 weak subkeys. The estimated weak-key class size is $2^{192-15 \cdot 12} = 2^{12}$, but the exact user keys that belong to this weak-key class are not known.

This linear analysis also allows us to study the diffusion power of MESH-96 with respect to modern tools such as the linear branch number. See Appendix, Sect. B.

The set of linear relations in Tables 3.100 and 3.101 means that the linear branch number *per round* of MESH-96 is only 4 (under weak-key assumptions). For instance, the 1-round relation $(\gamma, 0, 0, \gamma, 0, 0) \stackrel{1r}{\rightarrow} (\gamma, \gamma, 0, 0, 0, 0)$ has both the input and output masks with Hamming Weight 2 (under weak-key assumptions). So, even though MESH-96 provides full diffusion in a single round, a detailed analysis shows that its linear branch number is much lower than ideal. For a 6-word block, such as used in MESH-96, an MDS code using a 6×6 MDS matrix would rather provide a branch number of $1 + 6 = 7$. See Appendix, Sect. B.1.

3.13.6 LC of MESH-128 Under Weak-Key Assumptions

The following linear analyses of MESH-128 is based on [151], but are more extensive and more detailed.

The initial analysis obtains all linear relation patterns across the MA-box of MESH-128, under the corresponding weak-subkey restrictions. The strategy is to build linear relations in a bottom-up fashion, from the smallest components: $\odot$, $\boxplus$ and $\oplus$, up to an MA-box, up to a half-round, further to a full round and finally to multiple rounds.

Across unkeyed $\odot$ and $\boxplus$, the propagation of bitmasks follows from Table 3.98 and Table 3.99, respectively.

Table 3.102 lists all linear relation patterns across the MA-box of MESH-128. Only two 16-bit masks are used in the linear distinguishers: 0 and $\gamma = 1$. The bitmask 1 propagates across the modular addition with maximum bias, since it approximates only the least significant bit, which is not affected by carry bits.

The bias was approximated using the Piling-up lemma [131]. The biases for the approximations across unkeyed-$\odot$ follow from Table 3.98.

Table 3.102 Wordwise linear relations across the MA-box of MESH-128.

Bitmasks across the MA-box	# active $\odot$	Bias	Weak subkey(s)
$(0,0,0,0) \rightarrow (0,0,0,0)$	0	2^{-1}	—
$(0,0,0,\gamma) \rightarrow (0,\gamma,0,\gamma)$	2	$2^{-25.94}$	$Z_{10}^{(i)}$
$(0,0,\gamma,0) \rightarrow (\gamma,0,0,0)$	3	$2^{-38.41}$	$Z_{12}^{(i)}$
$(0,0,\gamma,\gamma) \rightarrow (\gamma,\gamma,0,\gamma)$	3	$2^{-38.41}$	$Z_{10}^{(i)}, Z_{12}^{(i)}$
$(0,\gamma,0,0) \rightarrow (\gamma,0,0,\gamma)$	3	$2^{-38.41}$	$Z_{11}^{(i)}$
$(0,\gamma,0,\gamma) \rightarrow (\gamma,\gamma,0,0)$	1	$2^{-13.47}$	$Z_{10}^{(i)}, Z_{11}^{(i)}$
$(0,\gamma,\gamma,0) \rightarrow (0,0,0,\gamma)$	2	$2^{-25.94}$	$Z_{11}^{(i)}, Z_{12}^{(i)}$
$(0,\gamma,\gamma,\gamma) \rightarrow (0,\gamma,0,0)$	2	$2^{-25.94}$	$Z_{10}^{(i)}, Z_{11}^{(i)}, Z_{12}^{(i)}$
$(\gamma,0,0,0) \rightarrow (0,\gamma,\gamma,0)$	1	$2^{-13.47}$	$Z_{9}^{(i)}, Z_{11}^{(i)}$
$(\gamma,0,0,\gamma) \rightarrow (0,0,\gamma,\gamma)$	1	$2^{-13.47}$	$Z_{9}^{(i)}, Z_{10}^{(i)}, Z_{11}^{(i)}$
$(\gamma,0,\gamma,0) \rightarrow (\gamma,\gamma,\gamma,0)$	2	$2^{-25.94}$	$Z_{9}^{(i)}, Z_{11}^{(i)}, Z_{12}^{(i)}$
$(\gamma,0,\gamma,\gamma) \rightarrow (\gamma,0,\gamma,\gamma)$	4	$2^{-50.88}$	$Z_{9}^{(i)}, Z_{10}^{(i)}, Z_{11}^{(i)}, Z_{12}^{(i)}$
$(\gamma,\gamma,0,0) \rightarrow (\gamma,\gamma,\gamma,\gamma)$	2	$2^{-25.94}$	$Z_{9}^{(i)}$
$(\gamma,\gamma,0,\gamma) \rightarrow (\gamma,0,\gamma,0)$	2	$2^{-25.94}$	$Z_{9}^{(i)}, Z_{10}^{(i)}$
$(\gamma,\gamma,\gamma,0) \rightarrow (0,\gamma,\gamma,\gamma)$	3	$2^{-38.41}$	$Z_{9}^{(i)}, Z_{12}^{(i)}$
$(\gamma,\gamma,\gamma,\gamma) \rightarrow (0,0,\gamma,0)$	1	$2^{-13.47}$	$Z_{9}^{(i)}, Z_{10}^{(i)}, Z_{12}^{(i)}$

The next step is to obtain from Table 3.102 all 1-round linear relations and the corresponding weak-subkeys conditions.

Tables 3.103 through 3.110 list all 255 nontrivial 1-round linear relations of MESH-128, with wordwise bitmasks either 0 or $\gamma = 1$. These tables provide an exhaustive analysis of how linear relations propagate across a single round of MESH-128.

Multiple-round linear relations are obtained by concatenating 1-round linear relations such that the output bitmask from one relation matches the input bitmask of the next relation.

From Tables 3.103 through 3.110, the multiple-round linear relations are clustered into 23 relations covering 1, 2, 3, 5, 6, 10, 15 or 30 rounds:

(1) $(0,0,0,0,0,0,0,\gamma) \stackrel{1r}{\rightarrow} (0,0,0,\gamma,0,\gamma,0,\gamma) \stackrel{1r}{\rightarrow} (\gamma,\ \gamma,\ 0,\ \gamma,\ \gamma,\ \gamma,\ 0,\ 0) \stackrel{1r}{\rightarrow} (\gamma,\ \gamma,\ \gamma,\ \gamma,\ \gamma,\ \gamma,\ \gamma,\ 0) \stackrel{1r}{\rightarrow} (\gamma,\gamma,\gamma,0,\gamma,0,\gamma,0) \stackrel{1r}{\rightarrow} (0,0,\gamma,0,0,0,\gamma,\gamma) \stackrel{1r}{\rightarrow}$ (0,0,0,0,0,0,0,γ), which is a 6-round iterative linear relation with bias $2^{-175.58}$.

Starting from an odd-numbered round i, the following 21 subkeys have to be weak: $Z_8^{(i)}$, $Z_{12}^{(i)}$, $Z_4^{(i+1)}$, $Z_9^{(i+1)}$, $Z_{10}^{(i+1)}$, $Z_{12}^{(i+1)}$, $Z_1^{(i+2)}$, $Z_6^{(i+2)}$, $Z_{12}^{(i+2)}$, $Z_2^{(i+3)}$, $Z_4^{(i+3)}$, $Z_5^{(i+3)}$, $Z_7^{(i+3)}$, $Z_{12}^{(i+3)}$, $Z_1^{(i+4)}$, $Z_3^{(i+4)}$, $Z_9^{(i+4)}$, $Z_{10}^{(i+4)}$, $Z_{12}^{(i+4)}$, $Z_7^{(i+5)}$ and $Z_{12}^{(i+5)}$.

Starting from an even-numbered round i, the following 23 subkeys have to be weak: $Z_{12}^{(i)}$, $Z_6^{(i+1)}$, $Z_8^{(i+1)}$, $Z_9^{(i+1)}$, $Z_{10}^{(i+1)}$, $Z_{12}^{(i+1)}$, $Z_2^{(i+2)}$, $Z_4^{(i+2)}$,

Table 3.103 Nontrivial 1-round linear relations for MESH-128 (part I).

1-round linear relations	Weak subkeys $Z_j^{(i)}$ (ith round)		Bias
	j value (odd i)	j value (even i)	
$(0,0,0,0,0,0,0,\gamma) \stackrel{1r}{\to} (0,0,0,\gamma,0,\gamma,0,\gamma)$	8; 12	12	$2^{-38.41}$
$(0,0,0,0,0,0,\gamma,0) \stackrel{1r}{\to} (0,0,\gamma,0,\gamma,\gamma,\gamma,\gamma)$	10; 11; 12	7; 10; 11; 12	$2^{-25.94}$
$(0,0,0,0,0,0,\gamma,\gamma) \stackrel{1r}{\to} (0,0,\gamma,\gamma,\gamma,0,\gamma,0)$	8; 10; 11	7; 10; 11	$2^{-13.47}$
$(0,0,0,0,0,\gamma,0,0) \stackrel{1r}{\to} (\gamma,\gamma,0,\gamma,\gamma,\gamma,\gamma,\gamma)$	6; 9; 10; 12	9; 10; 12	$2^{-13.47}$
$(0,0,0,0,0,\gamma,0,\gamma) \stackrel{1r}{\to} (\gamma,\gamma,0,0,\gamma,0,\gamma,0)$	6; 8; 9; 10	9; 10	$2^{-25.94}$
$(0,0,0,0,0,\gamma,\gamma,0) \stackrel{1r}{\to} (\gamma,\gamma,\gamma,\gamma,0,0,0,0)$	6; 9; 11	7; 9; 11	$2^{-13.47}$
$(0,0,0,0,0,\gamma,\gamma,\gamma) \stackrel{1r}{\to} (\gamma,\gamma,\gamma,0,0,\gamma,0,\gamma)$	6; 8; 9; 11; 12	7; 9; 11; 12	$2^{-25.94}$
$(0,0,0,0,\gamma,0,0,0) \stackrel{1r}{\to} (0,\gamma,\gamma,\gamma,\gamma,\gamma,0,0)$	11; 12	5; 11; 12	$2^{-25.94}$
$(0,0,0,0,\gamma,0,0,\gamma) \stackrel{1r}{\to} (0,\gamma,\gamma,0,\gamma,0,0,\gamma)$	8; 11	5; 11	$2^{-38.41}$
$(0,0,0,0,\gamma,0,\gamma,0) \stackrel{1r}{\to} (0,\gamma,0,\gamma,0,0,\gamma,\gamma)$	10	5; 7; 10	$2^{-25.94}$
$(0,0,0,0,\gamma,0,\gamma,\gamma) \stackrel{1r}{\to} (0,\gamma,0,0,0,\gamma,\gamma,0)$	8; 10; 12	5; 7; 10; 12	$2^{-38.41}$
$(0,0,0,0,\gamma,\gamma,0,0) \stackrel{1r}{\to} (\gamma,0,\gamma,0,0,0,\gamma,\gamma)$	6; 9; 10; 11	5; 9; 10; 11	$2^{-13.47}$
$(0,0,0,0,\gamma,\gamma,0,\gamma) \stackrel{1r}{\to} (\gamma,0,\gamma,\gamma,0,\gamma,\gamma,0)$	6; 8; 9; 10; 11; 12	5; 9; 10; 11; 12	$2^{-50.88}$
$(0,0,0,0,\gamma,\gamma,\gamma,0) \stackrel{1r}{\to} (\gamma,0,0,0,\gamma,\gamma,0,0)$	6; 9; 12	5; 7; 9; 12	$2^{-38.41}$
$(0,0,0,0,\gamma,\gamma,\gamma,\gamma) \stackrel{1r}{\to} (\gamma,0,0,\gamma,\gamma,0,0,\gamma)$	6; 8; 9	5; 7; 9	$2^{-25.94}$
$(0,0,0,\gamma,0,0,0,0) \stackrel{1r}{\to} (0,0,0,\gamma,0,\gamma,\gamma,0)$	12	4; 12	$2^{-38.41}$
$(0,0,0,\gamma,0,0,0,\gamma) \stackrel{1r}{\to} (0,0,0,0,0,0,\gamma,\gamma)$	8;	4;	2^{-1}
$(0,0,0,\gamma,0,0,\gamma,0) \stackrel{1r}{\to} (0,0,\gamma,\gamma,\gamma,0,0,\gamma)$	10; 11	4; 7; 10; 11	$2^{-13.47}$
$(0,0,0,\gamma,0,0,\gamma,\gamma) \stackrel{1r}{\to} (0,0,\gamma,0,\gamma,\gamma,0,0)$	8; 10; 11; 12	4; 7; 10; 11; 12	$2^{-25.94}$
$(0,0,0,\gamma,0,\gamma,0,0) \stackrel{1r}{\to} (\gamma,\gamma,0,0,\gamma,0,0,\gamma)$	6; 9; 10	4; 9; 10	$2^{-25.94}$
$(0,0,0,\gamma,0,\gamma,0,\gamma) \stackrel{1r}{\to} (\gamma,\gamma,0,\gamma,\gamma,\gamma,0,0)$	6; 8; 9; 10; 12	4; 9; 10; 12	$2^{-13.47}$
$(0,0,0,\gamma,0,\gamma,\gamma,0) \stackrel{1r}{\to} (\gamma,\gamma,\gamma,0,0,\gamma,\gamma,0)$	6; 9; 11; 12	4; 7; 9; 11; 12	$2^{-25.94}$
$(0,0,0,\gamma,0,\gamma,\gamma,\gamma) \stackrel{1r}{\to} (\gamma,\gamma,\gamma,\gamma,0,0,\gamma,\gamma)$	6; 8; 9; 11	4; 7; 9; 11	$2^{-13.47}$
$(0,0,0,\gamma,\gamma,0,0,0) \stackrel{1r}{\to} (0,\gamma,\gamma,0,\gamma,0,\gamma,0)$	11	4; 5; 11	$2^{-38.41}$
$(0,0,0,\gamma,\gamma,0,0,\gamma) \stackrel{1r}{\to} (0,\gamma,\gamma,\gamma,\gamma,\gamma,\gamma,\gamma)$	8; 11; 12	4; 5; 11; 12	$2^{-25.94}$
$(0,0,0,\gamma,\gamma,0,\gamma,0) \stackrel{1r}{\to} (0,\gamma,0,0,0,\gamma,0,\gamma)$	10; 12	4; 5; 7; 10; 12	$2^{-38.41}$
$(0,0,0,\gamma,\gamma,0,\gamma,\gamma) \stackrel{1r}{\to} (0,\gamma,0,\gamma,0,0,0,0)$	8; 10	4; 5; 7; 10	$2^{-25.94}$
$(0,0,0,\gamma,\gamma,\gamma,0,0) \stackrel{1r}{\to} (\gamma,0,\gamma,\gamma,0,\gamma,0,\gamma)$	6; 9; 10; 11; 12	4; 5; 9; 10; 11; 12	$2^{-50.88}$
$(0,0,0,\gamma,\gamma,\gamma,0,\gamma) \stackrel{1r}{\to} (\gamma,0,\gamma,0,0,0,0,0)$	6; 8; 9; 10; 11	4; 5; 9; 10; 11	$2^{-13.47}$
$(0,0,0,\gamma,\gamma,\gamma,\gamma,0) \stackrel{1r}{\to} (\gamma,0,0,\gamma,\gamma,0,\gamma,0)$	6; 9	4; 5; 7; 9	$2^{-25.94}$
$(0,0,0,\gamma,\gamma,\gamma,\gamma,\gamma) \stackrel{1r}{\to} (\gamma,0,0,0,\gamma,\gamma,\gamma,\gamma)$	6; 8; 9; 12	4; 5; 7; 9; 12	$2^{-38.41}$
$(0,0,\gamma,0,0,0,0,0) \stackrel{1r}{\to} (0,0,\gamma,\gamma,\gamma,0,\gamma,\gamma)$	3; 10; 11; 12	10; 11; 12	$2^{-25.94}$

Table 3.104 Nontrivial 1-round linear relations for MESH-128 (part II).

1-round linear relations	Weak subkeys $Z_j^{(i)}$ (ith round)		Bias
	j value (odd i)	j value (even i)	
$(0,0,\gamma,0,0,0,0,\gamma) \xrightarrow{1r} (0,0,\gamma,0,\gamma,\gamma,\gamma,0)$	3; 8; 10; 11	10; 11	$2^{-13.47}$
$(0,0,\gamma,0,0,0,\gamma,0) \xrightarrow{1r} (0,0,0,\gamma,0,\gamma,0,0)$	3	7	2^{-1}
$(0,0,\gamma,0,0,0,\gamma,\gamma) \xrightarrow{1r} (0,0,0,0,0,0,0,\gamma)$	3; 8; 12	7; 12	$2^{-38.41}$
$(0,0,\gamma,0,0,\gamma,0,0) \xrightarrow{1r} (\gamma,\gamma,\gamma,0,0,\gamma,0,0)$	3; 6; 9; 11	9; 11	$2^{-13.47}$
$(0,0,\gamma,0,0,\gamma,0,\gamma) \xrightarrow{1r} (\gamma,\gamma,\gamma,\gamma,0,0,0,\gamma)$	3; 6; 8; 9; 11; 12	9; 11; 12	$2^{-25.94}$
$(0,0,\gamma,0,0,\gamma,\gamma,0) \xrightarrow{1r} (\gamma,\gamma,0,0,\gamma,0,\gamma,\gamma)$	3; 6; 9; 10; 12	7; 9; 10; 12	$2^{-13.47}$
$(0,0,\gamma,0,0,\gamma,\gamma,\gamma) \xrightarrow{1r} (\gamma,\gamma,0,\gamma,\gamma,\gamma,\gamma,0)$	3; 6; 8; 9; 10	7; 9; 10	$2^{-25.94}$
$(0,0,\gamma,0,\gamma,0,0,0) \xrightarrow{1r} (0,\gamma,0,0,0,\gamma,\gamma,\gamma)$	3; 10	5; 10	$2^{-25.94}$
$(0,0,\gamma,0,\gamma,0,0,\gamma) \xrightarrow{1r} (0,\gamma,0,\gamma,0,0,\gamma,0)$	3; 8; 10; 12	5; 10; 12	$2^{-38.41}$
$(0,0,\gamma,0,\gamma,0,\gamma,0) \xrightarrow{1r} (0,\gamma,\gamma,0,\gamma,0,0,0)$	3; 11; 12	5; 7; 11; 12	$2^{-25.94}$
$(0,0,\gamma,0,\gamma,0,\gamma,\gamma) \xrightarrow{1r} (0,\gamma,\gamma,\gamma,\gamma,\gamma,0,\gamma)$	3; 8; 11	5; 7; 11	$2^{-38.41}$
$(0,0,\gamma,0,\gamma,\gamma,0,0) \xrightarrow{1r} (\gamma,0,0,\gamma,\gamma,0,0,0)$	3; 6; 9; 12	5; 9; 12	$2^{-38.41}$
$(0,0,\gamma,0,\gamma,\gamma,0,\gamma) \xrightarrow{1r} (\gamma,0,0,0,\gamma,\gamma,0,\gamma)$	3; 6; 8; 9	5; 9	$2^{-25.94}$
$(0,0,\gamma,0,\gamma,\gamma,\gamma,0) \xrightarrow{1r} (\gamma,0,\gamma,\gamma,0,\gamma,\gamma,\gamma)$	3; 6; 9; 10; 11	5; 7; 9; 10; 11	$2^{-13.47}$
$(0,0,\gamma,0,\gamma,\gamma,\gamma,\gamma) \xrightarrow{1r} (\gamma,0,\gamma,0,0,0,\gamma,0)$	3; 6; 8; 9; 10; 11; 12	5; 7; 9; 10; 11; 12	$2^{-50.88}$
$(0,0,\gamma,\gamma,0,0,0,0) \xrightarrow{1r} (0,0,\gamma,0,\gamma,\gamma,0,\gamma)$	3; 10; 11	4; 10; 11	$2^{-13.47}$
$(0,0,\gamma,\gamma,0,0,0,\gamma) \xrightarrow{1r} (0,0,\gamma,\gamma,\gamma,0,0,0)$	3; 8; 10; 11; 12	4; 10; 11; 12	$2^{-25.94}$
$(0,0,\gamma,\gamma,0,0,\gamma,0) \xrightarrow{1r} (0,0,0,0,0,0,\gamma,0)$	3; 12	4; 7; 12	$2^{-38.41}$
$(0,0,\gamma,\gamma,0,0,\gamma,\gamma) \xrightarrow{1r} (0,0,0,\gamma,0,\gamma,\gamma,\gamma)$	3; 8	4; 7	2^{-1}
$(0,0,\gamma,\gamma,0,\gamma,0,0) \xrightarrow{1r} (\gamma,\gamma,\gamma,\gamma,0,0,\gamma,0)$	3; 6; 9; 11; 12	4; 9; 11; 12	$2^{-25.94}$
$(0,0,\gamma,\gamma,0,\gamma,0,\gamma) \xrightarrow{1r} (\gamma,\gamma,\gamma,0,0,\gamma,\gamma,\gamma)$	3; 6; 8; 9; 11	4; 9; 11	$2^{-13.47}$
$(0,0,\gamma,\gamma,0,\gamma,\gamma,0) \xrightarrow{1r} (\gamma,\gamma,0,\gamma,\gamma,\gamma,0,\gamma)$	3; 6; 9; 10	4; 7; 9; 10	$2^{-25.94}$
$(0,0,\gamma,\gamma,0,\gamma,\gamma,\gamma) \xrightarrow{1r} (\gamma,\gamma,0,0,\gamma,0,0,0)$	3; 6; 8; 9; 10; 12	4; 7; 9; 10; 12	$2^{-13.47}$
$(0,0,\gamma,\gamma,\gamma,0,0,0) \xrightarrow{1r} (0,\gamma,0,\gamma,0,0,0,\gamma)$	3; 10; 12	4; 5; 10; 12	$2^{-38.41}$
$(0,0,\gamma,\gamma,\gamma,0,0,\gamma) \xrightarrow{1r} (0,\gamma,0,0,0,\gamma,0,0)$	3; 8; 10	4; 5; 10	$2^{-25.94}$
$(0,0,\gamma,\gamma,\gamma,0,\gamma,0) \xrightarrow{1r} (0,\gamma,\gamma,\gamma,\gamma,\gamma,\gamma,0)$	3; 11	4; 5; 7; 11	$2^{-38.41}$
$(0,0,\gamma,\gamma,\gamma,0,\gamma,\gamma) \xrightarrow{1r} (0,\gamma,\gamma,0,\gamma,0,\gamma,\gamma)$	3; 8; 11; 12	4; 5; 7; 11; 12	$2^{-25.94}$
$(0,0,\gamma,\gamma,\gamma,\gamma,0,0) \xrightarrow{1r} (\gamma,0,0,0,\gamma,\gamma,\gamma,0)$	3; 6; 9	4; 5; 9	$2^{-25.94}$
$(0,0,\gamma,\gamma,\gamma,\gamma,0,\gamma) \xrightarrow{1r} (\gamma,0,0,\gamma,\gamma,0,\gamma,\gamma)$	3; 6; 8; 9; 12	4; 5; 9; 12	$2^{-38.41}$
$(0,0,\gamma,\gamma,\gamma,\gamma,\gamma,0) \xrightarrow{1r} (\gamma,0,\gamma,0,0,0,0,\gamma)$	3; 6; 9; 10; 11; 12	4; 5; 7; 9; 10; 11; 12	$2^{-50.88}$
$(0,0,\gamma,\gamma,\gamma,\gamma,\gamma,\gamma) \xrightarrow{1r} (\gamma,0,\gamma,\gamma,0,\gamma,0,0)$	3; 6; 8; 9; 10; 11	4; 5; 7; 9; 10; 11	$2^{-13.47}$
$(0,\gamma,0,0,0,0,0,0) \xrightarrow{1r} (\gamma,\gamma,\gamma,\gamma,0,\gamma,\gamma,\gamma)$	9; 10; 12	2; 9; 10; 12	$2^{-13.47}$

Table 3.105 Nontrivial 1-round linear relations for MESH-128 (part III).

1-round linear relations	Weak subkeys $Z_j^{(i)}$ (ith round)		Bias
	j value (odd i)	j value (even i)	
$(0,\gamma,0,0,0,0,0,\gamma) \xrightarrow{1r} (\gamma,\gamma,\gamma,0,0,0,\gamma,0)$	8; 9; 10	2; 9; 10	$2^{-25.94}$
$(0,\gamma,0,0,0,0,\gamma,0) \xrightarrow{1r} (\gamma,\gamma,0,\gamma,\gamma,0,0,0)$	9; 11	2; 7; 9; 11	$2^{-13.47}$
$(0,\gamma,0,0,0,0,\gamma,\gamma) \xrightarrow{1r} (\gamma,\gamma,0,0,\gamma,\gamma,0,\gamma)$	8; 9; 11; 12	2; 7; 9; 11; 12	$2^{-25.94}$
$(0,\gamma,0,0,0,\gamma,0,0) \xrightarrow{1r} (0,0,\gamma,0,\gamma,0,0,0)$	6	2	2^{-1}
$(0,\gamma,0,0,0,\gamma,0,\gamma) \xrightarrow{1r} (0,0,\gamma,\gamma,\gamma,\gamma,0,\gamma)$	6; 8; 12	2; 12	$2^{-38.41}$
$(0,\gamma,0,0,0,\gamma,\gamma,0) \xrightarrow{1r} (0,0,0,0,0,\gamma,\gamma,\gamma)$	6; 10; 11; 12	2; 7; 10; 11; 12	$2^{-25.94}$
$(0,\gamma,0,0,0,\gamma,\gamma,\gamma) \xrightarrow{1r} (0,0,0,\gamma,0,0,\gamma,0)$	6; 8; 10; 11	2; 7; 10; 11	$2^{-13.47}$
$(0,\gamma,0,0,\gamma,0,0,0) \xrightarrow{1r} (\gamma,0,0,0,\gamma,0,\gamma,\gamma)$	9; 10; 11	2; 5; 9; 10; 11	$2^{-13.47}$
$(0,\gamma,0,0,\gamma,0,0,\gamma) \xrightarrow{1r} (\gamma,0,0,\gamma,\gamma,\gamma,\gamma,0)$	8; 9; 10; 11; 12	2; 5; 9; 10; 11; 12	$2^{-50.88}$
$(0,\gamma,0,0,\gamma,0,\gamma,0) \xrightarrow{1r} (\gamma,0,\gamma,0,0,\gamma,0,0)$	9; 12	2; 5; 7; 9; 12	$2^{-38.41}$
$(0,\gamma,0,0,\gamma,0,\gamma,\gamma) \xrightarrow{1r} (\gamma,0,\gamma,\gamma,0,0,0,\gamma)$	8; 9	2; 5; 7; 9	$2^{-25.94}$
$(0,\gamma,0,0,\gamma,\gamma,0,0) \xrightarrow{1r} (0,\gamma,0,\gamma,0,\gamma,0,0)$	6; 11; 12	2; 5; 11; 12	$2^{-25.94}$
$(0,\gamma,0,0,\gamma,\gamma,0,\gamma) \xrightarrow{1r} (0,\gamma,0,0,0,0,0,\gamma)$	6; 8; 11	2; 5; 11	$2^{-38.41}$
$(0,\gamma,0,0,\gamma,\gamma,\gamma,0) \xrightarrow{1r} (0,\gamma,\gamma,\gamma,\gamma,0,\gamma,\gamma)$	6; 10	2; 5; 7; 10	$2^{-25.94}$
$(0,\gamma,0,0,\gamma,\gamma,\gamma,\gamma) \xrightarrow{1r} (0,\gamma,\gamma,0,\gamma,\gamma,\gamma,0)$	6; 8; 10; 12	2; 5; 7; 10; 12	$2^{-38.41}$
$(0,\gamma,0,\gamma,0,0,0,0) \xrightarrow{1r} (\gamma,\gamma,\gamma,0,0,0,0,\gamma)$	9; 10	2; 4; 9; 10	$2^{-25.94}$
$(0,\gamma,0,\gamma,0,0,0,\gamma) \xrightarrow{1r} (\gamma,\gamma,\gamma,\gamma,0,\gamma,0,0)$	8; 9; 10; 12	2; 4; 9; 10; 12	$2^{-13.47}$
$(0,\gamma,0,\gamma,0,0,\gamma,0) \xrightarrow{1r} (\gamma,\gamma,0,0,\gamma,\gamma,\gamma,0)$	9; 11; 12	2; 4; 7; 9; 11; 12	$2^{-25.94}$
$(0,\gamma,0,\gamma,0,0,\gamma,\gamma) \xrightarrow{1r} (\gamma,\gamma,0,\gamma,\gamma,0,\gamma,\gamma)$	8; 9; 11	2; 4; 7; 9; 11	$2^{-13.47}$
$(0,\gamma,0,\gamma,0,\gamma,0,0) \xrightarrow{1r} (0,0,\gamma,\gamma,\gamma,\gamma,\gamma,0)$	6; 12	2; 4; 12	$2^{-38.41}$
$(0,\gamma,0,\gamma,0,\gamma,0,\gamma) \xrightarrow{1r} (0,0,\gamma,0,\gamma,0,\gamma,\gamma)$	6; 8	2; 4	2^{-1}
$(0,\gamma,0,\gamma,0,\gamma,\gamma,0) \xrightarrow{1r} (0,0,0,\gamma,0,0,0,\gamma)$	6; 10; 11	2; 4; 7; 10; 11	$2^{-13.47}$
$(0,\gamma,0,\gamma,0,\gamma,\gamma,\gamma) \xrightarrow{1r} (0,0,0,0,0,\gamma,0,0)$	6; 8; 10; 11; 12	2; 4; 7; 10; 11; 12	$2^{-25.94}$
$(0,\gamma,0,\gamma,\gamma,0,0,0) \xrightarrow{1r} (\gamma,0,0,\gamma,\gamma,\gamma,0,\gamma)$	9; 10; 11; 12	2; 4; 5; 9; 10; 11; 12	$2^{-50.88}$
$(0,\gamma,0,\gamma,\gamma,0,0,\gamma) \xrightarrow{1r} (\gamma,0,0,0,\gamma,0,0,0)$	8; 9; 10; 11	2; 4; 5; 9; 10; 11	$2^{-13.47}$
$(0,\gamma,0,\gamma,\gamma,0,\gamma,0) \xrightarrow{1r} (\gamma,0,\gamma,\gamma,0,0,\gamma,0)$	9	2; 4; 5; 7; 9	$2^{-25.94}$
$(0,\gamma,0,\gamma,\gamma,0,\gamma,\gamma) \xrightarrow{1r} (\gamma,0,\gamma,0,0,\gamma,\gamma,\gamma)$	8; 9; 12	2; 4; 5; 7; 9; 12	$2^{-38.41}$
$(0,\gamma,0,\gamma,\gamma,\gamma,0,0) \xrightarrow{1r} (0,\gamma,0,0,0,0,\gamma,0)$	6; 11	2; 4; 5; 11	$2^{-38.41}$
$(0,\gamma,0,\gamma,\gamma,\gamma,0,\gamma) \xrightarrow{1r} (0,\gamma,0,\gamma,0,\gamma,\gamma,\gamma)$	6; 8; 11; 12	2; 4; 5; 11; 12	$2^{-25.94}$
$(0,\gamma,0,\gamma,\gamma,\gamma,\gamma,0) \xrightarrow{1r} (0,\gamma,\gamma,0,\gamma,\gamma,0,\gamma)$	6; 10; 12	2; 4; 5; 7; 10; 12	$2^{-38.41}$
$(0,\gamma,0,\gamma,\gamma,\gamma,\gamma,\gamma) \xrightarrow{1r} (0,\gamma,\gamma,\gamma,\gamma,0,0,0)$	6; 8; 10	2; 4; 5; 7; 10	$2^{-25.94}$
$(0,\gamma,\gamma,0,0,0,0,0) \xrightarrow{1r} (\gamma,\gamma,0,0,\gamma,\gamma,0,0)$	3; 9; 11	2; 9; 11	$2^{-13.47}$

Table 3.106 Nontrivial 1-round linear relations for MESH-128 (part IV).

1-round linear relations	Weak subkeys $Z_j^{(i)}$ (ith round)		Bias
	j value (odd i)	j value (even i)	
$(0,\gamma,\gamma,0,0,0,0,\gamma) \stackrel{1r}{\rightarrow} (\gamma,\gamma,0,\gamma,\gamma,0,0,\gamma)$	3; 8; 9; 11; 12	2; 9; 11; 12	$2^{-25.94}$
$(0,\gamma,\gamma,0,0,0,\gamma,0) \stackrel{1r}{\rightarrow} (\gamma,\gamma,\gamma,0,0,0,\gamma,\gamma)$	3; 9; 10; 12	2; 7; 9; 10; 12	$2^{-13.47}$
$(0,\gamma,\gamma,0,0,0,\gamma,\gamma) \stackrel{1r}{\rightarrow} (\gamma,\gamma,\gamma,\gamma,0,\gamma,\gamma,0)$	3; 8; 9; 10	2; 7; 9; 10	$2^{-25.94}$
$(0,\gamma,\gamma,0,0,\gamma,0,0) \stackrel{1r}{\rightarrow} (0,0,0,\gamma,0,0,\gamma,\gamma)$	3; 6; 10; 11; 12	2; 10; 11; 12	$2^{-25.94}$
$(0,\gamma,\gamma,0,0,\gamma,0,\gamma) \stackrel{1r}{\rightarrow} (0,0,0,0,0,\gamma,\gamma,0)$	3; 6; 8; 10; 11	2; 10; 11	$2^{-13.47}$
$(0,\gamma,\gamma,0,0,\gamma,\gamma,0) \stackrel{1r}{\rightarrow} (0,0,\gamma,\gamma,\gamma,\gamma,0,0)$	3; 6;	2; 7;	2^{-1}
$(0,\gamma,\gamma,0,0,\gamma,\gamma,\gamma) \stackrel{1r}{\rightarrow} (0,0,\gamma,0,\gamma,0,0,\gamma)$	3; 6; 8; 12	2; 7; 12	$2^{-38.41}$
$(0,\gamma,\gamma,0,\gamma,0,0,0) \stackrel{1r}{\rightarrow} (\gamma,0,\gamma,\gamma,0,0,0,0)$	3; 9; 12	2; 5; 9; 12	$2^{-38.41}$
$(0,\gamma,\gamma,0,\gamma,0,0,\gamma) \stackrel{1r}{\rightarrow} (\gamma,0,\gamma,0,0,\gamma,0,\gamma)$	3; 8; 9	2; 5; 9	$2^{-25.94}$
$(0,\gamma,\gamma,0,\gamma,0,\gamma,0) \stackrel{1r}{\rightarrow} (\gamma,0,0,\gamma,\gamma,\gamma,\gamma,\gamma)$	3; 9; 10; 11	2; 5; 7; 9; 10; 11	$2^{-13.47}$
$(0,\gamma,\gamma,0,\gamma,0,\gamma,\gamma) \stackrel{1r}{\rightarrow} (\gamma,0,0,0,\gamma,0,\gamma,0)$	3; 8; 9; 10; 11; 12	2; 5; 7; 9; 10; 11; 12	$2^{-50.88}$
$(0,\gamma,\gamma,0,\gamma,\gamma,0,0) \stackrel{1r}{\rightarrow} (0,\gamma,\gamma,0,\gamma,\gamma,\gamma,\gamma)$	3; 6; 10	2; 5; 10	$2^{-25.94}$
$(0,\gamma,\gamma,0,\gamma,\gamma,0,\gamma) \stackrel{1r}{\rightarrow} (0,\gamma,\gamma,\gamma,\gamma,0,\gamma,0)$	3; 6; 8; 10; 12	2; 5; 10; 12	$2^{-38.41}$
$(0,\gamma,\gamma,0,\gamma,\gamma,\gamma,0) \stackrel{1r}{\rightarrow} (0,\gamma,0,0,0,0,0,0)$	3; 6; 11; 12	2; 5; 7; 11; 12	$2^{-25.94}$
$(0,\gamma,\gamma,0,\gamma,\gamma,\gamma,\gamma) \stackrel{1r}{\rightarrow} (0,\gamma,0,\gamma,0,\gamma,0,\gamma)$	3; 6; 8; 11	2; 5; 7; 11	$2^{-38.41}$
$(0,\gamma,\gamma,\gamma,0,0,0,0) \stackrel{1r}{\rightarrow} (\gamma,\gamma,0,\gamma,\gamma,0,\gamma,0)$	3; 9; 11; 12	2; 4; 9; 11; 12	$2^{-25.94}$
$(0,\gamma,\gamma,\gamma,0,0,0,\gamma) \stackrel{1r}{\rightarrow} (\gamma,\gamma,0,0,\gamma,\gamma,\gamma,\gamma)$	3; 8; 9; 11	2; 4; 9; 11	$2^{-13.47}$
$(0,\gamma,\gamma,\gamma,0,0,\gamma,0) \stackrel{1r}{\rightarrow} (\gamma,\gamma,\gamma,\gamma,0,\gamma,0,\gamma)$	3; 9; 10	2; 4; 7; 9; 10	$2^{-25.94}$
$(0,\gamma,\gamma,\gamma,0,0,\gamma,\gamma) \stackrel{1r}{\rightarrow} (\gamma,\gamma,\gamma,0,0,0,0,0)$	3; 8; 9; 10; 12	2; 4; 7; 9; 10; 12	$2^{-13.47}$
$(0,\gamma,\gamma,\gamma,0,\gamma,0,0) \stackrel{1r}{\rightarrow} (0,0,0,0,0,\gamma,0,\gamma)$	3; 6; 10; 11	2; 4; 10; 11	$2^{-13.47}$
$(0,\gamma,\gamma,\gamma,0,\gamma,0,\gamma) \stackrel{1r}{\rightarrow} (0,0,0,\gamma,0,0,0,0)$	3; 6; 8; 10; 11; 12	2; 4; 10; 11; 12	$2^{-25.94}$
$(0,\gamma,\gamma,\gamma,0,\gamma,\gamma,0) \stackrel{1r}{\rightarrow} (0,0,\gamma,0,\gamma,0,\gamma,0)$	3; 6; 12	2; 4; 7; 12	$2^{-38.41}$
$(0,\gamma,\gamma,\gamma,0,\gamma,\gamma,\gamma) \stackrel{1r}{\rightarrow} (0,0,\gamma,\gamma,\gamma,\gamma,\gamma,\gamma)$	3; 6; 8	2; 4; 7	2^{-1}
$(0,\gamma,\gamma,\gamma,\gamma,0,0,0) \stackrel{1r}{\rightarrow} (\gamma,0,\gamma,0,0,\gamma,\gamma,0)$	3; 9	2; 4; 5; 9	$2^{-25.94}$
$(0,\gamma,\gamma,\gamma,\gamma,0,0,\gamma) \stackrel{1r}{\rightarrow} (\gamma,0,\gamma,\gamma,0,0,\gamma,\gamma)$	3; 8; 9; 12	2; 4; 5; 9; 12	$2^{-38.41}$
$(0,\gamma,\gamma,\gamma,\gamma,0,\gamma,0) \stackrel{1r}{\rightarrow} (\gamma,0,0,0,\gamma,0,0,\gamma)$	3; 9; 10; 11; 12	2; 4; 5; 7; 9; 10; 11; 12	$2^{-50.88}$
$(0,\gamma,\gamma,\gamma,\gamma,0,\gamma,\gamma) \stackrel{1r}{\rightarrow} (\gamma,0,0,\gamma,\gamma,\gamma,0,0)$	3; 8; 9; 10; 11	2; 4; 5; 7; 9; 10; 11	$2^{-13.47}$
$(0,\gamma,\gamma,\gamma,\gamma,\gamma,0,0) \stackrel{1r}{\rightarrow} (0,\gamma,\gamma,\gamma,\gamma,0,0,\gamma)$	3; 6; 10; 12	2; 4; 5; 10; 12	$2^{-38.41}$
$(0,\gamma,\gamma,\gamma,\gamma,\gamma,0,\gamma) \stackrel{1r}{\rightarrow} (0,\gamma,\gamma,0,\gamma,\gamma,0,0)$	3; 6; 8; 10	2; 4; 5; 10	$2^{-25.94}$
$(0,\gamma,\gamma,\gamma,\gamma,\gamma,\gamma,0) \stackrel{1r}{\rightarrow} (0,\gamma,0,\gamma,0,\gamma,\gamma,0)$	3; 6; 11	2; 4; 5; 7; 11	$2^{-38.41}$
$(0,\gamma,\gamma,\gamma,\gamma,\gamma,\gamma,\gamma) \stackrel{1r}{\rightarrow} (0,\gamma,0,0,0,0,\gamma,\gamma)$	3; 6; 8; 11; 12	2; 4; 5; 7; 11; 12	$2^{-25.94}$
$(\gamma,0,0,0,0,0,0,0) \stackrel{1r}{\rightarrow} (\gamma,0,\gamma,\gamma,\gamma,\gamma,0,0)$	1; 11; 12	11; 12	$2^{-25.94}$

Table 3.107 Nontrivial 1-round linear relations for MESH-128 (part V).

1-round linear relations	Weak subkeys $Z_j^{(i)}$ (ith round)		Bias
	j value (odd i)	j value (even i)	
$(\gamma,0,0,0,0,0,0,\gamma) \xrightarrow{1r} (\gamma,0,\gamma,0,\gamma,0,0,\gamma)$	1; 8; 11	11	$2^{-38.41}$
$(\gamma,0,0,0,0,0,\gamma,0) \xrightarrow{1r} (\gamma,0,0,\gamma,0,0,\gamma,\gamma)$	1; 10	7; 10	$2^{-25.94}$
$(\gamma,0,0,0,0,0,\gamma,\gamma) \xrightarrow{1r} (\gamma,0,0,0,0,\gamma,\gamma,0)$	1; 8; 10; 12	7; 10; 12	$2^{-38.41}$
$(\gamma,0,0,0,0,\gamma,0,0) \xrightarrow{1r} (0,\gamma,\gamma,0,0,0,\gamma,\gamma)$	1; 6; 9; 10; 11	9; 10; 11	$2^{-13.47}$
$(\gamma,0,0,0,0,\gamma,0,\gamma) \xrightarrow{1r} (0,\gamma,\gamma,\gamma,0,\gamma,\gamma,0)$	1; 6; 8; 9; 10; 11; 12	9; 10; 11; 12	$2^{-50.88}$
$(\gamma,0,0,0,0,\gamma,\gamma,0) \xrightarrow{1r} (0,\gamma,0,0,\gamma,\gamma,0,0)$	1; 6; 9; 12	7; 9; 12	$2^{-38.41}$
$(\gamma,0,0,0,0,\gamma,\gamma,\gamma) \xrightarrow{1r} (0,\gamma,0,\gamma,\gamma,0,0,\gamma)$	1; 6; 8; 9	7; 9	$2^{-25.94}$
$(\gamma,0,0,0,\gamma,0,0,0) \xrightarrow{1r} (\gamma,\gamma,0,0,0,0,0,0)$	1	5	2^{-1}
$(\gamma,0,0,0,\gamma,0,0,\gamma) \xrightarrow{1r} (\gamma,\gamma,0,\gamma,0,\gamma,0,\gamma)$	1; 8; 12	5; 12	$2^{-38.41}$
$(\gamma,0,0,0,\gamma,0,\gamma,0) \xrightarrow{1r} (\gamma,\gamma,\gamma,0,\gamma,\gamma,\gamma,\gamma)$	1; 10; 11; 12	5; 7; 10; 11; 12	$2^{-25.94}$
$(\gamma,0,0,0,\gamma,0,\gamma,\gamma) \xrightarrow{1r} (\gamma,\gamma,\gamma,\gamma,\gamma,0,\gamma,0)$	1; 8; 10; 11	5; 7; 10; 11	$2^{-13.47}$
$(\gamma,0,0,0,\gamma,\gamma,0,0) \xrightarrow{1r} (0,0,0,\gamma,\gamma,\gamma,\gamma,\gamma)$	1; 6; 9; 10; 12	5; 9; 10; 12	$2^{-13.47}$
$(\gamma,0,0,0,\gamma,\gamma,0,\gamma) \xrightarrow{1r} (0,0,0,0,\gamma,0,\gamma,0)$	1; 6; 8; 9; 10	5; 9; 10	$2^{-25.94}$
$(\gamma,0,0,0,\gamma,\gamma,\gamma,0) \xrightarrow{1r} (0,0,\gamma,\gamma,0,0,0,0)$	1; 6; 9; 11	5; 7; 9; 11	$2^{-13.47}$
$(\gamma,0,0,0,\gamma,\gamma,\gamma,\gamma) \xrightarrow{1r} (0,0,\gamma,0,0,\gamma,0,\gamma)$	1; 6; 8; 9; 11; 12	5; 7; 9; 11; 12	$2^{-25.94}$
$(\gamma,0,0,\gamma,0,0,0,0) \xrightarrow{1r} (\gamma,0,\gamma,0,\gamma,0,\gamma,0)$	1; 11	4; 11	$2^{-38.41}$
$(\gamma,0,0,\gamma,0,0,0,\gamma) \xrightarrow{1r} (\gamma,0,\gamma,\gamma,\gamma,\gamma,\gamma,\gamma)$	1; 8; 11; 12	4; 11; 12	$2^{-25.94}$
$(\gamma,0,0,\gamma,0,0,\gamma,0) \xrightarrow{1r} (\gamma,0,0,0,0,\gamma,0,\gamma)$	1; 10; 12	4; 7; 10; 12	$2^{-38.41}$
$(\gamma,0,0,\gamma,0,0,\gamma,\gamma) \xrightarrow{1r} (\gamma,0,0,\gamma,0,0,0,0)$	1; 8; 10	4; 7; 10	$2^{-25.94}$
$(\gamma,0,0,\gamma,0,\gamma,0,0) \xrightarrow{1r} (0,\gamma,\gamma,\gamma,0,\gamma,0,\gamma)$	1; 6; 9; 10; 11; 12	4; 9; 10; 11; 12	$2^{-50.88}$
$(\gamma,0,0,\gamma,0,\gamma,0,\gamma) \xrightarrow{1r} (0,\gamma,\gamma,0,0,0,0,0)$	1; 6; 8; 9; 10; 11	4; 9; 10; 11	$2^{-13.47}$
$(\gamma,0,0,\gamma,0,\gamma,\gamma,0) \xrightarrow{1r} (0,\gamma,0,\gamma,\gamma,0,\gamma,0)$	1; 6; 9	4; 7; 9	$2^{-25.94}$
$(\gamma,0,0,\gamma,0,\gamma,\gamma,\gamma) \xrightarrow{1r} (0,\gamma,0,0,\gamma,\gamma,\gamma,\gamma)$	1; 6; 8; 9; 12	4; 7; 9; 12	$2^{-38.41}$
$(\gamma,0,0,\gamma,\gamma,0,0,0) \xrightarrow{1r} (\gamma,\gamma,0,\gamma,0,\gamma,\gamma,0)$	1; 12	4; 5; 12	$2^{-38.41}$
$(\gamma,0,0,\gamma,\gamma,0,0,\gamma) \xrightarrow{1r} (\gamma,\gamma,0,0,0,0,\gamma,\gamma)$	1; 8;	4; 5;	2^{-1}
$(\gamma,0,0,\gamma,\gamma,0,\gamma,0) \xrightarrow{1r} (\gamma,\gamma,\gamma,\gamma,\gamma,0,0,\gamma)$	1; 10; 11	4; 5; 7; 10; 11	$2^{-13.47}$
$(\gamma,0,0,\gamma,\gamma,0,\gamma,\gamma) \xrightarrow{1r} (\gamma,\gamma,\gamma,0,\gamma,\gamma,0,0)$	1; 8; 10; 11; 12	4; 5; 7; 10; 11; 12	$2^{-25.94}$
$(\gamma,0,0,\gamma,\gamma,\gamma,0,0) \xrightarrow{1r} (0,0,0,0,\gamma,0,0,\gamma)$	1; 6; 9; 10	4; 5; 9; 10	$2^{-25.94}$
$(\gamma,0,0,\gamma,\gamma,\gamma,0,\gamma) \xrightarrow{1r} (0,0,0,\gamma,\gamma,\gamma,0,0)$	1; 6; 8; 9; 10; 12	4; 5; 9; 10; 12	$2^{-13.47}$
$(\gamma,0,0,\gamma,\gamma,\gamma,\gamma,0) \xrightarrow{1r} (0,0,\gamma,0,0,\gamma,\gamma,0)$	1; 6; 9; 11; 12	4; 5; 7; 9; 11; 12	$2^{-25.94}$
$(\gamma,0,0,\gamma,\gamma,\gamma,\gamma,\gamma) \xrightarrow{1r} (0,0,\gamma,\gamma,0,0,\gamma,\gamma)$	1; 6; 8; 9; 11	4; 5; 7; 9; 11	$2^{-13.47}$
$(\gamma,0,\gamma,0,0,0,0,0) \xrightarrow{1r} (\gamma,0,0,0,0,\gamma,\gamma,\gamma)$	1; 3; 10	10	$2^{-25.94}$

Table 3.108 Nontrivial 1-round linear relations for MESH-128 (part VI).

1-round linear relations	Weak subkeys $Z_j^{(i)}$ (ith round)		Bias
	j value (odd i)	j value (even i)	
$(\gamma,0,\gamma,0,0,0,0,\gamma) \xrightarrow{1r} (\gamma,0,0,\gamma,0,0,\gamma,0)$	1; 3; 8; 10; 12	10; 12	$2^{-38.41}$
$(\gamma,0,\gamma,0,0,0,\gamma,0) \xrightarrow{1r} (\gamma,0,\gamma,0,\gamma,0,0,0)$	1; 3; 11; 12	7; 11; 12	$2^{-25.94}$
$(\gamma,0,\gamma,0,0,0,\gamma,\gamma) \xrightarrow{1r} (\gamma,0,\gamma,\gamma,\gamma,\gamma,0,\gamma)$	1; 3; 8; 11	7; 11	$2^{-38.41}$
$(\gamma,0,\gamma,0,0,\gamma,0,0) \xrightarrow{1r} (0,\gamma,0,\gamma,\gamma,0,0,0)$	1; 3; 6; 9; 12	9; 12	$2^{-38.41}$
$(\gamma,0,\gamma,0,0,\gamma,0,\gamma) \xrightarrow{1r} (0,\gamma,0,0,\gamma,\gamma,0,\gamma)$	1; 3; 6; 8; 9	9	$2^{-25.94}$
$(\gamma,0,\gamma,0,0,\gamma,\gamma,0) \xrightarrow{1r} (0,\gamma,\gamma,\gamma,0,\gamma,\gamma,\gamma)$	1; 3; 6; 9; 10; 11	7; 9; 10; 11	$2^{-13.47}$
$(\gamma,0,\gamma,0,0,\gamma,\gamma,\gamma) \xrightarrow{1r} (0,\gamma,\gamma,0,0,0,\gamma,0)$	1; 3; 6; 8; 9; 10; 11; 12	7; 9; 10; 11; 12	$2^{-50.88}$
$(\gamma,0,\gamma,0,\gamma,0,0,0) \xrightarrow{1r} (\gamma,\gamma,\gamma,\gamma,\gamma,0,\gamma,\gamma)$	1; 3; 10; 11; 12	5; 10; 11; 12	$2^{-25.94}$
$(\gamma,0,\gamma,0,\gamma,0,0,\gamma) \xrightarrow{1r} (\gamma,\gamma,\gamma,0,\gamma,\gamma,\gamma,0)$	1; 3; 8; 10; 11	5; 10; 11	$2^{-13.47}$
$(\gamma,0,\gamma,0,\gamma,0,\gamma,0) \xrightarrow{1r} (\gamma,\gamma,0,\gamma,0,\gamma,0,0)$	1; 3;	5; 7;	2^{-1}
$(\gamma,0,\gamma,0,\gamma,0,\gamma,\gamma) \xrightarrow{1r} (\gamma,\gamma,0,0,0,0,0,\gamma)$	1; 3; 8; 12	5; 7; 12	$2^{-38.41}$
$(\gamma,0,\gamma,0,\gamma,\gamma,0,0) \xrightarrow{1r} (0,0,\gamma,0,0,\gamma,0,0)$	1; 3; 6; 9; 11	5; 9; 11	$2^{-13.47}$
$(\gamma,0,\gamma,0,\gamma,\gamma,0,\gamma) \xrightarrow{1r} (0,0,\gamma,\gamma,0,0,0,\gamma)$	1; 3; 6; 8; 9; 11; 12	5; 9; 11; 12	$2^{-25.94}$
$(\gamma,0,\gamma,0,\gamma,\gamma,\gamma,0) \xrightarrow{1r} (0,0,0,0,\gamma,0,\gamma,\gamma)$	1; 3; 6; 9; 10; 12	5; 7; 9; 10; 12	$2^{-13.47}$
$(\gamma,0,\gamma,0,\gamma,\gamma,\gamma,\gamma) \xrightarrow{1r} (0,0,0,\gamma,\gamma,\gamma,\gamma,0)$	1; 3; 6; 8; 9; 10	5; 7; 9; 10	$2^{-25.94}$
$(\gamma,0,\gamma,\gamma,0,0,0,0) \xrightarrow{1r} (\gamma,0,0,\gamma,0,0,0,\gamma)$	1; 3; 10; 12	4; 10; 12	$2^{-38.41}$
$(\gamma,0,\gamma,\gamma,0,0,0,\gamma) \xrightarrow{1r} (\gamma,0,0,0,0,\gamma,0,0)$	1; 3; 8; 10	4; 10	$2^{-25.94}$
$(\gamma,0,\gamma,\gamma,0,0,\gamma,0) \xrightarrow{1r} (\gamma,0,\gamma,\gamma,\gamma,\gamma,\gamma,0)$	1; 3; 11	4; 7; 11	$2^{-38.41}$
$(\gamma,0,\gamma,\gamma,0,0,\gamma,\gamma) \xrightarrow{1r} (\gamma,0,\gamma,0,\gamma,0,\gamma,\gamma)$	1; 3; 8; 11; 12	4; 7; 11; 12	$2^{-25.94}$
$(\gamma,0,\gamma,\gamma,0,\gamma,0,0) \xrightarrow{1r} (0,\gamma,0,0,\gamma,\gamma,\gamma,0)$	1; 3; 6; 9	4; 9	$2^{-25.94}$
$(\gamma,0,\gamma,\gamma,0,\gamma,0,\gamma) \xrightarrow{1r} (0,\gamma,0,\gamma,\gamma,0,\gamma,\gamma)$	1; 3; 6; 8; 9; 12	4; 9; 12	$2^{-38.41}$
$(\gamma,0,\gamma,\gamma,0,\gamma,\gamma,0) \xrightarrow{1r} (0,\gamma,\gamma,0,0,0,0,\gamma)$	1; 3; 6; 9; 10; 11; 12	4; 7; 9; 10; 11; 12	$2^{-50.88}$
$(\gamma,0,\gamma,\gamma,0,\gamma,\gamma,\gamma) \xrightarrow{1r} (0,\gamma,\gamma,\gamma,0,\gamma,0,0)$	1; 3; 6; 8; 9; 10; 11	4; 7; 9; 10; 11	$2^{-13.47}$
$(\gamma,0,\gamma,\gamma,\gamma,0,0,0) \xrightarrow{1r} (\gamma,\gamma,\gamma,0,\gamma,\gamma,0,\gamma)$	1; 3; 10; 11	4; 5; 10; 11	$2^{-13.47}$
$(\gamma,0,\gamma,\gamma,\gamma,0,0,\gamma) \xrightarrow{1r} (\gamma,\gamma,\gamma,\gamma,\gamma,0,0,0)$	1; 3; 8; 10; 11; 12	4; 5; 10; 11; 12	$2^{-25.94}$
$(\gamma,0,\gamma,\gamma,\gamma,0,\gamma,0) \xrightarrow{1r} (\gamma,\gamma,0,0,0,0,\gamma,0)$	1; 3; 12	4; 5; 7; 12	$2^{-38.41}$
$(\gamma,0,\gamma,\gamma,\gamma,0,\gamma,\gamma) \xrightarrow{1r} (\gamma,\gamma,0,\gamma,0,\gamma,\gamma,\gamma)$	1; 3; 8	4; 5; 7	2^{-1}
$(\gamma,0,\gamma,\gamma,\gamma,\gamma,0,0) \xrightarrow{1r} (0,0,\gamma,\gamma,0,0,\gamma,0)$	1; 3; 6; 9; 11; 12	4; 5; 9; 11; 12	$2^{-25.94}$
$(\gamma,0,\gamma,\gamma,\gamma,\gamma,0,\gamma) \xrightarrow{1r} (0,0,\gamma,0,0,\gamma,\gamma,\gamma)$	1; 3; 6; 8; 9; 11	4; 5; 9; 11	$2^{-13.47}$
$(\gamma,0,\gamma,\gamma,\gamma,\gamma,\gamma,0) \xrightarrow{1r} (0,0,0,\gamma,\gamma,\gamma,0,\gamma)$	1; 3; 6; 9; 10	4; 5; 7; 9; 10	$2^{-25.94}$
$(\gamma,0,\gamma,\gamma,\gamma,\gamma,\gamma,\gamma) \xrightarrow{1r} (0,0,0,0,\gamma,0,0,0)$	1; 3; 6; 8; 9; 10; 12	4; 5; 7; 9; 10; 12	$2^{-13.47}$
$(\gamma,\gamma,0,0,0,0,0,0) \xrightarrow{1r} (0,\gamma,0,0,\gamma,0,\gamma,\gamma)$	1; 9; 10; 11	2; 9; 10; 11	$2^{-13.47}$

Table 3.109 Nontrivial 1-round linear relations for MESH-128 (part VII).

1-round linear relations	Weak subkeys $Z_j^{(i)}$ (ith round)		Bias
	j value (odd i)	j value (even i)	
$(\gamma,\gamma,0,0,0,0,0,\gamma) \stackrel{1r}{\to} (0,\gamma,0,\gamma,\gamma,\gamma,\gamma,0)$	1; 8; 9; 10; 11; 12	2; 9; 10; 11; 12	$2^{-50.88}$
$(\gamma,\gamma,0,0,0,0,\gamma,0) \stackrel{1r}{\to} (0,\gamma,\gamma,0,0,\gamma,0,0)$	1; 9; 12	2; 7; 9; 12	$2^{-38.41}$
$(\gamma,\gamma,0,0,0,0,\gamma,\gamma) \stackrel{1r}{\to} (0,\gamma,\gamma,\gamma,0,0,0,\gamma)$	1; 8; 9	2; 7; 9	$2^{-25.94}$
$(\gamma,\gamma,0,0,0,\gamma,0,0) \stackrel{1r}{\to} (\gamma,0,0,\gamma,0,\gamma,0,0)$	1; 6; 11; 12	2; 11; 12	$2^{-25.94}$
$(\gamma,\gamma,0,0,0,\gamma,0,\gamma) \stackrel{1r}{\to} (\gamma,0,0,0,0,0,0,\gamma)$	1; 6; 8; 11	2; 11	$2^{-38.41}$
$(\gamma,\gamma,0,0,0,\gamma,\gamma,0) \stackrel{1r}{\to} (\gamma,0,\gamma,\gamma,\gamma,0,\gamma,\gamma)$	1; 6; 10	2; 7; 10	$2^{-25.94}$
$(\gamma,\gamma,0,0,0,\gamma,\gamma,\gamma) \stackrel{1r}{\to} (\gamma,0,\gamma,0,\gamma,\gamma,\gamma,0)$	1; 6; 8; 10; 12	2; 7; 10; 12	$2^{-38.41}$
$(\gamma,\gamma,0,0,\gamma,0,0,0) \stackrel{1r}{\to} (0,0,\gamma,\gamma,0,\gamma,\gamma,\gamma)$	1; 9; 10; 12	2; 5; 9; 10; 12	$2^{-13.47}$
$(\gamma,\gamma,0,0,\gamma,0,0,\gamma) \stackrel{1r}{\to} (0,0,\gamma,0,0,0,\gamma,0)$	1; 8; 9; 10	2; 5; 9; 10	$2^{-25.94}$
$(\gamma,\gamma,0,0,\gamma,0,\gamma,0) \stackrel{1r}{\to} (0,0,0,\gamma,\gamma,0,0,0)$	1; 9; 11	2; 5; 7; 9; 11	$2^{-13.47}$
$(\gamma,\gamma,0,0,\gamma,0,\gamma,\gamma) \stackrel{1r}{\to} (0,0,0,0,\gamma,\gamma,0,\gamma)$	1; 8; 9; 11; 12	2; 5; 7; 9; 11; 12	$2^{-25.94}$
$(\gamma,\gamma,0,0,\gamma,\gamma,0,0) \stackrel{1r}{\to} (\gamma,\gamma,\gamma,0,\gamma,0,0,0)$	1; 6;	2; 5;	2^{-1}
$(\gamma,\gamma,0,0,\gamma,\gamma,0,\gamma) \stackrel{1r}{\to} (\gamma,\gamma,\gamma,\gamma,\gamma,\gamma,0,\gamma)$	1; 6; 8; 12	2; 5; 12	$2^{-38.41}$
$(\gamma,\gamma,0,0,\gamma,\gamma,\gamma,0) \stackrel{1r}{\to} (\gamma,\gamma,0,0,0,\gamma,\gamma,\gamma)$	1; 6; 10; 11; 12	2; 5; 7; 10; 11; 12	$2^{-25.94}$
$(\gamma,\gamma,0,0,\gamma,\gamma,\gamma,\gamma) \stackrel{1r}{\to} (\gamma,\gamma,0,\gamma,0,0,\gamma,0)$	1; 6; 8; 10; 11	2; 5; 7; 10; 11	$2^{-13.47}$
$(\gamma,\gamma,0,\gamma,0,0,0,0) \stackrel{1r}{\to} (0,\gamma,0,\gamma,\gamma,\gamma,0,\gamma)$	1; 9; 10; 11; 12	2; 4; 9; 10; 11; 12	$2^{-50.88}$
$(\gamma,\gamma,0,\gamma,0,0,0,\gamma) \stackrel{1r}{\to} (0,\gamma,0,0,\gamma,0,0,0)$	1; 8; 9; 10; 11	2; 4; 9; 10; 11	$2^{-13.47}$
$(\gamma,\gamma,0,\gamma,0,0,\gamma,0) \stackrel{1r}{\to} (0,\gamma,\gamma,\gamma,0,0,\gamma,0)$	1; 9	2; 4; 7; 9	$2^{-25.94}$
$(\gamma,\gamma,0,\gamma,0,0,\gamma,\gamma) \stackrel{1r}{\to} (0,\gamma,\gamma,0,0,\gamma,\gamma,\gamma)$	1; 8; 9; 12	2; 4; 7; 9; 12	$2^{-38.41}$
$(\gamma,\gamma,0,\gamma,0,\gamma,0,0) \stackrel{1r}{\to} (\gamma,0,0,0,0,0,\gamma,0)$	1; 6; 11	2; 4; 11	$2^{-38.41}$
$(\gamma,\gamma,0,\gamma,0,\gamma,0,\gamma) \stackrel{1r}{\to} (\gamma,0,0,\gamma,0,\gamma,\gamma,\gamma)$	1; 6; 8; 11; 12	2; 4; 11; 12	$2^{-25.94}$
$(\gamma,\gamma,0,\gamma,0,\gamma,\gamma,0) \stackrel{1r}{\to} (\gamma,0,\gamma,0,\gamma,\gamma,0,\gamma)$	1; 6; 10; 12	2; 4; 7; 10; 12	$2^{-38.41}$
$(\gamma,\gamma,0,\gamma,0,\gamma,\gamma,\gamma) \stackrel{1r}{\to} (\gamma,0,\gamma,\gamma,\gamma,0,0,0)$	1; 6; 8; 10	2; 4; 7; 10	$2^{-25.94}$
$(\gamma,\gamma,0,\gamma,\gamma,0,0,0) \stackrel{1r}{\to} (0,0,\gamma,0,0,0,0,\gamma)$	1; 9; 10	2; 4; 5; 9; 10	$2^{-25.94}$
$(\gamma,\gamma,0,\gamma,\gamma,0,0,\gamma) \stackrel{1r}{\to} (0,0,\gamma,\gamma,0,\gamma,0,0)$	1; 8; 9; 10; 12	2; 4; 5; 9; 10; 12	$2^{-13.47}$
$(\gamma,\gamma,0,\gamma,\gamma,0,\gamma,0) \stackrel{1r}{\to} (0,0,0,0,\gamma,\gamma,\gamma,0)$	1; 9; 11; 12	2; 4; 5; 7; 9; 11; 12	$2^{-25.94}$
$(\gamma,\gamma,0,\gamma,\gamma,0,\gamma,\gamma) \stackrel{1r}{\to} (0,0,0,\gamma,\gamma,0,\gamma,\gamma)$	1; 8; 9; 11	2; 4; 5; 7; 9; 11	$2^{-13.47}$
$(\gamma,\gamma,0,\gamma,\gamma,\gamma,0,0) \stackrel{1r}{\to} (\gamma,\gamma,\gamma,\gamma,\gamma,\gamma,\gamma,0)$	1; 6; 12	2; 4; 5; 12	$2^{-38.41}$
$(\gamma,\gamma,0,\gamma,\gamma,\gamma,0,\gamma) \stackrel{1r}{\to} (\gamma,\gamma,\gamma,0,\gamma,0,\gamma,\gamma)$	1; 6; 8	2; 4; 5	2^{-1}
$(\gamma,\gamma,0,\gamma,\gamma,\gamma,\gamma,0) \stackrel{1r}{\to} (\gamma,\gamma,0,\gamma,0,0,0,\gamma)$	1; 6; 10; 11	2; 4; 5; 7; 10; 11	$2^{-13.47}$
$(\gamma,\gamma,0,\gamma,\gamma,\gamma,\gamma,\gamma) \stackrel{1r}{\to} (\gamma,\gamma,0,0,0,\gamma,0,0)$	1; 6; 8; 10; 11; 12	2; 4; 5; 7; 10; 11; 12	$2^{-25.94}$
$(\gamma,\gamma,\gamma,0,0,0,0,0) \stackrel{1r}{\to} (0,\gamma,\gamma,\gamma,0,0,0,0)$	1; 3; 9; 12	2; 9; 12	$2^{-38.41}$

Table 3.110 Nontrivial 1-round linear relations for MESH-128 (part VIII).

1-round linear relations	Weak subkeys $Z_j^{(i)}$ (ith round)		Bias
	j value (odd i)	j value (even i)	
$(\gamma,\gamma,\gamma,0,0,0,0,\gamma) \xrightarrow{1r} (0,\gamma,\gamma,0,0,\gamma,0,\gamma)$	1; 3; 8; 9	2; 9	$2^{-25.94}$
$(\gamma,\gamma,\gamma,0,0,0,\gamma,0) \xrightarrow{1r} (0,\gamma,0,\gamma,\gamma,\gamma,\gamma,\gamma)$	1; 3; 9; 10; 11	2; 7; 9; 10; 11	$2^{-13.47}$
$(\gamma,\gamma,\gamma,0,0,0,\gamma,\gamma) \xrightarrow{1r} (0,\gamma,0,0,\gamma,0,\gamma,0)$	1; 3; 8; 9; 10; 11; 12	2; 7; 9; 10; 11; 12	$2^{-50.88}$
$(\gamma,\gamma,\gamma,0,0,\gamma,0,0) \xrightarrow{1r} (\gamma,0,\gamma,0,\gamma,\gamma,\gamma,\gamma)$	1; 3; 6; 10	2; 10	$2^{-25.94}$
$(\gamma,\gamma,\gamma,0,0,\gamma,0,\gamma) \xrightarrow{1r} (\gamma,0,\gamma,\gamma,\gamma,0,\gamma,0)$	1; 3; 6; 8; 10; 12	2; 10; 12	$2^{-38.41}$
$(\gamma,\gamma,\gamma,0,0,\gamma,\gamma,0) \xrightarrow{1r} (\gamma,0,0,0,0,0,0,0)$	1; 3; 6; 11; 12	2; 7; 11; 12	$2^{-25.94}$
$(\gamma,\gamma,\gamma,0,0,\gamma,\gamma,\gamma) \xrightarrow{1r} (\gamma,0,0,\gamma,0,\gamma,0,\gamma)$	1; 3; 6; 8; 11	2; 7; 11	$2^{-38.41}$
$(\gamma,\gamma,\gamma,0,\gamma,0,0,0) \xrightarrow{1r} (0,0,0,0,\gamma,\gamma,0,0)$	1; 3; 9; 11	2; 5; 9; 11	$2^{-13.47}$
$(\gamma,\gamma,\gamma,0,\gamma,0,0,\gamma) \xrightarrow{1r} (0,0,0,\gamma,\gamma,0,0,\gamma)$	1; 3; 8; 9; 11; 12	2; 5; 9; 11; 12	$2^{-25.94}$
$(\gamma,\gamma,\gamma,0,\gamma,0,\gamma,0) \xrightarrow{1r} (0,0,\gamma,0,0,0,\gamma,\gamma)$	1; 3; 9; 10; 12	2; 5; 7; 9; 10; 12	$2^{-13.47}$
$(\gamma,\gamma,\gamma,0,\gamma,0,\gamma,\gamma) \xrightarrow{1r} (0,0,\gamma,\gamma,0,\gamma,\gamma,0)$	1; 3; 8; 9; 10	2; 5; 7; 9; 10	$2^{-25.94}$
$(\gamma,\gamma,\gamma,0,\gamma,\gamma,0,0) \xrightarrow{1r} (\gamma,\gamma,0,\gamma,0,0,\gamma,\gamma)$	1; 3; 6; 10; 11; 12	2; 5; 10; 11; 12	$2^{-25.94}$
$(\gamma,\gamma,\gamma,0,\gamma,\gamma,0,\gamma) \xrightarrow{1r} (\gamma,\gamma,0,0,0,\gamma,\gamma,0)$	1; 3; 6; 8; 10; 11	2; 5; 10; 11	$2^{-13.47}$
$(\gamma,\gamma,\gamma,0,\gamma,\gamma,\gamma,0) \xrightarrow{1r} (\gamma,\gamma,\gamma,\gamma,\gamma,\gamma,0,0)$	1; 3; 6	2; 5; 7	2^{-1}
$(\gamma,\gamma,\gamma,0,\gamma,\gamma,\gamma,\gamma) \xrightarrow{1r} (\gamma,\gamma,\gamma,0,\gamma,0,0,\gamma)$	1; 3; 6; 8; 12	2; 5; 7; 12	$2^{-38.41}$
$(\gamma,\gamma,\gamma,\gamma,0,0,0,0) \xrightarrow{1r} (0,\gamma,\gamma,0,0,\gamma,\gamma,0)$	1; 3; 9	2; 4; 9	$2^{-25.94}$
$(\gamma,\gamma,\gamma,\gamma,0,0,0,\gamma) \xrightarrow{1r} (0,\gamma,\gamma,\gamma,0,0,\gamma,\gamma)$	1; 3; 8; 9; 12	2; 4; 9; 12	$2^{-38.41}$
$(\gamma,\gamma,\gamma,\gamma,0,0,\gamma,0) \xrightarrow{1r} (0,\gamma,0,0,\gamma,0,0,\gamma)$	1; 3; 9; 10; 11; 12	2; 4; 7; 9; 10; 11; 12	$2^{-50.88}$
$(\gamma,\gamma,\gamma,\gamma,0,0,\gamma,\gamma) \xrightarrow{1r} (0,\gamma,0,\gamma,\gamma,\gamma,0,0)$	1; 3; 8; 9; 10; 11	2; 4; 7; 9; 10; 11	$2^{-13.47}$
$(\gamma,\gamma,\gamma,\gamma,0,\gamma,0,0) \xrightarrow{1r} (\gamma,0,\gamma,\gamma,\gamma,0,0,\gamma)$	1; 3; 6; 10; 12	2; 4; 10; 12	$2^{-38.41}$
$(\gamma,\gamma,\gamma,\gamma,0,\gamma,0,\gamma) \xrightarrow{1r} (\gamma,0,\gamma,0,\gamma,\gamma,0,0)$	1; 3; 6; 8; 10	2; 4; 10	$2^{-25.94}$
$(\gamma,\gamma,\gamma,\gamma,0,\gamma,\gamma,0) \xrightarrow{1r} (\gamma,0,0,\gamma,0,\gamma,\gamma,0)$	1; 3; 6; 11	2; 4; 7; 11	$2^{-38.41}$
$(\gamma,\gamma,\gamma,\gamma,0,\gamma,\gamma,\gamma) \xrightarrow{1r} (\gamma,0,0,0,0,0,\gamma,\gamma)$	1; 3; 6; 8; 11; 12	2; 4; 7; 11; 12	$2^{-25.94}$
$(\gamma,\gamma,\gamma,\gamma,\gamma,0,0,0) \xrightarrow{1r} (0,0,0,\gamma,\gamma,0,\gamma,0)$	1; 3; 9; 11; 12	2; 4; 5; 9; 11; 12	$2^{-25.94}$
$(\gamma,\gamma,\gamma,\gamma,\gamma,0,0,\gamma) \xrightarrow{1r} (0,0,0,0,\gamma,\gamma,\gamma,\gamma)$	1; 3; 8; 9; 11	2; 4; 5; 9; 11	$2^{-13.47}$
$(\gamma,\gamma,\gamma,\gamma,\gamma,0,\gamma,0) \xrightarrow{1r} (0,0,\gamma,\gamma,0,\gamma,0,\gamma)$	1; 3; 9; 10	2; 4; 5; 7; 9; 10	$2^{-25.94}$
$(\gamma,\gamma,\gamma,\gamma,\gamma,0,\gamma,\gamma) \xrightarrow{1r} (0,0,\gamma,0,0,0,0,0)$	1; 3; 8; 9; 10; 12	2; 4; 5; 7; 9; 10; 12	$2^{-13.47}$
$(\gamma,\gamma,\gamma,\gamma,\gamma,\gamma,0,0) \xrightarrow{1r} (\gamma,\gamma,0,0,0,\gamma,0,\gamma)$	1; 3; 6; 10; 11	2; 4; 5; 10; 11	$2^{-13.47}$
$(\gamma,\gamma,\gamma,\gamma,\gamma,\gamma,0,\gamma) \xrightarrow{1r} (\gamma,\gamma,0,\gamma,0,0,0,0)$	1; 3; 6; 8; 10; 11; 12	2; 4; 5; 10; 11; 12	$2^{-25.94}$
$(\gamma,\gamma,\gamma,\gamma,\gamma,\gamma,\gamma,0) \xrightarrow{1r} (\gamma,\gamma,\gamma,0,\gamma,0,\gamma,0)$	1; 3; 6; 12	2; 4; 5; 7; 12	$2^{-38.41}$
$(\gamma,\gamma,\gamma,\gamma,\gamma,\gamma,\gamma,\gamma) \xrightarrow{1r} (\gamma,\gamma,\gamma,\gamma,\gamma,\gamma,\gamma,\gamma)$	1; 3; 6; 8	2; 4; 5; 7	2^{-1}

$Z_5^{(i+2)}$, $Z_{12}^{(i+2)}$, $Z_1^{(i+3)}$, $Z_3^{(i+3)}$, $Z_6^{(i+3)}$, $Z_{12}^{(i+3)}$, $Z_2^{(i+4)}$, $Z_5^{(i+4)}$, $Z_7^{(i+4)}$, $Z_9^{(i+4)}$, $Z_{10}^{(i+4)}$, $Z_{12}^{(i+4)}$, $Z_3^{(i+5)}$, $Z_8^{(i+5)}$ and $Z_{12}^{(i+5)}$.

(2) $(0,0,0,0,0,0,\gamma,0) \xrightarrow{1r} (0,0,\gamma,0,\gamma,\gamma,\gamma,\gamma) \xrightarrow{1r} (\gamma,\ 0,\ \gamma,\ 0,\ 0,\ 0,\ \gamma,\ 0) \xrightarrow{1r} (\gamma,\ 0,\ \gamma,\ 0,\ \gamma,\ 0,\ 0,\ 0) \xrightarrow{1r} (\gamma,\gamma,\gamma,\gamma,\gamma,0,\gamma,\gamma) \xrightarrow{1r} (0,0,\gamma,0,0,0,0,0) \xrightarrow{1r} (0,\ 0,\ \gamma,\ \gamma,\ \gamma,\ 0,\ \gamma,\ \gamma) \xrightarrow{1r} (0,\gamma,\gamma,0,\gamma,0,\gamma,\gamma) \xrightarrow{1r} (\gamma,0,0,0,\gamma,0,\gamma,0) \xrightarrow{1r} (\gamma,\gamma,\gamma,0,\gamma,\gamma,\gamma,\gamma) \xrightarrow{1r} (\gamma,\ \gamma,\ \gamma,\ 0,\ \gamma,\ 0,\ 0,\ \gamma) \xrightarrow{1r} (0,0,0,\gamma,\gamma,0,0,\gamma) \xrightarrow{1r}$

$(0, \gamma, \gamma, \gamma, \gamma, \gamma, \gamma, \gamma) \stackrel{1r}{\to} (0,\gamma,0,0,0,0,\gamma,\gamma) \stackrel{1r}{\to} (\gamma, \gamma, 0, 0, \gamma, \gamma, 0, \gamma) \stackrel{1r}{\to} (\gamma,\gamma,\gamma,\gamma,\gamma,\gamma,0,\gamma) \stackrel{1r}{\to} (\gamma,\gamma,0,\gamma,0,0,0,0) \stackrel{1r}{\to} (0, \gamma, 0, \gamma, \gamma, \gamma, 0, \gamma) \stackrel{1r}{\to} (0,\gamma,0,\gamma,0,\gamma,\gamma,\gamma) \stackrel{1r}{\to} (0,0,0,0,0,\gamma,0,0) \stackrel{1r}{\to} (\gamma, \gamma, 0, \gamma, \gamma, \gamma, \gamma, \gamma) \stackrel{1r}{\to} (\gamma, \gamma, 0, 0, 0, \gamma, 0, 0) \stackrel{1r}{\to} (\gamma,0,0,\gamma,0,\gamma,0,0) \stackrel{1r}{\to} (0, \gamma, \gamma, \gamma, 0, \gamma, 0, \gamma) \stackrel{1r}{\to} (0, 0, 0, \gamma, 0, 0, 0, 0) \stackrel{1r}{\to} (0,0,0,\gamma,0,\gamma,\gamma,0) \stackrel{1r}{\to} (\gamma, \gamma, \gamma, 0, 0, \gamma, \gamma, 0) \stackrel{1r}{\to} (\gamma, 0, 0, 0, 0, 0, 0, 0) \stackrel{1r}{\to} (\gamma,0,\gamma,\gamma,\gamma,\gamma,0,0) \stackrel{1r}{\to} (0, 0, \gamma, \gamma, 0, 0, \gamma, 0) \stackrel{1r}{\to} (0, 0, 0, 0, 0, 0, \gamma, 0)$, which is a 30-round iterative linear relation with bias $2^{-873.90}$.

Starting from an odd-numbered round i, the following 135 subkeys have to be weak: $Z_{10}^{(i)}$, $Z_{11}^{(i)}$, $Z_{12}^{(i)}$, $Z_5^{(i+1)}$, $Z_7^{(i+1)}$, $Z_9^{(i+1)}$, $Z_{10}^{(i+1)}$, $Z_{11}^{(i+1)}$, $Z_{12}^{(i+1)}$, $Z_1^{(i+2)}$, $Z_3^{(i+2)}$, $Z_{11}^{(i+2)}$, $Z_{12}^{(i+2)}$, $Z_5^{(i+3)}$, $Z_{10}^{(i+3)}$, $Z_{11}^{(i+3)}$, $Z_{12}^{(i+3)}$, $Z_1^{(i+4)}$, $Z_3^{(i+4)}$, $Z_8^{(i+4)}$, $Z_9^{(i+4)}$, $Z_{10}^{(i+4)}$, $Z_{12}^{(i+4)}$, $Z_{10}^{(i+5)}$, $Z_{11}^{(i+5)}$, $Z_{12}^{(i+5)}$, $Z_3^{(i+6)}$, $Z_8^{(i+6)}$, $Z_{11}^{(i+6)}$, $Z_{12}^{(i+6)}$, $Z_2^{(i+7)}$, $Z_5^{(i+7)}$, $Z_7^{(i+7)}$, $Z_9^{(i+7)}$, $Z_{10}^{(i+7)}$, $Z_{11}^{(i+7)}$, $Z_{12}^{(i+7)}$, $Z_1^{(i+8)}$, $Z_{10}^{(i+8)}$, $Z_{11}^{(i+8)}$, $Z_{12}^{(i+8)}$, $Z_2^{(i+9)}$, $Z_5^{(i+9)}$, $Z_7^{(i+9)}$, $Z_{12}^{(i+9)}$, $Z_1^{(i+10)}$, $Z_3^{(i+10)}$, $Z_8^{(i+10)}$, $Z_9^{(i+10)}$, $Z_{11}^{(i+10)}$, $Z_{12}^{(i+10)}$, $Z_4^{(i+11)}$, $Z_5^{(i+11)}$, $Z_{11}^{(i+11)}$, $Z_{12}^{(i+11)}$, $Z_3^{(i+12)}$, $Z_6^{(i+12)}$, $Z_8^{(i+12)}$, $Z_{11}^{(i+12)}$, $Z_{12}^{(i+12)}$, $Z_2^{(i+13)}$, $Z_7^{(i+13)}$, $Z_9^{(i+13)}$, $Z_{11}^{(i+13)}$, $Z_{12}^{(i+13)}$, $Z_1^{(i+14)}$, $Z_6^{(i+14)}$, $Z_8^{(i+14)}$, $Z_{12}^{(i+14)}$, $Z_2^{(i+15)}$, $Z_4^{(i+15)}$, $Z_5^{(i+15)}$, $Z_{10}^{(i+15)}$, $Z_{11}^{(i+15)}$, $Z_{12}^{(i+15)}$, $Z_1^{(i+16)}$, $Z_9^{(i+16)}$, $Z_{10}^{(i+16)}$, $Z_{11}^{(i+16)}$, $Z_{12}^{(i+16)}$, $Z_2^{(i+17)}$, $Z_4^{(i+17)}$, $Z_5^{(i+17)}$, $Z_{11}^{(i+17)}$, $Z_{12}^{(i+17)}$, $Z_6^{(i+18)}$, $Z_8^{(i+18)}$, $Z_{10}^{(i+18)}$, $Z_{11}^{(i+18)}$, $Z_{12}^{(i+18)}$, $Z_9^{(i+19)}$, $Z_{10}^{(i+19)}$, $Z_{12}^{(i+19)}$, $Z_1^{(i+20)}$, $Z_6^{(i+20)}$, $Z_8^{(i+20)}$, $Z_{10}^{(i+20)}$, $Z_{11}^{(i+20)}$, $Z_{12}^{(i+20)}$, $Z_2^{(i+21)}$, $Z_{11}^{(i+21)}$, $Z_{12}^{(i+21)}$, $Z_1^{(i+22)}$, $Z_6^{(i+22)}$, $Z_9^{(i+22)}$, $Z_{10}^{(i+22)}$, $Z_{11}^{(i+22)}$, $Z_{12}^{(i+22)}$, $Z_2^{(i+23)}$, $Z_4^{(i+23)}$, $Z_{10}^{(i+23)}$, $Z_{11}^{(i+23)}$, $Z_{12}^{(i+23)}$, $Z_{12}^{(i+24)}$, $Z_4^{(i+25)}$, $Z_7^{(i+25)}$, $Z_9^{(i+25)}$, $Z_{11}^{(i+25)}$, $Z_{12}^{(i+25)}$, $Z_1^{(i+26)}$, $Z_3^{(i+26)}$, $Z_6^{(i+26)}$, $Z_{11}^{(i+26)}$, $Z_{12}^{(i+26)}$, $Z_{11}^{(i+27)}$, $Z_{12}^{(i+27)}$, $Z_1^{(i+28)}$, $Z_3^{(i+28)}$, $Z_6^{(i+28)}$, $Z_9^{(i+28)}$, $Z_{11}^{(i+28)}$, $Z_{12}^{(i+28)}$, $Z_4^{(i+29)}$, $Z_7^{(i+29)}$ and $Z_{12}^{(i+29)}$.

Starting from an even-numbered round i, the following 141 subkeys have to be weak: $Z_7^{(i)}$, $Z_{10}^{(i)}$, $Z_{11}^{(i)}$, $Z_{12}^{(i)}$, $Z_3^{(i+1)}$, $Z_6^{(i+1)}$, $Z_8^{(i+1)}$, $Z_9^{(i+1)}$, $Z_{10}^{(i+1)}$, $Z_{11}^{(i+1)}$, $Z_{12}^{(i+1)}$, $Z_7^{(i+2)}$, $Z_{11}^{(i+2)}$, $Z_{12}^{(i+2)}$, $Z_1^{(i+3)}$, $Z_3^{(i+3)}$, $Z_{10}^{(i+3)}$, $Z_{11}^{(i+3)}$, $Z_{12}^{(i+3)}$, $Z_2^{(i+4)}$, $Z_4^{(i+4)}$, $Z_5^{(i+4)}$, $Z_7^{(i+4)}$, $Z_9^{(i+4)}$, $Z_{10}^{(i+4)}$, $Z_{12}^{(i+4)}$, $Z_3^{(i+5)}$, $Z_{10}^{(i+5)}$, $Z_{11}^{(i+5)}$, $Z_{12}^{(i+5)}$, $Z_4^{(i+6)}$, $Z_5^{(i+6)}$, $Z_7^{(i+6)}$, $Z_{11}^{(i+6)}$, $Z_{12}^{(i+6)}$, $Z_3^{(i+7)}$, $Z_8^{(i+7)}$, $Z_9^{(i+7)}$, $Z_{10}^{(i+7)}$, $Z_{11}^{(i+7)}$, $Z_{12}^{(i+7)}$, $Z_5^{(i+8)}$, $Z_7^{(i+8)}$, $Z_{10}^{(i+8)}$, $Z_{11}^{(i+8)}$, $Z_{12}^{(i+8)}$, $Z_1^{(i+9)}$, $Z_3^{(i+9)}$, $Z_6^{(i+9)}$, $Z_8^{(i+9)}$, $Z_{12}^{(i+9)}$, $Z_2^{(i+10)}$, $Z_5^{(i+10)}$, $Z_9^{(i+10)}$, $Z_{11}^{(i+10)}$, $Z_{12}^{(i+10)}$, $Z_8^{(i+11)}$, $Z_{11}^{(i+11)}$, $Z_{12}^{(i+11)}$, $Z_2^{(i+12)}$, $Z_4^{(i+12)}$, $Z_5^{(i+12)}$, $Z_7^{(i+12)}$, $Z_{11}^{(i+12)}$, $Z_{12}^{(i+12)}$, $Z_8^{(i+13)}$, $Z_9^{(i+13)}$, $Z_{11}^{(i+13)}$, $Z_{12}^{(i+13)}$, $Z_2^{(i+14)}$, $Z_5^{(i+14)}$, $Z_{12}^{(i+14)}$, $Z_1^{(i+15)}$, $Z_3^{(i+15)}$, $Z_6^{(i+15)}$, $Z_8^{(i+15)}$, $Z_{10}^{(i+15)}$, $Z_{11}^{(i+15)}$, $Z_{12}^{(i+15)}$, $Z_2^{(i+16)}$, $Z_4^{(i+16)}$, $Z_9^{(i+16)}$, $Z_{10}^{(i+16)}$, $Z_{11}^{(i+16)}$, $Z_{12}^{(i+16)}$, $Z_6^{(i+17)}$, $Z_8^{(i+17)}$, $Z_{11}^{(i+17)}$, $Z_{12}^{(i+17)}$, $Z_2^{(i+18)}$, $Z_4^{(i+18)}$, $Z_7^{(i+18)}$, $Z_{10}^{(i+18)}$, $Z_{11}^{(i+18)}$, $Z_{12}^{(i+18)}$, $Z_6^{(i+19)}$, $Z_9^{(i+19)}$, $Z_{10}^{(i+19)}$, $Z_{12}^{(i+19)}$, $Z_2^{(i+20)}$, $Z_4^{(i+20)}$, $Z_5^{(i+20)}$, $Z_7^{(i+20)}$, $Z_{10}^{(i+20)}$, $Z_{11}^{(i+20)}$, $Z_{12}^{(i+20)}$, $Z_1^{(i+21)}$, $Z_6^{(i+21)}$, $Z_{11}^{(i+21)}$, $Z_{12}^{(i+21)}$, $Z_4^{(i+22)}$, $Z_9^{(i+22)}$, $Z_{10}^{(i+22)}$, $Z_{11}^{(i+22)}$, $Z_{12}^{(i+22)}$, $Z_3^{(i+23)}$, $Z_6^{(i+23)}$,

$Z_8^{(i+23)}$, $Z_{10}^{(i+23)}$, $Z_{11}^{(i+23)}$, $Z_{12}^{(i+23)}$, $Z_4^{(i+24)}$, $Z_{12}^{(i+24)}$, $Z_6^{(i+25)}$, $Z_9^{(i+25)}$, $Z_{11}^{(i+25)}$, $Z_{12}^{(i+25)}$, $Z_2^{(i+26)}$, $Z_7^{(i+26)}$, $Z_{11}^{(i+26)}$, $Z_{12}^{(i+26)}$, $Z_1^{(i+27)}$, $Z_{11}^{(i+27)}$, $Z_{12}^{(i+27)}$, $Z_4^{(i+28)}$, $Z_5^{(i+28)}$, $Z_9^{(i+28)}$, $Z_{11}^{(i+28)}$, $Z_{12}^{(i+28)}$, $Z_3^{(i+29)}$ and $Z_{12}^{(i+29)}$.

(3) $(0,0,0,0,0,0,\gamma,\gamma) \stackrel{1r}{\rightarrow} (0,0,\gamma,\gamma,\gamma,0,\gamma,0) \stackrel{1r}{\rightarrow} (0, \gamma, \gamma, \gamma, \gamma, \gamma, \gamma, 0) \stackrel{1r}{\rightarrow} (0, \gamma, 0, \gamma, 0, \gamma, \gamma, 0) \stackrel{1r}{\rightarrow} (0,0,0,\gamma,0,0,0,\gamma) \stackrel{1r}{\rightarrow} (0,0,0,0,0,0,\gamma,\gamma)$, which is a 5-round iterative linear relation with bias $2^{-100.76}$.
Starting from an odd-numbered round i, the following 16 subkeys have to be weak: $Z_8^{(i)}$, $Z_{10}^{(i)}$, $Z_{11}^{(i)}$, $Z_4^{(i+1)}$, $Z_5^{(i+1)}$, $Z_7^{(i+1)}$, $Z_{11}^{(i+1)}$, $Z_3^{(i+2)}$, $Z_6^{(i+2)}$, $Z_{11}^{(i+2)}$, $Z_2^{(i+3)}$, $Z_4^{(i+3)}$, $Z_7^{(i+3)}$, $Z_{10}^{(i+3)}$, $Z_{11}^{(i+3)}$ and $Z_8^{(i+4)}$.
Starting from an even-numbered round i, the following 14 subkeys have to be weak: $Z_7^{(i)}$, $Z_{10}^{(i)}$, $Z_{11}^{(i)}$, $Z_3^{(i+1)}$, $Z_{11}^{(i+1)}$, $Z_2^{(i+2)}$, $Z_4^{(i+2)}$, $Z_5^{(i+2)}$, $Z_7^{(i+2)}$, $Z_{11}^{(i+2)}$, $Z_6^{(i+3)}$, $Z_{10}^{(i+3)}$, $Z_{11}^{(i+3)}$ and $Z_4^{(i+4)}$.

(4) $(0,0,0,0,0,\gamma,0,\gamma) \stackrel{1r}{\rightarrow} (\gamma,\gamma,0,0,\gamma,0,\gamma,0) \stackrel{1r}{\rightarrow} (0, 0, 0, \gamma, \gamma, 0, 0, 0) \stackrel{1r}{\rightarrow} (0, \gamma, \gamma, 0, \gamma, 0, \gamma, 0) \stackrel{1r}{\rightarrow} (\gamma,0,0,\gamma,\gamma,\gamma,\gamma,\gamma) \stackrel{1r}{\rightarrow} (0,0,\gamma,\gamma,0,0,\gamma,\gamma) \stackrel{1r}{\rightarrow} (0, 0, 0, \gamma, 0, \gamma, \gamma, \gamma) \stackrel{1r}{\rightarrow} (\gamma,\gamma,\gamma,\gamma,0,0,\gamma,\gamma) \stackrel{1r}{\rightarrow} (0,\gamma,0,\gamma,\gamma,\gamma,0,0) \stackrel{1r}{\rightarrow} (0, \gamma, 0, 0, 0, 0, \gamma, 0) \stackrel{1r}{\rightarrow} (\gamma, \gamma, 0, \gamma, \gamma, 0, 0, 0) \stackrel{1r}{\rightarrow} (0,0,\gamma,0,0,0,0,\gamma) \stackrel{1r}{\rightarrow} (0, 0, \gamma, 0, \gamma, \gamma, \gamma, 0) \stackrel{1r}{\rightarrow} (\gamma,0,\gamma,\gamma,0,\gamma,\gamma,\gamma) \stackrel{1r}{\rightarrow} (0,\gamma,\gamma,\gamma,0,\gamma,0,0) \stackrel{1r}{\rightarrow} (0, 0, 0, 0, 0, \gamma, 0, \gamma)$, which is a 15-round iterative linear relation with bias $2^{-250.40}$.
Starting from an odd-numbered round i, the following 58 subkeys have to be weak: $Z_6^{(i)}$, $Z_8^{(i)}$, $Z_9^{(i)}$, $Z_{10}^{(i)}$, $Z_2^{(i+1)}$, $Z_5^{(i+1)}$, $Z_7^{(i+1)}$, $Z_9^{(i+1)}$, $Z_{11}^{(i+1)}$, $Z_{11}^{(i+2)}$, $Z_2^{(i+3)}$, $Z_5^{(i+3)}$, $Z_7^{(i+3)}$, $Z_9^{(i+3)}$, $Z_{10}^{(i+3)}$, $Z_{11}^{(i+3)}$, $Z_1^{(i+4)}$, $Z_6^{(i+4)}$, $Z_8^{(i+4)}$, $Z_9^{(i+4)}$, $Z_{11}^{(i+4)}$, $Z_4^{(i+5)}$, $Z_7^{(i+5)}$, $Z_6^{(i+6)}$, $Z_8^{(i+6)}$, $Z_9^{(i+6)}$, $Z_{11}^{(i+6)}$, $Z_2^{(i+7)}$, $Z_4^{(i+7)}$, $Z_7^{(i+7)}$, $Z_9^{(i+7)}$, $Z_{10}^{(i+7)}$, $Z_{11}^{(i+7)}$, $Z_6^{(i+8)}$, $Z_{11}^{(i+8)}$, $Z_2^{(i+9)}$, $Z_7^{(i+9)}$, $Z_9^{(i+9)}$, $Z_{11}^{(i+9)}$, $Z_1^{(i+10)}$, $Z_9^{(i+10)}$, $Z_{10}^{(i+10)}$, $Z_{10}^{(i+11)}$, $Z_{11}^{(i+11)}$, $Z_3^{(i+12)}$, $Z_6^{(i+12)}$, $Z_9^{(i+12)}$, $Z_{10}^{(i+12)}$, $Z_{11}^{(i+12)}$, $Z_4^{(i+13)}$, $Z_7^{(i+13)}$, $Z_9^{(i+13)}$, $Z_{10}^{(i+13)}$, $Z_{11}^{(i+13)}$, $Z_3^{(i+14)}$, $Z_6^{(i+14)}$, $Z_{10}^{(i+14)}$ and $Z_{11}^{(i+14)}$.
Starting from an even-numbered round i, the following 60 subkeys have to be weak: $Z_9^{(i)}$, $Z_{10}^{(i)}$, $Z_1^{(i+1)}$, $Z_9^{(i+1)}$, $Z_{11}^{(i+1)}$, $Z_4^{(i+2)}$, $Z_5^{(i+2)}$, $Z_{11}^{(i+2)}$, $Z_3^{(i+3)}$, $Z_9^{(i+3)}$, $Z_{10}^{(i+3)}$, $Z_{11}^{(i+3)}$, $Z_4^{(i+4)}$, $Z_5^{(i+4)}$, $Z_7^{(i+4)}$, $Z_9^{(i+4)}$, $Z_{11}^{(i+4)}$, $Z_3^{(i+5)}$, $Z_8^{(i+5)}$, $Z_4^{(i+6)}$, $Z_7^{(i+6)}$, $Z_9^{(i+6)}$, $Z_{11}^{(i+6)}$, $Z_1^{(i+7)}$, $Z_3^{(i+7)}$, $Z_8^{(i+7)}$, $Z_9^{(i+7)}$, $Z_{10}^{(i+7)}$, $Z_{11}^{(i+7)}$, $Z_2^{(i+8)}$, $Z_4^{(i+8)}$, $Z_5^{(i+8)}$, $Z_{11}^{(i+8)}$, $Z_9^{(i+9)}$, $Z_{11}^{(i+9)}$, $Z_2^{(i+10)}$, $Z_4^{(i+10)}$, $Z_5^{(i+10)}$, $Z_9^{(i+10)}$, $Z_{10}^{(i+10)}$, $Z_3^{(i+11)}$, $Z_8^{(i+11)}$, $Z_{10}^{(i+11)}$, $Z_{11}^{(i+11)}$, $Z_5^{(i+12)}$, $Z_7^{(i+12)}$, $Z_9^{(i+12)}$, $Z_{10}^{(i+12)}$, $Z_{11}^{(i+12)}$, $Z_1^{(i+13)}$, $Z_3^{(i+13)}$, $Z_6^{(i+13)}$, $Z_8^{(i+13)}$, $Z_9^{(i+13)}$, $Z_{10}^{(i+13)}$, $Z_{11}^{(i+13)}$, $Z_2^{(i+14)}$, $Z_4^{(i+14)}$, $Z_{10}^{(i+14)}$ and $Z_{11}^{(i+14)}$.

(5) $(0,0,0,0,0,\gamma,\gamma,0) \stackrel{1r}{\rightarrow} (\gamma,\gamma,\gamma,\gamma,0,0,0,0) \stackrel{1r}{\rightarrow} (0, \gamma, \gamma, 0, 0, \gamma, \gamma, 0) \stackrel{1r}{\rightarrow} (0, 0, \gamma, \gamma, \gamma, \gamma, 0, 0) \stackrel{1r}{\rightarrow} (\gamma,0,0,0,\gamma,\gamma,\gamma,0) \stackrel{1r}{\rightarrow} (0,0,\gamma,\gamma,0,0,0,0) \stackrel{1r}{\rightarrow} (0, 0, \gamma, 0, \gamma, \gamma, 0, \gamma) \stackrel{1r}{\rightarrow} (\gamma,0,0,0,\gamma,\gamma,0,\gamma) \stackrel{1r}{\rightarrow} (0,0,0,0,\gamma,0,\gamma,0) \stackrel{1r}{\rightarrow} (0,\gamma,0,\gamma,0,0,\gamma,\gamma) \stackrel{1r}{\rightarrow} (\gamma, \gamma, 0, \gamma, \gamma, 0, \gamma, \gamma) \stackrel{1r}{\rightarrow} (0,0,0,\gamma,\gamma,0,\gamma,\gamma) \stackrel{1r}{\rightarrow} (0,\gamma,0,\gamma,0,0,0,0) \stackrel{1r}{\rightarrow} (\gamma, \gamma, \gamma, 0, 0, 0, 0, \gamma) \stackrel{1r}{\rightarrow} (0,\gamma,\gamma,0,0,\gamma,0,\gamma) \stackrel{1r}{\rightarrow} (0,$

0, 0, 0, 0, γ, γ, 0), which is a 15-round iterative linear relation with bias $2^{-275.34}$.

Starting from an odd-numbered round i, the following 48 subkeys have to be weak: $Z_6^{(i)}$, $Z_9^{(i)}$, $Z_{11}^{(i)}$, $Z_2^{(i+1)}$, $Z_4^{(i+1)}$, $Z_9^{(i+1)}$, $Z_3^{(i+2)}$, $Z_6^{(i+2)}$, $Z_4^{(i+3)}$, $Z_5^{(i+3)}$, $Z_9^{(i+3)}$, $Z_1^{(i+4)}$, $Z_6^{(i+4)}$, $Z_9^{(i+4)}$, $Z_{11}^{(i+4)}$, $Z_4^{(i+5)}$, $Z_{10}^{(i+5)}$, $Z_{11}^{(i+5)}$, $Z_3^{(i+6)}$, $Z_6^{(i+6)}$, $Z_8^{(i+6)}$, $Z_9^{(i+6)}$, $Z_5^{(i+7)}$, $Z_9^{(i+7)}$, $Z_{10}^{(i+7)}$, $Z_{10}^{(i+8)}$, $Z_2^{(i+9)}$, $Z_4^{(i+9)}$, $Z_7^{(i+9)}$, $Z_9^{(i+9)}$, $Z_{11}^{(i+9)}$, $Z_1^{(i+10)}$, $Z_8^{(i+10)}$, $Z_9^{(i+10)}$, $Z_{11}^{(i+10)}$, $Z_4^{(i+11)}$, $Z_5^{(i+11)}$, $Z_7^{(i+11)}$, $Z_{10}^{(i+11)}$, $Z_9^{(i+12)}$, $Z_{10}^{(i+12)}$, $Z_2^{(i+13)}$, $Z_9^{(i+13)}$, $Z_3^{(i+14)}$, $Z_6^{(i+14)}$, $Z_8^{(i+14)}$, $Z_{10}^{(i+14)}$ and $Z_{11}^{(i+14)}$.

Starting from an even-numbered round i, the following 50 subkeys have to be weak: $Z_7^{(i)}$, $Z_9^{(i)}$, $Z_{11}^{(i)}$, $Z_1^{(i+1)}$, $Z_3^{(i+1)}$, $Z_9^{(i+1)}$, $Z_2^{(i+2)}$, $Z_7^{(i+2)}$, $Z_3^{(i+3)}$, $Z_6^{(i+3)}$, $Z_9^{(i+3)}$, $Z_5^{(i+4)}$, $Z_7^{(i+4)}$, $Z_9^{(i+4)}$, $Z_{11}^{(i+4)}$, $Z_3^{(i+5)}$, $Z_{10}^{(i+5)}$, $Z_{11}^{(i+5)}$, $Z_5^{(i+6)}$, $Z_9^{(i+6)}$, $Z_1^{(i+7)}$, $Z_6^{(i+7)}$, $Z_8^{(i+7)}$, $Z_9^{(i+7)}$, $Z_{10}^{(i+7)}$, $Z_5^{(i+8)}$, $Z_7^{(i+8)}$, $Z_{10}^{(i+8)}$, $Z_8^{(i+9)}$, $Z_9^{(i+9)}$, $Z_{11}^{(i+9)}$, $Z_2^{(i+10)}$, $Z_4^{(i+10)}$, $Z_5^{(i+10)}$, $Z_7^{(i+10)}$, $Z_9^{(i+10)}$, $Z_{11}^{(i+10)}$, $Z_8^{(i+11)}$, $Z_{10}^{(i+11)}$, $Z_2^{(i+12)}$, $Z_4^{(i+12)}$, $Z_9^{(i+12)}$, $Z_{10}^{(i+12)}$, $Z_1^{(i+13)}$, $Z_3^{(i+13)}$, $Z_8^{(i+13)}$, $Z_9^{(i+13)}$, $Z_2^{(i+14)}$, $Z_{10}^{(i+14)}$ and $Z_{11}^{(i+14)}$.

(6) $(0,0,0,0,0,\gamma,\gamma,\gamma) \stackrel{1r}{\to} (\gamma,\gamma,\gamma,0,0,\gamma,0,\gamma) \stackrel{1r}{\to} (\gamma, 0, \gamma, \gamma, \gamma, 0, \gamma, 0) \stackrel{1r}{\to} (\gamma, \gamma, 0, 0, 0, 0, \gamma, 0) \stackrel{1r}{\to} (0,\gamma,\gamma,0,0,\gamma,0,0) \stackrel{1r}{\to} (0,0,0,\gamma,0,0,\gamma,\gamma) \stackrel{1r}{\to} (0, 0, \gamma, 0, \gamma, \gamma, 0, 0) \stackrel{1r}{\to} (\gamma,0,0,\gamma,\gamma,0,0,0) \stackrel{1r}{\to} (\gamma,\gamma,0,\gamma,0,\gamma,\gamma,0) \stackrel{1r}{\to} (\gamma, 0, \gamma, 0, \gamma, \gamma, 0, \gamma) \stackrel{1r}{\to} (0,0,\gamma,\gamma,0,0,0,\gamma) \stackrel{1r}{\to} (0,0,\gamma,\gamma,\gamma,0,0,0) \stackrel{1r}{\to} (0, \gamma, 0, \gamma, 0, 0, 0, \gamma) \stackrel{1r}{\to} (\gamma, \gamma, \gamma, \gamma, 0, \gamma, 0, 0) \stackrel{1r}{\to} (\gamma,0,\gamma,\gamma,\gamma,0,0,\gamma) \stackrel{1r}{\to} (\gamma,\gamma,\gamma,\gamma,\gamma,0,0,0) \stackrel{1r}{\to} (0, 0, 0, \gamma, \gamma, 0, \gamma, 0) \stackrel{1r}{\to} (0,\gamma,0,0,0,\gamma,0,\gamma) \stackrel{1r}{\to} (0,0,\gamma,\gamma,\gamma,\gamma,0,\gamma) \stackrel{1r}{\to} (\gamma,0,0,\gamma,\gamma,0,\gamma,\gamma) \stackrel{1r}{\to} (\gamma,\gamma,\gamma,0,\gamma,\gamma,0,0) \stackrel{1r}{\to} (\gamma, \gamma, 0, \gamma, 0, 0, \gamma, \gamma) \stackrel{1r}{\to} (0,\gamma,\gamma,0,0,\gamma,\gamma,\gamma) \stackrel{1r}{\to} (0, 0, \gamma, 0, \gamma, 0, 0, \gamma) \stackrel{1r}{\to} (0, \gamma, 0, \gamma, 0, 0, \gamma, 0) \stackrel{1r}{\to} (\gamma,\gamma,0,0,\gamma,\gamma,\gamma,0) \stackrel{1r}{\to} (\gamma, \gamma, 0, 0, 0, \gamma, \gamma, \gamma) \stackrel{1r}{\to} (\gamma,0,\gamma,0,\gamma,\gamma,\gamma,0) \stackrel{1r}{\to} (0,0,0,0,\gamma,0,\gamma,\gamma) \stackrel{1r}{\to} (0,\gamma,0,0,0,\gamma,\gamma,0) \stackrel{1r}{\to} (0, 0, 0, 0, 0, \gamma, \gamma, \gamma)$, which is a 30-round iterative linear relation with bias $2^{-923.78}$.

Starting from an odd-numbered round i, the following 129 subkeys have to be weak: $Z_6^{(i)}$, $Z_8^{(i)}$, $Z_9^{(i)}$, $Z_{11}^{(i)}$, $Z_{12}^{(i)}$, $Z_2^{(i+1)}$, $Z_{10}^{(i+1)}$, $Z_{12}^{(i+1)}$, $Z_1^{(i+2)}$, $Z_3^{(i+2)}$, $Z_{12}^{(i+2)}$, $Z_2^{(i+3)}$, $Z_7^{(i+3)}$, $Z_9^{(i+3)}$, $Z_{12}^{(i+3)}$, $Z_3^{(i+4)}$, $Z_6^{(i+4)}$, $Z_{10}^{(i+4)}$, $Z_{11}^{(i+4)}$, $Z_{12}^{(i+4)}$, $Z_4^{(i+5)}$, $Z_7^{(i+5)}$, $Z_{10}^{(i+5)}$, $Z_{11}^{(i+5)}$, $Z_{12}^{(i+5)}$, $Z_3^{(i+6)}$, $Z_6^{(i+6)}$, $Z_9^{(i+6)}$, $Z_{12}^{(i+6)}$, $Z_4^{(i+7)}$, $Z_5^{(i+7)}$, $Z_{12}^{(i+7)}$, $Z_1^{(i+8)}$, $Z_6^{(i+8)}$, $Z_{10}^{(i+8)}$, $Z_{12}^{(i+8)}$, $Z_5^{(i+9)}$, $Z_9^{(i+9)}$, $Z_{11}^{(i+9)}$, $Z_{12}^{(i+9)}$, $Z_3^{(i+10)}$, $Z_8^{(i+10)}$, $Z_{10}^{(i+10)}$, $Z_{11}^{(i+10)}$, $Z_{12}^{(i+10)}$, $Z_4^{(i+11)}$, $Z_5^{(i+11)}$, $Z_{10}^{(i+11)}$, $Z_{12}^{(i+11)}$, $Z_8^{(i+12)}$, $Z_9^{(i+12)}$, $Z_{10}^{(i+12)}$, $Z_{12}^{(i+12)}$, $Z_2^{(i+13)}$, $Z_4^{(i+13)}$, $Z_{10}^{(i+13)}$, $Z_{12}^{(i+13)}$, $Z_1^{(i+14)}$, $Z_3^{(i+14)}$, $Z_8^{(i+14)}$, $Z_{10}^{(i+14)}$, $Z_{11}^{(i+14)}$, $Z_{12}^{(i+14)}$, $Z_2^{(i+15)}$, $Z_4^{(i+15)}$, $Z_5^{(i+15)}$, $Z_9^{(i+15)}$, $Z_{11}^{(i+15)}$, $Z_{12}^{(i+15)}$, $Z_{10}^{(i+16)}$, $Z_{12}^{(i+16)}$, $Z_2^{(i+17)}$, $Z_{12}^{(i+17)}$, $Z_3^{(i+18)}$, $Z_6^{(i+18)}$, $Z_8^{(i+18)}$, $Z_9^{(i+18)}$, $Z_{12}^{(i+18)}$, $Z_4^{(i+19)}$, $Z_5^{(i+19)}$, $Z_7^{(i+19)}$, $Z_{10}^{(i+19)}$, $Z_{11}^{(i+19)}$, $Z_{12}^{(i+19)}$, $Z_1^{(i+20)}$, $Z_3^{(i+20)}$, $Z_6^{(i+20)}$, $Z_{10}^{(i+20)}$, $Z_{11}^{(i+20)}$, $Z_{12}^{(i+20)}$, $Z_2^{(i+21)}$,

$Z_4^{(i+21)}$, $Z_7^{(i+21)}$, $Z_9^{(i+21)}$, $Z_{12}^{(i+21)}$, $Z_3^{(i+22)}$, $Z_6^{(i+22)}$, $Z_8^{(i+22)}$, $Z_{12}^{(i+22)}$, $Z_5^{(i+23)}$, $Z_{10}^{(i+23)}$, $Z_{12}^{(i+23)}$, $Z_9^{(i+24)}$, $Z_{11}^{(i+24)}$, $Z_{12}^{(i+24)}$, $Z_2^{(i+25)}$, $Z_5^{(i+25)}$, $Z_7^{(i+25)}$, $Z_{10}^{(i+25)}$, $Z_{11}^{(i+25)}$, $Z_{12}^{(i+25)}$, $Z_1^{(i+26)}$, $Z_6^{(i+26)}$, $Z_8^{(i+26)}$, $Z_{10}^{(i+26)}$, $Z_{12}^{(i+26)}$, $Z_5^{(i+27)}$, $Z_7^{(i+27)}$, $Z_9^{(i+27)}$, $Z_{10}^{(i+27)}$, $Z_{12}^{(i+27)}$, $Z_8^{(i+28)}$, $Z_{10}^{(i+28)}$, $Z_{12}^{(i+28)}$, $Z_2^{(i+29)}$, $Z_7^{(i+29)}$, $Z_{10}^{(i+29)}$, $Z_{11}^{(i+29)}$ and $Z_{12}^{(i+29)}$.

Starting from an even-numbered round i, the following 131 subkeys have to be weak: $Z_7^{(i)}$, $Z_9^{(i)}$, $Z_{11}^{(i)}$, $Z_{12}^{(i)}$, $Z_1^{(i+1)}$, $Z_3^{(i+1)}$, $Z_6^{(i+1)}$, $Z_8^{(i+1)}$, $Z_{10}^{(i+1)}$, $Z_{12}^{(i+1)}$, $Z_4^{(i+2)}$, $Z_5^{(i+2)}$, $Z_7^{(i+2)}$, $Z_{12}^{(i+2)}$, $Z_1^{(i+3)}$, $Z_9^{(i+3)}$, $Z_{12}^{(i+3)}$, $Z_2^{(i+4)}$, $Z_{10}^{(i+4)}$, $Z_{11}^{(i+4)}$, $Z_{12}^{(i+4)}$, $Z_8^{(i+5)}$, $Z_{10}^{(i+5)}$, $Z_{11}^{(i+5)}$, $Z_{12}^{(i+5)}$, $Z_5^{(i+6)}$, $Z_9^{(i+6)}$, $Z_{12}^{(i+6)}$, $Z_1^{(i+7)}$, $Z_{12}^{(i+7)}$, $Z_2^{(i+8)}$, $Z_4^{(i+8)}$, $Z_7^{(i+8)}$, $Z_{10}^{(i+8)}$, $Z_{12}^{(i+8)}$, $Z_1^{(i+9)}$, $Z_3^{(i+9)}$, $Z_6^{(i+9)}$, $Z_8^{(i+9)}$, $Z_9^{(i+9)}$, $Z_{11}^{(i+9)}$, $Z_{12}^{(i+9)}$, $Z_4^{(i+10)}$, $Z_{10}^{(i+10)}$, $Z_{11}^{(i+10)}$, $Z_{12}^{(i+10)}$, $Z_3^{(i+11)}$, $Z_{10}^{(i+11)}$, $Z_{12}^{(i+11)}$, $Z_2^{(i+12)}$, $Z_4^{(i+12)}$, $Z_9^{(i+12)}$, $Z_{10}^{(i+12)}$, $Z_{12}^{(i+12)}$, $Z_1^{(i+13)}$, $Z_3^{(i+13)}$, $Z_6^{(i+13)}$, $Z_{10}^{(i+13)}$, $Z_{12}^{(i+13)}$, $Z_4^{(i+14)}$, $Z_5^{(i+14)}$, $Z_{10}^{(i+14)}$, $Z_{11}^{(i+14)}$, $Z_{12}^{(i+14)}$, $Z_1^{(i+15)}$, $Z_3^{(i+15)}$, $Z_9^{(i+15)}$, $Z_{11}^{(i+15)}$, $Z_{12}^{(i+15)}$, $Z_4^{(i+16)}$, $Z_5^{(i+16)}$, $Z_7^{(i+16)}$, $Z_{10}^{(i+16)}$, $Z_{12}^{(i+16)}$, $Z_6^{(i+17)}$, $Z_8^{(i+17)}$, $Z_{12}^{(i+17)}$, $Z_4^{(i+18)}$, $Z_5^{(i+18)}$, $Z_9^{(i+18)}$, $Z_{12}^{(i+18)}$, $Z_1^{(i+19)}$, $Z_8^{(i+19)}$, $Z_{10}^{(i+19)}$, $Z_{11}^{(i+19)}$, $Z_{12}^{(i+19)}$, $Z_2^{(i+20)}$, $Z_5^{(i+20)}$, $Z_{10}^{(i+20)}$, $Z_{11}^{(i+20)}$, $Z_{12}^{(i+20)}$, $Z_1^{(i+21)}$, $Z_8^{(i+21)}$, $Z_9^{(i+21)}$, $Z_{12}^{(i+21)}$, $Z_2^{(i+22)}$, $Z_7^{(i+22)}$, $Z_{12}^{(i+22)}$, $Z_3^{(i+23)}$, $Z_8^{(i+23)}$, $Z_{10}^{(i+23)}$, $Z_{12}^{(i+23)}$, $Z_2^{(i+24)}$, $Z_4^{(i+24)}$, $Z_7^{(i+24)}$, $Z_9^{(i+24)}$, $Z_{11}^{(i+24)}$, $Z_{12}^{(i+24)}$, $Z_1^{(i+25)}$, $Z_6^{(i+25)}$, $Z_{10}^{(i+25)}$, $Z_{11}^{(i+25)}$, $Z_{12}^{(i+25)}$, $Z_2^{(i+26)}$, $Z_7^{(i+26)}$, $Z_{10}^{(i+26)}$, $Z_{12}^{(i+26)}$, $Z_1^{(i+27)}$, $Z_3^{(i+27)}$, $Z_6^{(i+27)}$, $Z_9^{(i+27)}$, $Z_{10}^{(i+27)}$, $Z_{12}^{(i+27)}$, $Z_5^{(i+28)}$, $Z_7^{(i+28)}$, $Z_{10}^{(i+28)}$, $Z_{12}^{(i+28)}$, $Z_6^{(i+29)}$, $Z_{10}^{(i+29)}$, $Z_{11}^{(i+29)}$ and $Z_{12}^{(i+29)}$.

(7) $(0,0,0,0,\gamma,0,0,0) \stackrel{1r}{\to} (0,\gamma,\gamma,\gamma,\gamma,\gamma,0,0) \stackrel{1r}{\to} (0,\gamma,\gamma,\gamma,\gamma,0,0,\gamma) \stackrel{1r}{\to} (\gamma,0,\gamma,\gamma,0,0,\gamma,\gamma) \stackrel{1r}{\to} (\gamma,0,\gamma,0,\gamma,0,\gamma,\gamma) \stackrel{1r}{\to} (\gamma,\gamma,0,0,0,0,0,\gamma) \stackrel{1r}{\to} (0,\gamma,0,\gamma,\gamma,\gamma,\gamma,0) \stackrel{1r}{\to} (0,\gamma,\gamma,0,\gamma,\gamma,0,\gamma) \stackrel{1r}{\to} (0,\gamma,\gamma,\gamma,\gamma,0,\gamma,0) \stackrel{1r}{\to} (\gamma,0,0,0,\gamma,0,0,\gamma) \stackrel{1r}{\to} (\gamma,\gamma,0,\gamma,0,\gamma,0,\gamma) \stackrel{1r}{\to} (\gamma,0,0,\gamma,0,\gamma,\gamma,\gamma) \stackrel{1r}{\to} (0,\gamma,0,0,\gamma,\gamma,\gamma,\gamma) \stackrel{1r}{\to} (0,\gamma,\gamma,0,\gamma,\gamma,\gamma,0) \stackrel{1r}{\to} (0,\gamma,0,0,0,0,0,0) \stackrel{1r}{\to} (\gamma,\gamma,\gamma,\gamma,0,\gamma,\gamma,\gamma) \stackrel{1r}{\to} (\gamma,0,0,0,0,0,\gamma,\gamma) \stackrel{1r}{\to} (\gamma,0,0,0,0,\gamma,\gamma,0) \stackrel{1r}{\to} (0,\gamma,0,0,\gamma,\gamma,0,0) \stackrel{1r}{\to} (0,\gamma,0,\gamma,0,\gamma,0,0) \stackrel{1r}{\to} (0,0,\gamma,\gamma,\gamma,\gamma,\gamma,0) \stackrel{1r}{\to} (\gamma,0,\gamma,0,0,0,0,\gamma) \stackrel{1r}{\to} (\gamma,0,0,\gamma,0,0,\gamma,0) \stackrel{1r}{\to} (\gamma,0,0,0,0,\gamma,0,\gamma) \stackrel{1r}{\to} (0,\gamma,\gamma,\gamma,0,\gamma,\gamma,0) \stackrel{1r}{\to} (0,0,\gamma,0,\gamma,0,\gamma,0) \stackrel{1r}{\to} (0,\gamma,\gamma,0,\gamma,0,0,0) \stackrel{1r}{\to} (\gamma,0,\gamma,\gamma,0,0,0,0) \stackrel{1r}{\to} (\gamma,0,0,\gamma,0,0,0,\gamma) \stackrel{1r}{\to} (\gamma,0,\gamma,\gamma,\gamma,\gamma,\gamma,\gamma) \stackrel{1r}{\to} (0,0,0,0,\gamma,0,0,0)$, which is a 30-round iterative linear relation with bias $2^{-1023.54}$.

Starting from an odd-numbered round i, the following 115 subkeys have to be weak: $Z_{11}^{(i)}$, $Z_{12}^{(i)}$, $Z_2^{(i+1)}$, $Z_4^{(i+1)}$, $Z_5^{(i+1)}$, $Z_{10}^{(i+1)}$, $Z_{12}^{(i+1)}$, $Z_3^{(i+2)}$, $Z_8^{(i+2)}$, $Z_9^{(i+2)}$, $Z_{12}^{(i+2)}$, $Z_4^{(i+3)}$, $Z_7^{(i+3)}$, $Z_{11}^{(i+3)}$, $Z_{12}^{(i+3)}$, $Z_1^{(i+4)}$, $Z_3^{(i+4)}$, $Z_8^{(i+4)}$, $Z_{12}^{(i+4)}$, $Z_2^{(i+5)}$, $Z_9^{(i+5)}$, $Z_{10}^{(i+5)}$, $Z_{11}^{(i+5)}$, $Z_{12}^{(i+5)}$, $Z_6^{(i+6)}$, $Z_{10}^{(i+6)}$, $Z_{12}^{(i+6)}$, $Z_2^{(i+7)}$, $Z_5^{(i+7)}$, $Z_{10}^{(i+7)}$, $Z_{12}^{(i+7)}$, $Z_3^{(i+8)}$, $Z_9^{(i+8)}$, $Z_{10}^{(i+8)}$, $Z_{11}^{(i+8)}$, $Z_{12}^{(i+8)}$, $Z_5^{(i+9)}$, $Z_{12}^{(i+9)}$, $Z_1^{(i+10)}$, $Z_6^{(i+10)}$, $Z_8^{(i+10)}$, $Z_{11}^{(i+10)}$, $Z_{12}^{(i+10)}$, $Z_4^{(i+11)}$, $Z_7^{(i+11)}$,

$Z_9^{(i+11)}, Z_{12}^{(i+11)}, Z_6^{(i+12)}, Z_8^{(i+12)}, Z_{10}^{(i+12)}, Z_{12}^{(i+12)}, Z_2^{(i+13)}, Z_5^{(i+13)}, Z_7^{(i+13)}, Z_{11}^{(i+13)}, Z_{12}^{(i+13)}, Z_9^{(i+14)}, Z_{10}^{(i+14)}, Z_{12}^{(i+14)}, Z_2^{(i+15)}, Z_4^{(i+15)}, Z_7^{(i+15)}, Z_{11}^{(i+15)}, Z_{12}^{(i+15)}, Z_1^{(i+16)}, Z_8^{(i+16)}, Z_{10}^{(i+16)}, Z_{12}^{(i+16)}, Z_7^{(i+17)}, Z_9^{(i+17)}, Z_{12}^{(i+17)}, Z_6^{(i+18)}, Z_{11}^{(i+18)}, Z_{12}^{(i+18)}, Z_2^{(i+19)}, Z_4^{(i+19)}, Z_{12}^{(i+19)}, Z_3^{(i+20)}, Z_6^{(i+20)}, Z_9^{(i+20)}, Z_{10}^{(i+20)}, Z_{11}^{(i+20)}, Z_{12}^{(i+20)}, Z_{10}^{(i+21)}, Z_{12}^{(i+21)}, Z_1^{(i+22)}, Z_{10}^{(i+22)}, Z_{12}^{(i+22)}, Z_9^{(i+23)}, Z_{10}^{(i+23)}, Z_{11}^{(i+23)}, Z_{12}^{(i+23)}, Z_3^{(i+24)}, Z_6^{(i+24)}, Z_{12}^{(i+24)}, Z_5^{(i+25)}, Z_7^{(i+25)}, Z_{11}^{(i+25)}, Z_{12}^{(i+25)}, Z_3^{(i+26)}, Z_9^{(i+26)}, Z_{12}^{(i+26)}, Z_4^{(i+27)}, Z_{10}^{(i+27)}, Z_{12}^{(i+27)}, Z_1^{(i+28)}, Z_8^{(i+28)}, Z_{11}^{(i+28)}, Z_{12}^{(i+28)}, Z_4^{(i+29)}, Z_5^{(i+29)}, Z_7^{(i+29)}, Z_9^{(i+29)}, Z_{10}^{(i+29)}$ and $Z_{12}^{(i+29)}$.

Starting from an even-numbered round i, the following 137 subkeys have to be weak: $Z_5^{(i)}, Z_{11}^{(i)}, Z_{12}^{(i)}, Z_3^{(i+1)}, Z_6^{(i+1)}, Z_{10}^{(i+1)}, Z_{12}^{(i+1)}, Z_2^{(i+2)}, Z_4^{(i+2)}, Z_5^{(i+2)}, Z_9^{(i+2)}, Z_{12}^{(i+2)}, Z_1^{(i+3)}, Z_3^{(i+3)}, Z_8^{(i+3)}, Z_{11}^{(i+3)}, Z_{12}^{(i+3)}, Z_5^{(i+4)}, Z_7^{(i+4)}, Z_{12}^{(i+4)}, Z_1^{(i+5)}, Z_8^{(i+5)}, Z_9^{(i+5)}, Z_{10}^{(i+5)}, Z_{11}^{(i+5)}, Z_{12}^{(i+5)}, Z_2^{(i+6)}, Z_4^{(i+6)}, Z_5^{(i+6)}, Z_7^{(i+6)}, Z_{10}^{(i+6)}, Z_{12}^{(i+6)}, Z_3^{(i+7)}, Z_6^{(i+7)}, Z_8^{(i+7)}, Z_{10}^{(i+7)}, Z_{12}^{(i+7)}, Z_2^{(i+8)}, Z_4^{(i+8)}, Z_5^{(i+8)}, Z_7^{(i+8)}, Z_9^{(i+8)}, Z_{10}^{(i+8)}, Z_{11}^{(i+8)}, Z_{12}^{(i+8)}, Z_1^{(i+9)}, Z_8^{(i+9)}, Z_{12}^{(i+9)}, Z_2^{(i+10)}, Z_4^{(i+10)}, Z_{11}^{(i+10)}, Z_{12}^{(i+10)}, Z_1^{(i+11)}, Z_6^{(i+11)}, Z_8^{(i+11)}, Z_9^{(i+11)}, Z_{12}^{(i+11)}, Z_2^{(i+12)}, Z_5^{(i+12)}, Z_7^{(i+12)}, Z_{10}^{(i+12)}, Z_{12}^{(i+12)}, Z_3^{(i+13)}, Z_6^{(i+13)}, Z_{11}^{(i+13)}, Z_{12}^{(i+13)}, Z_2^{(i+14)}, Z_9^{(i+14)}, Z_{10}^{(i+14)}, Z_{12}^{(i+14)}, Z_1^{(i+15)}, Z_3^{(i+15)}, Z_6^{(i+15)}, Z_8^{(i+15)}, Z_{11}^{(i+15)}, Z_{12}^{(i+15)}, Z_7^{(i+16)}, Z_{10}^{(i+16)}, Z_{12}^{(i+16)}, Z_1^{(i+17)}, Z_6^{(i+17)}, Z_9^{(i+17)}, Z_{12}^{(i+17)}, Z_2^{(i+18)}, Z_5^{(i+18)}, Z_{11}^{(i+18)}, Z_{12}^{(i+18)}, Z_6^{(i+19)}, Z_{12}^{(i+19)}, Z_4^{(i+20)}, Z_5^{(i+20)}, Z_7^{(i+20)}, Z_9^{(i+20)}, Z_{10}^{(i+20)}, Z_{11}^{(i+20)}, Z_{12}^{(i+20)}, Z_1^{(i+21)}, Z_3^{(i+21)}, Z_8^{(i+21)}, Z_{10}^{(i+21)}, Z_{12}^{(i+21)}, Z_4^{(i+22)}, Z_7^{(i+22)}, Z_{10}^{(i+22)}, Z_{12}^{(i+22)}, Z_1^{(i+23)}, Z_6^{(i+23)}, Z_8^{(i+23)}, Z_9^{(i+23)}, Z_{10}^{(i+23)}, Z_{11}^{(i+23)}, Z_{12}^{(i+23)}, Z_2^{(i+24)}, Z_4^{(i+24)}, Z_7^{(i+24)}, Z_{12}^{(i+24)}, Z_3^{(i+25)}, Z_{11}^{(i+25)}, Z_{12}^{(i+25)}, Z_2^{(i+26)}, Z_5^{(i+26)}, Z_9^{(i+26)}, Z_{12}^{(i+26)}, Z_1^{(i+27)}, Z_3^{(i+27)}, Z_{10}^{(i+27)}, Z_{12}^{(i+27)}, Z_4^{(i+28)}, Z_{11}^{(i+28)}, Z_{12}^{(i+28)}, Z_1^{(i+29)}, Z_3^{(i+29)}, Z_6^{(i+29)}, Z_8^{(i+29)}, Z_9^{(i+29)}, Z_{10}^{(i+29)}$ and $Z_{12}^{(i+29)}$.

(8) $(0,0,0,0,\gamma,0,0,\gamma) \xrightarrow{1r} (0,\gamma,\gamma,0,\gamma,0,0,\gamma) \xrightarrow{1r} (\gamma, 0, \gamma, 0, 0, \gamma, 0, \gamma) \xrightarrow{1r} (0, \gamma, 0, 0, \gamma, \gamma, 0, \gamma) \xrightarrow{1r} (0,\gamma,0,0,0,0,0,\gamma) \xrightarrow{1r} (\gamma,\gamma,\gamma,0,0,0,\gamma,0) \xrightarrow{1r} (0, \gamma, 0, \gamma, \gamma, \gamma, \gamma, \gamma) \xrightarrow{1r} (0,\gamma,\gamma,\gamma,\gamma,0,0,0) \xrightarrow{1r} (\gamma,0,\gamma,0,0,\gamma,\gamma,0) \xrightarrow{1r} (0, \gamma, \gamma, \gamma, 0, \gamma, \gamma, \gamma) \xrightarrow{1r} (0, 0, \gamma, \gamma, \gamma, \gamma, \gamma, \gamma) \xrightarrow{1r} (\gamma,0,\gamma,\gamma,0,\gamma,0,0) \xrightarrow{1r} (0, \gamma, 0, 0, \gamma, \gamma, \gamma, 0) \xrightarrow{1r} (0, \gamma, \gamma, \gamma, \gamma, 0, \gamma, \gamma) \xrightarrow{1r} (\gamma,0,0,\gamma,\gamma,\gamma,0,0) \xrightarrow{1r} (0,0,0,0,\gamma,0,0,\gamma)$, which is a 15-round iterative linear relation with bias $2^{-325.22}$.

Starting from an odd-numbered round i, the following 58 subkeys have to be weak: $Z_8^{(i)}, Z_{11}^{(i)}, Z_2^{(i+1)}, Z_5^{(i+1)}, Z_9^{(i+1)}, Z_1^{(i+2)}, Z_3^{(i+2)}, Z_6^{(i+2)}, Z_8^{(i+2)}, Z_9^{(i+2)}, Z_2^{(i+3)}, Z_5^{(i+3)}, Z_{11}^{(i+3)}, Z_8^{(i+4)}, Z_9^{(i+4)}, Z_{10}^{(i+4)}, Z_2^{(i+5)}, Z_7^{(i+5)}, Z_9^{(i+5)}, Z_{10}^{(i+5)}, Z_{11}^{(i+5)}, Z_6^{(i+6)}, Z_8^{(i+6)}, Z_{10}^{(i+6)}, Z_2^{(i+7)}, Z_4^{(i+7)}, Z_5^{(i+7)}, Z_9^{(i+7)}, Z_1^{(i+8)}, Z_3^{(i+8)}, Z_6^{(i+8)}, Z_9^{(i+8)}, Z_{10}^{(i+8)}, Z_{11}^{(i+8)}, Z_2^{(i+9)}, Z_4^{(i+9)}, Z_7^{(i+9)}, Z_3^{(i+10)}, Z_6^{(i+10)}, Z_8^{(i+10)}, Z_9^{(i+10)}, Z_{10}^{(i+10)}, Z_{11}^{(i+10)}, Z_4^{(i+11)},$

$Z_9^{(i+11)}$, $Z_6^{(i+12)}$, $Z_{10}^{(i+12)}$, $Z_2^{(i+13)}$, $Z_4^{(i+13)}$, $Z_5^{(i+13)}$, $Z_7^{(i+13)}$, $Z_9^{(i+13)}$, $Z_{10}^{(i+13)}$, $Z_{11}^{(i+13)}$, $Z_1^{(i+14)}$, $Z_6^{(i+14)}$, $Z_9^{(i+14)}$ and $Z_{10}^{(i+14)}$.

Starting from an even-numbered round i, the following 54 subkeys have to be weak: $Z_5^{(i)}$, $Z_{11}^{(i)}$, $Z_3^{(i+1)}$, $Z_8^{(i+1)}$, $Z_9^{(i+1)}$, $Z_9^{(i+2)}$, $Z_6^{(i+3)}$, $Z_8^{(i+3)}$, $Z_{11}^{(i+3)}$, $Z_2^{(i+4)}$, $Z_9^{(i+4)}$, $Z_{10}^{(i+4)}$, $Z_1^{(i+5)}$, $Z_3^{(i+5)}$, $Z_9^{(i+5)}$, $Z_{10}^{(i+5)}$, $Z_{11}^{(i+5)}$, $Z_2^{(i+6)}$, $Z_4^{(i+6)}$, $Z_5^{(i+6)}$, $Z_7^{(i+6)}$, $Z_{10}^{(i+6)}$, $Z_3^{(i+7)}$, $Z_9^{(i+7)}$, $Z_7^{(i+8)}$, $Z_9^{(i+8)}$, $Z_{10}^{(i+8)}$, $Z_{11}^{(i+8)}$, $Z_3^{(i+9)}$, $Z_6^{(i+9)}$, $Z_8^{(i+9)}$, $Z_4^{(i+10)}$, $Z_5^{(i+10)}$, $Z_7^{(i+10)}$, $Z_9^{(i+10)}$, $Z_{10}^{(i+10)}$, $Z_{11}^{(i+10)}$, $Z_1^{(i+11)}$, $Z_3^{(i+11)}$, $Z_6^{(i+11)}$, $Z_9^{(i+11)}$, $Z_2^{(i+12)}$, $Z_5^{(i+12)}$, $Z_7^{(i+12)}$, $Z_{10}^{(i+12)}$, $Z_3^{(i+13)}$, $Z_8^{(i+13)}$, $Z_9^{(i+13)}$, $Z_{10}^{(i+13)}$, $Z_{11}^{(i+13)}$, $Z_4^{(i+14)}$, $Z_5^{(i+14)}$, $Z_9^{(i+14)}$ and $Z_{10}^{(i+14)}$.

(9) $(0,0,0,0,\gamma,\gamma,0,0) \xrightarrow{1r} (\gamma,0,\gamma,0,0,0,\gamma,\gamma) \xrightarrow{1r} (\gamma, 0, \gamma, \gamma, \gamma, \gamma, 0, \gamma) \xrightarrow{1r} (0, 0, \gamma, 0, 0, \gamma, \gamma, \gamma) \xrightarrow{1r} (\gamma,\gamma,0,\gamma,\gamma,\gamma,\gamma,0) \xrightarrow{1r} (\gamma,\gamma,0,\gamma,0,0,0,\gamma) \xrightarrow{1r} (0, \gamma, 0, 0, \gamma, 0, 0, 0) \xrightarrow{1r} (\gamma,0,0,0,\gamma,0,\gamma,\gamma) \xrightarrow{1r} (\gamma,\gamma,\gamma,\gamma,\gamma,0,\gamma,0) \xrightarrow{1r} (0, 0, \gamma, \gamma, 0, \gamma, 0, \gamma) \xrightarrow{1r} (\gamma, \gamma, \gamma, 0, 0, \gamma, \gamma, \gamma) \xrightarrow{1r} (\gamma,0,0,\gamma,0,\gamma,0,\gamma) \xrightarrow{1r} (0,\gamma,\gamma,0,0,0,0,0) \xrightarrow{1r} (\gamma, \gamma, 0, 0, \gamma, \gamma, 0, 0) \xrightarrow{1r} (\gamma,\gamma,\gamma,0,\gamma,0,0,0) \xrightarrow{1r} (0, 0, 0, 0, \gamma, \gamma, 0, 0)$, which is a 15-round iterative linear relation with bias $2^{-250.40}$.

Starting from an odd-numbered round i, the following 56 subkeys have to be weak: $Z_6^{(i)}$, $Z_9^{(i)}$, $Z_{10}^{(i)}$, $Z_{11}^{(i)}$, $Z_7^{(i+1)}$, $Z_{11}^{(i+1)}$, $Z_1^{(i+2)}$, $Z_3^{(i+2)}$, $Z_6^{(i+2)}$, $Z_8^{(i+2)}$, $Z_9^{(i+2)}$, $Z_{11}^{(i+2)}$, $Z_7^{(i+3)}$, $Z_9^{(i+3)}$, $Z_{10}^{(i+3)}$, $Z_1^{(i+4)}$, $Z_6^{(i+4)}$, $Z_{10}^{(i+4)}$, $Z_{11}^{(i+4)}$, $Z_2^{(i+5)}$, $Z_4^{(i+5)}$, $Z_9^{(i+5)}$, $Z_{10}^{(i+5)}$, $Z_{11}^{(i+5)}$, $Z_9^{(i+6)}$, $Z_{10}^{(i+6)}$, $Z_{11}^{(i+6)}$, $Z_5^{(i+7)}$, $Z_7^{(i+7)}$, $Z_{10}^{(i+7)}$, $Z_{11}^{(i+7)}$, $Z_1^{(i+8)}$, $Z_3^{(i+8)}$, $Z_9^{(i+8)}$, $Z_{10}^{(i+8)}$, $Z_4^{(i+9)}$, $Z_9^{(i+9)}$, $Z_{11}^{(i+9)}$, $Z_1^{(i+10)}$, $Z_3^{(i+10)}$, $Z_6^{(i+10)}$, $Z_8^{(i+10)}$, $Z_{11}^{(i+10)}$, $Z_4^{(i+11)}$, $Z_9^{(i+11)}$, $Z_{10}^{(i+11)}$, $Z_{11}^{(i+11)}$, $Z_3^{(i+12)}$, $Z_9^{(i+12)}$, $Z_{11}^{(i+12)}$, $Z_2^{(i+13)}$, $Z_5^{(i+13)}$, $Z_1^{(i+14)}$, $Z_3^{(i+14)}$, $Z_9^{(i+14)}$ and $Z_{11}^{(i+14)}$.

Starting from an even-numbered round i, the following 66 subkeys have to be weak: $Z_5^{(i)}$, $Z_9^{(i)}$, $Z_{10}^{(i)}$, $Z_{11}^{(i)}$, $Z_1^{(i+1)}$, $Z_3^{(i+1)}$, $Z_8^{(i+1)}$, $Z_{11}^{(i+1)}$, $Z_4^{(i+2)}$, $Z_5^{(i+2)}$, $Z_9^{(i+2)}$, $Z_{11}^{(i+2)}$, $Z_3^{(i+3)}$, $Z_6^{(i+3)}$, $Z_8^{(i+3)}$, $Z_9^{(i+3)}$, $Z_{10}^{(i+3)}$, $Z_2^{(i+4)}$, $Z_4^{(i+4)}$, $Z_5^{(i+4)}$, $Z_7^{(i+4)}$, $Z_{10}^{(i+4)}$, $Z_{11}^{(i+4)}$, $Z_1^{(i+5)}$, $Z_8^{(i+5)}$, $Z_9^{(i+5)}$, $Z_{10}^{(i+5)}$, $Z_{11}^{(i+5)}$, $Z_2^{(i+6)}$, $Z_5^{(i+6)}$, $Z_9^{(i+6)}$, $Z_{10}^{(i+6)}$, $Z_{11}^{(i+6)}$, $Z_1^{(i+7)}$, $Z_8^{(i+7)}$, $Z_{10}^{(i+7)}$, $Z_{11}^{(i+7)}$, $Z_2^{(i+8)}$, $Z_4^{(i+8)}$, $Z_5^{(i+8)}$, $Z_7^{(i+8)}$, $Z_9^{(i+8)}$, $Z_{10}^{(i+8)}$, $Z_3^{(i+9)}$, $Z_6^{(i+9)}$, $Z_8^{(i+9)}$, $Z_9^{(i+9)}$, $Z_{11}^{(i+9)}$, $Z_2^{(i+10)}$, $Z_7^{(i+10)}$, $Z_{11}^{(i+10)}$, $Z_1^{(i+11)}$, $Z_6^{(i+11)}$, $Z_8^{(i+11)}$, $Z_9^{(i+11)}$, $Z_{10}^{(i+11)}$, $Z_{11}^{(i+11)}$, $Z_2^{(i+12)}$, $Z_9^{(i+12)}$, $Z_{11}^{(i+12)}$, $Z_1^{(i+13)}$, $Z_6^{(i+13)}$, $Z_2^{(i+14)}$, $Z_5^{(i+14)}$, $Z_9^{(i+14)}$ and $Z_{11}^{(i+14)}$.

(10) $(0,0,0,0,\gamma,\gamma,0,\gamma) \xrightarrow{1r} (\gamma,0,\gamma,\gamma,0,\gamma,\gamma,0) \xrightarrow{1r} (0, \gamma, \gamma, 0, 0, 0, 0, \gamma) \xrightarrow{1r} (\gamma, \gamma, 0, \gamma, \gamma, 0, 0, \gamma) \xrightarrow{1r} (0,0,\gamma,\gamma,0,\gamma,0,0) \xrightarrow{1r} (\gamma,\gamma,\gamma,\gamma,0,0,\gamma,0) \xrightarrow{1r} (0, \gamma, 0, 0, \gamma, 0, 0, \gamma) \xrightarrow{1r} (\gamma,0,0,\gamma,\gamma,\gamma,\gamma,0) \xrightarrow{1r} (0,0,\gamma,0,0,\gamma,\gamma,0) \xrightarrow{1r} (\gamma, \gamma, 0, 0, \gamma, 0, \gamma, \gamma) \xrightarrow{1r} (0,0,0,0,\gamma,\gamma,0,\gamma)$, which is a 10-round iterative linear relation with bias $2^{-325.22}$.

Starting from an odd-numbered round i, the following 57 subkeys have to be weak: $Z_6^{(i)}$, $Z_8^{(i)}$, $Z_9^{(i)}$, $Z_{10}^{(i)}$, $Z_{11}^{(i)}$, $Z_{12}^{(i)}$, $Z_4^{(i+1)}$, $Z_7^{(i+1)}$, $Z_9^{(i+1)}$, $Z_{10}^{(i+1)}$, $Z_{11}^{(i+1)}$, $Z_{12}^{(i+1)}$, $Z_3^{(i+2)}$, $Z_8^{(i+2)}$, $Z_9^{(i+2)}$, $Z_{11}^{(i+2)}$, $Z_{12}^{(i+2)}$, $Z_2^{(i+3)}$, $Z_4^{(i+3)}$, $Z_5^{(i+3)}$, $Z_9^{(i+3)}$, $Z_{10}^{(i+3)}$, $Z_{12}^{(i+3)}$, $Z_3^{(i+4)}$, $Z_6^{(i+4)}$, $Z_9^{(i+4)}$, $Z_{11}^{(i+4)}$, $Z_{12}^{(i+4)}$, $Z_2^{(i+5)}$, $Z_4^{(i+5)}$, $Z_7^{(i+5)}$, $Z_9^{(i+5)}$, $Z_{10}^{(i+5)}$, $Z_{11}^{(i+5)}$, $Z_{12}^{(i+5)}$, $Z_8^{(i+6)}$, $Z_9^{(i+6)}$, $Z_{10}^{(i+6)}$, $Z_{11}^{(i+6)}$, $Z_{12}^{(i+6)}$, $Z_4^{(i+7)}$, $Z_5^{(i+7)}$, $Z_7^{(i+7)}$, $Z_9^{(i+7)}$, $Z_{11}^{(i+7)}$, $Z_{12}^{(i+7)}$, $Z_3^{(i+8)}$, $Z_6^{(i+8)}$, $Z_9^{(i+8)}$, $Z_{10}^{(i+8)}$, $Z_{12}^{(i+8)}$, $Z_2^{(i+9)}$, $Z_5^{(i+9)}$, $Z_7^{(i+9)}$, $Z_9^{(i+9)}$, $Z_{11}^{(i+9)}$ and $Z_{12}^{(i+9)}$.

Starting from an even-numbered round i, the following 51 subkeys have to be weak: $Z_5^{(i)}$, $Z_9^{(i)}$, $Z_{10}^{(i)}$, $Z_{11}^{(i)}$, $Z_{12}^{(i)}$, $Z_1^{(i+1)}$, $Z_3^{(i+1)}$, $Z_6^{(i+1)}$, $Z_9^{(i+1)}$, $Z_{10}^{(i+1)}$, $Z_{11}^{(i+1)}$, $Z_{12}^{(i+1)}$, $Z_2^{(i+2)}$, $Z_9^{(i+2)}$, $Z_{11}^{(i+2)}$, $Z_{12}^{(i+2)}$, $Z_1^{(i+3)}$, $Z_8^{(i+3)}$, $Z_9^{(i+3)}$, $Z_{10}^{(i+3)}$, $Z_{12}^{(i+3)}$, $Z_4^{(i+4)}$, $Z_9^{(i+4)}$, $Z_{11}^{(i+4)}$, $Z_{12}^{(i+4)}$, $Z_1^{(i+5)}$, $Z_3^{(i+5)}$, $Z_9^{(i+5)}$, $Z_{10}^{(i+5)}$, $Z_{11}^{(i+5)}$, $Z_{12}^{(i+5)}$, $Z_2^{(i+6)}$, $Z_5^{(i+6)}$, $Z_9^{(i+6)}$, $Z_{10}^{(i+6)}$, $Z_{11}^{(i+6)}$, $Z_{12}^{(i+6)}$, $Z_1^{(i+7)}$, $Z_6^{(i+7)}$, $Z_9^{(i+7)}$, $Z_{11}^{(i+7)}$, $Z_{12}^{(i+7)}$, $Z_7^{(i+8)}$, $Z_9^{(i+8)}$, $Z_{10}^{(i+8)}$, $Z_{12}^{(i+8)}$, $Z_1^{(i+9)}$, $Z_8^{(i+9)}$, $Z_9^{(i+9)}$, $Z_{11}^{(i+9)}$ and $Z_{12}^{(i+9)}$.

(11) $(0,0,0,0,\gamma,\gamma,\gamma,0) \stackrel{1r}{\to} (\gamma,0,0,0,\gamma,\gamma,0,0) \stackrel{1r}{\to} (0,0,0,\gamma,\gamma,\gamma,\gamma,\gamma) \stackrel{1r}{\to} (\gamma,0,0,0,\gamma,\gamma,\gamma,\gamma) \stackrel{1r}{\to} (0,0,\gamma,0,0,\gamma,0,\gamma) \stackrel{1r}{\to} (\gamma,\gamma,\gamma,\gamma,0,0,0,\gamma) \stackrel{1r}{\to} (0,\gamma,\gamma,\gamma,0,0,\gamma,\gamma) \stackrel{1r}{\to} (\gamma,\gamma,\gamma,0,0,0,0,0) \stackrel{1r}{\to} (0,\gamma,\gamma,\gamma,0,0,0,0) \stackrel{1r}{\to} (\gamma,\gamma,0,\gamma,\gamma,0,\gamma,0) \stackrel{1r}{\to} (0,0,0,0,\gamma,\gamma,\gamma,0)$, which is a 10-round iterative linear relation with bias $2^{-275.34}$.

Starting from an odd-numbered round i, the following 45 subkeys have to be weak: $Z_6^{(i)}$, $Z_9^{(i)}$, $Z_{12}^{(i)}$, $Z_5^{(i+1)}$, $Z_9^{(i+1)}$, $Z_{10}^{(i+1)}$, $Z_{12}^{(i+1)}$, $Z_6^{(i+2)}$, $Z_8^{(i+2)}$, $Z_9^{(i+2)}$, $Z_{12}^{(i+2)}$, $Z_5^{(i+3)}$, $Z_7^{(i+3)}$, $Z_9^{(i+3)}$, $Z_{11}^{(i+3)}$, $Z_{12}^{(i+3)}$, $Z_3^{(i+4)}$, $Z_6^{(i+4)}$, $Z_8^{(i+4)}$, $Z_9^{(i+4)}$, $Z_{11}^{(i+4)}$, $Z_{12}^{(i+4)}$, $Z_2^{(i+5)}$, $Z_4^{(i+5)}$, $Z_9^{(i+5)}$, $Z_{12}^{(i+5)}$, $Z_3^{(i+6)}$, $Z_8^{(i+6)}$, $Z_9^{(i+6)}$, $Z_{10}^{(i+6)}$, $Z_{12}^{(i+6)}$, $Z_2^{(i+7)}$, $Z_9^{(i+7)}$, $Z_{12}^{(i+7)}$, $Z_3^{(i+8)}$, $Z_9^{(i+8)}$, $Z_{11}^{(i+8)}$, $Z_{12}^{(i+8)}$, $Z_2^{(i+9)}$, $Z_4^{(i+9)}$, $Z_5^{(i+9)}$, $Z_7^{(i+9)}$, $Z_9^{(i+9)}$, $Z_{11}^{(i+9)}$ and $Z_{12}^{(i+9)}$.

Starting from an even-numbered round i, the following 47 subkeys have to be weak: $Z_5^{(i)}$, $Z_7^{(i)}$, $Z_9^{(i)}$, $Z_{12}^{(i)}$, $Z_1^{(i+1)}$, $Z_6^{(i+1)}$, $Z_9^{(i+1)}$, $Z_{10}^{(i+1)}$, $Z_{12}^{(i+1)}$, $Z_4^{(i+2)}$, $Z_5^{(i+2)}$, $Z_7^{(i+2)}$, $Z_9^{(i+2)}$, $Z_{12}^{(i+2)}$, $Z_1^{(i+3)}$, $Z_6^{(i+3)}$, $Z_8^{(i+3)}$, $Z_9^{(i+3)}$, $Z_{11}^{(i+3)}$, $Z_{12}^{(i+3)}$, $Z_9^{(i+4)}$, $Z_{11}^{(i+4)}$, $Z_{12}^{(i+4)}$, $Z_1^{(i+5)}$, $Z_3^{(i+5)}$, $Z_8^{(i+5)}$, $Z_9^{(i+5)}$, $Z_{12}^{(i+5)}$, $Z_2^{(i+6)}$, $Z_4^{(i+6)}$, $Z_7^{(i+6)}$, $Z_9^{(i+6)}$, $Z_{10}^{(i+6)}$, $Z_{12}^{(i+6)}$, $Z_1^{(i+7)}$, $Z_3^{(i+7)}$, $Z_9^{(i+7)}$, $Z_{12}^{(i+7)}$, $Z_2^{(i+8)}$, $Z_4^{(i+8)}$, $Z_9^{(i+8)}$, $Z_{11}^{(i+8)}$, $Z_{12}^{(i+8)}$, $Z_1^{(i+9)}$, $Z_9^{(i+9)}$, $Z_{11}^{(i+9)}$ and $Z_{12}^{(i+9)}$.

(12) $(0,0,0,0,\gamma,\gamma,\gamma,\gamma) \stackrel{1r}{\to} (\gamma,0,0,\gamma,\gamma,0,0,\gamma) \stackrel{1r}{\to} (\gamma,\gamma,0,0,0,0,\gamma,\gamma) \stackrel{1r}{\to} (0,\gamma,\gamma,\gamma,0,0,0,\gamma) \stackrel{1r}{\to} (\gamma,\gamma,0,0,\gamma,\gamma,\gamma,\gamma) \stackrel{1r}{\to} (\gamma,\gamma,0,\gamma,0,0,\gamma,0) \stackrel{1r}{\to} (0,\gamma,\gamma,\gamma,0,0,\gamma,0) \stackrel{1r}{\to} (\gamma,\gamma,\gamma,\gamma,0,\gamma,0,\gamma) \stackrel{1r}{\to} (\gamma,0,\gamma,0,\gamma,\gamma,0,0) \stackrel{1r}{\to} (0,0,\gamma,0,0,\gamma,0,0) \stackrel{1r}{\to} (\gamma,\gamma,\gamma,0,0,\gamma,0,0) \stackrel{1r}{\to} (\gamma,0,\gamma,0,\gamma,\gamma,\gamma,\gamma) \stackrel{1r}{\to} (0,0,0,\gamma,\gamma,\gamma,\gamma,0) \stackrel{1r}{\to} (\gamma,0,0,\gamma,\gamma,0,\gamma,0) \stackrel{1r}{\to} (\gamma,\gamma,\gamma,\gamma,\gamma,0,0,\gamma) \stackrel{1r}{\to}$

$(0,0,0,0,\gamma,\gamma,\gamma,\gamma)$, which is a 15-round iterative linear relation with bias $2^{-275.34}$.

Starting from an odd-numbered round i, the following 54 subkeys have to be weak: $Z_6^{(i)}$, $Z_8^{(i)}$, $Z_9^{(i)}$, $Z_4^{(i+1)}$, $Z_5^{(i+1)}$, $Z_1^{(i+2)}$, $Z_8^{(i+2)}$, $Z_9^{(i+2)}$, $Z_2^{(i+3)}$, $Z_4^{(i+3)}$, $Z_9^{(i+3)}$, $Z_{11}^{(i+3)}$, $Z_1^{(i+4)}$, $Z_6^{(i+4)}$, $Z_8^{(i+4)}$, $Z_{10}^{(i+4)}$, $Z_{11}^{(i+4)}$, $Z_2^{(i+5)}$, $Z_4^{(i+5)}$, $Z_7^{(i+5)}$, $Z_9^{(i+5)}$, $Z_3^{(i+6)}$, $Z_9^{(i+6)}$, $Z_{10}^{(i+6)}$, $Z_2^{(i+7)}$, $Z_4^{(i+7)}$, $Z_{10}^{(i+7)}$, $Z_1^{(i+8)}$, $Z_3^{(i+8)}$, $Z_6^{(i+8)}$, $Z_9^{(i+8)}$, $Z_{11}^{(i+8)}$, $Z_9^{(i+9)}$, $Z_{11}^{(i+9)}$, $Z_1^{(i+10)}$, $Z_3^{(i+10)}$, $Z_6^{(i+10)}$, $Z_{10}^{(i+10)}$, $Z_5^{(i+11)}$, $Z_7^{(i+11)}$, $Z_9^{(i+11)}$, $Z_{10}^{(i+11)}$, $Z_6^{(i+12)}$, $Z_9^{(i+12)}$, $Z_4^{(i+13)}$, $Z_5^{(i+13)}$, $Z_7^{(i+13)}$, $Z_{10}^{(i+13)}$, $Z_{11}^{(i+13)}$, $Z_1^{(i+14)}$, $Z_3^{(i+14)}$, $Z_8^{(i+14)}$, $Z_9^{(i+14)}$ and $Z_{11}^{(i+14)}$.

Starting from an even-numbered round i, the following 56 subkeys have to be weak: $Z_5^{(i)}$, $Z_7^{(i)}$, $Z_9^{(i)}$, $Z_1^{(i+1)}$, $Z_8^{(i+1)}$, $Z_2^{(i+2)}$, $Z_7^{(i+2)}$, $Z_9^{(i+2)}$, $Z_3^{(i+3)}$, $Z_8^{(i+3)}$, $Z_9^{(i+3)}$, $Z_{11}^{(i+3)}$, $Z_2^{(i+4)}$, $Z_5^{(i+4)}$, $Z_7^{(i+4)}$, $Z_{10}^{(i+4)}$, $Z_{11}^{(i+4)}$, $Z_1^{(i+5)}$, $Z_9^{(i+5)}$, $Z_2^{(i+6)}$, $Z_4^{(i+6)}$, $Z_7^{(i+6)}$, $Z_9^{(i+6)}$, $Z_{10}^{(i+6)}$, $Z_1^{(i+7)}$, $Z_3^{(i+7)}$, $Z_6^{(i+7)}$, $Z_8^{(i+7)}$, $Z_{10}^{(i+7)}$, $Z_5^{(i+8)}$, $Z_9^{(i+8)}$, $Z_{11}^{(i+8)}$, $Z_3^{(i+9)}$, $Z_6^{(i+9)}$, $Z_9^{(i+9)}$, $Z_{11}^{(i+9)}$, $Z_2^{(i+10)}$, $Z_{10}^{(i+10)}$, $Z_1^{(i+11)}$, $Z_3^{(i+11)}$, $Z_6^{(i+11)}$, $Z_8^{(i+11)}$, $Z_9^{(i+11)}$, $Z_{10}^{(i+11)}$, $Z_4^{(i+12)}$, $Z_5^{(i+12)}$, $Z_7^{(i+12)}$, $Z_9^{(i+12)}$, $Z_1^{(i+13)}$, $Z_{10}^{(i+13)}$, $Z_{11}^{(i+13)}$, $Z_2^{(i+14)}$, $Z_4^{(i+14)}$, $Z_5^{(i+14)}$, $Z_9^{(i+14)}$ and $Z_{11}^{(i+14)}$.

(13) $(0,0,0,\gamma,0,0,\gamma,0) \stackrel{1r}{\to} (0,0,\gamma,\gamma,\gamma,0,0,\gamma) \stackrel{1r}{\to} (0,\gamma,0,0,0,\gamma,0,0) \stackrel{1r}{\to} (0,0,\gamma,0,\gamma,0,0,0) \stackrel{1r}{\to} (0,\gamma,0,0,0,\gamma,\gamma,\gamma) \stackrel{1r}{\to} (0,0,0,\gamma,0,0,\gamma,0)$, which is a 5-round iterative linear relation with bias $2^{-75.82}$.

Starting from an odd-numbered round i, the following 12 subkeys have to be weak: $Z_{10}^{(i)}$, $Z_{11}^{(i)}$, $Z_4^{(i+1)}$, $Z_5^{(i+1)}$, $Z_{10}^{(i+1)}$, $Z_6^{(i+2)}$, $Z_5^{(i+3)}$, $Z_{10}^{(i+3)}$, $Z_6^{(i+4)}$, $Z_8^{(i+4)}$, $Z_{10}^{(i+4)}$ and $Z_{11}^{(i+4)}$.

Starting from an even-numbered round i, the following 14 subkeys have to be weak: $Z_4^{(i)}$, $Z_7^{(i)}$, $Z_{10}^{(i)}$, $Z_{11}^{(i)}$, $Z_3^{(i+1)}$, $Z_8^{(i+1)}$, $Z_{10}^{(i+1)}$, $Z_2^{(i+2)}$, $Z_3^{(i+3)}$, $Z_{10}^{(i+3)}$, $Z_2^{(i+4)}$, $Z_7^{(i+4)}$, $Z_{10}^{(i+4)}$ and $Z_{11}^{(i+4)}$.

(14) $(0,0,0,\gamma,0,\gamma,0,0) \stackrel{1r}{\to} (\gamma,\gamma,0,0,\gamma,0,0,\gamma) \stackrel{1r}{\to} (0,0,\gamma,0,0,0,\gamma,0) \stackrel{1r}{\to} (0,0,0,\gamma,0,\gamma,0,0)$, which is a 3-round iterative linear relation with bias $2^{-50.88}$.

Starting from an odd-numbered round i, the following seven subkeys have to be weak: $Z_6^{(i)}$, $Z_9^{(i)}$, $Z_{10}^{(i)}$, $Z_2^{(i+1)}$, $Z_5^{(i+1)}$, $Z_9^{(i+1)}$, $Z_{10}^{(i+1)}$ and $Z_3^{(i+2)}$.

Starting from an even-numbered round i, the following eight subkeys have to be weak: $Z_4^{(i)}$, $Z_9^{(i)}$, $Z_{10}^{(i)}$, $Z_1^{(i+1)}$, $Z_8^{(i+1)}$, $Z_9^{(i+1)}$, $Z_{10}^{(i+1)}$ and $Z_7^{(i+2)}$.

(15) $(0,0,0,\gamma,\gamma,\gamma,0,0) \stackrel{1r}{\to} (\gamma,0,\gamma,\gamma,0,\gamma,0,\gamma) \stackrel{1r}{\to} (0,\gamma,0,\gamma,\gamma,0,\gamma,\gamma) \stackrel{1r}{\to} (\gamma,0,\gamma,0,0,\gamma,\gamma,\gamma) \stackrel{1r}{\to} (0,\gamma,\gamma,0,0,0,\gamma,0) \stackrel{1r}{\to} (\gamma,\gamma,\gamma,0,0,0,\gamma,\gamma) \stackrel{1r}{\to} (0,\gamma,0,0,\gamma,0,\gamma,0) \stackrel{1r}{\to} (\gamma,0,\gamma,0,0,\gamma,0,0) \stackrel{1r}{\to} (0,\gamma,0,\gamma,\gamma,0,0,0) \stackrel{1r}{\to} (\gamma,0,0,\gamma,\gamma,\gamma,0,\gamma) \stackrel{1r}{\to} (0,0,0,\gamma,\gamma,\gamma,0,0)$, which is a 10-round iterative linear relation with bias $2^{-375.10}$.

Starting from an odd-numbered round i, the following 38 subkeys have to be weak: $Z_6^{(i)}$, $Z_9^{(i)}$, $Z_{10}^{(i)}$, $Z_{11}^{(i)}$, $Z_{12}^{(i)}$, $Z_4^{(i+1)}$, $Z_9^{(i+1)}$, $Z_{12}^{(i+1)}$, $Z_8^{(i+2)}$, $Z_9^{(i+2)}$, $Z_{12}^{(i+2)}$, $Z_7^{(i+3)}$, $Z_9^{(i+3)}$, $Z_{10}^{(i+3)}$, $Z_{11}^{(i+3)}$, $Z_{12}^{(i+3)}$, $Z_3^{(i+4)}$, $Z_9^{(i+4)}$, $Z_{10}^{(i+4)}$, $Z_{12}^{(i+4)}$, $Z_2^{(i+5)}$, $Z_7^{(i+5)}$, $Z_9^{(i+5)}$, $Z_{10}^{(i+5)}$, $Z_{11}^{(i+5)}$, $Z_{12}^{(i+5)}$, $Z_9^{(i+6)}$, $Z_{12}^{(i+6)}$, $Z_9^{(i+7)}$, $Z_{12}^{(i+7)}$, $Z_9^{(i+8)}$, $Z_{10}^{(i+8)}$, $Z_{11}^{(i+8)}$, $Z_{12}^{(i+8)}$, $Z_4^{(i+9)}$, $Z_5^{(i+9)}$, $Z_9^{(i+9)}$, $Z_{10}^{(i+9)}$ and $Z_{12}^{(i+9)}$.

Starting from an even-numbered round i, the following 61 subkeys have to be weak: $Z_4^{(i)}$, $Z_5^{(i)}$, $Z_9^{(i)}$, $Z_{10}^{(i)}$, $Z_{11}^{(i)}$, $Z_{12}^{(i)}$, $Z_1^{(i+1)}$, $Z_3^{(i+1)}$, $Z_6^{(i+1)}$, $Z_8^{(i+1)}$, $Z_9^{(i+1)}$, $Z_{12}^{(i+1)}$, $Z_2^{(i+2)}$, $Z_4^{(i+2)}$, $Z_5^{(i+2)}$, $Z_7^{(i+2)}$, $Z_9^{(i+2)}$, $Z_{12}^{(i+2)}$, $Z_1^{(i+3)}$, $Z_3^{(i+3)}$, $Z_6^{(i+3)}$, $Z_8^{(i+3)}$, $Z_9^{(i+3)}$, $Z_{10}^{(i+3)}$, $Z_{11}^{(i+3)}$, $Z_{12}^{(i+3)}$, $Z_2^{(i+4)}$, $Z_7^{(i+4)}$, $Z_9^{(i+4)}$, $Z_{10}^{(i+4)}$, $Z_{12}^{(i+4)}$, $Z_1^{(i+5)}$, $Z_3^{(i+5)}$, $Z_8^{(i+5)}$, $Z_9^{(i+5)}$, $Z_{10}^{(i+5)}$, $Z_{11}^{(i+5)}$, $Z_{12}^{(i+5)}$, $Z_2^{(i+6)}$, $Z_5^{(i+6)}$, $Z_7^{(i+6)}$, $Z_9^{(i+6)}$, $Z_{12}^{(i+6)}$, $Z_1^{(i+7)}$, $Z_3^{(i+7)}$, $Z_6^{(i+7)}$, $Z_9^{(i+7)}$, $Z_{12}^{(i+7)}$, $Z_2^{(i+8)}$, $Z_4^{(i+8)}$, $Z_5^{(i+8)}$, $Z_9^{(i+8)}$, $Z_{10}^{(i+8)}$, $Z_{11}^{(i+8)}$, $Z_{12}^{(i+8)}$, $Z_1^{(i+9)}$, $Z_6^{(i+9)}$, $Z_8^{(i+9)}$, $Z_9^{(i+9)}$, $Z_{10}^{(i+9)}$ and $Z_{12}^{(i+9)}$.

(16) $(0,0,0,\gamma,\gamma,\gamma,0,\gamma) \stackrel{1r}{\rightarrow} (\gamma,0,\gamma,0,0,0,0,0) \stackrel{1r}{\rightarrow} (\gamma,0,0,0,0,\gamma,\gamma,\gamma) \stackrel{1r}{\rightarrow} (0,\gamma,0,\gamma,\gamma,0,0,\gamma) \stackrel{1r}{\rightarrow} (\gamma,0,0,0,\gamma,0,0,0) \stackrel{1r}{\rightarrow} (\gamma,\gamma,0,0,0,0,0,0) \stackrel{1r}{\rightarrow} (0,\gamma,0,0,\gamma,0,\gamma,\gamma) \stackrel{1r}{\rightarrow} (\gamma,0,\gamma,\gamma,0,0,0,\gamma) \stackrel{1r}{\rightarrow} (\gamma,0,0,0,0,\gamma,0,0) \stackrel{1r}{\rightarrow} (0,\gamma,\gamma,0,0,0,\gamma,\gamma) \stackrel{1r}{\rightarrow} (\gamma,\gamma,\gamma,\gamma,0,\gamma,\gamma,0) \stackrel{1r}{\rightarrow} (\gamma,0,0,\gamma,0,\gamma,\gamma,0) \stackrel{1r}{\rightarrow} (0,\gamma,0,\gamma,\gamma,0,\gamma,0) \stackrel{1r}{\rightarrow} (\gamma,0,\gamma,\gamma,0,0,\gamma,0) \stackrel{1r}{\rightarrow} (\gamma,0,\gamma,\gamma,\gamma,\gamma,\gamma,0) \stackrel{1r}{\rightarrow} (0,0,0,\gamma,\gamma,\gamma,0,\gamma)$, which is a 15-round iterative linear relation with bias $2^{-315.22}$.

Starting from an odd-numbered round i, the following 50 subkeys have to be weak: $Z_6^{(i)}$, $Z_8^{(i)}$, $Z_9^{(i)}$, $Z_{10}^{(i)}$, $Z_{11}^{(i)}$, $Z_{10}^{(i+1)}$, $Z_1^{(i+2)}$, $Z_6^{(i+2)}$, $Z_8^{(i+2)}$, $Z_9^{(i+2)}$, $Z_2^{(i+3)}$, $Z_4^{(i+3)}$, $Z_5^{(i+3)}$, $Z_9^{(i+3)}$, $Z_{10}^{(i+3)}$, $Z_{11}^{(i+3)}$, $Z_1^{(i+4)}$, $Z_2^{(i+5)}$, $Z_9^{(i+5)}$, $Z_{10}^{(i+5)}$, $Z_{11}^{(i+5)}$, $Z_8^{(i+6)}$, $Z_9^{(i+6)}$, $Z_4^{(i+7)}$, $Z_{10}^{(i+7)}$, $Z_1^{(i+8)}$, $Z_6^{(i+8)}$, $Z_9^{(i+8)}$, $Z_{10}^{(i+8)}$, $Z_{11}^{(i+8)}$, $Z_2^{(i+9)}$, $Z_7^{(i+9)}$, $Z_9^{(i+9)}$, $Z_{10}^{(i+9)}$, $Z_1^{(i+10)}$, $Z_3^{(i+10)}$, $Z_6^{(i+10)}$, $Z_{11}^{(i+10)}$, $Z_4^{(i+11)}$, $Z_7^{(i+11)}$, $Z_9^{(i+11)}$, $Z_9^{(i+12)}$, $Z_4^{(i+13)}$, $Z_7^{(i+13)}$, $Z_{11}^{(i+13)}$, $Z_1^{(i+14)}$, $Z_3^{(i+14)}$, $Z_6^{(i+14)}$, $Z_9^{(i+14)}$ and $Z_{10}^{(i+14)}$.

Starting from an even-numbered round i, the following 54 subkeys have to be weak: $Z_4^{(i)}$, $Z_5^{(i)}$, $Z_9^{(i)}$, $Z_{10}^{(i)}$, $Z_{11}^{(i)}$, $Z_1^{(i+1)}$, $Z_3^{(i+1)}$, $Z_{10}^{(i+1)}$, $Z_7^{(i+2)}$, $Z_9^{(i+2)}$, $Z_8^{(i+3)}$, $Z_9^{(i+3)}$, $Z_{10}^{(i+3)}$, $Z_{11}^{(i+3)}$, $Z_5^{(i+4)}$, $Z_1^{(i+5)}$, $Z_9^{(i+5)}$, $Z_{10}^{(i+5)}$, $Z_{11}^{(i+5)}$, $Z_2^{(i+6)}$, $Z_5^{(i+6)}$, $Z_7^{(i+6)}$, $Z_9^{(i+6)}$, $Z_1^{(i+7)}$, $Z_3^{(i+7)}$, $Z_8^{(i+7)}$, $Z_{10}^{(i+7)}$, $Z_9^{(i+8)}$, $Z_{10}^{(i+8)}$, $Z_{11}^{(i+8)}$, $Z_3^{(i+9)}$, $Z_8^{(i+9)}$, $Z_9^{(i+9)}$, $Z_{10}^{(i+9)}$, $Z_2^{(i+10)}$, $Z_4^{(i+10)}$, $Z_7^{(i+10)}$, $Z_{11}^{(i+10)}$, $Z_1^{(i+11)}$, $Z_6^{(i+11)}$, $Z_9^{(i+11)}$, $Z_2^{(i+12)}$, $Z_4^{(i+12)}$, $Z_5^{(i+12)}$, $Z_7^{(i+12)}$, $Z_9^{(i+12)}$, $Z_1^{(i+13)}$, $Z_3^{(i+13)}$, $Z_{11}^{(i+13)}$, $Z_4^{(i+14)}$, $Z_5^{(i+14)}$, $Z_7^{(i+14)}$, $Z_9^{(i+14)}$ and $Z_{10}^{(i+14)}$.

(17) $(0,0,\gamma,0,\gamma,0,\gamma,\gamma) \stackrel{1r}{\rightarrow} (0,\gamma,\gamma,\gamma,\gamma,\gamma,0,\gamma) \stackrel{1r}{\rightarrow} (0,\gamma,\gamma,0,\gamma,\gamma,0,0) \stackrel{1r}{\rightarrow} (0,\gamma,\gamma,0,\gamma,\gamma,\gamma,\gamma) \stackrel{1r}{\rightarrow} (0,\gamma,0,\gamma,0,\gamma,0,\gamma) \stackrel{1r}{\rightarrow} (0,0,\gamma,0,\gamma,0,\gamma,\gamma)$, which is a 5-round iterative linear relation with bias $2^{-125.70}$.

Starting from an odd-numbered round i, the following 16 subkeys have to be weak: $Z_3^{(i)}$, $Z_8^{(i)}$, $Z_{11}^{(i)}$, $Z_2^{(i+1)}$, $Z_4^{(i+1)}$, $Z_5^{(i+1)}$, $Z_{10}^{(i+1)}$, $Z_3^{(i+2)}$, $Z_6^{(i+2)}$, $Z_{10}^{(i+2)}$, $Z_2^{(i+3)}$, $Z_5^{(i+3)}$, $Z_7^{(i+3)}$, $Z_{11}^{(i+3)}$, $Z_6^{(i+4)}$ and $Z_8^{(i+4)}$.
Starting from an even-numbered round i, the following 16 subkeys have to be weak: $Z_5^{(i)}$, $Z_7^{(i)}$, $Z_{11}^{(i)}$, $Z_3^{(i+1)}$, $Z_6^{(i+1)}$, $Z_8^{(i+1)}$, $Z_{10}^{(i+1)}$, $Z_2^{(i+2)}$, $Z_5^{(i+2)}$, $Z_{10}^{(i+2)}$, $Z_3^{(i+3)}$, $Z_6^{(i+3)}$, $Z_8^{(i+3)}$, $Z_{11}^{(i+3)}$, $Z_2^{(i+4)}$ and $Z_4^{(i+4)}$.

(18) $(0,0,\gamma,\gamma,0,\gamma,\gamma,0) \stackrel{1r}{\rightarrow} (\gamma,\gamma,0,\gamma,\gamma,\gamma,0,\gamma) \stackrel{1r}{\rightarrow} (\gamma,\gamma,\gamma,0,\gamma,0,\gamma,\gamma) \stackrel{1r}{\rightarrow} (0,0,\gamma,\gamma,0,\gamma,\gamma,0)$, which is a 3-round iterative linear relation with bias $2^{-50.88}$.
Starting from an odd-numbered round i, the following 12 subkeys have to be weak: $Z_3^{(i)}$, $Z_6^{(i)}$, $Z_9^{(i)}$, $Z_{10}^{(i)}$, $Z_2^{(i+1)}$, $Z_4^{(i+1)}$, $Z_5^{(i+1)}$, $Z_1^{(i+2)}$, $Z_3^{(i+2)}$, $Z_8^{(i+2)}$, $Z_9^{(i+2)}$ and $Z_{10}^{(i+2)}$.
Starting from an even-numbered round i, the following 12 subkeys have to be weak: $Z_4^{(i)}$, $Z_7^{(i)}$, $Z_9^{(i)}$, $Z_{10}^{(i)}$, $Z_1^{(i+1)}$, $Z_6^{(i+1)}$, $Z_8^{(i+1)}$, $Z_2^{(i+2)}$, $Z_5^{(i+2)}$, $Z_7^{(i+2)}$, $Z_9^{(i+2)}$ and $Z_{10}^{(i+2)}$.

(19) $(0,0,\gamma,\gamma,0,\gamma,\gamma,\gamma) \stackrel{1r}{\rightarrow} (\gamma,\gamma,0,0,\gamma,0,0,0) \stackrel{1r}{\rightarrow} (0,0,\gamma,\gamma,0,\gamma,\gamma,\gamma)$, which is a 2-round iterative linear relation with bias $2^{-25.94}$.
Starting from an odd-numbered round i, the following 11 subkeys have to be weak: $Z_3^{(i)}$, $Z_6^{(i)}$, $Z_8^{(i)}$, $Z_9^{(i)}$, $Z_{10}^{(i)}$, $Z_{12}^{(i)}$, $Z_2^{(i+1)}$, $Z_5^{(i+1)}$, $Z_9^{(i+1)}$, $Z_{10}^{(i+1)}$ and $Z_{12}^{(i+1)}$.
Starting from an even-numbered round i, the following nine subkeys have to be weak: $Z_4^{(i)}$, $Z_7^{(i)}$, $Z_9^{(i)}$, $Z_{10}^{(i)}$, $Z_{12}^{(i)}$, $Z_1^{(i+1)}$, $Z_9^{(i+1)}$, $Z_{10}^{(i+1)}$ and $Z_{12}^{(i+1)}$.

(20) $(\gamma,0,0,0,0,0,0,\gamma) \stackrel{1r}{\rightarrow} (\gamma,0,\gamma,0,\gamma,0,0,\gamma) \stackrel{1r}{\rightarrow} (\gamma,\gamma,\gamma,0,\gamma,\gamma,\gamma,0) \stackrel{1r}{\rightarrow} (\gamma,\gamma,\gamma,\gamma,\gamma,\gamma,0,0) \stackrel{1r}{\rightarrow} (\gamma,\gamma,0,0,0,\gamma,0,\gamma) \stackrel{1r}{\rightarrow} (\gamma,0,0,0,0,0,0,\gamma)$, which is a 5-round iterative linear relation with bias $2^{-100.76}$.
Starting from an odd-numbered round i, the following 18 subkeys have to be weak: $Z_1^{(i)}$, $Z_8^{(i)}$, $Z_{11}^{(i)}$, $Z_5^{(i+1)}$, $Z_{10}^{(i+1)}$, $Z_{11}^{(i+1)}$, $Z_1^{(i+2)}$, $Z_3^{(i+2)}$, $Z_6^{(i+2)}$, $Z_2^{(i+3)}$, $Z_4^{(i+3)}$, $Z_5^{(i+3)}$, $Z_{10}^{(i+3)}$, $Z_{11}^{(i+3)}$, $Z_1^{(i+4)}$, $Z_6^{(i+4)}$, $Z_8^{(i+4)}$ and $Z_{11}^{(i+4)}$.
Starting from an even-numbered round i, the following 16 subkeys have to be weak: $Z_{11}^{(i)}$, $Z_1^{(i+1)}$, $Z_3^{(i+1)}$, $Z_8^{(i+1)}$, $Z_{10}^{(i+1)}$, $Z_{11}^{(i+1)}$, $Z_2^{(i+2)}$, $Z_5^{(i+2)}$, $Z_7^{(i+2)}$, $Z_1^{(i+3)}$, $Z_3^{(i+3)}$, $Z_6^{(i+3)}$, $Z_{10}^{(i+3)}$, $Z_{11}^{(i+3)}$, $Z_2^{(i+4)}$ and $Z_{11}^{(i+4)}$.

(21) $(\gamma,0,0,0,0,0,\gamma,0) \stackrel{1r}{\rightarrow} (\gamma,0,0,\gamma,0,0,\gamma,\gamma) \stackrel{1r}{\rightarrow} (\gamma,0,0,\gamma,0,0,0,0) \stackrel{1r}{\rightarrow} (\gamma,0,\gamma,0,\gamma,0,\gamma,0) \stackrel{1r}{\rightarrow} (\gamma,\gamma,0,\gamma,0,\gamma,0,0) \stackrel{1r}{\rightarrow} (\gamma,0,0,0,0,0,\gamma,0)$, which is a 5-round iterative linear relation with bias $2^{-125.70}$.
Starting from an odd-numbered round i, the following 12 subkeys have to be weak: $Z_1^{(i)}$, $Z_{10}^{(i)}$, $Z_4^{(i+1)}$, $Z_7^{(i+1)}$, $Z_{10}^{(i+1)}$, $Z_1^{(i+2)}$, $Z_{11}^{(i+2)}$, $Z_5^{(i+3)}$, $Z_7^{(i+3)}$, $Z_1^{(i+4)}$, $Z_6^{(i+4)}$ and $Z_{11}^{(i+4)}$.
Starting from an even-numbered round i, the following 12 subkeys have to be weak: $Z_7^{(i)}$, $Z_{10}^{(i)}$, $Z_1^{(i+1)}$, $Z_8^{(i+1)}$, $Z_{10}^{(i+1)}$, $Z_4^{(i+2)}$, $Z_{11}^{(i+2)}$, $Z_1^{(i+3)}$, $Z_3^{(i+3)}$, $Z_2^{(i+4)}$, $Z_4^{(i+4)}$ and $Z_{11}^{(i+4)}$.

(22) $(\gamma, 0, \gamma, \gamma, \gamma, 0, 0, 0) \stackrel{1r}{\to} (\gamma, \gamma, \gamma, 0, \gamma, \gamma, 0, \gamma) \stackrel{1r}{\to} (\gamma, \gamma, 0, 0, 0, \gamma, \gamma, 0) \stackrel{1r}{\to} (\gamma, 0, \gamma, \gamma, \gamma, 0, \gamma, \gamma) \stackrel{1r}{\to} (\gamma, \gamma, 0, \gamma, 0, \gamma, \gamma, \gamma) \stackrel{1r}{\to} (\gamma, 0, \gamma, \gamma, \gamma, 0, 0, 0)$, which is a 5-round iterative linear relation with bias $2^{-75.82}$.
Starting from an odd-numbered round i, the following 18 subkeys have to be weak: $Z_1^{(i)}$, $Z_3^{(i)}$, $Z_{10}^{(i)}$, $Z_{11}^{(i)}$, $Z_2^{(i+1)}$, $Z_5^{(i+1)}$, $Z_{10}^{(i+1)}$, $Z_{11}^{(i+1)}$, $Z_1^{(i+2)}$, $Z_6^{(i+2)}$, $Z_{10}^{(i+2)}$, $Z_4^{(i+3)}$, $Z_5^{(i+3)}$, $Z_7^{(i+3)}$, $Z_1^{(i+4)}$, $Z_6^{(i+4)}$, $Z_8^{(i+4)}$ and $Z_{10}^{(i+4)}$.
Starting from an even-numbered round i, the following 20 subkeys have to be weak: $Z_4^{(i)}$, $Z_5^{(i)}$, $Z_{10}^{(i)}$, $Z_{11}^{(i)}$, $Z_1^{(i+1)}$, $Z_3^{(i+1)}$, $Z_6^{(i+1)}$, $Z_8^{(i+1)}$, $Z_{10}^{(i+1)}$, $Z_{11}^{(i+1)}$, $Z_2^{(i+2)}$, $Z_7^{(i+2)}$, $Z_{10}^{(i+2)}$, $Z_1^{(i+3)}$, $Z_3^{(i+3)}$, $Z_8^{(i+3)}$, $Z_2^{(i+4)}$, $Z_4^{(i+4)}$, $Z_7^{(i+4)}$ and $Z_{10}^{(i+4)}$.

(23) $(\gamma,\gamma,\gamma,\gamma,\gamma,\gamma,\gamma,\gamma) \stackrel{1r}{\to} (\gamma, \gamma, \gamma, \gamma, \gamma, \gamma, \gamma, \gamma)$, which is a 1-round iterative linear relation with bias 2^{-1}.
Starting from an odd-numbered round i, the following four subkeys have to be weak: $Z_1^{(i)}$, $Z_3^{(i)}$, $Z_6^{(i)}$ and $Z_8^{(i)}$.
Starting from an even-numbered round i, the following four subkeys have to be weak: $Z_2^{(i)}$, $Z_4^{(i)}$, $Z_5^{(i)}$ and $Z_7^{(i)}$.

Similar to the linear analysis of MESH-64, the key schedule of MESH-128 does not consist of a simple bit permutation like in IDEA. The fraction of keys that satisfies a weak-subkey restriction in MESH-128 was estimated (like in the case of MESH-64) to be about 2^{-15} per subkey. This estimate is in line with the assumption that a weak subkey is either 0 or 1, since the fifteen most significant bits have to be zero. Under this assumption, and since MESH-128 operates under a 256-bit key, at most $\lfloor 256/15 \rfloor = 17$ weak subkeys are allowed (for a nonempty weak-key class).

Under this estimate, linear relation (14) requires the minimum number of weak subkeys: only seven weak subkeys every three rounds (starting from an odd-numbered round) or eight weak subkeys every three rounds (starting from an even-numbered round). But, it holds with a bias of $2^{-50.88}$. To attack six rounds would require 15 weak subkeys. The data complexity is $c/(2 \cdot (2^{-50.88})^2)^2 = c \cdot 2^{201.52}$ known-plaintexts are needed (for a fixed constant $c \geq 1$). Extending it to seven round, requires 18 weak subkeys, which would imply an empty weak-key class.

Alternatively, linear relation (23) would allow to distinguish up to 4-round MESH-128 from a random permutation with only $c \cdot (2^{-1})^2 = 4c$ known plaintexts. Nonetheless, it would require 16 weak subkeys. The estimated weak-key class size is $2^{256-15 \cdot 16} = 2^{16}$, but the exact user keys that belong to this weak-key class are not known.

This linear analysis also allows us to study the diffusion power of MESH-128 with respect to modern tools such as the linear branch number. See Appendix, Sect. B.

The set of linear relations in Tables 3.103 through 3.110 means that the linear branch number *per round* of MESH-128 is only 4 (under weak-key assumptions). For instance, the 1-round relation $(0,0,0,0,0,0,0,\gamma) \stackrel{1r}{\to}$

$(0, 0, 0, \gamma, 0, \gamma, 0, \gamma)$ has total Hamming Weight 4. So, even though MESH-128 provides full diffusion in a single round, a detailed analysis shows that its linear branch number is much lower than ideal. For a 16-word block, such as used in MESH-128, an MDS code using a 16×16 MDS matrix would rather provide a branch number of $1 + 16 = 17$. See Appendix, Sect. B.1.

3.13.7 LC of MESH-64(8) Under Weak-Key Assumptions

The following linear analysis of MESH-64(8) are novel, that is, they have not been published before, although the analysis are based on techniques already used against other Lai-Massey cipher designs.

The attacks operate under weak-key assumptions.

The strategy is to build linear relations in a bottom-up fashion from the smallest components: $\odot$, $\boxplus$ and $\oplus$, up to an MA-box, up to a half-round, further to a full round, and finally to multiple rounds.

Across unkeyed $\odot$ and $\boxplus$, the propagation of bitmasks follows from Table 3.98 and Table 3.99, respectively.

Further, we obtain all linear relation patterns across the MA-box of MESH-64(8), under the corresponding weak-subkey restrictions. Table 3.111 lists all fifteen nontrivial linear relation patterns across the MA-box of MESH-64(8). Only two 16-bit masks are used in the linear distinguishers: 0 and $\gamma = 1$.

Table 3.111 All wordwise linear relations across the MA-box of MESH-64(8).

Bitmasks across the MA-box	# active $\odot$	Bias	Weak subkey(s)
$(0,0,0,0) \rightarrow (0,0,0,0)$	0	2^{-1}	—
$(0,0,0,\gamma) \rightarrow (0,0,\gamma,0)$	0	2^{-1}	$Z_{10}^{(i)}$
$(0,0,\gamma,0) \rightarrow (0,\gamma,0,\gamma)$	1	$2^{-6.5}$	—
$(0,0,\gamma,\gamma) \rightarrow (0,\gamma,\gamma,\gamma)$	2	2^{-12}	$Z_{10}^{(i)}$
$(0,\gamma,0,0) \rightarrow (\gamma,0,\gamma,0)$	1	$2^{-6.5}$	—
$(0,\gamma,0,\gamma) \rightarrow (\gamma,0,0,0)$	1	$2^{-6.5}$	$Z_{10}^{(i)}$
$(0,\gamma,\gamma,0) \rightarrow (\gamma,\gamma,\gamma,\gamma)$	1	$2^{-6.5}$	—
$(0,\gamma,\gamma,\gamma) \rightarrow (\gamma,\gamma,0,\gamma)$	1	$2^{-6.5}$	$Z_{10}^{(i)}$
$(\gamma,0,0,0) \rightarrow (\gamma,\gamma,0,0)$	0	2^{-1}	$Z_9^{(i)}$
$(\gamma,0,0,\gamma) \rightarrow (\gamma,\gamma,\gamma,0)$	0	2^{-1}	$Z_9^{(i)}, Z_{10}^{(i)}$
$(\gamma,0,\gamma,0) \rightarrow (\gamma,0,0,\gamma)$	2	2^{-12}	$Z_9^{(i)}$
$(\gamma,0,\gamma,\gamma) \rightarrow (\gamma,0,\gamma,\gamma)$	2	2^{-12}	$Z_9^{(i)}, Z_{10}^{(i)}$
$(\gamma,\gamma,0,0) \rightarrow (0,\gamma,\gamma,0)$	1	$2^{-6.5}$	$Z_9^{(i)}$
$(\gamma,\gamma,0,\gamma) \rightarrow (0,\gamma,0,0)$	1	$2^{-6.5}$	$Z_9^{(i)}, Z_{10}^{(i)}$
$(\gamma,\gamma,\gamma,0) \rightarrow (0,0,\gamma,\gamma)$	1	$2^{-6.5}$	$Z_9^{(i)}$
$(\gamma,\gamma,\gamma,\gamma) \rightarrow (0,0,0,\gamma)$	1	$2^{-6.5}$	$Z_9^{(i)}, Z_{10}^{(i)}$

The next step is to obtain from Table 3.111 all 1-round linear relations and the corresponding weak subkeys conditions for the linear relations to hold with highest possible bias.

Tables 3.112 through 3.119 list all 255 nontrivial 1-round linear relations of MESH-64(8), with wordwise bitmasks either 0 or $\gamma = 1$. These tables provide an exhaustive analysis of how linear relations propagate across a single round of MESH-64(8).

The bias was calculated using the Piling-up lemma [131].

Multiple-round linear relations are obtained by concatenating 1-round linear relations such that the output bitmask from one relation matches the input bitmask of the next relation.

From Tables 3.112 and 3.119, the multiple-round linear relations are clustered into 65 relations:

(1) $(0, 0, 0, 0, 0, 0, 0, \gamma) \stackrel{1r}{\to} (\gamma, \gamma, \gamma, \gamma, \gamma, \gamma, \gamma, 0) \stackrel{1r}{\to} (0, 0, 0, 0, 0, 0, 0, \gamma)$, which is 2-round iterative and holds with bias 2^{-12}. Assuming i is odd, the following nine subkeys have to be weak: $Z_8^{(i)}$, $Z_9^{(i)}$, $Z_{10}^{(i)}$, $Z_2^{(i+1)}$, $Z_4^{(i+1)}$, $Z_5^{(i+1)}$, $Z_7^{(i+1)}$, $Z_9^{(i+1)}$ and $Z_{10}^{(i+1)}$. Assuming i is even, the following seven subkeys have to be weak: $Z_9^{(i)}$, $Z_{10}^{(i)}$, $Z_1^{(i+1)}$, $Z_3^{(i+1)}$, $Z_6^{(i+1)}$, $Z_9^{(i+1)}$ and $Z_{10}^{(i+1)}$.

(2) $(0, 0, 0, 0, 0, 0, \gamma, 0) \stackrel{1r}{\to} (0, 0, 0, \gamma, 0, 0, \gamma, \gamma) \stackrel{1r}{\to} (0, 0, 0, \gamma, 0, 0, 0, 0) \stackrel{1r}{\to} (\gamma, \gamma, \gamma, \gamma, \gamma, \gamma, 0, \gamma) \stackrel{1r}{\to} (\gamma, \gamma, \gamma, 0, \gamma, \gamma, 0, 0) \stackrel{1r}{\to} (\gamma, \gamma, \gamma, 0, \gamma, \gamma, \gamma, \gamma) \stackrel{1r}{\to} (0, 0, 0, 0, 0, 0, \gamma, 0)$, which is 6-round iterative and holds with bias 2^{-12}. Assuming i is odd, the following 19 subkeys have to be weak: $Z_{10}^{(i)}$, $Z_4^{(i+1)}$, $Z_7^{(i+1)}$, $Z_{10}^{(i+1)}$, $Z_9^{(i+2)}$, $Z_{10}^{(i+2)}$, $Z_2^{(i+3)}$, $Z_4^{(i+3)}$, $Z_5^{(i+3)}$, $Z_{10}^{(i+3)}$, $Z_1^{(i+4)}$, $Z_3^{(i+4)}$, $Z_6^{(i+4)}$, $Z_{10}^{(i+4)}$, $Z_2^{(i+5)}$, $Z_5^{(i+5)}$, $Z_7^{(i+5)}$, $Z_9^{(i+5)}$ and $Z_{10}^{(i+5)}$. Assuming i is even, the following 21 subkeys have to be weak: $Z_7^{(i)}$, $Z_{10}^{(i)}$, $Z_8^{(i+1)}$, $Z_{10}^{(i+1)}$, $Z_4^{(i+2)}$, $Z_9^{(i+2)}$, $Z_{10}^{(i+2)}$, $Z_1^{(i+3)}$, $Z_3^{(i+3)}$, $Z_6^{(i+3)}$, $Z_8^{(i+3)}$, $Z_{10}^{(i+3)}$, $Z_2^{(i+4)}$, $Z_5^{(i+4)}$, $Z_{10}^{(i+4)}$, $Z_1^{(i+5)}$, $Z_3^{(i+5)}$, $Z_6^{(i+5)}$, $Z_8^{(i+5)}$, $Z_9^{(i+5)}$ and $Z_{10}^{(i+5)}$.

(3) $(0,0,0,0,0,0,\gamma,\gamma) \stackrel{1r}{\to} (\gamma, \gamma, \gamma, 0, \gamma, \gamma, 0, \gamma) \stackrel{1r}{\to} (0,0,0,\gamma,0,0,0,\gamma) \stackrel{1r}{\to} (0, 0, 0, 0, 0, 0, \gamma, \gamma)$, which is 3-round iterative and holds with bias 2^{-12}. Assuming i is odd, the following 6 subkeys have to be weak: $Z_8^{(i)}$, $Z_9^{(i)}$, $Z_2^{(i+1)}$, $Z_5^{(i+1)}$, $Z_9^{(i+1)}$ and $Z_8^{(i+2)}$. Assuming i is even, the following 8 subkeys have to be weak: $Z_7^{(i)}$, $Z_9^{(i)}$, $Z_1^{(i+1)}$, $Z_3^{(i+1)}$, $Z_6^{(i+1)}$, $Z_8^{(i+1)}$, $Z_9^{(i+1)}$ and $Z_4^{(i+2)}$.

(4) $(0,0,0,0,0,\gamma,0,0) \stackrel{1r}{\to} (\gamma,\gamma,0,0,\gamma,0,\gamma,\gamma) \stackrel{1r}{\to} (\gamma, \gamma, 0, 0, \gamma, \gamma, \gamma, 0) \stackrel{1r}{\to} (\gamma, \gamma, \gamma, \gamma, \gamma, 0, \gamma, \gamma) \stackrel{1r}{\to} (0,0,\gamma,\gamma,0,\gamma,0,0) \stackrel{1r}{\to} (0,0,\gamma,\gamma,0,0,0,\gamma) \stackrel{1r}{\to} (0,0,0,0,0,\gamma,0,0)$, which is 6-round iterative and holds with bias 2^{-34}. Assuming i is odd, the following 21 subkeys have to be weak: $Z_6^{(i)}$, $Z_9^{(i)}$, $Z_{10}^{(i)}$, $Z_2^{(i+1)}$, $Z_5^{(i+1)}$, $Z_7^{(i+1)}$, $Z_{10}^{(i+1)}$, $Z_1^{(i+2)}$, $Z_6^{(i+2)}$, $Z_{10}^{(i+2)}$, $Z_2^{(i+3)}$, $Z_4^{(i+3)}$,

Table 3.112 Nontrivial 1-round linear relations for MESH-64(8) (part I).

1-round linear relation	Weak subkeys $Z_j^{(i)}$ (ith round)		Bias
	j value (odd i)	j value (even i)	
$(0,0,0,0,0,0,0,\gamma) \xrightarrow{1r} (\gamma,\gamma,\gamma,\gamma,\gamma,\gamma,\gamma,0)$	8;9;10	9;10	$2^{-6.50}$
$(0,0,0,0,0,0,\gamma,0) \xrightarrow{1r} (0,0,0,\gamma,0,0,\gamma,\gamma)$	10	7;10	2^{-1}
$(0,0,0,0,0,0,\gamma,\gamma) \xrightarrow{1r} (\gamma,\gamma,\gamma,0,\gamma,\gamma,0,\gamma)$	8;9	7;9	$2^{-6.50}$
$(0,0,0,0,0,\gamma,0,0) \xrightarrow{1r} (\gamma,\gamma,0,0,\gamma,0,\gamma,\gamma)$	6;9;10	9;10	$2^{-6.50}$
$(0,0,0,0,0,\gamma,0,\gamma) \xrightarrow{1r} (0,0,\gamma,\gamma,0,\gamma,0,\gamma)$	6;8	—	$2^{-6.50}$
$(0,0,0,0,0,\gamma,\gamma,0) \xrightarrow{1r} (\gamma,\gamma,0,\gamma,\gamma,0,0,0)$	6;9	7;9	$2^{-6.50}$
$(0,0,0,0,0,\gamma,\gamma,\gamma) \xrightarrow{1r} (0,0,\gamma,0,0,\gamma,\gamma,0)$	6;8;10	7;10	2^{-12}
$(0,0,0,0,\gamma,0,0,0) \xrightarrow{1r} (0,\gamma,\gamma,0,\gamma,0,\gamma,\gamma)$	10	5;10	$2^{-6.50}$
$(0,0,0,0,\gamma,0,0,\gamma) \xrightarrow{1r} (\gamma,0,0,\gamma,0,\gamma,0,\gamma)$	8;9	5;9	2^{-12}
$(0,0,0,0,\gamma,0,\gamma,0) \xrightarrow{1r} (0,\gamma,\gamma,\gamma,\gamma,0,0,0)$	—	5;7	$2^{-6.50}$
$(0,0,0,0,\gamma,0,\gamma,\gamma) \xrightarrow{1r} (\gamma,0,0,0,0,\gamma,\gamma,0)$	8;9;10	5;7;9;10	2^{-12}
$(0,0,0,0,\gamma,\gamma,0,0) \xrightarrow{1r} (\gamma,0,\gamma,0,0,0,0,0)$	6;9	5;9	2^{-1}
$(0,0,0,0,\gamma,\gamma,0,\gamma) \xrightarrow{1r} (0,\gamma,0,\gamma,\gamma,\gamma,\gamma,0)$	6;8;10	5;10	$2^{-6.50}$
$(0,0,0,0,\gamma,\gamma,\gamma,0) \xrightarrow{1r} (\gamma,0,\gamma,\gamma,0,0,\gamma,\gamma)$	6;9;10	5;7;9;10	2^{-1}
$(0,0,0,0,\gamma,\gamma,\gamma,\gamma) \xrightarrow{1r} (0,\gamma,0,0,\gamma,\gamma,0,\gamma)$	6;8	5;7	$2^{-6.50}$
$(0,0,0,\gamma,0,0,0,0) \xrightarrow{1r} (\gamma,\gamma,\gamma,\gamma,\gamma,\gamma,0,\gamma)$	9;10	4;9;10	$2^{-6.50}$
$(0,0,0,\gamma,0,0,0,\gamma) \xrightarrow{1r} (0,0,0,0,0,0,\gamma,\gamma)$	8	4	2^{-1}
$(0,0,0,\gamma,0,0,\gamma,0) \xrightarrow{1r} (\gamma,\gamma,\gamma,0,\gamma,\gamma,\gamma,0)$	9	4;7;9	$2^{-6.50}$
$(0,0,0,\gamma,0,0,\gamma,\gamma) \xrightarrow{1r} (0,0,0,\gamma,0,0,0,0)$	8;10	4;7;10	2^{-1}
$(0,0,0,\gamma,0,\gamma,0,0) \xrightarrow{1r} (0,0,\gamma,\gamma,0,\gamma,\gamma,0)$	6	4	$2^{-6.50}$
$(0,0,0,\gamma,0,\gamma,0,\gamma) \xrightarrow{1r} (\gamma,\gamma,0,0,\gamma,0,0,0)$	6;8;9;10	4;9;10	$2^{-6.50}$
$(0,0,0,\gamma,0,\gamma,\gamma,0) \xrightarrow{1r} (0,0,\gamma,0,0,\gamma,0,\gamma)$	6;10	4;7;10	2^{-12}
$(0,0,0,\gamma,0,\gamma,\gamma,\gamma) \xrightarrow{1r} (\gamma,\gamma,0,\gamma,\gamma,0,\gamma,\gamma)$	6;8;9	4;7;9	$2^{-6.50}$
$(0,0,0,\gamma,\gamma,0,0,0) \xrightarrow{1r} (\gamma,0,0,\gamma,0,\gamma,\gamma,0)$	9	4;5;9	2^{-12}
$(0,0,0,\gamma,\gamma,0,0,\gamma) \xrightarrow{1r} (0,\gamma,\gamma,0,\gamma,0,0,0)$	8;10	4;5;10	$2^{-6.50}$
$(0,0,0,\gamma,\gamma,0,\gamma,0) \xrightarrow{1r} (\gamma,0,0,0,0,\gamma,0,\gamma)$	9;10	4;5;7;9;10	2^{-12}
$(0,0,0,\gamma,\gamma,0,\gamma,\gamma) \xrightarrow{1r} (0,\gamma,\gamma,\gamma,\gamma,0,\gamma,\gamma)$	8	4;5;7	$2^{-6.50}$
$(0,0,0,\gamma,\gamma,\gamma,0,0) \xrightarrow{1r} (0,\gamma,0,\gamma,\gamma,\gamma,0,\gamma)$	6;10	4;5;10	$2^{-6.50}$
$(0,0,0,\gamma,\gamma,\gamma,0,\gamma) \xrightarrow{1r} (\gamma,0,\gamma,0,0,0,\gamma,\gamma)$	6;8;9	4;5;9	2^{-1}
$(0,0,0,\gamma,\gamma,\gamma,\gamma,0) \xrightarrow{1r} (0,\gamma,0,0,\gamma,\gamma,\gamma,0)$	6	4;5;7	$2^{-6.50}$
$(0,0,0,\gamma,\gamma,\gamma,\gamma,\gamma) \xrightarrow{1r} (\gamma,0,\gamma,\gamma,0,0,0,0)$	6;8;9;10	4;5;7;9;10	2^{-1}
$(0,0,\gamma,0,0,0,0,0) \xrightarrow{1r} (0,0,0,0,0,\gamma,\gamma,\gamma)$	3;10	10	2^{-1}

Table 3.113 Nontrivial 1-round linear relations for MESH-64(8) (part II).

1-round linear relation	Weak subkeys $Z_j^{(i)}$ (ith round)		Bias
	j value (odd i)	j value (even i)	
$(0,0,\gamma,0,0,0,0,\gamma) \xrightarrow{1r} (\gamma,\gamma,\gamma,\gamma,\gamma,0,0,\gamma)$	3;8;9	9	$2^{-6.50}$
$(0,0,\gamma,0,0,0,\gamma,0) \xrightarrow{1r} (0,0,0,\gamma,0,\gamma,0,0)$	3	7	2^{-1}
$(0,0,\gamma,0,0,0,\gamma,\gamma) \xrightarrow{1r} (\gamma,\gamma,\gamma,0,\gamma,0,\gamma,0)$	3;8;9;10	7;9;10	$2^{-6.50}$
$(0,0,\gamma,0,0,\gamma,0,0) \xrightarrow{1r} (\gamma,\gamma,0,0,\gamma,\gamma,0,0)$	3;6;9	9	$2^{-6.50}$
$(0,0,\gamma,0,0,\gamma,0,\gamma) \xrightarrow{1r} (0,0,\gamma,\gamma,0,0,\gamma,0)$	3;6;8;10	10	2^{-12}
$(0,0,\gamma,0,0,\gamma,\gamma,0) \xrightarrow{1r} (\gamma,\gamma,0,\gamma,\gamma,\gamma,\gamma,\gamma)$	3;6;9;10	7;9;10	$2^{-6.50}$
$(0,0,\gamma,0,0,\gamma,\gamma,\gamma) \xrightarrow{1r} (0,0,\gamma,0,0,0,0,\gamma)$	3;6;8	7	$2^{-6.50}$
$(0,0,\gamma,0,\gamma,0,0,0) \xrightarrow{1r} (0,\gamma,\gamma,0,\gamma,\gamma,0,0)$	3	5	$2^{-6.50}$
$(0,0,\gamma,0,\gamma,0,0,\gamma) \xrightarrow{1r} (\gamma,0,0,\gamma,0,0,\gamma,0)$	3;8;9;10	5;9;10	2^{-12}
$(0,0,\gamma,0,\gamma,0,\gamma,0) \xrightarrow{1r} (0,\gamma,\gamma,\gamma,\gamma,\gamma,\gamma,\gamma)$	3;10	5;7;10	$2^{-6.50}$
$(0,0,\gamma,0,\gamma,0,\gamma,\gamma) \xrightarrow{1r} (\gamma,0,0,0,0,0,0,\gamma)$	3;8;9	5;7;9	2^{-12}
$(0,0,\gamma,0,\gamma,\gamma,0,0) \xrightarrow{1r} (\gamma,0,\gamma,0,0,\gamma,\gamma,\gamma)$	3;6;9;10	5;9;10	2^{-1}
$(0,0,\gamma,0,\gamma,\gamma,0,\gamma) \xrightarrow{1r} (0,\gamma,0,\gamma,\gamma,0,0,\gamma)$	3;6;8	5	$2^{-6.50}$
$(0,0,\gamma,0,\gamma,\gamma,\gamma,0) \xrightarrow{1r} (\gamma,0,\gamma,\gamma,0,\gamma,0,0)$	3;6;9	5;7;9	2^{-1}
$(0,0,\gamma,0,\gamma,\gamma,\gamma,\gamma) \xrightarrow{1r} (0,\gamma,0,0,\gamma,0,\gamma,0)$	3;6;8;10	5;7;10	$2^{-6.50}$
$(0,0,\gamma,\gamma,0,0,0,0) \xrightarrow{1r} (\gamma,\gamma,\gamma,\gamma,\gamma,0,\gamma,0)$	3;9	4;9	$2^{-6.50}$
$(0,0,\gamma,\gamma,0,0,0,\gamma) \xrightarrow{1r} (0,0,0,0,0,\gamma,0,0)$	3;8;10	4;10	2^{-1}
$(0,0,\gamma,\gamma,0,0,\gamma,0) \xrightarrow{1r} (\gamma,\gamma,\gamma,0,\gamma,0,0,\gamma)$	3;9;10	4;7;9;10	$2^{-6.50}$
$(0,0,\gamma,\gamma,0,0,\gamma,\gamma) \xrightarrow{1r} (0,0,0,\gamma,0,\gamma,\gamma,\gamma)$	3;8	4;7	2^{-1}
$(0,0,\gamma,\gamma,0,\gamma,0,0) \xrightarrow{1r} (0,0,\gamma,\gamma,0,0,0,\gamma)$	3;6;10	4;10	2^{-12}
$(0,0,\gamma,\gamma,0,\gamma,0,\gamma) \xrightarrow{1r} (\gamma,\gamma,0,0,\gamma,\gamma,\gamma,\gamma)$	3;6;8;9	4;9	$2^{-6.50}$
$(0,0,\gamma,\gamma,0,\gamma,\gamma,0) \xrightarrow{1r} (0,0,\gamma,0,0,0,\gamma,0)$	3;6	4;7	$2^{-6.50}$
$(0,0,\gamma,\gamma,0,\gamma,\gamma,\gamma) \xrightarrow{1r} (\gamma,\gamma,0,\gamma,\gamma,\gamma,0,0)$	3;6;8;9;10	4;7;9;10	$2^{-6.50}$
$(0,0,\gamma,\gamma,\gamma,0,0,0) \xrightarrow{1r} (\gamma,0,0,\gamma,0,0,0,\gamma)$	3;9;10	4;5;9;10	2^{-12}
$(0,0,\gamma,\gamma,\gamma,0,0,\gamma) \xrightarrow{1r} (0,\gamma,\gamma,0,\gamma,\gamma,\gamma,\gamma)$	3;8	4;5	$2^{-6.50}$
$(0,0,\gamma,\gamma,\gamma,0,\gamma,0) \xrightarrow{1r} (\gamma,0,0,0,0,0,\gamma,0)$	3;9	4;5;7;9	2^{-12}
$(0,0,\gamma,\gamma,\gamma,0,\gamma,\gamma) \xrightarrow{1r} (0,\gamma,\gamma,\gamma,\gamma,\gamma,0,0)$	3;8;10	4;5;7;10	$2^{-6.50}$
$(0,0,\gamma,\gamma,\gamma,\gamma,0,0) \xrightarrow{1r} (0,\gamma,0,\gamma,\gamma,0,\gamma,0)$	3;6	4;5	$2^{-6.50}$
$(0,0,\gamma,\gamma,\gamma,\gamma,0,\gamma) \xrightarrow{1r} (\gamma,0,\gamma,0,0,\gamma,0,0)$	3;6;8;9;10	4;5;9;10	2^{-1}
$(0,0,\gamma,\gamma,\gamma,\gamma,\gamma,0) \xrightarrow{1r} (0,\gamma,0,0,\gamma,0,0,\gamma)$	3;6;10	4;5;7;10	$2^{-6.50}$
$(0,0,\gamma,\gamma,\gamma,\gamma,\gamma,\gamma) \xrightarrow{1r} (\gamma,0,\gamma,\gamma,0,\gamma,\gamma,\gamma)$	3;6;8;9	4;5;7;9	2^{-1}
$(0,\gamma,0,0,0,0,0,0) \xrightarrow{1r} (\gamma,\gamma,\gamma,0,0,0,\gamma,\gamma)$	9;10	2;9;10	$2^{-6.50}$

Table 3.114 Nontrivial 1-round linear relations for MESH-64(8) (part III).

1-round linear relation	Weak subkeys $Z_j^{(i)}$ (ith round) j value (odd i)	j value (even i)	Bias
$(0,\gamma,0,0,0,0,0,\gamma) \stackrel{1r}{\rightarrow} (0,0,0,\gamma,\gamma,\gamma,0,\gamma)$	8	2	$2^{-6.50}$
$(0,\gamma,0,0,0,0,\gamma,0) \stackrel{1r}{\rightarrow} (\gamma,\gamma,\gamma,\gamma,0,0,0,0)$	9	2;7;9	$2^{-6.50}$
$(0,\gamma,0,0,0,0,\gamma,\gamma) \stackrel{1r}{\rightarrow} (0,0,0,0,\gamma,\gamma,\gamma,0)$	8;10	2;7;10	2^{-12}
$(0,\gamma,0,0,0,\gamma,0,0) \stackrel{1r}{\rightarrow} (0,0,\gamma,0,\gamma,0,0,0)$	6	2	2^{-1}
$(0,\gamma,0,0,0,\gamma,0,\gamma) \stackrel{1r}{\rightarrow} (\gamma,\gamma,0,\gamma,0,\gamma,\gamma,0)$	6;8;9;10	2;9;10	$2^{-6.50}$
$(0,\gamma,0,0,0,\gamma,\gamma,0) \stackrel{1r}{\rightarrow} (0,0,\gamma,\gamma,\gamma,0,\gamma,\gamma)$	6;10	2;7;10	2^{-1}
$(0,\gamma,0,0,0,\gamma,\gamma,\gamma) \stackrel{1r}{\rightarrow} (\gamma,\gamma,0,0,0,\gamma,0,\gamma)$	6;8;9	2;7;9	$2^{-6.50}$
$(0,\gamma,0,0,\gamma,0,0,0) \stackrel{1r}{\rightarrow} (\gamma,0,0,0,\gamma,0,0,0)$	9	2;5;9	2^{-1}
$(0,\gamma,0,0,\gamma,0,0,\gamma) \stackrel{1r}{\rightarrow} (0,\gamma,\gamma,\gamma,0,\gamma,\gamma,0)$	8;10	2;5;10	$2^{-6.50}$
$(0,\gamma,0,0,\gamma,0,\gamma,0) \stackrel{1r}{\rightarrow} (\gamma,0,0,\gamma,\gamma,0,\gamma,\gamma)$	9;10	2;5;7;9;10	2^{-1}
$(0,\gamma,0,0,\gamma,0,\gamma,\gamma) \stackrel{1r}{\rightarrow} (0,\gamma,\gamma,0,0,\gamma,0,\gamma)$	8	2;5;7	$2^{-6.50}$
$(0,\gamma,0,0,\gamma,\gamma,0,0) \stackrel{1r}{\rightarrow} (0,\gamma,0,0,0,0,\gamma,\gamma)$	6;10	2;5;10	$2^{-6.50}$
$(0,\gamma,0,0,\gamma,\gamma,0,\gamma) \stackrel{1r}{\rightarrow} (\gamma,0,\gamma,\gamma,\gamma,\gamma,0,\gamma)$	6;8;9	2;5;9	2^{-12}
$(0,\gamma,0,0,\gamma,\gamma,\gamma,0) \stackrel{1r}{\rightarrow} (0,\gamma,0,\gamma,0,0,0,0)$	6	2;5;7	$2^{-6.50}$
$(0,\gamma,0,0,\gamma,\gamma,\gamma,\gamma) \stackrel{1r}{\rightarrow} (\gamma,0,\gamma,0,\gamma,\gamma,\gamma,0)$	6;8;9;10	2;5;7;9;10	2^{-12}
$(0,\gamma,0,\gamma,0,0,0,0) \stackrel{1r}{\rightarrow} (0,0,0,\gamma,\gamma,\gamma,\gamma,0)$	–	2;4	$2^{-6.50}$
$(0,\gamma,0,\gamma,0,0,0,\gamma) \stackrel{1r}{\rightarrow} (\gamma,\gamma,\gamma,0,0,0,0,0)$	8;9;10	2;4;9;10	$2^{-6.50}$
$(0,\gamma,0,\gamma,0,0,\gamma,0) \stackrel{1r}{\rightarrow} (0,0,0,0,\gamma,\gamma,0,\gamma)$	10	2;4;7;10	2^{-12}
$(0,\gamma,0,\gamma,0,0,\gamma,\gamma) \stackrel{1r}{\rightarrow} (\gamma,\gamma,\gamma,\gamma,0,0,\gamma,\gamma)$	8;9	2;4;7;9	$2^{-6.50}$
$(0,\gamma,0,\gamma,0,\gamma,0,0) \stackrel{1r}{\rightarrow} (\gamma,\gamma,0,\gamma,0,\gamma,0,\gamma)$	6;9;10	2;4;9;10	$2^{-6.50}$
$(0,\gamma,0,\gamma,0,\gamma,0,\gamma) \stackrel{1r}{\rightarrow} (0,0,\gamma,0,\gamma,0,\gamma,\gamma)$	6;8	2;4	2^{-1}
$(0,\gamma,0,\gamma,0,\gamma,\gamma,0) \stackrel{1r}{\rightarrow} (\gamma,\gamma,0,0,0,\gamma,\gamma,0)$	6;9	2;4;7;9	$2^{-6.50}$
$(0,\gamma,0,\gamma,0,\gamma,\gamma,\gamma) \stackrel{1r}{\rightarrow} (0,0,\gamma,\gamma,\gamma,0,0,0)$	6;8;10	2;4;7;10	2^{-1}
$(0,\gamma,0,\gamma,\gamma,0,0,0) \stackrel{1r}{\rightarrow} (0,\gamma,\gamma,\gamma,0,\gamma,0,\gamma)$	10	2;4;5;10	$2^{-6.50}$
$(0,\gamma,0,\gamma,\gamma,0,0,\gamma) \stackrel{1r}{\rightarrow} (\gamma,0,0,0,\gamma,0,\gamma,\gamma)$	8;9	2;4;5;9	2^{-1}
$(0,\gamma,0,\gamma,\gamma,0,\gamma,0) \stackrel{1r}{\rightarrow} (0,\gamma,\gamma,0,0,\gamma,\gamma,0)$	—	2;4;5;7	$2^{-6.50}$
$(0,\gamma,0,\gamma,\gamma,0,\gamma,\gamma) \stackrel{1r}{\rightarrow} (\gamma,0,0,\gamma,\gamma,0,0,0)$	8;9;10	2;4;5;7;9;10	2^{-1}
$(0,\gamma,0,\gamma,\gamma,\gamma,0,0) \stackrel{1r}{\rightarrow} (\gamma,0,\gamma,\gamma,\gamma,\gamma,\gamma,0)$	6;9	2;4;5;9	2^{-12}
$(0,\gamma,0,\gamma,\gamma,\gamma,0,\gamma) \stackrel{1r}{\rightarrow} (0,\gamma,0,0,0,0,0,0)$	6;8;10	2;4;5;10	$2^{-6.50}$
$(0,\gamma,0,\gamma,\gamma,\gamma,\gamma,0) \stackrel{1r}{\rightarrow} (\gamma,0,\gamma,0,\gamma,\gamma,0,\gamma)$	6;9;10	2;4;5;7;9;10	2^{-12}
$(0,\gamma,0,\gamma,\gamma,\gamma,\gamma,\gamma) \stackrel{1r}{\rightarrow} (0,\gamma,0,\gamma,0,0,\gamma,\gamma)$	6;8	2;4;5;7	$2^{-6.50}$
$(0,\gamma,\gamma,0,0,0,0,0) \stackrel{1r}{\rightarrow} (\gamma,\gamma,\gamma,0,0,\gamma,0,0)$	3;9	2;9	$2^{-6.50}$

Table 3.115 Nontrivial 1-round linear relations for MESH-64(8) (part IV).

1-round linear relation	Weak subkeys $Z_j^{(i)}$ (ith round)		Bias
	j value (odd i)	j value (even i)	
$(0,\gamma,\gamma,0,0,0,0,\gamma) \xrightarrow{1r} (0,0,0,\gamma,\gamma,0,\gamma,0)$	3;8;10	2;10	2^{-12}
$(0,\gamma,\gamma,0,0,0,\gamma,0) \xrightarrow{1r} (\gamma,\gamma,\gamma,\gamma,0,\gamma,\gamma,\gamma)$	3;9;10	2;7;9;10	$2^{-6.50}$
$(0,\gamma,\gamma,0,0,0,\gamma,\gamma) \xrightarrow{1r} (0,0,0,0,\gamma,0,0,\gamma)$	3;8	2;7	$2^{-6.50}$
$(0,\gamma,\gamma,0,0,\gamma,0,0) \xrightarrow{1r} (0,0,\gamma,0,\gamma,\gamma,\gamma,\gamma)$	3;6;10	2;10	2^{-1}
$(0,\gamma,\gamma,0,0,\gamma,0,\gamma) \xrightarrow{1r} (\gamma,\gamma,0,\gamma,0,0,0,\gamma)$	3;6;8;9	2;9	$2^{-6.50}$
$(0,\gamma,\gamma,0,0,\gamma,\gamma,0) \xrightarrow{1r} (0,0,\gamma,\gamma,\gamma,\gamma,0,0)$	3;6	2;7	2^{-1}
$(0,\gamma,\gamma,0,0,\gamma,\gamma,\gamma) \xrightarrow{1r} (\gamma,\gamma,0,0,0,0,\gamma,0)$	3;6;8;9;10	2;7;9;10	$2^{-6.50}$
$(0,\gamma,\gamma,0,\gamma,0,0,0) \xrightarrow{1r} (\gamma,0,0,0,\gamma,\gamma,\gamma,\gamma)$	3;9;10	2;5;9;10	2^{-1}
$(0,\gamma,\gamma,0,\gamma,0,0,\gamma) \xrightarrow{1r} (0,\gamma,\gamma,\gamma,0,0,0,\gamma)$	3;8;	2;5;	$2^{-6.50}$
$(0,\gamma,\gamma,0,\gamma,0,\gamma,0) \xrightarrow{1r} (\gamma,0,0,\gamma,\gamma,\gamma,0,0)$	3;9	2;5;7;9	2^{-1}
$(0,\gamma,\gamma,0,\gamma,0,\gamma,\gamma) \xrightarrow{1r} (0,\gamma,\gamma,0,0,0,\gamma,0)$	3;8;10	2;5;7;10	$2^{-6.50}$
$(0,\gamma,\gamma,0,\gamma,\gamma,0,0) \xrightarrow{1r} (0,\gamma,0,0,0,\gamma,0,0)$	3;6	2;5	$2^{-6.50}$
$(0,\gamma,\gamma,0,\gamma,\gamma,0,\gamma) \xrightarrow{1r} (\gamma,0,\gamma,\gamma,\gamma,0,\gamma,0)$	3;6;8;9;10	2;5;9;10	2^{-12}
$(0,\gamma,\gamma,0,\gamma,\gamma,\gamma,0) \xrightarrow{1r} (0,\gamma,0,\gamma,0,\gamma,\gamma,\gamma)$	3;6;10	2;5;7;10	$2^{-6.50}$
$(0,\gamma,\gamma,0,\gamma,\gamma,\gamma,\gamma) \xrightarrow{1r} (\gamma,0,\gamma,0,\gamma,0,0,\gamma)$	3;6;8;9	2;5;7;9	2^{-12}
$(0,\gamma,\gamma,\gamma,0,0,0,0) \xrightarrow{1r} (0,0,0,\gamma,\gamma,0,0,\gamma)$	3;10	2;4;10	2^{-12}
$(0,\gamma,\gamma,\gamma,0,0,0,\gamma) \xrightarrow{1r} (\gamma,\gamma,\gamma,0,0,\gamma,\gamma,\gamma)$	3;8;9	2;4;9	$2^{-6.50}$
$(0,\gamma,\gamma,\gamma,0,0,\gamma,0) \xrightarrow{1r} (0,0,0,0,\gamma,0,\gamma,0)$	3	2;4;7	$2^{-6.50}$
$(0,\gamma,\gamma,\gamma,0,0,\gamma,\gamma) \xrightarrow{1r} (\gamma,\gamma,\gamma,\gamma,0,\gamma,0,0)$	3;8;9;10	2;4;7;9;10	$2^{-6.50}$
$(0,\gamma,\gamma,\gamma,0,\gamma,0,0) \xrightarrow{1r} (\gamma,\gamma,0,\gamma,0,0,\gamma,0)$	3;6;9	2;4;9	$2^{-6.50}$
$(0,\gamma,\gamma,\gamma,0,\gamma,0,\gamma) \xrightarrow{1r} (0,0,\gamma,0,\gamma,\gamma,0,0)$	3;6;8;10	2;4;10	2^{-1}
$(0,\gamma,\gamma,\gamma,0,\gamma,\gamma,0) \xrightarrow{1r} (\gamma,\gamma,0,0,0,0,0,\gamma)$	3;6;9;10	2;4;7;9;10	$2^{-6.50}$
$(0,\gamma,\gamma,\gamma,0,\gamma,\gamma,\gamma) \xrightarrow{1r} (0,0,\gamma,\gamma,\gamma,\gamma,\gamma,\gamma)$	3;6;8	2;4;7	2^{-1}
$(0,\gamma,\gamma,\gamma,\gamma,0,0,0) \xrightarrow{1r} (0,\gamma,\gamma,\gamma,0,0,\gamma,0)$	3	2;4;5	$2^{-6.50}$
$(0,\gamma,\gamma,\gamma,\gamma,0,0,\gamma) \xrightarrow{1r} (\gamma,0,0,0,\gamma,\gamma,0,0)$	3;8;9;10	2;4;5;9;10	2^{-1}
$(0,\gamma,\gamma,\gamma,\gamma,0,\gamma,0) \xrightarrow{1r} (0,\gamma,\gamma,0,0,0,0,\gamma)$	3;10	2;4;5;7;10	$2^{-6.50}$
$(0,\gamma,\gamma,\gamma,\gamma,0,\gamma,\gamma) \xrightarrow{1r} (\gamma,0,0,\gamma,\gamma,\gamma,\gamma,\gamma)$	3;8;9	2;4;5;7;9	2^{-1}
$(0,\gamma,\gamma,\gamma,\gamma,\gamma,0,0) \xrightarrow{1r} (\gamma,0,\gamma,\gamma,\gamma,0,0,\gamma)$	3;6;9;10	2;4;5;9;10	2^{-12}
$(0,\gamma,\gamma,\gamma,\gamma,\gamma,0,\gamma) \xrightarrow{1r} (0,\gamma,0,0,0,\gamma,\gamma,\gamma)$	3;6;8	2;4;5	$2^{-6.50}$
$(0,\gamma,\gamma,\gamma,\gamma,\gamma,\gamma,0) \xrightarrow{1r} (\gamma,0,\gamma,0,\gamma,0,\gamma,0)$	3;6;9	2;4;5;7;9	2^{-12}
$(0,\gamma,\gamma,\gamma,\gamma,\gamma,\gamma,\gamma) \xrightarrow{1r} (0,\gamma,0,\gamma,0,\gamma,0,0)$	3;6;8;10	2;4;5;7;10	$2^{-6.50}$
$(\gamma,0,0,0,0,0,0,0) \xrightarrow{1r} (\gamma,0,\gamma,0,\gamma,0,\gamma,\gamma)$	1;10	10	$2^{-6.50}$

Table 3.116 Nontrivial 1-round linear relations for MESH-64(8) (part V).

1-round linear relation	Weak subkeys $Z_j^{(i)}$ (ith round) j value (odd i)	Weak subkeys $Z_j^{(i)}$ (ith round) j value (even i)	Bias
$(\gamma,0,0,0,0,0,0,\gamma) \xrightarrow{1r} (0,\gamma,0,\gamma,0,\gamma,0,\gamma)$	1;8;9	9	2^{-12}
$(\gamma,0,0,0,0,0,\gamma,0) \xrightarrow{1r} (\gamma,0,\gamma,\gamma,\gamma,0,0,0)$	1	7	$2^{-6.50}$
$(\gamma,0,0,0,0,0,\gamma,\gamma) \xrightarrow{1r} (0,\gamma,0,0,0,\gamma,\gamma,0)$	1;8;9;10	7;9;10	2^{-12}
$(\gamma,0,0,0,0,\gamma,0,0) \xrightarrow{1r} (0,\gamma,\gamma,0,0,0,0,0)$	1;6;9	9	2^{-1}
$(\gamma,0,0,0,0,\gamma,0,\gamma) \xrightarrow{1r} (\gamma,0,0,\gamma,\gamma,\gamma,\gamma,0)$	1;6;8;10	10	$2^{-6.50}$
$(\gamma,0,0,0,0,\gamma,\gamma,0) \xrightarrow{1r} (0,\gamma,\gamma,\gamma,0,0,\gamma,\gamma)$	1;6;9;10	7;9;10	2^{-1}
$(\gamma,0,0,0,0,\gamma,\gamma,\gamma) \xrightarrow{1r} (\gamma,0,0,0,\gamma,\gamma,0,\gamma)$	1;6;8	7	$2^{-6.50}$
$(\gamma,0,0,0,\gamma,0,0,0) \xrightarrow{1r} (\gamma,\gamma,0,0,0,0,0,0)$	1	5	2^{-1}
$(\gamma,0,0,0,\gamma,0,0,\gamma) \xrightarrow{1r} (0,0,\gamma,\gamma,\gamma,\gamma,\gamma,0)$	1;8;9;10	5;9;10	$2^{-6.50}$
$(\gamma,0,0,0,\gamma,0,\gamma,0) \xrightarrow{1r} (\gamma,\gamma,0,\gamma,0,0,\gamma,\gamma)$	1;10	5;7;10	2^{-1}
$(\gamma,0,0,0,\gamma,0,\gamma,\gamma) \xrightarrow{1r} (0,0,\gamma,0,\gamma,\gamma,0,\gamma)$	1;8;9	5;7;9	$2^{-6.50}$
$(\gamma,0,0,0,\gamma,\gamma,0,0) \xrightarrow{1r} (0,0,0,0,\gamma,0,\gamma,\gamma)$	1;6;9;10	5;9;10	$2^{-6.50}$
$(\gamma,0,0,0,\gamma,\gamma,0,\gamma) \xrightarrow{1r} (\gamma,\gamma,\gamma,\gamma,0,\gamma,0,\gamma)$	1;6;8	5	$2^{-6.50}$
$(\gamma,0,0,0,\gamma,\gamma,\gamma,0) \xrightarrow{1r} (0,0,0,\gamma,\gamma,0,0,0)$	1;6;9	5;7;9	$2^{-6.50}$
$(\gamma,0,0,0,\gamma,\gamma,\gamma,\gamma) \xrightarrow{1r} (\gamma,\gamma,\gamma,0,0,\gamma,\gamma,0)$	1;6;8;10	5;7;10	2^{-12}
$(\gamma,0,0,\gamma,0,0,0,0) \xrightarrow{1r} (0,\gamma,0,\gamma,0,\gamma,\gamma,0)$	1;9	4;9	2^{-12}
$(\gamma,0,0,\gamma,0,0,0,\gamma) \xrightarrow{1r} (\gamma,0,\gamma,0,\gamma,0,0,0)$	1;8;10	4;10	$2^{-6.50}$
$(\gamma,0,0,\gamma,0,0,\gamma,0) \xrightarrow{1r} (0,\gamma,0,0,0,\gamma,0,\gamma)$	1;9;10	4;7;9;10	2^{-12}
$(\gamma,0,0,\gamma,0,0,\gamma,\gamma) \xrightarrow{1r} (\gamma,0,\gamma,\gamma,\gamma,0,\gamma,\gamma)$	1;8	4;7	$2^{-6.50}$
$(\gamma,0,0,\gamma,0,\gamma,0,0) \xrightarrow{1r} (\gamma,0,0,\gamma,\gamma,\gamma,0,\gamma)$	1;6;10	4;10	$2^{-6.50}$
$(\gamma,0,0,\gamma,0,\gamma,0,\gamma) \xrightarrow{1r} (0,\gamma,\gamma,0,0,0,\gamma,\gamma)$	1;6;8;9	4;9	2^{-1}
$(\gamma,0,0,\gamma,0,\gamma,\gamma,0) \xrightarrow{1r} (\gamma,0,0,0,\gamma,\gamma,\gamma,0)$	1;6	4;7	$2^{-6.50}$
$(\gamma,0,0,\gamma,0,\gamma,\gamma,\gamma) \xrightarrow{1r} (0,\gamma,\gamma,\gamma,0,0,0,0)$	1;6;8;9;10	4;7;9;10	2^{-1}
$(\gamma,0,0,\gamma,\gamma,0,0,0) \xrightarrow{1r} (0,0,\gamma,\gamma,\gamma,\gamma,0,\gamma)$	1;9;10	4;5;9;10	$2^{-6.50}$
$(\gamma,0,0,\gamma,\gamma,0,0,\gamma) \xrightarrow{1r} (\gamma,\gamma,0,0,0,0,\gamma,\gamma)$	1;8	4;5	2^{-1}
$(\gamma,0,0,\gamma,\gamma,0,\gamma,0) \xrightarrow{1r} (0,0,\gamma,0,\gamma,\gamma,\gamma,0)$	1;9	4;5;7;9	$2^{-6.50}$
$(\gamma,0,0,\gamma,\gamma,0,\gamma,\gamma) \xrightarrow{1r} (\gamma,\gamma,0,\gamma,0,0,0,0)$	1;8;10	4;5;7;10	2^{-1}
$(\gamma,0,0,\gamma,\gamma,\gamma,0,0) \xrightarrow{1r} (\gamma,\gamma,\gamma,\gamma,0,\gamma,\gamma,0)$	1;6	4;5	$2^{-6.50}$
$(\gamma,0,0,\gamma,\gamma,\gamma,0,\gamma) \xrightarrow{1r} (0,0,0,0,\gamma,0,0,0)$	1;6;8;9;10	4;5;9;10	$2^{-6.50}$
$(\gamma,0,0,\gamma,\gamma,\gamma,\gamma,0) \xrightarrow{1r} (\gamma,\gamma,\gamma,0,0,\gamma,0,\gamma)$	1;6;10	4;5;7;10	2^{-12}
$(\gamma,0,0,\gamma,\gamma,\gamma,\gamma,\gamma) \xrightarrow{1r} (0,0,0,\gamma,\gamma,0,\gamma,\gamma)$	1;6;8;9	4;5;7;9	$2^{-6.50}$
$(\gamma,0,\gamma,0,0,0,0,0) \xrightarrow{1r} (\gamma,0,\gamma,0,\gamma,\gamma,0,0)$	1;3	—	$2^{-6.50}$

Table 3.117 Nontrivial 1-round linear relations for MESH-64(8) (part VI).

1-round linear relation	Weak subkeys $Z_j^{(i)}$ (ith round)		Bias
	j value (odd i)	j value (even i)	
$(\gamma,0,\gamma,0,0,0,0,\gamma) \stackrel{1r}{\to} (0,\gamma,0,\gamma,0,0,\gamma,0)$	1;3;8;9; 10	9; 10	2^{-12}
$(\gamma,0,\gamma,0,0,0,\gamma,0) \stackrel{1r}{\to} (\gamma,0,\gamma,\gamma,\gamma,\gamma,\gamma,\gamma)$	1;3;10	7;10	$2^{-6.50}$
$(\gamma,0,\gamma,0,0,0,\gamma,\gamma) \stackrel{1r}{\to} (0,\gamma,0,0,0,0,0,\gamma)$	1;3;8;9	7;9	2^{-12}
$(\gamma,0,\gamma,0,0,\gamma,0,0) \stackrel{1r}{\to} (0,\gamma,\gamma,0,0,\gamma,\gamma,\gamma)$	1;3;6;9; 10	9; 10	2^{-1}
$(\gamma,0,\gamma,0,0,\gamma,0,\gamma) \stackrel{1r}{\to} (\gamma,0,0,\gamma,\gamma,0,0,\gamma)$	1;3;6;8	—	$2^{-6.50}$
$(\gamma,0,\gamma,0,0,\gamma,\gamma,0) \stackrel{1r}{\to} (0,\gamma,\gamma,\gamma,0,\gamma,0,0)$	1;3;6;9	7;9	2^{-1}
$(\gamma,0,\gamma,0,0,\gamma,\gamma,\gamma) \stackrel{1r}{\to} (\gamma,0,0,0,\gamma,0,\gamma,0)$	1;3;6;8;10	7;10	$2^{-6.50}$
$(\gamma,0,\gamma,0,\gamma,0,0,0) \stackrel{1r}{\to} (\gamma,\gamma,0,0,0,\gamma,\gamma,\gamma)$	1;3;10	5;10	2^{-1}
$(\gamma,0,\gamma,0,\gamma,0,0,\gamma) \stackrel{1r}{\to} (0,0,\gamma,\gamma,\gamma,0,0,\gamma)$	1;3;8;9	5;9	$2^{-6.50}$
$(\gamma,0,\gamma,0,\gamma,0,\gamma,0) \stackrel{1r}{\to} (\gamma,\gamma,0,\gamma,0,\gamma,0,0)$	1;3	5;7	2^{-1}
$(\gamma,0,\gamma,0,\gamma,0,\gamma,\gamma) \stackrel{1r}{\to} (0,0,\gamma,0,\gamma,0,\gamma,0)$	1;3;8;9;10	5;7;9;10	$2^{-6.50}$
$(\gamma,0,\gamma,0,\gamma,\gamma,0,0) \stackrel{1r}{\to} (0,0,0,0,\gamma,\gamma,0,0)$	1;3;6;9	5;9	$2^{-6.50}$
$(\gamma,0,\gamma,0,\gamma,\gamma,0,\gamma) \stackrel{1r}{\to} (\gamma,\gamma,\gamma,\gamma,0,0,\gamma,0)$	1;3;6;8;10	5;10	2^{-12}
$(\gamma,0,\gamma,0,\gamma,\gamma,\gamma,0) \stackrel{1r}{\to} (0,0,0,\gamma,\gamma,\gamma,\gamma,\gamma)$	1;3;6;9;10	5;7;9;10	$2^{-6.50}$
$(\gamma,0,\gamma,0,\gamma,\gamma,\gamma,\gamma) \stackrel{1r}{\to} (\gamma,\gamma,\gamma,0,0,0,0,\gamma)$	1;3;6;8	5;7	$2^{-6.50}$
$(\gamma,0,\gamma,\gamma,0,0,0,0) \stackrel{1r}{\to} (0,\gamma,0,\gamma,0,0,0,\gamma)$	1;3;9;10	4;9;10	2^{-12}
$(\gamma,0,\gamma,\gamma,0,0,0,\gamma) \stackrel{1r}{\to} (\gamma,0,\gamma,0,\gamma,\gamma,\gamma,\gamma)$	1;3;8	4	$2^{-6.50}$
$(\gamma,0,\gamma,\gamma,0,0,\gamma,0) \stackrel{1r}{\to} (0,\gamma,0,0,0,0,\gamma,0)$	1;3;9	4;7;9	2^{-12}
$(\gamma,0,\gamma,\gamma,0,0,\gamma,\gamma) \stackrel{1r}{\to} (\gamma,0,\gamma,\gamma,\gamma,\gamma,0,0)$	1;3;8;10	4;7;10	$2^{-6.50}$
$(\gamma,0,\gamma,\gamma,0,\gamma,0,0) \stackrel{1r}{\to} (\gamma,0,0,\gamma,\gamma,0,\gamma,0)$	1;3;6	4	$2^{-6.50}$
$(\gamma,0,\gamma,\gamma,0,\gamma,0,\gamma) \stackrel{1r}{\to} (0,\gamma,\gamma,0,0,\gamma,0,0)$	1;3;6;8;9;10	4;9;10	2^{-1}
$(\gamma,0,\gamma,\gamma,0,\gamma,\gamma,0) \stackrel{1r}{\to} (\gamma,0,0,0,\gamma,0,0,\gamma)$	1;3;6;10	4;7;10	$2^{-6.50}$
$(\gamma,0,\gamma,\gamma,0,\gamma,\gamma,\gamma) \stackrel{1r}{\to} (0,\gamma,\gamma,\gamma,0,\gamma,\gamma,\gamma)$	1;3;6;8;9	4;7;9	2^{-1}
$(\gamma,0,\gamma,\gamma,\gamma,0,0,0) \stackrel{1r}{\to} (0,0,\gamma,\gamma,\gamma,0,\gamma,0)$	1;3;9	4;5;9	$2^{-6.50}$
$(\gamma,0,\gamma,\gamma,\gamma,0,0,\gamma) \stackrel{1r}{\to} (\gamma,\gamma,0,0,0,\gamma,0,0)$	1;3;8;10	4;5;10	2^{-1}
$(\gamma,0,\gamma,\gamma,\gamma,0,\gamma,0) \stackrel{1r}{\to} (0,0,\gamma,0,\gamma,0,0,\gamma)$	1;3;9;10	4;5;7;9;10	$2^{-6.50}$
$(\gamma,0,\gamma,\gamma,\gamma,0,\gamma,\gamma) \stackrel{1r}{\to} (\gamma,\gamma,0,\gamma,0,\gamma,\gamma,\gamma)$	1;3;8	4;5;7	2^{-1}
$(\gamma,0,\gamma,\gamma,\gamma,\gamma,0,0) \stackrel{1r}{\to} (\gamma,\gamma,\gamma,\gamma,0,0,0,\gamma)$	1;3;6;10	4;5;10	2^{-12}
$(\gamma,0,\gamma,\gamma,\gamma,\gamma,0,\gamma) \stackrel{1r}{\to} (0,0,0,0,\gamma,\gamma,\gamma,\gamma)$	1;3;6;8;9	4;5;9	$2^{-6.50}$
$(\gamma,0,\gamma,\gamma,\gamma,\gamma,\gamma,0) \stackrel{1r}{\to} (\gamma,\gamma,\gamma,0,0,0,\gamma,0)$	1;3;6	4;5;7	$2^{-6.50}$
$(\gamma,0,\gamma,\gamma,\gamma,\gamma,\gamma,\gamma) \stackrel{1r}{\to} (0,0,0,\gamma,\gamma,\gamma,0,0)$	1;3;6;8;9;10	4;5;7;9;10	$2^{-6.50}$
$(\gamma,\gamma,0,0,0,0,0,0) \stackrel{1r}{\to} (0,\gamma,0,0,\gamma,0,0,0)$	1;9	2;9	2^{-1}

Table 3.118 Nontrivial 1-round linear relations for MESH-64(8) (part VII).

1-round linear relation	Weak subkeys $Z_j^{(i)}$ (ith round)		Bias
	j value (odd i)	j value (even i)	
$(\gamma,\gamma,0,0,0,0,0,\gamma) \stackrel{1r}{\rightarrow} (\gamma,0,\gamma,\gamma,0,\gamma,\gamma,0)$	1;8;10	2;10	$2^{-6.50}$
$(\gamma,\gamma,0,0,0,0,\gamma,0) \stackrel{1r}{\rightarrow} (0,\gamma,0,\gamma,\gamma,0,\gamma,\gamma)$	1;9;10	2;7;9;10	2^{-1}
$(\gamma,\gamma,0,0,0,0,\gamma,\gamma) \stackrel{1r}{\rightarrow} (\gamma,0,\gamma,0,0,\gamma,0,\gamma)$	1;8	2;7	$2^{-6.50}$
$(\gamma,\gamma,0,0,0,\gamma,0,0) \stackrel{1r}{\rightarrow} (\gamma,0,0,0,0,0,\gamma,\gamma)$	1;6;10	2;10	$2^{-6.50}$
$(\gamma,\gamma,0,0,0,\gamma,0,\gamma) \stackrel{1r}{\rightarrow} (0,\gamma,\gamma,\gamma,\gamma,\gamma,0,\gamma)$	1;6;8;9	2;9	2^{-12}
$(\gamma,\gamma,0,0,0,\gamma,\gamma,0) \stackrel{1r}{\rightarrow} (\gamma,0,0,\gamma,0,0,0,0)$	1;6	2;7	$2^{-6.50}$
$(\gamma,\gamma,0,0,0,\gamma,\gamma,\gamma) \stackrel{1r}{\rightarrow} (0,\gamma,\gamma,0,\gamma,\gamma,\gamma,0)$	1;6;8;9;10	2;7;9;10	2^{-12}
$(\gamma,\gamma,0,0,\gamma,0,0,0) \stackrel{1r}{\rightarrow} (0,0,\gamma,0,0,0,\gamma,\gamma)$	1;9;10	2;5;9;10	$2^{-6.50}$
$(\gamma,\gamma,0,0,\gamma,0,0,\gamma) \stackrel{1r}{\rightarrow} (\gamma,\gamma,0,\gamma,\gamma,\gamma,0,\gamma)$	1;8	2;5	$2^{-6.50}$
$(\gamma,\gamma,0,0,\gamma,0,\gamma,0) \stackrel{1r}{\rightarrow} (0,0,\gamma,\gamma,0,0,0,0)$	1;9	2;5;7;9	$2^{-6.50}$
$(\gamma,\gamma,0,0,\gamma,0,\gamma,\gamma) \stackrel{1r}{\rightarrow} (\gamma,\gamma,0,0,\gamma,\gamma,\gamma,0)$	1;8;10	2;5;7;10	2^{-12}
$(\gamma,\gamma,0,0,\gamma,\gamma,0,0) \stackrel{1r}{\rightarrow} (\gamma,\gamma,\gamma,0,\gamma,0,0,0)$	1;6	2;5	2^{-1}
$(\gamma,\gamma,0,0,\gamma,\gamma,0,\gamma) \stackrel{1r}{\rightarrow} (0,0,0,\gamma,0,\gamma,\gamma,0)$	1;6;8;9; 10	2;5;9; 10	$2^{-6.50}$
$(\gamma,\gamma,0,0,\gamma,\gamma,\gamma,0) \stackrel{1r}{\rightarrow} (\gamma,\gamma,\gamma,\gamma,\gamma,0,\gamma,\gamma)$	1;6;10	2;5;7;10	2^{-1}
$(\gamma,\gamma,0,0,\gamma,\gamma,\gamma,\gamma) \stackrel{1r}{\rightarrow} (0,0,0,0,0,\gamma,0,\gamma)$	1;6;8;9	2;5;7;9	$2^{-6.50}$
$(\gamma,\gamma,0,\gamma,0,0,0,0) \stackrel{1r}{\rightarrow} (\gamma,0,\gamma,\gamma,0,\gamma,0,\gamma)$	1;10	2;4;10	$2^{-6.50}$
$(\gamma,\gamma,0,\gamma,0,0,0,\gamma) \stackrel{1r}{\rightarrow} (0,\gamma,0,0,\gamma,0,\gamma,\gamma)$	1;8;9	2;4;9	2^{-1}
$(\gamma,\gamma,0,\gamma,0,0,\gamma,0) \stackrel{1r}{\rightarrow} (\gamma,0,\gamma,0,0,\gamma,\gamma,0)$	1	2;4;7	$2^{-6.50}$
$(\gamma,\gamma,0,\gamma,0,0,\gamma,\gamma) \stackrel{1r}{\rightarrow} (0,\gamma,0,\gamma,\gamma,0,0,0)$	1;8;9;10	2;4;7;9;10	2^{-1}
$(\gamma,\gamma,0,\gamma,0,\gamma,0,0) \stackrel{1r}{\rightarrow} (0,\gamma,\gamma,\gamma,\gamma,\gamma,\gamma,0)$	1;6;9	2;4;9	2^{-12}
$(\gamma,\gamma,0,\gamma,0,\gamma,0,\gamma) \stackrel{1r}{\rightarrow} (\gamma,0,0,0,0,0,0,0)$	1;6;8;10	2;4;10	$2^{-6.50}$
$(\gamma,\gamma,0,\gamma,0,\gamma,\gamma,0) \stackrel{1r}{\rightarrow} (0,\gamma,\gamma,0,\gamma,\gamma,0,\gamma)$	1;6;9;10	2;4;7;9;10	2^{-12}
$(\gamma,\gamma,0,\gamma,0,\gamma,\gamma,\gamma) \stackrel{1r}{\rightarrow} (\gamma,0,0,\gamma,0,0,\gamma,\gamma)$	1;6;8	2;4;7	$2^{-6.50}$
$(\gamma,\gamma,0,\gamma,\gamma,0,0,0) \stackrel{1r}{\rightarrow} (\gamma,\gamma,0,\gamma,\gamma,\gamma,\gamma,0)$	1	2;4;5	$2^{-6.50}$
$(\gamma,\gamma,0,\gamma,\gamma,0,0,\gamma) \stackrel{1r}{\rightarrow} (0,0,\gamma,0,0,0,0,0)$	1;8;9;10	2;4;5;9;10	$2^{-6.50}$
$(\gamma,\gamma,0,\gamma,\gamma,0,\gamma,0) \stackrel{1r}{\rightarrow} (\gamma,\gamma,0,0,\gamma,\gamma,0,\gamma)$	1;10	2;4;5;7;10	2^{-12}
$(\gamma,\gamma,0,\gamma,\gamma,0,\gamma,\gamma) \stackrel{1r}{\rightarrow} (0,0,\gamma,\gamma,0,0,\gamma,\gamma)$	1;8;9	2;4;5;7;9	$2^{-6.50}$
$(\gamma,\gamma,0,\gamma,\gamma,\gamma,0,0) \stackrel{1r}{\rightarrow} (0,0,0,\gamma,0,\gamma,0,\gamma)$	1;6;9;10	2;4;5;9;10	$2^{-6.50}$
$(\gamma,\gamma,0,\gamma,\gamma,\gamma,0,\gamma) \stackrel{1r}{\rightarrow} (\gamma,\gamma,\gamma,0,\gamma,0,\gamma,\gamma)$	1;6;8	2;4;5	2^{-1}
$(\gamma,\gamma,0,\gamma,\gamma,\gamma,\gamma,0) \stackrel{1r}{\rightarrow} (0,0,0,0,0,\gamma,\gamma,0)$	1;6;9	2;4;5;7;9	$2^{-6.50}$
$(\gamma,\gamma,0,\gamma,\gamma,\gamma,\gamma,\gamma) \stackrel{1r}{\rightarrow} (\gamma,\gamma,\gamma,\gamma,\gamma,0,0,0)$	1;6;8;10	2;4;5;7;10	2^{-1}
$(\gamma,\gamma,\gamma,0,0,0,0,0) \stackrel{1r}{\rightarrow} (0,\gamma,0,0,\gamma,\gamma,\gamma,\gamma)$	1;3;9;10	2;9;10	2^{-1}

Table 3.119 Nontrivial 1-round linear relations for MESH-64(8) (part VIII).

1-round linear relation	Weak subkeys $Z_j^{(i)}$ (*i*th round) *j* value (odd *i*)	*j* value (even *i*)	Bias
$(\gamma,\gamma,\gamma,0,0,0,0,\gamma) \stackrel{1r}{\to} (\gamma,0,\gamma,\gamma,0,0,0,\gamma)$	1;3;8	2	$2^{-6.50}$
$(\gamma,\gamma,\gamma,0,0,0,\gamma,0) \stackrel{1r}{\to} (0,\gamma,0,\gamma,\gamma,\gamma,0,0)$	1;3;9	2;7;9	2^{-1}
$(\gamma,\gamma,\gamma,0,0,0,\gamma,\gamma) \stackrel{1r}{\to} (\gamma,0,\gamma,0,0,0,\gamma,0)$	1;3;8;10	2;7;10	$2^{-6.50}$
$(\gamma,\gamma,\gamma,0,0,\gamma,0,0) \stackrel{1r}{\to} (\gamma,0,0,0,0,\gamma,0,0)$	1;3;6	2	$2^{-6.50}$
$(\gamma,\gamma,\gamma,0,0,\gamma,0,\gamma) \stackrel{1r}{\to} (0,\gamma,\gamma,\gamma,\gamma,0,\gamma,0)$	1;3;6;8;9;10	2;9;10	2^{-12}
$(\gamma,\gamma,\gamma,0,0,\gamma,\gamma,0) \stackrel{1r}{\to} (\gamma,0,0,\gamma,0,\gamma,\gamma,\gamma)$	1;3;6;10	2;7;10	$2^{-6.50}$
$(\gamma,\gamma,\gamma,0,0,\gamma,\gamma,\gamma) \stackrel{1r}{\to} (0,\gamma,\gamma,0,\gamma,0,0,\gamma)$	1;3;6;8;9	2;7;9	2^{-12}
$(\gamma,\gamma,\gamma,0,\gamma,0,0,0) \stackrel{1r}{\to} (0,0,\gamma,0,0,\gamma,0,0)$	1;3;9	2;5;9	$2^{-6.50}$
$(\gamma,\gamma,\gamma,0,\gamma,0,0,\gamma) \stackrel{1r}{\to} (\gamma,\gamma,0,\gamma,\gamma,0,\gamma,0)$	1;3;8;10	2;5;10	2^{-12}
$(\gamma,\gamma,\gamma,0,\gamma,0,\gamma,0) \stackrel{1r}{\to} (0,0,\gamma,\gamma,0,\gamma,\gamma,\gamma)$	1;3;9;10	2;5;7;9;10	$2^{-6.50}$
$(\gamma,\gamma,\gamma,0,\gamma,0,\gamma,\gamma) \stackrel{1r}{\to} (\gamma,\gamma,0,0,\gamma,0,0,\gamma)$	1;3;8	2;5;7	$2^{-6.50}$
$(\gamma,\gamma,\gamma,0,\gamma,\gamma,0,0) \stackrel{1r}{\to} (\gamma,\gamma,\gamma,0,\gamma,\gamma,\gamma,\gamma)$	1;3;6;10	2;5;10	2^{-1}
$(\gamma,\gamma,\gamma,0,\gamma,\gamma,0,\gamma) \stackrel{1r}{\to} (0,0,0,\gamma,0,0,0,\gamma)$	1;3;6;8;9	2;5;9	$2^{-6.50}$
$(\gamma,\gamma,\gamma,0,\gamma,\gamma,\gamma,0) \stackrel{1r}{\to} (\gamma,\gamma,\gamma,\gamma,\gamma,\gamma,0,0)$	1;3;6	2;5;7	2^{-1}
$(\gamma,\gamma,\gamma,0,\gamma,\gamma,\gamma,\gamma) \stackrel{1r}{\to} (0,0,0,0,0,0,\gamma,0)$	1;3;6;8;9;10	2;5;7;9;10	$2^{-6.50}$
$(\gamma,\gamma,\gamma,\gamma,0,0,0,0) \stackrel{1r}{\to} (\gamma,0,\gamma,\gamma,0,0,\gamma,0)$	1;3	2;4	$2^{-6.50}$
$(\gamma,\gamma,\gamma,\gamma,0,0,0,\gamma) \stackrel{1r}{\to} (0,\gamma,0,0,\gamma,\gamma,0,0)$	1;3;8;9;10	2;4;9;10	2^{-1}
$(\gamma,\gamma,\gamma,\gamma,0,0,\gamma,0) \stackrel{1r}{\to} (\gamma,0,\gamma,0,0,0,0,\gamma)$	1;3;10	2;4;7;10	$2^{-6.50}$
$(\gamma,\gamma,\gamma,\gamma,0,0,\gamma,\gamma) \stackrel{1r}{\to} (0,\gamma,0,\gamma,\gamma,\gamma,\gamma,\gamma)$	1;3;8;9	2;4;7;9	2^{-1}
$(\gamma,\gamma,\gamma,\gamma,0,\gamma,0,0) \stackrel{1r}{\to} (0,\gamma,\gamma,\gamma,\gamma,0,0,\gamma)$	1;3;6;9;10	2;4;9;10	2^{-12}
$(\gamma,\gamma,\gamma,\gamma,0,\gamma,0,\gamma) \stackrel{1r}{\to} (\gamma,0,0,0,0,\gamma,\gamma,\gamma)$	1;3;6;8	2;4	$2^{-6.50}$
$(\gamma,\gamma,\gamma,\gamma,0,\gamma,\gamma,0) \stackrel{1r}{\to} (0,\gamma,\gamma,0,\gamma,0,\gamma,0)$	1;3;6;9	2;4;7;9	2^{-12}
$(\gamma,\gamma,\gamma,\gamma,0,\gamma,\gamma,\gamma) \stackrel{1r}{\to} (\gamma,0,0,\gamma,0,\gamma,0,0)$	1;3;6;8;10	2;4;7;10	$2^{-6.50}$
$(\gamma,\gamma,\gamma,\gamma,\gamma,0,0,0) \stackrel{1r}{\to} (\gamma,\gamma,0,\gamma,\gamma,0,0,\gamma)$	1;3;10	2;4;5;10	2^{-12}
$(\gamma,\gamma,\gamma,\gamma,\gamma,0,0,\gamma) \stackrel{1r}{\to} (0,0,\gamma,0,0,\gamma,\gamma,\gamma)$	1;3;8;9	2;4;5;9	$2^{-6.50}$
$(\gamma,\gamma,\gamma,\gamma,\gamma,0,\gamma,0) \stackrel{1r}{\to} (\gamma,\gamma,0,0,\gamma,0,\gamma,0)$	1;3	2;4;5;7	$2^{-6.50}$
$(\gamma,\gamma,\gamma,\gamma,\gamma,0,\gamma,\gamma) \stackrel{1r}{\to} (0,0,\gamma,\gamma,0,\gamma,0,0)$	1;3;8;9;10	2;4;5;7;9;10	$2^{-6.50}$
$(\gamma,\gamma,\gamma,\gamma,\gamma,\gamma,0,0) \stackrel{1r}{\to} (0,0,0,\gamma,0,0,\gamma,0)$	1;3;6;9	2;4;5;9	$2^{-6.50}$
$(\gamma,\gamma,\gamma,\gamma,\gamma,\gamma,0,\gamma) \stackrel{1r}{\to} (\gamma,\gamma,\gamma,0,\gamma,\gamma,0,0)$	1;3;6;8;10	2;4;5;10	2^{-1}
$(\gamma,\gamma,\gamma,\gamma,\gamma,\gamma,\gamma,0) \stackrel{1r}{\to} (0,0,0,0,0,0,0,\gamma)$	1;3;6;9;10	2;4;5;7;9;10	$2^{-6.50}$
$(\gamma,\gamma,\gamma,\gamma,\gamma,\gamma,\gamma,\gamma) \stackrel{1r}{\to} (\gamma,\gamma,\gamma,\gamma,\gamma,\gamma,\gamma,\gamma)$	1;3;6;8	2;4;5;7	2^{-1}

$Z_5^{(i+3)}$, $Z_7^{(i+3)}$, $Z_9^{(i+3)}$, $Z_{10}^{(i+3)}$, $Z_3^{(i+4)}$, $Z_6^{(i+4)}$, $Z_{10}^{(i+4)}$, $Z_4^{(i+5)}$ and $Z_{10}^{(i+5)}$. Assuming i is even, the following 19 subkeys have to be weak: $Z_9^{(i)}$, $Z_{10}^{(i)}$, $Z_1^{(i+1)}$, $Z_8^{(i+1)}$, $Z_{10}^{(i+1)}$, $Z_2^{(i+2)}$, $Z_5^{(i+2)}$, $Z_7^{(i+2)}$, $Z_{10}^{(i+2)}$, $Z_1^{(i+3)}$, $Z_3^{(i+3)}$, $Z_8^{(i+3)}$, $Z_9^{(i+3)}$, $Z_{10}^{(i+3)}$, $Z_4^{(i+4)}$, $Z_{10}^{(i+4)}$, $Z_3^{(i+5)}$, $Z_8^{(i+5)}$ and $Z_{10}^{(i+5)}$.

(5) $(0,0,0,0,0,\gamma,0,\gamma) \stackrel{1r}{\to} (0,0,\gamma,\gamma,0,\gamma,0,\gamma) \stackrel{1r}{\to} (\gamma,\gamma,0,0,\gamma,\gamma,\gamma,\gamma) \stackrel{1r}{\to} (0,0,0,0,0,\gamma,0,\gamma)$, which is 3-round iterative and holds with bias

$2^{-17.50}$. Assuming i is odd, the following eight subkeys have to be weak: $Z_6^{(i)}$, $Z_8^{(i)}$, $Z_4^{(i+1)}$, $Z_9^{(i+1)}$, $Z_1^{(i+2)}$, $Z_6^{(i+2)}$, $Z_8^{(i+2)}$ and $Z_9^{(i+2)}$. Assuming i is even, the following eight subkeys have to be weak: $Z_3^{(i+1)}$, $Z_6^{(i+1)}$, $Z_8^{(i+1)}$, $Z_9^{(i+1)}$, $Z_2^{(i+2)}$, $Z_5^{(i+2)}$, $Z_7^{(i+2)}$ and $Z_9^{(i+2)}$.

(6) $(0,0,0,0,0,\gamma,\gamma,0) \stackrel{1r}{\rightarrow} (\gamma,\gamma,0,\gamma,\gamma,0,0,0) \stackrel{1r}{\rightarrow} (\gamma, \gamma, 0, \gamma, \gamma, \gamma, \gamma, 0) \stackrel{1r}{\rightarrow} (0, 0, 0, 0, 0, \gamma, \gamma, 0)$, which is 3-round iterative and holds with bias $2^{-17.50}$. Assuming i is odd, the following eight subkeys have to be weak: $Z_6^{(i)}$, $Z_9^{(i)}$, $Z_2^{(i+1)}$, $Z_4^{(i+1)}$, $Z_5^{(i+1)}$, $Z_1^{(i+2)}$, $Z_6^{(i+2)}$ and $Z_9^{(i+2)}$. Assuming i is even, the following eight subkeys have to be weak: $Z_7^{(i)}$, $Z_9^{(i)}$, $Z_1^{(i+1)}$, $Z_2^{(i+2)}$, $Z_4^{(i+2)}$, $Z_5^{(i+2)}$, $Z_7^{(i+2)}$ and $Z_9^{(i+2)}$.

(7) $(0,0,0,0,0,\gamma,\gamma,\gamma) \stackrel{1r}{\rightarrow} (0,0,\gamma,0,0,\gamma,\gamma,0) \stackrel{1r}{\rightarrow} (\gamma, \gamma, 0, \gamma, \gamma, \gamma, \gamma, \gamma) \stackrel{1r}{\rightarrow} (\gamma, \gamma, \gamma, \gamma, \gamma, 0, 0, 0) \stackrel{1r}{\rightarrow} (\gamma,\gamma,0,\gamma,\gamma,0,0,\gamma) \stackrel{1r}{\rightarrow} (0,0,\gamma,0,0,0,0,0) \stackrel{1r}{\rightarrow} (0,0,0,0,0,\gamma,\gamma,\gamma)$, which is 6-round iterative and holds with bias 2^{-34} Assuming i is odd, the following 19 subkeys have to be weak: $Z_6^{(i)}$, $Z_8^{(i)}$, $Z_{10}^{(i)}$, $Z_7^{(i+1)}$, $Z_9^{(i+1)}$, $Z_{10}^{(i+1)}$, $Z_1^{(i+2)}$, $Z_6^{(i+2)}$, $Z_8^{(i+2)}$, $Z_{10}^{(i+2)}$, $Z_2^{(i+3)}$, $Z_4^{(i+3)}$, $Z_5^{(i+3)}$, $Z_{10}^{(i+3)}$, $Z_1^{(i+4)}$, $Z_8^{(i+4)}$, $Z_9^{(i+4)}$, $Z_{10}^{(i+4)}$ and $Z_{10}^{(i+5)}$. Assuming i is even, the following 21 subkeys have to be weak: $Z_7^{(i)}$, $Z_{10}^{(i)}$, $Z_3^{(i+1)}$, $Z_6^{(i+1)}$, $Z_9^{(i+1)}$, $Z_{10}^{(i+1)}$, $Z_2^{(i+2)}$, $Z_4^{(i+2)}$, $Z_5^{(i+2)}$, $Z_7^{(i+2)}$, $Z_{10}^{(i+2)}$, $Z_1^{(i+3)}$, $Z_3^{(i+3)}$, $Z_{10}^{(i+3)}$, $Z_2^{(i+4)}$, $Z_4^{(i+4)}$, $Z_5^{(i+4)}$, $Z_9^{(i+4)}$, $Z_{10}^{(i+4)}$, $Z_3^{(i+5)}$ and $Z_{10}^{(i+5)}$.

(8) $(0,0,0,0,\gamma,0,0,0) \stackrel{1r}{\rightarrow} (0,\gamma,\gamma,0,\gamma,0,\gamma,\gamma) \stackrel{1r}{\rightarrow} (0, \gamma, \gamma, 0, 0, 0, \gamma, 0) \stackrel{1r}{\rightarrow} (\gamma, \gamma, \gamma, \gamma, 0, \gamma, \gamma, \gamma) \stackrel{1r}{\rightarrow} (\gamma,0,0,\gamma,0,\gamma,0,0) \stackrel{1r}{\rightarrow} (\gamma,0,0,\gamma,\gamma,\gamma,0,\gamma) \stackrel{1r}{\rightarrow} (0,0,0,0,\gamma,0,0,0)$, which is 6-round iterative and holds with bias 2^{-34}. Assuming i is odd, the following 19 subkeys have to be weak: $Z_{10}^{(i)}$, $Z_2^{(i+1)}$, $Z_5^{(i+1)}$, $Z_7^{(i+1)}$, $Z_{10}^{(i+1)}$, $Z_3^{(i+2)}$, $Z_9^{(i+2)}$, $Z_{10}^{(i+2)}$, $Z_2^{(i+3)}$, $Z_4^{(i+3)}$, $Z_7^{(i+3)}$, $Z_{10}^{(i+3)}$, $Z_1^{(i+4)}$, $Z_6^{(i+4)}$, $Z_{10}^{(i+4)}$, $Z_4^{(i+5)}$, $Z_5^{(i+5)}$, $Z_9^{(i+5)}$ and $Z_{10}^{(i+5)}$. Assuming i is even, the following 20 subkeys have to be weak: $Z_5^{(i)}$, $Z_{10}^{(i)}$, $Z_3^{(i+1)}$, $Z_8^{(i+1)}$, $Z_{10}^{(i+1)}$, $Z_2^{(i+2)}$, $Z_7^{(i+2)}$, $Z_9^{(i+2)}$, $Z_{10}^{(i+2)}$, $Z_1^{(i+3)}$, $Z_3^{(i+3)}$, $Z_6^{(i+3)}$, $Z_8^{(i+3)}$, $Z_{10}^{(i+3)}$, $Z_4^{(i+4)}$, $Z_{10}^{(i+4)}$, $Z_1^{(i+5)}$, $Z_6^{(i+5)}$, $Z_8^{(i+5)}$, $Z_9^{(i+5)}$ and $Z_{10}^{(i+5)}$.

(9) $(0, 0, 0, 0, \gamma, 0, 0, \gamma) \stackrel{1r}{\rightarrow} (\gamma, 0, 0, \gamma, 0, \gamma, 0, \gamma) \stackrel{1r}{\rightarrow} (0, \gamma, \gamma, 0, 0, 0, \gamma, \gamma) \stackrel{1r}{\rightarrow} (0, 0, 0, 0, \gamma, 0, 0, \gamma)$, which is 3-round iterative and holds with bias $2^{-17.5}$. Assuming i is odd, the following six subkeys have to be weak: $Z_8^{(i)}$, $Z_9^{(i)}$, $Z_4^{(i+1)}$, $Z_9^{(i+1)}$, $Z_3^{(i+2)}$ and $Z_8^{(i+2)}$. Assuming i is even, the following eight subkeys have to be weak: $Z_5^{(i)}$, $Z_9^{(i)}$, $Z_1^{(i+1)}$, $Z_6^{(i+1)}$, $Z_8^{(i+1)}$, $Z_9^{(i+1)}$, $Z_2^{(i+2)}$ and $Z_7^{(i+2)}$.

(10) $(0,0,0,0,\gamma,0,\gamma,0) \stackrel{1r}{\rightarrow} (0,\gamma,\gamma,\gamma,\gamma,0,0,0) \stackrel{1r}{\rightarrow} (0,\gamma,\gamma,\gamma,0,0,\gamma,0) \stackrel{1r}{\rightarrow} (0, 0, 0, 0, \gamma, 0, \gamma, 0)$, which is 3-round iterative and holds with bias $2^{-17.50}$. Assuming i is odd, the following four subkeys have to be weak: $Z_2^{(i+1)}$,

$Z_4^{(i+1)}$, $Z_5^{(i+1)}$ and $Z_3^{(i+2)}$. Assuming i is even, the following six subkeys have to be weak: $Z_5^{(i)}$, $Z_7^{(i)}$, $Z_3^{(i+1)}$, $Z_2^{(i+2)}$, $Z_4^{(i+2)}$ and $Z_7^{(i+2)}$.

(11) $(0,0,0,0,\gamma,0,\gamma,\gamma) \stackrel{1r}{\to} (\gamma,0,0,0,0,\gamma,\gamma,0) \stackrel{1r}{\to} (0,\gamma,\gamma,\gamma,0,0,\gamma,\gamma) \stackrel{1r}{\to} (\gamma,\gamma,\gamma,\gamma,0,\gamma,0,0) \stackrel{1r}{\to} (0,\gamma,\gamma,\gamma,\gamma,0,0,\gamma) \stackrel{1r}{\to} (\gamma,0,0,0,\gamma,\gamma,0,0) \stackrel{1r}{\to} (0,0,0,0,\gamma,0,\gamma,\gamma)$, which is 6-round iterative and holds with bias 2^{-34}. Assuming i is odd, the following 21 subkeys have to be weak: $Z_8^{(i)}$, $Z_9^{(i)}$, $Z_{10}^{(i)}$, $Z_7^{(i+1)}$, $Z_9^{(i+1)}$, $Z_{10}^{(i+1)}$, $Z_3^{(i+2)}$, $Z_8^{(i+2)}$, $Z_9^{(i+2)}$, $Z_{10}^{(i+2)}$, $Z_2^{(i+3)}$, $Z_4^{(i+3)}$, $Z_9^{(i+3)}$, $Z_{10}^{(i+3)}$, $Z_3^{(i+4)}$, $Z_8^{(i+4)}$, $Z_9^{(i+4)}$, $Z_{10}^{(i+4)}$, $Z_5^{(i+5)}$, $Z_9^{(i+5)}$ and $Z_{10}^{(i+5)}$. Assuming i is even, the following 27 subkeys have to be weak: $Z_5^{(i)}$, $Z_7^{(i)}$, $Z_9^{(i)}$, $Z_{10}^{(i)}$, $Z_1^{(i+1)}$, $Z_6^{(i+1)}$, $Z_9^{(i+1)}$, $Z_{10}^{(i+1)}$, $Z_2^{(i+2)}$, $Z_4^{(i+2)}$, $Z_7^{(i+2)}$, $Z_9^{(i+2)}$, $Z_{10}^{(i+2)}$, $Z_1^{(i+3)}$, $Z_3^{(i+3)}$, $Z_6^{(i+3)}$, $Z_9^{(i+3)}$, $Z_{10}^{(i+3)}$, $Z_2^{(i+4)}$, $Z_4^{(i+4)}$, $Z_5^{(i+4)}$, $Z_9^{(i+4)}$, $Z_{10}^{(i+4)}$, $Z_1^{(i+5)}$, $Z_6^{(i+5)}$, $Z_9^{(i+5)}$ and $Z_{10}^{(i+5)}$.

(12) $(0,0,0,0,\gamma,\gamma,0,0) \stackrel{1r}{\to} (\gamma,0,\gamma,0,0,0,0,0) \stackrel{1r}{\to} (\gamma,0,\gamma,0,\gamma,\gamma,0,0) \stackrel{1r}{\to} (0,0,0,0,\gamma,\gamma,0,0)$, which is 3-round iterative and holds with bias 2^{-12}. Assuming i is odd, the following six subkeys have to be weak: $Z_6^{(i)}$, $Z_9^{(i)}$, $Z_1^{(i+2)}$, $Z_3^{(i+2)}$, $Z_6^{(i+2)}$ and $Z_9^{(i+2)}$. Assuming i is even, the following six subkeys have to be weak: $Z_5^{(i)}$, $Z_9^{(i)}$, $Z_1^{(i+1)}$, $Z_3^{(i+1)}$, $Z_5^{(i+2)}$ and $Z_9^{(i+2)}$.

(13) $(0,0,0,0,\gamma,\gamma,0,\gamma) \stackrel{1r}{\to} (0,\gamma,0,\gamma,\gamma,\gamma,\gamma,0) \stackrel{1r}{\to} (\gamma,0,\gamma,0,\gamma,\gamma,0,\gamma) \stackrel{1r}{\to} (\gamma,\gamma,\gamma,\gamma,0,0,\gamma,0) \stackrel{1r}{\to} (\gamma,0,\gamma,0,0,0,0,\gamma) \stackrel{1r}{\to} (0,\gamma,0,\gamma,0,0,\gamma,0) \stackrel{1r}{\to} (0,0,0,0,\gamma,\gamma,0,\gamma)$, which is 6-round iterative and holds with bias 2^{-56}. Assuming i is odd, the following 27 subkeys have to be weak: $Z_6^{(i)}$, $Z_8^{(i)}$, $Z_{10}^{(i)}$, $Z_2^{(i+1)}$, $Z_4^{(i+1)}$, $Z_5^{(i+1)}$, $Z_7^{(i+1)}$, $Z_9^{(i+1)}$, $Z_{10}^{(i+1)}$, $Z_1^{(i+2)}$, $Z_3^{(i+2)}$, $Z_6^{(i+2)}$, $Z_8^{(i+2)}$, $Z_{10}^{(i+2)}$, $Z_2^{(i+3)}$, $Z_4^{(i+3)}$, $Z_7^{(i+3)}$, $Z_{10}^{(i+3)}$, $Z_1^{(i+4)}$, $Z_3^{(i+4)}$, $Z_8^{(i+4)}$, $Z_9^{(i+4)}$, $Z_{10}^{(i+4)}$, $Z_2^{(i+5)}$, $Z_4^{(i+5)}$, $Z_7^{(i+5)}$ and $Z_{10}^{(i+5)}$. Assuming i is even, the following 13 subkeys have to be weak: $Z_5^{(i)}$, $Z_{10}^{(i)}$, $Z_6^{(i+1)}$, $Z_9^{(i+1)}$, $Z_{10}^{(i+1)}$, $Z_5^{(i+2)}$, $Z_{10}^{(i+2)}$, $Z_1^{(i+3)}$, $Z_3^{(i+3)}$, $Z_{10}^{(i+3)}$, $Z_9^{(i+4)}$, $Z_{10}^{(i+4)}$ and $Z_{10}^{(i+5)}$.

(14) $(0,0,0,0,\gamma,\gamma,\gamma,0) \stackrel{1r}{\to} (\gamma,0,\gamma,\gamma,0,0,\gamma,\gamma) \stackrel{1r}{\to} (\gamma,0,\gamma,\gamma,\gamma,\gamma,0,0) \stackrel{1r}{\to} (\gamma,\gamma,\gamma,\gamma,0,0,0,\gamma) \stackrel{1r}{\to} (0,\gamma,0,0,\gamma,\gamma,0,0) \stackrel{1r}{\to} (0,\gamma,0,0,0,0,\gamma,\gamma) \stackrel{1r}{\to} (0,0,0,0,\gamma,\gamma,\gamma,0)$, which is 6-round iterative and holds with bias 2^{-34}. Assuming i is odd, the following 19 subkeys have to be weak: $Z_6^{(i)}$, $Z_9^{(i)}$, $Z_{10}^{(i)}$, $Z_4^{(i+1)}$, $Z_7^{(i+1)}$, $Z_{10}^{(i+1)}$, $Z_1^{(i+2)}$, $Z_3^{(i+2)}$, $Z_6^{(i+2)}$, $Z_{10}^{(i+2)}$, $Z_2^{(i+3)}$, $Z_4^{(i+3)}$, $Z_9^{(i+3)}$, $Z_{10}^{(i+3)}$, $Z_6^{(i+4)}$, $Z_{10}^{(i+4)}$, $Z_2^{(i+5)}$, $Z_7^{(i+5)}$ and $Z_{10}^{(i+5)}$. Assuming i is even, the following 21 subkeys have to be weak: $Z_5^{(i)}$, $Z_7^{(i)}$, $Z_9^{(i)}$, $Z_{10}^{(i)}$, $Z_1^{(i+1)}$, $Z_3^{(i+1)}$, $Z_8^{(i+1)}$, $Z_{10}^{(i+1)}$, $Z_4^{(i+2)}$, $Z_5^{(i+2)}$, $Z_{10}^{(i+2)}$, $Z_1^{(i+3)}$, $Z_3^{(i+3)}$, $Z_8^{(i+3)}$, $Z_9^{(i+3)}$, $Z_{10}^{(i+3)}$, $Z_2^{(i+4)}$, $Z_5^{(i+4)}$, $Z_{10}^{(i+4)}$, $Z_8^{(i+5)}$ and $Z_{10}^{(i+5)}$.

(15) $(0,0,0,0,\gamma,\gamma,\gamma,\gamma) \stackrel{1r}{\to} (0,\gamma,0,0,\gamma,\gamma,0,\gamma) \stackrel{1r}{\to} (\gamma,0,\gamma,\gamma,\gamma,\gamma,0,\gamma) \stackrel{1r}{\to} (0,0,0,0,\gamma,\gamma,\gamma,\gamma)$, which is 3-round iterative and holds with bias 2^{-23}. Assuming i is odd, the following ten subkeys have to be weak: $Z_6^{(i)}$, $Z_8^{(i)}$, $Z_2^{(i+1)}$, $Z_5^{(i+1)}$, $Z_9^{(i+1)}$, $Z_1^{(i+2)}$, $Z_3^{(i+2)}$, $Z_6^{(i+2)}$, $Z_8^{(i+2)}$ and $Z_9^{(i+2)}$. Assum-

ing i is even, the following eight subkeys have to be weak: $Z_5^{(i)}$, $Z_7^{(i)}$, $Z_6^{(i+1)}$, $Z_8^{(i+1)}$, $Z_9^{(i+1)}$, $Z_4^{(i+2)}$, $Z_5^{(i+2)}$ and $Z_9^{(i+2)}$.

(16) $(0,0,0,\gamma,0,0,\gamma,0) \stackrel{1r}{\rightarrow} (\gamma,\gamma,\gamma,0,\gamma,\gamma,\gamma,0) \stackrel{1r}{\rightarrow} (\gamma,\gamma,\gamma,\gamma,\gamma,\gamma,0,0) \stackrel{1r}{\rightarrow} (0, 0, 0, \gamma, 0, 0, \gamma, 0)$, which is 3-round iterative and holds with bias 2^{-12}. Assuming i is odd, the following eight subkeys have to be weak: $Z_9^{(i)}$, $Z_2^{(i+1)}$, $Z_5^{(i+1)}$, $Z_7^{(i+1)}$, $Z_1^{(i+2)}$, $Z_3^{(i+2)}$, $Z_6^{(i+2)}$ and $Z_9^{(i+2)}$. Assuming i is even, the following ten subkeys have to be weak: $Z_4^{(i)}$, $Z_7^{(i)}$, $Z_9^{(i)}$, $Z_1^{(i+1)}$, $Z_3^{(i+1)}$, $Z_6^{(i+1)}$, $Z_2^{(i+2)}$, $Z_4^{(i+2)}$, $Z_5^{(i+2)}$ and $Z_9^{(i+2)}$.

(17) $(0,0,0,\gamma,0,\gamma,0,0) \stackrel{1r}{\rightarrow} (0,0,\gamma,\gamma,0,\gamma,\gamma,0) \stackrel{1r}{\rightarrow} (0,0,\gamma,0,0,0,\gamma,0) \stackrel{1r}{\rightarrow} (0, 0, 0, \gamma, 0, \gamma, 0, 0)$, which is 3-round iterative and holds with bias 2^{-12}. Assuming i is odd, the following four subkeys have to be weak: $Z_6^{(i)}$, $Z_4^{(i+1)}$, $Z_7^{(i+1)}$ and $Z_3^{(i+2)}$. Assuming i is even, the following four subkeys have to be weak: $Z_4^{(i)}$, $Z_3^{(i+1)}$, $Z_6^{(i+1)}$ and $Z_7^{(i+2)}$.

(18) $(0,0,0,\gamma,0,\gamma,0,\gamma) \stackrel{1r}{\rightarrow} (\gamma,\gamma,0,0,\gamma,0,0,0) \stackrel{1r}{\rightarrow} (0,0,\gamma,0,0,0,\gamma,\gamma) \stackrel{1r}{\rightarrow} (\gamma, \gamma, \gamma, 0, \gamma, 0, \gamma, 0) \stackrel{1r}{\rightarrow} (0,0,\gamma,\gamma,0,\gamma,\gamma,\gamma) \stackrel{1r}{\rightarrow} (\gamma,\gamma,0,\gamma,\gamma,\gamma,0,0) \stackrel{1r}{\rightarrow} (0,0,0,\gamma,0,\gamma,0,\gamma)$, which is 6-round iterative and holds with bias 2^{-34}. Assuming i is odd, the following 27 subkeys have to be weak: $Z_6^{(i)}$, $Z_8^{(i)}$, $Z_9^{(i)}$, $Z_{10}^{(i)}$, $Z_2^{(i+1)}$, $Z_5^{(i+1)}$, $Z_9^{(i+1)}$, $Z_{10}^{(i+1)}$, $Z_3^{(i+2)}$, $Z_8^{(i+2)}$, $Z_9^{(i+2)}$, $Z_{10}^{(i+2)}$, $Z_2^{(i+3)}$, $Z_5^{(i+3)}$, $Z_7^{(i+3)}$, $Z_9^{(i+3)}$, $Z_{10}^{(i+3)}$, $Z_3^{(i+4)}$, $Z_6^{(i+4)}$, $Z_8^{(i+4)}$, $Z_9^{(i+4)}$, $Z_{10}^{(i+4)}$, $Z_2^{(i+5)}$, $Z_4^{(i+5)}$, $Z_5^{(i+5)}$, $Z_9^{(i+5)}$ and $Z_{10}^{(i+5)}$. Assuming i is even, the following 21 subkeys have to be weak: $Z_4^{(i)}$, $Z_9^{(i)}$, $Z_{10}^{(i)}$, $Z_1^{(i+1)}$, $Z_9^{(i+1)}$, $Z_{10}^{(i+1)}$, $Z_7^{(i+2)}$, $Z_9^{(i+2)}$, $Z_{10}^{(i+2)}$, $Z_1^{(i+3)}$, $Z_3^{(i+3)}$, $Z_9^{(i+3)}$, $Z_{10}^{(i+3)}$, $Z_4^{(i+4)}$, $Z_7^{(i+4)}$, $Z_9^{(i+4)}$, $Z_{10}^{(i+4)}$, $Z_1^{(i+5)}$, $Z_6^{(i+5)}$, $Z_9^{(i+5)}$ and $Z_{10}^{(i+5)}$.

(19) $(0,0,0,\gamma,0,\gamma,\gamma,0) \stackrel{1r}{\rightarrow} (0,0,\gamma,0,0,\gamma,0,\gamma) \stackrel{1r}{\rightarrow} (0,0,\gamma,\gamma,0,0,\gamma,0) \stackrel{1r}{\rightarrow} (\gamma, \gamma, \gamma, 0, \gamma, 0, 0, \gamma) \stackrel{1r}{\rightarrow} (\gamma,\gamma,0,\gamma,\gamma,0,\gamma,0) \stackrel{1r}{\rightarrow} (\gamma,\gamma,0,0,\gamma,\gamma,0,\gamma) \stackrel{1r}{\rightarrow} (0,0,0,\gamma,0,\gamma,\gamma,0)$, which is 6-round iterative and holds with bias 2^{-56}. Assuming i is odd, the following 15 subkeys have to be weak: $Z_6^{(i)}$, $Z_{10}^{(i)}$, $Z_{10}^{(i+1)}$, $Z_3^{(i+2)}$, $Z_9^{(i+2)}$, $Z_{10}^{(i+2)}$, $Z_2^{(i+3)}$, $Z_5^{(i+3)}$, $Z_{10}^{(i+3)}$, $Z_1^{(i+4)}$, $Z_{10}^{(i+4)}$, $Z_2^{(i+5)}$, $Z_5^{(i+5)}$, $Z_9^{(i+5)}$ and $Z_{10}^{(i+5)}$. Assuming i is even, the following 25 subkeys have to be weak: $Z_4^{(i)}$, $Z_7^{(i)}$, $Z_{10}^{(i)}$, $Z_3^{(i+1)}$, $Z_6^{(i+1)}$, $Z_8^{(i+1)}$, $Z_{10}^{(i+1)}$, $Z_4^{(i+2)}$, $Z_7^{(i+2)}$, $Z_9^{(i+2)}$, $Z_{10}^{(i+2)}$, $Z_1^{(i+3)}$, $Z_3^{(i+3)}$, $Z_8^{(i+3)}$, $Z_{10}^{(i+3)}$, $Z_2^{(i+4)}$, $Z_4^{(i+4)}$, $Z_5^{(i+4)}$, $Z_7^{(i+4)}$, $Z_{10}^{(i+4)}$, $Z_1^{(i+5)}$, $Z_6^{(i+5)}$, $Z_8^{(i+5)}$, $Z_9^{(i+5)}$ and $Z_{10}^{(i+5)}$.

(20) $(0,0,0,\gamma,0,\gamma,\gamma,\gamma) \stackrel{1r}{\rightarrow} (\gamma,\gamma,0,\gamma,\gamma,0,\gamma,\gamma) \stackrel{1r}{\rightarrow} (0, 0, \gamma, \gamma, 0, 0, \gamma, \gamma) \stackrel{1r}{\rightarrow} (0, 0, 0, \gamma, 0, \gamma, \gamma, \gamma)$, which is 3-round iterative and holds with bias 2^{-12}. Assuming i is odd, the following ten subkeys have to be weak: $Z_6^{(i)}$, $Z_8^{(i)}$, $Z_9^{(i)}$, $Z_2^{(i+1)}$, $Z_4^{(i+1)}$, $Z_5^{(i+1)}$, $Z_7^{(i+1)}$, $Z_9^{(i+1)}$, $Z_3^{(i+2)}$ and $Z_8^{(i+2)}$. Assuming i is even, the following eight subkeys have to be weak: $Z_4^{(i)}$, $Z_7^{(i)}$, $Z_9^{(i)}$, $Z_1^{(i+1)}$, $Z_8^{(i+1)}$, $Z_9^{(i+1)}$, $Z_4^{(i+2)}$ and $Z_7^{(i+2)}$.

(21) $(0,0,0,\gamma,\gamma,0,0,0) \overset{1r}{\to} (\gamma,0,0,\gamma,0,\gamma,\gamma,0) \overset{1r}{\to} (\gamma,0,0,0,\gamma,\gamma,\gamma,0) \overset{1r}{\to} (0,0,0,\gamma,\gamma,0,0,0)$, which is 3-round iterative and holds with bias 2^{-23}. Assuming i is odd, the following six subkeys have to be weak: $Z_9^{(i)}$, $Z_4^{(i+1)}$, $Z_7^{(i+1)}$, $Z_1^{(i+2)}$, $Z_6^{(i+2)}$ and $Z_9^{(i+2)}$. Assuming i is even, the following eight subkeys have to be weak: $Z_4^{(i)}$, $Z_5^{(i)}$, $Z_9^{(i)}$, $Z_1^{(i+1)}$, $Z_6^{(i+1)}$, $Z_5^{(i+2)}$, $Z_7^{(i+2)}$ and $Z_9^{(i+2)}$.

(22) $(0,0,0,\gamma,\gamma,0,0,\gamma) \overset{1r}{\to} (0,\gamma,\gamma,0,\gamma,0,0,0) \overset{1r}{\to} (\gamma,0,0,0,\gamma,\gamma,\gamma,\gamma) \overset{1r}{\to} (\gamma,\gamma,\gamma,0,0,\gamma,\gamma,0) \overset{1r}{\to} (\gamma,0,0,\gamma,0,\gamma,\gamma,\gamma) \overset{1r}{\to} (0,\gamma,\gamma,\gamma,0,0,0,0) \overset{1r}{\to} (0,0,0,\gamma,\gamma,0,0,\gamma)$, which is 6-round iterative and holds with bias 2^{-34}. Assuming i is odd, the following 21 subkeys have to be weak: $Z_8^{(i)}$, $Z_{10}^{(i)}$, $Z_2^{(i+1)}$, $Z_5^{(i+1)}$, $Z_9^{(i+1)}$, $Z_{10}^{(i+1)}$, $Z_1^{(i+2)}$, $Z_6^{(i+2)}$, $Z_8^{(i+2)}$, $Z_{10}^{(i+2)}$, $Z_2^{(i+3)}$, $Z_7^{(i+3)}$, $Z_{10}^{(i+3)}$, $Z_1^{(i+4)}$, $Z_6^{(i+4)}$, $Z_8^{(i+4)}$, $Z_9^{(i+4)}$, $Z_{10}^{(i+4)}$, $Z_2^{(i+5)}$, $Z_4^{(i+5)}$ and $Z_{10}^{(i+5)}$. Assuming i is even, the following 19 subkeys have to be weak: $Z_4^{(i)}$, $Z_5^{(i)}$, $Z_{10}^{(i)}$, $Z_3^{(i+1)}$, $Z_9^{(i+1)}$, $Z_{10}^{(i+1)}$, $Z_5^{(i+2)}$, $Z_7^{(i+2)}$, $Z_{10}^{(i+2)}$, $Z_1^{(i+3)}$, $Z_3^{(i+3)}$, $Z_6^{(i+3)}$, $Z_{10}^{(i+3)}$, $Z_4^{(i+4)}$, $Z_7^{(i+4)}$, $Z_9^{(i+4)}$, $Z_{10}^{(i+4)}$, $Z_3^{(i+5)}$ and $Z_{10}^{(i+5)}$.

(23) $(0,0,0,\gamma,\gamma,0,\gamma,0) \overset{1r}{\to} (\gamma,0,0,0,0,\gamma,0,\gamma) \overset{1r}{\to} (\gamma,0,0,\gamma,\gamma,\gamma,\gamma,0) \overset{1r}{\to} (\gamma,\gamma,\gamma,0,0,\gamma,0,\gamma) \overset{1r}{\to} (0,\gamma,\gamma,\gamma,\gamma,0,\gamma,0) \overset{1r}{\to} (0,\gamma,\gamma,0,0,0,0,\gamma) \overset{1r}{\to} (0,0,0,\gamma,\gamma,0,\gamma,0)$, which is 6-round iterative and holds with bias 2^{-56}. Assuming i is odd, the following 13 subkeys have to be weak: $Z_9^{(i)}$, $Z_{10}^{(i)}$, $Z_{10}^{(i+1)}$, $Z_1^{(i+2)}$, $Z_6^{(i+2)}$, $Z_{10}^{(i+2)}$, $Z_2^{(i+3)}$, $Z_9^{(i+3)}$, $Z_{10}^{(i+3)}$, $Z_3^{(i+4)}$, $Z_{10}^{(i+4)}$, $Z_2^{(i+5)}$ and $Z_{10}^{(i+5)}$. Assuming i is even, the following 27 subkeys have to be weak: $Z_4^{(i)}$, $Z_5^{(i)}$, $Z_7^{(i)}$, $Z_9^{(i)}$, $Z_{10}^{(i)}$, $Z_1^{(i+1)}$, $Z_6^{(i+1)}$, $Z_8^{(i+1)}$, $Z_{10}^{(i+1)}$, $Z_4^{(i+2)}$, $Z_5^{(i+2)}$, $Z_7^{(i+2)}$, $Z_{10}^{(i+2)}$, $Z_1^{(i+3)}$, $Z_3^{(i+3)}$, $Z_6^{(i+3)}$, $Z_8^{(i+3)}$, $Z_9^{(i+3)}$, $Z_{10}^{(i+3)}$, $Z_2^{(i+4)}$, $Z_4^{(i+4)}$, $Z_5^{(i+4)}$, $Z_7^{(i+4)}$, $Z_{10}^{(i+4)}$, $Z_3^{(i+5)}$, $Z_8^{(i+5)}$ and $Z_{10}^{(i+5)}$.

(24) $(0,0,0,\gamma,\gamma,0,\gamma,\gamma) \overset{1r}{\to} (0,\gamma,\gamma,\gamma,\gamma,0,\gamma,\gamma) \overset{1r}{\to} (\gamma,0,0,\gamma,\gamma,\gamma,\gamma,\gamma) \overset{1r}{\to} (0,0,0,\gamma,\gamma,0,\gamma,\gamma)$, which is 3-round iterative and holds with bias 2^{-12}. Assuming i is odd, the following ten subkeys have to be weak: $Z_8^{(i)}$, $Z_2^{(i+1)}$, $Z_4^{(i+1)}$, $Z_5^{(i+1)}$, $Z_7^{(i+1)}$, $Z_9^{(i+1)}$, $Z_1^{(i+2)}$, $Z_6^{(i+2)}$, $Z_8^{(i+2)}$ and $Z_9^{(i+2)}$. Assuming i is even, the following ten subkeys have to be weak: $Z_4^{(i)}$, $Z_5^{(i)}$, $Z_7^{(i)}$, $Z_3^{(i+1)}$, $Z_8^{(i+1)}$, $Z_9^{(i+1)}$, $Z_4^{(i+2)}$, $Z_5^{(i+2)}$, $Z_7^{(i+2)}$ and $Z_9^{(i+2)}$.

(25) $(0,0,0,\gamma,\gamma,\gamma,0,0) \overset{1r}{\to} (0,\gamma,0,\gamma,\gamma,\gamma,0,\gamma) \overset{1r}{\to} (0,\gamma,0,0,0,0,0,0) \overset{1r}{\to} (\gamma,\gamma,\gamma,0,0,0,\gamma,\gamma) \overset{1r}{\to} (\gamma,0,\gamma,0,0,0,\gamma,0) \overset{1r}{\to} (\gamma,0,\gamma,\gamma,\gamma,\gamma,\gamma,\gamma) \overset{1r}{\to} (0,0,0,\gamma,\gamma,\gamma,0,0)$, which is 6-round iterative and holds with bias 2^{-34}. Assuming i is odd, the following 19 subkeys have to be weak: $Z_6^{(i)}$, $Z_{10}^{(i)}$, $Z_2^{(i+1)}$, $Z_4^{(i+1)}$, $Z_5^{(i+1)}$, $Z_{10}^{(i+1)}$, $Z_9^{(i+2)}$, $Z_{10}^{(i+2)}$, $Z_2^{(i+3)}$, $Z_7^{(i+3)}$, $Z_{10}^{(i+3)}$, $Z_1^{(i+4)}$, $Z_3^{(i+4)}$, $Z_{10}^{(i+4)}$, $Z_4^{(i+5)}$, $Z_5^{(i+5)}$, $Z_7^{(i+5)}$, $Z_9^{(i+5)}$ and $Z_{10}^{(i+5)}$. Assuming i is even, the following 21 subkeys have to be weak: $Z_4^{(i)}$, $Z_5^{(i)}$, $Z_{10}^{(i)}$, $Z_6^{(i+1)}$, $Z_8^{(i+1)}$, $Z_{10}^{(i+1)}$, $Z_2^{(i+2)}$, $Z_9^{(i+2)}$, $Z_{10}^{(i+2)}$, $Z_1^{(i+3)}$, $Z_3^{(i+3)}$, $Z_8^{(i+3)}$, $Z_{10}^{(i+3)}$, $Z_7^{(i+4)}$, $Z_{10}^{(i+4)}$, $Z_1^{(i+5)}$, $Z_3^{(i+5)}$, $Z_6^{(i+5)}$, $Z_8^{(i+5)}$, $Z_9^{(i+5)}$ and $Z_{10}^{(i+5)}$.

(26) $(0,0,0,\gamma,\gamma,\gamma,0,\gamma) \xrightarrow{1r} (\gamma,0,\gamma,0,0,0,\gamma,\gamma) \xrightarrow{1r} (0,\gamma,0,0,0,0,0,\gamma) \xrightarrow{1r} (0, 0, 0, \gamma, \gamma, \gamma, 0, \gamma)$, which is 3-round iterative and holds with bias $2^{-17.5}$. Assuming i is odd, the following six subkeys have to be weak: $Z_6^{(i)}$, $Z_8^{(i)}$, $Z_9^{(i)}$, $Z_7^{(i+1)}$, $Z_9^{(i+1)}$ and $Z_8^{(i+2)}$. Assuming i is even, the following eight subkeys have to be weak: $Z_4^{(i)}$, $Z_5^{(i)}$, $Z_9^{(i)}$, $Z_1^{(i+1)}$, $Z_3^{(i+1)}$, $Z_8^{(i+1)}$, $Z_9^{(i+1)}$ and $Z_2^{(i+2)}$.

(27) $(0,0,0,\gamma,\gamma,\gamma,\gamma,0) \xrightarrow{1r} (0,\gamma,0,0,\gamma,\gamma,\gamma,0) \xrightarrow{1r} (0,\gamma,0,\gamma,0,0,0,0) \xrightarrow{1r} (0, 0, 0, \gamma, \gamma, \gamma, \gamma, 0)$, which is 3-round iterative and holds with bias $2^{-17.5}$. Assuming i is odd, the following four subkeys have to be weak: $Z_6^{(i)}$, $Z_2^{(i+1)}$, $Z_5^{(i+1)}$ and $Z_7^{(i+1)}$. Assuming i is even, the following six subkeys have to be weak: $Z_4^{(i)}$, $Z_5^{(i)}$, $Z_7^{(i)}$, $Z_6^{(i+1)}$, $Z_2^{(i+2)}$ and $Z_4^{(i+2)}$.

(28) $(0,0,0,\gamma,\gamma,\gamma,\gamma,\gamma) \xrightarrow{1r} (\gamma,0,\gamma,\gamma,0,0,0,0) \xrightarrow{1r} (0,\gamma,0,\gamma,0,0,0,\gamma) \xrightarrow{1r} (\gamma, \gamma, \gamma, 0, 0, 0, 0, 0) \xrightarrow{1r} (0,\gamma,0,0,\gamma,\gamma,\gamma,\gamma) \xrightarrow{1r} (\gamma,0,\gamma,0,\gamma,\gamma,\gamma,0) \xrightarrow{1r} (0,0,0,\gamma,\gamma,\gamma,\gamma,\gamma)$, which is 6-round iterative and holds with bias 2^{-34}. Assuming i is odd, the following 21 subkeys have to be weak: $Z_6^{(i)}$, $Z_8^{(i)}$, $Z_9^{(i)}$, $Z_{10}^{(i)}$, $Z_4^{(i+1)}$, $Z_9^{(i+1)}$, $Z_{10}^{(i+1)}$, $Z_8^{(i+2)}$, $Z_9^{(i+2)}$, $Z_{10}^{(i+2)}$, $Z_2^{(i+3)}$, $Z_9^{(i+3)}$, $Z_{10}^{(i+3)}$, $Z_6^{(i+4)}$, $Z_8^{(i+4)}$, $Z_9^{(i+4)}$, $Z_{10}^{(i+4)}$, $Z_5^{(i+5)}$, $Z_7^{(i+5)}$, $Z_9^{(i+5)}$ and $Z_{10}^{(i+5)}$. Assuming i is even, the following 27 subkeys have to be weak: $Z_4^{(i)}$, $Z_5^{(i)}$, $Z_7^{(i)}$, $Z_9^{(i)}$, $Z_{10}^{(i)}$, $Z_1^{(i+1)}$, $Z_3^{(i+1)}$, $Z_9^{(i+1)}$, $Z_{10}^{(i+1)}$, $Z_2^{(i+2)}$, $Z_4^{(i+2)}$, $Z_9^{(i+2)}$, $Z_{10}^{(i+2)}$, $Z_1^{(i+3)}$, $Z_3^{(i+3)}$, $Z_9^{(i+3)}$, $Z_{10}^{(i+3)}$, $Z_2^{(i+4)}$, $Z_5^{(i+4)}$, $Z_7^{(i+4)}$, $Z_9^{(i+4)}$, $Z_{10}^{(i+4)}$, $Z_1^{(i+5)}$, $Z_3^{(i+5)}$, $Z_6^{(i+5)}$, $Z_9^{(i+5)}$ and $Z_{10}^{(i+5)}$.

(29) $(0,0,\gamma,0,0,0,0,\gamma) \xrightarrow{1r} (\gamma,\gamma,\gamma,\gamma,\gamma,0,0,\gamma) \xrightarrow{1r} (0,0,\gamma,0,0,\gamma,\gamma,\gamma) \xrightarrow{1r} (0, 0, \gamma, 0, 0, 0, 0, \gamma)$, which is 3-round iterative and holds with bias $2^{-17.5}$. Assuming i is odd, the following ten subkeys have to be weak: $Z_3^{(i)}$, $Z_8^{(i)}$, $Z_9^{(i)}$, $Z_2^{(i+1)}$, $Z_4^{(i+1)}$, $Z_5^{(i+1)}$, $Z_9^{(i+1)}$, $Z_3^{(i+2)}$, $Z_6^{(i+2)}$ and $Z_8^{(i+2)}$;. Assuming i is even, the following six subkeys have to be weak: $Z_9^{(i)}$, $Z_1^{(i+1)}$, $Z_3^{(i+1)}$, $Z_8^{(i+1)}$, $Z_9^{(i+1)}$ and $Z_7^{(i+2)}$.

(30) $(0,0,\gamma,0,0,\gamma,0,0) \xrightarrow{1r} (\gamma,\gamma,0,0,\gamma,\gamma,0,0) \xrightarrow{1r} (\gamma,\gamma,\gamma,0,\gamma,0,0,0) \xrightarrow{1r} (0, 0, \gamma, 0, 0, \gamma, 0, 0)$, which is 3-round iterative and holds with bias 2^{-12}. Assuming i is odd, the following eight subkeys have to be weak: $Z_3^{(i)}$, $Z_6^{(i)}$, $Z_9^{(i)}$, $Z_2^{(i+1)}$, $Z_5^{(i+1)}$, $Z_1^{(i+2)}$, $Z_3^{(i+2)}$ and $Z_9^{(i+2)}$. Assuming i is even, the following six subkeys have to be weak: $Z_9^{(i)}$, $Z_1^{(i+1)}$, $Z_6^{(i+1)}$, $Z_2^{(i+2)}$, $Z_5^{(i+2)}$ and $Z_9^{(i+2)}$.

(31) $(0,0,\gamma,0,\gamma,0,0,0) \xrightarrow{1r} (0,\gamma,\gamma,0,\gamma,\gamma,0,0) \xrightarrow{1r} (0,\gamma,0,0,0,\gamma,0,0) \xrightarrow{1r} (0, 0, \gamma, 0, \gamma, 0, 0, 0)$, which is 3-round iterative and holds with bias 2^{-12}. Assuming i is odd, the following four subkeys have to be weak: $Z_3^{(i)}$, $Z_2^{(i+1)}$, $Z_5^{(i+1)}$ and $Z_6^{(i+2)}$. Assuming i is even, the following four subkeys have to be weak: $Z_5^{(i)}$, $Z_3^{(i+1)}$, $Z_6^{(i+1)}$ and $Z_2^{(i+2)}$.

(32) $(0,0,\gamma,0,\gamma,0,0,\gamma) \stackrel{1r}{\to} (\gamma,0,0,\gamma,0,0,\gamma,0) \stackrel{1r}{\to} (0,\gamma,0,0,0,\gamma,0,\gamma) \stackrel{1r}{\to} (\gamma,\gamma,0,\gamma,0,\gamma,\gamma,0) \stackrel{1r}{\to} (0,\gamma,\gamma,0,\gamma,\gamma,0,\gamma) \stackrel{1r}{\to} (\gamma,0,\gamma,\gamma,\gamma,0,\gamma,0) \stackrel{1r}{\to} (0,0,\gamma,0,\gamma,0,0,\gamma)$, which is 6-round iterative and holds with bias 2^{-56}. Assuming i is odd, the following 27 subkeys have to be weak: $Z_3^{(i)}$, $Z_8^{(i)}$, $Z_9^{(i)}$, $Z_{10}^{(i)}$, $Z_4^{(i+1)}$, $Z_7^{(i+1)}$, $Z_9^{(i+1)}$, $Z_{10}^{(i+1)}$, $Z_6^{(i+2)}$, $Z_8^{(i+2)}$, $Z_9^{(i+2)}$, $Z_{10}^{(i+2)}$, $Z_2^{(i+3)}$, $Z_4^{(i+3)}$, $Z_7^{(i+3)}$, $Z_9^{(i+3)}$, $Z_{10}^{(i+3)}$, $Z_3^{(i+4)}$, $Z_6^{(i+4)}$, $Z_8^{(i+4)}$, $Z_9^{(i+4)}$, $Z_{10}^{(i+4)}$, $Z_4^{(i+5)}$, $Z_5^{(i+5)}$, $Z_7^{(i+5)}$, $Z_9^{(i+5)}$ and $Z_{10}^{(i+5)}$. Assuming i is even, the following 21 subkeys have to be weak: $Z_5^{(i)}$, $Z_9^{(i)}$, $Z_{10}^{(i)}$, $Z_1^{(i+1)}$, $Z_9^{(i+1)}$, $Z_{10}^{(i+1)}$, $Z_2^{(i+2)}$, $Z_9^{(i+2)}$, $Z_{10}^{(i+2)}$, $Z_1^{(i+3)}$, $Z_6^{(i+3)}$, $Z_9^{(i+3)}$, $Z_{10}^{(i+3)}$, $Z_2^{(i+4)}$, $Z_5^{(i+4)}$, $Z_9^{(i+4)}$, $Z_{10}^{(i+4)}$, $Z_1^{(i+5)}$, $Z_3^{(i+5)}$, $Z_9^{(i+5)}$ and $Z_{10}^{(i+5)}$.

(33) $(0,0,\gamma,0,\gamma,0,\gamma,0) \stackrel{1r}{\to} (0,\gamma,\gamma,\gamma,\gamma,\gamma,\gamma,\gamma) \stackrel{1r}{\to} (0,\gamma,0,\gamma,0,\gamma,0,0) \stackrel{1r}{\to} (\gamma,\gamma,0,\gamma,0,\gamma,0,\gamma) \stackrel{1r}{\to} (\gamma,0,0,0,0,0,0,0) \stackrel{1r}{\to} (\gamma,0,\gamma,0,\gamma,0,\gamma,\gamma) \stackrel{1r}{\to} (0,0,\gamma,0,\gamma,0,\gamma,0)$, which is 6-round iterative and holds with bias 2^{-34}. Assuming i is odd, the following 19 subkeys have to be weak: $Z_3^{(i)}$, $Z_{10}^{(i)}$, $Z_2^{(i+1)}$, $Z_4^{(i+1)}$, $Z_5^{(i+1)}$, $Z_7^{(i+1)}$, $Z_{10}^{(i+1)}$, $Z_6^{(i+2)}$, $Z_9^{(i+2)}$, $Z_{10}^{(i+2)}$, $Z_2^{(i+3)}$, $Z_4^{(i+3)}$, $Z_{10}^{(i+3)}$, $Z_1^{(i+4)}$, $Z_{10}^{(i+4)}$, $Z_5^{(i+5)}$, $Z_7^{(i+5)}$, $Z_9^{(i+5)}$ and $Z_{10}^{(i+5)}$. Assuming i is even, the following 21 subkeys have to be weak: $Z_5^{(i)}$, $Z_7^{(i)}$, $Z_{10}^{(i)}$, $Z_3^{(i+1)}$, $Z_6^{(i+1)}$, $Z_8^{(i+1)}$, $Z_{10}^{(i+1)}$, $Z_2^{(i+2)}$, $Z_4^{(i+2)}$, $Z_9^{(i+2)}$, $Z_{10}^{(i+2)}$, $Z_1^{(i+3)}$, $Z_6^{(i+3)}$, $Z_8^{(i+3)}$, $Z_{10}^{(i+3)}$, $Z_{10}^{(i+4)}$, $Z_1^{(i+5)}$, $Z_3^{(i+5)}$, $Z_8^{(i+5)}$, $Z_9^{(i+5)}$ and $Z_{10}^{(i+5)}$.

(34) $(0,0,\gamma,0,\gamma,0,\gamma,\gamma) \stackrel{1r}{\to} (\gamma,0,0,0,0,0,0,\gamma) \stackrel{1r}{\to} (0,\gamma,0,\gamma,0,\gamma,0,\gamma) \stackrel{1r}{\to} (0,0,\gamma,0,\gamma,0,\gamma,\gamma)$, which is 3-round iterative and holds with bias 2^{-23}. Assuming i is odd, the following six subkeys have to be weak: $Z_3^{(i)}$, $Z_8^{(i)}$, $Z_9^{(i)}$, $Z_9^{(i+1)}$, $Z_6^{(i+2)}$ and $Z_8^{(i+2)}$. Assuming i is even, the following eight subkeys have to be weak: $Z_5^{(i)}$, $Z_7^{(i)}$, $Z_9^{(i)}$, $Z_1^{(i+1)}$, $Z_8^{(i+1)}$, $Z_9^{(i+1)}$, $Z_2^{(i+2)}$ and $Z_4^{(i+2)}$.

(35) $(0,0,\gamma,0,\gamma,\gamma,0,0) \stackrel{1r}{\to} (\gamma,0,\gamma,0,0,\gamma,\gamma,\gamma) \stackrel{1r}{\to} (\gamma,0,0,0,\gamma,0,\gamma,0) \stackrel{1r}{\to} (\gamma,\gamma,0,\gamma,0,0,\gamma,\gamma) \stackrel{1r}{\to} (0,\gamma,0,\gamma,\gamma,0,0,0) \stackrel{1r}{\to} (0,\gamma,\gamma,\gamma,0,\gamma,0,\gamma) \stackrel{1r}{\to} (0,0,\gamma,0,\gamma,\gamma,0,0)$, which is 6-round iterative and holds with bias 2^{-12}. Assuming i is odd, the following 17 subkeys have to be weak: $Z_3^{(i)}$, $Z_6^{(i)}$, $Z_9^{(i)}$, $Z_{10}^{(i)}$, $Z_7^{(i+1)}$, $Z_{10}^{(i+1)}$, $Z_1^{(i+2)}$, $Z_{10}^{(i+2)}$, $Z_2^{(i+3)}$, $Z_4^{(i+3)}$, $Z_7^{(i+3)}$, $Z_9^{(i+3)}$, $Z_{10}^{(i+3)}$, $Z_{10}^{(i+4)}$, $Z_2^{(i+5)}$, $Z_4^{(i+5)}$ and $Z_{10}^{(i+5)}$. Assuming i is even, the following 23 subkeys have to be weak: $Z_5^{(i)}$, $Z_9^{(i)}$, $Z_{10}^{(i)}$, $Z_1^{(i+1)}$, $Z_3^{(i+1)}$, $Z_6^{(i+1)}$, $Z_8^{(i+1)}$, $Z_{10}^{(i+1)}$, $Z_5^{(i+2)}$, $Z_7^{(i+2)}$, $Z_{10}^{(i+2)}$, $Z_1^{(i+3)}$, $Z_8^{(i+3)}$, $Z_9^{(i+3)}$, $Z_{10}^{(i+3)}$, $Z_2^{(i+4)}$, $Z_4^{(i+4)}$, $Z_5^{(i+4)}$, $Z_{10}^{(i+4)}$, $Z_3^{(i+5)}$, $Z_6^{(i+5)}$, $Z_8^{(i+5)}$ and $Z_{10}^{(i+5)}$.

(36) $(0,0,\gamma,0,\gamma,\gamma,0,\gamma) \stackrel{1r}{\to} (0,\gamma,0,\gamma,\gamma,0,0,\gamma) \stackrel{1r}{\to} (\gamma,0,0,0,\gamma,0,\gamma,\gamma) \stackrel{1r}{\to} (0,0,\gamma,0,\gamma,\gamma,0,\gamma)$, which is 3-round iterative and holds with bias 2^{-12}. Assuming i is odd, the following ten subkeys have to be weak: $Z_3^{(i)}$, $Z_6^{(i)}$, $Z_8^{(i)}$, $Z_2^{(i+1)}$, $Z_4^{(i+1)}$, $Z_5^{(i+1)}$, $Z_9^{(i+1)}$, $Z_1^{(i+2)}$, $Z_8^{(i+2)}$ and $Z_9^{(i+2)}$. Assuming

i is even, the following six subkeys have to be weak: $Z_5^{(i)}$, $Z_8^{(i+1)}$, $Z_9^{(i+1)}$, $Z_5^{(i+2)}$, $Z_7^{(i+2)}$ and $Z_9^{(i+2)}$.

(37) $(0,0,\gamma,0,\gamma,\gamma,\gamma,0) \stackrel{1r}{\rightarrow} (\gamma,0,\gamma,\gamma,0,\gamma,0,0) \stackrel{1r}{\rightarrow} (\gamma, 0, 0, \gamma, \gamma, 0, \gamma, 0) \stackrel{1r}{\rightarrow} (0, 0, \gamma, 0, \gamma, \gamma, \gamma, 0)$, which is 3-round iterative and holds with bias 2^{-12}. Assuming i is odd, the following six subkeys have to be weak: $Z_3^{(i)}$, $Z_6^{(i)}$, $Z_9^{(i)}$, $Z_4^{(i+1)}$, $Z_1^{(i+2)}$ and $Z_9^{(i+2)}$. Assuming i is even, the following ten subkeys have to be weak: $Z_5^{(i)}$, $Z_7^{(i)}$, $Z_9^{(i)}$, $Z_1^{(i+1)}$, $Z_3^{(i+1)}$, $Z_6^{(i+1)}$, $Z_4^{(i+2)}$, $Z_5^{(i+2)}$, $Z_7^{(i+2)}$ and $Z_9^{(i+2)}$.

(38) $(0,0,\gamma,0,\gamma,\gamma,\gamma,\gamma) \stackrel{1r}{\rightarrow} (0,\gamma,0,0,\gamma,0,\gamma,0) \stackrel{1r}{\rightarrow} (\gamma, 0, 0, \gamma, \gamma, 0, \gamma, \gamma) \stackrel{1r}{\rightarrow} (\gamma, \gamma, 0, \gamma, 0, 0, 0, 0) \stackrel{1r}{\rightarrow} (\gamma,0,\gamma,\gamma,0,\gamma,0,\gamma) \stackrel{1r}{\rightarrow} (0,\gamma,\gamma,0,0,\gamma,0,0) \stackrel{1r}{\rightarrow} (0,0,\gamma,0,\gamma,\gamma,\gamma,\gamma)$, which is 6-round iterative and holds with bias 2^{-12}. Assuming i is odd, the following 23 subkeys have to be weak: $Z_3^{(i)}$, $Z_6^{(i)}$, $Z_8^{(i)}$, $Z_{10}^{(i)}$, $Z_2^{(i+1)}$, $Z_5^{(i+1)}$, $Z_7^{(i+1)}$, $Z_9^{(i+1)}$, $Z_{10}^{(i+1)}$, $Z_1^{(i+2)}$, $Z_8^{(i+2)}$, $Z_{10}^{(i+2)}$, $Z_2^{(i+3)}$, $Z_4^{(i+3)}$, $Z_{10}^{(i+3)}$, $Z_1^{(i+4)}$, $Z_3^{(i+4)}$, $Z_6^{(i+4)}$, $Z_8^{(i+4)}$, $Z_9^{(i+4)}$, $Z_{10}^{(i+4)}$, $Z_2^{(i+5)}$ and $Z_{10}^{(i+5)}$. Assuming i is even, the following 17 subkeys have to be weak: $Z_5^{(i)}$, $Z_7^{(i)}$, $Z_{10}^{(i)}$, $Z_9^{(i+1)}$, $Z_{10}^{(i+1)}$, $Z_4^{(i+2)}$, $Z_5^{(i+2)}$, $Z_7^{(i+2)}$, $Z_{10}^{(i+2)}$, $Z_1^{(i+3)}$, $Z_{10}^{(i+3)}$, $Z_4^{(i+4)}$, $Z_9^{(i+4)}$, $Z_{10}^{(i+4)}$, $Z_3^{(i+5)}$, $Z_6^{(i+5)}$ and $Z_{10}^{(i+5)}$.

(39) $(0,0,\gamma,\gamma,0,0,0,0) \stackrel{1r}{\rightarrow} (\gamma,\gamma,\gamma,\gamma,\gamma,0,\gamma,0) \stackrel{1r}{\rightarrow} (\gamma,\gamma,0,0,\gamma,0,\gamma,0) \stackrel{1r}{\rightarrow} (0, 0, \gamma, \gamma, 0, 0, 0, 0)$, which is 3-round iterative and holds with bias $2^{-17.5}$. Assuming i is odd, the following eight subkeys have to be weak: $Z_3^{(i)}$, $Z_9^{(i)}$, $Z_2^{(i+1)}$, $Z_4^{(i+1)}$, $Z_5^{(i+1)}$, $Z_7^{(i+1)}$, $Z_1^{(i+2)}$ and $Z_9^{(i+2)}$. Assuming i is even, the following eight subkeys have to be weak: $Z_4^{(i)}$, $Z_9^{(i)}$, $Z_1^{(i+1)}$, $Z_3^{(i+1)}$, $Z_2^{(i+2)}$, $Z_5^{(i+2)}$, $Z_7^{(i+2)}$ and $Z_9^{(i+2)}$.

(40) $(0, 0, \gamma, \gamma, \gamma, 0, 0, 0) \stackrel{1r}{\rightarrow} (\gamma, 0, 0, \gamma, 0, 0, 0, \gamma) \stackrel{1r}{\rightarrow} (\gamma, 0, \gamma, 0, \gamma, 0, 0, 0) \stackrel{1r}{\rightarrow} (\gamma, \gamma, 0, 0, 0, \gamma, \gamma, \gamma) \stackrel{1r}{\rightarrow} (0, \gamma, \gamma, 0, \gamma, \gamma, \gamma, 0) \stackrel{1r}{\rightarrow} (0, \gamma, 0, \gamma, 0, \gamma, \gamma, \gamma) \stackrel{1r}{\rightarrow} (0, 0, \gamma, \gamma, \gamma, 0, 0, 0)$, which is 6-round iterative and holds with bias 2^{-34}. Assuming i is odd, the following 19 subkeys have to be weak: $Z_3^{(i)}$, $Z_9^{(i)}$, $Z_{10}^{(i)}$, $Z_4^{(i+1)}$, $Z_{10}^{(i+1)}$, $Z_1^{(i+2)}$, $Z_3^{(i+2)}$, $Z_{10}^{(i+2)}$, $Z_2^{(i+3)}$, $Z_7^{(i+3)}$, $Z_9^{(i+3)}$, $Z_{10}^{(i+3)}$, $Z_3^{(i+4)}$, $Z_6^{(i+4)}$, $Z_{10}^{(i+4)}$, $Z_2^{(i+5)}$, $Z_4^{(i+5)}$, $Z_7^{(i+5)}$ and $Z_{10}^{(i+5)}$. Assuming i is even, the following 21 subkeys have to be weak: $Z_4^{(i)}$, $Z_5^{(i)}$, $Z_9^{(i)}$, $Z_{10}^{(i)}$, $Z_1^{(i+1)}$, $Z_8^{(i+1)}$, $Z_{10}^{(i+1)}$, $Z_5^{(i+2)}$, $Z_{10}^{(i+2)}$, $Z_1^{(i+3)}$, $Z_6^{(i+3)}$, $Z_8^{(i+3)}$, $Z_9^{(i+3)}$, $Z_{10}^{(i+3)}$, $Z_2^{(i+4)}$, $Z_5^{(i+4)}$, $Z_7^{(i+4)}$, $Z_{10}^{(i+4)}$, $Z_6^{(i+5)}$, $Z_8^{(i+5)}$ and $Z_{10}^{(i+5)}$.

(41) $(0,0,\gamma,\gamma,\gamma,0,0,\gamma) \stackrel{1r}{\rightarrow} (0,\gamma,\gamma,0,\gamma,\gamma,\gamma,\gamma) \stackrel{1r}{\rightarrow} (\gamma,0,\gamma,0,\gamma,0,0,\gamma) \stackrel{1r}{\rightarrow} (0, 0, \gamma, \gamma, \gamma, 0, 0, \gamma)$, which is 3-round iterative and holds with bias 2^{-23}. Assuming i is odd, the following ten subkeys have to be weak: $Z_3^{(i)}$, $Z_8^{(i)}$, $Z_2^{(i+1)}$, $Z_5^{(i+1)}$, $Z_7^{(i+1)}$, $Z_9^{(i+1)}$, $Z_1^{(i+2)}$, $Z_3^{(i+2)}$, $Z_8^{(i+2)}$ and $Z_9^{(i+2)}$. Assuming i is even, the following eight subkeys have to be weak: $Z_4^{(i)}$, $Z_5^{(i)}$, $Z_3^{(i+1)}$, $Z_6^{(i+1)}$, $Z_8^{(i+1)}$, $Z_9^{(i+1)}$, $Z_5^{(i+2)}$ and $Z_9^{(i+2)}$.

(42) $(0,0,\gamma,\gamma,\gamma,0,\gamma,0) \stackrel{1r}{\to} (\gamma,0,0,0,0,0,\gamma,0) \stackrel{1r}{\to} (\gamma,0,\gamma,\gamma,\gamma,0,0,0) \stackrel{1r}{\to} (0, 0, \gamma, \gamma, \gamma, 0, \gamma, 0)$, which is 3-round iterative and holds with bias 2^{-23}. Assuming i is odd, the following six subkeys have to be weak: $Z_3^{(i)}$, $Z_9^{(i)}$, $Z_7^{(i+1)}$, $Z_1^{(i+2)}$, $Z_3^{(i+2)}$ and $Z_9^{(i+2)}$. Assuming i is even, the following eight subkeys have to be weak: $Z_4^{(i)}$, $Z_5^{(i)}$, $Z_7^{(i)}$, $Z_9^{(i)}$, $Z_1^{(i+1)}$, $Z_4^{(i+2)}$, $Z_5^{(i+2)}$ and $Z_9^{(i+2)}$.

(43) $(0,0,\gamma,\gamma,\gamma,0,\gamma,\gamma) \stackrel{1r}{\to} (0,\gamma,\gamma,\gamma,\gamma,\gamma,0,0) \stackrel{1r}{\to} (\gamma,0,\gamma,\gamma,\gamma,0,0,\gamma) \stackrel{1r}{\to} (\gamma, \gamma, 0, 0, 0, \gamma, 0, 0) \stackrel{1r}{\to} (\gamma,0,0,0,0,0,\gamma,\gamma) \stackrel{1r}{\to} (0,\gamma,0,0,0,\gamma,\gamma,0) \stackrel{1r}{\to} (0,0,\gamma,\gamma,\gamma,0,\gamma,\gamma)$, which is 6-round iterative and holds with bias 2^{-34}. Assuming i is odd, the following 21 subkeys have to be weak: $Z_3^{(i)}$, $Z_8^{(i)}$, $Z_{10}^{(i)}$, $Z_2^{(i+1)}$, $Z_4^{(i+1)}$, $Z_5^{(i+1)}$, $Z_9^{(i+1)}$, $Z_{10}^{(i+1)}$, $Z_1^{(i+2)}$, $Z_3^{(i+2)}$, $Z_8^{(i+2)}$, $Z_{10}^{(i+2)}$, $Z_2^{(i+3)}$, $Z_{10}^{(i+3)}$, $Z_1^{(i+4)}$, $Z_8^{(i+4)}$, $Z_9^{(i+4)}$, $Z_{10}^{(i+4)}$, $Z_2^{(i+5)}$, $Z_7^{(i+5)}$ and $Z_{10}^{(i+5)}$. Assuming i is even, the following 19 subkeys have to be weak: $Z_4^{(i)}$, $Z_5^{(i)}$, $Z_7^{(i)}$, $Z_{10}^{(i)}$, $Z_3^{(i+1)}$, $Z_6^{(i+1)}$, $Z_9^{(i+1)}$, $Z_{10}^{(i+1)}$, $Z_4^{(i+2)}$, $Z_5^{(i+2)}$, $Z_{10}^{(i+2)}$, $Z_1^{(i+3)}$, $Z_6^{(i+3)}$, $Z_{10}^{(i+3)}$, $Z_7^{(i+4)}$, $Z_9^{(i+4)}$, $Z_{10}^{(i+4)}$, $Z_6^{(i+5)}$ and $Z_{10}^{(i+5)}$.

(44) $(0,0,\gamma,\gamma,\gamma,\gamma,0,0) \stackrel{1r}{\to} (0,\gamma,0,\gamma,\gamma,0,\gamma,0) \stackrel{1r}{\to} (0,\gamma,\gamma,0,0,\gamma,\gamma,0) \stackrel{1r}{\to} (0, 0, \gamma, \gamma, \gamma, \gamma, 0, 0)$, which is 3-round iterative and holds with bias 2^{-12}. Assuming i is odd, the following eight subkeys have to be weak: $Z_3^{(i)}$, $Z_6^{(i)}$, $Z_2^{(i+1)}$, $Z_4^{(i+1)}$, $Z_5^{(i+1)}$, $Z_7^{(i+1)}$, $Z_3^{(i+2)}$ and $Z_6^{(i+2)}$. Assuming i is even, the following four subkeys have to be weak: $Z_4^{(i)}$, $Z_5^{(i)}$, $Z_2^{(i+2)}$ and $Z_7^{(i+2)}$.

(45) $(0,0,\gamma,\gamma,\gamma,\gamma,0,\gamma) \stackrel{1r}{\to} (\gamma,0,\gamma,0,0,\gamma,0,0) \stackrel{1r}{\to} (0,\gamma,\gamma,0,0,\gamma,\gamma,\gamma) \stackrel{1r}{\to} (\gamma, \gamma, 0, 0, 0, 0, \gamma, 0) \stackrel{1r}{\to} (0,\gamma,0,\gamma,\gamma,0,\gamma,\gamma) \stackrel{1r}{\to} (\gamma,0,0,\gamma,\gamma,0,0,0) \stackrel{1r}{\to} (0,0,\gamma,\gamma,\gamma,\gamma,0,\gamma)$, which is 6-round iterative and holds with bias 2^{-12}. Assuming i is odd, the following 23 subkeys have to be weak: $Z_3^{(i)}$, $Z_6^{(i)}$, $Z_8^{(i)}$, $Z_9^{(i)}$, $Z_{10}^{(i)}$, $Z_9^{(i+1)}$, $Z_{10}^{(i+1)}$, $Z_3^{(i+2)}$, $Z_6^{(i+2)}$, $Z_8^{(i+2)}$, $Z_9^{(i+2)}$, $Z_{10}^{(i+2)}$, $Z_2^{(i+3)}$, $Z_7^{(i+3)}$, $Z_9^{(i+3)}$, $Z_{10}^{(i+3)}$, $Z_8^{(i+4)}$, $Z_9^{(i+4)}$, $Z_{10}^{(i+4)}$, $Z_4^{(i+5)}$, $Z_5^{(i+5)}$, $Z_9^{(i+5)}$ and $Z_{10}^{(i+5)}$. Assuming i is even, the following 25 subkeys have to be weak: $Z_4^{(i)}$, $Z_5^{(i)}$, $Z_9^{(i)}$, $Z_{10}^{(i)}$, $Z_1^{(i+1)}$, $Z_3^{(i+1)}$, $Z_6^{(i+1)}$, $Z_9^{(i+1)}$, $Z_{10}^{(i+1)}$, $Z_2^{(i+2)}$, $Z_7^{(i+2)}$, $Z_9^{(i+2)}$, $Z_{10}^{(i+2)}$, $Z_1^{(i+3)}$, $Z_9^{(i+3)}$, $Z_{10}^{(i+3)}$, $Z_2^{(i+4)}$, $Z_4^{(i+4)}$, $Z_5^{(i+4)}$, $Z_7^{(i+4)}$, $Z_9^{(i+4)}$, $Z_{10}^{(i+4)}$, $Z_1^{(i+5)}$, $Z_9^{(i+5)}$ and $Z_{10}^{(i+5)}$.

(46) $(0,0,\gamma,\gamma,\gamma,\gamma,\gamma,0) \stackrel{1r}{\to} (0,\gamma,0,0,\gamma,0,0,\gamma) \stackrel{1r}{\to} (0,\gamma,\gamma,\gamma,0,\gamma,\gamma,0) \stackrel{1r}{\to} (\gamma, \gamma, 0, 0, 0, 0, 0, \gamma) \stackrel{1r}{\to} (\gamma,0,\gamma,\gamma,0,\gamma,\gamma,0) \stackrel{1r}{\to} (\gamma,0,0,0,\gamma,0,0,\gamma) \stackrel{1r}{\to} (0,0,\gamma,\gamma,\gamma,\gamma,\gamma,0)$, which is 6-round iterative and holds with bias 2^{-34}. Assuming i is odd, the following 19 subkeys have to be weak: $Z_3^{(i)}$, $Z_6^{(i)}$, $Z_{10}^{(i)}$, $Z_2^{(i+1)}$, $Z_5^{(i+1)}$, $Z_{10}^{(i+1)}$, $Z_3^{(i+2)}$, $Z_6^{(i+2)}$, $Z_9^{(i+2)}$, $Z_{10}^{(i+2)}$, $Z_2^{(i+3)}$, $Z_{10}^{(i+3)}$, $Z_1^{(i+4)}$, $Z_3^{(i+4)}$, $Z_6^{(i+4)}$, $Z_{10}^{(i+4)}$, $Z_5^{(i+5)}$, $Z_9^{(i+5)}$ and $Z_{10}^{(i+5)}$. Assuming i is even, the following 21 subkeys have to be weak: $Z_4^{(i)}$, $Z_5^{(i)}$, $Z_7^{(i)}$, $Z_{10}^{(i)}$, $Z_8^{(i+1)}$, $Z_{10}^{(i+1)}$, $Z_2^{(i+2)}$, $Z_4^{(i+2)}$, $Z_7^{(i+2)}$, $Z_9^{(i+2)}$, $Z_{10}^{(i+2)}$, $Z_1^{(i+3)}$, $Z_8^{(i+3)}$, $Z_{10}^{(i+3)}$, $Z_4^{(i+4)}$, $Z_7^{(i+4)}$, $Z_{10}^{(i+4)}$, $Z_1^{(i+5)}$, $Z_8^{(i+5)}$, $Z_9^{(i+5)}$ and $Z_{10}^{(i+5)}$.

(47) $(0, 0, \gamma, \gamma, \gamma, \gamma, \gamma, \gamma) \stackrel{1r}{\to} (\gamma, 0, \gamma, \gamma, 0, \gamma, \gamma, \gamma) \stackrel{1r}{\to} (0, \gamma, \gamma, \gamma, 0, \gamma, \gamma, \gamma) \stackrel{1r}{\to} (0, 0, \gamma, \gamma, \gamma, \gamma, \gamma, \gamma)$, which is 3-round iterative and holds with bias 2^{-1}. Assuming i is odd, the following ten subkeys have to be weak: $Z_3^{(i)}$, $Z_6^{(i)}$, $Z_8^{(i)}$, $Z_9^{(i)}$, $Z_4^{(i+1)}$, $Z_7^{(i+1)}$, $Z_9^{(i+1)}$, $Z_3^{(i+2)}$, $Z_6^{(i+2)}$ and $Z_8^{(i+2)}$. Assuming i is even, the following twelve subkeys have to be weak: $Z_4^{(i)}$, $Z_5^{(i)}$, $Z_7^{(i)}$, $Z_9^{(i)}$, $Z_1^{(i+1)}$, $Z_3^{(i+1)}$, $Z_6^{(i+1)}$, $Z_8^{(i+1)}$, $Z_9^{(i+1)}$, $Z_2^{(i+2)}$, $Z_4^{(i+2)}$ and $Z_7^{(i+2)}$.

(48) $(0,\gamma,0,0,0,0,\gamma,0) \stackrel{1r}{\to} (\gamma,\gamma,\gamma,\gamma,0,0,0,0) \stackrel{1r}{\to} (\gamma,0,\gamma,\gamma,0,0,\gamma,0) \stackrel{1r}{\to} (0, \gamma, 0, 0, 0, 0, \gamma, 0)$, which is 3-round iterative and holds with bias 2^{-23}. Assuming i is odd, the following six subkeys have to be weak: $Z_9^{(i)}$, $Z_2^{(i+1)}$, $Z_4^{(i+1)}$, $Z_1^{(i+2)}$, $Z_3^{(i+2)}$ and $Z_9^{(i+2)}$. Assuming i is even, the following eight subkeys have to be weak: $Z_2^{(i)}$, $Z_7^{(i)}$, $Z_9^{(i)}$, $Z_1^{(i+1)}$, $Z_3^{(i+1)}$, $Z_4^{(i+2)}$, $Z_7^{(i+2)}$ and $Z_9^{(i+2)}$.

(49) $(0,\gamma,0,0,0,\gamma,\gamma,\gamma) \stackrel{1r}{\to} (\gamma,\gamma,0,0,0,\gamma,0,\gamma) \stackrel{1r}{\to} (0,\gamma,\gamma,\gamma,\gamma,\gamma,0,\gamma) \stackrel{1r}{\to} (0, \gamma, 0, 0, 0, \gamma, \gamma, \gamma)$, which is 3-round iterative and holds with bias 2^{-23}. Assuming i is odd, the following eight subkeys have to be weak: $Z_6^{(i)}$, $Z_8^{(i)}$, $Z_9^{(i)}$, $Z_2^{(i+1)}$, $Z_9^{(i+1)}$, $Z_3^{(i+2)}$, $Z_6^{(i+2)}$ and $Z_8^{(i+2)}$. Assuming i is even, the following ten subkeys have to be weak: $Z_2^{(i)}$, $Z_7^{(i)}$, $Z_9^{(i)}$, $Z_1^{(i+1)}$, $Z_6^{(i+1)}$, $Z_8^{(i+1)}$, $Z_9^{(i+1)}$, $Z_2^{(i+2)}$, $Z_4^{(i+2)}$ and $Z_5^{(i+2)}$.

(50) $(0,\gamma,0,0,\gamma,0,0,0) \stackrel{1r}{\to} (\gamma,0,0,0,\gamma,0,0,0) \stackrel{1r}{\to} (\gamma,\gamma,0,0,0,0,0,0) \stackrel{1r}{\to} (0, \gamma, 0, 0, \gamma, 0, 0, 0)$, which is 3-round iterative and holds with bias 2^{-1}. Assuming i is odd, the following four subkeys have to be weak: $Z_9^{(i)}$, $Z_5^{(i+1)}$, $Z_1^{(i+2)}$ and $Z_9^{(i+2)}$. Assuming i is even, the following six subkeys have to be weak: $Z_2^{(i)}$, $Z_5^{(i)}$, $Z_9^{(i)}$, $Z_1^{(i+1)}$, $Z_2^{(i+2)}$ and $Z_9^{(i+2)}$.

(51) $(0,\gamma,0,0,\gamma,0,\gamma,\gamma) \stackrel{1r}{\to} (0,\gamma,\gamma,0,0,\gamma,0,\gamma) \stackrel{1r}{\to} (\gamma,\gamma,0,\gamma,0,0,0,\gamma) \stackrel{1r}{\to} (0, \gamma, 0, 0, \gamma, 0, \gamma, \gamma)$, which is 3-round iterative and holds with bias 2^{-12}. Assuming i is odd, the following six subkeys have to be weak: $Z_8^{(i)}$, $Z_2^{(i+1)}$, $Z_9^{(i+1)}$, $Z_1^{(i+2)}$, $Z_8^{(i+2)}$ and $Z_9^{(i+2)}$. Assuming i is even, the following ten subkeys have to be weak: $Z_2^{(i)}$, $Z_5^{(i)}$, $Z_7^{(i)}$, $Z_3^{(i+1)}$, $Z_6^{(i+1)}$, $Z_8^{(i+1)}$, $Z_9^{(i+1)}$, $Z_2^{(i+2)}$, $Z_4^{(i+2)}$ and $Z_9^{(i+2)}$.

(52) $(0,\gamma,0,\gamma,0,0,\gamma,\gamma) \stackrel{1r}{\to} (\gamma,\gamma,\gamma,\gamma,0,0,\gamma,\gamma) \stackrel{1r}{\to} (0,\gamma,0,\gamma,\gamma,\gamma,\gamma,\gamma) \stackrel{1r}{\to} (0, \gamma, 0, \gamma, 0, 0, \gamma, \gamma)$, which is 3-round iterative and holds with bias 2^{-12}. Assuming i is odd, the following eight subkeys have to be weak: $Z_8^{(i)}$, $Z_9^{(i)}$, $Z_2^{(i+1)}$, $Z_4^{(i+1)}$, $Z_7^{(i+1)}$, $Z_9^{(i+1)}$, $Z_6^{(i+2)}$ and $Z_8^{(i+2)}$. Assuming i is even, the following twelve subkeys have to be weak: $Z_2^{(i)}$, $Z_4^{(i)}$, $Z_7^{(i)}$, $Z_9^{(i)}$, $Z_1^{(i+1)}$, $Z_3^{(i+1)}$, $Z_8^{(i+1)}$, $Z_9^{(i+1)}$, $Z_2^{(i+2)}$, $Z_4^{(i+2)}$, $Z_5^{(i+2)}$ and $Z_7^{(i+2)}$.

(53) $(0,\gamma,0,\gamma,0,\gamma,\gamma,0) \stackrel{1r}{\to} (\gamma,\gamma,0,0,0,\gamma,\gamma,0) \stackrel{1r}{\to} (\gamma,0,0,\gamma,0,0,0,0) \stackrel{1r}{\to} (0, \gamma, 0, \gamma, 0, \gamma, \gamma, 0)$, which is 3-round iterative and holds with bias 2^{-23}. Assuming i is odd, the following six subkeys have to be weak: $Z_6^{(i)}$, $Z_9^{(i)}$,

$Z_2^{(i+1)}$, $Z_7^{(i+1)}$, $Z_1^{(i+2)}$ and $Z_9^{(i+2)}$. Assuming i is even, the following eight subkeys have to be weak: $Z_2^{(i)}$, $Z_4^{(i)}$, $Z_7^{(i)}$, $Z_9^{(i)}$, $Z_1^{(i+1)}$, $Z_6^{(i+1)}$, $Z_4^{(i+2)}$ and $Z_9^{(i+2)}$.

(54) $(0,\gamma,0,\gamma,\gamma,\gamma,0,0) \stackrel{1r}{\to} (\gamma,0,\gamma,\gamma,\gamma,\gamma,\gamma,0) \stackrel{1r}{\to} (\gamma,\gamma,\gamma,0,0,0,\gamma,0) \stackrel{1r}{\to} (0, \gamma, 0, \gamma, \gamma, \gamma, 0, 0)$, which is 3-round iterative and holds with bias $2^{-17.5}$. Assuming i is odd, the following eight subkeys have to be weak: $Z_6^{(i)}$, $Z_9^{(i)}$, $Z_4^{(i+1)}$, $Z_5^{(i+1)}$, $Z_7^{(i+1)}$, $Z_1^{(i+2)}$, $Z_3^{(i+2)}$ and $Z_9^{(i+2)}$. Assuming i is even, the following ten subkeys have to be weak: $Z_2^{(i)}$, $Z_4^{(i)}$, $Z_5^{(i)}$, $Z_9^{(i)}$, $Z_1^{(i+1)}$, $Z_3^{(i+1)}$, $Z_6^{(i+1)}$, $Z_2^{(i+2)}$, $Z_7^{(i+2)}$ and $Z_9^{(i+2)}$.

(55) $(0, \gamma, \gamma, 0, 0, 0, 0, 0) \stackrel{1r}{\to} (\gamma, \gamma, \gamma, 0, 0, \gamma, 0, 0) \stackrel{1r}{\to} (\gamma, 0, 0, 0, 0, \gamma, 0, 0) \stackrel{1r}{\to} (0, \gamma, \gamma, 0, 0, 0, 0, 0)$, which is 3-round iterative and holds with bias 2^{-12}. Assuming i is odd, the following six subkeys have to be weak: $Z_3^{(i)}$, $Z_9^{(i)}$, $Z_2^{(i+1)}$, $Z_1^{(i+2)}$, $Z_6^{(i+2)}$ and $Z_9^{(i+2)}$. Assuming i is even, the following six subkeys have to be weak: $Z_2^{(i)}$, $Z_9^{(i)}$, $Z_1^{(i+1)}$, $Z_3^{(i+1)}$, $Z_6^{(i+1)}$ and $Z_9^{(i+1)}$.

(56) $(0,\gamma,\gamma,0,\gamma,0,0,\gamma) \stackrel{1r}{\to} (0,\gamma,\gamma,\gamma,0,0,0,\gamma) \stackrel{1r}{\to} (\gamma,\gamma,\gamma,0,0,\gamma,\gamma,\gamma) \stackrel{1r}{\to} (0, \gamma, \gamma, 0, \gamma, 0, 0, \gamma)$, which is 3-round iterative and holds with bias 2^{-23}. Assuming i is odd, the following ten subkeys have to be weak: $Z_3^{(i)}$, $Z_8^{(i)}$, $Z_2^{(i+1)}$, $Z_4^{(i+1)}$, $Z_9^{(i+1)}$, $Z_1^{(i+2)}$, $Z_3^{(i+2)}$, $Z_6^{(i+2)}$, $Z_8^{(i+2)}$ and $Z_9^{(i+2)}$. Assuming i is even, the following eight subkeys have to be weak: $Z_2^{(i)}$, $Z_5^{(i)}$, $Z_3^{(i+1)}$, $Z_8^{(i+1)}$, $Z_9^{(i+1)}$, $Z_2^{(i+2)}$, $Z_7^{(i+2)}$ and $Z_9^{(i+2)}$.

(57) $(0,\gamma,\gamma,0,\gamma,0,\gamma,0) \stackrel{1r}{\to} (\gamma,0,0,\gamma,\gamma,\gamma,0,0) \stackrel{1r}{\to} (\gamma,\gamma,\gamma,\gamma,0,\gamma,\gamma,0) \stackrel{1r}{\to} (0, \gamma, \gamma, 0, \gamma, 0, \gamma, 0)$, which is 3-round iterative and holds with bias $2^{-17.5}$. Assuming i is odd, the following eight subkeys have to be weak: $Z_3^{(i)}$, $Z_9^{(i)}$, $Z_4^{(i+1)}$, $Z_5^{(i+1)}$, $Z_1^{(i+2)}$, $Z_3^{(i+2)}$, $Z_6^{(i+2)}$ and $Z_9^{(i+2)}$. Assuming i is even, the following ten subkeys have to be weak: $Z_2^{(i)}$, $Z_5^{(i)}$, $Z_7^{(i)}$, $Z_9^{(i)}$, $Z_1^{(i+1)}$, $Z_6^{(i+1)}$, $Z_2^{(i+2)}$, $Z_4^{(i+2)}$, $Z_7^{(i+2)}$ and $Z_9^{(i+2)}$.

(58) $(0,\gamma,\gamma,\gamma,0,\gamma,0,0) \stackrel{1r}{\to} (\gamma,\gamma,0,\gamma,0,0,\gamma,0) \stackrel{1r}{\to} (\gamma,0,\gamma,0,0,\gamma,\gamma,0) \stackrel{1r}{\to} (0, \gamma, \gamma, \gamma, 0, \gamma, 0, 0)$, which is 3-round iterative and holds with bias 2^{-12}. Assuming i is odd, the following ten subkeys have to be weak: $Z_3^{(i)}$, $Z_6^{(i)}$, $Z_9^{(i)}$, $Z_2^{(i+1)}$, $Z_4^{(i+1)}$, $Z_7^{(i+1)}$, $Z_1^{(i+2)}$, $Z_3^{(i+2)}$, $Z_6^{(i+2)}$ and $Z_9^{(i+2)}$. Assuming i is even, the following six subkeys have to be weak: $Z_2^{(i)}$, $Z_4^{(i)}$, $Z_9^{(i)}$, $Z_1^{(i+1)}$, $Z_7^{(i+2)}$ and $Z_9^{(i+2)}$.

(59) $(0,\gamma,\gamma,\gamma,\gamma,\gamma,\gamma,0) \stackrel{1r}{\to} (\gamma,0,\gamma,0,\gamma,0,\gamma,0) \stackrel{1r}{\to} (\gamma,\gamma,0,\gamma,0,\gamma,0,0) \stackrel{1r}{\to} (0, \gamma, \gamma, \gamma, \gamma, \gamma, \gamma, 0)$, which is 3-round iterative and holds with bias 2^{-23}. Assuming i is odd, the following eight subkeys have to be weak: $Z_3^{(i)}$, $Z_6^{(i)}$, $Z_9^{(i)}$, $Z_5^{(i+1)}$, $Z_7^{(i+1)}$, $Z_1^{(i+2)}$, $Z_6^{(i+2)}$ and $Z_9^{(i+2)}$. Assuming i is even, the following ten subkeys have to be weak: $Z_2^{(i)}$, $Z_4^{(i)}$, $Z_5^{(i)}$, $Z_7^{(i)}$, $Z_9^{(i)}$, $Z_1^{(i+1)}$, $Z_3^{(i+1)}$, $Z_2^{(i+2)}$, $Z_4^{(i+2)}$ and $Z_9^{(i+2)}$.

(60) $(\gamma,0,0,0,0,\gamma,\gamma,\gamma) \stackrel{1r}{\rightarrow} (\gamma,0,0,0,\gamma,\gamma,0,\gamma) \stackrel{1r}{\rightarrow} (\gamma,\gamma,\gamma,\gamma,0,\gamma,0,\gamma) \stackrel{1r}{\rightarrow} (\gamma,$ $0, 0, 0, 0, \gamma, \gamma, \gamma)$, which is 3-round iterative and holds with bias $2^{-17.5}$. Assuming i is odd, the following eight subkeys have to be weak: $Z_1^{(i)}$, $Z_6^{(i)}$, $Z_8^{(i)}$, $Z_5^{(i+1)}$, $Z_1^{(i+2)}$, $Z_3^{(i+2)}$, $Z_6^{(i+2)}$ and $Z_8^{(i+2)}$. Assuming i is even, the following six subkeys have to be weak: $Z_7^{(i)}$, $Z_1^{(i+1)}$, $Z_6^{(i+1)}$, $Z_8^{(i+1)}$, $Z_2^{(i+2)}$ and $Z_4^{(i+2)}$.

(61) $(\gamma,0,0,\gamma,0,0,\gamma,\gamma) \stackrel{1r}{\rightarrow} (\gamma,0,\gamma,\gamma,\gamma,0,\gamma,\gamma) \stackrel{1r}{\rightarrow} (\gamma,\gamma,0,\gamma,0,\gamma,\gamma,\gamma) \stackrel{1r}{\rightarrow} (\gamma,$ $0, 0, \gamma, 0, 0, \gamma, \gamma)$, which is 3-round iterative and holds with bias 2^{-12}. Assuming i is odd, the following eight subkeys have to be weak: $Z_1^{(i)}$, $Z_8^{(i)}$, $Z_4^{(i+1)}$, $Z_5^{(i+1)}$, $Z_7^{(i+1)}$, $Z_1^{(i+2)}$, $Z_6^{(i+2)}$ and $Z_8^{(i+2)}$. Assuming i is even, the following eight subkeys have to be weak: $Z_4^{(i)}$, $Z_7^{(i)}$, $Z_1^{(i+1)}$, $Z_3^{(i+1)}$, $Z_8^{(i+1)}$, $Z_2^{(i+2)}$, $Z_4^{(i+2)}$ and $Z_7^{(i+2)}$.

(62) $(\gamma,0,0,\gamma,\gamma,0,0,\gamma) \stackrel{1r}{\rightarrow} (\gamma,\gamma,0,0,0,0,\gamma,\gamma) \stackrel{1r}{\rightarrow} (\gamma,0,\gamma,0,0,\gamma,0,\gamma) \stackrel{1r}{\rightarrow} (\gamma,$ $0, 0, \gamma, \gamma, 0, 0, \gamma)$, which is 3-round iterative and holds with bias 2^{-12}. Assuming i is odd, the following eight subkeys have to be weak: $Z_1^{(i)}$, $Z_8^{(i)}$, $Z_2^{(i+1)}$, $Z_7^{(i+1)}$, $Z_1^{(i+2)}$, $Z_3^{(i+2)}$, $Z_6^{(i+2)}$ and $Z_8^{(i+2)}$. Assuming i is even, the following four subkeys have to be weak: $Z_4^{(i)}$, $Z_5^{(i)}$, $Z_1^{(i+1)}$ and $Z_8^{(i+1)}$.

(63) $(\gamma,0,\gamma,0,\gamma,\gamma,\gamma,\gamma) \stackrel{1r}{\rightarrow} (\gamma,\gamma,\gamma,0,0,0,0,\gamma) \stackrel{1r}{\rightarrow} (\gamma,0,\gamma,\gamma,0,0,0,\gamma) \stackrel{1r}{\rightarrow} (\gamma,$ $0, \gamma, 0, \gamma, \gamma, \gamma, \gamma)$, which is 3-round iterative and holds with bias $2^{-17.5}$. Assuming i is odd, the following eight subkeys have to be weak: $Z_1^{(i)}$, $Z_3^{(i)}$, $Z_6^{(i)}$, $Z_8^{(i)}$, $Z_2^{(i+1)}$, $Z_1^{(i+2)}$, $Z_3^{(i+2)}$ and $Z_8^{(i+2)}$. Assuming i is even, the following six subkeys have to be weak: $Z_5^{(i)}$, $Z_7^{(i)}$, $Z_1^{(i+1)}$, $Z_3^{(i+1)}$, $Z_8^{(i+1)}$ and $Z_4^{(i+2)}$.

(64) $(\gamma,\gamma,0,0,\gamma,0,0,\gamma) \stackrel{1r}{\rightarrow} (\gamma,\gamma,0,\gamma,\gamma,\gamma,0,\gamma) \stackrel{1r}{\rightarrow} (\gamma,\gamma,\gamma,0,\gamma,0,\gamma,\gamma) \stackrel{1r}{\rightarrow} (\gamma,$ $\gamma, 0, 0, \gamma, 0, 0, \gamma)$, which is 3-round iterative and holds with bias 2^{-12}. Assuming i is odd, the following eight subkeys have to be weak: $Z_1^{(i)}$, $Z_8^{(i)}$, $Z_2^{(i+1)}$, $Z_4^{(i+1)}$, $Z_5^{(i+1)}$, $Z_1^{(i+2)}$, $Z_3^{(i+2)}$ and $Z_8^{(i+2)}$. Assuming i is even, the following eight subkeys have to be weak: $Z_2^{(i)}$, $Z_5^{(i)}$, $Z_1^{(i+1)}$, $Z_6^{(i+1)}$, $Z_8^{(i+1)}$, $Z_2^{(i+2)}$, $Z_5^{(i+2)}$ and $Z_7^{(i+2)}$.

(65) $(\gamma,\gamma,\gamma,\gamma,\gamma,\gamma,\gamma,\gamma) \stackrel{1r}{\rightarrow} (\gamma,\gamma,\gamma,\gamma,\gamma,\gamma,\gamma,\gamma)$, which is 1-round iterative and holds with bias 2^{-1}. Assuming i is odd, the following four subkeys have to be weak: $Z_1^{(i)}$, $Z_3^{(i)}$, $Z_6^{(i)}$ and $Z_8^{(i)}$. Assuming i is even, the following four subkeys have to be weak: $Z_2^{(i)}$, $Z_4^{(i)}$, $Z_5^{(i)}$ and $Z_7^{(i)}$.

Similarly to the linear analysis of MESH-64, the key schedule of MESH-64(8) does not consist of a simple bit permutation like in IDEA. The fraction of keys that satisfy a weak-subkey restriction in MESH-64(8) was estimated to be about 2^{-7} per subkey. This estimate is in line with the assumption that a weak subkey has value 0 or 1, since the seven most significant bits of a subkey have to be zero. Under this assumption, and since MESH-64(8) operates under a 128-bit key, at most $\lfloor 128/7 \rfloor = 18$ weak subkeys are allowed (for a nonempty weak-key class).

Under this estimate, the linear relations (10), (17), (27), (31) and (50) require the least number of weak subkeys: about 1.33 per round on average. They are all 3-round iterative linear relations. Among them, the one holding with the largest bias is relation (50): 2^{-1}.

Overall, relations (47) and (65) also holds with bias 2^{-1}, but (47) is 3-round iterative and require ten weak subkeys, while (65) is 1-round iterative and require four weak subkeys.

Therefore, repeating relation (50) to cover 8.5-round MESH-64(8) will require 11 weak subkeys and a data complexity of $c/(2^{-1})^2 = 4c$ known plaintexts (for a fixed constant c). Therefore, it is possible to distinguish the full MESH-64(8) from a random permutation with low attack complexity. The cost is essentially $4c$ encryptions and the same amount of known plaintexts and memory. Consequently, MESH-64(8) is not a PRP. See Sect. 1.7.

The weak-key class size is estimated as $2^{128-7\cdot 11} = 2^{51}$. The exact user keys belonging to this weak-key class are not known. Identifying them is left as an open problem.

This linear analysis also allows us to study the diffusion power of MESH-64(8) with respect to modern tools such as the linear branch number. See Appendix, Sect. B.

The set of linear relations in Tables 3.112 through 3.119 indicates that the linear branch number *per round* of MESH-64(8) is only 4 (under weak-key assumptions). For instance, the 1-round relation $(\gamma,\gamma,0,0,0,0,0,0) \stackrel{1r}{\rightarrow} (0,\gamma,0,0,\gamma,0,0,0)$ has both the input and output masks with Hamming Weight 2. So, even though MESH-64(8) provides full diffusion in a single round, a detailed analysis shows that its linear branch number is much lower than ideal. For an 8-word block, such as in MESH-64(8), an MDS code using a 8×8 MDS matrix would rather provide a branch number of $1+8=9$. See Appendix, Sect. B.1.

3.13.8 LC of MESH-128(8) Under Weak-Key Assumptions

The following linear analysis of MESH-128(8) is more extensive than [151] and operates under weak-key assumptions.

The analysis follows a bottom-up approach, starting from the smallest components: $\odot$, $\boxplus$ and $\oplus$, up to an MA-box, up to a half-round, further to a full round, and finally to multiple rounds.

Across unkeyed $\odot$ and $\boxplus$, the propagation of bitmasks follows from Table 3.98 and Table 3.99, respectively.

Further, we obtain all linear relation patterns across the MA-box of MESH-128(8), under the corresponding weak-subkey restrictions.

Tables 3.120 through 3.127 lists all 255 nontrivial linear relation patterns across the MA-box of MESH-128(8). Only two 8-bit masks are used: 0 and $\gamma = 1$.

Table 3.120 Nontrivial bitmask propagation across the MA-box of MESH-128(8) (part I).

Bitmask propagation across the MA-box	Weak subkeys	Bias
$(\gamma,\gamma,\gamma,\gamma,\gamma,\gamma,\gamma,\gamma) \rightarrow (0,0,0,0,0,0,0,\gamma)$	$Z_{17}^{(i)}, Z_{18}^{(i)}$	$2^{-17.5}$
$(0,0,0,0,0,0,0,\gamma) \rightarrow (0,0,0,0,0,0,\gamma,0)$	$Z_{18}^{(i)}$	2^{-1}
$(\gamma,\gamma,\gamma,\gamma,\gamma,\gamma,\gamma,0) \rightarrow (0,0,0,0,0,0,\gamma,\gamma)$	$Z_{17}^{(i)}$	$2^{-17.5}$
$(\gamma,\gamma,\gamma,\gamma,\gamma,\gamma,0,\gamma) \rightarrow (0,0,0,0,0,\gamma,0,0)$	$Z_{17}^{(i)}, Z_{18}^{(i)}$	$2^{-17.5}$
$(0,0,0,0,0,0,\gamma,0) \rightarrow (0,0,0,0,0,\gamma,0,\gamma)$	—	2^{-12}
$(\gamma,\gamma,\gamma,\gamma,\gamma,\gamma,0,0) \rightarrow (0,0,0,0,0,\gamma,\gamma,0)$	$Z_{17}^{(i)}$	$2^{-17.5}$
$(0,0,0,0,0,0,\gamma,\gamma) \rightarrow (0,0,0,0,0,\gamma,\gamma,\gamma)$	$Z_{18}^{(i)}$	2^{-12}
$(0,0,0,0,0,\gamma,0,\gamma) \rightarrow (0,0,0,0,\gamma,0,0,0)$	$Z_{18}^{(i)}$	$2^{-6.5}$
$(\gamma,\gamma,\gamma,\gamma,\gamma,0,\gamma,0) \rightarrow (0,0,0,0,\gamma,0,0,\gamma)$	$Z_{17}^{(i)}$	2^{-23}
$(0,0,0,0,0,\gamma,0,0) \rightarrow (0,0,0,0,\gamma,0,\gamma,0)$	—	$2^{-6.5}$
$(\gamma,\gamma,\gamma,\gamma,\gamma,0,\gamma,\gamma) \rightarrow (0,0,0,0,\gamma,0,\gamma,\gamma)$	$Z_{17}^{(i)}, Z_{18}^{(i)}$	2^{-23}
$(\gamma,\gamma,\gamma,\gamma,\gamma,0,0,0) \rightarrow (0,0,0,0,\gamma,\gamma,0,0)$	$Z_{17}^{(i)}$	2^{-12}
$(0,0,0,0,0,\gamma,\gamma,\gamma) \rightarrow (0,0,0,0,\gamma,\gamma,0,\gamma)$	$Z_{18}^{(i)}$	$2^{-6.5}$
$(\gamma,\gamma,\gamma,\gamma,\gamma,0,0,\gamma) \rightarrow (0,0,0,0,\gamma,\gamma,\gamma,0)$	$Z_{17}^{(i)}, Z_{18}^{(i)}$	2^{-12}
$(0,0,0,0,0,\gamma,\gamma,0) \rightarrow (0,0,0,0,\gamma,\gamma,\gamma,\gamma)$	—	$2^{-6.5}$
$(\gamma,\gamma,\gamma,\gamma,0,\gamma,0,\gamma) \rightarrow (0,0,0,\gamma,0,0,0,0)$	$Z_{17}^{(i)}, Z_{18}^{(i)}$	$2^{-17.5}$
$(0,0,0,0,\gamma,0,\gamma,0) \rightarrow (0,0,0,\gamma,0,0,0,\gamma)$	—	2^{-23}
$(\gamma,\gamma,\gamma,\gamma,0,\gamma,0,0) \rightarrow (0,0,0,\gamma,0,0,\gamma,0)$	$Z_{17}^{(i)}$	$2^{-17.5}$
$(0,0,0,0,\gamma,0,\gamma,\gamma) \rightarrow (0,0,0,\gamma,0,0,\gamma,\gamma)$	$Z_{18}^{(i)}$	2^{-23}
$(0,0,0,0,\gamma,0,0,0) \rightarrow (0,0,0,\gamma,0,\gamma,0,0)$	—	2^{-12}
$(\gamma,\gamma,\gamma,\gamma,0,\gamma,\gamma,\gamma) \rightarrow (0,0,0,\gamma,0,\gamma,0,\gamma)$	$Z_{17}^{(i)}, Z_{18}^{(i)}$	$2^{-17.5}$
$(0,0,0,0,\gamma,0,0,\gamma) \rightarrow (0,0,0,\gamma,0,\gamma,\gamma,0)$	$Z_{18}^{(i)}$	2^{-12}
$(\gamma,\gamma,\gamma,\gamma,0,\gamma,\gamma,0) \rightarrow (0,0,0,\gamma,0,\gamma,\gamma,\gamma)$	$Z_{17}^{(i)}$	$2^{-17.5}$
$(\gamma,\gamma,\gamma,\gamma,0,0,0,0) \rightarrow (0,0,0,\gamma,\gamma,0,0,0)$	$Z_{17}^{(i)}$	2^{-12}
$(0,0,0,0,\gamma,\gamma,\gamma,\gamma) \rightarrow (0,0,0,\gamma,\gamma,0,0,\gamma)$	$Z_{18}^{(i)}$	$2^{-17.5}$
$(\gamma,\gamma,\gamma,\gamma,0,0,0,\gamma) \rightarrow (0,0,0,\gamma,\gamma,0,\gamma,0)$	$Z_{17}^{(i)}, Z_{18}^{(i)}$	2^{-12}
$(0,0,0,0,\gamma,\gamma,\gamma,0) \rightarrow (0,0,0,\gamma,\gamma,0,\gamma,\gamma)$	—	$2^{-17.5}$
$(0,0,0,0,\gamma,\gamma,0,\gamma) \rightarrow (0,0,0,\gamma,\gamma,\gamma,0,0)$	$Z_{18}^{(i)}$	$2^{-17.5}$
$(\gamma,\gamma,\gamma,\gamma,0,0,\gamma,0) \rightarrow (0,0,0,\gamma,\gamma,\gamma,0,\gamma)$	$Z_{17}^{(i)}$	2^{-23}
$(0,0,0,0,\gamma,\gamma,0,0) \rightarrow (0,0,0,\gamma,\gamma,\gamma,\gamma,0)$	—	$2^{-17.5}$
$(\gamma,\gamma,\gamma,\gamma,0,0,\gamma,\gamma) \rightarrow (0,0,0,\gamma,\gamma,\gamma,\gamma,\gamma)$	$Z_{17}^{(i)}, Z_{18}^{(i)}$	2^{-23}
$(0,0,0,\gamma,0,\gamma,0,\gamma) \rightarrow (0,0,\gamma,0,0,0,0,0)$	$Z_{18}^{(i)}$	2^{-12}

The next step is to obtain from Tables 3.120 through 3.127 all 1-round linear relations and the corresponding weak-subkey conditions. There are 65535 such 1-round linear relations in total[15].

[15] They are too many to be listed here. These 1-round linear relations for MESH-128(8) are available from the book's website http://www.springer.com/978-3-319-68272-3.

Table 3.121 Nontrivial bitmask propagation across the MA-box of MESH-128(8) (part II).

Bitmask propagation across the MA-box	Weak subkeys	Bias
$(\gamma,\gamma,\gamma,0,\gamma,0,\gamma,0) \rightarrow (0,0,\gamma,0,0,0,0,\gamma)$	$Z_{17}^{(i)}$	$2^{-28.5}$
$(0,0,0,\gamma,0,\gamma,0,0) \rightarrow (0,0,\gamma,0,0,0,\gamma,0)$	—	2^{-12}
$(\gamma,\gamma,\gamma,0,\gamma,0,\gamma,\gamma) \rightarrow (0,0,\gamma,0,0,0,\gamma,\gamma)$	$Z_{17}^{(i)}, Z_{18}^{(i)}$	$2^{-28.5}$
$(\gamma,\gamma,\gamma,0,\gamma,0,0,0) \rightarrow (0,0,\gamma,0,0,\gamma,0,0)$	$Z_{17}^{(i)}$	$2^{-17.5}$
$(0,0,0,\gamma,0,\gamma,\gamma,\gamma) \rightarrow (0,0,\gamma,0,0,\gamma,0,\gamma)$	$Z_{18}^{(i)}$	2^{-12}
$(\gamma,\gamma,\gamma,0,\gamma,0,0,\gamma) \rightarrow (0,0,\gamma,0,0,\gamma,\gamma,0)$	$Z_{17}^{(i)}, Z_{18}^{(i)}$	$2^{-17.5}$
$(0,0,0,\gamma,0,\gamma,\gamma,0) \rightarrow (0,0,\gamma,0,0,\gamma,\gamma,\gamma)$	—	2^{-12}
$(0,0,0,\gamma,0,0,0,0) \rightarrow (0,0,\gamma,0,\gamma,0,0,0)$	—	$2^{-6.5}$
$(\gamma,\gamma,\gamma,0,\gamma,\gamma,\gamma,\gamma) \rightarrow (0,0,\gamma,0,\gamma,0,0,\gamma)$	$Z_{17}^{(i)}, Z_{18}^{(i)}$	2^{-23}
$(0,0,0,\gamma,0,0,0,\gamma) \rightarrow (0,0,\gamma,0,\gamma,0,\gamma,0)$	$Z_{18}^{(i)}$	$2^{-6.5}$
$(\gamma,\gamma,\gamma,0,\gamma,\gamma,\gamma,0) \rightarrow (0,0,\gamma,0,\gamma,0,\gamma,\gamma)$	$Z_{17}^{(i)}$	2^{-23}
$(\gamma,\gamma,\gamma,0,\gamma,\gamma,0,\gamma) \rightarrow (0,0,\gamma,0,\gamma,\gamma,0,0)$	$Z_{17}^{(i)}, Z_{18}^{(i)}$	2^{-23}
$(0,0,0,\gamma,0,0,\gamma,0) \rightarrow (0,0,\gamma,0,\gamma,\gamma,0,\gamma)$	—	$2^{-17.5}$
$(\gamma,\gamma,\gamma,0,\gamma,\gamma,0,0) \rightarrow (0,0,\gamma,0,\gamma,\gamma,\gamma,0)$	$Z_{17}^{(i)}$	2^{-23}
$(0,0,0,\gamma,0,0,\gamma,\gamma) \rightarrow (0,0,\gamma,0,\gamma,\gamma,\gamma,\gamma)$	$Z_{18}^{(i)}$	$2^{-17.5}$
$(\gamma,\gamma,\gamma,0,0,0,0,0) \rightarrow (0,0,\gamma,\gamma,0,0,0,0)$	$Z_{17}^{(i)}$	$2^{-6.5}$
$(0,0,0,\gamma,\gamma,\gamma,\gamma,\gamma) \rightarrow (0,0,\gamma,\gamma,0,0,0,\gamma)$	$Z_{18}^{(i)}$	2^{-12}
$(\gamma,\gamma,\gamma,0,0,0,0,\gamma) \rightarrow (0,0,\gamma,\gamma,0,0,\gamma,0)$	$Z_{17}^{(i)}, Z_{18}^{(i)}$	$2^{-6.5}$
$(0,0,0,\gamma,\gamma,\gamma,\gamma,0) \rightarrow (0,0,\gamma,\gamma,0,0,\gamma,\gamma)$	—	2^{-12}
$(0,0,0,\gamma,\gamma,\gamma,0,\gamma) \rightarrow (0,0,\gamma,\gamma,0,\gamma,0,0)$	$Z_{18}^{(i)}$	2^{-12}
$(\gamma,\gamma,\gamma,0,0,0,\gamma,0) \rightarrow (0,0,\gamma,\gamma,0,\gamma,0,\gamma)$	$Z_{17}^{(i)}$	$2^{-17.5}$
$(0,0,0,\gamma,\gamma,\gamma,0,0) \rightarrow (0,0,\gamma,\gamma,0,\gamma,\gamma,0)$	—	2^{-12}
$(\gamma,\gamma,\gamma,0,0,0,\gamma,\gamma) \rightarrow (0,0,\gamma,\gamma,0,\gamma,\gamma,\gamma)$	$Z_{17}^{(i)}, Z_{18}^{(i)}$	$2^{-17.5}$
$(\gamma,\gamma,\gamma,0,0,\gamma,0,\gamma) \rightarrow (0,0,\gamma,\gamma,\gamma,0,0,0)$	$Z_{17}^{(i)}, Z_{18}^{(i)}$	2^{-12}
$(0,0,0,\gamma,\gamma,0,\gamma,0) \rightarrow (0,0,\gamma,\gamma,\gamma,0,0,\gamma)$	—	$2^{-17.5}$
$(\gamma,\gamma,\gamma,0,0,\gamma,0,0) \rightarrow (0,0,\gamma,\gamma,\gamma,0,\gamma,0)$	$Z_{17}^{(i)}$	2^{-12}
$(0,0,0,\gamma,\gamma,0,\gamma,\gamma) \rightarrow (0,0,\gamma,\gamma,\gamma,0,\gamma,\gamma)$	$Z_{18}^{(i)}$	$2^{-17.5}$
$(0,0,0,\gamma,\gamma,0,0,0) \rightarrow (0,0,\gamma,\gamma,\gamma,\gamma,0,0)$	—	$2^{-6.5}$
$(\gamma,\gamma,\gamma,0,0,\gamma,\gamma,\gamma) \rightarrow (0,0,\gamma,\gamma,\gamma,\gamma,0,\gamma)$	$Z_{17}^{(i)}, Z_{18}^{(i)}$	2^{-12}
$(0,0,0,\gamma,\gamma,0,0,\gamma) \rightarrow (0,0,\gamma,\gamma,\gamma,\gamma,\gamma,0)$	$Z_{18}^{(i)}$	$2^{-6.5}$
$(\gamma,\gamma,\gamma,0,0,\gamma,\gamma,0) \rightarrow (0,0,\gamma,\gamma,\gamma,\gamma,\gamma,\gamma)$	$Z_{17}^{(i)}$	2^{-12}
$(\gamma,\gamma,0,\gamma,0,\gamma,0,\gamma) \rightarrow (0,\gamma,0,0,0,0,0,0)$	$Z_{17}^{(i)}, Z_{18}^{(i)}$	$2^{-17.5}$

Starting from the first round, $i = 1$, the linear relations that require the least number of weak subkeys are:

(1) $(0,0,0,0,0,0,0,0,0,0,0,0,0,\gamma,\gamma,0) \stackrel{1r}{\rightarrow} (0, 0, 0, \gamma, \gamma, \gamma, 0, 0, 0, \gamma, \gamma, \gamma, \gamma, \gamma, \gamma, \gamma) \stackrel{1r}{\rightarrow} (0,0,0,0,0,0,0,\gamma,\gamma,\gamma,0,0,0,0,0,\gamma) \stackrel{1r}{\rightarrow} (0, \gamma, \gamma, 0, 0, 0, 0, 0, 0, 0, 0, 0, 0, 0, 0, 0) \stackrel{1r}{\rightarrow} (0,0,0,0,0,0,0,\gamma,\gamma,\gamma,0,0,0,\gamma,\gamma,\gamma) \stackrel{1r}{\rightarrow} (0, \gamma, \gamma, \gamma, \gamma, \gamma, 0, 0, 0, \gamma, \gamma, \gamma, \gamma, \gamma, \gamma, \gamma) \stackrel{1r}{\rightarrow} (0, 0, 0, 0, 0, 0, 0, 0, 0, 0, 0, 0, 0, \gamma, \gamma, 0)$, which is 6-round iterative and holds with a bias of 2^{-78}.

Table 3.122 Nontrivial bitmask propagation across the MA-box of MESH-128(8) (part III).

Bitmask propagation across the MA-box	Weak subkeys	Bias
$(0,0,\gamma,0,\gamma,0,\gamma,0) \rightarrow (0,\gamma,0,0,0,0,0,\gamma)$	—	2^{-34}
$(\gamma,\gamma,0,\gamma,0,\gamma,0,0) \rightarrow (0,\gamma,0,0,0,0,\gamma,0)$	$Z_{17}^{(i)}$	$2^{-17.5}$
$(0,0,\gamma,0,\gamma,0,\gamma,\gamma) \rightarrow (0,\gamma,0,0,0,0,\gamma,\gamma)$	$Z_{18}^{(i)}$	2^{-34}
$(0,0,\gamma,0,\gamma,0,0,0) \rightarrow (0,\gamma,0,0,0,\gamma,0,0)$	—	2^{-23}
$(\gamma,\gamma,0,\gamma,0,\gamma,\gamma,\gamma) \rightarrow (0,\gamma,0,0,0,\gamma,0,\gamma)$	$Z_{17}^{(i)}, Z_{18}^{(i)}$	$2^{-17.5}$
$(0,0,\gamma,0,\gamma,0,0,\gamma) \rightarrow (0,\gamma,0,0,0,\gamma,\gamma,0)$	$Z_{18}^{(i)}$	2^{-23}
$(\gamma,\gamma,0,\gamma,0,\gamma,\gamma,0) \rightarrow (0,\gamma,0,0,0,\gamma,\gamma,\gamma)$	$Z_{17}^{(i)}$	$2^{-17.5}$
$(\gamma,\gamma,0,\gamma,0,0,0,0) \rightarrow (0,\gamma,0,0,\gamma,0,0,0)$	$Z_{17}^{(i)}$	2^{-12}
$(0,0,\gamma,0,\gamma,\gamma,\gamma,\gamma) \rightarrow (0,\gamma,0,0,\gamma,0,0,\gamma)$	$Z_{18}^{(i)}$	$2^{-28.5}$
$(\gamma,\gamma,0,\gamma,0,0,0,\gamma) \rightarrow (0,\gamma,0,0,\gamma,0,\gamma,0)$	$Z_{17}^{(i)}, Z_{18}^{(i)}$	2^{-12}
$(0,0,\gamma,0,\gamma,\gamma,\gamma,0) \rightarrow (0,\gamma,0,0,\gamma,0,\gamma,\gamma)$	—	$2^{-28.5}$
$(0,0,\gamma,0,\gamma,\gamma,0,\gamma) \rightarrow (0,\gamma,0,0,\gamma,\gamma,0,0)$	$Z_{18}^{(i)}$	$2^{-28.5}$
$(\gamma,\gamma,0,\gamma,0,0,\gamma,0) \rightarrow (0,\gamma,0,0,\gamma,\gamma,0,\gamma)$	$Z_{17}^{(i)}$	2^{-23}
$(0,0,\gamma,0,\gamma,\gamma,0,0) \rightarrow (0,\gamma,0,0,\gamma,\gamma,\gamma,0)$	—	$2^{-28.5}$
$(\gamma,\gamma,0,\gamma,0,0,\gamma,\gamma) \rightarrow (0,\gamma,0,0,\gamma,\gamma,\gamma,\gamma)$	$Z_{17}^{(i)}, Z_{18}^{(i)}$	2^{-23}
$(0,0,\gamma,0,0,0,0,0) \rightarrow (0,\gamma,0,\gamma,0,0,0,0)$	—	2^{-12}
$(\gamma,\gamma,0,\gamma,\gamma,\gamma,\gamma,\gamma) \rightarrow (0,\gamma,0,\gamma,0,0,0,\gamma)$	$Z_{17}^{(i)}, Z_{18}^{(i)}$	$2^{-17.5}$
$(0,0,\gamma,0,0,0,0,\gamma) \rightarrow (0,\gamma,0,\gamma,0,0,\gamma,0)$	$Z_{18}^{(i)}$	2^{-12}
$(\gamma,\gamma,0,\gamma,\gamma,\gamma,\gamma,0) \rightarrow (0,\gamma,0,\gamma,0,0,\gamma,\gamma)$	$Z_{17}^{(i)}$	$2^{-17.5}$
$(\gamma,\gamma,0,\gamma,\gamma,\gamma,0,\gamma) \rightarrow (0,\gamma,0,\gamma,0,\gamma,0,0)$	$Z_{17}^{(i)}, Z_{18}^{(i)}$	$2^{-17.5}$
$(0,0,\gamma,0,0,0,\gamma,0) \rightarrow (0,\gamma,0,\gamma,0,\gamma,0,\gamma)$	—	2^{-23}
$(\gamma,\gamma,0,\gamma,\gamma,\gamma,0,0) \rightarrow (0,\gamma,0,\gamma,0,\gamma,\gamma,0)$	$Z_{17}^{(i)}$	$2^{-17.5}$
$(0,0,\gamma,0,0,0,\gamma,\gamma) \rightarrow (0,\gamma,0,\gamma,0,\gamma,\gamma,\gamma)$	$Z_{18}^{(i)}$	2^{-23}
$(0,0,\gamma,0,0,\gamma,0,\gamma) \rightarrow (0,\gamma,0,\gamma,\gamma,0,0,0)$	$Z_{18}^{(i)}$	$2^{-17.5}$
$(\gamma,\gamma,0,\gamma,\gamma,0,\gamma,0) \rightarrow (0,\gamma,0,\gamma,\gamma,0,0,\gamma)$	$Z_{17}^{(i)}$	2^{-23}
$(0,0,\gamma,0,0,\gamma,0,0) \rightarrow (0,\gamma,0,\gamma,\gamma,0,\gamma,0)$	—	$2^{-17.5}$
$(\gamma,\gamma,0,\gamma,\gamma,0,\gamma,\gamma) \rightarrow (0,\gamma,0,\gamma,\gamma,0,\gamma,\gamma)$	$Z_{17}^{(i)}, Z_{18}^{(i)}$	2^{-23}
$(\gamma,\gamma,0,\gamma,\gamma,0,0,0) \rightarrow (0,\gamma,0,\gamma,\gamma,\gamma,0,0)$	$Z_{17}^{(i)}$	2^{-12}
$(0,0,\gamma,0,0,\gamma,\gamma,\gamma) \rightarrow (0,\gamma,0,\gamma,\gamma,\gamma,0,\gamma)$	$Z_{18}^{(i)}$	$2^{-17.5}$
$(\gamma,\gamma,0,\gamma,\gamma,0,0,\gamma) \rightarrow (0,\gamma,0,\gamma,\gamma,\gamma,\gamma,0)$	$Z_{17}^{(i)}, Z_{18}^{(i)}$	2^{-12}
$(0,0,\gamma,0,0,\gamma,\gamma,0) \rightarrow (0,\gamma,0,\gamma,\gamma,\gamma,\gamma,\gamma)$	—	$2^{-17.5}$
$(\gamma,\gamma,0,0,0,0,0,0) \rightarrow (0,\gamma,\gamma,0,0,0,0,0)$	$Z_{17}^{(i)}$	$2^{-6.5}$

Assuming i is odd, the required weak subkeys are: $Z_{14}^{(i)}$, $Z_{17}^{(i)}$, $Z_{4}^{(i+1)}$, $Z_{6}^{(i+1)}$, $Z_{11}^{(i+1)}$, $Z_{13}^{(i+1)}$, $Z_{15}^{(i+1)}$, $Z_{10}^{(i+2)}$, $Z_{16}^{(i+2)}$, $Z_{17}^{(i+2)}$, $Z_{2}^{(i+3)}$, $Z_{17}^{(i+3)}$, $Z_{10}^{(i+4)}$, $Z_{14}^{(i+4)}$, $Z_{16}^{(i+4)}$, $Z_{2}^{(i+5)}$, $Z_{4}^{(i+5)}$, $Z_{6}^{(i+5)}$, $Z_{11}^{(i+5)}$, $Z_{13}^{(i+5)}$, $Z_{15}^{(i+5)}$ and $Z_{17}^{(i+5)}$.

Assuming i is even, the required weak subkeys are: $Z_{15}^{(i)}$, $Z_{17}^{(i)}$, $Z_{5}^{(i+1)}$, $Z_{10}^{(i+1)}$, $Z_{12}^{(i+1)}$, $Z_{14}^{(i+1)}$, $Z_{16}^{(i+1)}$, $Z_{8}^{(i+2)}$, $Z_{9}^{(i+2)}$, $Z_{17}^{(i+2)}$, $Z_{3}^{(i+3)}$, $Z_{17}^{(i+3)}$, $Z_{8}^{(i+4)}$, $Z_{9}^{(i+4)}$, $Z_{15}^{(i+4)}$, $Z_{3}^{(i+5)}$, $Z_{5}^{(i+5)}$, $Z_{10}^{(i+5)}$, $Z_{12}^{(i+5)}$, $Z_{14}^{(i+5)}$, $Z_{16}^{(i+5)}$ and $Z_{17}^{(i+5)}$.

Table 3.123 Nontrivial bitmask propagation across the MA-box of MESH-128(8) (part IV).

Bitmask propagation across the MA-box	Weak subkeys	Bias
$(0,0,\gamma,\gamma,\gamma,\gamma,\gamma,\gamma) \to (0,\gamma,\gamma,0,0,0,0,\gamma)$	$Z_{18}^{(i)}$	2^{-23}
$(\gamma,\gamma,0,0,0,0,0,\gamma) \to (0,\gamma,\gamma,0,0,0,\gamma,0)$	$Z_{17}^{(i)}, Z_{18}^{(i)}$	$2^{-6.5}$
$(0,0,\gamma,\gamma,\gamma,\gamma,\gamma,0) \to (0,\gamma,\gamma,0,0,0,\gamma,\gamma)$	—	2^{-23}
$(0,0,\gamma,\gamma,\gamma,\gamma,0,\gamma) \to (0,\gamma,\gamma,0,0,\gamma,0,0)$	$Z_{18}^{(i)}$	2^{-23}
$(\gamma,\gamma,0,0,0,0,\gamma,0) \to (0,\gamma,\gamma,0,0,\gamma,0,\gamma)$	$Z_{17}^{(i)}$	$2^{-17.5}$
$(0,0,\gamma,\gamma,\gamma,\gamma,0,0) \to (0,\gamma,\gamma,0,0,\gamma,\gamma,0)$	—	2^{-23}
$(\gamma,\gamma,0,0,0,0,\gamma,\gamma) \to (0,\gamma,\gamma,0,0,\gamma,\gamma,\gamma)$	$Z_{17}^{(i)}, Z_{18}^{(i)}$	$2^{-17.5}$
$(\gamma,\gamma,0,0,0,\gamma,0,\gamma) \to (0,\gamma,\gamma,0,\gamma,0,0,0)$	$Z_{17}^{(i)}, Z_{18}^{(i)}$	2^{-12}
$(0,0,\gamma,\gamma,\gamma,0,\gamma,0) \to (0,\gamma,\gamma,0,\gamma,0,0,\gamma)$	—	$2^{-28.5}$
$(\gamma,\gamma,0,0,0,\gamma,0,0) \to (0,\gamma,\gamma,0,\gamma,0,\gamma,0)$	$Z_{17}^{(i)}$	2^{-12}
$(0,0,\gamma,\gamma,\gamma,0,\gamma,\gamma) \to (0,\gamma,\gamma,0,\gamma,0,\gamma,\gamma)$	$Z_{18}^{(i)}$	$2^{-28.5}$
$(0,0,\gamma,\gamma,\gamma,0,0,0) \to (0,\gamma,\gamma,0,\gamma,\gamma,0,0)$	—	$2^{-17.5}$
$(\gamma,\gamma,0,0,0,\gamma,\gamma,\gamma) \to (0,\gamma,\gamma,0,\gamma,\gamma,0,\gamma)$	$Z_{17}^{(i)}, Z_{18}^{(i)}$	2^{-12}
$(0,0,\gamma,\gamma,\gamma,0,0,\gamma) \to (0,\gamma,\gamma,0,\gamma,\gamma,\gamma,0)$	$Z_{18}^{(i)}$	$2^{-17.5}$
$(\gamma,\gamma,0,0,0,\gamma,\gamma,0) \to (0,\gamma,\gamma,0,\gamma,\gamma,\gamma,\gamma)$	$Z_{17}^{(i)}$	2^{-12}
$(0,0,\gamma,\gamma,0,\gamma,0,\gamma) \to (0,\gamma,\gamma,\gamma,0,0,0,0)$	$Z_{18}^{(i)}$	2^{-23}
$(\gamma,\gamma,0,0,\gamma,0,\gamma,0) \to (0,\gamma,\gamma,\gamma,0,0,0,\gamma)$	$Z_{17}^{(i)}$	$2^{-28.5}$
$(0,0,\gamma,\gamma,0,\gamma,0,0) \to (0,\gamma,\gamma,\gamma,0,0,\gamma,0)$	—	2^{-23}
$(\gamma,\gamma,0,0,\gamma,0,\gamma,\gamma) \to (0,\gamma,\gamma,\gamma,0,0,\gamma,\gamma)$	$Z_{17}^{(i)}, Z_{18}^{(i)}$	$2^{-28.5}$
$(\gamma,\gamma,0,0,\gamma,0,0,0) \to (0,\gamma,\gamma,\gamma,0,\gamma,0,0)$	$Z_{17}^{(i)}$	$2^{-17.5}$
$(0,0,\gamma,\gamma,0,\gamma,\gamma,\gamma) \to (0,\gamma,\gamma,\gamma,0,\gamma,0,\gamma)$	$Z_{18}^{(i)}$	2^{-23}
$(\gamma,\gamma,0,0,\gamma,0,0,\gamma) \to (0,\gamma,\gamma,\gamma,0,\gamma,\gamma,0)$	$Z_{17}^{(i)}, Z_{18}^{(i)}$	$2^{-17.5}$
$(0,0,\gamma,\gamma,0,\gamma,\gamma,0) \to (0,\gamma,\gamma,\gamma,0,\gamma,\gamma,\gamma)$	—	2^{-23}
$(0,0,\gamma,\gamma,0,0,0,0) \to (0,\gamma,\gamma,\gamma,\gamma,0,0,0)$	—	$2^{-17.5}$
$(\gamma,\gamma,0,0,\gamma,\gamma,\gamma,\gamma) \to (0,\gamma,\gamma,\gamma,\gamma,0,0,\gamma)$	$Z_{17}^{(i)}, Z_{18}^{(i)}$	2^{-23}
$(0,0,\gamma,\gamma,0,0,0,\gamma) \to (0,\gamma,\gamma,\gamma,\gamma,0,\gamma,0)$	$Z_{18}^{(i)}$	$2^{-17.5}$
$(\gamma,\gamma,0,0,\gamma,\gamma,\gamma,0) \to (0,\gamma,\gamma,\gamma,\gamma,0,\gamma,\gamma)$	$Z_{17}^{(i)}$	2^{-23}
$(\gamma,\gamma,0,0,\gamma,\gamma,0,\gamma) \to (0,\gamma,\gamma,\gamma,\gamma,\gamma,0,0)$	$Z_{17}^{(i)}, Z_{18}^{(i)}$	2^{-23}
$(0,0,\gamma,\gamma,0,0,\gamma,0) \to (0,\gamma,\gamma,\gamma,\gamma,\gamma,0,\gamma)$	—	$2^{-28.5}$
$(\gamma,\gamma,0,0,\gamma,\gamma,0,0) \to (0,\gamma,\gamma,\gamma,\gamma,\gamma,\gamma,0)$	$Z_{17}^{(i)}$	2^{-23}
$(0,0,\gamma,\gamma,0,0,\gamma,\gamma) \to (0,\gamma,\gamma,\gamma,\gamma,\gamma,\gamma,\gamma)$	$Z_{18}^{(i)}$	$2^{-28.5}$
$(0,\gamma,0,\gamma,0,\gamma,0,\gamma) \to (\gamma,0,0,0,0,0,0,0)$	$Z_{18}^{(i)}$	$2^{-17.5}$

It requires 22 weak subkeys over six round, that is, 3.66 weak subkeys per round on average.

(2) $(0,0,0,0,0,0,0,0,0,0,0,0,\gamma,0,\gamma,0) \stackrel{1r}{\to} (0,0,0,\gamma,0,\gamma,0,\gamma,0,\gamma,0,0,0,0,0,0) \stackrel{1r}{\to} (0,0,0,0,0,0,\gamma,0,\gamma,0,\gamma,0,0,0,\gamma,0) \stackrel{1r}{\to} (0,\gamma,0,\gamma,0,0,0,0,0,0,0,0,0,0,0,0) \stackrel{1r}{\to} (0,0,0,0,0,0,\gamma,0,\gamma,0,\gamma,0,\gamma,0,0,0) \stackrel{1r}{\to} (0,\gamma,0,0,0,\gamma,0,\gamma,0,\gamma,0,0,0,0,0,0) \stackrel{1r}{\to} (0,0,0,0,0,0,0,0,0,0,0,0,\gamma,0,\gamma,0)$, which is 6-round iterative and holds with a bias of 2^{-67}. Assuming i is odd, the required weak subkeys are: $Z_4^{(i+1)}$, $Z_6^{(i+1)}$, $Z_8^{(i+1)}$, $Z_7^{(i+2)}$, $Z_2^{(i+3)}$, $Z_4,^{(i+3)}$, $Z_7^{(i+4)}$, $Z_2^{(i+5)}$, $Z_6^{(i+5)}$ and $Z_8^{(i+5)}$.

Table 3.124 Nontrivial bitmask propagation across the MA-box of MESH-128(8) (part V).

Bitmask propagation across the MA-box	Weak subkeys	Bias
$(\gamma,0,\gamma,0,\gamma,0,\gamma,0) \rightarrow (\gamma,0,0,0,0,0,0,\gamma)$	$Z_{17}^{(i)}$	2^{-34}
$(0,\gamma,0,\gamma,0,\gamma,0,0) \rightarrow (\gamma,0,0,0,0,0,\gamma,0)$	—	$2^{-17.5}$
$(\gamma,0,\gamma,0,\gamma,0,\gamma,\gamma) \rightarrow (\gamma,0,0,0,0,0,\gamma,\gamma)$	$Z_{17}^{(i)}, Z_{18}^{(i)}$	2^{-34}
$(\gamma,0,\gamma,0,\gamma,0,0,0) \rightarrow (\gamma,0,0,0,0,\gamma,0,0)$	$Z_{17}^{(i)}$	2^{-23}
$(0,\gamma,0,\gamma,0,\gamma,\gamma,\gamma) \rightarrow (\gamma,0,0,0,0,\gamma,0,\gamma)$	$Z_{18}^{(i)}$	$2^{-17.5}$
$(\gamma,0,\gamma,0,\gamma,0,0,\gamma) \rightarrow (\gamma,0,0,0,0,\gamma,\gamma,0)$	$Z_{17}^{(i)}, Z_{18}^{(i)}$	2^{-23}
$(0,\gamma,0,\gamma,0,\gamma,\gamma,0) \rightarrow (\gamma,0,0,0,0,\gamma,\gamma,\gamma)$	—	$2^{-17.5}$
$(0,\gamma,0,\gamma,0,0,0,0) \rightarrow (\gamma,0,0,0,\gamma,0,0,0)$	—	2^{-12}
$(\gamma,0,\gamma,0,\gamma,\gamma,\gamma,\gamma) \rightarrow (\gamma,0,0,0,\gamma,0,0,\gamma)$	$Z_{17}^{(i)}, Z_{18}^{(i)}$	$2^{-28.5}$
$(0,\gamma,0,\gamma,0,0,0,\gamma) \rightarrow (\gamma,0,0,0,\gamma,0,\gamma,0)$	$Z_{18}^{(i)}$	2^{-12}
$(\gamma,0,\gamma,0,\gamma,\gamma,\gamma,0) \rightarrow (\gamma,0,0,0,\gamma,0,\gamma,\gamma)$	$Z_{17}^{(i)}$	$2^{-28.5}$
$(\gamma,0,\gamma,0,\gamma,\gamma,0,\gamma) \rightarrow (\gamma,0,0,0,\gamma,\gamma,0,0)$	$Z_{17}^{(i)}, Z_{18}^{(i)}$	$2^{-28.5}$
$(0,\gamma,0,\gamma,0,0,\gamma,0) \rightarrow (\gamma,0,0,0,\gamma,\gamma,0,\gamma)$	—	2^{-23}
$(\gamma,0,\gamma,0,\gamma,\gamma,0,0) \rightarrow (\gamma,0,0,0,\gamma,\gamma,\gamma,0)$	$Z_{17}^{(i)}$	$2^{-28.5}$
$(0,\gamma,0,\gamma,0,0,\gamma,\gamma) \rightarrow (\gamma,0,0,0,\gamma,\gamma,\gamma,\gamma)$	$Z_{18}^{(i)}$	2^{-23}
$(\gamma,0,\gamma,0,0,0,0,0) \rightarrow (\gamma,0,0,\gamma,0,0,0,0)$	$Z_{17}^{(i)}$	2^{-12}
$(0,\gamma,0,\gamma,\gamma,\gamma,\gamma,\gamma) \rightarrow (\gamma,0,0,\gamma,0,0,0,\gamma)$	$Z_{18}^{(i)}$	$2^{-17.5}$
$(\gamma,0,\gamma,0,0,0,0,\gamma) \rightarrow (\gamma,0,0,\gamma,0,0,\gamma,0)$	$Z_{17}^{(i)}, Z_{18}^{(i)}$	2^{-12}
$(0,\gamma,0,\gamma,\gamma,\gamma,\gamma,0) \rightarrow (\gamma,0,0,\gamma,0,0,\gamma,\gamma)$	—	$2^{-17.5}$
$(0,\gamma,0,\gamma,\gamma,\gamma,0,\gamma) \rightarrow (\gamma,0,0,\gamma,0,\gamma,0,0)$	$Z_{18}^{(i)}$	$2^{-17.5}$
$(\gamma,0,\gamma,0,0,0,\gamma,0) \rightarrow (\gamma,0,0,\gamma,0,\gamma,0,\gamma)$	$Z_{17}^{(i)}$	2^{-23}
$(0,\gamma,0,\gamma,\gamma,\gamma,0,0) \rightarrow (\gamma,0,0,\gamma,0,\gamma,\gamma,0)$	—	$2^{-17.5}$
$(\gamma,0,\gamma,0,0,0,\gamma,\gamma) \rightarrow (\gamma,0,0,\gamma,0,\gamma,\gamma,\gamma)$	$Z_{17}^{(i)}, Z_{18}^{(i)}$	2^{-23}
$(\gamma,0,\gamma,0,0,\gamma,0,\gamma) \rightarrow (\gamma,0,0,\gamma,\gamma,0,0,0)$	$Z_{17}^{(i)}, Z_{18}^{(i)}$	$2^{-17.5}$
$(0,\gamma,0,\gamma,\gamma,0,\gamma,0) \rightarrow (\gamma,0,0,\gamma,\gamma,0,0,\gamma)$	—	2^{-23}
$(\gamma,0,\gamma,0,0,\gamma,0,0) \rightarrow (\gamma,0,0,\gamma,\gamma,0,\gamma,0)$	$Z_{17}^{(i)}$	$2^{-17.5}$
$(0,\gamma,0,\gamma,\gamma,0,\gamma,\gamma) \rightarrow (\gamma,0,0,\gamma,\gamma,0,\gamma,\gamma)$	$Z_{18}^{(i)}$	2^{-23}
$(0,\gamma,0,\gamma,\gamma,0,0,0) \rightarrow (\gamma,0,0,\gamma,\gamma,\gamma,0,0)$	—	2^{-12}
$(\gamma,0,\gamma,0,0,\gamma,\gamma,\gamma) \rightarrow (\gamma,0,0,\gamma,\gamma,\gamma,0,\gamma)$	$Z_{17}^{(i)}, Z_{18}^{(i)}$	$2^{-17.5}$
$(0,\gamma,0,\gamma,\gamma,0,0,\gamma) \rightarrow (\gamma,0,0,\gamma,\gamma,\gamma,\gamma,0)$	$Z_{18}^{(i)}$	2^{-12}
$(\gamma,0,\gamma,0,0,\gamma,\gamma,0) \rightarrow (\gamma,0,0,\gamma,\gamma,\gamma,\gamma,\gamma)$	$Z_{17}^{(i)}$	$2^{-17.5}$
$(0,\gamma,0,0,0,0,0,0) \rightarrow (\gamma,0,\gamma,0,0,0,0,0)$	—	$2^{-6.5}$

Assuming i is even, the required weak subkeys are: $Z_{13}^{(i)}$, $Z_{15}^{(i)}$, $Z_{10}^{(i+1)}$, $Z_{9}^{(i+2)}$, $Z_{11}^{(i+2)}$, $Z_{15}^{(i+2)}$, $Z_{9}^{(i+4)}$, $Z_{11}^{(i+4)}$, $Z_{13}^{(i+4)}$ and $Z_{10}^{(i+5)}$.
It requires 10 weak subkeys over six rounds, that is, 1.66 weak subkeys per round on average.

(3) $(0,0,0,0,0,0,0,0,0,0,0,\gamma,0,\gamma,0,0) \stackrel{1r}{\rightarrow} (0,0,0,0,0,0,\gamma,0,0,0,\gamma,0,0,0,0,0) \stackrel{1r}{\rightarrow} (0,0,0,0,0,\gamma,0,0,0,\gamma,0,\gamma,0,\gamma,0,0) \stackrel{1r}{\rightarrow} (0,0,\gamma,0,\gamma,0,0,0,0,0,0,0,0,0,0,0) \stackrel{1r}{\rightarrow} (0,0,0,0,0,\gamma,0,0,0,\gamma,0,0,0,0,0,0) \stackrel{1r}{\rightarrow} (0,0,\gamma,0,\gamma,0,\gamma,0,0,0,\gamma,0,0,0,0,0) \stackrel{1r}{\rightarrow} (0,0,0,0,0,0,0,0,0,0,0,\gamma,0,\gamma,0,0)$, which is 6-round iterative and holds with bias 2^{-67}.

Table 3.125 Nontrivial bitmask propagation across the MA-box of MESH-128(8) (part VI).

Bitmask propagation across the MA-box	Weak subkeys	Bias
$(\gamma,0,\gamma,\gamma,\gamma,\gamma,\gamma,\gamma) \rightarrow (\gamma,0,\gamma,0,0,0,0,\gamma)$	$Z_{17}^{(i)}, Z_{18}^{(i)}$	2^{-23}
$(0,\gamma,0,0,0,0,0,\gamma) \rightarrow (\gamma,0,\gamma,0,0,0,\gamma,0)$	$Z_{18}^{(i)}$	$2^{-6.5}$
$(\gamma,0,\gamma,\gamma,\gamma,\gamma,\gamma,0) \rightarrow (\gamma,0,\gamma,0,0,0,\gamma,\gamma)$	$Z_{17}^{(i)}$	2^{-23}
$(\gamma,0,\gamma,\gamma,\gamma,\gamma,0,\gamma) \rightarrow (\gamma,0,\gamma,0,0,\gamma,0,0)$	$Z_{17}^{(i)}, Z_{18}^{(i)}$	2^{-23}
$(0,\gamma,0,0,0,0,\gamma,0) \rightarrow (\gamma,0,\gamma,0,0,\gamma,0,\gamma)$	—	$2^{-17.5}$
$(\gamma,0,\gamma,\gamma,\gamma,\gamma,0,0) \rightarrow (\gamma,0,\gamma,0,0,\gamma,\gamma,0)$	$Z_{17}^{(i)}$	2^{-23}
$(0,\gamma,0,0,0,0,\gamma,\gamma) \rightarrow (\gamma,0,\gamma,0,0,\gamma,\gamma,\gamma)$	$Z_{18}^{(i)}$	$2^{-17.5}$
$(0,\gamma,0,0,0,\gamma,0,\gamma) \rightarrow (\gamma,0,\gamma,0,\gamma,0,0,0)$	$Z_{18}^{(i)}$	2^{-12}
$(\gamma,0,\gamma,\gamma,\gamma,0,\gamma,0) \rightarrow (\gamma,0,\gamma,0,\gamma,0,0,\gamma)$	$Z_{17}^{(i)}$	$2^{-28.5}$
$(0,\gamma,0,0,0,\gamma,0,0) \rightarrow (\gamma,0,\gamma,0,\gamma,0,\gamma,0)$	—	2^{-12}
$(\gamma,0,\gamma,\gamma,\gamma,0,\gamma,\gamma) \rightarrow (\gamma,0,\gamma,0,\gamma,0,\gamma,\gamma)$	$Z_{17}^{(i)}, Z_{18}^{(i)}$	$2^{-28.5}$
$(\gamma,0,\gamma,\gamma,\gamma,0,0,0) \rightarrow (\gamma,0,\gamma,0,\gamma,\gamma,0,0)$	$Z_{17}^{(i)}$	$2^{-17.5}$
$(0,\gamma,0,0,0,\gamma,\gamma,\gamma) \rightarrow (\gamma,0,\gamma,0,\gamma,\gamma,0,\gamma)$	$Z_{18}^{(i)}$	2^{-12}
$(\gamma,0,\gamma,\gamma,\gamma,0,0,\gamma) \rightarrow (\gamma,0,\gamma,0,\gamma,\gamma,\gamma,0)$	$Z_{17}^{(i)}, Z_{18}^{(i)}$	$2^{-17.5}$
$(0,\gamma,0,0,0,\gamma,\gamma,0) \rightarrow (\gamma,0,\gamma,0,\gamma,\gamma,\gamma,\gamma)$	—	2^{-12}
$(\gamma,0,\gamma,\gamma,0,\gamma,0,\gamma) \rightarrow (\gamma,0,\gamma,\gamma,0,0,0,0)$	$Z_{17}^{(i)}, Z_{18}^{(i)}$	2^{-23}
$(0,\gamma,0,0,\gamma,0,\gamma,0) \rightarrow (\gamma,0,\gamma,\gamma,0,0,0,\gamma)$	—	$2^{-28.5}$
$(\gamma,0,\gamma,\gamma,0,\gamma,0,0) \rightarrow (\gamma,0,\gamma,\gamma,0,0,\gamma,0)$	$Z_{17}^{(i)}$	2^{-23}
$(0,\gamma,0,0,\gamma,0,\gamma,\gamma) \rightarrow (\gamma,0,\gamma,\gamma,0,0,\gamma,\gamma)$	$Z_{18}^{(i)}$	$2^{-28.5}$
$(0,\gamma,0,0,\gamma,0,0,0) \rightarrow (\gamma,0,\gamma,\gamma,0,\gamma,0,0)$	—	$2^{-17.5}$
$(\gamma,0,\gamma,\gamma,0,\gamma,\gamma,\gamma) \rightarrow (\gamma,0,\gamma,\gamma,0,\gamma,0,\gamma)$	$Z_{17}^{(i)}, Z_{18}^{(i)}$	2^{-23}
$(0,\gamma,0,0,\gamma,0,0,\gamma) \rightarrow (\gamma,0,\gamma,\gamma,0,\gamma,\gamma,0)$	$Z_{18}^{(i)}$	$2^{-17.5}$
$(\gamma,0,\gamma,\gamma,0,\gamma,\gamma,0) \rightarrow (\gamma,0,\gamma,\gamma,0,\gamma,\gamma,\gamma)$	$Z_{17}^{(i)}$	2^{-23}
$(\gamma,0,\gamma,\gamma,0,0,0,0) \rightarrow (\gamma,0,\gamma,\gamma,\gamma,0,0,0)$	$Z_{17}^{(i)}$	$2^{-17.5}$
$(0,\gamma,0,0,\gamma,\gamma,\gamma,\gamma) \rightarrow (\gamma,0,\gamma,\gamma,\gamma,0,0,\gamma)$	$Z_{18}^{(i)}$	2^{-23}
$(\gamma,0,\gamma,\gamma,0,0,0,\gamma) \rightarrow (\gamma,0,\gamma,\gamma,\gamma,0,\gamma,0)$	$Z_{17}^{(i)}, Z_{18}^{(i)}$	$2^{-17.5}$
$(0,\gamma,0,0,\gamma,\gamma,\gamma,0) \rightarrow (\gamma,0,\gamma,\gamma,\gamma,0,\gamma,\gamma)$	—	2^{-23}
$(0,\gamma,0,0,\gamma,\gamma,0,\gamma) \rightarrow (\gamma,0,\gamma,\gamma,\gamma,\gamma,0,0)$	$Z_{18}^{(i)}$	2^{-23}
$(\gamma,0,\gamma,\gamma,0,0,\gamma,0) \rightarrow (\gamma,0,\gamma,\gamma,\gamma,\gamma,0,\gamma)$	$Z_{17}^{(i)}$	$2^{-28.5}$
$(0,\gamma,0,0,\gamma,\gamma,0,0) \rightarrow (\gamma,0,\gamma,\gamma,\gamma,\gamma,\gamma,0)$	—	2^{-23}
$(\gamma,0,\gamma,\gamma,0,0,\gamma,\gamma) \rightarrow (\gamma,0,\gamma,\gamma,\gamma,\gamma,\gamma,\gamma)$	$Z_{17}^{(i)}, Z_{18}^{(i)}$	$2^{-28.5}$
$(\gamma,0,0,0,0,0,0,0) \rightarrow (\gamma,\gamma,0,0,0,0,0,0)$	$Z_{17}^{(i)}$	2^{-1}

Assuming i is odd, the required weak subkeys are: $Z_{12}^{(i)}$, $Z_{14}^{(i)}$, $Z_{11}^{(i+1)}$, $Z_{10}^{(i+2)}$, $Z_{12}^{(i+2)}$, $Z_{14}^{(i+2)}$, $Z_{10}^{(i+4)}$ and $Z_{11}^{(i+5)}$.

Assuming i is even, the required weak subkeys are: $Z_7^{(i+1)}$, $Z_6^{(i+2)}$, $Z_3^{(i+3)}$, $Z_5^{(i+3)}$, $Z_6^{(i+4)}$, $Z_3^{(i+5)}$, $Z_5^{(i+5)}$ and $Z_7^{(i+5)}$.

It requires 8 weak subkeys over six round, that is, 1.33 weak subkeys per round on average.

(4) $(0,0,0,0,0,0,0,0,0,0,\gamma,0,\gamma,0,0,0) \stackrel{1r}{\rightarrow} (0, 0, 0, \gamma, 0, 0, 0, 0, 0, 0, \gamma, 0, 0, 0, 0) \stackrel{1r}{\rightarrow} (0,0,0,0,\gamma,0,0,0,0,0,\gamma,0,0,0,0,0) \stackrel{1r}{\rightarrow} (0, 0, 0, \gamma, 0, \gamma, 0,$

Table 3.126 Nontrivial bitmask propagation across the MA-box of MESH-128(8) (part VII).

Bitmask propagation across the MA-box	Weak subkeys	Bias
$(0,\gamma,\gamma,\gamma,\gamma,\gamma,\gamma,\gamma) \to (\gamma,\gamma,0,0,0,0,0,\gamma)$	$Z_{18}^{(i)}$	$2^{-17.5}$
$(\gamma,0,0,0,0,0,0,\gamma) \to (\gamma,\gamma,0,0,0,0,\gamma,0)$	$Z_{17}^{(i)}, Z_{18}^{(i)}$	2^{-1}
$(0,\gamma,\gamma,\gamma,\gamma,\gamma,\gamma,0) \to (\gamma,\gamma,0,0,0,0,\gamma,\gamma)$	—	$2^{-17.5}$
$(0,\gamma,\gamma,\gamma,\gamma,\gamma,0,\gamma) \to (\gamma,\gamma,0,0,0,\gamma,0,0)$	$Z_{18}^{(i)}$	$2^{-17.5}$
$(\gamma,0,0,0,0,0,\gamma,0) \to (\gamma,\gamma,0,0,0,\gamma,0,\gamma)$	$Z_{17}^{(i)}$	2^{-12}
$(0,\gamma,\gamma,\gamma,\gamma,\gamma,0,0) \to (\gamma,\gamma,0,0,0,\gamma,\gamma,0)$	—	$2^{-17.5}$
$(\gamma,0,0,0,0,0,\gamma,\gamma) \to (\gamma,\gamma,0,0,0,\gamma,\gamma,\gamma)$	$Z_{17}^{(i)}, Z_{18}^{(i)}$	2^{-12}
$(\gamma,0,0,0,0,\gamma,0,\gamma) \to (\gamma,\gamma,0,0,\gamma,0,0,0)$	$Z_{17}^{(i)}, Z_{18}^{(i)}$	$2^{-6.5}$
$(0,\gamma,\gamma,\gamma,\gamma,0,\gamma,0) \to (\gamma,\gamma,0,0,\gamma,0,0,\gamma)$	—	2^{-23}
$(\gamma,0,0,0,0,\gamma,0,0) \to (\gamma,\gamma,0,0,\gamma,0,\gamma,0)$	$Z_{17}^{(i)}$	$2^{-6.5}$
$(0,\gamma,\gamma,\gamma,\gamma,0,\gamma,\gamma) \to (\gamma,\gamma,0,0,\gamma,0,\gamma,\gamma)$	$Z_{18}^{(i)}$	2^{-23}
$(0,\gamma,\gamma,\gamma,\gamma,0,0,0) \to (\gamma,\gamma,0,0,\gamma,\gamma,0,0)$	—	2^{-12}
$(\gamma,0,0,0,0,\gamma,\gamma,\gamma) \to (\gamma,\gamma,0,0,\gamma,\gamma,0,\gamma)$	$Z_{17}^{(i)}, Z_{18}^{(i)}$	$2^{-6.5}$
$(0,\gamma,\gamma,\gamma,\gamma,0,0,\gamma) \to (\gamma,\gamma,0,0,\gamma,\gamma,\gamma,0)$	$Z_{18}^{(i)}$	2^{-12}
$(\gamma,0,0,0,0,\gamma,\gamma,0) \to (\gamma,\gamma,0,0,\gamma,\gamma,\gamma,\gamma)$	$Z_{17}^{(i)}$	$2^{-6.5}$
$(0,\gamma,\gamma,\gamma,0,\gamma,0,\gamma) \to (\gamma,\gamma,0,\gamma,0,0,0,0)$	$Z_{18}^{(i)}$	$2^{-17.5}$
$(\gamma,0,0,0,\gamma,0,\gamma,0) \to (\gamma,\gamma,0,\gamma,0,0,0,\gamma)$	$Z_{17}^{(i)}$	2^{-23}
$(0,\gamma,\gamma,\gamma,0,\gamma,0,0) \to (\gamma,\gamma,0,\gamma,0,0,\gamma,0)$	—	$2^{-17.5}$
$(\gamma,0,0,0,\gamma,0,\gamma,\gamma) \to (\gamma,\gamma,0,\gamma,0,0,\gamma,\gamma)$	$Z_{17}^{(i)}, Z_{18}^{(i)}$	2^{-23}
$(\gamma,0,0,0,\gamma,0,0,0) \to (\gamma,\gamma,0,\gamma,0,\gamma,0,0)$	$Z_{17}^{(i)}$	2^{-12}
$(0,\gamma,\gamma,\gamma,0,\gamma,\gamma,\gamma) \to (\gamma,\gamma,0,\gamma,0,\gamma,0,\gamma)$	$Z_{18}^{(i)}$	$2^{-17.5}$
$(\gamma,0,0,0,\gamma,0,0,\gamma) \to (\gamma,\gamma,0,\gamma,0,\gamma,\gamma,0)$	$Z_{17}^{(i)}, Z_{18}^{(i)}$	2^{-12}
$(0,\gamma,\gamma,\gamma,0,\gamma,\gamma,0) \to (\gamma,\gamma,0,\gamma,0,\gamma,\gamma,\gamma)$	—	$2^{-17.5}$
$(0,\gamma,\gamma,\gamma,0,0,0,0) \to (\gamma,\gamma,0,\gamma,\gamma,0,0,0)$	—	2^{-12}
$(\gamma,0,0,0,\gamma,\gamma,\gamma,\gamma) \to (\gamma,\gamma,0,\gamma,\gamma,0,0,\gamma)$	$Z_{17}^{(i)}, Z_{18}^{(i)}$	$2^{-17.5}$
$(0,\gamma,\gamma,\gamma,0,0,0,\gamma) \to (\gamma,\gamma,0,\gamma,\gamma,0,\gamma,0)$	$Z_{18}^{(i)}$	2^{-12}
$(\gamma,0,0,0,\gamma,\gamma,\gamma,0) \to (\gamma,\gamma,0,\gamma,\gamma,0,\gamma,\gamma)$	$Z_{17}^{(i)}$	$2^{-17.5}$
$(\gamma,0,0,0,\gamma,\gamma,0,\gamma) \to (\gamma,\gamma,0,\gamma,\gamma,\gamma,0,0)$	$Z_{17}^{(i)}, Z_{18}^{(i)}$	$2^{-17.5}$
$(0,\gamma,\gamma,\gamma,0,0,\gamma,0) \to (\gamma,\gamma,0,\gamma,\gamma,\gamma,0,\gamma)$	—	2^{-23}
$(\gamma,0,0,0,\gamma,\gamma,0,0) \to (\gamma,\gamma,0,\gamma,\gamma,\gamma,\gamma,0)$	$Z_{17}^{(i)}$	$2^{-17.5}$
$(0,\gamma,\gamma,\gamma,0,0,\gamma,\gamma) \to (\gamma,\gamma,0,\gamma,\gamma,\gamma,\gamma,\gamma)$	$Z_{18}^{(i)}$	2^{-23}
$(\gamma,0,0,\gamma,0,\gamma,0,\gamma) \to (\gamma,\gamma,\gamma,0,0,0,0,0)$	$Z_{17}^{(i)}, Z_{18}^{(i)}$	2^{-12}

$0, 0, 0, 0, 0, 0, 0, 0, 0) \stackrel{1r}{\to} (0,0,0,0,\gamma,0,0,0,0,0,0,0,\gamma,0,0,0) \stackrel{1r}{\to} (0, 0, 0, 0, 0, \gamma, 0, 0, 0, 0, 0, \gamma, 0, 0, 0, 0) \stackrel{1r}{\to} (0,0,0,0,0,0,0,0,0,0,\gamma,0,\gamma,0,0,0)$, which is 6-round iterative and holds with bias 2^{-34}.

Assuming i is odd, the required weak subkeys are: $Z_4^{(i+1)}$, $Z_5^{(i+2)}$, $Z_4^{(i+3)}$, $Z_6^{(i+3)}$, $Z_5^{(i+4)}$ and $Z_6^{(i+5)}$.

Assuming i is even, the required weak subkeys are: $Z_{11}^{(i)}$, $Z_{13}^{(i)}$, $Z_{12}^{(i+1)}$, $Z_{11}^{(i+2)}$, $Z_{13}^{(i+4)}$ and $Z_{12}^{(i+5)}$.

It requires six weak subkeys over six rounds, that is, one weak subkey per round on average.

Table 3.127 Nontrivial bitmask propagation across the MA-box of MESH-128(8) (part VIII).

Bitmask propagation across the MA-box	Weak subkeys	Bias
$(0,\gamma,\gamma,0,\gamma,0,\gamma,0) \rightarrow (\gamma,\gamma,\gamma,0,0,0,0,\gamma)$	—	$2^{-28.5}$
$(\gamma,0,0,\gamma,0,\gamma,0,0) \rightarrow (\gamma,\gamma,\gamma,0,0,0,\gamma,0)$	$Z_{17}^{(i)}$	2^{-12}
$(0,\gamma,\gamma,0,\gamma,0,\gamma,\gamma) \rightarrow (\gamma,\gamma,\gamma,0,0,0,\gamma,\gamma)$	$Z_{18}^{(i)}$	$2^{-28.5}$
$(0,\gamma,\gamma,0,\gamma,0,0,0) \rightarrow (\gamma,\gamma,\gamma,0,0,\gamma,0,0)$	—	$2^{-17.5}$
$(\gamma,0,0,\gamma,0,\gamma,\gamma,\gamma) \rightarrow (\gamma,\gamma,\gamma,0,0,\gamma,0,\gamma)$	$Z_{17}^{(i)}, Z_{18}^{(i)}$	2^{-12}
$(0,\gamma,\gamma,0,\gamma,0,0,\gamma) \rightarrow (\gamma,\gamma,\gamma,0,0,\gamma,\gamma,0)$	$Z_{18}^{(i)}$	$2^{-17.5}$
$(\gamma,0,0,\gamma,0,\gamma,\gamma,0) \rightarrow (\gamma,\gamma,\gamma,0,0,\gamma,\gamma,\gamma)$	$Z_{17}^{(i)}$	2^{-12}
$(\gamma,0,0,\gamma,0,0,0,0) \rightarrow (\gamma,\gamma,\gamma,0,\gamma,0,0,0)$	$Z_{17}^{(i)}$	$2^{-6.5}$
$(0,\gamma,\gamma,0,\gamma,\gamma,\gamma,\gamma) \rightarrow (\gamma,\gamma,\gamma,0,\gamma,0,0,\gamma)$	$Z_{18}^{(i)}$	2^{-23}
$(\gamma,0,0,\gamma,0,0,0,\gamma) \rightarrow (\gamma,\gamma,\gamma,0,\gamma,0,\gamma,0)$	$Z_{17}^{(i)}, Z_{18}^{(i)}$	$2^{-6.5}$
$(0,\gamma,\gamma,0,\gamma,\gamma,\gamma,0) \rightarrow (\gamma,\gamma,\gamma,0,\gamma,0,\gamma,\gamma)$	—	2^{-23}
$(0,\gamma,\gamma,0,\gamma,\gamma,0,\gamma) \rightarrow (\gamma,\gamma,\gamma,0,\gamma,\gamma,0,0)$	$Z_{18}^{(i)}$	2^{-23}
$(\gamma,0,0,\gamma,0,0,\gamma,0) \rightarrow (\gamma,\gamma,\gamma,0,\gamma,\gamma,0,\gamma)$	$Z_{17}^{(i)}$	$2^{-17.5}$
$(0,\gamma,\gamma,0,\gamma,\gamma,0,0) \rightarrow (\gamma,\gamma,\gamma,0,\gamma,\gamma,\gamma,0)$	—	2^{-23}
$(\gamma,0,0,\gamma,0,0,\gamma,\gamma) \rightarrow (\gamma,\gamma,\gamma,0,\gamma,\gamma,\gamma,\gamma)$	$Z_{17}^{(i)}, Z_{18}^{(i)}$	$2^{-17.5}$
$(0,\gamma,\gamma,0,0,0,0,0) \rightarrow (\gamma,\gamma,\gamma,\gamma,0,0,0,0)$	—	$2^{-6.5}$
$(\gamma,0,0,\gamma,\gamma,\gamma,\gamma,\gamma) \rightarrow (\gamma,\gamma,\gamma,\gamma,0,0,0,\gamma)$	$Z_{17}^{(i)}, Z_{18}^{(i)}$	2^{-12}
$(0,\gamma,\gamma,0,0,0,0,\gamma) \rightarrow (\gamma,\gamma,\gamma,\gamma,0,0,\gamma,0)$	$Z_{18}^{(i)}$	$2^{-6.5}$
$(\gamma,0,0,\gamma,\gamma,\gamma,\gamma,0) \rightarrow (\gamma,\gamma,\gamma,\gamma,0,0,\gamma,\gamma)$	$Z_{17}^{(i)}$	2^{-12}
$(\gamma,0,0,\gamma,\gamma,\gamma,0,\gamma) \rightarrow (\gamma,\gamma,\gamma,\gamma,0,\gamma,0,0)$	$Z_{17}^{(i)}, Z_{18}^{(i)}$	2^{-12}
$(0,\gamma,\gamma,0,0,0,\gamma,0) \rightarrow (\gamma,\gamma,\gamma,\gamma,0,\gamma,0,\gamma)$	—	$2^{-17.5}$
$(\gamma,0,0,\gamma,\gamma,\gamma,0,0) \rightarrow (\gamma,\gamma,\gamma,\gamma,0,\gamma,\gamma,0)$	$Z_{17}^{(i)}$	2^{-12}
$(0,\gamma,\gamma,0,0,0,\gamma,\gamma) \rightarrow (\gamma,\gamma,\gamma,\gamma,0,\gamma,\gamma,\gamma)$	$Z_{18}^{(i)}$	$2^{-17.5}$
$(0,\gamma,\gamma,0,0,\gamma,0,\gamma) \rightarrow (\gamma,\gamma,\gamma,\gamma,\gamma,0,0,0)$	$Z_{18}^{(i)}$	2^{-12}
$(\gamma,0,0,\gamma,\gamma,0,\gamma,0) \rightarrow (\gamma,\gamma,\gamma,\gamma,\gamma,0,0,\gamma)$	$Z_{17}^{(i)}$	$2^{-17.5}$
$(0,\gamma,\gamma,0,0,\gamma,0,0) \rightarrow (\gamma,\gamma,\gamma,\gamma,\gamma,0,\gamma,0)$	—	2^{-12}
$(\gamma,0,0,\gamma,\gamma,0,\gamma,\gamma) \rightarrow (\gamma,\gamma,\gamma,\gamma,\gamma,0,\gamma,\gamma)$	$Z_{17}^{(i)}, Z_{18}^{(i)}$	$2^{-17.5}$
$(\gamma,0,0,\gamma,\gamma,0,0,0) \rightarrow (\gamma,\gamma,\gamma,\gamma,\gamma,\gamma,0,0)$	$Z_{17}^{(i)}$	$2^{-6.5}$
$(0,\gamma,\gamma,0,0,\gamma,\gamma,\gamma) \rightarrow (\gamma,\gamma,\gamma,\gamma,\gamma,\gamma,0,\gamma)$	$Z_{18}^{(i)}$	2^{-12}
$(\gamma,0,0,\gamma,\gamma,0,0,\gamma) \rightarrow (\gamma,\gamma,\gamma,\gamma,\gamma,\gamma,\gamma,0)$	$Z_{17}^{(i)}, Z_{18}^{(i)}$	$2^{-6.5}$
$(0,\gamma,\gamma,0,0,\gamma,\gamma,0) \rightarrow (\gamma,\gamma,\gamma,\gamma,\gamma,\gamma,\gamma,\gamma)$	—	2^{-12}

Similarly to the linear analysis of MESH-64, the key schedule of MESH-128(8) does not consist of a simple bit permutation like in IDEA. The fraction of keys that satisfy a weak subkey restriction in MESH-128(8) was estimated to be about 2^{-7} per subkey. This estimate is in line with the assumption that a weak subkey has value 0 or 1, since the seven most significant bits of a subkey have to be zero. Under this assumption, and since MESH-128(8) operates under a 256-bit key, at most $\lfloor 256/7 \rfloor = 36$ weak subkeys are allowed (for a nonempty weak-key class).

Under this estimate, linear relation (4) requires the minimum number of weak subkeys on average among the linear relations listed previously. It also holds with the largest bias (per round). Iterating it over 8.5 rounds of MESH-

128(8) leads to a bias of 2^{-45}, and eight weak subkeys: $Z_4^{(2)}$, $Z_5^{(3)}$, $Z_4^{(4)}$, $Z_6^{(4)}$, $Z_5^{(5)}$, $Z_6^{(6)}$, $Z_4^{(8)}$ and $Z_5^{(9)}$. This leads to an estimated weak-key class size of $2^{256-7\cdot 8} = 2^{200}$. Note that the exact user keys that belong to this weak-key class are not known. Identifying them is left as an open problem.

It means that the full 8.5-round MESH-128(8) can be distinguished from a random permutation with $c/(2^{-45})^2 = c \cdot 2^{90}$ known plaintexts and an equivalent number of encryptions (for a fixed constant c).

Although the full 8.5-round MESH-128(8) cipher can be covered by a linear relation, the attack complexity is not polynomial in the cipher parameters. See Sect. 1.7.

This linear analysis also allows us to study the diffusion power of MESH-128(8) with respect to modern tools such as the linear branch number. See Appendix, Sect. B.

Analyzing the set of 65535 1-round linear relations for MESH-128(8), we found out that the linear branch number *per round* of MESH-128(8) is only 4 (under weak-key assumptions). For instance, the sum of the (wordwise) Hamming Weight of the input and output bitmasks of the 1-round linear relation

$$(0,0,0,0,0,0,0,0,0,0,0,0,0,0,\gamma,0) \stackrel{1r}{\rightarrow} (\gamma,\gamma,0,0,0,0,0,\gamma,0,0,0,0,0,0,0,0)$$

is 4. So, even though MESH-128(8) provides full diffusion in a single round, a detailed analysis shows that its linear branch number is much lower than ideal. For a 16-word block, such as in MESH-128(8), an MDS code using a 16×16 MDS matrix would rather provide a branch number of $1 + 16 = 17$. See Appendix, Sect. B.1.

3.13.9 LC of WIDEA-n Under Weak-Key Assumptions

The analyses in this section were adapted from [145, 113]. Background on the WIDEA-n ciphers are provided in Sect. 2.5.

As a preliminary step, we listed exhaustively all 1-round linear relations of IDEA [57] in Table 3.128 since WIDEA-n contains n copies of IDEA. All of these linear relations hold with maximum bias 2^{-1} under weak-subkey restrictions.

From Table 3.128 the best linear relations shall: (i) minimize the number of weak-subkey assumptions per round; (ii) maximize the bias and (iii) be iterative. Under these conditions, the best choices include the 3-round relation

$$(0,\gamma,\gamma,0) \stackrel{1r}{\rightarrow} (\gamma,0,\gamma,0) \stackrel{1r}{\rightarrow} (\gamma,\gamma,0,0) \stackrel{1r}{\rightarrow} (0,\gamma,\gamma,0), \qquad (3.172)$$

Table 3.128 Nontrivial 1-round linear relations of IDEA with $\gamma = 1$.

1-round linear relations	weak subkeys j-th round	masks in MA-box
$(0,0,0,\gamma) \stackrel{1r}{\rightarrow} (0,0,\gamma,0)$	$Z_{6(j-1)+3}, Z_{6(j-1)+5}$	$(0,\gamma) \rightarrow (\gamma,0)$
$(0,0,\gamma,0) \stackrel{1r}{\rightarrow} (\gamma,0,\gamma,\gamma)$	$Z_{6(j-1)+4}, Z_{6(j-1)+5}$	$(\gamma,\gamma) \rightarrow (0,\gamma)$
$(0,0,\gamma,\gamma) \stackrel{1r}{\rightarrow} (\gamma,0,0,\gamma)$	$Z_{6(j-1)+3}, Z_{6(j-1)+4}$	$(\gamma,0) \rightarrow (\gamma,\gamma)$
$(0,\gamma,0,0) \stackrel{1r}{\rightarrow} (0,0,0,\gamma)$	$Z_{6(j-1)+5}$	$(0,\gamma) \rightarrow (\gamma,1)$
$(0,\gamma,0,\gamma) \stackrel{1r}{\rightarrow} (0,0,\gamma,\gamma)$	$Z_{6(j-1)+3}$	**(0,0)→(0,0)**
$(0,\gamma,\gamma,0) \stackrel{1r}{\rightarrow} (\gamma,0,\gamma,0)$	$Z_{6(j-1)+4}$	$(\gamma,0) \rightarrow (\gamma,\gamma)$
$(0,\gamma,\gamma,\gamma) \stackrel{1r}{\rightarrow} (\gamma,0,0,0)$	$Z_{6(j-1)+3}, Z_{6(j-1)+4}, Z_{6(j-1)+5}$	$(\gamma,\gamma) \rightarrow (0,\gamma)$
$(\gamma,0,0,0) \stackrel{1r}{\rightarrow} (0,\gamma,\gamma,\gamma)$	$Z_{6(j-1)}, Z_{6(j-1)+4}, Z_{6(j-1)+5}$	$(\gamma,\gamma) \rightarrow (0,\gamma)$
$(\gamma,0,0,\gamma) \stackrel{1r}{\rightarrow} (0,\gamma,0,\gamma)$	$Z_{6(j-1)}, Z_{6(j-1)+3}, Z_{6(j-1)+4}$	$(\gamma,0) \rightarrow (\gamma,\gamma)$
$(\gamma,0,\gamma,0) \stackrel{1r}{\rightarrow} (\gamma,\gamma,0,0)$	$Z_{6(j-1)}$	**(0,0)→(0,0)**
$(\gamma,0,\gamma,\gamma) \stackrel{1r}{\rightarrow} (\gamma,\gamma,\gamma,0)$	$Z_{6(j-1)}, Z_{6(j-1)+3}, Z_{6(j-1)+5}$	$(0,\gamma) \rightarrow (\gamma,0)$
$(\gamma,\gamma,0,0) \stackrel{1r}{\rightarrow} (0,\gamma,\gamma,0)$	$Z_{6(j-1)}, Z_{6(j-1)+4}$	$(\gamma,0) \rightarrow (\gamma,\gamma)$
$(\gamma,\gamma,0,\gamma) \stackrel{1r}{\rightarrow} (0,\gamma,0,0)$	$Z_{6(j-1)}, Z_{6(j-1)+3}, Z_{6(j-1)+4}, Z_{6(j-1)+5}$	$(\gamma,\gamma) \rightarrow (0,\gamma)$
$(\gamma,\gamma,\gamma,0) \stackrel{1r}{\rightarrow} (\gamma,\gamma,0,\gamma)$	$Z_{6(j-1)}, Z_{6(j-1)+5}$	$(0,\gamma) \rightarrow (\gamma,0)$
$(\gamma,\gamma,\gamma,\gamma) \stackrel{1r}{\rightarrow} (\gamma,\gamma,\gamma,\gamma)$	$Z_{6(j-1)}, Z_{6(j-1)+3}$	**(0,0)→(0,0)**

with only four weak subkeys: $Z_{6(j-1)+4}, Z_{6j}, Z_{6(j+1)}, Z_{6(j+1)+4}$ starting from round j. All rotations of (3.172), for instance starting from $(\gamma,0,\gamma,0)$ instead of $(0,\gamma,\gamma,0)$, also fulfill the same criteria.

Another relevant choice is the 1-round iterative linear relation:

$$(\gamma,\gamma,\gamma,\gamma) \stackrel{1r}{\rightarrow} (\gamma,\gamma,\gamma,\gamma), \tag{3.173}$$

with only two weak subkeys per round: $Z_{6(j-1)}, Z_{6(j-1)+3}$ starting from round j.

Using the linear relation (3.173) we attack a single IDEA instance in WIDEA-4, resulting in the following iterative linear relation:

$$\begin{pmatrix} 0\,0\,0\,0 \\ 0\,0\,0\,0 \\ 0\,0\,0\,0 \\ \gamma\,\gamma\,\gamma\,\gamma \end{pmatrix} \rightarrow \begin{pmatrix} 0\,0\,0\,0 \\ 0\,0\,0\,0 \\ 0\,0\,0\,0 \\ \gamma\,\gamma\,\gamma\,\gamma \end{pmatrix}. \tag{3.174}$$

Note that there are *no* active MAD-boxes along (3.174) because all bitmasks at the input to the MAD-boxes are trivial: $(0,0,0,0)$. Thus, the fixed-point relation for the MDS matrix

$$\text{MDS}(0,0,0,0)^T = (0,0,0,0)^T$$

is exploited.

Concatenating (3.174) with itself across 8.5-round WIDEA-4, this linear relation holds with maximum bias 2^{-1} as long as the following eighteen subkeys are weak: $Z_{6(j-1)}, Z_{6(j-1)+3}$ for $0 \leq j \leq 9$. The weak-key class for this linear relation has size $2^{512-15\cdot 18} = 2^{242}$ keys. Using relation (3.172) instead of (3.173) would require all IDEA instance to be attacked at once. This means using the fixed point for the MDS matrix

$$\text{MDS}(\gamma, \gamma, \gamma, \gamma)^T = (\gamma, \gamma, \gamma, \gamma)^T,$$

but then, there are too many weak subkeys because some MAD-boxes become active.

The linear relation (3.174) can be used to distinguish the full 8.5-round WIDEA-4 from a random permutation using $c \cdot (2^{-1})^{-1} = 4c$ WIDEA-4 encryptions and $4c$ known plaintexts, where c is a fixed constant.

Applying (3.174) to WIDEA-8 gives even better results, since the later has key size of 1024 bits, and the same weak-subkey conditions lead to an estimated weak-key class of $2^{1024-15\cdot 18} = 2^{754}$ keys.

A (partial) key-recovery attack on the full 8.5-round WIDEA-4, using (3.174), can recover subkeys $Z_{48,0}$ and $Z_{51,0}$. In this case, we attack the last half-round and only sixteen subkeys need to be weak: $Z_{6(j-1)}, Z_{6(j-1)+3}$ for $0 \leq j \leq 8$, which implies a weak key class of about $2^{512-15\cdot 16} = 2^{272}$ keys. Using Matsui's estimation for a high-success rate attack, the data complexity is $8(2^{-1})^{-2} = 32$ known plaintexts. The attack effort is equivalent to 2^{32} multiplications per subkey, which is equivalent to fraction of $\frac{1}{17\cdot 2\cdot 4}$ of a full WIDEA-4 computation, or $2^{32}/(17 \cdot 2 \cdot 4) \approx 2^{25}$ WIDEA-4 encryptions. The memory needed is 32 counters.

The time complexity for WIDEA-8 becomes 2^{24} WIDEA-8 encryptions since there are eight IDEA instances. The weak-key class size is $2^{1024-15\cdot 16} = 2^{784}$.

3.13.10 LC of RIDEA Under Weak-Key Assumptions

The following linear analysis of RIDEA was not published before, although based on known techniques already used against other Lai-Massey cipher designs.

The strategy is to build linear relations in a bottom-up fashion, from the smallest components: $\odot$, $\boxplus$ and $\oplus$, up to an MA-box, up to a half-round, further to a full round, and finally to multiple rounds.

Across unkeyed $\odot$ and $\boxplus$, the propagation of bitmasks follows from Table 3.98 and Table 3.99, respectively.

Further, we obtain 1-round linear relations and the corresponding weak subkeys for the relations to hold with maximum possible bias.

Table 3.129 lists all fifteen nontrivial 1-round linear relations of RIDEA, under weak-key assumptions and wordwise linear bitmasks 0 and $\gamma = 1$.

Table 3.129 Nontrivial 1-round linear relations for RIDEA.

1-round linear relation	Weak subkeys	Bias	Linear relation across MA-box
$(0,0,0,\gamma) \xrightarrow{1r} (0,0,\gamma,0)$	$Z_4^{(i)}$	$2^{-13.47}$	$(0,\gamma) \rightarrow (\gamma,0)$
$(0,0,\gamma,0) \xrightarrow{1r} (\gamma,0,\gamma,\gamma)$	$Z_5^{(i)}$	$2^{-13.47}$	$(\gamma,\gamma) \rightarrow (0,\gamma)$
$(0,0,\gamma,\gamma) \xrightarrow{1r} (\gamma,0,0,\gamma)$	$Z_4^{(i)}, Z_5^{(i)}$	$2^{-13.47}$	$(\gamma,0) \rightarrow (\gamma,\gamma)$
$(0,\gamma,0,0) \xrightarrow{1r} (0,0,0,\gamma)$	—	$2^{-13.47}$	$(0,\gamma) \rightarrow (\gamma,0)$
$(0,\gamma,0,\gamma) \xrightarrow{1r} (0,0,\gamma,\gamma)$	$Z_4^{(i)}$	2^{-1}	$(0,0) \rightarrow (0,0)$
$(0,\gamma,\gamma,0) \xrightarrow{1r} (\gamma,0,\gamma,0)$	$Z_5^{(i)}$	2^{-1}	$(\gamma,0) \rightarrow (\gamma,\gamma)$
$(0,\gamma,\gamma,\gamma) \xrightarrow{1r} (\gamma,0,0,0)$	$Z_4^{(i)}, Z_5^{(i)}$	$2^{-13.47}$	$(\gamma,\gamma) \rightarrow (0,\gamma)$
$(\gamma,0,0,0) \xrightarrow{1r} (0,\gamma,\gamma,\gamma)$	$Z_1^{(i)}, Z_5^{(i)}$	$2^{-13.47}$	$(\gamma,\gamma) \rightarrow (0,\gamma)$
$(\gamma,0,0,\gamma) \xrightarrow{1r} (0,\gamma,0,\gamma)$	$Z_1^{(i)}, Z_4^{(i)}, Z_5^{(i)}$	2^{-1}	$(\gamma,0) \rightarrow (\gamma,\gamma)$
$(\gamma,0,\gamma,0) \xrightarrow{1r} (\gamma,\gamma,0,0)$	$Z_1^{(i)}$	2^{-1}	$(0,0) \rightarrow (0,0)$
$(\gamma,0,\gamma,\gamma) \xrightarrow{1r} (\gamma,\gamma,\gamma,0)$	$Z_1^{(i)}, Z_4^{(i)}$	$2^{-13.47}$	$(0,\gamma) \rightarrow (\gamma,0)$
$(\gamma,\gamma,0,0) \xrightarrow{1r} (0,\gamma,\gamma,0)$	$Z_1^{(i)}, Z_5^{(i)}$	2^{-1}	$(\gamma,0) \rightarrow (\gamma,\gamma)$
$(\gamma,\gamma,0,\gamma) \xrightarrow{1r} (0,\gamma,0,0)$	$Z_1^{(i)}, Z_4^{(i)}, Z_5^{(i)}$	$2^{-13.47}$	$(\gamma,\gamma) \rightarrow (0,\gamma)$
$(\gamma,\gamma,\gamma,0) \xrightarrow{1r} (\gamma,\gamma,0,\gamma)$	$Z_1^{(i)}$	$2^{-13.47}$	$(0,\gamma) \rightarrow (\gamma,0)$
$(\gamma,\gamma,\gamma,\gamma) \xrightarrow{1r} (\gamma,\gamma,\gamma,\gamma)$	$Z_1^{(i)}, Z_4^{(i)}$	2^{-1}	$(0,0) \rightarrow (0,0)$

Multiple-round linear relations are obtained by concatenating 1-round linear relations, such that the output mask from one relation matches the input mask to the next relation.

From Table 3.129, the multiple-round linear relations are clustered into five relations covering 1, 2, 3 or 6 rounds:

(1) $(0,0,0,\gamma) \xrightarrow{1r} (0,0,\gamma,0) \xrightarrow{1r} (\gamma,0,\gamma,\gamma) \xrightarrow{1r} (\gamma,\gamma,\gamma,0) \xrightarrow{1r} (\gamma,\gamma,0,\gamma) \xrightarrow{1r} (0,\gamma,0,0) \xrightarrow{1r} (0,0,0,\gamma)$, which is 6-round iterative and holds with bias $2^{5-13.47\cdot 6} = 2^{-75.82}$. Starting from round i, the weak subkeys required across six rounds are: $Z_4^{(i)}$, $Z_5^{(i+1)}$, $Z_1^{(i+2)}$, $Z_4^{(i+2)}$, $Z_1^{(i+3)}$, $Z_1^{(i+4)}$, $Z_4^{(i+4)}$ and $Z_5^{(i+4)}$.

(2) $(0,0,\gamma,\gamma) \xrightarrow{1r} (\gamma,0,0,\gamma) \xrightarrow{1r} (0,\gamma,0,\gamma) \xrightarrow{1r} (0,0,\gamma,\gamma)$, which is 3-round iterative and holds with bias $2^{-13.47}$. Starting from round i, the weak subkeys required are: $Z_4^{(i)}$, $Z_5^{(i)}$, $Z_1^{(i+1)}$, $Z_4^{(i+1)}$, $Z_5^{(i+1)}$ and $Z_4^{(i+2)}$.

(3) $(0,\gamma,\gamma,0) \xrightarrow{1r} (\gamma,0,\gamma,0) \xrightarrow{1r} (\gamma,\gamma,0,0) \xrightarrow{1r} (0,\gamma,\gamma,0)$, which is 3-round iterative and holds with bias 2^{-1}. Starting from round i, the weak subkeys required are: $Z_5^{(i)}$, $Z_1^{(i+1)}$, $Z_1^{(i+2)}$ and $Z_5^{(i+2)}$.

(4) $(0,\gamma,\gamma,\gamma) \xrightarrow{1r} (\gamma,0,0,0) \xrightarrow{1r} (0,\gamma,\gamma,\gamma)$ which is 2-round iterative and holds with bias $2^{1-2\cdot 13.47} = 2^{25.94}$. Starting from round i, the weak subkeys required are: $Z_4^{(i)}$, $Z_5^{(i)}$, $Z_1^{(i+1)}$ and $Z_5^{(i+1)}$.

(5) $(\gamma,\gamma,\gamma,\gamma) \stackrel{1r}{\to} (\gamma,\gamma,\gamma,\gamma)$ which is 1-round iterative and holds with bias 2^{-1}. Starting from round i, the weak subkeys required are: $Z_1^{(i)}$ and $Z_4^{(i)}$.

The linear relations (1) and (3) require the least number of weak subkeys per round, while relations (3) and (5) hold with the largest bias. The best trade-off is (3) which can cover the full 8.5-round RIDEA with bias 2^{-1} and requiring eleven weak subkeys: $Z_5^{(1)}$, $Z_1^{(2)}$, $Z_1^{(3)}$, $Z_5^{(3)}$, $Z_5^{(4)}$, $Z_1^{(5)}$, $Z_1^{(6)}$, $Z_5^{(6)}$, $Z_5^{(7)}$, $Z_1^{(8)}$ and $Z_1^{(9)}$.

According to the key schedule algorithm of RIDEA in Sec. 2.4.2, Table 3.130 lists the evolution of the weak-key class for linear relation (3) per round of RIDEA. Overall, relation (3) holds for a weak-key class of size 2^{14}

Table 3.130 Evolution of weak-key class size for linear relation (3) for RIDEA.

Weak subkeys	Key bits	Weak-key Class Size (bits)
$Z_5^{(1)}$	64–78	2^{113}
$Z_1^{(2)}$	96–110	2^{98}
$Z_1^{(3)}$, $Z_5^{(3)}$	64–78, 0–15	2^{77}
$Z_5^{(4)}$	96–110	2^{62}
$Z_1^{(5)}$	0–14	2^{52}
$Z_1^{(6)}$, $Z_5^{(6)}$	96–110, 32–46	2^{31}
$Z_5^{(7)}$	0–14	2^{24}
$Z_1^{(8)}$	32–46	2^{14}
$Z_1^{(9)}$	0–14	2^{14}

for the full 8.5-round RIDEA. Therefore, (3) can be used to distinguish the full RIDEA from a random permutation with $c/(2^{-1})^2 = 4c$ known plaintexts and equivalent encryption effort (for a fixed constant c).

Note that even though the weak-key class for linear relation (5) is smaller than that for relation (3), the former allows us to bypass the MA-box of RIDEA altogether (in every round). So, the new MA-box of RIDEA is not enough to strengthen RIDEA against linear cryptanalysis.

Table 3.131 depicts the evolution of the weak-key class for relation (5) per round of RIDEA.

Consequently, RIDEA is not an ideal primitive, that is, it is not a PRP. See Sect. 1.7.

Similarly to the case of PES and IDEA, these linear relations allow to study the diffusion power of RIDEA with respect to modern tools such as the linear branch number. See Appendix, Sect. B.

The set of linear relations in Table 3.129 indicates that the linear branch number of RIDEA *per round* is only 2 (under weak-key assumptions). For instance, the 1-round linear relation

Table 3.131 Evolution of weak-key class size for linear relation (5) for RIDEA.

Weak subkeys	Key bits	Weak-key Class Size
$Z_1^{(1)}, Z_1^{(4)}$	0–14, 48–62	2^{98}
$Z_2^{(1)}, Z_2^{(4)}$	96–110, 41–55	2^{76}
$Z_3^{(1)}, Z_3^{(4)}$	89–103, 9–23	2^{60}
$Z_4^{(1)}, Z_4^{(4)}$	82–96, 2–16	2^{53}
$Z_5^{(1)}, Z_5^{(4)}$	75–89, 123–9	2^{41}
$Z_6^{(1)}, Z_6^{(4)}$	43–57, 116–2	2^{34}
$Z_7^{(1)}, Z_7^{(4)}$	36–50, 84–98	2^{29}
$Z_8^{(1)}, Z_8^{(4)}$	29–43, 77–91	2^{22}
$Z_9^{(1)}, Z_9^{(4)}$	22–36, 70–84	2^{12}

$$(0,0,0,\gamma) \stackrel{1r}{\rightarrow} (0,0,\gamma,0)$$

has both the input and output bitmasks with Hamming Weight one. So, even though the design of RIDEA provides full diffusion in a single round, a detailed analysis shows that its linear branch number is 2, which is much lower than ideal. For a 4-word block, such as in RIDEA, an MDS code using a 4×4 MDS matrix would rather provide a linear branch number of 5. See Appendix, Sect. B.1.

3.13.11 LC of REESSE3+ Under Weak-Key Assumptions

The following linear analysis of REESSE3+ is based on [146], but the presented analyses are more extensive and more detailed. The linear attacks operate under weak-key assumptions.

The strategy is to build linear relations in a bottom-up fashion, from the smallest components: $\odot$, $\boxplus$ and $\oplus$, up to an MA-box, up to a half-round, further to a full round, and finally to multiple rounds.

Across unkeyed $\odot$ and $\boxplus$, the propagation of bitmasks follows from Table 3.98 and Table 3.99, respectively.

Let $(X \odot Z) \cdot \Gamma$ denote a linear approximation to $\odot$ using mask Γ. Under a weak subkey $Z \in \{0, 1\}$, this approximation becomes either:

- $(X \odot 1) \cdot 1 = X \cdot 1$,
- $(X \odot 0) \cdot 1 = (X \odot (-1)) \cdot 1 = (-X) \cdot 1 = (X \oplus 8000_x) \cdot 1 = X \cdot 1$.

So, in both cases, the mask 1 only concerns the least significant bit of X. It means that the multiplication by a weak subkey does not affect linear approximations of the least significant bit of its operand.

Further, we proceed to construct all linear relation patterns across the MA-box of REESSE3+, under the corresponding weak-subkey restrictions. Table 3.132 lists all linear relation patterns across the MA-box of REESSE3+. Only two 16-bit masks are used in the linear distinguishers: 0 and $\gamma = 1$. The bitmask 1 propagates across the modular addition operation with maximum bias, since it approximates only the least significant bit position, which is not affected by carry bits.

Table 3.132 All wordwise linear relations across the MA-box of REESSE3+.

Bitmasks across the MA-box	# active $\odot$	Bias
$(0,0,0,0) \rightarrow (0,0,0,0)$	0	2^{-1}
$(0,0,0,\gamma) \rightarrow (\gamma,0,\gamma,0)$	1	$2^{-13.47}$
$(0,0,\gamma,0) \rightarrow (0,0,\gamma,0)$	3	$2^{-38.41}$
$(0,0,\gamma,\gamma) \rightarrow (\gamma,0,0,0)$	1	$2^{-13.47}$
$(0,\gamma,0,0) \rightarrow (\gamma,\gamma,0,0)$	1	$2^{-13.47}$
$(0,\gamma,0,\gamma) \rightarrow (0,\gamma,\gamma,0)$	2	$2^{-25.94}$
$(0,\gamma,\gamma,0) \rightarrow (\gamma,\gamma,\gamma,0)$	3	$2^{-38.41}$
$(0,\gamma,\gamma,\gamma) \rightarrow (0,\gamma,0,0)$	2	$2^{-25.94}$
$(\gamma,0,0,0) \rightarrow (0,\gamma,0,\gamma)$	0	2^{-1}
$(\gamma,0,0,\gamma) \rightarrow (\gamma,\gamma,\gamma,\gamma)$	1	$2^{-13.47}$
$(\gamma,0,\gamma,0) \rightarrow (0,\gamma,\gamma,\gamma)$	2	$2^{-25.94}$
$(\gamma,0,\gamma,\gamma) \rightarrow (\gamma,\gamma,0,\gamma)$	1	$2^{-13.47}$
$(\gamma,\gamma,0,0) \rightarrow (\gamma,0,0,\gamma)$	1	$2^{-13.47}$
$(\gamma,\gamma,0,\gamma) \rightarrow (0,0,\gamma,\gamma)$	2	$2^{-25.94}$
$(\gamma,\gamma,\gamma,0) \rightarrow (\gamma,0,\gamma,\gamma)$	3	$2^{-38.41}$
$(\gamma,\gamma,\gamma,\gamma) \rightarrow (0,0,0,\gamma)$	2	$2^{-25.94}$

The next step is to obtain from Table 3.132 all 1-round linear relations (under wordwise bitmasks 0 and 1) and the corresponding weak-subkeys conditions.

Tables 3.133 through 3.140 list all 255 nontrivial 1-round linear relations of REESSE3+ under wordwise bitmasks either 0 or $\gamma = 1$. These tables provide an exhaustive analysis of how linear relations propagate across a single round of REESSE3+.

The bias was calculated using the Piling-up lemma [131].

Multiple-round linear relations are obtained by concatenating 1-round linear relations, such that the output mask from one relation matches the input mask to the next relation.

From Tables 3.133 up to 3.140, the multiple-round linear relations are clustered into 29 relations covering 1, 2, 3, 6 or 12 rounds:

(1) $(0,0,0,0,0,0,0,\gamma) \stackrel{1r}{\rightarrow} (\gamma,\gamma,\gamma,\gamma,\gamma,\gamma,\gamma,0) \stackrel{1r}{\rightarrow} (0,\ 0,\ 0,\ 0,\ 0,\ 0,\ 0,\ \gamma)$ is a 2-round iterative linear relation with bias $2^{-50.88}$. Starting at the i round, the following 4 subkeys are required to be weak: $Z_1^{(i+1)}$, $Z_4^{(i+1)}$, $Z_6^{(i+1)}$ and $Z_7^{(i+1)}$.

Table 3.133 Nontrivial 1-round linear relations for REESSE3+ (part I).

1-round linear relations	Weak subkeys $Z_j^{(i)}$ (ith round) j values	Bias
$(0,0,0,0,0,0,0,\gamma) \stackrel{1r}{\to} (\gamma,\gamma,\gamma,\gamma,\gamma,\gamma,\gamma,0)$	—	$2^{-25.94}$
$(0,0,0,0,0,0,\gamma,0) \stackrel{1r}{\to} (0,0,0,\gamma,0,0,0,0)$	7	$2^{-38.41}$
$(0,0,0,0,0,0,\gamma,\gamma) \stackrel{1r}{\to} (\gamma,\gamma,\gamma,0,\gamma,\gamma,\gamma,0)$	7	$2^{-25.94}$
$(0,0,0,0,0,\gamma,0,0) \stackrel{1r}{\to} (\gamma,\gamma,\gamma,\gamma,\gamma,\gamma,0,\gamma)$	6	$2^{-25.94}$
$(0,0,0,0,0,\gamma,0,\gamma) \stackrel{1r}{\to} (0,0,0,0,0,0,\gamma,\gamma)$	6	2^{-1}
$(0,0,0,0,0,\gamma,\gamma,0) \stackrel{1r}{\to} (\gamma,\gamma,\gamma,0,\gamma,\gamma,0,\gamma)$	6; 7	$2^{-25.94}$
$(0,0,0,0,0,\gamma,\gamma,\gamma) \stackrel{1r}{\to} (0,0,0,\gamma,0,0,\gamma,\gamma)$	6; 7	$2^{-38.41}$
$(0,0,0,0,\gamma,0,0,0) \stackrel{1r}{\to} (0,0,0,0,0,\gamma,0,0)$	—	$2^{-38.41}$
$(0,0,0,0,\gamma,0,0,\gamma) \stackrel{1r}{\to} (\gamma,\gamma,\gamma,\gamma,\gamma,0,\gamma,0)$	—	$2^{-25.94}$
$(0,0,0,0,\gamma,0,\gamma,0) \stackrel{1r}{\to} (0,0,0,\gamma,0,\gamma,0,0)$	7	2^{-1}
$(0,0,0,0,\gamma,0,\gamma,\gamma) \stackrel{1r}{\to} (\gamma,\gamma,\gamma,0,\gamma,0,\gamma,0)$	7	$2^{-25.94}$
$(0,0,0,0,\gamma,\gamma,0,0) \stackrel{1r}{\to} (\gamma,\gamma,\gamma,\gamma,\gamma,0,0,\gamma)$	6	$2^{-25.94}$
$(0,0,0,0,\gamma,\gamma,0,\gamma) \stackrel{1r}{\to} (0,0,0,0,0,\gamma,\gamma,\gamma)$	6	$2^{-38.41}$
$(0,0,0,0,\gamma,\gamma,\gamma,0) \stackrel{1r}{\to} (\gamma,\gamma,\gamma,0,\gamma,0,0,\gamma)$	6; 7	$2^{-25.94}$
$(0,0,0,0,\gamma,\gamma,\gamma,\gamma) \stackrel{1r}{\to} (0,0,0,\gamma,0,\gamma,\gamma,\gamma)$	6; 7	2^{-1}
$(0,0,0,\gamma,0,0,0,0) \stackrel{1r}{\to} (0,0,\gamma,\gamma,0,\gamma,\gamma,\gamma)$	4	$2^{-25.94}$
$(0,0,0,\gamma,0,0,0,\gamma) \stackrel{1r}{\to} (\gamma,\gamma,0,0,\gamma,0,0,\gamma)$	4	2^{-1}
$(0,0,0,\gamma,0,0,\gamma,0) \stackrel{1r}{\to} (0,0,\gamma,0,0,\gamma,\gamma,\gamma)$	4; 7	$2^{-25.94}$
$(0,0,0,\gamma,0,0,\gamma,\gamma) \stackrel{1r}{\to} (\gamma,\gamma,0,\gamma,\gamma,0,0,\gamma)$	4; 7	$2^{-25.94}$
$(0,0,0,\gamma,0,\gamma,0,0) \stackrel{1r}{\to} (\gamma,\gamma,0,0,\gamma,0,\gamma,0)$	4; 6	2^{-1}
$(0,0,0,\gamma,0,\gamma,0,\gamma) \stackrel{1r}{\to} (0,0,\gamma,\gamma,0,\gamma,0,0)$	4; 6	$2^{-25.94}$
$(0,0,0,\gamma,0,\gamma,\gamma,0) \stackrel{1r}{\to} (\gamma,\gamma,0,\gamma,\gamma,0,\gamma,0)$	4; 6; 7	$2^{-25.94}$
$(0,0,0,\gamma,0,\gamma,\gamma,\gamma) \stackrel{1r}{\to} (0,0,\gamma,0,0,\gamma,0,0)$	4; 6; 7	$2^{-25.94}$
$(0,0,0,\gamma,\gamma,0,0,0) \stackrel{1r}{\to} (0,0,\gamma,\gamma,0,0,\gamma,\gamma)$	4	$2^{-25.94}$
$(0,0,0,\gamma,\gamma,0,0,\gamma) \stackrel{1r}{\to} (\gamma,\gamma,0,0,\gamma,\gamma,0,\gamma)$	4	$2^{-25.94}$
$(0,0,0,\gamma,\gamma,0,\gamma,0) \stackrel{1r}{\to} (0,0,\gamma,0,0,0,\gamma,\gamma)$	4; 7	$2^{-25.94}$
$(0,0,0,\gamma,\gamma,0,\gamma,\gamma) \stackrel{1r}{\to} (\gamma,\gamma,0,\gamma,\gamma,\gamma,0,\gamma)$	4; 7	2^{-1}
$(0,0,0,\gamma,\gamma,\gamma,0,0) \stackrel{1r}{\to} (\gamma,\gamma,0,0,\gamma,\gamma,\gamma,0)$	4; 6	$2^{-25.94}$
$(0,0,0,\gamma,\gamma,\gamma,0,\gamma) \stackrel{1r}{\to} (0,0,\gamma,\gamma,0,0,0,0)$	4; 6	$2^{-25.94}$
$(0,0,0,\gamma,\gamma,\gamma,\gamma,0) \stackrel{1r}{\to} (\gamma,\gamma,0,\gamma,\gamma,\gamma,\gamma,0)$	4; 6; 7	2^{-1}
$(0,0,0,\gamma,\gamma,\gamma,\gamma,\gamma) \stackrel{1r}{\to} (0,0,\gamma,0,0,0,0,0)$	4; 6; 7	$2^{-25.94}$
$(0,0,\gamma,0,0,0,0,0) \stackrel{1r}{\to} (0,\gamma,0,\gamma,0,\gamma,\gamma,\gamma)$	—	$2^{-13.47}$

Table 3.134 Nontrivial 1-round linear relations for REESSE3+ (part II).

1-round linear relations	Weak subkeys $Z_j^{(i)}$ (ith round) j values	Bias
$(0,0,\gamma,0,0,0,0,\gamma) \stackrel{1r}{\to} (\gamma,0,\gamma,0,\gamma,0,0,\gamma)$	—	$2^{-13.47}$
$(0,0,\gamma,0,0,0,\gamma,0) \stackrel{1r}{\to} (0,\gamma,0,0,0,\gamma,\gamma,\gamma)$	7	$2^{-13.47}$
$(0,0,\gamma,0,0,0,\gamma,\gamma) \stackrel{1r}{\to} (\gamma,0,\gamma,\gamma,\gamma,0,0,\gamma)$	7	$2^{-38.41}$
$(0,0,\gamma,0,0,\gamma,0,0) \stackrel{1r}{\to} (\gamma,0,\gamma,0,\gamma,0,\gamma,0)$	6	$2^{-13.47}$
$(0,0,\gamma,0,0,\gamma,0,\gamma) \stackrel{1r}{\to} (0,\gamma,0,\gamma,0,\gamma,0,0)$	6	$2^{-13.47}$
$(0,0,\gamma,0,0,\gamma,\gamma,0) \stackrel{1r}{\to} (\gamma,0,\gamma,\gamma,\gamma,0,\gamma,0)$	6; 7	$2^{-38.41}$
$(0,0,\gamma,0,0,\gamma,\gamma,\gamma) \stackrel{1r}{\to} (0,\gamma,0,0,0,\gamma,0,0)$	6; 7	$2^{-13.47}$
$(0,0,\gamma,0,\gamma,0,0,0) \stackrel{1r}{\to} (0,\gamma,0,\gamma,0,0,\gamma,\gamma)$	—	$2^{-13.47}$
$(0,0,\gamma,0,\gamma,0,0,\gamma) \stackrel{1r}{\to} (\gamma,0,\gamma,0,\gamma,\gamma,0,\gamma)$	—	$2^{-38.41}$
$(0,0,\gamma,0,\gamma,0,\gamma,0) \stackrel{1r}{\to} (0,\gamma,0,0,0,0,\gamma,\gamma)$	7	$2^{-13.47}$
$(0,0,\gamma,0,\gamma,0,\gamma,\gamma) \stackrel{1r}{\to} (\gamma,0,\gamma,\gamma,\gamma,\gamma,0,\gamma)$	7	$2^{-13.47}$
$(0,0,\gamma,0,\gamma,\gamma,0,0) \stackrel{1r}{\to} (\gamma,0,\gamma,0,\gamma,\gamma,\gamma,0)$	6	$2^{-38.41}$
$(0,0,\gamma,0,\gamma,\gamma,0,\gamma) \stackrel{1r}{\to} (0,\gamma,0,\gamma,0,0,0,0)$	6	$2^{-13.47}$
$(0,0,\gamma,0,\gamma,\gamma,\gamma,0) \stackrel{1r}{\to} (\gamma,0,\gamma,\gamma,\gamma,\gamma,\gamma,0)$	6; 7	$2^{-13.47}$
$(0,0,\gamma,0,\gamma,\gamma,\gamma,\gamma) \stackrel{1r}{\to} (0,\gamma,0,0,0,0,0,0)$	6; 7	$2^{-13.47}$
$(0,0,\gamma,\gamma,0,0,0,0) \stackrel{1r}{\to} (0,\gamma,\gamma,0,0,0,0,0)$	4	$2^{-13.47}$
$(0,0,\gamma,\gamma,0,0,0,\gamma) \stackrel{1r}{\to} (\gamma,0,0,\gamma,\gamma,\gamma,\gamma,0)$	4	$2^{-13.47}$
$(0,0,\gamma,\gamma,0,0,\gamma,0) \stackrel{1r}{\to} (0,\gamma,\gamma,\gamma,0,0,0,0)$	4; 7	$2^{-38.41}$
$(0,0,\gamma,\gamma,0,0,\gamma,\gamma) \stackrel{1r}{\to} (\gamma,0,0,0,\gamma,\gamma,\gamma,0)$	4; 7	$2^{-13.47}$
$(0,0,\gamma,\gamma,0,\gamma,0,0) \stackrel{1r}{\to} (\gamma,0,0,\gamma,\gamma,\gamma,0,\gamma)$	4; 6	$2^{-13.47}$
$(0,0,\gamma,\gamma,0,\gamma,0,\gamma) \stackrel{1r}{\to} (0,\gamma,\gamma,0,0,0,\gamma,\gamma)$	4; 6	$2^{-13.47}$
$(0,0,\gamma,\gamma,0,\gamma,\gamma,0) \stackrel{1r}{\to} (\gamma,0,0,0,\gamma,\gamma,0,\gamma)$	4; 6; 7	$2^{-13.47}$
$(0,0,\gamma,\gamma,0,\gamma,\gamma,\gamma) \stackrel{1r}{\to} (0,\gamma,\gamma,\gamma,0,0,\gamma,\gamma)$	4; 6; 7	$2^{-38.41}$
$(0,0,\gamma,\gamma,\gamma,0,0,0) \stackrel{1r}{\to} (0,\gamma,\gamma,0,0,\gamma,0,0)$	4	$2^{-38.41}$
$(0,0,\gamma,\gamma,\gamma,0,0,\gamma) \stackrel{1r}{\to} (\gamma,0,0,\gamma,\gamma,0,\gamma,0)$	4	$2^{-13.47}$
$(0,0,\gamma,\gamma,\gamma,0,\gamma,0) \stackrel{1r}{\to} (0,\gamma,\gamma,\gamma,0,\gamma,0,0)$	4; 7	$2^{-13.47}$
$(0,0,\gamma,\gamma,\gamma,0,\gamma,\gamma) \stackrel{1r}{\to} (\gamma,0,0,0,\gamma,0,\gamma,0)$	4; 7	$2^{-13.47}$
$(0,0,\gamma,\gamma,\gamma,\gamma,0,0) \stackrel{1r}{\to} (\gamma,0,0,\gamma,\gamma,0,0,\gamma)$	4; 6	$2^{-13.47}$
$(0,0,\gamma,\gamma,\gamma,\gamma,0,\gamma) \stackrel{1r}{\to} (0,\gamma,\gamma,0,0,\gamma,\gamma,\gamma)$	4; 6	$2^{-38.41}$
$(0,0,\gamma,\gamma,\gamma,\gamma,\gamma,0) \stackrel{1r}{\to} (\gamma,0,0,0,\gamma,0,0,\gamma)$	4; 6; 7	$2^{-13.47}$
$(0,0,\gamma,\gamma,\gamma,\gamma,\gamma,\gamma) \stackrel{1r}{\to} (0,\gamma,\gamma,\gamma,0,\gamma,\gamma,\gamma)$	4; 6; 7	$2^{-13.47}$
$(0,\gamma,0,0,0,0,0,0) \stackrel{1r}{\to} (0,0,0,\gamma,\gamma,\gamma,\gamma,\gamma)$	—	$2^{-25.94}$

Table 3.135 Nontrivial 1-round linear relations for REESSE3+ (part III).

1-round linear relations	Weak subkeys $Z_j^{(i)}$ (ith round) j values	Bias
$(0,\gamma,0,0,0,0,0,\gamma) \xrightarrow{1r} (\gamma,\gamma,\gamma,0,0,0,0,\gamma)$	—	2^{-1}
$(0,\gamma,0,0,0,0,\gamma,0) \xrightarrow{1r} (0,0,0,0,\gamma,\gamma,\gamma,\gamma)$	7	$2^{-25.94}$
$(0,\gamma,0,0,0,0,\gamma,\gamma) \xrightarrow{1r} (\gamma,\gamma,\gamma,\gamma,0,0,0,\gamma)$	7	$2^{-25.94}$
$(0,\gamma,0,0,0,\gamma,0,0) \xrightarrow{1r} (\gamma,\gamma,\gamma,0,0,0,\gamma,0)$	6	2^{-1}
$(0,\gamma,0,0,0,\gamma,0,\gamma) \xrightarrow{1r} (0,0,0,\gamma,\gamma,\gamma,0,0)$	6	$2^{-25.94}$
$(0,\gamma,0,0,0,\gamma,\gamma,0) \xrightarrow{1r} (\gamma,\gamma,\gamma,\gamma,0,0,\gamma,0)$	6; 7	$2^{-25.94}$
$(0,\gamma,0,0,0,\gamma,\gamma,\gamma) \xrightarrow{1r} (0,0,0,0,\gamma,\gamma,0,0)$	6; 7	$2^{-25.94}$
$(0,\gamma,0,0,\gamma,0,0,0) \xrightarrow{1r} (0,0,0,\gamma,\gamma,0,\gamma,\gamma)$	—	$2^{-25.94}$
$(0,\gamma,0,0,\gamma,0,0,\gamma) \xrightarrow{1r} (\gamma,\gamma,\gamma,0,0,\gamma,0,\gamma)$	—	$2^{-25.94}$
$(0,\gamma,0,0,\gamma,0,\gamma,0) \xrightarrow{1r} (0,0,0,0,\gamma,0,\gamma,\gamma)$	7	$2^{-25.94}$
$(0,\gamma,0,0,\gamma,0,\gamma,\gamma) \xrightarrow{1r} (\gamma,\gamma,\gamma,\gamma,0,\gamma,0,\gamma)$	7	2^{-1}
$(0,\gamma,0,0,\gamma,\gamma,0,0) \xrightarrow{1r} (\gamma,\gamma,\gamma,0,0,\gamma,\gamma,0)$	6	$2^{-25.94}$
$(0,\gamma,0,0,\gamma,\gamma,0,\gamma) \xrightarrow{1r} (0,0,0,\gamma,\gamma,0,0,0)$	6	$2^{-25.94}$
$(0,\gamma,0,0,\gamma,\gamma,\gamma,0) \xrightarrow{1r} (\gamma,\gamma,\gamma,\gamma,0,\gamma,\gamma,0)$	6; 7	2^{-1}
$(0,\gamma,0,0,\gamma,\gamma,\gamma,\gamma) \xrightarrow{1r} (0,0,0,0,\gamma,0,0,0)$	6; 7	$2^{-25.94}$
$(0,\gamma,0,\gamma,0,0,0,0) \xrightarrow{1r} (0,0,\gamma,0,\gamma,0,0,0)$	4	2^{-1}
$(0,\gamma,0,\gamma,0,0,0,\gamma) \xrightarrow{1r} (\gamma,\gamma,0,\gamma,0,\gamma,\gamma,0)$	4	$2^{-25.94}$
$(0,\gamma,0,\gamma,0,0,\gamma,0) \xrightarrow{1r} (0,0,\gamma,\gamma,\gamma,0,0,0)$	4; 7	$2^{-38.41}$
$(0,\gamma,0,\gamma,0,0,\gamma,\gamma) \xrightarrow{1r} (\gamma,\gamma,0,0,0,\gamma,\gamma,0)$	4; 7	$2^{-25.94}$
$(0,\gamma,0,\gamma,0,\gamma,0,0) \xrightarrow{1r} (\gamma,\gamma,0,\gamma,0,\gamma,0,\gamma)$	4; 6	$2^{-25.94}$
$(0,\gamma,0,\gamma,0,\gamma,0,\gamma) \xrightarrow{1r} (0,0,\gamma,0,\gamma,0,\gamma,\gamma)$	4; 6	2^{-1}
$(0,\gamma,0,\gamma,0,\gamma,\gamma,0) \xrightarrow{1r} (\gamma,\gamma,0,0,0,\gamma,0,\gamma)$	4; 6; 7	$2^{-25.94}$
$(0,\gamma,0,\gamma,0,\gamma,\gamma,\gamma) \xrightarrow{1r} (0,0,\gamma,\gamma,\gamma,0,\gamma,\gamma)$	4; 6; 7	$2^{-38.41}$
$(0,\gamma,0,\gamma,\gamma,0,0,0) \xrightarrow{1r} (0,0,\gamma,0,\gamma,\gamma,0,0)$	4	$2^{-38.41}$
$(0,\gamma,0,\gamma,\gamma,0,0,\gamma) \xrightarrow{1r} (\gamma,\gamma,0,\gamma,0,0,\gamma,0)$	4	$2^{-25.94}$
$(0,\gamma,0,\gamma,\gamma,0,\gamma,0) \xrightarrow{1r} (0,0,\gamma,\gamma,\gamma,\gamma,0,0)$	4; 7	2^{-1}
$(0,\gamma,0,\gamma,\gamma,0,\gamma,\gamma) \xrightarrow{1r} (\gamma,\gamma,0,0,0,0,\gamma,0)$	4; 7	$2^{-25.94}$
$(0,\gamma,0,\gamma,\gamma,\gamma,0,0) \xrightarrow{1r} (\gamma,\gamma,0,\gamma,0,0,0,\gamma)$	4; 6	$2^{-25.94}$
$(0,\gamma,0,\gamma,\gamma,\gamma,0,\gamma) \xrightarrow{1r} (0,0,\gamma,0,\gamma,\gamma,\gamma,\gamma)$	4; 6	$2^{-38.41}$
$(0,\gamma,0,\gamma,\gamma,\gamma,\gamma,0) \xrightarrow{1r} (\gamma,\gamma,0,0,0,0,0,\gamma)$	4; 6; 7	$2^{-25.94}$
$(0,\gamma,0,\gamma,\gamma,\gamma,\gamma,\gamma) \xrightarrow{1r} (0,0,\gamma,\gamma,\gamma,\gamma,\gamma,\gamma)$	4; 6; 7	2^{-1}
$(0,\gamma,\gamma,0,0,0,0,0) \xrightarrow{1r} (0,\gamma,0,0,\gamma,0,0,0)$	—	$2^{-13.47}$

Table 3.136 Nontrivial 1-round linear relations for REESSE3+ (part IV).

1-round linear relations	Weak subkeys $Z_j^{(i)}$ (ith round) j values	Bias
$(0,\gamma,\gamma,0,0,0,0,\gamma) \stackrel{1r}{\to} (\gamma,0,\gamma,\gamma,0,\gamma,\gamma,0)$	—	$2^{-13.47}$
$(0,\gamma,\gamma,0,0,0,\gamma,0) \stackrel{1r}{\to} (0,\gamma,0,\gamma,\gamma,0,0,0)$	7	$2^{-38.41}$
$(0,\gamma,\gamma,0,0,0,\gamma,\gamma) \stackrel{1r}{\to} (\gamma,0,\gamma,0,0,\gamma,\gamma,0)$	7	$2^{-13.47}$
$(0,\gamma,\gamma,0,0,\gamma,0,0) \stackrel{1r}{\to} (\gamma,0,\gamma,\gamma,0,\gamma,0,\gamma)$	6	$2^{-13.47}$
$(0,\gamma,\gamma,0,0,\gamma,0,\gamma) \stackrel{1r}{\to} (0,\gamma,0,0,\gamma,0,\gamma,\gamma)$	6	$2^{-13.47}$
$(0,\gamma,\gamma,0,0,\gamma,\gamma,0) \stackrel{1r}{\to} (\gamma,0,\gamma,0,0,\gamma,0,\gamma)$	6; 7	$2^{-13.47}$
$(0,\gamma,\gamma,0,0,\gamma,\gamma,\gamma) \stackrel{1r}{\to} (0,\gamma,0,\gamma,\gamma,0,\gamma,\gamma)$	6; 7	$2^{-38.41}$
$(0,\gamma,\gamma,0,\gamma,0,0,0) \stackrel{1r}{\to} (0,\gamma,0,0,\gamma,\gamma,0,0)$	—	$2^{-38.41}$
$(0,\gamma,\gamma,0,\gamma,0,0,\gamma) \stackrel{1r}{\to} (\gamma,0,\gamma,\gamma,0,0,\gamma,0)$	—	$2^{-13.47}$
$(0,\gamma,\gamma,0,\gamma,0,\gamma,0) \stackrel{1r}{\to} (0,\gamma,0,\gamma,\gamma,\gamma,0,0)$	7	$2^{-13.47}$
$(0,\gamma,\gamma,0,\gamma,0,\gamma,\gamma) \stackrel{1r}{\to} (\gamma,0,\gamma,0,0,0,\gamma,0)$	7	$2^{-13.47}$
$(0,\gamma,\gamma,0,\gamma,\gamma,0,0) \stackrel{1r}{\to} (\gamma,0,\gamma,\gamma,0,0,0,\gamma)$	6	$2^{-13.47}$
$(0,\gamma,\gamma,0,\gamma,\gamma,0,\gamma) \stackrel{1r}{\to} (0,\gamma,0,0,\gamma,\gamma,\gamma,\gamma)$	6	$2^{-38.41}$
$(0,\gamma,\gamma,0,\gamma,\gamma,\gamma,0) \stackrel{1r}{\to} (\gamma,0,\gamma,0,0,0,0,\gamma)$	6; 7	$2^{-13.47}$
$(0,\gamma,\gamma,0,\gamma,\gamma,\gamma,\gamma) \stackrel{1r}{\to} (0,\gamma,0,\gamma,\gamma,\gamma,\gamma,\gamma)$	6; 7	$2^{-13.47}$
$(0,\gamma,\gamma,\gamma,0,0,0,0) \stackrel{1r}{\to} (0,\gamma,\gamma,\gamma,\gamma,\gamma,\gamma,\gamma)$	4	$2^{-13.47}$
$(0,\gamma,\gamma,\gamma,0,0,0,\gamma) \stackrel{1r}{\to} (\gamma,0,0,0,0,0,0,\gamma)$	4	$2^{-13.47}$
$(0,\gamma,\gamma,\gamma,0,0,\gamma,0) \stackrel{1r}{\to} (0,\gamma,\gamma,0,\gamma,\gamma,\gamma,\gamma)$	4; 7	$2^{-13.47}$
$(0,\gamma,\gamma,\gamma,0,0,\gamma,\gamma) \stackrel{1r}{\to} (\gamma,0,0,\gamma,0,0,0,\gamma)$	4; 7	$2^{-38.41}$
$(0,\gamma,\gamma,\gamma,0,\gamma,0,0) \stackrel{1r}{\to} (\gamma,0,0,0,0,0,\gamma,0)$	4; 6	$2^{-13.47}$
$(0,\gamma,\gamma,\gamma,0,\gamma,0,\gamma) \stackrel{1r}{\to} (0,\gamma,\gamma,\gamma,\gamma,\gamma,0,0)$	4; 6	$2^{-13.47}$
$(0,\gamma,\gamma,\gamma,0,\gamma,\gamma,0) \stackrel{1r}{\to} (\gamma,0,0,\gamma,0,0,\gamma,0)$	4; 6; 7	$2^{-38.41}$
$(0,\gamma,\gamma,\gamma,0,\gamma,\gamma,\gamma) \stackrel{1r}{\to} (0,\gamma,\gamma,0,\gamma,\gamma,0,0)$	4; 6; 7	$2^{-13.47}$
$(0,\gamma,\gamma,\gamma,\gamma,0,0,0) \stackrel{1r}{\to} (0,\gamma,\gamma,\gamma,\gamma,0,\gamma,\gamma)$	4	$2^{-13.47}$
$(0,\gamma,\gamma,\gamma,\gamma,0,0,\gamma) \stackrel{1r}{\to} (\gamma,0,0,0,0,\gamma,0,\gamma)$	4	$2^{-38.41}$
$(0,\gamma,\gamma,\gamma,\gamma,0,\gamma,0) \stackrel{1r}{\to} (0,\gamma,\gamma,0,\gamma,0,\gamma,\gamma)$	4; 7	$2^{-13.47}$
$(0,\gamma,\gamma,\gamma,\gamma,0,\gamma,\gamma) \stackrel{1r}{\to} (\gamma,0,0,\gamma,0,\gamma,0,\gamma)$	4; 7	$2^{-13.47}$
$(0,\gamma,\gamma,\gamma,\gamma,\gamma,0,0) \stackrel{1r}{\to} (\gamma,0,0,0,0,\gamma,\gamma,0)$	4; 6	$2^{-38.41}$
$(0,\gamma,\gamma,\gamma,\gamma,\gamma,0,\gamma) \stackrel{1r}{\to} (0,\gamma,\gamma,\gamma,\gamma,0,0,0)$	4; 6	$2^{-13.47}$
$(0,\gamma,\gamma,\gamma,\gamma,\gamma,\gamma,0) \stackrel{1r}{\to} (\gamma,0,0,\gamma,0,\gamma,\gamma,0)$	4; 6; 7	$2^{-13.47}$
$(0,\gamma,\gamma,\gamma,\gamma,\gamma,\gamma,\gamma) \stackrel{1r}{\to} (0,\gamma,\gamma,0,\gamma,0,0,0)$	4; 6; 7	$2^{-13.47}$
$(\gamma,0,0,0,0,0,0,0) \stackrel{1r}{\to} (\gamma,0,0,\gamma,0,\gamma,\gamma,\gamma)$	1	$2^{-13.47}$

Table 3.137 Nontrivial 1-round linear relations for REESSE3+ (part V).

1-round linear relations	Weak subkeys $Z_j^{(i)}$ (ith round) j values	Bias
$(\gamma,0,0,0,0,0,0,\gamma) \stackrel{1r}{\rightarrow} (0,\gamma,\gamma,0,\gamma,0,0,\gamma)$	1	$2^{-13.47}$
$(\gamma,0,0,0,0,0,\gamma,0) \stackrel{1r}{\rightarrow} (\gamma,0,0,0,0,\gamma,\gamma,\gamma)$	1; 7	$2^{-13.47}$
$(\gamma,0,0,0,0,0,\gamma,\gamma) \stackrel{1r}{\rightarrow} (0,\gamma,\gamma,\gamma,\gamma,0,0,\gamma)$	1; 7	$2^{-38.41}$
$(\gamma,0,0,0,0,\gamma,0,0) \stackrel{1r}{\rightarrow} (0,\gamma,\gamma,0,\gamma,0,\gamma,0)$	1; 6	$2^{-13.47}$
$(\gamma,0,0,0,0,\gamma,0,\gamma) \stackrel{1r}{\rightarrow} (\gamma,0,0,\gamma,0,\gamma,0,0)$	1; 6	$2^{-13.47}$
$(\gamma,0,0,0,0,\gamma,\gamma,0) \stackrel{1r}{\rightarrow} (0,\gamma,\gamma,\gamma,\gamma,0,\gamma,0)$	1; 6; 7	$2^{-38.41}$
$(\gamma,0,0,0,0,\gamma,\gamma,\gamma) \stackrel{1r}{\rightarrow} (\gamma,0,0,0,0,\gamma,0,0)$	1; 6; 7	$2^{-13.47}$
$(\gamma,0,0,0,\gamma,0,0,0) \stackrel{1r}{\rightarrow} (\gamma,0,0,\gamma,0,0,\gamma,\gamma)$	1	$2^{-13.47}$
$(\gamma,0,0,0,\gamma,0,0,\gamma) \stackrel{1r}{\rightarrow} (0,\gamma,\gamma,0,\gamma,\gamma,0,\gamma)$	1	$2^{-38.41}$
$(\gamma,0,0,0,\gamma,0,\gamma,0) \stackrel{1r}{\rightarrow} (\gamma,0,0,0,0,0,\gamma,\gamma)$	1; 7	$2^{-13.47}$
$(\gamma,0,0,0,\gamma,0,\gamma,\gamma) \stackrel{1r}{\rightarrow} (0,\gamma,\gamma,\gamma,\gamma,\gamma,0,\gamma)$	1; 7	$2^{-13.47}$
$(\gamma,0,0,0,\gamma,\gamma,0,0) \stackrel{1r}{\rightarrow} (0,\gamma,\gamma,0,\gamma,\gamma,\gamma,0)$	1; 6	$2^{-38.41}$
$(\gamma,0,0,0,\gamma,\gamma,0,\gamma) \stackrel{1r}{\rightarrow} (\gamma,0,0,\gamma,0,0,0,0)$	1; 6	$2^{-13.47}$
$(\gamma,0,0,0,\gamma,\gamma,\gamma,0) \stackrel{1r}{\rightarrow} (0,\gamma,\gamma,\gamma,\gamma,\gamma,\gamma,0)$	1; 6; 7	$2^{-13.47}$
$(\gamma,0,0,0,\gamma,\gamma,\gamma,\gamma) \stackrel{1r}{\rightarrow} (\gamma,0,0,0,0,0,0,0)$	1; 6; 7	$2^{-13.47}$
$(\gamma,0,0,\gamma,0,0,0,0) \stackrel{1r}{\rightarrow} (\gamma,0,\gamma,0,0,0,0,0)$	1; 4	$2^{-13.47}$
$(\gamma,0,0,\gamma,0,0,0,\gamma) \stackrel{1r}{\rightarrow} (0,\gamma,0,\gamma,\gamma,\gamma,\gamma,0)$	1; 4	$2^{-13.47}$
$(\gamma,0,0,\gamma,0,0,\gamma,0) \stackrel{1r}{\rightarrow} (\gamma,0,\gamma,\gamma,0,0,0,0)$	1; 4; 7	$2^{-38.41}$
$(\gamma,0,0,\gamma,0,0,\gamma,\gamma) \stackrel{1r}{\rightarrow} (0,\gamma,0,0,\gamma,\gamma,\gamma,0)$	1; 4; 7	$2^{-13.47}$
$(\gamma,0,0,\gamma,0,\gamma,0,0) \stackrel{1r}{\rightarrow} (0,\gamma,0,\gamma,\gamma,\gamma,0,\gamma)$	1; 4; 6	$2^{-13.47}$
$(\gamma,0,0,\gamma,0,\gamma,0,\gamma) \stackrel{1r}{\rightarrow} (\gamma,0,\gamma,0,0,0,\gamma,\gamma)$	1; 4; 6	$2^{-13.47}$
$(\gamma,0,0,\gamma,0,\gamma,\gamma,0) \stackrel{1r}{\rightarrow} (0,\gamma,0,0,\gamma,\gamma,0,\gamma)$	1; 4; 6; 7	$2^{-13.47}$
$(\gamma,0,0,\gamma,0,\gamma,\gamma,\gamma) \stackrel{1r}{\rightarrow} (\gamma,0,\gamma,\gamma,0,0,\gamma,\gamma)$	1; 4; 6; 7	$2^{-38.41}$
$(\gamma,0,0,\gamma,\gamma,0,0,0) \stackrel{1r}{\rightarrow} (\gamma,0,\gamma,0,0,\gamma,0,0)$	1; 4	$2^{-38.41}$
$(\gamma,0,0,\gamma,\gamma,0,0,\gamma) \stackrel{1r}{\rightarrow} (0,\gamma,0,\gamma,\gamma,0,\gamma,0)$	1; 4	$2^{-13.47}$
$(\gamma,0,0,\gamma,\gamma,0,\gamma,0) \stackrel{1r}{\rightarrow} (\gamma,0,\gamma,\gamma,0,\gamma,0,0)$	1; 4; 7	$2^{-13.47}$
$(\gamma,0,0,\gamma,\gamma,0,\gamma,\gamma) \stackrel{1r}{\rightarrow} (0,\gamma,0,0,\gamma,0,\gamma,0)$	1; 4; 7	$2^{-13.47}$
$(\gamma,0,0,\gamma,\gamma,\gamma,0,0) \stackrel{1r}{\rightarrow} (0,\gamma,0,\gamma,\gamma,0,0,\gamma)$	1; 4; 6	$2^{-13.47}$
$(\gamma,0,0,\gamma,\gamma,\gamma,0,\gamma) \stackrel{1r}{\rightarrow} (\gamma,0,\gamma,0,0,\gamma,\gamma,\gamma)$	1; 4; 6	$2^{-38.41}$
$(\gamma,0,0,\gamma,\gamma,\gamma,\gamma,0) \stackrel{1r}{\rightarrow} (0,\gamma,0,0,\gamma,0,0,\gamma)$	1; 4; 6; 7	$2^{-13.47}$
$(\gamma,0,0,\gamma,\gamma,\gamma,\gamma,\gamma) \stackrel{1r}{\rightarrow} (\gamma,0,\gamma,\gamma,0,\gamma,\gamma,\gamma)$	1; 4; 6; 7	$2^{-13.47}$
$(\gamma,0,\gamma,0,0,0,0,0) \stackrel{1r}{\rightarrow} (\gamma,\gamma,0,0,0,0,0,0)$	1	2^{-1}

Table 3.138 Nontrivial 1-round linear relations for REESSE3+ (part VI).

1-round linear relations	Weak subkeys $Z_j^{(i)}$ (ith round) j values	Bias
$(\gamma,0,\gamma,0,0,0,0,\gamma) \xrightarrow{1r} (0,0,\gamma,\gamma,\gamma,\gamma,\gamma,0)$	1	$2^{-25.94}$
$(\gamma,0,\gamma,0,0,0,\gamma,0) \xrightarrow{1r} (\gamma,\gamma,0,\gamma,0,0,0,0)$	1; 7	$2^{-38.41}$
$(\gamma,0,\gamma,0,0,0,\gamma,\gamma) \xrightarrow{1r} (0,0,\gamma,0,\gamma,\gamma,\gamma,0)$	1; 7	$2^{-25.94}$
$(\gamma,0,\gamma,0,0,\gamma,0,0) \xrightarrow{1r} (0,0,\gamma,\gamma,\gamma,\gamma,0,\gamma)$	1; 6	$2^{-25.94}$
$(\gamma,0,\gamma,0,0,\gamma,0,\gamma) \xrightarrow{1r} (\gamma,\gamma,0,0,0,0,\gamma,\gamma)$	1; 6	2^{-1}
$(\gamma,0,\gamma,0,0,\gamma,\gamma,0) \xrightarrow{1r} (0,0,\gamma,0,\gamma,\gamma,0,\gamma)$	1; 6; 7	$2^{-25.94}$
$(\gamma,0,\gamma,0,0,\gamma,\gamma,\gamma) \xrightarrow{1r} (\gamma,\gamma,0,\gamma,0,0,\gamma,\gamma)$	1; 6; 7	$2^{-38.41}$
$(\gamma,0,\gamma,0,\gamma,0,0,0) \xrightarrow{1r} (\gamma,\gamma,0,0,0,\gamma,0,0)$	1	$2^{-38.41}$
$(\gamma,0,\gamma,0,\gamma,0,0,\gamma) \xrightarrow{1r} (0,0,\gamma,\gamma,\gamma,0,\gamma,0)$	1	$2^{-25.94}$
$(\gamma,0,\gamma,0,\gamma,0,\gamma,0) \xrightarrow{1r} (\gamma,\gamma,0,\gamma,0,\gamma,0,0)$	1; 7	2^{-1}
$(\gamma,0,\gamma,0,\gamma,0,\gamma,\gamma) \xrightarrow{1r} (0,0,\gamma,0,\gamma,0,\gamma,0)$	1; 7	$2^{-25.94}$
$(\gamma,0,\gamma,0,\gamma,\gamma,0,0) \xrightarrow{1r} (0,0,\gamma,\gamma,\gamma,0,0,\gamma)$	1; 6	$2^{-25.94}$
$(\gamma,0,\gamma,0,\gamma,\gamma,0,\gamma) \xrightarrow{1r} (\gamma,\gamma,0,0,0,\gamma,\gamma,\gamma)$	1; 6	$2^{-38.41}$
$(\gamma,0,\gamma,0,\gamma,\gamma,\gamma,0) \xrightarrow{1r} (0,0,\gamma,0,\gamma,0,0,\gamma)$	1; 6; 7	$2^{-25.94}$
$(\gamma,0,\gamma,0,\gamma,\gamma,\gamma,\gamma) \xrightarrow{1r} (\gamma,\gamma,0,\gamma,0,\gamma,\gamma,\gamma)$	1; 6; 7	2^{-1}
$(\gamma,0,\gamma,\gamma,0,0,0,0) \xrightarrow{1r} (\gamma,\gamma,\gamma,\gamma,0,\gamma,\gamma,\gamma)$	1; 4	$2^{-25.94}$
$(\gamma,0,\gamma,\gamma,0,0,0,\gamma) \xrightarrow{1r} (0,0,0,0,\gamma,0,0,\gamma)$	1; 4	2^{-1}
$(\gamma,0,\gamma,\gamma,0,0,\gamma,0) \xrightarrow{1r} (\gamma,\gamma,\gamma,0,0,\gamma,\gamma,\gamma)$	1; 4; 7	$2^{-25.94}$
$(\gamma,0,\gamma,\gamma,0,0,\gamma,\gamma) \xrightarrow{1r} (0,0,0,\gamma,\gamma,0,0,\gamma)$	1; 4; 7	$2^{-25.94}$
$(\gamma,0,\gamma,\gamma,0,\gamma,0,0) \xrightarrow{1r} (0,0,0,0,\gamma,0,\gamma,0)$	1; 4; 6	2^{-1}
$(\gamma,0,\gamma,\gamma,0,\gamma,0,\gamma) \xrightarrow{1r} (\gamma,\gamma,\gamma,\gamma,0,\gamma,0,0)$	1; 4; 6	$2^{-25.94}$
$(\gamma,0,\gamma,\gamma,0,\gamma,\gamma,0) \xrightarrow{1r} (0,0,0,\gamma,\gamma,0,\gamma,0)$	1; 4; 6; 7	$2^{-25.94}$
$(\gamma,0,\gamma,\gamma,0,\gamma,\gamma,\gamma) \xrightarrow{1r} (\gamma,\gamma,\gamma,0,0,\gamma,0,0)$	1; 4; 6; 7	$2^{-25.94}$
$(\gamma,0,\gamma,\gamma,\gamma,0,0,0) \xrightarrow{1r} (\gamma,\gamma,\gamma,\gamma,0,0,\gamma,\gamma)$	1; 4	$2^{-25.94}$
$(\gamma,0,\gamma,\gamma,\gamma,0,0,\gamma) \xrightarrow{1r} (0,0,0,0,\gamma,\gamma,0,\gamma)$	1; 4	$2^{-25.94}$
$(\gamma,0,\gamma,\gamma,\gamma,0,\gamma,0) \xrightarrow{1r} (\gamma,\gamma,\gamma,0,0,0,\gamma,\gamma)$	1; 4; 7	$2^{-25.94}$
$(\gamma,0,\gamma,\gamma,\gamma,0,\gamma,\gamma) \xrightarrow{1r} (0,0,0,\gamma,\gamma,\gamma,0,\gamma)$	1; 4; 7	2^{-1}
$(\gamma,0,\gamma,\gamma,\gamma,\gamma,0,0) \xrightarrow{1r} (0,0,0,0,\gamma,\gamma,\gamma,0)$	1; 4; 6	$2^{-25.94}$
$(\gamma,0,\gamma,\gamma,\gamma,\gamma,0,\gamma) \xrightarrow{1r} (\gamma,\gamma,\gamma,\gamma,0,0,0,0)$	1; 4; 6	$2^{-25.94}$
$(\gamma,0,\gamma,\gamma,\gamma,\gamma,\gamma,0) \xrightarrow{1r} (0,0,0,\gamma,\gamma,\gamma,\gamma,0)$	1; 4; 6; 7	2^{-1}
$(\gamma,0,\gamma,\gamma,\gamma,\gamma,\gamma,\gamma) \xrightarrow{1r} (\gamma,\gamma,\gamma,0,0,0,0,0)$	1; 4; 6; 7	$2^{-25.94}$
$(\gamma,\gamma,0,0,0,0,0,0) \xrightarrow{1r} (\gamma,0,0,0,\gamma,0,0,0)$	1	$2^{-13.47}$

Table 3.139 Nontrivial 1-round linear relations for REESSE3+ (part VII).

1-round linear relations	Weak subkeys $Z_j^{(i)}$ (ith round) j values	Bias
$(\gamma,\gamma,0,0,0,0,0,\gamma) \stackrel{1r}{\to} (0,\gamma,\gamma,\gamma,0,\gamma,\gamma,0)$	1	$2^{-13.47}$
$(\gamma,\gamma,0,0,0,0,\gamma,0) \stackrel{1r}{\to} (\gamma,0,0,\gamma,\gamma,0,0,0)$	1; 7	$2^{-38.41}$
$(\gamma,\gamma,0,0,0,0,\gamma,\gamma) \stackrel{1r}{\to} (0,\gamma,\gamma,0,0,\gamma,\gamma,0)$	1; 7	$2^{-13.47}$
$(\gamma,\gamma,0,0,0,\gamma,0,0) \stackrel{1r}{\to} (0,\gamma,\gamma,\gamma,0,\gamma,0,\gamma)$	1; 6	$2^{-13.47}$
$(\gamma,\gamma,0,0,0,\gamma,0,\gamma) \stackrel{1r}{\to} (\gamma,0,0,0,\gamma,0,\gamma,\gamma)$	1; 6	$2^{-13.47}$
$(\gamma,\gamma,0,0,0,\gamma,\gamma,0) \stackrel{1r}{\to} (0,\gamma,\gamma,0,0,\gamma,0,\gamma)$	1; 6; 7	$2^{-13.47}$
$(\gamma,\gamma,0,0,0,\gamma,\gamma,\gamma) \stackrel{1r}{\to} (\gamma,0,0,\gamma,\gamma,0,\gamma,\gamma)$	1; 6; 7	$2^{-38.41}$
$(\gamma,\gamma,0,0,\gamma,0,0,0) \stackrel{1r}{\to} (\gamma,0,0,0,\gamma,\gamma,0,0)$	1	$2^{-38.41}$
$(\gamma,\gamma,0,0,\gamma,0,0,\gamma) \stackrel{1r}{\to} (0,\gamma,\gamma,\gamma,0,0,\gamma,0)$	1	$2^{-13.47}$
$(\gamma,\gamma,0,0,\gamma,0,\gamma,0) \stackrel{1r}{\to} (\gamma,0,0,\gamma,\gamma,\gamma,0,0)$	1; 7	$2^{-13.47}$
$(\gamma,\gamma,0,0,\gamma,0,\gamma,\gamma) \stackrel{1r}{\to} (0,\gamma,\gamma,0,0,0,\gamma,0)$	1; 7	$2^{-13.47}$
$(\gamma,\gamma,0,0,\gamma,\gamma,0,0) \stackrel{1r}{\to} (0,\gamma,\gamma,\gamma,0,0,0,\gamma)$	1; 6	$2^{-13.47}$
$(\gamma,\gamma,0,0,\gamma,\gamma,0,\gamma) \stackrel{1r}{\to} (\gamma,0,0,0,\gamma,\gamma,\gamma,\gamma)$	1; 6	$2^{-38.41}$
$(\gamma,\gamma,0,0,\gamma,\gamma,\gamma,0) \stackrel{1r}{\to} (0,\gamma,\gamma,0,0,0,0,\gamma)$	1; 6; 7	$2^{-13.47}$
$(\gamma,\gamma,0,0,\gamma,\gamma,\gamma,\gamma) \stackrel{1r}{\to} (\gamma,0,0,\gamma,\gamma,\gamma,\gamma,\gamma)$	1; 6; 7	$2^{-13.47}$
$(\gamma,\gamma,0,\gamma,0,0,0,0) \stackrel{1r}{\to} (\gamma,0,\gamma,\gamma,\gamma,\gamma,\gamma,\gamma)$	1; 4	$2^{-13.47}$
$(\gamma,\gamma,0,\gamma,0,0,0,\gamma) \stackrel{1r}{\to} (0,\gamma,0,0,0,0,0,\gamma)$	1; 4	$2^{-13.47}$
$(\gamma,\gamma,0,\gamma,0,0,\gamma,0) \stackrel{1r}{\to} (\gamma,0,\gamma,0,\gamma,\gamma,\gamma,\gamma)$	1; 4; 7	$2^{-13.47}$
$(\gamma,\gamma,0,\gamma,0,0,\gamma,\gamma) \stackrel{1r}{\to} (0,\gamma,0,\gamma,0,0,0,\gamma)$	1; 4; 7	$2^{-38.41}$
$(\gamma,\gamma,0,\gamma,0,\gamma,0,0) \stackrel{1r}{\to} (0,\gamma,0,0,0,0,\gamma,0)$	1; 4; 6	$2^{-13.47}$
$(\gamma,\gamma,0,\gamma,0,\gamma,0,\gamma) \stackrel{1r}{\to} (\gamma,0,\gamma,\gamma,\gamma,\gamma,0,0)$	1; 4; 6	$2^{-13.47}$
$(\gamma,\gamma,0,\gamma,0,\gamma,\gamma,0) \stackrel{1r}{\to} (0,\gamma,0,\gamma,0,0,\gamma,0)$	1; 4; 6; 7	$2^{-38.41}$
$(\gamma,\gamma,0,\gamma,0,\gamma,\gamma,\gamma) \stackrel{1r}{\to} (\gamma,0,\gamma,0,\gamma,\gamma,0,0)$	1; 4; 6; 7	$2^{-13.47}$
$(\gamma,\gamma,0,\gamma,\gamma,0,0,0) \stackrel{1r}{\to} (\gamma,0,\gamma,\gamma,\gamma,0,\gamma,\gamma)$	1; 4	$2^{-13.47}$
$(\gamma,\gamma,0,\gamma,\gamma,0,0,\gamma) \stackrel{1r}{\to} (0,\gamma,0,0,0,\gamma,0,\gamma)$	1; 4	$2^{-38.41}$
$(\gamma,\gamma,0,\gamma,\gamma,0,\gamma,0) \stackrel{1r}{\to} (\gamma,0,\gamma,0,\gamma,0,\gamma,\gamma)$	1; 4; 7	$2^{-13.47}$
$(\gamma,\gamma,0,\gamma,\gamma,0,\gamma,\gamma) \stackrel{1r}{\to} (0,\gamma,0,\gamma,0,\gamma,0,\gamma)$	1; 4; 7	$2^{-13.47}$
$(\gamma,\gamma,0,\gamma,\gamma,\gamma,0,0) \stackrel{1r}{\to} (0,\gamma,0,0,0,\gamma,\gamma,0)$	1; 4; 6	$2^{-38.41}$
$(\gamma,\gamma,0,\gamma,\gamma,\gamma,0,\gamma) \stackrel{1r}{\to} (\gamma,0,\gamma,\gamma,\gamma,0,0,0)$	1; 4; 6	$2^{-13.47}$
$(\gamma,\gamma,0,\gamma,\gamma,\gamma,\gamma,0) \stackrel{1r}{\to} (0,\gamma,0,\gamma,0,\gamma,\gamma,0)$	1; 4; 6; 7	$2^{-13.47}$
$(\gamma,\gamma,0,\gamma,\gamma,\gamma,\gamma,\gamma) \stackrel{1r}{\to} (\gamma,0,\gamma,0,\gamma,0,0,0)$	1; 4; 6; 7	$2^{-13.47}$
$(\gamma,\gamma,\gamma,0,0,0,0,0) \stackrel{1r}{\to} (\gamma,\gamma,0,\gamma,\gamma,\gamma,\gamma,\gamma)$	1	$2^{-25.94}$

Table 3.140 Nontrivial 1-round linear relations for REESSE3+ (part VIII).

1-round linear relations	Weak subkeys $Z_j^{(i)}$ (ith round) j values	Bias
$(\gamma,\gamma,\gamma,0,0,0,0,\gamma) \stackrel{1r}{\rightarrow} (0,0,\gamma,0,0,0,0,\gamma)$	1	2^{-1}
$(\gamma,\gamma,\gamma,0,0,0,\gamma,0) \stackrel{1r}{\rightarrow} (\gamma,\gamma,0,0,\gamma,\gamma,\gamma,\gamma)$	1; 7	$2^{-25.94}$
$(\gamma,\gamma,\gamma,0,0,0,\gamma,\gamma) \stackrel{1r}{\rightarrow} (0,0,\gamma,\gamma,0,0,0,\gamma)$	1; 7	$2^{-25.94}$
$(\gamma,\gamma,\gamma,0,0,\gamma,0,0) \stackrel{1r}{\rightarrow} (0,0,\gamma,0,0,0,\gamma,0)$	1; 6	2^{-1}
$(\gamma,\gamma,\gamma,0,0,\gamma,0,\gamma) \stackrel{1r}{\rightarrow} (\gamma,\gamma,0,\gamma,\gamma,\gamma,0,0)$	1; 6	$2^{-25.94}$
$(\gamma,\gamma,\gamma,0,0,\gamma,\gamma,0) \stackrel{1r}{\rightarrow} (0,0,\gamma,\gamma,0,0,\gamma,0)$	1; 6; 7	$2^{-25.94}$
$(\gamma,\gamma,\gamma,0,0,\gamma,\gamma,\gamma) \stackrel{1r}{\rightarrow} (\gamma,\gamma,0,0,\gamma,\gamma,0,0)$	1; 6; 7	$2^{-25.94}$
$(\gamma,\gamma,\gamma,0,\gamma,0,0,0) \stackrel{1r}{\rightarrow} (\gamma,\gamma,0,\gamma,\gamma,0,\gamma,\gamma)$	1	$2^{-25.94}$
$(\gamma,\gamma,\gamma,0,\gamma,0,0,\gamma) \stackrel{1r}{\rightarrow} (0,0,\gamma,0,0,\gamma,0,\gamma)$	1	$2^{-25.94}$
$(\gamma,\gamma,\gamma,0,\gamma,0,\gamma,0) \stackrel{1r}{\rightarrow} (\gamma,\gamma,0,0,\gamma,0,\gamma,\gamma)$	1; 7	$2^{-25.94}$
$(\gamma,\gamma,\gamma,0,\gamma,0,\gamma,\gamma) \stackrel{1r}{\rightarrow} (0,0,\gamma,\gamma,0,\gamma,0,\gamma)$	1; 7	2^{-1}
$(\gamma,\gamma,\gamma,0,\gamma,\gamma,0,0) \stackrel{1r}{\rightarrow} (0,0,\gamma,0,0,\gamma,\gamma,0)$	1; 6	$2^{-25.94}$
$(\gamma,\gamma,\gamma,0,\gamma,\gamma,0,\gamma) \stackrel{1r}{\rightarrow} (\gamma,\gamma,0,\gamma,\gamma,0,0,0)$	1; 6	$2^{-25.94}$
$(\gamma,\gamma,\gamma,0,\gamma,\gamma,\gamma,0) \stackrel{1r}{\rightarrow} (0,0,\gamma,\gamma,0,\gamma,\gamma,0)$	1; 6; 7	2^{-1}
$(\gamma,\gamma,\gamma,0,\gamma,\gamma,\gamma,\gamma) \stackrel{1r}{\rightarrow} (\gamma,\gamma,0,0,\gamma,0,0,0)$	1; 6; 7	$2^{-25.94}$
$(\gamma,\gamma,\gamma,\gamma,0,0,0,0) \stackrel{1r}{\rightarrow} (\gamma,\gamma,\gamma,0,\gamma,0,0,0)$	1; 4	2^{-1}
$(\gamma,\gamma,\gamma,\gamma,0,0,0,\gamma) \stackrel{1r}{\rightarrow} (0,0,0,\gamma,0,\gamma,\gamma,0)$	1; 4	$2^{-25.94}$
$(\gamma,\gamma,\gamma,\gamma,0,0,\gamma,0) \stackrel{1r}{\rightarrow} (\gamma,\gamma,\gamma,\gamma,\gamma,0,0,0)$	1; 4; 7	$2^{-38.41}$
$(\gamma,\gamma,\gamma,\gamma,0,0,\gamma,\gamma) \stackrel{1r}{\rightarrow} (0,0,0,0,0,\gamma,\gamma,0)$	1; 4; 7	$2^{-25.94}$
$(\gamma,\gamma,\gamma,\gamma,0,\gamma,0,0) \stackrel{1r}{\rightarrow} (0,0,0,\gamma,0,\gamma,0,\gamma)$	1; 4; 6	$2^{-25.94}$
$(\gamma,\gamma,\gamma,\gamma,0,\gamma,0,\gamma) \stackrel{1r}{\rightarrow} (\gamma,\gamma,\gamma,0,\gamma,0,\gamma,\gamma)$	1; 4; 6	2^{-1}
$(\gamma,\gamma,\gamma,\gamma,0,\gamma,\gamma,0) \stackrel{1r}{\rightarrow} (0,0,0,0,0,\gamma,0,\gamma)$	1; 4; 6; 7	$2^{-25.94}$
$(\gamma,\gamma,\gamma,\gamma,0,\gamma,\gamma,\gamma) \stackrel{1r}{\rightarrow} (\gamma,\gamma,\gamma,\gamma,\gamma,0,\gamma,\gamma)$	1; 4; 6; 7	$2^{-38.41}$
$(\gamma,\gamma,\gamma,\gamma,\gamma,0,0,0) \stackrel{1r}{\rightarrow} (\gamma,\gamma,\gamma,0,\gamma,\gamma,0,0)$	1; 4	$2^{-38.41}$
$(\gamma,\gamma,\gamma,\gamma,\gamma,0,0,\gamma) \stackrel{1r}{\rightarrow} (0,0,0,\gamma,0,0,\gamma,0)$	1; 4	$2^{-25.94}$
$(\gamma,\gamma,\gamma,\gamma,\gamma,0,\gamma,0) \stackrel{1r}{\rightarrow} (\gamma,\gamma,\gamma,\gamma,\gamma,\gamma,0,0)$	1; 4; 7	2^{-1}
$(\gamma,\gamma,\gamma,\gamma,\gamma,0,\gamma,\gamma) \stackrel{1r}{\rightarrow} (0,0,0,0,0,0,\gamma,0)$	1; 4; 7	$2^{-25.94}$
$(\gamma,\gamma,\gamma,\gamma,\gamma,\gamma,0,0) \stackrel{1r}{\rightarrow} (0,0,0,\gamma,0,0,0,\gamma)$	1; 4; 6	$2^{-25.94}$
$(\gamma,\gamma,\gamma,\gamma,\gamma,\gamma,0,\gamma) \stackrel{1r}{\rightarrow} (\gamma,\gamma,\gamma,0,\gamma,\gamma,\gamma,\gamma)$	1; 4; 6	$2^{-38.41}$
$(\gamma,\gamma,\gamma,\gamma,\gamma,\gamma,\gamma,0) \stackrel{1r}{\rightarrow} (0,0,0,0,0,0,0,\gamma)$	1; 4; 6; 7	$2^{-25.94}$
$(\gamma,\gamma,\gamma,\gamma,\gamma,\gamma,\gamma,\gamma) \stackrel{1r}{\rightarrow} (\gamma,\gamma,\gamma,\gamma,\gamma,\gamma,\gamma,\gamma)$	1; 4; 6; 7	2^{-1}

(2) $(0,0,0,0,0,0,\gamma,0) \stackrel{1r}{\rightarrow} (0,0,0,\gamma,0,0,0,0) \stackrel{1r}{\rightarrow} (0,0,\gamma,\gamma,0,\gamma,\gamma,\gamma) \stackrel{1r}{\rightarrow} (0,\gamma,\gamma,\gamma,0,0,\gamma,\gamma) \stackrel{1r}{\rightarrow} (\gamma,0,0,\gamma,0,0,0,\gamma) \stackrel{1r}{\rightarrow} (0,\gamma,0,\gamma,\gamma,\gamma,\gamma,0) \stackrel{1r}{\rightarrow} (\gamma,\gamma,0,0,0,0,0,\gamma) \stackrel{1r}{\rightarrow} (0,\gamma,\gamma,\gamma,0,\gamma,\gamma,0) \stackrel{1r}{\rightarrow} (\gamma,0,0,\gamma,0,0,\gamma,0) \stackrel{1r}{\rightarrow} (\gamma,0,\gamma,\gamma,0,0,0,0) \stackrel{1r}{\rightarrow} (\gamma,\gamma,\gamma,\gamma,0,\gamma,\gamma,\gamma) \stackrel{1r}{\rightarrow} (\gamma,\gamma,\gamma,\gamma,\gamma,0,\gamma,\gamma) \stackrel{1r}{\rightarrow} (0,0,0,0,0,0,\gamma,0)$, which is 12-round iterative and holds with bias $2^{-348.56}$.

Starting at the i round, the following 28 subkeys are required to be weak: $Z_7^{(i)}$, $Z_4^{(i+1)}$, $Z_4^{(i+2)}$, $Z_6^{(i+2)}$, $Z_7^{(i+2)}$, $Z_4^{(i+3)}$, $Z_7^{(i+3)}$, $Z_1^{(i+4)}$, $Z_4^{(i+4)}$, $Z_4^{(i+5)}$, $Z_6^{(i+5)}$, $Z_7^{(i+5)}$, $Z_1^{(i+6)}$, $Z_4^{(i+7)}$, $Z_6^{(i+7)}$, $Z_7^{(i+7)}$, $Z_1^{(i+8)}$, $Z_4^{(i+8)}$, $Z_7^{(i+8)}$, $Z_1^{(i+9)}$, $Z_4^{(i+9)}$, $Z_1^{(i+10)}$, $Z_4^{(i+10)}$, $Z_6^{(i+10)}$, $Z_7^{(i+10)}$, $Z_1^{(i+11)}$, $Z_4^{(i+11)}$ and $Z_7^{(i+11)}$.

(3) $(0,0,0,0,0,0,\gamma,\gamma) \xrightarrow{1r} (\gamma,\gamma,\gamma,0,\gamma,\gamma,\gamma,0) \xrightarrow{1r} (0,0,\gamma,\gamma,0,\gamma,\gamma,0) \xrightarrow{1r} (\gamma,0,0,0,\gamma,\gamma,0,\gamma) \xrightarrow{1r} (\gamma,0,0,\gamma,0,0,0,0) \xrightarrow{1r} (\gamma,0,\gamma,0,0,0,0,0) \xrightarrow{1r} (\gamma,\gamma,0,0,0,0,0,0) \xrightarrow{1r} (\gamma,0,0,0,\gamma,0,0,0) \xrightarrow{1r} (\gamma,0,0,\gamma,0,0,\gamma,\gamma) \xrightarrow{1r} (0,\gamma,0,0,\gamma,\gamma,\gamma,0) \xrightarrow{1r} (\gamma,\gamma,\gamma,\gamma,0,\gamma,\gamma,0) \xrightarrow{1r} (0,0,0,0,0,\gamma,0,\gamma) \xrightarrow{1r} (0,0,0,0,0,0,\gamma,\gamma)$, which is 12-round iterative and holds with bias $2^{-125.70}$.

Starting at the i round, the following 24 subkeys are required to be weak: $Z_7^{(i)}$, $Z_1^{(i+1)}$, $Z_6^{(i+1)}$, $Z_7^{(i+1)}$, $Z_4^{(i+2)}$, $Z_6^{(i+2)}$, $Z_7^{(i+2)}$, $Z_1^{(i+3)}$, $Z_6^{(i+3)}$, $Z_1^{(i+4)}$, $Z_4^{(i+4)}$, $Z_1^{(i+5)}$, $Z_1^{(i+6)}$, $Z_1^{(i+7)}$, $Z_1^{(i+8)}$, $Z_4^{(i+8)}$, $Z_7^{(i+8)}$, $Z_6^{(i+9)}$, $Z_7^{(i+9)}$, $Z_1^{(i+10)}$, $Z_4^{(i+10)}$, $Z_6^{(i+10)}$, $Z_7^{(i+10)}$ and $Z_6^{(i+11)}$.

(4) $(0,0,0,0,0,\gamma,0,0) \xrightarrow{1r} (\gamma,\gamma,\gamma,\gamma,\gamma,\gamma,0,\gamma) \xrightarrow{1r} (\gamma,\gamma,\gamma,0,\gamma,\gamma,\gamma,\gamma) \xrightarrow{1r} (\gamma,\gamma,0,0,\gamma,0,0,0) \xrightarrow{1r} (\gamma,0,0,0,\gamma,\gamma,0,0) \xrightarrow{1r} (0,\gamma,\gamma,0,\gamma,\gamma,\gamma,0) \xrightarrow{1r} (\gamma,0,\gamma,0,0,0,0,\gamma) \xrightarrow{1r} (0,0,\gamma,\gamma,\gamma,\gamma,\gamma,0) \xrightarrow{1r} (\gamma,0,0,0,\gamma,0,0,\gamma) \xrightarrow{1r} (0,\gamma,\gamma,0,\gamma,\gamma,0,\gamma) \xrightarrow{1r} (0,\gamma,0,0,\gamma,\gamma,\gamma,\gamma) \xrightarrow{1r} (0,0,0,0,\gamma,0,0,0) \xrightarrow{1r} (0,0,0,0,0,\gamma,0,0)$, which is 12-round iterative and holds with bias $2^{-350.16}$.

Starting at the i round, the following 20 subkeys are required to be weak: $Z_6^{(i)}$, $Z_1^{(i+1)}$, $Z_4^{(i+1)}$, $Z_6^{(i+1)}$, $Z_1^{(i+2)}$, $Z_6^{(i+2)}$, $Z_7^{(i+2)}$, $Z_1^{(i+3)}$, $Z_1^{(i+4)}$, $Z_6^{(i+4)}$, $Z_6^{(i+5)}$, $Z_7^{(i+5)}$, $Z_1^{(i+6)}$, $Z_4^{(i+7)}$, $Z_6^{(i+7)}$, $Z_7^{(i+7)}$, $Z_1^{(i+8)}$, $Z_6^{(i+9)}$, $Z_6^{(i+10)}$ and $Z_7^{(i+10)}$.

(5) $(0,0,0,0,0,\gamma,\gamma,0) \xrightarrow{1r} (\gamma,\gamma,\gamma,0,\gamma,\gamma,0,\gamma) \xrightarrow{1r} (\gamma,\gamma,0,\gamma,\gamma,0,0,0) \xrightarrow{1r} (\gamma,0,\gamma,\gamma,\gamma,0,\gamma,\gamma) \xrightarrow{1r} (0,0,0,\gamma,\gamma,\gamma,0,\gamma) \xrightarrow{1r} (0,0,\gamma,\gamma,0,0,0,0) \xrightarrow{1r} (0,\gamma,\gamma,0,0,0,0,0) \xrightarrow{1r} (0,\gamma,0,0,\gamma,0,0,0) \xrightarrow{1r} (0,0,0,\gamma,\gamma,0,\gamma,\gamma) \xrightarrow{1r} (\gamma,\gamma,0,\gamma,\gamma,\gamma,0,\gamma) \xrightarrow{1r} (\gamma,0,\gamma,\gamma,\gamma,0,0,0) \xrightarrow{1r} (\gamma,\gamma,\gamma,\gamma,0,0,\gamma,\gamma) \xrightarrow{1r} (0,0,0,0,0,\gamma,\gamma,0)$, which is 12-round iterative and holds with bias $2^{-190.52}$.

Starting at the i round, the following 22 subkeys are required to be weak: $Z_6^{(i)}$, $Z_7^{(i)}$, $Z_1^{(i+1)}$, $Z_6^{(i+1)}$, $Z_1^{(i+2)}$, $Z_4^{(i+2)}$, $Z_1^{(i+3)}$, $Z_4^{(i+3)}$, $Z_7^{(i+3)}$, $Z_4^{(i+4)}$, $Z_6^{(i+4)}$, $Z_4^{(i+5)}$, $Z_4^{(i+8)}$, $Z_7^{(i+8)}$, $Z_1^{(i+9)}$, $Z_4^{(i+9)}$, $Z_6^{(i+9)}$, $Z_1^{(i+10)}$, $Z_4^{(i+10)}$, $Z_1^{(i+11)}$, $Z_4^{(i+11)}$ and $Z_7^{(i+11)}$.

(6) $(0,0,0,0,0,\gamma,\gamma,\gamma) \xrightarrow{1r} (0,0,0,\gamma,0,0,\gamma,\gamma) \xrightarrow{1r} (\gamma,\gamma,0,\gamma,\gamma,0,0,\gamma) \xrightarrow{1r} (0,\gamma,0,0,0,\gamma,0,\gamma) \xrightarrow{1r} (0,0,0,\gamma,\gamma,\gamma,0,0) \xrightarrow{1r} (\gamma,\gamma,0,0,\gamma,\gamma,\gamma,0) \xrightarrow{1r} (0,\gamma,\gamma,0,0,0,0,\gamma) \xrightarrow{1r} (\gamma,0,\gamma,\gamma,0,\gamma,\gamma,0) \xrightarrow{1r} (0,0,0,\gamma,\gamma,0,\gamma,0) \xrightarrow{1r} (0,0,\gamma,0,0,0,\gamma,\gamma) \xrightarrow{1r} (\gamma,0,\gamma,\gamma,\gamma,0,0,\gamma) \xrightarrow{1r} (0,0,0,0,\gamma,\gamma,0,\gamma) \xrightarrow{1r} (0,0,0,0,0,\gamma,\gamma,\gamma)$, which is 12-round iterative and holds with bias $2^{-325.22}$.

Starting at the i round, the following 22 subkeys are required to be weak: $Z_6^{(i)}$, $Z_7^{(i)}$, $Z_4^{(i+1)}$, $Z_7^{(i+1)}$, $Z_1^{(i+2)}$, $Z_4^{(i+2)}$, $Z_6^{(i+3)}$, $Z_4^{(i+4)}$, $Z_6^{(i+4)}$, $Z_1^{(i+5)}$, $Z_6^{(i+5)}$, $Z_7^{(i+5)}$, $Z_1^{(i+7)}$, $Z_4^{(i+7)}$, $Z_6^{(i+7)}$, $Z_7^{(i+7)}$, $Z_4^{(i+8)}$, $Z_7^{(i+8)}$, $Z_7^{(i+9)}$, $Z_1^{(i+10)}$, $Z_4^{(i+10)}$ and $Z_6^{(i+11)}$.

(7) $(0,0,0,0,\gamma,0,0,\gamma) \stackrel{1r}{\to} (\gamma,\gamma,\gamma,\gamma,\gamma,0,\gamma,0) \stackrel{1r}{\to} (\gamma,\gamma,\gamma,\gamma,\gamma,\gamma,0,0) \stackrel{1r}{\to} (0,0,0,\gamma,0,0,0,\gamma) \stackrel{1r}{\to} (\gamma,\gamma,0,0,\gamma,0,0,\gamma) \stackrel{1r}{\to} (0,\gamma,\gamma,\gamma,0,0,\gamma,0) \stackrel{1r}{\to} (0,\gamma,\gamma,0,\gamma,\gamma,\gamma,\gamma) \stackrel{1r}{\to} (0,\gamma,0,\gamma,\gamma,\gamma,\gamma,\gamma) \stackrel{1r}{\to} (0,0,\gamma,\gamma,\gamma,\gamma,\gamma,\gamma) \stackrel{1r}{\to} (0,\gamma,\gamma,\gamma,0,\gamma,\gamma,\gamma) \stackrel{1r}{\to} (0,\gamma,\gamma,0,\gamma,\gamma,0,0) \stackrel{1r}{\to} (\gamma,0,\gamma,\gamma,0,0,0,\gamma) \stackrel{1r}{\to} (0,0,0,0,\gamma,0,0,\gamma)$, which is 12-round iterative and holds with bias $2^{-125.70}$. Starting at the i round, the following 24 subkeys are required to be weak: $Z_1^{(i+1)}$, $Z_4^{(i+1)}$, $Z_7^{(i+1)}$, $Z_1^{(i+2)}$, $Z_4^{(i+2)}$, $Z_6^{(i+2)}$, $Z_4^{(i+3)}$, $Z_1^{(i+4)}$, $Z_4^{(i+5)}$, $Z_7^{(i+5)}$, $Z_6^{(i+6)}$, $Z_7^{(i+6)}$, $Z_4^{(i+7)}$, $Z_6^{(i+7)}$, $Z_7^{(i+7)}$, $Z_4^{(i+8)}$, $Z_6^{(i+8)}$, $Z_7^{(i+8)}$, $Z_4^{(i+9)}$, $Z_6^{(i+9)}$, $Z_7^{(i+9)}$, $Z_6^{(i+10)}$, $Z_1^{(i+11)}$ and $Z_4^{(i+11)}$.

(8) $(0,0,0,0,\gamma,0,\gamma,0) \stackrel{1r}{\to} (0,0,0,\gamma,0,\gamma,0,0) \stackrel{1r}{\to} (\gamma,\gamma,0,0,\gamma,0,\gamma,0) \stackrel{1r}{\to} (\gamma,0,0,\gamma,\gamma,\gamma,0,0) \stackrel{1r}{\to} (0,\gamma,0,\gamma,\gamma,0,0,\gamma) \stackrel{1r}{\to} (\gamma,\gamma,0,\gamma,0,0,\gamma,0) \stackrel{1r}{\to} (\gamma,0,\gamma,0,\gamma,\gamma,\gamma,\gamma) \stackrel{1r}{\to} (\gamma,\gamma,0,\gamma,0,\gamma,\gamma,\gamma) \stackrel{1r}{\to} (\gamma,0,\gamma,0,\gamma,\gamma,0,0) \stackrel{1r}{\to} (0,0,\gamma,\gamma,\gamma,0,0,\gamma) \stackrel{1r}{\to} (\gamma,0,0,\gamma,\gamma,0,\gamma,0) \stackrel{1r}{\to} (\gamma,0,\gamma,\gamma,0,\gamma,0,0) \stackrel{1r}{\to} (0,0,0,0,\gamma,0,\gamma,0)$, which is 12-round iterative and holds with bias $2^{-125.70}$. Starting at the i round, the following 28 subkeys are required to be weak: $Z_7^{(i)}$, $Z_4^{(i+1)}$, $Z_6^{(i+1)}$, $Z_1^{(i+2)}$, $Z_7^{(i+2)}$, $Z_1^{(i+3)}$, $Z_4^{(i+3)}$, $Z_6^{(i+3)}$, $Z_4^{(i+4)}$, $Z_1^{(i+5)}$, $Z_4^{(i+5)}$, $Z_7^{(i+5)}$, $Z_1^{(i+6)}$, $Z_6^{(i+6)}$, $Z_7^{(i+6)}$, $Z_1^{(i+7)}$, $Z_4^{(i+7)}$, $Z_6^{(i+7)}$, $Z_7^{(i+7)}$, $Z_1^{(i+8)}$, $Z_6^{(i+8)}$, $Z_4^{(i+9)}$, $Z_1^{(i+10)}$, $Z_4^{(i+10)}$, $Z_7^{(i+10)}$, $Z_1^{(i+11)}$, $Z_4^{(i+11)}$ and $Z_6^{(i+11)}$.

(9) $(0,0,0,0,\gamma,0,\gamma,\gamma) \stackrel{1r}{\to} (\gamma,\gamma,\gamma,0,\gamma,0,\gamma,0) \stackrel{1r}{\to} (\gamma,\gamma,0,0,\gamma,0,\gamma,\gamma) \stackrel{1r}{\to} (0,\gamma,\gamma,0,0,0,\gamma,0) \stackrel{1r}{\to} (0,\gamma,0,\gamma,\gamma,0,0,0) \stackrel{1r}{\to} (0,0,\gamma,0,\gamma,\gamma,0,0) \stackrel{1r}{\to} (\gamma,0,\gamma,0,\gamma,\gamma,\gamma,0) \stackrel{1r}{\to} (0,0,\gamma,0,\gamma,0,0,\gamma) \stackrel{1r}{\to} (\gamma,0,\gamma,0,\gamma,\gamma,0,\gamma) \stackrel{1r}{\to} (\gamma,\gamma,0,0,0,\gamma,\gamma,\gamma) \stackrel{1r}{\to} (\gamma,0,0,\gamma,\gamma,0,\gamma,\gamma) \stackrel{1r}{\to} (0,\gamma,0,0,\gamma,0,\gamma,0) \stackrel{1r}{\to} (0,0,0,0,\gamma,0,\gamma,\gamma)$, which is 12-round iterative and holds with bias $2^{-350.16}$. Starting at the i round, the following 20 subkeys are required to be weak: $Z_7^{(i)}$, $Z_1^{(i+1)}$, $Z_7^{(i+1)}$, $Z_1^{(i+2)}$, $Z_7^{(i+2)}$, $Z_7^{(i+3)}$, $Z_4^{(i+4)}$, $Z_6^{(i+5)}$, $Z_1^{(i+6)}$, $Z_6^{(i+6)}$, $Z_7^{(i+6)}$, $Z_1^{(i+8)}$, $Z_6^{(i+8)}$, $Z_1^{(i+9)}$, $Z_6^{(i+9)}$, $Z_7^{(i+9)}$, $Z_1^{(i+10)}$, $Z_4^{(i+10)}$, $Z_7^{(i+10)}$ and $Z_7^{(i+11)}$.

(10) $(0,0,0,0,\gamma,\gamma,0,0) \stackrel{1r}{\to} (\gamma,\gamma,\gamma,\gamma,\gamma,0,0,\gamma) \stackrel{1r}{\to} (0,0,0,\gamma,0,0,\gamma,0) \stackrel{1r}{\to} (0,0,\gamma,0,0,\gamma,\gamma,\gamma) \stackrel{1r}{\to} (0,\gamma,0,0,0,\gamma,0,0) \stackrel{1r}{\to} (\gamma,\gamma,\gamma,0,0,0,\gamma,0) \stackrel{1r}{\to} (\gamma,\gamma,0,0,\gamma,\gamma,\gamma,\gamma) \stackrel{1r}{\to} (\gamma,0,0,\gamma,\gamma,\gamma,\gamma,\gamma) \stackrel{1r}{\to} (\gamma,0,\gamma,\gamma,0,\gamma,\gamma,\gamma) \stackrel{1r}{\to} (\gamma,\gamma,\gamma,0,0,\gamma,0,0) \stackrel{1r}{\to} (0,0,\gamma,0,0,0,\gamma,0) \stackrel{1r}{\to} (0,\gamma,0,0,0,\gamma,\gamma,\gamma) \stackrel{1r}{\to} (0,0,0,0,\gamma,\gamma,0,0)$, which is 12-round iterative and holds with bias $2^{-202.52}$. Starting at the i round, the following 26 subkeys are required to be weak: $Z_6^{(i)}$, $Z_1^{(i+1)}$, $Z_4^{(i+1)}$, $Z_4^{(i+2)}$, $Z_7^{(i+2)}$, $Z_6^{(i+3)}$, $Z_7^{(i+3)}$, $Z_6^{(i+4)}$, $Z_1^{(i+5)}$, $Z_7^{(i+5)}$,

$Z_1^{(i+6)}$, $Z_6^{(i+6)}$, $Z_7^{(i+6)}$, $Z_1^{(i+7)}$, $Z_4^{(i+7)}$, $Z_6^{(i+7)}$, $Z_7^{(i+7)}$, $Z_1^{(i+8)}$, $Z_4^{(i+8)}$, $Z_6^{(i+8)}$, $Z_7^{(i+8)}$, $Z_1^{(i+9)}$, $Z_6^{(i+9)}$, $Z_7^{(i+10)}$, $Z_6^{(i+11)}$ and $Z_7^{(i+11)}$.

(11) $(0,0,0,0,\gamma,\gamma,\gamma,0) \xrightarrow{1r} (\gamma,\gamma,\gamma,0,\gamma,0,0,\gamma) \xrightarrow{1r} (0,0,\gamma,0,0,\gamma,0,\gamma) \xrightarrow{1r} (0, \gamma, 0, \gamma, 0, \gamma, 0, 0) \xrightarrow{1r} (\gamma,\gamma,0,\gamma,0,\gamma,0,\gamma) \xrightarrow{1r} (\gamma,0,\gamma,\gamma,\gamma,\gamma,0,0) \xrightarrow{1r} (0, 0, 0, 0, \gamma, \gamma, \gamma, 0)$, which is 6-round iterative and holds with bias $2^{-125.70}$. Starting at the i round, the following 12 subkeys are required to be weak: $Z_6^{(i)}$, $Z_7^{(i)}$, $Z_1^{(i+1)}$, $Z_6^{(i+2)}$, $Z_4^{(i+3)}$, $Z_6^{(i+3)}$, $Z_1^{(i+4)}$, $Z_4^{(i+4)}$, $Z_6^{(i+4)}$, $Z_1^{(i+5)}$, $Z_4^{(i+5)}$ and $Z_6^{(i+5)}$.

(12) $(0,0,0,0,\gamma,\gamma,\gamma,\gamma) \xrightarrow{1r} (0,0,0,\gamma,0,\gamma,\gamma,\gamma) \xrightarrow{1r} (0, 0, \gamma, 0, 0, \gamma, 0, 0) \xrightarrow{1r} (\gamma, 0, \gamma, 0, \gamma, 0, \gamma, 0) \xrightarrow{1r} (\gamma,\gamma,0,\gamma,0,\gamma,0,0) \xrightarrow{1r} (0, \gamma, 0, 0, 0, 0, \gamma, 0) \xrightarrow{1r} (0, 0, 0, 0, \gamma, \gamma, \gamma, \gamma)$, which is 6-round iterative and holds with bias $2^{-75.82}$. Starting at the i round, the following 12 subkeys are required to be weak: $Z_6^{(i)}$, $Z_7^{(i)}$, $Z_4^{(i+1)}$, $Z_6^{(i+1)}$, $Z_7^{(i+1)}$, $Z_6^{(i+2)}$, $Z_1^{(i+3)}$, $Z_7^{(i+3)}$, $Z_1^{(i+4)}$, $Z_4^{(i+4)}$, $Z_6^{(i+4)}$ and $Z_7^{(i+5)}$.

(13) $(0,0,0,\gamma,0,\gamma,0,\gamma) \xrightarrow{1r} (0,0,\gamma,\gamma,0,\gamma,0,0) \xrightarrow{1r} (\gamma,0,0,\gamma,\gamma,\gamma,0,\gamma) \xrightarrow{1r} (\gamma, 0, \gamma, 0, 0, \gamma, \gamma, \gamma) \xrightarrow{1r} (\gamma,\gamma,0,\gamma,0,0,\gamma,\gamma) \xrightarrow{1r} (0,\gamma,0,\gamma,0,0,0,\gamma) \xrightarrow{1r} (\gamma, \gamma, 0, \gamma, 0, \gamma, \gamma, 0) \xrightarrow{1r} (0,\gamma,0,\gamma,0,0,\gamma,0) \xrightarrow{1r} (0,0,\gamma,\gamma,\gamma,0,0,0) \xrightarrow{1r} (0, \gamma, \gamma, 0, 0, \gamma, 0, 0) \xrightarrow{1r} (\gamma,0,\gamma,\gamma,0,\gamma,0,\gamma) \xrightarrow{1r} (\gamma,\gamma,\gamma,\gamma,0,\gamma,0,0) \xrightarrow{1r} (0, 0, 0, \gamma, 0, \gamma, 0, \gamma)$, which is 12-round iterative and holds with bias $2^{-350.16}$. Starting at the i round, the following 28 subkeys are required to be weak: $Z_4^{(i)}$, $Z_6^{(i)}$, $Z_4^{(i+1)}$, $Z_6^{(i+1)}$, $Z_1^{(i+2)}$, $Z_4^{(i+2)}$, $Z_6^{(i+2)}$, $Z_1^{(i+3)}$, $Z_6^{(i+3)}$, $Z_7^{(i+3)}$, $Z_1^{(i+4)}$, $Z_4^{(i+4)}$, $Z_7^{(i+4)}$, $Z_4^{(i+5)}$, $Z_1^{(i+6)}$, $Z_4^{(i+6)}$, $Z_6^{(i+6)}$, $Z_7^{(i+6)}$, $Z_4^{(i+7)}$, $Z_7^{(i+7)}$, $Z_4^{(i+8)}$, $Z_6^{(i+9)}$, $Z_1^{(i+10)}$, $Z_4^{(i+10)}$, $Z_6^{(i+10)}$, $Z_1^{(i+11)}$, $Z_4^{(i+11)}$ and $Z_6^{(i+11)}$.

(14) $(0,0,0,\gamma,0,\gamma,\gamma,0) \xrightarrow{1r} (\gamma,\gamma,0,\gamma,\gamma,0,\gamma,0) \xrightarrow{1r} (\gamma,0,\gamma,0,\gamma,0,\gamma,\gamma) \xrightarrow{1r} (0, 0, \gamma, 0, \gamma, 0, \gamma, 0) \xrightarrow{1r} (0,\gamma,0,0,0,0,\gamma,\gamma) \xrightarrow{1r} (\gamma,\gamma,\gamma,\gamma,0,0,0,\gamma) \xrightarrow{1r} (0, 0, 0, \gamma, 0, \gamma, \gamma, 0)$, which is 6-round iterative and holds with bias $2^{-125.70}$. Starting at the i round, the following 12 subkeys are required to be weak: $Z_4^{(i)}$, $Z_6^{(i)}$, $Z_7^{(i)}$, $Z_1^{(i+1)}$, $Z_4^{(i+1)}$, $Z_7^{(i+1)}$, $Z_1^{(i+2)}$, $Z_7^{(i+2)}$, $Z_7^{(i+3)}$, $Z_7^{(i+4)}$, $Z_1^{(i+5)}$ and $Z_4^{(i+5)}$.

(15) $(0,0,0,\gamma,\gamma,0,0,0) \xrightarrow{1r} (0,0,\gamma,\gamma,0,0,\gamma,\gamma) \xrightarrow{1r} (\gamma,0,0,0,\gamma,\gamma,\gamma,0) \xrightarrow{1r} (0, \gamma, \gamma, \gamma, \gamma, \gamma, \gamma, 0) \xrightarrow{1r} (\gamma,0,0,\gamma,0,\gamma,\gamma,0) \xrightarrow{1r} (0,\gamma,0,0,\gamma,\gamma,0,\gamma) \xrightarrow{1r} (0, 0, 0, \gamma, \gamma, 0, 0, 0)$, which is 6-round iterative and holds with bias $2^{-100.76}$. Starting at the i round, the following 14 subkeys are required to be weak: $Z_4^{(i)}$, $Z_4^{(i+1)}$, $Z_7^{(i+1)}$, $Z_1^{(i+2)}$, $Z_6^{(i+2)}$, $Z_7^{(i+2)}$, $Z_4^{(i+3)}$, $Z_6^{(i+3)}$, $Z_7^{(i+3)}$, $Z_1^{(i+4)}$, $Z_4^{(i+4)}$, $Z_6^{(i+4)}$, $Z_7^{(i+4)}$ and $Z_6^{(i+5)}$.

(16) $(0,0,0,\gamma,\gamma,0,0,\gamma) \xrightarrow{1r} (\gamma,\gamma,0,0,\gamma,\gamma,0,\gamma) \xrightarrow{1r} (\gamma,0,0,0,\gamma,\gamma,\gamma,\gamma) \xrightarrow{1r} (\gamma, 0, 0, 0, 0, 0, 0, 0) \xrightarrow{1r} (\gamma,0,0,\gamma,0,\gamma,\gamma,\gamma) \xrightarrow{1r} (\gamma,0,\gamma,\gamma,0,0,\gamma,\gamma) \xrightarrow{1r} (0, 0, 0, \gamma, \gamma, 0, 0, \gamma)$, which is 6-round iterative and holds with bias $2^{-150.64}$.

Starting at the i round, the following 14 subkeys are required to be weak: $Z_4^{(i)}$, $Z_1^{(i+1)}$, $Z_6^{(i+1)}$, $Z_1^{(i+2)}$, $Z_6^{(i+2)}$, $Z_7^{(i+2)}$, $Z_1^{(i+3)}$, $Z_1^{(i+4)}$, $Z_4^{(i+4)}$, $Z_6^{(i+4)}$, $Z_7^{(i+4)}$, $Z_1^{(i+5)}$, $Z_4^{(i+5)}$ and $Z_7^{(i+5)}$.

(17) $(0,0,0,\gamma,\gamma,\gamma,\gamma,0) \xrightarrow{1r} (\gamma,\gamma,0,\gamma,\gamma,\gamma,\gamma,0) \xrightarrow{1r} (0,\gamma,0,\gamma,0,\gamma,\gamma,0) \xrightarrow{1r} (\gamma, \gamma, 0, 0, 0, \gamma, 0, \gamma) \xrightarrow{1r} (\gamma,0,0,0,\gamma,0,\gamma,\gamma) \xrightarrow{1r} (0,\gamma,\gamma,\gamma,\gamma,\gamma,0,\gamma) \xrightarrow{1r} (0, \gamma, \gamma, \gamma, \gamma, 0, 0, 0) \xrightarrow{1r} (0,\gamma,\gamma,\gamma,\gamma,0,\gamma,\gamma) \xrightarrow{1r} (\gamma,0,0,\gamma,0,\gamma,0,\gamma) \xrightarrow{1r} (\gamma, 0, \gamma, 0, 0, 0, \gamma, \gamma) \xrightarrow{1r} (0,0,\gamma,0,\gamma,\gamma,\gamma,0) \xrightarrow{1r} (\gamma,0,\gamma,\gamma,\gamma,\gamma,\gamma,0) \xrightarrow{1r} (0, 0, 0, \gamma, \gamma, \gamma, \gamma, 0)$, which is 12-round iterative and holds with bias $2^{-150.64}$.
Starting at the i round, the following 30 subkeys are required to be weak: $Z_4^{(i)}$, $Z_6^{(i)}$, $Z_7^{(i)}$, $Z_1^{(i+1)}$, $Z_4^{(i+1)}$, $Z_6^{(i+1)}$, $Z_7^{(i+1)}$, $Z_4^{(i+2)}$, $Z_6^{(i+2)}$, $Z_7^{(i+2)}$, $Z_1^{(i+3)}$, $Z_6^{(i+3)}$, $Z_1^{(i+4)}$, $Z_7^{(i+4)}$, $Z_4^{(i+5)}$, $Z_6^{(i+5)}$, $Z_4^{(i+6)}$, $Z_4^{(i+7)}$, $Z_7^{(i+7)}$, $Z_1^{(i+8)}$, $Z_4^{(i+8)}$, $Z_6^{(i+8)}$, $Z_1^{(i+9)}$, $Z_7^{(i+9)}$, $Z_6^{(i+10)}$, $Z_7^{(i+10)}$, $Z_1^{(i+11)}$, $Z_4^{(i+11)}$, $Z_6^{(i+11)}$ and $Z_7^{(i+11)}$.

(18) $(0,0,0,\gamma,\gamma,\gamma,\gamma,\gamma) \xrightarrow{1r} (0,0,\gamma,0,0,0,0,0) \xrightarrow{1r} (0,\gamma,0,\gamma,0,\gamma,\gamma,\gamma) \xrightarrow{1r} (0, 0, \gamma, \gamma, \gamma, 0, \gamma, \gamma) \xrightarrow{1r} (\gamma,0,0,0,\gamma,0,\gamma,0) \xrightarrow{1r} (\gamma,0,0,0,0,0,\gamma,\gamma) \xrightarrow{1r} (0, \gamma, \gamma, \gamma, \gamma, 0, 0, \gamma) \xrightarrow{1r} (\gamma,0,0,0,0,\gamma,0,\gamma) \xrightarrow{1r} (\gamma,0,0,\gamma,0,\gamma,0,0) \xrightarrow{1r} (0, \gamma, 0, \gamma, \gamma, \gamma, 0, \gamma) \xrightarrow{1r} (0,0,\gamma,0,\gamma,\gamma,\gamma,\gamma) \xrightarrow{1r} (0,\gamma,0,0,0,0,0,0) \xrightarrow{1r} (0, 0, 0, \gamma, \gamma, \gamma, \gamma, \gamma)$, which is 12-round iterative and holds with bias $2^{-275.34}$.
Starting at the i round, the following 22 subkeys are required to be weak: $Z_4^{(i)}$, $Z_6^{(i)}$, $Z_7^{(i)}$, $Z_4^{(i+2)}$, $Z_6^{(i+2)}$, $Z_7^{(i+2)}$, $Z_4^{(i+3)}$, $Z_7^{(i+3)}$, $Z_1^{(i+4)}$, $Z_7^{(i+4)}$, $Z_1^{(i+5)}$, $Z_7^{(i+5)}$, $Z_4^{(i+6)}$, $Z_1^{(i+7)}$, $Z_6^{(i+7)}$, $Z_1^{(i+8)}$, $Z_4^{(i+8)}$, $Z_6^{(i+8)}$, $Z_4^{(i+9)}$, $Z_6^{(i+9)}$, $Z_6^{(i+10)}$ and $Z_7^{(i+10)}$.

(19) $(0,0,\gamma,0,0,0,0,\gamma) \xrightarrow{1r} (\gamma,0,\gamma,0,\gamma,0,0,\gamma) \xrightarrow{1r} (0,0,\gamma,\gamma,\gamma,0,\gamma,0) \xrightarrow{1r} (0, \gamma, \gamma, \gamma, 0, \gamma, 0, 0) \xrightarrow{1r} (\gamma,0,0,0,0,0,\gamma,0) \xrightarrow{1r} (\gamma,0,0,0,0,\gamma,\gamma,\gamma) \xrightarrow{1r} (\gamma, 0, 0, 0, 0, \gamma, 0, 0) \xrightarrow{1r} (0,\gamma,\gamma,0,\gamma,0,\gamma,0) \xrightarrow{1r} (0,\gamma,0,\gamma,\gamma,\gamma,0,0) \xrightarrow{1r} (\gamma, \gamma, 0, \gamma, 0, 0, 0, \gamma) \xrightarrow{1r} (0,\gamma,0,0,0,0,0,\gamma) \xrightarrow{1r} (\gamma,\gamma,\gamma,0,0,0,0,\gamma) \xrightarrow{1r} (0,0,\gamma,0,0,0,0,\gamma)$, which is 12-round iterative and holds with bias $2^{-150.64}$.
Starting at the i round, the following 18 subkeys are required to be weak: $Z_1^{(i+1)}$, $Z_4^{(i+2)}$, $Z_7^{(i+2)}$, $Z_4^{(i+3)}$, $Z_6^{(i+3)}$, $Z_1^{(i+4)}$, $Z_7^{(i+4)}$, $Z_1^{(i+5)}$, $Z_6^{(i+5)}$, $Z_7^{(i+5)}$, $Z_1^{(i+6)}$, $Z_6^{(i+6)}$, $Z_7^{(i+7)}$, $Z_4^{(i+8)}$, $Z_6^{(i+8)}$, $Z_1^{(i+9)}$, $Z_4^{(i+9)}$ and $Z_1^{(i+11)}$.

(20) $(0,0,\gamma,0,0,\gamma,\gamma,0) \xrightarrow{1r} (\gamma,0,\gamma,\gamma,\gamma,0,\gamma,0) \xrightarrow{1r} (\gamma,\gamma,\gamma,0,0,0,\gamma,\gamma) \xrightarrow{1r} (0, 0, \gamma, \gamma, 0, 0, 0, \gamma) \xrightarrow{1r} (\gamma,0,0,\gamma,\gamma,\gamma,\gamma,0) \xrightarrow{1r} (0,\gamma,0,0,\gamma,0,0,\gamma) \xrightarrow{1r} (\gamma, \gamma, \gamma, 0, 0, \gamma, 0, \gamma) \xrightarrow{1r} (\gamma,\gamma,0,\gamma,\gamma,\gamma,0,0) \xrightarrow{1r} (0,\gamma,0,0,0,\gamma,\gamma,0) \xrightarrow{1r} (\gamma, \gamma, \gamma, \gamma, 0, 0, \gamma, 0) \xrightarrow{1r} (\gamma,\gamma,\gamma,\gamma,\gamma,0,0,0) \xrightarrow{1r} (\gamma,\gamma,\gamma,0,\gamma,\gamma,0,0) \xrightarrow{1r} (0, 0, \gamma, 0, 0, \gamma, \gamma, 0)$, which is 12-round iterative and holds with bias $2^{-325.22}$.
Starting at the i round, the following 26 subkeys are required to be weak: $Z_6^{(i)}$, $Z_7^{(i)}$, $Z_1^{(i+1)}$, $Z_4^{(i+1)}$, $Z_7^{(i+1)}$, $Z_1^{(i+2)}$, $Z_7^{(i+2)}$, $Z_4^{(i+3)}$, $Z_1^{(i+4)}$, $Z_4^{(i+4)}$,

$Z_6^{(i+4)}$, $Z_7^{(i+4)}$, $Z_1^{(i+6)}$, $Z_6^{(i+6)}$, $Z_1^{(i+7)}$, $Z_4^{(i+7)}$, $Z_6^{(i+7)}$, $Z_6^{(i+8)}$, $Z_7^{(i+8)}$, $Z_1^{(i+9)}$, $Z_4^{(i+9)}$, $Z_7^{(i+9)}$, $Z_1^{(i+10)}$, $Z_4^{(i+10)}$, $Z_1^{(i+11)}$ and $Z_6^{(i+11)}$.

(21) $(0,0,\gamma,0,\gamma,0,0,0) \xrightarrow{1r} (0,\gamma,0,\gamma,0,0,\gamma,\gamma) \xrightarrow{1r} (\gamma,\gamma,0,0,0,\gamma,\gamma,0) \xrightarrow{1r} (0,\gamma,\gamma,0,0,\gamma,0,\gamma) \xrightarrow{1r} (0,\gamma,0,0,\gamma,0,\gamma,\gamma) \xrightarrow{1r} (\gamma,\gamma,\gamma,\gamma,0,\gamma,0,\gamma) \xrightarrow{1r} (\gamma,\gamma,\gamma,0,\gamma,0,\gamma,\gamma) \xrightarrow{1r} (0,0,\gamma,\gamma,0,\gamma,0,\gamma) \xrightarrow{1r} (0,\gamma,\gamma,0,0,0,\gamma,\gamma) \xrightarrow{1r} (\gamma,0,\gamma,0,0,\gamma,\gamma,0) \xrightarrow{1r} (0,0,\gamma,0,\gamma,\gamma,0,\gamma) \xrightarrow{1r} (0,\gamma,0,\gamma,0,0,0,0) \xrightarrow{1r} (0,0,\gamma,0,\gamma,0,0,0)$, which is 12-round iterative and holds with bias $2^{-125.70}$. Starting at the i round, the following 20 subkeys are required to be weak: $Z_4^{(i+1)}$, $Z_7^{(i+1)}$, $Z_1^{(i+2)}$, $Z_6^{(i+2)}$, $Z_7^{(i+2)}$, $Z_6^{(i+3)}$, $Z_7^{(i+4)}$, $Z_1^{(i+5)}$, $Z_4^{(i+5)}$, $Z_6^{(i+5)}$, $Z_1^{(i+6)}$, $Z_7^{(i+6)}$, $Z_4^{(i+7)}$, $Z_6^{(i+7)}$, $Z_7^{(i+8)}$, $Z_1^{(i+9)}$, $Z_6^{(i+9)}$, $Z_7^{(i+9)}$, $Z_6^{(i+10)}$ and $Z_4^{(i+11)}$.

(22) $(0,0,\gamma,0,\gamma,0,\gamma,\gamma) \xrightarrow{1r} (\gamma,0,\gamma,\gamma,\gamma,\gamma,0,\gamma) \xrightarrow{1r} (\gamma,\gamma,\gamma,\gamma,0,0,0,0) \xrightarrow{1r} (\gamma,\gamma,\gamma,0,\gamma,0,0,0) \xrightarrow{1r} (\gamma,\gamma,0,\gamma,\gamma,0,\gamma,\gamma) \xrightarrow{1r} (0,\gamma,0,\gamma,0,\gamma,0,\gamma) \xrightarrow{1r} (0,0,\gamma,0,\gamma,0,\gamma,\gamma)$, which is 6-round iterative and holds with bias $2^{-75.82}$. Starting at the i round, the following 12 subkeys are required to be weak: $Z_7^{(i)}$, $Z_1^{(i+1)}$, $Z_4^{(i+1)}$, $Z_6^{(i+1)}$, $Z_1^{(i+2)}$, $Z_4^{(i+2)}$, $Z_1^{(i+3)}$, $Z_1^{(i+4)}$, $Z_4^{(i+4)}$, $Z_7^{(i+4)}$, $Z_4^{(i+5)}$ and $Z_6^{(i+5)}$.

(23) $(0,0,\gamma,\gamma,0,0,\gamma,0) \xrightarrow{1r} (0,\gamma,\gamma,\gamma,0,0,0,0) \xrightarrow{1r} (0,\gamma,\gamma,\gamma,\gamma,\gamma,\gamma,\gamma) \xrightarrow{1r} (0,\gamma,\gamma,0,\gamma,0,0,0) \xrightarrow{1r} (0,\gamma,0,0,\gamma,\gamma,0,0) \xrightarrow{1r} (\gamma,\gamma,\gamma,0,0,\gamma,\gamma,0) \xrightarrow{1r} (0,0,\gamma,\gamma,0,0,\gamma,0)$, which is 6-round iterative and holds with bias $2^{-150.64}$. Starting at the i round, the following 10 subkeys are required to be weak: $Z_4^{(i)}$, $Z_7^{(i)}$, $Z_4^{(i+1)}$, $Z_4^{(i+2)}$, $Z_6^{(i+2)}$, $Z_7^{(i+2)}$, $Z_6^{(i+4)}$, $Z_1^{(i+5)}$, $Z_6^{(i+5)}$ and $Z_7^{(i+5)}$.

(24) $(0,0,\gamma,\gamma,\gamma,\gamma,0,0) \xrightarrow{1r} (\gamma,0,0,\gamma,\gamma,0,0,\gamma) \xrightarrow{1r} (0,\gamma,0,\gamma,\gamma,0,\gamma,0) \xrightarrow{1r} (0,0,\gamma,\gamma,\gamma,\gamma,0,0)$, which is 3-round iterative and holds with bias $2^{-25.94}$. Starting at the i round, the following 6 subkeys are required to be weak: $Z_4^{(i)}$, $Z_6^{(i)}$, $Z_1^{(i+1)}$, $Z_4^{(i+1)}$, $Z_4^{(i+2)}$ and $Z_7^{(i+2)}$.

(25) $(0,0,\gamma,\gamma,\gamma,\gamma,0,\gamma) \xrightarrow{1r} (0,\gamma,\gamma,0,0,\gamma,\gamma,\gamma) \xrightarrow{1r} (0,\gamma,0,\gamma,\gamma,0,\gamma,\gamma) \xrightarrow{1r} (\gamma,\gamma,0,0,0,0,\gamma,0) \xrightarrow{1r} (\gamma,0,0,\gamma,\gamma,0,0,0) \xrightarrow{1r} (\gamma,0,\gamma,0,0,\gamma,0,0) \xrightarrow{1r} (0,0,\gamma,\gamma,\gamma,\gamma,0,\gamma)$, which is 6-round iterative and holds with bias $2^{-206.52}$. Starting at the i round, the following 12 subkeys are required to be weak: $Z_4^{(i)}$, $Z_6^{(i)}$, $Z_6^{(i+1)}$, $Z_7^{(i+1)}$, $Z_4^{(i+2)}$, $Z_7^{(i+2)}$, $Z_1^{(i+3)}$, $Z_7^{(i+3)}$, $Z_1^{(i+4)}$, $Z_4^{(i+4)}$, $Z_1^{(i+5)}$ and $Z_6^{(i+5)}$.

(26) $(0,\gamma,\gamma,0,0,\gamma,\gamma,0) \xrightarrow{1r} (\gamma,0,\gamma,0,0,\gamma,0,\gamma) \xrightarrow{1r} (\gamma,\gamma,0,0,0,0,\gamma,\gamma) \xrightarrow{1r} (0,\gamma,\gamma,0,0,\gamma,\gamma,0)$, which is 3-round iterative and holds with bias $2^{-25.94}$. Starting at the i round, the following 6 subkeys are required to be weak: $Z_6^{(i)}$, $Z_7^{(i)}$, $Z_1^{(i+1)}$, $Z_6^{(i+1)}$, $Z_1^{(i+2)}$ and $Z_7^{(i+2)}$.

(27) $(0,\gamma,\gamma,0,\gamma,0,0,\gamma) \xrightarrow{1r} (\gamma,0,\gamma,\gamma,0,0,\gamma,0) \xrightarrow{1r} (\gamma,\gamma,\gamma,0,0,\gamma,\gamma,\gamma) \xrightarrow{1r} (\gamma,\gamma,0,0,\gamma,\gamma,0,0) \xrightarrow{1r} (0,\gamma,\gamma,\gamma,0,0,0,\gamma) \xrightarrow{1r} (\gamma,0,0,0,0,0,0,\gamma) \xrightarrow{1r} (0,\gamma,\gamma,0,\gamma,0,0,\gamma)$, which is 6-round iterative and holds with bias $2^{-206.52}$.

Starting at the i round, the following 10 subkeys are required to be weak: $Z_1^{(i+1)}$, $Z_4^{(i+1)}$, $Z_7^{(i+1)}$, $Z_1^{(i+2)}$, $Z_6^{(i+2)}$, $Z_7^{(i+2)}$, $Z_1^{(i+3)}$, $Z_6^{(i+3)}$, $Z_4^{(i+4)}$ and $Z_1^{(i+5)}$.

(28) $(0,\gamma,\gamma,0,\gamma,0,\gamma,\gamma) \xrightarrow{1r} (\gamma,0,\gamma,0,0,0,\gamma,0) \xrightarrow{1r} (\gamma,\gamma,0,\gamma,0,0,0,0) \xrightarrow{1r} (\gamma, 0, \gamma, \gamma, \gamma, \gamma, \gamma, \gamma) \xrightarrow{1r} (\gamma,\gamma,\gamma,0,0,0,0,0) \xrightarrow{1r} (\gamma,\gamma,0,\gamma,\gamma,\gamma,\gamma,\gamma) \xrightarrow{1r} (\gamma, 0, \gamma, 0, \gamma, 0, 0, 0) \xrightarrow{1r} (\gamma,\gamma,0,0,0,\gamma,0,0) \xrightarrow{1r} (0,\gamma,\gamma,\gamma,0,\gamma,0,\gamma) \xrightarrow{1r} (0, \gamma, \gamma, \gamma, \gamma, \gamma, 0, 0) \xrightarrow{1r} (\gamma,0,0,0,0,\gamma,\gamma,0) \xrightarrow{1r} (0,\gamma,\gamma,\gamma,\gamma,0,\gamma,0) \xrightarrow{1r} (0, \gamma, \gamma, 0, \gamma, 0, \gamma, \gamma)$, which is 12-round iterative and holds with bias $2^{-275.34}$. Starting at the i round, the following 26 subkeys are required to be weak: $Z_7^{(i)}$, $Z_1^{(i+1)}$, $Z_7^{(i+1)}$, $Z_1^{(i+2)}$, $Z_4^{(i+2)}$, $Z_1^{(i+3)}$, $Z_4^{(i+3)}$, $Z_6^{(i+3)}$, $Z_7^{(i+3)}$, $Z_1^{(i+4)}$, $Z_1^{(i+5)}$, $Z_4^{(i+5)}$, $Z_6^{(i+5)}$, $Z_7^{(i+5)}$, $Z_1^{(i+6)}$, $Z_1^{(i+7)}$, $Z_6^{(i+7)}$, $Z_4^{(i+8)}$, $Z_6^{(i+8)}$, $Z_4^{(i+9)}$, $Z_6^{(i+9)}$, $Z_1^{(i+10)}$, $Z_6^{(i+10)}$, $Z_7^{(i+10)}$, $Z_4^{(i+11)}$ and $Z_7^{(i+11)}$.

(29) $(\gamma,\gamma,\gamma,\gamma,\gamma,\gamma,\gamma,\gamma) \xrightarrow{1r} (\gamma,\gamma,\gamma,\gamma,\gamma,\gamma,\gamma,\gamma)$, which is 1-round iterative and holds with bias 2^{-1}. Starting at the i round, the following 4 subkeys are required to be weak: $Z_1^{(i)}$, $Z_4^{(i)}$, $Z_6^{(i)}$ and $Z_7^{(i)}$.

The linear relation (19) accounts for the least amount of weak subkeys per round, only 1.5, but across 8.5-round REESSE3+, the corresponding bias is below 2^{-64}.

Concerning the bias, relations (24), (26) and (29) achieve the highest average bias per round: $2^{-8.64}$, $2^{-8.64}$ and 2^{-1}, respectively.

Overall, the best trade-off is the linear relation (29). This is a 1-round iterative relation, which means it can be concatenated with itself. The exact number of rounds is a trade-off between the number of weak-key conditions and the existence of user keys that satisfy those conditions. Coincidentally, the weak subkeys needed for (29) are in the exact same positions and in the exact same quantity as the ones in DC in Sect. 3.5.10. Thus, the very same reasoning applies concerning the size of weak-key classes for increasing number of rounds as in Sect. 3.5.10.

Note that (29) is a wide linear trail, just like characteristic (29) in Sect. 3.5.10 is a wide differential trail.

Summarizing, linear attacks under weak-key conditions are possible for the full 8.5-round REESSE3+ and even for 16-round REESSE3+.

Independent of the trail (be it differential or linear), at least one special user key exists that satisfies any trail, $K = 0^{256}$, because this key makes every multiplicative subkey weak (whether the subkey is part of the trail or not).

As mentioned previously, another consequence of the key schedule design of REESSE3+ is that the full 8.5-round REESSE3+ can be distinguished from a random permutation with low attack complexity, since the linear relation (29) holds with maximum bias 2^{-1}. Apart from the weak-key assumptions, the attack cost is the same as that of the attacks on PES and IDEA: essentially $c \cdot (2^{-1})^2 = 4c$ known plaintexts and equivalent number of encryptions, for some constant c.

Consequently, REESSE3+ is not an ideal primitive, that is, it is not a PRP. See Sect. 1.7.

Similarly to the case of PES and IDEA, these linear relations allow to study the diffusion power of REESSE3+ with respect to modern tools such as the linear branch number. See Appendix, Sect. B.

The set of linear relations in Tables 3.133 up to 3.140 indicates that the linear branch number *per round* of REESSE3+ is only 6 (without any weak-key assumption). For instance, the 1-round linear relation

$$(0,0,\gamma,0,0,0,0,\gamma) \stackrel{1r}{\rightarrow} (\gamma,0,\gamma,0,\gamma,0,0,\gamma)$$

has the input and output bitmasks with Hamming Weight summing up to six. So, even though the design of REESSE+ provides text full diffusion in a single round, analysis shows that its linear branch number is much lower than ideal. For a 16-word block, such as in REESSE3+, an MDS code using a 16×16 MDS matrix would rather provide a branch number of 17. See Appendix, Sect. B.1.

3.13.12 LC of IDEA* Without Weak-Key Assumptions

The following linear analysis of IDEA* is based on [121].

The strategy is to build linear relations in a bottom-up fashion, from the smallest components: $\odot$, $\boxplus$ and $\oplus$, up to an MA-box, up to a half-round, further to a full round, and finally to multiple rounds.

Across unkeyed $\odot$ and $\boxplus$, the propagation of bitmasks follows from Table 3.98 and Table 3.99, respectively.

Further we proceed to obtain 1-round linear relations (under wordwise bitmasks 0 and 1) and the conditions for the linear relations to hold with highest possible bias while preserving only two 16-bit masks in the linear distinguishers: 0 and $\gamma = 1$. The bitmask 1 propagates across the modular addition with maximum bias, since it approximates only the least significant bit position, which is not affected by carry bits.

Table 3.141 lists all fifteen nontrivial 1-round linear relations of IDEA*, using only wordwise bitmasks 0 and $\gamma = 1$. There are **no weak-key nor weak-subkey assumptions**. Bias was approximated using the Piling-up lemma [131]. The approximations across unkeyed $\odot$ was $(1,1) \stackrel{\odot}{\rightarrow} 1$ with bias $2^{-13.47}$. The same approximation holds for $\boxdot$.

Multiple-round linear relations are obtained by concatenating 1-round linear relations, such that the output mask from one relation matches the input mask to the next relation.

From Table 3.141, there are only five multi-round iterative linear relations:

Table 3.141 Nontrivial 1-round linear relations of IDEA*.

1-round linear relations	Bias	# Active unkeyed $\odot$	# Active unkeyed $\boxplus$	Linear Rel. across the AX-box
$(0,0,0,\gamma) \overset{1r}{\to} (0,0,\gamma,0)$	$2^{-25.94}$	1	1	$(0,\gamma) \to (\gamma,0)$
$(0,0,\gamma,0) \overset{1r}{\to} (\gamma,0,\gamma,\gamma)$	$2^{-63.35}$	3	2	$(\gamma,\gamma) \to (0,\gamma)$
$(0,0,\gamma,\gamma) \overset{1r}{\to} (\gamma,0,0,\gamma)$	$2^{-38.41}$	2	1	$(\gamma,0) \to (\gamma,\gamma)$
$(0,\gamma,0,0) \overset{1r}{\to} (0,0,0,\gamma)$	$2^{-25.94}$	1	1	$(0,\gamma) \to (\gamma,0)$
$(0,\gamma,0,\gamma) \overset{1r}{\to} (0,0,\gamma,\gamma)$	$2^{-25.94}$	2	0	$(0,0) \to (0,0)$
$(0,\gamma,\gamma,0) \overset{1r}{\to} (\gamma,0,\gamma,0)$	$2^{-38.41}$	2	1	$(\gamma,0) \to (\gamma,\gamma)$
$(0,\gamma,\gamma,\gamma) \overset{1r}{\to} (\gamma,0,0,0)$	$2^{-38.41}$	1	2	$(\gamma,\gamma) \to (0,\gamma)$
$(\gamma,0,0,0) \overset{1r}{\to} (0,\gamma,\gamma,\gamma)$	$2^{-63.35}$	3	2	$(\gamma,\gamma) \to (0,\gamma)$
$(\gamma,0,0,\gamma) \overset{1r}{\to} (0,\gamma,0,\gamma)$	$2^{-38.41}$	2	1	$(\gamma,0) \to (\gamma,\gamma)$
$(\gamma,0,\gamma,0) \overset{1r}{\to} (\gamma,\gamma,0,0)$	$2^{-25.94}$	2	0	$(0,0) \to (0,0)$
$(\gamma,0,\gamma,\gamma) \overset{1r}{\to} (\gamma,\gamma,\gamma,0)$	$2^{-50.88}$	3	1	$(0,\gamma) \to (\gamma,0)$
$(\gamma,\gamma,0,0) \overset{1r}{\to} (0,\gamma,\gamma,0)$	$2^{-38.41}$	2	1	$(\gamma,0) \to (\gamma,\gamma)$
$(\gamma,\gamma,0,\gamma) \overset{1r}{\to} (0,\gamma,0,0)$	$2^{-38.41}$	1	2	$(\gamma,\gamma) \to (0,\gamma)$
$(\gamma,\gamma,\gamma,0) \overset{1r}{\to} (\gamma,\gamma,0,\gamma)$	$2^{-50.88}$	3	1	$(0,\gamma) \to (\gamma,0)$
$(\gamma,\gamma,\gamma,\gamma) \overset{1r}{\to} (\gamma,\gamma,\gamma,\gamma)$	$2^{-50.88}$	4	0	$(0,0) \to (0,0)$

(1) $(0,0,0,\gamma) \overset{1r}{\to} (0,0,\gamma,0) \overset{1r}{\to} (\gamma,0,\gamma,\gamma) \overset{1r}{\to} (\gamma,\gamma,\gamma,0) \overset{1r}{\to} (\gamma,\gamma,0,\gamma) \overset{1r}{\to} (0,\gamma,0,0) \overset{1r}{\to} (0,0,0,\gamma)$, which is 6-round iterative and holds with bias $2^{-260.40}$.
(2) $(0,0,\gamma,\gamma) \overset{1r}{\to} (\gamma,0,0,\gamma) \overset{1r}{\to} (0,\gamma,0,\gamma) \overset{1r}{\to} (0,0,\gamma,\gamma)$, which is 3-round iterative and holds with bias $2^{-100.76}$.
(3) $(0,\gamma,\gamma,0) \overset{1r}{\to} (\gamma,0,\gamma,0) \overset{1r}{\to} (\gamma,\gamma,0,0) \overset{1r}{\to} (0,\gamma,\gamma,0)$, which is 3-round iterative and holds with bias $2^{-100.76}$.
(4) $(0,\gamma,\gamma,\gamma) \overset{1r}{\to} (\gamma,0,0,0) \overset{1r}{\to} (0,\gamma,\gamma,\gamma)$, which is 2-round iterative and holds with bias $2^{-110.76}$.
(5) $(\gamma,\gamma,\gamma,\gamma) \overset{1r}{\to} (\gamma,\gamma,\gamma,\gamma)$, which is 1-round iterative and holds with bias $2^{-50.88}$.

All multi-round linear relation have too low biases even for a distinguish-from-random attack (the codebook size of IDEA* is only 2^{64} text blocks).

From Table 3.141, even linear relations covering two rounds already have too low bias. The best trade-offs are linear relations covering 1.5 rounds, such as $(0,0,0,\gamma) \overset{1.5r}{\to} (0,0,\gamma,0)$, $(0,\gamma,0,0) \overset{1.5r}{\to} (0,0,0,\gamma)$, $(0,\gamma,0,\gamma) \overset{1.5r}{\to} (0,0,\gamma,\gamma)$ or $(\gamma, 0, \gamma, 0) \overset{1.5r}{\to} (\gamma, \gamma, 0, 0)$, since an additional KM half-round contains only $\oplus$ and $\boxplus$. Even in these cases, the data complexity is estimated at $c/(2^{-25.94})^{-2} = c \cdot 2^{51.88}$ known plaintexts, where c is a constant.

A key-recovery attack on 2.5-round IDEA* using this 1.5-round distinguisher could guess the four subkeys in an additional AX half-rounds before and after the distinguisher. For instance, starting from the first round, the subkeys $Z_5^{(1)}$, $Z_6^{(1)}$, $Z_5^{(3)}$ and $Z_6^{(3)}$ would be recovered.

The attack effort would be about $2 \cdot c \cdot 2^{51.88} \cdot (2^{16})^4 = c \cdot 2^{116.88}$ AX half-round computations, but it would only recover 64 subkey bits, not the full key. Moreover, from the key schedule of IDEA* (Sect. 2.8.2), these four subkeys, which are 1.5-round apart, do not provide enough information to reconstruct the original 128-bit user key:

$$\begin{aligned}
Z_5^{(1)} &= ((((((K_{-3} \boxplus K_{-2}) \oplus K_{-1}) \boxplus Z_2^{(1)}) \oplus Z_3^{(1)}) \boxplus Z_4^{(1)}) \lll 7 \oplus c_{12},\\
Z_6^{(1)} &= ((((((K_{-2} \boxplus K_{-1}) \oplus K_0) \boxplus Z_3^{(1)}) \oplus Z_4^{(1)}) \boxplus Z_5^{(1)}) \lll 7 \oplus c_{13},\\
Z_5^{(3)} &= ((((((Z_3^{(1)} \boxplus Z_4^{(1)}) \oplus Z_5^{(1)}) \boxplus Z_2^{(2)}) \oplus Z_3^{(2)}) \boxplus Z_4^{(2)}) \lll 7 \oplus c_{18},\\
Z_6^{(3)} &= ((((((Z_4^{(1)} \boxplus Z_5^{(1)}) \oplus Z_6^{(1)}) \boxplus Z_3^{(2)}) \oplus Z_4^{(2)}) \boxplus Z_5^{(2)}) \lll 7 \oplus c_{19},
\end{aligned}$$

where the 128-bit user key is denoted by the 8-tuple:

$$(K_{-7}, K_{-6}, K_{-5}, K_{-4}, K_{-3}, K_{-2}, K_{-1}, K_0).$$

The set of linear relations in Tables 3.141 indicates that the linear branch number *per round* of IDEA* is only 2 (without any weak-key assumption). For instance, the 1-round linear relation

$$(0, 0, 0, \gamma) \xrightarrow{1r} (0, 0, \gamma, 0)$$

has the input and output bitmasks with Hamming Weight summing up to 2. So, even though the design of IDEA* provides full diffusion in a single round, a detailed analysis shows that its linear branch number is much lower than ideal. For a 4-word block, such as in IDEA*, an MDS code using a 4×4 MDS matrix would rather provide a branch number of 5. See Appendix, Sect. B.1.

3.13.13 LC of YBC Without Weak-Key Assumptions

The following linear analysis of YBC is unpublished. The strategy is to build linear relations in a bottom-up fashion, from the smallest components: $\odot$, $\boxplus$ and $\oplus$, up to a half-round, further to a full round and finally to multiple rounds.

Across unkeyed $\odot$ and $\boxplus$, the propagation of bitmasks follows from Table 3.98 and Table 3.99, respectively.

Further we proceed to obtain 1-round linear relations (under wordwise bitmasks 0 and 1) and the conditions for the linear relations to hold with highest possible bias while preserving only two 16-bit masks in the linear distinguishers: 0 and $\gamma = 1$. The bitmask 1 propagates across the modular addition with maximum bias, since it approximates only the least significant bit position, which is not affected by carry bits.

Table 3.142 lists all fifteen nontrivial 1-round linear relations of YBC, using only wordwise bitmasks 0 and $\gamma = 1$. There are **no weak-key nor weak-subkey assumptions**. Bias was approximated using the Piling-up lemma [131]. The approximations across unkeyed $\odot$ was $(1,1) \stackrel{\odot}{\rightarrow} 1$ with bias $2^{-13.47}$. The same approximation holds for $\boxdot$.

Table 3.142 Nontrivial 1-round linear relations of YBC.

1-round linear relations	Bias	# Active unkeyed $\odot$
$(0,\gamma,0,0) \stackrel{1r}{\rightarrow} (0,0,0,\gamma)$	$2^{-25.94}$	2
$(0,0,\gamma,0) \stackrel{1r}{\rightarrow} (0,0,\gamma,0)$	$2^{-25.94}$	2
$(0,\gamma,\gamma,0) \stackrel{1r}{\rightarrow} (0,0,\gamma,\gamma)$	$2^{-25.94}$	2
$(0,\gamma,0,\gamma) \stackrel{1r}{\rightarrow} (0,\gamma,0,0)$	$2^{-38.41}$	3
$(0,0,0,\gamma) \stackrel{1r}{\rightarrow} (0,\gamma,0,\gamma)$	$2^{-38.41}$	3
$(0,\gamma,\gamma,\gamma) \stackrel{1r}{\rightarrow} (0,\gamma,\gamma,0)$	$2^{-13.47}$	1
$(0,0,\gamma,\gamma) \stackrel{1r}{\rightarrow} (0,\gamma,\gamma,\gamma)$	$2^{-38.41}$	3
$(\gamma,0,0,0) \stackrel{1r}{\rightarrow} (\gamma,0,0,0)$	$2^{-25.94}$	2
$(\gamma,\gamma,0,0) \stackrel{1r}{\rightarrow} (\gamma,0,0,\gamma)$	$2^{-25.94}$	2
$(\gamma,0,\gamma,0) \stackrel{1r}{\rightarrow} (\gamma,0,\gamma,0)$	$2^{-50.88}$	4
$(\gamma,\gamma,\gamma,0) \stackrel{1r}{\rightarrow} (\gamma,0,\gamma,\gamma)$	$2^{-25.94}$	2
$(\gamma,\gamma,0,\gamma) \stackrel{1r}{\rightarrow} (\gamma,\gamma,0,0)$	$2^{-38.41}$	3
$(\gamma,0,0,\gamma) \stackrel{1r}{\rightarrow} (\gamma,\gamma,0,\gamma)$	$2^{-13.47}$	1
$(\gamma,\gamma,\gamma,\gamma) \stackrel{1r}{\rightarrow} (\gamma,\gamma,\gamma,0)$	$2^{-13.47}$	1
$(\gamma,\gamma,\gamma,0) \stackrel{1r}{\rightarrow} (\gamma,\gamma,\gamma,\gamma)$	$2^{-13.47}$	1

Multiple-round linear relations are obtained by concatenating 1-round linear relations, such that the output mask from one relation matches the input mask to the next relation.

From Table 3.142, there are only four multi-round iterative linear relations:

(1) $(0,\gamma,0,0) \stackrel{1r}{\rightarrow} (0,0,0,\gamma) \stackrel{1r}{\rightarrow} (0,\gamma,0,\gamma) \stackrel{1r}{\rightarrow} (0,\gamma,0,0)$, which is 3-round iterative and holds with bias $2^{-100.76}$.

(2) $(0,\gamma,\gamma,0) \stackrel{1r}{\rightarrow} (0,0,\gamma,\gamma) \stackrel{1r}{\rightarrow} (0,\gamma,\gamma,\gamma) \stackrel{1r}{\rightarrow} (0,\gamma,\gamma,0)$, which is 3-round iterative and holds with bias $2^{-75.82}$.

(3) $(\gamma,\gamma,0,0) \stackrel{1r}{\rightarrow} (\gamma,0,0,\gamma) \stackrel{1r}{\rightarrow} (\gamma,\gamma,0,\gamma) \stackrel{1r}{\rightarrow} (\gamma,\gamma,0,0)$, which is 3-round iterative and holds with bias $2^{-75.82}$.

(4) $(\gamma,\gamma,\gamma,0) \stackrel{1r}{\rightarrow} (\gamma,0,\gamma,\gamma) \stackrel{1r}{\rightarrow} (\gamma,\gamma,\gamma,\gamma) \stackrel{1r}{\rightarrow} (\gamma,\gamma,\gamma,0)$, which is 3-round iterative and holds with bias $2^{-50.88}$.

All multi-round linear relation have too low biases even for a distinguish-from-random attack (the codebook size of YBC is only 2^{64} text blocks).

From Table 3.142, the best 2-round linear relation is for instance $(\gamma, 0, \gamma, \gamma) \stackrel{1r}{\rightarrow} (\gamma, \gamma, \gamma, \gamma) \stackrel{1r}{\rightarrow} (\gamma, \gamma, \gamma, 0)$, which holds with bias $2^{-25.94}$. It allows a distinguish-from-random attack on up to 2.5-round YBC with about $c \cdot (2^{-25.94})^2 = c \cdot 2^{51.88}$ known plaintexts where c is a constant.

These linear relations allows us to study the diffusion power of YBC with respect to modern tools such as the linear branch number. See Appendix, Sect. B.

From Table 3.142, the linear branch number *per round* of YBC is only 2. For instance, the 1-round linear relation

$$(0, \gamma, 0, 0) \stackrel{1r}{\rightarrow} (0, 0, 0, \gamma)$$

has input and output bitmasks with Hamming Weight summing up to 2. So, even though the design of YBC provides full text diffusion is a single round, analysis shows that its linear branch number is much lower than ideal. For a 4-word block such as in YBC an MDS code using a 4×4 MDS matrix would rather provide a branch number of 5. See Appendix, Sect. B.1.

3.14 Differential-Linear Cryptanalysis

Previous analysis comparing differential (DC) and linear cryptanalysis (LC) was presented by Chabaud and Vaudenay in [51], but focusing on similarities and analogies between the two techniques separately.

The differential-linear cryptanalysis (DLC) technique was developed by S. Langford and M.E. Hellman [119, 118] and applied to 8-round DES. Their attack recovered 10 key bits with 80% success rate using 512 chosen plaintexts. The success rate increases to 95% using 768 chosen plaintexts.

Langford and Hellman showed that it is possible to combine both DC [30] and LC [131], or more specifically, a differential characteristic and a linear relation, in this order, to build a longer distinguisher combining both techniques. For their attack on 8-round DES, the differential characteristic held with certainty.

Later, in [18], Biham *et al.* generalized the DL technique to the case where the characteristic holds with probability p where $0 < p < 1$.

This attack has been applied to several block ciphers such as FEAL [4], IDEA [48, 81], Serpent [20], COCONUT98 [18] and also to stream ciphers such as Phelix [185].

Further developments of the DLC technique include [41, 126, 123].

The following description of the differential-linear technique comes from [21]. Let $E_K = E_1 \circ E_0$ be a block cipher parameterized by a key K. For a plaintext/ciphertext pair (P, C), it follows that $C = E_K(P) = E_1(E_0(P))$. Let us denote the partial encryption of P by E_0 as T, that is, $T = E_0(P) = E_1^{-1}(C)$.

Langford and Hellman [119] showed that a concatenation of a differential characteristic and a linear approximation is feasible. The main idea in that combination is to encrypt plaintext pairs and check if the corresponding ciphertext pair has the same parity for a given output mask (or not).

Let $\Delta_P \to \Delta_T$ denote a differential across E_0 with probability one. Let $\Gamma_T \to \Gamma_C$ denote a linear approximation across E_1 with bias q. Start with plaintexts $(P_1, P_2) = (P_1, P_1 \oplus \Delta_P)$. After partial encryption by E_0, the intermediate encrypted values are $(T_1, T_2) = (T_1, T_1 \oplus \Delta_T)$. For any intermediate value T and its corresponding ciphertext C, the relation $\Gamma_T \cdot T = \Gamma_C \cdot C$ holds with probability $\frac{1}{2} + q$. For (P_1, P_2), the ciphertexts are (C_1, C_2) and the intermediate values are (T_1, T_2). Therefore, each of the relations

$$\Gamma_c \cdot C_1 = \Gamma_T \cdot T_1$$

and

$$\Gamma_C \cdot C_2 = \Gamma_T \cdot T_2 = \Gamma_T \cdot T_1 \oplus \Gamma_T \cdot \Delta_T$$

is satisfied with probability $\frac{1}{2} \pm q$. Hence, with probability $\frac{1}{2} + 2q^2$, the relation

$$\Gamma_C \cdot C_1 = \Gamma_C \cdot C_2 \oplus \Gamma_T \cdot \Delta_T$$

holds.

Note that Γ_T and Δ_T are known and thus, the condition on C_1 and C_2 has probability $\frac{1}{2} + 2q^2$, while for a random ciphertext pairs, this condition is satisfied with probability $\frac{1}{2}$. This fact can be used in distinguishers and key-recovery attacks. Hellman and Langford noted that it is possible to use truncated differentials [100] as long as $\Gamma_T \cdot \Delta_T$ is predictable.

Later, it was shown [18] that it is possible to have $\Gamma_T \cdot \Delta_T$ unknown but fixed. Also, it was shown that it is possible to use the differential-linear technique when the differential has probability $p \neq 1$. In this case, the probability that

$$\Gamma_T \cdot T_1 = \Gamma_T \cdot T_2 \oplus \Gamma_T \cdot \Delta_T$$

is $\frac{1}{2} + \frac{p}{2}$, and thus, the event

$$\Gamma_C \cdot C_1 = \Gamma_C \cdot C_2 \oplus \Gamma_T \cdot \Delta_T$$

holds with probability $\frac{1}{2} + 2pq^2$.

Even if $\Gamma_T \cdot \Delta_T$ is unknown to the adversary but constant for a given key, the attack still succeeds. In that case, we know that the value $\Gamma_C \cdot C_1 \oplus \Gamma_C \cdot C_2$ is either 0 or 1 with a bias of $2q^2$. This case is similar to the setting of a linear cryptanalysis when $\Gamma_K \cdot K$ is unknown, and can be either 0 or 1.

3.14.1 DL Analysis of PES

This section is based on analyses described in [149].

The largest differential-linear distinguishers found for PES can cover the full cipher (under weak-key assumptions). For instance, Table 3.143 describes a differential-linear distinguisher combining a 3-round characteristic with a 5-round linear relation, where $\delta_2 \oplus \delta_4 = \nu = 8000_x$. Note that since the linear relation starts with the bitmasks $(0, 1, 0, 1)$, only the nonzero word differences δ_2 and δ_4 actually matter.

Table 3.143 Differential-linear distinguisher for the full 8.5-round PES.

Round i	Differential-linear relation	$\text{msb}_{15}(Z_1^{(i)})$	$\text{msb}_{15}(Z_2^{(i)})$
1	$(\nu, 0, \nu, 0) \stackrel{1r}{\to} (\nu, 0, \nu, 0)$	0–14	—
2	$(\nu, 0, \nu, 0) \stackrel{1r}{\to} (\nu, 0, \nu, 0)$	96–110	—
3	$(\nu, 0, \nu, 0) \stackrel{1r}{\to} (\delta_1, \delta_2, \delta_3, \delta_4)$	—	—
4	$(0, 1, 0, 1) \stackrel{1r}{\to} (0, 1, 0, 1)$	—	98–112
5	$(0, 1, 0, 1) \stackrel{1r}{\to} (0, 1, 0, 1)$	—	91–105
6	$(0, 1, 0, 1) \stackrel{1r}{\to} (0, 1, 0, 1)$	—	59–73
7	$(0, 1, 0, 1) \stackrel{1r}{\to} (0, 1, 0, 1)$	—	52–66
8	$(0, 1, 0, 1) \stackrel{1r}{\to} (0, 1, 0, 1)$	—	45–59

The 3-round characteristic holds with certainty for a differential-linear weak-key class of size 2^{62}, in which key bits 15–44, 74-90 and 113–127 can be arbitrary. The linear relation holds with maximum bias 2^{-1} under weak-key assumptions.

This differential-linear distinguisher can be used to distinguish the full 8.5-round PES from a random permutation given low attack complexity: essentially $c \cdot (2^{-1})^{-1} = 4c$ encryptions and equivalent data complexities.

Another differential-linear distinguisher for the full PES cipher is detailed in Table 3.144. It holds with certainty for a weak-key class of size 2^{62}. Key bits numbered 15–44, 74-88 and 111-127 can be arbitrary values.

3.14.2 DL Analysis of IDEA

The first reported differential-linear analysis of reduced-round variants of the IDEA cipher was by Borst in [47]. The following attack description comes from [47, 48].

Let (n_i, q_i) denote the inputs to the i-th MA-box of IDEA and let (r_i, s_i) denote the corresponding output, for $1 \leq i \leq 8$. Let t_i denote the output of the first $\odot$ in the i-th MA-box. Thus, $t_i = n_i \odot Z_5^{(i)}$.

Table 3.144 Differential-linear distinguisher for the full 8.5-round PES.

Round i	Differential-linear relation	$\text{msb}_{15}(Z_1^{(i)})$	$\text{msb}_{15}(Z_2^{(i)})$
1	$(\nu,0,\nu,0) \stackrel{1r}{\rightarrow} (\nu,0,\nu,0)$	0–14	—
2	$(\nu,0,\nu,0) \stackrel{1r}{\rightarrow} (\nu,0,\nu,0)$	96–110	—
3	$(\nu,0,\nu,0) \stackrel{1r}{\rightarrow} (\nu,0,\nu,0)$	89–103	—
4	$(\nu,0,\nu,0) \stackrel{1r}{\rightarrow} (\delta_1,\delta_2,\delta_3,\delta_4)$	—	—
5	$(0,1,0,1) \stackrel{1r}{\rightarrow} (0,1,0,1)$	—	91–105
6	$(0,1,0,1) \stackrel{1r}{\rightarrow} (0,1,0,1)$	—	59–73
7	$(0,1,0,1) \stackrel{1r}{\rightarrow} (0,1,0,1)$	—	52–66
8	$(0,1,0,1) \stackrel{1r}{\rightarrow} (0,1,0,1)$	—	45–59

Let μ_i denote a 16-bit value such that $\text{lsb}_i(\mu_i) = 1$ and $\text{lsb}_j(\mu_i) = 0$ for all $j \neq i$.

Initially, subkeys $Z_4^{(1)}$ and $Z_5^{(3)}$ were recovered (up to equivalent subkeys $2^{16} + 1 - Z_4^{(1)}$ and $2^{16} + 1 - Z_5^{(3)}$).

The input difference has the form $(0, \delta_1, 0, \delta_2)$, and the difference operator is bitwise exclusive-or. Encrypt plaintext pairs (P, P^*) under the given difference, where $P = (p_1, p_2, p_3, p_4)$ and $P^* = (p_1^*, p_2^*, p_3^*, p_4^*)$. Let $C = (c_1, c_2, c_3, c_4)$ and $C^* = (c_1^*, c_2^*, c_3^*, c_4^*)$ denote the corresponding ciphertexts.

With $\delta_1 = \mu_i$, one gets after the first KM half-round a difference μ_i after $Z_2^{(1)}$ with probability $\frac{1}{2}$ [167, 104]. Similarly, with probability $\frac{1}{2}$, the difference after μ_i appears after $Z_3^{(2)}$. This part of the differential holds with probability $\frac{1}{4}$. One can choose six plaintext pairs such that this part of the differential holds at least once [47].

The values of (p_4, p_4^*) are chosen such that

$$(Z_4^{(1)} \odot p_4) \oplus (Z_4^{(1)} \odot p_4^*) = (p_2 \boxplus Z_2^{(1)}) \oplus (p_2^* \boxplus Z_2^{(1)}).$$

This choice of difference ensures that the input difference to the first MA-box is $(0, 0)$.

For one of the six plaintext pairs, the difference after the second KM half-round will be

$$(0, 0, \mu, \alpha), \tag{3.175}$$

Now, concerning the first multiplication in the second MA-box:

$$\Delta t_2 = (n_2 \odot Z_5^{(2)}) \oplus ((n_2 \oplus \mu_i) \odot Z_5^{(2)}).$$

For every choice of $Z_5^{(2)}$, we observed that there are several possible values for μ_i such that

$$\Delta t_2 \cdot 1 = 0, \tag{3.176}$$

with probability p such that the bias $|p - \frac{1}{2}| > 0.166$ over all (p_2, p_2^*) [47]. Further, we observed that for all but 26 of the 2^{16} possible values of $Z_5^{(2)}$, there is at least one μ_i for which the bias is larger than $\frac{1}{4}$. This fact will be used in the linear part of the attack. Instead of using one linear relation that holds with an average probability for each key, we are going to use a set of relations. For each key, at least one of the relations has a large bias. This idea is central to the attack.

From now on, only the least significant bit will be used in the linear approximations of 16-bit words in all internal cipher states.

Using (3.175), the difference after the second round becomes

$$(\Delta s_2, \mu_i \oplus \Delta s_2, \Delta t_2 \oplus \Delta s_2, \alpha \oplus \Delta t_2 \oplus \Delta s_2),$$

where $r_2 = t_2 \boxplus s_2$.

Since there is no output transformation and no swap of middle words in the ciphertexts:

$$\Delta s_3 \cdot 1 = \Delta c_3 \cdot 1 \oplus \Delta t_2 \cdot 1 \oplus \Delta s_2 \cdot 1,$$

$$\Delta t_3 \cdot 1 = \Delta s_3 \cdot 1 \oplus \Delta c_2 \cdot 1 \oplus \mu_i \oplus \Delta s_2 \cdot 1.$$

It means that we are able to predict the least significant bit of the output difference of the first multiplication in the last MA-box. The inputs to this multiplication are $Z_5^{(3)}$ and $(c_1 \oplus c_3, c_1^* \oplus c_3^*)$. For every ciphertext pair, we can predict Δt_3 with high probability. Keep a counter for every candidate for $Z_5^{(3)}$ and increment the counter for subkey values compatible with the $(c_1 \oplus c_3, c_1^* \oplus c_3^*)$ and the computed Δt_3.

We do not know which μ_i makes (3.176) hold with large probability. Therefore, we repeat the attack for several μ_i values. Also, we guess the value of $Z_4^{(1)}$. Experiments suggest that for wrong $Z_4^{(1)}$ candidates, the attack fails to suggest a specific value for $Z_5^{(3)}$. Thus, we can recognize wrong subkey guesses. Still, it is not possible to distinguish $Z_4^{(1)}$ from $2^{16} + 1 - Z_4^{(1)}$.

Once $Z_4^{(1)}$ is recovered, tests have shown that we need at most 9000 pairs to determine $Z_5^{(3)}$. On average, we guess correctly after 2^{15} trials. Therefore, we need about 2^{29} plaintext pairs. Examining one plaintext pair takes a few $\oplus$ operations and 2^{16} table look-ups, one for each $Z_5^{(3)}$ value. Since we examine 16 differentials, the attack needs in total 2^{20} additions and exclusive-ors for each pair. The total workload is about 2^{49} simple operations, which is estimated at 2^{44} 3-round IDEA computations.

According to the key schedule of IDEA, $Z_4^{(1)}$ account for the user key bits 48–63, while $Z_5^{(3)}$ represent user key bits 50–65. But, this attack assumed independent round subkeys [47], which means the original key schedule of IDEA was not used. Therefore, the key overlapping property could not be used to reduce the attack complexity. Also, this assumption means that the key space increased from 2^{128} (in the original IDEA cipher) to 2^{288} (in IDEA with independent subkeys).

Two additional subkeys $Z_1^{(3)}$ and $Z_6^{(3)}$ were also aimed during the analysis of [47, 48], but no further information was provided about the recovery of the remaining $288 - 4 \cdot 30 = 168$ subkey bits.

Differential-linear analysis of IDEA variants were also described in [79, 81].

A differential-linear (DL) distinguisher will be denoted by a triple

$$(\delta_1 \overset{t}{\to} \delta_2, \gamma_1 \overset{m}{\to} \gamma_2, \eta),$$

where $\delta_1 \overset{t}{\to} \delta_2$ denotes a t-round differential characteristic (holding with high, nonzero probability), $\gamma_1 \overset{m}{\to} \gamma_2$ denotes a m-round linear relation (holding with high, nonzero bias) and $\eta \in \{0, 1\}$.

Let $X^{(i)} = (X_1^{(i)}, X_2^{(i)}, X_3^{(i)}, X_4^{(i)})$ denote the input to the i-th round of IDEA. A DL distinguisher predicts that there exists plaintext pairs (P, P^*) such that $P \oplus P^* = \delta_1$, $X^{(t)} \oplus X^{*(t)} = \delta_2$, and $\gamma_1 \cdot (X^{(t)} \oplus X^{*(t)}) = \eta$. This last expression connects the differential characteristic with the linear relation.

The probability of a DL distinguisher is

$$\Pr(\gamma_1 \cdot (X^{(t)} \oplus X^{*(t)}) = \eta | P \oplus P^* = \delta_1).$$

For IDEA, a DL distinguisher under weak-key assumptions leads to a membership test for a weak-key class W. W is a set of 128-bit user keys for which some DL distinguisher $(\delta_1 \overset{t}{\to} \delta_2, \gamma_1 \overset{m}{\to} \gamma_2, \eta)$ holds with certainty.

Hawkes in [79, 81], describes the following DL distinguisher across the full 8.5-round IDEA, where $\nu = 8000_x$:

$$((0, \nu, 0, \nu) \overset{4r}{\to} (0, 0, \nu, \nu), (1, 1, 0, 0) \overset{3r}{\to} (1, 1, 0, 0), 0) \tag{3.177}$$

for which the key bit positions 0–18, 29–71 and 125–127 are zero. These key bit position imply that the following subkeys are weak: $Z_4^{(1)}$, $Z_4^{(2)}$, $Z_5^{(2)}$, $Z_5^{(3)}$, $Z_4^{(4)}$, $Z_1^{(6)}$, $Z_5^{(6)}$, $Z_5^{(7)}$ and $Z_1^{(8)}$.

For details on the 1-round characteristics for IDEA, see Table 3.9 in Sect. 3.5.2. For details on the 1-round linear relations for IDEA, see Table 3.94 in Sect. 3.13.2.

At the input to the 5th round, the text difference is $(0, 0, \nu, \nu)$. After the 5th KM half-round, the difference has the form $(0, 0, \nu, \delta)$ where δ is unknown since $Z_5^{(4)}$ is not restricted. The 5th MA-box input difference has the form (ν, δ). Since $Z_5^{(5)}$ is not restricted (not weak), the round output difference has the form $(\nabla_1, \nabla_2, \nabla_3, \nabla_4)$. But, using the involution property of the MA half-round, it follows that $\nabla_1 \oplus \nabla_2 = \nu$.

One round further, at the connecting point between the differential characteristic and the linear relation, the combination of the nonzero difference words and the nonzero bitmasks gives $\delta_2 \cdot \gamma_1 = \nu \cdot 1 = 0$. Note that the bits approximated by the linear mask have difference zero, that is, they are equal. So, the parity is $\eta = 0$.

The DL distinguisher (3.177) does not cover the output transformation of IDEA. In particular, it does not cover $Z_1^{(9)}$ for which the input bitmask is 1. The weak-key class W for (3.177) includes key bits in positions 29-37 (the nine least significant bits of $Z_1^{(9)}$). Thus, only seven bits of $Z_1^{(9)}$ need to be found.

Candidate values $z_{1,9}$ for $Z_1^{(9)}$ must satisfy

$$(X_1^{(9)} \odot z_{1,9}^{-1}) \cdot 1 \oplus (X_1^{*(9)} \odot z_{1,9}^{-1}) \cdot 1 \oplus X_2^{(9)} \cdot 1 \oplus X_2^{*(9)} \cdot 1 = 0, \quad (3.178)$$

where the last swap of middle words in undone since we cover the full 8.5-round IDEA. A ciphertext pair $(X^{(9)}, X^{(9)*})$ for which (3.178) holds is said to be a DL-right pair for $z_{1,9}$.

Table 3.145 lists the DL distinguishers for several reduced-round variants of IDEA (starting from the first round), and the membership test complexities.

Table 3.145 Differential-linear distinguishers for IDEA variants: parameters and complexities.

# Rounds	δ_1	λ_1	$\|WKC\|$	Weak Key bit positions	Average Data Complexity (CP)
4	$(0,0,\nu,\nu)$	$(1,1,0,0)$	2^{99}	50–78	18
4.5	$(0,\nu,0,\nu)$	$(1,0,1,0)$	2^{97}	41–71	38.4
5	$(0,0,\nu,\nu)$	$(1,0,1,0)$	2^{84}	50–71, 75–96	18
5.5	$(0,\nu,0,\nu)$	$(1,0,1,0)$	2^{82}	2–16, 41–71	4.5
6	$(0,\nu,0,\nu)$	$(0,1,1,0)$	2^{82}	2–16, 41–71	8.1
6.5	$(0,\nu,0,\nu)$	$(0,1,1,0)$	2^{80}	2–18, 41–71	4.5
7	$(0,\nu,0,\nu)$	$(1,1,0,0)$	2^{80}	2–18, 41–71	14.8
7.5	$(0,\nu,0,\nu)$	$(1,1,0,0)$	2^{75}	0–18, 41–71, 125–127	31.2
8	$(0,\nu,0,\nu)$	$(1,1,0,0)$	2^{66}	0–25, 41–71, 123–127	38.4
8.5	$(0,\nu,0,\nu)$	$(1,1,0,0)$	2^{63}	0–18, 29–71, 125–127	19.5

3.14.3 Higher-Order Differential-Linear Analysis of IDEA

The Higher-Order Differential-Linear (HODL) attack combines higher-order differentials [100, 115] with linear approximations [131, 132].

The following description is extracted from [21].

Let a target block cipher E_K be decomposed into $E_K = E_1 \circ E_0$, where K is a secret key. The idea of the attack is to use a higher-order differential covering E_0 to predict the difference of an intermediate cipher state T, and then use a (highly-biased) linear approximation to cover E_1.

Let S denote a plaintext structure $S = \{P_1, P_2, \ldots, P_m\}$ such that the higher-order differential (across E_0) predicts the value $\oplus_{i=1}^{m} T_i$ with some nonzero probability p, and where T_i, $1 \leq i \leq m$, are intermediate cipher states (after E_0). Under standard independent assumptions, this means that the parity of any subset of bits taken over all the T_i is biased with bias $p' = p/2$. Moreover, assume that there is a linear approximation covering E_1 that predicts the value $\Gamma_T \cdot T \oplus \Gamma_C \cdot C$ with probability $\frac{1}{2} + q$.

Lemma 3.2 gives the probability that the HODL relation holds for a structure S.

Lemma 3.2. *[21]*
Let the I be defined as

$$I = \{\Gamma_T \cdot (T_1 \oplus T_2 \ldots \oplus T_m) = \Gamma_C \cdot (C_1 \oplus C_2 \oplus \ldots \oplus C_m)\}.$$

Then, under standard independence assumptions, $Pr[I] = \frac{1}{2} + 2^{m-1} q^m$.

Lemma 3.3 gives the bias of a HODL relation.

Lemma 3.3. *[21]*
Given a plaintext structure S with the input requirements of the higher-order differential, the bias of the event that the exclusive-or of the output bit mask in all ciphertexts equals the value predicted by the linear approximation is $\epsilon = 2^{m-1} \cdot p \cdot q^m$.

An application of the HODL attack on IDEA uses the linear weak-key class of 2^{23} keys in [57]. It is depicted in Table 3.95. This linear approximation has bias 2^{-1} and is used to cover the last 5.5 rounds of IDEA.

In [21], it is mentioned that the higher-order differential covers the first 3 rounds of IDEA and uses structures of the form (A, P, A, P) like in a square attack: the terminology A means the corresponding 16-bit words are active, while P means that the 16-bit words are passive (constant). According to [21], after 3-round IDEA the exclusive-or sum of the least significant bit of the first and second words in a block are zero. But, following the propagation of active and passive words per round, $(A, P, A, P) \stackrel{1r}{\rightarrow} (A, A, P, P) \stackrel{1r}{\rightarrow} (?, ?, ?, ?)$, where the patterns after the first round is due to the fact that $Z_1^{(1)} = 0 = Z_3^{(1)}$ as part of the weak-key restrictions. Since $Z_1^{(2)} \neq 0$ and $Z_5^{(2)} \neq 0$, the pattern degenerates after 2 rounds.

Assuming the higher-order differential in fact covers the first two rounds of IDEA, note that the integral of the two least significant bits of the first and second output words of the second round, over the Λ-set of 2^{16} plaintexts, gives 0 since the exclusive-or of the first and second output words of the second round corresponds to an active word (the first input to the MA-box of the second round).

Using the linear relation depicted in Table 3.95, starting with the mask $(1,1,0,0)$ *in the third round* implies the linear trail in Table 3.146 covering the last 6.5 rounds of IDEA.

Table 3.146 Linear relation covering the last 6.5 rounds of IDEA for HODL attack.

i-th Round	Linear Relation	$\text{msb}_{15}(Z_1^{(i)})$	$\text{msb}_{15}(Z_5^{(i)})$
3	$(\gamma,\gamma,0,0) \stackrel{1r}{\rightarrow} (0,\gamma,\gamma,0)$	89–103	50–64
4	$(0,\gamma,\gamma,0) \stackrel{1r}{\rightarrow} (\gamma,0,\gamma,0)$	—	18–32
5	$(\gamma,0,\gamma,0) \stackrel{1r}{\rightarrow} (\gamma,\gamma,0,0)$	75–89	—
6	$(\gamma,\gamma,0,0) \stackrel{1r}{\rightarrow} (0,\gamma,\gamma,0)$	43–57	4–18
7	$(0,\gamma,\gamma,0) \stackrel{1r}{\rightarrow} (\gamma,0,\gamma,0)$	—	125–11
8	$(\gamma,0,\gamma,0) \stackrel{1r}{\rightarrow} (\gamma,\gamma,0,0)$	29–43	—
OT	$(\gamma,\gamma,0,0) \stackrel{1r}{\rightarrow} (\gamma,\gamma,0,0)$	22–36	—

For this weak-key class the output mask of all ciphertexts in a given structure is the same. Using this fact and about 100 structures, it is possible to identify if the unknown key is in the weak-key class.

The 97 key bits numbered 0–64, 75–103, 125–127 are zero. The remaining $128-97=31$ key bits can be arbitrary. The new weak-key class contains 2^{31} keys.

The membership tests require $2^{16}\cdot 100 \approx 2^{23}$ chosen plaintexts, and negligible computing time. The memory complexity is 2^{16} text blocks.

3.14.4 DL Analysis of MESH-64

The largest differential-linear distinguisher found for MESH-64 can cover 5 rounds, starting from the second round. Table 3.147 describes a differential-linear distinguisher combining a 2.5-round characteristic (holding with certainty) with a 2.5-round linear relation (holding with bias 2^{-1}), under weak-subkey assumptions. The characteristic and linear relation were chosen in order to minimize the number of weak subkeys needed.

The difference operator is bitwise exclusive-or and the difference $\nu = 8000_x$.

Note that the bitmask $(1,1,1,1)$ is applied after 2.5 rounds, where the output difference is $(\nu,0,0,0)$. Note that $\nu\cdot 1 = 0$.

Although weak subkeys are required for this distinguisher to hold, it is not known which user key(s) could actually lead to the eight subkeys in Table 3.147 to be weak, according to the key schedule algorithm of MESH-64. This question is left as an open problem.

Table 3.147 Differential-linear distinguisher for 5-round MESH-64.

Round i	Differential-linear relation	Weak subkeys
2	$(0,0,0,\nu) \stackrel{1r}{\to} (\nu,0,\nu,0)$	$Z_6^{(2)}$, $Z_7^{(2)}$
3	$(0,0,\nu,0) \stackrel{1r}{\to} (\nu,0,0,0)$	$Z_5^{(3)}$, $Z_6^{(3)}$
3.5	$(\nu,0,0,0) \stackrel{0.5r}{\to} (\nu,0,0,0)$	—
4	$(1,1,1,1) \stackrel{0.5r}{\to} (1,1,1,1)$	—
5	$(1,1,1,1) \stackrel{1r}{\to} (1,1,1,1)$	$Z_1^{(5)}$, $Z_4^{(5)}$
6	$(1,1,1,1) \stackrel{1r}{\to} (1,1,1,1)$	$Z_2^{(6)}$, $Z_3^{(6)}$

Assuming each weak subkey occurs with a probability of 2^{-15}, the estimated size of the weak-key class for which the differential-linear distinguisher holds is $2^{128-8\cdot 15} = 2^8$.

3.14.5 DL Analysis of MESH-96

The largest differential-linear distinguisher found for MESH-96 can cover 5 rounds, starting from the first round. Table 3.148 describes a differential-linear distinguisher combining a 1.5-round characteristic (holding with certainty) with a 3.5-round linear relation (holding with bias 2^{-1}), under weak-subkey assumptions. The characteristic and linear relation were chosen in order to minimize the number of weak subkeys needed.

The difference operator is bitwise exclusive-or and the difference $\nu = 8000_x$.

Table 3.148 Differential-linear distinguisher for 5-round MESH-96.

Round i	Differential-linear relation	Weak subkeys
1	$(\nu,0,0,\nu,0,0) \stackrel{1r}{\to} (\nu,\nu,0,0,0,0)$	$Z_1^{(1)}$
1.5	$(\nu,\nu,0,0,0,0) \stackrel{0.5r}{\to} (\nu,\nu,0,0,0,0)$	$Z_2^{(2)}$
2	$(1,1,1,1,1,1) \stackrel{0.5r}{\to} (1,1,1,1,1,1)$	—
3	$(1,1,1,1,1,1) \stackrel{1r}{\to} (1,1,1,1,1,1)$	$Z_1^{(3)}$, $Z_3^{(3)}$, $Z_5^{(3)}$
4	$(1,1,1,1,1,1) \stackrel{1r}{\to} (1,1,1,1,1,1)$	$Z_2^{(4)}$, $Z_4^{(4)}$, $Z_6^{(4)}$
5	$(1,1,1,1,1,1) \stackrel{1r}{\to} (1,1,1,1,1,1)$	$Z_1^{(5)}$, $Z_3^{(5)}$, $Z_5^{(5)}$

Note that the bitmask $(1,1,1,1,1,1)$ is applied after 1.5 rounds, where the output difference is $(\nu,\nu,0,0,0,0)$. Note that $\nu \cdot 1 = 0$.

Although weak subkeys are required for this distinguisher to hold, it is not known which user key(s) could actually lead to the eleven subkeys in Table 3.148 to be weak, according to the key schedule algorithm of MESH-96. This question is left as an open problem.

Assuming each weak subkey occurs with a probability of 2^{-15}, the estimated size of the weak-key class for which the differential-linear distinguisher holds is $2^{192-11\cdot 15} = 2^{27}$.

3.14.6 DL Analysis of MESH-128

The largest differential-linear distinguisher found for MESH-128 can cover 5 rounds, starting from the first round. Table 3.149 describes a differential-linear distinguisher combining a 1.5-round characteristic (holding with certainty) with a 3.5-round linear relation (holding with bias 2^{-1}), under weak-subkey assumptions. The characteristic and linear relation were chosen in order to minimize the number of weak subkeys needed.

The difference operator is bitwise exclusive-or and the difference $\nu = 8000_x$.

Table 3.149 Differential-linear distinguisher for 5-round MESH-128.

Round i	Differential-linear relation	Weak subkeys
1	$(\nu,0,0,0,\nu,0,0,0) \stackrel{1r}{\to} (\nu,\nu,0,0,0,0,0,0)$	$Z_1^{(1)}$
1.5	$(\nu,\nu,0,0,0,0,0,0) \stackrel{0.5r}{\to} (\nu,\nu,0,0,0,0,0,0)$	$Z_2^{(2)}$
2	$(1,1,1,1,1,1,1,1) \stackrel{0.5r}{\to} (1,1,1,1,1,1,1,1)$	—
3	$(1,1,1,1,1,1,1,1) \stackrel{1r}{\to} (1,1,1,1,1,1,1,1)$	$Z_1^{(3)}$, $Z_3^{(3)}$, $Z_6^{(3)}$, $Z_8^{(3)}$
4	$(1,1,1,1,1,1,1,1) \stackrel{1r}{\to} (1,1,1,1,1,1,1,1)$	$Z_2^{(4)}$, $Z_4^{(4)}$, $Z_5^{(4)}$, $Z_7^{(4)}$
5	$(1,1,1,1,1,1,1,1) \stackrel{1r}{\to} (1,1,1,1,1,1,1,1)$	$Z_1^{(5)}$, $Z_3^{(5)}$, $Z_6^{(5)}$, $Z_8^{(5)}$

Note that the bitmask $(1,1,1,1,1,1,1,1)$ is applied after 1.5 rounds, where the output difference is $(\nu,\nu,0,0,0,0,0,0)$. Note that $\nu \cdot 1 = 0$.

Although weak subkeys are required for this distinguisher to hold, it is not known which user key(s) could actually lead to the fourteen subkeys in Table 3.149 to be weak, according to the key schedule algorithm of MESH-128. Identifying these user keys is left as an open problem.

Assuming each weak subkey occurs with a probability of 2^{-15}, the estimated size of the weak-key class for which the differential-linear distinguisher holds is $2^{256-14\cdot 15} = 2^{46}$.

3.15 Square/Multiset Attacks

The Square attack was originally developed by Daemen *et al.* as a dedicated attack on the Square block cipher [58]. See Table 1.4 in Chap. 1 for the general parameters of this cipher.

This attack exploits the fact that the Square cipher has a byte-oriented design, that is, all its internal components are neatly organized bytewise, and all these operations are bijective mappings.

The Square attack was further extended and applied to several other modern block ciphers which also operate wordwise, such as, AES [71], SAFER [129], Skipjack [157], Misty1 [134] and IDEA [114].

The Square attack is also known as the *saturation* attack [127], the multiset attack [148], the integral cryptanalysis [107, 84] and the structural cryptanalysis [36].

The Square attack works in a chosen-plaintext setting, and shares similarities with the higher-order differential technique [115] in the sense that the former uses a kind of generalized (text) difference (under some group operation) as a distinguishing property. But, in [107], Knudsen and Wagner renamed the Square attack as *integral cryptanalysis*, setting it in a different framework from that of higher-order differentials.

A most prominent application of the Square attack is the result of Y. Todo on the full 8-round Misty1 block cipher [177].

The basic concepts in a Square analysis include the notion of a multiset, word status and integrals.

Definition 3.22. (Λ-set) [58]
A Λ-set is a multiset (a set where values can have multiplicities, that is, values can appear more than once) of 2^m n-bit text blocks. Each text block is partitioned into m-bit words (n is a multiple of m) which are either active, passive, balanced or garbled[16].

The choice of word size m is usually based on the target block cipher, which typically operates on well-defined m-bit words. For instance, $n = 128$ and $m = 8$ for the AES cipher, but $n = 64$ and $m = 16$ for IDEA.

Definition 3.23. (Integral) [107, 84]
Let x_i^j denote the j-th value of the m-bit word x_i in the i-th position (in left-to-right order) in the text blocks in a Λ-set. The sum

$$\otimes_{j=0}^{2^m-1} x_i^j$$

denotes the integral value of the word x_i. The operation $\otimes$ depends on the underlying algebraic group (in order to obtain an invariant sum). For instance, for $(\mathbb{Z}_2^{16}, \oplus)$ and $(\mathrm{GF}(2^{16}+1), \odot)$ the integral uses $\otimes = \oplus$. For $(\mathbb{Z}_{2^{16}}, \boxplus)$, the integral uses $\otimes = \boxplus$.

The integral is simply the sum of the values of a word in a fixed position in a text block, over a set of blocks (the Λ-set). The most convenient group operation for the sum depends on the group operation involved, but once set, the same group operation is used for the entire cipher.

[16] These word types are defined in the following paragraphs.

Definition 3.24. (Active word)
An m-bit word in a Λ-set that assumes all possible 2^m values exactly once (that is, a permutation of the 2^m values), is called *active* and denoted by the symbol A. Active words are also called *saturated.*

Active words contain just a permutation of 2^m values. Depending on the group operation used, their integral is invariant, that is, a constant.

Definition 3.25. (Passive word)
An m-bit word in a Λ-set that assumes a fixed (arbitrary) value, is called *passive* and denoted by P.

Passive words contain a fixed, constant value, and depending on the group operation used, their integral is invariant, that is, a constant.

Definition 3.26. (Balanced word)
An m-bit word in a Λ-set whose integral is a fixed, predictable value (depending on the underlying group operation) is called *balanced* and denoted by the symbol B. For instance, for $(\mathbb{Z}_2^{16}, \oplus)$ and $(\mathrm{GF}(2^{16}+1), \odot)$, the integral value for balanced words is 0. For $(\mathbb{Z}_{2^{16}}, \boxplus)$, the integral value is 2^{15}.

Unlike active and passive words (whose elements contain clearly identifiable patterns), balanced words can only be distinguished based on their integral value.

Definition 3.27. (Garbled word)
An m-bit word in a Λ-set, whose integral is unpredictable, is called *garbled* and denoted by the symbol ?.

An example may help clarify these concepts. Let $n = 15$, $m = 3$ and let $(\mathbb{Z}_2^m, \oplus)$ be the underlying algebraic group. Let $\Omega = \{(0,3,2,7,0),$ $(0,1,3,0,2)$, $(0,5,3,1,3)$, $(0,7,2,6,1)$, $(0,6,3,1,1)$, $(0,2,1,4,2)$, $(0,4,3,5,4)$, $(0,0,1,3,7)\}$ be a Λ-set where each n-bit text block contains $n/m = 5$ words.

We order words from left-to-right. The first (leftmost) 3-bit word in Ω is passive since this word position always contains the same constant value 0. The second word in Ω is active because it contains the values $\{3,1,5,7,6,2,4,0\}$, which represent a permutation of all possible 3-bit values. The order of these values in the text blocks does not matter, but their relative position (in each block) matters. The third word in Ω is balanced because its integral is $2 \oplus 3 \oplus 3 \oplus 2 \oplus 3 \oplus 1 \oplus 3 \oplus 1 = 0$. This is one peculiar type of balanced word (called *even*) where very value repeats an even number of times. The fourth word in Ω is garbled: its integral is $7 \oplus 0 \oplus 1 \oplus 6 \oplus 1 \oplus 4 \oplus 5 \oplus 3 = 3$. The fifth word in Ω can be represented as $\{0,2,3,1,1,2,4,7\}$ and is another kind of balanced word since $0 \oplus 2 \oplus 3 \oplus 1 \oplus 1 \oplus 2 \oplus 4 \oplus 7 = 0$. Notice, that though the integral is zero, its values do not appear in pairs. In summary, the Λ-set Ω can be represented by the type of each word: $(P, A, B, ?, B)$.

From their definition, active and passive words are necessarily balanced, but not the other way around. For instance, the first and second words in

Ω have integral equal to zero, but the third word does not contain either a permutation of m-bit values or a constant value.

Thus, the set of balanced words properly contain both the sets of active and passive words.

Let us study the propagation of each type of word across some usual internal cipher components.

Consider the mapping $Y = F(X, K) = X \oplus K$ for a fixed secret key K and variable X. From the definition, if X is an active word, then $X \oplus K$ still contains a permutation of values because K is fixed. The fact that K is secret means that Y contains an unknown permutation of the original values in X, and thus an active word. Therefore, active words propagate with certainty across F, independent of the unknown value K.

Likewise, if X is a passive word, then X contains a fixed, constant value. Therefore, $X \oplus K$ is still a constant (although probably a different constant), and so the passive word propagates across F with certainty, independent of the value of K.

Now, if X is a balanced word, neither active nor passive, then the integral (of 2^t values x_i) $\oplus_{i=1}^{2^t} x_i$ is a known quantity, for instance, zero. Consequently, the integral of the values in $X \oplus K$ will be $\oplus_{i=1}^{2^t}(x_i \oplus K) = \oplus_{i=1}^{2^t} K = 0$, independent of the value of K, since the same value K is xored an even number of times.

Now, consider $G(X, K) = S[X \oplus K]$, where K is a secret key and S is a bijective substitution box (S-box). If X is active, then $X \oplus K$ remains active (as described before), and $S[X \oplus K]$ will be a permutation of $X \oplus K$, since S is a bijective mapping. Therefore, $S[X \oplus K]$ remains active with certainty, although containing an unknown permutation of values from the original X.

Similarly, if X is passive, then $X \oplus K$ will remain passive. Further, $S[X \oplus K]$ will remain passive since the same value is input to S every time.

Finally, let X be balanced, but not active nor passive. Then, the integral (of 2^t values x_i) $\oplus_{i=1}^{2^t} x_i$ is a known quantity, for instance, zero. Further, for $X \oplus K$, the integral $\oplus_{i=1}^{2^t} x_i \oplus K = \oplus_{i=1}^{2^t} K = 0$, since K is a fixed value and exclusive-ored an even number of times. Now, if the values in $X \oplus K$ repeat in pairs (like in an *even* word). Then, each value $S[x_i \oplus K]$ will also appear in pairs (an even number of times). Therefore, the integral $\oplus_{i=1}^{2^t} S[x_i \oplus K] = 0$. Otherwise, the exact value of the integral $\oplus_{i=1}^{2^t} S[x_i \oplus K]$ will depend on the S-box.

Similar to DC and LC, a *multiset trail* consists of a sequence of balanced words across multiple components/rounds of a cipher.

Likewise, a *multiset distinguisher* can be characterized by Λ-sets covering multiple, consecutive rounds, where at least one word is balanced (in every round). In extreme cases, the word size can be reduced to a single bit.

The concepts of balanced words and integrals are fundamental in multiset distinguishers, just like differences in DC and linear relations in LC. When all words in a multiset degrade into garbled words, then, there is no more distinguishing capability in a multiset distinguisher. So, the existence of at least one balanced word delimits the beginning and the end of a multiset distinguisher.

Unlike in DC and LC, multiset distinguishers are not probabilistic, and they are not constructed by concatenating 1-round distinguishers, for instance, like in DC and LC. Rather, an r-round multiset distinguisher starts with a carefully chosen initial multiset which propagates freely across r rounds (of a target cipher) until there are no more balanced words in the Λ-set. Consequently, multiset distinguishers always hold with certainty.

Multiset distinguishers allow to identify a given block cipher from a set of random permutation (over the same plaintext space), in a so called *distinguish-from-random (DFR) attack.* This kind of distinguisher is a fundamental tool for key-recovery attacks that allow not only to extend the attacks to more rounds, but also to recover subkeys around the distinguisher. A terminology for key-recovery attacks is

Definition 3.28. (bR-attack)
An bR integral attack stands for an attack recovering key information from b rounds of a cipher. The usual strategy is to guess subkeys surrounding a t-round distinguisher, $t < b$, and use the integral property to identify the correct subkey from the wrong ones.

A generalization of the multiset, similar to the generalization of differences in [115], involves the concept of higher-order multisets.

Definition 3.29. (Higher-Order Multiset)
Let a multiset Ω consist of n-bit text blocks which can be partitioned into m-bit words, where n is a multiple of m. A higher-order multiset Ω^t (also containing n-bit text blocks) operates on words of tm bits, for some integer t.

Higher-order multisets exploit the fact that some cipher components typically group individual words into larger chunks. For instance, the MDS matrix is AES groups four bytes (of the cipher state) into a 32-bit word in order to perform matrix multiplication [71].

Thus, higher-order multisets exploit cipher operations that naturally merge smaller words into larger ones. The higher-order multisets usually requires a much higher data complexity than the original multiset attack.

An example of higher-order multiset attack is [69] on Rijndael, where $n = 128$, $m = 8$ and $t = 4$. Also, see the higher-order multisets in IDEA in [1].

The terminology used to denote a Λ-set from now on will simply use the status of the words in the Λ-set itself. For instance, $(A, B, ?, P)$ denotes a Λ-set whose words are active, balanced, garbled and passive (in left-to-right order).

Further, to denote the propagation of Λ-sets across rounds, the former will be simply separated by an arrow $\stackrel{1r}{\to}$ to indicate that the leftmost Λ-set leads to (or causes) the rightmost Λ-set after one full round of a given block cipher. This propagation across rounds holds with certainty.

For example, $(P, A, P, P) \stackrel{1r}{\to} (A, A, B, A)$ denotes that the input Λ-set (P, A, P, P) leads to the output Λ-set (A, A, B, A) across one full round with probability one.

Further details about the propagation of Λ-sets inside a full round will be provided depending on each cipher's internal structure. For instance, for PES and IDEA, the propagation of Λ-set words across the MA-box will be represented by its input/output words.

As an example, $(P, A, P, P) \stackrel{(P,A)\stackrel{\text{MA}}{\to}(A,A)}{\longrightarrow} (A, A, B, A)$ denotes the propagation of a Λ-set across 1-round IDEA, where $(P, A) \stackrel{\text{MA}}{\to} (A, A)$ denotes the propagation of the smaller Λ-set (P, A) across the MA-box.

A sequence of Λ-sets across consecutive rounds will provide a *multiset trail*, similar to differential and linear trails. As such, multiset trails indicate the path followed by balanced words across consecutive cipher components. The multiset trail ends when there are no more balanced words in the trail.

The notions of multiset trails and multiset distinguishers will sometimes be used interchangeably.

3.15.1 Square/Multiset Attacks on IDEA

The attacks in this section are based on the analyses of [153, 147].

For IDEA, multiset distinguishers can cover half-rounds because the former can either start or end in the middle of a round. Thus, the terminology $\stackrel{1r}{\to}$ may be replaced by $\stackrel{0.5r}{\to}$ to denote that a half-round is covered instead of a full round. If the distinguisher starts in the middle of a round, then it covers an MA half-round. If it ends in the middle of a round, then it covers a KM half-round.

Assuming integrals use the exclusive-or operator, $\oplus$, the behavior of the different kinds of words, A, P, B and ?, according to each operator in IDEA, is summarized in Table 3.150. All of these operators are binary. There are three tables, one for each operator: $\oplus$, $\boxplus$ and $\odot$. The leftmost column in each table indicates the status of one of the operands, while the topmost row indicates the status of the other operand. The tables are symmetric because these operators are commutative.

Table 3.150 Propagation of balanced/garbled words across the operators $\oplus$, $\boxplus$ and $\odot$.

$\oplus$	A	P	B	?
A	B	A	B	?
P	A	P	B	?
B	B	B	B	?
?	?	?	?	?

$\boxplus$	A	P	B	?
A	?	A	?	?
P	A	P	?	?
B	?	?	?	?
?	?	?	?	?

$\odot$	A	P	B	?
A	?	A	?	?
P	A	P	?	?
B	?	?	?	?
?	?	?	?	?

In the next paragraphs some arguments and counterexamples will be provided to support the content of Table 3.150.

Let $X = \{x_1, x_2, \ldots, x_{2^m}\}$ and $Y = \{y_1, y_2, \ldots, y_{2^m}\}$ denote two active m-bit words. If $Z = X \oplus Y$, then the integral $\oplus_{i=1}^{2^m} z_i = \oplus_{i=1}^{2^m}(x_i \oplus y_i) = (\oplus_{i=1}^{2^m} x_i) \oplus (\oplus_{i=1}^{2^m} y_i) = 0 \oplus 0 = 0$. Thus, Z is balanced, independent of how values in X are matched to values in Y. It may happen that the particular order in which each x_i is paired with y_i may lead to Z become passive, but it is not guaranteed. What can be stated in all cases is that Z will be balanced.

Now, let $Z = X \boxplus Y$. The integral becomes $\oplus_{i=1}^{2^m} z_i = \oplus_{i=1}^{2^m}(x_i \boxplus y_i) = (\oplus_{i=1}^{2^m} x_i) \oplus (\oplus_{i=1}^{2^m} y_i) \oplus (\oplus_{i=1}^{2^m} c_i) = 0 \oplus 0 \oplus \oplus_{i=1}^{2^m} c_i = \oplus_{i=1}^{2^m} c_i$, where $c_i = x_i \oplus y_i \oplus c_{i-1}$, with $c_0 = 0$, is the vector of carry bits in the modular addition of X and Y. The integral value in this case is unpredictable and depends on how each x_i is paired to each y_i. For instance, let $m = 3$, $X = Y = \{0, 1, 2, 3, 4, 5, 6, 7\}$. In this case, $Z = \{0, 2, 4, 6, 0, 2, 4, 6\}$, which means Z is balanced. If instead, $Y = \{7, 6, 5, 4, 3, 2, 1, 0\}$, then $Z = \{7, 7, 7, 7, 7, 7, 7, 7\}$, that is, a passive word. In general, though, for a random Y, such as $Y = \{2, 7, 1, 4, 6, 3, 0, 5\}$, the integral will be nonzero (and unpredictable), and Z will be garbled[17].

Finally, let $Z = X \odot Y$. The integral in this case is $\oplus_{i=1}^{2^m} z_i = (\oplus_{i=1}^{2^m} x_i) \odot (\oplus_{i=1}^{2^m} y_i)$, and the integral value depends on how the x_i are paired with the the y_i. For instance, let $X = \{0, 1, 2, 3\}$ and $Y = \{3, 2, 1, 0\}$ in GF(2^2+1)=GF(5), where $0 \equiv 2^2$. In this (exceptional) case, $Z = \{2, 2, 2, 2\}$, that is, a passive word. In general, the integral is unpredictable and Z becomes a garbled word.

Now, let X be active and Y be passive. If $Z = X \oplus Y$, then $\oplus_{i=1}^{2^m} z_i = (\oplus_{i=1}^{2^m}(x_i \oplus y_i) = \oplus_{i=1}^{2^m}(x_i \oplus y)$, where y is a constant. But, exclusive-oring a constant to a permutation of m-bit values just permutes it further. Thus, Z contains yet another permutation of the values in X, that is, Z is still an active word.

If $Z = X \boxplus Y$ then $\oplus_{i=1}^{2^m} z_i = \oplus_{i=1}^{2^m}(x_i \boxplus y_i) = \oplus_{i=1}^{2^m}(x_i \boxplus y)$. Similar to the previous paragraph, adding a constant y to a permutation of m-bit values, just leads to another permutation of the same values. Thus, Z is still active, although its values contain a different permutation of the values in X.

If $Z = X \odot Y$ then $\oplus_{i=1}^{2^m} z_i = \oplus_{i=1}^{2^m}(x_i \odot y_i) = \oplus_{i=1}^{2^m}(x_i \odot y)$ and once again, the same reasoning as in the previous paragraph applies: multiplying a constant to all elements of a permutation leads to another permutation. Thus, Z is still an active word.

Now, let X be active and Y be balanced (but not active nor passive). If $Z = X \oplus Y$, then $\oplus_{i=1}^{2^m} z_i = (\oplus_{i=1}^{2^m} x_i) \oplus (\oplus_{i=1}^{2^m} y_i) = 0 \oplus 0 = 0$. Thus, Z is balanced.

If $Z = X \boxplus Y$ then $\oplus_{i=1}^{2^m} z_i = \oplus_{i=1}^{2^m}(x_i \boxplus y_i)$ and the integral value will depend on how each x_i is paired to each y_i. For instance, let $X = \{1, 2, 3, 4, 5, 6, 7, 0\}$ and $Y = \{5, 6, 5, 6, 3, 2, 1, 0\}$. In this case (as well as in the general case), $Z = \{6, 0, 0, 2, 0, 0, 0, 0\}$ is garbled since the integral equals

[17] Although the reasoning is applied to small word sizes and to particular orderings of values inside the words, they exemplify the general case. Moreover, a single counterexample is enough to demonstrate that the integral value is unpredictable.

$6 \oplus 2 = 4$. In rare cases, though, such as if $Y = \{0, 1, 0, 1, 1, 0, 1, 0\}$, we have $Z = \{3, 2, 1, 0, 3, 2, 1, 0\}$ which is balanced.

If $Z = X \odot Y$ then $\oplus_{i=1}^{2^m} z_i = \oplus_{i=1}^{2^m}(x_i \odot y_i)$ and again the integral value will depend on how each x_i is paired to each y_i. For instance, let $X = \{0, 1, 2, 3, 4, 5, 6, 7, 8, 9, 10, 11, 12, 13, 14, 15\}$ and $Y = \{1, 2, 2, 1, 1, 1, 1, 2, 2, 1, 1, 1, 1, 1, 1, 1\}$ in GF($2^4 + 1$)=GF(17), where $0 \equiv 2^4$. Then, $Z = \{0, 2, 4, 3, 4, 5, 6, 14, 0, 9, 10, 11, 12, 13, 14, 15\}$, and the integral equals 4. In general, Z is a garbled word.

Now, let X be active and Y be garbled. If $Z = X \oplus Y$ then $\oplus_{i=1}^{2^m} z_i = (\oplus_{i=1}^{2^m} x_i) \oplus (\oplus_{i=1}^{2^m} y_i) = 0 \oplus (\oplus_{i=1}^{2^m} y_i)$, and the integral value is unpredictable since Y is garbled. Therefore, Z is garbled as well.

If $Z = X \boxplus Y$ then $\oplus_{i=1}^{2^m} z_i = \oplus_{i=1}^{2^m}(x_i \boxplus y_i)$ and the integral will depend on how each x_i is paired to each y_i. For instance, let $X = \{0, 1, 2, 3, 4, 5, 6, 7\}$ and $Y = \{2, 2, 0, 3, 1, 0, 3, 2\}$. Then, $Z = \{2, 3, 2, 6, 5, 5, 1, 1\}$, which is garbled. This is the general case. As an exceptional case, it is possible to have a garbled $Y = \{7, 6, 5, 4, 3, 2, 0, 7\}$ so that $Z = \{7, 7, 7, 7, 7, 7, 6, 6\}$, which is balanced.

If $Z = X \odot Y$ then $\oplus_{i=1}^{2^m} z_i = \oplus_{i=1}^{2^m}(x_i \odot y_i)$ and again the integral value will depend on how each x_i is paired to each y_i. In general, the result is a garbled word. For instance, let $X = \{0, 1, 2, 3\}$ and $Y = \{2, 2, 0, 3\}$ in GF($2^2 + 1$)=GF(5), where $0 \equiv 2^2$. Then, $Z = \{3, 2, 3, 4\}$, which is garbled.

Now, let X and Y be both passive words. If $Z = X \oplus Y$ then $\oplus_{i=1}^{2^m} z_i = (\oplus_{i=1}^{2^m} x_i) \oplus (\oplus_{i=1}^{2^m} y_i) = (\oplus_{i=1}^{2^m} x) \oplus (\oplus_{i=1}^{2^m} y)$, where x and y are constants. Therefore, Z contains only constants and is passive.

If $Z = X \boxplus Y$ then $\oplus_{i=1}^{2^m} z_i = \oplus_{i=1}^{2^m}(x_i \boxplus y_i) = \oplus_{i=1}^{2^m}(x \boxplus y)$, where x and y are constants. Therefore, Z contains only constants and is a passive word.

If $Z = X \odot Y$ then $\oplus_{i=1}^{2^m} z_i = \oplus_{i=1}^{2^m}(x_i \odot y_i) = \oplus_{i=1}^{2^m}(x \odot y)$, where x and y are constants. Therefore, Z contains only constants and is a passive word.

Now, let X be passive and Y be balanced (but not active nor passive). If $Z = X \oplus Y$ then $\oplus_{i=1}^{2^m} z_i = (\oplus_{i=1}^{2^m} x_i) \oplus (\oplus_{i=1}^{2^m} y_i) = 0 \oplus 0$, which means Z is a balanced word.

If $Z = X \boxplus Y$ then $\oplus_{i=1}^{2^m} z_i = \oplus_{i=1}^{2^m}(x_i \boxplus y_i)$. In general, the result is a garbled word due to carry bits in the modular addition. In exceptional cases, depending on how each x_i is paired to each y_i, the result may be balanced. For instance, let $X = \{1, 1, 1, 1, 1, 1, 1, 1\}$ and $Y = \{3, 2, 1, 0, 3, 2, 1, 0\}$. Then, $Z = \{4, 3, 2, 1, 4, 3, 2, 1\}$.

If $Z = X \odot Y$ then the integral is $\oplus_{i=1}^{2^m} z_i = \oplus_{i=1}^{2^m}(x_i \odot y_i) = \oplus_{i=1}^{2^m}(x \odot y_i)$, where x is a constant. In general, the result in a garbled word. It is not possible to factor the common constant x from the integral since $\odot$ does not commute with $\oplus$. As an example let $X = \{3, 3, 3, 3, 3, 3, 3, 3, 3, 3, 3, 3, 3, 3, 3, 3\}$ and $Y = \{3, 2, 2, 1, 2, 1, 1, 1, 1, 1, 1, 1, 1, 1, 1, 0\}$ in GF($2^4 + 1$) with $2^4 \equiv 0$. Then, $Z = \{9, 6, 6, 3, 6, 3, 3, 3, 3, 3, 3, 3, 3, 3, 3, 14\}$, which is garbled since the integral equals 2.

Now, let X be passive and Y be garbled. If $Z = X \oplus Y$ then $\oplus_{i=1}^{2^m} z_i = (\oplus_{i=1}^{2^m} x) \oplus (\oplus_{i=1}^{2^m} y_i) = \oplus_{i=1}^{2^m} y_i$, whose result is unpredictable since Y is garbled. Thus, Z is garbled as well.

If $Z = X \boxplus Y$ then $\oplus_{i=1}^{2^m} z_i = \oplus_{i=1}^{2^m}(x_i \boxplus y_i) = \oplus_{i=1}^{2^m}(x \boxplus y_i)$, where x is a constant. In general, the result is a garbled word due to carry bits in the modular addition. For instance, let $X = \{1,1,1,1,1,1,1,1\}$ and $Y = \{3,2,1,0,4,3,2,0\}$. Then, $Z = \{4,3,2,1,5,4,3,1\}$, whose integral equals $5 \oplus 2 = 7$.

If $Z = X \odot Y$ then the integral is $\oplus_{i=1}^{2^m} z_i = \oplus_{i=1}^{2^m}(x \odot y_i)$, where x is a constant. In general, the result is a garbled word.

Now, let X and Y be both balanced (but not active nor passive) words. If $Z = X \oplus Y$ then $\oplus_{i=1}^{2^m} z_i = \oplus_{i=1}^{2^m} x_i \oplus \oplus_{i=1}^{2^m} y_i = 0 \oplus 0$. Thus, Z is also balanced. In exceptional cases, it may happen that the result is passive such as when $X = \{0,1,2,3,0,1,2,3\}$ and $Y = \{3,2,1,0,3,2,1,0\}$. Other examples can be constructed by appropriately pairing x_i values with y_i values.

If $Z = X \boxplus Y$ then $\oplus_{i=1}^{2^m} z_i = \oplus_{i=1}^{2^m}(x_i \boxplus y_i)$. The result is usually a garbled word since $\oplus$ and $\boxplus$ are not associative: $\oplus_{i=1}^{2^m}(x_i \boxplus y_i) \neq (\oplus_{i=1}^{2^m} x_i) \boxplus (\oplus_{i=1}^{2^m} y_i)$. In exceptional cases, by pairing x_i with y_i values appropriately, the result may be balanced. For instance, if $X = \{0,1,2,3,0,1,2,3\}$ and $Y = \{3,2,1,0,3,2,1,0\}$, then $Z = \{3,3,3,3,3,3,3,3\}$.

If $Z = X \odot Y$ then the integral is $\oplus_{i=1}^{2^m} z_i = \oplus_{i=1}^{2^m}(x_i \odot y_i)$. The integral is usually a garbled word $\oplus_{i=1}^{2^m}(x_i \odot y_i) \neq (\oplus_{i=1}^{2^m} x_i) \odot (\oplus_{i=1}^{2^m} y_i)$.

Now, let X be balanced and Y be garbled. If $Z = X \oplus Y$ then $\oplus_{i=1}^{2^m} z_i = (\oplus_{i=1}^{2^m} x_i) \oplus (\oplus_{i=1}^{2^m} y_i) = 0 \oplus \oplus_{i=1}^{2^m} y_i$, which means Z is garbled.

If $Z = X \boxplus Y$ then $\oplus_{i=1}^{2^m} z_i = \oplus_{i=1}^{2^m}(x_i \boxplus y_i)$. In general, the result will be garbled due to the carry bits in modular addition. Even if it were possible to split the integral in two parts, for X and Y values, separately, Y is already garbled.

If $Z = X \odot Y$ then the integral is $\oplus_{i=1}^{2^m} z_i = \oplus_{i=1}^{2^m}(x_i \odot y_i)$. In general, this integral value will be unpredictable and Z will be garbled. Even if it were possible to split this integral in two parts, for X and Y, separately, Y is already garbled.

Now, let both X and Y be garbled. If $Z = X \oplus Y$ then $\oplus_{i=1}^{2^m} z_i = (\oplus_{i=1}^{2^m} x_i) \oplus (\oplus_{i=1}^{2^m} y_i)$, which is the exclusive-or of two garbled words. Depending on how the x_i are paired with the y_i, the integral value can be any value in the range $[0, \ldots, 2^m - 1]$. In general, the result is garbled.

If $Z = X \boxplus Y$ then $\oplus_{i=1}^{2^m} z_i = \oplus_{i=1}^{2^m}(x_i \boxplus y_i)$. In general, the result will be garbled. In exceptional cases, depending on how x_i values are paired with y_i, the result may be balanced. For instance, let $X = \{0,1,2,1,3,4,1,2\}$ and $Y = \{2,1,0,1,1,0,1,0\}$. Then, $Z = \{2,2,2,2,4,4,2,2\}$. This example demonstrate that even out of garbled words, structured words can emerge. But, this is very exceptional, and the order of values in the Λ-set is usually key dependent.

Finally, if $Z = X \odot Y$ then the integral is $\oplus_{i=1}^{2^m} z_i = \oplus_{i=1}^{2^m}(x_i \odot y_i)$. Again, in general, Z will be garbled, just as in the cases where only one of X and Y were garbled.

The construction of multiset distinguishers usually follows a bottom-up approach: from the smallest components up to larger ones. Table 3.150 showed the propagation of Λ-sets across a single binary operator: $\oplus$, $\boxplus$ and $\odot$.

In Lai-Massey cipher designs such as in IDEA, the next larger component to study is the MA-box. There are not many Λ-sets (with at least one balanced word) that can propagate across an MA-box and still retain a balanced word at the output. Table 3.151 lists all possible Λ-sets with at least one balanced word at the input.

Table 3.151 Propagation of multisets across the MA-box of IDEA.

$(P,P) \overset{\text{MA}}{\to} (P,P)$	$(P,A) \overset{\text{MA}}{\to} (A,A)$
$(A,P) \overset{\text{MA}}{\to} (?,A)$	$(A,A) \overset{\text{MA}}{\to} (?,?)$
$(P,B) \overset{\text{MA}}{\to} (?,?)$	$(B,P) \overset{\text{MA}}{\to} (?,?)$
$(A,B) \overset{\text{MA}}{\to} (?,?)$	$(B,A) \overset{\text{MA}}{\to} (?,?)$
$(B,B) \overset{\text{MA}}{\to} (?,?)$	$(P,?) \overset{\text{MA}}{\to} (?,?)$
$(?,P) \overset{\text{MA}}{\to} (?,?)$	$(A,?) \overset{\text{MA}}{\to} (?,?)$
$(?,A) \overset{\text{MA}}{\to} (?,?)$	$(B,?) \overset{\text{MA}}{\to} (?,?)$
$(?,B) \overset{\text{MA}}{\to} (?,?)$	

Table 3.151 shows that the MA-box of IDEA is effective in destroying the invariant property for the majority of the Λ-sets. The small set of balanced word patterns that can cross the MA-box serve to further construct 1-round multiset distinguishers.

An example of a multiset distinguisher covering 1.5-round IDEA, using 16-bit words and exclusive-or as integral operator, is the following:

$$(P,A,P,P) \overset{(P,A)\overset{\text{MA}}{\to}(A,A)}{\to} (A,A,B,A) \overset{(?,B)\overset{\text{MA}}{\to}((?,?)}{\to} (?,?,?,?). \quad (3.179)$$

Although, there is a balanced word at the input to the second MA-box, there is no balanced word at the end of the second round because the MA-box output becomes $(?,?)$. Due to this fact, (3.179) becomes a 1.5-round multiset distinguisher.

A dual 1.5-round multiset distinguisher to (3.179) is:

$$(P,P,P,A) \overset{(P,A)\overset{\text{MA}}{\to}(A,A)}{\to} (A,A,A,B) \overset{(B,?)\overset{\text{MA}}{\to}((?,?)}{\to} (?,?,?,?). \quad (3.180)$$

In [1], Akgun *et al.* did extensive analysis of IDEA and found (higher-order) square distinguishers on 2-, 3-, 4- and 5-round IDEA variants (the 5-round distinguisher requires some weak-subkey assumptions).

An attack on 2.5-round IDEA, using (3.179), can recover subkeys ($Z_3^{(3)}$, $Z_4^{(3)}$) using the fact that the rightmost input to the second MA-box is balanced.

Let $C_i = (c_1^i, c_2^i, c_3^i, c_4^i)$ denote the i-th ciphertext corresponding to a plaintext in a Λ-set. Thus,

$$(c_3^i \boxminus z_{3,3}) \oplus (c_4^i \odot (z_{4,3})^{-1}), \tag{3.181}$$

should be balanced for the correct subkey value. $z_{3,3}$ and $z_{4,3}$ are candidate values for $Z_3^{(3)}$ and $Z_4^{(3)}$. Expression (3.181) represents a 16-bit condition. The most significant bit of $Z_3^{(3)}$ cannot be uniquely identified, though. Thus, we recover 31 subkey bits.

Using two Λ-sets, the number of wrong subkeys surviving the test (3.181) is $2^{31}/(2^{16})^2 < 1$, and the correct subkey value can be uniquely identified.

A single computation of (3.181) costs one $\odot$, one $\oplus$ and one $\boxminus$. Using two Λ-sets means repeating the computation of (3.181) $2 \cdot 2^{16} = 2^{17}$ times. Assuming an $\odot$ costs about three times a $\oplus$, and that the latter costs about the same as a $\boxminus$, then the estimated cost of (3.181) over two Λ-sets is $2^{17} \cdot 5$ $\oplus$ operations. Similarly, a 2.5-round IDEA computation costs ten $\odot$, ten $\boxplus$ and twelve $\oplus$, or about 52 $\oplus$. Thus, this part of the attack costs about $2^{17} \cdot 5/52 \approx 2^{14}$ 2.5-round IDEA encryptions. The data complexity is $2 \cdot 2^{16} = 2^{17}$ chosen plaintexts. The memory complexity is 2^{16} text blocks since the same memory can be used to store each Λ-set separately.

The dual multiset distinguisher (3.180) can be used to recover two additional subkeys, $(Z_1^{(3)}, Z_2^{(3)})$, except for the most significant bit of $Z_2^{(3)}$, by testing the expression

$$(c_1^i \odot (z_{1,3})^{-1}) \oplus (c_2^i \boxminus z_{2,3}), \tag{3.182}$$

which should be a balanced word according to the leftmost input to the second MA-box in (3.180). $z_{1,3}$ and $z_{2,3}$ are candidate values for $Z_1^{(3)}$ and $Z_2^{(3)}$. This expression represents a 16-bit condition.

Using two Λ-sets is enough to identify 31 subkey bits. The complexity of this step is the same as the previous step to recover 31 bits of $(Z_3^{(3)}, Z_4^{(3)})$, that is, about 2^{14} 2.5-round IDEA encryptions.

The combined effort from the previous steps is about 2^{15} 2.5-round IDEA encryptions.

Let $Y_i = (y_{1,i}, y_{2,i}, y_{3,i}, y_{4,i})$ denote the output block of the i-th round.

Once we obtain $(Z_1^{(3)}, Z_4^{(3)})$ and the fifteen least significant bits of $Z_2^{(3)}$ and $Z_3^{(3)}$, we can partially decrypt the last KM half-round, and obtain the Y_2 text blocks over the Λ-sets.

Further, we can recover $(Z_5^{(2)}, Z_6^{(2)})$ and the most significant bits of $Z_2^{(3)}$ and $Z_3^{(3)}$ by decrypting the Y_2 blocks over the Λ-sets. The correct subkey value should lead to active words after the second KM half-round in all Λ-sets used.

In total, we guess 34 subkey bits. Each word in Y_2 gives a 16-bit condition. We can reuse the Λ-sets we used in the previous step since we recover at most 16 subkey bits at a time.

Using the key schedule of IDEA, $Z_1^{(3)}$ corresponds to bit 89–104 of the user key; $Z_2^{(3)}$ corresponds to bits 105–120; $Z_3^{(3)}$ corresponds to bits 121–8; $Z_4^{(3)}$ to bits 9–24; $Z_5^{(2)}$ to bits 57–72 and $Z_6^{(2)}$ to bits 73–88. In total, they account for 96 bits of the user key, numbered 0–24 and 57–127.

The remaining $128 - 96 = 32$ bits can be found by exhaustive search. The final attack complexity is 2^{32} 2.5-round encryptions. The data complexity is $3 \cdot 2^{16} = 2^{17.58}$ chosen plaintexts. The memory complexity is 2^{16} text blocks.

3.15.1.1 An Alternative Square/Multiset Attack on IDEA

Consider the following 1.5-round multiset distinguisher for IDEA:

$$(A, P, A, P) \overset{(P,P)\overset{\text{MA}}{\to}(P,P)}{\longrightarrow} (A, A, P, P) \overset{(A,A)\overset{\text{MA}}{\to}((?,?)}{\longrightarrow} (?, ?, ?, ?). \quad (3.183)$$

Although (3.183) is depicted across two rounds, balanced words can only be traced up to 1.5 rounds. In this distinguisher, the active words must contain the same permutation, so that the input to the first MA-box becomes (P, P).

A dual 1.5-round multiset distinguisher is:

$$(P, A, P, A) \overset{(P,P)\overset{\text{MA}}{\to}(P,P)}{\longrightarrow} (P, P, A, A) \overset{(A,A)\overset{\text{MA}}{\to}((?,?)}{\longrightarrow} (?, ?, ?, ?). \quad (3.184)$$

Although there are two active words in the beginning of both (3.183) and (3.184) the Λ-sets require only 2^{16} chosen plaintexts.

We use both distinguishers to attack 2.5-round IDEA. The approach is the following: (i) instead of independent active words, we input the same permutation in both active entries of the distinguisher; (ii) in (3.183), we recover $(Z_1^{(1)}, Z_3^{(1)})$. When the guessed subkey values are correct, then the active words (which contain the same permutation) will cancel out at the input to the MA-box, and result in a passive word. When this collision happens, the input to the first MA-box will have the form (P, P). Otherwise, the active words combining to form the input the MA-box will not contain the same permutation, and the result will have the form (B, P). In this case, the output from the MA-box will have the form $(?, ?)$, and the distinguisher will not be satisfied.

The attack continues at the end of 2.5-round IDEA: we recover $(Z_1^{(3)}, Z_2^{(3)})$ and $(Z_3^{(3)}, Z_4^{(3)})$ separately, by checking if

$$(c_1^i \odot (z_{1,3})^{-1}) \oplus (c_2^i \boxminus z_{2,3}), \quad (3.185)$$

and

$$(c_3^i \boxminus z_{3,3}) \oplus (c_4^i \odot (z_{4,3})^{-1}), \qquad (3.186)$$

are both active words, according to the pattern (A, A) at the input to the second MA-box in both distinguishers. $(z_{1,3},\ z_{2,3},\ z_{3,3},\ z_{4,3})$ are candidate values for $(Z_1^{(3)},\ Z_2^{(3)},\ Z_3^{(3)},\ Z_4^{(3)})$.

Expressions (3.185) and (3.186) are both 16-bit conditions.

The most significant bits of $Z_2^{(3)}$ and $Z_3^{(3)}$ cannot be uniquely identified.

The data complexity is $4 \cdot 2^{16} \cdot 2^{32} = 2^{50}$ chosen plaintexts because the guessed values of $(Z_1^{(1)}, Z_3^{(1)})$ are embedded in the initial Λ-set. The memory complexity is 2^{16} text blocks, since the same memory can be used to store each Λ-set separately.

We recover 31 bits from $(Z_1^{(3)}, Z_2^{(3)})$, as well as 31 bits from $(Z_1^{(1)}, Z_3^{(1)})$. The effort to compute (3.185) is one $\odot$, one $\oplus$ and one $\boxminus$, which is about five $\oplus$. The effort to compute 2.5-round IDEA is about 52 $\oplus$.

Assuming that one $\odot$ costs about 3 $\oplus$, then the cost of this step is $4 \cdot 2^{16}(2^{31} \cdot 4 + 5 \cdot 2^{16})/52 \approx 2^{43.29}$ 2.5-round computations. The number of wrong subkey candidates surviving after four Λ-sets is $2^{62}/(2^{16})^4 < 1$.

The time complexity for the attack using (3.186) is smaller since $(Z_1^{(1)},\ Z_3^{(1)})$ was already recovered.

The dual distinguisher (3.184) can be used to further recover $(Z_2^{(1)},\ Z_4^{(1)})$. The attack effort is also $2^{43.29}$ 2.5-round computations.

Using the key schedule of IDEA, $Z_1^{(1)}$ corresponds to bits 0–15 of the user key; $Z_2^{(1)}$ corresponds to bits 16–31; $Z_3^{(1)}$ corresponds to bits 32–47; $Z_4^{(1)}$ corresponds to bits 48–63; $Z_1^{(3)}$ corresponds to bits 89-104; the most significant fifteen bits of $Z_2^{(3)}$ correspond to bits 106–120; the most significant fifteen bits of $Z_3^{(3)}$ correspond to bits 122–8; $Z_4^{(3)}$ correspond to bits 9–24. In summary, we recover 101 user key bits, taking account of all bits that overlap. The remaining $128 - 101 = 27$ bits can be recovered by exhaustive search. The final time complexity is $2^{44.29}$ 2.5-round IDEA encryptions.

Higher-order square/multiset attacks, as reported in [147] do not reach more rounds nor have lower attack complexities than the square attacks described in this Section.

3.15.1.2 Square-Like Attacks on IDEA

The attacks described this section are based on [60].

Let $X^{(i)} = (X_1^{(i)}, X_2^{(i)}, X_3^{(i)}, X_4^{(i)})$ denote the i-th round input to IDEA. For $i = 1$, $X^{(1)}$ is the plaintext block. For $i > 1$, $X^{(i)}$ denotes also the $(i-1)$-th output block.

Let $Y^{(i)} = (Y_1^{(i)}, Y_2^{(i)}, Y_3^{(i)}, Y_4^{(i)})$ denote the output of the i-th KM half-round. For $i = 9$, $Y^{(9)}$ denotes the output of the full 8.5-round IDEA.

Lemma 3.4. *[60] Let (n_i, q_i) denote the inputs to the i-th MA-box of IDEA and let (r_i, s_i) denote the corresponding output, for $1 \leq i \leq 8$. If n_i is fixed (a passive word) and q_i is active, then both r_i and s_i are active.*

Proof. If n_i is passive and q_i is active, then $s_i = (n_i \odot Z_5^{(i)} \boxplus q_i) \odot Z_6^{(i)}$ is active, assuming $Z_5^{(i)}$ and $Z_6^{(i)}$ are fixed. Likewise, $r_i = n_i \odot Z_5^{(i)} \boxplus s_i$ is also passive, since s_i is active.

Corollary 3.1. *[60] Consider a Λ-set of the form (P, P, P, A), that is, the first three words are passive (fixed, arbitrary values) while the fourth word is active. Encrypt this Λ-set through 1- and 1.5-round IDEA. The ciphertext Λ-sets have the form (A, A, A, B) and $(A, A, A, ?)$, respectively.*

The proof is in [60].

The distribution of values in Corollary 3.1 can be used as a distinguisher of 1- and 1.5-round IDEA from a random permutation. The probability of such a distribution of values in a random permutation is $\frac{2^{16}!}{(2^{16})^{2^{16}}} \approx 2^{-281720}$.

Lemma 3.5. *[60] $lsb_1(r_i \oplus s_i) = lsb_1(n_i \odot Z_5^{(i)})$, for $1 \leq i \leq 8$.*

The proof is in [60].

Corollary 3.2. *[60] Consider a Λ-set of 2^{16} plaintext blocks of the form (P, P, P, A). Encrypt this Λ-set across 1-round IDEA. Then, the output Λ-set consists of text blocks $X^{(2)}$ satisfying: $X_1^{(2)} \oplus X_2^{(2)}$ and $lsb_1(X_2^{(2)} \oplus X_3^{(2)})$ are constants.*

Proof. Since $n_1 = (X_1^{(1)} \odot Z_1^{(1)}) \oplus (X_3^{(1)} \boxplus Z_3^{(1)})$, the leftmost input to the MA-box is a passive word. By Lemma 3.5, the least significant bit of the exclusive-or of the MA-box outputs is also passive. But, $X_2^{(2)} \oplus X_3^{(2)} = (X_3^{(1)} \boxplus Z_3^{(1)}) \oplus r_1 \oplus (X_2^{(1)} \boxplus Z_2^{(1)}) \oplus s_1$. Since the least significant bit of modular addition is not affected by carry bis, and the round subkeys are fixed, it follows that $\text{lsb}_1(X_2^{(2)} \oplus X_3^{(2)})$ is a constant.

Lemma 3.6. *[60] Consider a Λ-set of the form (P, P, P, A). Encrypt it through 2-round IDEA. Then, the output Λ-set consists of text blocks $X^{(3)}$ such that $lsb_1(X_2^{(3)} \oplus X_3^{(3)} \oplus Z_5^{(2)} \odot (X_1^{(3)} \oplus X_2^{(3)}))$ is constant.*

Proof. In the second round, $X_2^{(3)} = (X_3^{(2)} \boxplus Z_3^{(2)}) \oplus s_2$ and $X_3^{(3)} = (X_2^{(2)} \boxplus Z_2^{(2)}) \oplus r_2$. Then, $\text{lsb}_1(X_2^{(3)} \oplus s_2 \oplus X_3^{(3)} \oplus r_2) = \text{lsb}_1((X_3^{(2)} \boxplus Z_3^{(2)}) \oplus (X_2^{(2)} \boxplus Z_2^{(2)}))$. By Lemma 3.5, $\text{lsb}_1(r_2 \oplus s_2) = \text{lsb}_1(Z_5^{(2)} \odot (X_1^{(3)} \oplus X_2^{(3)}))$ and we have $\text{lsb}_1(X_2^{(3)} \oplus X_3^{(3)} \oplus Z_5^{(2)} \odot (X_1^{(3)} \oplus X_2^{(3)})) = \text{lsb}_1(X_2^{(2)} \oplus X_3^{(2)} \oplus Z_2^{(2)} \oplus Z_3^{(2)})$. Since the first three words of the input Λ-set are passive, by Corollary 3.2, $\text{lsb}_1(X_2^{(2)} \oplus X_3^{(2)})$ is constant and the result follows.

Another consequence of Lemma 3.5 is the following

Corollary 3.3. *[60] Consider a Λ-set of the form (P, A, P, P). Encrypt it through 1-round IDEA. Then, the output Λ-set consisting of text blocks $X^{(2)}$ satisfies: $lsb_1(X_2^{(2)} \oplus X_3^{(2)})$ takes the same value for all plaintexts such that $lsb_1(X_2^{(1)}) = 0$, and takes the complementary value for $lsb_1(X_2^{(1)}) = 1$.*

Proof. By Lemma 3.5, $r_1 \oplus s_1$ is a constant. But, $X_2^{(2)} \oplus X_3^{(2)} = (X_3^{(1)} \boxplus Z_3^{(1)}) \oplus s_1 \oplus (X_2^{(1)} \boxplus Z_2^{(1)}) \oplus r_1$. Since the subkeys are fixed and the least significant bit of modular addition is not affected by carry bits, $\text{lsb}_1(X_2^{(2)} \oplus X_3^{(2)})$ depends only of $\text{lsb}_1(X_2^{(1)})$, and the result follows.

Lemma 3.6 leads to an attack on 2-round IDEA. The correct $Z_5^{(2)}$ satisfies the 1-bit condition

$$\text{lsb}_1(X_2^{(3)} \oplus X_3^{(3)} \oplus Z_5^{(2)} \odot (X_1^{(3)} \oplus X_2^{(3)})) = \text{constant},$$

for any given Λ-set, whereas the wrong subkey candidates will give random values. Since $Z_5^{(2)}$ has 16 bits, 16 Λ-set will be needed. From [117], if $x, y \notin \{0, 1\}$, then $(x \odot y) + (2^{16} + 1 - y) = 2^{16} + 1$. Therefore, for $x, y \notin \{0, 1\}$, $\text{lsb}_1(x \odot y) = \text{lsb}_1(x \odot (2^{16} + 1 - y)) \oplus 1$. Consequently, for any value of $Z_5^{(2)}$, there is a dual subkey $2^{16} + 1 - Z_5^{(2)}$ that is indistinguishable from the correct value by this multiset attack. That means that the integral over a Λ-set cannot distinguish between $Z_5^{(2)}$ and $2^{16} + 1 - Z_5^{(2)}$.

If $Z_5^{(2)} \notin \{0, 1\}$, the attack eliminates all subkey candidates except $Z_5^{(2)}$ and $2^{16}+1-Z_5^{(2)}$. If $Z_5^{(2)} \in \{0, 1\}$, the attack eliminates all subkey candidates except 0 and 1. Therefore, it is enough to search half of the subkey space, but in the end, we will have two candidates for $Z_5^{(2)}$.

The effort to compute the 1-bit condition (once) is roughly equivalent to $\frac{7}{24}$ of 1-round IDEA, assuming the 1-bit condition requires one $\odot$ and three $\oplus$, which is approximated as seven $\oplus$. In comparison, 1-round IDEA requires four $\odot$, four $\boxplus$ and six $\oplus$, which is approximated as 22 $\oplus$.

We can use Corollary 3.2 to find $Z_6^{(2)}$, $Z_1^{(2)}$, $Z_2^{(2)}$ and $Z_5^{(2)}$. Since the first inputs to the first MA-box should be equal for all Λ-sets, we can guess the 48 bits of $Z_6^{(2)}$, $Z_1^{(2)}$, $Z_2^{(2)}$ (assuming the recovered $Z_5^{(2)}$). The first input to the MA-box gives a 16-bit condition. Thus, three Λ-sets will be needed, which means $3 \cdot 2^{16}$ chosen plaintexts. The attack effort is roughly equivalent to $2^{48} \cdot 3 \cdot 2^{16} \cdot \frac{7}{44} \approx 2^{63}$ 2-round IDEA computations.

In order to extend the attack to 2.5-round IDEA, consider the exclusive-or of the least significant bits of the two outputs of an MA-box: $\text{lsb}_1(X_2^{(3)} \oplus Z_2^{(3)} \oplus X_3^{(3)} \oplus Z_3^{(3)})$.

For any ciphertext Λ-set, there are two possible values for $\text{lsb}_1(X_2^{(3)} \oplus X_3^{(3)})$, one of which is the bitwise complement of the other.

We can also calculate $X_1^{(3)}$ and $X_2^{(3)}$ by guessing $Z_1^{(3)}$ and $Z_2^{(3)}$. Further, we can check if $\text{lsb}_1(X_2^{(3)} \oplus X_3^{(3)} \oplus Z_5^{(2)} \odot (X_1^{(3)} \oplus X_2^{(3)}))$ is constant, which serves as a filter to eliminate the wrong subkey values. In total, 48 subkey bits are recovered: $Z_1^{(3)}$, $Z_2^{(3)}$ and $Z_5^{(2)}$. The attack requires 48 Λ-sets, which means $48 \cdot 2^{16}$ chosen plaintexts. The attack effort is $48 \cdot 2^{16} \cdot \frac{7}{56} \approx 2^{18}$ 2.5-round IDEA computations (assuming 2.5-round IDEA costs 56 $\oplus$). The remaining $128 - 48 = 80$ key bits can be recovered by exhaustive key search. The final attack complexity becomes 2^{80} 2.5-round IDEA computations.

In order to extend the attack to 3-round IDEA, we need to guess $Z_5^{(3)}$ and $Z_6^{(3)}$. According to the key schedule of IDEA, $(Z_5^{(3)}, Z_6^{(3)})$ contains the key bits of $Z_5^{(2)}$. So, we need to guess 64 key bits, numbered 51-82, 90-105 and 106-121. From the 1-bit condition, we will need 64 Λ-sets, which gives a data complexity of $64 \cdot 2^{16} = 2^{20}$ chosen plaintexts.

The attack effort is $2^{20} \cdot \frac{7}{72} \approx 2^{18}$ 3-round IDEA computations (assuming 3-round IDEA costs 72 $\oplus$). The remaining 128 - 64 = 64 key bits can be recovered by exhaustive key search, and the final attack effort becomes 2^{64} 4-round IDEA computations.

For attacking 3.5-round IDEA, we need to additionally guess $Z_1^{(4)}$, $Z_2^{(4)}$, $Z_3^{(4)}$ and $Z_4^{(4)}$ in the last KM half-round. From the key schedule of IDEA, this means a total of 96 key bits, numbered 51-127 and 0-18. The data complexity is 96 Λ-sets or $96 \cdot 2^{16} = 2^{22.58}$ chosen plaintexts.

The attack effort is $96 \cdot 2^{16} \cdot \frac{7}{80} \approx 2^{19}$ 3.5-round IDEA computations (assuming 3.5-round IDEA costs 80 $\oplus$). The remaining 128 - 80 - 48 key bits can be recovered by exhaustive key search. The final attack effort is 2^{48} 3.5-round IDEA computations.

3.15.2 Square/Multiset Attacks on MESH-64

A square/multiset attack on MESH-64 can be done in a bottom-up fashion, in a similar way as done for IDEA. Table 3.150 showed the propagation of word patterns across individual operators such as $\oplus$, $\odot$ and $\boxplus$.

Further, Table 3.152 shows the propagation of word patterns across a bigger component: the MA-box of MESH-64, in a similar way as Table 3.151 showed it for the MA-box of IDEA. Comparatively, fewer multiset patterns can cross the MA-box of MESH-64 with at least one balanced word at the output than across the MA-box of IDEA. The reason for this behavior is the longer MA-box of MESH-64.

Proceeding with the bottom-up approach, the longest multiset distinguisher found for MESH-64 covers 1.5 rounds, for instance:

$$(A, P, A, P) \overset{(P,P)\overset{\text{MA}}{\to}(P,P)}{\longrightarrow} (A, A, P, P) \overset{(A,A)\overset{\text{MA}}{\to}((?,?)}{\longrightarrow} (?, ?, ?, ?). \quad (3.187)$$

The reasoning for (3.187) is the same as for (3.183): the initial active words must contain the same permutation, so that the input to the first MA-box becomes (P, P).

Table 3.152 Propagation of multisets across the MA-box of MESH-64.

$(P,P) \stackrel{\text{MA}}{\to} (P,P)$	$(P,A) \stackrel{\text{MA}}{\to} (A,?)$
$(A,P) \stackrel{\text{MA}}{\to} (?,?)$	$(A,A) \stackrel{\text{MA}}{\to} (?,?)$
$(P,B) \stackrel{\text{MA}}{\to} (?,?)$	$(B,P) \stackrel{\text{MA}}{\to} (?,?)$
$(A,B) \stackrel{\text{MA}}{\to} (?,?)$	$(B,A) \stackrel{\text{MA}}{\to} (?,?)$
$(B,B) \stackrel{\text{MA}}{\to} (?,?)$	$(P,?) \stackrel{\text{MA}}{\to} (?,?)$
$(?,P) \stackrel{\text{MA}}{\to} (?,?)$	$(A,?) \stackrel{\text{MA}}{\to} (?,?)$
$(?,A) \stackrel{\text{MA}}{\to} (?,?)$	$(B,?) \stackrel{\text{MA}}{\to} (?,?)$
$(?,B) \stackrel{\text{MA}}{\to} (?,?)$	

A dual 1.5-round multiset distinguisher is

$$(P,A,P,A) \stackrel{(P,P)\stackrel{\text{MA}}{\to}(P,P)}{\longrightarrow} (P,P,A,A) \stackrel{(A,A)\stackrel{\text{MA}}{\to}((?,?)}{\longrightarrow} (?,?,?,?). \quad (3.188)$$

Using distinguishers like (3.187) allows us to recover subkeys at both ends of the distinguisher. The most interesting subkeys to recover are at the top, because they lead directly to bits of the user key. Otherwise, if we use distinguishers like (3.179), then we could only recover subkeys at the bottom end of 2.5 rounds, which do not provide direct information on the user key bits.

In (3.187), the two active words contain the same permutation, that is, the elements are in the same order. Thus, the initial Λ-set has the form $(i \odot z_{1,1}^{-1}, d_1, i \boxminus z_{3,1}, d_2)$ where d_1, d_2 are arbitrary constants, $0 \leq i \leq 2^{16} - 1$, and $(z_{1,1}, z_{3,1})$ are candidate values for $(Z_1^{(1)}, Z_3^{(1)})$. The reasoning for mixing $(z_{1,1}, z_{3,1})$ in the Λ-set is to eliminate $(Z_1^{(1)}, Z_3^{(1)})$ when their values are guessed correctly. When this matching happens, the Λ-set at the input to the first MA half-round will have the form (i, d_3, i, d_4), where d_3, d_4 are constants. Note that i represent an active 16-bit word, and each block in the Λ-set has the same value for them. Therefore, at the input to the first MA-box, the word pattern becomes (P,P), as predicted in (3.187). The Λ-set at the end of the first round has the form (A,A,P,P), which becomes $(?,?,?,?)$ after a half-round.

Let $C_i = (c_1^i, c_2^i, c_3^i, c_4^i)$ denote the ciphertext of the i-th plaintext in the initial Λ-set. The subkeys $(Z_1^{(3)}, Z_2^{(3)})$ can be further recovered by checking if

$$(c_1^i \odot z_{1,3}^{-1}) \oplus (c_2^i \boxminus z_{2,3}), \quad (3.189)$$

is an active word, as predicted by (3.187), where $(z_{1,3}, z_{2,3})$ is a candidate pair for $(Z_1^{(3)}, Z_2^{(3)})$. Note that (3.189) represents a 16-bit condition.

The following issues were observed in this attack:

- multiplicative subkeys such as $Z_1^{(1)}$ and $Z_1^{(3)}$ cannot be uniquely identified. For instance, we cannot distinguish $Z_1^{(1)}$ from $Z_1^{(1)} \odot 0$ because

$(Z_1^{(1)} \odot 0)^{-1} \odot Z_1^{(1)} \odot 0 = 0^{-1} \odot (Z_1^{(1)})^{-1} \odot Z_1^{(1)} \odot 0 = (Z_1^{(1)})^{-1} \odot Z_1^{(1)}$. This phenomenon is due to the fact that $0^{-1} \odot 0 = 1$ over $\mathrm{GF}(2^{16}+1)$. Thus, both $Z_1^{(1)}$ from $Z_1^{(1)} \odot 0$ result in the same active word.

- additive subkeys such as $Z_3^{(1)}$ and $Z_2^{(3)}$ cannot be fully recovered. In particular, $Z_3^{(1)}$ and$Z_3^{(1)} \oplus 8000_x$ cannot be distinguished. The most significant bit can be flipped without altering the integral value since 8000_x is exclusive-ored an even number of times in an integral.

The data complexity to recover about 60 subkey bits from ($Z_1^{(1)}$, $Z_3^{(1)}$, $Z_1^{(3)}$, $Z_2^{(3)}$) is $4 \cdot 2^{16} \cdot 2^{32} = 2^{50}$ chosen plaintexts, because the guess for ($Z_1^{(1)}$, $Z_3^{(1)}$) is embedded in the initial Λ-sets. The memory complexity is only 2^{16} text blocks, since the same storage can be reused for each Λ-set.

The number of wrong subkey candidates surviving the filtering condition (3.189) after four Λ-sets is $2^{60}/(2^{16})^4 < 1$.

The time complexity accounts for the effort to recover ($Z_1^{(1)}$, $Z_3^{(1)}$) in the initial Λ-set and to recover ($Z_1^{(3)}$, $Z_2^{(3)}$) through (3.189). The former step costs 2^{16} $\odot$ and 2^{16} $\boxplus$ per Λ-set. The latter costs ones $\odot$, one $\boxminus$ and one $\oplus$ per Λ-set. Assuming each $\odot$ costs three $\oplus$ or three $\boxminus$, and that $\oplus$ and $\boxminus$ cost the same, then we have $4 \cdot 2^{16}(2^{16} \cdot 4 \cdot 2^{16} + 5) \approx 2^{52}$ $\oplus$.

Since 2.5-round MESH-64 contains 12 $\odot$, 12 $\boxplus$ and 12 $\oplus$, that is, about 60 $\oplus$, the attack effort is estimated at $2^{52}/60 \approx 2^{46}$ 2.5-round MESH-64 computations.

We can recover $(Z_3^{(3)}, Z_4^{(3)})$ using the fact that

$$(c_4^i \odot z_{4,3}^{-1}) \oplus (c_3^i \boxminus z_{3,3}), \tag{3.190}$$

should be an active word according to (3.187); $(z_{3,3}, z_{4,3})$ is a candidate pair for $(Z_3^{(3)}, Z_4^{(3)})$. Note that (3.190) represents a 16-bit condition.

This step has reduced complexity since we already know $(Z_1^{(1)}, Z_3^{(1)})$. The effort is only $2 \cdot 2^{16} \cdot 5/82 \approx 2^{13}$ 2.5-round MESH-64 computations.

The dual distinguisher (3.188) can be used to recover $(Z_2^{(1)}, Z_4^{(1)})$ since we already discovered $Z_j^{(3)}$ for $1 \leq j \leq 4$. The effort is $2 \cdot 2^{16}(2^{32} \cdot 4)/82 \approx 2^{45}$ 2.5-round MESH-64 computations.

Recovering (most of) $(Z_1^{(1)}, Z_2^{(1)}, Z_3^{(1)}, Z_4^{(1)})$ will translate into recovering an equal amount of bits from the user key. But, according to the key schedule of MESH-64, these subkeys do not help in recovering $(Z_1^{(3)}, Z_2^{(3)}, Z_3^{(3)}, Z_4^{(3)})$ because

$$Z_1^{(3)} = (((((Z_7^{(1)} \boxplus Z_1^{(2)}) \oplus Z_2^{(2)}) \boxplus Z_5^{(2)}) \oplus Z_6^{(2)}) \boxplus Z_7^{(2)}) \lll 7 \oplus c_{14},$$

$$Z_2^{(3)} = (((((Z_1^{(2)} \boxplus Z_2^{(2)}) \oplus Z_3^{(2)}) \boxplus Z_6^{(2)}) \oplus Z_7^{(2)}) \boxplus Z_1^{(3)}) \lll 7 \oplus c_{15},$$

$$Z_3^{(3)} = (((((Z_2^{(2)} \boxplus Z_3^{(2)}) \oplus Z_4^{(2)}) \boxplus Z_7^{(2)}) \oplus Z_1^{(3)}) \boxplus Z_2^{(3)}) \lll 7 \oplus c_{16},$$

and

$$Z_4^{(3)} = (((((Z_3^{(2)} \boxplus Z_4^{(2)}) \oplus Z_5^{(2)}) \boxplus Z_1^{(3)}) \oplus Z_2^{(3)}) \boxplus Z_3^{(3)}) \lll 7 \oplus c_{17}.$$

Therefore, the remaining $128 - 60 = 68$ bits of the user key can be recovered by exhaustive search. The final time complexity is 2^{68} 2.5-round MESH-64 computations. The data complexity is $2 \cdot 2^{50} = 2^{51}$ chosen plaintexts. The memory complexity is 2^{16} text blocks since the same storage can be reused for each Λ-set separately.

3.15.3 Square/Multiset Attacks on MESH-96

Following on the bottom-up approach as with the attack on IDEA in Sect. 3.15.1 and on MESH-64 in Sect. 3.15.2, the longest multiset distinguishers found for MESH-96 cover up to 1.5 rounds, for instance:

$$(A,P,P,A,P,P) \overset{(P,P,P)\overset{\text{MA}}{\to}(P,P,P)}{\longrightarrow} (A,A,P,P,P,P) \overset{(A,A,P)\overset{\text{MA}}{\to}((?,?,?)}{\longrightarrow} (?,?,?,?,?,?). \quad (3.191)$$

Each 6-tuple represent the status of the six 16-bit words making up one text block. The larger arrow denotes the transition of Λ-set between rounds. The 3-tuples on top of the arrow indicate the transition across the MA-box.

There are two dual 1.5-round multiset distinguishers:

$$(P,A,P,P,A,P) \overset{(P,P,P)\overset{\text{MA}}{\to}(P,P,P)}{\longrightarrow} (P,P,A,A,P,P) \overset{(A,P,A)\overset{\text{MA}}{\to}((?,?,?)}{\longrightarrow} (?,?,?,?,?,?), \quad (3.192)$$

and

$$(P,P,A,P,P,A) \overset{(P,P,P)\overset{\text{MA}}{\to}(P,P,P)}{\longrightarrow} (P,P,P,P,A,A) \overset{(P,A,A)\overset{\text{MA}}{\to}((?,?,?)}{\longrightarrow} (?,?,?,?,?,?). \quad (3.193)$$

The rationale of the construction of these distinguishers is the same as in (3.187) for MESH-64: the two active words in the initial Λ-set in (3.191) contain the same permutation. Therefore, the initial Λ-set in (3.191) consists of text blocks of the form $(i \odot (z_{1,1})^{-1}, d_1, d_2, i \boxminus z_{4,1}, d_3, d_4)$, where d_1, d_2, d_3, d_4 are constants, $0 \leq i \leq 2^{16} - 1$ and $(z_{1,1}, z_{4,1})$ are candidate values for $(Z_1^{(1)}, Z_4^{(1)})$.

The guessed values $(z_{1,1}, z_{4,1})$ are embedded in the text blocks of the initial Λ-set, so that when the correct subkey values are guessed, the Λ-set acquires the form $(i, e_1, e_2, i, e_3, e_4)$ after the first KM half-round, where e_1, e_2, e_3, e_4

are constants. From this point on, the input to the MA-box has the form (P, P, P), as predicted in (3.191). Further, the Λ-sets follow the distinguisher.

If the guessed subkey value is wrong then the Λ-set acquire the form (i, e_1, e_2, j, e_3, e_4), where $0 \leq i, j \leq 2^{16} - 1$ but i and j do not contain the same permutation. Consequently, the input to the first MA-box has the form (B, P, P). The output Λ-set from the MA-box will be $(?, ?, ?)$, and the distinguisher will collapse after the first round.

Let $C_i = (c_1^i, c_2^i, c_3^i, c_4^i, c_5^i, c_6^i)$ denote the i-th ciphertext block (after 2.5-round MESH-96) corresponding to the i-th plaintext block in a Λ-set.

The correct value of ($Z_1^{(1)}$, $Z_4^{(1)}$, $Z_1^{(3)}$, $Z_2^{(3)}$) can be found by checking if

$$(c_1^i \odot z_{1,3}^{-1}) \oplus (c_2^i \boxminus z_{2,3}), \tag{3.194}$$

is an active word, according to (3.191), where $(z_{1,3}, z_{2,3})$ are candidate values for $(Z_1^{(3)}, Z_2^{(3)})$. Similar to the multiset analysis of MESH-64, there is uncertainty on the correct value of multiplicative subkey Z and $Z \odot 0$; also, the most significant bit of additive subkeys Z cannot be uniquely identified, that is, Z and $Z \oplus 8000_x$.

Therefore, we recover about 60 bits of information from ($Z_1^{(1)}$, $Z_4^{(1)}$, $Z_1^{(3)}$, $Z_2^{(3)}$). The data complexity is $4 \cdot 2^{16} \cdot 2^{32} = 2^{50}$ chosen plaintexts, because the guessed values of ($Z_1^{(1)}$, $Z_4^{(1)}$) is embedded in the initial Λ-sets. Assuming one $\odot$ costs three $\oplus$, and that $\oplus$, $\boxplus$ and $\boxminus$ cost about the same, this step costs about $4 \cdot 2^{16} \cdot 2^{32} \cdot 5 = 5 \cdot 2^{50}$ $\oplus$s.

Computing (3.194) costs one $\odot$, one $\oplus$ and one $\boxminus$, which means about 5 $\odot$s per Λ-set.

Further, 2.5-round MESH-96 costs 19 $\odot$, 17 $\boxplus$ and 18 $\oplus$, which is about 82 $\oplus$. So, the time complexity of this step is about $(5 \cdot 2^{50} + 4 \cdot 2^{16} \cdot 5)/82 \approx 2^{44}$ 2.5-round MESH-96 computations.

Further, we can recover $(Z_3^{(3)}, Z_4^{(3)})$ using the fact that

$$(c_3^i \odot z_{3,3}^{-1}) \oplus (c_4^i \boxminus z_{4,3}), \tag{3.195}$$

should be a 16-bit passive word; $(z_{3,3}, z_{4,3})$ are candidate values for ($Z_3^{(3)}$, $Z_4^{(3)}$). The attack complexity of this step is reduced since ($Z_1^{(1)}$, $Z_4^{(1)}$) were already recovered in the previous step. The time complexity is $4 \cdot 2^{16} \cdot 5/82 \approx 2^{14}$ 2.5-round MESH-96 computations. The same data used in the previous step can be reused, as well as the memory.

Further, we can recover $(Z_5^{(3)}, Z_6^{(3)})$ using the fact that

$$(c_5^i \odot z_{5,3}^{-1}) \oplus (c_6^i \boxminus z_{6,3}), \tag{3.196}$$

should be a 16-bit passive word; $(z_{5,3}, z_{6,3})$ are candidate values for $(Z_5^{(3)}, Z_6^{(3)})$. The attack complexity of this step is about 2^{14} 2.5-round MESH-96 computations.

Now, the alternative distinguisher (3.192) can be used to recover only $(Z_2^{(1)}, Z_5^{(1)})$ because $Z_j^{(3)}$, for $1 \leq j \leq 6$, have already been recovered. The complexity is $2 \cdot 2^{16} \cdot 2^{32}/82 \approx 2^{43}$ 2.5-round MESH-96 computations.

Likewise, the distinguisher (3.193) can be used to recover only $(Z_3^{(1)}, Z_6^{(1)})$ since $Z_j^{(3)}$, for $1 \leq j \leq 6$, have already been recovered. The complexity is also $2 \cdot 2^{16} \cdot 2^{32}/82 \approx 2^{43}$ 2.5-round MESH-96 computations.

According to the key schedule of MESH-96, recovery of ($Z_1^{(1)}$, $Z_2^{(1)}$, $Z_3^{(1)}$, $Z_4^{(1)}$, $Z_5^{(1)}$, $Z_6^{(1)}$) will immediately allow recovery of the corresponding number of bits of the user key, but:

$$Z_1^{(3)} = (((((Z_7^{(1)} \boxplus Z_2^{(2)}) \oplus Z_4^{(2)}) \boxplus Z_6^{(2)}) \oplus Z_8^{(2)}) \boxplus Z_9^{(2)}) \lll 9 \oplus c_{12},$$

$$Z_2^{(3)} = (((((Z_8^{(1)} \boxplus Z_3^{(2)}) \oplus Z_5^{(2)}) \boxplus Z_7^{(2)}) \oplus Z_9^{(2)}) \boxplus Z_1^{(3)}) \lll 9 \oplus c_{13},$$

$$Z_3^{(3)} = (((((Z_9^{(1)} \boxplus Z_4^{(2)}) \oplus Z_6^{(2)}) \boxplus Z_8^{(2)}) \oplus Z_1^{(3)}) \boxplus Z_2^{(3)}) \lll 9 \oplus c_{14},$$

$$Z_4^{(3)} = (((((Z_1^{(2)} \boxplus Z_5^{(2)}) \oplus Z_7^{(2)}) \boxplus Z_9^{(2)}) \oplus Z_2^{(3)}) \boxplus Z_3^{(3)}) \lll 9 \oplus c_{15},$$

$$Z_5^{(3)} = (((((Z_2^{(2)} \boxplus Z_6^{(2)}) \oplus Z_8^{(2)}) \boxplus Z_1^{(3)}) \oplus Z_3^{(3)}) \boxplus Z_4^{(3)}) \lll 9 \oplus c_{16},$$

$$Z_6^{(3)} = (((((Z_3^{(2)} \boxplus Z_7^{(2)}) \oplus Z_9^{(2)}) \boxplus Z_2^{(3)}) \oplus Z_4^{(3)}) \boxplus Z_5^{(3)}) \lll 9 \oplus c_{17}.$$

Therefore, knowledge of ($Z_1^{(1)}$, $Z_2^{(1)}$, $Z_3^{(1)}$, $Z_4^{(1)}$), $Z_5^{(1)}$, $Z_6^{(1)}$) is not enough to uniquely recover ($Z_1^{(3)}$, $Z_2^{(3)}$, $Z_3^{(3)}$, $Z_4^{(3)}$), $Z_5^{(3)}$, $Z_6^{(3)}$).

In total, 90 user key bits are recovered. The remaining $192-90=102$ user key bits can be recovered by exhaustive search. The final time complexity is 2^{102} 2.5-round MESH-96 computations. The data complexity is $3 \cdot 2^{50} \approx 2^{51.5}$ chosen plaintexts. The memory complexity is 2^{16} text blocks.

To attack 3-round MESH-96, the subkeys $(Z_7^{(3)}, Z_8^{(3)}, Z_9^{(3)})$ in an additional MA half-round can be guessed, and the previous attack on 2.5-rounds can be applied. Only the time complexity changes to $2^{48} \cdot 2^{102} = 2^{150}$ 3-round MESH-96 computations.

3.15.4 Square/Multiset Attacks on MESH-128

A bottom-up approach is used to construct multiset distinguishers for MESH-128, in a similar fashion as used for the other MESH ciphers.

The longest multiset distinguishers found cover up to 1.5-round MESH-128:

$$(A, P, P, P, A, P, P, P) \overset{(P,P,P,P)\overset{\text{MA}}{\to}(P,P,P,P)}{\longrightarrow} (A, A, P, P, P, P, P, P)$$
$$\overset{(A,A,P,P)\overset{\text{MA}}{\to}(?,?,?,?)}{\longrightarrow} (?, ?, ?, ?, ?, ?, ?, ?). \quad (3.197)$$

Each 8-tuple represent the status of the eight 16-bit words making up one text block. The larger arrow denotes the transition of Λ-set between rounds. The 4-tuples on top of the arrow indicate the transition across the MA-box.

There are dual multiset distinguishers covering 1.5-round MESH-128, based on pairing active words at the input to the MA-box. They are:

$$(P, A, P, P, P, A, P, P) \overset{(P,P,P,P)\overset{\text{MA}}{\to}(P,P,P,P)}{\longrightarrow} (P, P, A, P, A, P, P, P)$$
$$\overset{(A,P,A,P)\overset{\text{MA}}{\to}(?,?,?,?)}{\longrightarrow} (?, ?, ?, ?, ?, ?, ?, ?), \quad (3.198)$$

and

$$(P, P, A, P, P, P, A, P) \overset{(P,P,P,P)\overset{\text{MA}}{\to}(P,P,P,P)}{\longrightarrow} (P, P, P, A, P, A, P, P)$$
$$\overset{(P,A,P,A)\overset{\text{MA}}{\to}(?,?,?,?)}{\longrightarrow} (?, ?, ?, ?, ?, ?, ?, ?), \quad (3.199)$$

and

$$(P, P, P, A, P, P, P, A) \overset{(P,P,P,P)\overset{\text{MA}}{\to}(P,P,P,P)}{\longrightarrow} (P, P, P, P, P, P, A, A)$$
$$\overset{(P,P,A,A)\overset{\text{MA}}{\to}(?,?,?,?)}{\longrightarrow} (?, ?, ?, ?, ?, ?, ?, ?). \quad (3.200)$$

Similar to previous multiset distinguisher constructions, the two active words in the initial Λ-set in (3.197) contain the same permutation. Therefore, the initial Λ-set contains plaintexts of the form $(i \odot z_{1,1}^{-1}, d_1, d_2, d_3, i \boxminus z_{5,1}, d_4, d_5, d_6)$, where d_j for $1 \leq j \leq 6$ are constants, $0 \leq i \leq 2^{16} - 1$, and $(z_{1,1}, z_{5,1})$ are candidate values for $(Z_1^{(1)}, Z_5^{(1)})$.

Therefore, the guessed values of $(Z_1^{(1)}, Z_5^{(1)})$ are embedded in the initial Λ-set. When the correct subkey values are guessed, the Λ-set after the first KM half-round will have the form $(i, e_1, e_2, e_3, i, e_4, e_5, e_6)$, where e_j, for $1 \leq j \leq 6$, are constants and $0 \leq i \leq 2^{16} - 1$.

Consequently, the input to the first MA-box will consist of passive words only: (P, P, P, P). After the second KM half-round, the Λ-set will have the form (A, A, P, P, P, P, P, P). The second MA-box will destroy any balanced word in the input, and the round output Λ-set will be $(?, ?, ?, ?, ?, ?, ?, ?)$.

When $(z_{1,1}, z_{5,1})$ are not the correct subkey value, the initial Λ-set will not be satisfied, and the distinguisher will collapse after the first MA-box because the input to MA-box will be (B, P, P, P) instead of (P, P, P, P). Thus, the output Λ-set of the MA-box after the first round will already have the form $(?, ?, ?, ?)$.

When (3.197) holds, subkeys can also be guessed in pairs at the bottom end of 2.5 rounds. Let $C_i = (c_1^i, c_2^i, c_3^i, c_4^i, c_5^i, c_6^i, c_7^i, c_8^i)$ denote the i-th ciphertext block corresponding to the i-th plaintext block in a Λ-set.

We can additionally recover $(Z_1^{(3)}, Z_2^{(3)})$ by checking candidate values $(z_{1,3}, z_{2,3})$ until

$$(c_1^i \odot (z_{1,3})^{-1}) \oplus (c_2^i \boxminus z_{2,3}), \tag{3.201}$$

is an active 16-bit word, according to (3.197).

We can additionally recover $(Z_3^{(3)}, Z_5^{(3)})$ by checking candidate values $(z_{3,3}, z_{5,3})$ until

$$(c_3^i \odot (z_{3,3})^{-1}) \oplus (c_5^i \boxminus z_{5,3}), \tag{3.202}$$

is an active 16-bit word, according to (3.197).

Alternatively, we can additionally recover $(Z_4^{(3)}, Z_6^{(3)})$ by checking candidate values $(z_{4,3}, z_{6,3})$ until

$$(c_6^i \odot (z_{6,3})^{-1}) \oplus (c_4^i \boxminus z_{4,3}), \tag{3.203}$$

is a passive 16-bit word, according to (3.197).

Finally, we can recover $(Z_7^{(3)}, Z_8^{(3)})$ by checking candidate values $(z_{7,3}, z_{8,3})$ until

$$(c_8^i \odot (z_{8,3})^{-1}) \oplus (c_7^i \boxminus z_{7,3}), \tag{3.204}$$

is a passive 16-bit word, according to (3.197).

Similar to previous multiset analysis of MESH ciphers, there is uncertainty over the recovered value of multiplicative subkeys, or more precisely, between z and $z \odot 0$; likewise, the most significant bit of additive subkeys cannot be uniquely identified.

Thus, using (3.201), we recover about 60 bits of information from $(Z_1^{(1)}, Z_5^{(1)}, Z_3^{(3)}, Z_5^{(3)})$

Embedding the guesses for $(Z_1^{(1)}, Z_5^{(1)})$ in the initial Λ-set costs 2^{32} combined $\odot$ and $\boxplus$ operations. Checking (3.201) once costs one $\odot$, one $\oplus$ and one $\boxminus$.

In comparison, 2.5-round MESH-128 costs 28 $\odot$, 28 $\boxplus$ and 24 $\oplus$. Assuming each $\odot$ costs three $\oplus$, and that $\oplus$, $\boxplus$ and $\boxminus$ cost about the same, then, this stage of the attack costs about $4 \cdot 2^{16}(2^{32} \cdot 4 + 5)/136 \approx 2^{45}$ 2.5-round MESH-128 computations.

The same attack can be repeated using (3.202) to recover only $(Z_3^{(3)}, Z_5^{(3)})$ because $(Z_1^{(1)}, Z_5^{(1)})$ were recovered in the previous step. The time complexity decreases to $2 \cdot 2^{16} \cdot 5/136 \approx 2^{12}$ 2.5-round MESH-128. Since less bits are recovered this time, we only need two Λ-sets.

Likewise, we can use (3.203) to recover $(Z_4^{(3)}, Z_6^{(3)})$ because $(Z_1^{(1)}, Z_5^{(1)})$ were already recovered. The time complexity is again 2^{12} 2.5-round MESH-128.

Finally, we can use (3.204) to recover $(Z_7^{(3)}, Z_8^{(3)})$ because $(Z_1^{(1)}, Z_5^{(1)})$ were already recovered. The time complexity is also 2^{12} 2.5-round MESH-128.

The overall time complexity is still dominated by the first step: 2^{45} 2.5-round MESH-128 computations.

Now that most bits of $(Z_1^{(3)}, Z_2^{(3)}, Z_3^{(3)}, Z_4^{(3)}, Z_5^{(3)}, Z_6^{(3)}, Z_7^{(3)}, Z_8^{(3)})$ have been recovered, we can use (3.198) to recover $(Z_2^{(1)}, Z_6^{(1)})$, with effort about $2 \cdot 2^{16} \cdot 2^{32} \cdot 4/136 = 2^{50}/68 \approx 2^{44}$ 2.5-round computations.

Likewise, we can use (3.199) to recover $(Z_3^{(1)}, Z_7^{(1)})$, with effort about 2^{44} 2.5-round computations.

Finally, we can use (3.200) to recover $Z_j^{(1)}$, for $1 \leq j \leq 8$, with effort about 2^{44} 2.5-round computations.

So, the effort add up to $2^{45} + 3 \cdot 2^{44} \approx 2^{46}$ 2.5-round MESH-128 computations. The data complexity is $4 \cdot 2^{50} = 2^{52}$ chosen plaintexts. The memory complexity is 2^{16} text blocks, since the same storage can be reused for each Λ-set.

According to the key schedule of MESH-128, the subkeys $Z_j^{(1)}$, for $1 \leq j \leq 8$, lead immediately to recovery of an equivalent amount of user key information. But, for the subkeys $Z_j^{(3)}$, for $1 \leq j \leq 8$:

$$Z_1^{(3)} = (((((Z_9^{(1)} \boxplus Z_{12}^{(1)}) \oplus Z_1^{(2)}) \boxplus Z_5^{(2)}) \oplus Z_{11}^{(2)}) \boxplus Z_{12}^{(2)}) \lll 11 \oplus c_{24},$$

$$Z_2^{(3)} = (((((Z_{10}^{(1)} \boxplus Z_1^{(2)}) \oplus Z_2^{(2)}) \boxplus Z_6^{(2)}) \oplus Z_{12}^{(2)}) \boxplus Z_1^{(3)}) \lll 11 \oplus c_{25},$$

$$Z_3^{(3)} = (((((Z_{11}^{(1)} \boxplus Z_2^{(2)}) \oplus Z_3^{(2)}) \boxplus Z_7^{(2)}) \oplus Z_1^{(3)}) \boxplus Z_2^{(3)}) \lll 11 \oplus c_{26},$$

$$Z_4^{(3)} = (((((Z_{12}^{(1)} \boxplus Z_3^{(2)}) \oplus Z_4^{(2)}) \boxplus Z_8^{(2)}) \oplus Z_2^{(3)}) \boxplus Z_3^{(3)}) \lll 11 \oplus c_{27},$$

$$Z_5^{(3)} = (((((Z_1^{(2)} \boxplus Z_4^{(2)}) \oplus Z_5^{(2)}) \boxplus Z_9^{(2)}) \oplus Z_3^{(3)}) \boxplus Z_4^{(3)}) \lll 11 \oplus c_{28},$$

$$Z_6^{(3)} = (((((Z_2^{(2)} \boxplus Z_5^{(2)}) \oplus Z_6^{(2)}) \boxplus Z_{10}^{(2)}) \oplus Z_4^{(3)}) \boxplus Z_5^{(3)}) \lll 11 \oplus c_{29},$$

$$Z_7^{(3)} = (((((Z_3^{(2)} \boxplus Z_6^{(2)}) \oplus Z_7^{(2)}) \boxplus Z_{11}^{(2)}) \oplus Z_5^{(3)}) \boxplus Z_6^{(3)}) \lll 11 \oplus c_{30},$$

and

$$Z_8^{(3)} = (((((Z_4^{(2)} \boxplus Z_7^{(2)}) \oplus Z_8^{(2)}) \boxplus Z_{12}^{(2)}) \oplus Z_6^{(3)}) \boxplus Z_7^{(3)}) \lll 11 \oplus c_{31}.$$

Thus, knowledge of $Z_j^{(1)}$, for $1 \leq j \leq 8$, is not sufficient to uniquely identify $Z_j^{(3)}$, for $1 \leq j \leq 8$. Further, the latter does not translate into user key bits like the former. Therefore, the remaining $256 - (16 \cdot 8 - 8) = 144$ user key bits can be recovered by exhaustive search. The final time complexity is 2^{114} 2.5-round MESH-128 computations.

To attack 3-round MESH-128, the subkeys $Z_9^{(3)}$, $Z_{10}^{(3)}$, $Z_{11}^{(3)}$, $Z_{12}^{(3)}$) of an additional MA half-round can be guessed, and the previous attack on 2.5 rounds can be applied. The time complexity increases to $2^{114+64} = 2^{178}$ 3-round MESH-128 computations.

3.15.5 Square/Multiset Attacks on MESH-64(8)

A bottom-up approach is used to construct multiset distinguishers for MESH-64(8), in a similar fashion as used for the other MESH ciphers.

The longest multiset distinguishers found cover up to 1.5-round MESH-64(8):

$$(A,P,P,P,A,P,P,P) \overset{(P,P,P,P)\overset{\text{MA}}{\to}(P,P,P,P)}{\to} (A,A,P,P,P,P,P,P)$$
$$\overset{(A,A,P,P)\overset{\text{MA}}{\to}(?,?,?,?)}{\to} (?,?,?,?,?,?,?,?). \quad (3.205)$$

Each 8-tuple represent the status of the eight 8-bit words making up one text block. The larger arrow denotes the transition of Λ-sets between rounds. The 4-tuples on top of the arrow indicate the transition across the MA-box of MESH-64(8).

There are dual distinguishers to (3.205) also covering 1.5 rounds. They are:

$$(P,A,P,P,P,A,P,P) \overset{(P,P,P,P)\overset{\text{MA}}{\to}(P,P,P,P)}{\to} (P,P,A,P,A,P,P,P)$$
$$\overset{(A,P,A,P)\overset{\text{MA}}{\to}(?,?,?,?)}{\to} (?,?,?,?,?,?,?,?), \quad (3.206)$$

and

$$(P, P, A, P, P, P, A, P) \overset{(P,P,P,P)\overset{\text{MA}}{\to}(P,P,P,P)}{\to} (P, P, P, A, P, A, P, P) \overset{(P,A,P,A)\overset{\text{MA}}{\to}(?,?,?,?)}{\to} (?, ?, ?, ?, ?, ?, ?, ?), \tag{3.207}$$

and

$$(P, P, P, A, P, P, P, A) \overset{(P,P,P,P)\overset{\text{MA}}{\to}(P,P,P,P)}{\to} (P, P, P, P, P, P, A, A) \overset{(P,P,A,A)\overset{\text{MA}}{\to}(?,?,?,?)}{\to} (?, ?, ?, ?, ?, ?, ?, ?). \tag{3.208}$$

Similar to previous multiset distinguisher constructions, the two active words in the initial Λ-set in (3.205) contain the same permutation. Therefore, the initial Λ-set contains plaintexts of the form $(i \odot z_{1,1}^{-1}, d_1, d_2, d_3, i \boxminus z_{5,1}, d_4, d_5, d_6)$, where d_j, for $1 \leq j \leq 6$, are constants, $0 \leq i \leq 2^{16} - 1$, and $(z_{1,1}, z_{5,1})$ are candidate values for $(Z_1^{(1)}, Z_5^{(1)})$.

Therefore, the guessed values of $(Z_1^{(1)}, Z_5^{(1)})$ are embedded in the initial Λ-set. When the correct subkey values are guessed, the Λ-set after the first KM half-round will have the form $(i, e_1, e_2, e_3, i, e_4, e_5, e_6)$, where e_j for $1 \leq j \leq 6$ are constants and $0 \leq i \leq 2^{16} - 1$.

Consequently, the input to the first MA-box will consist of passive words only: (P, P, P, P). After the second KM half-round, the Λ-set will have the form (A, A, P, P, P, P, P, P). The second MA-box will destroy any balanced word in the input, and the round output Λ-set will be $(?, ?, ?, ?, ?, ?, ?, ?)$.

When $(z_{1,1}, z_{5,1})$ are not the correct subkey value, the initial Λ-set will not be satisfied, and the distinguisher will collapse after the first MA-box because the input to MA-box will be (B, P, P, P) instead of (P, P, P, P). Thus, the output MA-box Λ-set after the first MA-box will already have the form $(?, ?, ?, ?)$.

When (3.205) holds, subkeys can also be guessed in pairs at the bottom end of 2.5 rounds. Let $C_i = (c_1^i, c_2^i, c_3^i, c_4^i, c_5^i, c_6^i, c_7^i, c_8^i)$ denote the i-th ciphertext block corresponding to the i-th plaintext block in a Λ-set.

We can additionally recover $(Z_1^{(3)}, Z_2^{(3)})$ by checking candidate values $(z_{1,3}, z_{2,3})$ until

$$(c_1^i \odot (z_{1,3})^{-1}) \oplus (c_2^i \boxminus z_{2,3}), \tag{3.209}$$

is an active 8-bit word, according to (3.205).

We can additionally recover $(Z_3^{(3)}, Z_5^{(3)})$ by checking candidate values $(z_{3,3}, z_{5,3})$ until

$$(c_3^i \odot (z_{3,3})^{-1}) \oplus (c_5^i \boxminus z_{5,3}), \tag{3.210}$$

is an active 8-bit word, according to (3.205).

We can additionally recover $(Z_4^{(3)}, Z_6^{(3)})$ by checking candidate values $(z_{4,3}, z_{6,3})$ until

$$(c_6^i \odot (z_{6,3})^{-1}) \oplus (c_4^i \boxminus z_{4,3}), \tag{3.211}$$

is a passive 8-bit word, according to (3.205).

Finally, we can additionally recover $(Z_7^{(3)}, Z_8^{(3)})$ by checking candidate values $(z_{7,3}, z_{8,3})$ until

$$(c_8^i \odot (z_{8,3})^{-1}) \oplus (c_7^i \boxminus Z_{7,3}), \tag{3.212}$$

is a passive 8-bit word, according to (3.205).

Similar to previous multiset analysis of MESH ciphers, there is uncertainty over the recovered value of multiplicative subkeys, or more precisely, between z and $z \odot 0$; likewise, the most significant bit of additive subkeys cannot be uniquely identified.

Thus, using (3.209), we can recover about 28 bits of information from $(Z_1^{(1)}, Z_5^{(1)}, Z_1^{(3)}, Z_2^{(3)})$.

Embedding the guesses for $(Z_1^{(1)}, Z_5^{(1)})$ in the initial Λ-set costs 2^{16} combined $\odot$ and $\boxplus$ operations. Checking (3.209) once costs one $\odot$, one $\oplus$ and one $\boxminus$.

In comparison, 2.5-round MESH-64(8) costs 20 $\odot$, 20 $\boxplus$ and 24 $\oplus$. Assuming each $\odot$ costs 3 $\oplus$, and that $\oplus$, $\boxplus$ and $\boxminus$ cost about the same, then, this stage of the attack costs about $4 \cdot 2^8(2^{16} \cdot 4 + 5)/104 \approx 2^{22}$ 2.5-round MESH-64(8) computations.

The same attack can be repeated using (3.210) to recover $(Z_3^{(3)}, Z_5^{(3)})$ because $(Z_1^{(1)}, Z_5^{(1)})$ was recovered in the previous step. The complexity decreases to $2 \cdot 2^8 \cdot 5/104 \approx 2^5$ 2.5-round MESH-64(8). Since less bits are recovered this time, we only need two Λ-sets.

Likewise, we use (3.211) to recover only $(Z_4^{(3)}, Z_6^{(3)})$ because $(Z_1^{(1)}, Z_5^{(1)})$ were already recovered. The time complexity is again 2^5 2.5-round MESH-64(8).

Finally, we use (3.212) to recover only $(Z_7^{(3)}, Z_8^{(3)})$ because $(Z_1^{(1)}, Z_5^{(1)})$ were already recovered. The time complexity is also 2^5 2.5-round MESH-64(8).

The attack on (3.205) can now be repeated for (3.206), (3.207) and (3.208) to recover subkeys of the first KM half-round, since the subkeys $Z_j^{(3)}$ for $1 \leq j \leq 8$ were already recovered. This strategy means that the attack complexities will be reduced. For instance, using (3.206), we can recover $(Z_2^{(1)}, Z_6^{(1)})$ and check its correctness using the previously recovered values of $Z_j^{(3)}$ for $1 \leq j \leq 8$. The complexity is $2 \cdot 2^8 \cdot 2^{16} \cdot 4/104 \approx 2^{18}$ 2.5-round MESH-64(8) computations.

The data complexity is $2 \cdot 2^8 \cdot 2^{16} = 2^{25}$ chosen plaintexts. The memory complexity is 2^8 text blocks since the same storage can be reused for each Λ-set.

The same approach can be used with (3.207) and (3.208) to recover $(Z_3^{(1)}, Z_7^{(1)})$ and $(Z_4^{(1)}, Z_8^{(1)})$, respectively.

The overall time complexity is still dominated by the first step: 2^{22} 2.5-round MESH-64(8) computations.

The subkeys $Z_j^{(1)}$, for $1 \leq j \leq 8$, lead immediately to recovery of an equivalent amount of user key information. But, the subkeys $Z_j^{(3)}$ for $1 \leq j \leq 8$ according to the key schedule of MESH-64(8) can be represented as:

$$Z_1^{(3)} = (((Z_5^{(1)} \boxplus Z_9^{(1)}) \oplus Z_8^{(2)}) \boxplus Z_9^{(2)}) \lll 1 \oplus c_{20},$$

$$Z_2^{(3)} = (((Z_6^{(1)} \boxplus Z_{10}^{(1)}) \oplus Z_9^{(2)}) \boxplus Z_{10}^{(2)}) \lll 1 \oplus c_{21},$$

$$Z_3^{(3)} = (((Z_7^{(1)} \boxplus Z_1^{(2)}) \oplus Z_{10}^{(2)}) \boxplus Z_1^{(3)}) \lll 1 \oplus c_{22},$$

$$Z_4^{(3)} = (((Z_8^{(1)} \boxplus Z_2^{(2)}) \oplus Z_1^{(3)}) \boxplus Z_2^{(3)}) \lll 1 \oplus c_{23},$$

$$Z_5^{(3)} = (((Z_9^{(1)} \boxplus Z_3^{(2)}) \oplus Z_2^{(3)}) \boxplus Z_1^{(3)}) \lll 1 \oplus c_{24},$$

$$Z_6^{(3)} = (((Z_{10}^{(1)} \boxplus Z_4^{(2)}) \oplus Z_3^{(3)}) \boxplus Z_4^{(3)}) \lll 1 \oplus c_{25},$$

$$Z_7^{(3)} = (((Z_1^{(2)} \boxplus Z_5^{(2)}) \oplus Z_4^{(3)}) \boxplus Z_5^{(3)}) \lll 1 \oplus c_{26},$$

and

$$Z_8^{(3)} = (((Z_2^{(2)} \boxplus Z_6^{(2)}) \oplus Z_5^{(3)}) \boxplus Z_6^{(3)}) \lll 1 \oplus c_{27}.$$

Thus, knowledge of $Z_j^{(1)}$ for $1 \leq j \leq 8$ is not sufficient to uniquely recover $Z_j^{(3)}$ for $1 \leq j \leq 8$. Furthermore, the latter does not translate into user key bits like the former. Therefore, the remaining $128 - 56 = 72$ key bits can be found by exhaustive search. The final time complexity becomes 2^{72} 2.5-round MESH-64(8) computations.

To attack 3-round MESH-64(8), the subkeys $(Z_9^{(3)}, Z_{10}^{(3)})$ of an additional MA half-round can be guessed and the previous attack on 2.5 rounds can be applied. The time complexity increases to $2^{72+16} = 2^{88}$ 3-round MESH-64(8) computations.

3.15.6 Square/Multiset Attacks on MESH-128(8)

A bottom-up approach is used to construct multiset distinguishers for MESH-128(8), in a similar fashion as used for the other MESH ciphers.

The longest multiset distinguishers found cover 1.5-round MESH-128(8):

$$(A,P,P,P,P,P,P,P,A,P,P,P,P,P,P,P) \overset{(P,P,P,P,P,P,P,P)\overset{\text{MA}}{\to}(P,P,P,P,P,P,P,P)}{\longrightarrow}$$
$$(A,A,P,P,P,P,P,P,P,P,P,P,P,P,P,P) \overset{(A,A,P,P,P,P,P,P)\overset{\text{MA}}{\to}(?,?,?,?,?,?,?,?)}{\longrightarrow}$$
$$(?,?,?,?,?,?,?,?,?,?,?,?,?,?,?,?). \quad (3.213)$$

Each 16-tuple represents the status of the sixteen 8-bit words making up one text block. The larger arrow denotes the transition of Λ-sets between rounds. The 8-tuples on top of the arrow indicate the transition across the MA-box of MESH-128(8).

There are dual distinguishers to (3.213) based on pairing inputs to the MA-box. They are:

$$(P,A,P,P,P,P,P,P,P,A,P,P,P,P,P,P) \overset{(P,P,P,P,P,P,P,P)\overset{\text{MA}}{\to}(P,P,P,P,P,P,P,P)}{\longrightarrow}$$
$$(P,P,A,P,P,P,P,P,A,P,P,P,P,P,P,P) \overset{(A,P,A,P,P,P,P,P)\overset{\text{MA}}{\to}(?,?,?,?,?,?,?,?)}{\longrightarrow}$$
$$(?,?,?,?,?,?,?,?,?,?,?,?,?,?,?,?), \quad (3.214)$$

and

$$(P,P,A,P,P,P,P,P,P,P,A,P,P,P,P,P) \overset{(P,P,P,P,P,P,P,P)\overset{\text{MA}}{\to}(P,P,P,P,P,P,P,P)}{\longrightarrow}$$
$$(P,P,P,A,P,P,P,P,P,A,P,P,P,P,P,P) \overset{(P,A,P,A,P,P,P,P)\overset{\text{MA}}{\to}(?,?,?,?,?,?,?,?)}{\longrightarrow}$$
$$(?,?,?,?,?,?,?,?,?,?,?,?,?,?,?,?), \quad (3.215)$$

and

$$(P,P,P,A,P,P,P,P,P,P,P,A,P,P,P,P) \overset{(P,P,P,P,P,P,P,P)\overset{\text{MA}}{\to}(P,P,P,P,P,P,P,P)}{\longrightarrow}$$
$$(P,P,P,P,A,P,P,P,P,P,A,P,P,P,P,P) \overset{(P,P,A,P,A,P,P,P)\overset{\text{MA}}{\to}(?,?,?,?,?,?,?,?)}{\longrightarrow}$$
$$(?,?,?,?,?,?,?,?,?,?,?,?,?,?,?,?), \quad (3.216)$$

and

$$(P,P,P,P,A,P,P,P,P,P,P,P,A,P,P,P) \overset{(P,P,P,P,P,P,P,P)\overset{\text{MA}}{\to}(P,P,P,P,P,P,P,P)}{\longrightarrow}$$
$$(P,P,P,P,P,A,P,P,P,P,P,A,P,P,P,P) \overset{(P,P,P,A,P,A,P,P)\overset{\text{MA}}{\to}(?,?,?,?,?,?,?,?)}{\longrightarrow}$$
$$(?,?,?,?,?,?,?,?,?,?,?,?,?,?,?,?), \quad (3.217)$$

and

$$(P,P,P,P,P,A,P,P,P,P,P,P,P,A,P,P) \overset{(P,P,P,P,P,P,P,P)\overset{\mathrm{MA}}{\longrightarrow}(P,P,P,P,P,P,P,P)}{}$$

$$(P,P,P,P,P,P,A,P,P,P,P,P,A,P,P,P) \overset{(P,P,P,P,A,P,A,P)\overset{\mathrm{MA}}{\longrightarrow}(?,?,?,?,?,?,?,?)}{}$$

$$(?,?,?,?,?,?,?,?,?,?,?,?,?,?,?,?), \tag{3.218}$$

and

$$(P,P,P,P,P,P,A,P,P,P,P,P,P,P,A,P) \overset{(P,P,P,P,P,P,P,P)\overset{\mathrm{MA}}{\longrightarrow}(P,P,P,P,P,P,P,P)}{}$$

$$(P,P,P,P,P,P,P,A,P,P,P,P,A,P,P,P) \overset{(P,P,P,P,P,A,P,A)\overset{\mathrm{MA}}{\longrightarrow}(?,?,?,?,?,?,?,?)}{}$$

$$(?,?,?,?,?,?,?,?,?,?,?,?,?,?,?,?), \tag{3.219}$$

and

$$(P,P,P,P,P,P,P,A,P,P,P,P,P,P,P,A) \overset{(P,P,P,P,P,P,P,P)\overset{\mathrm{MA}}{\longrightarrow}(P,P,P,P,P,P,P,P)}{}$$

$$(P,P,P,P,P,P,P,P,P,P,P,P,P,P,A,A) \overset{(P,P,P,P,P,P,A,A)\overset{\mathrm{MA}}{\longrightarrow}(?,?,?,?,?,?,?,?)}{}$$

$$(?,?,?,?,?,?,?,?,?,?,?,?,?,?,?,?). \tag{3.220}$$

Similar to previous multiset distinguisher constructions, the two active words in the initial Λ-set in (3.213) contain the same permutation. Therefore, the initial Λ-set contains plaintexts of the form ($i \odot z_{1,1}^{-1}$, d_1, d_2, d_3, d_4, d_5, d_6, d_7, $i \boxminus z_{9,1}$, d_8, d_9, d_{10}, d_{11}, d_{12}, d_{13}, d_{14}), where d_j, for $1 \leq j \leq 14$, are constants, $0 \leq i \leq 2^{16}-1$, and $(z_{1,1}, z_{9,1})$ are candidate values for $(Z_1^{(1)}, Z_9^{(1)})$.

Therefore, the guessed values of $(Z_1^{(1)}, Z_9^{(1)})$ are embedded in the initial Λ-set. When the correct subkey values are guessed, the Λ-set after the first KM half-round will have the form (i, e_1, e_2, e_3, e_4, e_5, e_6, e_7, i, e_8, e_9, e_{10}, e_{11}, e_{12}, e_{13}, e_{14}), where e_j, for $1 \leq j \leq 14$, are constants and $0 \leq i \leq 2^{16}-1$.

Consequently, the input to the first MA-box will be (P,P,P,P,P,P,P,P). After the second KM half-round, the Λ-set will have the form (A, A, P, P, P, P, P, P). The second MA-box will destroy any balanced word in the input, and the round output Λ-set will be $(?,?,?,?,?,?,?,?,?,?,?,?,?,?,?,?)$.

When $(z_{1,1}, z_{9,1})$ are not the correct subkey values, the distinguisher will collapse after the first MA-box because the input to MA-box will be (B,P,P,P,P,P,P,P) instead of (P,P,P,P,P,P,P,P). Thus, the output Λ-set after the first MA-box will already have the form $(?,?,?,?,?,?,?,?)$.

When (3.213) holds, subkeys can also be guessed in pairs at the bottom end of 2.5 rounds. Let $C_i = (c_1^i, c_2^i, c_3^i, c_4^i, c_5^i, c_6^i, c_7^i, c_8^i, c_9^i, c_{10}^i, c_{11}^i, c_{12}^i, c_{13}^i, c_{14}^i, c_{15}^i, c_{16}^i)$ denote the i-th ciphertext block corresponding to the i-th plaintext block in a Λ-set.

We can additionally recover $(Z_1^{(3)}, Z_2^{(3)})$ by checking candidate values $(z_{1,3}, z_{2,3})$ until

$$(c_1^i \odot (z_{1,3})^{-1}) \oplus (c_2^i \boxminus z_{2,3}), \tag{3.221}$$

is an active 8-bit word, according to (3.213).

We can additionally recover $(Z_3^{(3)}, Z_9^{(3)})$ by checking candidate values $(z_{3,3}, z_{9,3})$ until

$$(c_3^i \odot (z_{3,3})^{-1}) \oplus (c_9^i \boxminus z_{9,3}), \tag{3.222}$$

is an active 8-bit word, according to (3.213).

We can additionally recover $(Z_4^{(3)}, Z_{10}^{(3)})$ by checking candidate values $(z_{4,3}, z_{10,3})$ until

$$(c_{10}^i \odot (z_{10,3})^{-1}) \oplus (c_4^i \boxminus z_{4,3}), \tag{3.223}$$

is a passive 8-bit word, according to (3.213).

We can additionally recover $(Z_5^{(3)}, Z_{11}^{(3)})$ by checking candidate values $(z_{5,3}, z_{11,3})$ until

$$(c_5^i \odot (z_{5,3})^{-1}) \oplus (c_{11}^i \boxminus z_{11,3}), \tag{3.224}$$

is a passive 8-bit word, according to (3.213).

We can additionally recover $(Z_6^{(3)}, Z_{12}^{(3)})$ by checking candidate values $(z_{6,3}, z_{12,3})$ until

$$(c_{12}^i \odot (z_{12,3})^{-1}) \oplus (c_6^i \boxminus z_{6,3}), \tag{3.225}$$

is a passive 8-bit word, according to (3.213).

We can additionally recover $(Z_7^{(3)}, Z_{13}^{(3)})$ by checking candidate values $(z_{7,3}, z_{13,3})$ until

$$(c_7^i \odot (z_{7,3})^{-1}) \oplus (c_{13}^i \boxminus z_{13,3}), \tag{3.226}$$

is a passive 8-bit word, according to (3.213).

We can additionally recover $(Z_8^{(3)}, Z_{14}^{(3)})$ by checking candidate values $(z_{8,3}, z_{14,3})$ until

$$(c_{14}^i \odot (z_{14,3})^{-1}) \oplus (c_8^i \boxminus z_{8,3}), \tag{3.227}$$

is a passive 8-bit word, according to (3.213).

Finally, we can recover $(Z_{15}^{(3)}, Z_{16}^{(3)})$ by checking candidate values $(z_{15,3}, z_{16,3})$ until

$$(c_{16}^i \odot (z_{16,3})^{-1}) \oplus (c_{15}^i \boxminus z_{15,3}), \tag{3.228}$$

is a passive 8-bit word, according to (3.213).

Similar to previous multiset analysis of MESH ciphers, there is uncertainty over the recovered value of multiplicative subkeys, or more precisely, between

z and $z \odot 0$; likewise, the most significant bit of additive subkeys cannot be uniquely identified.

Thus, using (3.221), we can recover about 28 bits of information from $(Z_1^{(1)}, Z_9^{(1)}, Z_1^{(3)}, Z_2^{(3)})$.

Embedding the guesses for $(Z_1^{(1)}, Z_9^{(1)})$ in the initial Λ-set costs 2^{16} combined $\odot$ and $\boxplus$ operations. Checking (3.221) once costs one $\odot$, one $\oplus$ and one $\boxminus$.

In comparison, 2.5-round MESH-128(8) costs 40 $\odot$, 40 $\boxplus$ and 48 $\oplus$. Assuming each $\odot$ costs 3 $\oplus$, and that $\oplus$, $\boxplus$ and $\boxminus$ cost about the same, then, this stage of the attack costs about $4 \cdot 2^8(2^{16} \cdot 4 + 5)/208 \approx 2^{20}$ 2.5-round MESH-128(8) computations.

The same attack can be repeated using (3.222) to recover $(Z_3^{(3)}, Z_9^{(3)})$ because $(Z_1^{(1)}, Z_9^{(1)})$ were recovered in the previous step. The complexity decreases to $2 \cdot 2^8 \cdot 5/208 \approx 2^4$ 2.5-round MESH-128(8). Since less bits are recovered this time, we only need two Λ-sets.

Likewise, we can use (3.223) to recover only $(Z_4^{(3)}, Z_{10}^{(3)})$ because $(Z_1^{(1)}, Z_5^{(1)})$ were already recovered. The time complexity is again 2^4 2.5-round MESH-128(8).

The same reasoning holds concerning (3.224) to recover only $(Z_5^{(3)}, Z_{11}^{(3)})$, and (3.225) to recover only $(Z_6^{(3)}, Z_{12}^{(3)})$, and (3.226) to recover $(Z_7^{(3)}, Z_{13}^{(3)})$, and (3.227) to recover $(Z_8^{(3)}, Z_{14}^{(3)})$, and (3.228) to recover $(Z_{15}^{(3)}, Z_{16}^{(3)})$. The time complexity is in all these cases is about 2^4 2.5-round MESH-128(8) computations. The combined attack complexity of these additional steps sum up to $7 \cdot 2^4 \approx 2^7$ 2.5-round MESH-128(8) computations.

The attack on (3.213) can now be repeated for (3.214), (3.215) until (3.220) to recover subkeys of the first KM half-round, since the subkeys $Z_j^{(3)}$, for $1 \leq j \leq 16$, were already recovered. This strategy means that the attack complexities will be reduced. For instance, using (3.214), we can recover $(Z_2^{(1)}, Z_{10}^{(1)})$ and check its correctness using the previously recovered values of $Z_j^{(3)}$ for $1 \leq j \leq 16$. The complexity is $2 \cdot 2^8 \cdot 2^{16} \cdot 4/208 \approx 2^{18}$ 2.5-round MESH-128(8) computations.

The data complexity is $2 \cdot 2^8 \cdot 2^{16} = 2^{25}$ chosen plaintexts (because only two subkeys are recovered this time). The memory complexity is 2^8 text blocks since the same storage can be reused for each Λ-set.

The same approach can be used with (3.215) until (3.220) to recover $(Z_3^{(1)}, Z_{11}^{(1)})$, $(Z_4^{(1)}, Z_{12}^{(1)})$, $(Z_5^{(1)}, Z_{13}^{(1)})$, $(Z_6^{(1)}, Z_{14}^{(1)})$, $(Z_7^{(1)}, Z_{15}^{(1)})$ and $(Z_8^{(1)}, Z_{16}^{(1)})$, respectively.

The overall time complexity is about $2^{22} + 7 \cdot 2^{18} \approx 2^{22.5}$ 2.5-round MESH-128(8) computations.

The subkeys $Z_j^{(1)}$ for $1 \leq j \leq 16$, lead immediately to recovery of an equivalent amount of user key information. But, the subkeys $Z_j^{(3)}$ for $1 \leq j \leq 16$ according to the key schedule of MESH-128(8) can be represented as:

$$Z_1^{(3)} = (((Z_5^{(1)} \boxplus Z_{15}^{(1)}) \oplus Z_{17}^{(2)}) \boxplus Z_{18}^{(2)}) \lll 1 \oplus c_{36},$$

$$Z_2^{(3)} = (((Z_6^{(1)} \boxplus Z_{16}^{(1)}) \oplus Z_{18}^{(2)}) \boxplus Z_1^{(3)}) \lll 1 \oplus c_{37},$$

$$Z_3^{(3)} = (((Z_7^{(1)} \boxplus Z_{17}^{(1)}) \oplus Z_1^{(3)}) \boxplus Z_2^{(3)}) \lll 1 \oplus c_{38},$$

$$Z_4^{(3)} = (((Z_8^{(1)} \boxplus Z_{18}^{(1)}) \oplus Z_2^{(3)}) \boxplus Z_3^{(3)}) \lll 1 \oplus c_{39},$$

$$Z_5^{(3)} = (((Z_9^{(1)} \boxplus Z_1^{(2)}) \oplus Z_3^{(3)}) \boxplus Z_4^{(3)}) \lll 1 \oplus c_{40},$$

$$Z_6^{(3)} = (((Z_{10}^{(1)} \boxplus Z_2^{(2)}) \oplus Z_4^{(3)}) \boxplus Z_5^{(3)}) \lll 1 \oplus c_{41},$$

$$Z_7^{(3)} = (((Z_{11}^{(1)} \boxplus Z_3^{(2)}) \oplus Z_5^{(3)}) \boxplus Z_6^{(3)}) \lll 1 \oplus c_{42},$$

$$Z_8^{(3)} = (((Z_{12}^{(1)} \boxplus Z_4^{(2)}) \oplus Z_6^{(3)}) \boxplus Z_7^{(3)}) \lll 1 \oplus c_{43},$$

$$Z_9^{(3)} = (((Z_{13}^{(1)} \boxplus Z_5^{(2)}) \oplus Z_7^{(3)}) \boxplus Z_8^{(3)}) \lll 1 \oplus c_{44},$$

$$Z_{10}^{(3)} = (((Z_{14}^{(1)} \boxplus Z_6^{(2)}) \oplus Z_8^{(3)}) \boxplus Z_9^{(3)}) \lll 1 \oplus c_{45},$$

$$Z_{11}^{(3)} = (((Z_{15}^{(1)} \boxplus Z_7^{(2)}) \oplus Z_9^{(3)}) \boxplus Z_{10}^{(3)}) \lll 1 \oplus c_{46},$$

$$Z_{12}^{(3)} = (((Z_{16}^{(1)} \boxplus Z_8^{(2)}) \oplus Z_{10}^{(3)}) \boxplus Z_{11}^{(3)}) \lll 1 \oplus c_{47},$$

$$Z_{13}^{(3)} = (((Z_{17}^{(1)} \boxplus Z_9^{(2)}) \oplus Z_{11}^{(3)}) \boxplus Z_{12}^{(3)}) \lll 1 \oplus c_{48},$$

$$Z_{14}^{(3)} = (((Z_{18}^{(1)} \boxplus Z_{10}^{(2)}) \oplus Z_{12}^{(3)}) \boxplus Z_{13}^{(3)}) \lll 1 \oplus c_{49},$$

$$Z_{15}^{(3)} = (((Z_1^{(2)} \boxplus Z_{11}^{(2)}) \oplus Z_{13}^{(3)}) \boxplus Z_{14}^{(3)}) \lll 1 \oplus c_{50},$$

and

$$Z_{16}^{(3)} = (((Z_2^{(2)} \boxplus Z_{12}^{(2)}) \oplus Z_{14}^{(3)}) \boxplus Z_{15}^{(3)}) \lll 1 \oplus c_{51}.$$

Thus, knowledge of $Z_j^{(1)}$ for $1 \leq j \leq 16$ is not sufficient to uniquely recover $Z_j^{(3)}$, for $1 \leq j \leq 16$. Furthermore, the latter does not translate into user key bits like the former. Therefore, the remaining $256-120=136$ key bits can be found by exhaustive search. The final time complexity becomes 2^{136} 2.5-round MESH-128(8) computations.

To attack 3-round MESH-128(8), the subkeys $(Z_{17}^{(3)}, Z_{18}^{(3)})$ of an additional MA half-round can be guessed and the previous attack on 2.5 rounds can be applied. The time complexity increases to $2^{136+16}=2^{152}$ 3-round MESH-128(8) computations.

3.15.7 Square/Multiset Attacks on REESSE3+

All internal operations in REESSE3+ are wordwise. This neat wordwise cipher framework makes it an ideal target to square attacks.

Square attacks on REESSE3+ work with λ-sets, or more specifically, multisets of 2^{16t} plaintexts in which different words (in a block) contain values with multiplicities ($t \geq 1$).

The design of the MA-box of REESSE3+ imply that the most effective pattern that can propagate across the MA-box, with certainty, has the form $(P,P,A,P) \stackrel{\text{MA}}{\rightarrow} (A,A,A,A)$, that is, a single A word in the third input position makes all four outputs also A. Thus, the pattern (P,P,A,P) preserves the highest number of active words across the MA-box among all patterns A, P, B and ?.

Other patterns were not as successful across the MA-box: $(A,P,P,P) \stackrel{\text{MA}}{\rightarrow} (?,?,?,A)$, $(P,P,P,A) \stackrel{\text{MA}}{\rightarrow} (A,A,?,A)$ and $(P,A,P,P) \stackrel{\text{MA}}{\rightarrow} (?,A,?,A)$.

The pattern $(P,P,A,P) \stackrel{\text{MA}}{\rightarrow} (A,A,A,A)$ translates into the following 1-round distinguishers: $(P,P,P,P,A,P,P,P) \stackrel{1r}{\rightarrow} (A,A,A,B,A,A,A,A)$ and $(P,P,P,P,P,P,A,P) \stackrel{1r}{\rightarrow} (A,A,A,A,A,B,A,A)$.

Further, propagating across an additional MA-box, respectively:

- the pattern $(B,?,B,B) \stackrel{\text{MA}}{\rightarrow} (?,?,?,?)$ means that the previous 1-round square distinguisher can be extended across an additional KM half-round: $(P,P,P,P,A,P,P,P) \stackrel{1.5r}{\rightarrow} (A,A,A,?,A,A,A,A)$.
- the pattern $(B,B,B,?) \stackrel{\text{MA}}{\rightarrow} (?,?,?,?)$ means that the previous 1-round square distinguisher can be extended across an additional KM half-round: $(P,P,P,P,P,P,A,P) \stackrel{1.5r}{\rightarrow} (A,A,A,A,A,?,A,A)$.

Both of them can be used as distinguishers of 2-round REESSE3+: from the output ciphertext Λ-set pick pairs of words that reconstruct the input of the MA-boxes from the bottom up. For instance, let (a,b,c,d,e,f,g,h) denote the output Λ-set. Then, the $(a \oplus c,\ b \oplus d,\ e \oplus g,\ f \oplus h)$ should be (B,B,B,B).

This distinguishing attack can be extended to a key-recovery attack on 2.5 rounds.

Without loss of generality, such an attack works as follows:

- choose 2^{16} plaintexts with word patterns according to the Λ-set (P, P, P, P, A, P, P, P); the Λ-set after 1.5 rounds is $(A, A, A, ?, A, A, A, A)$, but after two full rounds the Λ-set contains only ? in all eight word positions.
- guess the subkeys $Z_1^{(3)}$ and $Z_2^{(3)}$ (considering the word swap), and partially decrypt the last KM and MA half-round for all texts in the output Λ-set.
- keep the subkey values for which the decrypted Λ-set has the pattern B, the leftmost entry to the MA-box of the second round; each non-? word pattern is used as a distinguisher to test for the correct key guess.
- for a wrong subkey pair there is a chance of 2^{-16} that the word B is satisfied: essentially the integral should be 0 in all sixteen bits. Using two Λ-sets, we get an error probability of 2^{-32}. Since there are $2^{32}-1$ wrong subkeys, the expected number of wrong subkeys that survive this filtering is $(2^{32}-1)\cdot 2^{-32} < 1$, and only the correct subkey pair is expected to survive. The same procedure is repeated for the other four subkeys of the third KM half-round. The time complexities are the same in each case, but the same data can be used for all subkeys and the same storage.

The attack complexity is $2 \cdot 2^{16}$ chosen plaintexts (CP), $4 \cdot 2^{32}$ KM half-round partial computations, which means $2^{34} \cdot 0.5/2.5 \approx 2^{32}$ 2.5-round REESSE3+ computations, and 2^{17} blocks for storage.

Note that this attack applies equally well to three full rounds of REESSE3+. This fact follows from the absence of subkeys in the MA half-round (which can be easily removed by the adversary).

3.15.8 Square/Multiset Attacks on FOX/IDEA-NXT

The attacks described in this section are based on [181], in which the authors detail a 3-round multiset distinguisher for FOX128.

Let $P = (LL^1, LR^1, RL^1, RR^1)$ denote a 128-bit plaintext block, where LL^1, LR^1, RL^1, $RR^1 \in \mathbb{Z}_2^{32}$. Consider the following multiset construction: $LL^1 = LR^1 = (c, c, c, c)$ and $RL^1 = RR^1 = (c, c, c, x)$, where x takes all possible 8-bit values and c is an arbitrary, fixed constant. The input of the first f64 round function is $X^1 = (0, 0, 0, 0, 0, 0, 0, 0)$ because $LL^1 = LR^1$ and $RL^1 = RR^1$.

Let $\text{f64}(0, 0, 0, 0, 0, 0, 0, 0) = (a_0, a_1, a_2, a_3, a_4, a_5, a_6, a_7, a_8)$, where a_i, for $0 \le i \le 7$, are determined by the round subkey $RK^1 = (RK_0^1, RK_1^1)$. So, a_i, for $0 \le i \le 7$, are constants when the user key is fixed. The first round output becomes $LL^2 = (a_2 \oplus c,\ a_3 \oplus c,\ a_0 \oplus a_2,\ a_1 \oplus a_3)$, $LR^2 = (a_0 \oplus c,$

$a_1 \oplus c$, $a_2 \oplus c$, $a_3 \oplus c)$, $RL^2 = (a_6 \oplus c$, $a_7 \oplus x$, $a_4 \oplus a_6$, $a_5 \oplus a_7 \oplus x \oplus c)$ and $RR^2 = (a_4 \oplus c$, $a_5 \oplus c$, $a_6 \oplus c$, $a_7 \oplus x)$.

Therefore, the second round input to f64 becomes $X^2 = (X_0^2, X_1^2, X_2^2, X_3^2, X_4^2, X_5^2, X_6^2, X_7^2)$, where $X_i^2 \in \mathbb{Z}_2^8$ for $0 \leq i \leq 7$: $X_0^2 = a_0 \oplus a_2$, $X_1^2 = a_1 \oplus a_3$, $X_2^2 = a_0 \oplus c$, $X_3^2 = a_1 \oplus c$, $X_4^2 = a_4 \oplus a_6$, $X_5^2 = a_5 \oplus a_7 \oplus c \oplus x$, $X_6^2 = a_4 \oplus c$ and $X_7^2 = a_5 \oplus c$.

Let $\text{f64}(X^2) = (y_0, y_1, y_2, y_3, y_4, y_5, y_6, y_7)$, where $y_i \in \mathbb{Z}_2^8$ for $0 \leq i \leq 7$. Then, the second round output becomes $LL^3 = (y_2 \oplus a_0 \oplus a_2$, $y_3 \oplus a_1 \oplus a_3$, $y_0 \oplus y_2 \oplus a_0 \oplus c$, $y_1 \oplus y_3 \oplus a_1 \oplus c)$, $LR^3 = (y_0 \oplus a_0 \oplus c$, $y_1 \oplus a_1 \oplus c$, $y_2 \oplus a_2 \oplus c$, $y_3 \oplus a_3 \oplus c)$, $RL^3 = (y_6 \oplus a_4 \oplus a_6$, $y_7 \oplus a_5 \oplus a_7 \oplus c \oplus x$, $y_4 \oplus y_6 \oplus a_4 \oplus c$, $y_5 \oplus y_7 \oplus a_5 \oplus c)$ and $RR^3 = (y_4 \oplus a_4 \oplus c$, $y_5 \oplus a_5 \oplus c$, $y_6 \oplus a_6 \oplus c$, $y_7 \oplus a_7 \oplus x)$.

Therefore, the input to f64 in the third round becomes $X^3 = (X_0^3, X_1^3, X_2^3, X_3^3, X_4^3, X_5^3, X_6^3, X_7^3)$, where $X_i^3 \in \mathbb{Z}_2^8$ for $0 \leq i \leq 7$:

$X_0^3 = y_0 \oplus y_2 \oplus a_2 \oplus c$, $X_1^3 = y_1 \oplus y_3 \oplus a_3 \oplus c$, $X_2^3 = y_0 \oplus a_0 \oplus a_2$, $X_3^3 = y_1 \oplus a_1 \oplus a_3$, $X_4^3 = y_4 \oplus y_6 \oplus a_6 \oplus c$, $X_5^3 = y_5 \oplus y_7 \oplus a_7 \oplus x$, $X_6^3 = y_4 \oplus a_4 \oplus a_6$, $X_7^3 = y_5 \oplus a_5 \oplus a_7 \oplus x$.

From the high-level structure of FOX128, it follows that

$$\text{io}(LL^4) \oplus LR^4 = (X_0^3, X_1^3, X_2^3, X_3^3),$$

and

$$\text{io}(RL^4) \oplus RR^4 = (X_4^3, X_5^3, X_6^3, X_7^3).$$

From the definition of $\text{io} = \text{or}^{-1}$, it follows that

$$\text{io}(LL^4) = (LL_0^4 \oplus LL_2^4, LL_1^4 \oplus LL_3^4, LL_0^4 \oplus LL_1^4),$$

and

$$\text{io}(RL^4) = (RL_0^4 \oplus RL_2^4, RL_1^4 \oplus RL_3^4, RL_0^4 \oplus RL_1^4).$$

Thus, we get the following relations:

$$LL_0^4 \oplus LL_2^4 \oplus LR_0^4 = y_0 \oplus y_2 \oplus a_2 \oplus c, \tag{3.229}$$

$$LL_1^4 \oplus LL_3^4 \oplus LR_1^4 = y_1 \oplus y_3 \oplus a_3 \oplus c, \tag{3.230}$$

$$LL_0^4 \oplus LR_2^4 = y_0 \oplus a_0 \oplus a_2, \tag{3.231}$$

$$LL_1^4 \oplus LR_3^4 = y_1 \oplus a_a \oplus a_3, \tag{3.232}$$

$$RL_0^4 \oplus RL_2^4 \oplus RR_0^4 = y_4 \oplus y_6 \oplus a_6 \oplus c, \tag{3.233}$$

$$RL_1^4 \oplus RL_3^4 \oplus RR_1^4 = y_5 \oplus y_7 \oplus a_7 \oplus X, \tag{3.234}$$

$$RL_0^4 \oplus RR_2^4 = y_4 \oplus a_4 \oplus a_6, \tag{3.235}$$

$$RL_1^4 \oplus RR_3^4 = y_5 \oplus a_5 \oplus a_7 \oplus x. \tag{3.236}$$

Further, the following relations also hold:

$$LL_2^4 \oplus LR_0^4 \oplus LR_2^4 = y_2 \oplus a_0 \oplus c, \tag{3.237}$$

$$LL_3^4 \oplus LR_3^4 \oplus LR_3^4 = y_3 \oplus a_1 \oplus c, \tag{3.238}$$

$$LL_0^4 \oplus LR_2^4 = y_0 \oplus a_0 \oplus a_2, \tag{3.239}$$

$$LL_1^4 \oplus LR_3^4 = y_1 \oplus a_1 \oplus a_3, \tag{3.240}$$

$$RL_2^4 \oplus RR_0^4 \oplus RR_2^4 = y_6 \oplus a_4 \oplus c, \tag{3.241}$$

$$RL_3^4 \oplus RR_1^4 \oplus RR_3^4 = y_7 \oplus a_5, \tag{3.242}$$

$$RL_0^4 \oplus RR_2^4 = y_4 \oplus a_4 \oplus a_6. \tag{3.243}$$

Now, we analyze the y_i values, $0 \leq i \leq 7$. Let $y = S[x \oplus a_5 \oplus a_7 \oplus c \oplus RK_0^2]$, where S denotes the bijective S-box used in FOX128. Then, $y_i = S[y \oplus b_i] \oplus RK_{0i}^2$, where b_i, for $0 \leq i \leq 7$, are entirely determined by a_i, c and RK^2. So, b_i are constants when the user key is fixed.

Since S is bijective, $y = S[x \oplus a_5 \oplus a_7 \oplus c \oplus RK_0^2]$ differs when x takes different values, and the user key is fixed. As a consequence, $y_i = S[y \oplus b_i] \oplus RK_{0i}^2$ will have different values when x varies and the user key is fixed.

Thus, we know that $LL_2^4 \oplus LR_0^4 \oplus LR_2^4$, $LL_3^4 \oplus LR_1^4 \oplus LR_3^4$, $LL_0^4 \oplus LR_2^4$, $LL_1^4 \oplus LR_3^4$, $RL_2^4 \oplus RR_0^4 \oplus RR_2^4$, $RL_3^4 \oplus RR_1^4 \oplus RR_3^4$ and $RL_0^4 \oplus RR_2^4$ will each have different values when x varies. Therefore, the following theorem holds:

Theorem 3.2. *[181] Let $P = (LL^1, LR^1, RL^1, RR^1)$ and $P^* = (LL^{1*}, LR^{1*}, RL^{1*}, RR^{1*})$ be two plaintext inputs to 3-round FOX128. Let $C = (LL^4, LR^4, RL^4, RR^4)$ and $C^* = (LL^{4*}, LR^{4*}, RL^{4*}, RR^{4*})$ be the corresponding ciphertexts. RR_i, for $0 \leq i \leq 7$, denotes the $(i+1)$-th byte of RR. If $LL^1 = LR^1 = LL^{1*} = LR^{1*}$, $RL^1 = RR^1$, $RL^{1*} = RR^{1*}$, $RR_i^1 = RR_i^{1*}$, for $i \in \{0, 1, 2\}$, $RR_3^1 \neq RR_3^{1*}$, then C and C^* satisfy the following inequalities:*

$$LL_2^4 \oplus LR_0^4 \oplus LR_2^4 \neq LL_2^{4*} \oplus LR_0^{4*} \oplus LR_2^{4*}, \tag{3.244}$$

$$LL_3^4 \oplus LR_1^4 \oplus LR_3^4 \neq LL_3^{4*} \oplus LR_1^{4*} \oplus LR_3^{4*}, \tag{3.245}$$

$$LL_0^4 \oplus LR_2^4 \neq LL_0^{4*} \oplus LR_2^{4*}, \tag{3.246}$$

$$LL_1^4 \oplus LR_3^4 \neq LL_1^{4*} \oplus LR_3^{4*}, \tag{3.247}$$

$$RL_2^4 \oplus RR_0^4 \oplus RR_2^4 \neq RL_2^{4*} \oplus RR_0^{4*} \oplus RR_2^{4*}, \tag{3.248}$$

$$RL_3^4 \oplus RR_1^4 \oplus RR_3^4 \neq RL_3^{4*} \oplus RR_1^{4*} \oplus RR_3^{4*}, \tag{3.249}$$

$$RL_0^4 \oplus RR_2^4 \neq RL_0^{4*} \oplus RR_2^{4*}. \tag{3.250}$$

From the previous arguments, $RL_1^4 \oplus RR_3^4 = y_5 \oplus x \oplus a_5 \oplus a_7$, and y_5 assumes different values when x varies. So, we obtain the following corollary, which is similar to the multiset distinguisher in [92, 139].

Corollary 3.4. *[181] Let $P_j = (LL_j^1, LR_j^1, RL_j^1, RR_j^1)$, for $0 \leq j \leq 255$, be 256 plaintexts. Let $C_j = (LL_j^4, LR_j^4, RL_j^4, RR_j^4)$ denote the corresponding ciphertexts after 3-round FOX128. If $LL_j^1 = LR_j^1$, $RL_j^1 = RR_j^1$, RL_{j_i} for $i \in \{0,1,2\}$ are constants, and LR_{j_3} take all possible byte values, then C_j for $0 \leq j \leq 255$ satisfy*

$$\bigoplus_{j=0}^{255}(RL_{j_1}^4 \oplus RR_{j_3}^4) = 0. \tag{3.251}$$

The first multiset attack is on 4-round FOX128 where the last orthomorphism transformation is omitted. Initially, 72 bits of subkeys RK_0^4 and $RK_{1_0}^4$ will be recovered, where $RK_{1_0}^4$ stands for the first, leftmost byte of RK_1^4.

Let $P = (LL^1, LR^1, RL^1, RR^1)$ denote a plaintext block and $C = (LL^5, LR^5, RL^5, RR^5)$ denote the corresponding ciphertext block. The input to the fourth f64 round function is $(LL^5 \oplus LR^5, RL^5 \oplus RR^5)$. Note that $LL_2^4 \oplus LR_2^4 = LL_2^5 \oplus LR_2^5$.

If we guess the value of LR_0^4 then we can guess $LL_2^4 \oplus LR_2^4 \oplus LR_0^4$. From the structure of the f64 function, the value of LR_0^4 is fully determined from the input $(LL^5 \oplus LR^5, RL^5 \oplus RR^5)$, RK_0^4 and $RK_{1_0}^4$.

Using inequality (3.244) of Theorem 3.2, the following algorithm will be used to recover RK_0^4 and $RK_{1_0}^4$:

Algorithm 1:

(1) choose 166 plaintexts $P_j = (LL_j^1, LR_j^1, RL_j^1, RR_j^1)$, for $0 \leq j \leq 165$, as follows: $LL_j^1 = (c,c,c,c)$, $LR_j^1 = (c,c,c,c)$, $RL_j^1 = (c,c,c,j)$, $RR_j^1 = (c,c,c,j)$, where c is a constant and $0 \leq j \leq 165$. The corresponding ciphertext blocks are $C_j = (LL_j^5, LR_j^5, RL_j^5, RR_j^5)$.

(2) for each possible value of $(RK_0^4, RK_{1_0}^4)$, compute the first byte Y_{j0}^4 of f64$(LL_j^5 \oplus LR_j^5, RL_j^5 \oplus RR_j^5)$ and then compute

$$\Delta_j = Y_{j0}^5 \oplus LL_{j_2}^5 \oplus LR_{j_2}^5 \oplus LR_{j_0}^5. \tag{3.252}$$

Check for a collision among the Δ_j. If a collision is found, then discard the value of $(RK_0^4, RK_{1_0}^4)$. Otherwise, output the current value of $(RK_0^4, RK_{1_0}^4)$.

(3) from the values output in step (2), choose some other plaintexts and repeat step (2). The values output in repeated trials are the correct ones.

The probability of at least one collision can be modeled as the following event: throw 166 balls into 256 bins at random, and look for a bin with more than one ball. This probability is larger than $1 - e^{-166\cdot 165/2^9} \geq 1 - 2^{-76}$. So, the probability of passing step (2) is less than 2^{-76}. Since the right subkey must pass step (2), the number of subkey candidates passing step (2) is about $1 + 2^{72} \cdot 2^{-76} \approx 1.06$. Then, only two plaintexts are needed in step (3).

The data complexity of the attack is about 168 chosen plaintexts.

The time complexity is mainly due to step (2), the computation of the Δ_j, which is less than 1-round FOX128 computation. So, the time complexity is less than $2^{72} \cdot 168/4 \approx 42 \cdot 2^{72}$ 4-round FOX128 encryptions.

For recovering $RK_{1_1}^4$, the steps are similar to Algorithm 1, except that RK_0^4 is already known. The number of subkey candidates is 2^8. Only 64 chosen plaintexts are needed since we can reuse the previous data in Algorithm 1 again.

Using inequality (3.245) in Theorem 3.2, compute

$$\Delta_j = Y_{j_1}^4 \oplus LL_{j_3}^5 \oplus LR_{j_3}^5 \oplus LR_{j_1}^5. \tag{3.253}$$

This part of the attack requires $2^8 \cdot 64/4 = 2^{12}$ 4-round FOX128 encryptions.

Using the recovered RK_0^4 and $RK_{1_0}^4$, inequality (3.246) in Theorem 3.2, and the chosen plaintexts in Algorithm 1, we can recover $RK_{1_2}^4$ by computing

$$\Delta_j = Y_{j_0}^4 \oplus Y_{j_2}^4 \oplus LL_{j_0}^5 \oplus LR_{j_2}^5. \tag{3.254}$$

This part of the attack requires 2^{12} 4-round FOX128 encryptions.

Similarly, from the recovered RK_0^4, $RK_{1_1}^4$, inequality (3.247) in Theorem 3.2, and the chosen plaintexts in Algorithm 1, we can recover $RK_{1_3}^4$ by computing

$$\Delta_j = Y_{j_1}^4 \oplus Y_{j_3}^4 \oplus LL_{j_1}^5 \oplus LR_{j_3}^5. \tag{3.255}$$

This part of the attack requires 2^{12} 4-round FOX128 encryptions.

Further, using inequality (3.248) in Theorem 3.2 and the chosen plaintexts in Algorithm 1, we can recover $RK_{1_4}^4$ by computing

$$\Delta_j = Y_{j_4}^4 \oplus RL_{j_2}^5 \oplus RR_{j_2}^5 \oplus RR_{j_0}^5. \tag{3.256}$$

This part of the attack requires 2^{12} 4-round FOX128 encryptions.

Using inequality (3.249) in Theorem 3.2 and the chosen plaintexts in Algorithm 1, we can recover $RK_{1_5}^4$ by computing

$$\Delta_j = Y^4_{j_5} \oplus RL^5_{j_3} \oplus RR^5_{j_3} \oplus RR^5_{j_1}. \tag{3.257}$$

This part of the attack requires 2^{12} 4-round FOX128 encryptions.

Using the recovered RK^4_0 $RK^4_{1_4}$, inequality (3.250) in Theorem 3.2 and the chosen plaintexts in Algorithm 1, we can recover $RK^4_{1_6}$ by computing

$$\Delta_j = Y^4_{j_4} \oplus RL^5_{j_0} \oplus Y^5_{j_6} \oplus RR^5_{j_2}. \tag{3.258}$$

This part of the attack requires 2^{12} 4-round FOX128 encryptions.

From RK^4_0, $RK^4_{1_5}$ and inequality (3.250) in Theorem 3.2, we can use the following algorithm to recover $RK^4_{1_7}$:

Algorithm 2

(1) choose 256 plaintexts $P_j = (LL^1_j, LR^1_j, RL^1_j, RR^1_j)$, for $0 \leq j \leq 255$ as follows: $LL^1_j = (c, c, c, c)$, $LR^1_j = (c, c, c, c)$, $RL^1_j = (c, c, c, j)$, $RR^1_j = (c, c, c, j)$, where c is a constant and $0 \leq j \leq 255$. The corresponding ciphertext blocks are $C_j = (LL^5_j, LR^5_j, RL^5_j, RR^5_j)$.

(2) for each possible value of $RK^4_{1_7}$, compute $Y^4_{j_7}$ and then

$$\Delta = \bigoplus_{j=0}^{255} (RL^5_1 \oplus RR^5_3 \oplus Y^4_{j_5} \oplus Y^4_{j_7}). \tag{3.259}$$

Check if $\Delta = 0$. If not, then discard the value of $RK^4_{1_7}$. Otherwise, output the value of $RK^4_{1_7}$.

(3) from the output values in step (2), choose another group of plaintexts, and repeat step (2) until the key candidate is uniquely determined.

Wrong subkey candidates will pass step (2) with probability 2^{-8}. Thus, Algorithm 2 requires about 2^9 chosen plaintexts. The time complexity is about $2^9 \cdot 2^8/4 = 2^{15}$ 4-round FOX128 encryptions. The data in Algorithm 1 can be repeatedly used again, so the data complexity for recovering RK^4 is about 2^9 and the time complexity is about $42 \cdot 2^{72} + 6 \cdot 2^{12} + 2^{15}$ 4-round FOX128 encryptions.

By decrypting the fourth round, we can recover RK^3. The time complexity is less than $2^{73} + 6 \cdot 2^{12} + 2^{15}$ 4-round FOX128 encryptions.

Similarly, we can recover RK^2 and RK^1. The time complexity for both subkeys is less than $2^{73} + 6 \cdot 2^{!2} + 2^{15}$ 4-round FOX128 encryptions.

In summary, the attack requires 2^9 chosen plaintexts and about $2^{77.6}$ 4-round FOX128 encryptions.

This attack can be extended to 5-round FOX128 by guessing RK^5. Only the time complexity changes to $2^{205.6}$ 5-round FOX128 encryptions.

To attack FOX64, the following theorem and corollary are given:

Theorem 3.3. *[181] Let $P = (L^1, R^1)$ and $P^* = (L^{1*}, R^{1*})$ denote two 64-bit plaintext blocks for 3-round FOX64. Let $C = (L^4, R^4)$ and $C^* = (L^{4*}, R^{4*})$ denote the corresponding ciphertext blocks. Let L_i denote the $(i+1)$-th byte of L. If $L^1 = R^1$, $L^{1*} = R^{1*}$ and $L^1_i = L^{1*}_i$ for $i \in \{0, 1, 2\}$, $L^1_3 \neq L^{1*}_3$, then C and C^* satisfy the following inequalities:*

$$L_2^4 \oplus R_2^4 \oplus R_0^4 \neq L_2^{4*} \oplus R_2^{4*} \oplus R_0^{4*}, \tag{3.260}$$

$$L_3^4 \oplus R_3^4 \oplus R_1^4 \neq L_3^{4*} \oplus R_3^{4*} \oplus R_1^{4*}, \tag{3.261}$$

$$L_0^4 \oplus R_2^4 \neq L_0^{4*} \oplus R_2^{4*}. \tag{3.262}$$

Corollary 3.5. *[181] Let $P_j = (L_j^1, R_j^1)$, for $0 \leq j \leq 255$, denote 256 plaintext. Let $C_j = (L_j^4, R_j^4)$ denote the corresponding ciphertext blocks after 3-round FOX64. If $L_j^1 = R_j^1$, $L_{j_i}^1$ for $i \in \{0,1,2\}$ are constants and L_{j_3} take all possible values between 0 and 255, then C_j, for $0 \leq j \leq 255$, satisfy*

$$\bigoplus_{j=0}^{255} (L_{j_1}^4 \oplus R_{j_3}^4) = 0. \tag{3.263}$$

Using Theorem 3.3 and Corollary 3.5, similar algorithms can be constructed to recover subkeys of 4-round FOX64. The attacks require about 2^9 chosen plaintexts, and the time complexity is about $2^{45.4}$ 4-round FOX64 encryptions. This attack can be extended to 5-, 6- and 7-round FOX64 by guessing additional round subkeys. The time complexities for 5-, 6- and 7-round FOX64 increase to $2^{109.4}$, $2^{173.4}$ and $2^{237.4}$ 5-, 6- and 7-round FOX64 encryptions, respectively.

3.16 Demirci Attack

In [60], Demirci presented a dedicated square attack on reduced-round variants of IDEA. See Sect. 3.15.1.2, where the Demirci relation (Lemma 3.5) was described.

In the following sections, we describe such attacks on other Lai-Massey cipher designs.

3.16.1 Demirci Attack on MESH-64

Demirci's attack using 1st-order integrals [58, 107] can be adapted to attack reduced-round MESH-64, but starting from the second round, or any other even-numbered round, because in these rounds the two middle words are combined with subkeys using modular addition operation. This fact is a consequence of the design of MESH-64. The use of $\odot$ in odd-numbered rounds invalidates the Demirci relation in Lemma 3.5.

Consider for instance, a 64-bit plaintext multiset of the form (P, P, P, A), that is, where the first three 16-bit words are constants (passive) while the fourth word is active.

Let $X^{(i)} = (X_1^{(i)}, X_2^{(i)}, X_3^{(i)}, X_4^{(i)})$ denote the i-th input text block in a Λ-set. So, for $i = 1$, $X^{(1)}$ denotes the plaintext blocks in a Λ-set. Let $Y^{(i)} = (Y_1^{(i)}, Y_2^{(i)}, Y_3^{(i)}, Y_4^{(i)})$ denote the internal cipher data in a Λ-set after the i-th KM half-round.

Let (n_i, q_i) denote the input to the i-th MA-box, and (r_i, s_i) denote the corresponding output. Moreover, let $s_i = r_i \boxplus t_i$, where $t_i = (n_i \odot Z_5^{(i)} \boxplus q_i) \odot Z_6^{(i)}$.

After 1-round of MESH-64, the output multiset has the form $(?, ?, A, *)$, that is, the first two words are garbled, the third word is active and the fourth word is balanced. The multiset after 1.5 rounds becomes $(?, ?, A, ?)$, but the least significant bit of $Y_2^{(3)}$ is constant because it is a combination of only active words from the MA-box of the previous round.

Demirci's attack exploits the property that the integral, restricted to the least significant bit, is constant along a path across 1-round MESH-64, that involves only modular addition and exclusive-or operations. These operations preserve the integral value of the least significant bit over a multiset, as long as only balanced words are added to the integral. This property can be used as a distinguisher to attack 2-round MESH-64, from the second to the third rounds.

Consider $X^{(2)}$, $X^{(3)}$ and $X^{(4)}$. From the ciphertext, $X^{(4)}$, we have

$$\mathrm{lsb}_1(X_2^{(4)} \oplus s_3 \oplus X_3^{(4)} \oplus r_3) = \mathrm{lsb}_1(X_2^{(3)} \oplus Z_2^{(3)} \oplus X_3^{(3)} \oplus Z_3^{(3)}), \quad (3.264)$$

which exploits the fact that the least significant bit position is linear across modular addition. Thus, (3.264) holds with certainty for any internal cipher data and round subkeys.

Further,

$$\mathrm{lsb}_1(r_3 \oplus s_3) = \mathrm{lsb}_1(t_3) = \mathrm{lsb}_1((n_3 \odot Z_5^{(3)} \boxplus q_3) \odot Z_6^{(3)}). \quad (3.265)$$

Note that n_3 and q_3 can be replaced by the ciphertext words as $n_3 = X_1^{(4)} \oplus X_2^{(4)}$ and $q_3 = X_3^{(4)} \oplus X_4^{(4)}$, assuming the swap of two middle words is still there.

Combining (3.264), (3.265) with the values of n_3 and q_3, we have

$$\mathrm{lsb}_1(X_2^{(4)} \oplus X_3^{(4)} \oplus X_2^{(3)} \oplus Z_2^{(3)} \oplus X_3^{(3)} \oplus Z_3^{(3)}) =$$
$$\mathrm{lsb}_1(((X_1^{(4)} \oplus X_2^{(4)}) \odot Z_5^{(3)} \boxplus (X_3^{(4)} \oplus X_4^{(4)})) \odot Z_6^{(3)}). \quad (3.266)$$

Note that (3.266) depends on the least significant bits of the ciphertext $X^{(4)}$ and the internal state $X^{(3)}$. In particular, the least significant bits of $X_2^{(3)}$ and $X_3^{(3)}$ are balanced, according to the multiset used for the attack.

Over the given multiset, the integral of (3.266) is zero because the subkeys are fixed and the intermediate values have the least significant bits balanced.

Therefore, (3.266) represents a 1-bit condition that holds with certainty. It can be used to recover $Z_5^{(3)}$ and $Z_6^{(3)}$.

The attack requires $32 \cdot 2^{16} = 2^{21}$ chosen plaintexts and an effort of $2^{32} \cdot 2^{16} + 2^{31} \cdot 2^{16} + \ldots + 2 \cdot 2^{16} \approx 2^{49}$ computations of (3.266). Assuming $\odot$ costs about three times an $\oplus$ and that an $\boxplus$ and an $\oplus$ cost about the same, then, computing 2-round MESH-64 costs about 52 $\oplus$, and one computation of (3.266) costs about 14 $\oplus$. Thus, the time complexity is estimated at 2^{47} 2-round MESH-64 computations.

The memory complexity is 2^{16} text blocks since the same storage can be reused for each new Λ-set.

Unfortunately, (3.266) does not provide information to recover the remaining subkeys of 2-round MESH-64. Moreover, according to the key schedule of MESH-64, the knowledge of $Z_5^{(3)}$ and $Z_6^{(3)}$ is not enough to uniquely recover the remaining subkeys. An exhaustive search for the remaining subkeys of 2-round MESH-64 would need to recover at least four subkeys. For instance, $Z_6^{(3)} = (((((Z_5^{(2)} \boxplus Z_6^{(2)}) \oplus Z_7^{(2)}) \boxplus Z_3^{(3)}) \oplus Z_4^{(3)}) \boxplus Z_5^{(3)}) \lll 7 \oplus c_{19}$ means that in order to recover $Z_5^{(2)}$, knowledge of $Z_6^{(2)}$, $Z_7^{(2)}$, $Z_3^{(3)}$ and $Z_4^{(3)}$ is needed.

3.16.2 Demirci Attack on MESH-96

Demirci's attack using 1st-order integrals [58, 107] can be adapted to attack reduced-round MESH-96, starting from the first round.

Consider a 96-bit plaintext multiset of the form (P, P, P, P, P, A), that is, where the first five 16-bit words are constants (passive) while the sixth word is active.

Let $X^{(i)} = (X_1^{(i)}, X_2^{(i)}, X_3^{(i)}, X_4^{(i)}, X_5^{(i)}, X_6^{(i)})$ denote the i-th input text block in a Λ-set. So, for $i = 1$, $X^{(1)}$ denotes the plaintext blocks in a Λ-set. Let $Y^{(i)} = (Y_1^{(i)}, Y_2^{(i)}, Y_3^{(i)}, Y_4^{(i)}, Y_5^{(i)}, Y_6^{(i)})$ denote the internal cipher data in a Λ-set after the i-th KM half-round.

Let (n_i, q_i, m_i) denote the input to the i-th MA-box, and (r_i, s_i, t_i) denote the corresponding output. Moreover, let $o_i = n_i \odot Z_7^{(i)}$, $w_i = o_i \boxplus q_i$, $j_i = w_i \odot m_i$, $l_i = j_i \boxplus Z_8^{(i)}$, $v_i = w_i \odot l_i$, $u_i = o_i \boxplus v_i$. Then, $r_i = u_i \odot Z_9^{(i)}$ and $s_i = r_i \boxplus v_i$.

After 1-round MESH-96, the output multiset has the form $(?, ?, ?, ?, A, *)$, that is, the first four words are garbled, the fifth word is active and the sixth word is balanced. The multiset after 1.5 rounds becomes $(?, ?, ?, ?, A, ?)$, but the least significant bit of $Y_6^{(2)}$ is constant because it is a combination of active words only from the MA-box of the previous round.

Consider $X^{(1)}$, $X^{(2)}$ and $X^{(3)}$. From the ciphertext, $X^{(3)}$, we have

$$\text{lsb}_1(X_3^{(3)} \oplus s_2 \oplus X_5^{(3)} \oplus r_2) = \text{lsb}_1(X_3^{(2)} \oplus Z_3^{(2)} \oplus X_5^{(2)} \oplus Z_5^{(2)}), \quad (3.267)$$

which exploits the fact that the least significant bit position is linear across modular addition. Thus, (3.267) holds with certainty for any internal cipher data and round subkeys.

Further,

$$\mathrm{lsb}_1(r_2 \oplus s_2) = \mathrm{lsb}_1(v_2) = \mathrm{lsb}_1(w_2 \odot l_2) = \mathrm{lsb}_1((o_2 \oplus q_2) \odot (j_2 \boxplus Z_8^{(2)})). \tag{3.268}$$

Note that n_3, q_3 and m_3 can be replaced by ciphertext words as $n_3 = X_1^{(3)} \oplus X_2^{(3)}$, $q_3 = X_3^{(3)} \oplus X_4^{(3)}$ and $m_3 = X_5^{(3)} \oplus X_6^{(3)}$, assuming the swap of four middle words is still there.

Combining (3.267), (3.268) with the values of n_3, q_3, m_3, we have

$$\mathrm{lsb}_1(X_3^{(3)} \oplus X_5^{(3)} \oplus X_3^{(2)} \oplus Z_3^{(2)} \oplus X_5^{(2)} \oplus Z_5^{(2)}) =$$
$$\mathrm{lsb}_1(((n_2 \odot Z_7^{(2)}) \boxplus q_2) \odot (Z_8^{(2)} \boxplus (m_2 \odot ((n_2 \odot Z_7^{(2)}) \boxplus q_2)))). \tag{3.269}$$

Note that (3.269) depends on the least significant bits of the ciphertext blocks and the internal state. In particular, the least significant bits of $X_3^{(2)}$ and $X_5^{(2)}$ are balanced, according to the multiset used for the attack.

Over the given multiset, the integral of (3.269) is zero because the subkeys are fixed and the intermediate values have the least significant bits balanced.

Therefore, (3.269) represents a 1-bit condition that holds with certainty. It can be used to recover $Z_7^{(2)}$ and $Z_8^{(2)}$. Note that the least significant bit of $Z_3^{(2)}$ and $Z_5^{(2)}$ vanish in the integral.

The attack requires $32 \cdot 2^{16} = 2^{21}$ chosen plaintexts and an effort of $2^{32} \cdot 2^{16} + 2^{31} \cdot 2^{16} + \ldots + 2 \cdot 2^{16} \approx 2^{49}$ computations of (3.269). Assuming $\odot$ costs about three times an $\oplus$ and that an $\boxplus$ and an $\oplus$ cost about the same, then, computing 2-round MESH-96 costs about 80 $\oplus$, and one computation of (3.269) costs about 16 $\oplus$. Thus, the time complexity is estimated at 2^{47} 2-round MESH-96 computations.

The memory complexity is 2^{16} text blocks since the same storage can be reused for each new Λ-set.

Unfortunately, (3.269) does not provide information to recover the remaining subkeys of 2-round MESH-96. Moreover, according to the key schedule of MESH-96, the knowledge of $Z_7^{(2)}$ and $Z_8^{(2)}$ is not enough to uniquely recover the remaining subkeys. The situation is similar to the attack on 2-round MESH-64 in Sect. 3.16.1.

3.16.3 Demirci Attack on MESH-128

Demirci's attack using 1st-order integrals [58, 107] can be adapted to attack reduced-round MESH-128, starting from the second round, or any other even-numbered round, because the different KM half-rounds for odd- and even-numbered rounds. This is a consequence of the design of MESH-128. Otherwise, the presence of $\odot$ along the path of the Demirci relation (Lemma 3.5) would invalidate it.

Consider a 128-bit plaintext multiset of the form (P, P, P, P, P, P, P, A), that is, where the first seven 16-bit words are constants (passive) while the eighth word is active.

Let $X^{(i)} = (X_1^{(i)}, X_2^{(i)}, X_3^{(i)}, X_4^{(i)}, X_5^{(i)}, X_6^{(i)}, X_7^{(i)}, X_8^{(i)})$ denote the i-th input text block in a Λ-set. So, for $i = 1$, $X^{(1)}$ denotes the plaintext blocks in a Λ-set. Let $Y^{(i)}$ =($Y_1^{(i)}$, $Y_2^{(i)}$, $Y_3^{(i)}$, $Y_4^{(i)}$, $Y_5^{(i)}$, $Y_6^{(i)}$, $Y_7^{(i)}$, $Y_8^{(i)}$) denote the internal cipher data in a Λ-set after the i-th KM half-round.

Let (n_i, q_i, m_i, u_i) denote the input to the i-th MA-box, and (r_i, s_i, t_i, v_i) denote the corresponding output. Moreover, let $h_i = m_i \odot q_i$, $g_i = u_i \boxplus h_i$, $l_i = g_i \odot Z_{10}^{(i)}$, $o_i = h_i \boxplus l_i$, $j_i = o_i \odot (q_i \boxplus n_i \odot Z_9^{(i)})$, $w_i = j_i \boxplus n_i \odot Z_9^{(i)}$, $y_i = w_i \odot Z_{11}^{(i)}$. Then, $r_i = s_i \boxplus y_i$.

After 1-round MESH-128, the output multiset is $(?, ?, ?, ?, ?, ?, ?, ?)$, that is, all words are garbled. The reason is that the MA-box of MESH-128 has four layers of $\boxplus$ and $\odot$. All further multisets have the form (?, ?, ?, ?, ?, ?, ?, ?). But, the least significant bit of $r_1 \oplus s_1$ is balanced since y_1 is an active word.

Consider $X^{(2)}$, $X^{(3)}$ and $X^{(4)}$. From the ciphertext, $X^{(4)}$, we have

$$\text{lsb}_1(X_4^{(4)} \oplus s_3 \oplus X_7^{(4)} \oplus r_3) = \text{lsb}_1(X_4^{(3)} \oplus Z_4^{(3)} \oplus X_7^{(3)} \oplus Z_7^{(3)}), \quad (3.270)$$

which exploits the fact that the least significant bit position is linear across modular addition. Thus, (3.270) holds with certainty for any internal cipher data and round subkeys.

Further,

$$\text{lsb}_1(r_3 \oplus s_3) = \text{lsb}_1(y_3) = \text{lsb}_1(w_3 \odot Z_{11}^{(3)}), \quad (3.271)$$

where w_3 can be expressed in terms of (n_3, q_3, m_3, u_3), which can be further replaced by ciphertext words such as $n_3 = X_1^{(4)} \oplus X_2^{(4)}$, $q_3 = X_3^{(4)} \oplus X_5^{(4)}$, $m_3 = X_4^{(4)} \oplus X_6^{(4)}$ and $u_3 = X_7^{(4)} \oplus X_8^{(4)}$, assuming the swap of four middle words is still there.

Combining (3.270), (3.271) with the values of n_3, q_3, m_3 and u_3, we have

$$\text{lsb}_1(X_4^{(4)} \oplus X_7^{(4)} \oplus X_4^{(3)} \oplus Z_4^{(3)} \oplus X_7^{(3)} \oplus Z_7^{(3)}) =$$
$$\text{lsb}_1(Z_{11}^{(3)} \odot (((n_3 \odot Z_9^{(3)}) \boxplus q_3) \odot ((m_3 \odot q_3) \boxplus Z_{10}^{(3)} \odot (u_3 \boxplus (m_3 \odot q_3))))). \quad (3.272)$$

Note that (3.272) depends on the least significant bits of the ciphertext blocks and the internal state. In particular, the least significant bits of $X_4^{(3)}$ and $X_7^{(3)}$ are balanced, according to the multiset used for the attack.

Over the given multiset, the integral of (3.272) is zero because the subkeys are fixed and the intermediate values have the least significant bits balanced.

Therefore, (3.272) represents a 1-bit condition that holds with certainty. It can be used to recover $Z_9^{(3)}$, $Z_{10}^{(3)}$ and $Z_{11}^{(3)}$.

The attack requires $48 \cdot 2^{16} l = 2^{21.58}$ chosen plaintexts and an effort of $2^{48} \cdot 2^{16} + 2^{48} \cdot 2^{16} + \ldots + 2 \cdot 2^{16} \approx 2^{66}$ computations of (3.272). Assuming $\odot$ costs about three times an $\oplus$ and that an $\boxplus$ and an $\oplus$ cost about the same, then, computing 2-round MESH-128 costs about 120 $\oplus$, and one computation of (3.272) costs about 27 $\oplus$. Thus, the time complexity is estimated at 2^{64} 2-round MESH-128 computations.

The memory complexity is 2^{16} text blocks since the same storage can be reused for each new Λ-set.

Unfortunately, (3.272) does not provide information to recover the remaining subkeys of 2-round MESH-128. Moreover, according to the key schedule of MESH-128, the knowledge of $Z_9^{(3)}$, $Z_{10}^{(3)}$ and $Z_{11}^{(3)}$ is not enough to uniquely recover the remaining subkeys. The situation is similar to the attack on 2-round MESH-64 in Sect. 3.16.1.

3.17 Biryukov-Demirci Attack

The attacks in this section are based on the analysis in [150].

The Biryukov-Demirci (BD) attack is a dedicated known-plaintext attack on Lai-Massey cipher designs that trades-off a small number of known data blocks for a larger time complexity.

The BD attack combines two ideas:

- an observation by A. Biryukov that in the computational graph of IDEA (Fig. 2.2) the two middle 16-bit words in a block are only combined with subkeys or internal cipher data via two group operations, $\boxplus$ and $\oplus$, across the full cipher. Therefore, restricted to the least significant bit only, their exclusive-or combination results in a linear trail that holds with certainty.
- an attack by Demirci in [61], in which a narrow linear trail exists across the MA-box of IDEA involving only one subkey and one input word.

This attack operates without any weak-key or weak-subkey assumptions.

Interestingly, this attack does not hold for the PES cipher, where all words in a block are combined the three group operations along the cipher.

3.17.1 Biryukov-Demirci Attack on IDEA

Let us focus on the IDEA block cipher. The analysis starts from the MA-box or the Multiplication-Addition box. It is composed of a sequence of multiplications and additions, alternated, and computed in a zigzag order. See Fig. 2.2.

Let (n_i, q_i) and (r_i, s_i) denote the inputs and outputs to the i-th MA-box of IDEA, respectively, and let (p_1, p_2, p_3, p_4) and (c_1, c_2, c_3, c_4) denote a plaintext and corresponding ciphertext after 8.5-round IDEA.

If one tracks the value of the input and output of only the two middle words in a block, then the results are equations (3.273) and (3.274). The linear trails are depicted (in red) in the computational graph of IDEA in Fig. 3.27.

$$((((((((p_2 \boxplus Z_2^{(1)}) \oplus r_1 \boxplus Z_3^{(2)}) \oplus s_2 \boxplus Z_2^{(3)}) \oplus r_3 \boxplus Z_3^{(4)}) \oplus s_4 \boxplus Z_2^{(5)}) \oplus r_5 \boxplus Z_3^{(6)}) \oplus s_6 \boxplus Z_2^{(7)}) \oplus r_7 \boxplus Z_3^{(8)}) \oplus s_8 \boxplus Z_3^{(9)} = c_3 \,, \quad (3.273)$$

and

$$((((((((p_3 \boxplus Z_3^{(1)}) \oplus s_1 \boxplus Z_2^{(2)}) \oplus r_2 \boxplus Z_3^{(3)}) \oplus s_3 \boxplus Z_2^{(4)}) \oplus r_4 \boxplus Z_3^{(5)}) \oplus s_5 \boxplus Z_2^{(6)}) \oplus r_6 \boxplus Z_3^{(7)}) \oplus s_7 \boxplus Z_2^{(8)}) \oplus r_8 \boxplus Z_2^{(9)} = c_2 \,. \quad (3.274)$$

Let $\text{lsb}_1(x)$ denote the least significant bit of x. This operator is linear in relation to $\oplus$, that is, $\text{lsb}_1(x \oplus y) = \text{lsb}_1(x) \oplus \text{lsb}_1(y)$.

Equations (3.273) and (3.274) still hold with certainty if restricted only to the least significant bit. In this case, all operators become exclusive-or:

$$\text{lsb}_1(p_2 \oplus Z_2^{(1)} \oplus r_1 \oplus Z_3^{(2)} \oplus s_2 \oplus Z_2^{(3)} \oplus r_3 \oplus Z_3^{(4)} \oplus s_4 \oplus Z_2^{(5)} \oplus r_5 \oplus Z_3^{(6)} \oplus s_6 \oplus Z_2^{(7)} \oplus r_7 \oplus Z_3^{(8)} \oplus s_8 \oplus Z_3^{(9)} \oplus c_3) = 0 \,, \quad (3.275)$$

and

$$\text{lsb}_1(p_3 \oplus Z_3^{(1)} \oplus s_1 \oplus Z_2^{(2)} \oplus r_2 \oplus Z_3^{(3)} \oplus s_3 \oplus Z_2^{(4)} \oplus r_4 \oplus Z_3^{(5)} \oplus s_5 \oplus Z_2^{(6)} \oplus r_6 \oplus Z_3^{(7)} \oplus s_7 \oplus Z_2^{(8)} \oplus r_8 \oplus Z_2^{(9)} \oplus c_2) = 0 \,. \quad (3.276)$$

If the exclusive-or of the least significant bits of r_i and s_i could be discovered, then two bits of information on the key could be derived for IDEA, in a know-plaintext setting, namely:

$$\text{lsb}_1(Z_2^{(1)} \oplus Z_3^{(2)} \oplus Z_2^{(3)} \oplus Z_3^{(4)} \oplus Z_2^{(5)} \oplus Z_3^{(6)} \oplus Z_2^{(7)} \oplus Z_3^{(8)} \oplus Z_3^{(9)})$$

and

$$\text{lsb}_1(Z_3^{(1)} \oplus Z_2^{(2)} \oplus Z_3^{(3)} \oplus Z_2^{(4)} \oplus Z_3^{(5)} \oplus Z_2^{(6)} \oplus Z_3^{(7)} \oplus Z_2^{(8)} \oplus Z_2^{(9)})$$

and similarly for any number of rounds.

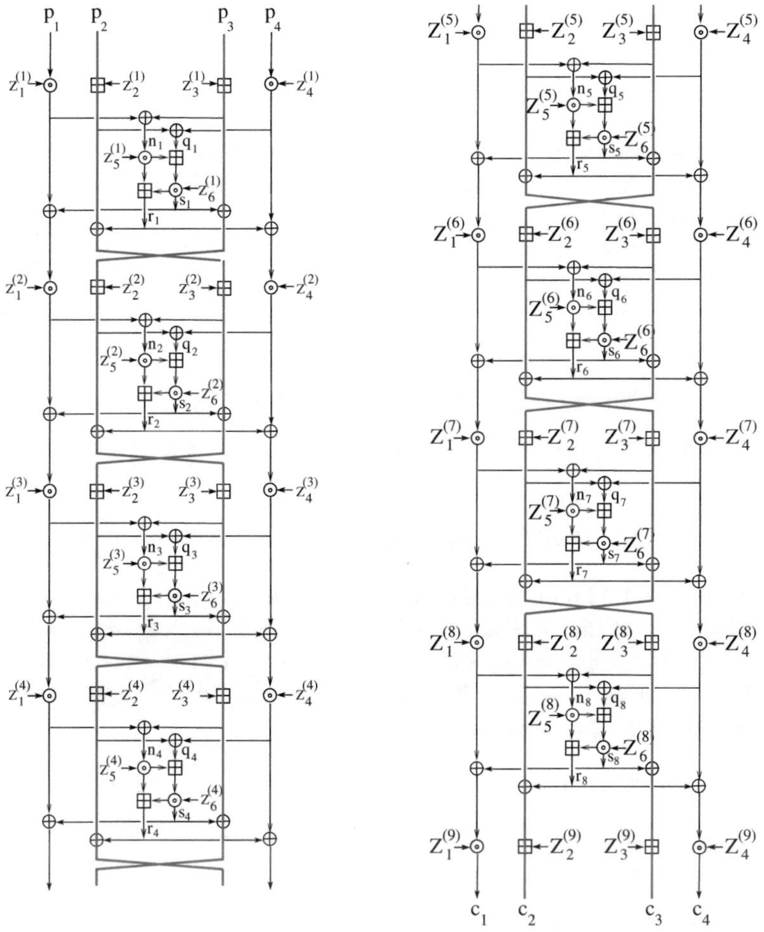

Fig. 3.27 Linear trails (depicted in red) across the full 8.5-round IDEA cipher.

The following analysis combines this observation by Biryukov, and the approach by Demirci in [61]. The analysis starts with the first 1.5 rounds of IDEA.

Let $(X_1^{(i)}, X_2^{(i)}, X_3^{(i)}, X_4^{(i)})$ denote the i-th round input, and $(Y_1^{(i)}, Y_2^{(i)}, Y_3^{(i)}, Y_4^{(i)})$ denote the output of the i-th KM half-round. The value n_1 can be computed as:

$$n_1 = (p_1 \odot Z_1^{(1)}) \oplus (p_3 \boxplus Z_3^{(1)}) = (Y_1^{(2)} \odot (Z_1^{(2)})^{-1}) \oplus (Y_2^{(2)} \boxminus Z_2^{(2)}) . \quad (3.277)$$

The other input to the MA-box, q_1, can be computed as

$$q_1 = (p_2 \boxplus Z_2^{(1)}) \oplus (p_4 \odot Z_4^{(1)}) = (Y_3^{(2)} \boxminus Z_3^{(2)}) \oplus (Y_4^{(2)} \odot (Z_4^{(2)})^{-1}) . \quad (3.278)$$

Demirci's relation for the least significant bits of (r_1, s_1) is:

$$\text{lsb}_1(r_1 \oplus s_1) = \text{lsb}_1(n_1 \odot Z_5^{(1)}) . \tag{3.279}$$

There is an alternative relation such as $\text{lsb}_1(s_1 \odot (Z_6^{(i)})^{-1}) = \text{lsb}(n_1 \odot Z_5^{(1)} \boxplus q_1)$, but it depends on both inputs (n_i, q_i) and both subkeys, and thus is less attractive than (3.279).

Following the input and output of the two middle words in a block gives:

$$\text{lsb}_1(p_2 \oplus Z_2^{(1)} \oplus r_1 \oplus Z_3^{(2)}) = \text{lsb}_1(Y_3^{(2)}) , \tag{3.280}$$

and

$$\text{lsb}_1(p_3 \oplus Z_3^{(1)} \oplus s_1 \oplus Z_2^{(2)}) = \text{lsb}_1(Y_2^{(2)}) . \tag{3.281}$$

From the exclusive-or-combination of (3.279), (3.280) and (3.281), we obtain:

$$\text{lsb}_1(Y_2^{(2)} \oplus Y_3^{(2)} \oplus Z_2^{(2)} \oplus Z_3^{(2)} \oplus p_2 \oplus Z_2^{(1)} \oplus p_3 \oplus Z_3^{(1)}) = \text{lsb}_1(n_1 \odot Z_5^{(1)}) . \tag{3.282}$$

Expressions (3.277) and (3.282), plus the key overlapping property in the key schedule algorithm of IDEA provide the one-bit distinguisher (3.283). This distinguisher holds with certainty and allows us to recover 49 bits of information about the key: $Z_5^{(1)}$, $Z_1^{(1)}$, $Z_3^{(1)}$ and $\text{lsb}_1(Z_2^{(2)} \oplus Z_3^{(2)} \oplus Z_2^{(1)} \oplus Z_3^{(1)})$, using only 49 known plaintext/ciphertext pairs:

$$\begin{aligned} \text{lsb}_1(Y_2^{(2)} \oplus Y_3^{(2)} \oplus Z_2^{(2)} \oplus Z_3^{(2)} \oplus p_2 \oplus Z_2^{(1)} \oplus p_3 \oplus Z_3^{(1)} \oplus \\ Z_5^{(1)} \odot ((p_1 \odot Z_1^{(1)}) \oplus (p_3 \boxplus Z_3^{(1)}))) = 0 . \end{aligned} \tag{3.283}$$

Equation (3.283) contains nine $\oplus$, two $\odot$ and one $\boxplus$. Assuming that one $\odot$ costs about three times one $\boxplus$ or $\oplus$, and that $\boxplus$ cost the same as $\oplus$, then computing (3.283) is equivalent to 16 $\oplus$, or $16/30 = 8/15$ of the cost of 1.5-round IDEA computation.

The BD attack recovers 49 bits of the key (numbered 0–15, 32–47, and the exclusive-or of bits numbered 31, 40 and 127).

After 49 known plaintext/ciphertext pairs, the number of candidate subkeys surviving is $2^{-49} \cdot 2^{49} = 1$, and thus, the correct one can be identified. The time complexity is $\frac{8}{15} \cdot 2^{49} \approx 2^{48}$ 1.5-round IDEA computations. This is a known-plaintext attack without weak-key assumptions. The remaining $128 - 49 = 79$ key bits can be recovered by exhaustive key search.

A BD attack on the first 2.5 rounds of IDEA follows a similar procedure. Note that the exclusive-or of the least significant bits of the input and output of the two middle words in a block gives:

$$\text{lsb}_1(p_2 \oplus Z_2^{(1)} \oplus r_1 \oplus Z_3^{(2)} \oplus s_2 \oplus Z_2^{(3)}) = \text{lsb}_1(Y_2^{(3)}) , \tag{3.284}$$

and

$$\text{lsb}_1(p_3 \oplus Z_3^{(1)} \oplus s_1 \oplus Z_2^{(2)} \oplus r_2 \oplus Z_3^{(3)}) = \text{lsb}_1(Y_3^{(3)}) . \quad (3.285)$$

Demirci's relation for 1-round IDEA results in:

$$\text{lsb}_1(r_1 \oplus s_1) = \text{lsb}_1(Z_5^{(1)} \odot ((p_1 \odot Z_1^{(1)}) \oplus (p_3 \boxplus Z_3^{(1)}))) , \quad (3.286)$$

$$\text{lsb}_1(r_2 \oplus s_2) = \text{lsb}_1(Z_5^{(2)} \odot ((Y_1^{(3)} \odot (Z_1^{(3)})^{-1}) \oplus (Y_2^{(3)} \boxminus Z_2^{(3)}))) . \quad (3.287)$$

Combining (3.284), (3.285), (3.286), and (3.287) gives (3.288) which is a 1-bit distinguisher holding with certainty across 2.5-round IDEA. According to the key schedule of IDEA, (3.287) involves 90 user key bits (numbered 64–79, 0–15, 32–47, 57–72, 89–104, 105–120, and the exclusive-or of key bits numbered 31, 47, 40, 120, 8):

$$\begin{aligned} \text{lsb}_1(p_2 \oplus Z_2^{(1)} \oplus p_3 \oplus Z_3^{(1)} \oplus Z_3^{(2)} \oplus Z_2^{(2)} \oplus Y_2^{(3)} \oplus Z_2^{(3)} \oplus Z_3^{(3)} \oplus Y_3^{(3)}) = \\ \text{lsb}_1(r_1 \oplus s_1) \oplus \text{lsb}_1(r_2 \oplus s_2) = \text{lsb}_1(Z_5^{(1)} \odot ((p_1 \odot Z_1^{(1)})\oplus \\ (p_3 \boxplus Z_3^{(1)}))) \oplus \text{lsb}_1(Z_5^{(2)} \odot ((Y_1^{(3)} \odot (Z_1^{(3)})^{-1}) \oplus (Y_2^{(3)} \boxminus Z_2^{(3)}))) . \end{aligned} \quad (3.288)$$

Equation (3.288) contains twelve $\oplus$, four $\odot$ and two $\boxplus$, equivalent to 26 $\oplus$ or 26/52 of the cost of 2.5-round IDEA. Equation (3.288) allows us to recover the 90 subkey bits: $Z_5^{(1)}$, $Z_1^{(1)}$, $Z_3^{(1)}$, $Z_5^{(2)}$, $Z_1^{(3)}$, $Z_2^{(3)}$ and $\text{lsb}_1(Z_2^{(1)} \oplus Z_3^{(1)} \oplus Z_3^{(2)} \oplus Z_2^{(3)} \oplus Z_2^{(2)} \oplus Z_3^{(3)})$. Using 90 known plaintexts/ciphertext pairs, the number of surviving subkey candidates is $2^{-90} \cdot 2^{90} = 1$, and thus, the correct subkey can be found. The time complexity is $\frac{26}{52} \cdot 2^{90} = 2^{89}$ 2.5-round IDEA computations. The remaining $128 - 90 = 38$ key bits can be recovered by exhaustive search.

A BD attack on the first 3.5 rounds of IDEA uses a linear relation involving the least significant bits of the input and output of the two middle words in a block:

$$\text{lsb}_1(p_2 \oplus Z_2^{(1)} \oplus r_1 \oplus Z_3^{(2)} \oplus s_2 \oplus Z_2^{(3)} \oplus r_3 \oplus Z_3^{(4)}) = \text{lsb}_1(Y_3^{(4)}) , \quad (3.289)$$

and

$$\text{lsb}_1(p_3 \oplus Z_3^{(1)} \oplus s_1 \oplus Z_2^{(2)} \oplus r_2 \oplus Z_3^{(3)} \oplus s_3 \oplus Z_2^{(4)}) = \text{lsb}_1(Y_2^{(4)}) . \quad (3.290)$$

Combining (3.289) and (3.290) results in:

$$\begin{aligned} \text{lsb}_1(p_2 \oplus Z_2^{(1)} \oplus Z_3^{(2)} \oplus Z_2^{(3)} \oplus Z_3^{(4)} \oplus p_3 \oplus Z_3^{(1)} \oplus Z_2^{(2)} \oplus Z_3^{(3)} \oplus Z_2^{(4)} \oplus \\ Y_3^{(4)} \oplus Y_2^{(4)}) = \text{lsb}_1(r_1 \oplus s_1) \oplus \text{lsb}_1(r_2 \oplus s_2) \oplus \text{lsb}_1(r_3 \oplus s_3) . \end{aligned} \quad (3.291)$$

Using Demirci's relation for 1-round IDEA, results in:

$$\mathrm{lsb}_1(r_2 \oplus s_2) = \mathrm{lsb}_1(Z_5^{(2)} \odot ((Y_1^{(3)} \odot (Z_1^{(3)})^{-1}) \oplus (Y_2^{(3)} \boxminus Z_2^{(3)}))) =$$
$$\mathrm{lsb}_1(Z_5^{(2)} \odot ((((Y_1^{(4)} \odot (Z_1^{(4)})^{-1} \oplus s_3) \odot (Z_1^{(3)})^{-1})) \oplus$$
$$((Y_3^{(4)} \boxminus Z_3^{(4)}) \oplus r_3 \boxminus Z_2^{(3)}))) \,, \qquad (3.292)$$

and

$$\mathrm{lsb}_1(r_3 \oplus s_3) = \mathrm{lsb}_1(Z_5^{(3)} \odot ((Y_1^{(4)} \odot (Z_1^{(4)})^{-1}) \oplus (Y_2^{(4)} \boxminus Z_2^{(4)}))) \,. \qquad (3.293)$$

The individual values of r_3 and s_3 are needed in (3.292):

$$s_3 = (((Y_1^{(4)} \odot (Z_1^{(4)})^{-1}) \oplus (Y_2^{(4)} \boxminus Z_2^{(4)})) \odot Z_5^{(3)} \boxplus$$
$$((Y_3^{(4)} \boxminus Z_3^{(4)}) \oplus (Y_4^{(4)} \odot (Z_4^{(4)})^{-1})) \odot Z_6^{(3)}) \,, \qquad (3.294)$$

and

$$r_3 = s_3 \boxplus (((Y_1^{(4)} \odot (Z_1^{(4)})^{-1}) \oplus (Y_2^{(4)} \boxminus Z_2^{(4)})) \odot Z_5^{(3)}) \,. \qquad (3.295)$$

Combining (3.286), (3.291), (3.292), (3.293), (3.294) and (3.295) totals 14 $\odot$, 10 $\boxplus$ and 20 $\oplus$, which is equivalent to 72 $\oplus$. This combination gives a 1-bit distinguisher that holds with certainty and allows us to recover 112 key bits (numbered 0–17, 32–47, 50–127), according to the key schedule of IDEA. The subkeys involved are $Z_5^{(1)}$, $Z_1^{(1)}$, $Z_3^{(1)}$, $Z_5^{(2)}$, $Z_1^{(3)}$, $Z_2^{(3)}$, $Z_1^{(4)}$, $Z_2^{(4)}$, $Z_5^{(3)}$, $Z_3^{(4)}$, $Z_4^{(4)}$, $Z_6^{(3)}$, and $\mathrm{lsb}_1(Z_2^{(1)} \oplus Z_3^{(2)} \oplus Z_2^{(3)} \oplus Z_3^{(4)} \oplus Z_3^{(1)} \oplus Z_2^{(2)} \oplus Z_3^{(3)} \oplus Z_2^{(4)})$.

Using 112 known plaintexts, the number of subkey candidates surviving this filtering is $2^{-112} \cdot 2^{112} = 1$, and thus, the correct one can be uniquely identified.

The time complexity is $\frac{72}{74} \cdot 2^{112} \approx 2^{112}$ 3.5-round IDEA computations. The remaining $128 - 112 = 16$ key bits can be recovered by exhaustive search.

A BD attack on the first 4 rounds of IDEA uses a linear relation involving the exclusive-or of the least significant bits of the input and output of the two middle words in a block:

$$\mathrm{lsb}_1(p_2 \oplus Z_2^{(1)} \oplus r_1 \oplus Z_3^{(2)} \oplus s_2 \oplus Z_2^{(3)} \oplus r_3 \oplus Z_3^{(4)} \oplus s_4) = \mathrm{lsb}_1(X_2^{(5)}) \,, \qquad (3.296)$$

and

$$\mathrm{lsb}_1(p_3 \oplus Z_3^{(1)} \oplus s_1 \oplus Z_2^{(2)} \oplus r_2 \oplus Z_3^{(3)} \oplus s_3 \oplus Z_2^{(4)} \oplus r_4) = \mathrm{lsb}_1(X_3^{(5)}) \,. \qquad (3.297)$$

Combining (3.296) and (3.297) results in:

$$\mathrm{lsb}_1(p_2 \oplus Z_2^{(1)} \oplus Z_3^{(2)} \oplus Z_2^{(3)} \oplus Z_3^{(4)} \oplus p_3 \oplus Z_3^{(1)} \oplus Z_2^{(2)} \oplus Z_3^{(3)} \oplus Z_2^{(4)} \oplus$$
$$X_3^{(5)} \oplus X_2^{(5)}) = \mathrm{lsb}_1(r_1 \oplus s_1) \oplus \mathrm{lsb}_1(r_2 \oplus s_2) \oplus \mathrm{lsb}_1(r_3 \oplus s_3) \oplus$$
$$\mathrm{lsb}_1(r_4 \oplus s_4) \,. \qquad (3.298)$$

Using Demirci's relation for 1-round IDEA results in:

$$\mathrm{lsb}_1(r_2 \oplus s_2) = \mathrm{lsb}_1(Z_5^{(2)} \odot ((p_1 \odot Z_1^{(1)} \oplus s_1) \odot Z_1^{(2)} \oplus (((p_2 \boxplus Z_2^{(1)}) \oplus r_1) \boxplus Z_3^{(2)}))), \tag{3.299}$$

$$\mathrm{lsb}_1(r_3 \oplus s_3) = \mathrm{lsb}_1(Z_5^{(3)} \odot ((Y_1^{(4)} \odot (Z_1^{(4)})^{-1}) \oplus (Y_2^{(4)} \boxminus Z_2^{(4)}))), \tag{3.300}$$

$$\mathrm{lsb}_1(r_4 \oplus s_4) = \mathrm{lsb}_1((X_1^{(5)} \oplus X_2^{(5)}) \odot Z_5^{(4)}). \tag{3.301}$$

For (3.299), the individual values of r_1 and s_1 are needed:

$$s_1 = (((p_1 \odot Z_1^{(1)}) \oplus (p_3 \boxplus Z_3^{(1)})) \odot Z_5^{(1)} \boxplus ((p_2 \boxplus Z_2^{(1)}) \oplus (p_4 \odot Z_4^{(1)}))) \odot Z_6^{(1)}, \tag{3.302}$$

$$r_1 = s_1 \boxplus ((p_1 \odot Z_1^{(1)}) \oplus (p_3 \boxplus Z_3^{(1)})) \odot Z_5^{(1)}. \tag{3.303}$$

For (3.300) the individual values of $Y_1^{(4)}$ and $Y_2^{(4)}$ are needed:

$$Y_1^{(4)} = X_1^{(5)} \oplus s_4 = X_1^{(5)} \oplus ((X_1^{(5)} \oplus X_2^{(5)}) \odot Z_5^{(4)} \boxplus (X_3^{(5)} \oplus X_4^{(5)})) \odot Z_6^{(4)}, \tag{3.304}$$

$$Y_2^{(4)} = X_3^{(5)} \oplus r_4 = X_3^{(5)} \oplus (s_4 \boxplus (X_1^{(5)} \oplus X_2^{(5)}) \odot Z_5^{(4)}). \tag{3.305}$$

Combining (3.286), (3.298), (3.299), (3.300), (3.301), (3.302), (3.303), (3.304), and (3.305) gives 17 $\odot$, 11 $\boxplus$ and 25 $\oplus$, which is equivalent to about 87 $\oplus$. This combination gives a 1-bit distinguisher that holds with certainty. This distinguisher allows us to recover 114 key bits (numbered 0-113 according to the key schedule of IDEA). The subkeys recovered are $Z_1^{(1)}$, $Z_3^{(1)}$, $Z_5^{(1)}$, $Z_5^{(2)}$, $Z_1^{(2)}$, $Z_2^{(1)}$, $Z_3^{(2)}$, $Z_4^{(1)}$, $Z_6^{(1)}$, $Z_5^{(3)}$, $Z_1^{(4)}$, $Z_2^{(4)}$, $Z_5^{(4)}$, $Z_6^{(4)}$, and $\mathrm{lsb}_1(Z_2^{(1)} \oplus Z_3^{(2)} \oplus Z_2^{(3)} \oplus Z_3^{(4)} \oplus Z_3^{(1)} \oplus Z_2^{(2)} \oplus Z_3^{(3)} \oplus Z_2^{(4)})$.

Using 114 known plaintexts, the number of surviving subkey candidates is $2^{-114}.2^{114} = 1$, and thus, the correct one can be uniquely identified. The time complexity is $\frac{87}{88} \cdot 2^{114} \approx 2^{114}$ 4-round IDEA computations. The remaining $128 - 114 = 14$ key bits can be recovered by exhaustive search.

A BD attack on the first 4.5 rounds of IDEA would involve all 128 key bits and would therefore not be more efficient than an exhaustive key search.

3.17.1.1 New Biryukov-Demirci Attacks on IDEA

Let $X^{(i)} = (X_1^{(i)}, X_2^{(i)}, X_3^{(i)}, X_4^{(i)})$ denote the input text block to the i-th round in IDEA; it also denotes the output from the $(i-1)$-th round for $i > 1$. Let $Y^{(i)} = (Y_1^{(i)}, Y_2^{(i)}, Y_3^{(i)}, Y_4^{(i)})$ denote the output text block from the i-th KM half-round.

The first attack described in [90] is on 1.5-round IDEA and uses the Biryukov-Demirci relation

$$\text{lsb}_1(X_2^{(1)} \oplus X_3^{(1)} \oplus X_2^{(2)} \oplus X_3^{(2)} \oplus Z_2^{(1)} \oplus Z_3^{(1)} \oplus Z_2^{(2)} \oplus Z_3^{(2)}) = \text{lsb}_1(Z_5^{(1)} \odot ((X_1^{(1)} \odot Z_1^{(1)}) \oplus (X_3^{(1)} \boxplus Z_3^{(1)}))), \quad (3.306)$$

which provides a 1-bit condition that holds with certainty.

The BD attack initially recovers the subkeys $(Z_1^{(1)}, Z_3^{(1)}, Z_5^{(1)})$ which represent 48 user key bits, which correspond to user key bits numbered 0-15, 32-47 and 64-79, respectively, according to the key schedule algorithm of IDEA. Additionally, the $\text{lsb}_1(Z_2^{(1)} \oplus Z_2^{(2)} \oplus Z_3^{(2)})$ can be recovered by checking if the equation (3.306) is satisfied. For the correct subkey candidate, (3.306) shall always provide the same (bit) result whatever the plaintext/ciphertext pair. Wrong subkey candidates satisfy (3.306) with probability 2^{-1} for any given plaintext/ciphertext pair.

After 49 different, random plaintext/ciphertext pairs, the number of wrong candidates remaining is $2^{48}/2^{49} < 1$, and the correct candidate can be uniquely identified.

Assuming a $\odot$ costs about three time an $\oplus$, that a single $\oplus$ or $\boxplus$ cost the same, and that 1.5-round IDEA costs about 30 $\oplus$, then the attack effort so far is about $2^{48} \cdot 49 \cdot \frac{15}{30} \approx 2^{52}$ 1.5-round IDEA computations.

Instead of recovering the remaining $128 - 48 = 80$ key bits by exhaustive search, equation (3.306) can be re-framed from the decryption direction as

$$\text{lsb}_1(X_2^{(1)} \oplus X_3^{(1)} \oplus X_2^{(2)} \oplus X_3^{(2)} \oplus Z_2^{(1)} \oplus Z_3^{(1)} \oplus Z_2^{(2)} \oplus Z_3^{(2)}) = \text{lsb}_1(Z_5^{(1)} \odot ((X_1^{(2)} \boxdot Z_1^{(2)}) \oplus (X_2^{(2)} \boxminus Z_2^{(2)}))), \quad (3.307)$$

which also provides a 1-bit condition that holds with certainty. Note that $\boxdot$ means modular division and $\boxminus$ means modular subtraction.

Equation (3.307) can be used to recover $(Z_1^{(2)}, Z_2^{(2)})$ which account for 32 user key bits (numbered 96–127). The effort so far is $2^{32} \cdot 33 \cdot \frac{15}{30} = 2^{36}$ 1.5-round IDEA computations.

The remaining $128 - 48 - 32 = 48$ user key bits can be recovered by exhaustive search. The overall effort is $2^{52} + 2^{36} + 2^{48} \approx 2^{52}$ 1.5-round IDEA computations. The data complexity is $49+33 = 82$ known plaintexts. The memory complexity accounts for the storage of these text blocks.

To extend the BD attack to 2-round IDEA, the Biryukov-Demirci relation is

$$\text{lsb}_1(X_2^{(1)} \oplus X_3^{(1)} \oplus X_2^{(3)} \oplus X_3^{(3)} \oplus Z_2^{(1)} \oplus Z_3^{(1)} \oplus Z_2^{(2)} \oplus Z_3^{(2)} \oplus Z_5^{(1)} \odot ((X_1^{(1)} \odot Z_1^{(1)}) \oplus (X_3^{(1)} \boxplus Z_3^{(1)})) \oplus Z_5^{(2)} \odot (X_1^{(3)} \oplus X_2^{(3)})) = 0, \quad (3.308)$$

which is a 1-bit condition that holds with certainty.

If we fix $(X_1^{(1)}, X_3^{(1)})$ to constant values in all plaintexts used in this attack, then $Z_5^{(1)} \odot ((X_1^{(1)} \odot Z_1^{(1)}) \oplus (X_3^{(1)} \boxplus Z_3^{(1)}))$ becomes a constant value in (3.308).

In this way, we can recover only $Z_5^{(2)}$ at a cost of $2^{16} \cdot 17 \cdot \frac{9}{60} \approx 2^{17}$ 2-round IDEA computations.

Once $Z_5^{(2)}$ is found, then $Z_6^{(2)}$ can be recovered checking equation (3.306) after partially decrypting the last MA half-round. For the correct $Z_6^{(2)}$, equation (3.306) might provide a constant (bit) value. The effort of this step is $2^{16} \cdot 17 \cdot \frac{15}{60} = 2^{18}$ 2-round IDEA computations.

Further, the previous attack on 1.5-round IDEA can be used to recover $(Z_1^{(1)}, Z_3^{(1)}, Z_5^{(1)})$ with an effort of $2^{48} \cdot 49 \cdot \frac{15}{60} = 2^{52}$ 2-round IDEA computations. Next, $(Z_1^{(2)}, Z_2^{(2)})$ can also be recovered with the time complexity $2^{32} \cdot 33 \cdot \frac{15}{60} = 2^{35}$ 2-round IDEA computations.

The remaining $128 - 96 = 32$ user key bits can be recovered by exhaustive search. The final attack effort is about 2^{52} 2-round IDEA computations. The data complexity is $17 + 50 + 33 = 90$ chosen plaintexts (CP). The memory complexity accounts for storing the plaintexts.

To extend the BD attack to 2.5-round IDEA, the Biryukov-Demirci relation becomes:

$$\begin{aligned}\mathrm{lsb}_1(X_2^{(1)} \oplus X_3^{(1)} \oplus Y_2^{(3)} \oplus Y_3^{(3)} \oplus Z_2^{(1)} \oplus Z_3^{(1)} \oplus Z_2^{(2)} \oplus Z_3^{(2)} \oplus Z_2^{(3)} \oplus Z_3^{(3)}) \oplus \\ \mathrm{lsb}_1(Z_5^{(1)} \odot ((X_1^{(1)} \odot Z_1^{(1)}) \oplus (X_3^{(1)} \boxplus Z_3^{(1)}))) \oplus \\ \mathrm{lsb}_1(Z_5^{(2)} \odot ((Y_1^{(3)} \boxdot Z_1^{(3)}) \oplus (Y_2^{(3)} \boxminus Z_2^{(3)}))) = 0,\end{aligned} \tag{3.309}$$

where $\boxdot$ represents modular division and $\boxminus$ stands for modular subtraction.

Fixing $(X_1^{(1)}, X_3^{(1)})$ to constant values allows us to recover $(Z_5^{(2)}, Z_1^{(3)}, Z_2^{(3)})$, which according to the key schedule of IDEA corresponds to 48 user key bits 57–72 and 89–120. The effort is about $2^{48} \cdot 49 \cdot \frac{18}{68} \approx 2^{52}$ 2.5-round IDEA computations.

Next, we can recover $(Z_1^{(1)}, Z_3^{(1)})$ and the remaining unknown bits of $Z_5^{(1)}$, using (3.309) but without fixing $(X_1^{(1)}, X_3^{(1)})$. These 39 user key bits (numbered 0–15, 32–47 and 73–79) require an effort of $2^{39} \cdot 40 \cdot \frac{18}{68} \approx 2^{42}$ 2.5-round IDEA computations, assuming 2.5-round IDEA costs about 68 $\oplus$.

The remaining unknown 41 user key bits can be recovered by exhaustive search. The data complexity is 50 chosen plaintexts plus 40 known plaintexts. The memory is the storage needed for 50 text blocks.

To extend the BD attack to 3-round IDEA, the Biryukov-Demirci relation is

$$\begin{aligned}\mathrm{lsb}_1(X_2^{(1)} \oplus X_3^{(1)} \oplus X_2^{(4)} \oplus X_3^{(4)} \oplus Z_2^{(1)} \oplus Z_3^{(1)} \oplus Z_2^{(2)} \oplus Z_3^{(2)} \oplus Z_2^{(3)} \oplus Z_3^{(3)}) \oplus \\ \mathrm{lsb}_1(Z_5^{(1)} \odot ((X_1^{(1)} \odot Z_1^{(1)}) \oplus (X_3^{(1)} \boxplus Z_3^{(1)}))) \oplus \\ \mathrm{lsb}_1(Z_5^{(2)} \odot ((Y_1^{(3)} \boxdot Z_1^{(3)}) \oplus (Y_2^{(3)} \boxminus Z_2^{(3)}))) \oplus \mathrm{lsb}_1(Z_5^{(3)} \odot (X_1^{(4)} \oplus X_2^{(4)})) = 0,\end{aligned}$$

where $\boxdot$ represents modular division and $\boxminus$ stands for modular subtraction.

If we fix $(X_1^{(1)}, X_3^{(1)})$ to constant values, then we can recover $(Z_5^{(2)}, Z_1^{(3)}, Z_2^{(3)}, Z_5^{(3)}, Z_6^{(3)})$, which corresponds to 64 user key bits (numbered 50–81 and

89–120), according to the key schedule of IDEA. Essentially, when we guess $(Z_5^{(3)}, Z_6^{(3)})$, we are decrypting the last MA half-round, and we are back to 2.5-round IDEA.

The attack effort is $2^{64} \cdot 65 \cdot \frac{22}{90} \approx 2^{68}$ 3-round IDEA computations, assuming 3-round IDEA costs about 90 $\oplus$.

Once these subkeys are recovered, we can apply the previous attack on 2.5-round IDEA to further recover 49 additional user key bits numbered 0–15, 32–47, 73–79 and 127, with complexity about $2^{49} \cdot 50 \cdot \frac{18}{90} \approx 2^{52}$ 3-round IDEA computations. The remaining 15 user key bits can be found by exhaustive search. The final time complexity is 2^{68} 3-round IDEA computations. The data complexity is $65 + 49 = 114$ chosen plaintexts. The memory complexity is essentially the storage for 65 text blocks.

To attack 3.5-round IDEA, the approach is to fix $(X_1^{(1)}, X_3^{(1)})$ to constant values, guess the four subkeys of the last MA half-round, and apply the previous attack on 3-round IDEA. The subkeys are $(Z_5^{(2)}$, $Z_1^{(3)}$, $Z_2^{(3)}$, $Z_5^{(3)}$, $Z_6^{(3)}$, $Z_1^{(4)}$, $Z_2^{(4)}$, $Z_3^{(4)}$, $Z_4^{(4)})$, which means 96 user key bits, except for the 32 bits numbered 18-49.

The attack effort is $2^{96} \cdot 97 \cdot \frac{22}{98} \approx 2^{100}$ 3.5-round IDEA computations. The remaining 32 user key bits can be found by exhaustive search. The data complexity is 97 chosen plaintexts. The memory complexity is essentially storage for these text blocks.

3.17.1.2 TMTO Biryukov-Demirci Attack on IDEA

A Time-Memory Trade-Off (TMTO) approach to the Biryukov-Demirci attack was presented in [90].

Consider 2.5-round IDEA, and the rewritten equation:

$$\begin{aligned} \mathrm{lsb}_1(X_2^{(1)} \oplus X_3^{(1)} \oplus Y_2^{(3)} \oplus Y_3^{(3)} \oplus Z_2^{(1)} \oplus Z_3^{(1)} \oplus Z_2^{(2)} \oplus Z_3^{(2)} \oplus Z_2^{(3)} \oplus Z_3^{(3)} \oplus \\ Z_5^{(1)} \odot ((X_1^{(1)} \odot Z_1^{(1)}) \oplus (X_3^{(1)} \boxplus Z_3^{(1)}))) = \\ \mathrm{lsb}_1(Z_5^{(2)} \odot ((Y_1^{(3)} \boxdot Z_1^{(3)}) \oplus (Y_2^{(3)} \boxminus Z_2^{(3)}))), \end{aligned} \tag{3.310}$$

where $\boxdot$ represents modular division and $\boxminus$ stands for modular subtraction. Note that the left-hand side of (3.310) depends on the plaintext (and key), while the right-hand side of (3.310) depends on the ciphertext (and key) of 2.5-round IDEA.

Note that the value $\mathrm{lsb}_1(Z_2^{(1)} \oplus Z_2^{(2)} \oplus Z_3^{(2)} \oplus Z_2^{(3)} \oplus Z_3^{(3)})$ is unknown but fixed. So, it accounts for a single, unknown bit.

For all 2^{48} possible values of $(Z_1^{(1)}, Z_3^{(1)}, Z_5^{(1)})$ and for 49 known plaintext/ciphertext pairs, compute the left-hand side of (3.310) and store the 49-bit result in a hash table. Such table has 2^{48} entries, each of which holds 49 bits, each bit corresponding to one evaluation of the left-hand side of (3.310). This step costs $2^{48} \cdot 49 \cdot \frac{13}{68} \approx 2^{51}$ 2.5-round IDEA computations, assuming each 2.5-round IDEA costs about 68 $\oplus$s. The memory complexity is $2^{48} \cdot 49/64 \approx 2^{48}$ text blocks. These subkeys correspond to user key bits numbered 0–15, 32–47 and 64–79, according to the key schedule algorithm of IDEA.

Similarly, for all 2^{48} possible values of $(Z_1^{(3)}, Z_2^{(3)}, Z_5^{(2)})$ and for the same 49 known plaintext/ciphertext pairs, compute the right-hand side of (3.310) and look for a matching between the 49-bit result and an entry in the hash table. This step costs $2^{48} \cdot 49 \cdot \frac{8}{68} \approx 2^{50}$ 2.5-round IDEA computations.

The probability of matching between two 49-bit strings is about 2^{-49} assuming each bit matching is independent. There are $2^{48} + 2^{48}$ values being compared, so the chances of a matching are high.

Any matching provides a potential candidate for ($Z_1^{(1)}$, $Z_3^{(1)}$, $Z_5^{(1)}$, $Z_1^{(3)}$, $Z_2^{(3)}$, $Z_5^{(2)}$). In total, 87 user key bits are recovered because $Z_5^{(2)}$ and $Z_5^{(1)}$ share nine bits. The remaining $128 - 87 = 41$ user key bits can be found by exhaustive search. The total attack effort is $2^{50} + 2^{51} \approx 2^{51.5}$ 2.5-round IDEA computations. The data complexity is 49 known plaintexts. The memory complexity is the storage needed for these plaintexts.

To attack 3-round IDEA, consider the rewritten Biryukov-Demirci relation:

$$\begin{aligned}\mathrm{lsb}_1(X_2^{(1)} \oplus X_3^{(1)} \oplus X_2^{(4)} \oplus X_3^{(4)} \oplus Z_2^{(1)} \oplus Z_3^{(1)} \oplus Z_2^{(2)} \oplus Z_3^{(2)} \oplus Z_2^{(3)} \oplus Z_3^{(3)}) \oplus \\ Z_5^{(1)} \odot ((X_1^{(1)} \odot Z_1^{(1)}) \oplus (X_3^{(1)} \boxplus Z_3^{(1)}))) = \\ \mathrm{lsb}_1(Z_5^{(2)} \odot ((Y_1^{(3)} \boxdot Z_1^{(3)}) \oplus (Y_2^{(3)} \boxminus Z_2^{(3)})) \oplus Z_5^{(3)} \odot (X_1^{(4)} \oplus X_2^{(4)})),\end{aligned} \tag{3.311}$$

where $\boxdot$ represents modular division and $\boxminus$ stands for modular subtraction. Note that the left-hand side of (3.311) depends on the plaintext (and key), while the right-hand side of (3.311) depends on the ciphertext (and key).

For all 2^{48} possible values of $(Z_1^{(1)}, Z_3^{(1)}, Z_5^{(1)})$ and for 56 known plaintext/ciphertext pairs, compute the left-hand side of (3.311) and store the 56-bit result in a hash table. This step costs $2^{48} \cdot 56 \cdot \frac{13}{90} \approx 2^{51}$ 3-round IDEA computations. The memory complexity is $2^{48} \cdot 56/64 \approx 2^{48}$ text blocks.

Similarly, for all 2^{55} possible values of $(Z_1^{(3)}, Z_2^{(3)}, Z_5^{(2)}, Z_5^{(3)})$ and for the same 56 known plaintext/ciphertext pairs, compute the right-hand side of (3.311) and try to find a match between the 56-bit result and an entry in the hash table. This step costs $2^{55} \cdot 56 \cdot \frac{12}{90} \approx 2^{58}$ 3-round IDEA computations.

Any such match provides a potential candidate for ($Z_1^{(1)}$, $Z_3^{(1)}$, $Z_5^{(1)}$, $Z_1^{(3)}$, $Z_2^{(3)}$, $Z_5^{(2)}$, $Z_5^{(3)}$).

Note that in the first step we only had 48 key bits to recover, but we had 55 key bits in the second step. That is the reason we used 56 plaintext/ciphertext pairs in both steps: we need a common bit-string size to compare both computations.

In total, $48+55 = 103$ user key bits were recovered. The remaining 25 user key bits can be found by exhaustive search. The final effort is 2^{58} 3-round IDEA computations. The data complexity is 56 known plaintexts, and the memory needed to store them.

To attack 3.5-round IDEA, the four subkeys of an additional MA half-round can be guessed: $(Z_1^{(4)}, Z_2^{(4)}, Z_3^{(4)}, Z_4^{(4)})$, on top of the 55 key bits $(Z_1^{(3)}, Z_2^{(3)}, Z_5^{(2)}, Z_5^{(3)})$, and the previous attack on 3-round IDEA can be applied. This means $55 + 25 = 80$ key bits, according to the key schedule of IDEA. So, 81 known plaintext/ciphertext pairs will be needed.

The effort in the first step to recover $(Z_1^{(1)}, Z_3^{(1)}, Z_5^{(1)})$ will be $2^{48} \cdot 81 \cdot \frac{13}{98} \approx 2^{51}$ 3.5-round IDEA computations. The effort of the second step becomes $2^{80} \cdot 81 \cdot \frac{12}{98} \approx 2^{83}$ 3.5-round IDEA computations. The remaining 25 user key bits can be found by exhaustive search. The data complexity is 81 known plaintext. The memory needed is the storage of these text blocks.

3.17.2 Biryukov-Demirci Attack on MESH-64

The Biryukov-Demirci attack to (reduced-round) IDEA can be adapted to (reduced-round) variants of MESH-64, as long as the total effort is less than that of an exhaustive key search. We use the same terminology as in the attack on IDEA. For instance, (n_i, q_i) denote the i-th MA-box input and (r_i, s_i) denote the i-th MA-box output.

For an attack on the first 1.5 rounds of MESH-64, there are two possible trails to follow.

First, consider the trails involving the two middle words in a block of MESH-64 (see Fig. 2.3(a)):

$$\text{lsb}_1(p_2 \oplus Z_2^{(1)} \oplus r_1) = \text{lsb}_1(Y_3^{(2)} \odot (Z_3^{(2)})^{-1}), \tag{3.312}$$

$$\text{lsb}_1(p_3 \oplus Z_3^{(1)} \oplus s_1) = \text{lsb}_1(Y_2^{(2)} \odot (Z_2^{(2)})^{-1}). \tag{3.313}$$

Combining (3.312) and (3.313) results in:

$$\text{lsb}_1(p_2 \oplus Z_2^{(1)} \oplus p_3 \oplus Z_3^{(1)} \oplus Y_3^{(2)} \odot (Z_3^{(2)})^{-1} \oplus Y_2^{(2)} \odot (Z_2^{(2)})^{-1}) = \text{lsb}_1(r_1 \oplus s_1). \tag{3.314}$$

The right-hand side of (3.314) can be represented as:

$$\begin{aligned} \text{lsb}_1(r_1 \oplus s_1) = \quad & \text{lsb}_1((((p_1 \odot Z_1^{(1)}) \oplus (p_3 \boxplus Z_3^{(1)})) \odot Z_5^{(1)} \boxplus \\ & ((p_2 \boxplus Z_2^{(1)}) \oplus (p_4 \odot Z_4^{(1)}))) \odot Z_6^{(1)}) \quad (3.315) \\ = & \text{lsb}_1((((Y_1^{(2)} \boxminus Z_1^{(2)}) \oplus (Y_2^{(2)} \odot (Z_2^{(2)})^{-1})) \odot Z_5^{(1)} \boxplus \\ & ((Y_3^{(2)} \odot (Z_3^{(2)})^{-1}) \oplus (Y_4^{(2)} \boxminus Z_4^{(2)}))) \odot Z_6^{(1)}). \quad (3.316) \end{aligned}$$

Equations (3.314) and (3.315) involve 7 $\oplus$, 6 $\odot$ and 3 $\boxplus$, which is equivalent to 28 $\oplus$ (assuming each $\odot$ costs three $\oplus$ and that the cost of a $\boxplus$ is the same as that of an $\oplus$). These equations provide a 1-bit condition that hold with certainty and allow to recover 128 bits of information on the key: $Z_1^{(1)}$, $Z_2^{(1)}$, $Z_3^{(1)}$, $Z_4^{(1)}$, $Z_5^{(1)}$, $Z_6^{(1)}$, $Z_2^{(2)}$ and $Z_3^{(2)}$. According to the key schedule of MESH-64:

$$Z_2^{(2)} = (((Z_1^{(1)} \boxplus Z_2^{(1)}) \oplus Z_3^{(1)} \boxplus Z_6^{(1)}) \oplus Z_7^{(1)} \boxplus Z_1^{(2)}) \lll 7 \oplus c_8 \,,$$
$$Z_3^{(2)} = (((Z_2^{(1)} \boxplus Z_3^{(1)}) \oplus Z_4^{(1)} \boxplus Z_7^{(1)}) \oplus Z_1^{(2)} \boxplus Z_2^{(2)}) \lll 7 \oplus c_9 \,,$$

which implies that the values of $Z_2^{(2)}$ and $Z_3^{(2)}$ cannot be deduced from the other subkeys to be recovered, namely there is no key bit overlap property such as in the key schedule of IDEA to help reduce the time complexity. The attack requires 128 known plaintexts and $\frac{28}{34} \cdot 2^{128} \approx 2^{128}$ 1.5-round MESH-64 computations.

Equations (3.314) and (3.316) provide a 1-bit condition to recover 128 bits of information on the key: $Z_1^{(1)}$, $Z_2^{(1)}$, $Z_3^{(1)}$, $Z_4^{(1)}$, $Z_5^{(1)}$, $Z_6^{(1)}$, $Z_3^{(2)}$ and $Z_2^{(2)}$. The data complexity is 128 known plaintexts. The memory is 128 text blocks. The time complexity is 2^{128} computations of (3.314) and (3.316).

Next, consider trails involving the first and fourth words in a block of MESH-64:

$$\text{lsb}_1(p_1 \odot Z_1^{(1)} \oplus s_1) = \text{lsb}_1(Y_1^{(2)} \oplus Z_1^{(2)}) \,, \quad (3.317)$$
$$\text{lsb}_1(p_4 \odot Z_4^{(1)} \oplus r_1) = \text{lsb}_1(Y_4^{(2)} \oplus Z_4^{(2)}) \,. \quad (3.318)$$

Combining (3.317) and (3.318) results in:

$$\text{lsb}_1(p_1 \odot Z_1^{(1)} \oplus p_4 \odot Z_4^{(1)} \oplus Y_1^{(2)} \oplus Z_1^{(2)} \oplus Y_4^{(2)} \oplus Z_4^{(2)}) = \text{lsb}_1(r_1 \oplus s_1) \,. \quad (3.319)$$

The right-hand side of (3.319) can be represented as (3.315) or (3.316).

Equations (3.319) and (3.315) provide a 1-bit condition that holds with certainty and allows us to recover 97 bits of information on the key: $Z_1^{(1)}$, $Z_2^{(1)}$, $Z_3^{(1)}$, $Z_4^{(1)}$, $Z_5^{(1)}$, $Z_6^{(1)}$ and $\text{lsb}_1(Z_1^{(2)} \oplus Z_4^{(2)})$. Using 97 known plaintexts, the number of surviving subkey candidates is $2^{-97} \cdot 2^{97} = 1$ and only the correct subkey is expected to survive. The time complexity is $\frac{28}{34} \cdot 2^{97} \approx 2^{97}$ 1.5-round MESH-64 computations.

Expressions (3.319) and (3.316) provide a 1-bit condition to recover 128 bits of information on the key: $Z_1^{(1)}$, $Z_4^{(1)}$, $Z_5^{(1)}$, $Z_6^{(1)}$, $Z_1^{(2)}$, $Z_2^{(2)}$, $Z_3^{(2)}$ and $Z_4^{(2)}$, using 128 known plaintexts, and $\frac{28}{34} \cdot 2^{128} \approx 2^{128}$ 1.5-round MESH-64 computations (therefore, it is *not* better than exhaustive key search).

To attack 2-round MESH-64, there are two possibilities: consider first the trail involving the two middle words in a block:

$$\text{lsb}_1((p_2 \boxplus Z_2^{(1)} \oplus r_1) \odot Z_3^{(2)} \oplus s_2) = \text{lsb}_1(X_2^{(3)}), \quad (3.320)$$
$$\text{lsb}_1((p_3 \boxplus Z_3^{(1)} \oplus s_1) \odot Z_2^{(2)} \oplus r_2) = \text{lsb}_1(X_3^{(3)}). \quad (3.321)$$

Combining (3.320) and (3.321) results in:

$$\text{lsb}_1((p_2 \oplus Z_2^{(1)} \oplus r_1) \odot Z_3^2 \oplus (p_3 \oplus Z_3^{(1)} \oplus s_1) \odot Z_2^{(2)} \oplus X_2^{(3)} \oplus X_3^{(3)}) = \text{lsb}_1(r_2 \oplus s_2). \quad (3.322)$$

The right-hand side of (3.322) can be represented as:

$$\text{lsb}_1(r_2 \oplus s_2) = \text{lsb}_1(((X_1^{(3)} \oplus X_2^{(3)}) \odot Z_5^{(2)} \boxplus (X_3^{(3)} \oplus X_4^{(3)})) \odot Z_6^{(2)}). \quad (3.323)$$

To solve (3.322), the individual values of r_1 and s_1 are needed:

$$r_1 = (n_1 \odot Z_5^{(1)} \boxplus (n_1 \odot Z_5^{(1)} \boxplus q_1) \odot Z_6^{(1)}) \odot Z_7^{(1)}, \quad (3.324)$$

and

$$s_1 = r_1 \boxplus (n_1 \odot Z_5^{(1)} \boxplus q_1) \odot Z_6^{(1)}. \quad (3.325)$$

Combining (3.322), (3.323), (3.324) and (3.325) requires ten $\odot$, seven $\oplus$ and seven $\boxplus$. This combination results in a 1-bit condition that holds with certainty and allows us to recover $Z_1^{(1)}$, $Z_2^{(1)}$, $Z_3^{(1)}$, $Z_4^{(1)}$, $Z_5^{(1)}$, $Z_6^{(1)}$, $Z_7^{(1)}$, $Z_2^{(2)}$, $Z_3^{(2)}$, $Z_5^{(2)}$ and $Z_6^{(2)}$, which according to the key schedule of MESH-64, correspond to the full 128-bit user key. The attack effort is about 75% that of an exhaustive key search.

Next, consider the trails involving the first and fourth words in a block across 2-round MESH-64:

$$\text{lsb}_1(p_1 \odot Z_1^{(1)} \oplus s_1 \oplus Z_1^{(2)} \oplus s_2) = \text{lsb}_1(X_1^{(3)}), \quad (3.326)$$
$$\text{lsb}_1(p_4 \odot Z_4^{(1)} \oplus r_1 \oplus Z_4^{(2)} \oplus r_2) = \text{lsb}_1(X_4^{(3)}). \quad (3.327)$$

Combining (3.326) and (3.327) results in:

$$\text{lsb}_1(p_1 \odot Z_1^{(1)} \oplus Z_1^{(2)} \oplus X_1^{(3)} \oplus p_4 \odot Z_4^{(1)} \oplus Z_4^{(2)} \oplus X_4^{(3)}) = \text{lsb}_1(r_1 \oplus s_1) \oplus \text{lsb}_1(r_2 \oplus s_2) = \text{lsb}_1(((p_1 \odot Z_1^{(1)} \oplus (p_3 \boxplus Z_3^{(1)}) \odot Z_5^{(1)} \boxplus ((p_2 \boxplus Z_2^{(1)}) \oplus (p_4 \odot Z_4^{(1)}))) \odot Z_6^{(1)}) \oplus \text{lsb}_1(((X_1^{(3)} \oplus X_2^{(3)}) \odot Z_5^2 \boxplus (X_3^{(3)} \oplus X_4^{(3)})) \odot Z_6^{(2)}). \quad (3.328)$$

Equation (3.328) contains six $\odot$, four $\boxplus$ and five $\oplus$, which is equivalent to 27 $\oplus$. This expression is a 1-bit condition that holds with certainty and allows us to recover 128 bits of information on the key: $Z_1^{(1)}$, $Z_2^{(1)}$, $Z_3^{(1)}$, $Z_4^{(1)}$, $Z_5^{(1)}$, $Z_6^{(1)}$, $Z_5^{(2)}$ and $Z_6^{(2)}$, with 128 known plaintexts and $\frac{27}{52} \cdot 2^{128} \approx 2^{127}$ 2-round MESH-64 computations. This attack is not much better than exhaustive key search.

3.17.3 Biryukov-Demirci Attack on MESH-96

Let $P = (p_1, p_2, p_3, p_4, p_5, p_6)$ denote a plaintext block, $(X_1^{(i)}, X_2^{(i)}, X_3^{(i)}, X_4^{(i)}, X_5^{(i)}, X_6^{(i)})$ denote the input to the i-th round, and $(Y_1^{(i)}, Y_2^{(i)}, Y_3^{(i)}, Y_4^{(i)}, Y_5^{(i)}, Y_6^{(i)})$ denote the output of the i-th KM half-round.

Let (n_i, q_i, m_i) denote the input to the i-th MA-box, and (r_i, s_i, t_i) denote the i-th MA-box output.

To attack 1.5-round MESH-96, consider the trails involving p_3 and p_5.

$$\text{lsb}_1(p_3 \odot Z_3^{(1)} \oplus r_1) = \text{lsb}_1(Y_5^{(2)} \boxminus Z_5^{(2)})\,, \tag{3.329}$$

$$\text{lsb}_1(p_5 \odot Z_5^{(1)} \oplus s_1) = \text{lsb}_1(Y_3^{(2)} \boxminus Z_3^{(2)})\,. \tag{3.330}$$

Combining (3.329) and (3.330) results in:

$$\text{lsb}_1(p_3 \odot Z_3^{(1)} \oplus p_5 \odot Z_5^{(1)} \oplus Y_5^{(2)} \oplus Z_5^{(2)} \oplus Y_3^{(2)} \oplus Z_3^{(2)}) = \text{lsb}_1(s_1 \oplus r_1)\,. \tag{3.331}$$

The right-hand side of (3.331) can be represented as:

$$\text{lsb}_1(s_1 \oplus r_1) = \text{lsb}_1((n_1 \odot Z_7^{(1)} \boxplus q_1) \odot ((n_1 \odot Z_7^{(1)} \boxplus q_1) \odot m_1 \boxplus Z_8^{(1)}))\,. \tag{3.332}$$

For (3.332), the values of n_1, m_1 and q_1 are required. They are: $n_1 = (p_1 \odot Z_1^{(1)}) \oplus (p_4 \boxplus Z_4^{(1)})$, $q_1 = (p_2 \boxplus Z_2^{(1)}) \oplus (p_5 \odot Z_5^{(1)})$ and $m_1 = (p_3 \odot Z_3^{(1)}) \oplus (p_6 \boxplus Z_6^{(1)})$.

Expressions (3.331) and (3.332) contain nine $\odot$, six $\boxplus$ and eight $\oplus$, which is equivalent to 41 $\oplus$. These expressions provide a 1-bit condition that holds with certainty and allows us to recover all 128 bits of information on the key: $Z_1^{(1)}$, $Z_2^{(1)}$, $Z_3^{(1)}$, $Z_4^{(1)}$, $Z_5^{(1)}$, $Z_6^{(1)}$, $Z_7^{(1)}$, $Z_8^{(1)}$ and $\text{lsb}_1(Z_5^{(2)} \oplus Z_3^{(2)})$. The effort needed is more than that of an exhaustive key search for 1.5-round MESH-96.

To attack 2-round MESH-96, consider the trails involving p_3 and p_5:

$$\text{lsb}_1(p_3 \odot Z_3^{(1)} \oplus r_1 \oplus Z_5^{(2)} \oplus s_2) = \text{lsb}_1(X_3^{(2)})\,, \tag{3.333}$$

$$\text{lsb}_1(p_5 \odot Z_5^{(1)} \oplus s_1 \oplus Z_3^{(2)} \oplus r_2) = \text{lsb}_1(X_5^{(2)})\,. \tag{3.334}$$

Combining (3.333) and (3.334) results in:

$$\text{lsb}_1(p_3 \odot Z_3^{(1)} \oplus Z_5^{(2)} \oplus X_3^{(2)} \oplus p_5 \odot Z_5^{(1)} \oplus Z_3^{(2)} \oplus X_5^{(2)}) = \text{lsb}_1(r_1 \oplus s_1) \oplus \text{lsb}_1(r_2 \oplus s_2)\,. \tag{3.335}$$

The terms in the right-hand side of (3.335) can be represented as:

$$\begin{aligned}&\mathrm{lsb}_1(r_1 \oplus s_1) = \\ &\mathrm{lsb}_1(((((n_1 \odot Z_7^{(1)}) \boxplus q_1) \odot m_1) \boxplus Z_8^{(1)}) \odot ((n_1 \odot Z_7^{(1)}) \boxplus q_1)) \oplus \\ &\mathrm{lsb}_1(((Y_1^{(1)} \oplus Y_4^{(1)}) \odot Z_7^{(1)} \boxplus (Y_2^{(1)} \oplus Y_5^{(1)})) \odot \\ &(Z_8^{(1)} \boxplus ((Y_1^{(1)} \oplus Y_4^{(1)}) \odot Z_7^{(1)} \boxplus (Y_2^{(1)} \oplus Y_5^{(1)})) \odot (Y_3^{(1)} \oplus Y_6^{(1)})))\end{aligned} \tag{3.336}$$

and

$$\begin{aligned}\mathrm{lsb}_1(r_2 \oplus s_2) = \mathrm{lsb}_1(((X_1^{(2)} \oplus X_2^{(2)}) \odot Z_7^{(2)} \boxplus (X_4^{(2)} \oplus X_3^{(2)})) \odot \\ (Z_8^{(2)} \boxplus ((X_1^{(2)} \oplus X_2^{(2)}) \odot Z_7^{(2)} \boxplus (X_4^{(2)} \oplus X_3^{(2)})) \odot (X_5^{(2)} \oplus Y_6^{(2)}))).\end{aligned} \tag{3.337}$$

Expressions (3.335), (3.336) and (3.337) contain thirteen $\odot$, sixteen $\oplus$ and nine $\boxplus$, which is equivalent to about 64 $\oplus$. This combination results in a 1-bit condition that holds with certainty, and allows us to recover 161 bits of information on the key: $Z_i^{(1)}$, for $1 \leq i \leq 8$, $Z_7^{(2)}$, $Z_8^{(2)}$, and $\mathrm{lsb}_1(Z_5^{(2)} \oplus Z_3^{(2)})$. Using 161 known plaintexts, the number of surviving subkey candidates is $2^{-161} \cdot 2^{161} = 1$ and only the correct subkey is expected to survive. The time complexity is $\frac{64}{80} \cdot 2^{161} \approx 2^{161}$ 2-round MESH-96 computations.

To attack 2.5-round MESH-96, consider the trails involving p_3, and p_5. The procedure is similar to the attack on 2 rounds, but involves more than 256 bits of information on the key: $Z_i^{(1)}, 1 \leq i \leq 8$, $Z_7^{(2)}$, $Z_8^{(2)}$, $Z_i^{(3)}, 1 \leq i \leq 6$, and $\mathrm{lsb}_1(Z_5^{(2)} \oplus Z_3^{(2)})$. This is more effort than an exhaustive key search.

3.17.4 Biryukov-Demirci Attack on MESH-128

Let $P = (p_1, p_2, p_3, p_4, p_5, p_6, p_7, p_8)$ denote a plaintext block, $(X_1^{(i)}, X_2^{(i)}, X_3^{(i)}, X_4^{(i)}, X_5^{(i)}, X_6^{(i)}, X_7^{(i)}, X_8^{(i)})$ denote the input to the i-th round, and $(Y_1^{(i)}, Y_2^{(i)}, Y_3^{(i)}, Y_4^{(i)}, Y_5^{(i)}, Y_6^{(i)}, Y_7^{(i)}, Y_8^{(i)})$ denote the output of the i-th KM half-round.

Let (n_i, q_i, m_i, u_i) denote the input to the i-th MA-box, and (r_i, s_i, t_i, v_i) denote the i-th MA-box output.

To attack 1.5-round MESH-128, consider the trails starting at p_4 and p_7:

$$\mathrm{lsb}_1(p_4 \boxplus Z_4^{(1)} \oplus r_1) = \mathrm{lsb}_1(Y_7^{(2)} \odot (Z_7^{(2)})^{-1}), \tag{3.338}$$

$$\mathrm{lsb}_1(p_7 \boxplus Z_7^{(1)} \oplus s_1) = \mathrm{lsb}_1(Y_4^{(2)} \odot (Z_4^{(2)})^{-1}). \tag{3.339}$$

Combining (3.338) and (3.339) results in:

$$\mathrm{lsb}_1(p_4 \oplus Z_4^{(1)} \oplus p_7 \oplus Z_7^{(1)} \oplus Y_7^{(2)} \odot (Z_7^{(2)})^{-1} \oplus Y_4^{(2)} \odot (Z_4^{(2)})^{-1}) = \mathrm{lsb}_1(r_1 \oplus s_1)\,. \quad (3.340)$$

The right-hand side of (3.340) can be expressed as

$$\mathrm{lsb}_1(r_1 \oplus s_1) = \mathrm{lsb}_1((((((((n_1 \odot Z_9^{(1)}) \boxplus q_1) \odot m_1) \boxplus u_1) \odot Z_{10}^{(1)} \boxplus (((n_1 \odot Z_9^{(1)}) \boxplus q_1) \odot m_1)) \odot ((n_1 \odot Z_9^{(1)}) \boxplus q_1) \boxplus (n_1 \odot Z_9^{(1)})) \odot Z_{11}^{(1)})\,, \quad (3.341)$$

where $n_1 = (p_1 \odot Z_1^{(1)}) \oplus (p_5 \boxplus Z_5^{(1)})$, $q_1 = (p_2 \boxplus Z_2^{(1)}) \oplus (p_6 \odot Z_6^{(1)})$, $m_1 = (p_3 \odot Z_3^{(1)}) \oplus (p_7 \boxplus Z_7^{(1)})$ and $u_1 = (p_4 \boxplus Z_4^{(1)}) \oplus (p_8 \odot Z_8^{(1)})$.

The combination of (3.340), (3.341) as well as the values of n_1, q_1, m_1 and u_1, involves fifteen $\odot$, ten $\boxplus$ and nine $\oplus$, which is equivalent to about 64 $\oplus$. In total, this combination provides a 1-bit condition that holds with certainty. It allows us to recover 208 bits of information on the key: $Z_i^{(1)}$, for $1 \leq i \leq 11$, $Z_4^{(2)}$ and $Z_7^{(2)}$. Using 208 known plaintexts and $\frac{64}{76} \cdot 2^{208} \approx 2^{208}$ 1.5-round MESH-128 computations.

To attack 2-round MESH-128, consider the extended trails used previously for 1.5 rounds:

$$\mathrm{lsb}_1((p_4 \boxplus Z_4^{(1)} \oplus r_1) \odot Z_7^{(2)} \oplus s_2) = \mathrm{lsb}_1(X_4^{(2)})\,, \quad (3.342)$$

$$\mathrm{lsb}_1((p_7 \boxplus Z_7^{(1)} \oplus s_1) \odot Z_4^{(2)} \oplus r_2) = \mathrm{lsb}_1(X_7^{(2)})\,. \quad (3.343)$$

Combining (3.342) and (3.343) results in:

$$\mathrm{lsb}_1((p_4 \boxplus Z_4^{(1)} \oplus r_1) \odot Z_7^{(2)} \oplus (p_7 \boxplus Z_7^{(1)} \oplus s_1) \odot Z_4^{(2)} \oplus X_4^{(2)} \oplus X_7^{(2)} = \mathrm{lsb}_1(r_2 \oplus s_2)\,. \quad (3.344)$$

Equation (3.344) depends on $r_2 \oplus s_2$, which can be expressed as

$$\mathrm{lsb}_1(r_2 \oplus s_2) = \mathrm{lsb}_1((((((((n_2 \odot Z_9^{(2)}) \boxplus q_2) \odot m_2) \boxplus u_2) \odot Z_{10}^{(2)} \boxplus (((n_2 \odot Z_9^{(2)}) \boxplus q_2) \odot m_2)) \odot ((n_2 \odot Z_9^{(2)}) \boxplus q_2) \boxplus (n_2 \odot Z_9^{(2)})) \odot Z_{11}^{(2)})\,, \quad (3.345)$$

where $n_2 = X_1^{(2)} \oplus X_2^{(2)}$, $q_2 = X_3^{(2)} \oplus X_5^{(2)}$, $m_2 = X_4^{(2)} \oplus X_6^{(2)}$ and $u_2 = X_7^{(2)} \oplus X_8^{(2)}$.

The individual values of r_1 and s_1 are needed in (3.344) as well. Let

$$h_1 = ((n_1 \odot Z_9^{(1)}) \boxplus q_1) \odot m_1\,, \quad (3.346)$$

$$g_1 = h_1 \boxplus u_1\,, \quad (3.347)$$

$$l_1 = g_1 \odot Z_{10}^{(1)}\,, \tag{3.348}$$

$$i_1 = l_1 \boxplus h_1\,, \tag{3.349}$$

$$j_1 = i_1 \odot ((n_1 \odot Z_9^{(1)}) \boxplus q_1)\,, \tag{3.350}$$

and

$$w_1 = (j_1 \boxplus (n_1 \odot Z_9^{(1)})) \odot Z_{11}^{(1)}\,. \tag{3.351}$$

Then,

$$s_1 = ((((w_1 \boxplus j_1) \odot i_1) \boxplus l_1) \odot Z_{12}^{(1)} \boxplus ((w_1 \boxplus j_1) \odot i_1)) \odot (w_1 \boxplus j_1)\,, \tag{3.352}$$

and finally, $r_1 = w_1 \boxplus s_1$.

Taking into account equations (3.345) until (3.352), equation (3.344) requires 22 $\odot$, 19 $\boxplus$ and 9 $\oplus$, which is equivalent to about 49 $\oplus$. The subkeys recovered are $Z_i^{(1)}$, for $1 \leq i \leq 12$, $Z_j^{(2)}$ for $j \in \{4, 7, 9, 10, 11\}$, which means 272 subkey bits, which includes the full 256-bit user key (in the subkeys of the first round). Thus, the effort is not less than that of an exhaustive key search.

3.17.5 Biryukov-Demirci Attack on MESH-64(8)

Let $P = (p_1, p_2, p_3, p_4, p_5, p_6, p_7, p_8)$ denote a plaintext block of MESH-64(8), $(X_1^{(i)}, X_2^{(i)}, X_3^{(i)}, X_4^{(i)}, X_5^{(i)}, X_6^{(i)}, X_7^{(i)}, X_8^{(i)})$ denote the input to the i-th round, and $(Y_1^{(i)}, Y_2^{(i)}, Y_3^{(i)}, Y_4^{(i)}, Y_5^{(i)}, Y_6^{(i)}, Y_7^{(i)}, Y_8^{(i)})$ denote the output of the i-th KM half-round. The word size is 8 bits.

Let (n_i, q_i, m_i, u_i) denote the input to the i-th MA-box, and (r_i, s_i, t_i, v_i) denote the i-th MA-box output.

To attack 1.5-round MESH-64(8), consider the trails starting at p_2 and p_5:

$$\text{lsb}_1(p_2 \boxplus Z_2^{(1)} \oplus s_1) = \text{lsb}_1(Y_5^{(2)} \odot (Z_5^{(2)})^{-1})\,, \tag{3.353}$$

$$\text{lsb}_1(p_5 \boxplus Z_5^{(1)} \oplus r_1) = \text{lsb}_1(Y_2^{(2)} \odot (Z_2^{(2)})^{-1})\,. \tag{3.354}$$

Combining (3.353) and (3.354) results in:

$$\text{lsb}_1(p_2 \oplus Z_2^{(1)} \oplus p_5 \oplus Z_5^{(1)} \oplus Y_5^{(2)} \odot (Z_5^{(2)})^{-1} \oplus Y_2^{(2)} \odot (Z_2^{(2)})^{-1} = \text{lsb}_1(r_1 \oplus s_1)\,. \quad (3.355)$$

The right-hand side of (3.355) can be expressed as

$$\text{lsb}_1(r_1 \oplus s_1) = \text{lsb}_1(n_1 \odot Z_9^{(1)})\,, \quad (3.356)$$

where $n_1 = (p_1 \odot Z_1^{(1)}) \oplus (p_5 \boxplus Z_5^{(1)})$.

The combination of (3.355), (3.356), as well as the value of n_1, involves four $\odot$, one $\boxplus$ and six $\oplus$, which is equivalent to 19 $\oplus$. In total, this combination provides a 1-bit condition that holds with certainty. It allows us to recover 40 bits of information on the key: $Z_1^{(1)}$, $Z_5^{(1)}$, $Z_9^{(1)}$, $Z_2^{(2)}$ and $Z_5^{(2)}$, using 40 known plaintexts and $\frac{19}{60} \cdot 2^{40} \approx 2^{38.34}$ 1.5-round MESH-64(8) computations. The remaining $128 - 40 = 88$ key bits can be recovered by exhaustive key search.

To attack 2-round MESH-64(8), consider the extended trails used previously for 1.5 rounds:

$$\text{lsb}_1((p_2 \boxplus Z_2^{(1)} \oplus s_1) \odot Z_5^{(2)} \oplus r_2) = \text{lsb}_1(X_2^{(2)})\,, \quad (3.357)$$

$$\text{lsb}_1((p_5 \boxplus Z_5^{(1)} \oplus r_1) \odot Z_2^{(2)} \oplus s_2) = \text{lsb}_1(X_5^{(2)})\,. \quad (3.358)$$

Combining (3.357) and (3.358) results in:

$$\text{lsb}_1((p_2 \boxplus Z_2^{(1)} \oplus s_1) \odot Z_5^{(2)} \oplus (p_5 \boxplus Z_5^{(1)} \oplus r_1) \odot Z_2^{(2)} \oplus X_2^{(2)} \oplus X_5^{(2)} = \text{lsb}_1(r_2 \oplus s_2)\,. \quad (3.359)$$

Equation (3.359) depends on $r_2 \oplus s_2$, which can be expressed as

$$\text{lsb}_1(r_2 \oplus s_2) = \text{lsb}_1(n_2 \odot Z_9^{(2)})\,, \quad (3.360)$$

where $n_2 = X_1^{(2)} \oplus X_2^{(2)}$.

The individual values of r_1 and s_1 are needed in (3.359) as well. Let

$$h_1 = ((n_1 \odot Z_9^{(1)}) \boxplus q_1) \odot m_1\,, \quad (3.361)$$

$$g_1 = h_1 \boxplus u_1\,, \quad (3.362)$$

$$v_1 = g_1 \odot Z_{10}^{(1)}\,, \quad (3.363)$$

$$t_1 = h_1 \boxplus v_1\,, \quad (3.364)$$

$$s_1 = t_1 \odot ((n_1 \odot Z_9^{(1)}) \boxplus q_1)\,, \tag{3.365}$$

and

$$r_1 = s_1 \boxplus (n_1 \odot Z_9^{(1)})\,, \tag{3.366}$$

where $n_1 = (p_1 \odot Z_1^{(1)}) \oplus (p_5 \boxplus Z_5^{(1)})$, $q_1 = (p_2 \boxplus Z_2^{(1)}) \oplus (p_6 \odot Z_6^{(1)})$, $m_1 = (p_3 \odot Z_3^{(1)}) \oplus (p_7 \boxplus Z_7^{(1)})$ and $u_1 = (p_4 \boxplus Z_4^{(1)}) \oplus (p_8 \odot Z_8^{(1)})$.

Taking into account equations (3.360) until (3.366), equation (3.359) requires fourteen $\odot$, eleven $\boxplus$ and eight $\oplus$, which is equivalent to about 61 $\oplus$. The subkeys recovered are $Z_i^{(1)}$, for $1 \leq i \leq 10$, $Z_2^{(2)}$ and $Z_5^{(2)}$, which means 96 subkey bits. Using 96 known plaintexts and $\frac{61}{88} \cdot 2^{96} \approx 2^{96}$ 2-round MESH-64(8) computations.

Attacking 2.5-round MESH-64(8) would include one additional KM half-round on top of the previous attack, but the complexity would be larger than that of an exhaustive key search.

3.17.6 Biryukov-Demirci Attack on MESH-128(8)

Let $P = (p_1, p_2, p_3, p_4, p_5, p_6, p_7, p_8, p_9, p_{10}, p_{11}, p_{12}, p_{13}, p_{14}, p_{15}, p_{16})$ denote a plaintext block of MESH-128(8), $(X_1^{(i)}, X_2^{(i)}, X_3^{(i)}, X_4^{(i)}, X_5^{(i)}, X_6^{(i)}, X_7^{(i)}, X_8^{(i)}, X_9^{(i)}, X_{10}^{(i)}, X_{11}^{(i)}, X_{12}^{(i)}, X_{13}^{(i)}, X_{14}^{(i)}, X_{15}^{(i)}, X_{16}^{(i)})$ denote the output of the i-th round, and $(Y_1^{(i)}, Y_2^{(i)}, Y_3^{(i)}, Y_4^{(i)}, Y_5^{(i)}, Y_6^{(i)}, Y_7^{(i)}, Y_8^{(i)}, Y_9^{(i)}, Y_{10}^{(i)}, Y_{11}^{(i)}, Y_{12}^{(i)}, Y_{13}^{(i)}, Y_{14}^{(i)}, Y_{15}^{(i)}, Y_{16}^{(i)})$ denote the output of the i-th KM half-round. The word size is 8 bits.

Let $(n_i, q_i, m_i, u_i, a_i, b_i, d_i, e_i)$ and $(r_i, s_i, t_i, v_i, f_i, g_i, h_i, l_i)$ denote the i-th MA-box input and output, respectively.

To attack 1.5-round MESH-128(8), consider the trails starting at p_8 and p_{15}:

$$\text{lsb}_1(p_8 \boxplus Z_8^{(1)} \oplus r_1) = \text{lsb}_1(Y_{15}^{(2)} \odot (Z_{15}^{(2)})^{-1})\,, \tag{3.367}$$
$$\text{lsb}_1(p_{15} \boxplus Z_{15}^{(1)} \oplus s_1) = \text{lsb}_1(Y_8^{(2)} \odot (Z_8^{(2)})^{-1})\,. \tag{3.368}$$

Combining (3.367) and (3.368) results in:

$$\text{lsb}_1(p_8 \oplus Z_8^{(1)} \oplus p_{15} \oplus Z_{15}^{(1)} \oplus Y_{15}^{(2)} \odot (Z_{15}^{(2)})^{-1} \oplus Y_8^{(2)} \odot (Z_8^{(2)})^{-1}) = \text{lsb}_1(r_1 \oplus s_1)\,. \tag{3.369}$$

The right-hand side of (3.369) can be expressed as

$$\text{lsb}_1(r_1 \oplus s_1) = \text{lsb}_1(n_1 \odot Z_{17}^{(1)})\,, \tag{3.370}$$

where $n_1 = (p_1 \odot Z_1^{(1)}) \oplus (p_9 \boxplus Z_9^{(1)})$.

The combination of (3.369), (3.370) as well as the values of n_1, q_1, m_1 and u_1, involves four $\odot$, one $\boxplus$ and six $\oplus$, which is equivalent to 19 $\oplus$. In total, this combination provides a 1-bit condition that holds with certainty. It allows us to recover 40 bits of information on the key: $Z_8^{(1)}$, $Z_{15}^{(1)}$, $Z_{17}^{(1)}$, $Z_8^{(2)}$ and $Z_{15}^{(2)}$. Using 40 known plaintexts and $\frac{19}{120} \cdot 2^{40} \approx 2^{37.34}$ 1.5-round MESH-128(8) computations.

To attack 2-round MESH-128(8), consider the extended trails used previously for 1.5 rounds:

$$\mathrm{lsb}_1((p_8 \boxplus Z_8^{(1)} \oplus r_1) \odot Z_{15}^{(2)} \oplus s_2) = \mathrm{lsb}_1(X_8^{(2)})\,, \tag{3.371}$$

$$\mathrm{lsb}_1((p_{15} \boxplus Z_{15}^{(1)} \oplus s_1) \odot Z_8^{(2)} \oplus r_2) = \mathrm{lsb}_1(X_{15}^{(2)})\,. \tag{3.372}$$

Combining (3.371) and (3.372) results in:

$$\mathrm{lsb}_1((p_8 \boxplus Z_8^{(1)} \oplus r_1) \odot Z_{15}^{(2)} \oplus (p_{15} \boxplus Z_{15}^{(1)} \oplus s_1) \odot Z_8^{(2)} \oplus X_8^{(2)} \oplus X_{15}^{(2)}) = \mathrm{lsb}_1(r_2 \oplus s_2)\,. \tag{3.373}$$

Equation (3.373) depends on $r_2 \oplus s_2$, which can be expressed as

$$\mathrm{lsb}_1(r_2 \oplus s_2) = \mathrm{lsb}_1(n_2 \odot Z_{17}^{(2)})\,, \tag{3.374}$$

where $n_2 = X_1^{(2)} \oplus X_2^{(2)}$.

The individual values of r_1 and s_1 are needed in (3.373) as well. Let

$$w_1 = ((n_1 \odot Z_{17}^{(1)}) \boxplus q_1) \odot m_1\,, \tag{3.375}$$

$$c_1 = w_1 \boxplus u_1\,, \tag{3.376}$$

$$x_1 = c_1 \odot a_1\,, \tag{3.377}$$

$$y_1 = x_1 \boxplus b_1\,, \tag{3.378}$$

$$o_1 = y_1 \odot d_1\,, \tag{3.379}$$

$$j_1 = o_1 \boxplus e_1\,, \tag{3.380}$$

$$l_1 = j_1 \odot Z_{18}^{(1)}\,, \tag{3.381}$$

$$h_1 = o_1 \boxplus l_1\,, \tag{3.382}$$

$$g_1 = h_1 \odot y_1\,, \tag{3.383}$$

$$f_1 = x_1 \boxplus g_1\,, \tag{3.384}$$

$$v_1 = f_1 \odot c_1\,, \tag{3.385}$$

$$t_1 = v_1 \boxplus w_1\,, \tag{3.386}$$

$$s_1 = t_1 \odot ((n_1 \odot Z_{17}^{(1)}) \boxplus q_1)\,, \tag{3.387}$$

and

$$r_1 = s_1 \boxplus (n_1 \odot Z_{17}^{(1)})\,, \tag{3.388}$$

where $n_1 = (p_1 \odot Z_1^{(1)}) \oplus (p_9 \boxplus Z_9^{(1)})$, $q_1 = (p_2 \boxplus Z_2^{(1)}) \oplus (p_{10} \odot Z_{10}^{(1)})$, $m_1 = (p_3 \odot Z_3^{(1)}) \oplus (p_{11} \boxplus Z_{11}^{(1)})$, $u_1 = (p_4 \boxplus Z_4^{(1)}) \oplus (p_{12} \odot Z_{12}^{(1)})$, $a_1 = (p_5 \odot Z_5^{(1)}) \oplus (p_{13} \boxplus Z_{13}^{(1)})$, $b_1 = (p_6 \boxplus Z_6^{(1)}) \oplus (p_{14} \odot Z_{14}^{(1)})$, $d_1 = (p_7 \odot Z_7^{(1)}) \oplus (p_{15} \boxplus Z_{15}^{(1)})$ and $e_1 = (p_8 \boxplus Z_8^{(1)}) \oplus (p_{16} \odot Z_{16}^{(1)})$.

Taking into account equations (3.374) until (3.388), equation (3.373) requires eighteen $\odot$, nineteen $\boxplus$ and thirteen $\oplus$, which is equivalent to about 86 $\oplus$. The subkeys recovered are $Z_i^{(1)}$, for $1 \leq i \leq 18$, $Z_8^{(2)}$ and $Z_{15}^{(2)}$, which means 160 subkey bits. Using 160 known plaintexts and $\frac{86}{176} \cdot 2^{160} \approx 2^{159}$ 2-round MESH-128(8) computations.

Attacking 2.5-round MESH-128(8) would include one additional KM half-round on top of the previous attack on 2 rounds, but the complexity would be larger than that of an exhaustive key search.

3.18 Key-Dependent Distribution Attack

This section is based on [174]. Let us assume some randomly encrypted texts are provided. Some intermediate values calculated from the given text blocks and the correct key/subkey should comply with some expected key-dependent distribution. Otherwise, if the wrong key/subkey value is used to calculate the intermediate value, then the result should rather fit a random distribution. This generic reasoning is employed in Differential and Linear cryptanalysis [89].

Using this key-dependent distribution, the authors of [174] formalize a key-recovery attack using statistical hypothesis testing and named it as the *key-dependent attack*. For a given key, the null hypothesis of the test is that the intermediate value conforms to a key-dependent distribution, determined by the key. The text samples are the intermediate values calculated from the given plaintext/ciphertext blocks. If the test succeeds, the key/subkey used is considered to be the correct one; otherwise, it is discarded.

For keys that share the same key-dependent distribution and the same intermediate value computed, the corresponding hypothesis tests can be merged to reduce the time complexity of the attack. By this reasoning, the whole key space is divided into several key-dependent classes. Consequently, the time complexity of the attack depends on the time taken to distinguish between the key-dependent and the random distributions.

The attack effort depends on the entropy of the key-dependent distribution: the closer the key-dependent distribution is to a random distribution, the more encryptions are needed. For each key-dependent class, the number of encryptions and the criteria for rejecting hypothesis can be chosen to optimize the attack effort. The expected attack time for each class is also obtained. The total expected time complexity can be computed from the expected time for each key-dependent class. Different orderings of the key-dependent classes lead to different expected time complexities.

The total expected time complexity is minimized if the correct key is assumed to be chosen uniformly from the key space.

Previous key-dependent attacks in the literature include [10] on the Lucifer cipher and [105] on the DFC cipher.

An instantiation of the key-dependent attack on reduced-round variants of the IDEA cipher is presented in the following paragraphs, using the Biryukov-Demirci relation [150] in a differential cryptanalysis setting.

Definition 3.30. [174]
For a block cipher, if the probability distribution of an intermediate value varies for different keys under some specific constraints, then this probability distribution is a key-dependent distribution.

Definition 3.31. [174]
A key-dependent class is a tuple (P, U), where P is a fixed key-dependent distribution of intermediate values (internal cipher data), and U is a set of keys that share the same key-dependent distribution P and the same intermediate value calculations.

Definition 3.32. [174]
The key fraction f of a key-dependent class is the ratio between the size of U and the size of the key space.

The key-dependent attack determines which class contains the correct key by conducting hypothesis testings on each class. Such procedure on a key-dependent class (P, U) is called *individual attack* and is composed of four phases:

- Parameter-determining phase: determine the sample size and the criteria for rejecting the hypothesis that the intermediate values conform to P.

- Data-collecting phase: randomly choose encrypted data according to the specific constraints. Though each individual attack chooses encrypted data randomly, one such data can be used by many individual attacks in order to reduce the overall data complexity.
- Judgment phase: calculate the intermediate values from the collected encrypted data. If the results satisfy the rejection criteria, then discard the key-dependent class. Otherwise, proceed to the next phase.
- Exhaustive-search phase: exhaustively search U to find the key. If the search does not find the actual key, then start another individual attack on the next key-dependent class.

The time complexity of the key-dependent attack is determined by the time complexity of each individual attack and the order in which they are performed. The time complexity of each individual attack is determined by the corresponding key-dependent distribution P. For each key-dependent class, the number of encryptions and the criteria for rejecting the hypothesis are chosen in order to minimize the time complexity of each individual attack. To minimize this time complexity, the attack should consider the probability of two types of error, denoted Type-I and Type-II.

A Type-I error occurs when the hypothesis is rejected for a key-dependent class while in fact the correct key is in U, and the attack will fail to find the actual key in this case. The probability of a Type-I error is also called the significance level of the test and is denoted α.

A Type-II error occurs when the test succeeds but in fact it should fail because the correct key is not in U. In this case, the attack will arrive at the exhaustive search phase, but will not find the actual key. The probability of a Type-II error is denoted β.

With a fixed sample size, denoted N, and a significance level α, the criteria of rejecting the hypothesis is determined, and the probability β is also fixed. For a fixed sample size, it is impossible to reduce both α and β simultaneously. In order to reduce both α and β, the attack would have to use a larger sample size, but then the time and data complexities would increase. Therefore, an individual attack has to balance the sample size and the probability of making wrong decisions.

For a key-dependent class (P, U), if the actual key is not in this class, the expected time complexity (measured in number of encryptions) of an individual attack on this class is

$$W = N + \beta|U|. \tag{3.389}$$

If the actual key is in this class, the expected time of an individual attack is

$$R = N + (1-\alpha)\frac{|U|}{2}. \tag{3.390}$$

Since the time complexity is dominated by the time taken attacking the wrong key-dependent classes (there is only one class containing the correct key), the attack only needs to minimize the time spent attacking the wrong classes to minimize the total time complexity.

Although α does not appear in (3.389), α affects the success probability of the attack, so α should also be considered.

An upper-bound is set on α to guarantee the success probability is above a fixed threshold, and then the sample size is chosen in order to minimize equation (3.389), which minimizes the time complexity of individual attacks. It is possible that some key-dependent distribution is too close to a random distribution causing the expected time for performing hypothesis testing longer than directly searching the class. For these classes, the attack exhaustively searches the class directly instead of using statistical hypothesis testing.

The time complexity depends on the order the individual attacks are performed on different classes. Since the expected time complexity of individual attacks are not the same, different orderings of the individual attacks result in different total time complexities.

Assume that a key-dependent attack performs individual attacks on m key-dependent classes in the order (P_1, U_1), (P_2, U_2), ..., (P_m, U_m). Let R_i denote the expected time for (P_i, U_i) if the actual key is in U_i, and let W_i denote the expected time if the correct key is not in U_i.

Theorem 3.4. *[174]*
The expected time for the whole key-dependent attack is minimized if the following condition is satisfied:

$$\frac{f_1}{W_1} \geq \frac{f_2}{W_2} \geq \ldots \frac{f_m}{W_m}. \tag{3.391}$$

The proof of this theorem is in [174].

3.18.1 Key-Dependent Attacks on IDEA

Let $X^{(i)} = (X_1^{(i)}, X_2^{(i)}, X_3^{(i)}, X_4^{(i)})$ denote the 64-bit input block to the i-th round in IDEA, $1 \leq i \leq 9$. Let $Y^{(i)} = (Y_1^{(i)}, Y_2^{(i)}, Y_3^{(i)}, Y_4^{(i)})$ denote the 64-bit output block from the i-th KM half-round. Let $P = (P_1, P_2, P_3, P_4)$ denote a plaintext block, and $C = (C_1, C_2, C_3, C_4)$ denote the corresponding ciphertext block after 8.5-round IDEA.

A concrete instantiation of the key-dependent attack will use the Biryukov-Demirci relation [150] in a differential cryptanalysis setting [24].

Let the inputs to the i-th MA-box be denoted (n_i, q_i), and the corresponding outputs be (r_i, s_i), for $1 \leq i \leq 8$.

The Biryukov-Demirci relation across the full 8.5-round IDEA is

$$\text{lsb}_1(C_2 \oplus C_3) = \text{lsb}_1(P_2 \oplus P_3 \oplus Z_2^{(1)} \oplus Z_3^{(1)} \oplus s_1 \oplus r_1 \oplus Z_2^{(2)} \oplus Z_3^{(2)} \oplus s_2 \oplus r_2 \oplus Z_2^{(3)} \oplus Z_3^{(3)} \oplus s_3 \oplus r_3 \oplus Z_2^{(4)} \oplus Z_3^{(4)} \oplus s_4 \oplus r_4 \oplus Z_2^{(5)} \oplus Z_3^{(5)} \oplus s_5 \oplus r_5 \oplus Z_2^{(6)} \oplus Z_3^{(6)} \oplus s_6 \oplus r_6 \oplus Z_2^{(7)} \oplus Z_3^{(7)} \oplus s_7 \oplus r_7 \oplus Z_2^{(8)} \oplus Z_3^{(8)} \oplus s_8 \oplus r_8 \oplus Z_2^{(9)} \oplus Z_3^{(9)}). \quad (3.392)$$

Demirci's relation [60] for the least significant bit of $r_i \oplus s_i$ (denote it as t_i) is:

$$\text{lsb}_1(r_i \oplus s_i) = \text{lsb}_1(n_i \odot Z_5^{(i)}) = \text{lsb}_1(t_i). \quad (3.393)$$

Let us assuming the difference operator as exclusive-or ($\oplus$). So, $\Delta P = P \oplus P^*$ denotes the difference between plaintext blocks (P, P^*), where $P = (P_1, P_2, P_3, P_4)$ and $P^* = (P_1^*, P_2^*, P_3^*, P_4^*)$. The corresponding ciphertexts blocks are denoted $C = (C_1, C_2, C_3, C_4)$ and $C^* = (C_1^*, C_2^*, C_3^*, C_4^*)$, respectively.

It was shown in [60] that for a chosen plaintext pair (P, P^*), the difference of the contents in the Biryukov-Demirci relation (3.392) results in the *keyless Biryukov-Demirci relation*:

$$\text{lsb}_1(C_2 \oplus C_2^* \oplus C_3 \oplus C_3^*) = \text{lsb}_1(P_2 \oplus P_2^* \oplus P_3 \oplus P_3^* \oplus \bigoplus_{i=1}^{8} \Delta t_i), \quad (3.394)$$

where $\bigoplus_{i=1}^{8} \Delta t_i$ denoted the exclusive-or sum of the differences Δt_i, for $1 \leq i \leq 8$.

Theorem 3.5. *[174]*
Consider the i-th round in IDEA. If one pair of intermediate values (n_i, n_i^*) satisfies $\Delta n_i = 8000_x$ then

$$Pr(lsb_1(\Delta n_i) = lsb_1(8000_x \odot Z_5^{(i)})) = \frac{|W|}{2^{15}}, \quad (3.395)$$

where W is the set of all 16-bit words w such that $1 \leq w \leq 8000_x$ and $(w * Z_5^{(i)}) + (8000_x * Z_5^{(i)}) < 2^{16} + 1$, with $*$ defined as follows

$$a * b = \begin{cases} a \odot b, & \text{if } a \odot b \neq 0 \\ 2^{16} & \text{if } a \odot b = 0 \end{cases}$$

The proof of Theorem 3.5 is in [174].

In general, there are four cases for the $\Pr(\text{lsb}_1(\Delta n_i) = 1)$ as the value of $Z_5^{(i)}$ ranges from 0 to $2^{16} - 1$.

$$\Pr(\text{lsb}_1(\Delta n_i) = 1) \approx \begin{cases} \frac{Z_5^{(i)}}{2^{17}}, & \text{if } \text{lsb}_2(Z_5^{(i)}) = 00 \\ 0.5 - \frac{Z_5^{(i)}}{2^{17}}, & \text{if } \text{lsb}_2(Z_5^{(i)}) = 01 \\ 1.0 - \frac{Z_5^{(i)}}{2^{17}}, & \text{if } \text{lsb}_2(Z_5^{(i)}) = 10 \\ 0.5 + \frac{Z_5^{(i)}}{2^{17}}, & \text{if } \text{lsb}_2(Z_5^{(i)}) = 11 \end{cases} \quad (3.396)$$

From equation (3.396), the following approximation holds for most values of $Z_5^{(i)}$:

$$\min\{\Pr(\text{lsb}_1(\Delta n_i) = 0), \Pr(\text{lsb}_1(\Delta n_i) = 1)\} \approx \begin{cases} \frac{Z_5^{(i)}}{2^{17}}, & \text{if } \text{lsb}_1(Z_5^{(i)}) = 0 \\ 0.5 - \frac{Z_5^{(i)}}{2^{17}}, & \text{if } \text{lsb}_1(Z_5^{(i)}) = 1 \end{cases}$$

Calculations by the authors in [174] showed that for only 219 out of the 2^{16} possible values of $Z_5^{(i)}$, the difference between the values in (3.396), (3.397) and the real probability is larger than 0.01.

Equation (3.397) indicates that we can approximate the left-hand side of (3.397) by fixing several most and least significant bits of $Z_5^{(i)}$.

An attack on 5.5-round IDEA starts from the third round until after the 8th KM half-round. The idea is to perform a key-dependent attack based on the distribution of Δn_4.

Consider the following differential equation of the keyless Biryukov-Demirci relation for a 5.5-round IDEA variant:

$$\text{lsb}_1(\Delta t_4) = \text{lsb}_1(\Delta X_2^{(3)} \oplus \Delta X_3^{(3)} \oplus \Delta Y_2^{(8)} \oplus \Delta Y_3^{(8)} \oplus \Delta t_3 \oplus \Delta t_5 \oplus \Delta t_6 \oplus \Delta t_7). \quad (3.397)$$

We first need to construct a plaintext pair satisfying the constraint $\Delta n_4 = 8000_x$. This construction is based on the following lemma.

Lemma 3.7. *[174]*
For any α, if two 16-bit words x and x^* have the same fifteen least significant bits, then
- *$x \oplus \alpha$ and $x^* \oplus \alpha$ have the same fifteen least significant bits*
- *$x \boxplus \alpha$ and $x^* \boxplus \alpha$ have the same fifteen least significant bits.*

Proposition 3.2. *[174]*
If a pair of intermediate values $(Y^{(3)}, Y^{(3)*})$ satisfy the following conditions:
(a) $\Delta Y_1^{(3)} = \Delta Y_3^{(3)} = 0$
(b) $\Delta Y_2^{(3)} = 8000_x$
(c) $Y_2^{(3)} \oplus Y_4^{(3)} = Y_2^{(3)*} \oplus Y_4^{(3)*}$
then $\Delta n_3 = 0$ and the Pr(lsb$_1(\Delta n_4) = 0$) can be determined from Theorem 3.5.

The proof of Proposition 3.2 is in [174].

The attack on 5.5-round IDEA uses plaintext pairs satisfying Proposition 3.2. Condition (a) is satisfied by letting $\Delta X_1^{(3)} = \Delta X_3^{(3)} = 0$. By Lemma 3.7, (P_2, P_2^*) are fixed to have the same 15 least significant bits, and hence $\Delta Y_2^{(1)} = 8000_x$. In order to fulfill condition (c), we have to guess $Z_4^{(3)}$ and, based on it, choose $(X_1^{(3)}, X_1^{(3)*})$ which satisfy $\Delta Y_4^{(3)} = 8000_x$.

By Proposition 3.2, $\Delta t_3 = 0$. In order to get the right-hand side of (3.397), we still need to get Δt_5, Δt_6 and Δt_7. We need to guess $Z_5^{(5)}$, $Z_1^{(6)}$, $Z_2^{(6)}$, $Z_5^{(6)}$, $Z_6^{(6)}$, $Z_1^{(7)}$, $Z_2^{(7)}$, $Z_3^{(7)}$, $Z_4^{(7)}$, $Z_5^{(7)}$, $Z_6^{(7)}$, $Z_1^{(8)}$, $Z_2^{(8)}$, $Z_3^{(8)}$ and $Z_4^{(8)}$.

According to [24], one can partially decrypt one ciphertext pair using these fifteen subkeys to calculate the values of Δt_5, Δt_6 and Δt_7. These subkeys represent 103 user key bits, according to the key schedule of IDEA. For each guess, we calculate the value of Δt_4 from a special encryption pair. Note that these 103 key bits also include the bits of $Z_5^{(4)}$, which determines the key-dependent distribution of Δt_4, according to Theorem 3.5.

The key space can be divided into 2^{103} key-dependent classes by the 103 key bits, each containing 2^{25} keys.

For a key-dependent class (P, U), let $p = \Pr(\text{lsb}_1(\Delta t_4) = \text{lsb}_1(8000_x \odot Z_5^{(4)}))$.

For simplicity, in the following analysis, we assume that $p \leq 0.5$. The case $p > 0.5$ is similar. Assume that the sample size is n encryption pairs, that satisfy the specific constraint on this key-dependent class, and m of them satisfy $\text{lsb}_1(\Delta t_4) = \text{lsb}_1(8000_x \odot Z_5^{(4)})$. The criteria for not rejecting the hypothesis is that $m \leq k$, for a fixed k.

The probability of a Type-I error is $\alpha = \sum_{i=k+1}^{n} p^i (1-p)^{n-i}$, while the probability of a Type-II error is $\beta = \sum_{i=0}^{k} \binom{n}{i} 0.5^n$.

If (P, U) is a wrong class, the expected time complexity of checking this class is

$$W = 2n + 2^{25}\beta. \tag{3.398}$$

The attack sets $\alpha \leq 0.01$ to ensure that the probability of false rejection will not exceed 0.01. Under this precondition, the attack chooses n and β so that $\alpha < 0.01$, which minimizes (3.398) in order to minimize the time complexity of each key-dependent class (P, U). We minimize the total expected time complexity with this method. Since this choice depends only on $Z_5^{(4)}$, we only need to get n and k for 2^{16} different values.

For example, for a key-dependent class (P, U), with $Z_5^{(4)} = 8000_x$, $p \approx 0.666687$. The attack checks every possible n and k to find the minimum expected time complexity of the individual attacks for this class. The expected time complexity for each class is upper-bounded by the exhaustive search effort on the class, which is 2^{25}. Hence, the attack only checks all the n and k smaller than 2^{25}. The expected time is minimized under $\alpha < 0.01$

when $n = 425$ and $k = 164$. In this case, $\alpha = 0.009970$, $\beta = 0.000001$ and $W = 899.094678$.

Since all key-dependent classes have the same key fraction, the order of execution of individual attacks with minimal expected time complexity becomes the ascending order of W for all key-dependent classes, due to Theorem 3.4.

The total expected time complexity of the attack becomes (according to Theorem 3.4):

$$\sum_{i+1}^{m} f_i R_i + \sum_{i=1}^{m} (f_i \sum_{j=1}^{i-1} W_j) + \alpha \sum_{i=1}^{m} (f_i \sum_{j=i+1}^{m} W_j) =$$
$$\frac{1}{2^{103}} (\sum_{i=1}^{2^{103}} R_i + \sum_{i=1}^{2^{103}} \sum_{j=1}^{i-1} W_j + 0.01 \sum_{i=1}^{2^{103}} \sum_{j=i+1}^{2^{103}} W_j)$$
$$\leq \frac{1}{2^{103}} (\sum_{i=1}^{2^{103}} 2^{26} + \sum_{i=1}^{2^{103}} \sum_{j=1}^{i-1} W_j + 0.01 \sum_{i=1}^{2^{103}} \sum_{j=i+1}^{2^{103}} W_j) =$$
$$\frac{1}{2^{103}} (2^{103} \cdot 2^{26} + \sum_{i=1}^{2^{103}} (2^{103} - i + 0.01 \cdot i) W_i) \approx 2^{112.1}.$$

The success rate of the attack is 99% if n and β are chosen for each key-dependent class and the order of the individual attacks is determined as described previously. The number of text pairs needed in one test is about 2^{19} in the worst case. The attack uses a set of about 2^{21} plaintexts which can provide 2^{20} plaintext pairs satisfying the conditions of Proposition 3.2 for each key-dependent class.

The attack can be summarized as follows:

- for every possible $Z_5^{(4)}$, calculate the number n of plaintext pairs and the criteria for not rejecting the hypothesis k.
- let S be an empty set. Randomly enumerate a 16-bit word s, insert s and $s \oplus 8000_x$ in S. Repeat this enumeration until S contains 2^5 different words. Ask for the encryption of all plaintexts of the form (A, B, C, D), where A and C are fixed to arbitrary constants, B takes all values in S and D takes all possible 16-bit values.
- enumerate the key-dependent classes in ascending order of W:
  - randomly choose a set of plaintext pairs with cardinality n from the known encryptions. The plaintext pairs must satisfy the requirements of Proposition 3.2.
  - partially decrypt all the selected encryption pairs and count the number of occurrences of $\text{lsb}_1(\Delta t_4) = 1$.
  - test the hypothesis: if the hypotheses is not rejected, perform exhaustive search for the remaining 25 key bits.

We can extend the previous attack to 6-round IDEA, starting from the MA half-round of the second round, until after the KM half-round of the eighth round. The data complexity is 2^{49} chosen plaintexts, and the time complexity is $2^{112.1}$ 6-round IDEA computations.

Note that $Z_5^{(2)}$ and $Z_6^{(2)}$ are included in the 103 key bits recovered in the attack on 5.5-round IDEA. Thus, we can add this MA half-round to the 5.5-round attack without affecting the time complexity.

But, it is more difficult to construct right pairs satisfying Proposition 3.2. Consider a pair of intermediate values X^3 and X^{3*} before the third round, which satisfy Proposition 3.2. If we partially decrypt X^3 and X^{3*} using any possible $Z_5^{(2)}$ and $Z_6^{(2)}$, then all the results have the same exclusive-or of the first and third words. Thus, the attack selects all the plaintexts P where the 15 least significant bits of $P_1 \oplus P_3$ are fixed to an arbitrary 15-bit constant. The total number of selected plaintexts is 2^{49}. Is it possible to provide 2^{48} plaintext pairs satisfying the conditions in Proposition 3.2 in the test for $(Z_5^{(2)}, Z_6^{(2)}, Z_4^{(3)})$.

Further, we describe an attack on 5-round IDEA starting from the first round. The differential version of the Biryukov-Demirci relation is the following:

$$\text{lsb}_1(\Delta t_2) = \text{lsb}_1(\Delta P_2 \oplus \Delta P_3 \oplus \Delta C_2 \oplus \Delta C_3 \oplus \Delta t_1 \oplus \Delta t_3 \oplus \Delta t_4 \oplus \Delta t_5). \tag{3.399}$$

Plaintext pairs are chosen according to Proposition 3.2 before the first round by guessing $Z_4^{(1)}$ and then $\Delta t_1 = 0$ as in the attack on 5.5-round IDEA. In order to determine the right-hand side of (3.399), we need to discover $Z_5^{(3)}$, $Z_1^{(4)}$, $Z_2^{(4)}$, $Z_5^{(4)}$, $Z_6^{(4)}$, $Z_1^{(5)}$, $Z_2^{(5)}$, $Z_3^{(5)}$, $Z_4^{(5)}$, $Z_5^{(5)}$ and $Z_6^{(5)}$. These eleven subkeys cover the bits numbered 75–65 of the user key. These 119 key bits only cover the nine most significant bits of $Z_5^{(2)}$, which determines the probability distribution of $\text{lsb}_1(\Delta t_2)$. It is not necessary to guess the full $Z_5^{(2)}$. The attack continues to guess the least significant bit of $Z_5^{(2)}$, the 72nd bit of the user key, and estimates the probability that $\text{lsb}_1(\Delta t_2) = 1$ according to (3.396).

Hence, the attack divides the key space into 2^{120} key-dependent classes by the 120 key bits, and performs the individual attacks on each class. The attack uses statistical hypothesis testing to determine which class the actual key is in. The subkey $Z_5^{(2)}$, which determines the $\Pr(\text{lsb}_1(\Delta t_1) = 1)$, cannot be approximated by (3.396), and the attack exhaustively searches the remaining key bits.

In this attack, it is possible that the expected time of individual attacks is larger than exhaustive search of some classes, which means that $2n + \beta \cdot 2^8 \geq 2^8$.

Under this condition, the attack also uses exhaustive key search to determine the remaining 8 key bits to make sure the time needed does not exceed the exhaustive search effort. This attack also choose $\alpha < 0.01$ to ensure a success rate of 99%. In this case, the total expected time complexity is $2^{125.5}$ 5-round IDEA encryptions.

Experiments suggest that the attack needs at most 75 pairs for one test. We ask for 2^{17} encryptions, which can provide 2^{16} pairs, and this is sufficient for the test.

In the second attack on 5-round IDEA, we try to obtain the plaintext pairs satisfying Proposition 3.2 before the second round. In order to determine $\text{lsb}_1(\Delta t_3)$ we need to know the least significant bits of Δt_1, Δt_2, Δt_4 and Δt_5. Hence, we need to know the subkeys $Z_1^{(1)}$, $Z_2^{(1)}$, $Z_3^{(1)}$, $Z_4^{(1)}$, $Z_5^{(1)}$, $Z_6^{(1)}$, $Z_4^{(2)}$, $Z_5^{(3)}$, $Z_5^{(4)}$, $Z_1^{(5)}$, $Z_2^{(5)}$, $Z_5^{(5)}$ and $Z_6^{(5)}$. These thirteen subkeys correspond to 107 user key bits, numbered 0–106. For every guessed 106 key candidate, we use similar techniques as before. The expected time complexity is $2^{115.3}$ 5-round IDEA encryptions.

Since it is not possible to predict the plaintext pairs which produce the intermediate pairs satisfying Proposition 3.2 before the second round, the whole codebook is required for this attack.

3.19 BDK Attacks

In [23], Biham *et al.* presented new attacks on reduced-round variants of IDEA based on nontrivial relations involving all three arithmetic operations: $\odot$, $\boxplus$ and $\oplus$.

The acronym *BDK* stands for the surname of the authors of these attacks: Eli Biham, Orr Dunkelman and Nathan Keller.

Let $X^{(i)} = (X_1^{(i)}, X_2^{(i)}, X_3^{(i)}, X_4^{(i)})$ denote the 64-bit input block to the i-th round in IDEA, for $1 \leq i \leq 9$. Further, let $Y^{(i)} = (Y_1^{(i)}, Y_2^{(i)}, Y_3^{(i)}, Y_4^{(i)})$ denote the 64-bit output block from the i-th KM half-round.

Let the inputs to the i-th MA-box be denoted (n_i, q_i), and the corresponding outputs be (r_i, s_i), for $1 \leq i \leq 8$. Further, let $t_i = n_i \odot Z_5^{(i)}$.

Let the difference operator of two bit-strings (X, X^*) be $\oplus$ and their difference be denoted $\Delta X = X \oplus X^*$.

The following proposition was stated and proved in [23]:

Proposition 3.3. *[23]*
Assume that the exclusive-or difference $(\Delta n_i, \Delta q_i) = (0, \alpha)$ holds for some value $\alpha \neq 0$. Also, assume that there is no key difference in $Z_5^{(i)}$, but there is no assumption whether there is a nonzero difference in $Z_6^{(i)}$. Then:
(1) the $\text{lsb}_1(\Delta r_i \oplus \Delta s_i) = 0$.
(2) the average probability of the event $(\Delta r_i, \Delta s_i) = (8000_x, 8000_x)$ over all the possible keys is 2^{-16} (if $\alpha \neq 0$ or if there is a nonzero difference in $Z_6^{(i)}$).
(3) if $\alpha \neq 0$ of if there a nonzero difference in $Z_6^{(i)}$, then

$$\sum_{\nu,\tau} Pr^2[(\Delta r_i, \Delta s_i) = (\nu, \tau)] = 2^{-23.72}.$$

If the MA-box were a truly random mapping, then the probability of the event $(\Delta r_i, \Delta s_i) = (8000_x, 8000_x)$ in item (2) would be 2^{-32}. This same probability would be expected in item (3).

The Biryukov-Demirci relation (3.392) is another important tool in the attacks of [23]. In order to eliminate the dependence of (3.392) on the subkeys, consider the exclusive-or difference of (3.392) for the plaintext pair (P, P^*):

$$\text{lsb}_1(P_2 \oplus P_2^* \oplus P_3 \oplus P_3^* \oplus \bigoplus_{i=1}^{8} \Delta t_i) = \text{lsb}_1(C_2 \oplus C_2^* \oplus C_3 \oplus C_3^*), \tag{3.400}$$

which is known as the *keyless Biryukov-Demirci relation.*

A distinguishing attack on the first 2.5-round IDEA (which also holds for any 2.5 consecutive rounds starting and ending in a KM half-round) uses (3.400) restricted to 2.5 rounds:

$$\text{lsb}_1(P_2 \oplus P_2^* \oplus P_3 \oplus P_3^* \oplus \Delta t_1 \oplus \Delta t_2) = \text{lsb}_1(C_2 \oplus C_2^* \oplus C_3 \oplus C_3^*). \tag{3.401}$$

From item (1) in Proposition 3.3: if $(\Delta n_i, \Delta q_i) = (0, \alpha)$ then $\Delta t_i = 0$. In order to satisfy this property, consider plaintext pairs (P, P^*) such that $\Delta X^{(1)} = (0, \beta, 0, \gamma)$ for arbitrary nonzero β and γ. For these pairs, $\Delta Y_1^{(1)} = \Delta Y_3^{(1)} = 0$, independent of the values of $Z_1^{(1)}$ and $Z_3^{(1)}$, and hence $\Delta n_1 = 0$. Since the difference in $Z_5^{(i)}$ is zero, it follows that $\Delta t_1 = 0$.

Similarly, if we take only ciphertext pairs satisfying $\Delta Y^{(3)} = (0, 0, \beta^*, \gamma^*)$ for arbitrary nonzero β^* and γ^*, then $(\Delta n_2, \Delta q_2) = (0, \alpha^*)$ for some nonzero α^*, and hence $\Delta t_2 = 0$.

If the plaintext/ciphertext pairs (P, C) and (P^*, C^*) satisfy both $\Delta t_1 = 0$ and $\Delta t_2 = 0$ then, equation (3.401) becomes

$$\text{lsb}_1(P_2 \oplus P_2^* \oplus P_3 \oplus P_3^*) = \text{lsb}_1(C_2 \oplus C_2^* \oplus C_3 \oplus C_3^*), \tag{3.402}$$

which is a 1-bit condition that can be verified using only the least significant bits of the plaintext/ciphertext pairs.

Therefore, a simple distinguishing attack on 2.5-round IDEA is

(1) ask for the encryption of 2^{18} plaintexts of the form (U, V, W, X) where U and W are fixed, while V and X assume arbitrary random values.
(2) insert the ciphertexts into a hash table sorted by the first two words.
(3) for every ciphertext pair in the same entry of the hash table, check whether equation (3.402) holds for the corresponding plaintext/ciphertext pair.
(4) if there is a pair for which equation (3.402) does not hold, conclude that the cipher is not 2.5-round IDEA. If there is no such pair satisfying (3.402), conclude that the cipher is 2.5-round IDEA.

The 2^{18} plaintexts can be combined into $2^{18}(2^{18} - 1)/2 \approx 2^{35}$ possible pairs, and a fraction of about 2^{-32} of them is expected to have ciphertext difference of the form $(0, 0, \beta^*, \gamma^*)$. Hence, the expected number of pairs in item (3) is $2^{35} \cdot 2^{-32} = 8$ If there is a pair for which equation (3.402) does not hold, we know for sure that the cipher is not 2.5-round IDEA. On the other hand, for a random permutation, the probability that this equation holds for all the eight pairs is 2^{-8}. Therefore, the distinguisher succeeds with probability greater than 99.5%.

Since the steps (2) and (3) of the attack are implemented using a hash table, the time complexity is dominated by the time complexity of the encryptions in the first part of the attack. The data complexity is 2^{18} chosen plaintexts, the memory complexity is 2^{18} blocks, and time complexity is 2^{18} 2.5-round IDEA encryptions.

The 2.5-round distinguisher can be used for a key-recovery attack on 3 full rounds (starting in a KM half-round and ending after an MA half-round). The attack operates as follows:

(1) ask for the encryption of 2^{19} plaintexts of the form (A, X, B, Y) where A and B are fixed while X and Y assume arbitrary, random values.
(2) for each guess of the 32-bit subkeys of the last MA half-round:
    (a) partially decrypt the last MA half-round for all the ciphertexts and insert the $Y^{(3)}$ values into a hash table sorted by the first 32 bits.
    (b) for every pair of values in the same bin of the hash table, check if (3.402) holds for the corresponding plaintext/ciphertext pair.
    (c) if there is a pair for which this equation does not hold, discard the guess for the subkeys. Otherwise, keep the guessed value.
(3) output all the subkey guesses that were not discarded.

Since there are 2^{19} plaintexts, there are about $2^{19}(2^{19}-1)/2 \approx 2^{37}$ possible pairs, and about 32 pairs are examined in step 2(b). Hence, for a wrong key guess, the probability that equation (3.402) holds for all the pairs is $(2^{-1})^{32} = 2^{-32}$. Therefore, only few possible subkey candidates remain, including the right one.

The time complexity of the attack is dominated by step 2(b) in which decryption of all ciphertexts are performed under all subkeys of the MA half-round. The data complexity is 2^{19} chosen plaintext. Memory complexity is 2^{19} 64-bit blocks. The time complexity is equivalent to $2^{19} \cdot 2^{32}/6 \approx 2^{48.5}$ 3-round IDEA encryptions. This attack only recovers 32 user key bits. The remaining $128 - 32 = 96$ key bits can be recovered by exhaustive search, resulting in a time complexity of 2^{96} 3-round IDEA encryptions.

Consider now a 4.5-round IDEA variant starting at round 4. We adapt equation (3.400) to

$$\begin{aligned} \text{lsb}_1(P_2 \oplus P_2^* \oplus P_3 \oplus P_3^* \oplus \Delta t_4 \oplus \Delta t_5 \oplus \Delta t_6 \oplus \Delta t_7) = \\ \text{lsb}_1(C_2 \oplus C_2^* \oplus C_3 \oplus C_3^*). \end{aligned} \quad (3.403)$$

This attack uses plaintext pairs with input difference $\Delta X^{(4)} = (0, \beta, 0, \gamma)$. Thus, $\Delta t_4 = 0$. In order to compute Δt_j, for $5 \leq j \leq 7$, we guess part of the user key and partially decrypt the last three rounds.

Subkeys $Z_1^{(8)}$, $Z_2^{(8)}$, $Z_3^{(8)}$, $Z_4^{(8)}$, $Z_1^{(7)}$, $Z_2^{(7)}$, $Z_3^{(8)}$, $Z_4^{(7)}$, $Z_5^{(7)}$, $Z_6^{(7)}$, $Z_6^{(6)}$ and $Z_5^{(6)}$ are guessed to compute the Δt_j values and decrypt two rounds. Further, $Z_1^{(6)}$, $Z_2^{(6)}$, $Z_5^{(5)}$ are guessed to compute Δt_5. These 15 subkeys amount to 103 user key bits. Hence, these 103 key bits are guessed and for each guess we check if equation (3.403) holds for the plaintext/ciphertext pair. Finding the right subkeys requires about 128 pairs for the analysis. These pairs can be constructed from about 16 chosen plaintexts. Starting the attack in another round would require guessing more than 103 key bits.

Extending the attack to 5 rounds requires guessing the subkeys in the third MA half-round, which does not increase the time complexity since $(Z_5^{(3)}, Z_6^{(3)})$ correspond to bits 50–81 of the user key, and those bits are already included in the 103 bits guessed in the 4.5-round attack. However, the additional MA half-round affects the data complexity.

The issue is how to get plaintext pairs with difference of the form $\Delta X^{(4)} = (0, \beta, 0, \gamma)$. Since for every guess of $(Z_5^{(3)}, Z_6^{(3)})$, different plaintext pairs are needed to fulfill this differential requirement, this attack uses known plaintexts instead of chosen plaintexts. Start with 2^{19} known plaintexts that can provide $2^{19}(2^{19}-1)/2 \approx 2^{37}$ possible pairs. For each guess of $(Z_5^{(3)}, Z_6^{(3)})$, partially encrypt all the plaintexts and choose the pairs that have difference $\Delta X^{(4)} = (0, \beta, 0, \gamma)$. We expect $2^{37} \cdot 2^{-32} = 32$ such pairs.

The time complexity of this step is negligible compared to that of other steps.

The attack can be described as follows:

(1) ask for the encryption of 2^{19} known plaintexts.
(2) for each guess of $(Z_5^{(3)}, Z_6^{(3)})$, i.e. key bits 50–81, perform the following:
    (a) partially encrypt the plaintext through the third MA half-round and insert the resulting $X^{(4)}$ values into a hash table indexed by the first and third words.
    (b) for each guess of key bits 0–49, 82–99 and 125–127, and for all the colliding pairs, perform the following:
        (i) partially decrypt all pairs through rounds 6 and 7, and the 5th MA half-round.
        (ii) verify that equation (3.403) holds for all the pairs. If not, discard the key guess.
    (c) if the key guess passed the filtering, perform exhaustive search on the remaining 25 key bits.

For every guess of key bits 50–81, we expect that 32 pairs are analyzed in step 2(b). Hence, the probability that a wrong key guess passes the filtering is 2^{-32}. Thus, we expect that about $2^{103} \cdot 2^{-32} = 2^{71}$ key guesses enter step 2(c). The time complexity of step 2(c) is equivalent to $2^{25} \cdot 2^{71} = 2^{96}$ encryptions.

The time complexity of the attack is dominated by the partial decryptions of step 2(b). Note that half of the key guesses are discarded after the first pair, half of the remaining key guesses are discarded after the second pair, and so on. Therefore, instead of decrypting all pairs at once, the adversary can decrypt the first pair and check whether equation (3.403) holds, then it if the key guess was not discarded, decrypt the second pair and check the equation again for it. With this approach, the time complexity of this step becomes $2^{103} + 2^{102} + 2^{101} + \ldots \approx 2^{104}$ partial decryptions which are roughly equivalent to 2^{103} 5-round IDEA encryptions. The data complexity is 2^{19} known plaintexts. Memory complexity is 2^{19} text blocks.

The attacks in this section are summarized in Table 3.153.

Due to the extensive use of the *keyless Biryukov-Demirci relation* (3.400), these attacks are called *Keyless-BD* attacks.

Table 3.153 BDK attacks (in the single-key model) on reduced-round IDEA variants in [25].

#Rounds	Attack	Time	Data	Memory	Comments
2.5	Keyless-BD	2^{18}	2^{18} CP	2^{18}	distinguish-from-random attack
3	Keyless-BD	2^{96}	2^{19} CP	2^{19}	
4.5	Keyless-BD	2^{103}	16 CP	16	starting from the 4th KM half-round
5	Keyless-BD	2^{103}	2^{19} KP	2^{19}	starting from the 3rd MA half-round

3.20 Meet-in-the-Middle Attacks

The following explanation of the Meet-in-the-Middle (MITM) attack is based on [25].

The standard MITM attack on a block cipher uses the observation that some intermediate state V can be computed both from the plaintext end given only part of the user key, denoted K_t (where the subscript t stands for *top*), and from the ciphertext end given a part of the user key denoted K_b (where the subscript b stands for *bottom*).

The adversary considers several plaintext/ciphertext pairs and for each guess of K_t, the V values are computed from the plaintexts and stored in a hash table H. For each guess of K_b, the V values are computed from the ciphertexts and looked-up for a match in H. If $|K_t| > |K_b|$, the roles of K_t and K_b are swapped to reduce the storage needed for H, where $|K|$ is the size of K.

The memory complexity of the attack is $2^{|K_{\min}|}$ $(|V|+|K_{\min}|)$-bit values, where $|K_{\min}| = \min(|K_t|, |K_b|)$.

The time complexity is $2^{\max(|K_t|,|K_b|)}$ partial encryptions/decryptions, assuming that computing V from both the plaintext and ciphertext ends costs about the same.

The data complexity is $(|K_t| + |K_b|)/|V|$ plaintext/ciphertext pairs, required for discarding the wrong values of (K_t, K_b). Note that the matching value V provides a $|V|$-bit condition, that is, it allows us to discard a fraction $1/|V|$ of the candidates for each plaintext/ciphertext pair.

3.20.1 Meet-in-the-Middle Attacks on IDEA

The following MITM attacks are based on [25].

Let $X^{(i)} = (X_1^{(i)}, X_2^{(i)}, X_3^{(i)}, X_4^{(i)})$ denote the 64-bit input block to the i-th round of IDEA, for $1 \leq i \leq 9$. Let $Y^{(i)} = (Y_1^{(i)}, Y_2^{(i)}, Y_3^{(i)}, Y_4^{(i)})$ denote the 64-bit output block from the i-th KM half-round. Let the inputs to the i-th MA-box be denoted (n_i, q_i), while the corresponding outputs are (r_i, s_i), for $1 \leq i \leq 8$.

Consider 3.5-round IDEA, and the 16-bit value n_2, the leftmost input to the MA-box in the second round. Recall that $n_2 = (X_1^{(2)} \odot Z_1^{(2)}) \oplus (X_3^{(2)} \boxplus Z_3^{(2)})$, where $X_1^{(2)} = (X_1^{(1)} \odot Z_1^{(1)}) \oplus s_1$, $X_3^{(2)} = (X_2^{(1)} \boxplus Z_2^{(1)}) \oplus r_1$ and (r_1, s_1) can be computed from the input plaintext $X^{(1)}$ and the first round subkeys.

Therefore, $|K_t| = 112$ since n_2 can be computed from $X^{(1)}$ and user key bits 0–111, according to the key schedule of IDEA.

Analogously, $n_2 = X_1^{(3)} \oplus X_2^{(3)}$ and developing $X_1^{(3)}$ and $X_2^{(3)}$ further down, n_2 can be computed from the ciphertext and from the user key bits 50–17. Thus, $|K_b| = 96$. In this case, $|V| = |n_2| = 16$ bits and a standard MITM attack on 3.5-round IDEA has data complexity $(112 + 96)/16 = 13$ known plaintexts. The memory complexity is 2^{96} (16+96)-bit values and the time complexity is 2^{112} partial IDEA decryptions.

In [25], several improvements are presented to reduce the attack complexities:

- reducing the memory complexity: note that K_t and K_b share 80 user key bits, numbered 0–17, 50–111. Instead of guessing these bits twice in K_t and K_b, the adversary starts by guessing these 80 key bits once, and for each guess, perform the standard MITM attack assuming these 80 key bits are fixed. This approach reduces $|K_t|$ to 32 bits, and $|K_b|$ to 16 bits. As a result, for each guess of the common 80 key bits, the memory complexity of the attack drops to 2^{16}, while the time complexity drops to 2^{32} partial encryptions/decryptions. The total time complexity remains $2^{80} \cdot 2^{32} = 2^{112}$ partial encryptions/decryptions, but the memory complexity is only 2^{16} because the same memory slots can be reused by both K_t and K_b.
- reducing the time complexity: note that both the $\boxplus$ and the $\oplus$ operations share the property that the k least significant bits of the result depend only on the k least significant bits of the inputs. This simple observation allows us to reduce $|K_t|$ by restricting the analysis to the $\text{lsb}_k(V)$. Instead

of analyzing $V = n_2$, we consider $V = q_2$ the rightmost input to the MA-box in the second round. Note that q_2 can be computed from the plaintexts and key bits 0–95 and 112–127; likewise, q_2 can be computed from the ciphertexts and user key bits 50–24. Thus, $|K_t| = 112$ and $|K_b| = 103$.

If we restrict attention to $\text{lsb}_k(q_2)$, then we need only the $\text{lsb}_k(Z_2^{(2)})$, which corresponds to user key bits 112–127, and $|K_t|$ is reduced to $96+k$. Taking $k = 7$ results in $|K_t| = |K_b| = 103$ and the time complexity becomes 2^{103} partial encryptions/decryptions. However, the memory complexity increases to 2^{25} since K_t and K_b share only 78 key bits. The data complexity also increases to $\lceil (103 + 103)/7 \rceil = 30$ known plaintexts.

- reducing the data complexity: note that there is no need to discard all possible (K_t, K_b) values during the MITM attack. It is sufficient to discard all but 2^{103} values since exhaustively searching the remaining values can be done within an additional time complexity of at most 2^{103} partial encryptions/decryptions. If the adversary considers only 4 known plaintexts, which provide a filtering of $4 \cdot |V| = 28$ bits, then for each guess of the 78 common user key bits, only $2^{128-78} \cdot 2^{-28} = 2^{22}$ suggestions for the remaining bits of K_t and K_b are expected to remain. Each such suggestion, along with the 78 common key bits, yields a suggestion for the full user key, which can be checked with a couple of trial encryptions. Thus, the complexity of discarding the remaining suggestions is $2^{78} \cdot 2^{22} = 2^{100}$ partial encryption.

By combining all the three improvements, the MITM attack on 3.5-round IDEA requires 2^{103} partial encryptions, 4 known plaintexts and 2^{25} memory.

This attack on the first 3.5-round IDEA can be improved further to use only two known plaintexts by using as V the third and fourth words after the second KM half-round: $Y_3^{(2)}$ and $Y_4^{(2)}$. Thus, $|V| = 32$.

This approach follows from the fact that the 103 key bits guessed at the bottom half cover the bits required for a full decryption of two rounds, including the second MA half-round. From the top side it is enough to guess 96 key bits to obtain the values of these two words after the second KM half-round.

The MITM attack can be extended to 4.5-round IDEA, but from the 4th round up to the 8th KM half-round. Then, $V = n_5$ can be computed from $X^{(4)}$ and user key bits 75–49. Also, V can be computed from $Y^{(8)}$ and user key bits 125–99. Hence, $|K_t| = |K_b| = 103$, which means a time complexity of 2^{103} partial encryptions. The memory complexity is 2^{25} since K_t and K_b share 78 key bits. However, the data complexity decreases to just two known plaintexts since each value of $V = n_5$ supplies a 16-bit filtering condition. Thus, for each 78-bit key guess only $2^{50} \cdot 2^{-32} = 2^{18}$ suggestions are left after the MITM phase. The total number of suggested keys is $2^{78} \cdot 2^{18} = 2^{96}$ and they can be checked by trial encryption.

3.20.1.1 Meet-in-the-Middle Biryukov-Demirci Attacks on IDEA

This section describes attacks that combine the (keyless) Biryukov-Demirci relation (3.394) with the Meet-in-the-Middle technique [25] to attack 6-round IDEA from the input to the second MA half-round until after the 8th KM half-round. This combination of techniques received the acronym *MITM BD* attack.

Let (P, P') denote a plaintext pair and (C, C') denote the corresponding ciphertext pair. In particular, $P = (P_1, P_2, P_3, P_4)$ and analogously $P' = (P'_1, P'_2, P'_3, P'_4)$. Likewise, $C = (C_1, C_2, C_3, C_4)$ and analogously $C' = (C'_1, C'_2, C'_3, C'_4)$.

The keyless Biryukov-Demirci relation for the specific 6-round IDEA variant is

$$\begin{aligned} \text{lsb}_1(P_2 \oplus P'_2 \oplus P_3 \oplus P'_3 \oplus \Delta t_2 \oplus \Delta t_3 \oplus \Delta t_4) = \\ \text{lsb}_1(C_2 \oplus C'_2 \oplus C_3 \oplus C'_3 \oplus \Delta t_5 \oplus \Delta t_6 \oplus \Delta t_7). \end{aligned} \quad (3.404)$$

The adversary chooses two sets as follows:

- the first set consists of: P_2, P_3, P'_2, P'_3, Δt_2, Δt_3, Δt_4.
- the second set consists of: C_2, C'_2, C_3, C'_3, Δt_5, Δt_6, Δt_7.

In the standard MITM attack, the adversary has to partially encrypt/decrypt from the plaintext and ciphertext ends until a common value V is obtained. The use of the Biryukov-Demirci relation allows us to *jump over* one round in the middle: the adversary computes only up to Δt_4 in the encryption direction and only up to Δt_5 in the decryption direction. The meet-in-the-middle effect is achieved using (3.404) to connect both ends.

The linear key schedule of IDEA is essential in allowing to extend the previous attack on 4.5 rounds to an efficient attack on 6 rounds. The terms of the first set can be computed from the plaintexts and the user key bits 50–63: all subkeys of the second MA half-round, of the full third round, plus $Z_1^{(4)}$, $Z_3^{(4)}$, $Z_5^{(4)}$. The terms of the second set can be computer from the ciphertexts and the user key bits 125–99: all subkeys of the 8th KM half-round, the full 7th rounds, plus $Z_1^{(6)}$, $Z_2^{(6)}$, $Z_5^{(6)}$. Hence, $|K_t| = 112$, $|K_b| = 103$ and $|V| = 1$ due to (3.404).

Note that K_t and K_b share 87 key bits: 125–33 and 50–99. Thus, using the improvements in Sect. 3.20.1, the data complexity is 16 known plaintexts. The memory complexity is 2^{16} 32-bit blocks, that is, 2^{15} 64-bit blocks. The time complexity is 2^{112} partial encryptions of 16 plaintexts, which is equivalent to 2^{115} 6-round IDEA encryptions.

3.20.1.2 TMTO Meet-in-the-Middle Biryukov-Demirci Attacks on IDEA

The most time-consuming part of the attack on 6-round IDEA in Sect. 3.20.1.1 is computing Δt_3 and Δt_4 for 16 plaintexts, which requires knowledge of 112 key bits, numbered 50–33, according to the key schedule of IDEA. Biham *et al.* observed that bits 25–33 are required only for $Z_5^{(4)}$, which is used only in the last multiplication operation in the computation of Δt_4. Hence, it seems that the adversary can guess the 103 key bits 50–24 and perform all operations except for the last multiplication and then guess the remaining nine key bits and perform a single multiplication operation for the 16 plaintexts. However, this is not possible since the user key bits 25–33 are also part of K_b, and hence their value should be guessed and fixed in advance, before the beginning of the MITM attack.

This problem can be solved at the expense of increasing the memory complexity. The adversary simply ignores the fact that key bits 25–33 are shared by K_t and K_b and treats them as independent parts of K_t and K_b. Consequently, the number of shared key bits decreases to 78, and the memory complexity increases to 2^{25} 40-bit blocks: 15 bits for the value of keyless Biryukov-Demirci relation in the 15 pairs, and 25 bits for the value of the key bits 25–49.

On the other hand, the computation time of Δt_3 and Δt_4 decreases since it is now possible to postpone the guessing of bits 25–33 until the last multiplication. As a result, this phase of the attack requires $2^{112} \cdot 16 = 2^{116}$ multiplications. Since each computation of 6-round IDEA contains 24 $\odot$ (and other operations), the time complexity is less than $2^{111.42}$ 6-round IDEA encryptions.

After reducing the time complexity of the MITM step, discarding the 2^{113} remaining subkey candidates becomes the most time consuming phase of the attack. However, this part can also be performed more efficiently: when generating the hash table, the adversary also computes $n_5 = X_1^{(6)} \oplus X_2^{(6)}$ for one of the plaintext/ciphertext pairs, and stores it in the hash table. For the remaining subkey guess, the adversary only computer n_5 for that plaintext/ciphertext pair from the plaintext side and checks whether it matches the value in the corresponding entry of the hash table. This value is a 16-bit filtering condition, and only 2^{97} key candidates remain after this stage (which can be easily checked by trial encryption).

During the computation of Δt_3 and Δt_4, the adversary already performs full encryption through the third round and partial encryption of the 4th round, obtaining the value n_5, which requires only 3 $\odot$, that are roughly equivalent to 1/8 of 6-round IDEA encryption. Thus, the time complexity of this phase is $2^{113}/8 = 2^{110}$ 6-round IDEA encryptions. The total time complexity is $2^{111.42} + 2^{110} = 2^{111.9}$ 6-round IDEA encryptions. The memory complexity increases to 2^{25} 56-bit blocks, which is less than 2^{25} 64-bit blocks.

This attack on 6-round IDEA can be further optimized to use only two known plaintexts. First, the adversary constructs the tables and performs the MITM phase. With two known plaintexts, the adversary can check the validity of the keyless Biryukov-Demirci relation only once, and thus, 2^{127} key suggestions remain after this phase. Most of these suggestions can be discarded efficiently by storing n_5 in the table in one of the encryptions, and computing it from the plaintext end for each subkey suggestion. In order to make this step even more efficient, the adversary can make a small change in the MITM phase: in addition to computing Δt_3 and Δt_4, the adversary computes the intermediate values until the multiplication with $Z_6^{(4)}$ in the 4th MA half-round. With these intermediate values, n_5 can be computed with only two $\odot$, two $\boxplus$ and two $\oplus$, which mean less than 1/12 of 6-round IDEA encryptions.

The time complexity is dominated by the second phase: $2^{127}/12 \approx 2^{123.42}$ 6-round IDEA encryptions.

Other attacks using the Splice-and-Cut technique [5] combined with the MITM BD attack (termed SaC MITM BD attack) on reduced-round variants of IDEA are described in [25].

Table 3.154 summarizes the attacks in [25].

Table 3.154 Attacks on reduced-round IDEA variants in [25].

#Rounds	Attack	Time	Data	Memory	Comments
4.5	MITM	2^{103}	2 KP	2^{16}	Data complex. equals unicity distance
5	MITM	2^{119}	10 KP	2^{24}	first 5 rounds
5	MITM	2^{119}	10 KP	2^{24}	last 5.5 rounds
6	MITM BD	$2^{123.4}$	2 KP	2^{25}	Data complex. equals unicity distance
6	MITM BD	$2^{111.9}$	16 KP	2^{25}	
6.5	SaC MITM BD	2^{122}	2^{10} CP	2^{10}	
6.5	SaC MITM BD	2^{113}	2^{23} CP	2^{23}	starting at second round
6.5	SaC MITM BD	$2^{111.9}$	2^{32} CP	2^{32}	
7	SaC MITM BD	2^{123}	2^{38} CP	2^{38}	
7	SaC MITM BD	2^{112}	2^{48} CP	2^{48}	
7.5	SaC MITM BD	$2^{125.9}$	16 CP	16	
7.5	SaC MITM BD	2^{114}	2^{63} CP	2^{63}	
8.5	SaC MITM BD	$2^{126.8}$	16 CP	16	

3.20.2 More Meet-in-the-Middle Attacks on IDEA

In [61], Demirci *et al.* described a meet-in-the-middle attack on reduced-round versions of IDEA. This attack uses the Demirci relation (3.393).

Let $X^{(i)} = (X_1^{(i)}, X_2^{(i)}, X_3^{(i)}, X_4^{(i)})$ denote the 64-bit input block to the i-th round in IDEA. Let $Y^{(i)} = (Y_1^{(i)}, Y_2^{(i)}, Y_3^{(i)}, Y_4^{(i)})$ denote the 64-bit output block from the i-th KM half-round. Let the inputs to the i-th MA-box be denoted (n_i, q_i), while the corresponding outputs are denoted (r_i, s_i).

A preliminary analysis involves a series of lemmas and theorems.

Lemma 3.8. *(Demirci relation) [60]*
$lsb_1(r_i \oplus s_i) = lsb_1(n_i \odot Z_5^{(i)})$.

Extending the result of Lemma 3.8 to 1-round IDEA, we have Corollary 3.6.

Corollary 3.6. *[61]*

$$lsb_1(X_2^{(i)} \oplus X_3^{(i)} \oplus Z_5^{(i-1)} \odot (X_1^{(i)} \oplus X_2^{(i)})) = lsb_1(X_2^{(i-1)} \oplus X_3^{(i-1)} \oplus Z_2^{(i-1)} \oplus Z_3^{(i-1)}).$$

Proof. From Lemma 3.8, $\text{lsb}_1(r_{i-1} \oplus s_{i-1}) = \text{lsb}_1(Z_5^{(i-1)} \odot (X_2^{(i)} \oplus X_3^{(i)}))$. Consider the paths connecting the inputs $X^{(i-1)}$ and the outputs $X^{(i)}$ of the $(i-1)$-th round: $X_2^{(i)} = (X_3^{(i-1)} \boxplus Z_3^{(i-1)}) \oplus s_{i-1}$ and $X_3^{(i)} = (X_2^{(i-1)} \boxplus Z_2^{(i-1)}) \oplus r_{i-1}$. Since the least significant bit is not affect by carry bits, all $\boxplus$ become $\oplus$, and the expression (3.6) follows.

Extending Lemma 3.8 further to 2-round IDEA results in Corollary 3.7:

Corollary 3.7. *[61]*

$$\begin{aligned} lsb_1(X_2^{(i)} \oplus X_3^{(i)} \oplus Z_5^{(i-2)} \odot (X_1^{(i-2)} \oplus X_2^{(i-2)}) \oplus Z_5^{(i-1)} \odot (X_1^{(i)} \oplus X_2^{(i)})) = \\ lsb_1(X_2^{(i-2)} \oplus X_3^{(i-2)} \oplus Z_2^{(i-1)} \oplus Z_3^{(i-1)} \oplus Z_2^{(i-2)} \oplus Z_3^{(i-2)}). \quad (3.405) \end{aligned}$$

The proof is similar to that of Corollary 3.6 and can be found in [61].

Theorem 3.6. *[61]*
Consider a set of 256 plaintexts $P = \{(p_1, p_2, p_3, p_4)\}$ where p_1, p_3 and the $lsb_8(p_2)$ are fixed; the $msb_8(p_2)$ take on every possible 8-bit value once; p_4 varies according to p_2 such that $(p_2 \boxplus Z_2^{(1)}) \oplus (p_4 \odot Z_4^{(1)})$ is a constant value. After these plaintexts are encrypted through 2-round IDEA, the following properties hold for the 256 plaintexts in P:
(i) the $lsb_8(n_2)$ are fixed, but the $msb_8(n_2)$ take every possible value once.
(ii) the $lsb_1(n_2 \odot Z_5^{(2)})$ equals either $lsb_1(X_2^{(3)} \oplus X_3^{(3)})$ or $lsb_1(X_2^{(3)} \oplus X_3^{(3)}) \oplus 1$.

The proof of Theorem 3.6 can be found in [61].

Corollary 3.8. *[61]*
Consider the plaintexts (p_1, p_2, p_3, p_4) in the set P in Theorem 3.6 encrypted through 2-round IDEA. Ordering the 256 values n_2 according to the $msb_8(p_2)$ in P, beginning with $msb_8(p_2) = 0$, will lead to the sequence

$$(x \oplus y_0|z), (x \oplus y_1|z), \ldots, (x \oplus y_{255}|z),$$

for some fixed 8-bit values x and z, for $y_i = (((i \boxplus a) \oplus b) \boxplus c) \oplus d$ where $0 \leq i \leq 255$ and fixed 8-bit values a, b, c and d.

The proof of Corollary 3.8 can be found in [61].

An attack on 3-round IDEA comprises the following steps:

(1) Pre-computation: prepare a sieving set of 256 strings:

$$S = \{f(a, b, c, Z_5^{(2)}, x, z) : 0 \leq a, b, c, x, z < 2^8, 0 \leq Z_5^{(2)} < 2^{16}\}$$

where f is a function mapping a given $(a, b, c, Z_5^{(2)}, x, z)$ to a 256-bit string defined by

$$f(a, b, c, Z_5^{(2)}, x, z)[i] = \text{lsb}_1(Z_5^{(2)} \odot (x \oplus y_i | z))$$

for $y_i = (((i \boxplus a) \oplus b) \boxplus c)$, and $0 \leq i \leq 255$. So, the memory needed for this step is 2^{56} 256-bit strings, or 2^{58} 64-bit text blocks.

(2) take a set of 2^{24} plaintexts $P = \{(p_1, p_2, p_3, p_4)\}$ such that p_1, p_3 and the $\text{lsb}_8(p_2)$ are fixed, while p_4 and the $\text{msb}_8(p_2)$ take each possible 24-bit value once. Encrypt this set with 3-round IDEA.

(3) for each value of $Z_2^{(1)}$ and $Z_4^{(1)}$, take 256 plaintexts from P such that the $\text{msb}_8(p_2)$ change from 0 to 255 and $(p_2 \boxplus Z_2^{(1)}) \oplus (p_4 \odot Z_4^{(1)})$ are constant. For each candidate value for $Z_5^{(3)}$, compute

$$\text{lsb}_1(X_2^{(4)} \oplus X_3^{(4)} \oplus Z_5^{(3)} \odot (X_1^{(4)} \oplus X_2^{(4)})) \tag{3.406}$$

over the selected 256 plaintexts. At this point, if $Z_5^{(3)}$ in (3.406) is correct, then $\text{lsb}_1(X_2^{(4)} \oplus X_3^{(4)} \oplus Z_5^{(3)} \odot (X_1^{(4)} \oplus X_2^{(4)}))$ are all equal either to $\text{lsb}_1(X_2^{(3)} \oplus X_3^{(3)})$ or to $\text{lsb}_1(X_2^{(3)} \oplus X_3^{(3)}) \oplus 1$ by Theorem 3.6.

(4) sort the 256 bits in step (3) according to the plaintexts $\text{msb}_8(p_2)$ for $0 \leq \text{msb}_8(p_2) \leq 255$. From Theorem 3.6(i), $\text{lsb}_1(X_2^{(3)} \oplus X_3^{(3)})$ equals either $\text{lsb}_1(n_2 \odot Z_5^{(2)})$ or $\text{lsb}_1(n_2 \odot Z_5^{(2)}) \oplus 1$, and the n_2 values after been sorted according to the $\text{msb}_8(p_2)$, follow the pattern in Corollary 3.8. Therefore, the sorted 256-bit sequence that corresponds to the right choice of $(Z_2^{(1)}, Z_4^{(1)}, Z_5^{(3)})$ must be present in the sieving set S. Check whether the sorted 256-bit sequence is present in S. If not, eliminate the corresponding candidates $(Z_2^{(1)}, Z_4^{(1)}, Z_5^{(3)})$. When the correct subkey values are tried, $\text{lsb}_1(X_2^{(4)} \oplus X_3^{(4)} \oplus Z_5^{(3)} \odot (X_1^{(4)} \oplus X_2^{(4)}))$ equals either $\text{lsb}_1(n_2 \odot Z_5^{(2)})$ or $\text{lsb}_1(n_2 \odot Z_5^{(2)}) \oplus 1$. In the former case, the 256-bit sequence has to be in S. In the latter case, $\text{lsb}_1(n_2 \odot Z_5^{(2)}) \oplus 1 = \text{lsb}_1(n_2 \odot (2^{16} + 1 - Z_5^{(2)}))$. Hence, the 256-bit sequence again has to be present in S.

(5) if more than two key combinations survive, return to (2) and change the plaintext set S. Continue until only two subkey combinations remain: $(Z_2^{(1)}, Z_4^{(1)}, Z_5^{(3)})$ and $(Z_2^{(1)}, Z_4^{(1)}, 2^{16} + 1 - Z_5^{(3)})$.

The attack finds subkeys $(Z_2^{(1)}, Z_4^{(1)}, Z_5^{(3)})$. The correct $Z_5^{(2)}$ is found from the elements of S that matches the remaining 256-bit string. According to the key schedule of IDEA, these four subkeys correspond to the 41 key bits numbered 16–31 and 48–72 of the user key. The remaining $128 - 41 = 87$ key bits can be found by exhaustive key search. The final time complexity is 2^{88} 3-round IDEA computations, according to [61]. The data complexity is 2^{24} chosen plaintexts.

To attack 3.5-round IDEA, steps (1) and (2) are identical to the attack on 3-round IDEA, except that the plaintexts are encrypted through 3.5 rounds. The attack proceeds as follows:

- for every value of $Z_2^{(1)}$ and $Z_4^{(1)}$ take the 256-bit plaintext blocks from P that keep $(p_2 \boxplus Z_2^{(1)}) \oplus (p_4 \odot Z_4^{(1)})$ constant. For every value of $Z_1^{(4)}$ and $Z_2^{(4)}$, partially decrypt the ciphertexts to obtain $X_1^{(4)}$ and $X_2^{(4)}$. Further, compute for each $Z_5^{(3)}$ the value:

$$\mathrm{lsb}_1(Y_2^{(4)} \oplus Y_3^{(4)} \oplus Z_5^{(3)} \odot (X_1^{(4)} \oplus X_2^{(4)})). \tag{3.407}$$

  Note that $\mathrm{lsb}_1(Y_2^{(4)} \oplus Y_3^{(4)}) = \mathrm{lsb}_1(X_2^{(4)} \oplus X_3^{(4)})$ because the KM half-round for the middle words uses only $\boxplus$, and the bit computed in (3.407) equals either $\mathrm{lsb}_1(X_2^{(3)} \oplus X_3^{(3)})$ or $\mathrm{lsb}_1(X_2^{(3)} \oplus X_3^{(3)}) \oplus 1$ for all ciphertexts. If the choices for $Z_2^{(1)}$, $Z_4^{(1)}$, $Z_1^{(4)}$, $Z_2^{(4)}$ and $Z_5^{(3)}$ are correct, the derived 256-bit sequence must exist in S. Steps (4) and (5) are executed as in the 3-round attack.

According to the key schedule of IDEA, subkeys $Z_2^{(1)}$, $Z_4^{(1)}$, $Z_5^{(2)}$, $Z_1^{(4)}$, $Z_2^{(4)}$ and $Z_5^{(3)}$ correspond to 73 user key bits. The remaining $128 - 73 = 55$ key bits can be recovered by exhaustive search.

The final time complexity is 2^{73} 3.5-round IDEA computations, according to [61]. The data complexity is 2^{24} chosen plaintexts.

For an attack on 4-round IDEA, the main difference is the partial decryption in step (3): first, partially decrypt to find $X_1^{(4)}$ and $X_2^{(4)}$ using $Z_1^{(4)}$, $Z_2^{(4)}$, $Z_5^{(4)}$ and $Z_6^{(4)}$. Further, calculate the 256 values:

$$\mathrm{lsb}_1(X_2^{(5)} \oplus X_3^{(5)} \oplus Z_5^{(4)} \odot (X_1^{(5)} \oplus X_2^{(5)}) \oplus Z_5^{(3)} \odot (X_1^{(4)} \oplus X_2^{(4)})). \tag{3.408}$$

From Corollary 3.7, (3.408) equals either $\mathrm{lsb}_1(X_2^{(3)} \oplus X_3^{(3)})$ or $\mathrm{lsb}_1(X_2^{(3)} \oplus X_3^{(3)}) \oplus 1$ for all the 256 ciphertexts. From these bits, the 256-bit sequence is produced by sorting the bits according to their plaintexts $\mathrm{msb}_8(p_2)$. The key elimination is carried out as in the previous attacks.

The recovered subkeys are $Z_1^{(2)}$, $Z_4^{(1)}$, $Z_5^{(3)}$, $Z_1^{(4)}$, $Z_2^{(4)}$, $Z_5^{(4)}$ and $Z_6^{(4)}$. According to the key schedule of IDEA, they correspond to 82 user key bits. The remaining $128 - 82 = 46$ key bits can be recovered by exhaustive search.

The time complexity is 2^{89} 4-round computations, according to [61]. The data complexity is 2^{24} chosen plaintexts.

3.20.3 Improved Meet-in-the-Middle Attacks on IDEA

In [7], Ayaz and Selçuk revisited the Meet-in-the-Middle attack in [61] described in Sect. 3.20.2.

In [61], the memory complexity and pre-computation time are independent of the number of rounds, while the key search time varies depending on the number of rounds attacked.

The memory complexity is determined by the size of the sieving set, which consists of 2^{56} 256-bit strings, or 2^{58} 64-bit text blocks.

Pre-computation time is dedicated to prepare the sieving set. The f function is calculated once for each bit of the sieving set. There are 2^{56} 256-bit strings. Therefore, the pre-computation time is 2^{64} f computations.

The key search time for attacking 3, 3.5, 4, 4.5 and 5 rounds depends on the number of key bits searched. In each of these attacks, a look-up string is computed over 256 ciphertexts for each key candidate, contributing a factor of 2^8.

In the attack on 3-round IDEA, the key searched has 34 bits, making the search time complexity 2^{42} partial decryptions. The attack on 3.5 rounds searches 32 more key bits, making the time complexity 2^{74}. The 4-round attack searches 16 more key bits, raising the time complexity to 2^{90}. For 4.5 rounds, 114 key bits are searched, with complexity 2^{122}. For 5 rounds, 119 key bits are searched with complexity 2^{127}.

The original attack in [61] partitioned p_2 into an 8-bit fixed and an 8-bit variable part, where the later took all possible 2^8 values over the set P. It is not necessary to have a balanced partition of p_2. The attack works just as well with an unbalanced partition. Accordingly, one can obtain significant savings by reducing the size of the variable part. Let v denote the number of most significant bits in the variable part of p_2. The sieving set for the attack becomes

$$S = \{f(a, b, c, d, z, Z_5^{(2)} : 0 \leq a, b, c, d < 2^v, 0 \leq z < 2^{16-v}, 0 \leq Z_5^{(2)} < 2^{16}\}. \tag{3.409}$$

Shortening v narrows the sieving set both for each element in S and for the whole set S. With a variable v-bit part, the sieving set entries will be 2^v bits each instead of 2^8 bits. Furthermore, the number of entries in the sieving set will be reduced by a factor of $2^{3(8-v)}$. This change also decreases the key search time by a factor 2^{8-v} since for each key candidate we encrypt 2^v plaintexts to form the bit-string to be searched in the sieving set instead of 2^8.

Therefore, using an unbalanced partition of p_2, we obtain an improvement by a factor of 2^9 in pre-computation time, 2^3 in key search time and 2^{12} in memory.

Further reduction in the sieving set size comes from *collisions* in the entries for different (a, b, c, d) tuples. In the original attack, all elements of the sieving set were thought to be distinct. But, a significant number of collisions exist among the sieving set entries.

The analytical findings were obtained according to the y_i values:

Definition 3.33. [7]
Let two 4-tuples (a, b, c, d), for $0 \leq a, b, c, d < 2^v$, be called *equivalent* if they result in the same $y_i = (((i \boxplus a) \oplus b) \boxplus c) \oplus d$ value for all $0 \leq i < 2^v$.

Lemma 3.9. *[7]*
For any 4-tuple (a, b, c, d), complementing the most significant bit of any two or four of a, b, c, d yields an equivalent 4-tuple.

The proof of Lemma 3.9 can be found in [7]. Lemma 3.9 gives eight equivalent (a, b, c, d) 4-tuples. Further equivalence is provided by the complement operation.

Lemma 3.10. *[7]*
(a, b, c, d) is equivalent to $(a, \overline{b}, \overline{c} \boxplus 1, \overline{d})$ for $0 \leq a, b, c, d < 2^v$.

The proof of Lemma 3.10 can be found in [7].

Lemma 3.11. *[7]*
(a, b, c, d) is equivalent to (i) $(a \boxplus 2^{v-2}, b, c \boxplus 2^{v-2}, d)$ if the second most significant bit of b is 1, and to (ii) $(a \boxplus 2^{v-2}, b, c \boxminus 2^{v-2}, d)$ if the second most significant bit of b is 0.

The proof of Lemma 3.11 can be found in [7].

When Lemma 3.11 is applied to all 16 equivalent 4-tuples, the size of the equivalence class is doubled, yielding 32 equivalent 4-tuples.

If the two most significant bits of a and one most significant bit of b, c, d are discarded, we will find exactly one of these 32 equivalent 4-tuples, since the equivalent 4-tuples take all possible values over these five bits. Therefore, in the sieving set formation step, we do not have to search all combinations of (a, b, c, d). Conducting the search on $\mathrm{lsb}_{v-2}(a)$, $\mathrm{lsb}_{v-1}(b)$, $\mathrm{lsb}_{v-1}(c)$, $\mathrm{lsb}_{v-1}(d)$ suffices. This reduction in effort decreases both the pre-computation time and the sieving set size by a factor of 2^5.

The collisions are exclusively based on equivalent (a, b, c, d) 4-tuples, but there are other collisions as well, and the actual collision rate is assumed to be 2^6 or higher.

The effectiveness of the attack in [61] can be significantly improved using the described elimination power from the sieving set. When a look-up string is matched with a sieving set entry, we can do a further correctness test on the key by checking whether the key values used to obtain the set entry are consistent with the round subkeys used to obtain the look-up string.

First, we can check the $Z_5^{(2)}$ value found in the sieving set hit for consistency with the key used in the partial decryption. The 3-round attack searches for $Z_2^{(1)}$, $Z_4^{(1)}$ and $Z_5^{(3)}$, which overlaps with $Z_5^{(2)}$ in 9 bits, numbered 58–66, according to the key schedule of IDEA. If we store these 9 bits of $Z_5^{(2)}$ for each sieving set entry and compare them to the corresponding bits of the key candidate used in the partial decryption in case of a hit, a wrong key's chances of passing the sieving test will be reduced by a factor of 2^9.

The subkeys found in attacks on 3.5 rounds: $Z_1^{(4)}$, $Z_2^{(4)}$; on 4 rounds: $Z_5^{(4)}$, $Z_6^{(4)}$; or on 4.5 rounds: $Z_1^{(5)}$, $Z_2^{(5)}$, $Z_3^{(5)}$, $Z_4^{(5)}$ do not overlap with $Z_5^{(2)}$ in more bits.

The seven bits of $Z_5^{(2)}$ that do not overlap with the searched subkeys can be used to deduce the corresponding seven bits of the user key. Moreover, in attacks that use multiple elimination rounds, a check on these bits can be carried out to test the consistency of the sieving set hits across different elimination rounds. Either way, these seven bits can be used to reduce the set of key candidates by a factor of 2^7 per elimination round.

A similar consistency check can be applied to the a values of the sieving set entries. The 32 equivalent 4-tuples have the same $\text{lsb}_{v-2}(a)$ values. Hence, in case of a sieving set hit, the a value of the sieving set entry matched can be compared to the $v-2$ low order bits of the a value of the partial decryption

$$a = \text{msb}_v(Z_2^{(1)}) + \text{carry}(\text{lsb}_{16-v}(p_2) \boxplus \text{lsb}_{16-v}(Z_2^{(1)})),$$

which is fixed and known over the plaintext set P. This extension brings an extra elimination power of 2^{v-2} to the attack while costing $v-2$ bits of storage per sieving set entry.

A similar check can be carried out over the c values. The 32 equivalent 4-tuples are equal to $\pm c \bmod 2^{v-2}$ over $\text{lsb}_{v-2}(c)$ while $\text{msb}_2(c)$ takes all possible four values. Moreover, for every c value, there are two possible values of $\text{msb}_v(Z_3^{(2)})$ since

$$c = \text{msb}_v(Z_3^{(2)}) + \text{carry}(((\text{lsb}_{16-v}(p_2) \boxplus \text{lsb}_{16-v}(Z_2^{(1)})) \oplus \text{lsb}_{16-v}(r_1)) \boxplus \text{lsb}_{16-v}(Z_3^{(2)})),$$

where the carry bit is unknown. The $\text{msb}_v(Z_3^{(2)})$ are covered by $Z_2^{(1)}$ for $v \leq 7$, which is the case in our attacks. Therefore, by conducting a consistency check between the key candidate tried and the c value of the sieving set entry matched, we can reduce the number of keys by an additional factor of 2^{v-4}. As in the case of a, this check on c costs an extra $v-2$ bits of storage per sieving set entry.

The improvements in the attacks in [61] mean significant reductions in the memory, pre-computation time and key search time complexities. Memory complexity is mainly the size of the sieving set. Each sieving set entry contains a 2^v-bit look-up string. Additionally, we need to store $Z_5^{(2)}$, $\text{lsb}_{v-2}(a)$ and $\text{lsb}_{v-2}(c)$ values to have the extra elimination power, which costs an extra $12+2v$ bits per entry. The number of entries in the set is about 2^{3v+26}. Thus, the overall memory complexity of the sieving set is $2^{3v+26} \cdot (2^v + 2v + 12)$ bits. In terms of IDEA blocks, this is less than 2^{41} text blocks for $v = 5$.

Pre-computations time complexity is the time required to compute the sieving set. We need to compute f 2^v times for each sieving set entry. The number of entries computer for the sieving set is $2^{3v+32-5}$ since the most significant bits of a, b, c, d and the second most significant bits of a need not be searched. Thus, the pre-computation complexity is 2^{4v+27} f computations, which is roughly equivalent to 2^{4v+26} IDEA rounds. The pre-computation time is the dominant time complexity only for the 3-round attack.

Key search time complexity depends on both the number of rounds attacked and the number of variable bits in p_2. For each candidate key set, we take 2^v values for $\text{msb}_v(p_2)$ and calculate the look-up string by partial decryptions. This procedure may need to be repeated several times if the attack requires multiple elimination rounds. The effect of elimination rounds on the attack complexity is twofold. First, a different plaintext set would be needed for each elimination round, making the data complexity of the attack become $t \cdot 2^{16+v}$ chosen plaintext for t the number of elimination rounds.

Second, the complexity of the key search step would increase due to multiple repetitions of the elimination procedure, However, this increase can be expected to be relatively marginal, since the additional elimination rounds will be applied only to the keys that have passed the previous tests. Since each elimination round will remove the vast majority of the wrong keys, the additional time complexity from the extra elimination rounds will be negligible. The attack complexities of the improved attacks (for $v = 5$) compared to those of [61] are listed in Table 3.155. Time complexity is measured in number of rounds attacked. Memory complexity is measured in number of 64-bit blocks.

Table 3.155 Improved and previous MITM attack complexities on reduced-round IDEA.

#Rounds	[7], $v = 5$			[61]		
	Time	Data (CP)	Memory	Time	Data (CP)	Memory
3	$2^{44.41}$	2^{23}	2^{41}	$2^{61.41}$	2^{24}	2^{58}
3.5	$2^{68.19}$	$2^{23.6}$	2^{41}	$2^{71.19}$	2^{24}	2^{58}
4	2^{82}	2^{24}	2^{41}	2^{87}	2^{24}	2^{58}
4.5	$2^{115.83}$	$2^{24.6}$	2^{41}	$2^{118.83}$	2^{24}	2^{58}
5	$2^{120.67}$	$2^{24.6}$	2^{41}	$2^{123.67}$	2^{24}	2^{58}

3.20.4 Meet-in-the-Middle Attack on FOX128

An attack described in [85], using the MITM technique was called *All-Subkeys Recovery* (ASR) attack, and was applied to reduced-round variants of FOX128.

The basic concept of the ASR attack is to convert the user-key searching procedure to another procedure of finding all round subkeys, which are regarded as independent variables in this attack.

Any adversary who knows the round subkeys can encrypt and decrypt any plaintext/ciphertext even if the user key is unknown or not recoverable. In addition, if the key schedule algorithm is invertible, then the user key can in fact be recovered from the round subkeys.

The ASR attack depends on the following parameters: (i) the size of the user key; (ii) the size of the round subkeys; (iii) the structure of the data processing part (balanced Feistel Network, SPN or Lai-Massey).

The ASR attack works even if the key schedule were an ideal function. Nonetheless, concrete block ciphers currently employ key schedule algorithms which are much weaker than ideal functions. Consequently, the number of attacked rounds may increase by thoroughly analyzing the key schedule algorithms.

The ASR attack uses a Time-Memory-Trade-Off (TMTO) approach and operates in the single-key setting.

Suppose an n-bit block cipher E operating under a k-bit key K and consisting of R rounds, where each round has an l-bit subkey.

Initially, the adversary must find an s-bit match in an intermediate cipher state denoted S. The state S can be computed from a plaintext P and a set of subkey bits K_1 by a function F_1 as $F_1(P, K_1)$. Similarly, S can be computed from the corresponding ciphertext C and another set of subkey bits K_2 by a function F_2 as $S = F_2^{-1}(C, K_2)$. The remaining independent subkeys bits are denoted K_3. Thus, $|K_1| + |K_2| + |K_3| = R \cdot l$. Using K_1 and K_2, the adversary can compute $F_1(P, K_1)$ and $F_2^{-1}(C, K_2)$ independently to find an s-bit match $F_1(P, K_1) = F_2^{-1}(C, K_2)$ when the correct values of the bits of K_1 and K_2 are guessed.

For instance, suppose $|K_1| \leq |K_2|$. Compute $F_1(P, K_1)$ exhaustively for K_1 and store the result in a hash table H indexed by $F_1(P, K_1)$, that is, $H(F_1(P, K_1)) = K_1$. Further, compute $F_2^{-1}(C, K_2)$ exhaustively for K_2 and look for a match in H, that is, if $H(F_2^{-1}(C, K_2))$ is not empty, then a pair (k_1, k_2) of candidates for (K_1, K_2) is suggested.

The matching state provides an s-bit condition per known (P, C) pair, and there are $R \cdot l$ subkey bits in total. After a single (P, C) pair, it is expected that there will be $2^{R \cdot l - s}$ subkey candidates remaining. This number can be reduced by using additional (P, C) pairs. Using N plaintext/ciphertext pairs, the number of subkey candidates suggested becomes $2^{R \cdot l - N \cdot s}$ as long as $N \leq (|K_1| + |K_2|)/s$.

Further, the adversary exhaustively searches the correct remaining subkey bits from the surviving subkey candidates in K_3.

The computational effort in number of calls to E is estimated as

$$\max(2^{|K_1|}, 2^{|K_2|}) \cdot N + 2^{R \cdot l - N \cdot s}. \qquad (3.410)$$

The data complexity is $\max(N, \lceil (R \cdot l - N \cdot s)/n \rceil)$ known plaintexts. The memory complexity is about $\min(2^{|K_1|}, 2^{|K_2|}) \cdot N$ text blocks, which is the size of the hash table H.

The ASR attack works faster than exhaustive key search when (3.410) is less than 2^K calls to E. The cost of memory accesses is not taken into account in (3.410), assuming a table look-up has negligible cost compared to a computation of F_1 or F_2. However, strictly speaking those costs should be considered, which is $\max(2^{|K_1|}, 2^{|K_2|}) \cdot N$ memory accesses.

If R and s are fixed, then the time and memory complexities are dominated by $|K_1|$ and $|K_2|$. Thus, the smaller K_1 and K_2 are, the more efficient the attack becomes with respect to time and memory complexities.

Therefore, the objective of an ASR attack is to find a matching state S by the smallest $\max(|K_1|, |K_2|)$.

Let $X^{(i)} = (X_1^{(i)}, X_2^{(i)}, X_3^{(i)}, X_4^{(i)})$ denote the input text block to the i-th round in FOX128, where $|X_j^{(i)}| = 32$ bits. We refer to Sect. 2.6.1 for further terminology on the components of FOX128 cipher.

An ASR attack on 5-round FOX128 uses the following one-round relation for the matching state:

$$X_1^{(i+1)} \oplus \mathrm{io}(X_2^{(i+1)}) = X_1^{(i)} \oplus X_2^{(i)}. \tag{3.411}$$

If the adversary knows $X_1^{(1)}$ and $X_2^{(1)}$, then $X_1^{(2)} \oplus \mathrm{io}(X_2^{(2)})$ can be computed.

Thus, $(X_1^{(1)}, X_2^{(1)})$ are chosen as the matching state in the forward (encryption) direction. The 32-bit matching value (3.411) is computed from a 128-bit round subkey RK_1^2, a 64-bit subkey LK_2^2 and the leftmost 32 bits of RK_2^2, that is, $(X_1^{(1)}, X_2^{(1)}) = F_1(P, K_1)$ and $|K_1| = 224 = 128 + 64 + 32$.

Similarly, choose $(X_3^{(2)}, X_4^{(2)})$ as the matching state in the backward (decryption) direction. In this case, $(X_3^{(2)}, X_4^{(2)}) = F_2(C, K_2)$, where

$$K_2 \in \{K_5, \mathrm{LK}_4^2, \mathrm{msb}_{32}(\mathrm{RK}_4^2)\}$$

and $|K_2| = 224$.

Using $N = 13$ plaintext/ciphertext pairs, since $13 \leq (224 + 224)/32$, the time complexity for finding all round subkeys is estimated as $\max(2^{224}, 2^{224}) \cdot 13 + 2^{128 \cdot 5 - 13 \cdot 32} = 2^{228}$ encryptions of 5-round FOX128. The data complexity is 13 known plaintexts. The memory complexity is $\min(2^{224}, 2^{224}) \cdot 13$, which is about 2^{228} text blocks.

The ASR attack assumes that there is no equivalent keys, which is a reasonable assumption for modern block ciphers. Otherwise, there is more than one key mapping a given plaintext to a ciphertext, and the ASR attack recovers only one of these equivalent keys.

In the ASR attack, it is *not* mandatory to analyze the key schedule algorithm, which makes the attack simpler and more generic. Therefore, this attack works even if the key schedule can be modeled as an ideal function.

While all round subkeys are regarded as independent variables in the ASR attacks, there may be mathematical relations between these subkey bits in actual block ciphers. If an adversary exploits these relations in the underlying key schedule algorithm, then the attack can be improved. For instance:

- recovering the user key from the round subkeys.
- reducing the search space and the attack complexities by using mathematical relations that involve the round subkeys.

3.20.5 Improved ASR Attack on FOX Ciphers

In the All-Subkey Recovery (ASR) attack in [87], the number of round subkey bits required to compute the internal cipher state S from a given plaintext/ciphertext (P/C) pair is denoted K_1 and K_2, respectively[18]. The sizes of K_1 and K_2 are the dominant parameters for the attack complexities. Thus, reducing $|K_1|$ and $|K_2|$ could allow to extend the ASR attack to more rounds of a given target cipher.

The following techniques are aimed at reducing the sizes of K_1 and K_2.

Function Reduction technique: the basic concept of this technique is to fix some plaintext or ciphertext bits by exploiting some degrees of freedom of the P/C pairs. This approach allows an adversary to consider a key-dependent variable as a new subkey. It is expected that this reasoning will reduce the amount of subkeys required to compute the matching state.

Using the function reduction technique, a lower bound on the security of several Feistel Network ciphers was established against generic key-recovery attacks in [86].

Suppose that the i-th round state S_i is computed from the $(i-1)$-th round state S_{i-1} and the i-th round function G_i as $S_i = G_i(S_{i-1} \oplus K_i)$. The function reduction technique consists of two parts:

- Key Linearization: since G_i is a nonlinear function, $G_i(S_{i-1} \oplus K_i) \neq G_i(S_{i-1}) \oplus G_i(K_i)$. The key linearization step exploits the degrees of freedom of the P/C pairs to express S_i as a linear function of S_{i-1} and K_i as $S_i = L_i(S_{i-1}, K_i)$, where L_i is a linear mapping. Once L_i is found, K_i can be moved to the next nonlinear function by an equivalent transformation.
- Equivalent transformation: after the key linearization step, K_i is replaced by K_i', an equivalent subkey value. In order to reduce the number of involved subkey bits on the trails to the matching state, all subkeys on the

trails affected by K_i' are also replaced by new variables by an equivalent transformation. This is how the number of subkey bits required to compute the matching state can be reduced.

Since the function reduction technique exploits the degrees of freedom of plaintext/ciphertext (P/C) pairs, sometimes an attack becomes unfeasible due to shortage of enough data blocks. In such cases, a variant of the ASR attack is used. It is called Improved ASR (IASR) attack, and it simply repeatedly applies the ASR attack to detect the correct subkey value. This variant attack can reduce the required data for each individual ASR phase, though the total amount of data remains unchanged. The repetitive ASR attack is as follows:

- mount the ASR attack with N P/C pairs, with $N < (|K_1| + |K_2|)/s$. Then, put the remaining subkey candidates in a table T_1. The number of expected candidates is $|K_1| + |K_2| - N \cdot s$.
- repeatedly apply the ASR attack with different N P/C pairs. If the remaining subkey candidates match (i.e. overlap) with the ones in T_1, then put these candidates in another table T_2. The expected number of candidates is $|K_1| + |K_2| - 2 \cdot N \cdot s$.
- repeat the previous steps until the correct key is found, that is, $M = (|K_1| + |K_2|)/(N \cdot s)$ times.

After this attack is repeated M (≥ 2) times, the time complexity to recover K_1 and K_2 is estimated as $\max(2^{|K_1|}, 2^{|K_2|}) \cdot N \cdot M + 2^{|K_1|+|K_2|-N\cdot s} + \ldots + 2^{|K_1|+|K_2|-(M-1)\cdot N\cdot s}$.

The total data complexity is $(|K_1|+|K_2|)/s$, while each phase is done with $N = (|K_1|+|K_2|)/M \cdot s)$ P/C pairs, which is M times less data than required in the basic ASR attack.

The memory complexity is about $\max(2^{|K_1|+|K_2|-N\cdot s}, \min(2^{|K_1|}, 2^{|K_|}) \cdot N)$ text blocks, which includes the storage needed for the tables used in the matching steps.

3.20.6 Improved ASR Attack on FOX64

Let the 64-bit i-th round input to FOX64 be denoted (L_{i-1}, R_{i-1}), for $i \geq 1$, where $L_{i-1}, R_{i-1} \in \mathbb{Z}_2^{32}$. The i-th round bijective functions f32: $\mathbb{Z}_2^{32} \rightarrow \mathbb{Z}_2^{32}$ are parameterized by the 64-bit i-th round subkey $K_i = (LK_i, RK_i)$. The orthomorphism function is denoted $\text{or}(x_0, x_1) = (x_1, x_0 \oplus x_1)$, with $x_0, x_1 \in \mathbb{Z}_2^{16}$.

The key linearization step starts as follows: let $L_0 \oplus R_0 = c_0$ denote a constant value fixed at the input to the first f32 function. Now regarding $\text{f32}(c_0, K_1)$ as a 32-bit new key K_1', the state after f32 is $(L_0 \oplus K_1', R_0 \oplus K_1')$. The round output becomes $(L_1, R_1) = (\text{or}(L_0) \oplus \text{or}(K_1'),\ R_0 \oplus K_1')$. Let us denote $OK_1' = \text{or}(K_1')$. This way, K_1 linearly affects L_1 and R_1.

Now, the equivalent transformation step: in the second round, OK_1' and K_1' are exclusive-ored to LK_2 in the first and in the last operations in the f32 function. Let $LK_2' = LK_2 \oplus K_1' \oplus OK_1'$, $K_1'' = K_1' \oplus LK_2$ and $OK_1''=$ or$(OK_1' \oplus LK_2)$ be new subkeys. Then, f32 contains $K_2' = (LK_2', RL_2)$ and K_1'' and OK_1'' linearly affect the outputs of the second round.

In the third round, OK_1'' and K_1'' are exclusive-ored to LK_3 in the first and the last operations of the f32 function. Let $LK_3' = LK_3 \oplus K_1'' \oplus OK_1''$, $K_1'' = K_1' \oplus LK_2$ and $OK_1''=$ or$(OK_1'' \oplus LK_2)$ be new subkeys.

The attack on 6-round FOX64 will use the following 1-round keyless linear relation that holds with certainty:

$$\text{io}(L_{i+1}) \oplus R_{i+1} = L_i \oplus R_i. \qquad (3.412)$$

Note that (3.412) implies a 32-bit condition.

From (3.412), the following 16-bit linear relation can be derived:

$$(L_4^{(1)} \oplus L_4^{(3)}, L_4^{(3)}) \oplus (R_4^{(3)}, R_4^{(1)}) = (L_3^{(3)}, L_3^{(1)}) \oplus (R_3^{(3)}, R_3^{(1)}), \qquad (3.413)$$

where $L_i = (L_i^{(3)}, L_i^{(2)}, L_i^{(1)}, L_i^{(0)})$ and $R_i = (R_i^{(3)}, R_i^{(2)}, R_i^{(1)}, R_i^{(0)})$.

The forward computation stage is as follows: for given K_2', LK_3', $RK_3^{'(3)}$, $RK_3^{'(1)}$, $K_1^{'''(3)}$, $K_1^{'''(1)}$, $OK_1^{'''(3)}$ and $OK_1^{'''(1)}$, the value $(L_3^{(3)}, L_3^{(1)}) \oplus (R_3^{(3)}, R_3^{(1)})$ is computable. Since $(K_1^{'''(3)}, K_1^{'''(1)})$ and $(OK_1^{'''(3)}, OK_1^{'''(1)})$ linearly affect $(L_3^{(3)}, L_3^{(1)})$ and $(R_3^{(3)}, R_3^{(1)})$, respectively, we can regard $(K_1^{'''(3)}, K_1^{'''(1)}) \oplus (OK_1^{'''(3)}, OK_1^{'''(1)})$ as a new 16-bit key XK_1. Then, $(L_3^{(3)}, L_3^{(1)}) \oplus (R_3^{(3)}, R_3^{(1)})$ is obtained from the 112 bits of key K_2', LK_3', $RK_3^{'(3)}$, $RK_3^{'(1)}$ and linearly dependent 16-bit key XK_1.

The backward computation stage is as follows: $(L_4^{(1)} \oplus L_4^{(3)}, L_4^{(3)}) \oplus (R_4^{(3)}, R_4^{(1)})$ is obtained from the 112 bits of key K_6, LK_5, $RK_5^{(1)}$, $RK_5^{(3)}$. Using the indirect matching technique in [3], 8 bits out of the 16 bits of XK_1 are moved to the left half of the matching equation. Then, the left and right halves of the equation contain 120 bits of the key each, that is, $|K_1| = |K_2| = 120$.

Using $N = 15$, the time complexity for finding 240 key bits is estimated as $\max(2^{120}, 2^{120}) \cdot 15 + 2^{240-15\cdot 16} \approx 2^{124}$ 6-round FOX64 computations. The data complexity is 15 chosen plaintexts. The memory complexity is about 2^{124} text blocks.

To extend the attack to 7-round FOX64, the function reduction technique is applied in the backwards direction, that is, the value $L_7 \oplus R_7$ is fixed to a constant c_7. Using the involution property of the round construction of FOX64, $(L_4^{(1)} \oplus L_4^{(3)}, L_4^{(3)}) \oplus (R_4^{(3)}, R_4^{(1)})$ is also obtained from 112 key bits and linearly-dependent 16-bit key XK_2.

In this attack, $XK_1 \oplus XK_2$ is regarded as a new 16-bit key value. Similar to the attack on 6-round FOX64, the left and right halves of the matching equation contain 120 key bits each, that is, $|K_1| = |K_2| = 120$.

Repetitive ASR attack: plaintexts and ciphertexts have to satisfy the 32-bit conditions $L_0 \oplus R_0 = c_0$ and $L_7 \oplus R_7 = c_7$ for fixed constants c_0 and c_7. The data required for finding such pairs is estimated by a game that an adversary plays to find 32-bit multi-collisions with 32-bit restricted inputs. An n-bit t-multi-collision is found using $t!^{1/t} \cdot 2^{n(t-1)/t}$ random data with high probability [175].

In the basic ASR attack, at least 15 (=240/16) multi-collisions are necessary to detect the 240-bit key. To obtain such pairs with high probability, it requires $2^{32.55}$ (=$15!^{1/15} \cdot 2^{23 \cdot 14/15}$) plaintext/ciphertext pairs. However, it is unfeasible, since the degree of freedom of plaintexts is only 32 bits. In order to overcome this problem, the repetitive ASR variant is used with $M = 2$. In each ASR phase, the data required is reduced to 8 and 7. Then, eight 32-bit multi-collisions are obtained from $2^{29.9}$ plaintext/ciphertext pairs with high probability. Thus, we can obtain the required data by exploiting 32 free bits.

The estimated time complexity for finding 240 key bits is $\max(2^{120}, 2^{120}) \cdot 8 \cdot 2 + 2^{240-8\cdot 16} = 2^{124}$.

The remaining $448 - 240 = 208$ key bits are obtained by recursively applying ASR attacks. The time complexity of this phase is estimated as 2^{106} using 4 (=$\lfloor 208/6 \rfloor$) plaintext/ciphertext pairs.

The data complexity is $2^{30.9}$ (=$2^{29.9} \cdot 2$) plaintext/ciphertext pairs. The memory complexity is about 2^{123} (=$\max(2^{240-128}, \min(2^{120}, 2^{120}) \cdot 8$) text blocks.

3.20.7 Improved ASR Attack on FOX128

Let the 128-bit i-th round input to FOX128 be denoted (LL_{i-1}, LR_{i-1}, RL_{i-1}, RR_{i-1}), for $i \geq 1$, where LL_{i-1}, LR_{i-1}, RL_{i-1}, $RR_{i-1} \in \mathbb{Z}_2^{32}$. The i-th round bijective mapping f64: $\mathbb{Z}_2^{64} \to \mathbb{Z}_2^{64}$ is parameterized by the 128-bit i-th round subkey $K_i = (LK_i, RK_i)$, with $LK_i, RK_i \in \mathbb{Z}_2^{64}$. The orthomorphism function is denoted $\mathrm{or}(x_0, x_1) = (x_1, x_0 \oplus x_1)$, with $x_0, x_1 \in \mathbb{Z}_2^{16}$.

The i-th round function updates the input state as

$$(LL_i, LR_i) = (\mathrm{or}(LL_{i-1} \oplus \phi_L), LR_{i-1} \oplus \phi_L),$$

and

$$(RL_i, RR_i) = (\mathrm{or}(RL_{i-1} \oplus \phi_R), RR_{i-1} \oplus \phi_R),$$

where the f64 output is denoted

$$(\phi_L, \phi_R) = \mathrm{f64}((L_{i-1} \oplus LR_{i-}, RL_{i-1} \oplus RR_{i-1}, K_i).$$

The key linearization step starts as follows: let us fix two 16-bit values input to f64 as

$$LL_0 \oplus LR_0 = c_1,$$
$$RL_0 \oplus RR_0 = c_2.$$

The input to the first f64 function becomes $((c_1, c_2), K_1)$ and the output result, $\text{f64}((c_1, c_2), K_1)$, is considered a new 64-bit key $K_1' = (KL_1', KR_1')$. Further, KL_1' and KR_1' are exclusive-ored to LR_0 and RR_0, respectively.

The state after the first round becomes

$$(LL_1, LR_1, RL_1, RR_1) =$$

$$(\text{or}(LL_0) \oplus OKL_1', LR_0 \oplus KL_1', \text{or}(R_0) \oplus OKR_1', RR_0 \oplus KR_1'),$$

where $OKL_1' = \text{or}(KL_1')$ and $OKR_1' = \text{or}(KR_1')$. Therefore, the first round subkeys linearly affect LL_1, LR_1, RL_1 and RR_1.

The equivalent transformation is performed similarly to that on FOX64.

The following 1-round keyless relation holding with certainty is used to attack FOX128:

$$\text{io}(LL_{i+1} \oplus LR_{i+1} = LL_i \oplus LR_i, \tag{3.414}$$

which is a 32-bit condition.

From (3.414), the following 16-bit relation is obtained:

$$(LL_4^{(1)} \oplus LL_4^{(3)}, LL_4^{(3)}) \oplus (LR_4^{(3)}, LR_4^{(1)}) = (LL_3^{(3)}, LL_3^{(1)}) \oplus (LR_3^{(3)}, LR_3^{(1)}) \tag{3.415}$$

The forward computation stage works as follows: for given (K_2', LK_3', $RKL_3^{'(3)}$, $RKL_3^{'(1)}$, $KL_1^{'''(3)}$, $KL_1^{'''(1)}$, $OKL_1^{'''(3)}$, $OKL_1^{'''(1)}$), the value ($LL_3^{(3)}$, $LL_3^{(1)}$) $\oplus$ ($LR_3^{(3)}$, $LR_3^{(1)}$) can be computed. Since ($KL_1^{'''(3)}$, $KL_1^{'''(1)}$) and ($OKL_1^{'''(3)}$, $OKL_1^{'''(1)}$) linearly affect the matching states ($LL_3^{(3)}$, $LL_3^{(1)}$) and ($LR_3^{(3)}$, $LR_3^{(1)}$), respectively, we can regard ($LK_1^{'''(3)}$, $LK_1^{'''(1)}$) $\oplus$ ($OKL_1^{'''(3)}$, $OKL_1^{'''(1)}$) as a new 16-bit key XK_1. Then, ($LL_3^{(3)}$, $LL_3^{(1)}$) $\oplus$ ($LR_3^{(3)}$, $LR_3^{(1)}$) is obtained from the 208 key bits K_2', LK_3', $RKL_3^{'(3)}$, $RKL_3^{'(1)}$ and the 16-bit linearly-dependent XK_1.

The backwards computation stage is: ($LL_4^{(1)} \oplus LL_4^{(3)}$, $LL_4^{(3)}$) $\oplus$ ($LR_4^{(3)}$, $LR_4^{(1)}$) is obtained from the 208 key bits (K_6, LK_5, $RKL_5^{(1)}$, $RKL_5^{(3)}$). Using the indirect matching technique in [3], 8 bits out of the 16 bits of XK_1 can be moved to the left half of the matching equation. Then, the left and right halves of this equation contain 216 bits of the key, that is, $|K_1| = |K_2| = 216$.

With $N = 26$, the time complexity for recovering 432 key bits is estimated as $\max(2^{216}, 2^{216}) \cdot 26 = 2^{221}$.

The remaining $768 - 416 = 352$ key bits can be obtained by recursively applying the ASR attack. The time complexity for recovering the remaining subkeys is estimated as $2^{352/2+1.6} = 2^{177.6}$ 6-round FOX128 computations, using $\lceil 352/128 \rceil = 2$ plaintext/ciphertext pairs.

The data complexity is only 26 chosen plaintexts, and the memory complexity is $\min(2^{216}, 2^{216}) \cdot 26 \approx 2^{221}$ text blocks.

If the function reduction technique is also used in the backwards direction, the attack can be extended to 7 rounds. It uses two additional 16-bit relations: $LL_7 \oplus RL_7 = c_3$ and $RL_7 \oplus RR_7 = c_4$.

Using the involutory property of the round function of FOX128:

$$(LL_4^{(1)} \oplus LL_4^{(3)}, L_4^{(3)}) \oplus (LR_4^{(3)}, LR_4^{(1)})$$

is obtained from 208 key bits and the 16-bit linearly-dependent key XK_2. In this attack, $XK_1 \oplus XK_2$ is further regarded as a 16-bit new key. Similar to the attack on 6-round FOX128, the left and right halves of the equation contain 216 bits of the key.

The repetitive ASR approach is employed as follows: recall that plaintexts and ciphertexts need to satisfy 64-bit relations $LL_0 \oplus LR_0$, $RL_0 \oplus RR_0$, $LL_7 \oplus LR_7$ and $RL_7 \oplus RR_7$. The cost is estimated by a game that an adversary plays to find 64-bit multi-collisions with 64-bit restricted inputs.

In the basic ASR attack, at least $432/16 = 27$ multi-collisions are needed to detect the 432 key bits. To obtain such pairs with a high probability, $27!^{1/27} \cdot 2^{64 \cdot 26/27} \approx 2^{65.1}$ pairs are required. However, it is unfeasible since the degrees of freedom of plaintexts is only 64 bits.

We use the repetitive ASR approach with $M = 2$. In each ASR recovery phase, the required data is reduced to 13 and 14. The fourteen 64-bit multi-collisions are obtained given 2^{62} plaintext/ciphertext pairs with high probability.

The time complexity for finding 432 key bits is estimated as $\max(2^{216}, 2^{216}) \cdot 14 \cdot 2 + 2^{432-16\cdot 14} = 2^{224}$ 7-round FOX128 computations. The remaining $896 - 432 = 480$ key bits are obtained by recursively applying the ASR attack. The time complexity for this phase is estimated as $2^{480/2+2} = 2^{242}$ using $\lceil 480/128 \rceil = 4$ plaintext/ciphertext pairs.

The data complexity is $2^{62} \cdot 2 = 2^{63}$ plaintext/ciphertext pairs. The memory complexity is about 2^{242} text blocks.

3.21 Related-Key Attacks

Related-key (RK) attacks exploit the behavior of a block cipher under different keys. Examples of RK attacks include [96, 95, 12].

RK attacks test the limits of the security of block ciphers by allowing the key channel to be exploited by the adversary. Since block ciphers serve as building blocks in several cryptographic constructions, RK attacks become relevant in settings where block ciphers become components in larger cryptographic functions, such as hash functions [136], where the key input can be manipulated by the adversary [180]. In some of these cases, the key input can be under the control of the adversary, and the key itself may not be secret anymore.

A usual setting for related-key attacks is in combination with other techniques such as differential cryptanalysis.

A conventional (non-related-key) differential for an n-bit block cipher E_K uses plaintext pairs with a chosen difference ΔP and obtains a ciphertext pair with difference ΔC such that

$$\Pr_{P,K}(E_K(P) \oplus E_K(P \oplus \Delta P) = \Delta C)$$

holds with high probability (compared to 2^{-n}) or is zero (for the case of impossible differentials [102, 15]).

A related-key differential for an n-bit block cipher E_K uses a 3-tuple of differences: $(\Delta P, \Delta C, \Delta K)$ for the plaintext, the ciphertext and the key, respectively, such that

$$\Pr_{P,K}(E_K(P) \oplus E_{K \oplus \Delta K}(P \oplus \Delta P) = \Delta C)$$

holds with high probability (compared to 2^{-n}) or is zero (for the case of impossible differentials [102, 15]).

In both cases, the common assumption is that the probability of the differential is independent of P and K and is uniformly distributed over all plaintexts and keys.

3.21.1 RK Differential-Linear Attacks on IDEA

In [79, 81], Hawkes described related-key differential-linear (RKDL) distinguishers for IDEA. The approach is to use plaintext pairs (P, P^*), with a particular difference ΔP, encrypted under two related keys (K, K^*), which differ in specific bit positions. The plaintexts are chosen in order to cancel the difference in the subkeys in the first rounds, so that afterwards the intermediate ciphertext words are the same in both texts.

A RKDL attack on 4-round IDEA was described in [79, 81]. It is an extension of the attack in [95].

Let $X^{(i)} = (X_1^{(i)}, X_2^{(i)}, X_3^{(i)}, X_4^{(i)})$ denote the input to the i-th round of IDEA. Let $Y^{(i)} = (Y_1^{(i)}, Y_2^{(i)}, Y_3^{(i)}, Y_4^{(i)})$ denote the 64-bit output block from the i-th KM half-round.

A RKDL distinguisher is denoted by a tuple (k_Δ, γ, η), where $0 \leq k_\Delta \leq 127$, $\gamma \in \mathbb{Z}_2^{64} - \{0\}$ and $\eta \in \{0, 1\}$. The RKDL distinguisher predicts that there exists plaintext pairs (P, P^*) such that $\gamma \cdot (C \oplus C^*) = \eta$, where P

encrypts to C under K and P^* encrypts to C^* under K^*, where K and K^* differ in bit positions given by k_Δ.

The probability of the RKDL distinguisher is

$$\Pr(\gamma \cdot (C \oplus C^*) = \eta | C = E_K(P), C^* = E_K(P^*)),$$

where K and K^* differ in bit positions given by k_Δ.

A RKDL weak-key class W is a set of user keys for which some RKDL distinguisher holds with certainty.

Consider the first 4 rounds of IDEA. Suppose K and K^* are two related keys that differ in bit position $k_\Delta = 16$. Then, $Z_i^{(j)} = Z_i^{(j)*}$ for $1 \leq i \leq 6$ and $1 \leq j \leq 4$ except that $Z_2^{(1)} = Z_2^{(1)*} \oplus \nu$, $Z_4^{(3)} = Z_4^{(3)*} \oplus 2^8$ and $Z_4^{(3)} = Z_4^{(3)*} \oplus 2$.

Further, if $\Delta P = P \oplus P^* = (0, \nu, 0, 0)$, then $X^{(2)} = X^{(2)*}$ and $X^{(3)} = X^{(3)*}$ since ΔP cancels the difference due to $(Z_2^{(1)}, Z_2^{(1)*})$, and thus there is no difference along 2.5 rounds. Thus, the difference in $(\Delta Y_1^{(3)}, \Delta Y_2^{(3)}, \Delta Y_3^{(3)})$ is also zero, but $\Delta Y_4^{(3)} \neq 0$ because of the difference in $(Z_4^{(3)}, Z_4^{(3)*})$.

Using the linear approximation $(1, 0, 1, 0)$ at $Y^{(3)}$ results in zero parity, since the difference in the leftmost three words is zero. The linear relation covers the third MA half-round and has output bitmask $(1, 1, 0, 0)$.

For the 4th round, $(1, 1, 0, 0) \stackrel{1r}{\rightarrow} (0, 1, 1, 0)$.

Therefore, the differential-linear relation becomes

$$(X_2^{(5)} \oplus X_2^{(5)*}) \cdot 1 \oplus (X_3^{(5)} \oplus X_3^{(5)*}) \cdot 1 = 0.$$

This linear relation involves the exclusive-or of the least significant bis of the two middle words in a block, which can be represented as $(X_2^{(5)} \oplus X_3^{(5)}) \cdot 1 = ((X_1^{(5)} \oplus X_2^{(5)}) \odot Z_5^{(4)}) \cdot 1$, using the Demirci relation (3.279).

Therefore, it is possible to recover $Z_5^{(4)}$. Note that no weak subkeys were needed.

Table 3.156 lists some of the parameters for RKDL attacks on several IDEA variants.

Table 3.156 Parameters and complexity figures for RKDL attacks on IDEA variants [79].

#Rounds	k_Δ	γ	Weak key bit positions	$\|WKC\|$	Average Data Complexity (CP)
4	16	$(0,1,1,0)$	—	2^{128}	38.4
4.5	16	$(1,1,0,0)$	82–96	2^{113}	19.5
5	16	$(1,1,0,0)$	75–96	2^{98}	38.4
5.5	16	$(1,1,0,0)$	18–32, 75–89	2^{99}	38.4
6	0	$(0,1,1,0)$	11–25, 75–96	2^{91}	$2^{13.2}$
6.5	0	$(1,0,1,0)$	4–25, 75–89	2^{84}	614
8	111	$(1,1,0,0)$	50–71, 75–110	2^{70}	$2^{19.3}$

3.21.2 Related-Key Keyless-BD Attack

In [23], Biham *et al.* presented a related-key attack on the first 7.5 rounds of IDEA. This attack uses the keyless Biryukov-Demirci relation (3.400), and the context (including the terminology) of Sect. 3.19.

Let K and K^* be two user keys that are equal in all key bits except bit 34 and in any non-empty subset of the nine bits $\{41, 42, \ldots, 49\}$. Let P and P^* denote two plaintexts such that $Y^{(2)}$ and $Y^{(2)*}$, the intermediate states after the second KM half-round, satisfy:

$$Y_1^{(2)} = Y_1^{(2)*},\ Y_2^{(2)} = Y_2^{(2)*},\ Y_3^{(2)} = Y_3^{(2)*},\ Y_4^{(2)} = Y_4^{(2)*}. \quad (3.416)$$

In such pair, the intermediate values are equal until the 4th MA half-round. In that MA half-round, the input difference is $(\Delta n_4, \Delta q_4) = (0, 0)$ and the key difference affects only $Z_6^{(4)}$. Hence, by Proposition 3.3 in Sect. 3.19, $\Delta t_2 = \Delta t_3 = \Delta t_4 = 0$.

For such pairs, equation (3.400) becomes

$$\begin{aligned} \mathrm{lsb}_1(P_2 \oplus P_2^* \oplus P_3 \oplus P_3^* \oplus \Delta t_1 \oplus \Delta t_5 \oplus \Delta t_6 \oplus \Delta t_7) = \\ \mathrm{lsb}_1(C_2 \oplus C_2^* \oplus C_3 \oplus C_3^*). \end{aligned} \quad (3.417)$$

If the adversary is able to construct plaintext pairs that satisfy (3.416), he can partially encrypt/decrypt the plaintext/ciphertext pairs through rounds 1, 5, 6 and 7, and check if (3.417) is satisfied. Therefore, the adversary has to guess $Z_1^{(1)}$, $Z_3^{(1)}$ and $Z_5^{(1)}$ to partially encrypt the first round, and $Z_5^{(5)}$, $Z_1^{(6)}$, $Z_2^{(6)}$, $Z_5^{(6)}$, $Z_6^{(6)}$, $Z_1^{(7)} - Z_6^{(7)}$, $Z_1^{(8)} - Z_4^{(8)}$ for the partial decryption.

These 18 subkeys represent 103 user key bits, and hence guessing them and checking if (3.416) holds for some plaintext/ciphertext pairs satisfying (3.416) yields an attack faster than exhaustive key search.

The attack algorithm on 7.5-round IDEA is the following:

(1) ask for the encryption of $2^{42.5}$ known plaintexts under key K; denote this set of plaintexts/ciphertexts by Q.
(2) ask for the encryption of $2^{42.5}$ known plaintexts under key K^*; denote this set of plaintexts/ciphertexts by Q^*.
(3) for each guess of $(Z_1^{(1)}, Z_2^{(1)}, Z_3^{(1)}, Z_4^{(1)})$:
    (a) partially encrypt all plaintexts in Q and Q^* through the first KM half-round.
    (b) find all pairs of $Y^{(1)}$ encrypted under K and $Y^{(1)*}$ encrypted under K^* such that $Y^{(1)} \oplus Y^{(1)*} = (0000_x, 0040_x, 0000_x, 0040_x)$.
    (c) for each such pair and each guess of $(Z_5^{(1)}, Z_6^{(1)}, Z_3^{(2)}, Z_4^{(2)})$:
        (i) if the pair satisfies (3.416), then guess $Z_5^{(5)}$, $Z_1^{(6)}$, $Z_2^{(6)}$, $Z_5^{(6)}$, $Z_6^{(6)}$, $Z_1^{(7)} - Z_6^{(7)}$ and $Z_1^{(8)} - Z_4^{(8)}$ and check if (3.417) is satisfied.

(ii) if (3.417) is not satisfied, discard the subkeys guessed.

(4) for each remaining subkey, exhaustively try all 25 remaining key bits.

There are $2^{42.5}(2^{42.5}-1)/2 = 2^{85}$ plaintexts pairs, of which $2^{85} \cdot 2^{-64} = 2^{21}$ have difference $(0000_x, 0040_x, 0000_x, 0040_x)$ after the first KM half-round. For each guess of $(Z_5^{(1)}, Z_6^{(1)}, Z_3^{(2)}, Z_4^{(2)})$ about $2^{21} \cdot 2^{-17} = 16$ pairs have a zero difference after the second KM half-round satisfying (3.416).

For the correct subkeys, all these pairs should satisfy (3.417). For wrong subkeys, the probability that (3.417) is satisfied for all these pairs is 2^{-16}. There are 2^{103} key candidates and hence the number of subkeys that enter step (4) is expected to be $2^{103} \cdot 2^{-16} = 2^{87}$.

Steps (1) and (2) have time complexity $2^{42.5}$ encryptions each. Step (4) has time complexity $2^{87} \cdot 2^{25} = 2^{112}$ trial encryptions. The time complexity of the attack is dominated by step (3). Step 3(a) is repeated 2^{64} times, and each time $2^{43.5}$ values are partially encrypted through one KM half-round. Hence, the time complexity is $2^{64} \cdot 2^{43.5} = 2^{107.5}$ partial encryptions. Step 3(b) can be executed efficiently using a hash table. In step 3(c)(i) only 2^{21} pairs (or 2^{22} values) are analyzed but this step requires guessing $Z_3^{(2)}$ and $Z_4^{(2)}$, or 32 bits not covered by the bits in step 3(a). Thus, finding the pairs satisfying (3.416) has time complexity $2^{64} \cdot 2^{22} \cdot 2^{32} = 2^{118}$ 1-round IDEA decryptions.

The second part of step 3(c)(i), checking if (3.417) is satisfied, costs less. Even though 9 more key bits are guessed, there are only 32 pairs (64 values) that enter this step. The total time complexity is about $2^{118} \cdot \frac{1}{7 \cdot 5} = 2^{115.1}$ 7.5-round IDEA encryptions.

3.21.3 Related-Key Boomerang Attack on IDEA

Let E_K denote the encryption operation of a target block cipher E under an unknown key K. Let us assume that E_K can be decomposed into two parts as $E_K = E^1 \circ E^0 = E^1(E^0)$, where each of E^1 and E^0 does not necessarily correspond to half of E_K. From the decryption direction, we have $D_K = E_K^{-1}$, and therefore, $D_K = (E^0)^{-1} \circ (E^1)^{-1} = D^0 \circ D^1$.

The difference operator is bitwise exclusive-or.

Consider a related-key differential $\Delta \rightarrow \Delta^*$ across E^0, holding with probability p, and under a key difference ΔK_0. For instance, a plaintext pair $(P_1, P_2) = (P_1, P_1 \oplus \Delta)$ is encrypted as $X_1 = E_K^0(P_1)$ and $X_2 = E^0_{K \oplus \Delta K_0}(P_2)$, respectively, and $X_1 \oplus X_2 = \Delta^*$ holds with probability p.

Consider another related-key differential $\nabla \rightarrow \nabla^*$ across $(E^1)^{-1}$, holding with probability q, and under a key difference ΔK_1. For instance, a ciphertext pair $(C_1, C_3) = (C_1, C_1 \oplus \nabla)$ is decrypted as $X_1 = D_K^0(C_1)$ and $X_3 = D^0_{K \oplus \Delta K_1}(C_3)$, respectively, and $X_1 \oplus X_3 = \nabla^*$ holds with probability q. These relations are depicted in Fig. 3.28.

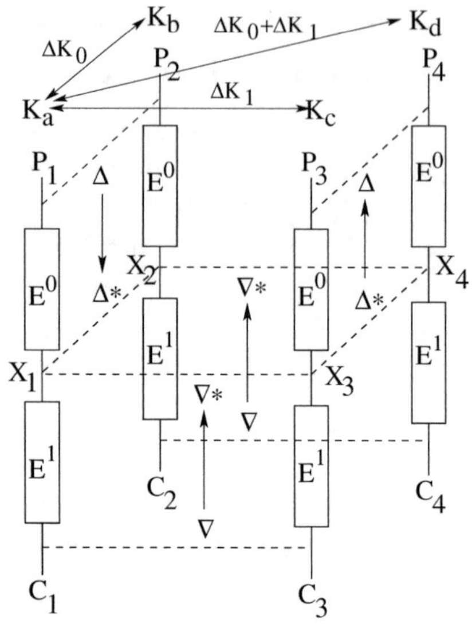

Fig. 3.28 Pictorial representation of a related-key boomerang. The vertical arrows indicate the direction differences propagate.

A related-key boomerang distinguisher uses four different unknown related keys: K_a, $K_b = K_a \oplus \Delta K_0$, $K_c = K_a \oplus \Delta K_1$ and $K_d = K_a \oplus \Delta K_0 \oplus \Delta K_1$.

The attack algorithm is as follows [22]:

(1) choose a random plaintext block P_1 and compute $P_2 = P_1 \oplus \Delta$.
(2) ask for the ciphertexts $C_1 = E_{k_a}(P_1)$ and $C_2 = E_{K_b}(P_2)$ to the encryption oracle.
(3) compute $C_3 = C_1 \oplus \nabla$ and $C_4 = C_2 \oplus \nabla$.
(4) ask for the plaintexts $P_3 = D_{K_c}(C_3)$ and $P_4 = D_{K_d}(C_4)$ to the decryption oracle.
(5) test whether $P_3 \oplus P_4 = \Delta$.

For a random permutation operating on n-bit text blocks, the probability that the test in (5) is satisfied is 2^{-n}. For the target n-bit block cipher E_K, the probability that the test in (5) is satisfied is $(pq)^2$, just like for a conventional boomerang.

The related-key boomerang distinguisher can use multiple differentials across E^0 and E^1 [17], under the condition that all related-key differentials across E^0 use the same key difference ΔK_0 and the same input difference Δ. Also, all related-key differentials across E^1 shall use the same key difference ΔK_1 and the same output difference ∇. The final probability of a quartet to be a right quartet becomes $(\hat{p}\hat{q})^2$, where $\hat{p}$ is defined in (3.130) and $\hat{q}$ is defined in (3.131) taking into account the key difference as well.

In the appendix of [23], Biham *et al.* presented related-key distinguishers and attacks on 7.5-round variants of IDEA that improve the analyses in [22]. Results in Sect. 3.19 are used as well.

First, they presented a 5.5-round related-key boomerang distinguisher starting from the first MA half-round until the end of the 6th round.

The first related-key differential starts after the first KM half-round with difference $(0000_x, 0040_x, 0000_x, 0040_x)$ and ends after the 4th MA half-round. The key difference is in bit 34 of the user key and in any non-empty subset of the nine bits $\{41, 42, 43, \ldots, 49\}$.

The second related-key differential starts at the beginning of round 5 with the difference $(0000_x, 8000_x, 0000_x, 0000_x)$ and key difference in key bit 91. Then, zero difference is preserved until key bit 91 is used again in $Z_4^{(7)}$.

For the first differential, a pair with input difference $(0000_x$, 0040_x, 0000_x, $0040_x)$ in the input to the first MA half-round causes difference $(0000_x$, 0000_x, 0040_x, $0040_x)$ after the MA half-round with certainty. With probability 2^{-1}, the key difference cancels the data difference in the third word, and with probability 2^{-16}, the key difference cancels the data difference in the fourth word. Thus, with probability 2^{-17}, the pair has zero difference after the second KM half-round. This zero difference is preserved until the last multiplication in the 4th MA half-round. Therefore, in that MA half-round, both Δn_4 and the key difference in $Z_5^{(4)}$ are zero. Applying the observation in Proposition 3.3, we obtain $\hat{p} = 2^{-17} \cdot 2^{-11.86} = 2^{-28.86}$. Therefore, taking two pairs with difference $(0000_x$, 0040_x, 0000_x, $0040_x)$ before the first MA half-round, they are expected to have the same difference after round 4 with probability at least $\hat{p}^2 = 2^{-57.76}$. The key difference ΔK_0 can be any of 511 possible values. For instance, the difference used in ΔK_0 is in bits 34 and 49 only. The key difference between two related user keys K_a and K_b is denoted $\Delta K_0 = K_a \oplus K_b$.

Using these differentials results in a 5.5-round related-key boomerang distinguisher that requires $2^{59.32}$ adaptively chosen-plaintexts and ciphertext (ACPC): $2^{57.32}$ values are encrypted under four different keys.

A related-key rectangle attack on the first 6.5 rounds of IDEA uses the 5.5-round related-key boomerang distinguisher described previously.

Let K_a, K_b, K_c and K_d denote the related keys such that $K_b = K_a \oplus \Delta K_0$, $K_c = K_a \oplus \Delta K_1$, and $K_d = K_c \oplus \Delta K_0$.

The attack algorithm consists of

(1) Data Collection Phase

(a) generate 2^{35} structures denoted $S_1^a, S_2^a, \ldots, S_{2^{35}}^a$ of 2^{28} plaintexts each. In each structure, the first word, the six least significant bits of the second word and the 14 least significant bits of the third word are fixed. Ask for the encryption of these structures under K_a to the encryption oracle.

(b) flip bit 6 of the second word and bit 13 of the third word of any plaintext encrypted under K_a, and ask for the encryption of the resulting plaintexts under K_b, to the encryption oracle. These structures are denoted $S_1^b, S_2^b \ldots, S_{2^{35}}^b$.

(c) generate 2^{35} structures denoted $S_1^c, S_2^c \ldots, S_{2^{35}}^c$ of 2^{28} plaintexts each, where the first word, the six least significant bits of the second word, and the 14 least significant bits of the third word are fixed. Ask for the encryption of these structures under key K_c, to the encryption oracle.

(d) flip bit 6 of the second word and bit 13 of the third word of any plaintext encrypted under K_c, and ask for the encryption of the resulting plaintexts under key K_d. These structures are denoted $S_1^d, S_2^d \ldots, S_{2^{35}}^d$.

(2) Finding Candidate Quartets

(a) find all plaintext pairs $C_1 \in S_i^a$ and $C_3 \in S_j^c$ such that they have the same value in the first, the second and the third words.

(b) for each such pair, check whether there are ciphertext pairs $C_2 \in S_i^b$ and $C_4 \in S_j^d$ such that they have the same value in the first, the second and the third words. If such a pair exists, transfer the corresponding plaintexts (P_1, P_2, P_3, P_4) to analysis.

(3) Analysis of Candidate Quartets

(a) initialize 2^{64} counters, each corresponding to a different guess of $Z_1^{(2)}$, $Z_1^{(3)}$, $Z_1^{(4)}$, $Z_7^{(4)}$.

(b) for each guess of these four subkeys, and each candidate quartet, check if the partial encryption and the partial decryption of the pairs of the quartet lead to the required differences. If they do, then increment the respective counter.

(4) Output: output all subkey guesses whose counter has value greater than 8.

The attack is expected to encounter more than 16 right quartets. Thus, the counter that corresponds to the right subkey is expected to be greater than 8. The probability that a quartet satisfies the differences for a wrong subkey is 2^{-128}, and as for each subkey guess there are 2^{126} quartets, then the expected value of the counters that correspond to wrong subkeys is $2^{126} \cdot 2^{-128} = 2^{-2}$. With probability 95.7%, the right subkey is in the list of subkeys returned in step (4), and only one out of $2^{42.1}$ wrong subkeys is in the list. The remaining key bits can be found by exhaustive search. The time complexity is $2^{85.9}$ 6.5-round IDEA encryptions.

The time complexity of Step (1) is 2^{65} encryptions, while Step (2) requires the time to insert 2^{64} ciphertexts into a hash table indexed by the first 48 bits, i.e. 2^{64} memory accesses. The expected number of collisions is $2^{63} \cdot 2^{63} \cdot 2^{-48} = 2^{78}$, and for each such collision, two more memory accesses are needed in Step 2(b). The total time complexity of Step 2(b) is 2^{79} memory accesses. Step 2(b) usually finds about 256 ciphertext pairs from S_i^b and S_j^d as there are 2^{28}

ciphertexts in each ciphertext set. Therefore, 2^{96} candidate quartets enter Step (3). The analysis of each subkey is independent of the other subkeys, and thus, can be efficiently done using table look-up. For example, consider the 4th words in (C_1, C_3) and (C_2, C_4). For the right subkey and a right quartet, the difference before the multiplication with $Z_4^{(7)}$ should be zero. It is possible to generate a table of 2^{32} values corresponding to the two words (the 4th word of C_1 and the 4th word of C_3, or the 4th word of C_2 and the 4th word of C_4), and the subkey values for which the pair has zero difference before the multiplication. Two accesses to the table is enough to check that both (C_1, C_3) and (C_2, C_4) suggest the same value for $Z_4^{(7)}$.

As multiplication has a relatively random behavior, then it is expected that for any subkey, it decrypts the pairs of values correctly with probability 2^{-16}. Therefore, it is expected that each entry of the table contains one subkey on average, and the probability that the pairs (C_1, C_3) and (C_2, C_4) suggest the same subkey, for a wrong quartet, is 2^{-16}. If there is no subkey suggested by the two pairs, then the quartet is wrong, and hence, the quartet can be discarded from further analysis. Thus, by a careful ordering of the memory accesses and using four tables, one for each of the four keys and appropriate differences, it is possible to find the subkeys suggested by the quartet in approximately two memory accesses per quartet, and the total time complexity of Step (3) is about 2^{87} memory accesses.

The time complexity of Step (4) is 2^{64} memory accesses. Thus, the attack requires 2^{87} memory accesses and data complexity of 2^{65} related-key chosen plaintexts (RK-CP). Recovering the remaining key bits requires $2^{85.9}$ additional encryptions.

This attack can be extended to 7-round IDEA by guessing the subkeys and partially decrypting the last MA half-round, and further applying the attack on 6.5-round IDEA.

The complexity would increase to $2^{32} \cdot 2^{87} = 2^{119}$ memory accesses, while the data complexity would remain 2^{65} related-key chosen plaintexts.

But, $Z_6^{(7)}$ and $Z_2^{(1)}$ share 12 key bits. This fact allows us to filter most of the wrong quartets by evaluating after the addition in the first KM half-round. We have to consider the cases where there is a carry from bit 3 and where there is no such carry. Since it is not known the correct case, we discard all pairs whose difference is affected by the carry from bit 3.

The attack starts with Step (1) of the 6.5-round attack. For each guess of $Z_5^{(7)}$, $Z_6^{(7)}$, all the ciphertexts are partially decrypted one half-round. Then, Step 2(a) of the 6.5-round attack is applied.

Step 2(b) changes: the $\text{msb}_{12}(Z_1^{(2)})$ are known from the key bits guessed, and the adversary can use them to find for each plaintext $P_a \in S_i^a$ encrypted under K_a only 2^{18} plaintext $P_b \in S_i^b$ that may have the required difference with P_a after the first KM half-round. A similar procedure can be done with P_c and P_d.

The number of pairs $(P_b, P_d) \in S_i^b \times S_j^d$ satisfying the differences with P_a ad P_c, such that the corresponding ciphertexts have zero difference in the first, second and third words for a given (P_a, P_c) pair that enters Step 2(c) is reduced to $2^{18} \cdot 2^{18} \cdot 2^{-48} = 2^{-12}$. Thus, it is expected that only $2^{78} \cdot 2^{-12} = 2^{66}$ quartets enter Step (3) for each guess of $Z_5^{(7)}$, $Z_6^{(7)}$. This approach leads to discarding 2^{-2} of the right pairs, that is, pairs with difference $(0000_x, 0040_x, 0000_x, 0040_x)$ after the first KM half-round. Therefore, 9 right quartets are expected. On the other hand, this approach also reduces the number of quartets suggesting a wrong subkey by a factor of $\frac{9}{16}$.

The total time complexity (without the data collection and the partial decryption steps) is $2^{32}(2^{64}+2^{79}+2^{67}+2^{64}) = 2^{111}$ memory accesses. Assuming 3 clock cycles for one memory access, this complexity becomes equivalent to $2^{104.2}$ 7-round IDEA encryptions.

3.21.4 Further Related-Key Boomerang and Rectangle Attack on IDEA

If the key schedule algorithm is a linear function, such as in DES and IDEA, a given difference in the key determines the difference in the round subkeys [22]. Otherwise, if the key schedule algorithm is nonlinear, such as in the MESH ciphers, the exact difference in the subkeys, needed to match the internal data difference in a related-key differential might be unknown. In the latter case, the adversary needs to examine the design of the key schedule in order to determine the probability of difference propagation, and to harmonize these differences with those in the encryption framework. This strategy implies that the adversary will have to repeat the attack with several keys (under the needed key difference) until obtaining the necessary subkey difference(s) for the related-key differential to hold.

A 5.5-round boomerang distinguisher for IDEA starts from the second round until after the seventh MA half-round. This 5.5-round IDEA variant is decomposed as $E_1 \circ E_0$, where E_0 covers three rounds, while E_1 covers the following 2.5 rounds.

The input difference to E_0 has the form $\Delta = (0000_x, 0000_x, 8000_x, 0000_x)$. The key difference is $\Delta K_0 = e_{25}$, where e_i denotes an exclusive-or difference only in the i-th key bit of the user key.

The input (text) difference in Δ is canceled by the subkey difference and the resulting zero difference remains up to the 4th MA half-round.

The key difference further reappears in $Z_5^{(4)}$, leading to an unknown difference Δ^* after the MA half-round. There are at most 2^{32} Δ^* difference values after the MA half-round, and in the worst case, all of them are equiprobable with probability 2^{-32}. Since all of them are used, $\hat{p} = \sqrt{2^{32} \cdot (2^{-32})^2} = 2^{-16}$.

The second differential for E_1^{-1} has output difference

$$\nabla = (0000_x, 0000_x, 0100_x, 0000_x)$$

and key difference $\Delta K_1 = e_{75}$. In the decryption of a ciphertext pair with difference ∇, the key difference ΔK_1 cancels the text difference with probability 2^{-1}. If this difference cancellation happens, the zero difference remains through decryption until the beginning of the 5th MA half-round. This time, there are 2^{16} possible ∇^* values and in the worst case, all of them are equiprobable with probability 2^{-17}. All of them are used simultaneously, and thus, $\hat{q} = \sqrt{2^{16} \cdot 2^{-34}} = 2^{-9}$. In summary, this distinguishing attack on 5.5-round IDEA uses $2^{51.6}$ quartets of adaptively chosen plaintexts and ciphertexts (ACPC) or $2^{49.6}$ texts encrypted/decrypted under four different keys.

3.21.4.1 A Related-Key Boomerang Attack on 6-round IDEA

Let K_a be an unknown key, $K_b = K_a \oplus \Delta K_0 = K_a \oplus e_{25}$, $K_c = K_a \oplus \Delta K_1 = K_a \oplus e_{75}$ and $K_d = K_a \oplus \Delta K_0 \oplus \Delta K_1$. The following paragraphs describe a related-key boomerang attack on 6-round IDEA starting at the first MA half-round and reaching the end of the 7th KM half-round.

The related-key boomerang attack is as follows:

(1) for each guess of key bits 64-95 of K_a, set a counter initialized to zero.
(2) choose $2^{17.6}$ 32-bit values (r, t) and for each such value:

- choose a structure A of 2^{32} plaintexts of the form (x, y, z, w) such that $x \oplus z = t$ and $y \oplus w = r$.
- choose a structure B of 2^{32} plaintexts of the form $(x, y \oplus 8000_x, z, w)$ such that $x \oplus z = t$ and $y \oplus w = r$.
- ask for the encryption of the plaintexts in A under K_a, and obtain A'. Similarly, ask for the encryption of the plaintexts in B under K_b and obtain B'.
- for any ciphertext in A', compute its exclusive-or with the output difference ∇ and obtain C'. Ask for the decryption of C' under K_c to obtain C.
- for any ciphertext in B', compute its exclusive-or with the output difference ∇ and obtain D'. Ask for the decryption of D' under K_d to obtain D.
- insert all the plaintexts in C and D into a hash table indexed by the exclusive-or of the first and third words, and the exclusive-or of the second and fourth words.
- examine a pair of colliding plaintexts (P_3, P_4). Let P_1 be the plaintext that was encrypted, exclusive-ored to ∇ and decrypted to P_3. Let P_2 be the plaintext that was encrypted, exclusive-ored to ∇ and decrypted to P_4. For any guess of bits 64-95 of K_a:

(a) partially encrypt (P_1, P_2, P_3, P_4) across the first MA half-round. If the differences of the partial encryptions of (P_1, P_2) and (P_3, P_4) are both Δ, then continue the analysis. Otherwise, try the next subkey candidate.
(b) verify that the difference after a partial decryption of the respective ciphertext pair is zero (bits 64-95 of the user key cover the subkeys that interact with the third word of the ciphertext). If this is the case, increment the counter of the subkey.

(3) output the subkey whose counter is maximal.

The structures were chosen so that in each pair (A, B) of structures, there are 2^{32} pairs with input Δ after the first MA half-round. For each such pair of structures, 2^{32} plaintext pairs (P_3, P_4) are expected to be analyzed. Under random distribution assumptions, 2^{32} quartets from each pair of structures are encountered. However, most of them are discarded, and wrong quartets have probability 2^{-32} to agree with the subkey of the first MA half-round.

Thus, we have $2^{17.6}$ quartets in total, each of which suggests 32-bit subkey values. The second filtering reduces this number by a factor of 4.

The attack suggests $2^{15.6}$ possible values of 32 bits of K_a. We expect four right quartets. The right value is expected to be the one with the maximal counter with high probability.

The data complexity of the attack is $2^{51.6}$ adaptive chosen plaintexts and ciphertexts (ACPC). The time complexity is $2^{51.6}$ MA half-round evaluations, equivalent to about 2^{48} encryptions. The remaining $128 - 32 = 96$ key bits can be recovered by exhaustive search, leading to a final time complexity of 2^{96} 6-round IDEA computations.

3.21.4.2 A Related-Key Rectangle Attack on 6.5-round IDEA

The previous related-key boomerang attack on 6-round IDEA in Sect. 3.21.4.1 can be extended to a related-key rectangle attack on 6.5-round IDEA, starting after the first KM half-round, until the end of the 7th round.

This attack uses the same differentials as in Sect. 3.21.4.1, and can be described as follows:

- choose $2^{25.8}$ pairs (r, t). Generate two plaintext structures for each (r, t) pair, like in the boomerang attack.
- ask for the encryption of the structures under K_a and K_b as in the boomerang attack.
- ask for encryption under K_c of each plaintext encrypted under K_a
- ask for the encryption under K_d of each plaintext encrypted under K_b.
- for each guess of the subkey of the 7th MA half-round, partially decrypt all the ciphertexts under the guessed subkeys, and perform the related-key boomerang attack on 6-round IDEA.

Out of the $2^{57.8}$ plaintexts encrypted under each key, we get about $2^{51.6}$ pairs with differences required for the attack on 6-round IDEA.

The time complexity of the attack is $2^{32} \cdot (4 \cdot 2^{57.8}/13 + 2^{51.6}/13) \approx 2^{88.1}$ 6.5-round encryptions, and data complexity of $2^{59.8}$ chosen plaintexts (CP) under four related keys.

The remaining $128 - 32 - 32 = 64$ key bits can be recovered by exhaustive search.

References

1. Akgun, M., Demirci, H., Sagiroglu, M.S., Kavak, P.: Improved Square Properties of IDEA. Turkish Journal Elec. Engineering and Computer Science **20**(4), 493–506 (2012)
2. Amirazizi, H.R., Hellman, M.E.: Time-Memory-Processor Trade-Offs. IEEE Transactions on Information Theory **34**(3), 505–512 (1988)
3. Aoki, K., Guo, J., Matusiewicz, K., Sasaki, Y., Wang, L.: Preimages for Step-Reduced SHA-2. In: M. Matsui (ed.) Advances in Cryptology, Asiacrypt, LNCS 5912, pp. 578–597. Springer (2009)
4. Aoki, K., Ohta, K.: Differential-Linear Attack on FEAL. IEICE Transactions on Fundamentals **E79-A**(1), 20–27 (1996)
5. Aoki, K., Sasaki, Y.: Preimage Attacks on One-Block MD4, 63-step MD5 and more. In: R.M. Avanzi, L. Keliher, S. Sica (eds.) Selected Areas in Cryptography (SAC), LNCS 5381, pp. 103–119. Springer (2008)
6. Aoki, K., Sasaki, Y.: Meet-in-the-Middle Preimage Attacks Against Reduced SHA-0 and SHA-1. In: S. Halevi (ed.) Advances in Cryptology, Crypto, LNCS 5677, pp. 70–89. Springer (2009)
7. Ayaz, E.S., Selçuk, A.A.: Improved DST Cryptanalysis of IDEA. In: E. Biham, A.M. Youssef (eds.) Selected Areas in Cryptography (SAC), LNCS 4356, pp. 1–14. Springer (2006)
8. Babbage, S.: A Space/Time Tradeoff in Exhaustive Search Attacks on Stream Ciphers. European Convention on Security and Detection, IEE Conference Publication no. 408 (1995)
9. Bard, G.V.: Algebraic Cryptanalysis. Springer (2009)
10. Ben-Aroya, I., Biham, E.: Differential Cryptanalysis of Lucifer. Journal of Cryptology **9**(1), 21–34 (1996)
11. Biere, A., Heule, M., Van Maaren, H., Walsh, T.: Handbook of Satisfiability - Frontiers in Artificial Intelligence and Applications, vol. 185. IOS Press (2009)
12. Biham, E.: New Types of Cryptanalytic Attacks using Related Keys. In: T. Helleseth (ed.) Advances in Cryptology, Eurocrypt, LNCS 765, pp. 398–409. Springer (1993)
13. Biham, E.: How to Decrypt or Even Substitute DES-encrypted Messages in 2^{28} Steps. Information Processing Letters **84**(3), 117–124 (2002)
14. Biham, E., Anderson, R.J., Knudsen, L.R.: Serpent: a New Block Cipher Proposal. In: S. Vaudenay (ed.) Fast Software Encryption (FSE), LNCS 1372, pp. 222–238. Springer (1998)
15. Biham, E., Biryukov, A., Shamir, A.: Cryptanalysis of Skipjack Reduced to 31 rounds using Impossible Differentials. In: J. Stern (ed.) Advances in Cryptology, Eurocrypt, LNCS 1592, pp. 12–23. Springer (1999)
16. Biham, E., Biryukov, A., Shamir, A.: Miss-in-the-Middle Attacks on IDEA, Khufu and Khafre. In: L.R. Knudsen (ed.) Fast Software Encryption (FSE), LNCS 1636, pp. 124–138. Springer (1999)

17. Biham, E., Dunkelman, O., Keller, N.: The Rectangle Attack - Rectangling the Serpent. In: B. Pfitzmann (ed.) Advances in Cryptology, Eurocrypt, LNCS 2045, pp. 340–357. Springer (2001)
18. Biham, E., Dunkelman, O., Keller, N.: Enhancing Differential-Linear Cryptanalysis. In: Y. Zheng (ed.) Advances in Cryptology, Asiacrypt, LNCS 2501, pp. 254–266. Springer (2002)
19. Biham, E., Dunkelman, O., Keller, N.: New Results on Boomerang and Rectangle Attacks. In: J. Daemen, V. Rijmen (eds.) Fast Software Encryption (FSE), LNCS 2365, pp. 1–16. Springer (2002)
20. Biham, E., Dunkelman, O., Keller, N.: Differential-Linear Cryptanalysis of Serpent. In: T. Johansson (ed.) Fast Software Encryption (FSE), LNCS 2887, pp. 9–21. Springer (2003)
21. Biham, E., Dunkelman, O., Keller, N.: New Combined Attacks on Block Ciphers. In: H. Gilbert, H. Handschuh (eds.) Fast Software Encryption (FSE), LNCS 3557, pp. 126–144. Springer (2005)
22. Biham, E., Dunkelman, O., Keller, N.: Related-key Boomerang and Rectangle Attacks. In: R. Cramer (ed.) Advances in Cryptology, Eurocrypt, LNCS 3494, pp. 507–525. Springer (2005)
23. Biham, E., Dunkelman, O., Keller, N.: New Cryptanalytic Results on IDEA. In: X. Lai, K. Chen (eds.) Advances in Cryptology, Asiacrypt, LNCS 4284, pp. 412–427. Springer (2006)
24. Biham, E., Dunkelman, O., Keller, N.: A New Attack on 6-round IDEA. In: A. Biryukov (ed.) Fast Software Encryption (FSE), LNCS 4593, pp. 211–224. Springer (2007)
25. Biham, E., Dunkelman, O., Keller, N., Shamir, A.: New Data-Efficient Attacks on Reduced-Round IDEA. IACR ePrint Archive 2011/417 (2011)
26. Biham, E., Furman, V.: Improved Impossible Differentials on Twofish. In: B. Roy, E. Okamoto (eds.) Progress in Cryptology, Indocrypt, LNCS 1977, pp. 80–92 (2000)
27. Biham, E., Keller, N.: Cryptanalysis of Reduced-Round Variants of Rijndael. Third AES Conference (2000)
28. Biham, E., Shamir, A.: Differential Cryptanalysis of DES-like Cryptosystems (extended abstract). In: A.J. Menezes, S.A. Vanstone (eds.) Advances in Cryptology, Crypto, LNCS 537, pp. 1–19. Springer (1990)
29. Biham, E., Shamir, A.: Differential Cryptanalysis of DES-like Cryptosystems. Journal of Cryptology **4**(1), 3–72 (1991)
30. Biham, E., Shamir, A.: Differential Cryptanalysis of the Data Encryption Standard. Springer (1993)
31. Biham, E., Shamir, A.: Differential Cryptanalysis of the Full 16-round DES. In: E.F. Brickell (ed.) Advances in Cryptology, Crypto, LNCS 740, pp. 487–496. Springer (1993)
32. Biryukov, A.: Some Thoughts on Time-Memory-Data Tradeoffs. IACR ePrint archive, 2005/207 (2005)
33. Biryukov, A., Mukhopadhyay, S., Sarkar, P.: Improved Time-Memory Tradeoff with Multiple Data. In: B. Preneel, S. Tavares (eds.) Selected Areas in Cryptography (SAC), LNCS 3897, pp. 110–127. Springer (2006)
34. Biryukov, A., Nakahara Jr, J., Preneel, B., Vandewalle, J.: New Weak-Key Classes of IDEA. In: R.H. Deng, S. Qing, F. Bao, J. Zhou (eds.) Information and Communications Security (ICICS), LNCS 2513, pp. 315–326. Springer (2002)
35. Biryukov, A., Shamir, A.: Cryptanalytic Time/Memory/Data Tradeoffs for Stream Ciphers. In: T. Okamoto (ed.) Advances in Cryptology, Asiacrypt, LNCS 1976, pp. 1–13. Springer (2000)
36. Biryukov, A., Shamir, A.: Structural Cryptanalysis of SASAS. In: B. Pfitzmann (ed.) Advances in Cryptology, Eurocrypt, LNCS 2045, pp. 394–405. Springer (2001)

37. Biryukov, A., Wagner, D.: Slide Attacks. In: L.R. Knudsen (ed.) Fast Software Encryption (FSE), LNCS 1636, pp. 245–259. Springer (1999)
38. Biryukov, A., Wagner, D.: Advanced Slide Attacks. In: B. Preneel (ed.) Advances in Cryptology, Eurocrypt, LNCS 1807, pp. 586–606. Springer (2000)
39. Black, J., Rogaway, P., Shrimpton, T.: Black-Box Analysis of the Block-Cipher-based Hash-Function Constructions from PGV. In: M. Yung (ed.) Advances in Cryptology, Crypto, LNCS 2442, pp. 320–335. Springer (2002)
40. Blocher, U., Dichtl, M.: Problems with the Linear Cryptanalysis of DES using More than One Active S-box per Round. In: R. Anderson (ed.) Fast Software Encryption (FSE), LNCS 809, pp. 256–274. Springer (1994)
41. Blondeau, C., Leander, G., Nyberg, K.: Differential-Linear Cryptanalysis Revisited. In: S. Moriai (ed.) Fast Software Encryption (FSE), LNCS 8540, pp. 411–430. Springer (2014)
42. Bogdanov, A., Chang, D., Ghosh, M., Sanadhya, S.K.: Bicliques with Minimal Data and Time Complexity for AES. In: J. Lee, J. Kim (eds.) Information Security and Cryptology, ICISC, LNCS 8949, pp. 160–174. Springer (2015)
43. Bogdanov, A., Khovratovich, D., Rechberger, C.: Biclique Cryptanalysis of the Full AES. In: D.H. Lee, X. Wang (eds.) Advances in Cryptology, Asiacrypt, LNCS 7073, pp. 344–371. Springer (2011)
44. Bogdanov, A., Rijmen, V.: Linear Hulls with Correlation Zero and Linear Cryptanalysis of Block Ciphers. IACR ePrint archive 2011/123 (2011)
45. Bogdavov, A., Rechberger, C.: A 3-Subset Meet-in-the-Middle Attack: Cryptanalysis of the Lightweight Block Cipher KTANTAN. In: A. Biryukov, G. Gong, D.R. Stinson (eds.) Selected Areas in Cryptography (SAC), LNCS 6544, pp. 229–240. Springer (2010)
46. Borisov, N., Chew, M., Johnson, R., Wagner, D.: Multiplicative Differentials. In: J. Daemen, V. Rijmen (eds.) Fast Software Encryption (FSE), LNCS 2365, pp. 17–33. Springer (2002)
47. Borst, J.: Differential-Linear Cryptanalysis of IDEA. ESAT-COSIC Tech report 96/2, 14 pgs (1996)
48. Borst, J., Knudsen, L.R., Rijmen, V.: Two Attacks on Reduced IDEA (extended abstract). In: W. Fumy (ed.) Advances in Cryptology, Eurocrypt, LNCS 1233, pp. 1–13. Springer (1997)
49. Brickell, E., Pointcheval, D., Vaudenay, S., Yung, M.: Design Validations for Discrete Logarithm based Signature Schemes. In: H. Imai, Y. Zheng (eds.) Public-Key Cryptography (PKP), LNCS 1751, pp. 276–292. Springer (2000)
50. Brown, L., Kwan, M., Pieprzyk, J., Seberry, J.: Improving Resistance to Differential Cryptanalysis and the Redesign of LOKI. In: H. Imai, R.L. Rivest, T. Matsumoto (eds.) Advances in Cryptology, Asiacrypt, LNCS 739, pp. 36–50. Springer (1991)
51. Chabaud, F., Vaudenay, S.: Links Between Differential and Linear Cryptanalysis. Tech. rep., Laboratoire d'Informatique de l'ENS (1994)
52. Chen, J., Xue, D., Lai, X.: An Analysis of International Data Encryption Algorithm (IDEA) Security against Differential Cryptanalysis. China Crypt (2008)
53. Clapp, C.: Joint Hardware/Software Design of a Fast Stream Cipher. In: S. Vaudenay (ed.) Fast Software Encryption (FSE), LNCS 1372, pp. 75–92 (1998)
54. Coppersmith, D.: The Data Encryption Standard (DES) and its Strength against Attacks. IBM Journal of Research and Development **38**(3), 243–250 (1994)
55. Cormen, T.H., Leiserson, C.E., Rivest, R.L., Stein, C.: Introduction to Algorithms. MIT Press (2009)
56. Daemen, J.: Cipher and Hash Function Design – Strategies based on Linear and Differential Cryptanalysis. Ph.D. thesis, Dept. Elektrotechniek, ESAT, Katholieke Universiteit Leuven, Belgium (1995)
57. Daemen, J., Govaerts, R., Vandewalle, J.: Weak Keys for IDEA. In: D.R. Stinson (ed.) Advances in Cryptology, Crypto, LNCS 773, pp. 224–231. Springer (1993)

58. Daemen, J., Knudsen, L.R., Rijmen, V.: The Block Cipher SQUARE. In: E. Biham (ed.) Fast Software Encryption (FSE), LNCS 1267, pp. 149–165. Springer (1997)
59. Daemen, J., Rijmen, V.: AES Proposal: Rijndael. First AES Conference, California, USA, http://www.nist.gov/aes (1998)
60. Demirci, H.: Square-like Attacks on Reduced Rounds of IDEA. In: K. Nyberg, H. Heys (eds.) Selected Areas in Cryptography (SAC), LNCS 2595, pp. 147–159. Springer (2002)
61. Demirci, H., Selçuk, A.A., Türe, E.: A New Meet-in-the-Middle Attack on the IDEA Block Cipher. In: M. Matsui, R. Zuccherato (eds.) Selected Areas in Cryptography (SAC), LNCS 3006, pp. 117–129. Springer (2003)
62. Denning, D.E.: Cryptography and Data Security. Addison-Wesley (1982)
63. Dj Golic, J.: Cryptanalysis of Alleged A5 Stream Cipher. In: W. Fumy (ed.) Advances in Cryptology, Eurocrypt, LNCS 1233, pp. 239–255. Springer (1997)
64. Dj Golic, J., Bagini, V., Morgari, G.: Linear Cryptanalysis of Bluetooth Stream Cipher. In: L.R. Knudsen (ed.) Advances in Cryptology, Eurocrypt, LNCS 2332, pp. 238–255. Springer (2002)
65. Dunkelman, O., Keller, N.: The Effect of the Omission of Last Round's MixColumns on AES. Information Processing Letters **110**(8-9), 304–308 (2010)
66. Electronic-Frontier-Foundation: Cracking DES - Secrets of Encryption Research Wiretap Politics and Chip Design. O'Reilly and Associates Inc. (1998)
67. Feistel, H.: Cryptography and Computer Privacy. Scientific American **228**(5), 15–23 (1973)
68. Feller, W.: An Introduction to Probability Theory and Its Applications, vol. 1. Wiley (1968)
69. Ferguson, N., Kelsey, J., Lucks, S., Schneier, B., Stay, M., Wagner, D., Whiting, D.: Improved Cryptanalysis of Rijndael. In: B. Schneier (ed.) Fast Software Encryption (FSE), LNCS 1978, pp. 213–230. Springer (2000)
70. Fiat, A., Naor, M.: Rigourous Time/Space Tradeoffs for Inverting Functions. SIAM Journal on Computing **29**(3), 790–803 (1999)
71. FIPS197: Advanced Encryption Standard (AES). FIPS PUB 197 Federal Information Processing Standard Publication 197, United States Department of Commerce (2001)
72. Flajolet, P., Odlyzko, A.: Random Mapping Statistics. In: J.J. Quisquater, J. Vandewalle (eds.) Advances in Cryptology, Eurocrypt, LNCS 434, pp. 329–354. Springer (1989)
73. Forré, R.: The Strict Avalanche Criterion: Spectral Properties of Boolean Functions and an Extended Definition. In: S. Goldwasser (ed.) Advances in Cryptology, Crypto, LNCS 403, pp. 450–468. Springer (1990)
74. Girault, M., Stern, J.: On the Length of Cryptographic Hash-values used in Identification Schemes. In: Y.G. Desmedt (ed.) Advances in Cryptology, Crypto, LNCS 839, pp. 202–215. Springer (1994)
75. Grover, L.K.: A Fast Quantum Mechanical Algorithm for Database Search. In: G.L. Miller (ed.) Proceedings of the 20th Annual ACM Symposium on the Theory of Computing, pp. 22–24. ACM (1996)
76. Guo, R., Jin, C.: Impossible Differential Cryptanalysis on Lai-Massey Scheme. ETRI Journal **36**(6), 1032–1040 (2014)
77. Harpes, C.: Cryptanalysis of Iterated Block Ciphers, *ETH Series in Information Processing*, vol. 7. Hartung-Gorre Verlag, Konstanz (1996)
78. Harpes, C., Kramer, G.G., Massey, J.L.: A Generalization of Linear Cryptanalysis and the Applicability of Matsui's Piling-Up Lemma. In: L.C. Guillou, J.J. Quisquater (eds.) Advances in Cryptology, Eurocrypt, LNCS 921, pp. 24–38. Springer (1995)
79. Hawkes, P.: Differential-Linear Weak Key Classes of IDEA. In: K. Nyberg (ed.) Advances in Cryptology, Eurocrypt, LNCS 1403, pp. 112–126. Springer (1998)

80. Hawkes, P., O'Connor, L.: On Applying Linear Cryptanalysis to IDEA. In: K. Kim, T. Matsumoto (eds.) Advances in Cryptology, Asiacrypt, LNCS 1163, pp. 105–115. Springer (1996)
81. Hawkes, P.M.: Asymptotic Bounds on Differential Probabilities and an Analysis of the Block Cipher IDEA. Ph.D. thesis, The University of Queensland, St. Lucia, Australia (1998)
82. Hellman, M.E.: A Cryptanalytic Time-Memory Trade-Off. IEEE Transactions on Information Theory **IT-26**(4), 401–406 (1980)
83. Hong, J., Sarkar, P.: New Applications of Time Memory Data Tradeoffs. In: B. Roy (ed.) Advances in Cryptology, Asiacrypt, LNCS 3788, pp. 353–372. Springer (2005)
84. Hu, Y., Zhang, Y., Xiao, G.: Integral Cryptanalysis of SAFER+. Electronic Letters **35**(17), 1458–1459 (1999)
85. Isobe, T., Shibutani, K.: All Subkeys Recovery Attack on Block Ciphers: Extending Meet-in-the-Middle Approach. In: L.R. Knudsen, H. Wu (eds.) Selected Areas in Cryptography (SAC), LNCS 7707, pp. 202–221. Springer (2013)
86. Isobe, T., Shibutani, K.: Generic Key-Recovery Attack on Feistel Scheme. In: K. Sako, P. Sarkar (eds.) Advances in Cryptology, Asiacrypt, LNCS 8269, pp. 464–485. Springer (2013)
87. Isobe, T., Shibutani, K.: Improved All-Subkeys Recovery Attack on FOX, KATAN and SHACAL-2 Block Ciphers. In: C. Cid, C. Rechberger (eds.) Fast Software Encryption (FSE), LNCS 8540, pp. 104–126. Springer (2015)
88. Joux, A.: Multicollisions in Iterated Hash Functions. Application to Cascade Constructions. In: M. Franklin (ed.) Advances in Cryptology, Crypto, LNCS 3152, pp. 306–316. Springer (2004)
89. Junod, P.: On the Optimality of Linear, Differential and Sequential Distinguishers. In: E. Biham (ed.) Advances in Cryptology, Eurocrypt, LNCS 2656, pp. 17–32. Springer (2003)
90. Junod, P.: New Attacks against Reduced-Round Versions of IDEA. In: H. Gilbert, H. Handschuh (eds.) Fast Software Encryption (FSE), LNCS 3557, pp. 384–397. Springer (2005)
91. Junod, P., Macchetti, M.: Revisiting the IDEA Philosophy. In: O. Dunkelman (ed.) Fast Software Encryption (FSE), LNCS 5665, pp. 277–295. Springer (2009)
92. Junod, P., Vaudenay, S.: FOX: a New Family of Block Ciphers. In: H. Handschuh, M.A. Hasan (eds.) Selected Areas in Cryptography (SAC), LNCS 3357, pp. 114–129. Springer (2004)
93. Kaliski Jr, B.S., Rivest, R.L., Sherman, A.T.: Is the Data Encryption Standard a Group? Journal of Cryptology **1**, 3–36 (1988)
94. Kelsey, J., Kohno, T., Schneier, B.: Amplified Boomerang Attacks against Reduced-Round MARS and Serpent. In: B. Schneier (ed.) Fast Software Encryption (FSE), LNCS 1978, pp. 75–93. Springer (2000)
95. Kelsey, J., Schneier, B., Wagner, D.: Key-Schedule Cryptanalysis of IDEA, G-DES, GOST, SAFER and triple-DES. In: N. Koblitz (ed.) Advances in Cryptology, Crypto, LNCS 1109, pp. 237–251. Springer (1996)
96. Kelsey, J., Schneier, B., Wagner, D.: Related-Key Cryptanalysis of 3-Way, Biham-DES, CAST, DES-X, NewDES, RC2, and TEA. In: Y. Han, T. Okamoto, S. Qing (eds.) Information and Communication Security (ICICS), LNCS 1334, pp. 233–246. Springer (1997)
97. Khovratovich, D., Leurent, G., Rechberger, C.: Narrow-Bicliques: Cryptanalysis of Full IDEA. In: D. Pointcheval, T. Johansson (eds.) Advances in Cryptology, Eurocrypt, LNCS 7237, pp. 392–410. Springer (2012)
98. Kim, K., Park, S., Lee, S.: Reconstruction of s^2-DES S-boxes and Their Immunity to Differential Cryptanalysis. JW-ISC, 1993 Korean-Japan Joint Workshop on Information Security and Cryptology (1993)

99. Knudsen, L.R.: Block Ciphers: Analysis, Design and Applications. Ph.D. thesis, Aarhus University, Denmark (1994)
100. Knudsen, L.R.: Truncated and Higher Order Differentials. In: B. Preneel (ed.) Fast Software Encryption (FSE), LNCS 1008, pp. 196–211. Springer (1995)
101. Knudsen, L.R.: Block Ciphers - a Survey. In: B. Preneel, V. Rijmen (eds.) COSIC Course, LNCS 1528, pp. 18–48. Springer (1998)
102. Knudsen, L.R.: DEAL – a 128-bit Block Cipher. Technical Report #151, University of Bergen, Dept. of Informatics, Norway (1998)
103. Knudsen, L.R., Berson, T.A.: Truncated Differentials of SAFER. In: D. Gollmann (ed.) Fast Software Encryption (FSE), LNCS 1039, pp. 15–26. Springer (2005)
104. Knudsen, L.R., Meier, W.: Improved Differential Attack on RC5. In: N. Koblitz (ed.) Advances in Cryptology, Crypto, LNCS 1109, pp. 216–228. Springer (1996)
105. Knudsen, L.R., Rijmen, V.: On the Decorrelated Fast Cipher (DFC) and its Theory. In: L.R. Knudsen (ed.) Fast Software Encryption (FSE), LNCS 1636, pp. 81–94 (1999)
106. Knudsen, L.R., Robshaw, M.J.B., Wagner, D.: Truncated Differentials and Skipjack. In: M.J. Wiener (ed.) Advances in Cryptology, CRYPTO, LNCS 1666, pp. 165–180. Springer (1999)
107. Knudsen, L.R., Wagner, D.: Integral Cryptanalysis. In: J. Daemen, V. Rijmen (eds.) Fast Software Encryption (FSE), LNCS 2365, pp. 112–127. Springer (2002)
108. Koyama, T., Wang, L., Sasaki, Y., Sakiyama, K., Ohta, K.: New Truncated Differential Cryptanalysis on 3D Block Cipher. In: M.D. Ryan, B. Smyth, G. Wang (eds.) 8th International Conference on Information Security Practice and Experience (ISPEC), LNCS 7232, pp. 109–125. Springer (2012)
109. Kuhn, U.: Cryptanalysis of Reduced-Round MISTY. In: B. Pfitzmann (ed.) Advances in Cryptology, Eurocrypt, LNCS 2045, pp. 325–339 (2001)
110. Kumar, S., Paar, C., Pelzl, J., Pfeiffer, G., Schimmler, M.: Breaking cipher with COPACOBANA - a Cost-Optimized Parallel Code Breaker. In: L. Goubin, M. Matsui (eds.) Cryptographic Hardware and Embedded Systems (CHES), LNCS 4249, pp. 101–118. Springer (2006)
111. Kusuda, K., Matsumoto, T.: Optimization of Time-Memory Trade-Off Cryptanalysis and its Application to DES, FEAL-32 and Skipjack. IEICE Transactions on Fundamentals **E79-A**(1), 35–48 (1996)
112. Kusuda, K., Matsumoto, T.: A Strength Evaluation of the Data Encryption Standard. IMES discussion paper series, Institute for Monetary and Economic Studies, Bank of Japan, 97E-5 (1997)
113. Lafitte, F., Nakahara Jr, J., Van Heule, D.: Applications of SAT Solvers in Cryptanalysis: Finding Weak Keys and Preimages. Journal of Satisfiability, Boolean Modeling and Computation **9**(1), 1–25 (2014)
114. Lai, X.: On the Design and Security of Block Ciphers. Ph.D. thesis, ETH no. 9752, Swiss Federal Institute of Technology, Zurich (1992)
115. Lai, X.: Higher Order Derivatives and Differential Cryptanalysis. In: Communications and Cryptography: Two Sides of One Tapestry, pp. 227–233. Kluwer Academic Publishers (1994)
116. Lai, X.: On the Design and Security of Block Ciphers, *ETH Series in Information Processing*, vol. 1. Hartung-Gorre Verlag, Konstanz (1995)
117. Lai, X., Massey, J.L., Murphy, S.: Markov Ciphers and Differential Cryptanalysis. In: D.W. Davies (ed.) Advances in Cryptology, Eurocrypt, LNCS 547, pp. 17–38. Springer (1991)
118. Langford, S.K.: Differential-Linear Cryptanalysis and Threshold Signatures. Ph.D. thesis, Stanford University, USA (1995)
119. Langford, S.K., Hellman, M.E.: Differential-Linear Cryptanalysis. In: Y.G. Desmedt (ed.) Advances in Cryptology, Crypto, LNCS 839, pp. 17–25. Springer (1994)
120. Lee, W., Nandi, M., Sarkar, P., Chang, D., Lee, S., Sakurai, K.: A Generalization of PGV-Hash Functions and Security Analysis in Black-Box Model. In: H. Wang,

J. Pieprzyk, V. Varadharajan (eds.) 9th Information Security and Privacy (ACISP), LNCS 3108, pp. 212–223. Springer (2004)
121. Lerman, L., Nakahara Jr, J., Veshchikov, N.: Improving Block Cipher Design by Rearranging Internal Operations. In: P. Samarati (ed.) 10th International Conference on Security and Cryptography (SECRYPT), pp. 27–38. IEEE (2013)
122. Leurent, G.: Cryptanalysis of WIDEA. In: S. Moriai (ed.) Fast Software Encryption (FSE), LNCS 8424, pp. 39–51. Springer (2014)
123. Liu, Z., Gu, D., Zhang, J., Li, W.: Differential-Multiple Lnear Cryptanalysis. In: F. Bao, M. Yung, D. Lin, J. Jiang (eds.) 5th Internatonal Conference on Information Security and Cryptology (Inscrypt), LNCS 6151, pp. 35–49. Springer (2009)
124. Lu, J.: Cryptanalysis of Block Ciphers. Royal Holloway, University of London (2008)
125. Lu, J.: The (Related-key) Impossible Boomerang Attack and its Application to the AES Block Cipher. Designs, Codes and Cryptography **60**, 123–143 (2011)
126. Lu, J.: A Methodology for Differential-Linear Cryptanalysis and its Applications (extended abstract). In: A. Canteaut (ed.) Fast Software Encryption (FSE), LNCS 7549, pp. 69–89. Springer (2012)
127. Lucks, S.: The Saturation Attack - a Bait for Twofish. In: M. Matsui (ed.) Fast Software Encryption (FSE), LNCS 2355, pp. 1–5. Springer (2001)
128. Massacci, F., Marraro, L.: Logical Cryptanalysis as a SAT Problem. Journal of Automated Reasoning **24**(1), 165–203 (2000)
129. Massey, J.L.: SAFER K–64: a Byte-Oriented Block-Ciphering Algorithm. In: R. Anderson (ed.) Fast Software Encryption (FSE), LNCS 809, pp. 1–17. Springer (1994)
130. Matsui, M.: Linear Cryptanalysis of DES Cipher (I), ver. 1.03
131. Matsui, M.: Linear Cryptanalysis Method for DES Cipher. In: T. Helleseth (ed.) Advances in Cryptology, Eurocrypt, LNCS 765, pp. 386–397. Springer (1993)
132. Matsui, M.: The First Experimental Cryptanalysis of the Data Encryption Standard. In: Y.G. Desmedt (ed.) Advances in Cryptology, Crypto, LNCS 839, pp. 1–11. Springer (1994)
133. Matsui, M.: On Correlation between the Order of S-boxes and the Strength of DES. In: A. De Santis (ed.) Advances in Cryptology, Eurocrypt, LNCS 950, pp. 366–375. Springer (1994)
134. Matsui, M.: New Block Encryption Algorithm MISTY. In: E. Biham (ed.) Fast Software Encryption (FSE), LNCS 1267, pp. 54–68. Springer (1997)
135. Matsui, M., Yamagishi, A.: A New Method for Known-Plaintext Attack of FEAL Cipher. In: R.A. Rueppel (ed.) Advances in Cryptology, Eurocrypt, LNCS 658, pp. 81–91. Springer (1993)
136. Menezes, A.J., Van Oorschot, P., Vanstone, S.A.: Handbook of Applied Cryptography. CRC Press (1997)
137. Merkle, R.C.: Fast Software Encryption Functions. In: A.J. Menezes, S.A. Vanstone (eds.) Advances in Cryptology, Crypto, LNCS 537, pp. 476–501. Springer (1991)
138. Meyer, C.H., Matyas, S.M.: Cryptography: a New Dimension in Computer Data Security. Wiley (1982)
139. Minier, M.: An Integral Cryptanalysis against a Five Rounds Version of FOX. Western European Workshop on Research in Cryptography, WEWoRC (2005)
140. Moore, J.H., Simmons, G.J.: Cycle Structure of the DES for Keys having Palindromic (or Antipalindromic) Sequences of Round Keys. IEEE Transactions on Software Engineering **SE**(2), 262–273 (1987)
141. Muller, F., Peyrin, T.: Linear Cryptanalysis of the TSC Family of Stream Ciphers. In: B. Roy (ed.) Advances in Cryptology, Asiacrypt, LNCS 3788, pp. 373–394. Springer (2005)
142. Murphy, S.: The Cryptanalysis of FEAL-4 with 20 Chosen Plaintexts. Journal of Cryptology **2**(3), 145–154 (1990)
143. Nakahara Jr, J.: Cryptanalysis and Design of Block Ciphers. Ph.D. thesis, Dept. Elektrotechniek, ESAT, Katholieke Universiteit Leuven, Belgium (2003)

144. Nakahara Jr, J.: A Linear Analysis of Blowfish and Khufu. In: E. Dawson, S. Duncan (eds.) Information Security Practice and Experience Conference (ISPEC), LNCS 4464, pp. 20–32. Springer (2007)
145. Nakahara Jr, J.: Differential and Linear Attacks on the Full WIDEA-n Block Ciphers (under Weak Keys). In: J. Pieprzyk, A.R. Sadeghi, M. Manulis (eds.) International Conference on Cryptology and Network Security (CANS), LNCS 7712, pp. 56–71. Springer (2012)
146. Nakahara Jr, J.: Cryptanalysis of the Full 8.5-round REESSE3+ Block Cipher. In: K. Lauter, F. Rodriguez-Henriquez (eds.) Progress in Cryptology, Latincrypt, LNCS 9230, pp. 170–186. Springer (2015)
147. Nakahara Jr, J., Barreto, P.S.L.M., Preneel, B., Vandewalle, J., Kim, H.Y.: Square Attacks on Reduced-Round PES and IDEA Block Ciphers. In: B. Macq, J.J. Quisquater (eds.) 23rd Symposium on Information Theory in the Benelux, pp. 187–195 (2002)
148. Nakahara Jr, J., de Freitas, D.S., Phan, R.C.W.: New Multiset Attacks on Rijndael with Large Blocks. In: E. Dawson, S. Vaudenay (eds.) MYCRYPT, LNCS 3715, pp. 277–295. Springer (2005)
149. Nakahara Jr, J., Preneel, B., Vandewalle, J.: A Note on Weak-Keys of PES, IDEA and some Extended Variants. In: C. Boyd, W. Mao (eds.) 6th Information Security Conference (ISC), LNCS 2851, pp. 267–279. Springer (2003)
150. Nakahara Jr, J., Preneel, B., Vandewalle, J.: The Biryukov-Demirci Attack on Reduced-Round Version of IDEA and MESH Ciphers. In: H. Wang, J. Pieprzyk, V. Varadharajan (eds.) 9th Australasian Conference on Information Security and Privacy (ACISP), LNCS 3108, pp. 98–109. Springer (2004)
151. Nakahara Jr, J., Rijmen, V., Preneel, B., Vandewalle, J.: The MESH Block Ciphers. In: K. Chae, M. Yung (eds.) Information Security Applications (WISA), LNCS 2908, pp. 458–473. Springer (2003)
152. Nakahara Jr, J., Sepehrdad, P., Zhang, B., Wang, M.: Linear (Hull) and Algebraic Cryptanalysis of the Block Cipher PRESENT. In: J.A. Garay, A. Otsuka (eds.) 8th International Conference on Cryptology and Network Security (CANS), LNCS 5888, pp. 58–75. Springer (2009)
153. Nakahara, J.J., Preneel, B., Vandewalle, J.: Square Attacks on Reduced-Round PES and IDEA Block Ciphers. 23rd Symposium on Information Theory in the Benelux (2002)
154. NBS: Data Encryption Standard (DES). FIPS PUB 46, Federal Information Processing Standards Publication 46, U.S. Department of Commerce (1977)
155. NBS: Data Encryption Standard (DES). FIPS PUB 46–3, Federal Information Processing Standards Publication 46–3 (1999)
156. Nishimura, K., Sibuya, M.: Probability to Meet-in-the-Middle. Journal of Cryptology **2**(1), 13–22 (1990)
157. NIST: Skipjack and KEA Specification, version 2.0. http://csrc.nist.gov/ (1998)
158. NIST: Secure Hash Standard. Federal Information Processing Standards, FIPS PUB 180-3 (2008)
159. Nyberg, K.: Differentially Uniform Mappings for Cryptography. In: T. Helleseth (ed.) Advances in Cryptology, Eurocrypt, LNCS 765, pp. 55–64. Springer (1993)
160. Nyberg, K.: Linear Approximation of Block Ciphers. In: A. De Santis (ed.) Advances in Cryptology, Eurocrypt, LNCS 950, pp. 439–444. Springer (1995)
161. Preneel, B.: Analysis and Design of Cryptographic Hash Functions. Ph.D. thesis, Dept. Elektrotechniek, ESAT, Katholieke Universiteit Leuven, Belgium (1993)
162. Preneel, B., Govaerts, R., Vandewalle, J.: Hash Functions based on Block Ciphers: a Synthetic Approach. In: D.R. Stinson (ed.) Advances in Cryptology, Crypto, LNCS 773, pp. 368–378. Springer (1993)
163. Quisquater, J.J., Delescaille, J.P.: Other Cycling Tests for DES. In: C. Pomerance (ed.) Advances in Cryptology, Crypto, LNCS 293, pp. 255–256. Springer (1987)

164. Rabin, M.O.: Digitalized Signatures. In: R.A. DeMillo (ed.) Foundations of Secure Computation, pp. 155–166. Academic Press (1978)
165. Raddum, H.: Cryptanalysis of IDEA-X/2. In: T. Johansson (ed.) Fast Software Encryption (FSE), LNCS 2887, pp. 1–8. Springer (2003)
166. Rivest, R.L., Shamir, A.: Payword and Micromint: Two Simple Micropayment Schemes. In: Security Protocols Workshop, pp. 69–87 (1996)
167. Rueppel, R.A.: Analysis and Design of Stream Ciphers. Springer (1986)
168. Sasaki, Y., Aoki, K.: Finding Preimages in Full MD5 Faster than Exhaustive Search. In: A. Joux (ed.) Advances in Cryptology, Eurocrypt, LNCS 5479, pp. 134–152. Springer (2009)
169. Schneier, B.: Description of a New Variable-Length Key, 64-bit Block Cipher (Blowfish). In: R. Anderson (ed.) Fast Software Encryption (FSE), LNCS 809, pp. 191–204. Springer (1994)
170. Shamir, A.: On the Security of DES. In: H.C. Williams (ed.) Advances in Cryptology, Crypto, LNCS 218, pp. 280–281. Springer (1985)
171. Shirai, T., Shibutani, K.: On the Diffusion Matrix Employed in the Whirlpool Hashing Function. The NESSIE project (2003)
172. Soos, M.: CryptoMiniSat 2.5.0. SAT Race competitive event booklet (2010)
173. Su, S., Lu, S.: A 128-bit Block Cipher based on Three Group Arithmetics. IACR ePrint archive 2014/704 (2014)
174. Sun, X., Lai, X.: The Key-Dependent Attack on Block Ciphers. In: M. Matsui (ed.) Advances in Cryptology, Asiacrypt, LNCS 5912, pp. 19–36. Springer (2009)
175. Suzuki, K., Tonien, D., Kurosawa, K., Toyota, K.: Birthday Paradox for Multi-Collisions. In: M.S. Rhee, B. Lee (eds.) Information Security and Cryptology (ICISC), LNCS 4296, pp. 29–40. SPringer (2006)
176. Tardy-Corfdir, A., Gilbert, H.: A Known-Plaintext Attack of FEAL-4 and FEAL-6. In: J. Feigenbaum (ed.) Advances in Cryptology, Crypto, LNCS 576, pp. 172–182. Springer (1992)
177. Todo, Y.: Integral Cryptanalysis on Full Misty1. In: R. Gennaro, M. Robshaw (eds.) Advances in Cryptology, Crypto, LNCS 9215, pp. 413–432. Springer (2015)
178. Vaudenay, S.: On the Weak Keys of Blowfish. In: D. Gollmann (ed.) Fast Software Encryption (FSE), LNCS 1039, pp. 27–32. Springer (2005)
179. Wagner, D.: The Boomerang Attack. In: L.R. Knudsen (ed.) Fast Software Encryption (FSE), LNCS 1636, pp. 156–170. Springer (1999)
180. Wei, L., Peyrin, T., Sokolowski, P., Ling, S., Pieprzyk, J., Wang, H.: On the (In)Security of IDEA in Various Hashing Modes. IACR ePrint archive 2012/264 (2012)
181. Wenling, W., Wentao, Z., Dengguo, F.: Integral Cryptanalysis of Reduced FOX Block Cipher. In: D. Won, S. Kim (eds.) ICISC, LNCS 3935, pp. 229–241. Springer (2006)
182. Wheeler, D.J., Needham, R.M.: TEA, a Tiny Encryption Algorithm. In: B. Preneel (ed.) Fast Software Encryption (FSE), LNCS 1008, pp. 363–366. Springer (1995)
183. Winternitz, R., Hellman, M.E.: Chosen-Key Attacks on a Block Cipher. Cryptologia **11**(1), 16–20 (1987)
184. Wu, H., Bao, F., Deng, R.H., Ye, Q.Z.: Improved Truncated Differential Attacks on SAFER. In: K. Ohta, D. Pei (eds.) Advances in Cryptology, Asiacrypt, LNCS 1514, pp. 133–147. Springer (1998)
185. Wu, H., Preneel, B.: Differential-Linear Attacks against the Stream Cipher Phelix. In: A. Biryukov (ed.) Fast Software Encryption (FSE), LNCS 4593, pp. 87–100 (2007)
186. Wu, Z., Lai, X., Zhu, B., Luo, Y.: Impossible Differential Cryptanalysis of FOX. IACR ePrint archive, 2009/357 (2009)
187. Yıldırım, H.M.: Some Linear Relations for Block Cipher IDEA. Ph.D. thesis, The Middle East Technical University, Turkey (2002)

188. Yıldırım, H.M.: Nonlinearity Properties of the Mixing Operations of the Block Cipher IDEA. In: T. Johansson, S. Maitra (eds.) Progress in Cryptology, Indocrypt, LNCS 2904, pp. 68–81. Springer (2003)
189. Yuval, G.: How to Swindle Rabin. Cryptologia **3**(3), 187–189 (1979)

Chapter 4
New Cipher Designs

Life can only be understood backwards, but it must be lived forwards.
– Søren Kierkegaard

Abstract

This chapter describes new Lai-Massey cipher designs which incorporate new insights and lessons learned from previous attacks including third-party cryptanalyses. Also, preliminary differential and linear cryptanalyses are presented and are used to motivate the minimum number of rounds. Further analyses and attacks are left as open problems.

All new ciphers are iterated designs operating on well-defined w-bit words. The allowed word sizes are w bits with $w \in \{4, 8, 16\}$.

To identify each new design, the ciphers will be designated as XC_i, for $i \geq 1$, meaning the i-th *eXperimental Cipher* design.

4.1 XC_1: Encryption and Decryption Frameworks

The first new design is denoted as XC_1 and operates on $6w$-bit text blocks under a $12w$-bit key and iterates r rounds plus an output transformation, where $r \geq 8$. For $w = 16$ bits, XC_1 operates on 96-bit text blocks and under a 192-bit key. The smaller words size $w \in \{4, 8\}$ correspond to mini-versions of XC_1, which may be suitable for cryptanalysis experiments and attack simulations.

Each full round of XC_1 for encryption can be split into two parts: a key-mixing (KM) and an Xor-Add-Multiply (XAM) half-round, in this order.

XC_1 uses only the three original group operations from PES [2], IDEA [1, 3] and the MESH [5] ciphers, but no S-boxes nor MDS codes. Also, like IDEA* [4], the $\boxdot$ operation is used at the input to the XAM-box.

Let $X^{(i)} = (X_1^{(i)}, X_2^{(i)}, X_3^{(i)}, X_4^{(i)}, X_5^{(i)}, X_6^{(i)})$ denote the input block to the i-th round, $1 \leq i \leq r+1$. Therefore, $X^{(1)}$ denotes the plaintext (block).

The output of the i-th KM half-round is denoted

J. Nakahara Jr., *Lai-Massey Cipher Designs*, https://doi.org/10.1007/978-3-319-68273-0_4

$$Y^{(i)} = (Y_1^{(i)}, Y_2^{(i)}, Y_3^{(i)}, Y_4^{(i)}, Y_5^{(i)}, Y_6^{(i)})$$

where

$$Y^{(i)} = (X_1^{(i)} \oplus Z_1^{(i)}, X_2^{(i)} \boxplus Z_2^{(i)}, X_3^{(i)} \odot Z_3^{(i)}, X_4^{(i)} \odot Z_4^{(i)}, X_5^{(i)} \oplus Z_5^{(i)}, X_5^{(i)} \boxplus Z_6^{(i)})$$

which also becomes the input to the i-th XAM half-round.

The input to the i-th XAM-box is the 3-tuple

$$(Y_1^{(i)} \boxminus Y_4^{(i)}, Y_2^{(i)} \boxdot Y_5^{(i)}, Y_3^{(i)} \oplus Y_6^{(i)}) = (n_i, q_i, m_i).$$

The output of this XAM-box is denoted (r_i, s_i, t_i), where $a_i = n_i \oplus Z_7^{(i)}$; $b_i = a_i \boxplus q_i$; $c_i = b_i \odot m_i$; $d_i = c_i \oplus Z_8^{(i)}$; $e_i = b_i \odot d_i$; $f_i = a_i \boxplus e_i$; $r_i = f_i \oplus Z_9^{(i)}$; $s_i = r_i \boxplus e_i$; $t_i = s_i \odot d_i$.

Note that r_i ultimately depends on all words of (n_i, q_i, m_i). Consequently, r_i depends on all words in $Y^{(i)}$ and in $X^{(i)}$. Therefore, s_i and t_i also depend on all words in $Y^{(i)}$ and in $X^{(i)}$.

Finally, the XAM-box output is combined with the original input words in $Y^{(i)}$ resulting in $(Y_1^{(i)} \boxplus r_i, Y_2^{(i)} \odot s_i, Y_3^{(i)} \oplus t_i, Y_4^{(i)} \boxplus r_i, Y_5^{(i)} \odot s_i, Y_6^{(i)} \oplus t_i)$. Note that each word depends on at least one word of (r_i, s_i, t_i). Thus, complete text diffusion is achieved in every single round.

Note that all three group operations, $\oplus$, $\boxplus$ and $\odot$, are used in all places: both in the KM and XAM half-rounds, both at the input and the output of the XAM-box, and in the XAM-box itself. In fact, there are exactly eight instances of each of these group operations (or their inverses) in every full round.

As a last operation in a full round, there is a fixed word permutation denoted σ. The syntax is $\sigma(a, b, c, d, e, f) = (d, e, f, a, b, c)$. The i-th round output is denoted

$$X^{(i+1)} = (Y_4^{(i)} \boxplus r_i, Y_5^{(i)} \odot s_i, Y_6^{(i)} \oplus t_i, Y_1^{(i)} \boxplus r_i, Y_2^{(i)} \odot s_i, Y_3^{(i)} \oplus t_i).$$

This (full) round transformation is repeated r times. The output transformation consists of σ followed by a KM half-round. Therefore, the 6-tuple

$$Y^{(r+1)} = (X_1^{(r+1)} \oplus Z_1^{(r+1)}, X_2^{(r+1)} \boxplus Z_2^{(r+1)}, X_3^{(r+1)} \odot Z_3^{(r+1)}, X_4^{(r+1)} \odot Z_4^{(r+1)},$$
$$X_5^{(r+1)} \oplus Z_5^{(r+1)}, X_6^{(r+1)} \boxplus Z_6^{(r+1)})$$

denotes the ciphertext (block) of XC_1.

For decryption, the inverse of the i-th KM half-round consists of the reverse order of round subkeys appropriately transformed by the inverse of $\odot$ or $\boxplus$ or $\oplus$.

The inverse of the i-th XAM half-round is similar to the original i-th XAM half-round: the input to the XAM-box is also (n_i, q_i, m_i), but the output (r_i, s_i, t_i) is combined via the inverses of $\boxplus$ and $\odot$ ($\oplus$ is its own inverse).

Therefore, the output of the XAM-box is combined as follows:

$$(Y_1^{(i)} \boxminus r_i, Y_2^{(i)} \boxdot s_i, Y_3^{(i)} \oplus t_i, Y_4^{(i)} \boxminus r_i, Y_5^{(i)} \boxdot s_i, Y_6^{(i)} \oplus t_i).$$

Note that the second and fifth words use $\boxdot$ to invert the original multiplication performed during encryption. Likewise, the first and third words use $\boxplus$. See Fig. 4.1.

If the original input to the i-th XAM-box is (n_i, q_i, m_i), this design will guarantee that this same input reappears for proper decryption:

$$(Y_1^{(i)} \boxplus r_i \boxminus r_i \boxminus (Y_4^{(i)} \boxplus r_i \boxminus r_i), ((Y_2^{(i)} \odot s_i) \boxdot s_i) \boxdot (Y_5^{(i)} \odot s_i \boxdot s_i),$$

$$Y_3^{(i)} \oplus t_i \oplus t_i \oplus Y_6^{(i)} \oplus t_i \oplus t_i) = (n_i, q_i, m_i).$$

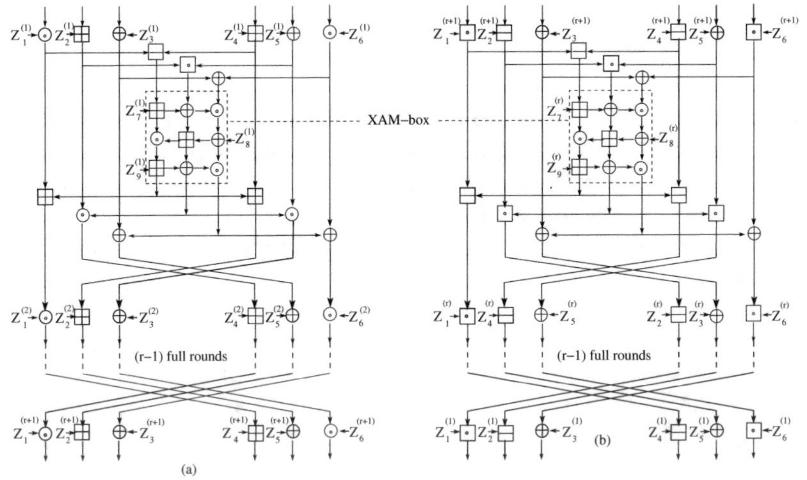

Fig. 4.1 Pictorial representation of: (a) encryption and (b) decryption frameworks of XC_1.

New features of XC_1 include:

- use of all three group operations along the whole cipher framework: in each half-round and both at the input and at the output of the XAM-box.
- no keyed-$\odot$ operations in the XAM-box, so it is not easier to traverse it by assuming weak subkeys. In fact, it is not possible to have weak subkeys in the XAM-box.
- the XAM-box is composed only of $\oplus$, $\boxplus$ and $\odot$ operations. But, subkeys are combined via $\oplus$.
- one type of KM half-round is used for all rounds, unlike in MESH-96 [5] which uses two types of KM half-rounds.
- there are $\oplus$, $\boxplus$ and $\odot$ operations combining the output of the XAM-box at the end of the XAM half-round. This means that the inverse

operations have to be used for decryption. See Fig. 4.1(b). Therefore, the same framework cannot be used interchangeably for encryption and decryption by just changing the order and value of round subkeys. Thus, XC_1 shares this same feature with IDEA* [4].

4.1.1 Key Schedule Algorithm

The key schedule of MESH-96 (Sect. 2.3.4) can be used to generate subkeys for XC_1 since both ciphers operate under 192-bit keys. Moreover, both ciphers require exactly six 16-bit subkeys per round. In general, as many subkeys can be generated for an arbitrary number of rounds r.

4.1.2 Differential Analysis

This section provides preliminary differential analysis of XC_1 (for a variable number of rounds). The analysis approach is bottom-up, from the smallest components $\odot$, $\boxplus$ and $\oplus$, up to larger ones.

The difference operator is bitwise exclusive-or.

Across unkeyed $\odot$ and $\boxplus$, the propagation of 16-bit difference words follows from Tables 3.12 and 3.13, respectively.

The next step is to obtain all difference patterns across the XAM-box of XC_1.

The only 16-bit difference values used are $\delta = 8000_x$ and 0000_x. The difference 8000_x propagates across the modular addition with probability one, since it affects only the most significant bit (and the last carry-out bit is truncated due to modular reduction.

Table 4.1 lists the propagation of wordwise differences across the XAM-box, the corresponding probability and the number of active $\odot$ operations in the differential trail.

Table 4.1 Difference propagation across the XAM-box of XC_1.

Difference propagation across the XAM-box	# active $\odot$	Probability
$(0,0,0) \to (0,0,0)$	0	1
$(0,0,\delta) \to (\delta,0,\delta)$	3	2^{-45}
$(0,\delta,0) \to (0,0,\delta)$	2	2^{-30}
$(0,\delta,\delta) \to (\delta,0,0)$	2	2^{-30}
$(\delta,0,0) \to (\delta,\delta,0)$	3	2^{-45}
$(\delta,0,\delta) \to (0,\delta,\delta)$	3	2^{-45}
$(\delta,\delta,0) \to (\delta,\delta,\delta)$	2	2^{-30}
$(\delta,\delta,\delta) \to (0,\delta,0)$	3	2^{-45}

The next step is to construct 1-round characteristics based on Table 4.1. Tables 4.2 and 4.3 list all nontrivial 1-round differential characteristics of XC_1, under weak-key assumptions, using only wordwise differences 0000_x and 8000_x.

Table 4.2 Nontrivial 1-round characteristics of XC_1 using xor differences and $\delta = 8000_x$.

1-round characteristics	Weak Subkeys (i-th round)	Difference across the XAM-box	Prob.
$(0,0,0,0,0,\delta) \xrightarrow{1r} (\delta,\delta,0,0,\delta,0)$	$Z_6^{(i)}$	$(0,0,\delta) \rightarrow (\delta,0,\delta)$	2^{-45}
$(0,0,0,0,\delta,0) \xrightarrow{1r} (0,0,\delta,0,\delta,\delta)$		$(0,\delta,0) \rightarrow (0,0,\delta)$	2^{-60}
$(0,0,0,0,\delta,\delta) \xrightarrow{1r} (\delta,\delta,\delta,0,0,\delta)$	$Z_6^{(i)}$	$(0,\delta,\delta) \rightarrow (\delta,0,0)$	2^{-60}
$(0,0,0,\delta,0,0) \xrightarrow{1r} (\delta,0,\delta,\delta,0,0)$		$(\delta,0,0) \rightarrow (\delta,\delta,0)$	2^{-75}
$(0,0,0,\delta,0,\delta) \xrightarrow{1r} (0,\delta,\delta,\delta,\delta,0)$	$Z_6^{(i)}$	$(\delta,0,\delta) \rightarrow (0,\delta,\delta)$	2^{-75}
$(0,0,0,\delta,\delta,0) \xrightarrow{1r} (\delta,0,0,\delta,\delta,\delta)$		$(\delta,\delta,0) \rightarrow (\delta,\delta,\delta)$	2^{-75}
$(0,0,0,\delta,\delta,\delta) \xrightarrow{1r} (0,\delta,0,\delta,0,\delta)$	$Z_6^{(i)}$	$(\delta,\delta,\delta) \rightarrow (0,\delta,0)$	2^{-90}
$(0,0,\delta,0,0,0) \xrightarrow{1r} (\delta,\delta,0,0,0,\delta)$		$(0,0,\delta) \rightarrow (\delta,0,\delta)$	2^{-45}
$(0,0,\delta,0,0,\delta) \xrightarrow{1r} (0,0,0,0,\delta,\delta)$	$Z_6^{(i)}$	$(0,0,0) \rightarrow (0,0,0)$	1
$(0,0,\delta,0,\delta,0) \xrightarrow{1r} (\delta,\delta,\delta,0,\delta,0)$		$(0,\delta,\delta) \rightarrow (\delta,0,0)$	2^{-60}
$(0,0,\delta,0,\delta,\delta) \xrightarrow{1r} (0,0,\delta,0,0,0)$	$Z_6^{(i)}$	$(0,\delta,0) \rightarrow (0,0,\delta)$	2^{-60}
$(0,0,\delta,\delta,0,0) \xrightarrow{1r} (0,\delta,\delta,\delta,0,\delta)$		$(\delta,0,\delta) \rightarrow (0,\delta,\delta)$	2^{-75}
$(0,0,\delta,\delta,0,\delta) \xrightarrow{1r} (\delta,0,\delta,\delta,\delta,\delta)$	$Z_6^{(i)}$	$(\delta,0,0) \rightarrow (\delta,\delta,0)$	2^{-75}
$(0,0,\delta,\delta,\delta,0) \xrightarrow{1r} (0,\delta,0,\delta,\delta,0)$		$(\delta,\delta,\delta) \rightarrow (0,\delta,0)$	2^{-90}
$(0,0,\delta,\delta,\delta,\delta) \xrightarrow{1r} (\delta,0,0,\delta,0,0)$	$Z_6^{(i)}$	$(\delta,\delta,0) \rightarrow (\delta,\delta,\delta)$	2^{-75}
$(0,\delta,0,0,0,0) \xrightarrow{1r} (0,0,0,\delta,\delta,\delta)$		$(0,\delta,0) \rightarrow (0,0,\delta)$	2^{-60}
$(0,\delta,0,0,0,\delta) \xrightarrow{1r} (\delta,\delta,0,\delta,0,\delta)$	$Z_6^{(i)}$	$(0,\delta,\delta) \rightarrow (\delta,0,0)$	2^{-60}
$(0,\delta,0,0,\delta,0) \xrightarrow{1r} (0,0,\delta,\delta,0,0)$		$(0,0,0) \rightarrow (0,0,0)$	2^{-45}
$(0,\delta,0,0,\delta,\delta) \xrightarrow{1r} (\delta,\delta,\delta,\delta,\delta,0)$	$Z_6^{(i)}$	$(0,0,\delta) \rightarrow (\delta,0,\delta)$	2^{-90}
$(0,\delta,0,\delta,0,0) \xrightarrow{1r} (\delta,0,\delta,0,\delta,\delta)$		$(\delta,\delta,0) \rightarrow (\delta,\delta,\delta)$	2^{-75}
$(0,\delta,0,\delta,0,\delta) \xrightarrow{1r} (0,\delta,\delta,0,0,\delta)$	$Z_6^{(i)}$	$(\delta,\delta,\delta) \rightarrow (0,\delta,0)$	2^{-90}
$(0,\delta,0,\delta,\delta,0) \xrightarrow{1r} (\delta,0,0,0,0,0)$		$(\delta,0,0) \rightarrow (\delta,\delta,0)$	2^{-90}
$(0,\delta,0,\delta,\delta,\delta) \xrightarrow{1r} (0,\delta,0,0,\delta,0)$	$Z_6^{(i)}$	$(\delta,0,\delta) \rightarrow (0,\delta,\delta)$	2^{-90}
$(0,\delta,\delta,0,0,0) \xrightarrow{1r} (\delta,\delta,0,\delta,\delta,0)$		$(0,\delta,\delta) \rightarrow (\delta,0,0)$	2^{-60}
$(0,\delta,\delta,0,0,\delta) \xrightarrow{1r} (0,0,0,\delta,0,0)$	$Z_6^{(i)}$	$(0,\delta,0) \rightarrow (0,0,\delta)$	2^{-60}
$(0,\delta,\delta,0,\delta,0) \xrightarrow{1r} (\delta,\delta,\delta,\delta,0,\delta)$		$(0,0,\delta) \rightarrow (\delta,0,\delta)$	2^{-90}
$(0,\delta,\delta,0,\delta,\delta) \xrightarrow{1r} (0,0,\delta,\delta,\delta,\delta)$	$Z_6^{(i)}$	$(0,0,0) \rightarrow (0,0,0)$	2^{-45}
$(0,\delta,\delta,\delta,0,0) \xrightarrow{1r} (0,\delta,\delta,0,\delta,0)$		$(\delta,\delta,\delta) \rightarrow (0,\delta,0)$	2^{-90}
$(0,\delta,\delta,\delta,0,\delta) \xrightarrow{1r} (\delta,0,\delta,0,0,0)$	$Z_6^{(i)}$	$(\delta,\delta,0) \rightarrow (\delta,\delta,\delta)$	2^{-75}
$(0,\delta,\delta,\delta,\delta,0) \xrightarrow{1r} (0,\delta,0,0,0,\delta)$		$(\delta,0,\delta) \rightarrow (0,\delta,\delta)$	2^{-90}
$(0,\delta,\delta,\delta,\delta,\delta) \xrightarrow{1r} (\delta,0,0,0,\delta,\delta)$	$Z_6^{(i)}$	$(\delta,0,0) \rightarrow (\delta,\delta,0)$	2^{-90}
$(\delta,0,0,0,0,0) \xrightarrow{1r} (0,\delta,\delta,\delta,0,0)$	$Z_1^{(i)}$	$(\delta,0,0) \rightarrow (\delta,\delta,0)$	2^{-75}

Table 4.3 Nontrivial 1-round characteristics of XC_1 using xor differences and $\delta = 8000_x$.

1-round characteristics	Weak Subkeys (i-th round)	Difference across the MA-box	Prob.
$(\delta,0,0,0,0,\delta) \stackrel{1r}{\rightarrow} (\delta,0,\delta,\delta,\delta,0)$	$Z_1^{(i)}, Z_6^{(i)}$	$(\delta,0,\delta) \rightarrow (0,\delta,\delta)$	2^{-75}
$(\delta,0,0,0,\delta,0) \stackrel{1r}{\rightarrow} (0,\delta,0,\delta,\delta,\delta)$	$Z_1^{(i)}$	$(\delta,\delta,0) \rightarrow (\delta,\delta,\delta)$	2^{-75}
$(\delta,0,0,0,\delta,\delta) \stackrel{1r}{\rightarrow} (\delta,0,0,\delta,0,\delta)$	$Z_1^{(i)}, Z_6^{(i)}$	$(\delta,\delta,\delta) \rightarrow (0,\delta,0)$	2^{-90}
$(\delta,0,0,\delta,0,0) \stackrel{1r}{\rightarrow} (\delta,\delta,0,0,0,0)$	$Z_1^{(i)}$	$(0,0,0) \rightarrow (0,0,0)$	1
$(\delta,0,0,\delta,0,\delta) \stackrel{1r}{\rightarrow} (0,0,0,0,\delta,0)$	$Z_1^{(i)}, Z_6^{(i)}$	$(0,0,\delta) \rightarrow (\delta,0,\delta)$	2^{-45}
$(\delta,0,0,\delta,\delta,0) \stackrel{1r}{\rightarrow} (\delta,\delta,\delta,0,\delta,\delta)$	$Z_1^{(i)}$	$(0,\delta,0) \rightarrow (0,0,\delta)$	2^{-60}
$(\delta,0,0,\delta,\delta,\delta) \stackrel{1r}{\rightarrow} (0,0,\delta,0,0,\delta)$	$Z_1^{(i)}, Z_6^{(i)}$	$(0,\delta,\delta) \rightarrow (\delta,0,0)$	2^{-60}
$(\delta,0,\delta,0,0,0) \stackrel{1r}{\rightarrow} (\delta,0,\delta,\delta,0,\delta)$	$Z_1^{(i)}$	$(\delta,0,\delta) \rightarrow (0,\delta,\delta)$	2^{-75}
$(\delta,0,\delta,0,0,\delta) \stackrel{1r}{\rightarrow} (0,\delta,\delta,\delta,\delta,\delta)$	$Z_1^{(i)}, Z_6^{(i)}$	$(\delta,0,0) \rightarrow (\delta,\delta,0)$	2^{-75}
$(\delta,0,\delta,0,\delta,0) \stackrel{1r}{\rightarrow} (\delta,0,0,\delta,\delta,0)$	$Z_1^{(i)}$	$(\delta,\delta,\delta) \rightarrow (0,\delta,0)$	2^{-90}
$(\delta,0,\delta,0,\delta,\delta) \stackrel{1r}{\rightarrow} (0,\delta,0,\delta,0,0)$	$Z_1^{(i)}, Z_6^{(i)}$	$(\delta,\delta,0) \rightarrow (\delta,\delta,\delta)$	2^{-75}
$(\delta,0,\delta,\delta,0,0) \stackrel{1r}{\rightarrow} (0,0,0,0,0,\delta)$	$Z_1^{(i)}$	$(0,0,\delta) \rightarrow (\delta,0,\delta)$	2^{-45}
$(\delta,0,\delta,\delta,0,\delta) \stackrel{1r}{\rightarrow} (\delta,\delta,0,0,\delta,\delta)$	$Z_1^{(i)}, Z_6^{(i)}$	$(0,0,0) \rightarrow (0,0,0)$	1
$(\delta,0,\delta,\delta,\delta,0) \stackrel{1r}{\rightarrow} (0,0,\delta,0,\delta,0)$	$Z_1^{(i)}$	$(0,\delta,\delta) \rightarrow (\delta,0,0)$	2^{-60}
$(\delta,0,\delta,\delta,\delta,\delta) \stackrel{1r}{\rightarrow} (\delta,\delta,\delta,0,0,0)$	$Z_1^{(i)}, Z_6^{(i)}$	$(0,\delta,0) \rightarrow (0,0,\delta)$	2^{-60}
$(\delta,\delta,0,0,0,0) \stackrel{1r}{\rightarrow} (0,\delta,\delta,0,\delta,\delta)$	$Z_1^{(i)}$	$(\delta,\delta,0) \rightarrow (\delta,\delta,\delta)$	2^{-75}
$(\delta,\delta,0,0,0,\delta) \stackrel{1r}{\rightarrow} (\delta,0,\delta,0,0,\delta)$	$Z_1^{(i)}, Z_6^{(i)}$	$(\delta,\delta,\delta) \rightarrow (0,\delta,0)$	2^{-90}
$(\delta,\delta,0,0,\delta,0) \stackrel{1r}{\rightarrow} (0,\delta,0,0,0,0)$	$Z_1^{(i)}$	$(\delta,0,0) \rightarrow (\delta,\delta,0)$	2^{-90}
$(\delta,\delta,0,0,\delta,\delta) \stackrel{1r}{\rightarrow} (\delta,0,0,0,\delta,0)$	$Z_1^{(i)}, Z_6^{(i)}$	$(\delta,0,\delta) \rightarrow (0,\delta,\delta)$	2^{-90}
$(\delta,\delta,0,\delta,0,0) \stackrel{1r}{\rightarrow} (\delta,\delta,0,\delta,\delta,\delta)$	$Z_1^{(i)}$	$(0,\delta,0) \rightarrow (0,0,\delta)$	2^{-60}
$(\delta,\delta,0,\delta,0,\delta) \stackrel{1r}{\rightarrow} (0,0,0,\delta,0,\delta)$	$Z_1^{(i)}, Z_6^{(i)}$	$(0,\delta,\delta) \rightarrow (\delta,0,0)$	2^{-60}
$(\delta,\delta,0,\delta,\delta,0) \stackrel{1r}{\rightarrow} (\delta,\delta,\delta,\delta,0,0)$	$Z_1^{(i)}$	$(0,0,0) \rightarrow (0,0,0)$	2^{-45}
$(\delta,\delta,0,\delta,\delta,\delta) \stackrel{1r}{\rightarrow} (0,0,\delta,\delta,\delta,0)$	$Z_1^{(i)}, Z_6^{(i)}$	$(0,0,\delta) \rightarrow (\delta,0,\delta)$	2^{-90}
$(\delta,\delta,\delta,0,0,0) \stackrel{1r}{\rightarrow} (\delta,0,\delta,0,\delta,0)$	$Z_1^{(i)}$	$(\delta,\delta,\delta) \rightarrow (0,\delta,0)$	2^{-90}
$(\delta,\delta,\delta,0,0,\delta) \stackrel{1r}{\rightarrow} (0,\delta,\delta,0,0,0)$	$Z_1^{(i)}, Z_6^{(i)}$	$(\delta,\delta,0) \rightarrow (\delta,\delta,\delta)$	2^{-75}
$(\delta,\delta,\delta,0,\delta,0) \stackrel{1r}{\rightarrow} (\delta,0,0,0,0,\delta)$	$Z_1^{(i)}$	$(\delta,0,\delta) \rightarrow (0,\delta,\delta)$	2^{-90}
$(\delta,\delta,\delta,0,\delta,\delta) \stackrel{1r}{\rightarrow} (0,\delta,0,0,\delta,\delta)$	$Z_1^{(i)}, Z_6^{(i)}$	$(\delta,0,0) \rightarrow (\delta,\delta,0)$	2^{-90}
$(\delta,\delta,\delta,\delta,0,0) \stackrel{1r}{\rightarrow} (0,0,0,\delta,\delta,0)$	$Z_1^{(i)}$	$(0,\delta,\delta) \rightarrow (\delta,0,0)$	2^{-60}
$(\delta,\delta,\delta,\delta,0,\delta) \stackrel{1r}{\rightarrow} (\delta,\delta,0,\delta,0,0)$	$Z_1^{(i)}, Z_6^{(i)}$	$(0,\delta,0) \rightarrow (0,0,\delta)$	2^{-60}
$(\delta,\delta,\delta,\delta,\delta,0) \stackrel{1r}{\rightarrow} (0,0,\delta,\delta,0,\delta)$	$Z_1^{(i)}$	$(0,0,\delta) \rightarrow (\delta,0,\delta)$	2^{-90}
$(\delta,\delta,\delta,\delta,\delta,\delta) \stackrel{1r}{\rightarrow} (\delta,\delta,\delta,\delta,\delta,\delta)$	$Z_1^{(i)}, Z_6^{(i)}$	$(0,0,0) \rightarrow (0,0,0)$	2^{-45}

Multiple-round characteristics are obtained by concatenating 1-round characteristics such that the output difference from one characteristic matches the input difference of the next characteristic.

From Tables 4.2 and 4.3, the following characteristics can be derived.

(1) $(0,0,0,0,0,\delta) \xrightarrow{1r} (\delta,\delta,0,0,\delta,0) \xrightarrow{1r} (0,\delta,0,0,0,0) \xrightarrow{1r} (0, 0, 0, \delta, \delta, \delta) \xrightarrow{1r} (0, \delta, 0, \delta, 0, \delta) \xrightarrow{1r} (0,\delta,\delta,0,0,\delta) \xrightarrow{1r} (0,0,0,\delta,0,0) \xrightarrow{1r} (\delta,0,\delta,\delta,0,0) \xrightarrow{1r} (0,0,0,0,0,\delta)$, which is 8-round iterative and holds with probability 2^{-555}.
The following six subkeys have to be weak: $Z_6^{(i)}$, $Z_1^{(i+1)}$, $Z_6^{(i+3)}$, $Z_6^{(i+4)}$, $Z_6^{(i+5)}$ and $Z_1^{(i+7)}$.

(2) $(0,0,0,0,\delta,0) \xrightarrow{1r} (0,0,\delta,0,\delta,\delta) \xrightarrow{1r} (0,0,\delta,0,0,0) \xrightarrow{1r} (\delta, \delta, 0, 0, 0, \delta) \xrightarrow{1r} (\delta, 0, \delta, 0, 0, \delta) \xrightarrow{1r} (0,\delta,\delta,\delta,\delta,\delta) \xrightarrow{1r} (\delta,0,0,0,\delta,\delta) \xrightarrow{1r} (\delta,0,0,\delta,0,\delta) \xrightarrow{1r} (0,0,0,0,\delta,0)$, which is 8-round iterative and holds with probability 2^{-555}.
The following ten subkeys have to be weak: $Z_6^{(i+1)}$, $Z_1^{(i+3)}$, $Z_6^{(i+3)}$, $Z_1^{(i+4)}$, $Z_6^{(i+4)}$, $Z_6^{(i+5)}$, $Z_1^{(i+6)}$, $Z_6^{(i+6)}$, $Z_1^{(i+7)}$ and $Z_6^{(i+7)}$.

(3) $(0,0,0,0,\delta,\delta) \xrightarrow{1r} (\delta,\delta,\delta,0,0,\delta) \xrightarrow{1r} (0,\delta,\delta,0,0,0) \xrightarrow{1r} (\delta,\delta,0,\delta,\delta,0) \xrightarrow{1r} (\delta, \delta, \delta, \delta, 0, 0) \xrightarrow{1r} (0,0,0,\delta,\delta,0) \xrightarrow{1r} (\delta,0,0,\delta,\delta,\delta) \xrightarrow{1r} (0,0,\delta,0,0,\delta) \xrightarrow{1r} (0,0,0,0,\delta,\delta)$, which is 8-round iterative and holds with probability 2^{-435}.
The following eight subkeys have to be weak: $Z_6^{(i)}$, $Z_1^{(i+1)}$, $Z_6^{(i+1)}$, $Z_1^{(i+3)}$, $Z_1^{(i+4)}$, $Z_1^{(i+6)}$, $Z_6^{(i+6)}$ and $Z_6^{(i+7)}$.

(4) $(0,0,0,\delta,0,\delta) \xrightarrow{1r} (0,\delta,\delta,\delta,\delta,0) \xrightarrow{1r} (0,\delta,0,0,0,\delta) \xrightarrow{1r} (\delta,\delta,0,\delta,0,\delta) \xrightarrow{1r} (0,0,0,\delta,0,\delta)$, which is 4-round iterative and holds with probability 2^{-285}.
The following four subkeys have to be weak: $Z_6^{(i)}$, $Z_6^{(i+2)}$, $Z_1^{(i+3)}$ and $Z_6^{(i+3)}$.

(5) $(0,0,\delta,0,\delta,0) \xrightarrow{1r} (\delta,\delta,\delta,0,\delta,0) \xrightarrow{1r} (\delta,0,0,0,0,\delta) \xrightarrow{1r} (\delta,0,\delta,\delta,\delta,0) \xrightarrow{1r} (0, 0, \delta, 0, \delta, 0)$, which is 4-round iterative and holds with probability 2^{-285}.
The following four subkeys have to be weak: $Z_1^{(i+1)}$, $Z_1^{(i+2)}$, $Z_6^{(i+2)}$ and $Z_1^{(i+3)}$.

(6) $(0,0,\delta,\delta,0,0) \xrightarrow{1r} (0,\delta,\delta,\delta,0,\delta) \xrightarrow{1r} (\delta,0,\delta,0,0,0) \xrightarrow{1r} (\delta,0,\delta,\delta,0,\delta) \xrightarrow{1r} (\delta, \delta, 0, 0, \delta, \delta) \xrightarrow{1r} (\delta,0,0,0,\delta,0) \xrightarrow{1r} (0,\delta,0,\delta,\delta,\delta) \xrightarrow{1r} (0, \delta, 0, 0, \delta, 0) \xrightarrow{1r} (0,0,\delta,\delta,0,0)$, which is 8-round iterative and holds with probability 2^{-525}.
The following eight subkeys have to be weak: $Z_6^{(i+1)}$, $Z_1^{(i+2)}$, $Z_1^{(i+3)}$, $Z_6^{(i+3)}$, $Z_1^{(i+4)}$, $Z_6^{(i+4)}$, $Z_1^{(i+5)}$ and $Z_6^{(i+6)}$.

(7) $(0,0,\delta,\delta,0,\delta) \xrightarrow{1r} (\delta,0,\delta,\delta,\delta,\delta) \xrightarrow{1r} (\delta,\delta,\delta,0,0,0) \xrightarrow{1r} (\delta,0,\delta,0,\delta,0) \xrightarrow{1r} (\delta, 0, 0, \delta, \delta, 0) \xrightarrow{1r} (\delta,\delta,\delta,0,\delta,\delta) \xrightarrow{1r} (0,\delta,0,0,\delta,\delta) \xrightarrow{1r} (\delta,\delta,\delta,\delta,\delta,0) \xrightarrow{1r} (0, 0, \delta, \delta, 0, \delta)$, which is 8-round iterative and holds with probability 2^{-645}.

The following ten subkeys have to be weak: $Z_6^{(i)}$, $Z_1^{(i+1)}$, $Z_6^{(i+1)}$, $Z_1^{(i+2)}$, $Z_1^{(i+3)}$, $Z_1^{(i+4)}$, $Z_1^{(i+5)}$, $Z_6^{(i+5)}$, $Z_6^{(i+6)}$ and $Z_1^{(i+7)}$.

(8) $(0,0,\delta,\delta,\delta,0) \stackrel{1r}{\to} (0,\delta,0,\delta,\delta,0) \stackrel{1r}{\to} (\delta,0,0,0,0,0) \stackrel{1r}{\to} (0,\delta,\delta,\delta,0,0) \stackrel{1r}{\to} (0,\ \delta,\ \delta,\ 0,\ \delta,\ 0) \stackrel{1r}{\to} (\delta,\delta,\delta,\delta,0,\delta) \stackrel{1r}{\to} (\delta,\delta,0,\delta,0,0) \stackrel{1r}{\to} (\delta,\delta,0,\delta,\delta,\delta) \stackrel{1r}{\to} (0,0,\delta,\delta,\delta,0)$, which is 8-round iterative and holds with probability 2^{-645}.
The following six subkeys have to be weak: $Z_1^{(i+2)}$, $Z_1^{(i+5)}$, $Z_6^{(i+5)}$, $Z_1^{(i+6)}$, $Z_1^{(i+7)}$ and $Z_6^{(i+7)}$.

(9) $(0,0,\delta,\delta,\delta,\delta) \stackrel{1r}{\to} (\delta,0,0,\delta,0,0) \stackrel{1r}{\to} (\delta,\delta,0,0,0,0) \stackrel{1r}{\to} (0,\delta,\delta,0,\delta,\delta) \stackrel{1r}{\to} (0,\ 0,\ \delta,\ \delta,\ \delta,\ \delta)$, which is 4-round iterative and holds with probability 2^{-195}.
The following four subkeys have to be weak: $Z_6^{(i)}$, $Z_1^{(i+1)}$, $Z_1^{(i+2)}$ and $Z_6^{(i+3)}$.

(10) $(0,\delta,0,\delta,0,0) \stackrel{1r}{\to} (\delta,0,\delta,0,\delta,\delta) \stackrel{1r}{\to} (0,\delta,0,\delta,0,0)$, which is 2-round iterative and holds with probability 2^{-150}.
The following two subkeys have to be weak: $Z_1^{(i+1)}$ and $Z_6^{(i+1)}$.

(11) $(\delta,\delta,\delta,\delta,\delta,\delta) \stackrel{1r}{\to} (\delta,\delta,\delta,\delta,\delta,\delta)$, which is 1-round iterative and holds with probability 2^{-45}.
The following two subkeys have to be weak: $Z_1^{(i)}$ and $Z_6^{(i)}$.

The characteristics requiring the least number of weak-subkey assumptions per round are (1) and (8), but their probabilities are too low. Even the whole codebook could not provide enough pairs to identify any of them.

For the key schedule of XC_1, using a 192-bit key, assuming each weak-subkey assumption holds with probability 2^{-15}, then at most $192/15 = 13$ weak subkeys will be allowed.

The best trade-off involving the highest (average) probability per round and the lowest number of weak subkeys per round is (11). It can cover up to 4-round XC_1 with probability 2^{-180}, and requiring eight weak subkeys. The weak-key class is estimated as $2^{192-15\cdot 8} = 2^{72}$. The data complexity to distinguish 4-round XC_1 from a random permutation is estimated as $c \cdot 2^{180}$ CP, where c is a fixed constant.

4.1.3 Linear Analysis

This section provides preliminary linear cryptanalysis of XC_1. The analysis approach is bottom-up, from the smallest components $\odot$, $\boxplus$ and $\oplus$, up to larger ones.

Table 3.98 lists the bitmask propagation and the corresponding biases across the unkeyed-$\odot$ operation (under wordwise bitmasks 0 and $\gamma = 1$).

Table 3.99 lists the bitmask propagation and the corresponding biases across $\boxplus$ (under wordwise bitmasks 0 and 1).

The next step is to obtain linear relations across the XAM-box. Table 4.4 lists the bitmasks, the corresponding bias and the number of active $\odot$ along the linear trail.

Table 4.4 Linear mask propagation across the XAM-box of XC_1.

Bitmask propagation across the XAM-box	# active $\odot$	Bias
$(0,0,0) \rightarrow (0,0,0)$	0	2^{-1}
$(0,\gamma,\gamma) \rightarrow (0,0,\gamma)$	3	$2^{-38.41}$
$(\gamma,0,0) \rightarrow (0,\gamma,0)$	1	$2^{-13.47}$
$(\gamma,\gamma,\gamma) \rightarrow (0,\gamma,\gamma)$	2	$2^{-25.94}$
$(\gamma,0,\gamma) \rightarrow (\gamma,0,0)$	2	$2^{-25.94}$
$(\gamma,\gamma,0) \rightarrow (\gamma,0,\gamma)$	1	$2^{-13.47}$
$(0,0,\gamma) \rightarrow (\gamma,\gamma,0)$	1	$2^{-13.47}$
$(0,\gamma,0) \rightarrow (\gamma,\gamma,\gamma)$	2	$2^{-25.94}$

The next step is to construct 1-round linear relations based on Table 4.4. Tables 4.5 and 4.6 list all nontrivial 1-round linear relations for XC_1 under weak-key assumptions and using only bitmasks 0 and $\gamma = 1$.

From Tables 4.5 and 4.6, the following eleven linear relations can be derived:

(1) $(0,0,0,0,0,\gamma) \stackrel{1r}{\rightarrow} (0,0,\gamma,\gamma,\gamma,0) \stackrel{1r}{\rightarrow} (0,\gamma,0,\gamma,\gamma,0) \stackrel{1r}{\rightarrow} (\gamma,0,\gamma,\gamma,\gamma,\gamma) \stackrel{1r}{\rightarrow} (0,0,\gamma,0,\gamma,\gamma) \stackrel{1r}{\rightarrow} (\gamma,\gamma,\gamma,0,\gamma,\gamma) \stackrel{1r}{\rightarrow} (0,\gamma,\gamma,\gamma,0,0) \stackrel{1r}{\rightarrow} (0,\gamma,\gamma,0,\gamma,0) \stackrel{1r}{\rightarrow} (0,0,0,0,0,\gamma)$, which is 8-round iterative and holds with bias $2^{-375.10}$. The following six subkeys have to be weak: $Z_6^{(i)}$, $Z_1^{(i+3)}$, $Z_6^{(i+3)}$, $Z_6^{(i+4)}$, $Z_1^{(i+5)}$ and $Z_6^{(i+5)}$.

(2) $(0,0,0,0,\gamma,0) \stackrel{1r}{\rightarrow} (\gamma,\gamma,\gamma,0,0,0) \stackrel{1r}{\rightarrow} (\gamma,0,\gamma,0,\gamma,0) \stackrel{1r}{\rightarrow} (\gamma,0,0,\gamma,\gamma,0) \stackrel{1r}{\rightarrow} (0,0,\gamma,0,0,0) \stackrel{1r}{\rightarrow} (0,0,\gamma,\gamma,0,\gamma) \stackrel{1r}{\rightarrow} (\gamma,0,0,0,0,0) \stackrel{1r}{\rightarrow} (0,\gamma,0,0,\gamma,\gamma) \stackrel{1r}{\rightarrow} (0,0,0,0,\gamma,0)$, which is 8-round iterative and holds with bias $2^{-325.22}$. The following six subkeys have to be weak: $Z_1^{(i+1)}$, $Z_1^{(i+2)}$, $Z_1^{(i+3)}$, $Z_6^{(i+5)}$, $Z_1^{(i+6)}$ and $Z_6^{(i+7)}$.

(3) $(0,0,0,0,\gamma,\gamma) \stackrel{1r}{\rightarrow} (\gamma,\gamma,0,\gamma,\gamma,0) \stackrel{1r}{\rightarrow} (\gamma,\gamma,\gamma,\gamma,0,0) \stackrel{1r}{\rightarrow} (0,0,\gamma,0,0,\gamma) \stackrel{1r}{\rightarrow} (0,0,0,0,\gamma,\gamma)$, which is 4-round iterative and holds with bias $2^{-125.70}$. The following four subkeys have to be weak: $Z_6^{(i)}$, $Z_1^{(i+2)}$, $Z_1^{(i+3)}$ and $Z_6^{(i+4)}$.

(4) $(0,0,0,\gamma,0,0) \stackrel{1r}{\rightarrow} (\gamma,0,0,0,\gamma,\gamma) \stackrel{1r}{\rightarrow} (\gamma,0,0,\gamma,0,\gamma) \stackrel{1r}{\rightarrow} (\gamma,\gamma,\gamma,\gamma,\gamma,0) \stackrel{1r}{\rightarrow} (\gamma,\gamma,0,0,0,\gamma) \stackrel{1r}{\rightarrow} (\gamma,0,\gamma,0,0,\gamma) \stackrel{1r}{\rightarrow} (0,\gamma,0,0,0,0) \stackrel{1r}{\rightarrow} (\gamma,\gamma,0,\gamma,0,0) \stackrel{1r}{\rightarrow} (0,0,0,\gamma,0,0)$, which is 8-round iterative and holds with bias $2^{-325.22}$. The following ten subkeys have to be weak: $Z_1^{(i+1)}$, $Z_6^{(i+1)}$, $Z_1^{(i+2)}$, $Z_6^{(i+2)}$, $Z_1^{(i+3)}$, $Z_1^{(i+4)}$, $Z_6^{i+4)}$, $Z_1^{(i+5)}$, $Z_6^{(i+5)}$ and $Z_1^{(i+7)}$.

Table 4.5 Nontrivial 1-round linear relations of XC_1 using bitmasks 0 and $\gamma = 1$.

1-round linear relation	Weak Subkeys (i-th round)	Bitmasks across the XAM-box	Bias
$(0,0,0,0,0,\gamma) \stackrel{1r}{\rightarrow} (0,0,\gamma,\gamma,\gamma,0)$	$Z_6^{(i)}$	$(0,\gamma,\gamma) \rightarrow (0,0,\gamma)$	$2^{-75.82}$
$(0,0,0,0,\gamma,0) \stackrel{1r}{\rightarrow} (\gamma,\gamma,\gamma,0,0,0)$		$(\gamma,0,0) \rightarrow (0,\gamma,0)$	$2^{-25.94}$
$(0,0,0,0,\gamma,\gamma) \stackrel{1r}{\rightarrow} (\gamma,\gamma,0,\gamma,\gamma,0)$	$Z_6^{(i)}$	$(\gamma,\gamma,\gamma) \rightarrow (0,\gamma,\gamma)$	$2^{-50.88}$
$(0,0,0,\gamma,0,0) \stackrel{1r}{\rightarrow} (\gamma,0,0,0,\gamma,\gamma)$		$(\gamma,0,\gamma) \rightarrow (\gamma,0,0)$	$2^{-25.94}$
$(0,0,0,\gamma,0,\gamma) \stackrel{1r}{\rightarrow} (\gamma,0,\gamma,\gamma,0,\gamma)$	$Z_6^{(i)}$	$(\gamma,\gamma,0) \rightarrow (\gamma,0,\gamma)$	$2^{-50.88}$
$(0,0,0,\gamma,\gamma,0) \stackrel{1r}{\rightarrow} (0,\gamma,\gamma,0,\gamma,\gamma)$		$(0,0,\gamma) \rightarrow (\gamma,\gamma,0)$	$2^{-25.94}$
$(0,0,0,\gamma,\gamma,\gamma) \stackrel{1r}{\rightarrow} (0,\gamma,0,\gamma,0,\gamma)$	$Z_6^{(i)}$	$(0,\gamma,0) \rightarrow (\gamma,\gamma,\gamma)$	$2^{-50.88}$
$(0,0,\gamma,0,0,0) \stackrel{1r}{\rightarrow} (0,0,\gamma,\gamma,0,\gamma)$		$(0,\gamma,\gamma) \rightarrow (0,0,\gamma)$	$2^{-75.82}$
$(0,0,\gamma,0,0,\gamma) \stackrel{1r}{\rightarrow} (0,0,0,0,\gamma,\gamma)$	$Z_6^{(i)}$	$(0,0,0) \rightarrow (0,0,0)$	2^{-1}
$(0,0,\gamma,0,\gamma,0) \stackrel{1r}{\rightarrow} (\gamma,\gamma,0,\gamma,0,\gamma)$		$(\gamma,\gamma,\gamma) \rightarrow (0,\gamma,\gamma)$	$2^{-50.88}$
$(0,0,\gamma,0,\gamma,\gamma) \stackrel{1r}{\rightarrow} (\gamma,\gamma,\gamma,0,\gamma,\gamma)$	$Z_6^{(i)}$	$(\gamma,0,0) \rightarrow (0,\gamma,0)$	$2^{-25.94}$
$(0,0,\gamma,\gamma,0,0) \stackrel{1r}{\rightarrow} (\gamma,0,\gamma,\gamma,\gamma,0)$		$(\gamma,\gamma,0) \rightarrow (\gamma,0,\gamma)$	$2^{-50.88}$
$(0,0,\gamma,\gamma,0,\gamma) \stackrel{1r}{\rightarrow} (\gamma,0,0,0,0,0)$	$Z_6^{(i)}$	$(\gamma,0,\gamma) \rightarrow (\gamma,0,0)$	$2^{-25.94}$
$(0,0,\gamma,\gamma,\gamma,0) \stackrel{1r}{\rightarrow} (0,\gamma,0,\gamma,\gamma,0)$		$(0,\gamma,0) \rightarrow (\gamma,\gamma,\gamma)$	$2^{-50.88}$
$(0,0,\gamma,\gamma,\gamma,\gamma) \stackrel{1r}{\rightarrow} (0,\gamma,\gamma,0,0,0)$	$Z_6^{(i)}$	$(0,0,\gamma) \rightarrow (\gamma,\gamma,0)$	$2^{-25.94}$
$(0,\gamma,0,0,0,0) \stackrel{1r}{\rightarrow} (\gamma,\gamma,0,\gamma,0,0)$		$(\gamma,0,0) \rightarrow (0,\gamma,0)$	$2^{-25.94}$
$(0,\gamma,0,0,0,\gamma) \stackrel{1r}{\rightarrow} (\gamma,\gamma,\gamma,0,\gamma,0)$	$Z_6^{(i)}$	$(\gamma,\gamma,\gamma) \rightarrow (0,\gamma,\gamma)$	$2^{-50.88}$
$(0,\gamma,0,0,\gamma,0) \stackrel{1r}{\rightarrow} (0,0,\gamma,\gamma,0,0)$		$(0,0,0) \rightarrow (0,0,0)$	$2^{-25.94}$
$(0,\gamma,0,0,\gamma,\gamma) \stackrel{1r}{\rightarrow} (0,0,0,0,\gamma,0)$	$Z_6^{(i)}$	$(0,\gamma,\gamma) \rightarrow (0,0,\gamma)$	$2^{-50.88}$
$(0,\gamma,0,\gamma,0,0) \stackrel{1r}{\rightarrow} (0,\gamma,0,\gamma,\gamma,\gamma)$		$(0,0,\gamma) \rightarrow (\gamma,\gamma,0)$	$2^{-25.94}$
$(0,\gamma,0,\gamma,0,\gamma) \stackrel{1r}{\rightarrow} (0,\gamma,\gamma,0,0,\gamma)$	$Z_6^{(i)}$	$(0,\gamma,0) \rightarrow (\gamma,\gamma,\gamma)$	$2^{-50.88}$
$(0,\gamma,0,\gamma,\gamma,0) \stackrel{1r}{\rightarrow} (\gamma,0,\gamma,\gamma,\gamma,\gamma)$		$(\gamma,0,\gamma) \rightarrow (\gamma,0,0)$	$2^{-50.88}$
$(0,\gamma,0,\gamma,\gamma,\gamma) \stackrel{1r}{\rightarrow} (\gamma,0,0,0,0,\gamma)$	$Z_6^{(i)}$	$(\gamma,\gamma,0) \rightarrow (\gamma,0,\gamma)$	$2^{-25.94}$
$(0,\gamma,\gamma,0,0,0) \stackrel{1r}{\rightarrow} (\gamma,\gamma,\gamma,0,0,\gamma)$		$(\gamma,\gamma,\gamma) \rightarrow (0,\gamma,\gamma)$	$2^{-50.88}$
$(0,\gamma,\gamma,0,0,\gamma) \stackrel{1r}{\rightarrow} (\gamma,\gamma,0,\gamma,\gamma,\gamma)$	$Z_6^{(i)}$	$(\gamma,0,0) \rightarrow (0,\gamma,0)$	$2^{-25.94}$
$(0,\gamma,\gamma,0,\gamma,0) \stackrel{1r}{\rightarrow} (0,0,0,0,0,\gamma)$		$(0,\gamma,\gamma) \rightarrow (0,0,\gamma)$	$2^{-50.88}$
$(0,\gamma,\gamma,0,\gamma,\gamma) \stackrel{1r}{\rightarrow} (0,0,\gamma,\gamma,\gamma,\gamma)$	$Z_6^{(i)}$	$(0,0,0) \rightarrow (0,0,0)$	$2^{-25.94}$
$(0,\gamma,\gamma,\gamma,0,0) \stackrel{1r}{\rightarrow} (0,\gamma,\gamma,0,\gamma,0)$		$(0,\gamma,0) \rightarrow (\gamma,\gamma,\gamma)$	$2^{-50.88}$
$(0,\gamma,\gamma,\gamma,0,\gamma) \stackrel{1r}{\rightarrow} (0,\gamma,0,\gamma,0,0)$	$Z_6^{(i)}$	$(0,0,\gamma) \rightarrow (\gamma,\gamma,0)$	$2^{-25.94}$
$(0,\gamma,\gamma,\gamma,\gamma,0) \stackrel{1r}{\rightarrow} (\gamma,0,0,0,\gamma,0)$		$(\gamma,\gamma,0) \rightarrow (\gamma,0,\gamma)$	$2^{-25.94}$
$(0,\gamma,\gamma,\gamma,\gamma,\gamma) \stackrel{1r}{\rightarrow} (\gamma,0,\gamma,\gamma,0,0)$	$Z_6^{(i)}$	$(\gamma,0,\gamma) \rightarrow (\gamma,0,0)$	$2^{-50.88}$
$(\gamma,0,0,0,0,0) \stackrel{1r}{\rightarrow} (0,\gamma,0,0,\gamma,\gamma)$	$Z_1^{(i)}$	$(\gamma,0,\gamma) \rightarrow (\gamma,0,0)$	$2^{-25.94}$

Table 4.6 Nontrivial 1-round linear relations of XC_1 using bitmasks 0 and $\gamma = 1$.

1-round linear relation	Weak Subkeys (i-th round)	Bitmasks across the XAM-box	Bias
$(\gamma,0,0,0,0,\gamma) \stackrel{1r}{\to} (0,\gamma,\gamma,\gamma,0,\gamma)$	$Z_1^{(i)}, Z_6^{(i)}$	$(\gamma,\gamma,0) \to (\gamma,0,\gamma)$	$2^{-50.88}$
$(\gamma,0,0,0,\gamma,0) \stackrel{1r}{\to} (\gamma,0,\gamma,0,\gamma,\gamma)$	$Z_1^{(i)}$	$(0,0,\gamma) \to (\gamma,\gamma,0)$	$2^{-25.94}$
$(\gamma,0,0,0,\gamma,\gamma) \stackrel{1r}{\to} (\gamma,0,0,\gamma,0,\gamma)$	$Z_1^{(i)}, Z_6^{(i)}$	$(0,\gamma,0) \to (\gamma,\gamma,\gamma)$	$2^{-50.88}$
$(\gamma,0,0,\gamma,0,0) \stackrel{1r}{\to} (\gamma,\gamma,0,0,0,0)$	$Z_1^{(i)}$	$(0,0,0) \to (0,0,0)$	2^{-1}
$(\gamma,0,0,\gamma,0,\gamma) \stackrel{1r}{\to} (\gamma,\gamma,\gamma,\gamma,\gamma,0)$	$Z_1^{(i)}, Z_6^{(i)}$	$(0,\gamma,\gamma) \to (0,0,\gamma)$	$2^{-75.82}$
$(\gamma,0,0,\gamma,\gamma,0) \stackrel{1r}{\to} (0,0,\gamma,0,0,0)$	$Z_1^{(i)}$	$(\gamma,0,0) \to (0,\gamma,0)$	$2^{-25.94}$
$(\gamma,0,0,\gamma,\gamma,\gamma) \stackrel{1r}{\to} (0,0,0,\gamma,\gamma,0)$	$Z_1^{(i)}, Z_6^{(i)}$	$(\gamma,\gamma,\gamma) \to (0,\gamma,\gamma)$	$2^{-50.88}$
$(\gamma,0,\gamma,0,0,0) \stackrel{1r}{\to} (0,\gamma,\gamma,\gamma,\gamma,0)$	$Z_1^{(i)}$	$(\gamma,\gamma,0) \to (\gamma,0,\gamma)$	$2^{-50.88}$
$(\gamma,0,\gamma,0,0,\gamma) \stackrel{1r}{\to} (0,\gamma,0,0,0,0)$	$Z_1^{(i)}, Z_6^{(i)}$	$(\gamma,0,\gamma) \to (\gamma,0,0)$	$2^{-25.94}$
$(\gamma,0,\gamma,0,\gamma,0) \stackrel{1r}{\to} (\gamma,0,0,\gamma,\gamma,0)$	$Z_1^{(i)}$	$(0,\gamma,0) \to (\gamma,\gamma,\gamma)$	$2^{-50.88}$
$(\gamma,0,\gamma,0,\gamma,\gamma) \stackrel{1r}{\to} (\gamma,0,\gamma,0,0,0)$	$Z_1^{(i)}, Z_6^{(i)}$	$(0,0,\gamma) \to (\gamma,\gamma,0)$	$2^{-25.94}$
$(\gamma,0,\gamma,\gamma,0,0) \stackrel{1r}{\to} (\gamma,\gamma,\gamma,\gamma,0,\gamma)$	$Z_1^{(i)}$	$(0,\gamma,\gamma) \to (0,0,\gamma)$	$2^{-75.82}$
$(\gamma,0,\gamma,\gamma,0,\gamma) \stackrel{1r}{\to} (\gamma,\gamma,0,0,\gamma,\gamma)$	$Z_1^{(i)}, Z_6^{(i)}$	$(0,0,0) \to (0,0,0)$	2^{-1}
$(\gamma,0,\gamma,\gamma,\gamma,0) \stackrel{1r}{\to} (0,0,0,\gamma,0,\gamma)$	$Z_1^{(i)}$	$(\gamma,\gamma,\gamma) \to (0,\gamma,\gamma)$	$2^{-50.88}$
$(\gamma,0,\gamma,\gamma,\gamma,\gamma) \stackrel{1r}{\to} (0,0,\gamma,0,\gamma,\gamma)$	$Z_1^{(i)}, Z_6^{(i)}$	$(\gamma,0,0) \to (0,\gamma,0)$	$2^{-25.94}$
$(\gamma,\gamma,0,0,0,0) \stackrel{1r}{\to} (\gamma,0,0,\gamma,\gamma,\gamma)$	$Z_1^{(i)}$	$(0,0,\gamma) \to (\gamma,\gamma,0)$	$2^{-25.94}$
$(\gamma,\gamma,0,0,0,\gamma) \stackrel{1r}{\to} (\gamma,0,\gamma,0,0,\gamma)$	$Z_1^{(i)}, Z_6^{(i)}$	$(0,\gamma,0) \to (\gamma,\gamma,\gamma)$	$2^{-50.88}$
$(\gamma,\gamma,0,0,\gamma,0) \stackrel{1r}{\to} (0,\gamma,\gamma,\gamma,\gamma,\gamma)$	$Z_1^{(i)}$	$(\gamma,0,\gamma) \to (\gamma,0,0)$	$2^{-50.88}$
$(\gamma,\gamma,0,0,\gamma,\gamma) \stackrel{1r}{\to} (0,\gamma,0,0,0,\gamma)$	$Z_1^{(i)}, Z_6^{(i)}$	$(\gamma,\gamma,0) \to (\gamma,0,\gamma)$	$2^{-25.94}$
$(\gamma,\gamma,0,\gamma,0,0) \stackrel{1r}{\to} (0,0,0,\gamma,0,0)$	$Z_1^{(i)}$	$(\gamma,0,0) \to (0,\gamma,0)$	$2^{-25.94}$
$(\gamma,\gamma,0,\gamma,0,\gamma) \stackrel{1r}{\to} (0,0,\gamma,0,\gamma,0)$	$Z_1^{(i)}, Z_6^{(i)}$	$(\gamma,\gamma,\gamma) \to (0,\gamma,\gamma)$	$2^{-50.88}$
$(\gamma,\gamma,0,\gamma,\gamma,0) \stackrel{1r}{\to} (\gamma,\gamma,\gamma,\gamma,0,0)$	$Z_1^{(i)}$	$(0,0,0) \to (0,0,0)$	$2^{-25.94}$
$(\gamma,\gamma,0,\gamma,\gamma,\gamma) \stackrel{1r}{\to} (\gamma,\gamma,0,0,\gamma,0)$	$Z_1^{(i)}, Z_6^{(i)}$	$(0,\gamma,\gamma) \to (0,0,\gamma)$	$2^{-50.88}$
$(\gamma,\gamma,\gamma,0,0,0) \stackrel{1r}{\to} (\gamma,0,\gamma,0,\gamma,0)$	$Z_1^{(i)}$	$(0,\gamma,0) \to (\gamma,\gamma,\gamma)$	$2^{-50.88}$
$(\gamma,\gamma,\gamma,0,0,\gamma) \stackrel{1r}{\to} (\gamma,0,0,\gamma,0,0)$	$Z_1^{(i)}, Z_6^{(i)}$	$(0,0,\gamma) \to (\gamma,\gamma,0)$	$2^{-25.94}$
$(\gamma,\gamma,\gamma,0,\gamma,0) \stackrel{1r}{\to} (0,\gamma,0,0,\gamma,0)$	$Z_1^{(i)}$	$(\gamma,\gamma,0) \to (\gamma,0,\gamma)$	$2^{-25.94}$
$(\gamma,\gamma,\gamma,0,\gamma,\gamma) \stackrel{1r}{\to} (0,\gamma,\gamma,\gamma,0,0)$	$Z_1^{(i)}, Z_6^{(i)}$	$(\gamma,0,\gamma) \to (\gamma,0,0)$	$2^{-50.88}$
$(\gamma,\gamma,\gamma,\gamma,0,0) \stackrel{1r}{\to} (0,0,\gamma,0,0,\gamma)$	$Z_1^{(i)}$	$(\gamma,\gamma,\gamma) \to (0,\gamma,\gamma)$	$2^{-50.88}$
$(\gamma,\gamma,\gamma,\gamma,0,\gamma) \stackrel{1r}{\to} (0,0,0,\gamma,\gamma,\gamma)$	$Z_1^{(i)}, Z_6^{(i)}$	$(\gamma,0,0) \to (0,\gamma,0)$	$2^{-25.94}$
$(\gamma,\gamma,\gamma,\gamma,\gamma,0) \stackrel{1r}{\to} (\gamma,\gamma,0,0,0,\gamma)$	$Z_1^{(i)}$	$(0,\gamma,\gamma) \to (0,0,\gamma)$	$2^{-50.88}$
$(\gamma,\gamma,\gamma,\gamma,\gamma,\gamma) \stackrel{1r}{\to} (\gamma,\gamma,\gamma,\gamma,\gamma,\gamma)$	$Z_1^{(i)}, Z_6^{(i)}$	$(0,0,0) \to (0,0,0)$	$2^{-25.94}$

(5) $(0,0,0,\gamma,0,\gamma) \stackrel{1r}{\to} (\gamma,0,\gamma,\gamma,0,\gamma) \stackrel{1r}{\to} (\gamma,\gamma,0,0,\gamma,\gamma) \stackrel{1r}{\to} (0,\gamma,0,0,0,\gamma) \stackrel{1r}{\to} (\gamma,\gamma,\gamma,0,\gamma,0) \stackrel{1r}{\to} (0,\gamma,0,0,\gamma,0) \stackrel{1r}{\to} (0,0,\gamma,\gamma,0,0) \stackrel{1r}{\to} (\gamma,0,\gamma,\gamma,\gamma,0) \stackrel{1r}{\to} (0,0,0,\gamma,0,\gamma)$, which is 8-round iterative and holds with bias $2^{-275.34}$. The following eight subkeys have to be weak: $Z_6^{(i)}$, $Z_1^{(i+1)}$, $Z_6^{(i+1)}$, $Z_1^{(i+2)}$, $Z_6^{(i+2)}$, $Z_6^{(i+3)}$, $Z_1^{(i+4)}$ and $Z_1^{(i+7)}$.

(6) $(0,0,0,\gamma,\gamma,0) \stackrel{1r}{\rightarrow} (0,\gamma,\gamma,0,\gamma,\gamma) \stackrel{1r}{\rightarrow} (0,0,\gamma,\gamma,\gamma,\gamma) \stackrel{1r}{\rightarrow} (0,\gamma,\gamma,0,0,0) \stackrel{1r}{\rightarrow} (\gamma,\gamma,\gamma,0,0,\gamma) \stackrel{1r}{\rightarrow} (\gamma,0,0,\gamma,0,0) \stackrel{1r}{\rightarrow} (\gamma,\gamma,0,0,0,0) \stackrel{1r}{\rightarrow} (\gamma,0,0,\gamma,\gamma,\gamma) \stackrel{1r}{\rightarrow} (0,0,0,\gamma,\gamma,0)$, which is 8-round iterative and holds with bias $2^{-225.46}$.
The following eight subkeys have to be weak: $Z_6^{(i+1)}$, $Z_6^{(i+2)}$, $Z_1^{(i+4)}$, $Z_6^{(i+4)}$, $Z_1^{(i+5)}$, $Z_1^{(i+6)}$, $Z_1^{(i+7)}$ and $Z_6^{(i+7)}$.

(7) $(0,0,0,\gamma,\gamma,\gamma) \stackrel{1r}{\rightarrow} (0,\gamma,0,\gamma,0,\gamma) \stackrel{1r}{\rightarrow} (0,\gamma,\gamma,0,0,\gamma) \stackrel{1r}{\rightarrow} (\gamma,\gamma,0,\gamma,\gamma,\gamma) \stackrel{1r}{\rightarrow} (\gamma,\gamma,0,0,\gamma,0) \stackrel{1r}{\rightarrow} (0,\gamma,\gamma,\gamma,\gamma,\gamma) \stackrel{1r}{\rightarrow} (\gamma,0,\gamma,\gamma,0,0) \stackrel{1r}{\rightarrow} (\gamma,\gamma,\gamma,\gamma,0,\gamma) \stackrel{1r}{\rightarrow} (0,0,0,\gamma,\gamma,\gamma)$, which is 8-round iterative and holds with bias $2^{-375.10}$.
The following ten subkeys have to be weak: $Z_6^{(i)}$, $Z_6^{(i+1)}$, $Z_6^{(i+2)}$, $Z_1^{(i+3)}$, $Z_6^{(i+3)}$, $Z_1^{(i+4)}$, $Z_6^{(i+5)}$, $Z_1^{(i+6)}$, $Z_1^{(i+7)}$ and $Z_6^{(i+7)}$.

(8) $(0,0,\gamma,0,\gamma,0) \stackrel{1r}{\rightarrow} (\gamma,\gamma,0,\gamma,0,\gamma) \stackrel{1r}{\rightarrow} (0,0,\gamma,0,\gamma,0)$, which is 2-round iterative and holds with bias $2^{-100.76}$.
The following two subkeys have to be weak: $Z_1^{(i)}$ and $Z_6^{(i)}$.

(9) $(0,\gamma,0,\gamma,0,0) \stackrel{1r}{\rightarrow} (0,\gamma,0,\gamma,\gamma,\gamma) \stackrel{1r}{\rightarrow} (\gamma,0,0,0,0,\gamma) \stackrel{1r}{\rightarrow} (0,\gamma,\gamma,\gamma,0,\gamma) \stackrel{1r}{\rightarrow} (0,\gamma,0,\gamma,0,0)$, which is 4-round iterative and holds with bias $2^{-125.70}$.
The following four subkeys have to be weak: $Z_6^{(i+1)}$, $Z_1^{(i+1)}$, $Z_6^{(i+1)}$ and $Z_6^{(i+2)}$.

(10) $(0,\gamma,\gamma,\gamma,\gamma,0) \stackrel{1r}{\rightarrow} (\gamma,0,0,0,\gamma,0) \stackrel{1r}{\rightarrow} (\gamma,0,\gamma,0,\gamma,\gamma) \stackrel{1r}{\rightarrow} (\gamma,0,\gamma,0,0,0) \stackrel{1r}{\rightarrow} (0,\gamma,\gamma,\gamma,\gamma,0)$, which is 4-round iterative and holds with bias $2^{-125.70}$.
The following four subkeys have to be weak: $Z_1^{(i)}$, $Z_1^{(i+1)}$, $Z_6^{(i+1)}$ and $Z_1^{(i+2)}$.

(11) $(\gamma,\gamma,\gamma,\gamma,\gamma,\gamma) \stackrel{1r}{\rightarrow} (\gamma,\gamma,\gamma,\gamma,\gamma,\gamma)$, which is 1-round iterative and holds with bias $2^{-25.94}$.
The following two subkeys have to be weak: $Z_1^{(i)}$ and $Z_6^{(i)}$.

Among the eleven linear relations listed, (1) and (2) require the least average number of weak subkeys per round: about 0.75, but the bias is too low. Even the full codebook available would not be enough.

Considering the key schedule of XC_1 using a 192-bit key, and assuming each weak-subkey assumption holds with probability 2^{-15}, then at most $192/15 = 13$ weak subkeys will be allowed.

The best trade-off involving the highest (average) bias per round and the lowest number of weak subkeys per round is linear relation (11). It can cover up to 3-round XC_1 with bias $2^{-75.82}$ and requiring six weak subkeys. The weak-key class size is estimated as $2^{192-15\cdot 6} = 2^{102}$. The data complexity to distinguish 3-round XC_1 from a random permutation is estimated as $c \cdot (2^{-75.82})^2 = c \cdot 2^{151.64}$ KP where c is a fixed constant.

These preliminary differential and linear analyses of XC_1 help corroborate the minimum number of rounds for XC_1 as $r \geq 8$.

4.2 XC_2: Encryption and Decryption Frameworks

Another new design, XC_2, operates on $5w$-bit text blocks under a $10w$-bit key and iterates r rounds plus an output transformation, where $r \geq 8$. For $w = 16$ bits, XC_2 operates on 80-bit text blocks and under a 160-bit key.

Each full round of XC_2 for encryption can be split into two parts: a key-mixing (KM) and an Xor-Add (XA) half-round, in this order.

XC_2 uses only the three original group operations from PES [2], IDEA [1, 3] and the MESH [5] ciphers, that is, no S-boxes nor MDS codes. Also, like IDEA* [4], the $\boxdot$ operation is used to form (part of) the input to the XA-box.

Let $X^{(i)} = (X_1^{(i)}, X_2^{(i)}, X_3^{(i)}, X_4^{(i)}, X_5^{(i)})$ denote the input block to the i-th round for $1 \leq i \leq r+1$. Therefore, $X^{(1)}$ denotes the plaintext (block).

The output of the i-th KM half-round is denoted

$$Y^{(i)} = (Y_1^{(i)}, Y_2^{(i)}, Y_3^{(i)}, Y_4^{(i)}, Y_5^{(i)}).$$

where

$$Y^{(i)} = (X_1^{(i)} \oplus Z_1^{(i)}, X_2^{(i)} \boxplus Z_2^{(i)}, X_3^{(i)} \oplus Z_3^{(i)}, X_4^{(i)} \boxplus Z_4^{(i)}, X_5^{(i)} \oplus Z_5^{(i)})$$

which also becomes the input to the i-th XA half-round. This half-round contains a so called XA-box, which consists of $\oplus$ and $\boxplus$ operations, some of which are keyed.

The input to the i-th XA-box is the 5-tuple

$$(Y_1^{(i)} \boxdot Y_3^{(i)}, Y_2^{(i)} \boxdot Y_4^{(i)}, Y_3^{(i)} \boxdot Y_5^{(i)}, Y_4^{(i)} \boxdot Y_1^{(i)}, Y_5^{(i)} \boxdot Y_2^{(i)}) = (n_i, q_i, m_i, u_i, v_i).$$

The output of the i-th XA-box is denoted p_i, where $p_i = ((((((u_i \oplus v_i) \boxplus Z_6^{(i)}) \oplus m_i) \boxplus Z_7^{(i)}) \oplus q_i) \boxplus Z_8^{(i)}) \oplus n_i$.

Note that p_i depends on all words of $(n_i, q_i, m_i, u_i, v_i)$. Consequently, p_i depends on all words in $Y^{(i)}$ and in $X^{(i)}$.

Finally, the XA-box output is combined with the original input words in $Y^{(i)}$ resulting in

$$(Y_1^{(i)} \odot p_i, Y_2^{(i)} \odot p_i, Y_3^{(i)} \odot p_i, Y_4^{(i)} \odot p_i, Y_5^{(i)} \odot p_i).$$

Note that each round output word depends on all words of $(n_i, q_i, m_i, u_i, v_i)$. Thus, complete text diffusion is achieved in every single round.

Note that no subkey is combined via $\odot$. Thus, the design of XC_2 is similar to that of IDEA*.

As a last encryption operation in a full round, there is a fixed (involutory) permutation denoted σ. The syntax is $\sigma(a, b, c, d, e) = (d, e, c, a, b)$. The i-th round output is denoted

$$X^{(i+1)} = (Y_4^{(i)} \odot p_i, Y_5^{(i)} \odot p_i, Y_3^{(i)} \odot p_i, Y_1^{(i)} \odot p_i, Y_2^{(i)} \odot p_i).$$

This (full) round transformation is repeated r times. The output transformation consists of σ followed by a KM half-round. Therefore, the 5-tuple

$$Y^{(r+1)} = (X_1^{(r+1)} \oplus Z_1^{(r+1)}, X_2^{(r+1)} \boxplus Z_2^{(r+1)}, X_3^{(r+1)} \oplus Z_3^{(r+1)},$$
$$X_4^{(r+1)} \boxplus Z_4^{(r+1)}, X_5^{(r+1)} \oplus Z_5^{(r+1)})$$

denotes the ciphertext (block).

For decryption, the inverse of the i-th KM half-round consists of the reverse order of round subkeys appropriately transformed by the inverse of $\boxplus$ or $\oplus$.

The inverse of the i-th XA half-round is similar to the original i-th XA half-round: the input to the XA-box is also $(n_i, q_i, m_i, u_i, v_i)$, but the output p_i is combined via $\boxdot$. Therefore, the output of the XA-box is as follows:

$$(Y_1^{(i)} \boxdot p_i, Y_2^{(i)} \boxdot p_i, Y_3^{(i)} \boxdot p_i, Y_4^{(i)} \boxdot p_i, Y_5^{(i)} \boxdot p_i).$$

Note that all five words in $Y^{(i)}$ are combined pairwise to form the 5-tuple input to the XA-box. See Fig. 4.2.

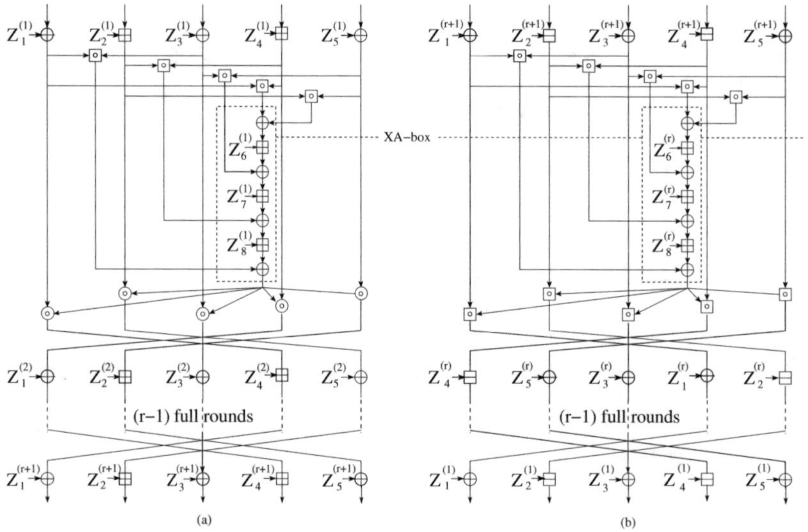

Fig. 4.2 Pictorial representation of: (a) encryption and (b) decryption frameworks of XC_2.

New features of XC_2 include:

- the block size contains an odd number of words: five.
- these five words in a block are combined in pairs to form a 5-tuple input to the XA-box.

- overall, no round subkey is combined via $\odot$ to avoid weak subkeys. This design is similar to that adopted in IDEA* [4].
- the XA-box itself is a non-bijective mapping, unlike all previous Lai-Massey designs presented so far. This XA-box is a mapping XA: $\mathbb{Z}_2^{5w} \rightarrow \mathbb{Z}_2^w$. Due to this design feature, the same XA-box output is combined to all five words in $Y^{(i)}$. This is actually necessary because of the pairing of $Y^{(i)}$ values at the input to the XA-box.
- like XC_1, the same framework cannot be used for both encryption and decryption in XC_2 by just altering the value and order of round subkeys because at the output of the XA-box the $\odot$ is used, and it is not keyed. The inverse of $\odot$ is $\boxdot$ for decryption.

4.2.1 Key Schedule Algorithm

The key schedule of XC_2 generates eight 16-bit subkeys per round, for a total of $8 \cdot r + 5$ subkeys for r rounds plus the output transformation.

The key schedule of XC_2 uses a nonlinear feedback shift register (NLFSR) design and consists of the following steps:

- a preliminary step is to define 16-bit constants c_i as follows: $c_0 = 1$ and $c_i = 3 \cdot c_{i-1}$ for $i \geq 1$. Multiplication in $\mathrm{GF}(2)[x]/p(x)$ is denoted by $\cdot$ with $p(x)$ a primitive polynomial over GF(2): $p(x) = x^{16} + x^5 + x^3 + x^2 + 1$. The constant 3 in the construction of c_i represents the polynomial $x + 1$ in GF(2).
- the 160-bit user key $K = (k_0, k_1, \ldots, k_{159})$ is partitioned into ten 16-bit words K_l for $0 \leq l \leq 9$ and assigned to the subkeys as follows: $Z_{j+1}^{(1)} = K_j \oplus c_j$ for $0 \leq j \leq 7$, $Z_1^{(2)} = K_8 \oplus c_8$ and $Z_2^{(2)} = K_9 \oplus c_9$.
- each subsequent 16-bit subkey is constructed iteratively as follows:

$$Z_{l(i)}^{h(i)} = (((Z_{l(i-10)}^{h(i-10)} \oplus Z_{l(i-8)}^{h(i-8)}) \boxplus Z_{l(i-7)}^{h(i-7)}) \oplus Z_{l(i-2)}^{h(i-2)}) \lll 7 \boxplus c_i, \quad (4.1)$$

  with $10 \leq i \leq 8 \cdot r + 4$; $x \lll 7$ means x left rotated by seven bits; $h(i) = \lfloor i/8 \rfloor + 1$ and $l(i) = i \bmod 8 + 1$.

This key schedule was designed to achieve fast key diffusion due to the use of the irreducible polynomial $r(x) = x^{10} + x^8 + x^3 + x^2 + 1$ in $l(i)$ and $h(i)$, the bitwise rotation and the interleaving of $\oplus$ and $\boxplus$.

Let us deduce the first subkey that depends on all eight user key words. Initially, $Z_1^{(1)} = K_0 \oplus c_0$, $Z_2^{(1)} = K_1 \oplus c_1$, $Z_3^{(1)} = K_2 \oplus c_2$, $Z_4^{(1)} = K_3 \oplus c_3$, $Z_5^{(1)} = K_4 \oplus c_4$, $Z_6^{(1)} = K_5 \oplus c_5$, $Z_7^{(1)} = K_6 \oplus c_6$, $Z_8^{(1)} = K_7 \oplus c_7$, $Z_1^{(2)} = K_8 \oplus c_8$ and $Z_2^{(2)} = K_9 \oplus c_9$.

According to (4.1), for $i = 10$: $Z_3^{(2)} = (((Z_1^{(1)} \oplus Z_3^{(1)}) \boxplus Z_4^{(1)}) \oplus Z_1^{(2)}) \lll 7 \oplus c_{10}$.

Further, for $i = 11$: $Z_4^{(2)} = (((Z_2^{(1)} \oplus Z_4^{(1)}) \boxplus Z_5^{(1)}) \oplus Z_2^{(2)}) \lll 7 \oplus c_{11}$.

Further, for $i = 12$: $Z_5^{(2)} = (((Z_3^{(1)} \oplus Z_5^{(1)}) \boxplus Z_6^{(1)}) \oplus Z_3^{(2)}) \lll 7 \oplus c_{12}$.
Further, for $i = 13$: $Z_6^{(2)} = (((Z_4^{(1)} \oplus Z_6^{(1)}) \boxplus Z_7^{(1)}) \oplus Z_4^{(2)}) \lll 7 \oplus c_{13}$.
Further, for $i = 14$: $Z_7^{(2)} = (((Z_5^{(1)} \oplus Z_7^{(1)}) \boxplus Z_8^{(1)}) \oplus Z_5^{(2)}) \lll 7 \oplus c_{14}$.

Note that $Z_7^{(2)}$ is the first subkey that depends nonlinearly on all words K_l, $0 \leq 9$. Consequently, complete key diffusion is achieved after 2 rounds.

4.2.2 Differential Analysis

This section provides preliminary differential analysis of XC_2 (for a variable number of rounds). The analysis approach is bottom-up, from the smallest components such as $\odot$, $\boxplus$ and $\oplus$, up to larger ones.

The difference operator is bitwise exclusive-or.

Across unkeyed $\odot$ and $\boxplus$, the propagation of 16-bit difference words follows from Tables 3.12 and 3.13, respectively.

The next step is to obtain all difference patterns across the XA-box of XC_2.

The only 16-bit difference values used are $\delta = 8000_x$ and 0000_x. The difference 8000_x propagates across the modular addition with probability one, since it affects only the most significant bit (and the last carry-out bit is truncated due to modular reduction.

Table 4.7 lists the propagation of wordwise differences across the XA-box. These differences propagate with certainty.

Note that due to the construction of the XA-box and the fact that the only nonzero difference is $\delta = 8000_x$, the output difference depends only on the parity of the nonzero wordwise differences: an odd number of δ differences leads to an output difference of δ, otherwise, the output difference is zero: $\Delta p_i = \Delta u_i \oplus \Delta v_i \oplus \Delta m_i \oplus \Delta q_i \oplus \Delta n_i$.

The next step is to construct 1-round characteristics based on Table 4.7. Table 4.8 lists all nontrivial 1-round differential characteristics for XC_2 using only wordwise differences 0000_x and $\delta = 8000_x$.

Multiple-round characteristics are obtained by concatenating 1-round characteristics such that the output difference from one characteristic matches the input difference of the next characteristic.

From Table 4.8, the following characteristics can be derived.

(1) $(0, 0, 0, 0, \delta) \stackrel{1r}{\rightarrow} (0, \delta, 0, 0, 0) \stackrel{1r}{\rightarrow} (0, 0, 0, 0, \delta)$, which is 2-round iterative and holds with probability 2^{-120}.
(2) $(0, 0, 0, \delta, 0) \stackrel{1r}{\rightarrow} (\delta, 0, 0, 0, 0) \stackrel{1r}{\rightarrow} (0, 0, 0, \delta, 0)$, which is 2-round iterative and holds with probability 2^{-90}.
(3) $(0, 0, 0, \delta, \delta) \stackrel{1r}{\rightarrow} (\delta, \delta, 0, 0, 0) \stackrel{1r}{\rightarrow} (0, 0, 0, \delta, \delta)$, which is 2-round iterative and holds with probability 2^{-210}.

Table 4.7 Difference propagation across the XA-box of XC$_2$.

Wordwise difference across the XA-box	Wordwise difference across the XA-box
$(0,0,0,0,0) \to 0$	$(\delta,0,0,0,0) \to \delta$
$(0,0,0,0,\delta) \to \delta$	$(\delta,0,0,0,\delta) \to 0$
$(0,0,0,\delta,0) \to \delta$	$(\delta,0,0,\delta,0) \to 0$
$(0,0,0,\delta,\delta) \to 0$	$(\delta,0,0,\delta,\delta) \to \delta$
$(0,0,\delta,0,0) \to \delta$	$(\delta,0,\delta,0,0) \to 0$
$(0,0,\delta,0,\delta) \to 0$	$(\delta,0,\delta,0,\delta) \to \delta$
$(0,0,\delta,\delta,0) \to 0$	$(\delta,0,\delta,\delta,0) \to \delta$
$(0,0,\delta,\delta,\delta) \to \delta$	$(\delta,0,\delta,\delta,\delta) \to 0$
$(0,\delta,0,0,0) \to \delta$	$(\delta,\delta,0,0,0) \to 0$
$(0,\delta,0,0,\delta) \to 0$	$(\delta,\delta,0,0,\delta) \to \delta$
$(0,\delta,0,\delta,0) \to 0$	$(\delta,\delta,0,\delta,0) \to \delta$
$(0,\delta,0,\delta,\delta) \to \delta$	$(\delta,\delta,0,\delta,\delta) \to 0$
$(0,\delta,\delta,0,0) \to 0$	$(\delta,\delta,\delta,0,0) \to \delta$
$(0,\delta,\delta,0,\delta) \to \delta$	$(\delta,\delta,\delta,0,\delta) \to 0$
$(0,\delta,\delta,\delta,0) \to \delta$	$(\delta,\delta,\delta,\delta,0) \to 0$
$(0,\delta,\delta,\delta,\delta) \to 0$	$(\delta,\delta,\delta,\delta,\delta) \to \delta$

(4) $(0,0,\delta,0,0) \overset{1r}{\to} (0,0,\delta,0,0)$, which is 1-round iterative and holds with probability 2^{-45}.

(5) $(0,0,\delta,0,\delta) \overset{1r}{\to} (0,\delta,\delta,0,0) \overset{1r}{\to} (0,0,\delta,0,\delta)$, which is 2-round iterative and holds with probability 2^{-195}.

(6) $(0,0,\delta,\delta,0) \overset{1r}{\to} (\delta,0,\delta,0,0) \overset{1r}{\to} (0,0,\delta,\delta,0)$, which is 2-round iterative and holds with probability 2^{-165}.

(7) $(0,0,\delta,\delta,\delta) \overset{1r}{\to} (\delta,\delta,\delta,0,0) \overset{1r}{\to} (0,0,\delta,\delta,\delta)$, which is 2-round iterative and holds with probability 2^{-270}.

(8) $(0,\delta,0,0,\delta) \overset{1r}{\to} (0,\delta,0,0,\delta)$, which is 1-round iterative and holds with probability 2^{-90}.

(9) $(0,\delta,0,\delta,0) \overset{1r}{\to} (\delta,0,0,0,\delta) \overset{1r}{\to} (0,\delta,0,\delta,0)$, which is 2-round iterative and holds with probability 2^{-195}.

(10) $(0,\delta,0,\delta,\delta) \overset{1r}{\to} (\delta,\delta,0,0,\delta) \overset{1r}{\to} (0,\delta,0,\delta,\delta)$, which is 2-round iterative and holds with probability 2^{-255}.

(11) $(0,\delta,\delta,0,\delta) \overset{1r}{\to} (0,\delta,\delta,0,\delta)$, which is 1-round iterative and holds with probability 2^{-120}.

(12) $(0,\delta,\delta,\delta,0) \overset{1r}{\to} (\delta,0,\delta,0,\delta) \overset{1r}{\to} (0,\delta,\delta,\delta,0)$, which is 2-round iterative and holds with probability 2^{-255}.

(13) $(0,\delta,\delta,\delta,\delta) \overset{1r}{\to} (\delta,\delta,\delta,0,\delta) \overset{1r}{\to} (0,\delta,\delta,\delta,\delta)$, which is 2-round iterative and holds with probability 2^{-300}.

(14) $(\delta,0,0,\delta,0) \overset{1r}{\to} (\delta,0,0,\delta,0)$, which is 1-round iterative and holds with probability 2^{-75}.

(15) $(\delta,0,0,\delta,\delta) \overset{1r}{\to} (\delta,\delta,0,\delta,0) \overset{1r}{\to} (\delta,0,0,\delta,\delta)$, which is 2-round iterative and holds with probability 2^{-255}.

Table 4.8 Nontrivial 1-round characteristics of XC_2 using xor differences and $\delta = 8000_x$.

1-round characteristics	Difference across XA-box	Prob.
$(0,0,0,0,\delta) \xrightarrow{1r} (0,\delta,0,0,0)$	$(0,0,\delta,0,\delta) \to 0$	2^{-60}
$(0,0,0,\delta,0) \xrightarrow{1r} (\delta,0,0,0,0)$	$(0,\delta,0,\delta,0) \to 0$	2^{-45}
$(0,0,0,\delta,\delta) \xrightarrow{1r} (\delta,\delta,0,0,0)$	$(0,\delta,\delta,\delta,\delta) \to 0$	2^{-105}
$(0,0,\delta,0,0) \xrightarrow{1r} (0,0,\delta,0,0)$	$(\delta,0,\delta,0,0) \to 0$	2^{-45}
$(0,0,\delta,0,\delta) \xrightarrow{1r} (0,\delta,\delta,0,0)$	$(\delta,0,0,0,\delta) \to 0$	2^{-90}
$(0,0,\delta,\delta,0) \xrightarrow{1r} (\delta,0,\delta,0,0)$	$(\delta,\delta,\delta,\delta,0) \to 0$	2^{-90}
$(0,0,\delta,\delta,\delta) \xrightarrow{1r} (\delta,\delta,\delta,0,0)$	$(\delta,\delta,0,\delta,\delta) \to 0$	2^{-135}
$(0,\delta,0,0,0) \xrightarrow{1r} (0,0,0,0,\delta)$	$(0,\delta,0,0,\delta) \to 0$	2^{-60}
$(0,\delta,0,0,\delta) \xrightarrow{1r} (0,\delta,0,0,\delta)$	$(0,\delta,\delta,0,0) \to 0$	2^{-90}
$(0,\delta,0,\delta,0) \xrightarrow{1r} (\delta,0,0,0,\delta)$	$(0,0,0,\delta,\delta) \to 0$	2^{-90}
$(0,\delta,0,\delta,\delta) \xrightarrow{1r} (\delta,\delta,0,0,\delta)$	$(0,0,\delta,\delta,0) \to 0$	2^{-120}
$(0,\delta,\delta,0,0) \xrightarrow{1r} (0,0,\delta,0,\delta)$	$(\delta,\delta,\delta,0,\delta) \to 0$	2^{-105}
$(0,\delta,\delta,0,\delta) \xrightarrow{1r} (0,\delta,\delta,0,\delta)$	$(\delta,\delta,0,0,0) \to 0$	2^{-120}
$(0,\delta,\delta,\delta,0) \xrightarrow{1r} (\delta,0,\delta,0,\delta)$	$(\delta,0,\delta,\delta,\delta) \to 0$	2^{-135}
$(0,\delta,\delta,\delta,\delta) \xrightarrow{1r} (\delta,\delta,\delta,0,\delta)$	$(\delta,0,0,\delta,0) \to 0$	2^{-150}
$(\delta,0,0,0,0) \xrightarrow{1r} (0,0,0,\delta,0)$	$(\delta,0,0,\delta,0) \to 0$	2^{-45}
$(\delta,0,0,0,\delta) \xrightarrow{1r} (0,\delta,0,\delta,0)$	$(\delta,0,\delta,\delta,\delta) \to 0$	2^{-105}
$(\delta,0,0,\delta,0) \xrightarrow{1r} (\delta,0,0,\delta,0)$	$(\delta,\delta,0,0,0) \to 0$	2^{-75}
$(\delta,0,0,\delta,\delta) \xrightarrow{1r} (\delta,\delta,0,\delta,0)$	$(\delta,\delta,\delta,0,\delta) \to 0$	2^{-135}
$(\delta,0,\delta,0,0) \xrightarrow{1r} (0,0,\delta,\delta,0)$	$(0,0,\delta,\delta,0) \to 0$	2^{-75}
$(\delta,0,\delta,0,\delta) \xrightarrow{1r} (0,\delta,\delta,\delta,0)$	$(0,0,0,\delta,\delta) \to 0$	2^{-120}
$(\delta,0,\delta,\delta,0) \xrightarrow{1r} (\delta,0,\delta,\delta,0)$	$(0,\delta,\delta,0,0) \to 0$	2^{-105}
$(\delta,0,\delta,\delta,\delta) \xrightarrow{1r} (\delta,\delta,\delta,\delta,0)$	$(0,\delta,0,0,\delta) \to 0$	2^{-150}
$(\delta,\delta,0,0,0) \xrightarrow{1r} (0,0,0,\delta,\delta)$	$(\delta,\delta,0,\delta,\delta) \to 0$	2^{-105}
$(\delta,\delta,0,0,\delta) \xrightarrow{1r} (0,\delta,0,\delta,\delta)$	$(\delta,\delta,\delta,\delta,0) \to 0$	2^{-135}
$(\delta,\delta,0,\delta,0) \xrightarrow{1r} (\delta,0,0,\delta,\delta)$	$(\delta,0,0,0,\delta) \to 0$	2^{-120}
$(\delta,\delta,0,\delta,\delta) \xrightarrow{1r} (\delta,\delta,0,\delta,\delta)$	$(\delta,0,\delta,0,0) \to 0$	2^{-150}
$(\delta,\delta,\delta,0,0) \xrightarrow{1r} (0,0,\delta,\delta,\delta)$	$(0,\delta,\delta,\delta,\delta) \to 0$	2^{-135}
$(\delta,\delta,\delta,0,\delta) \xrightarrow{1r} (0,\delta,\delta,\delta,\delta)$	$(0,\delta,0,\delta,0) \to 0$	2^{-150}
$(\delta,\delta,\delta,\delta,0) \xrightarrow{1r} (\delta,0,\delta,\delta,\delta)$	$(0,0,\delta,0,\delta) \to 0$	2^{-150}
$(\delta,\delta,\delta,\delta,\delta) \xrightarrow{1r} (\delta,\delta,\delta,\delta,\delta)$	$(0,0,0,0,0) \to 0$	2^{-165}

(16) $(\delta,0,\delta,\delta,0) \xrightarrow{1r} (\delta,0,\delta,\delta,0)$, which is 1-round iterative and holds with probability 2^{-105}.

(17) $(\delta,0,\delta,\delta,\delta) \xrightarrow{1r} (\delta,\delta,\delta,\delta,0) \xrightarrow{1r} (\delta,0,\delta,\delta,\delta)$, which is 2-round iterative and holds with probability 2^{-300}.

(18) $(\delta,\delta,0,\delta,\delta) \xrightarrow{1r} (\delta,\delta,0,\delta,\delta)$, which is 1-round iterative and holds with probability 2^{-150}.

(19) $(\delta,\delta,\delta,\delta,\delta) \xrightarrow{1r} (\delta,\delta,\delta,\delta,\delta)$, which is 1-round iterative and holds with probability 2^{-165}.

The best trade-off is achieved by characteristics (4), (8) and (9) each of which holds with probability 2^{-45} per round. Moreover, (1) is 1-round iterative. There are *no weak-subkey assumptions* for any of the characteristics.

Characteristic (1) can be used to distinguish up to 3-round XC_2 from a random permutation with probability 2^{-135}. Since no weak subkey are required, this attack applies to any key. The data complexity is estimated as $c \cdot 2^{135}$ CP, where c is a constant.

4.2.3 Linear Analysis

This section provides preliminary linear cryptanalysis of XC_2. The analysis approach is bottom-up, from the smallest components $\odot$, $\boxplus$ and $\oplus$, up to larger ones.

Table 3.98 lists the bitmask propagation and the corresponding biases across the unkeyed-$\odot$ operation (under wordwise bitmasks 0 and $\gamma = 1$).

Table 3.99 lists the bitmask propagation and the corresponding biases across $\boxplus$ (under wordwise bitmasks 0 and 1).

The next step is to obtain linear relations across the XA-box. There are only two possible linear trails across the XA-box, both holding with maximum bias: $(0,0,0,0,0) \rightarrow 0$ and $(1,1,1,1,1) \rightarrow 1$.

The next step is to construct 1-round linear relations based on the trails across the XA-box.

Table 4.9 lists all nontrivial 1-round linear relations for XC_2 using only bitmasks 0 and $\gamma = 1$.

From Table 4.9, the following chains of linear relations can be derived:

(1) $(0,0,0,0,\gamma) \xrightarrow{1r} (0,\gamma,0,0,0) \xrightarrow{1r} (0,0,0,0,\gamma)$, which is 2-round iterative and holds with bias $2^{-150.64}$.
(2) $(0,0,0,\gamma,0) \xrightarrow{1r} (\gamma,0,0,0,0) \xrightarrow{1r} (0,0,0,\gamma,0)$, which is 2-round iterative and holds with bias $2^{-150.64}$.
(3) $(0,0,0,\gamma,\gamma) \xrightarrow{1r} (\gamma,\gamma,0,0,0) \xrightarrow{1r} (0,0,0,\gamma,\gamma)$, which is 2-round iterative and holds with bias $2^{-50.88}$.
(4) $(0,0,\gamma,0,0) \xrightarrow{1r} (0,0,\gamma,0,0)$, which is 1-round iterative and holds with bias $2^{-75.82}$.
(5) $(0,0,\gamma,0,\gamma) \xrightarrow{1r} (0,\gamma,\gamma,0,0) \xrightarrow{1r} (0,0,\gamma,0,\gamma)$, which is 2-round iterative and holds with bias $2^{-50.88}$.
(6) $(0,0,\gamma,\gamma,0) \xrightarrow{1r} (\gamma,0,\gamma,0,0) \xrightarrow{1r} (0,0,\gamma,\gamma,0)$, which is 2-round iterative and holds with bias $2^{-50.88}$.
(7) $(0,0,\gamma,\gamma,\gamma) \xrightarrow{1r} (\gamma,\gamma,\gamma,0,0) \xrightarrow{1r} (0,0,0,\gamma,\gamma,\gamma)$, which is 2-round iterative and holds with bias $2^{-200.52}$.

Table 4.9 Nontrivial 1-round linear relations of XC_2 using bitmasks 0 and $\gamma = 1$.

1-round linear relation	Bitmasks across XA-box	Bias
$(0,0,0,0,\gamma) \stackrel{1r}{\to} (0,\gamma,0,0,0)$	$(\gamma,\gamma,\gamma,\gamma,\gamma) \to 1$	$2^{-75.82}$
$(0,0,0,\gamma,0) \stackrel{1r}{\to} (\gamma,0,0,0,0)$	$(\gamma,\gamma,\gamma,\gamma,\gamma) \to 1$	$2^{-75.82}$
$(0,0,0,\gamma,\gamma) \stackrel{1r}{\to} (\gamma,\gamma,0,0,0)$	$(0,0,0,0,0) \to 0$	$2^{-25.94}$
$(0,0,\gamma,0,0) \stackrel{1r}{\to} (0,0,\gamma,0,0)$	$(\gamma,\gamma,\gamma,\gamma,\gamma) \to 1$	$2^{-75.82}$
$(0,0,\gamma,0,\gamma) \stackrel{1r}{\to} (0,\gamma,\gamma,0,0)$	$(0,0,0,0,0) \to 0$	$2^{-25.94}$
$(0,0,\gamma,\gamma,0) \stackrel{1r}{\to} (\gamma,0,\gamma,0,0)$	$(0,0,0,0,0) \to 0$	$2^{-25.94}$
$(0,0,\gamma,\gamma,\gamma) \stackrel{1r}{\to} (\gamma,\gamma,\gamma,0,0)$	$(\gamma,\gamma,\gamma,\gamma,\gamma) \to 1$	$2^{-100.76}$
$(0,\gamma,0,0,0) \stackrel{1r}{\to} (0,0,0,0,\gamma)$	$(\gamma,\gamma,\gamma,\gamma,\gamma) \to 1$	$2^{-75.82}$
$(0,\gamma,0,0,\gamma) \stackrel{1r}{\to} (0,\gamma,0,0,\gamma)$	$(0,0,0,0,0) \to 0$	$2^{-25.94}$
$(0,\gamma,0,\gamma,0) \stackrel{1r}{\to} (\gamma,0,0,0,\gamma)$	$(0,0,0,0,0) \to 0$	$2^{-25.94}$
$(0,\gamma,0,\gamma,\gamma) \stackrel{1r}{\to} (\gamma,\gamma,0,0,\gamma)$	$(\gamma,\gamma,\gamma,\gamma,\gamma) \to 1$	$2^{-100.76}$
$(0,\gamma,\gamma,0,0) \stackrel{1r}{\to} (0,0,\gamma,0,\gamma)$	$(0,0,0,0,0) \to 0$	$2^{-25.94}$
$(0,\gamma,\gamma,0,\gamma) \stackrel{1r}{\to} (0,\gamma,\gamma,0,\gamma)$	$(\gamma,\gamma,\gamma,\gamma,\gamma) \to 1$	$2^{-100.76}$
$(0,\gamma,\gamma,\gamma,0) \stackrel{1r}{\to} (\gamma,0,\gamma,0,\gamma)$	$(\gamma,\gamma,\gamma,\gamma,\gamma) \to 1$	$2^{-100.76}$
$(0,\gamma,\gamma,\gamma,\gamma) \stackrel{1r}{\to} (\gamma,\gamma,\gamma,0,\gamma)$	$(0,0,0,0,0) \to 0$	$2^{-50.88}$
$(\gamma,0,0,0,0) \stackrel{1r}{\to} (0,0,0,\gamma,0)$	$(\gamma,\gamma,\gamma,\gamma,\gamma) \to 1$	$2^{-75.82}$
$(\gamma,0,0,0,\gamma) \stackrel{1r}{\to} (0,\gamma,0,\gamma,0)$	$(0,0,0,0,0) \to 0$	$2^{-25.94}$
$(\gamma,0,0,\gamma,0) \stackrel{1r}{\to} (\gamma,0,0,\gamma,0)$	$(0,0,0,0,0) \to 0$	$2^{-25.94}$
$(\gamma,0,0,\gamma,\gamma) \stackrel{1r}{\to} (\gamma,\gamma,0,\gamma,0)$	$(\gamma,\gamma,\gamma,\gamma,\gamma) \to 1$	$2^{-100.76}$
$(\gamma,0,\gamma,0,0) \stackrel{1r}{\to} (0,0,\gamma,\gamma,0)$	$(0,0,0,0,0) \to 0$	$2^{-25.94}$
$(\gamma,0,\gamma,0,\gamma) \stackrel{1r}{\to} (0,\gamma,\gamma,\gamma,0)$	$(\gamma,\gamma,\gamma,\gamma,\gamma) \to 1$	$2^{-100.76}$
$(\gamma,0,\gamma,\gamma,0) \stackrel{1r}{\to} (\gamma,0,\gamma,\gamma,0)$	$(\gamma,\gamma,\gamma,\gamma,\gamma) \to 1$	$2^{-100.76}$
$(\gamma,0,\gamma,\gamma,\gamma) \stackrel{1r}{\to} (\gamma,\gamma,\gamma,\gamma,0)$	$(0,0,0,0,0) \to 0$	$2^{-50.88}$
$(\gamma,\gamma,0,0,0) \stackrel{1r}{\to} (0,0,0,\gamma,\gamma)$	$(0,0,0,0,0) \to 0$	$2^{-25.94}$
$(\gamma,\gamma,0,0,\gamma) \stackrel{1r}{\to} (0,\gamma,0,\gamma,\gamma)$	$(\gamma,\gamma,\gamma,\gamma,\gamma) \to 1$	$2^{-100.76}$
$(\gamma,\gamma,0,\gamma,0) \stackrel{1r}{\to} (\gamma,0,0,\gamma,\gamma)$	$(\gamma,\gamma,\gamma,\gamma,\gamma) \to 1$	$2^{-100.76}$
$(\gamma,\gamma,0,\gamma,\gamma) \stackrel{1r}{\to} (\gamma,\gamma,0,\gamma,\gamma)$	$(0,0,0,0,0) \to 0$	$2^{-50.88}$
$(\gamma,\gamma,\gamma,0,0) \stackrel{1r}{\to} (0,0,\gamma,\gamma,\gamma)$	$(\gamma,\gamma,\gamma,\gamma,\gamma) \to 1$	$2^{-100.76}$
$(\gamma,\gamma,\gamma,0,\gamma) \stackrel{1r}{\to} (0,\gamma,\gamma,\gamma,\gamma)$	$(0,0,0,0,0) \to 0$	$2^{-50.88}$
$(\gamma,\gamma,\gamma,\gamma,0) \stackrel{1r}{\to} (\gamma,0,\gamma,\gamma,\gamma)$	$(0,0,0,0,0) \to 0$	$2^{-50.88}$
$(\gamma,\gamma,\gamma,\gamma,\gamma) \stackrel{1r}{\to} (\gamma,\gamma,\gamma,\gamma,\gamma)$	$(\gamma,\gamma,\gamma,\gamma,\gamma) \to 1$	$2^{-125.70}$

(8) $(0,\gamma,0,0,\gamma) \stackrel{1r}{\to} (0,\gamma,0,0,\gamma)$, which is 1-round iterative and holds with bias $2^{-25.94}$.

(9) $(0,\gamma,0,\gamma,0) \stackrel{1r}{\to} (\gamma,0,0,0,\gamma) \stackrel{1r}{\to} (0,\gamma,0,\gamma,0)$, which is 2-round iterative and holds with bias $2^{-50.88}$.

(10) $(0,\gamma,0,\gamma,\gamma) \stackrel{1r}{\to} (\gamma,\gamma,0,0,\gamma) \stackrel{1r}{\to} (0,\gamma,0,\gamma,\gamma)$, which is 2-round iterative and holds with bias $2^{200.52}$.

(11) $(0,\gamma,\gamma,0,\gamma) \stackrel{1r}{\to} (0,\gamma,\gamma,0,\gamma)$, which is 1-round iterative and holds with bias $2^{-100.76}$.

(12) $(0,\gamma,\gamma,\gamma,0) \stackrel{1r}{\rightarrow} (\gamma,0,\gamma,0,\gamma) \stackrel{1r}{\rightarrow} (0,\gamma,\gamma,\gamma,0)$, which is 2-round iterative and holds with bias $2^{-200.52}$.
(13) $(0,\gamma,\gamma,\gamma,\gamma) \stackrel{1r}{\rightarrow} (\gamma,\gamma,\gamma,0,\gamma) \stackrel{1r}{\rightarrow} (0,\gamma,\gamma,\gamma,\gamma)$, which is 2-round iterative and holds with bias $2^{-100.76}$.
(14) $(\gamma,0,0,\gamma,0) \stackrel{1r}{\rightarrow} (\gamma,0,0,\gamma,0)$, which is 1-round iterative and holds with bias $2^{-25.94}$.
(15) $(\gamma,0,0,\gamma,\gamma) \stackrel{1r}{\rightarrow} (\gamma,\gamma,0,\gamma,0) \stackrel{1r}{\rightarrow} (\gamma,0,0,\gamma,\gamma)$, which is 2-round iterative and holds with bias $2^{-200.52}$.
(16) $(\gamma,0,\gamma,\gamma,0) \stackrel{1r}{\rightarrow} (\gamma,0,\gamma,\gamma,0)$, which is 1-round iterative and holds with bias $2^{-100.76}$.
(17) $(\gamma,0,\gamma,\gamma,\gamma) \stackrel{1r}{\rightarrow} (\gamma,\gamma,\gamma,\gamma,0) \stackrel{1r}{\rightarrow} (\gamma,0,\gamma,\gamma,\gamma)$, which is 2-round iterative and holds with bias $2^{-100.76}$.
(18) $(\gamma,\gamma,0,\gamma,\gamma) \stackrel{1r}{\rightarrow} (\gamma,\gamma,0,\gamma,\gamma)$, which is 1-round iterative and holds with bias $2^{-50.88}$.
(19) $(\gamma,\gamma,\gamma,\gamma,\gamma) \stackrel{1r}{\rightarrow} (\gamma,\gamma,\gamma,\gamma,\gamma)$, which is 1-round iterative and holds with bias $2^{-125.70}$.

The best trade-off is achieved by linear relation (8) which is 1-round iterative and holds with bias $2^{-25.94}$ per round. There are no weak-subkey assumptions for any linear relation.

This linear relation can be used to distinguish up to 3-round XC_2 from a random permutation using about $c \cdot 2^{-2} \cdot (2^{-25.94})^{-3} = c \cdot 2^{75.82}$ KP and equivalent encryption effort. c is a constant.

So far, the preliminary differential and linear of XC_2 analyses indicate that the lower bound $r \geq 8$ is adequate.

4.3 XC_3: Encryption and Decryption Frameworks

The third new design, XC_3, operates on $5w$-bit text blocks under a $10w$-bit key and iterates r rounds plus an output transformation, where $r \geq 8$. For $w = 16$ bits, XC_3 operates on 80-bit text blocks, under a 160-bit key.

XC_3 uses only the three original group operations from PES [2], IDEA [1, 3] and the MESH [5] ciphers, that is, no S-boxes nor MDS codes.

Each full round of XC_3 for encryption can be split into two parts: a key-mixing (KM) and an Multiplication-Addition-Xor (MAX) half-round, in this order.

Let $X^{(i)} = (X_1^{(i)}, X_2^{(i)}, X_3^{(i)}, X_4^{(i)}, X_5^{(i)})$ denote the input block to the i-th round, $1 \leq i \leq r+1$. Therefore, $X^{(1)}$ denotes the plaintext (block).

The output of the i-th KM half-round is denoted

$$Y^{(i)} = (Y_1^{(i)}, Y_2^{(i)}, Y_3^{(i)}, Y_4^{(i)}, Y_5^{(i)})$$

where

$$Y^{(i)} = (X_1^{(i)} \boxplus Z_1^{(i)}, X_2^{(i)} \boxplus Z_2^{(i)}, X_3^{(i)} \odot Z_3^{(i)}, X_4^{(i)} \boxplus Z_4^{(i)}, X_5^{(i)} \boxplus Z_5^{(i)})$$

which also becomes the input to the i-th MAX half-round.

The input to the i-th MAX-box is the 3-tuple

$$(Y_1^{(i)} \oplus Y_3^{(i)}, Y_2^{(i)} \boxdot Y_4^{(i)}, Y_3^{(i)} \oplus Y_5^{(i)}) = (n_i, q_i, m_i).$$

The output of the i-th MAX-box is denoted (r_i, s_i, t_i), where $a_i = n_i \boxplus Z_6^{(i)}$; $b_i = a_i \oplus q_i$; $c_i = b_i \odot m_i$; $d_i = c_i \oplus Z_7^{(i)}$; $e_i = b_i \boxplus d_i$; $f_i = a_i \odot e_i$; $r_i = f_i \boxplus Z_8^{(i)}$, $s_i = e_i \oplus r_i$ and $t_i = d_i \odot s_i$.

Note that r_i ultimately depends on all words of (n_i, q_i, m_i). Consequently, r_i depends on all words in $Y^{(i)}$ and in $X^{(i)}$. Therefore, s_i and t_i also depend on all words in $Y^{(i)}$ and in $X^{(i)}$.

Finally, the MAX-box output is combined with the original input words in $Y^{(i)}$ resulting in

$$(Y_1^{(i)} \oplus r_i \oplus t_i, Y_2^{(i)} \odot s_i, Y_3^{(i)} \oplus r_i \oplus t_i, Y_4^{(i)} \odot s_i, Y_5^{(i)} \oplus r_i \oplus t_i).$$

Note that each word depends on at least one word of (r_i, s_i, t_i). Thus, complete text diffusion is achieved in every single round.

Note that $\oplus$ operators are repeated at the output of the MAX-box, which violates one of the original design criterion in PES, IDEA and MESH ciphers, but it is necessary in order to perform proper decryption.

As a last encryption operation in a full round, there is a fixed word permutation denoted σ. The syntax is $\sigma(a, b, c, d, e) = (d, e, c, a, b)$. The i-th round output is denoted

$$X^{(i+1)} = (Y_4^{(i)} \odot s_i, Y_5^{(i)} \oplus r_i \oplus t_i, Y_3^{(i)} \oplus r_i \oplus t_i, Y_1^{(i)} \oplus r_i \oplus t_i, Y_2^{(i)} \odot s_i).$$

This (full) round transformation is repeated r times. The output transformation consists of σ followed by a KM half-round. Therefore, the 5-tuple

$$Y^{(r+1)} = (X_1^{(r+1)} \boxplus Z_1^{(r+1)}, X_2^{(r+1)} \boxplus Z_2^{(r+1)}, X_3^{(r+1)} \odot Z_3^{(r+1)},$$
$$X_4^{(r+1)} \boxplus Z_4^{(r+1)}, X_5^{(r+1)} \boxplus Z_5^{(r+1)})$$

denotes the ciphertext (block) of XC_3.

For decryption, the inverse of the i-th KM half-round consists of the reverse order of round subkeys appropriately transformed by the inverse of $\odot$ or $\boxplus$.

The inverse of the i-th MAX half-round is similar to the original i-th MAX half-round: the input to the MAX-box is also (n_i, q_i, m_i), but the output (r_i, s_i, t_i) is combined via the inverses of $\oplus$ and $\odot$. Therefore, the output of the MAX-box is as follows:

$$(Y_1^{(i)} \oplus r_i \oplus t_i, Y_2^{(i)} \odot s_i, Y_3^{(i)} \oplus r_i \oplus t_i, Y_4^{(i)} \odot s_i, Y_5^{(i)} \oplus r_i \oplus t_i).$$

If the original input to the i-th MAX-box is (n_i, q_i, m_i), then the same input reappears for proper decryption:

$$(Y_1^{(i)} \oplus r_i \oplus t_i \oplus (Y_3^{(i)} \oplus r_i \oplus t_i), ((Y_2^{(i)} \odot s_i) \boxdot (Y_4^{(i)} \odot s_i),$$

$$Y_3^{(i)} \oplus r_i \oplus t_i \oplus (Y_5^{(i)} \oplus r_i \oplus t_i)) = (n_i, q_i, m_i).$$

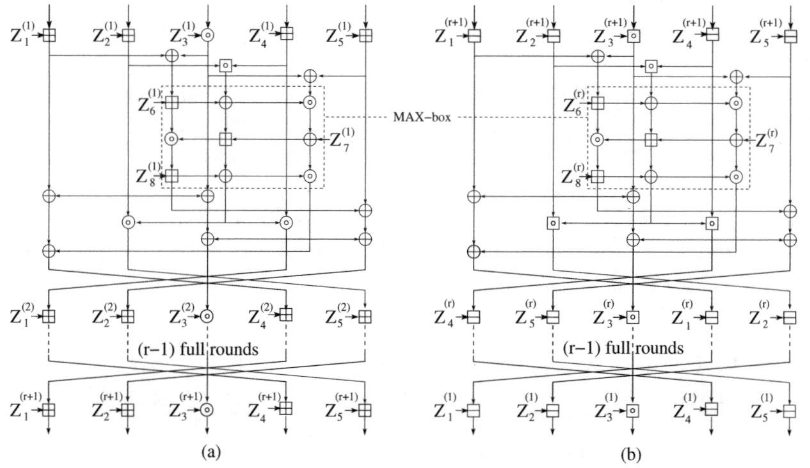

Fig. 4.3 Pictorial representation of the (a) encryption and (b) decryption frameworks of XC_3.

New features of XC_3 include:

- the block size contains an odd number of words, like in XC_2.
- $Y_3^{(i)}$ participates twice in the pairing of words at the input to the i-th MAX-box. This means that $Y_3^{(i)}$ affects more than one input to the MAX-box.
- the KM half-round is the same for all rounds, unlike in MESH ciphers [5] which use two types of KM half-rounds.
- the operators in the input to the MAX-box are $\odot$ and $\oplus$.

4.3.1 Key Schedule Algorithm

The key schedule of XC_3 generates eight 16-bit sukkeys per round, for a total of $8 \cdot r + 5$ subkeys for r rounds plus the output transformation. This is the same number of subkeys as in XC_2.

XC_3 can use the same key schedule as XC_2 in Sect. 4.2.1, inheriting the same properties and security guarantees.

4.3.2 Differential Analysis

This section provides preliminary differential analysis of XC_3 (for a variable number of rounds). The analysis approach is bottom-up, from the smallest components $\odot$, $\boxplus$ and $\oplus$, up to larger ones.

The difference operator is bitwise exclusive-or.

Across unkeyed $\odot$ and $\boxplus$, the propagation of 16-bit difference words follows from Tables 3.12 and 3.13, respectively.

The next step is to obtain all difference patterns across the MAX-box of XC_3.

The only 16-bit difference values used are $\delta = 8000_x$ and 0000_x. The difference 8000_x propagates across the modular addition with certainty, since it affects only the most significant bit (and the last carry-out bit is truncated due to modular reduction.

Table 4.10 lists the propagation of wordwise differences across the MAX-box, the corresponding probability and the number of active $\odot$ operations in the differential trail.

Table 4.10 Difference propagation across the MAX-box of XC_3.

Wordwise differences across the MAX-box	# active $\odot$	Probability
$(0,0,0) \to (0,0,0)$	0	1
$(0,0,\delta) \to (\delta,0,\delta)$	3	2^{-45}
$(0,\delta,0) \to (0,0,\delta)$	2	2^{-30}
$(0,\delta,\delta) \to (\delta,0,0)$	2	2^{-30}
$(\delta,0,0) \to (\delta,\delta,0)$	3	2^{-45}
$(\delta,0,\delta) \to (0,\delta,\delta)$	3	2^{-45}
$(\delta,\delta,0) \to (\delta,\delta,\delta)$	2	2^{-30}
$(\delta,\delta,\delta) \to (0,\delta,0)$	3	2^{-45}

The next step is to construct 1-round characteristics based on Table 4.10.

Table 4.11 lists all nontrivial 1-round differential characteristics of XC_3 under weak-key assumptions, using only wordwise differences 0000_x and 8000_x.

Multiple-round characteristics are obtained by concatenating 1-round characteristics such that the output difference from one characteristic matches the input difference of the next characteristic.

From Table 4.11, the following characteristics can be derived.

(1) $(0,0,0,0,\delta) \xrightarrow{1r} (0,\delta,0,0,0) \xrightarrow{1r} (0,\delta,\delta,\delta,\delta) \xrightarrow{1r} (0,0,0,\delta,0) \xrightarrow{1r} (\delta,\delta,\delta,\delta,0) \xrightarrow{1r} (\delta,0,\delta,\delta,\delta) \xrightarrow{1r} (\delta,0,0,0,0) \xrightarrow{1r} (\delta,\delta,\delta,0,\delta) \xrightarrow{1r} (0,0,0,0,\delta)$, which is 8-round iterative and holds with probability 2^{-540}. The required weak subkeys are: $Z_3^{(i+2)}$, $Z_3^{(i+4)}$, $Z_3^{(i+5)}$ and $Z_3^{(i+7)}$.

(2) $(0,0,0,\delta,\delta) \xrightarrow{1r} (\delta,0,\delta,\delta,0) \xrightarrow{1r} (\delta,\delta,0,0,0) \xrightarrow{1r} (\delta,0,0,\delta,0) \xrightarrow{1r} (0,0,0,\delta,\delta)$, which is 4-round iterative and holds with probability 2^{-270}. The required weak subkey is: $Z_3^{(i+1)}$.

Table 4.11 Nontrivial 1-round characteristics of XC_3 using xor differences and $\delta = 8000_x$.

1-round characteristics	Weak Subkeys	Difference across MAX-box	Prob.
$(0,0,0,0,\delta) \stackrel{1r}{\to} (0,\delta,0,0,0)$	—	$(0,0,\delta) \to (\delta,0,\delta)$	2^{-45}
$(0,0,0,\delta,0) \stackrel{1r}{\to} (\delta,\delta,\delta,\delta,0)$	—	$(0,\delta,0) \to (0,0,\delta)$	2^{-60}
$(0,0,0,\delta,\delta) \stackrel{1r}{\to} (\delta,0,\delta,\delta,0)$	—	$(0,\delta,\delta) \to (\delta,0,0)$	2^{-60}
$(0,0,\delta,0,0) \stackrel{1r}{\to} (\delta,\delta,0,\delta,\delta)$	$Z_3^{(i)}$	$(\delta,0,\delta) \to (0,\delta,\delta)$	2^{-75}
$(0,0,\delta,0,\delta) \stackrel{1r}{\to} (\delta,0,0,\delta,\delta)$	$Z_3^{(i)}$	$(\delta,0,0) \to (\delta,\delta,0)$	2^{-75}
$(0,0,\delta,\delta,0) \stackrel{1r}{\to} (0,0,\delta,0,\delta)$	$Z_3^{(i)}$	$(\delta,\delta,\delta) \to (0,\delta,0)$	2^{-90}
$(0,0,\delta,\delta,\delta) \stackrel{1r}{\to} (0,\delta,\delta,0,\delta)$	$Z_3^{(i)}$	$(\delta,\delta,0) \to (\delta,\delta,\delta)$	2^{-75}
$(0,\delta,0,0,0) \stackrel{1r}{\to} (0,\delta,\delta,\delta,\delta)$	—	$(0,\delta,0) \to (0,0,\delta)$	2^{-60}
$(0,\delta,0,0,\delta) \stackrel{1r}{\to} (0,0,\delta,\delta,\delta)$	—	$(0,\delta,\delta) \to (\delta,0,0)$	2^{-60}
$(0,\delta,0,\delta,0) \stackrel{1r}{\to} (\delta,0,0,0,\delta)$	—	$(0,0,0) \to (0,0,0)$	2^{-45}
$(0,\delta,0,\delta,\delta) \stackrel{1r}{\to} (\delta,\delta,0,0,\delta)$	—	$(0,0,\delta) \to (\delta,0,\delta)$	2^{-90}
$(0,\delta,\delta,0,0) \stackrel{1r}{\to} (\delta,0,\delta,0,0)$	$Z_3^{(i)}$	$(\delta,\delta,\delta) \to (0,\delta,0)$	2^{-90}
$(0,\delta,\delta,0,\delta) \stackrel{1r}{\to} (\delta,\delta,\delta,0,0)$	$Z_3^{(i)}$	$(\delta,\delta,0) \to (\delta,\delta,\delta)$	2^{-75}
$(0,\delta,\delta,\delta,0) \stackrel{1r}{\to} (0,\delta,0,\delta,0)$	$Z_3^{(i)}$	$(\delta,0,\delta) \to (0,\delta,\delta)$	2^{-90}
$(0,\delta,\delta,\delta,\delta) \stackrel{1r}{\to} (0,0,0,\delta,0)$	$Z_3^{(i)}$	$(\delta,0,0) \to (\delta,\delta,0)$	2^{-90}
$(\delta,0,0,0,0) \stackrel{1r}{\to} (\delta,\delta,\delta,0,\delta)$	—	$(\delta,0,0) \to (\delta,\delta,0)$	2^{-75}
$(\delta,0,0,0,\delta) \stackrel{1r}{\to} (\delta,0,\delta,0,\delta)$	—	$(\delta,0,\delta) \to (0,\delta,\delta)$	2^{-75}
$(\delta,0,0,\delta,0) \stackrel{1r}{\to} (0,0,0,\delta,\delta)$	—	$(\delta,\delta,0) \to (\delta,\delta,\delta)$	2^{-75}
$(\delta,0,0,\delta,\delta) \stackrel{1r}{\to} (0,\delta,0,\delta,\delta)$	—	$(\delta,\delta,\delta) \to (0,\delta,0)$	2^{-90}
$(\delta,0,\delta,0,0) \stackrel{1r}{\to} (0,0,\delta,\delta,0)$	$Z_3^{(i)}$	$(0,0,\delta) \to (\delta,0,\delta)$	2^{-45}
$(\delta,0,\delta,0,\delta) \stackrel{1r}{\to} (0,\delta,\delta,\delta,0)$	$Z_3^{(i)}$	$(0,0,0) \to (0,0,0)$	1
$(\delta,0,\delta,\delta,0) \stackrel{1r}{\to} (\delta,\delta,0,0,0)$	$Z_3^{(i)}$	$(0,\delta,\delta) \to (\delta,0,0)$	2^{-60}
$(\delta,0,\delta,\delta,\delta) \stackrel{1r}{\to} (\delta,0,0,0,0)$	$Z_3^{(i)}$	$(0,\delta,0) \to (0,0,\delta)$	2^{-60}
$(\delta,\delta,0,0,0) \stackrel{1r}{\to} (\delta,0,0,\delta,0)$	—	$(\delta,\delta,0) \to (\delta,\delta,\delta)$	2^{-75}
$(\delta,\delta,0,0,\delta) \stackrel{1r}{\to} (\delta,\delta,0,\delta,0)$	—	$(\delta,\delta,\delta) \to (0,\delta,0)$	2^{-90}
$(\delta,\delta,0,\delta,0) \stackrel{1r}{\to} (0,\delta,\delta,0,0)$	—	$(\delta,0,0) \to (\delta,\delta,0)$	2^{-90}
$(\delta,\delta,0,\delta,\delta) \stackrel{1r}{\to} (0,0,\delta,0,0)$	—	$(\delta,0,\delta) \to (0,\delta,\delta)$	2^{-90}
$(\delta,\delta,\delta,0,0) \stackrel{1r}{\to} (0,\delta,0,0,\delta)$	$Z_3^{(i)}$	$(0,\delta,\delta) \to (\delta,0,0)$	2^{-60}
$(\delta,\delta,\delta,0,\delta) \stackrel{1r}{\to} (0,0,0,0,\delta)$	$Z_3^{(i)}$	$(0,\delta,0) \to (0,0,\delta)$	2^{-60}
$(\delta,\delta,\delta,\delta,0) \stackrel{1r}{\to} (\delta,0,\delta,\delta,\delta)$	$Z_3^{(i)}$	$(0,0,\delta) \to (\delta,0,\delta)$	2^{-90}
$(\delta,\delta,\delta,\delta,\delta) \stackrel{1r}{\to} (\delta,\delta,\delta,\delta,\delta)$	$Z_3^{(i)}$	$(0,0,0) \to (0,0,0)$	2^{-45}

(3) $(0,0,\delta,0,0) \stackrel{1r}{\to} (\delta,\delta,0,\delta,\delta) \stackrel{1r}{\to} (0,0,\delta,0,0)$, which is 2-round iterative and holds with probability 2^{-165}. The required weak subkey is $Z_3^{(i)}$.

(4) $(0,0,\delta,0,\delta) \stackrel{1r}{\to} (\delta,0,0,\delta,\delta) \stackrel{1r}{\to} (0,\delta,0,\delta,\delta) \stackrel{1r}{\to} (\delta,\delta,0,0,\delta) \stackrel{1r}{\to} (\delta,\delta,0,\delta,0) \stackrel{1r}{\to} (0,\delta,\delta,0,0) \stackrel{1r}{\to} (\delta,0,\delta,0,0) \stackrel{1r}{\to} (0,0,\delta,\delta,0) \stackrel{1r}{\to} (0,0,\delta,0,\delta)$, which is 8-round iterative and holds with probability 2^{-660}. The required weak subkeys are: $Z_3^{(i)}$, $Z_3^{(i+5)}$, $Z_3^{(i+6)}$ and $Z_3^{(i+7)}$.

(5) $(0,0,\delta,\delta,\delta) \stackrel{1r}{\to} (0,\delta,\delta,0,\delta) \stackrel{1r}{\to} (\delta,\delta,\delta,0,0) \stackrel{1r}{\to} (0,\delta,0,0,\delta) \stackrel{1r}{\to} (0,0,\delta,\delta,\delta)$, which is 4-round iterative and holds with probability 2^{-270}. The required weak subkeys are: $Z_3^{(i)}$, $Z_3^{(i+1)}$ and $Z_3^{(i+2)}$.

(6) $(0,\delta,0,\delta,0) \stackrel{1r}{\to} (\delta,0,0,0,\delta) \stackrel{1r}{\to} (\delta,0,\delta,0,\delta) \stackrel{1r}{\to} (0,\delta,\delta,\delta,0) \stackrel{1r}{\to} (0,\delta,0,\delta,0)$, which is 4-round iterative and holds with probability 2^{-210}. The required weak subkeys are: $Z_3^{(i+2)}$ and $Z_3^{(i+3)}$.

(7) $(\delta,\delta,\delta,\delta,\delta) \stackrel{1r}{\to} (\delta,\delta,\delta,\delta,\delta)$, which is 1-round iterative and holds with probability 2^{-45}. The required weak subkey is $Z_3^{(i)}$.

Characteristic (7) provides the best trade-off in terms of largest probability and least number of weak subkeys per round. Considering XC_3 has 80-bit blocks, (7) can be used to distinguish up to 3-round XC_3 from a random permutation, using about 2^{68} chosen plaintexts and an equivalent encryption effort.

4.3.3 Linear Analysis

This section provides preliminary linear cryptanalysis of XC_3. The analysis approach is bottom-up, from the smallest components $\odot$, $\boxplus$ and $\oplus$, up to larger ones.

Table 3.98 lists the bitmask propagation and the corresponding biases across the unkeyed-$\odot$ operation (under wordwise bitmasks 0 and $\gamma = 1$).

Table 3.99 lists the bitmask propagation and the corresponding biases across $\boxplus$ (under wordwise bitmasks 0 and 1).

The next step is to obtain linear relations across the MAX-box. Table 4.12 lists the bitmasks, the corresponding bias and the number of active $\odot$ along the linear trail.

Table 4.12 Linear bitmask propagation across the MAX-box of XC_3.

Bitmask propagation across the MAX-box	# active $\odot$	Bias
$(0,0,0) \to (0,0,0)$	0	2^{-1}
$(0,\gamma,\gamma) \to (0,0,\gamma)$	3	$2^{-38.41}$
$(\gamma,0,0) \to (0,\gamma,0)$	1	$2^{-13.47}$
$(\gamma,\gamma,\gamma) \to (0,\gamma,\gamma)$	2	$2^{-25.94}$
$(\gamma,0,\gamma) \to (\gamma,0,0)$	2	$2^{-25.94}$
$(\gamma,\gamma,0) \to (\gamma,0,\gamma)$	1	$2^{-13.47}$
$(0,0,\gamma) \to (\gamma,\gamma,0)$	1	$2^{-13.47}$
$(0,\gamma,0) \to (\gamma,\gamma,\gamma)$	2	$2^{-25.94}$

The next step is to construct 1-round linear relations based on Table 4.12. Table 4.13 lists all nontrivial 1-round linear relations for XC_3 under weak-key assumptions and using only bitmasks 0 and $\gamma = 1$.

From Table 4.13, the following linear relations can be derived:

Table 4.13 Nontrivial 1-round linear relations of XC3 using bitmasks 0 and $\gamma = 1$.

1-round linear relation	Weak subkeys	Bitmasks across XA-box	Bias
$(0,0,0,0,\gamma) \stackrel{1r}{\to} (\gamma,\gamma,\gamma,\gamma,\gamma)$	—	$(\gamma,\gamma,0) \to (\gamma,0,\gamma)$	$2^{-50.88}$
$(0,0,0,\gamma,0) \stackrel{1r}{\to} (\gamma,0,\gamma,\gamma,0)$	—	$(\gamma,0,0) \to (0,\gamma,0)$	$2^{-25.94}$
$(0,0,0,\gamma,\gamma) \stackrel{1r}{\to} (0,\gamma,0,0,\gamma)$	—	$(0,\gamma,0) \to (\gamma,\gamma,\gamma)$	$2^{-50.88}$
$(0,0,\gamma,0,0) \stackrel{1r}{\to} (\gamma,0,0,\gamma,\gamma)$	$Z_3^{(i)}$	$(\gamma,\gamma,0) \to (\gamma,0,\gamma)$	$2^{-50.88}$
$(0,0,\gamma,0,\gamma) \stackrel{1r}{\to} (0,\gamma,\gamma,0,0)$	$Z_3^{(i)}$	$(0,0,0) \to (0,0,0)$	2^{-1}
$(0,0,\gamma,\gamma,0) \stackrel{1r}{\to} (0,0,\gamma,0,\gamma)$	$Z_3^{(i)}$	$(0,\gamma,0) \to (\gamma,\gamma,\gamma)$	$2^{-50.88}$
$(0,0,\gamma,\gamma,\gamma) \stackrel{1r}{\to} (\gamma,\gamma,0,\gamma,0)$	$Z_3^{(i)}$	$(\gamma,0,0) \to (0,\gamma,0)$	$2^{-25.94}$
$(0,\gamma,0,0,0) \stackrel{1r}{\to} (0,0,\gamma,\gamma,\gamma)$	—	$(\gamma,0,0) \to (0,\gamma,0)$	$2^{-25.94}$
$(0,\gamma,0,0,\gamma) \stackrel{1r}{\to} (\gamma,\gamma,0,0,0)$	—	$(0,\gamma,0) \to (\gamma,\gamma,\gamma)$	$2^{-50.88}$
$(0,\gamma,0,\gamma,0) \stackrel{1r}{\to} (\gamma,0,0,0,\gamma)$	—	$(0,0,0) \to (0,0,0)$	$2^{-25.94}$
$(0,\gamma,0,\gamma,\gamma) \stackrel{1r}{\to} (0,\gamma,\gamma,\gamma,0)$	—	$(\gamma,\gamma,0) \to (\gamma,0,\gamma)$	$2^{-25.94}$
$(0,\gamma,\gamma,0,0) \stackrel{1r}{\to} (\gamma,0,\gamma,0,0)$	$Z_3^{(i)}$	$(0,\gamma,0) \to (\gamma,\gamma,\gamma)$	$2^{-50.88}$
$(0,\gamma,\gamma,0,\gamma) \stackrel{1r}{\to} (0,\gamma,0,\gamma,\gamma)$	$Z_3^{(i)}$	$(\gamma,0,0) \to (0,\gamma,0)$	$2^{-25.94}$
$(0,\gamma,\gamma,\gamma,0) \stackrel{1r}{\to} (0,0,0,\gamma,0)$	$Z_3^{(i)}$	$(\gamma,\gamma,0) \to (\gamma,0,\gamma)$	$2^{-25.94}$
$(0,\gamma,\gamma,\gamma,\gamma) \stackrel{1r}{\to} (\gamma,\gamma,\gamma,0,\gamma)$	$Z_3^{(i)}$	$(0,0,0) \to (0,0,0)$	$2^{-25.94}$
$(\gamma,0,0,0,0) \stackrel{1r}{\to} (\gamma,0,\gamma,0,\gamma)$	—	$(\gamma,\gamma,0) \to (\gamma,0,\gamma)$	$2^{-50.88}$
$(\gamma,0,0,0,\gamma) \stackrel{1r}{\to} (0,\gamma,0,\gamma,0)$	—	$(0,0,0) \to (0,0,0)$	2^{-1}
$(\gamma,0,0,\gamma,0) \stackrel{1r}{\to} (0,0,0,\gamma,\gamma)$	—	$(0,\gamma,0) \to (\gamma,\gamma,\gamma)$	$2^{-50.88}$
$(\gamma,0,0,\gamma,\gamma) \stackrel{1r}{\to} (\gamma,\gamma,\gamma,0,0)$	—	$(\gamma,0,0) \to (0,\gamma,0)$	$2^{-25.94}$
$(\gamma,0,\gamma,0,0) \stackrel{1r}{\to} (0,0,\gamma,\gamma,0)$	$Z_3^{(i)}$	$(0,0,0) \to (0,0,0)$	2^{-1}
$(\gamma,0,\gamma,0,\gamma) \stackrel{1r}{\to} (\gamma,\gamma,0,0,\gamma)$	$Z_3^{(i)}$	$(\gamma,\gamma,0) \to (\gamma,0,\gamma)$	$2^{-50.88}$
$(\gamma,0,\gamma,\gamma,0) \stackrel{1r}{\to} (\gamma,0,0,0,0)$	$Z_3^{(i)}$	$(\gamma,0,0) \to (0,\gamma,0)$	$2^{-25.94}$
$(\gamma,0,\gamma,\gamma,\gamma) \stackrel{1r}{\to} (0,\gamma,\gamma,\gamma,\gamma)$	$Z_3^{(i)}$	$(0,\gamma,0) \to (\gamma,\gamma,\gamma)$	$2^{-50.88}$
$(\gamma,\gamma,0,0,0) \stackrel{1r}{\to} (\gamma,0,0,\gamma,0)$	—	$(0,\gamma,0) \to (\gamma,\gamma,\gamma)$	$2^{-50.88}$
$(\gamma,\gamma,0,0,\gamma) \stackrel{1r}{\to} (0,\gamma,\gamma,0,\gamma)$	—	$(\gamma,0,0) \to (0,\gamma,0)$	$2^{-25.94}$
$(\gamma,\gamma,0,\gamma,0) \stackrel{1r}{\to} (0,0,\gamma,0,0)$	—	$(\gamma,\gamma,0) \to (\gamma,0,\gamma)$	$2^{-25.94}$
$(\gamma,\gamma,0,\gamma,\gamma) \stackrel{1r}{\to} (\gamma,\gamma,0,\gamma,\gamma)$	—	$(0,0,0) \to (0,0,0)$	$2^{-25.94}$
$(\gamma,\gamma,\gamma,0,0) \stackrel{1r}{\to} (0,0,0,0,\gamma)$	$Z_3^{(i)}$	$(\gamma,0,0) \to (0,\gamma,0)$	$2^{-25.94}$
$(\gamma,\gamma,\gamma,0,\gamma) \stackrel{1r}{\to} (\gamma,\gamma,\gamma,\gamma,0)$	$Z_3^{(i)}$	$(0,\gamma,0) \to (\gamma,\gamma,\gamma)$	$2^{-50.88}$
$(\gamma,\gamma,\gamma,\gamma,0) \stackrel{1r}{\to} (\gamma,0,\gamma,\gamma,\gamma)$	$Z_3^{(i)}$	$(0,0,0) \to (0,0,0)$	$2^{-25.94}$
$(\gamma,\gamma,\gamma,\gamma,\gamma) \stackrel{1r}{\to} (0,\gamma,0,0,0)$	$Z_3^{(i)}$	$(\gamma,\gamma,0) \to (\gamma,0,\gamma)$	$2^{-25.94}$

(1) $(0,0,0,0,\gamma) \stackrel{1r}{\to} (\gamma,\gamma,\gamma,\gamma,\gamma) \stackrel{1r}{\to} (0,\gamma,0,0,0) \stackrel{1r}{\to} (0,0,\gamma,\gamma,\gamma) \stackrel{1r}{\to} (\gamma,\gamma,0,\gamma,0) \stackrel{1r}{\to} (0,0,\gamma,0,0) \stackrel{1r}{\to} (\gamma,0,0,\gamma,\gamma) \stackrel{1r}{\to} (\gamma,\gamma,\gamma,0,0) \stackrel{1r}{\to} (0,0,0,0,\gamma)$, which is 8-round iterative and holds with bias $2^{-250.40}$.
The required weak subkeys are: $Z_3^{(i+1)}$, $Z_3^{(i+3)}$, $Z_3^{(i+5)}$ and $Z_3^{(i+7)}$.

(2) $(0,0,0,\gamma,0) \stackrel{1r}{\to} (\gamma,0,\gamma,\gamma,0) \stackrel{1r}{\to} (\gamma,0,0,0,0) \stackrel{1r}{\to} (\gamma,0,\gamma,0,\gamma) \stackrel{1r}{\to} (\gamma,\gamma,0,0,\gamma) \stackrel{1r}{\to} (0,\gamma,\gamma,0,\gamma) \stackrel{1r}{\to} (0,\gamma,0,\gamma,\gamma) \stackrel{1r}{\to} (0,\gamma,\gamma,\gamma,0) \stackrel{1r}{\to} (0,0,0,\gamma,0)$, which is 8-round iterative and holds with bias $2^{-250.40}$.

The required weak subkeys are: $Z_3^{(i+1)}$, $Z_3^{(i+3)}$, $Z_3^{(i+5)}$ and $Z_3^{(i+7)}$.

(3) $(0,0,0,\gamma,\gamma) \stackrel{1r}{\rightarrow} (0,\gamma,0,0,\gamma) \stackrel{1r}{\rightarrow} (\gamma,\gamma,0,0,0) \stackrel{1r}{\rightarrow} (\gamma,0,0,\gamma,0) \stackrel{1r}{\rightarrow} (0, 0, 0, \gamma, \gamma)$, which is 4-round iterative and holds with bias $2^{-200.52}$.
No weak subkeys are required.

(4) $(0,0,\gamma,0,\gamma) \stackrel{1r}{\rightarrow} (0,\gamma,\gamma,0,0) \stackrel{1r}{\rightarrow} (\gamma,0,\gamma,0,0) \stackrel{1r}{\rightarrow} (0,0,\gamma,\gamma,0) \stackrel{1r}{\rightarrow} (0, 0, \gamma, 0, \gamma)$, which is 4-round iterative and holds with bias $2^{-100.76}$.
The required weak subkeys are: $Z_3^{(i)}$, $Z_3^{(i+1)}$, $Z_3^{(i+2)}$ and $Z_3^{(i+3)}$.

(5) $(0,\gamma,0,\gamma,0) \stackrel{1r}{\rightarrow} (\gamma,0,0,0,\gamma) \stackrel{1r}{\rightarrow} (0,\gamma,0,\gamma,0)$, which is 2-round iterative and holds with bias $2^{-25.94}$. No weak subkeys are required.

(6) $(0,\gamma,\gamma,\gamma,\gamma) \stackrel{1r}{\rightarrow} (\gamma,\gamma,\gamma,0,\gamma) \stackrel{1r}{\rightarrow} (\gamma,\gamma,\gamma,\gamma,0) \stackrel{1r}{\rightarrow} (\gamma,0,\gamma,\gamma,\gamma) \stackrel{1r}{\rightarrow} (0, \gamma, \gamma, \gamma, \gamma)$, which is 4-round iterative and holds with bias $2^{-149.64}$. The required weak subkeys are: $Z_3^{(i)}$, $Z_3^{(i+1)}$, $Z_3^{(i+2)}$ and $Z_3^{(i+3)}$.

(7) $(\gamma,\gamma,0,\gamma,\gamma) \stackrel{1r}{\rightarrow} (\gamma,\gamma,0,\gamma,\gamma)$, which is 1-round iterative and holds with bias $2^{-25.94}$. No weak subkeys are required.

The best trade-off is achieved by linear relation (7) which is 1-round iterative, requires no weak subkeys and holds with bias $2^{-25.94}$ per round.

This linear relation can be used to distinguish up to 3-round XC_3 from a random permutation using about $c \cdot 2^{-2} \cdot (2^{-25.94})^{-3} = c \cdot 2^{75.82}$ KP and equivalent encryption effort. c is a constant.

Therefore, these preliminary differential and linear analyses of XC_3 indicate that the lower bound $r \geq 8$ is adequate.

4.4 XC_4: Encryption and Decryption Frameworks

The next new design, XC_4, operates on $4w$-bit text blocks under an $8w$-bit key and iterates r rounds plus an output transformation, where $r \geq 8$. For $w = 16$ bits, XC_4 operates on 64-bit text blocks, under a 128-bit key.

XC_4 uses only the three original group operations from PES [2], IDEA [1, 3] and the MESH [5] ciphers, that is, no S-boxes nor MDS codes.

Each full round of XC_4 for encryption can be split into two parts: a key-mixing (KM) and an Addition-Multiplication (AM) half-round in this order.

Let $X^{(i)} = (X_1^{(i)}, X_2^{(i)}, X_3^{(i)}, X_4^{(i)})$ denote the input block to the i-th round, $1 \leq i \leq r+1$. Therefore, $X^{(1)}$ denotes the plaintext (block).

The output of the i-th KM half-round is denoted

$$Y^{(i)} = (Y_1^{(i)}, Y_2^{(i)}, Y_3^{(i)}, Y_4^{(i)})$$

where

$$Y^{(i)} = (X_1^{(i)} \odot Z_1^{(i)}, X_2^{(i)} \boxplus Z_2^{(i)}, X_3^{(i)} \boxplus Z_3^{(i)}, X_4^{(i)} \odot Z_4^{(i)})$$

which also becomes the input to the i-th AM half-round.

The input to the i-th AM-box is the 2-tuple

$$(Y_1^{(i)} \oplus Y_2^{(i)} \oplus Y_4^{(i)}, Y_1^{(i)} \oplus Y_3^{(i)} \oplus Y_4^{(i)}) = (n_i, q_i).$$

The output of the i-th AM-box is denoted (r_i, s_i), where $a_i = n_i \boxplus Z_5^{(i)}$; $b_i = a_i \odot q_i$; $s_i = b_i \boxplus Z_6^{(i)}$ and $r_i = a_i \odot s_i$.

Note that r_i ultimately depends on all words of (n_i, q_i). Consequently, r_i depends on all words in $Y^{(i)}$ and in $X^{(i)}$. Therefore, s_i also depends on all words in $Y^{(i)}$ and in $X^{(i)}$.

Finally, the AM-box output is combined with the original input words in $Y^{(i)}$ resulting in

$$(Y_1^{(i)} \oplus r_i Y_2^{(i)} \oplus r_i \oplus s_i, Y_3^{(i)} \oplus s_i \oplus r_i, Y_4^{(i)} \oplus s_i).$$

Note that each word depends on at least one word of (r_i, s_i). Thus, complete text diffusion is achieved in every single round.

Note that $\oplus$ operators are repeated after the AM-box, which violates one of the original design criterion in PES, IDEA and MESH ciphers, but it is necessary in order to perform proper decryption.

As a last encryption operation in a full round, there is a fixed word permutation denoted σ. The syntax is $\sigma(a, b, c, d) = (c, d, a, b)$. The i-th round output is denoted

$$X^{(i+1)} = (Y_3^{(i)} \oplus r_i \oplus s_i, Y_4^{(i)} \oplus s_i, Y_1^{(i)} \oplus r_i, Y_2^{(i)} \oplus r_i \oplus s_i).$$

This (full) round transformation is repeated r times. The output transformation consists of σ followed by a KM half-round. Therefore, the 5-tuple

$$Y^{(r+1)} = (X_1^{(r+1)} \odot Z_1^{(r+1)}, X_2^{(r+1)} \boxplus Z_2^{(r+1)}, X_3^{(r+1)} \boxplus Z_3^{(r+1)}, X_4^{(r+1)} \odot Z_4^{(r+1)})$$

denotes the ciphertext (block) of XC$_4$.

For decryption, the inverse of the i-th KM half-round consists of the reverse order of round subkeys appropriately transformed by the inverse of $\odot$ or $\boxplus$.

The inverse of the i-th AM half-round is similar to the original i-th AM half-round: the input to the AM-box is also (n_i, q_i) and the output (r_i, s_i) is combined via $\oplus$.

If the original input to the i-th AM-box is (n_i, q_i), then the same input reappears for proper decryption:

$$((Y_1^{(i)} \oplus r_i) \oplus (Y_2^{(i)} \oplus r_i \oplus s_i) \oplus (Y_4^{(i)} \oplus s_i), ((Y_1^{(i)} \oplus r_i) \oplus (Y_3^{(i)} \oplus r_i \oplus s_i) \oplus (Y_4^{(i)} \oplus s_i)) =$$
$$(n_i, q_i).$$

New features of XC$_4$ include:

- the inputs to the AM-box contain triplets of the four words in a block. One objective is to avoid iterative characteristics such as $(\delta, \delta, \delta, \delta)$, where

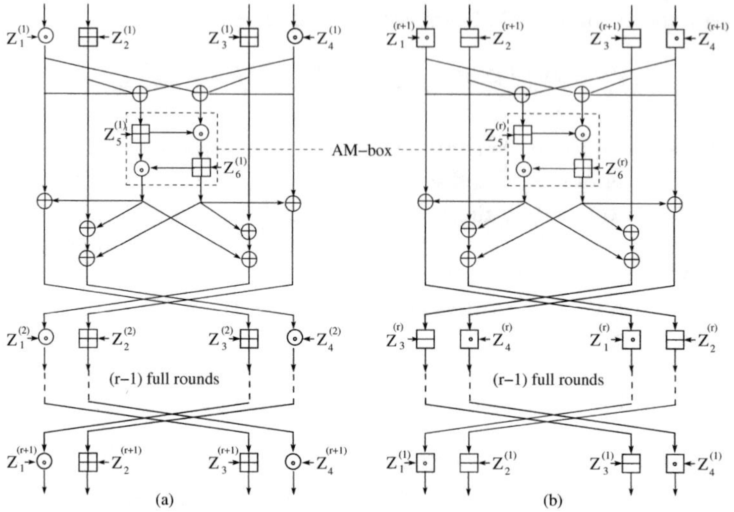

Fig. 4.4 Pictorial representation of the (a) encryption and (b) decryption frameworks of XC_4.

$\delta \neq 0$ is a wordwise difference, from propagating across a round and bypassing the AM-box.
- the KM half-round is the same for all rounds, unlike in MESH ciphers which use two types of KM half-rounds.
- the $\oplus$ operators are repeated at the output of the AM-box.

4.4.1 Key Schedule Algorithm

The key schedule of XC_4 generates six 16-bit sukkeys per round, for a total of $6 \cdot r + 4$ subkeys for r rounds plus the output transformation.

Due to the same block size, XC_4 will use the same key schedule as MESH-64, inheriting the same properties and security guarantees.

4.4.2 Differential Analysis

This section provides preliminary differential analysis of XC_4 (for a variable number of rounds). The analysis approach is bottom-up, from the smallest components $\odot$, $\boxplus$ and $\oplus$, up to larger ones.

The difference operator is bitwise exclusive-or.

Across unkeyed $\odot$ and $\boxplus$, the propagation of 16-bit difference words follows from Tables 3.12 and 3.13, respectively.

The next step is to obtain all difference patterns across the AM-box of XC_4.

The only 16-bit difference values used are $\delta = 8000_x$ and 0000_x. The difference 8000_x propagates across the modular addition with certainty, since it affects only the most significant bit (and the last carry-out bit is truncated due to modular reduction.

Table 4.14 lists the propagation of wordwise differences across the AM-box, the corresponding probability and the number of active $\odot$ operations in the differential trail.

Table 4.14 Difference propagation across the AM-box of XC_4.

Wordwise differences across the AM-box	# active $\odot$	Probability
$(0,0) \rightarrow (0,0)$	0	1
$(0,\delta) \rightarrow (\delta,\delta)$	2	2^{-30}
$(\delta,0) \rightarrow (0,\delta)$	2	2^{-30}
$(\delta,\delta) \rightarrow (\delta,0)$	2	2^{-30}

The next step is to construct 1-round characteristics based on Table 4.14.

Table 4.15 lists all nontrivial 1-round differential characteristics of XC_4 under weak-key assumptions, using only wordwise differences 0000_x and 8000_x.

Table 4.15 Nontrivial 1-round characteristics of XC_4 using xor differences and $\delta = 8000_x$.

1-round characteristics	Weak Subkeys	Difference across AM-box	Prob.
$(0,0,0,\delta) \xrightarrow{1r} (\delta,\delta,\delta,\delta)$	$Z_4^{(i)}$	$(\delta,\delta) \rightarrow (\delta,0)$	2^{-30}
$(0,0,\delta,0) \xrightarrow{1r} (\delta,\delta,\delta,0)$	—	$(0,\delta) \rightarrow (\delta,\delta)$	2^{-30}
$(0,0,\delta,\delta) \xrightarrow{1r} (0,0,0,\delta)$	$Z_4(i)$	$(\delta,0) \rightarrow (0,\delta)$	2^{-30}
$(0,\delta,0,0) \xrightarrow{1r} (\delta,\delta,0,0)$	—	$(\delta,0) \rightarrow (0,\delta)$	2^{-30}
$(0,\delta,0,\delta) \xrightarrow{1r} (0,0,\delta,\delta)$	$Z_4^{(i)}$	$(0,\delta) \rightarrow (\delta,\delta)$	2^{-30}
$(0,\delta,\delta,0) \xrightarrow{1r} (0,0,\delta,0)$	—	$(\delta,\delta) \rightarrow (\delta,0)$	2^{-30}
$(0,\delta,\delta,\delta) \xrightarrow{1r} (\delta,\delta,0,\delta)$	$Z_4^{(i)}$	$(0,0) \rightarrow (0,0)$	1
$(\delta,0,0,0) \xrightarrow{1r} (\delta,0,0,\delta)$	$Z_1^{(i)}$	$(\delta,\delta) \rightarrow (\delta,0)$	2^{-30}
$(\delta,0,0,\delta) \xrightarrow{1r} (0,\delta,\delta,0)$	$Z_1^{(i)}, Z_4^{(i)}$	$(0,0) \rightarrow (0,0)$	1
$(\delta,0,\delta,0) \xrightarrow{1r} (0,\delta,\delta,\delta)$	$Z_1^{(i)}$	$(\delta,0) \rightarrow (0,\delta)$	2^{-30}
$(\delta,0,\delta,\delta) \xrightarrow{1r} (\delta,0,0,0)$	$Z_1^{(i)}, Z_4^{(i)}$	$(0,\delta) \rightarrow (\delta,\delta)$	2^{-30}
$(\delta,\delta,0,0) \xrightarrow{1r} (0,\delta,0,\delta)$	$Z_1^{(i)}$	$(0,\delta) \rightarrow (\delta,\delta)$	2^{-30}
$(\delta,\delta,0,\delta) \xrightarrow{1r} (\delta,0,\delta,0)$	$Z_1^{(i)}, Z_4^{(i)}$	$(\delta,0) \rightarrow (0,\delta)$	2^{-30}
$(\delta,\delta,\delta,0) \xrightarrow{1r} (\delta,0,\delta,\delta)$	$Z_1^{(i)}$	$(0,0) \rightarrow (0,0)$	1
$(\delta,\delta,\delta,\delta) \xrightarrow{1r} (0,\delta,0,0)$	$Z_1^{(i)}, Z_4^{(i)}$	$(\delta,\delta) \rightarrow (\delta,0)$	2^{-30}

Multiple-round characteristics are obtained by concatenating 1-round characteristics such that the output difference from one characteristic matches the input difference of the next characteristic.

From Table 4.15, the following characteristics can be derived.

(1) $(0,0,0,\delta) \stackrel{1r}{\rightarrow} (\delta,\delta,\delta,\delta) \stackrel{1r}{\rightarrow} (0,\delta,0,0) \stackrel{1r}{\rightarrow} (\delta,\delta,0,0) \stackrel{1r}{\rightarrow} (0,\delta,0,\delta) \stackrel{1r}{\rightarrow} (0,0,\delta,\delta) \stackrel{1r}{\rightarrow} (0,0,0,\delta)$, which is 6-round iterative and holds with probability 2^{-180}. The required weak subkeys are: $Z_4^{(i)}$, $Z_1^{(i+1)}$, $Z_4^{(i+1)}$, $Z_1^{(i+3)}$, $Z_4^{(i+4)}$ and $Z_4^{(i+5)}$.

(2) $(0,0,\delta,0) \stackrel{1r}{\rightarrow} (\delta,\delta,\delta,0) \stackrel{1r}{\rightarrow} (\delta,0,\delta,\delta) \stackrel{1r}{\rightarrow} (\delta,0,0,0) \stackrel{1r}{\rightarrow} (\delta,0,0,\delta) \stackrel{1r}{\rightarrow} (0,\delta,\delta,0) \stackrel{1r}{\rightarrow} (0,0,\delta,0)$, which is 6-round iterative and holds with probability 2^{-120}. The required weak subkeys are: $Z_1^{(i+1)}$, $Z_1^{(i+2)}$, $Z_4^{(i+2)}$, $Z_1^{(i+3)}$, $Z_1^{(i+4)}$ and $Z_4^{(i+4)}$.

(3) $(0,\delta,\delta,\delta) \stackrel{1r}{\rightarrow} (\delta,\delta,0,\delta) \stackrel{1r}{\rightarrow} (\delta,0,\delta,0) \stackrel{1r}{\rightarrow} (0,\delta,\delta,\delta)$, which is 3-round iterative and holds with probability 2^{-60}. The required weak subkeys are: $Z_4^{(i)}$, $Z_1^{(i+1)}$, $Z_4^{(i+1)}$ and $Z_1^{(i+2)}$.

(4) $(\delta,\delta,\delta,\delta) \stackrel{1r}{\rightarrow} (\delta,\delta,\delta,\delta)$, which is 1-round iterative and holds with probability 2^{-30}. The required weak subkeys are: $Z_1^{(i)}$ and $Z_4^{(i)}$.

Characteristic (4) provides the best trade-off in terms of largest probability and least number of weak subkeys per round. Considering XC_4 has 64-bit blocks, (4) can be used to distinguish up to 3-round XC_4 from a random permutation, with about 2^{46} chosen plaintexts and an equivalent encryption effort.

4.4.3 Linear Analysis

This section provides preliminary linear cryptanalysis of XC_4. The analysis approach is bottom-up, from the smallest components$\odot$, $\boxplus$ and $\oplus$, up to larger ones.

Table 3.98 lists the bitmask propagation and the corresponding biases across the unkeyed-$\odot$ operation (under wordwise bitmasks 0 and $\gamma = 1$).

Table 3.99 lists the bitmask propagation and the corresponding biases across $\boxplus$ (under wordwise bitmasks 0 and 1).

The next step is to obtain linear relations across the AM-box. Table 4.12 lists the bitmasks, the corresponding bias and the number of active $\odot$ along the linear trail.

The next step is to construct 1-round linear relations based on Table 4.16. Table 4.17 lists all nontrivial 1-round linear relations for XC_4 under weak-key assumptions and using only bitmasks 0 and $\gamma = 1$.

From Table 4.17, the following linear relations can be derived:

Table 4.16 Linear bitmask propagation across the AM-box of XC_4.

Bitmask propagation across the AM-box	# active $\odot$	Bias
$(0,0) \rightarrow (0,0)$	0	2^{-1}
$(0,\gamma) \rightarrow (\gamma,0)$	1	$2^{-13.47}$
$(\gamma,0) \rightarrow (\gamma,\gamma)$	2	$2^{-25.94}$
$(\gamma,\gamma) \rightarrow (0,\gamma)$	1	$2^{-13.47}$

Table 4.17 Nontrivial 1-round linear relations of XC_4 using bitmasks 0 and $\gamma = 1$.

1-round linear relation	Weak subkeys	Bitmasks across AM-box	Bias
$(0,0,0,\gamma) \overset{1r}{\rightarrow} (\gamma,\gamma,0,\gamma)$	—	$(\gamma,\gamma) \rightarrow (0,\gamma)$	$2^{-13.47}$
$(0,0,\gamma,0) \overset{1r}{\rightarrow} (\gamma,\gamma,\gamma,\gamma)$	$Z_4^{(i)}$	$(\gamma,0) \rightarrow (\gamma,\gamma)$	$2^{-13.47}$
$(0,0,\gamma,\gamma) \overset{1r}{\rightarrow} (0,0,\gamma,0)$	—	$(0,\gamma) \rightarrow (\gamma,0)$	$2^{-25.94}$
$(0,\gamma,0,0) \overset{1r}{\rightarrow} (0,\gamma,\gamma,0)$	$Z_4^{(i)}$	$(\gamma,0) \rightarrow (\gamma,\gamma)$	$2^{-13.47}$
$(0,\gamma,0,\gamma) \overset{1r}{\rightarrow} (\gamma,0,\gamma,\gamma)$	$Z_4^{(i)}$	$(0,\gamma) \rightarrow (\gamma,0)$	$2^{-25.94}$
$(0,\gamma,\gamma,0) \overset{1r}{\rightarrow} (\gamma,0,0,\gamma)$	—	$(0,0) \rightarrow (0,0)$	2^{-1}
$(0,\gamma,\gamma,\gamma) \overset{1r}{\rightarrow} (0,\gamma,0,0)$	$Z_4^{(i)}$	$(\gamma,\gamma) \rightarrow (0,\gamma)$	$2^{-13.47}$
$(\gamma,0,0,0) \overset{1r}{\rightarrow} (\gamma,\gamma,0,0)$	$Z_1^{(i)}$	$(0,\gamma) \rightarrow (\gamma,0)$	$2^{-25.94}$
$(\gamma,0,0,\gamma) \overset{1r}{\rightarrow} (0,0,0,\gamma)$	$Z_1^{(i)}, Z_4^{(i)}$	$(\gamma,0) \rightarrow (\gamma,\gamma)$	$2^{-13.47}$
$(\gamma,0,\gamma,0) \overset{1r}{\rightarrow} (0,0,\gamma,\gamma)$	$Z_1^{(i)}$	$(\gamma,\gamma) \rightarrow (0,\gamma)$	$2^{-13.47}$
$(\gamma,0,\gamma,\gamma) \overset{1r}{\rightarrow} (\gamma,\gamma,\gamma,0)$	$Z_1^{(i)}, Z_4^{(i)}$	$(0,0) \rightarrow (0,0)$	2^{-1}
$(\gamma,\gamma,0,0) \overset{1r}{\rightarrow} (\gamma,0,\gamma,0)$	$Z_1^{(i)}$	$(\gamma,\gamma) \rightarrow (0,\gamma)$	$2^{-13.47}$
$(\gamma,\gamma,0,\gamma) \overset{1r}{\rightarrow} (0,\gamma,\gamma,\gamma)$	$Z_1^{(i)}, Z_4^{(i)}$	$(0,0) \rightarrow (0,0)$	2^{-1}
$(\gamma,\gamma,\gamma,0) \overset{1r}{\rightarrow} (0,\gamma,0,\gamma)$	$Z_1^{(i)}$	$(0,\gamma) \rightarrow (\gamma,0)$	$2^{-25.94}$
$(\gamma,\gamma,\gamma,\gamma) \overset{1r}{\rightarrow} (\gamma,0,0,0)$	$Z_1^{(i)}, Z_4^{(i)}$	$(\gamma,0) \rightarrow (\gamma,\gamma)$	$2^{-13.47}$

(1) $(0,0,0,\gamma) \overset{1r}{\rightarrow} (\gamma,\gamma,0,\gamma) \overset{1r}{\rightarrow} (0,\gamma,\gamma,\gamma) \overset{1r}{\rightarrow} (0,\gamma,0,0) \overset{1r}{\rightarrow} (0,\gamma,\gamma,0) \overset{1r}{\rightarrow} (\gamma,0,0,\gamma) \overset{1r}{\rightarrow} (0,0,0,\gamma)$, which is 6-round iterative and holds with bias $2^{-50.88}$.
The required weak subkeys are: $Z_4^{(i)}$, $Z_1^{(i+1)}$, $Z_4^{(i+1)}$, $Z_4^{(i+2)}$, $Z_1^{(i+5)}$ and $Z_4^{(i+5)}$.

(2) $(0,0,\gamma,0) \overset{1r}{\rightarrow} (\gamma,\gamma,\gamma,\gamma) \overset{1r}{\rightarrow} (\gamma,0,0,0) \overset{1r}{\rightarrow} (\gamma,\gamma,0,0) \overset{1r}{\rightarrow} (\gamma,0,\gamma,0) \overset{1r}{\rightarrow} (0,0,\gamma,\gamma) \overset{1r}{\rightarrow} (0,0,\gamma,0)$, which is 6-round iterative and holds with bias $2^{-100.76}$.
The required weak subkeys are: $Z_1^{(i+1)}$, $Z_4^{(i+1)}$, $Z_1^{(i+2)}$, $Z_1^{(i+3)}$, $Z_1^{(i+4)}$ and $Z_4^{(i+5)}$.

(3) $(0,\gamma,0,\gamma) \overset{1r}{\rightarrow} (\gamma,0,\gamma,\gamma) \overset{1r}{\rightarrow} (\gamma,\gamma,\gamma,0) \overset{1r}{\rightarrow} (0,\gamma,0,\gamma)$, which is 3-round iterative and holds with bias $2^{-50.88}$.
The required weak subkeys are: $Z_4^{(i)}$, $Z_1^{(i+1)}$, $Z_4^{(i+1)}$ and $Z_1^{(i+2)}$.

The best trade-off is achieved by the first three rounds of linear relation (1), which can be used to distinguish up to 3-round XC_4 from a random permutation using $c \cdot 2^{-1} \cdot (2^{-25.94})^{-2} = c \cdot 2^{50.88}$ known plaintexts (KP) and equivalent encryption effort (where c is a constant).

So far, the preliminary differential and linear analyses of XC_4 indicate that the lower bound $r \geq 8$ is adequate.

References

1. Lai, X.: On the Design and Security of Block Ciphers, *ETH Series in Information Processing*, vol. 1. Hartung-Gorre Verlag, Konstanz (1995)
2. Lai, X., Massey, J.L.: A Proposal for a New Block Encryption Standard. In: I.B. Damgård (ed.) Advances in Cryptology, Eurocrypt, LNCS 473, pp. 389–404. Springer (1990)
3. Lai, X., Massey, J.L., Murphy, S.: Markov Ciphers and Differential Cryptanalysis. In: D.W. Davies (ed.) Advances in Cryptology, Eurocrypt, LNCS 547, pp. 17–38. Springer (1991)
4. Lerman, L., Nakahara Jr, J., Veshchikov, N.: Improving Block Cipher Design by Rearranging Internal Operations. In: P. Samarati (ed.) 10th International Conference on Security and Cryptography (SECRYPT), pp. 27–38. IEEE (2013)
5. Nakahara Jr, J., Rijmen, V., Preneel, B., Vandewalle, J.: The MESH Block Ciphers. In: K. Chae, M. Yung (eds.) Information Security Applications (WISA), LNCS 2908, pp. 458–473. Springer (2003)

Chapter 5
Conclusions

A pessimist sees the difficulty in every opportunity; an optimist sees the opportunity in every difficulty. – Winston S. Churchill

Abstract

This chapter summarizes attack results, draws our final conclusions and comments on the Lai-Massey cipher designs analyzed in previous chapters.

5.1 The Lai-Massey Design Paradigm

This book focuses on a specific type of cryptographic algorithm: block ciphers. More specifically, we discuss and study block ciphers that follow the design paradigm known as *Lai-Massey*.

The terminology *Lai-Massey cipher designs* used in this book was meant to refer to those block ciphers that follow, in one way or another, the design criteria originally set for the PES and IDEA ciphers created by X. Lai and J.L. Massey. For the IDEA cipher, there was also the contribution of S. Murphy [15].

Chap. 1 presented a non-exhaustive survey of known block ciphers, their main parameters and design paradigms. The majority of the existing block ciphers follows either a Feistel Network or a Substitution-Permutation Network (SPN) paradigm. But, there is a third paradigm called *Lai-Massey*, that although accounting for a minor fraction of the existing block ciphers, still present relevant design contributions to cryptography, such as:

- round functions performing full text diffusion in a single round.
- originally, use of only three (incompatible) group operations: bitwise exclusive-or, modular addition in $Z_{2^{16}}$ and modular multiplication in $GF(2^{16}+1)$, with $0 \equiv 2^{16}$. Moreover, none of these operations were ever used twice in sequence along the encryption or decryption frameworks.
- small number of rounds compared to Feistel Network and SPN designs.
- originally, no S-boxes nor MDS codes were used.

J. Nakahara Jr., *Lai-Massey Cipher Designs*, https://doi.org/10.1007/978-3-319-68273-0_5

Chap. 2 provided detailed descriptions of sixteen ciphers (or cipher families) that adopt somehow the Lai-Massey design paradigm. See Table 5.1. For FOX ciphers, k is a multiple of 8.

Table 5.1 Lai-Massey cipher (and hybrid) designs in Chap. 2 in chronological order.

Cipher	Block Size (bits)	Key Size (bits)	# Rounds	Word size (bits)	Reference	**Year**
PES	64	128	8.5	16	[14]	1990
IDEA	64	128	8.5	16	[15]	1991
YBC	64	128	6.5	16	[27]	1996
RIDEA	64	128	8.5	16	[28]	2002
MESH-64	64	128	8.5	16	[23]	2003
MESH-96	96	192	10.5	16	[23]	2003
MESH-128	128	256	12.5	16	[23]	2003
FOX-64/k/r	64	$0 \leq k \leq 256$	$12 \leq r \leq 255$	8	[11]	2004
FOX-128/k/r	128	$0 \leq k \leq 256$	$12 \leq r \leq 255$	8	[11]	2004
MESH-64(8)	64	128	8.5	8	[21]	2005
MESH-128(8)	128	256	8.5	8	[21]	2005
Bel-T	128	128, 192, 256	8	32	[1]	2007
WIDEA-4	256	512	8.5	16	[10]	2009
WIDEA-8	512	1024	8.5	16	[10]	2009
IDEA*	64	128	6.5	16	[16]	2013
REESSE3+	128	256	8.5	16	[24]	2014

Moreover, this chapter also showed some interesting design contributions of each cipher.

- in PES and IDEA, all multiplications are keyed, as well as half of all modular addition operations.
- YBC was the first design to use the unkeyed-$\odot$ and keyed-$\oplus$ operations (see Table 5.2). Also, YBC did not have a clearly defined MA-box. Moreover, only the KM half-round was keyed. In fact, the YBC framework has a hybrid design combining an unbalanced Feistel Network with some features from Lai-Massey ciphers. No analysis of YBC was found in public records. Finally, YBC was the first design to iterate less than 8.5 rounds.
- the FOX ciphers were the first designs to drop the $\boxplus$ and $\odot$ operations and to use S-boxes and MDS codes. The absence of $\boxplus$ and $\odot$ helped remove the issue of weak keys (that is a concern for PES and IDEA). FOX ciphers operate on 8-bit words (instead of 16 bits), and use a fourth group operation: multiplication in GF(2^{16}) (see Table 5.2).
- WIDEA-n were the first designs to operate on block sizes larger than 128 bits, and to adopt a 3-dimensional design.
- the MESH ciphers were the first designs to use two different types of KM half-rounds; MESH-96 and MESH-128 were the first designs with block sizes larger than 64 bits, and also to adopt the unkeyed $\odot$ operation.

- REESSE3+, like YBC, also used the unkeyed $\odot$ operation, although the former had a well-defined MA-box.
- IDEA* was the first design to use unkeyed $\boxdot$ in the encryption framework. Also, all $\oplus$ operations are keyed in IDEA*. Due to the use of $\odot$ in the output of the AX half-round, the encryption and decryption frameworks are more dissimilar than previous Lai-Massey designs: the decryption framework has to use $\boxdot$ instead of $\odot$. Finally, IDEA* contains a nonlinear diffusion layer: although the AX-box is a linear transformation because the unkeyed-$\odot$ (or unkeyed-$\boxdot$) used to combined its output are nonlinear operations.
- Bel-T, the Belarussian block cipher standard, is also a hybrid design like YBC, combining features from Feistel and Lai-Massey frameworks.

Table 5.2 lists and compares the different internal components of the fifteen Lai-Massey cipher designs described in Chap. 2. This table shows quantitatively the type and number of elementary operations such as $\oplus$, $\boxplus$, $\odot$ and S-boxes used by each cipher.

Table 5.2 Operators used in the cipher designs in Chapter 2.

Cipher	# Unkeyed			# Keyed			# S-boxes	MDS codes
	$\oplus$	$\boxplus$	$\odot$	$\oplus$	$\boxplus$	$\odot$		
PES	48	16	—	—	18	34	—	No
IDEA	48	16	—	—	18	34	—	No
YBC	36	24	24	28	—	—	—	No
Bel-T	40	40	—	—	56	—	28	No
RIDEA	48	8	8	—	26	26	—	No
MESH-64	48	24	—	—	18	42	—	No
MESH-96	90	30	30	—	43	53	—	No
MESH-128	144	96	48	—	52	100	—	No
FOX-64/k/r†	112	—	—	96	—	—	128	Yes
FOX-128/k/r†	192	—	—	224	—	—	256	Yes
MESH-64(8)†	96	32	16	—	36	52	—	No
MESH-128(8)†	192	64	48	—	72	88	—	No
WIDEA-4	192	64	—	—	72	136	—	Yes
WIDEA-8	384	128	—	—	144	272	—	Yes
IDEA*	—	12	36	26	14	—	—	No
REESSE3+	96	24	24	—	32	32	—	No

†: these ciphers use 8-bit words, but the number of operations are counted as if words were 16 bits wide; moreover, for FOX ciphers, the number of $\oplus$ in the MDS matrix computations are not accounted for. All figures correspond to the nomimal number of rounds. See Table 5.1. For FOX ciphers, the figures correspond to $r = 16$ rounds.

Chap. 3 provided a cryptanalytic survey with detailed descriptions of both known and unpublished attacks on the Lai-Massey cipher designs in Chap 2. More specifically, Chap. 3 covered twenty different analysis techniques (both general and dedicated attacks): exhaustive search, dictionary attack, Time-Memory Trade-off (TMTO), Differential Cryptanalysis, Truncated Differentials, Multiplicative Differential analysis, Impossible Differentials, Slide, Advanced slide, Biclique, Boomerang, Linear, Differential-Linear, Square, Demirci, Biryukov-Demirci, Key-Dependent, BDK, Meet-in-the-Middle and related-key attacks.

The attacks detailed in this book cover research results published during the period 1990-2016, that is, during the last 27 years (since PES was released).

Many of the described attacks follow a top-down approach, which allows not only independent verification of the attack complexities (reproducibility) by third parties, but also serves an instructive purpose for readers willing to better understand these cryptanalytic attacks.

The attack coverage in Chap. 3 is not uniform across all Lai-Massey cipher designs. The majority of the analyses results in the cryptographic literature are on IDEA (for a variable number of rounds), since it is the oldest (together with PES) and most well-known design. Moreover, due to its redesign from PES, in order to withstand differential cryptanalysis, IDEA became a benchmark against which all new analysis techniques were tested (other benchmark ciphers include the AES and the DES).

Tables 5.3 to 5.28 summarize the attack complexities of the ciphers in Chap. 3. The list of acronyms (in the beginning of this book) contains the terminology used in these tables.

Unless described otherwise, the time complexity is measured in number of encryption rounds indicated in the leftmost column. Sometimes, another unit of measurement is used: number memory accesses (MA). In these cases, data is computed (off-line) and stored in hash tables to save computations that are repeated many times during an attack. It is a time-memory trade-off approach. The problem is how to translate the cost of a table lookup into an equivalent number of t-round encryptions, where the cipher under analysis has t rounds. Unless the authors explicitly made such a conversion, the two units of measure are listed.

It is important to note that sometimes, the attack complexities listed for each attack may differ from the original source because: (i) different units of measures of complexity were used; (ii) not the whole user key was recovered, but only part of it. In order to put all the attacks on the same footing (full user key recovery, and the same units of measure), some attack complexities had to be recomputed, which means the attack complexities increased compared to the original source. In these cases, the source mentioned is a section in this book, where the attacks are described in detail.

The data complexity is measured in number of text blocks needed under different setting, such as KP, CP, CC, ACPCC and so on. No attacks operating in a ciphertext-only (CO) setting were found or reported in the literature on any Lai-Massey cipher design.

The memory complexity is the storage needed for the attack and is measured in number of full text blocks.

Table 5.3 summarizes the complexities of distinguish-from-random (DFR) attacks on the PES cipher, under weak-key assumptions, for a variable number of rounds. The aim of DFR attacks is not to recover key material but rather provide tools and evidence to distinguish a given cipher from a random permutation (over the same text domain). Beyond seventeen rounds of PES there is only one (weak) user key for which PES can be distinguished from a random permutation: the all-zero key 0^{128}.

The largest weak-key class (for 8.5 rounds) is identified by a differential-linear attack, and it contains 2^{62} keys (out of 2^{128}).

Several block ciphers have been reported with weak keys for the full number of rounds, for example, DES [20], TEA [12], PES [22], MISTY1 [17], LOKI89 [3] and Crypton [6].

The existence of weak keys for PES is a persistence concern for the encryption/decryption operations since the key schedule of PES is so weak that there is at least one weak (user) key for any number of rounds [22] (way beyond the original 8.5 rounds): the all-zero key 0^{128}.

Although for a small number of rounds, the fraction of weak keys is small compared to the key space, weak keys become a real security issue if PES is used as a building block in other cryptographic constructions such as hash functions [25], in which the key input may be under the control of an adversary. In this case, a single weak key is enough to compromise security because the key input can be manipulated to be weak, affecting the behavior of the encryption scheme in a predictable way that can be exploited by the adversary.

The fact that the full 8.5-round PES can be distinguished from a random permutation with low attack complexities means that PES is not a pseudorandom permutation (PRP), according to Sect. 1.7. Consequently, PES is not an ideal cryptographic primitive, which becomes evident from the discussion of the use of PES as a cryptographic component in a hash function setting.

Table 5.3 DFR attack complexities on PES (under weak-key assumptions); c is a constant.

#Rounds	Attack	Time	Data/Memory	Source	$\|WKC\|$
8.5	Differential	2	2 CP	Table 3.7	2^{41}
8.5	Linear	$4c$	$4c$ KP	Table 3.93	2^{48}
8.5	Diff.-Linear	$4c$	$4c$ CP	Sect. 3.14.1	2^{62}
> 17	Linear	2	2 KP	Sect. 3.13.1	1

Table 5.4 summarizes the complexities of key-recovery attacks in Chap. 3 on the PES cipher for a variable number of rounds. These attacks hold for any key, that is, there is no weak-key assumption.

The differential attack in [15] uses the whole codebook. It is arguable whether key-recovery attacks that require the whole codebook can be justified. Once an adversary is in possession of the whole codebook (for a fixed, unknown) key, he/she can encrypt or decrypt any plaintext/ciphertext of his/her choice while the key is still in use. In fact, the codebook discloses the whole encryption mapping for the unknown key only[1]. There is no clear justification for recovering the user key in this case. Moreover, if the whole codebook is available (to the adversary), then the unknown key is quite probably not in use anymore.

Table 5.4 Key-recovery attack complexities on PES (without weak-key assumptions).

#Rounds	Time	Data	Memory	Attack	Source	Comments
8.5	2^{64}	2^{64} CP	2^{64}	Differential	[15]	
any ‡	2^{64}	2^{32} †	2^{64}	TMDTO	[4, 5]	2^{96} OPP
any ‡	2^{64}	2^{64} †	2^{64}	Key Collision	[26]	
any ‡	2^{64}	2^{64} †	2^{64}	Biham's	[2]	2^{64} OPP
any ‡	$2^{85.33}$	1 †	$2^{85.33}$	Hellman's TMTO	[9]	2^{128} OPP
any ‡	$2^{85.33}$	$2^{42.66}$ †	$2^{42.66}$	TMDTO	[4, 5]	$2^{85.33}$ OPP

†: Data complexity means the number of different keys.
‡: the number of rounds is such that all user key bits are used. For PES, it means at least 1.5 rounds.

Table 5.5 summarizes distinguish-from-random (DFR) attack complexities on the IDEA cipher, under weak-key assumptions, for a variable number of rounds. The largest weak-key class (for 8.5 rounds) is identified by a boomerang attack and it contains 2^{64} keys (out of 2^{128}).

Since IDEA inherited the same key schedule algorithm of PES, similar weak-key problems exist for IDEA, under different attack settings.

The fact that the full 8.5-round IDEA (and even variants with many more rounds) can be distinguished from a random permutation with low attack complexities means that IDEA is not a pseudorandom permutation (PRP), according to Sect. 1.7. Consequently, IDEA is not an ideal cryptographic primitive, similar to the situation of PES when used as a component in a hash function setting [25].

There are many other Differential-Linear distinguishers listed in Table 3.145 for IDEA variants but with less than 8.5 rounds. Also, Table 3.156 lists other related-key Differential-Linear distinguishers for IDEA variants with less than 8.5 rounds. Finally, there are also many Boomerang distinguishers for IDEA variants with less than 8.5 round in Sect. 3.12.

Table 5.6 summarizes the complexities of key-recovery attacks on several reduced-round variants of IDEA, *without* weak-key assumptions. This means that these attacks hold for any key. The attacks are ordered by increasing number of rounds. For a fixed number of rounds, attacks are order by increasing time complexity. Time complexity is used as reference because these

[1] Except for the case of equivalent keys

Table 5.5 DFR attack complexities on IDEA (under weak-key assumptions); c is a constant.

#Rounds	Attack	Time	Data	Memory	Source	$\|WKC\|$
8	Rel.-Key Diff.Linear	2	2 CP	2	Sect. 3.21.1	2^{70}
8.5	HO Diff.-Linear	1	2^{23} CP	2^{16}	Sect. 3.14.3	2^{31}
8.5	Differential	2	2 CP	2	Table 3.10	2^{35}
8.5	Linear	$4c$	$4c$ KP	$4c$	Table 3.95	2^{23}
8.5	Diff.-Linear	19.5	19.5 CP	19.5	Sect. 3.14.2	2^{63}
8.5	Boomerang	2^{16}	2^{16} ACPC	2^{16}	Table 3.87	2^{64}
14	Linear	$4c$	$4c$ KP	$4c$	Sect. 3.13.2	16
> 14	Linear	2	2 KP	2	Sect. 3.13.2	1

attacks are compared to the exhaustive key search (EKS) attack. The memory complexity is measured in number of full text blocks.

The attack figures related to *any* number of rounds are derived from Table 3.2 in Chap. 3.

Based only on the time complexity, the Biclique attack reaches the full 8.5-round IDEA with effort less than that of an EKS. But, taking into account the Hellman's and TMDTO settings at the end of Table 5.7, the Biclique attack does not achieve a better complexity trade-off.

Table 5.8 summarizes distinguish-from-random attack complexities on MESH-64, under weak-key assumptions, for a variable number of rounds. The largest weak-key class (for the largest number of rounds) is identified by boomerang attack and it contains 2^8 keys (out of 2^{128}).

Table 5.9 summarizes the complexities of key-recovery attacks on MESH-64, for a variable number of rounds, *without* weak-key assumptions. This means that these attacks hold for any key. The attacks are ordered by increasing number of rounds. For a fixed number of rounds, attacks are ordered by increasing time complexity.

The DFR attacks on up to 5.5 rounds and key-recovery attacks on up to 4 rounds so far corroborate the design choices for MESH-64, including its number of rounds.

Taking into account only the time complexity, the best known key-recovery attack uses the Impossible-Differential technique and reaches 4-round MESH-64. The other attack on 4-round MESH-64 requires the whole codebook and is not faster than EKS.

The best generic attack (for any number of rounds) is TMDTO, requiring an offline pre-processing complexity of 2^{96} MESH-64 computations, and recovering at least one of 2^{32} keys. The Impossible Differential attack reaches about half of MESH-64, but it does not offer a better complexity trade-off than the TMDTO approach.

Table 5.10 summarizes distinguish-from-random attack complexities on MESH-96, for a variable number of rounds, under weak-key assumptions.

Table 5.6 Key-recovery attack complexities on IDEA (without weak-key assumptions).

#Rounds	Time	Data	Memory	Attack	Source
1.5	2^{52}	82 KP	82	Biryukov-Demirci	Sect. 3.17.1.1
2	2^{42}	2^{10} CP	2^{10}	Differential	[19]
2	2^{52}	90 KP	90	Biryukov-Demirci	Sect. 3.17.1.1
2.5	2^{18}	2^{18} CP	2^{18}	Keyless-BD	Sect. 3.19
2.5	2^{32}	$2^{17.58}$ CP	2^{16}	Square/Multiset	Sect. 3.15.1
2.5	2^{42}	50 CP + 40 KP	90	Biryukov-Demirci	Sect. 3.17.1.1
2.5	$2^{44.29}$	2^{50} CP	2^{16}	Square/Multiset	Sect. 3.15.1.1
2.5	$2^{51.5}$	49 KP	49	TMTO Biryukov-Demirci	Sect. 3.17.1.2
2.5	2^{89}	90 KP	90	Biryukov-Demirci	Sect. 3.17.1
2.5	2^{106}	2^{10} CP	2^{10}	Differential	[19]
3	$2^{44.41}$	2^{23} CP	2^{41}	MITM	Sect. 3.20.3
3	2^{58}	56 KP	56	TMTO Biryukov-Demirci	Sect. 3.17.1.2
3	2^{68}	114 KP	114	Biryukov-Demirci	Sect. 3.17.1.1
3	2^{88}	2^{24} CP	2^{58}	MITM	Sect. 3.20.2
3	2^{96}	2^{19} cp	2^{19}	Keyless-BD	Sect. 3.19
3.5	2^{45}	2^{30} CP	2^{29}	Diff.-Linear	[7]
3.5	2^{48}	$2^{22.58}$ CP	2^{16}	Square/Multiset	Sect. 3.15.1.2
3.5	$2^{52.59}$	$2^{38.5}$ CP	$2^{52.49}$	Imp. Diff.	Sect. 3.8.1.1
3.5	2^{67}	2^{56} CP	2^{32}	Trunc. Diff.	Sect. 3.6.1.1
3.5	$2^{68.19}$	$2^{23.6}$ CP	2^{41}	MITM	Sect. 3.20.3
3.5	2^{73}	2^{24} CP	2^{58}	MITM	Sect. 3.20.2
3.5	2^{83}	81 KP	81	TMTO Biryukov-Demirci	Sect. 3.17.1.2
3.5	2^{100}	97 CP	97	Biryukov-Demirci	Sect. 3.17.1.1
3.5	2^{112}	112 KP	112	Biryukov-Demirci	Sect. 3.17.1
4	2^{70}	2^{38} CP	2^{63}	Imp. Diff.	Sect. 3.8.1.2
4	2^{82}	2^{24} CP	2^{41}	MITM	Sect. 3.20.3
4	2^{89}	2^{24} CP	2^{58}	MITM	Sect. 3.20.2
4	2^{114}	114 KP	114	Biryukov-Demirci	Sect. 3.17.1
4.5	2^{103}	2 KP	2^{16}	MITM	Sect. 3.20.1.2
4.5	2^{103}	16 CP	16	Keyless-BD	Sect. 3.19
4.5	2^{108}	2^{64} CP	2^{32}	Imp. Diff.	Sect. 3.8.1.3
4.5	$2^{115.83}$	$2^{24.6}$ CP	2^{41}	MITM	Sect. 3.20.3
4.5	$2^{120.9}$	$2^{58.5}$ KP	$2^{43.6}$	Statistical Integral	[8]
5	2^{103}	2^{19} KP	2^{19}	Keyless-BD	Sect. 3.19
5	$2^{101.5} + 2^{112}$ MA	2^{25} CP	2^{110}	Biclique	[13]
5	2^{110}	2^{25} CP	2^{16}	Biclique	[13]
5	$2^{115.3}$	2^{64} KP	2^{64}	Key-Dependent	Sect. 3.18.1
5	2^{119}	10 KP	2^{24}	MITM	Sect. 3.20.1.2
5	2^{119}	10 KP	2^{24}	MITM	Sect. 3.20.1.2
5	$2^{120.67}$	$2^{24.6}$ CP	2^{41}	MITM	Sect. 3.20.3
5	$2^{125.5}$	2^{17} CP	2^{17}	Key-Dependent	Sect. 3.18.1
5.5	$2^{112.1}$	2^{21} CP	2^{21}	Key-Dependent	Sect. 3.18.1
6	2^{96}	$2^{51.6}$ RK-ACPC	$2^{51.6}$	Rel.-Key Boomerang	Sect. 3.21.4.1
6	$2^{111.9}$	16 KP	2^{25}	MITM	Sect. 3.20.1.2
6	$2^{112.1}$	2^{49} CP	2^{49}	Key-Dependent	Sect. 3.18.1
6	$2^{118.9}$	2^{25} CP	2^{12}	Biclique	[13]
6	$2^{123.4}$	2 KP	2^{25}	MITM	Sect. 3.20.1.2
6.5	$2^{85.9}$	2^{65} RK-CP	2^{65}	Rel.-Key Rectangle	Sect. 3.21.3
6.5	$2^{88.1}$	$2^{59.8}$ RK-CP	259.8	Rel.-Key Rectangle	Sect. 3.21.4.2
6.5	$2^{111.9}$	2^{32} CP	2^{32}	SaC MITM BD	Sect. 3.20.1.2
6.5	2^{113}	2^{23} CP	2^{23}	SaC MITM BD	Sect. 3.20.1.2
6.5	2^{122}	2^{10} CP	2^{10}	SaC MITM BD	Sect. 3.20.1.2
7	$2^{104.2}$	2^{65} RK-CP	2^{65}	Rel.-Key Rectangle	Sect. 3.21.3
7	2^{112}	2^{48} CP	2^{48}	SaC MITM BD	Sect. 3.20.1.2
7	2^{123}	2^{38} CP	2^{38}	SaC MITM BD	Sect. 3.20.1.2

Table 5.7 Key-recovery attack complexities on IDEA (without weak-key assumptions).

#Rounds	Time	Data	Memory	Attack	Source	Comments
7.5	2^{114}	2^{63} CP	2^{63}	SaC MITM BD	Sect. 3.20.1.2	
7.5	$2^{115.1}$	$2^{43.5}$ RK-KP	$2^{43.5}$	Rel.-Key Keyless-BD	Sect. 3.21.2	
7.5	$2^{123.9}$	2^{52} CP	2^{7}	Biclique	Sect. 3.11.1.2	
7.5	$2^{125.9}$	16 CP	16	SaC MITM BD	Sect. 3.20.1.2	
7.5	$2^{126.5}$	2^{18} CP	2^{3}	Biclique	Sect. 3.11.1.2	
8.5	$2^{125.97}$	2^{59} CP	2^{3}	Biclique	Sect. 3.11.1.1	
8.5	$2^{126.06}$	2^{52} CP	2^{3}	Biclique	Sect. 3.11.1.1	
8.5	$2^{126.8}$	16 CP	16	SaC MITM BD	Sect. 3.20.1.2	
any ‡	2^{64}	2^{32} †	2^{64}	TMDTO	[4, 5]	2^{96} OPP
any ‡	2^{64}	2^{64} †	2^{64}	Key Collision	[26]	
any ‡	2^{64}	2^{64} †	2^{64}	Biham's	[2]	2^{64} OPP
any ‡	$2^{85.33}$	1 †	$2^{85.33}$	Hellman's TMTO	[9]	2^{128} OPP
any ‡	$2^{85.33}$	$2^{42.66}$ †	$2^{42.66}$	TMDTO	[4, 5]	$2^{85.33}$ OPP

†: Data complexity means the number of different keys.
‡: the number of rounds is such that all user key bits are used. For IDEA, it means at least 1.5 rounds.

Table 5.8 DFR attack complexities on the MESH-64 cipher (under weak-key assumptions); c is a constant.

#Rounds	Time	Data/Memory	Attack	Source	$\|WKC\|$
4	2	2 CP	Differential	Sect. 3.5.3	2^{8}
4	$4c$	$4c$ KP	Linear	Sect 3.13.4	2^{8}
5	$4c$	$4c$ CP	Diff.-Linear	Sect. 3.14.4	2^{8}
5.5	2^{10}	2^{10} ACPC	Boomerang	Sect. 3.12.2	2^{23}

Table 5.9 Key-recovery attack complexities on the MESH-64 cipher (without weak-key assumptions).

#Rounds	Time	Data	Memory	Attack	Source	Comments
1.5	2^{97}	97 KP	97	Biryukov-Demirci	Sect. 3.17.2	
1.5	2^{128}	128 KP	128	Biryukov-Demirci	Sect. 3.17.2	
2.5	2^{68}	2^{51} CP	2^{16}	Square/Multiset	Sect. 3.15.2	
3.5	2^{64}	$2^{39.5}$ CP	2^{58}	Imp. Diff.	Sect. 3.8.2	
3.5	2^{77}	2^{64} CP	2^{32}	Trunc. Diff.	Sect. 3.6.2	WCU
4	2^{112}	$2^{39.5}$ CP	2^{58}	Imp. Diff.	Sect. 3.8.2	
4	2^{128}	2^{64} CP	2^{32}	Trunc. Diff.	Sect. 3.6.2	WCU
any ‡	2^{64}	2^{32} †	2^{64}	TMDTO	[4, 5]	2^{96} OPP
any ‡	2^{64}	2^{64} †	2^{64}	Key Collision	[26]	
any ‡	2^{64}	2^{64} †	2^{64}	Biham's	[2]	2^{64} OPP
any ‡	$2^{85.33}$	1 †	$2^{85.33}$	Hellman's TMTO	[9]	2^{128} OPP
any ‡	$2^{85.33}$	$2^{42.66}$ †	$2^{42.66}$	TMDTO	[4, 5]	$2^{85.33}$ OPP

†: Data complexity means the number of different keys.
‡: the number of rounds is such that all user key bits are used. For MESH-64, it means at least 1.5 rounds.

The largest weak-key class (for the largest number of rounds) is identified by a differential attack and it contains 2^{72} keys (out of 2^{192}).

Table 5.10 DFR attack complexities on MESH-96 cipher (under weak-key assumptions); c is a constant.

#Rounds	Time	Data/Memory	Attack	Source	$\|WKC\|$
4	$4c$	$4c$ KP	Linear	Sect. 3.13.5	2^{12}
5	$4c$	$4c$ CP	Diff.-Linear	Sect. 3.14.5	2^{27}
5.5	2^{26}	2^{26} ACPC	Boomerang	Sect. 3.12.3	2^{27}
6	2^{92}	2^{92} CP	Differential	Sect. 3.5.4	2^{72}

Table 5.11 summarizes the complexities of key-recovery attacks on MESH-96, for a variable number of rounds, *without* weak-key assumptions. This means that these attacks hold for any key. The attacks are ordered by increasing number of rounds. For a fixed number of rounds, attacks are ordered by increasing time complexity.

Table 5.11 Key-recovery attack complexities on MESH-96 cipher (without weak-key assumptions).

#Rounds	Time	Data	Memory	Attack	Source	Comments
2	2^{161}	161 KP	161	Biryukov-Demirci	Sect. 3.17.3	
3	2^{150}	$2^{51.5}$ CP	2^{16}	Square/Multiset	Sect. 3.15.3	
3.5	2^{96}	2^{56} CP	$2^{89.41}$	Imp. Diff.	Sect. 3.8.3.1	
3.5	2^{109}	2^{96} CP	2^{64}	Trunc. Diff.	Sect. 3.6.3	WCU
3.5	2^{144}	2^{56} CP	$2^{89.41}$	Imp. Diff.	Sect. 3.8.3.1	
4	2^{157}	2^{96} CP	2^{64}	Trunc. Diff.	Sect. 3.6.3	WCU
any ‡	2^{96}	2^{48} †	2^{96}	TMDTO	[4, 5]	2^{144} OPP
any ‡	2^{96}	2^{96} †	2^{96}	Key Collision	[26]	
any ‡	2^{96}	2^{96} †	2^{96}	Biham's	[2]	2^{96} OPP
any ‡	2^{128}	2^{64} †	2^{64}	TMDTO	[4, 5]	2^{128} OPP
any ‡	2^{128}	1 †	2^{128}	Hellman's TMTO	[9]	2^{192} OPP

†: Data complexity means the number of different keys.
‡: the number of rounds is such that all user key bits are used. For MESH-96, it means at least 1.5 rounds.

Taking into account only the time complexity, the best key-recovery attack not using the whole codebook is the Impossible-Differential technique reaching up to 3.5-round MESH-96.

The best generic attack (for any number of rounds) is TMDTO, requiring an offline pre-processing complexity of 2^{144} MESH-96 computations and recovering at least one of 2^{48} keys.

Table 5.12 summarizes distinguish-from-random attack complexities on MESH-128, for a variable number of rounds, under weak-key assumptions.

The largest weak-key class (for the largest number of rounds) is identified by a boomerang attack and it contains 2^{46} keys (out of 2^{256}).

Table 5.12 DFR attack complexities on MESH-128 cipher (under weak-key assumptions); c is a constant.

#Rounds	Time	Data/Memory	Attack	Source	$\|WKC\|$
4	$4c$	$4c$ KP	Linear	Sect. 3.13.6	2^{16}
4	2	2 CP	Differential	Sect. 3.5.5	2^{16}
5	$4c$	$4c$ CP	Diff.-Linear	Sect. 3.14.6	2^{46}
5.5	2^{34}	2^{34} ACPC	Boomerang	Sect. 3.12.4	2^{46}

Table 5.13 summarizes the complexities of key-recovery attacks on MESH-128, for a variable number of rounds, *without* weak-key assumptions. This means that these attacks hold for any key. The attacks are ordered by increasing number of rounds. For a fixed number of rounds, attacks are ordered by increasing time complexity.

Table 5.13 Key-recovery attack complexities on MESH-128 cipher (without weak-key assumptions).

#Rounds	Time	Data	Memory	Attack	Source	Comments
1.5	2^{208}	208 KP	208	Biryukov-Demirci	Sect. 3.17.4	
3	2^{178}	2^{52} CP	2^{16}	Square/Multiset	Sect. 3.15.4	
3.5	2^{128}	2^{64} CP	2^{153}	Imp. Diff.	Sect. 3.8.4.1	
3.5	2^{141}	2^{128} CP	2^{64}	Trunc. Diff.	Sect. 3.6.4	WCU
4	2^{192}	2^{64} CP	2^{153}	Imp. Diff.	Sect. 3.8.4.1	
4	2^{206}	2^{128} CP	2^{64}	Trunc. DIff.	Sect. 3.6.4	WCU
any ‡	2^{128}	2^{64} †	2^{128}	TMDTO	[4, 5]	2^{192} OPP
any ‡	2^{128}	2^{128} †	2^{128}	Key Collision	[26]	
any ‡	2^{128}	2^{128} †	2^{128}	Biham's	[2]	2^{128} OPP
any ‡	$2^{170.66}$	$2^{85.33}$ †	$2^{185.33}$	TMDTO	[4, 5]	$2^{170.66}$ OPP
any ‡	$2^{170.66}$	1 †	$2^{170.66}$	Hellman's TMTO	[9]	2^{256} OPP

†: Data complexity means the number of different keys.
‡: the number of rounds is such that all user key bits are used. For MESH-128, it means at least 1.5 rounds.

Taking into account only the time complexity, the best known key-recovery attack is the impossible differential reaching up to 4-round MESH-128 (but it does not achieve a better balance than Hellman's or TMDTO attacks). The other attack on 4 rounds required the whole codebook.

The best generic attack (for any number of rounds) is TMDTO, requiring an offline pre-processing complexity of 2^{192} MESH-128 computations, and recovering at least one of 2^{64} keys.

Table 5.14 summarizes distinguish-from-random attack complexities on MESH-64(8), for a variable number of rounds, under weak-key assumptions. The largest weak-key class (for 8.5 rounds) is identified by a linear attack and it contains 2^{51} keys (out of 2^{128}).

The fact that the full 8.5-round MESH-64(8) can be distinguished from a random permutation with low attack complexities means that MESH-64(8) is not a pseudorandom permutation (PRP), according to Sect. 1.7. Consequently, MESH-64(8) is not an ideal cryptographic primitive, similar to the situation of PES when used as a component in a hash function setting.

Table 5.14 DFR attack complexities on MESH-64(8) cipher (under weak-key assumptions); c is a constant.

#Rounds	Time	Data/Memory	Attack	Source	$\|WKC\|$
8.5	$4c$	$4c$ KP	Linear	Sect. 3.13.7	2^{51}
8.5	2	2 CP	Differential	Sect. 3.5.6	2^{44}

Table 5.15 summarizes the complexities of key-recovery attacks on MESH-64(8) for a variable number of rounds, *without* weak-key assumptions. This means that these attacks hold for any key. The attacks are ordered by increasing number of rounds. For a fixed number of rounds, attacks are ordered by increasing time complexity.

Table 5.15 Key-recovery attack complexities on MESH-64(8) cipher (without weak-key assumptions).

#Rounds	Time	Data	Memory	Attack	Source	Comments
1.5	2^{88}	40 KP	40	Biryukov-Demirci	Sect. 3.17.5	
2	2^{96}	96 KP	96	Biryukov-Demirci	Sect. 3.17.5	
2.5	2^{72}	2^{25} CP	2^{16}	Square/Multiset	Sect. 3.15.5	
3	2^{88}	2^{25} CP	2^{16}	Square/Multiset	Sect. 3.15.5	
3.5	2^{64}	$2^{39.5}$ CP	2^{58}	Imp. Diff.	Sect. 3.8.5.1	
4	2^{80}	$2^{39.5}$ CP	2^{58}	Imp. Diff.	Sect. 3.8.5.1	
any ‡	2^{64}	2^{32} †	2^{64}	TMDTO	[4, 5]	2^{96} OPP
any ‡	2^{64}	2^{64} †	2^{64}	Key Collision	[26]	
any ‡	2^{64}	2^{64} †	2^{64}	Biham's	[2]	2^{64} OPP
any ‡	$2^{85.33}$	1 †	$2^{85.33}$	Hellman's TMTO	[9]	2^{128} OPP
any ‡	$2^{85.33}$	$2^{42.66}$ †	$2^{42.66}$	TMDTO	[4, 5]	$2^{85.33}$ OPP

†: Data complexity means the number of different keys.
‡: the number of rounds is such that all user key bits are used. For MESH-64(8), it means at least 1.5 rounds.

Taking into account only the time complexity, the best known key-recovery attack on MESH-64(8) is the Impossible-Differential technique reaching up to 4 rounds.

The best generic attack (for any number of rounds) is TMDTO, requiring offline pre-processing complexity of 2^{96} MESH-64(8) computations, and recovering at least one of 2^{64} keys.

Table 5.16 summarizes distinguish-from-random attack complexities on MESH-128(8) for a variable number of rounds, under weak-key assumptions. The largest weak-key class (for 8.5 rounds) is identified by a linear attack and it contains 2^{200} keys (out of 2^{256}).

Even though the full 8.5-round MESH-128(8) can be distinguished from a random permutation for large weak-key classes, the attack complexities are not a polynomial in the cipher parameters. This means that MESH-128(8) still cannot be discarded as a pseudorandom permutation (PRP) according to Sect. 1.7.

Table 5.16 DFR attack complexities on MESH-128(8) cipher (under weak-key assumptions); c is a constant.

#Rounds	Time	Data/Memory	Attack	Source	$\|WKC\|$
8.5	$c \cdot 2^{90}$	$c \cdot 2^{90}$ KP	Linear	Sect. 3.13.8	2^{200}
8.5	2^{99}	2^{99} CP	Differential	Sect. 3.5.7	2^{193}

Table 5.17 summarizes the complexities of key-recovery attacks on MESH-128(8), for a variable number of rounds, *without* weak-key assumptions. This means that these attacks hold for any key. The attacks are ordered by increasing number of rounds. For a fixed number of rounds, attacks are ordered by increasing time complexity.

Taking into account only the time complexity, the best known key-recovery attack is the Impossible-Differential attack reaching up to 4-round MESH-128(8).

The best generic attack (for any number of rounds) is TMDTO, requiring an offline pre-processing complexity of 2^{192} MESH-128(8) computations, and recovering at least one of 2^{64} keys.

Table 5.18 summarizes the complexities of key-recovery attacks on IDEA*, for a variable number of rounds, *without* weak-key assumptions. This means that these attacks hold for any key. So far, only generic attacks are listed.

Table 5.19 summarizes the complexities of distinguish-from-random attacks on FOX64, for a variable number of rounds, *without* weak-key assumptions. This means that these attacks hold for any key. The attacks are ordered by increasing number of rounds. For a fixed number of rounds, attacks are ordered by increasing time complexity.

Table 5.17 Key-recovery attack complexities on MESH-128(8) cipher (without weak-key assumptions).

#Rounds	Time	Data	Memory	Attack	Source	Comments
1.5	$2^{37.34}$	40 KP	40	Biryukov-Demirci	Sect. 3.17.6	
2	2^{159}	160 KP	160	Biryukov-Demirci	Sect. 3.17.6	
2.5	2^{136}	2^{25} CP	2^8	Square/Multiset	Sect. 3.15.6	
3	2^{152}	2^{25} CP	2^8	Square/Multiset	Sect. 3.15.6	
3.5	2^{128}	$2^{72.5}$ CP	2^{121}	Imp. Diff.	Sect. 3.8.6.1	
4	2^{144}	$2^{72.5}$ CP	2^{121}	Imp. Diff.	Sect. 3.8.6.1	
any ‡	2^{128}	2^{64} †	2^{128}	TMDTO	[4, 5]	2^{192} OPP
any ‡	2^{128}	2^{128} †	2^{128}	Key Collision	[26]	
any ‡	2^{128}	2^{128} †	2^{128}	Biham's	[2]	2^{128} OPP
any ‡	$2^{170.66}$	$2^{85.33}$ †	$2^{185.33}$	TMDTO	[4, 5]	$2^{170.66}$ OPP
any ‡	$2^{170.66}$	1 †	$2^{170.66}$	Hellman's TMTO	[9]	2^{256} OPP

†: Data complexity means the number of different keys.
‡: the number of rounds is such that all user key bits are used. For MESH-128(8), it means at least 1.5 rounds.

Table 5.18 Key-recovery attack complexities on IDEA* cipher.

#Rounds	Time	Data	Memory	Attack	Source	Comments
any ‡	2^{64}	2^{32} †	2^{64}	TMDTO	[4, 5]	2^{96} preprocessing
any ‡	2^{64}	2^{64} †	2^{64}	Key Collision	[26]	
any ‡	2^{64}	2^{64} †	2^{64}	Biham's	[2]	2^{64} preprocessing
any ‡	$2^{85.33}$	1 †	$2^{85.33}$	Hellman's TMTO	[9]	2^{128} preprocessing
any ‡	$2^{85.33}$	$2^{42.66}$ †	$2^{42.66}$	TMDTO	[4, 5]	$2^{85.33}$ preprocessing

†: Data complexity means the number of different keys.
‡: the number of rounds is such that all user key bits are used. For IDEA*, it means at least 1.5 rounds.

Table 5.19 DFR attack complexities on FOX64/k/r cipher.

#Rounds	Time	Data	Memory	Attack	Source
1	1	1 KP	1	Generic	[18]
2	2^{16}	2^{16} KP	2^{16}	Generic	[18]
3	2^{32}	2^{32} KP	2^{32}	Generic	[18]
4	2^{48}	2^{48} KP	2^{48}	Generic	[18]
5	2^{56}	2^{56} KP	2^{56}	Generic	[18]

Table 5.20 summarizes key-recovery attacks on variants of FOX64/k/r. The default values are $(k, r) = (128, 16)$ for which FOX64/k/r is denoted simply FOX64. If the time complexity exceeds 2^{128}, then the attacks apply to other variants of FOX64/k/r for the key range stated in the comments column.

Table 5.20 Key-recovery attack complexities on FOX64/k/r cipher.

#Rounds	Time	Data	Memory	Attack	Source	Comments
4	$2^{45.4}$	2^9 CP	2^9	Square/Multiset	Sect. 3.15.8	
5	2^{73}	2^{40} CP	2^{58}	Imp. Diff.	Sect. 3.8.7.1	
5	$2^{109.4}$	2^9 CP	2^9	Square/Multiset	Sect. 3.15.8	
6	2^{124}	15 KP	2^{124}	Improved ASR	Sect. 3.20.6	
6	2^{137}	2^{40} CP	2^{58}	Imp. Diff.	Sect. 3.8.7.1	$k > 137$, $r > 6$
6	$2^{173.4}$	2^9 CP	2^9	Square/Multiset	Sect. 3.15.8	$k > 173$, $r > 6$
7	2^{186}	$2^{30.9}$ KP	2^{123}	Improved ASR	Sect. 3.20.6	$k > 186$, $r > 7$
7	$2^{237.4}$	2^9 CP	2^9	Square/Multiset	Sect. 3.15.8	$k > 237$, $r > 7$
any ‡	2^{64}	2^{32} †	2^{64}	TMDTO	[4, 5]	2^{96} OPP
any ‡	2^{64}	2^{64} †	2^{64}	Key Collision	[26]	
any ‡	2^{64}	2^{64} †	2^{64}	Biham's	[2]	2^{64} OPP
any ‡	$2^{85.33}$	1 †	$2^{85.33}$	Hellman's TMTO	[9]	2^{128} OPP
any ‡	$2^{85.33}$	$2^{42.66}$ †	$2^{42.66}$	TMDTO	[4, 5]	$2^{85.33}$ OPP

†: Data complexity means the number of different keys.
‡: the number of rounds is such that all user key bits are used. For FOX64/128/r, it means at least 1.5 rounds.

Table 5.21 summarizes key-recovery attacks on variants of FOX128/k/r. The default values are $(k, r) = (256, 16)$, for which FOX128/k/r is denoted simply FOX128. If the time complexity exceeds 2^{256}, then the attack applies to other variants of FOX128/k/r for the key range stated in the comments column.

Table 5.22 summarizes distinguish-from-random attack complexities on RIDEA under weak-key assumptions. The largest weak-key class (for 8.5 rounds) is identified by a linear attack and it contains 2^{14} keys (out of 2^{128}).

Table 5.23 summarizes distinguish-from-random (DFR) attack complexities for REESSE3+ for a variable number of rounds, under weak-key assumptions. The largest weak-key class (for 8.5 rounds) is identified by a differential attack and it contains 2^{15} keys (out of 2^{256}).

The fact that the full 8.5-round REESSE3+ (and even variants with many more rounds) can be distinguished from a random permutation with low attack complexities means that REESSE3+ is not a pseudorandom permutation (PRP), according to Sect. 1.7. Consequently, REESSE3+ is not an ideal cryptographic primitive, similar to the situation about PES when used as a component in a hash function setting.

Table 5.24 summarizes complexities of key-recovery attacks on RESSE3+, for a variable number of rounds *without* weak-key assumptions. This means

Table 5.21 Key-recovery attack complexities on FOX128/k/r cipher.

#Rounds	Time	Data	Memory	Attack	Source	Comments
4	$2^{77.6}$	2^9 CP	2^9	Square/Multiset	Sect. 3.15.8	
5	2^{133}	2^{72} CP	2^{121}	Imp. Diff.	Sect. 3.8.7.2	
5	$2^{205.6}$	2^9 CP	2^9	Square/Multiset	Sect. 3.15.8	
5	2^{228}	13 KP	2^{228}	ASR	Sect. 3.20.4	
6	$2^{177.6}$	26 CP	2^{221}	Improved ASR	Sect. 3.20.7	
6	2^{261}	2^{72} CP	2^{121}	Imp. Diff.	Sect. 3.8.7.2	$k > 261$, $r > 6$
7	2^{242}	2^{63} CP	2242	Improved ASR	Sect. 3.20.7	
any ‡	2^{128}	2^{64} †	2^{128}	TMDTO	[4, 5]	2^{192} OPP
any ‡	2^{128}	2^{128} †	2^{128}	Key Collision	[26]	
any ‡	2^{128}	2^{128} †	2^{128}	Biham's	[2]	2^{128} OPP
any ‡	$2^{170.66}$	1 †	$2^{270.66}$	Hellman's TMTO	[9]	2^{256} OPP
any ‡	$2^{170.66}$	$2^{85.33}$ †	$2^{85.33}$	TMDTO	[4, 5]	$2^{170.66}$ OPP

†: Data complexity means the number of different keys.
‡: the number of rounds is such that all user key bits are used. For FOX128/256/r, it means at least 2 rounds.

Table 5.22 DFR attack complexities on RIDEA cipher (under weak-key assumptions); c is a constant.

#Rounds	Time	Data/Memory	Attack	Source	$\|WKC\|$
8.5	2	2 CP	Differential	Sect. 3.5.9	2^{12}
8.5	$4c$	$4c$ KP	Linear	Sect. 3.13.10	2^{14}

Table 5.23 DFR attack complexities on REESSE3+ cipher (under weak-key assumptions); c is a constant.

#Rounds	Time	Data/Memory	Attack	Source	$\|WKC\|$
8.5	$4c$	$4c$ KP	Linear	Sect. 3.13.11	2^{15}
8.5	2	2 CP	Differential	Sect. 3.5.10	2^{15}
16	2	2 CP	Differential	Sect. 3.5.10	2
> 16	2	2 CP	Differential	Sect. 3.5.10	1

that these attacks hold for any key. The attacks are ordered by increasing number of rounds. For a fixed number of rounds, attacks are ordered by increasing time complexity.

Table 5.25 summarizes distinguish-from-random attack complexities on WIDEA-4 under weak-key assumptions. Both of these attacks have practical attack complexities. The largest weak-key class (for largest number of rounds) is identified by a differential (or linear) attack and it contains 2^{242} keys (out of 2^{512}).

The fact that the full 8.5-round WIDEA-4 can be distinguished from a random permutation with low attack complexities means that WIDEA-4 is not

Table 5.24 Key-recovery attack complexities on REESSE3+ cipher (without weak-key assumptions).

#Rounds	Time	Data	Memory	Attack	Source	Comments
2.5	2^{32}	2^{17} CP	2^{17}	Square/Multiset	Sect. 3.15.7	
3.5–4	2^{79}	2^{72} CP	2^{57}	Imp. Diff.	Sect. 3.8.8	
any ‡	2^{128}	2^{64} †	2^{128}	TMDTO	[4, 5]	2^{192} OPP
any ‡	2^{128}	2^{128} †	2^{128}	Key Collision	[26]	
any ‡	2^{128}	2^{128} †	2^{128}	Biham's	[2]	2^{128} OPP
any ‡	$2^{170.66}$	1 †	$2^{170.66}$	Hellman's TMTO	[9]	2^{256} OPP
any ‡	$2^{170.66}$	$2^{85.33}$ †	$2^{85.33}$	TMDTO	[4, 5]	$2^{170.66}$ OPP

†: Data complexity means the number of different keys.
‡: the number of rounds is such that all user key bits are used. For REESSE3+, it means at least 1.5 rounds.

a pseudorandom permutation (PRP), according to Sect. 1.7. Consequently, WIDEA-4 is not an ideal cryptographic primitive, similar to the situation about PES when used as a component in a hash function setting.

Table 5.25 DFR attack complexities on WIDEA-4 cipher (under weak-key assumptions); c is a constant.

#Rounds	Time	Data/ Memory	Attack	Source	$\|WKC\|$
8.5	2	2 CP	Differential	Sect. 3.5.8.1	2^{242}
8.5	$4c$	$4c$ KP	Linear	Sect. 3.13.9	2^{242}

Table 5.26 summarizes the complexities of key-recovery attacks on WIDEA-4, for a variable number of rounds, *without* weak-key assumptions. This means that these attacks hold for any key. The attacks are ordered by increasing number of rounds. For a fixed number of rounds, attacks are ordered by increasing time complexity.

Table 5.26 Summary of attack complexities on WIDEA-4 cipher (without weak-key assumptions).

#Rounds	Time	Data	Memory	Attack	Source	Comments
8.5	2^{48}	2^{70} CP	2^{64}	Trunc. Diff.	Sect. 3.6.5.1	KR
8.5	2^{65}	2^{65} CP	2^{64}	Trunc. Diff.	Sect. 3.6.5	DFR

Table 5.27 summarizes distinguish-from-random attack complexities on WIDEA-8 under weak-key assumptions. Both of these attacks have practical attack complexities. The largest weak-key class (for the largest number of

rounds) is identified by a differential (or linear) attack and it contains 2^{754} keys (out of 2^{1024}).

The fact that the full 8.5-round WIDEA-8 can be distinguished from a random permutation with low attack complexities means that WIDEA-8 is not a pseudorandom permutation (PRP), according to Sect. 1.7. Consequently, WIDEA-8 is not an ideal cryptographic primitive, similar to the situation about PES when used as a component in a hash function setting.

Table 5.27 DFR attack complexities on WIDEA-8 cipher (under weak-key assumptions); c is a constant.

#Rounds	Time	Data/Memory	Attack	Source	$\|WKC\|$
8.5	2	2 CP	Differential	Sect. 3.5.8.1	2^{754}
8.5	$4c$	$4c$ KP	Linear	Sect. 3.13.9	2^{754}

Table 5.28 summarizes the complexities of key-recovery attacks on WIDEA-8, for a variable number of rounds, *without* weak-key assumptions. This means that these attacks hold for any key. The attacks are ordered by increasing number of rounds. For a fixed number of rounds, attacks are ordered by increasing time complexity.

Table 5.28 Summary of attack complexities on WIDEA-8 cipher (without weak-key assumptions).

#Rounds	Time	Data	Memory	Attack	Source	Comments
8.5	2^{48}	2^{74} CP	2^{64}	Trunc. Diff.	Sect. 3.6.5.1	KR
8.5	2^{65}	2^{65} CP	2^{64}	Trunc. Diff.	Sect. 3.6.5	DFR

References

1. Belarus: Data Encryption and Integrity Algorithms. Preliminary State Standard of Republic of Belarus (STB P 34.101.31-2007). http://apmi.bsu.by/assets/files/std/belt-spec27.pdf (2007)
2. Biham, E.: How to Decrypt or Even Substitute DES-encrypted Messages in 2^{28} Steps. Information Processing Letters **84**(3), 117–124 (2002)
3. Biham, E., Shamir, A.: Differential Cryptanalysis of Snefru, Khafre, Redoc-III, Loki and Lucifer. In: J. Feigenbaum (ed.) Advances in Cryptology, Crypto, LNCS 576, pp. 156–171. Springer (1991)
4. Biryukov, A.: Some Thoughts on Time-Memory-Data Tradeoffs. IACR ePrint archive, 2005/207 (2005)

5. Biryukov, A., Mukhopadhyay, S., Sarkar, P.: Improved Time-Memory Tradeoff with Multiple Data. In: B. Preneel, S. Tavares (eds.) Selected Areas in Cryptography (SAC), LNCS 3897, pp. 110–127. Springer (2006)
6. Borst, J.: Weak Keys of Crypton. AES Competition, First round (1998)
7. Borst, J., Knudsen, L.R., Rijmen, V.: Two Attacks on Reduced IDEA (extended abstract). In: W. Fumy (ed.) Advances in Cryptology, Eurocrypt, LNCS 1233, pp. 1–13. Springer (1997)
8. Cui, T., Chen, H., Wen, L., Wang, M.: Statistical Integral Attack on CAST-256 and IDEA. Cryptography and Communications - Discrete Structures, Boolean Functions and Sequences **10**(1), 195–209 (2018)
9. Hellman, M.E.: A Cryptanalytic Time-Memory Trade-Off. IEEE Transactions on Information Theory **IT-26**(4), 401–406 (1980)
10. Junod, P., Macchetti, M.: Revisiting the IDEA Philosophy. In: O. Dunkelman (ed.) Fast Software Encryption (FSE), LNCS 5665, pp. 277–295. Springer (2009)
11. Junod, P., Vaudenay, S.: FOX: a New Family of Block Ciphers. In: H. Handschuh, M.A. Hasan (eds.) Selected Areas in Cryptography (SAC), LNCS 3357, pp. 114–129. Springer (2004)
12. Kelsey, J., Schneier, B., Wagner, D.: Key-Schedule Cryptanalysis of IDEA, G-DES, GOST, SAFER and triple-DES. In: N. Koblitz (ed.) Advances in Cryptology, Crypto, LNCS 1109, pp. 237–251. Springer (1996)
13. Khovratovich, D., Leurent, G., Rechberger, C.: Narrow-Bicliques: Cryptanalysis of Full IDEA. In: D. Pointcheval, T. Johansson (eds.) Advances in Cryptology, Eurocrypt, LNCS 7237, pp. 392–410. Springer (2012)
14. Lai, X., Massey, J.L.: A Proposal for a New Block Encryption Standard. In: I.B. Damgård (ed.) Advances in Cryptology, Eurocrypt, LNCS 473, pp. 389–404. Springer (1990)
15. Lai, X., Massey, J.L., Murphy, S.: Markov Ciphers and Differential Cryptanalysis. In: D.W. Davies (ed.) Advances in Cryptology, Eurocrypt, LNCS 547, pp. 17–38. Springer (1991)
16. Lerman, L., Nakahara Jr, J., Veshchikov, N.: Improving Block Cipher Design by Rearranging Internal Operations. In: P. Samarati (ed.) 10th International Conference on Security and Cryptography (SECRYPT), pp. 27–38. IEEE (2013)
17. Lu, J., Yap, W.S., Wei, Y.: Weak Keys of the Full MISTY1 Block Cipher for Related-Key Cryptanalysis. IACR ePrint archive 2012/066
18. Luo, Y., Lai, X., Zhou, Y.: Generic Attacks on the Lai-Massey Scheme. Designs, Codes and Cryptography (2016)
19. Meier, W.: On the Security of the IDEA Block Cipher. In: T. Helleseth (ed.) Advances in Cryptology, Eurocrypt, LNCS 765, pp. 371–385. Springer (1994)
20. Moore, J.H., Simmons, G.J.: Cycle Structure of the DES for Keys having Palindromic (or Antipalindromic) Sequences of Round Keys. IEEE Transactions on Software Engineering **SE**(2), 262–273 (1987)
21. Nakahara Jr, J.: Faster Variants of the MESH Block Ciphers. In: A. Canteaut, K. Viswanathan (eds.) International Conference on Cryptology in India, Indocrypt, LNCS 3348, pp. 162–174. Springer (2004)
22. Nakahara Jr, J., Preneel, B., Vandewalle, J.: A Note on Weak-Keys of PES, IDEA and some Extended Variants. In: C. Boyd, W. Mao (eds.) 6th Information Security Conference (ISC), LNCS 2851, pp. 267–279. Springer (2003)
23. Nakahara Jr, J., Rijmen, V., Preneel, B., Vandewalle, J.: The MESH Block Ciphers. In: K. Chae, M. Yung (eds.) Information Security Applications (WISA), LNCS 2908, pp. 458–473. Springer (2003)
24. Su, S., Lu, S.: A 128-bit Block Cipher based on Three Group Arithmetics. IACR ePrint archive 2014/704 (2014)
25. Wei, L., Peyrin, T., Sokolowski, P., Ling, S., Pieprzyk, J., Wang, H.: On the (In)Security of IDEA in Various Hashing Modes. IACR ePrint archive 2012/264 (2012)

26. Winternitz, R., Hellman, M.E.: Chosen-Key Attacks on a Block Cipher. Cryptologia **11**(1), 16–20 (1987)
27. Yi, X.: On Design and Analysis of a New Block Cipher. In: J. Jaffar, R.H.C. Yap (eds.) Concurrency and Parallelism, Programming, Networking and Security, ASIAN 1996, LNCS 1179, pp. 213–222. Springer (1996)
28. Yıldırım, H.M.: Nonlinearity Properties of the Mixing Operations of the Block Cipher IDEA. In: T. Johansson, S. Maitra (eds.) Progress in Cryptology, Indocrypt, LNCS 2904, pp. 68–81. Springer (2003)

Appendix A
Monoids, Groups, Rings and Fields

Information theory is the science of communication in the presence of noise.
- C.E. Shannon
Cryptology is the science of communication in the presence of adversaries.
- R.L. Rivest

A *monoid*, denoted $(S, \cdot)$, consists of a non-empty set S and a binary operation $\cdot$ defined on S satisfying the following properties:

(1) associativity: $(a \cdot b) \cdot c = a \cdot (b \cdot c)$ for all $a, b, c \in S$
(2) identity: there exist an element $1 \in S$ such that $a \cdot 1 = 1 \cdot a = a$ for all $a \in S$

An example of a monoid is the set $\mathbf{N}$ of natural numbers, with the addition operation. The neutral element is 0. This same set under multiplication has neutral element 1.

A *group*, denoted $(S, +)$, is composed of a non-empty set S and a binary operation, denoted $+$, defined on S. Groups can be either finite or infinite depending on the number of elements in S. The number of distinct elements in a group is called the *order of the group*.

The fundamental properties of a group are:

(3) *closure* under $+$: $a + b \in S$ for all $a, b \in S$.
(4) $(S, +)$ is a monoid.
(5) additive *inverse*: for each $x \in S$, there is an element $-x \in S$ such that $x + (-x) = (-x) + x = 0$.

If the following property holds, then $(S, +)$ is called a commutative or Abelian group, otherwise it is non-Abelian:

(6) *commutativity* of $+$: $a + b = b + a$ for all $a, b \in S$.

An example of a (finite) Abelian group is $(\mathbb{Z}_n, +)$, where

$$\mathbb{Z}_n = \{x \in \mathbb{Z} | 0 \leq x \leq n - 1\},$$

where $n > 1$ is a positive integer. The set $\mathbb{Z}$ of integers under multiplication, denoted $(\mathbb{Z}, *)$, is an example of an infinite Abelian group.

An example of a non-Abelian group is the set of invertible square matrices with entries in $\mathbb{Z}$, under the multiplication operation. It is an infinite non-Abelian group.

J. Nakahara Jr., *Lai-Massey Cipher Designs*, https://doi.org/10.1007/978-3-319-68273-0

In contrast, the group of permutations of n elements, for fixed $n > 1$, under composition of mappings is a finite non-Abelian group.

A *ring*, denoted $(S, +, *)$, is a group under two operations usually denoted $+$ (addition) and $*$ (multiplication). Besides properties (1)-(5), a ring satisfies the following properties:

(7) *closure* under $*$: $a * b \in S$ for all $a, b \in S$.
(8) *associativity* of $*$: $a * (b * c) = (a * b) * c$ for all $a, b, c \in S$.
(9) left- and right-*distributive* laws: $a * (b + c) = a * b + a * c$ and $(a + b) * c = a * c + b * c$ for all $a, b, c \in S$.

If both operations in a ring are commutative, then the ring is a commutative or an Abelian ring.

Examples of Abelian rings: $\mathbb{Z}[x]$, the set of polynomials in x with integer coefficients, is an infinite ring; the set $\mathbb{Z}_n^d[x]$ of polynomials of degree d with coefficients in $\mathbb{Z}_n$ (residue class ring or factor ring) for fixed integers d and n is a finite ring.

Examples of non-Abelian rings: the set $\mathrm{GL}(k, \mathbb{Z})$, the *general linear group* of invertible $k \times k$ matrices with entries from $\mathbb{Z}$, is an infinite ring; $\mathrm{GL}(k, \mathbb{Z}_n)$, the set of invertible $k \times k$ matrices with values in Z_n, for fixed n, is a finite ring.

A *field*, denoted $(F, +, *)$, consists of a non-empty set F and two binary operations, usually denoted $+$ (addition) and $*$ (multiplication) defined on F with the following properties:

- $(F, +, *)$ is a ring.
- commutativity of $*$: $a * b = b * a$ for all $a, b \in F$.
- multiplicative identity: there exist $1 \in F$ such that $a * 1 = 1 * a = a$ for all $a \in F$.
- multiplicative inverse: if $a \in F$ and $a \neq 0$, then there exist $b \in F$ such that $a * b = b * a = 1$.

Examples of infinite fields: $\mathbb{Q}$, $\mathbb{R}$, $\mathbb{C}$ with the addition $(+)$ and multiplication $(*)$ operations in each field.

Examples of finite fields: (i) Z_q with q a prime number; (ii) $\mathrm{GF}(2^n) = \mathrm{GF}(2)[\mathrm{x}]/(p(x))$ where n is a positive integer and $p(x)$ is a primitive polynomial of degree n over $\mathrm{GF}(2)$.

The question whether all elements are invertible is the distinguishing feature between a *group* and a *monoid*, or between a *field* and a *ring*.

Appendix B
Differential and Linear Branch Number

Research is to see what everybody else has seen, and to think what nobody else has thought.
–Albert Szent-Györgi

Definition B.1. [55]
The *differential branch number* of a linear transformation θ is defined as

$$B(\theta) = \min_{a \neq 0}\{\mathrm{HW}(a) + \mathrm{HW}(\theta(a))\},$$

where $\mathrm{HW}(a)$ is the Hamming Weight of the vector a in the underlying vector space.

The Hamming distance between two vectors u and v from the n-dimensional vector space $\mathrm{GF}(2^m)^n$ is the number of m-bit coordinates where u and v differ.

The Hamming weight, $\mathrm{HW}(u)$ of an element $u \in \mathrm{GF}(2^m)$ is the Hamming Distance between u and the null vector of $\mathrm{GF}(2^m)^n$, that is, the number of non-zero components of u.

Definition B.2. [55]
A linear $[n, k, d]$-code $\mathcal{C}$ over $\mathrm{GF}(2^m)$ is a k-dimensional subspace of the vector space $\mathrm{GF}(2^m)^n$, where two different vectors of the subspace have a Hamming distance of at least d, and d is the largest number with this property.

A linear code $\mathcal{C}$ can be described by a generator matrix G, which is a $k \times n$ matrix whose rows form a vector space basis for the code. Since the choice of a basis in a vector space is not unique, a code has many different generator matrices that can be reduced to one another by elementary row operations and column permutations. The *echelon (or standard) form* of the generator matrix is the following:

$$G_e = [I_{k\times k} \| A_{k\times(n-k)}],$$

where $I_{k\times k}$ is the identity matrix of order k.

The dual code $\mathcal{C}^t$ of a code $\mathcal{C}$ is defined as the set of vectors that are orthogonal to all the vectors of $\mathcal{C}$: $\mathcal{C}^t = \{x| < x, y >= 0, \forall y \in \mathcal{C}\}$, where $< x, y >$ is the dot product between the vectors x and y.

J. Nakahara Jr., *Lai-Massey Cipher Designs*, https://doi.org/10.1007/978-3-319-68273-0

The *differential branch number* of a linear mapping can be related to the minimal distance of the associated linear code.

Definition B.3. [55]
Let θ be a linear mapping from $GF(2^m)^n$ to $GF(2^m)^n$. The associated code of θ, $\mathcal{C}_\theta$, is the linear code that has codewords given by the vectors $(x|\theta(x))$. The code $\mathcal{C}_\theta$ has 2^n codewords and has length $2n$.

It follows from the definition that the *differential branch number* of a mapping equals the minimal distance between two different codewords of its associated code. The upper bound on the differential branch number of a mapping corresponds to the Singleton bound on the minimal distance of a linear code. This relation between linear transformations and linear codes allows to efficiently construct mappings with a high differential branch number. Given a linear code $\mathcal{C}_\theta$, the associated mapping θ is given by $\theta(x) = x \cdot A$, where A is found from the echelon form of the generator matrix of $\mathcal{C}$. It can be proved that the *linear branch number* of a mapping θ is equal to the minimal distance of the dual code of $\mathcal{C}_\theta$.

From a linear cryptanalysis perspective, an analogous quantity can be defined:

Definition B.4. (Linear Branch Number)
The *linear branch number* of a transformation ϕ with respect to x is given by

$$L(\phi, x) = \min_{x \neq 0}\{HW(x) + HW(\phi(x))\}.$$

B.1 MDS Codes

A linear code over the Galois Field GF(2^m) is denoted by an $[n, k, d]$-code, where n is the symbol length of the encoded message, k is the length of the original message and d is the minimal symbol distance between any two encoded messages.

Definition B.5. [55][chap.11]
Linear $[n, k, d]$-codes obey the Singleton bound: $d \leq n - k + 1$. A code that meets the bound, $d = n - k + 1$, is called a *Maximum Distance Separable (MDS)* code. A linear $[n, k, d]$-code $\mathcal{C}$ with generator matrix $G = [I_{k\times k} \| A_{k\times(n-k)}]$ is MDS if and only if every square submatrix formed from i rows and i columns of A, for $i \in \{1, \ldots, \min(k, n-k)\}$, is nonsingular.

Consequently, by definition, each square submatrix of an MDS matrix is MDS as well. Therefore, a single $n \times n$ MDS matrix automatically provides several $m \times m$ MDS matrices for all $1 < m < n$. These $m \times m$ MDS matrices are not customized though, that is, their coefficients may not be small nor lead necessarily to an efficient implementation. On the other hand, this simple observation allows one to find several MDS matrices as submatrices whose dimensions are not powers of 2 (the vast majority of MDS matrices used in cipher constructions in the literature have dimensions which are powers of 2).

Both the AES and the Twofish ciphers use [8, 4, 5] MDS codes that is: $n = 8$, $k = 4$ and $d = n - k + 1 = 8 - 4 + 1 = 5$.

Known MDS codes are Reed-Solomon codes, (3,1) Hamming codes, (4,1) extended Hamming codes and dual MDS codes [55].

MDS codes are used to provide perfect diffusion. An earlier concept, closely related to MDS codes, is that of a multipermutation.

Definition B.6. [80]
A permutation $f : V^2 \to V^2$ defined as $f(x, y) = (f_1(x, y), f_2(x, y))$, is a *multipermutation* if for every $x, y \in V$, the mappings $f_i(a, *)$, $f_i(*, b)$ for $i \in \{1, 2\}$ are permutations on V.

Vaudenay in [92] generalized the concept of multipermutation to that of an (r, t)-multipermutation.

Definition B.7. [92]
A function $f : V^r \to V^t$ is an (r, t)-multipermutation if any two distinct $(r + t)$-tuples of the form $(x_1, \ldots, x_r, f(x_1, \ldots, x_r))$ cannot collide in any r positions.

The perfect diffusion with (r, t)-multipermutations is achieved because for any $x_1, \ldots, x_r \in V$ and any integer i such that $1 \leq i \leq r$, changing any i input values will change at least $t - i + 1$ output values of f.

From [64] and using Def. B.5, the set of all words of the form $(x_1, \ldots, x_r, f(x_1, \ldots, x_r))$ can be seen as a systematic error-correcting code of $|V|^r$ words of length $r + t$ with minimal distance $t + 1$, which matches the Singleton bound. The connection between multipermutations and MDS codes comes from the fact that if V is a finite field, a linear (r, t)-multipermutation is a $[r+t, r, t+1]$ MDS code in the sense that any word of length r is coded by the concatenation of the word and its multipermutation image. Equivalence holds only when the MDS code minimal distance is at least r, with $t+1 \geq r$. Linear (r, t)-multipermutations f can be represented using a $r \times t$ MDS matrix M as $f : x \to M \cdot x$.

MDS matrices form a pervasive component not only in modern block ciphers such as AES [33], BKSQ [27] and SHARK [74], but also in stream ciphers such as MUGI [94] as well as in hash functions such as Grostl [34] and Maelstrom [32].

Although MDS codes appear in many cipher designs, PES [49], IDEA [50], MESH [67], Keccak [10], PRESENT [15] and Serpent [3] are examples of modern cryptographic algorithms that do not use MDS matrices at all in their design.

MDS matrices do not need to be square matrices necessarily. Their use in SPN cipher designs though, requires that they be invertible, and thus, be square matrices, so that decryption is possible.

A convenient design criteria (for SPN cipher designs) is to impose that such matrices also be *involutory*. A square matrix X is *involutory* is $X^2 = I$ that is, $X = X^{-1}$, where I is the identity matrix. Otherwise, the matrix is non-involutory. If an involutory matrix is used as a cipher component in SPN designs, it means that both encryption and decryption will have the same (implementation) cost concerning this diffusion component.

For Lai-Massey cipher designs that incorporate MDS matrices, there is no need for the involutory property. For instance, the FOX cipher family uses non-involutory MDS matrices [40].

A square matrix S of order d is called *symmetric* if $A = A^T$, that is, A is identical to its transpose matrix, or equivalently, $a_{i,j} = a_{j,i}$ for all $0 \leq i, j < d$.

B.1.1 How to Construct MDS Matrices?

There are several known techniques to construct MDS matrices:

- **circulant matrices**: each row of a circulant matrix forms a cyclically rotated version of the other rows of the matrix. Similarly for the columns. An $m \times m$ circulant matrix C can be denoted

$$C = \begin{pmatrix} a_0 & a_1 \ldots a_{m-1} \\ a_{m-1} & a_0 \ldots a_{m-2} \\ \ldots & \ldots\ldots \ldots \\ a_1 & a_2 \ldots \; a_0 \end{pmatrix}.$$

  Thus, a circulant matrix can be characterized by any of its rows. Without loss of generality, it is usual to denote $C = \text{circ}(a_0, a_1, \ldots, a_{m-1})$ by its first, topmost row. If $C = [c_{i,j}]$, with $0 \leq i, j \leq m-1$, then $\text{circ}(a_0, a_1, \ldots, a_{m-1}) = C$ means that $c_{i,j} = a_{(j-i) \bmod m}$.
  M_1 is an example of circulant MDS matrix used in AES/Rijndael [33], GRAND CRU [16] and Rainbow [52]. M_1 has entries in the finite field $\text{GF}(2^8) = \text{GF}(2)[x]/(x^8 + x^4 + x^3 + x + 1)$:

$$M_1 = \begin{pmatrix} \texttt{02}_\texttt{x} & \texttt{03}_\texttt{x} & \texttt{01}_\texttt{x} & \texttt{01}_\texttt{x} \\ \texttt{01}_\texttt{x} & \texttt{02}_\texttt{x} & \texttt{03}_\texttt{x} & \texttt{01}_\texttt{x} \\ \texttt{01}_\texttt{x} & \texttt{01}_\texttt{x} & \texttt{02}_\texttt{x} & \texttt{03}_\texttt{x} \\ \texttt{03}_\texttt{x} & \texttt{01}_\texttt{x} & \texttt{01}_\texttt{x} & \texttt{02}_\texttt{x} \end{pmatrix}$$

  and can be simply denoted $\text{circ}(\texttt{02}_\texttt{x}, \texttt{03}_\texttt{x}, \texttt{01}_\texttt{x}, \texttt{01}_\texttt{x})$. In this case, the entries of the MDS matrix were carefully chosen to be small in order to minimize the implementation cost for encryption (number of finite field multiplications).

In block ciphers with 2-dimensional states such as the AES, the decision to multiply a row vector (of the state) on the left of an MDS matrix, or a column vector (of the state) on the right of the MDS matrix depends on the intended diffusion effect. Multiplying on the left of the MDS matrix means diffusion across the elements in each row of the state, while multiplying on the right means diffusion across the elements in each column of the state. In AES, each column of the state is multiplied on the right of a 4×4 MDS matrix:

$$\begin{pmatrix} b_0 \\ b_1 \\ b_2 \\ b_3 \end{pmatrix} = M_1 \cdot \begin{pmatrix} a_0 \\ a_1 \\ a_2 \\ a_3 \end{pmatrix} .$$

M_1 is not involutory, and its inverse matrix has entries with higher Hamming Weight than those in M_1:

$$M_1^{-1} = \begin{pmatrix} \texttt{0E}_\texttt{x} & \texttt{09}_\texttt{x} & \texttt{0D}_\texttt{x} & \texttt{0B}_\texttt{x} \\ \texttt{0B}_\texttt{x} & \texttt{0E}_\texttt{x} & \texttt{09}_\texttt{x} & \texttt{0D}_\texttt{x} \\ \texttt{0D}_\texttt{x} & \texttt{0B}_\texttt{x} & \texttt{0E}_\texttt{x} & \texttt{09}_\texttt{x} \\ \texttt{09}_\texttt{x} & \texttt{0D}_\texttt{x} & \texttt{0B}_\texttt{x} & \texttt{0E}_\texttt{x} \end{pmatrix} .$$

M_1^{-1} can be denoted simply as circ($\texttt{0E}_\texttt{x}, \texttt{09}_\texttt{x}, \texttt{0D}_\texttt{x}, \texttt{0B}_\texttt{x}$).

The BKSQ block cipher [27] uses a non-involutory, but circulant 3×3 MDS matrix with entries in $\mathrm{GF}(2^8) = \mathrm{GF}(2)[x]/(x^8+x^7+x^6+x^5+x^4+x^2+1)$:

$$M_2 = \begin{pmatrix} \texttt{03}_\texttt{x} & \texttt{02}_\texttt{x} & \texttt{02}_\texttt{x} \\ \texttt{02}_\texttt{x} & \texttt{03}_\texttt{x} & \texttt{02}_\texttt{x} \\ \texttt{02}_\texttt{x} & \texttt{02}_\texttt{x} & \texttt{03}_\texttt{x} \end{pmatrix} .$$

M_2 is a rare example of an MDS matrix whose order is not a power of 2. It can be denoted $M_2 = $ circ($\texttt{03}_\texttt{x}, \texttt{02}_\texttt{x}, \texttt{02}_\texttt{x}$).

The Hierocrypt-3 block cipher [88] uses a 4×4 circulant, non-involutory MDS matrix, over $\mathrm{GF}(2^8) = \mathrm{GF}(2)[x]/(x^8 + x^6 + x^5 + x + 1)$:

$$M_3 = \begin{pmatrix} \texttt{C4}_\texttt{x} & \texttt{65}_\texttt{x} & \texttt{C8}_\texttt{x} & \texttt{8B}_\texttt{x} \\ \texttt{8B}_\texttt{x} & \texttt{C4}_\texttt{x} & \texttt{65}_\texttt{x} & \texttt{C8}_\texttt{x} \\ \texttt{C8}_\texttt{x} & \texttt{8B}_\texttt{x} & \texttt{C4}_\texttt{x} & \texttt{65}_\texttt{x} \\ \texttt{65}_\texttt{x} & \texttt{C8}_\texttt{x} & \texttt{8B}_\texttt{x} & \texttt{C4}_\texttt{x} \end{pmatrix} ,$$

whose inverse is also a circulant MDS matrix:

$$M_3^{-1} = \begin{pmatrix} \texttt{82}_\texttt{x} & \texttt{C4}_\texttt{x} & \texttt{34}_\texttt{x} & \texttt{F6}_\texttt{x} \\ \texttt{F6}_\texttt{x} & \texttt{82}_\texttt{x} & \texttt{C4}_\texttt{x} & \texttt{34}_\texttt{x} \\ \texttt{34}_\texttt{x} & \texttt{F6}_\texttt{x} & \texttt{82}_\texttt{x} & \texttt{C4}_\texttt{x} \\ \texttt{C4}_\texttt{x} & \texttt{34}_\texttt{x} & \texttt{F6}_\texttt{x} & \texttt{82}_\texttt{x} \end{pmatrix}.$$

Another example of a circulant MDS matrix is:

$$M_4 = \begin{pmatrix} \texttt{05}_\texttt{x} & \texttt{19}_\texttt{x} & \texttt{06}_\texttt{x} & \texttt{1B}_\texttt{x} \\ \texttt{1B}_\texttt{x} & \texttt{05}_\texttt{x} & \texttt{19}_\texttt{x} & \texttt{06}_\texttt{x} \\ \texttt{06}_\texttt{x} & \texttt{1B}_\texttt{x} & \texttt{05}_\texttt{x} & \texttt{19}_\texttt{x} \\ \texttt{19}_\texttt{x} & \texttt{06}_\texttt{x} & \texttt{1B}_\texttt{x} & \texttt{61}_\texttt{x} \end{pmatrix}.$$

The WIDEA-8 block cipher [39] uses an 8×8 circulant MDS matrix based on a $[16, 8, 9]$-code, whose entries are in $GF(2^{16}) = GF(2)[x]/(x^{16} + x^5 + x^3 + x^2 + 1)$ since the word size of WIDEA-8 is 16 bits:

$$M_5 = \begin{pmatrix} \texttt{01}_\texttt{x} & \texttt{01}_\texttt{x} & \texttt{04}_\texttt{x} & \texttt{01}_\texttt{x} & \texttt{08}_\texttt{x} & \texttt{05}_\texttt{x} & \texttt{02}_\texttt{x} & \texttt{09}_\texttt{x} \\ \texttt{09}_\texttt{x} & \texttt{01}_\texttt{x} & \texttt{01}_\texttt{x} & \texttt{04}_\texttt{x} & \texttt{01}_\texttt{x} & \texttt{08}_\texttt{x} & \texttt{05}_\texttt{x} & \texttt{02}_\texttt{x} \\ \texttt{02}_\texttt{x} & \texttt{09}_\texttt{x} & \texttt{01}_\texttt{x} & \texttt{01}_\texttt{x} & \texttt{04}_\texttt{x} & \texttt{01}_\texttt{x} & \texttt{08}_\texttt{x} & \texttt{05}_\texttt{x} \\ \texttt{05}_\texttt{x} & \texttt{02}_\texttt{x} & \texttt{09}_\texttt{x} & \texttt{01}_\texttt{x} & \texttt{01}_\texttt{x} & \texttt{04}_\texttt{x} & \texttt{01}_\texttt{x} & \texttt{08}_\texttt{x} \\ \texttt{08}_\texttt{x} & \texttt{05}_\texttt{x} & \texttt{02}_\texttt{x} & \texttt{09}_\texttt{x} & \texttt{01}_\texttt{x} & \texttt{01}_\texttt{x} & \texttt{04}_\texttt{x} & \texttt{01}_\texttt{x} \\ \texttt{01}_\texttt{x} & \texttt{08}_\texttt{x} & \texttt{05}_\texttt{x} & \texttt{02}_\texttt{x} & \texttt{09}_\texttt{x} & \texttt{01}_\texttt{x} & \texttt{01}_\texttt{x} & \texttt{04}_\texttt{x} \\ \texttt{04}_\texttt{x} & \texttt{01}_\texttt{x} & \texttt{08}_\texttt{x} & \texttt{05}_\texttt{x} & \texttt{02}_\texttt{x} & \texttt{09}_\texttt{x} & \texttt{01}_\texttt{x} & \texttt{01}_\texttt{x} \\ \texttt{01}_\texttt{x} & \texttt{04}_\texttt{x} & \texttt{01}_\texttt{x} & \texttt{08}_\texttt{x} & \texttt{05}_\texttt{x} & \texttt{02}_\texttt{x} & \texttt{09}_\texttt{x} & \texttt{01}_\texttt{x} \end{pmatrix}.$$

It can be denoted $M_5 = \text{circ}(\texttt{01}_\texttt{x}, \texttt{01}_\texttt{x}, \texttt{04}_\texttt{x}, \texttt{01}_\texttt{x}, \texttt{08}_\texttt{x}, \texttt{05}_\texttt{x}, \texttt{02}_\texttt{x}, \texttt{09}_\texttt{x})$. Curiously, M_5 is also used in the Whirlpool hash function [7], but under the finite field $GF(2^8)$=$GF(2)[x]/(x^8 + x^4 + x^3 + x^2 + 1)$. So, the same matrix remains MDS under different finite fields.

- **Algebraic construction**: an example of an MDS matrix constructed algebraically is the one used in the Curupira block cipher [8]. It is an involutory 3×3 MDS matrix with entries in $GF(2^8) = GF(2)[x]/(x^8 + x^6 + x^3 + x^2 + 1)$:

$$M_6 = \begin{pmatrix} \texttt{03}_\texttt{x} & \texttt{02}_\texttt{x} & \texttt{02}_\texttt{x} \\ \texttt{04}_\texttt{x} & \texttt{05}_\texttt{x} & \texttt{04}_\texttt{x} \\ \texttt{06}_\texttt{x} & \texttt{06}_\texttt{x} & \texttt{07}_\texttt{x} \end{pmatrix}.$$

M_6 is another rare example of an MDS matrix whose order is not a power of 2.

Due to its small size, the construction followed the definition of MDS matrices (B.5): every square submatrix of an MDS matrix should be nonsingular. By defining a generic 3×3 matrix

$$X = \begin{pmatrix} a & b & c \\ d & e & f \\ g & h & i \end{pmatrix}$$

one can set conditions on the nine coefficients $\{a, b, c, \ldots, i\}$ such that all square submatrices of X are nonsingular. Moreover, to make X involutory, an additional condition is imposed: $X^2 = I$ or $X = X^{-1}$. On top of those conditions, one can have some degree of freedom in choosing the coefficients to be small in order to minimize the implementation costs. This approach is feasible only for small square matrices. It requires computing the determinant of all square submatrices of the 3×3 matrix. The number of 1×1 square submatrices of X is $\binom{3}{1}^2 = 9$, which means all individual coefficients should be nonzero; the number of 2×2 square submatrices is $\binom{3}{2}^2 = 9$. In general, for an $n \times n$ matrix, there are $\binom{n}{i}^2$ possible $i \times i$ square submatrices. So, the total number of determinants computed for all submatrices of order 1 up to $n-1$ is $S = \sum_{i=1}^{n-1} \binom{n}{i}^2$. That means a system of S equations in n^2 variables. For the particular case of X, that means 18 conditions on 9 variables.

- **Cauchy matrices** [97]: given $x_0, \ldots, x_{n-1}$ and $y_0, \ldots, y_{n-1}$, the matrix $A = [a_{i,j}]$, for $0 \leq i, j \leq n-1$, where $a_{i,j} = \frac{1}{x_i + y_j}$ is called a Cauchy matrix. It is known that

$$\det(A) = \frac{\prod_{0 \leq i < j \leq n-1} (x_j - x_i)(y_j - y_i)}{\prod_{0 \leq i, j \leq n-1} (x_i + y_j)}.$$

Thus, provided the x_i are distinct, the y_j are distinct and $x_i + y_j \neq 0$ for all i, j, it follows that any square submatrix of a Cauchy matrix is nonsingular over any field.
An example provided in [97] is the involutory, symmetric 8×8 MDS matrix M_7 over the finite field $\mathrm{GF}(2^8) = \mathrm{GF}(2)/(x^8 + x^4 + x^3 + x^2 + 1)$:

$$M_7 = \begin{pmatrix} 93_x & 13_x & 57_x & DA_x & 58_x & 47_x & 0C_x & 1F_x \\ 13_x & 93_x & DA_x & 57_x & 47_x & 58_x & 1F_x & 0C_x \\ 57_x & DA_x & 93_x & 13_x & 0C_x & 1F_x & 58_x & 47_x \\ DA_x & 57_x & 13_x & 93_x & 1F_x & 0C_x & 47_x & 58_x \\ 58_x & 47_x & 0C_x & 1F_x & 93_x & 13_x & 57_x & DA_x \\ 47_x & 58_x & 1F_x & 0C_x & 13_x & 93_x & DA_x & 57_x \\ 0C_x & 1F_x & 58_x & 47_x & 57_x & DA_x & 93_x & 13_x \\ 1F_x & 0C_x & 47_x & 58_x & DA_x & 57_x & 13_x & 93_x \end{pmatrix}.$$

In M_7, the coefficients $a_{i,j}$ are quite large which imply a high implementation cost.
In [36], Gupta and Ray provided more efficient constructions of Cauchy matrices that are MDS and whose coefficients have low Hamming Weight.

- **Vandermonde matrices** [48, 77]: a Vandermonde matrix V of order d is defined by successive powers of its entries:

$$V = \begin{pmatrix} 1 & m_0 & m_0^2 & \ldots & m_0^{d-1} \\ 1 & m_1 & m_1^2 & \ldots & m_1^{d-1} \\ \ldots & \ldots & \ldots & \ldots & \ldots \\ 1 & m_i & m_i^2 & \ldots & m_i^{d-1} \\ \ldots & \ldots & \ldots & \ldots & \ldots \\ 1 & m_{d-1} & m_{d-1}^2 & \ldots & m_{d-1}^{d-1} \end{pmatrix}$$

where multiplication is defined in some finite field. Thus, the Vandermonde matrix V can be denoted by d entries: $V = \mathrm{van}(m_0, m_1, \ldots, m_{d-1})$. The Anubis block cipher [5] uses a non-involutory 4×4 MDS matrix derived from a Vandermonde matrix, in its key schedule algorithm:

$$M_8 = \begin{pmatrix} \mathtt{01_x} & \mathtt{01_x} & \mathtt{01_x}^2 & \mathtt{01_x}^3 \\ \mathtt{01_x} & \mathtt{02_x} & \mathtt{02_x}^2 & \mathtt{02_x}^3 \\ \mathtt{01_x} & \mathtt{06_x} & \mathtt{06_x}^2 & \mathtt{06_x}^3 \\ \mathtt{01_x} & \mathtt{08_x} & \mathtt{08_x}^2 & \mathtt{08_x}^3 \end{pmatrix} = \begin{pmatrix} \mathtt{01_x} & \mathtt{01_x} & \mathtt{01_x} & \mathtt{01_x} \\ \mathtt{01_x} & \mathtt{02_x} & \mathtt{04_x} & \mathtt{08_x} \\ \mathtt{01_x} & \mathtt{06_x} & \mathtt{14_x} & \mathtt{78_x} \\ \mathtt{01_x} & \mathtt{08_x} & \mathtt{40_x} & \mathtt{5D_x} \end{pmatrix}.$$

- **Hadamard matrices** [36, 77]: in [77], Sajadieh *et al.* constructed MDS matrices called Finite Field Hadamard (FFHadamard), which is defined as follows: a $2^m \times 2^m$ matrix H is a Finite Field Hadamard matrix in $\mathrm{GF}(2^n)$ if it can be represented as

$$H = \begin{pmatrix} U & V \\ V & U \end{pmatrix},$$

where the submatrices $U_{2^{m/2} \times 2^{m/2}}$ and $V_{2^{m/2} \times 2^{m/2}}$ are also FFHadamard. By definition, all FFHadamard matrices are symmetric.
From the symmetry in the FFHadamard matrix construction H, it is possible to characterize an FFHadamard matrix from any of its rows.
Without loss of generality, usually the first, topmost row is used [77]: let $H = [h_{i,j}]$ be a $2^m \times 2^m$ matrix whose topmost row is $(x_0, x_1, \ldots, x_{2^m-1})$ and $h_{i,j} = x_{i \oplus j}$, for $0 \leq i, j < 2^m$. Then, H is FFHadamard, and can be uniquely characterized by 2^m elements and is denoted as $H = \mathrm{had}(x_0, x_1, \ldots, x_{2^m-1})$.
In [36], detailed algorithms and constructions for FFHadamard matrices of dimensions which are powers of 2 were presented, along with concrete examples.
An example of a 16×16 involutory FFHadamard MDS matrix is

$$M_9 = \begin{pmatrix}
\mathtt{01_x} & \mathtt{03_x} & \mathtt{08_x} & \mathtt{B2_x} & \mathtt{0D_x} & \mathtt{60_x} & \mathtt{E8_x} & \mathtt{1C_x} & \mathtt{0F_x} & \mathtt{2C_x} & \mathtt{A2_x} & \mathtt{8B_x} & \mathtt{C9_x} & \mathtt{7A_x} & \mathtt{AC_x} & \mathtt{35_x} \\
\mathtt{03_x} & \mathtt{01_x} & \mathtt{B2_x} & \mathtt{08_x} & \mathtt{60_x} & \mathtt{0D_x} & \mathtt{1C_x} & \mathtt{E8_x} & \mathtt{2C_x} & \mathtt{0F_x} & \mathtt{8B_x} & \mathtt{A2_x} & \mathtt{7A_x} & \mathtt{C9_x} & \mathtt{35_x} & \mathtt{AC_x} \\
\mathtt{08_x} & \mathtt{B2_x} & \mathtt{01_x} & \mathtt{03_x} & \mathtt{E8_x} & \mathtt{1C_x} & \mathtt{0D_x} & \mathtt{60_x} & \mathtt{A2_x} & \mathtt{8B_x} & \mathtt{0F_x} & \mathtt{2C_x} & \mathtt{AC_x} & \mathtt{35_x} & \mathtt{C9_x} & \mathtt{7A_x} \\
\mathtt{B2_x} & \mathtt{08_x} & \mathtt{03_x} & \mathtt{01_x} & \mathtt{1C_x} & \mathtt{E8_x} & \mathtt{60_x} & \mathtt{0D_x} & \mathtt{8B_x} & \mathtt{A2_x} & \mathtt{2C_x} & \mathtt{0F_x} & \mathtt{35_x} & \mathtt{AC_x} & \mathtt{7A_x} & \mathtt{C9_x} \\
\mathtt{0D_x} & \mathtt{60_x} & \mathtt{E8_x} & \mathtt{1C_x} & \mathtt{01_x} & \mathtt{03_x} & \mathtt{08_x} & \mathtt{B2_x} & \mathtt{C9_x} & \mathtt{7A_x} & \mathtt{AC_x} & \mathtt{35_x} & \mathtt{0F_x} & \mathtt{2C_x} & \mathtt{A2_x} & \mathtt{8B_x} \\
\mathtt{60_x} & \mathtt{0D_x} & \mathtt{1C_x} & \mathtt{E8_x} & \mathtt{03_x} & \mathtt{01_x} & \mathtt{B2_x} & \mathtt{08_x} & \mathtt{7A_x} & \mathtt{C9_x} & \mathtt{35_x} & \mathtt{AC_x} & \mathtt{2C_x} & \mathtt{0F_x} & \mathtt{8B_x} & \mathtt{A2_x} \\
\mathtt{E8_x} & \mathtt{1C_x} & \mathtt{0D_x} & \mathtt{60_x} & \mathtt{08_x} & \mathtt{B2_x} & \mathtt{01_x} & \mathtt{03_x} & \mathtt{AC_x} & \mathtt{35_x} & \mathtt{C9_x} & \mathtt{7A_x} & \mathtt{A2_x} & \mathtt{8B_x} & \mathtt{0F_x} & \mathtt{2C_x} \\
\mathtt{1C_x} & \mathtt{E8_x} & \mathtt{60_x} & \mathtt{0D_x} & \mathtt{B2_x} & \mathtt{08_x} & \mathtt{03_x} & \mathtt{01_x} & \mathtt{35_x} & \mathtt{AC_x} & \mathtt{7A_x} & \mathtt{C9_x} & \mathtt{8B_x} & \mathtt{A2_x} & \mathtt{2C_x} & \mathtt{0F_x} \\
\mathtt{0F_x} & \mathtt{2C_x} & \mathtt{A2_x} & \mathtt{8B_x} & \mathtt{C9_x} & \mathtt{7A_x} & \mathtt{AC_x} & \mathtt{35_x} & \mathtt{01_x} & \mathtt{03_x} & \mathtt{08_x} & \mathtt{B2_x} & \mathtt{0D_x} & \mathtt{60_x} & \mathtt{E8_x} & \mathtt{1C_x} \\
\mathtt{2C_x} & \mathtt{0F_x} & \mathtt{8B_x} & \mathtt{A2_x} & \mathtt{7A_x} & \mathtt{C9_x} & \mathtt{35_x} & \mathtt{AC_x} & \mathtt{03_x} & \mathtt{01_x} & \mathtt{B2_x} & \mathtt{08_x} & \mathtt{60_x} & \mathtt{0D_x} & \mathtt{1C_x} & \mathtt{E8_x} \\
\mathtt{A2_x} & \mathtt{8B_x} & \mathtt{0F_x} & \mathtt{2C_x} & \mathtt{AC_x} & \mathtt{35_x} & \mathtt{C9_x} & \mathtt{7A_x} & \mathtt{08_x} & \mathtt{B2_x} & \mathtt{01_x} & \mathtt{03_x} & \mathtt{E8_x} & \mathtt{1C_x} & \mathtt{0D_x} & \mathtt{60_x} \\
\mathtt{8B_x} & \mathtt{A2_x} & \mathtt{2C_x} & \mathtt{0F_x} & \mathtt{35_x} & \mathtt{AC_x} & \mathtt{7A_x} & \mathtt{C9_x} & \mathtt{B2_x} & \mathtt{08_x} & \mathtt{03_x} & \mathtt{01_x} & \mathtt{1C_x} & \mathtt{E8_x} & \mathtt{60_x} & \mathtt{0D_x} \\
\mathtt{C9_x} & \mathtt{7A_x} & \mathtt{AC_x} & \mathtt{35_x} & \mathtt{0F_x} & \mathtt{2C_x} & \mathtt{A2_x} & \mathtt{8B_x} & \mathtt{0D_x} & \mathtt{60_x} & \mathtt{E8_x} & \mathtt{1C_x} & \mathtt{01_x} & \mathtt{03_x} & \mathtt{08_x} & \mathtt{B2_x} \\
\mathtt{7A_x} & \mathtt{C9_x} & \mathtt{35_x} & \mathtt{AC_x} & \mathtt{2C_x} & \mathtt{0F_x} & \mathtt{8B_x} & \mathtt{A2_x} & \mathtt{60_x} & \mathtt{0D_x} & \mathtt{1C_x} & \mathtt{E8_x} & \mathtt{03_x} & \mathtt{01_x} & \mathtt{B2_x} & \mathtt{08_x} \\
\mathtt{AC_x} & \mathtt{35_x} & \mathtt{C9_x} & \mathtt{7A_x} & \mathtt{A2_x} & \mathtt{8B_x} & \mathtt{0F_x} & \mathtt{2C_x} & \mathtt{E8_x} & \mathtt{1C_x} & \mathtt{0D_x} & \mathtt{60_x} & \mathtt{08_x} & \mathtt{B2_x} & \mathtt{01_x} & \mathtt{03_x} \\
\mathtt{35_x} & \mathtt{AC_x} & \mathtt{7A_x} & \mathtt{C9_x} & \mathtt{8B_x} & \mathtt{A2_x} & \mathtt{2C_x} & \mathtt{0F_x} & \mathtt{1C_x} & \mathtt{E8_x} & \mathtt{60_x} & \mathtt{0D_x} & \mathtt{B2_x} & \mathtt{08_x} & \mathtt{03_x} & \mathtt{01_x}
\end{pmatrix}.$$

Notice the symmetric pattern of the submatrices in M_9 like in H. Note also that M_9 can be denoted simply had($\mathtt{01_x}$, $\mathtt{03_x}$, $\mathtt{08_x}$, $\mathtt{B2_x}$, $\mathtt{0D_x}$, $\mathtt{60_x}$, $\mathtt{E8_x}$, $\mathtt{1C_x}$, $\mathtt{0F_x}$, $\mathtt{2C_x}$, $\mathtt{A2_x}$, $\mathtt{8B_x}$, $\mathtt{C9_x}$, $\mathtt{7A_x}$, $\mathtt{AC_x}$, $\mathtt{35_x}$), from its topmost row.

The CLEFIA block cipher [86] uses two 4×4 involutory, FFHadamard MDS matrices with entries in $\mathrm{GF}(2^8) = \mathrm{GF}(2)[x]/(x^8+x^4+x^3+x^2+1)$. The matrix M_{10} is also used in the Anubis block cipher [5]:

$$M_{10} = \begin{pmatrix}
\mathtt{01_x} & \mathtt{02_x} & \mathtt{04_x} & \mathtt{06_x} \\
\mathtt{02_x} & \mathtt{01_x} & \mathtt{06_x} & \mathtt{04_x} \\
\mathtt{04_x} & \mathtt{06_x} & \mathtt{01_x} & \mathtt{02_x} \\
\mathtt{06_x} & \mathtt{04_x} & \mathtt{02_x} & \mathtt{01_x}
\end{pmatrix}$$

and

$$M_{11} = \begin{pmatrix}
\mathtt{01_x} & \mathtt{08_x} & \mathtt{02_x} & \mathtt{0A_x} \\
\mathtt{08_x} & \mathtt{01_x} & \mathtt{0A_x} & \mathtt{02_x} \\
\mathtt{02_x} & \mathtt{0A_x} & \mathtt{01_x} & \mathtt{08_x} \\
\mathtt{0A_x} & \mathtt{02_x} & \mathtt{08_x} & \mathtt{01_x}
\end{pmatrix}.$$

Thus, M_{10} =had($\mathtt{01_x}, \mathtt{02_x}, \mathtt{04_x}, \mathtt{06_x}$), M_{11} =had($\mathtt{01_x}, \mathtt{08_x}, \mathtt{02_x}, \mathtt{0A_x}$).

The Khazad block cipher [6] uses an 8×8 involutory FFHadamard MDS matrix, with entries in $\mathrm{GF}(2^8) = \mathrm{GF}(2)[x]/(x^8+x^4+x^3+x^2+1)$:

$$M_{12} = \begin{pmatrix}
\mathtt{01_x} & \mathtt{03_x} & \mathtt{04_x} & \mathtt{05_x} & \mathtt{06_x} & \mathtt{08_x} & \mathtt{0B_x} & \mathtt{07_x} \\
\mathtt{03_x} & \mathtt{01_x} & \mathtt{05_x} & \mathtt{04_x} & \mathtt{08_x} & \mathtt{06_x} & \mathtt{07_x} & \mathtt{0B_x} \\
\mathtt{04_x} & \mathtt{05_x} & \mathtt{01_x} & \mathtt{03_x} & \mathtt{0B_x} & \mathtt{07_x} & \mathtt{06_x} & \mathtt{08_x} \\
\mathtt{05_x} & \mathtt{04_x} & \mathtt{03_x} & \mathtt{01_x} & \mathtt{07_x} & \mathtt{0B_x} & \mathtt{08_x} & \mathtt{06_x} \\
\mathtt{06_x} & \mathtt{08_x} & \mathtt{0B_x} & \mathtt{07_x} & \mathtt{01_x} & \mathtt{03_x} & \mathtt{04_x} & \mathtt{05_x} \\
\mathtt{08_x} & \mathtt{06_x} & \mathtt{07_x} & \mathtt{0B_x} & \mathtt{03_x} & \mathtt{01_x} & \mathtt{05_x} & \mathtt{04_x} \\
\mathtt{0B_x} & \mathtt{07_x} & \mathtt{06_x} & \mathtt{08_x} & \mathtt{04_x} & \mathtt{05_x} & \mathtt{01_x} & \mathtt{03_x} \\
\mathtt{07_x} & \mathtt{0B_x} & \mathtt{08_x} & \mathtt{06_x} & \mathtt{05_x} & \mathtt{04_x} & \mathtt{03_x} & \mathtt{01_x}
\end{pmatrix}.$$

Thus, M_{12} = had($\mathtt{01_x}, \mathtt{03_x}, \mathtt{04_x}, \mathtt{05_x}, \mathtt{06_x}, \mathtt{08_x}, \mathtt{0B_x}, \mathtt{07_x}$).

The Grostl hash function [34] uses the 8×8 FFHadamard matrix M_{13} = had($\mathtt{02_x}$, $\mathtt{02_x}$, $\mathtt{03_x}$, $\mathtt{04_x}$, $\mathtt{05_x}$, $\mathtt{03_x}$, $\mathtt{05_x}$, $\mathtt{07_x}$).

The Whirlwind hash function [4] uses two FFHadamard matrices:

$$M_{14} = \text{had}(\texttt{5}_\text{x}, \texttt{4}_\text{x}, \texttt{A}_\text{x}, \texttt{6}_\text{x}, \texttt{2}_\text{x}, \texttt{D}_\text{x}, \texttt{8}_\text{x}, \texttt{3}_\text{x})$$

and

$$M_{15} = \text{had}(\texttt{5}_\text{x}, \texttt{E}_\text{x}, \texttt{4}_\text{x}, \texttt{7}_\text{x}, \texttt{1}_\text{x}, \texttt{3}_\text{x}, \texttt{F}_\text{x}, \texttt{8}_\text{x}),$$

both with entries in $\text{GF}(2^4) = \text{GF}(2)[x]/(x^4 + x + 1)$.
Although the 1024-bit state of Whirlwind contains 16-bit words, each of these words is further partitioned into four nibbles, and individual nibbles are diffused using the two FFHadamard matrices.

- *ad hoc* **designs**: in [41], Junod and Vaudenay described a procedure to construct MDS matrices aimed at maximizing the number of entries equal to 1 in such matrices, and minimizing the number of larger entries. The objective is to minimize the implementation cost of the resulting MDS matrices in $\text{GF}(2^m)$, since multiplying by 1 costs nothing. One drawback is that there is no guarantee that such matrices will be involutory. Thus, there is no guarantee about the size of the entries in the inverse matrix. But, the authors intended to use such MDS matrices in ciphers whose design did not require the inverse matrix (for decryption).
An example of such a construction is the 4×4 matrix denoted M_{16} with nine entries equal to 1 and only two other nonzero entries:

$$M_{16} = \begin{pmatrix} a & 1 & 1 & 1 \\ 1 & 1 & b & a \\ 1 & a & 1 & b \\ 1 & b & a & 1 \end{pmatrix},$$

where a and b are nonzero constants in $\text{GF}(q)$, q a prime or a prime power, $a \neq 1$, $a \neq 0$. For $a = 2$ and $b = 3$, the matrix M_{16} has nine 1s compared with eight 1s in the AES MDS matrix M_1.
The FOX64 block cipher [40] uses a 4×4 MDS matrix denoted M_{17} following this approach:

$$M_{17} = \begin{pmatrix} 1 & 1 & 1 & a \\ 1 & z & a & 1 \\ z & a & 1 & 1 \\ a & 1 & z & 1 \end{pmatrix},$$

where $z = a^{-1} + 1 = a^7 + a^6 + a^5 + a^4 + a^3 + a^2 + 1$, with a a root of the irreducible polynomial $a^8 + a^7 + a^6 + a^5 + a^4 + a^3 + 1$. In fact, $a = \texttt{02}_\text{x}$ and $z = \texttt{FD}_\text{x}$.
The FOX128 block cipher [40], uses the 8×8 MDS matrix:

$$M_{18} = \begin{pmatrix} 1 & 1 & 1 & 1 & 1 & 1 & 1 & a \\ 1 & a & b & c & d & e & f & 1 \\ a & b & c & d & e & f & 1 & 1 \\ b & c & d & e & f & 1 & a & 1 \\ c & d & e & f & 1 & a & b & 1 \\ d & e & f & 1 & a & b & c & 1 \\ e & f & 1 & a & b & c & d & 1 \\ f & 1 & a & b & c & d & e & 1 \end{pmatrix},$$

where $a = x+1$, $b = x^7+x$, $c = x$, $d = x^2$, $e = x^7+x^6+x^5+x^4+x^3+x^2$, $f = x^6+x^5+x^4+x^3+x^2+x$, and x is a root of the irreducible polynomial $x^8+x^7+x^6+x^5+x^4+x^3+1$. In fact, $a = 03_x$, $b = 82_x$, $c = 02_x$, $d = 04_x$, $e = fc_x$ and $f = 79_x$.

M_{19} is an example of a 3×3 involutory MDS matrix with unknown construction method:

$$M_{19} = \begin{pmatrix} \texttt{02}_{\texttt{x}} & \texttt{07}_{\texttt{x}} & \texttt{04}_{\texttt{x}} \\ \texttt{03}_{\texttt{x}} & \texttt{06}_{\texttt{x}} & \texttt{04}_{\texttt{x}} \\ \texttt{03}_{\texttt{x}} & \texttt{07}_{\texttt{x}} & \texttt{05}_{\texttt{x}} \end{pmatrix}.$$

The Hierocrypt-L1 block cipher [89] uses a 4×4 MDS matrix, denoted M_{20}, with entries from $\mathrm{GF}(2^8) = \mathrm{GF}(2)[x]/(x^8 + x^6 + x^5 + x + 1)$:

$$M_{20} = \begin{pmatrix} \texttt{6C}_{\texttt{x}} & \texttt{25}_{\texttt{x}} & \texttt{9B}_{\texttt{x}} & \texttt{03}_{\texttt{x}} \\ \texttt{6D}_{\texttt{x}} & \texttt{06}_{\texttt{x}} & \texttt{C8}_{\texttt{x}} & \texttt{18}_{\texttt{x}} \\ \texttt{75}_{\texttt{x}} & \texttt{78}_{\texttt{x}} & \texttt{9E}_{\texttt{x}} & \texttt{1F}_{\texttt{x}} \\ \texttt{42}_{\texttt{x}} & \texttt{78}_{\texttt{x}} & \texttt{EB}_{\texttt{x}} & \texttt{61}_{\texttt{x}} \end{pmatrix}.$$

The Twofish block cipher [79] uses a non-involutory, non-symmetric 4×4 MDS matrix, denoted M_{21}, with entries from $\mathrm{GF}(2^8) = \mathrm{GF}(2)[x]/(x^8 + x^6 + x^5 + x^3 + 1)$:

$$M_{21} = \begin{pmatrix} \texttt{01}_{\texttt{x}} & \texttt{EF}_{\texttt{x}} & \texttt{5B}_{\texttt{x}} & \texttt{5B}_{\texttt{x}} \\ \texttt{5B}_{\texttt{x}} & \texttt{EF}_{\texttt{x}} & \texttt{EF}_{\texttt{x}} & \texttt{01}_{\texttt{x}} \\ \texttt{EF}_{\texttt{x}} & \texttt{5B}_{\texttt{x}} & \texttt{01}_{\texttt{x}} & \texttt{EF}_{\texttt{x}} \\ \texttt{EF}_{\texttt{x}} & \texttt{01}_{\texttt{x}} & \texttt{EF}_{\texttt{x}} & \texttt{5B}_{\texttt{x}} \end{pmatrix}.$$

The SHARK block cipher [74] uses a non-circulant, non-involutory, non-symmetric, 8×8 MDS matrix, denoted M_{22}, with entries from $\mathrm{GF}(2^8) = \mathrm{GF}(2)[x]/(x^8 + x^7 + x^6 + x^5 + x^4 + x^2 + 1)$:

$$M_{22} = \begin{pmatrix} \texttt{CE}_{\texttt{x}} & \texttt{95}_{\texttt{x}} & \texttt{57}_{\texttt{x}} & \texttt{82}_{\texttt{x}} & \texttt{8A}_{\texttt{x}} & \texttt{19}_{\texttt{x}} & \texttt{B0}_{\texttt{x}} & \texttt{01}_{\texttt{x}} \\ \texttt{E7}_{\texttt{x}} & \texttt{FE}_{\texttt{x}} & \texttt{05}_{\texttt{x}} & \texttt{D2}_{\texttt{x}} & \texttt{52}_{\texttt{x}} & \texttt{C1}_{\texttt{x}} & \texttt{88}_{\texttt{x}} & \texttt{F1}_{\texttt{x}} \\ \texttt{B9}_{\texttt{x}} & \texttt{DA}_{\texttt{x}} & \texttt{4D}_{\texttt{x}} & \texttt{D1}_{\texttt{x}} & \texttt{9E}_{\texttt{x}} & \texttt{17}_{\texttt{x}} & \texttt{83}_{\texttt{x}} & \texttt{86}_{\texttt{x}} \\ \texttt{D0}_{\texttt{x}} & \texttt{9D}_{\texttt{x}} & \texttt{26}_{\texttt{x}} & \texttt{2C}_{\texttt{x}} & \texttt{5D}_{\texttt{x}} & \texttt{9F}_{\texttt{x}} & \texttt{6D}_{\texttt{x}} & \texttt{75}_{\texttt{x}} \\ \texttt{52}_{\texttt{x}} & \texttt{A9}_{\texttt{x}} & \texttt{07}_{\texttt{x}} & \texttt{6C}_{\texttt{x}} & \texttt{B9}_{\texttt{x}} & \texttt{8F}_{\texttt{x}} & \texttt{70}_{\texttt{x}} & \texttt{17}_{\texttt{x}} \\ \texttt{87}_{\texttt{x}} & \texttt{28}_{\texttt{x}} & \texttt{3A}_{\texttt{x}} & \texttt{5A}_{\texttt{x}} & \texttt{F4}_{\texttt{x}} & \texttt{33}_{\texttt{x}} & \texttt{0B}_{\texttt{x}} & \texttt{6C}_{\texttt{x}} \\ \texttt{74}_{\texttt{x}} & \texttt{51}_{\texttt{x}} & \texttt{15}_{\texttt{x}} & \texttt{CF}_{\texttt{x}} & \texttt{09}_{\texttt{x}} & \texttt{A4}_{\texttt{x}} & \texttt{62}_{\texttt{x}} & \texttt{09}_{\texttt{x}} \\ \texttt{0B}_{\texttt{x}} & \texttt{31}_{\texttt{x}} & \texttt{7F}_{\texttt{x}} & \texttt{86}_{\texttt{x}} & \texttt{BE}_{\texttt{x}} & \texttt{05}_{\texttt{x}} & \texttt{83}_{\texttt{x}} & \texttt{34}_{\texttt{x}} \end{pmatrix},$$

and its inverse is

$$M_{22}^{-1} = \begin{pmatrix} \texttt{E7}_\texttt{x} & \texttt{30}_\texttt{x} & \texttt{90}_\texttt{x} & \texttt{85}_\texttt{x} & \texttt{D0}_\texttt{x} & \texttt{4B}_\texttt{x} & \texttt{91}_\texttt{x} & \texttt{41}_\texttt{x} \\ \texttt{53}_\texttt{x} & \texttt{95}_\texttt{x} & \texttt{9B}_\texttt{x} & \texttt{A5}_\texttt{x} & \texttt{96}_\texttt{x} & \texttt{BC}_\texttt{x} & \texttt{A1}_\texttt{x} & \texttt{68}_\texttt{x} \\ \texttt{02}_\texttt{x} & \texttt{45}_\texttt{x} & \texttt{F7}_\texttt{x} & \texttt{65}_\texttt{x} & \texttt{5C}_\texttt{x} & \texttt{1F}_\texttt{x} & \texttt{B6}_\texttt{x} & \texttt{52}_\texttt{x} \\ \texttt{A2}_\texttt{x} & \texttt{CA}_\texttt{x} & \texttt{22}_\texttt{x} & \texttt{94}_\texttt{x} & \texttt{44}_\texttt{x} & \texttt{63}_\texttt{x} & \texttt{2A}_\texttt{x} & \texttt{A2}_\texttt{x} \\ \texttt{FC}_\texttt{x} & \texttt{67}_\texttt{x} & \texttt{8E}_\texttt{x} & \texttt{10}_\texttt{x} & \texttt{29}_\texttt{x} & \texttt{75}_\texttt{x} & \texttt{85}_\texttt{x} & \texttt{71}_\texttt{x} \\ \texttt{24}_\texttt{x} & \texttt{45}_\texttt{x} & \texttt{A2}_\texttt{x} & \texttt{CF}_\texttt{x} & \texttt{2F}_\texttt{x} & \texttt{22}_\texttt{x} & \texttt{C1}_\texttt{x} & \texttt{0E}_\texttt{x} \\ \texttt{A1}_\texttt{x} & \texttt{F1}_\texttt{x} & \texttt{71}_\texttt{x} & \texttt{40}_\texttt{x} & \texttt{91}_\texttt{x} & \texttt{27}_\texttt{x} & \texttt{18}_\texttt{x} & \texttt{A5}_\texttt{x} \\ \texttt{56}_\texttt{x} & \texttt{F4}_\texttt{x} & \texttt{AF}_\texttt{x} & \texttt{32}_\texttt{x} & \texttt{D2}_\texttt{x} & \texttt{A4}_\texttt{x} & \texttt{DC}_\texttt{x} & \texttt{71}_\texttt{x} \end{pmatrix}.$$

Note that the entries of both M_{22} and M_{22}^{-1} have quite heavy Hamming Weight.

Designing MDS matrices satisfying too many criteria at once may sometimes involve conflicts. Some attempts at designing such MDS matrices have failed, which means that the security claims coming from the assumption that the matrix is MDS cannot be upheld [36, 66, 84].

B.2 Implementation Costs

As discussed previously, involutory MDS matrices are useful when the inverse matrix is needed, for instance, in the decryption procedure of some blocks cipher e.g. those following the SPN structure.

Another practical criterion for efficient implementation is the effective size of each entry of the MDS matrix. One approach used in the FOX family of block ciphers [41] was to maximize the number of entries equal to 1, and minimize the number of other (larger) coefficients.

The cost of a matrix multiplication can be measured by the total number of two primitive operations: bitwise exclusive-or ($\oplus$) and xtime (which is multiplication by 2 in the underlying finite field). Note that every element in the finite field $\mathrm{GF}(2^m)$ can be decomposed in a binary form, for instance, $\texttt{0E}_\texttt{x} = 00001110_2 = 8 \oplus 4 \oplus 2$. Thus, if $\texttt{0E}_\texttt{x}$ is an entry of an MDS matrix, a multiplication by an element s means $\texttt{0E}_\texttt{x} \cdot s = (8 \cdot s) \oplus (4 \cdot s) \oplus (2 \cdot s)$ where $\cdot$ stands for multiplication in $\mathrm{GF}(2^m)$. Consequently, it can be summarized by a series of multiplications of s by powers of 2, or repeated multiplication by 2.

Let $\mathrm{xtime}(s) = 2 \cdot s$. Then,

$$\texttt{0E}_\texttt{x} \cdot s = \mathrm{xtime}(\mathrm{xtime}(\mathrm{xtime}(s))) \oplus \mathrm{xtime}(\mathrm{xtime}(s)) \oplus \mathrm{xtime}(s),$$

which means six xtimes calls and two xors. For higher efficiency, we can chain the calls since the same s is present in all cases. Thus,

$$\texttt{0E}_\texttt{x} \cdot s = \mathrm{xtime}(\mathrm{xtime}(\mathrm{xtime}(s \oplus 1) \oplus 1),$$

and this multiplication between two m-bit values in GF(2^m) cost only two xors and three calls to xtimes.

Since the coefficients of the MDS matrix are fixed, it becomes clear that the smaller those coefficients are the lower the total cost of a single matrix multiplication. The overall size of the entries is more important than its Hamming Weight. Suppose $\texttt{11}_\texttt{x}$, which has a lower Hamming Weight than $\texttt{0E}_\texttt{x}$. The multiplication $\texttt{11}_\texttt{x} \cdot s = (16 \cdot s) \oplus s = \text{xtime}(\text{xtime}(\text{xtime}(\text{xtime}((s))))) \oplus s$ costs four xtimes and one xor.

The implementation of xtimes can be performed by an explicit algorithm or by a table look-up (which costs memory for storage). In either case, a single xtime call costs more than a single xor, and the multiplication by $\texttt{11}_\texttt{x}$ is more expensive than by $\texttt{0E}_\texttt{x}$.

In Table B.1, we summarize and compare properties and costs of some of the MDS matrices described previously.

Table B.1 Summary of properties of MDS matrices. The polynomial $p(x)$ or finite field is listed in the 5th column.

Matrix	Order	Type	Invol.	Finite Field GF(2)[x]/($p(x)$)	#xors	# xtime	Ref.
M_1	4×4	circulant	No	$x^8+x^4+x^3+x+1$	4	8	[33]
M_1^{-1}	4×4	circulant	No	$x^8+x^4+x^3+x+1$	28	48	[33]
M_2	3×3	circulant	No	$x^8+x^7+x^6+x^5+x^4+x^2+1$	3	9	[27]
M_5	8×8	circulant	No	$x^{16}+x^5+x^3+x^2+1$	16	88	[39]
M_6	3×3	algebraic	Yes	$x^8+x^6+x^3+x^2+1$	6	15	[8]
M_7	8×8	Cauchy	Yes	$x^8+x^4+x^3+x^2+1$	184	344	[97]
M_8	4×4	Vandermonde	No	$x^8+x^4+x^3+x^2+1$	9	33	[5]
M_9	16×16	FFHadamard	Yes	GF(2^8)	544	1248	[36]
M_{10}	4×4	FFHadamard	Yes	$x^8+x^4+x^3+x^2+1$	4	20	[86]
M_{14}	8×8	FFHadamard	Yes	x^4+x+1	24	68	[4]
M_{17}	4×4	*ad hoc*	No	$x^8+x^7+x^6+x^5+x^4+x^3+1$	24	32	[40]
M_{18}	8×8	*ad hoc*	No	$x^8+x^7+x^6+x^5+x^4+x^3+1$	78	169	[40]

Notice the implementation cost of matrices of the same order varies because of the different entries. For instance, M_6 costs more xtime operations than M_2, but M_6 is involutory, which may be advantageous if the inverse matrix is needed, since M_2 is not involutory.

The same reasoning applies to M_1 and M_8, as well as between M_{17} and M_{10}. Although M_1 implementation costs are the lowest, its inverse costs more than M_{10} which is involutory.

The same holds for M_5, M_7, M_{14} and M_{18}. Among the involutory matrices, M_{14} has the lowest implementation cost.

Appendix C
Substitution Boxes (S-boxes)

Anything is possible if approached with conviction.
– Konosuke Matsushita

This is a brief survey of cryptographic Substitution boxes, or simply S-boxes, and a number of relevant properties that are expected from S-boxes used in a cryptographic setting, such as the differential and linear profiles, the algebraic normal form and the nonlinearity.

S-boxes are the main nonlinear components in several cryptographic primitives, such as block and stream ciphers, hash functions and Message Authentication Codes (MAC) [62]. S-boxes fulfill the role of confusion as stated by Shannon in [81] and additionally provide (local) diffusion in the sense that all of its output bits depend (nonlinearly) on all input bits.

Previous research works on S-boxes include [13, 72, 73, 76, 95].

C.1 S-box Definition

An $n \times m$-bit Substitution Box (S-box) is a nonlinear mapping formally denoted $S : GF(2)^n \rightarrow GF(2)^m$ that is typically used to provide the confusion property [81] in several cryptographic primitives such as block ciphers, stream ciphers and hash functions.

In Substitution-Permutation Network (SPN) ciphers, the S-boxes are bijective mappings due to the need for decryption [33]. In Feistel Network ciphers, the S-boxes do not need to be bijective [68], but this can lead to a number of attacks that could be averted with the use of bijective S-boxes [28, 75, 65]. Some attacks though, depend exactly on the fact that S-boxes are bijective mappings, such as the reflection attack [37] against the GOST cipher [98] and the SASAS attack [14].

The majority of known S-boxes published in the literature are fixed mappings. Some S-boxes, though, are key dependent such as in Blowfish [78], or secret and variable depending on the application environment, such as the 4×4-bit S-boxes of GOST cipher [98]. The latest trend of lightweight cipher designs use small, fixed, 4×4-bit S-boxes [96, 26, 15, 83, 3].

If the S-boxes have small dimensions then they are often implemented as a static truth table (or look-up tables).

J. Nakahara Jr., *Lai-Massey Cipher Designs*, https://doi.org/10.1007/978-3-319-68273-0

Some S-boxes are involutory mappings, that is, they are their own inverses. This concept was adopted in ciphers such as Khazad and Anubis (also in the tweaked versions), to save memory for decryption. But, most block ciphers nowadays either do not adopt this strategy or do not need them, such as ciphers using the Feistel [68] or the Lai-Massey designs [40] which do not need the inverse S-box. The AES S-box without the affine transformation would be involutory, that is, if $S[x] = 1/x$, then $S[S[x]] = 1/(1/x) = x$.

In some cases, like in the DES cipher [68], there are eight contiguous S-boxes, and not only the individual S-box contents but also their relative position matters for the security of the DES [58] because some input bits are shared by neighboring S-boxes.

For 4×4 bijective S-boxes, exhaustive analysis over the whole space of such S-boxes has already been performed in [76]. In [13], Bilgin *et al.* performed analysis also for all 3×3 S-boxes and for the DES 6×4 S-boxes. Note that there are only $2^3! = 40320 \approx 2^{15.2992}$ possible permutations (bijective 3×3 S-boxes), and $2^4! = 16! = 20922789888000 \approx 2^{44.2501}$ permutations (bijective 4×4 S-boxes). For DES S-boxes, considered simply as surjective mappings, there are $2^6!/(2^6 - 2^4)! = 10221346459144248675287040000 \approx 2^{93.045}$ such mappings. But, since each DES S-box consists of four 4×4-bit S-boxes, there are in fact $4!\binom{16!}{4} = \approx 2^{177.00056}$ such mappings.

Tables C.1 and C.2 provide a non-exhaustive enumeration of parameters and properties of S-boxes, listed in alphabetical order of the cryptographic algorithm they are embedded in.

An $n \times m$ S-box can be interpreted as a vectorial Boolean mapping $S : GF(2)^n \rightarrow GF(2)^m$. Therefore, it can be described in terms of its m component Boolean functions $s_i : GF(2)^n \rightarrow GF(2)$ for $0 \leq i < m$. Thus, $S[x]=(s_0(x), s_1(x), \ldots, s_{m-1}(x))$.

Concerning the differential cryptanalysis technique [12, 71], relevant properties for an S-box include the *differential profile* and the *differential uniformity*.

The differential profile is related to the distribution of the differences such as:

$$\delta_S(a, b) = \#\{x \in GF(2^n) : S[x \oplus a] \oplus S[x] = b\}.$$

The value $S[x \oplus a] \oplus S[x]$ is called the *output xor-difference* to the S-box S, where a is the input difference. The value

$$\delta_{\max} = \max_{a \neq 0, b} \delta_S(a, b)$$

is the *differential uniformity* of S, and identifies the most probable nontrivial input/output difference pair(s) (a, b) that can propagate across S.

An extensive listing of $\delta_S(a, b)$ for all $0 \leq a < 2^n$, for $b = S[x \oplus a] \oplus S[x]$ and for a given difference operator, such as $\oplus$, is called the *Difference Distribution Table (DDT)* of S. The S-box is said to be $\delta_{\max}$-differentially uniform, in the sense that all nontrivial entries in its DDT are less than or equal to $\delta_{\max}$.

Table C.1 General parameters and properties of S-boxes.

Algorithm	Dimensions (bits)	#S-box	Biject?	Fixed? ⋆	Invol.?	Ref./Year
3WAY	3×3	1	Yes	Yes	No	[24]/1993
AES/Rijndael	8×8	1	Yes	Yes	No	[33]/2001
ARIA	8×8	2	Yes	Yes	No	[47]/2003
BASEKING	3×3	1	Yes	Yes	No	[23]/1995
BKSQ	8×8	1	Yes	Yes	No	[27]/2000
BLOWFISH	8×32	1	No	No	No	[78]/1994
CAMELLIA	8×8	4	Yes	Yes	No	[61]/2004
CAST-128/256	8×32	4	No	Yes	No	[2]/1997
CIPHERUNICORN-A	8×8	4	Yes	Yes	No	[91]/2000
CIPHERUNICORN-E	8×8	4	Yes	Yes	No	[90]/1998
CLEFIA	8×8	2	Yes	Yes	No	[85]/2007
COCONUT98	8×24	1	No	Yes	No	[93]/1998
CRYPTON	8×8	2	Yes	Yes	No	[54]/1998
CS	8×8	1	Yes	Yes	No	[87]/1998
CTC	3×3	1	Yes	Yes	No	[22]/2006
DES	6×4	8	No	Yes	No	[68]/1999
s^2-DES	6×4	8	No	Yes	No	[42]/1991
s^3-DES	6×4	8	No	Yes	No	[44]/1993
s^5-DES	6×4	8	No	Yes	No	[43]/1995
DRAGON	8×32	2	No	Yes	No	[29]/2006
E2	8×8	1	Yes	Yes	No	[70]/1998
FOX/IDEA-NXT	8×8	1	Yes	Yes	No	[40]/2004
HAMSI	4×4	1	Yes	Yes	No	[46]/1998
HIEROCRYPT-L1/3	8×8	1	Yes	Yes	No	[88, 89]/2000
JH	4×4	2	Yes	Yes	No	[96]/2011
KASUMI	7×7	1	Yes	Yes	No	[1]/1999
KASUMI	9×9	1	Yes	Yes	No	[1]/1999
KECCAK	5×5	1	Yes	Yes	No	[23]/1995
KHAZAD/ANUBIS	8×8	1	Yes	Yes	Yes	[6]/2000
KHUFU	8×32	‡	No	No	No	[63]/1991
LED	4×4	1	Yes	Yes	No	[35]/2011
LOKI89	12×8	1	No	Yes	No	[19]/1990
LOKI91	12×8	1	No	Yes	No	[17]/1991
LOKI97	11×8	1	No	Yes	No	[18]/1998
LOKI97	13×8	1	No	Yes	No	[18]/1998
MAGENTA	8×8	1	Yes	Yes	No	[38]/1998
MARS	9×32	1	No	Yes	No	[20]/1999
MISTY1	7×7	1	Yes	Yes	No	[60]/1997
MISTY1	9×9	1	Yes	Yes	No	[60]/1997
NOEKEON	4×4	1	Yes	Yes	Yes	[26]/2000
PANAMA	3×3	1	Yes	Yes	No	[23]/1995
PRESENT	4×4	1	Yes	Yes	No	[15]/2007

⋆: if the S-box is not static/fixed, then it is key dependent.

‡: key dependent S-boxes; the number of S-boxes depends on the number of rounds

Table C.2 General parameters and properties of S-boxes (cont.).

Algorithms	Dimensions (bits)	#S-box	Biject?	Fixed?	Invol.?	Ref./Year
PRINTcipher	3×3	1	Yes	Yes	No	[45]/2010
RADIOGATÚN	3×3	1	Yes	Yes	No	[9]/2006
RAINBOW	8×8	1	Yes	Yes	No	[52]/1998
SAFER K/SK/+/++	8×8	2	Yes	Yes	No	[56]/1994
SC2000	4×4	1	Yes	Yes	No	[83]/2002
SC2000	5×5	1	Yes	Yes	No	[83]/2002
SC2000	6×6	1	Yes	Yes	No	[83]/2002
SEED	8×8	2	Yes	Yes	No	[53]/2005
SERPENT-0	4×4	32	Yes	Yes	No	[11]/1998
SERPENT	4×4	8	Yes	Yes	No	[3]/1998
SHARK	8×8	1	Yes	Yes	No	[74]/1996
SKIPJACK	8×8	1	Yes	Yes	No	[69]/1998
SMS4	8×8	1	Yes	Yes	No	[30]/2008
SQUARE	8×8	1	Yes	Yes	No	[25]/1997
TWOFISH	8×8	1	Yes	Yes	No	[79]/1997
WHIRLPOOL	8×8	1	Yes	Yes	No	[7]/2003
WHIRLWIND	16×16	1	Yes	Yes	Yes	[4]/2010
ZODIAC	8×8	2	Yes	Yes	No	[51]/2000
ZUC	8×8	2	Yes	Yes	No	[31]/2011

Concerning the linear cryptanalysis technique [57], relevant properties for an S-box include the *linear profile* and the *linear uniformity*. The linear profile means the distribution of the linear bias values. Let $< a, x >$ denote the dot product of two bit strings $a, x \in GF(2)^t$, that is, $< a, x > = < x, a > = \oplus_{j=0}^{t-1} x_j \cdot a_j$, where $a = (a_0, a_1, \ldots, a_{t-1})$. Let $\gamma_S(a, b) = \#\{x : 0 \leq x < 2^n, < x, a > = < S[x], b >\} - 2^{n/2}$, where $a \in GF(2)^n$ and $b \in GF(2)^m$. When $\gamma_S(a, b)$ is nonzero, there is a nonzero correlation between a linear combination of a input bits and b output bits. The value $\gamma_{\max} = \max_{a \neq 0, b \neq 0} \gamma_S(a, b)$ indicates the most biased nontrivial linear relation(s) across S for all bitmasks a and b. The value $\gamma_{\max}$ is the linear uniformity (the counterpart of the differential uniformity).

An extensive listing of $\gamma_S(a, b) \cdot 2^n$ values for all possible input/output bitmask pairs (a, b) is called the *Linear Distribution Table (LAT)* of S, following [57]. The S-box S is said to be $\gamma_{\max}$-linearly uniform in the sense that all nontrivial entries in its LAT are less than or equal to $\gamma_{\max}$.

Table C.3 gives a non-exhaustive listing of some cryptographic parameters of some S-boxes. We use the value $\delta_{\max}/2^n$ since it gives the probability of the best nontrivial difference propagating across each S-box. Similarly, we use $|\gamma_{\max}/2^n|$ since it gives the largest bias among the nontrivial bitmasks propagating across the S-box, where n is the input size of S.

Table C.3 Cryptographic parameters of S-boxes.

Cipher	Dimensions ($n \times m$ bits)	$\delta\max/2^n$ ⋆	$\|\gamma\max\|/2^n$
3WAY/BASEKING	3×3	2^{-2}	2^{-2}
CTC	3×3	2^{-2}	2^{-2}
PANAMA/RADIOGATÚN	3×3	2^{-2}	2^{-2}
PRINTcipher	3×3	2^{-2}	2^{-2}
HAMSI	4×4	2^{-2}	2^{-2}
NOEKEON	4×4	2^{-2}	2^{-2}
PRESENT	4×4	2^{-2}	2^{-2}
JH (S0,S1) †	4×4	2^{-2}	2^{-2}
SC2000	4×4	2^{-2}	2^{-2}
SERPENT ‡	4×4	2^{-2}	2^{-2}
SC2000	5×5	2^{-4}	2^{-3}
SC2000	6×6	2^{-4}	2^{-3}
DES (S1)	6×4	2^{-2}	$2^{-1.83}$
DES (S2)	6×4	2^{-2}	2^{-2}
DES (S3)	6×4	2^{-2}	2^{-2}
DES (S4)	6×4	2^{-2}	2^{-2}
DES (S5)	6×4	2^{-2}	$2^{-1.67}$
DES (S6)	6×4	2^{-2}	$2^{-2.19}$
DES (S7)	6×4	2^{-2}	$2^{-1.83}$
DES (S8)	6×4	2^{-2}	2^{-2}
s^2-DES (S1)	6×4	$2^{-2.19}$	$2^{-2.19}$
s^2-DES (S2)	6×4	$2^{-2.19}$	$2^{-2.19}$
s^2-DES (S3)	6×4	$2^{-2.19}$	$2^{-2.19}$
s^2-DES (S4)	6×4	2^{-2}	$2^{-2.19}$
s^2-DES (S5)	6×4	2^{-2}	$2^{-1.83}$
s^2-DES (S6)	6×4	2^{-2}	$2^{-2.19}$
s^2-DES (S7)	6×4	2^{-2}	2^{-2}
s^2-DES (S8)	6×4	2^{-2}	$2^{-2.19}$
s^3-DES (S1)	6×4	$2^{-1.67}$	2^{-2}
s^3-DES (S2)	6×4	$2^{-1.83}$	2^{-2}
s^3-DES (S3)	6×4	$2^{-1.67}$	2^{-2}
s^3-DES (S4)	6×4	$2^{-1.67}$	$2^{-1.41}$
s^3-DES (S5)	6×4	$2^{-1.67}$	$2^{-1.41}$
s^3-DES (S6)	6×4	$2^{-1.67}$	$2^{-1.67}$
s^3-DES (S7)	6×4	$2^{-1.67}$	$2^{-1.67}$
s^3-DES (S8)	6×4	$2^{-1.67}$	2^{-2}
s^5-DES (S1)	6×4	$2^{-1.83}$	2^{-2}
s^5-DES (S2)	6×4	$2^{-2.67}$	2^{-2}
s^5-DES (S3)	6×4	$2^{-2.67}$	2^{-2}
s^5-DES (S4)	6×4	$2^{-2.67}$	2^{-2}
s^5-DES (S5)	6×4	$2^{-2.67}$	2^{-2}
s^5-DES (S6)	6×4	$2^{-1.83}$	2^{-2}
s^5-DES (S7)	6×4	$2^{-1.83}$	2^{-2}
s^5-DES (S8)	6×4	$2^{-1.83}$	2^{-2}

⋆: the inverse S-box, when it exists, has the same differential and linear profiles.

†: both s-boxes have the same differential and linear profiles.

‡: all eight S-boxes (S1 to S8) have the same δ_S and $|\gamma_S|$.

Table C.4 Cryptographic parameters of S-boxes (cont.).

Cipher	Dimensions ($n \times m$ bits)	$\delta\text{max}/2^n$	$\|\gamma\text{max}\|/2^n$
KASUMI (S7)	7×7	2^{-6}	2^{-4}
MISTY1 (S7)	7×7	2^{-6}	2^{-4}
AES/Rijndael	8×8	2^{-6}	2^{-4}
ANUBIS (original)	8×8	2^{-5}	$2^{-2.91}$
ANUBIS (tweak)	8×8	2^{-5}	2^{-3}
ARIA (S1)	8×8	2^{-6}	2^{-4}
ARIA (S2)	8×8	2^{-6}	2^{-4}
BKSQ	8×8	2^{-6}	2^{-4}
CAMELLIA	8×8	2^{-6}	2^{-4}
CIPHERUNICORN-A	8×8	2^{-6}	2^{-4}
CIPHERUNICORN-E	8×8	2^{-6}	2^{-4}
CLEFIA (S0)	8×8	$2^{-4.68}$	$2^{-3.19}$
CLEFIA (S1)	8×8	2^{-6}	2^{-4}
CRYPTON	8×8	2^{-5}	2^{-3}
CS	8×8	2^{-4}	2^{-3}
E2	8×8	$2^{-4.68}$	$2^{-3.19}$
FOX/IDEANXT	8×8	2^{-4}	2^{-3}
HIEROCRYPT-L1/3	8×8	2^{-6}	2^{-4}
KHAZAD (tweak)	8×8	2^{-5}	2^{-3}
MAGENTA	8×8	2^{-5}	$2^{-3.29}$
RAINBOW	8×8	$2^{-5.41}$	2^{-3}
SAFER K/SK/+/++	8×8	2^{-1}	$2^{-2.47}$
SEED	8×8	2^{-6}	2^{-4}
SHARK	8×8	2^{-6}	2^{-4}
SKIPJACK	8×8	$2^{-4.41}$	$2^{-3.19}$
SMS4	8×8	2^{-6}	2^{-4}
SQUARE	8×8	2^{-6}	2^{-4}
TWOFISH	8×8	$2^{-4.67}$	2^{-3}
WHIRLPOOL	8×8	2^{-5}	$2^{-3.19}$
ZODIAC (S1)	8×8	$2^{-4.67}$	$2^{-2.91}$
ZODIAC (S2)	8×8	$2^{-4.41}$	$2^{-2.87}$
ZUC (S0)	8×8	2^{-5}	2^{-3}
ZUC (S1)	8×8	2^{-6}	2^{-4}
COCONUT98	8×24	2^{-6}	$2^{-2.24}$
CAST-128/256	8×32	2^{-7}	2^{-1}
DRAGON	8×32	2^{-7}	2^{-1}
KASUMI (S9)	9×9	2^{-8}	2^{-5}
MISTY1 (S9)	9×9	2^{-8}	2^{-5}
MARS	9×32	2^{-7}	$2^{-2.61}$
LOKI97	11×8	2^{-7}	2^{-6}
LOKI89	12×8	2^{-4}	$2^{-4.67}$
LOKI91	12×8	$2^{-4.95}$	$2^{-4.64}$
LOKI97	13×8	2^{-7}	2^{-7}
WHIRLWIND	16×16	2^{-14}	2^{-8}

C.2 S-box Representations

There are several possible representations for an S-box:

- **Table Look-up** or *Truth Table*: an S-box $S : GF(2)^n \rightarrow GF(2)^m$ requires $m \cdot 2^n$ bits of storage. Depending on the application environment and the dimensions n, m, this representation may provide very fast access at an affordable storage space.
- **Algebraic Normal Form (ANF)** is a method of standardizing and normalizing logical formulas. The ANF of a Boolean mapping $f : GF(2)^n \rightarrow GF(2)$ is the unique representation of f as a polynomial over the polynomial ring $F_2[x_{n-1}, \ldots, x_0]/(x_{n-1}^2 + x_{n-1}, \ldots, x_0^2 + x_0)$. Assume the following bit numbering $(x_{n-1}, \ldots, x_1, x_0)$ for the S-box input and $(y_{n-1}, \ldots, y_1, y_0)$ for the S-box output, each of which has its own ANF. In general, the ANF of f has the form $f(x_{n-1}, \ldots, x_0) = a_0 + a_1.x_0 + \ldots + a_n.x_{n-1} + a_{0,1}.x_0.x_1 + \ldots + a_{n-2,n-1}.x_{n-2}.x_{n-1} + \ldots + a_{0,1,\ldots,n-1}.x_0.x_1 \ldots x_{n-1}$, where the coefficients $a_{i_1,i_2,\ldots} \in GF(2)$, $+$ denotes bitwise exclusive-or and . denotes the bitwise-AND operation.
  As an example, consider the 5-tuple input $(x_0, x_1, x_2, x_3, x_4)$ and the 5-tuple output $(y_0, y_1, y_2, y_3, y_4)$. The ANF of the 5×5-bit S-box of KECCAK (see Table C.8) is: $y_0 = x_0 + \overline{x_1}x_2 = x_0 + x_2 + x_1x_2$, $y_1 = x_1 + x_3 + x_2x_3$, $y_2 = x_2 + x_4 + x_3x_4$, $y_3 = x_0 + x_3 + x_0x_4$ and $y_4 = x_1 + x_4 + x_0x_1$.
  There are also many other normal forms. A propositional formula in Conjunctive Normal Form (CNF) is a formula in the form

$$\wedge_{i=1}^{n}(\vee_{j=1}^{m_i} A_{ij}) = (A_{11} \vee \ldots \vee A_{1m_1}) \wedge \ldots \wedge (A_{n1} \vee \ldots \vee A_{nm_n}),$$

  where each A_{ij}, with $1 \leq i \leq n$, $1 \leq j \leq m_i$, is either a symbol x or its negation $\overline{x}$; $\vee$ denotes bitwise-OR; $\wedge$ denotes bitwise-AND. A propositional formula in Disjunctive Normal Form (DNF) is a formula in the form

$$\vee_{i=1}^{n}(\wedge_{j=1}^{m_i} A_{ij}) = (A_{11} \wedge \ldots \wedge A_{1m_1}) \vee \ldots \vee (A_{n1} \wedge \ldots \wedge A_{nm_n}),$$

  where each A_{ij}, with $1 \leq i \leq n$, $1 \leq j \leq m_i$, is either a symbol or the negation of a symbol.
  To transform a formula in ANF to CNF, one has to account for the exclusive-or: $a + b \equiv (a \wedge \overline{b}) \vee (\overline{a} \wedge b)$, which is in CNF. Analogously, $a + b \equiv (\overline{a} \vee \overline{b}) \wedge (a \vee b)$ which is in DNF.
- **Polynomial expression over a finite field**: for instance, the AES [33] S-box relation $y = S[x]$, in $GF(2^8)$ can be represented as the polynomial (with only nine nonzero coefficients) $y = 63_x + 8f_x \cdot x^{127} + b5_x \cdot x^{191} + x^{223} + f4_x \cdot x^{239} + 25_x \cdot x^{247} + f9_x \cdot x^{251} + 09_x \cdot x^{253} + 05_x \cdot x^{254}$.
  SHARK S-box [74], which is also based on the inversion operation in $GF(2^8)$, plus an affine transformation, has a similarly sparse polynomial

expression (with only nine nonzero coefficients): $y = b1_x + bb_x.x^{127} + 1f_x.x^{191} + 99_x.x^{223} + 98_x.x^{239} + da_x.x^{247} + 9d_x.x^{251} + a3_x.x^{253} + 3e_x.x^{254}$. The sparsity becomes evident compared to the polynomial expression for the S-box S0 of CLEFIA (with 247 nonzero coefficients):

$y = 57_x + 43_x.x + 13_x.x^2 + 73_x.x^3 + 0b_x.x^4 + 7e_x.x^5 + bf_x.x^6 + 7b_x.x^7 + 96_x.x^8 + 4e_x.x^9 + 87_x.x^{10} + de_x.x^{11} + fd_x.x^{12} + 7b_x.x^{13} + c4_x.x^{14} + ca_x.x^{15} + f2_x.x^{16} + 27_x.x^{17} + e7_x.x^{18} + 54_x.x^{19} + 82_x.x^{20} + 86_x.x^{21} + 5e_x.x^{22} + d4_x.x^{23} + 9a_x.x^{24} + ae_x.x^{25} + f1_x.x^{26} + d6_x.x^{27} + 12_x.x^{28} + 6a_x.x^{29} + 28_x.x^{30} + 6c_x.x^{31} + fa_x.x^{32} + 05_x.x^{33} + 3a_x.x^{34} + a9_x.x^{35} + 4d_x.x^{36} + e6_x.x^{37} + 7b_x.x^{38} + f6_x.x^{39} + 2a_x.x^{40} + b0_x.x^{41} + 37_x.x^{42} + ff_x.x^{43} + 23_x.x^{44} + 65_x.x^{45} + ef_x.x^{46} + 49_x.x^{47} + 68_x.x^{48} + 8f_x.x^{49} + 45_x.x^{50} + ba_x.x^{51} + f2_x.x^{52} + 67_x.x^{53} + cb_x.x^{54} + 7c_x.x^{55} + b6_x.x^{56} + 71_x.x^{57} + 9c_x.x^{58} + c6_x.x^{59} + 3f_x.x^{60} + 3c_x.x^{61} + 72_x.x^{62} + 25_x.x^{63} + f7_x.x^{64} + 0a_x.x^{65} + 9e_x.x^{66} + 9c_x.x^{67} + 71_x.x^{68} + e8_x.x^{69} + c8_x.x^{70} + 69_x.x^{71} + ad_x.x^{72} + 0a_x.x^{73} + 5b_x.x^{74} + ad_x.x^{75} + d1_x.x^{76} + 36_x.x^{77} + 6e_x.x^{78} + f1_x.x^{79} + 4f_x.x^{80} + 14_x.x^{81} + 7f_x.x^{82} + f8_x.x^{83} + c9_x.x^{84} + a8_x.x^{85} + ef_x.x^{86} + 7c_x.x^{87} + 03_x.x^{88} + 92_x.x^{89} + e4_x.x^{90} + 5d_x.x^{91} + 69_x.x^{92} + 9c_x.x^{93} + 49_x.x^{94} + e2_x.x^{95} + af_x.x^{96} + 83_x.x^{97} + af_x.x^{98} + 48_x.x^{99} + 88_x.x^{100} + d2_x.x^{101} + 8e_x.x^{102} + 5d_x.x^{103} + f1_x.x^{104} + 70_x.x^{105} + eb_x.x^{106} + 8e_x.x^{107} + 4c_x.x^{108} + e7_x.x^{109} + 0f_x.x^{110} + 06_x.x^{111} + ec_x.x^{112} + fa_x.x^{113} + 82_x.x^{114} + 31_x.x^{115} + 98_x.x^{116} + 40_x.x^{117} + bb_x.x^{118} + d2_x.x^{119} + e9_x.x^{120} + af_x.x^{121} + eb_x.x^{122} + 36_x.x^{123} + 0f_x.x^{124} + da_x.x^{125} + 51_x.x^{126} + a5_x.x^{128} + 85_x.x^{129} + 76_x.x^{130} + f4_x.x^{131} + 52_x.x^{132} + bd_x.x^{133} + 4c_x.x^{134} + 7d_x.x^{135} + 27_x.x^{136} + 78_x.x^{137} + a2_x.x^{138} + 53_x.x^{139} + fb_x.x^{140} + 68_x.x^{141} + ef_x.x^{142} + 07_x.x^{143} + da_x.x^{144} + d1_x.x^{145} + a7_x.x^{146} + 6b_x.x^{147} + de_x.x^{148} + b9_x.x^{149} + 4e_x.x^{150} + fb_x.x^{151} + 8a_x.x^{152} + 44_x.x^{153} + 0b_x.x^{154} + cc_x.x^{155} + 9b_x.x^{156} + 5f_x.x^{157} + 8d_x.x^{158} + a2_x.x^{159} + 71_x.x^{160} + dc_x.x^{161} + ce_x.x^{162} + eb_x.x^{163} + a7_x.x^{164} + 07_x.x^{165} + 28_x.x^{166} + dd_x.x^{167} + 1e_x.x^{168} + 9e_x.x^{169} + 12_x.x^{170} + 78_x.x^{171} + 04_x.x^{172} + 8a_x.x^{173} + 83_x.x^{174} + e2_x.x^{175} + 3d_x.x^{176} + 0e_x.x^{177} + 3b_x.x^{178} + 0a_x.x^{179} + 77_x.x^{180} + e6_x.x^{181} + 15_x.x^{182} + 57_x.x^{183} + 57_x.x^{184} + 76_x.x^{185} + 50_x.x^{186} + 1a_x.x^{187} + 9b_x.x^{188} + a6_x.x^{189} + bb_x.x^{190} + 32_x.x^{192} + 15_x.x^{193} + 1f_x.x^{194} + 39_x.x^{195} + 4c_x.x^{196} + 35_x.x^{197} + 8e_x.x^{198} + 73_x.x^{199} + f4_x.x^{200} + 92_x.x^{201} + 02_x.x^{202} + bd_x.x^{203} + 2b_x.x^{204} + cb_x.x^{205} + 40_x.x^{206} + 92_x.x^{207} + da_x.x^{208} + 5b_x.x^{209} + 7a_x.x^{210} + e0_x.x^{211} + 0b_x.x^{212} + 1e_x.x^{213} + 12_x.x^{214} + e6_x.x^{215} + 59_x.x^{216} + 50_x.x^{217} + 5d_x.x^{218} + 02_x.x^{219} + 20_x.x^{220} + e5_x.x^{221} + 6b_x.x^{222} + 79_x.x^{224} + 9a_x.x^{225} + a7_x.x^{226} + 3d_x.x^{227} + 21_x.x^{228} + 5b_x.x^{229} + 13_x.x^{230} + 21_x.x^{231} + b1_x.x^{232} + 1a_x.x^{233} + 0a_x.x^{234} + 24_x.x^{235} + c9_x.x^{236} + 1c_x.x^{237} + 58_x.x^{238} + 20_x.x^{240} + 2e_x.x^{241} + 7e_x.x^{242} + 15_x.x^{243} + 60_x.x^{244} + b9_x.x^{245} + 2f_x.x^{246} + 42_x.x^{248} + 71_x.x^{249} + 64_x.x^{250} + db_x.x^{252}$

Analogously, for the S-box S1 of CLEFIA (with 253 nonzero coefficients):

$y = 6c_x + d3_x.x + 47_x.x^2 + c8_x.x^3 + ca_x.x^4 + 5b_x.x^5 + 0b_x.x^6 + 9f_x.x^7$

$+ 05_x.x^8 + 8b_x.x^9 + f4_x.x^{10} + b9_x.x^{11} + d9_x.x^{12} + eb_x.x^{13} + 0d_x.x^{14} + a9_x.x^{15} + a7_x.x^{16} + f5_x.x^{17} + da_x.x^{18} + 14_x.x^{19} + 19_x.x^{20} + ce_x.x^{21} + 2c_x.x^{22} + cf_x.x^{23} + 3d_x.x^{24} + 0c_x.x^{25} + 99_x.x^{26} + fd_x.x^{27} + aa.x^{28} + 65_x.x^{29} + f4_x.x^{30} + 96_x.x^{31} + 32_x.x^{32} + cd_x.x^{33} + 56_x.x^{34} + 97_x.x^{35} + ba_x.x^{36} + 39.x^{37} + 84_x.x^{38} + 9b_x.x^{39} + 1a_x.x^{40} + 6a_x.x^{41} + a5_x.x^{42} + 1c_x.x^{43} + c8_x.x^{44} + 8e_x.x^{45} + 90_x.x^{46} + 4c_x.x^{47} + 96_x.x^{48} + 8a_x.x^{49} + e1_x.x^{50} + d5_x.x^{51} + 46_x.x^{52} + 1e_x.x^{53} + 04_x.x^{54} + 16_x.x^{55} + 30_x.x^{56} + c1_x.x^{58} + 90_x.x^{59} + 90_x.x^{60} + 10_x.x^{61} + 06_x.x^{62} + b2_x.x^{63} + 12_x.x^{64} + 69_x.x^{65} + 3e_x.x^{66} + 76_x.x^{67} + 73_x.x^{68} + 4e_x.x^{69} + c8_x.x^{70} + 78_x.x^{71} + 67_x.x^{72} + fa_x.x^{73} + 4c_x.x^{74} + 3c_x.x^{75} + e1_x.x^{76} + a5_x.x^{77} + 4d_x.x^{78} + 9a_x.x^{79} + df_x.x^{80} + 95_x.x^{81} + a3_x.x^{82} + 87_x.x^{83} + 2c_x.x^{84} + 91_x.x^{85} + ac_x.x^{86} + 4a_x.x^{87} + bc_x.x^{88} + ba_x.x^{89} + 57_x.x^{90} + f7_x.x^{91} + e5_x.x^{92} + 3c_x.x^{93} + c2_x.x^{94} + f8_x.x^{95} + c7_x.x^{96} + ce_x.x^{97} + e0_x.x^{98} + 28_x.x^{99} + 4b_x.x^{100} + 53_x.x^{101} + cb_x.x^{102} + 71_x.x^{103} + 26_x.x^{104} + e1_x.x^{105} + e3_x.x^{106} + a7_x.x^{107} + eb_x.x^{108} + 03_x.x^{109} + ff_x.x^{110} + 70_x.x^{111} + a9_x.x^{112} + 30_x.x^{113} + 69_x.x^{114} + 94_x.x^{115} + 4c_x.x^{116} + 6c_x.x^{117} + 08_x.x^{118} + 2f_x.x^{119} + 5a_x.x^{120} + 86_x.x^{121} + a3_x.x^{122} + 27_x.x^{123} + dd_x.x^{124} + e9_x.x^{125} + c5_x.x^{126} + 46_x.x^{127} + 50_x.x^{128} + 95_x.x^{129} + 66_x.x^{130} + 31_x.x^{131} + c1_x.x^{132} + 1b_x.x^{133} + 40_x.x^{134} + f9_x.x^{135} + af_x.x^{136} + 5b_x.x^{137} + 10_x.x^{138} + 58_x.x^{139} + de_x.x^{140} + 5b_x.x^{141} + 18_x.x^{142} + a7_x.x^{143} + 99_x.x^{144} + be_x.x^{145} + 78_x.x^{146} + 20_x.x^{147} + 86_x.x^{148} + e2_x.x^{149} + 11_x.x^{150} + 1b_x.x^{151} + 2e_x.x^{152} + f1_x.x^{154} + cc_x.x^{155} + e7_x.x^{156} + a1_x.x^{157} + b0_x.x^{158} + bd_x.x^{159} + 93_x.x^{160} + 08_x.x^{161} + 9b_x.x^{162} + 1a_x.x^{163} + b2_x.x^{164} + 6c_x.x^{165} + ed_x.x^{166} + 96_x.x^{167} + df_x.x^{168} + d1_x.x^{169} + 7e_x.x^{170} + e8_x.x^{171} + ac_x.x^{172} + 10_x.x^{173} + 4d_x.x^{174} + ab_x.x^{175} + 3e_x.x^{176} + e4_x.x^{177} + 42_x.x^{178} + 28_x.x^{179} + 92_x.x^{180} + e7_x.x^{181} + b9_x.x^{182} + 61_x.x^{183} + 20_x.x^{184} + 7a_x.x^{185} + f6_x.x^{186} + e6_x.x^{187} + 29_x.x^{188} + 7b_x.x^{189} + f9_x.x^{190} + 34_x.x^{191} + 1f_x.x^{192} + f5_x.x^{193} + b0_x.x^{194} + 8b_x.x^{195} + a6_x.x^{196} + 09_x.x^{197} + 8f_x.x^{198} + bb_x.x^{199} + 35_x.x^{200} + bb_x.x^{201} + f4_x.x^{202} + b3_x.x^{203} + a9_x.x^{204} + a3_x.x^{205} + 74_x.x^{206} + 3b_x.x^{207} + 1c_x.x^{208} + 25_x.x^{209} + 28_x.x^{210} + 64_x.x^{211} + 2c_x.x^{212} + be_x.x^{213} + 9f_x.x^{214} + b6_x.x^{215} + 8f_x.x^{216} + 0e_x.x^{217} + fa_x.x^{218} + 12_x.x^{219} + 84_x.x^{220} + 70_x.x^{221} + 64_x.x^{222} + ca_x.x^{223} + 36_x.x^{224} + 17_x.x^{225} + 78_x.x^{226} + e1_x.x^{227} + 32_x.x^{228} + 4b_x.x^{229} + 18_x.x^{230} + 68_x.x^{231} + a2_x.x^{232} + fc_x.x^{233} + 38_x.x^{234} + fb_x.x^{235} + 64_x.x^{236} + db_x.x^{237} + 85_x.x^{238} + 1c_x.x^{239} + 11_x.x^{240} + d1_x.x^{241} + 6d_x.x^{242} + d7_x.x^{243} + d5_x.x^{244} + 02_x.x^{245} + 95_x.x^{246} + 44_x.x^{247} + a6_x.x^{248} + 39_x.x^{249} + 06_x.x^{250} + 5f_x.x^{251} + f1_x.x^{252} + 78_x.x^{253} + 1e_x.x^{254}$

Additionally, algebraic expressions of low algebraic degree such as quadratic expressions are of particular interest [21, 82]. Courtois and Pieprzyk in [21] demonstrated how to check the existence of quadratic expressions for S-boxes.

For Serpent S-box S_0, 21 linearly independent quadratic algebraic relations were obtained (as predicted for 4×4-bit S-boxes). These algebraic relations are multivariate equations holding with certainty and mixing both the input and the output variables. A common notation for all of Serpent S-boxes: the input is denoted (x_3, x_2, x_1, x_0) and the output is denoted (y_3, y_2, y_1, y_0). The following 21 multivariate quadratic relations for Serpent S_0 were obtained[1]:

- $x_0 + x_1 + x_2 + x_3 + y_3 + x_0.x_3 = 0,$
- $x_2 + x_3 + y_0 + y_1 + x_0.x_1 + x_1.x_3 = 0,$
- $x_0 + x_0.x_1 + x_0.x_2 + x_0.y_3 = 0,$
- $x_0.x_1 + x_1.x_2 + x_1.y_0 + x_1.y_1 = 0,$
- $1 + y_0 + y_3 + x_0.x_2 + x_0.y_1 + x_0.y_2 + x_1.x_2 + x_1.y_2 = 0,$
- $x_0 + x_3 + y_0 + y_1 + y_2 + y_3 + x_0.x_1 + x_0.y_2 + x_1.x_2 + x_1.y_0 + x_1.y_3 = 0,$
- $x_3 + y_2 + x_0.x_2 + x_0.y_0 + x_0.y_1 + x_0.y_2 + x_1.y_0 = 0,$
- $x_0 + x_3 + y_0 + y_1 + y_2 + y_3 + x_0.x_2 + x_0.y_1 + x_0.y_2 + x_1.x_2 + x_2.x_3 + x_2.y_0 = 0,$
- $x_0 + x_1 + x_3 + y_3 + x_0.y_1 + x_0.y_2 + x_1.x_2 + x_2.y_1 = 0,$
- $x_0.x_2 + x_1.x_2 + x_2.x_3 + x_2.y_2 = 0,$
- $1 + x_0 + x_1 + x_2 + x_3 + y_1 + y_2 + x_0.x_1 + x_0.x_2 + x_2.x_3 + x_2.y_3 = 0,$
- $1 + x_2 + y_1 + y_2 + y_3 + x_0.x_1 + x0.x_2 + x_0.y_1 + x_1.x_2 + x_2.x_3 + x_3.y_0 = 0,$
- $1 + x_0 + x_1 + x_2 + y_0 + x_0.x_1 + x0.x_2 + x_0.y_1 + x_0.y_2 + x_1.x_2 + x_1.y_0 + x_3.y_1 = 0,$
- $1 + x_1 + x_3 + y_0 + y_3 + x_0.x_1 + x_0.y_2 + x_1.x_2 + x_3.y_2 = 0,$
- $x_2 + y_0 + y_1 + x_0.x_1 + x_2.x_3 + x_3.y_3 = 0,$
- $1 + x_2 + y_0 + y_1 + y_2 + y_3 + y_0.y_1 = 0,$
- $1 + x_0 + x_1 + y_1 + x_0.x_1 + x_0.x_2 + x_0.y_1 + x_0.y_2 + x_1.x_2 + x_2.x_3 = 0,$
- $1 + x_1 + x_3 + y_0 + y_2 + y_3 + x_0.x_1 = 0,$
- $x_2 + x_3 + y_0 + y_1 + y_2 + x_0.x_1 + x_0.y_1 + x_0.y_2 + x_1.x_2 = 0,$

[1] Each of them holds with certainty and they are all linearly independent relations. These same properties apply to the quadratic relations of all the other S-boxes of Serpent.

- $x_0 + x_1 + x_2 + y_0 + y_1 + y_2 + y_3 + x_0.x_1 = 0$,
- $1 + x_0 + x_3 + y_1 + x_0.x_1 + x_0.y_1 + x_1.y_0 + x_2.x_3 = 0$.

For Serpent S-box S_1, the following 21 quadratic expressions were obtained:

- $1 + x_1 + x_2 + x_3 + y_2 + x_0.x_1 = 0$,
- $x_0 + x_1 + x_2 + y_1 + y_3 + x_0.x_1 + x_0.x_3 + x_1.x_3 = 0$,
- $x_0 + x_0.x_2 + x_0.x_3 + x_0.y_2 = 0$,
- $x_0 + x_1 + x_2 + y_2 + y_3 + x_0.y_0 + x_0.y_3 + x_1.y_0 + x_1.y_1 = 0$,
- $x_0 + x_1 + x_2 + y_1 + y_3 + x_0.x_3 + x_1.x_2 + x_1.y_2 = 0$,
- $x_0 + x_1 + y_1 + y_2 + x_0.x_1 + x_0.x_2 + x_0.y_0 + x_0.y_1 + x_1.x_2 + x_1.y_0 + x_1.y_3 = 0$,
- $1 + x_1 + x_2 + y_0 + y_2 + y_3 + x_0.x_1 + x_0.x_3 + x_0.y_1 + x_0.y_3 + x_1.x_2 + x_2.x_3 = 0$,
- $1 + x_1 + x_2 + y_0 + x_0.x_3 + x_0.y_3 + x_1.y_0 + x_2.y_0 = 0$,
- $1 + x_0 + x_1 + x_2 + y_0 + y_2 + y_3 + x_0.x_2 + x_0.y_0 + x_0.y_1 + x_1.y_0 + x_2.y_1 = 0$,
- $1 + x_0 + x_1 + x_2 + y_0 + y_2 + y_3 + x_0.x_1 + x_0.x_2 + x_0.x_3 + x_0.y_0 + x_1.x_2 + x_1.y_0 + x_2.y_2 = 0$,
- $1 + x_0 + x_1 + x_2 + y_0 + x_0.x_1 + x_0.x_2 + x_0.y_1 + x_2.y_3 = 0$,
- $x_1 + x_2 + y_0 + y_1 + x_0.x_1 + x_0.x_2 + x_0.y_1 + x_1.y_0 + x_3.y_0 = 0$,
- $1 + x_0 + x_2 + y_0 + y_1 + y_3 + x_0.x_1 + x_0.x_2 + x_0.x_3 + x_0.y_3 + x_1.y_0 + x_3.y_1 = 0$,
- $1 + y_0 + y_1 + y_2 + x_0.x_2 + x_0.x_3 + x_1.x_2 + x_3.y_2 = 0$,
- $x_1 + y_1 + y_2 + x_0.x_3 + x_0.y_3 + x_1.x_2 + x_1.y_0 + x_3.y_3 = 0$,
- $1 + x_0 + y_0 + y_1 + x_0.x_1 + x_0.x_3 + y_0.y_1 = 0$,
- $1 + x_0 + y_0 + y_1 + x_0.x_1 + x_0.x_3 + x_0.y_0 + x_1.y_0 = 0$,
- $x_1 + y_1 + y_2 + y_3 + x_0.x_1 + x_0.x_3 = 0$,
- $x_2 + y_3 + x_0.x_2 + x_0.y_3 + x_1.y_0 = 0$,

- $1 + x_1 + x_2 + y_0 + y_3 + x_0.x_1 = 0$,
- $1 + x_1 + x_2 + y_0 + y_3 + x_0.y_3 + x_1.x_2 + x_1.y_0 = 0$.

For Serpent S-box S_2, the following 21 quadratic expressions were obtained:

- $1 + x_0 + x_1 + x_3 + y_1 + y_2 + y_3 + x_0.x_1 + x_0.y_0 = 0$,
- $x_0 + x_2 + x_3 + y_1 + y_2 + x_0.x_1 + x_0.y_2 + x_1.x_3 = 0$,
- $1 + x_0 + x_2 + y_3 + x_1.x_2 + x_1.y_0 = 0$,
- $x_1 + x_2 + x_3 + y_0 + x_0.x_2 = 0$,
- $x_1 + x_2 + x_3 + y_0 + x_0.y_1 + x_0.y_2 + x_1.x_2 + x_1.y_1 + x_1.y_2 = 0$,
- $1 + x_0 + x_1 + x_2 + y_3 + x_0.x_1 + x_1.x_2 + x_1.y_3 = 0$,
- $1 + x_0 + x_1 + x_3 + y_1 + y_2 + y_3 + x_0.x_3 = 0$,
- $x_1 + x_3 + y_0 + x_1.x_2 + x_2.x_3 + x_2.y_0 = 0$,
- $x_0.x_1 + x_0.y_1 + x_0.y_2 + x_0.y_3 = 0$,
- $x_2 + y_1 + x_0.y_1 + x_1.x_2 + x_1.y_1 + x_2.x_3 + x_2.y_2 = 0$,
- $1+x_2+x_3+y_1+y_2+y_3+x_0.y_2+x_1.x_2+x_1.y_1+x_2.x_3+x_2.y_1+x_2.y_3 = 0$,
- $1 + x_0 + x_1 + x_2 + y_0 + y_1 + y_3 + x_0.x_1 + x_0.y_1 + x_1.x_2 + x_3.y_0 = 0$,
- $1 + x_0 + x_1 + y_1 + y_3 + x_0.y_2 + x_1.x_2 + x_1.y_1 + x_2.x_3 + x_3.y_1 = 0$,
- $1+x_0+x_2+x_3+y_0+y_2+y_3+x_0.y_2+x_1.x_2+x_1.y_1+x_2.x_3+x_2.y_1+x_3.y_2 = 0$,
- $x_0 + x_1 + y_0 + x_0.x_1 + x_0.y_2 + x_2.y_1 + x_3.y_3 = 0$,
- $x_0 + x_1 + x_3 + y_1 + y_2 + x_0.y_1 + x_2.x_3 + y_0.y_1 = 0$,
- $x_0 + x_1 + x_3 + y_1 + x_0.y_1 + x_0.y_2 + x_1.y_1 + x_2.x_3 + x_2.y_1 = 0$,
- $1 + x_1 + x_2 + y_0 + y_3 + x_0.x_1 + x_0.y_2 + x_1.y_1 + x_2.y_1 = 0$,
- $x_1 + x_3 + y_0 + y_1 + x_2.x_3 = 0$,

- $x_0 + x_1 + x_3 + y_2 + x_2.x_3 = 0$,

- $1 + x_0 + x_2 + x_3 + y_0 + y_1 + y_2 + y_3 + x_2.x_3 = 0$.

For Serpent S-box S_3, the following 21 quadratic expressions were obtained:

- $x_0.x_1 + x_0.x_2 + x_0.y_2 + x_0.y_3 = 0$,

- $x_0 + y_1 + y_2 + y_3 + x_0.x_3 + x_1.x_3 = 0$,

- $x_1 + x_2 + x_3 + y_2 + x_0.x_1 + x_0.y_0 + x_1.x_2 + x_1.y_0 = 0$,

- $x_1 + x_2 + y_0 + y_3 + x_0.x_1 + x_0.x_2 + x_0.y_0 + x_0.y_3 + x_1.x_2 + x_1.y_1 = 0$,

- $x_2 + x_3 + y_1 + y_3 + x_0.x_1 + x_0.x_3 + x_1.x_2 + x_1.y_2 = 0$,

- $x_1 + x_2 + y_0 + y_1 + y_2 + x_0.x_2 + x_0.y_0 + x_1.y_3 = 0$,

- $x_2 + x_3 + y_1 + y_3 + x_0.x_2 + x_0.x_3 + x_0.y_1 + x_0.y_3 = 0$,

- $x_0 + x_2 + y_0 + y_3 + x_0.x_1 + x_0.y_3 + x_1.x_2 + x_2.y_0 = 0$,

- $x_0.x_1 + x_0.x_3 + x_0.y_0 + x_0.y_3 + x_1.x_2 + x_2.x_3 + x_2.y_1 = 0$,

- $x_0 + y_0 + y_3 + x_0.x_3 + x_0.y_0 + x_1.x_2 + x_2.x_3 + x_2.y_2 = 0$,

- $x_1 + x_3 + y_3 + x_0.x_1 + x_0.x_2 + x_0.x_3 + x_0.y_3 + x_1.x_2 + x_2.x_3 + x_2.y_3 = 0$,

- $x_0 + x_1 + x_2 + y_1 + y_2 + x_0.x_1 + x_0.x_2 + x_0.y_3 + x_3.y_0 = 0$,

- $x_2 + x_3 + y_2 + x_0.x_1 + x_0.x_3 + x_0.y_3 + x_2.x_3 + x_3.y_1 = 0$,

- $x_0 + x_3 + y_1 + y_2 + y_3 + x_0.x_1 + x_0.x_3 + x_0.y_0 + x_0.y_3 + x_2.x_3 + x_3.y_2 = 0$,

- $x_2 + y_2 + x_0.x_3 + x_0.y_0 + x_3.y_3 = 0$,

- $x_1 + x_2 + y_0 + y_1 + y_2 + x_0.x_1 + x_0.x_3 + y_0.y_1 = 0$,

- $x_0 + x_1 + x_2 + y_1 + y_2 + x_0.x_3 = 0$,

- $x_0 + x_3 + y_1 + x_0.x_2 + x_0.y_3 + x_1.x_2 = 0$,

- $x_0 + x_3 + y_0 + y_1 + x_0.x_1 + x_0.x_3 = 0$,

- $x_1 + y_1 + y_2 + y_3 + x_0.x_2 + x_1.x_2 + x_2.x_3 = 0,$
- $x_1 + y_1 + y_2 + x_0.x_2 + x_0.y_3 + x_1.x_2 + x_2.x_3 = 0.$

For Serpent S-box S_4, the following 21 quadratic expressions were obtained:

- $1 + x_3 + y_0 + y_3 + x_0.x_1 + x_0.x_2 + x_0.y_0 + x_1.x_2 = 0,$
- $1 + x_1 + x_2 + x_3 + y_0 + x_0.x_1 + x_0.x_3 + x_1.x_3 = 0,$
- $1 + x_0 + x_3 + y_0 + y_3 + x_1.y_0 = 0,$
- $1 + x_3 + y_0 + y_3 + x_0.x_2 + x_0.x_3 + x_0.y_1 + x_1.y_1 = 0,$
- $1 + x_1 + x_2 + x_3 + y_0 + x_0.y_0 + x_0.y_3 + x_1.y_2 = 0,$
- $1 + x_2 + x_3 + y_0 + x_0.x_3 + x_1.y_3 = 0,$
- $x_2 + x_3 + y_1 + y_2 + x_0.x_2 + x_0.x_3 + x_0.y_1 + x_0.y_2 = 0,$
- $1 + x_1 + x_2 + y_0 + y_1 + x_0.y_3 + x_2.y_0 = 0,$
- $1 + x_2 + y_0 + y_1 + y_2 + y_3 + x_0.x_1 + x_0.x_3 + x_0.y_3 + x_2.y_1 = 0,$
- $1 + x_0 + x_2 + x_3 + y_0 + y_3 + x_0.x_2 + x_0.y_1 + x_0.y_3 + x_2.x_3 + x_2.y_2 = 0,$
- $x_1 + x_3 + y_1 + y_3 + x_0.x_1 + x_0.x_3 + x_0.y_1 + x_0.y_3 + x_2.x_3 + x_2.y_3 = 0,$
- $x_0 + x_0.x_2 + x_0.x_3 + x_0.y_0 + x_2.x_3 + x_3.y_0 = 0,$
- $1 + x_1 + x_2 + y_0 + x_0.x_1 + x_2.x_3 + x_3.y_1 = 0,$
- $1 + x_0 + x_1 + x_2 + x_3 + y_0 + x_0.x_2 + x_0.y_0 + x_0.y_1 + x_0.y_3 + x_3.y_2 = 0,$
- $x_1 + y_2 + y_3 + x_0.y_0 + x_0.y_3 + x_3.y_3 = 0,$
- $1 + x_2 + y_0 + y_1 + y_2 + y_3 + x_0.x_3 + y_0.y_1 = 0,$
- $x_0 + x_1 + y_2 + y_3 + x_0.x_2 + x_0.x_3 + x_2.x_3 = 0,$
- $1 + x_3 + y_0 + y_3 + x_0.x_2 + x_0.x_3 + x_2.x_3 = 0,$
- $x_0 + x_1 + y_2 + y_3 + x_0.x_1 + x_0.x_2 + x_0.x_3 + x_0.y_0 + x_0.y_3 + x_2.x_3 = 0,$

- $1 + x_1 + x_3 + y_0 + y_2 + x_0.x_2 + x_0.x_3 + x_0.y_0 + x_0.y_3 + x_2.x_3 = 0$,
- $1 + x_0 + x_1 + y_0 + y_1 + x_0.y_0 = 0$.

For Serpent S-box S_5, the following 21 quadratic expressions were obtained:

- $1 + x_0 + x_1 + x_2 + x_3 + y_3 + x_0.x_1 + x_0.x_2 + x_0.x_3 + x_0.y_3 = 0$,
- $1 + x_1 + x_2 + x_3 + y_0 + x_0.x_1 + x_0.x_3 + x_1.x_3 = 0$,
- $x_0 + x_0.x_1 + x_0.x_2 + x_0.y_0 + x_1.x_2 + x_1.y_0 = 0$,
- $1+x_1+x_3+y_0+y_1+y_2+x_0.x_1+x_0.x_2+x_0.y_0+x_0.y_1+x_1.x_2+x_1.y_1 = 0$,
- $y_0 + y_3 + x_0.x_1 + x_0.y_1 + x_0.y_2 + x_1.y_2 = 0$,
- $y_0 + y_3 + x_0.y_1 + x_0.y_2 + x_1.x_2 + x_1.y_3 = 0$,
- $x_1 + y_0 + y_1 + x_0.x_2 + x_0.x_3 + x_0.y_0 + x_2.x_3 = 0$,
- $x_2 + y_2 + y_3 + x_0.x_1 + x_0.x_2 + x_0.y_0 + x_0.y_1 + x_1.x_2 + x_2.y_0 = 0$,
- $x_1 + x_2 + y_0 + y_1 + y_2 + y_3 + x_0.x_1 + x_0.x_3 + x_0.y_0 + x_0.y_2 + x_2.y_1 = 0$,
- $1+x_0+x_2+x_3+y_0+y_1+y_2+x_0.x_1+x_0.y_0+x_0.y_2+x_1.x_2+x_2.y_2 = 0$,
- $1 + x_0 + x_2 + x_3 + y_0 + y_1 + y_3 + x_0.x_1 + x_0.x_2 + x_0.x_3 + x_0.y_0 + x_0.y_1 + x_1.x_2 + x_2.y_3 = 0$,
- $x_0 + x_1 + y_0 + y_1 + x_3.y_0 = 0$,
- $1 + x_1 + x_2 + x_3 + y_0 + x_0.x_1 + x_3.y_1 = 0$,
- $x_2 + y_0 + y_2 + x_0.y_1 + x_3.y_2 = 0$,
- $1 + x_2 + x_3 + y_1 + x_0.x_1 + x_0.y_1 + x_0.y_2 + x_3.y_3 = 0$,
- $x_1 + y_1 + x_0.x_1 + x_0.x_2 + x_1.x_2 + y_0.y_1 = 0$,
- $1 + x_0 + x_3 + y_1 + y_2 + x_0.x_1 + x_0.x_2 + x_1.x_2 = 0$,
- $x_0 + x_1 + x_2 + y_1 + y_2 + y_3 + x_0.x_1 = 0$,
- $1 + x_0 + x_1 + x_3 + y_2 + y_3 + x_0.x_1 + x_0.x_2 + x_0.y_0 + x_1.x_2 = 0$,

- $x_0 + x_1 + y_0 + y_1 + y_3 + x_0.x_1 + x_0.x_2 + x_0.y_1 + x_1.x_2 = 0$,
- $x_1 + x_2 + y_0 + y_1 + y_2 + x_0.x_2 + x_0.x_3 + x_0.y_1 + x_1.x_2 = 0$.

For Serpent S-box S_6, the following 21 quadratic expressions were obtained:

- $1 + x_0 + x_1 + x_2 + y_1 + x_0.x_1 + x_0.x_2 + x_0.y_1 = 0$,
- $1 + x_1 + x_2 + y_1 + x_0.x_3 = 0$,
- $x_0 + x_3 + y_0 + y_1 + x_0.y_2 + x_0.y_3 + x_1.x_2 + x_1.x_3 + x_1.y_0 = 0$,
- $1 + x_0 + x_1 + x_3 + y_2 + y_3 + x_0.x_2 + x_1.x_3 + x_1.y_1 = 0$,
- $x_0 + x_1 + x_2 + x_3 + y_1 + y_2 + y_3 + x_0.x_1 + x_0.y_2 + x_0.y_3 + x_1.x_2 + x_1.y_2 = 0$,
- $x_1 + x_0.x_1 + x_1.x_2 + x_1.x_3 + x_1.y_3 = 0$,
- $x_1 + x_3 + y_0 + y_2 + x_0.x_1 + x_0.x_2 + x_1.x_3 + x_2.x_3 = 0$,
- $x_0 + x_1 + y_1 + y_2 + x_0.y_0 + x_1.x_2 + x_1.x_3 + x_2.y_0 = 0$,
- $x_0 + x_2 + x_3 + y_1 + y_2 + y_3 + x_0.y_0 + x_0.y_2 + x_1.x_3 + x_2.y_1 = 0$,
- $x_0 + x_1 + x_2 + y_0 + y_1 + y_3 + x_0.x_2 + x_0.y_0 + x_2.y_2 = 0$,
- $1 + x_1 + x_2 + y_0 + y_1 + y_2 + y_3 + x_0.x_1 + x_0.y_2 + x_0.y_3 + x_1.x_2 + x_1.x_3 + x_2.y_3 = 0$,
- $x_0 + x_1 + x_2 + y_0 + y_1 + y_3 + x_0.x_2 + x_0.y_2 + x_1.x_2 + x_1.x_3 + x_3.y_0 = 0$,
- $1 + x_2 + y_0 + y_1 + y_2 + x_0.x_1 + x_0.x_2 + x_3.y_1 = 0$,
- $1 + x_0 + x_1 + y_2 + y_3 + x_0.x_1 + x_0.x_2 + x_0.y_0 + x_1.x_2 + x_3.y_2 = 0$,
- $x_0 + y_0 + y_1 + x_0.x_2 + x_0.y_3 + x_1.x_2 + x_1.x_3 + x_3.y_3 = 0$,
- $1 + x_0 + y_3 + x_0.x_1 + x_0.x_2 + x_0.y_2 + x_0.y_3 + x_1.x_2 + x_1.x_3 + y_0.y_1 = 0$,
- $1 + x_1 + y_1 + y_2 + y_3 = 0$,
- $1 + x_3 + y_1 + y_3 + x_0.x_1 + x_0.x_2 = 0$,
- $1 + x_1 + x_2 + x_3 + y_0 + y_1 + x_0.x_2 + x_0.y_2 + x_0.y_3 + x_1.x_2 = 0$,

- $x_0 + x_1 + x_2 + x_3 + y_0 + y_1 + y_3 + x_0.y_0 + x_0.y_3 + x_1.x_2 + x_1.x_3 = 0$,
- $x_0 + x_1 + y_0 + y_1 + x_0.x_1 + x_0.x_2 = 0$.

For Serpent S-box S_7, the following 21 quadratic expressions were obtained:

- $x_0 + x_3 + y_1 + y_3 + x_0.x_1 + x_0.y_1 + x_1.x_2 = 0$,
- $x_1 + x_2 + y_3 + x_0.x_1 + x_0.x_2 + x_0.y_3 = 0$,
- $x_1 + x_3 + y_1 + y_3 + x_0.x_1 + x_0.x_2 + x_0.x_3 + x_0.y_0 + x_0.y_1 + x_1.x_3 + x_1.y_0 = 0$,
- $x_0 + x_1 + x_0.x_2 + x_0.y_2 + x_1.x_3 + x_1.y_1 = 0$,
- $x_0 + x_1 + y_1 + y_2 + x_0.x_2 + x_0.x_3 + x_0.y_0 + x_0.y_1 + x_0.y_2 + x_1.x_3 + x_1.y_2 = 0$,
- $x_2 + x_3 + y_1 + x_0.x_2 + x_0.y_0 + x_0.y_1 + x_0.y_2 + x_1.y_3 = 0$,
- $1 + x_0 + x_1 + x_3 + y_0 + y_2 + x_0.x_1 + x_0.y_1 + x_2.x_3 = 0$,
- $1 + x_2 + x_3 + y_0 + y_2 + y_3 + x_0.x_1 + x_0.x_2 + x_0.x_3 + x_0.y_1 + x_2.y_0 = 0$,
- $1 + x_1 + x_2 + x_3 + y_0 + y_2 + x_0.y_1 + x_0.y_2 + x_2.y_1 = 0$,
- $1 + x_0 + x_1 + x_2 + y_0 + y_1 + y_2 + y_3 + x_0.x_1 + x_0.x_3 + x_0.y_0 + x_0.y_1 + x_2.y_2 = 0$,
- $x_1 + x_3 + y_1 + x_0.x_1 + x_0.x_2 + x_0.y_0 + x_0.y_2 + x_2.y_3 = 0$,
- $y_2 + y_3 + x_0.x_2 + x_0.y_0 + x_0.y_1 + x_0.y_2 + x_3.y_0 = 0$,
- $1 + x_2 + x_3 + y_0 + x_0.x_1 + x_0.x_3 + x_0.y_0 + x_0.y_1 + x_0.y_2 + x_3.y_1 = 0$,
- $1 + x_0 + x_2 + x_3 + y_0 + x_0.x_2 + x_0.x_3 + x_0.y_1 + x_0.y_2 + x_1.x_3 + x_3.y_2 = 0$,
- $1 + x_0 + x_2 + x_3 + y_0 + y_2 + y_3 + x_0.y_0 + x_1.x_3 + x_3.y_3 = 0$,
- $1 + x_3 + y_0 + y_1 + y_2 + y_3 + x_0.x_1 + x_0.x_3 + y_0.y_1 = 0$,
- $1 + x_0 + y_0 + y_1 + x_1.x_3 = 0$,
- $1 + x_1 + x_2 + y_0 + y_1 + y_2 + x_1.x_3 = 0$,
- $x_0 + y_2 + x_0.x_1 + x_0.x_2 + x_0.x_3 + x_0.y_0 + x_0.y_2 + x_1.x_3 = 0$,

- $1 + x_1 + x_3 + y_0 + y_2 + y_3 + x_0.y_0 + x_0.y_2 + x_1.x_3 = 0$,
- $1 + x_0 + x_1 + y_0 + x_0.x_1 + x_0.y_1 + x_1.x_3 = 0$.

Concerning the ANF of S-boxes output bits (as Boolean mappings), the following are the ANFs of SERPENT S-boxes S_i, for $0 \leq i \leq 7$, where the input is denoted (x_3, x_2, x_1, x_0) and the output is denoted (y_3, y_2, y_1, y_0).

For S-box S_0:

$y_3 = x_0 + x_1 + x_2 + x_3 + x_0.x_3$,
$y_2 = x_1 + x_0.x_1 + x_0.x_2 + x_0.x_1.x_2 + x_3 + x_1.x_3 + x_1.x_2.x_3$,
$y_1 = 1 + x_0 + x_0.x_2 + x_1.x_2 + x_0.x_1.x_2 + x_1.x_3 + x_0.x_2.x_3 + x_1.x_2.x_3$,
$y_0 = 1 + x_0 + x_0.x_1 + x_2 + x_0.x_2 + x_1.x_2 + x_0.x_1.x_2 + x_3 + x_0.x_2.x_3 + x_1.x_2.x_3$.

For S-box S_1:

$y_3 = 1 + x_1 + x_0.x_2 + x_3 + x_0.x_3 + x_0.x_1.x_3 + x_0.x_2.x_3 + x_1.x_2.x_3$,
$y_2 = 1 + x_1 + x_0.x_1 + x_2 + x_3$,
$y_1 = 1 + x_0 + x_0.x_1 + x_2 + x_0.x_2 + x_3 + x_1.x_3 + x_0.x_1.x_3 + x_0.x_2.x_3 + x_1.x_2.x_3$,
$y_0 = 1 + x_0 + x_1 + x_1.x_2 + x_0.x_3 + x_2.x_3 + x_0.x_2.x_3 + x_1.x_2.x_3$.

For S-box S_2:

$y_3 = 1 + x_0 + x_1 + x_2 + x_0.x_1.x_2 + x_1.x_3$,
$y_2 = x_0 + x_1 + x_1.x_2 + x_3 + x_1.x_3 + x_0.x_1.x_3 + x_2.x_3 + x_0.x_2.x_3$,
$y_1 = x_0 + x_1 + x_2 + x_1.x_2 + x_0.x_1.x_2 + x_0.x_3 + x_0.x_1.x_3 + x_2.x_3 + x_0.x_2.x_3$,
$y_0 = x_1 + x_2 + x_0.x_2 + x_3$.

For S-box S_3:

$y_3 = x_0 + x_1 + x_0.x_1 + x_2 + x_0.x_2 + x_0.x_1.x_2 + x_3 + x_2.x_3 + x_0.x_2.x_3$,
$y_2 = x_0 + x_0.x_1 + x_2 + x_0.x_1.x_2 + x_3 + x_1.x_3 + x_0.x_1.x_3$,
$y_1 = x_0 + x_1 + x_0.x_2 + x_0.x_3 + x_0.x_1.x_3 + x_2.x_3 + x_0.x_2.x_3$,
$y_0 = x_0 + x_1 + x_1.x_2 + x_3 + x_0.x_3 + x_2.x_3 + x_0.x_2.x_3 + x_1.x_2.x_3$.

For S-box S_4:

$y_3 = x_0 + x_1 + x_2 + x_1.x_2 + x_0.x_3 + x_1.x_3 + x_0.x_1.x_3$,
$y_2 = x_0 + x_0.x_1 + x_2 + x_1.x_2 + x_0.x_1.x_2 + x_1.x_3 + x_0.x_1.x_3 + x_2.x_3 + x_1.x_2.x_3$,
$y_1 = x_0 + x_0.x_2 + x_1.x_2 + x_3 + x_1.x_3 + x_2.x_3 + x_0.x_2.x_3 + x_1.x_2.x_3$,
$y_0 = 1 + x_1 + x_0.x_1 + x_2 + x_3 + x_0.x_3 + x_1.x_3$.

For S-box S_5:

$y_3 = 1 + x_0 + x_1 + x_2 + x_0.x_1.x_2 + x_3 + x_0.x_3 + x_0.x_2.x_3$,
$y_2 = 1 + x_1 + x_0.x_2 + x_3 + x_0.x_1.x_3 + x_2.x_3 + x_0.x_2.x_3 + x_1.x_2.x_3$,
$y_1 = 1 + x_0 + x_0.x_1 + x_2 + x_3 + x_1.x_3 + x_0.x_1.x_3 + x_2.x_3$,
$y_0 = 1 + x_1 + x_0.x_1 + x_2 + x_3 + x_0.x_3 + x_1.x_3$.

For S-box S_6:

$y_3 = x_1 + x_0.x_1 + x_2 + x_0.x_2 + x_0.x_1.x_2 + x_3 + x_2.x_3 + x_1.x_2.x_3$,
$y_2 = 1 + x_0 + x_0.x_1 + x_2 + x_1.x_2 + x_0.x_1.x_2 + x_1.x_3 + x_0.x_1.x_3 + x_2.x_3 + x_1.x_2.x_3$,
$y_1 = 1 + x_1 + x_2 + x_0.x_3$,
$y_0 = 1 + x_0 + x_1 + x_2 + x_0.x_2 + x_1.x_2 + x_0.x_1.x_2 + x_3 + x_0.x_1.x_3 + x_1.x_2.x_3$.

For S-box S_7:

$y_3 = x_0 + x_1 + x_2 + x_0.x_2 + x_0.x_1.x_2 + x_0.x_3$,
$y_2 = x_0 + x_1 + x_2 + x_0.x_1.x_2 + x_3 + x_0.x_3 + x_1.x_3 + x_0.x_1.x_3 + x_1.x_2.x_3$,

$y_1 = x_1 + x_0.x_1 + x_2 + x_0.x_2 + x_1.x_2 + x_3 + x_0.x_3 + x_0.x_1.x_3 + x_0.x_2.x_3$,
$y_0 = 1 + x_0.x_1 + x_2 + x_0.x_3 + x_1.x_3 + x_2.x_3 + x_0.x_2.x_3 + x_1.x_2.x_3$.

The ANFs of the 4×4 S-box of the PRESENT cipher are:
$y_3 = x_0 + x_2 + x_1.x_2 + x_3$,
$y_2 = x_1 + x_0.x_1.x_2 + x_3 + x_1.x_3 + x_0.x_1.x_3 + x_2.x_3 + x_0.x_2.x_3$,
$y_1 = 1 + x_0.x_1 + x_2 + x_3 + x_0.x_3 + x_1.x_3 + x_0.x_1.x_3 + x_0.x_2.x_3$,
$y_0 = 1 + x_0 + x_1 + x_1.x_2 + x_0.x_1.x_2 + x_3 + x_0.x_1.x_3 + x_0.x_2.x_3$.

The ANFs of the 7×7-bit S-box of Misty1 [59] are:
$y_0 = x_0 + x_1.x_3 + x_0.x_3.x_4 + x_1.x_5 + x_0.x_2.x_5 + x_4.x_5 + x_0.x_1.x_6 + x_2.x_6 + x_0.x_5.x_6 + x_3.x_5.x_6 + 1$,
$y_1 = x_0.x_2 + x_0.x_4 + x_3.x_4 + x_1.x_5 + x_2.x_4.x_5 + x_6 + x_0.x_6 + x_3.x_6 + x_2.x_3.x_6 + x_1.x_4.x_6 + x_0.x_5.x_6 + 1$,
$y_2 = x_1.x_2 + x_0.x_2.x_3 + x_4 + x_1.x_4 + x_0.x_1.x_4 + x_0.x_5 + x_0.x_4.x_5 + x_3.x_4.x_5 + x_1.x_6 + x_3.x_6 + x_0.x_3.x6 + x_4.x_6 + x_2.x_4.x_6$,
$y_3 = x_0 + x_1 + x_0.x_1.x_2 + x_0.x_3 + x_2.x_4 + x_1.x_4.x_5 + x_2.x_6 + x_1.x_3.x_6 + x_0.x_4.x_6 + x_5.x_6 + 1$
$y_4 = x_2.x_3 + x_0.x_4 + x_1.x_3.x_4 + x_5 + x_2.x_5 + x_1.x_2.x_5 + x_0.x_3.x_5 + x_1.x_6 + x_1.x_5.x_6 + x_4.x_5.x_6 + 1$,
$y_5 = x_0 + x_1 + x_2 + x_0.x_1.x_2 + x_0.x_3 + x_1.x_2.x_3 + x_1.x_4 + x_0.x_2.x_4 + x_0.x_5 + x_0.x_1.x_5 + x_3.x_5 + x_0.x_6 + x_2.x_5.x_6$,
$y_6 = x_0.x_1 + x_3 + x_0.x_3 + x_2.x_3.x_4 + x_0.x_5 + x_2.x_5 + x_3.x_5 + x_1.x_3.x_5 + x_1.x_6 + x_1.x_2.x_6 + x_0.x_3.x_6 + x_4.x_6 + x_2.x_5.x_6$.

The ANFs of the 7×7-bit S-box of Kasumi [1] are:
$y_0 = x_1.x_3 + x_4 + x_0.x_1.x_4 + x_5 + x_2.x_5 + x_3.x_4.x_5 + x_6 + x_0.x_6 + x_1.x_6 + x_3.x_6 + x_2.x_4.x_6 + x_1.x_5.x_6 + x_4.x_5.x_6$,
$y_1 = x_0.x_1 + x_0.x_4 + x_2.x_4 + x_5 + x_1.x_2.x_5 + x_0.x_3.x_5 + x_6 + x_0.x_2.x_6 + x_3.x_6 + x_4.x_5.x_6 + 1$,
$y_2 = x_0 + x_0.x_3 + x_2.x_3 + x_1.x_2.x_4 + x_0.x_3.x_4 + x_1.x_5 + x_0.x_2.x_5 + x_0.x_6 + x_0.x_1.x_6 + x_2.x_6 + x_4.x_6 + 1$,
$y_3 = x_1 + x_0.x_1.x_2 + x_1.x_4 + x_3.x_4 + x_0.x_5 + x_0.x_1.x_5 + x_2.x_3.x_5 + x_1.x_4.x_5 + x_2.x_6 + x_1.x_3.x_6$,
$y_4 = x_0.x_2 + x_3 + x_1.x_3 + x_1.x_4 + x_0.x_1.x_4 + x_2.x_3.x_4 + x_0.x_5 + x_1.x_3.x_5 + x_0.x_4.x_5 + x_1.x_6 + x_3.x_6 + x_0.x_3.x_6 + x_5.x_6 + 1$,
$y_5 = x_2 + x_0.x_2 + x_0.x_3 + x_1.x_2.x_3 + x_0.x_2.x_4 + x_0.x_5 + x_2.x_5 + x_4.x_5 + x_1.x_6 + x_1.x_2.x_6 + x_0.x_3.x_6 + x_3.x_4.x_6 + x_2.x_5.x_6 + 1$,
$y_6 = x_1.x_2 + x_0.x_1.x_3 + x_0.x_4 + x_1.x_5 + x_3.x_5 + x_6 + x_0.x_1.x_6 + x_2.x_3.x_6 + x_1.x_4.x_6 + x_0.x_5.x_6$.

The ANFs of the 9×9-bit S-box of Misty1 [59] are:
$y_0 = x_0.x_4 + x_0.x_5 + x_1.x_5 + x_1.x_6 + x_2.x_6 + x_2.x_7 + x_3.x_7 + x_3.x_8 + x_4.x_8 + 1$,
$y_1 = x_0.x_2 + x_3 + x_1.x_3 + x_2.x_3 + x_3.x_4 + x_4.x_5 + x_0.x_6 + x_2.x_6 + x_7 + x_0.x_8 + x_3.x_8 + x_5.x_8 + 1$,
$y_2 = x_0.x_1 + x_1.x_3 + x_4 + x_0.x_4 + x_2.x_4 + x_3.x_4 + x_4.x_5 + x_0.x_6 + x_5.x_6 + x_1.x_7 + x_3.x_7 + x_8$,
$y_3 = x_0 + x_1.x_2 + x_2.x_4 + x_5 + x_1.x_5 + x_3.x_5 + x_4.x_5 + x_5.x_6 + x_1.x_7 + x_6.x_7 + x_2.x_8 + x_4.x_8$,

$y_4 = x_1 + x_0.x_3 + x_2.x_3 + x_0.x_5 + x_3.x_5 + x_6 + x_2.x_6 + x_4.x_6 + x_5.x_6 + x_6.x_7 + x_2.x_8 + x_7.x_8$,
$y_5 = x_2 + x_0.x_3 + x_1.x_4 + x_3.x_4 + x_1.x_6 + x_4.x_6 + x_7 + x_3.x_7 + x_5.x_7 + x_6.x_7 + x_0.x_8 + x_7.x_8$,
$y_6 = x_0.x_1 + x_3 + x_1.x_4 + x_2.x_5 + x_4.x_5 + x_2.x_7 + x_5.x_7 + x_8 + x_0.x_8 + x_4.x_8 + x_6.x_8 + x_7.x_8 + 1$,
$y_7 = x_1 + x_0.x_1 + x_1.x_2 + x_2.x_3 + x_0.x_4 + x_5 + x_1.x_6 + x_3.x_6 + x_0.x_7 + x_4.x_7 + x_6.x_7 + x_1.x_8 + 1$,
$y_8 = x_0 + x_0.x_1 + x_1.x_2 + x_4 + x_0.x_5 + x_2.x_5 + x_3.x_6 + x_5.x_6 + x_0.x_7 + x_0.x_8 + x_3.x_8 + x_6.x_8 + 1$.

For the ANFs of the AES S-box, the input is the ordered 8-tuple (x_0, x_1, x_2, x_3, x_4, x_5, x_6, x_7) and the output is the ordered 8-tuple (y_0, y_1, y_2, y_3, y_4, y_5, y_6, y_7):

- $y_7 = x_2 + x_0.x_2 + x_1.x_2 + x_0.x_2.x_3 + x_1.x_2.x_3 + x_0.x_1.x_2.x_3 + x_4 + x_0.x_1.x_4 + x_2.x_4 + x_0.x_1.x_2.x_4 + x_0.x_1.x_3.x_4 + x_1.x_2.x_3.x_4 + x_0.x_1.x_2.x_3.x_4 + x_5 + x_0.x_1.x_5 + x_0.x_2.x_5 + x_0.x_1.x_2.x_5 + x_3.x_5 + x_0.x_3.x_5 + x_1.x_3.x_5 + x_2.x_3.x_5 + x_0.x_2.x_3.x_5 + x_1.x_2.x_3.x_5 + x_0.x_4.x_5 + x_1.x_2.x_4.x_5 + x_3.x_4.x_5 + x_0.x_1.x_3.x_4.x_5 + x_0.x_2.x_3.x_4.x_5 + x_0.x_6 + x_0.x_1.x_6 + x_2.x_6 + x_1.x_2.x_6 + x_0.x_3.x_6 + x_1.x_3.x_6 + x_0.x_1.x_3.x_6 + x_0.x_2.x_3.x_6 + x_0.x_1.x_2.x_3.x_6 + x_4.x_6 + x_2.x_4.x_6 + x_0.x_2.x_4.x_6 + x_1.x_2.x_4.x_6 + x_0.x_1.x_2.x_4.x_6 + x_0.x_3.x_4.x_6 + x_0.x_2.x_3.x_4.x_6 + x_0.x_5.x_6 + x_2.x_5.x_6 + x_0.x_3.x_5.x_6 + x_1.x_2.x_3.x_5.x_6 + x_4.x_5.x_6 + x_0.x_1.x_4.x_5.x_6 + x_1.x_2.x_4.x_5.x_6 + x_3.x_4.x_5.x_6 + x_0.x_3.x_4.x_5.x_6 + x_2.x_3.x_4.x_5.x_6 + x_0.x_2.x_3.x_4.x_5.x_6 + x_7 + x_0.x_7 + x_1.x_7 + x_0.x_2.x_7 + x_0.x_1.x_2.x_7 + x_0.x_3.x_7 + x_0.x_2.x_3.x_7 + x_1.x_2.x_3.x_7 + x_1.x_4.x_7 + x_0.x_1.x_4.x_7 + x_0.x_2.x_4.x_7 + x_0.x_3.x_4.x_7 + x_1.x_3.x_4.x_7 + x_0.x_1.x_3.x_4.x_7 + x_2.x_3.x_4.x_7 + x_0.x_2.x_3.x_4.x_7 + x_5.x_7 + x_0.x_1.x_2.x_5.x_7 + x_3.x_5.x_7 + x_0.x_3.x_5.x_7 + x_4.x_5.x_7 + x_1.x_4.x_5.x_7 + x_0.x_1.x_4.x_5.x_7 + x_0.x_1.x_2.x_4.x_5.x_7 + x_0.x_1.x_3.x_4.x_5.x_7 + x_0.x_2.x_3.x_4.x_5.x_7 + x_0.x_6.x_7 + x_2.x_6.x_7 + x_0.x_1.x_2.x_6.x_7 + x_0.x_3.x_6.x_7 + x_1.x_3.x_6.x_7 + x_0.x_1.x_2.x_3.x_6.x_7 + x_4.x_6.x_7 + x_0.x_4.x_6.x_7 + x_1.x_4.x_6.x_7 + x_2.x_4.x_6.x_7 + x_1.x_2.x_4.x_6.x_7 + x_0.x_3.x_4.x_6.x_7 + x_0.x_1.x_3.x_4.x_6.x_7 + x_0.x_2.x_3.x_4.x_6.x_7 + x_1.x_5.x_6.x_7 + x_0.x_1.x_5.x_6.x_7 + x_2.x_5.x_6.x_7 + x_1.x_2.x_5.x_6.x_7 + x_0.x_3.x_5.x_6.x_7 + x_1.x_3.x_5.x_6.x_7 + x_0.x_1.x_3.x_5.x_6.x_7 + x_0.x_2.x_3.x_5.x_6.x_7 + x_4.x_5.x_6.x_7 + x_1.x_4.x_5.x_6.x_7 + x_0.x_2.x_4.x_5.x_6.x_7 + x_3.x_4.x_5.x_6.x_7 + x_1.x_3.x_4.x_5.x_6.x_7 + x_0.x_1.x_3.x_4.x_5.x_6.x_7 + x_0.x_2.x_3.x_4.x_5.x_6.x_7$,
- $y_6 = 1 + x_3 + x_1.x_3 + x_0.x_1.x_3 + x_2.x_3 + x_0.x_4 + x_0.x_1.x_4 + x_0.x_2.x_4 + x_0.x_1.x_2.x_4 + x_1.x_3.x_4 + x_0.x_1.x_3.x_4 + x_2.x_3.x_4 + x_1.x_2.x_3.x_4 + x_5 + x_0.x_5 + x_0.x_1.x_5 + x_0.x_2.x_5 + x_1.x_2.x_5 + x_0.x_1.x_2.x_5 + x_3.x_5 + x_0.x_3.x_5 + x_0.x_2.x_3.x_5 + x_1.x_2.x_3.x_5 + x_0.x_1.x_2.x_3.x_5 + x_0.x_4.x_5 + x_0.x_2.x_4.x_5 + x_1.x_2.x_4.x_5 + x_0.x_1.x_3.x_4.x_5 + x_2.x_3.x_4.x_5 + x_0.x_2.x_3.x_4.x_5 + x_1.x_2.x_3.x_4.x_5 + x_6 + x_0.x_2.x_6 + x_1.x_2.x_6 + x_0.x_3.x_6 + x_1.x_3.x_6 + x_0.x_1.x_3.x_6 + x_0.x_2.x_3.x_6 + x_1.x_2.x_3.x_6 + x_0.x_1.x_2.x_3.x_6 + x_4.x_6 + x_0.x_4.x_6 + x_1.x_4.x_6 + x_2.x_4.x_6 + x_0.x_2.x_4.x_6 + x_3.x_4.x_6 + x_0.x_3.x_4.x_6 + x_1.x_3.x_4.x_6 + x_2.x_3.x_4.x_6 + x_0.x_5.x_6 + x_1.x_5.x_6 + x_2.x_3.x_5.x_6 + x_0.x_2.x_3.x_5.x_6$

$+ x_0.x_1.x_2.x_3.x_5.x_6 + x_4.x_5.x_6 + x_1.x_2.x_4.x_5.x_6 + x_0.x_1.x_2.x_4.x_5.x_6 +$
$x_1.x_3.x_4.x_5.x_6 + x_0.x_1.x_3.x_4.x_5.x_6 + x_0.x_7 + x_1.x_7 + x_1.x_2.x_7 + x_3.x_7$
$+ x_0.x_1.x_3.x_7 + x_2.x_3.x_7 + x_0.x_2.x_3.x_7 + x_0.x_1.x_2.x_3.x_7 + x_0.x_4.x_7$
$+ x_1.x_4.x_7 + x_2.x_4.x_7 + x_1.x_2.x_4.x_7 + x_0.x_3.x_4.x_7 + x_1.x_3.x_4.x_7 +$
$x_1.x_2.x_3.x_4.x_7 + x_0.x_1.x_2.x_3.x_4.x_7 + x_5.x_7 + x_0.x_5.x_7 + x_0.x_2.x_5.x_7 +$
$x_3.x_5.x_7 + x_1.x_3.x_5.x_7 + x_2.x_3.x_5.x_7 + x_1.x_2.x_3.x_5.x_7 + x_1.x_4.x_5.x_7 +$
$x_0.x_1.x_4.x_5.x_7 + x_0.x_2.x_4.x_5.x_7 + x_0.x_1.x_2.x_4.x_5.x_7 + x_1.x_3.x_4.x_5.x_7 +$
$x_0.x_1.x_3.x_4.x_5.x_7 + x_1.x_6.x_7 + x_0.x_1.x_6.x_7 + x_3.x_6.x_7 + x_0.x_1.x_3.x_6.x_7$
$+ x_0.x_4.x_6.x_7 + x_1.x_4.x_6.x_7 + x_0.x_2.x_4.x_6.x_7 + x_0.x_1.x_2.x_4.x_6.x_7 +$
$x_0.x_3.x_4.x_6.x_7 + x_0.x_1.x_3.x_4.x_6.x_7 + x_2.x_3.x_4.x_6.x_7 + x_5.x_6.x_7 + x_0.x_2.$
$x_5.x_6.x_7 + x_1.x_2.x_5.x_6.x_7 + x_0.x_1.x_2.x_5.x_6.x_7 + x_0.x_1.x_3.x_5.x_6.x_7 +$
$x_2.x_3.x_5.x_6.x_7 + x_4.x_5.x_6.x_7 + x_0.x_4.x_5.x_6.x_7 + x_1.x_4.x_5.x_6.x_7 + x_0.x_1.$
$x_2.x_4.x_5.x_6.x_7 + x_3.x_4.x_5.x_6.x_7 + x_1.x_3.x_4.x_5.x_6.x_7 + x_0.x_1.x_3.x_4.x_5.x_6.x_7,$

- $y_5 = 1 + x_0.x_1.x_2 + x_0.x_3 + x_0.x_1.x_3 + x_0.x_1.x_2.x_3 + x_4 + x_0.x_1.x_4$
$+ x_2.x_4 + x_0.x_2.x_4 + x_1.x_2.x_4 + x_3.x_4 + x_0.x_1.x_3.x_4 + x_0.x_2.x_3.x_4 +$
$x_0.x_1.x_2.x_3.x_4 + x_1.x_5 + x_0.x_1.x_5 + x_1.x_2.x_5 + x_0.x_1.x_2.x_5 + x_0.x_3.x_5$
$+ x_1.x_3.x_5 + x_0.x_2.x_3.x_5 + x_1.x_2.x_3.x_5 + x_0.x_1.x_2.x_3.x_5 + x_0.x_1.x_4.x_5$
$+ x_2.x_4.x_5 + x_0.x_2.x_4.x_5 + x_1.x_2.x_4.x_5 + x_0.x_1.x_2.x_4.x_5 + x_3.x_4.x_5 +$
$x_2.x_3.x_4.x_5 + x_0.x_1.x_2.x_3.x_4.x_5 + x_6 + x_1.x_6 + x_0.x_1.x_6 + x_0.x_2.x_6 +$
$x_1.x_2.x_6 + x_1.x_3.x_6 + x_2.x_3.x_6 + x_1.x_2.x_3.x_6 + x_0.x_1.x_2.x_3.x_6 + x_4.x_6$
$+ x_1.x_4.x_6 + x_0.x_1.x_4.x_6 + x_0.x_1.x_2.x_4.x_6 + x_1.x_3.x_4.x_6 + x_2.x_3.x_4.x_6$
$+ x_1.x_2.x_3.x_4.x_6 + x_0.x_5.x_6 + x_1.x_5.x_6 + x_0.x_1.x_5.x_6 + x_1.x_3.x_5.x_6$
$+ x_2.x_3.x_5.x_6 + x_0.x_1.x_2.x_3.x_5.x_6 + x_0.x_1.x_4.x_5.x_6 + x_1.x_2.x_4.x_5.x_6$
$+ x_0.x_1.x_2.x_4.x_5.x_6 + x_3.x_4.x_5.x_6 + x_0.x_3.x_4.x_5.x_6 + x_1.x_3.x_4.x_5.x_6 +$
$x_2.x_3.x_4.x_5.x_6 + x_7 + x_0.x_1.x_7 + x_1.x_3.x_7 + x_0.x_1.x_3.x_7 + x_2.x_3.x_7$
$+ x_0.x_1.x_2.x_3.x_7 + x_0.x_4.x_7 + x_1.x_4.x_7 + x_0.x_1.x_4.x_7 + x_2.x_4.x_7 +$
$x_1.x_2.x_4.x_7 + x_0.x_3.x_4.x_7 + x_1.x_3.x_4.x_7 + x_2.x_3.x_4.x_7 + x_0.x_2.x_3.x_4.x_7$
$+ x_1.x_2.x_3.x_4.x_7 + x_5.x_7 + x_1.x_5.x_7 + x_0.x_1.x_5.x_7 + x_2.x_5.x_7 + x_3.x_5.x_7$
$+ x_1.x_3.x_5.x_7 + x_0.x_1.x_3.x_5.x_7 + x_0.x_2.x_3.x_5.x_7 + x_0.x_1.x_2.x_3.x_5.x_7 +$
$x_4.x_5.x_7 + x_1.x_4.x_5.x_7 + x_0.x_1.x_4.x_5.x_7 + x_2.x_4.x_5.x_7 + x_0.x_2.x_4.x_5.x_7$
$+ x_0.x_1.x_2.x_4.x_5.x_7 + x_3.x_4.x_5.x_7 + x_0.x_3.x_4.x_5.x_7 + x_1.x_6.x_7 + x_2.x_6.x_7$
$+ x_0.x_1.x_2.x_6.x_7 + x_0.x_2.x_3.x_6.x_7 + x_0.x_4.x_6.x_7 + x_0.x_1.x_4.x_6.x_7 +$
$x_3.x_4.x_6.x_7 + x_1.x_3.x_4.x_6.x_7 + x_0.x_2.x_3.x_4.x_6.x_7 + x_1.x_2.x_3.x_4.x_6.x_7 +$
$x_5.x_6.x_7 + x_0.x_5.x_6.x_7 + x_0.x_1.x_5.x_6.x_7 + x_2.x_3.x_5.x_6.x_7 + x_1.x_2.x_3.x_5.$
$x_6.x_7 + x_0.x_1.x_2.x_3.x_5.x_6.x_7 + x_0.x_4.x_5.x_6.x_7 + x_2.x_4.x_5.x_6.x_7 + x_0.x_2.$
$x_4.x_5.x_6.x_7 + x_1.x_2.x_4.x_5.x_6.x_7 + x_0.x_1.x_2.x_4.x_5.x_6.x_7,$
- $y_4 = x_0 + x_1 + x_0.x_1 + x_2 + x_3 + x_2.x_3 + x_0.x_2.x_3 + x_0.x_4 + x_1.x_4$
$+ x_1.x_2.x_4 + x_3.x_4 + x_0.x_3.x_4 + x_2.x_3.x_4 + x_0.x_2.x_3.x_4 + x_1.x_2.x_3.x_4$
$+ x_0.x_1.x_2.x_3.x_4 + x_5 + x_0.x_5 + x_1.x_5 + x_2.x_5 + x_3.x_5 + x_0.x_3.x_5$
$+ x_1.x_3.x_5 + x_0.x_1.x_3.x_5 + x_2.x_3.x_5 + x_1.x_2.x_3.x_5 + x_4.x_5 + x_0.x_4.x_5$
$+ x_1.x_4.x_5 + x_0.x_1.x_4.x_5 + x_2.x_4.x_5 + x_0.x_1.x_2.x_4.x_5 + x_3.x_4.x_5 +$
$x_0.x_3.x_4.x_5 + x_1.x_3.x_4.x_5 + x_0.x_2.x_3.x_4.x_5 + x_0.x_1.x_2.x_3.x_4.x_5 + x_0.x_6$
$+ x_1.x_6 + x_0.x_2.x_6 + x_0.x_1.x_2.x_6 + x_0.x_1.x_3.x_6 + x_2.x_3.x_6 + x_0.x_1.x_2.x_3.$
$x_6 + x_4.x_6 + x_0.x_4.x_6 + x_0.x_1.x_4.x_6 + x_1.x_2.x_4.x_6 + x_3.x_4.x_6 +$
$x_0.x_3.x_4.x_6 + x_0.x_1.x_3.x_4.x_6 + x_0.x_2.x_3.x_4.x_6 + x_1.x_2.x_3.x_4.x_6 + x_0.x_1.$

$x_5.x_6 + x_2.x_5.x_6 + x_3.x_5.x_6 + x_0.x_3.x_5.x_6 + x_2.x_3.x_5.x_6 + x_0.x_2.x_3.x_5.x_6$
$+ x_1.x_2.x_3.x_5.x_6 + x_0.x_1.x_2.x_3.x_5.x_6 + x_0.x_4.x_5.x_6 + x_0.x_1.x_4.x_5.x_6 +$
$x_1.x_2.x_4.x_5.x_6 + x_0.x_1.x_2.x_4.x_5.x_6 + x_1.x_3.x_4.x_5.x_6 + x_0.x_1.x_3.x_4.x_5.x_6$
$+ x_2.x_3.x_4.x_5.x_6 + x_1.x_2.x_3.x_4.x_5.x_6 + x_0.x_7 + x_3.x_7 + x_0.x_1.x_3.x_7$
$+ x_0.x_1.x_2.x_3.x_7 + x_0.x_4.x_7 + x_2.x_4.x_7 + x_3.x_4.x_7 + x_0.x_3.x_4.x_7 +$
$x_1.x_3.x_4.x_7 + x_2.x_3.x_4.x_7 + x_0.x_1.x_2.x_3.x_4.x_7 + x_5.x_7 + x_1.x_5.x_7 +$
$x_0.x_1.x_5.x_7 + x_0.x_2.x_5.x_7 + x_1.x_2.x_5.x_7 + x_0.x_1.x_2.x_5.x_7 + x_3.x_5.x_7 +$
$x_0.x_3.x_5.x_7 + x_0.x_1.x_3.x_5.x_7 + x_2.x_3.x_5.x_7 + x_0.x_2.x_3.x_5.x_7 + x_1.x_2.x_3.$
$x_5.x_7 + x_4.x_5.x_7 + x_1.x_2.x_4.x_5.x_7 + x_3.x_4.x_5.x_7 + x_1.x_3.x_4.x_5.x_7 +$
$x_2.x_3.x_4.x_5.x_7 + x_0.x_2.x_3.x_4.x_5.x_7 + x_1.x_2.x_3.x_4.x_5.x_7 + x_0.x_1.x_2.x_3.x_4.$
$x_5.x_7 + x_6.x_7 + x_0.x_6.x_7 + x_1.x_6.x_7 + x_0.x_1.x_6.x_7 + x_2.x_6.x_7 +$
$x_1.x_2.x_6.x_7 + x_0.x_1.x_2.x_6.x_7 + x_3.x_6.x_7 + x_0.x_3.x_6.x_7 + x_0.x_1.x_3.x_6.x_7$
$+ x_4.x_6.x_7 + x_0.x_4.x_6.x_7 + x_0.x_2.x_4.x_6.x_7 + x_3.x_4.x_6.x_7 + x_0.x_3.x_4.x_6.x_7$
$+ x_2.x_3.x_4.x_6.x_7 + x_1.x_2.x_3.x_4.x_6.x_7 + x_0.x_1.x_2.x_3.x_4.x_6.x_7 + x_5.x_6.x_7$
$+ x_0.x_1.x_5.x_6.x_7 + x_1.x_2.x_5.x_6.x_7 + x_3.x_5.x_6.x_7 + x_0.x_3.x_5.x_6.x_7 +$
$x_1.x_3.x_5.x_6.x_7 + x_1.x_2.x_3.x_5.x_6.x_7 + x_0.x_4.x_5.x_6.x_7 + x_1.x_4.x_5.x_6.x_7 +$
$x_2.x_4.x_5.x_6.x_7 + x_3.x_4.x_5.x_6.x_7 + x_0.x_3.x_4.x_5.x_6.x_7 + x_0.x_2.x_3.x_4.x_5.x_6.x_7,$

- $y_3 = x_0 + x_1.x_2 + x_0.x_3 + x_0.x_1.x_3 + x_2.x_3 + x_0.x_2.x_3 + x_1.x_2.x_3$
$+ x_0.x_1.x_2.x_3 + x_4 + x_0.x_1.x_4 + x_0.x_2.x_4 + x_0.x_1.x_2.x_4 + x_0.x_3.x_4 +$
$x_1.x_3.x_4 + x_2.x_3.x_4 + x_0.x_1.x_2.x_3.x_4 + x_0.x_1.x_5 + x_1.x_2.x_5 + x_0.x_1.x_2.x_5$
$+ x_2.x_3.x_5 + x_1.x_2.x_3.x_5 + x_0.x_1.x_2.x_3.x_5 + x_4.x_5 + x_0.x_4.x_5 + x_2.x_4.x_5$
$+ x_1.x_2.x_4.x_5 + x_1.x_3.x_4.x_5 + x_0.x_1.x_3.x_4.x_5 + x_2.x_3.x_4.x_5 + x_0.x_2.x_3.$
$x_4.x_5 + x_6 + x_0.x_1.x_6 + x_0.x_2.x_6 + x_1.x_2.x_6 + x_3.x_6 + x_0.x_3.x_6 +$
$x_0.x_1.x_3.x_6 + x_1.x_2.x_3.x_6 + x_0.x_4.x_6 + x_0.x_1.x_4.x_6 + x_0.x_2.x_4.x_6 +$
$x_0.x_3.x_4.x_6 + x_1.x_3.x_4.x_6 + x_0.x_1.x_3.x_4.x_6 + x_0.x_1.x_2.x_3.x_4.x_6 + x_5.x_6$
$+ x_1.x_5.x_6 + x_0.x_1.x_5.x_6 + x_0.x_2.x_5.x_6 + x_0.x_1.x_2.x_5.x_6 + x_3.x_5.x_6$
$+ x_1.x_3.x_5.x_6 + x_0.x_1.x_3.x_5.x_6 + x_0.x_2.x_3.x_5.x_6 + x_1.x_2.x_3.x_5.x_6 +$
$x_0.x_1.x_2.x_3.x_5.x_6 + x_0.x_4.x_5.x_6 + x_1.x_4.x_5.x_6 + x_2.x_4.x_5.x_6 + x_0.x_2.x_4.$
$x_5.x_6 + x_1.x_2.x_4.x_5.x_6 + x_3.x_4.x_5.x_6 + x_0.x_3.x_4.x_5.x_6 + x_2.x_3.x_4.x_5.x_6$
$+ x_1.x_2.x_3.x_4.x_5.x_6 + x_0.x_1.x_2.x_3.x_4.x_5.x_6 + x_7 + x_0.x_7 + x_1.x_7 +$
$x_0.x_1.x_7 + x_0.x_2.x_7 + x_1.x_2.x_7 + x_3.x_7 + x_0.x_3.x_7 + x_2.x_3.x_7 +$
$x_0.x_2.x_3.x_7 + x_1.x_2.x_3.x_7 + x_0.x_1.x_2.x_3.x_7 + x_0.x_2.x_4.x_7 + x_1.x_2.x_4.x_7$
$+ x_0.x_1.x_2.x_4.x_7 + x_3.x_4.x_7 + x_1.x_3.x_4.x_7 + x_0.x_1.x_3.x_4.x_7 + x_0.x_1.x_2.$
$x_3.x_4.x_7 + x_5.x_7 + x_2.x_5.x_7 + x_1.x_2.x_5.x_7 + x_0.x_1.x_2.x_5.x_7 + x_3.x_5.x_7$
$+ x_0.x_2.x_3.x_5.x_7 + x_0.x_1.x_2.x_3.x_5.x_7 + x_1.x_4.x_5.x_7 + x_0.x_1.x_4.x_5.x_7$
$+ x_2.x_4.x_5.x_7 + x_0.x_2.x_4.x_5.x_7 + x_0.x_1.x_2.x_4.x_5.x_7 + x_0.x_3.x_4.x_5.x_7$
$+ x_0.x_1.x_3.x_4.x_5.x_7 + x_2.x_3.x_4.x_5.x_7 + x_1.x_2.x_3.x_4.x_5.x_7 + x_6.x_7 +$
$x_0.x_1.x_6.x_7 + x_0.x_2.x_6.x_7 + x_1.x_2.x_6.x_7 + x_0.x_3.x_6.x_7 + x_0.x_2.x_3.x_6.x_7$
$+ x_0.x_1.x_2.x_3.x_6.x_7 + x_1.x_4.x_6.x_7 + x_2.x_4.x_6.x_7 + x_0.x_1.x_2.x_4.x_6.x_7$
$+ x_3.x_4.x_6.x_7 + x_0.x_3.x_4.x_6.x_7 + x_1.x_3.x_4.x_6.x_7 + x_0.x_2.x_3.x_4.x_6.x_7 +$
$x_1.x_2.x_3.x_4.x_6.x_7 + x_5.x_6.x_7 + x_0.x_5.x_6.x_7 + x_1.x_5.x_6.x_7 + x_0.x_1.x_5.x_6.x_7$
$+ x_0.x_2.x_5.x_6.x_7 + x_1.x_2.x_5.x_6.x_7 + x_3.x_5.x_6.x_7 + x_0.x_3.x_5.x_6.x_7 +$
$x_1.x_3.x_5.x_6.x_7 + x_0.x_2.x_3.x_5.x_6.x_7 + x_1.x_2.x_3.x_5.x_6.x_7 + x_0.x_1.x_2.x_3.x_5.$
$x_6.x_7 + x_4.x_5.x_6.x_7 + x_0.x_4.x_5.x_6.x_7 + x_2.x_4.x_5.x_6.x_7 + x_1.x_2.x_4.x_5.x_6.x_7$

$+ x_3.x_4.x_5.x_6.x_7 + x_0.x_1.x_3.x_4.x_5.x_6.x_7 + x_2.x_3.x_4.x_5.x_6.x_7 + x_0.x_2.x_3.$
$x_4.x_5.x_6.x_7,$

- $y_2 = x_0 + x_1 + x_0.x_2 + x_0.x_3 + x_0.x_1.x_3 + x_2.x_3 + x_0.x_2.x_3 + x_0.x_4 +$
$x_1.x_4 + x_0.x_1.x_4 + x_0.x_2.x_4 + x_1.x_2.x_4 + x_3.x_4 + x_0.x_3.x_4 + x_1.x_3.x_4 +$
$x_0.x_1.x_3.x_4 + x_2.x_3.x_4 + x_0.x_2.x_3.x_4 + x_1.x_2.x_3.x_4 + x_0.x_1.x_2.x_3.x_4 +$
$x_5 + x_0.x_5 + x_0.x_1.x_5 + x_0.x_2.x_5 + x_1.x_2.x_5 + x_0.x_1.x_2.x_5 + x_0.x_3.x_5 +$
$x_1.x_3.x_5 + x_0.x_1.x_3.x_5 + x_0.x_2.x_3.x_5 + x_1.x_2.x_3.x_5 + x_0.x_1.x_2.x_3.x_5 +$
$x_0.x_4.x_5 + x_1.x_4.x_5 + x_2.x_4.x_5 + x_0.x_1.x_2.x_4.x_5 + x_3.x_4.x_5 + x_1.x_2.x_3.$
$x_4.x_5 + x_0.x_1.x_2.x_3.x_4.x_5 + x_0.x_6 + x_0.x_1.x_6 + x_1.x_2.x_6 + x_0.x_3.x_6$
$+ x_0.x_1.x_3.x_6 + x_2.x_3.x_6 + x_1.x_2.x_3.x_6 + x_0.x_1.x_2.x_3.x_6 + x_0.x_4.x_6 +$
$x_0.x_1.x_4.x_6 + x_0.x_2.x_4.x_6 + x_3.x_4.x_6 + x_0.x_3.x_4.x_6 + x_0.x_1.x_3.x_4.x_6 +$
$x_2.x_3.x_4.x_6 + x_1.x_2.x_3.x_4.x_6 + x_0.x_1.x_2.x_3.x_4.x_6 + x_5.x_6 + x_1.x_5.x_6 +$
$x_0.x_1.x_5.x_6 + x_3.x_5.x_6 + x_0.x_3.x_5.x_6 + x_2.x_3.x_5.x_6 + x_0.x_1.x_2.x_3.x_5.x_6$
$+ x_2.x_4.x_5.x_6 + x_0.x_2.x_4.x_5.x_6 + x_1.x_2.x_4.x_5.x_6 + x_0.x_1.x_2.x_4.x_5.x_6$
$+ x_3.x_4.x_5.x_6 + x_1.x_3.x_4.x_5.x_6 + x_0.x_2.x_3.x_4.x_5.x_6 + x_7 + x_0.x_7 +$
$x_0.x_1.x_7 + x_0.x_2.x_7 + x_1.x_2.x_7 + x_0.x_1.x_2.x_7 + x_0.x_3.x_7 + x_1.x_3.x_7 +$
$x_2.x_3.x_7 + x_0.x_2.x_3.x_7 + x_0.x_1.x_2.x_3.x_7 + x_4.x_7 + x_0.x_4.x_7 + x_1.x_4.x_7 +$
$x_0.x_2.x_4.x_7 + x_1.x_2.x_4.x_7 + x_0.x_1.x_2.x_4.x_7 + x_0.x_3.x_4.x_7 + x_0.x_1.x_3.x_4.x_7$
$+ x_2.x_3.x_4.x_7 + x_0.x_2.x_3.x_4.x_7 + x_0.x_1.x_2.x_3.x_4.x_7 + x_0.x_5.x_7 + x_1.x_5.x_7$
$+ x_1.x_2.x_5.x_7 + x_0.x_1.x_2.x_5.x_7 + x_0.x_3.x_5.x_7 + x_1.x_3.x_5.x_7 + x_0.x_1.x_3.$
$x_5.x_7 + x_0.x_2.x_3.x_5.x_7 + x_1.x_4.x_5.x_7 + x_2.x_4.x_5.x_7 + x_1.x_2.x_4.x_5.x_7 +$
$x_0.x_1.x_2.x_4.x_5.x_7 + x_0.x_3.x_4.x_5.x_7 + x_0.x_1.x_3.x_4.x_5.x_7 + x_1.x_2.x_3.x_4.x_5.$
$x_7 + x_6.x_7 + x_0.x_6.x_7 + x_2.x_6.x_7 + x_1.x_2.x_6.x_7 + x_0.x_1.x_2.x_6.x_7 +$
$x_1.x_3.x_6.x_7 + x_0.x_1.x_3.x_6.x_7 + x_1.x_2.x_3.x_6.x_7 + x_4.x_6.x_7 + x_0.x_1.x_4.x_6.x_7$
$+ x_2.x_4.x_6.x_7 + x_1.x_2.x_4.x_6.x_7 + x_1.x_3.x_4.x_6.x_7 + x_0.x_1.x_3.x_4.x_6.x_7 +$
$x_2.x_3.x_4.x_6.x_7 + x_0.x_2.x_3.x_4.x_6.x_7 + x_1.x_2.x_3.x_4.x_6.x_7 + x_0.x_1.x_2.x_3.x_4.$
$x_6.x_7 + x_1.x_5.x_6.x_7 + x_2.x_5.x_6.x_7 + x_0.x_2.x_5.x_6.x_7 + x_3.x_5.x_6.x_7 +$
$x_0.x_3.x_5.x_6.x_7 + x_1.x_3.x_5.x_6.x_7 + x_0.x_1.x_3.x_5.x_6.x_7 + x_2.x_3.x_5.x_6.x_7 +$
$x_0.x_2.x_3.x_5.x_6.x_7 + x_0.x_1.x_2.x_3.x_5.x_6.x_7 + x_4.x_5.x_6.x_7 + x_0.x_4.x_5.x_6.x_7$
$+ x_1.x_4.x_5.x_6.x_7 + x_0.x_1.x_4.x_5.x_6.x_7 + x_0.x_2.x_4.x_5.x_6.x_7 + x_3.x_4.x_5.x_6.x_7$
$+ x_0.x_1.x_3.x_4.x_5.x_6.x_7 + x_2.x_3.x_4.x_5.x_6.x_7 + x_0.x_2.x_3.x_4.x_5.x_6.x_7 +$
$x_1.x_2.x_3.x_4.x_5.x_6.x_7,$
- $y_1 = 1 + x_0 + x_0.x_1 + x_0.x_2 + x_3 + x_0.x_3 + x_1.x_3 + x_0.x_1.x_3 +$
$x_2.x_3 + x_0.x_2.x_3 + x_1.x_2.x_3 + x_0.x_4 + x_1.x_4 + x_0.x_1.x_4 + x_1.x_2.x_4$
$+ x_0.x_3.x_4 + x_1.x_3.x_4 + x_0.x_1.x_3.x_4 + x_2.x_3.x_4 + x_0.x_1.x_2.x_3.x_4 +$
$x_1.x_3.x_5 + x_2.x_3.x_5 + x_1.x_2.x_3.x_5 + x_4.x_5 + x_0.x_4.x_5 + x_0.x_1.x_4.x_5 +$
$x_0.x_2.x_4.x_5 + x_0.x_1.x_2.x_4.x_5 + x_0.x_3.x_4.x_5 + x_1.x_3.x_4.x_5 + x_2.x_3.x_4.x_5$
$+ x_0.x_2.x_3.x_4.x_5 + x_6 + x_0.x_1.x_6 + x_2.x_6 + x_0.x_1.x_2.x_6 + x_1.x_3.x_6$
$+ x_0.x_1.x_3.x_6 + x_2.x_3.x_6 + x_0.x_2.x_3.x_6 + x_0.x_1.x_2.x_3.x_6 + x_4.x_6 +$
$x_0.x_4.x_6 + x_0.x_1.x_2.x_4.x_6 + x_3.x_4.x_6 + x_0.x_3.x_4.x_6 + x_0.x_1.x_3.x_4.x_6 +$
$x_1.x_2.x_3.x_4.x_6 + x_0.x_1.x_2.x_3.x_4.x_6 + x_1.x_5.x_6 + x_0.x_1.x_5.x_6 + x_0.x_2.x_5.x_6$
$+ x_1.x_2.x_5.x_6 + x_3.x_5.x_6 + x_1.x_3.x_5.x_6 + x_0.x_1.x_3.x_5.x_6 + x_1.x_2.x_3.x_5.x_6$
$+ x_0.x_1.x_2.x_3.x_5.x_6 + x_0.x_4.x_5.x_6 + x_0.x_1.x_4.x_5.x_6 + x_0.x_2.x_4.x_5.x_6 +$
$x_1.x_2.x_4.x_5.x_6 + x_3.x_4.x_5.x_6 + x_1.x_3.x_4.x_5.x_6 + x_0.x_1.x_3.x_4.x_5.x_6 +$
$x_0.x_2.x_3.x_4.x_5.x_6 + x_1.x_2.x_3.x_4.x_5.x_6 + x_7 + x_0.x_7 + x_1.x_7 + x_2.x_7$

$+ x_0.x_2.x_7 + x_3.x_7 + x_1.x_3.x_7 + x_0.x_1.x_2.x_3.x_7 + x_0.x_4.x_7 + x_1.x_4.x_7$
$+ x_0.x_1.x_4.x_7 + x_1.x_2.x_4.x_7 + x_0.x_1.x_2.x_4.x_7 + x_3.x_4.x_7 + x_1.x_3.x_4.x_7$
$+ x_0.x_1.x_3.x_4.x_7 + x_2.x_3.x_4.x_7 + x_0.x_2.x_3.x_4.x_7 + x_0.x_1.x_2.x_3.x_4.x_7 +$
$x_0.x_5.x_7 + x_2.x_5.x_7 + x_1.x_2.x_5.x_7 + x_3.x_5.x_7 + x_1.x_3.x_5.x_7 + x_0.x_1.x_3.$
$x_5.x_7 + x_2.x_3.x_5.x_7 + x_0.x_2.x_3.x_5.x_7 + x_0.x_4.x_5.x_7 + x_1.x_4.x_5.x_7 +$
$x_2.x_4.x_5.x_7 + x_0.x_1.x_2.x_4.x_5.x_7 + x_3.x_4.x_5.x_7 + x_0.x_3.x_4.x_5.x_7 + x_1.x_3.$
$x_4.x_5.x_7 + x_2.x_3.x_4.x_5.x_7 + x_0.x_2.x_3.x_4.x_5.x_7 + x_0.x_6.x_7 + x_2.x_6.x_7 +$
$x_1.x_2.x_6.x_7 + x_0.x_1.x_2.x_6.x_7 + x_3.x_6.x_7 + x_0.x_1.x_3.x_6.x_7 + x_0.x_2.x_3.x_6.x_7$
$+ x_0.x_4.x_6.x_7 + x_1.x_4.x_6.x_7 + x_2.x_4.x_6.x_7 + x_0.x_1.x_2.x_4.x_6.x_7 + x_3.x_4.$
$x_6.x_7 + x_0.x_1.x_3.x_4.x_6.x_7 + x_2.x_3.x_4.x_6.x_7 + x_0.x_2.x_3.x_4.x_6.x_7 + x_1.x_2.$
$x_3.x_4.x_6.x_7 + x_0.x_1.x_2.x_3.x_4.x_6.x_7 + x_5.x_6.x_7 + x_0.x_1.x_5.x_6.x_7 + x_0.x_2.$
$x_5.x_6.x_7 + x_0.x_1.x_2.x_5.x_6.x_7 + x_0.x_3.x_5.x_6.x_7 + x_0.x_2.x_3.x_5.x_6.x_7 +$
$x_1.x_2.x_3.x_5.x_6.x_7 + x_0.x_1.x_2.x_3.x_5.x_6.x_7 + x_0.x_4.x_5.x_6.x_7 + x_2.x_4.x_5.x_6.x_7$
$+ x_0.x_1.x_2.x_4.x_5.x_6.x_7 + x_1.x_3.x_4.x_5.x_6.x_7 + x_0.x_1.x_3.x_4.x_5.x_6.x_7,$

- $y_0 = 1 + x_0 + x_0.x_1 + x_2 + x_1.x_2 + x_3 + x_1.x_3 + x_2.x_3 + x_1.x_2.x_3$
$+ x_0.x_1.x_2.x_3 + x_4 + x_0.x_4 + x_1.x_4 + x_0.x_1.x_4 + x_2.x_4 + x_0.x_2.x_4$
$+ x_1.x_2.x_4 + x_0.x_3.x_4 + x_1.x_3.x_4 + x_0.x_1.x_2.x_3.x_4 + x_0.x_5 + x_0.x_2.x_5$
$+ x_0.x_1.x_2.x_5 + x_0.x_3.x_5 + x_2.x_3.x_5 + x_0.x_2.x_3.x_5 + x_1.x_2.x_3.x_5 +$
$x_0.x_1.x_4.x_5 + x_0.x_2.x_4.x_5 + x_0.x_1.x_2.x_4.x_5 + x_0.x_2.x_3.x_4.x_5 + x_0.x_6$
$+ x_1.x_6 + x_0.x_1.x_6 + x_2.x_6 + x_0.x_2.x_6 + x_1.x_2.x_6 + x_0.x_1.x_2.x_6 +$
$x_0.x_3.x_6 + x_1.x_2.x_3.x_6 + x_0.x_1.x_2.x_3.x_6 + x_4.x_6 + x_0.x_4.x_6 + x_1.x_4.x_6 +$
$x_1.x_2.x_4.x_6 + x_0.x_3.x_4.x_6 + x_1.x_3.x_4.x_6 + x_0.x_1.x_3.x_4.x_6 + x_0.x_2.x_3.x_4.x_6$
$+ x_0.x_1.x_2.x_3.x_4.x_6 + x_5.x_6 + x_1.x_5.x_6 + x_2.x_5.x_6 + x_0.x_2.x_5.x_6 +$
$x_0.x_3.x_5.x_6 + x_0.x_1.x_3.x_5.x_6 + x_1.x_2.x_3.x_5.x_6 + x_4.x_5.x_6 + x_0.x_4.x_5.x_6$
$+ x_1.x_4.x_5.x_6 + x_0.x_1.x_4.x_5.x_6 + x_2.x_4.x_5.x_6 + x_1.x_2.x_4.x_5.x_6 + x_0.x_3.$
$x_4.x_5.x_6 + x_2.x_3.x_4.x_5.x_6 + x_0.x_2.x_3.x_4.x_5.x_6 + x_0.x_1.x_7 + x_2.x_7 +$
$x_0.x_2.x_7 + x_0.x_1.x_2.x_7 + x_1.x_3.x_7 + x_2.x_3.x_7 + x_0.x_2.x_3.x_7 + x_1.x_2.x_3.x_7$
$+ x_0.x_1.x_2.x_3.x_7 + x_0.x_4.x_7 + x_0.x_1.x_4.x_7 + x_2.x_4.x_7 + x_0.x_2.x_4.x_7 +$
$x_1.x_2.x_4.x_7 + x_0.x_1.x_2.x_4.x_7 + x_3.x_4.x_7 + x_0.x_1.x_2.x_3.x_4.x_7 + x_5.x_7 +$
$x_2.x_5.x_7 + x_0.x_2.x_5.x_7 + x_0.x_1.x_2.x_5.x_7 + x_3.x_5.x_7 + x_0.x_1.x_3.x_5.x_7$
$+ x_2.x_3.x_5.x_7 + x_1.x_2.x_3.x_5.x_7 + x_0.x_1.x_2.x_3.x_5.x_7 + x_0.x_4.x_5.x_7 +$
$x_1.x_4.x_5.x_7 + x_2.x_4.x_5.x_7 + x_0.x_2.x_4.x_5.x_7 + x_0.x_1.x_2.x_4.x_5.x_7 + x_0.x_3.$
$x_4.x_5.x_7 + x_1.x_3.x_4.x_5.x_7 + x_2.x_3.x_4.x_5.x_7 + x_0.x_2.x_3.x_4.x_5.x_7 + x_6.x_7$
$+ x_2.x_6.x_7 + x_1.x_2.x_6.x_7 + x_0.x_1.x_2.x_6.x_7 + x_3.x_6.x_7 + x_1.x_3.x_6.x_7$
$+ x_0.x_1.x_3.x_6.x_7 + x_2.x_3.x_6.x_7 + x_0.x_1.x_2.x_3.x_6.x_7 + x_0.x_4.x_6.x_7 +$
$x_0.x_2.x_4.x_6.x_7 + x_1.x_2.x_4.x_6.x_7 + x_0.x_3.x_4.x_6.x_7 + x_0.x_1.x_3.x_4.x_6.x_7$
$+ x_1.x_2.x_3.x_4.x_6.x_7 + x_0.x_1.x_2.x_3.x_4.x_6.x_7 + x_5.x_6.x_7 + x_1.x_5.x_6.x_7$
$+ x_0.x_1.x_5.x_6.x_7 + x_1.x_2.x_5.x_6.x_7 + x_0.x_1.x_2.x_5.x_6.x_7 + x_3.x_5.x_6.x_7 +$
$x_0.x_3.x_5.x_6.x_7 + x_0.x_2.x_3.x_5.x_6.x_7 + x_0.x_1.x_4.x_5.x_6.x_7 + x_2.x_4.x_5.x_6.x_7$
$+ x_1.x_2.x_4.x_5.x_6.x_7 + x_0.x_1.x_2.x_4.x_5.x_6.x_7 + x_1.x_3.x_4.x_5.x_6.x_7 + x_2.x_3.$
$x_4.x_5.x_6.x_7 + x_0.x_2.x_3.x_4.x_5.x_6.x_7.$

For the ANFs of the SKIPJACK S-box [69], the input is denoted (x_0, x_1, x_2, x_3, x_4, x_5, x_6, x_7) and the output is (y_0, y_1, y_2, y_3, y_4, y_5, y_6, y_7):

- $y_7 = 1 + x_1 + x_0.x_1 + x_0.x_2 + x_1.x_2 + x_0.x_3 + x_0.x_1.x_2.x_3 + x_0.x_4 + x_0.x_1.x_4 + x_0.x_2.x_4 + x_3.x_4 + x_0.x_3.x_4 + x_0.x_2.x_3.x_4 + x_0.x_1.x_2.x_3.x_4 + x_5 + x_0.x_5 + x_1.x_5 + x_0.x_1.x_5 + x_0.x_2.x_5 + x_1.x_2.x_5 + x_0.x_1.x_2.x_5 + x_0.x_3.x_5 + x_2.x_3.x_5 + x_0.x_2.x_3.x_5 + x_1.x_2.x_3.x_5 + x_4.x_5 + x_1.x_4.x_5 + x_1.x_2.x_4.x_5 + x_0.x_2.x_3.x_4.x_5 + x_6 + x_0.x_6 + x_1.x_6 + x_2.x_6 + x_1.x_2.x_6 + x_3.x_6 + x_1.x_3.x_6 + x_2.x_3.x_6 + x_0.x_2.x_3.x_6 + x_0.x_1.x_2.x_3.x_6 + x_0.x_4.x_6 + x_1.x_4.x_6 + x_0.x_1.x_4.x_6 + x_2.x_4.x_6 + x_0.x_2.x_4.x_6 + x_0.x_1.x_2.x_4.x_6 + x_0.x_3.x_4.x_6 + x_1.x_3.x_4.x_6 + x_2.x_3.x_4.x_6 + x_0.x_2.x_3.x_4.x_6 + x_0.x_1.x_2.x_3.x_4.x_6 + x_5.x_6 + x_0.x_5.x_6 + x_0.x_1.x_5.x_6 + x_2.x_5.x_6 + x_0.x_2.x_5.x_6 + x_3.x_5.x_6 + x_0.x_3.x_5.x_6 + x_1.x_3.x_5.x_6 + x_0.x_4.x_5.x_6 + x_1.x_4.x_5.x_6 + x_0.x_1.x_4.x_5.x_6 + x_2.x_4.x_5.x_6 + x_0.x_2.x_4.x_5.x_6 + x_0.x_1.x_2.x_4.x_5.x_6 + x_3.x_4.x_5.x_6 + x_0.x_3.x_4.x_5.x_6 + x_1.x_3.x_4.x_5.x_6 + x_2.x_3.x_4.x_5.x_6 + x_7 + x_0.x_7 + x_0.x_1.x_7 + x_1.x_2.x_7 + x_0.x_1.x_2.x_7 + x_0.x_3.x_7 + x_1.x_3.x_7 + x_2.x_3.x_7 + x_1.x_2.x_3.x_7 + x_4.x_7 + x_0.x_1.x_4.x_7 + x_1.x_2.x_4.x_7 + x_0.x_1.x_2.x_4.x_7 + x_3.x_4.x_7 + x_0.x_1.x_3.x_4.x_7 + x_5.x_7 + x_0.x_5.x_7 + x_2.x_5.x_7 + x_0.x_1.x_2.x_5.x_7 + x_0.x_3.x_5.x_7 + x_1.x_3.x_5.x_7 + x_2.x_3.x_5.x_7 + x_1.x_2.x_3.x_5.x_7 + x_0.x_1.x_2.x_3.x_5.x_7 + x_1.x_4.x_5.x_7 + x_0.x_1.x_4.x_5.x_7 + x_0.x_2.x_4.x_5.x_7 + x_1.x_2.x_4.x_5.x_7 + x_0.x_1.x_2.x_4.x_5.x_7 + x_3.x_4.x_5.x_7 + x_0.x_3.x_4.x_5.x_7 + x_1.x_3.x_4.x_5.x_7 + x_1.x_2.x_3.x_4.x_5.x_7 + x_0.x_1.x_2.x_3.x_4.x_5.x_7 + x_0.x_6.x_7 + x_1.x_6.x_7 + x_0.x_3.x_6.x_7 + x_1.x_3.x_6.x_7 + x_1.x_2.x_3.x_6.x_7 + x_1.x_4.x_6.x_7 + x_0.x_1.x_4.x_6.x_7 + x_1.x_2.x_4.x_6.x_7 + x_1.x_3.x_4.x_6.x_7 + x_2.x_3.x_4.x_6.x_7 + x_0.x_1.x_2.x_3.x_4.x_6.x_7 + x_0.x_1.x_5.x_6.x_7 + x_1.x_2.x_5.x_6.x_7 + x_0.x_1.x_3.x_5.x_6.x_7 + x_0.x_2.x_3.x_5.x_6.x_7 + x_0.x_4.x_5.x_6.x_7 + x_2.x_4.x_5.x_6.x_7 + x_0.x_2.x_4.x_5.x_6.x_7 + x_1.x_2.x_4.x_5.x_6.x_7 + x_3.x_4.x_5.x_6.x_7 + x_1.x_3.x_4.x_5.x_6.x_7 + x_0.x_1.x_3.x_4.x_5.x_6.x_7$,
- $y_6 = x_0 + x_0.x_1 + x_2 + x_0.x_2 + x_0.x_1.x_2 + x_0.x_3 + x_2.x_3 + x_0.x_2.x_3 + x_0.x_1.x_2.x_3 + x_4 + x_0.x_1.x_4 + x_2.x_4 + x_0.x_3.x_4 + x_2.x_3.x_4 + x_2.x_5 + x_0.x_2.x_5 + x_1.x_2.x_5 + x_0.x_1.x_2.x_5 + x_1.x_3.x_5 + x_0.x_1.x_3.x_5 + x_0.x_2.x_3.x_5 + x_1.x_2.x_3.x_5 + x_4.x_5 + x_0.x_4.x_5 + x_1.x_4.x_5 + x_0.x_1.x_4.x_5 + x_1.x_2.x_4.x_5 + x_0.x_1.x_2.x_4.x_5 + x_3.x_4.x_5 + x_0.x_3.x_4.x_5 + x_0.x_1.x_3.x_4.x_5 + x_2.x_3.x_4.x_5 + x_1.x_2.x_3.x_4.x_5 + x_0.x_1.x_2.x_3.x_4.x_5 + x_0.x_6 + x_1.x_6 + x_0.x_1.x_2.x_6 + x_3.x_6 + x_0.x_1.x_3.x_6 + x_0.x_4.x_6 + x_1.x_4.x_6 + x_2.x_4.x_6 + x_0.x_2.x_4.x_6 + x_3.x_4.x_6 + x_0.x_3.x_4.x_6 + x_1.x_3.x_4.x_6 + x_2.x_3.x_4.x_6 + x_1.x_2.x_3.x_4.x_6 + x_0.x_5.x_6 + x_0.x_1.x_5.x_6 + x_0.x_2.x_5.x_6 + x_1.x_2.x_5.x_6 + x_0.x_1.x_2.x_5.x_6 + x_0.x_3.x_5.x_6 + x_1.x_3.x_5.x_6 + x_2.x_3.x_5.x_6 + x_0.x_2.x_3.x_5.x_6 + x_1.x_2.x_3.x_5.x_6 + x_0.x_1.x_2.x_3.x_5.x_6 + x_1.x_4.x_5.x_6 + x_0.x_1.x_2.x_4.x_5.x_6 + x_1.x_3.x_4.x_5.x_6 + x_0.x_1.x_3.x_4.x_5.x_6 + x_2.x_3.x_4.x_5.x_6 + x_0.x_2.x_3.x_4.x_5.x_6 + x_1.x_2.x_3.x_4.x_5.x_6 + x_7 + x_0.x_7 + x_1.x_7 + x_2.x_7 + x_1.x_2.x_7 + x_0.x_1.x_2.x_7 + x_3.x_7 + x_1.x_3.x_7 + x_0.x_1.x_3.x_7 + x_0.x_2.x_3.x_7 + x_1.x_2.x_3.x_7 + x_0.x_4.x_7 + x_1.x_4.x_7 + x_0.x_1.x_4.x_7 + x_2.x_4.x_7 + x_0.x_2.x_4.x_7 + x_1.x_2.x_4.x_7 + x_0.x_3.x_4.x_7 + x_1.x_3.x_4.x_7 + x_0.x_1.x_2.x_3.x_4.x_7 + x_0.x_5.x_7 + x_1.x_5.x_7 + x_0.x_1.x_5.x_7 + x_0.x_2.x_5.x_7 + x_0.x_1.x_2.x_5.x_7 + x_3.x_5.x_7 + x_0.x_3.x_5.x_7 + x_0.x_1.x_3.x_5.x_7 + x_2.x_3.x_5.x_7 + x_4.x_5.x_7 + x_0.x_1.x_4.x_5.x_7 + x_0.x_2.x_4.x_5.x_7 + x_1.x_3.x_4.x_5.x_7 + x_0.x_1.x_3.x_4.x_5.x_7 + x_2.x_3.x_4.x_5.x_7 + x_0.x_2.x_3.x_4.x_5.x_7 + x_1.x_2.x_3.x_4.x_5.x_7 + x_0.x_1.x_2.x_3.x_4.x_5.x_7 + x_6.x_7 + x_0.x_6.x_7 + x_0.x_1.x_6.$

$x_7 + x_1.x_2.x_6.x_7 + x_0.x_1.x_2.x_6.x_7 + x_0.x_1.x_3.x_6.x_7 + x_0.x_2.x_3.x_6.x_7 + x_1.x_2.x_3.x_6.x_7 + x_4.x_6.x_7 + x_0.x_4.x_6.x_7 + x_0.x_3.x_4.x_6.x_7 + x_2.x_3.x_4.x_6.x_7 + x_1.x_2.x_3.x_4.x_6.x_7 + x_0.x_5.x_6.x_7 + x_1.x_5.x_6.x_7 + x_0.x_1.x_5.x_6.x_7 + x_2.x_5.x_6.x_7 + x_3.x_5.x_6.x_7 + x_1.x_3.x_5.x_6.x_7 + x_0.x_1.x_3.x_5.x_6.x_7 + x_1.x_2.x_3.x_5.x_6.x_7 + x_0.x_1.x_2.x_3.x_5.x_6.x_7 + x_4.x_5.x_6.x_7 + x_0.x_4.x_5.x_6.x_7 + x_1.x_4.x_5.x_6.x_7 + x_0.x_1.x_4.x_5.x_6.x_7 + x_2.x_4.x_5.x_6.x_7 + x_0.x_1.x_2.x_4.x_5.x_6.x_7 + x_3.x_4.x_5.x_6.x_7 + x_1.x_3.x_4.x_5.x_6.x_7 + x_0.x_1.x_3.x_4.x_5.x_6.x_7 + x_2.x_3.x_4.x_5.x_6.x_7 + x_0.x_2.x_3.x_4.x_5.x_6.x_7$,

- $y_5 = 1 + x_0 + x_1 + x_0.x_1 + x_1.x_2 + x_0.x_3 + x_2.x_3 + x_0.x_2.x_3 + x_1.x_2.x_3 + x_0.x_4 + x_0.x_1.x_4 + x_2.x_4 + x_0.x_1.x_2.x_4 + x_3.x_4 + x_0.x_3.x_4 + x_1.x_3.x_4 + x_0.x_1.x_2.x_3.x_4 + x_5 + x_0.x_5 + x_1.x_5 + x_0.x_2.x_5 + x_1.x_3.x_5 + x_0.x_1.x_3.x_5 + x_0.x_2.x_3.x_5 + x_1.x_2.x_3.x_5 + x_0.x_1.x_2.x_3.x_5 + x_0.x_4.x_5 + x_1.x_4.x_5 + x_0.x_2.x_4.x_5 + x_1.x_2.x_4.x_5 + x_0.x_1.x_2.x_4.x_5 + x_1.x_3.x_4.x_5 + x_0.x_1.x_3.x_4.x_5 + x_0.x_2.x_3.x_4.x_5 + x_0.x_1.x_2.x_3.x_4.x_5 + x_0.x_6 + x_1.x_6 + x_2.x_6 + x_1.x_2.x_6 + x_0.x_1.x_2.x_6 + x_0.x_3.x_6 + x_1.x_3.x_6 + x_2.x_3.x_6 + x_0.x_2.x_3.x_6 + x_1.x_2.x_3.x_6 + x_4.x_6 + x_0.x_1.x_4.x_6 + x_2.x_4.x_6 + x_3.x_4.x_6 + x_0.x_1.x_3.x_4.x_6 + x_0.x_2.x_3.x_4.x_6 + x_1.x_2.x_3.x_4.x_6 + x_5.x_6 + x_0.x_1.x_5.x_6 + x_2.x_5.x_6 + x_0.x_2.x_5.x_6 + x_1.x_2.x_5.x_6 + x_3.x_5.x_6 + x_0.x_3.x_5.x_6 + x_1.x_3.x_5.x_6 + x_0.x_1.x_3.x_5.x_6 + x_2.x_3.x_5.x_6 + x_1.x_2.x_3.x_5.x_6 + x_1.x_4.x_5.x_6 + x_2.x_4.x_5.x_6 + x_0.x_3.x_4.x_5.x_6 + x_1.x_3.x_4.x_5.x_6 + x_0.x_1.x_3.x_4.x_5.x_6 + x_2.x_3.x_4.x_5.x_6 + x_0.x_1.x_2.x_3.x_4.x_5.x_6 + x_7 + x_1.x_7 + x_0.x_1.x_7 + x_1.x_2.x_7 + x_0.x_1.x_2.x_7 + x_3.x_7 + x_0.x_3.x_7 + x_1.x_3.x_7 + x_0.x_1.x_3.x_7 + x_1.x_2.x_3.x_7 + x_0.x_1.x_2.x_3.x_7 + x_1.x_4.x_7 + x_2.x_4.x_7 + x_1.x_2.x_4.x_7 + x_1.x_3.x_4.x_7 + x_0.x_1.x_3.x_4.x_7 + x_2.x_3.x_4.x_7 + x_0.x_1.x_2.x_3.x_4.x_7 + x_0.x_5.x_7 + x_1.x_5.x_7 + x_0.x_1.x_5.x_7 + x_2.x_5.x_7 + x_1.x_2.x_5.x_7 + x_0.x_1.x_2.x_5.x_7 + x_0.x_3.x_5.x_7 + x_2.x_3.x_5.x_7 + x_0.x_2.x_3.x_5.x_7 + x_1.x_2.x_3.x_5.x_7 + x_0.x_1.x_2.x_3.x_5.x_7 + x_1.x_4.x_5.x_7 + x_0.x_2.x_4.x_5.x_7 + x_1.x_2.x_4.x_5.x_7 + x_3.x_4.x_5.x_7 + x_1.x_3.x_4.x_5.x_7 + x_0.x_1.x_3.x_4.x_5.x_7 + x_2.x_3.x_4.x_5.x_7 + x_6.x_7 + x_0.x_6.x_7 + x_1.x_6.x_7 + x_0.x_2.x_6.x_7 + x_3.x_6.x_7 + x_0.x_1.x_3.x_6.x_7 + x_2.x_3.x_6.x_7 + x_1.x_2.x_3.x_6.x_7 + x_0.x_1.x_2.x_3.x_6.x_7 + x_0.x_4.x_6.x_7 + x_0.x_1.x_4.x_6.x_7 + x_2.x_4.x_6.x_7 + x_1.x_2.x_4.x_6.x_7 + x_3.x_4.x_6.x_7 + x_0.x_1.x_3.x_4.x_6.x_7 + x_0.x_2.x_3.x_4.x_6.x_7 + x_1.x_2.x_3.x_4.x_6.x_7 + x_5.x_6.x_7 + x_2.x_5.x_6.x_7 + x_1.x_3.x_5.x_6.x_7 + x_2.x_3.x_5.x_6.x_7 + x_0.x_2.x_3.x_5.x_6.x_7 + x_1.x_2.x_3.x_5.x_6.x_7 + x_0.x_1.x_2.x_3.x_5.x_6.x_7 + x_4.x_5.x_6.x_7 + x_0.x_4.x_5.x_6.x_7 + x_1.x_4.x_5.x_6.x_7 + x_0.x_1.x_4.x_5.x_6.x_7 + x_2.x_4.x_5.x_6.x_7 + x_0.x_2.x_4.x_5.x_6.x_7 + x_1.x_2.x_4.x_5.x_6.x_7 + x_1.x_3.x_4.x_5.x_6.x_7 + x_0.x_1.x_3.x_4.x_5.x_6.x_7 + x_1.x_2.x_3.x_4.x_5.x_6.x_7$,
- $y_4 = x_0 + x_0.x_1 + x_2 + x_3 + x_2.x_3 + x_0.x_2.x_3 + x_1.x_2.x_3 + x_0.x_4 + x_0.x_1.x_4 + x_2.x_4 + x_1.x_2.x_3.x_4 + x_0.x_1.x_2.x_3.x_4 + x_0.x_2.x_5 + x_1.x_2.x_5 + x_0.x_1.x_2.x_5 + x_0.x_3.x_5 + x_2.x_3.x_5 + x_0.x_2.x_3.x_5 + x_1.x_2.x_3.x_5 + x_0.x_1.x_2.x_3.x_5 + x_4.x_5 + x_0.x_4.x_5 + x_1.x_4.x_5 + x_0.x_2.x_4.x_5 + x_3.x_4.x_5 + x_0.x_3.x_4.x_5 + x_0.x_2.x_3.x_4.x_5 + x_0.x_1.x_2.x_3.x_4.x_5 + x_6 + x_0.x_6 + x_0.x_2.x_6 + x_0.x_1.x_2.x_6 + x_0.x_3.x_6 + x_1.x_3.x_6 + x_0.x_1.x_3.x_6 + x_2.x_3.x_6 + x_0.x_2.x_3.x_6 + x_1.x_2.x_3.x_6 + x_0.x_4.x_6 + x_0.x_1.x_4.x_6 + x_3.x_4.x_6 + x_0.x_3.x_4.x_6 + x_0.x_1.x_3.x_4.x_6 + x_2.x_3.x_4.x_6 + x_0.x_2.x_3.x_4.x_6 + x_1.x_2.x_3.x_4.x_6 + x_0.x_1.x_2.x_3.x_4.x_6 + x_1.x_5.x_6 + x_0.x_1.x_5.x_6 + x_2.x_5.x_6 + x_0.x_2.$

$x_5.x_6 + x_1.x_2.x_5.x_6 + x_0.x_1.x_2.x_5.x_6 + x_1.x_3.x_5.x_6 + x_0.x_1.x_3.x_5.x_6 + x_2.x_3.x_5.x_6 + x_0.x_2.x_3.x_5.x_6 + x_1.x_2.x_3.x_5.x_6 + x_0.x_1.x_2.x_3.x_5.x_6 + x_4.x_5.x_6 + x_0.x_4.x_5.x_6 + x_0.x_1.x_4.x_5.x_6 + x_2.x_4.x_5.x_6 + x_0.x_1.x_2.x_4.x_5.x_6 + x_1.x_3.x_4.x_5.x_6 + x_0.x_1.x_3.x_4.x_5.x_6 + x_0.x_2.x_3.x_4.x_5.x_6 + x_0.x_7 + x_1.x_7 + x_2.x_7 + x_0.x_2.x_7 + x_1.x_2.x_7 + x_3.x_7 + x_0.x_3.x_7 + x_1.x_3.x_7 + x_0.x_1.x_3.x_7 + x_1.x_2.x_3.x_7 + x_0.x_4.x_7 + x_2.x_4.x_7 + x_0.x_2.x_4.x_7 + x_1.x_3.x_4.x_7 + x_0.x_1.x_3.x_4.x_7 + x_1.x_2.x_3.x_4.x_7 + x_5.x_7 + x_0.x_5.x_7 + x_0.x_1.x_2.x_5.x_7 + x_2.x_3.x_5.x_7 + x_1.x_2.x_3.x_5.x_7 + x_4.x_5.x_7 + x_0.x_4.x_5.x_7 + x_0.x_1.x_4.x_5.x_7 + x_0.x_3.x_4.x_5.x_7 + x_1.x_3.x_4.x_5.x_7 + x_0.x_1.x_3.x_4.x_5.x_7 + x_2.x_3.x_4.x_5.x_7 + x_0.x_2.x_3.x_4.x_5.x_7 + x_0.x_1.x_2.x_3.x_4.x_5.x_7 + x_6.x_7 + x_0.x_6.x_7 + x_1.x_6.x_7 + x_2.x_6.x_7 + x_0.x_1.x_2.x_6.x_7 + x_0.x_3.x_6.x_7 + x_1.x_3.x_6.x_7 + x_0.x_4.x_6.x_7 + x_1.x_4.x_6.x_7 + x_0.x_1.x_4.x_6.x_7 + x_3.x_4.x_6.x_7 + x_0.x_3.x_4.x_6.x_7 + x_1.x_3.x_4.x_6.x_7 + x_0.x_1.x_3.x_4.x_6.x_7 + x_0.x_2.x_3.x_4.x_6.x_7 + x_5.x_6.x_7 + x_0.x_1.x_5.x_6.x_7 + x_2.x_5.x_6.x_7 + x_1.x_2.x_5.x_6.x_7 + x_0.x_1.x_2.x_5.x_6.x_7 + x_3.x_5.x_6.x_7 + x_0.x_3.x_5.x_6.x_7 + x_1.x_3.x_5.x_6.x_7 + x_0.x_1.x_3.x_5.x_6.x_7 + x_0.x_2.x_3.x_5.x_6.x_7 + x_0.x_1.x_2.x_3.x_5.x_6.x_7 + x_0.x_4.x_5.x_6.x_7 + x_1.x_2.x_4.x_5.x_6.x_7 + x_0.x_1.x_2.x_4.x_5.x_6.x_7 + x_0.x_3.x_4.x_5.x_6.x_7 + x_0.x_1.x_3.x_4.x_5.x_6.x_7$,

- $y_3 = x_1 + x_0.x_1 + x_2 + x_0.x_1.x_2 + x_1.x_3 + x_0.x_2.x_3 + x_0.x_1.x_2.x_3 + x_0.x_4 + x_0.x_2.x_4 + x_1.x_2.x_4 + x_0.x_3.x_4 + x_1.x_3.x_4 + x_0.x_1.x_3.x_4 + x_1.x_2.x_3.x_4 + x_0.x_1.x_2.x_3.x_4 + x_5 + x_0.x_1.x_5 + x_1.x_2.x_5 + x_3.x_5 + x_1.x_3.x_5 + x_0.x_1.x_3.x_5 + x_0.x_2.x_3.x_5 + x_4.x_5 + x_0.x_4.x_5 + x_2.x_4.x_5 + x_0.x_2.x_4.x_5 + x_0.x_1.x_2.x_4.x_5 + x_0.x_3.x_4.x_5 + x_1.x_3.x_4.x_5 + x_2.x_3.x_4.x_5 + x_1.x_2.x_3.x_4.x_5 + x_0.x_1.x_2.x_3.x_4.x_5 + x_6 + x_0.x_6 + x_1.x_6 + x_0.x_2.x_6 + x_0.x_1.x_2.x_6 + x_1.x_3.x_6 + x_0.x_2.x_3.x_6 + x_1.x_2.x_3.x_6 + x_4.x_6 + x_0.x_4.x_6 + x_1.x_4.x_6 + x_0.x_1.x_4.x_6 + x_3.x_4.x_6 + x_1.x_3.x_4.x_6 + x_0.x_2.x_3.x_4.x_6 + x_1.x_2.x_3.x_4.x_6 + x_0.x_1.x_2.x_3.x_4.x_6 + x_0.x_5.x_6 + x_2.x_5.x_6 + x_0.x_2.x_5.x_6 + x_1.x_2.x_5.x_6 + x_0.x_1.x_2.x_5.x_6 + x_3.x_5.x_6 + x_1.x_3.x_5.x_6 + x_0.x_1.x_3.x_5.x_6 + x_2.x_3.x_5.x_6 + x_0.x_2.x_3.x_5.x_6 + x_1.x_2.x_3.x_5.x_6 + x_0.x_1.x_2.x_3.x_5.x_6 + x_1.x_4.x_5.x_6 + x_2.x_4.x_5.x_6 + x_0.x_1.x_2.x_4.x_5.x_6 + x_3.x_4.x_5.x_6 + x_1.x_3.x_4.x_5.x_6 + x_0.x_1.x_3.x_4.x_5.x_6 + x_2.x_3.x_4.x_5.x_6 + x_0.x_2.x_3.x_4.x_5.x_6 + x_0.x_7 + x_0.x_2.x_7 + x_1.x_2.x_7 + x_0.x_1.x_2.x_7 + x_0.x_3.x_7 + x_0.x_1.x_3.x_7 + x_0.x_2.x_3.x_7 + x_0.x_1.x_2.x_3.x_7 + x_4.x_7 + x_2.x_4.x_7 + x_0.x_2.x_4.x_7 + x_1.x_2.x_4.x_7 + x_0.x_3.x_4.x_7 + x_1.x_3.x_4.x_7 + x_1.x_2.x_3.x_4.x_7 + x_5.x_7 + x_1.x_5.x_7 + x_0.x_1.x_2.x_5.x_7 + x_3.x_5.x_7 + x_2.x_3.x_5.x_7 + x_0.x_2.x_3.x_5.x_7 + x_1.x_4.x_5.x_7 + x_2.x_4.x_5.x_7 + x_0.x_2.x_4.x_5.x_7 + x_3.x_4.x_5.x_7 + x_0.x_3.x_4.x_5.x_7 + x_0.x_1.x_3.x_4.x_5.x_7 + x_2.x_3.x_4.x_5.x_7 + x_0.x_6.x_7 + x_1.x_6.x_7 + x_0.x_1.x_6.x_7 + x_0.x_2.x_6.x_7 + x_0.x_1.x_2.x_6.x_7 + x_2.x_3.x_6.x_7 + x_1.x_2.x_3.x_6.x_7 + x_0.x_4.x_6.x_7 + x_2.x_4.x_6.x_7 + x_1.x_2.x_4.x_6.x_7 + x_3.x_4.x_6.x_7 + x_1.x_3.x_4.x_6.x_7 + x_0.x_1.x_3.x_4.x_6.x_7 + x_0.x_2.x_3.x_4.x_6.x_7 + x_0.x_5.x_6.x_7 + x_1.x_5.x_6.x_7 + x_0.x_1.x_5.x_6.x_7 + x_2.x_5.x_6.x_7 + x_0.x_1.x_2.x_5.x_6.x_7 + x_0.x_1.x_3.x_5.x_6.x_7 + x_4.x_5.x_6.x_7 + x_0.x_4.x_5.x_6.x_7 + x_2.x_4.x_5.x_6.x_7 + x_3.x_4.x_5.x_6.x_7 + x_0.x_3.x_4.x_5.x_6.x_7$,
- $y_2 = x_0 + x_0.x_1 + x_0.x_2 + x_1.x_2 + x_0.x_1.x_2 + x_0.x_3 + x_1.x_3 + x_0.x_2.x_3 + x_1.x_2.x_3 + x_0.x_1.x_2.x_3 + x_4 + x_0.x_4 + x_0.x_2.x_4 + x_0.x_1.x_2.x_4 + x_3.x_4 + x_1.x_3.x_4 + x_0.x_1.x_3.x_4 + x_2.x_3.x_4 + x_0.x_2.x_3.x_4 + x_0.x_1.x_2.x_3.x_4 +$

$x_2.x_5 + x_0.x_2.x_5 + x_0.x_1.x_2.x_5 + x_0.x_3.x_5 + x_1.x_3.x_5 + x_2.x_3.x_5 + x_1.x_2.x_3.x_5 + x_0.x_1.x_2.x_3.x_5 + x_1.x_4.x_5 + x_0.x_1.x_4.x_5 + x_0.x_2.x_4.x_5 + x_0.x_1.x_2.x_4.x_5 + x_3.x_4.x_5 + x_0.x_1.x_3.x_4.x_5 + x_2.x_3.x_4.x_5 + x_0.x_1.x_2.x_3.x_4.x_5 + x_0.x_1.x_6 + x_1.x_2.x_6 + x_0.x_1.x_2.x_6 + x_3.x_6 + x_2.x_3.x_6 + x_1.x_2.x_3.x_6 + x_0.x_4.x_6 + x_1.x_4.x_6 + x_1.x_2.x_4.x_6 + x_0.x_1.x_2.x_4.x_6 + x_3.x_4.x_6 + x_1.x_3.x_4.x_6 + x_2.x_3.x_4.x_6 + x_0.x_2.x_3.x_4.x_6 + x_5.x_6 + x_0.x_5.x_6 + x_1.x_5.x_6 + x_0.x_1.x_5.x_6 + x_0.x_2.x_5.x_6 + x_3.x_5.x_6 + x_0.x_3.x_5.x_6 + x_1.x_3.x_5.x_6 + x_2.x_3.x_5.x_6 + x_4.x_5.x_6 + x_0.x_1.x_2.x_4.x_5.x_6 + x_0.x_3.x_4.x_5.x_6 + x_1.x_3.x_4.x_5.x_6 + x_1.x_7 + x_0.x_2.x_7 + x_1.x_2.x_7 + x_3.x_7 + x_0.x_3.x_7 + x_1.x_2.x_3.x_7 + x_4.x_7 + x_1.x_4.x_7 + x_2.x_4.x_7 + x_0.x_2.x_4.x_7 + x_0.x_3.x_4.x_7 + x_1.x_3.x_4.x_7 + x_2.x_3.x_4.x_7 + x_0.x_2.x_3.x_4.x_7 + x_0.x_1.x_2.x_3.x_4.x_7 + x_0.x_5.x_7 + x_1.x_5.x_7 + x_0.x_2.x_5.x_7 + x_1.x_2.x_5.x_7 + x_3.x_5.x_7 + x_0.x_3.x_5.x_7 + x_1.x_3.x_5.x_7 + x_0.x_1.x_3.x_5.x_7 + x_2.x_3.x_5.x_7 + x_0.x_2.x_3.x_5.x_7 + x_4.x_5.x_7 + x_2.x_4.x_5.x_7 + x_0.x_2.x_4.x_5.x_7 + x_0.x_3.x_4.x_5.x_7 + x_1.x_3.x_4.x_5.x_7 + x_0.x_1.x_3.x_4.x_5.x_7 + x_0.x_2.x_3.x_4.x_5.x_7 + x_1.x_2.x_3.x_4.x_5.x_7 + x_6.x_7 + x_0.x_6.x_7 + x_0.x_1.x_6.x_7 + x_0.x_2.x_6.x_7 + x_1.x_2.x_6.x_7 + x_3.x_6.x_7 + x_1.x_2.x_3.x_6.x_7 + x_0.x_1.x_2.x_3.x_6.x_7 + x_1.x_4.x_6.x_7 + x_2.x_4.x_6.x_7 + x_0.x_2.x_4.x_6.x_7 + x_1.x_2.x_4.x_6.x_7 + x_3.x_4.x_6.x_7 + x_2.x_3.x_4.x_6.x_7 + x_0.x_1.x_2.x_3.x_4.x_6.x_7 + x_0.x_5.x_6.x_7 + x_1.x_5.x_6.x_7 + x_2.x_5.x_6.x_7 + x_3.x_5.x_6.x_7 + x_0.x_3.x_5.x_6.x_7 + x_1.x_3.x_5.x_6.x_7 + x_1.x_2.x_3.x_5.x_6.x_7 + x_0.x_1.x_2.x_3.x_5.x_6.x_7 + x_4.x_5.x_6.x_7 + x_0.x_4.x_5.x_6.x_7 + x_1.x_4.x_5.x_6.x_7 + x_0.x_1.x_4.x_5.x_6.x_7 + x_2.x_4.x_5.x_6.x_7 + x_0.x_2.x_4.x_5.x_6.x_7 + x_0.x_1.x_3.x_4.x_5.x_6.x_7 + x_2.x_3.x_4.x_5.x_6.x_7,$

- $y_1 = 1 + x_1 + x_0.x_1 + x_2 + x_0.x_3 + x_0.x_2.x_3 + x_0.x_4 + x_0.x_1.x_4 + x_2.x_4 + x_1.x_2.x_4 + x_0.x_1.x_2.x_4 + x_0.x_3.x_4 + x_0.x_2.x_3.x_4 + x_1.x_2.x_3.x_4 + x_1.x_5 + x_2.x_5 + x_0.x_2.x_5 + x_1.x_2.x_5 + x_0.x_3.x_5 + x_0.x_1.x_3.x_5 + x_2.x_3.x_5 + x_0.x_2.x_3.x_5 + x_1.x_2.x_3.x_5 + x_0.x_1.x_4.x_5 + x_2.x_4.x_5 + x_0.x_1.x_2.x_4.x_5 + x_3.x_4.x_5 + x_0.x_3.x_4.x_5 + x_0.x_1.x_3.x_4.x_5 + x_0.x_2.x_3.x_4.x_5 + x_0.x_1.x_2.x_3.x_4.x_5 + x_6 + x_0.x_6 + x_2.x_6 + x_1.x_2.x_6 + x_0.x_1.x_2.x_6 + x_3.x_6 + x_0.x_3.x_6 + x_1.x_3.x_6 + x_2.x_3.x_6 + x_0.x_1.x_2.x_3.x_6 + x_0.x_1.x_4.x_6 + x_0.x_2.x_4.x_6 + x_1.x_2.x_4.x_6 + x_0.x_1.x_2.x_4.x_6 + x_1.x_3.x_4.x_6 + x_0.x_1.x_3.x_4.x_6 + x_0.x_5.x_6 + x_0.x_2.x_5.x_6 + x_1.x_2.x_5.x_6 + x_3.x_5.x_6 + x_0.x_3.x_5.x_6 + x_1.x_3.x_5.x_6 + x_0.x_1.x_3.x_5.x_6 + x_2.x_3.x_5.x_6 + x_1.x_2.x_3.x_5.x_6 + x_0.x_1.x_2.x_3.x_5.x_6 + x_4.x_5.x_6 + x_0.x_1.x_4.x_5.x_6 + x_0.x_2.x_4.x_5.x_6 + x_0.x_1.x_2.x_4.x_5.x_6 + x_0.x_1.x_3.x_4.x_5.x_6 + x_0.x_7 + x_1.x_7 + x_0.x_2.x_7 + x_2.x_3.x_7 + x_4.x_7 + x_0.x_4.x_7 + x_0.x_2.x_4.x_7 + x_0.x_3.x_4.x_7 + x_2.x_3.x_4.x_7 + x_0.x_1.x_2.x_3.x_4.x_7 + x_5.x_7 + x_0.x_5.x_7 + x_1.x_5.x_7 + x_0.x_1.x_5.x_7 + x_2.x_5.x_7 + x_0.x_2.x_5.x_7 + x_1.x_2.x_5.x_7 + x_0.x_1.x_2.x_5.x_7 + x_0.x_3.x_5.x_7 + x_0.x_2.x_3.x_5.x_7 + x_1.x_2.x_3.x_5.x_7 + x_4.x_5.x_7 + x_0.x_4.x_5.x_7 + x_2.x_4.x_5.x_7 + x_0.x_2.x_4.x_5.x_7 + x_1.x_2.x_4.x_5.x_7 + x_3.x_4.x_5.x_7 + x_1.x_3.x_4.x_5.x_7 + x_0.x_1.x_3.x_4.x_5.x_7 + x_0.x_2.x_3.x_4.x_5.x_7 + x_0.x_1.x_2.x_3.x_4.x_5.x_7 + x_1.x_6.x_7 + x_0.x_2.x_6.x_7 + x_1.x_2.x_6.x_7 + x_0.x_1.x_2.x_6.x_7 + x_3.x_6.x_7 + x_2.x_3.x_6.x_7 + x_0.x_1.x_2.x_3.x_6.x_7 + x_4.x_6.x_7 + x_0.x_4.x_6.x_7 + x_1.x_4.x_6.x_7 + x_0.x_1.x_4.x_6.x_7 + x_2.x_4.x_6.x_7 + x_1.x_2.x_4.x_6.x_7 + x_1.x_3.x_4.x_6.x_7 + x_2.x_3.x_4.x_6.x_7 + x_0.x_2.x_3.x_4.x_6.x_7 + x_5.x_6.x_7 + x_0.x_5.x_6.x_7 + x_1.x_5.x_6.x_7 + x_2.x_5.x_6.x_7 + x_0.x_1.x_2.x_5.x_6.x_7 + x_1.x_3.x_5.x_6.x_7 + x_0.x_1.x_3.x_5.x_6.x_7 + x_1.x_2.x_3.x_5.x_6.x_7 + x_0.x_1.x_2.x_3.x_5.x_6.x_7 +$

$x_4.x_5.x_6.x_7 + x_0.x_1.x_4.x_5.x_6.x_7 + x_2.x_4.x_5.x_6.x_7 + x_0.x_2.x_4.x_5.x_6.x_7 + x_1.x_2.x_4.x_5.x_6.x_7 + x_0.x_3.x_4.x_5.x_6.x_7 + x_2.x_3.x_4.x_5.x_6.x_7 + x_0.x_2.x_3.x_4.x_5.x_6.x_7 + x_1.x_2.x_3.x_4.x_5.x_6.x_7$,

- $y_0 = 1 + x_2 + x_0.x_1.x_3 + x_2.x_3 + x_0.x_1.x_2.x_3 + x_0.x_1.x_4 + x_0.x_1.x_2.x_4 + x_3.x_4 + x_0.x_3.x_4 + x_1.x_3.x_4 + x_0.x_2.x_3.x_4 + x_1.x_2.x_3.x_4 + x_5 + x_0.x_5 + x_0.x_1.x_5 + x_0.x_2.x_5 + x_1.x_2.x_5 + x_0.x_1.x_2.x_5 + x_2.x_3.x_5 + x_0.x_2.x_3.x_5 + x_1.x_2.x_3.x_5 + x_0.x_1.x_2.x_3.x_5 + x_0.x_4.x_5 + x_1.x_4.x_5 + x_0.x_1.x_4.x_5 + x_2.x_4.x_5 + x_0.x_1.x_2.x_4.x_5 + x_3.x_4.x_5 + x_0.x_3.x_4.x_5 + x_1.x_3.x_4.x_5 + x_2.x_3.x_4.x_5 + x_1.x_2.x_3.x_4.x_5 + x_0.x_6 + x_0.x_1.x_6 + x_2.x_6 + x_0.x_2.x_6 + x_0.x_1.x_2.x_6 + x_3.x_6 + x_0.x_3.x_6 + x_2.x_3.x_6 + x_1.x_2.x_3.x_6 + x_0.x_4.x_6 + x_1.x_4.x_6 + x_0.x_1.x_4.x_6 + x_0.x_2.x_4.x_6 + x_1.x_2.x_4.x_6 + x_3.x_4.x_6 + x_0.x_1.x_3.x_4.x_6 + x_1.x_2.x_3.x_4.x_6 + x_0.x_1.x_2.x_3.x_4.x_6 + x_5.x_6 + x_1.x_5.x_6 + x_0.x_1.x_5.x_6 + x_0.x_2.x_5.x_6 + x_3.x_5.x_6 + x_0.x_1.x_3.x_5.x_6 + x_2.x_3.x_5.x_6 + x_0.x_1.x_2.x_3.x_5.x_6 + x_0.x_4.x_5.x_6 + x_0.x_1.x_4.x_5.x_6 + x_0.x_2.x_4.x_5.x_6 + x_3.x_4.x_5.x_6 + x_0.x_2.x_3.x_4.x_5.x_6 + x_1.x_2.x_3.x_4.x_5.x_6 + x_0.x_1.x_2.x_3.x_4.x_5.x_6 + x_7 + x_0.x_7 + x_0.x_1.x_7 + x_3.x_7 + x_2.x_3.x_7 + x_1.x_2.x_3.x_7 + x_0.x_1.x_2.x_3.x_7 + x_4.x_7 + x_0.x_4.x_7 + x_1.x_4.x_7 + x_0.x_1.x_4.x_7 + x_2.x_4.x_7 + x_0.x_2.x_4.x_7 + x_1.x_2.x_4.x_7 + x_3.x_4.x_7 + x_0.x_3.x_4.x_7 + x_0.x_1.x_3.x_4.x_7 + x_2.x_3.x_4.x_7 + x_0.x_2.x_3.x_4.x_7 + x_1.x_2.x_3.x_4.x_7 + x_5.x_7 + x_1.x_5.x_7 + x_0.x_1.x_5.x_7 + x_0.x_2.x_5.x_7 + x_3.x_5.x_7 + x_1.x_3.x_5.x_7 + x_4.x_5.x_7 + x_0.x_4.x_5.x_7 + x_1.x_4.x_5.x_7 + x_0.x_2.x_4.x_5.x_7 + x_3.x_4.x_5.x_7 + x_0.x_3.x_4.x_5.x_7 + x_0.x_1.x_3.x_4.x_5.x_7 + x_1.x_2.x_3.x_4.x_5.x_7 + x_0.x_1.x_2.x_3.x_4.x_5.x_7 + x_6.x_7 + x_0.x_6.x_7 + x_2.x_6.x_7 + x_0.x_2.x_6.x_7 + x_0.x_2.x_3.x_6.x_7 + x_1.x_2.x_3.x_6.x_7 + x_0.x_1.x_2.x_3.x_6.x_7 + x_2.x_4.x_6.x_7 + x_0.x_2.x_4.x_6.x_7 + x_0.x_1.x_2.x_4.x_6.x_7 + x_3.x_4.x_6.x_7 + x_0.x_3.x_4.x_6.x_7 + x_1.x_3.x_4.x_6.x_7 + x_0.x_1.x_3.x_4.x_6.x_7 + x_1.x_2.x_3.x_4.x_6.x_7 + x_0.x_1.x_2.x_3.x_4.x_6.x_7 + x_0.x_5.x_6.x_7 + x_1.x_5.x_6.x_7 + x_0.x_1.x_5.x_6.x_7 + x_2.x_5.x_6.x_7 + x_0.x_2.x_5.x_6.x_7 + x_1.x_3.x_5.x_6.x_7 + x_0.x_2.x_3.x_5.x_6.x_7 + x_0.x_1.x_2.x_3.x_5.x_6.x_7 + x_0.x_1.x_4.x_5.x_6.x_7 + x_0.x_2.x_4.x_5.x_6.x_7 + x_0.x_1.x_2.x_4.x_5.x_6.x_7 + x_0.x_3.x_4.x_5.x_6.x_7 + x_1.x_3.x_4.x_5.x_6.x_7 + x_0.x_2.x_3.x_4.x_5.x_6.x_7$.

C.2.1 Examples of Real S-boxes

This section lists several examples of S-boxes used in cryptographic primitives. The S-boxes are depicted in tabular form. Most of the S-boxes are bijective mappings. The inverse S-boxes are not always presented but can be easily computed: for an S-box $S : \mathbb{Z}_2^n \to \mathbb{Z}_2^n$, if $S[x] = y$ then $S^{-1}[y] = x$ for all $x, y \in \mathbb{Z}_2^n$.

Table C.5 The 3 × 3 S-boxes of CTC, 3WAY (BASEKING), PANAMA (StepRightUp and RadioGatún), PRINTcipher.

Cipher	x							
	0	1	2	3	4	5	6	7
CTC[x]	7	6	0	4	2	5	1	3
3WAY[x]	7	2	4	5	1	6	3	0
PANAMA[x]	7	4	1	6	2	3	5	0
PRINTcipher[x]	0	1	3	6	7	4	5	2

Table C.6 The 4 × 4 S-boxes of HAMSI (S-box S_2 of SERPENT), PRESENT (and LED), JH (two S-boxes), NOEKEON and SC2000.

Cipher	x															
	0	1	2	3	4	5	6	7	8	9	10	11	12	13	14	15
HAMSI[x]	8	6	7	9	3	12	10	15	13	1	14	4	0	11	5	2
PRESENT[x]	12	5	6	11	9	0	10	13	3	14	15	8	4	7	1	2
$JH_1[x]$	9	0	4	11	13	12	3	15	1	10	2	6	7	5	8	14
$JH_2[x]$	3	12	6	13	5	7	1	9	15	2	0	4	11	10	14	8
NOEKEON[x]	7	10	2	12	4	8	15	0	5	9	1	14	3	13	11	6
SC2000[x]	2	5	10	12	7	15	1	11	13	6	0	9	4	8	3	14

Table C.7 The eight 4 × 4 SERPENT S-boxes.

S-box	x															
	0	1	2	3	4	5	6	7	8	9	10	11	12	13	14	15
$S_0[x]$	3	8	15	1	10	6	5	11	14	13	4	2	7	0	9	12
$S_1[x]$	15	12	2	7	9	0	5	10	1	11	14	8	6	13	3	4
$S_2[x]$	8	6	7	9	3	12	10	15	13	1	14	4	0	11	5	2
$S_3[x]$	0	15	11	8	12	9	6	3	13	1	2	4	10	7	5	14
$S_4[x]$	1	15	8	3	12	0	11	6	2	5	4	10	9	14	7	13
$S_5[x]$	15	5	2	11	4	10	9	12	0	3	14	8	13	6	7	1
$S_6[x]$	7	2	12	5	8	4	6	11	14	9	1	15	13	3	10	0
$S_7[x]$	1	13	15	0	14	8	2	11	7	4	12	10	9	3	5	6

Table C.8 The 5 × 5 KECCAK S-box and its inverse.

i	0	1	2	3	4	5	6	7	8	9	10	11	12	13	14	15
S[i]	0	5	10	11	20	17	22	23	9	12	3	2	13	8	15	14
i	16	17	18	19	20	21	22	23	24	25	26	27	28	29	30	31
S[i]	18	21	24	27	6	1	4	7	26	29	16	19	30	25	28	31
i	0	1	2	3	4	5	6	7	8	9	10	11	12	13	14	15
$S^{-1}[i]$	0	21	11	10	22	1	20	23	13	8	2	3	9	12	15	14
i	16	17	18	19	20	21	22	23	24	25	26	27	28	29	30	31
$S^{-1}[i]$	26	5	16	27	4	17	6	7	18	29	24	19	30	25	28	31

Table C.9 The 7×7 MISTY1 S-box. For $0 \leq i \leq 7$, $0 \leq j \leq 15$, the input is $(i \ll 3)|j \in \mathbb{Z}_2^7$ and the output is $S[(i \ll 3)|j] \in \mathbb{Z}_2^7$. Values in hexadecimal.

i\j	0	1	2	3	4	5	6	7	8	9	10	11	12	13	14	15
0	$1b_x$	32_x	33_x	$5a_x$	$3b_x$	10_x	17_x	54_x	$5b_x$	$1a_x$	72_x	73_x	$6b_x$	$2c_x$	66_x	49_x
1	$1f_x$	24_x	13_x	$6c_x$	37_x	$2e_x$	$3f_x$	$4a_x$	$5d_x$	$0f_x$	40_x	56_x	25_x	51_x	$1c_x$	04_x
2	$0b_x$	46_x	20_x	$0d_x$	$7b_x$	35_x	44_x	42_x	$2b_x$	$1e_x$	41_x	14_x	$4b_x$	79_x	15_x	$6f_x$
3	$0e_x$	55_x	09_x	36_x	74_x	$0c_x$	67_x	53_x	28_x	$0a_x$	$7e_x$	38_x	02_x	07_x	60_x	29_x
4	19_x	12_x	65_x	$2f_x$	30_x	39_x	08_x	68_x	$5f_x$	78_x	$2a_x$	$4c_x$	64_x	45_x	75_x	$3d_x$
5	59_x	48_x	03_x	57_x	$7c_x$	$4f_x$	62_x	$3c_x$	$1d_x$	21_x	$5e_x$	27_x	$6a_x$	70_x	$4d_x$	$3a_x$
6	01_x	$6d_x$	$6e_x$	63_x	18_x	77_x	23_x	05_x	26_x	76_x	00_x	31_x	$2d_x$	$7a_x$	$7f_x$	61_x
7	50_x	22_x	11_x	06_x	47_x	16_x	52_x	$4e_x$	71_x	$3e_x$	69_x	43_x	34_x	$5c_x$	58_x	$7d_x$

Table C.10 The 9 × 9 MISTY1 S-box. For $0 \leq i \leq 31$, $0 \leq j \leq 15$, the input is $(i \ll 5)|j \in \mathbb{Z}_2^9$ and the output is $S[(i \ll 5)|j] \in \mathbb{Z}_2^9$. Values in hexadecimal.

i\j	0	1	2	3	4	5	6	7	8	9	10	11	12	13	14	15
0	$1c3_x$	$0cb_x$	153_x	$19f_x$	$1e3_x$	$0e9_x$	$0fb_x$	035_x	181_x	$0b9_x$	117_x	$1eb_x$	133_x	009_x	$02d_x$	$0d3_x$
1	$0c7_x$	$14a_x$	037_x	$07e_x$	$0eb_x$	164_x	193_x	$1d8_x$	$0a3_x$	$11e_x$	055_x	$02c_x$	$01d_x$	$1a2_x$	163_x	118_x
2	$14b_x$	152_x	$1d2_x$	$00f_x$	$02b_x$	030_x	$13a_x$	$0e5_x$	111_x	138_x	$18e_x$	063_x	$0e3_x$	$0c8_x$	$1f4_x$	$01b_x$
3	001_x	$09d_x$	$0f8_x$	$1a0_x$	$16d_x$	$1f3_x$	$01c_x$	146_x	$07d_x$	$0d1_x$	082_x	$1ea_x$	183_x	$12d_x$	$0f4_x$	$19e_x$
4	$1d3_x$	$0dd_x$	$1e2_x$	128_x	$1e0_x$	$0ec_x$	059_x	091_x	011_x	$12f_x$	026_x	$0dc_x$	$0b0_x$	$18c_x$	$10f_x$	$1f7_x$
5	$0e7_x$	$16c_x$	$0b6_x$	$0f9_x$	$0d8_x$	151_x	101_x	$14c_x$	103_x	$0b8_x$	154_x	$12b_x$	$1ae_x$	017_x	071_x	$00c_x$
6	047_x	058_x	$07f_x$	$1a4_x$	134_x	129_x	084_x	$15d_x$	$19d_x$	$1b2_x$	$1a3_x$	048_x	$07c_x$	051_x	$1ca_x$	023_x
7	$13d_x$	$1a7_x$	165_x	$03b_x$	042_x	$0da_x$	192_x	$0ce_x$	$0c1_x$	$06b_x$	$09f_x$	$1f1_x$	$12c_x$	184_x	$0fa_x$	196_x
8	$1e1_x$	169_x	$17d_x$	031_x	180_x	$10a_x$	094_x	$1da_x$	186_x	$13e_x$	$11c_x$	060_x	175_x	$1cf_x$	067_x	119_x
9	065_x	068_x	099_x	150_x	008_x	007_x	$17c_x$	$0b7_x$	024_x	019_x	$0de_x$	127_x	$0db_x$	$0e4_x$	$1a9_x$	052_x
10	109_x	090_x	$19c_x$	$1c1_x$	028_x	$1b3_x$	135_x	$16a_x$	176_x	$0df_x$	$1e5_x$	188_x	$0c5_x$	$16e_x$	$1de_x$	$1b1_x$
11	$0c3_x$	$1df_x$	036_x	$0ee_x$	$1ee_x$	$0f0_x$	093_x	049_x	$09a_x$	$1b6_x$	069_x	081_x	125_x	$00b_x$	$05e_x$	$0b4_x$
12	149_x	$1c7_x$	174_x	$03e_x$	$13b_x$	$1b7_x$	$08e_x$	$1c6_x$	$0ae_x$	010_x	095_x	$1ef_x$	$04e_x$	$0f2_x$	$1fd_x$	085_x
13	$0fd_x$	$0f6_x$	$0a0_x$	$16f_x$	083_x	$08a_x$	156_x	$09b_x$	$13c_x$	107_x	167_x	098_x	$1d0_x$	$1e9_x$	003_x	$1fe_x$
14	$0bd_x$	122_x	089_x	$0d2_x$	$18f_x$	012_x	033_x	$06a_x$	142_x	$0ed_x$	170_x	$11b_x$	$0e2_x$	$14f_x$	158_x	131_x
15	147_x	$05d_x$	113_x	$1cd_x$	079_x	161_x	$1a5_x$	179_x	$09e_x$	$1b4_x$	$0cc_x$	022_x	132_x	$01a_x$	$0e8_x$	004_x
16	187_x	$1ed_x$	197_x	039_x	$1bf_x$	$1d7_x$	027_x	$18b_x$	$0c6_x$	$09c_x$	$0d0_x$	$14e_x$	$06c_x$	034_x	$1f2_x$	$06e_x$
17	$0ca_x$	025_x	$0ba_x$	191_x	$0fe_x$	013_x	106_x	$02f_x$	$1ad_x$	172_x	$1db_x$	$0c0_x$	$10b_x$	$1d6_x$	$0f5_x$	$1ec_x$
18	$10d_x$	076_x	114_x	$1ab_x$	075_x	$10c_x$	$1e4_x$	159_x	054_x	$11f_x$	$04b_x$	$0c4_x$	$1be_x$	$0f7_x$	029_x	$0a4_x$
19	$00e_x$	$1f0_x$	077_x	$04d_x$	$17a_x$	086_x	$08b_x$	$0b3_x$	171_x	$0bf_x$	$10e_x$	104_x	097_x	$15b_x$	160_x	168_x
20	$0d7_x$	$0bb_x$	066_x	$1ce_x$	$0fc_x$	092_x	$1c5_x$	$06f_x$	016_x	$04a_x$	$0a1_x$	139_x	$0af_x$	$0f1_x$	190_x	$00a_x$
21	$1aa_x$	143_x	$17b_x$	056_x	$18d_x$	166_x	$0d4_x$	$1fb_x$	$14d_x$	194_x	$19a_x$	087_x	$1f8_x$	123_x	$0a7_x$	$1b8_x$
22	141_x	$03c_x$	$1f9_x$	140_x	$02a_x$	155_x	$11a_x$	$1a1_x$	198_x	$0d5_x$	126_x	$1af_x$	061_x	$12e_x$	157_x	$1dc_x$
23	072_x	$18a_x$	$0aa_x$	096_x	115_x	$0ef_x$	045_x	$07b_x$	$08d_x$	145_x	053_x	$05f_x$	178_x	$0b2_x$	$02e_x$	020_x
24	$1d5_x$	$03f_x$	$1c9_x$	$1e7_x$	$1ac_x$	044_x	038_x	014_x	$0b1_x$	$16b_x$	$0ab_x$	$0b5_x$	$05a_x$	182_x	$1c8_x$	$1d4_x$
25	018_x	177_x	064_x	$0cf_x$	$06d_x$	100_x	199_x	130_x	$15a_x$	005_x	120_x	$1bb_x$	$1bd_x$	$0e0_x$	$04f_x$	$0d6_x$
26	$13f_x$	$1c4_x$	$12a_x$	015_x	006_x	$0ff_x$	$19b_x$	$0a6_x$	043_x	088_x	050_x	$15f_x$	$1e8_x$	121_x	073_x	$17e_x$
27	$0bc_x$	$0c2_x$	$0c9_x$	173_x	189_x	$1f5_x$	074_x	$1cc_x$	$1e6_x$	$1a8_x$	195_x	$01f_x$	041_x	$00d_x$	$1ba_x$	032_x
28	$03d_x$	$1d1_x$	080_x	$0a8_x$	057_x	$1b9_x$	162_x	148_x	$0d9_x$	105_x	062_x	$07a_x$	021_x	$1ff_x$	112_x	108_x
29	$1c0_x$	$0a9_x$	$11d_x$	$1b0_x$	$1a6_x$	$0cd_x$	$0f3_x$	$05c_x$	102_x	$05b_x$	$1d9_x$	144_x	$1f6_x$	$0ad_x$	$0a5_x$	$03a_x$
30	$1cb_x$	136_x	$17f_x$	046_x	$0e1_x$	$01e_x$	$1dd_x$	$0e6_x$	137_x	$1fa_x$	185_x	$08c_x$	$08f_x$	040_x	$1b5_x$	$0be_x$
31	078_x	000_x	$0ac_x$	110_x	$15e_x$	124_x	002_x	$1bc_x$	$0a2_x$	$0ea_x$	070_x	$1fc_x$	116_x	$15c_x$	$04c_x$	$1c2_x$

Table C.11 The eight 6×4 DES S-boxes. The input is denoted $(x_1x_2x_3x_4x_5x_6) \in \mathbb{Z}_2^6$. The output is $S_i[x_1x_6][x_2x_3x_4x_5] \in \mathbb{Z}_2^4$.

		$x_2x_3x_4x_5$															
S_1		0	1	2	3	4	5	6	7	8	9	10	11	12	13	14	15
	0	14	4	13	1	2	15	11	8	3	10	6	12	5	9	0	7
x_1x_6	1	0	15	7	4	14	2	13	1	10	6	12	11	9	5	3	8
	2	4	1	14	8	13	6	2	11	15	12	9	7	3	10	5	0
	3	15	12	8	2	4	9	1	7	5	11	3	14	10	0	6	13
S_2		0	1	2	3	4	5	6	7	8	9	10	11	12	13	14	15
	0	15	1	8	14	6	11	3	4	9	7	2	13	12	0	5	10
x_1x_6	1	3	13	4	7	15	2	8	14	12	0	1	10	6	9	11	5
	2	0	14	7	11	10	4	13	1	5	8	12	6	9	3	2	15
	3	13	8	10	1	3	15	4	2	11	6	7	12	0	5	14	9
S_3		0	1	2	3	4	5	6	7	8	9	10	11	12	13	14	15
	0	10	0	9	14	6	3	15	5	1	13	12	7	11	4	2	8
x_1x_6	1	13	7	0	9	3	4	6	10	2	8	5	14	12	11	15	1
	2	13	6	4	9	8	15	3	0	11	1	2	12	5	10	14	7
	3	1	10	13	0	6	9	8	7	4	15	14	3	11	5	2	12
S_4		0	1	2	3	4	5	6	7	8	9	10	11	12	13	14	15
	0	7	13	14	3	0	6	9	10	1	2	8	5	11	12	4	15
x_1x_6	1	13	8	11	5	6	15	0	3	4	7	2	12	1	10	14	9
	2	10	6	9	0	12	11	7	13	15	1	3	14	5	2	8	4
	3	3	15	0	6	10	1	13	8	9	4	5	11	12	7	2	14
S_5		0	1	2	3	4	5	6	7	8	9	10	11	12	13	14	15
	0	2	12	4	1	7	10	11	6	8	5	3	15	13	0	14	9
x_1x_6	1	14	11	2	12	4	7	13	1	5	0	15	10	3	9	8	6
	2	4	2	1	11	10	13	7	8	15	9	12	5	6	3	0	14
	3	11	8	12	7	1	14	2	13	6	15	0	9	10	4	5	3
S_6		0	1	2	3	4	5	6	7	8	9	10	11	12	13	14	15
	0	12	1	10	15	9	2	6	8	0	13	3	4	14	7	5	11
x_1x_6	1	10	15	4	2	7	12	9	5	6	1	13	14	0	11	3	8
	2	9	14	15	5	2	8	12	3	7	0	4	10	1	13	11	6
	3	4	3	2	12	9	5	15	10	11	14	1	7	6	0	8	13
S_7		0	1	2	3	4	5	6	7	8	9	10	11	12	13	14	15
	0	4	11	2	14	15	0	8	13	3	12	9	7	5	10	6	1
x_1x_6	1	13	0	11	7	4	9	1	10	14	3	5	12	2	15	8	6
	2	1	4	11	13	12	3	7	14	10	15	6	8	0	5	9	2
	3	6	11	13	8	1	4	10	7	9	5	0	15	14	2	3	12
S_8		0	1	2	3	4	5	6	7	8	9	10	11	12	13	14	15
	0	13	2	8	4	6	15	11	1	10	9	3	14	5	0	12	7
x_1x_6	1	1	15	13	8	10	3	7	4	12	5	6	11	0	14	9	2
	2	7	11	4	1	9	12	14	2	0	6	10	13	15	3	5	8
	3	2	1	14	7	4	10	8	13	15	12	9	0	3	5	6	11

Table C.12 The eight 6×4 s^2-DES S-boxes. The input is denoted $(x_1x_2x_3x_4x_5x_6) \in \mathbb{Z}_2^6$. The output is $S_i[x_1x_6][x_2x_3x_4x_5] \in \mathbb{Z}_2^4$.

		$x_2x_3x_4x_5$															
S_1		0	1	2	3	4	5	6	7	8	9	10	11	12	13	14	15
x_1x_6	0	12	14	1	15	11	10	8	4	7	9	5	0	3	2	13	6
	1	3	5	4	12	9	14	0	8	2	7	10	1	13	6	15	11
	2	10	7	9	11	15	13	2	5	14	6	1	4	12	3	8	0
	3	6	0	5	8	14	7	1	3	12	4	2	13	11	15	10	9
S_2		0	1	2	3	4	5	6	7	8	9	10	11	12	13	14	15
x_1x_6	0	2	12	4	6	3	0	8	5	10	11	15	7	13	1	14	9
	1	14	8	3	11	9	13	10	2	5	0	1	6	7	12	15	4
	2	4	13	14	3	1	10	5	7	9	15	8	12	11	2	0	6
	3	0	11	1	15	4	5	12	14	7	10	9	13	6	8	2	3
S_3		0	1	2	3	4	5	6	7	8	9	10	11	12	13	14	15
x_1x_6	0	5	10	7	12	13	2	0	4	6	14	11	15	3	1	9	8
	1	3	13	6	14	2	0	15	12	1	5	10	7	4	11	8	9
	2	9	2	11	6	1	13	10	15	14	12	3	5	0	8	4	7
	3	1	8	14	11	5	15	9	3	7	6	4	0	12	10	13	2
S_4		0	1	2	3	4	5	6	7	8	9	10	11	12	13	14	15
x_1x_6	0	13	2	12	11	3	1	5	9	15	6	8	0	14	10	4	7
	1	9	1	14	5	11	7	12	4	8	15	0	6	3	2	13	10
	2	14	5	15	13	7	2	9	6	0	12	10	11	4	8	1	3
	3	6	10	5	3	2	11	14	0	7	4	1	8	9	13	15	12
S_5		0	1	2	3	4	5	6	7	8	9	10	11	12	13	14	15
x_1x_6	0	10	2	9	11	8	7	6	3	5	13	12	15	0	4	14	1
	1	2	1	0	5	11	8	15	7	12	9	14	6	3	13	10	4
	2	0	5	8	6	4	3	13	14	9	1	15	11	2	10	7	12
	3	7	14	11	13	12	2	10	9	1	8	0	4	5	6	3	15
S_6		0	1	2	3	4	5	6	7	8	9	10	11	12	13	14	15
x_1x_6	0	0	11	7	10	12	9	14	6	1	3	5	15	2	4	8	13
	1	3	1	4	5	0	12	8	7	10	2	14	13	6	9	15	11
	2	5	12	15	6	7	11	10	13	0	8	9	14	4	2	1	3
	3	4	8	10	11	6	5	7	1	14	15	12	2	3	13	9	0
S_7		0	1	2	3	4	5	6	7	8	9	10	11	12	13	14	15
x_1x_6	0	5	0	11	14	10	2	9	8	13	3	12	6	4	7	1	15
	1	10	12	13	4	9	1	3	0	6	8	5	15	14	11	2	7
	2	6	11	12	9	0	3	4	14	1	7	8	13	10	2	15	5
	3	9	4	7	0	3	11	2	1	15	5	6	8	12	13	10	14
S_8		0	1	2	3	4	5	6	7	8	9	10	11	12	13	14	15
x_1x_6	0	7	13	4	14	10	15	8	2	11	9	6	3	1	12	5	0
	1	6	14	10	12	5	7	0	1	2	13	11	4	8	9	15	3
	2	10	6	12	1	11	9	14	3	13	15	4	5	0	2	8	7
	3	5	3	15	6	0	1	13	9	4	10	14	8	12	7	11	2

Table C.13 The eight 6×4 s^3-DES S-boxes. The input is denoted $(x_1x_2x_3x_4x_5x_6) \in \mathbb{Z}_2^6$. The output is $S_i[x_1x_6][x_2x_3x_4x_5] \in \mathbb{Z}_2^4$.

		$x_2x_3x_4x_5$															
S_1		0	1	2	3	4	5	6	7	8	9	10	11	12	13	14	15
	0	15	8	3	14	4	2	9	5	0	11	10	1	13	7	6	12
x_1x_6	1	6	15	9	5	3	12	10	0	13	8	4	11	14	2	1	7
	2	9	14	5	8	2	4	15	3	10	7	6	13	1	11	12	0
	3	10	5	3	15	12	9	0	6	1	2	8	4	11	14	7	13
S_2		0	1	2	3	4	5	6	7	8	9	10	11	12	13	14	15
	0	13	14	0	3	10	4	7	9	11	8	12	6	1	15	2	5
x_1x_6	1	8	2	11	13	4	1	14	7	5	15	0	3	10	6	9	12
	2	14	9	3	10	0	7	13	4	8	5	6	15	11	12	1	2
	3	1	4	14	7	11	13	8	2	6	3	5	10	12	0	15	9
S_3		0	1	2	3	4	5	6	7	8	9	10	11	12	13	14	15
	0	13	3	11	5	14	8	0	6	4	15	1	12	7	2	10	9
x_1x_6	1	4	13	1	8	7	2	14	11	15	10	12	3	9	5	0	6
	2	6	5	8	11	13	14	3	0	9	2	4	1	10	7	15	12
	3	1	11	7	2	8	13	4	14	6	12	10	15	3	0	9	5
S_4		0	1	2	3	4	5	6	7	8	9	10	11	12	13	14	15
	0	9	0	7	11	12	5	10	6	15	3	1	14	2	8	4	13
x_1x_6	1	5	10	12	6	0	15	3	9	8	13	11	1	7	2	14	4
	2	10	7	9	12	5	0	6	11	3	14	4	2	8	13	15	1
	3	3	9	15	0	6	10	5	12	14	2	1	7	13	4	8	11
S_5		0	1	2	3	4	5	6	7	8	9	10	11	12	13	14	15
	0	5	15	9	10	0	3	14	4	2	12	7	1	13	6	8	11
x_1x_6	1	6	9	3	15	5	12	0	10	8	7	13	4	2	11	14	1
	2	15	0	10	9	3	5	4	14	8	11	1	7	6	12	13	2
	3	12	5	0	6	15	10	9	3	7	2	14	11	8	1	4	13
S_6		0	1	2	3	4	5	6	7	8	9	10	11	12	13	14	15
	0	4	3	7	10	9	0	14	13	15	5	12	6	2	11	1	8
x_1x_6	1	14	13	11	4	2	7	1	8	9	10	5	3	15	0	12	6
	2	13	0	10	9	4	3	7	14	1	15	6	12	8	5	11	2
	3	1	7	4	14	11	8	13	2	10	12	3	5	6	15	0	9
S_7		0	1	2	3	4	5	6	7	8	9	10	11	12	13	14	15
	0	4	10	15	12	2	9	1	6	11	5	0	3	7	14	13	8
x_1x_6	1	10	15	6	0	5	3	12	9	1	8	11	13	14	4	7	2
	2	2	12	9	6	15	10	4	1	5	11	3	0	8	7	14	13
	3	12	6	3	9	0	5	10	15	2	13	4	14	7	11	1	8
S_8		0	1	2	3	4	5	6	7	8	9	10	11	12	13	14	15
	0	13	10	0	7	3	9	14	4	2	15	12	1	5	6	11	8
x_1x_6	1	2	7	13	1	4	14	11	8	15	12	6	10	9	5	0	3
	2	4	13	14	0	9	3	7	10	1	8	2	11	15	5	12	6
	3	8	11	7	14	2	4	13	1	6	5	9	0	12	15	3	10

Table C.14 The eight 6 × 4 s^5-DES S-boxes. The input is denoted $(x_1x_2x_3x_4x_5x_6) \in \mathbb{Z}_2^6$. The output is $S_i[x_1x_6][x_2x_3x_4x_5] \in \mathbb{Z}_2^4$.

		$x_2x_3x_4x_5$															
S_1		0	1	2	3	4	5	6	7	8	9	10	11	12	13	14	15
x_1x_6	0	9	10	15	1	4	7	2	12	6	5	3	14	8	11	13	0
	1	2	13	8	4	11	1	14	7	12	3	15	9	5	6	0	10
	2	10	12	4	7	9	2	15	1	3	6	13	8	14	5	0	11
	3	4	11	1	13	14	7	8	2	10	0	6	3	9	12	15	5
S_2		0	1	2	3	4	5	6	7	8	9	10	11	12	13	14	15
x_1x_6	0	6	3	5	0	8	14	11	13	9	10	12	7	15	4	2	1
	1	9	6	10	12	15	0	5	3	4	1	7	11	2	13	14	8
	2	5	8	3	14	6	13	0	11	10	15	9	2	12	1	7	4
	3	6	3	15	9	0	10	12	5	13	8	2	4	11	7	1	14
S_3		0	1	2	3	4	5	6	7	8	9	10	11	12	13	14	15
x_1x_6	0	11	5	8	2	6	12	1	15	7	14	13	4	0	9	10	3
	1	7	8	1	14	11	2	13	4	12	3	6	9	5	15	0	10
	2	8	11	1	12	15	6	2	5	4	7	10	9	3	0	13	14
	3	13	2	4	7	1	11	14	8	10	9	15	0	12	6	3	5
S_4		0	1	2	3	4	5	6	7	8	9	10	11	12	13	14	15
x_1x_6	0	13	11	8	14	3	0	6	5	4	7	2	9	15	12	1	10
	1	10	0	3	5	15	6	12	9	1	13	4	14	8	11	2	7
	2	6	5	11	8	0	14	13	3	9	12	7	2	10	1	4	15
	3	9	12	5	15	6	3	0	10	7	11	2	8	13	4	14	1
S_5		0	1	2	3	4	5	6	7	8	9	10	11	12	13	14	15
x_1x_6	0	12	6	2	11	5	8	15	1	3	13	9	14	0	7	10	4
	1	15	0	12	5	3	6	9	10	4	11	2	8	14	1	7	13
	2	1	12	15	5	6	11	8	2	4	7	10	9	13	0	3	14
	3	6	3	10	0	9	12	5	15	13	4	1	14	7	11	8	2
S_6		0	1	2	3	4	5	6	7	8	9	10	11	12	13	14	15
x_1x_6	0	14	8	2	5	9	15	4	3	7	1	12	6	0	10	11	13
	1	1	13	11	8	2	4	7	14	10	6	0	15	5	9	12	3
	2	4	2	9	15	14	8	3	5	10	7	0	12	13	1	6	11
	3	8	11	7	4	13	1	14	2	5	0	9	10	6	15	3	12
S_7		0	1	2	3	4	5	6	7	8	9	10	11	12	13	14	15
x_1x_6	0	4	13	10	3	7	0	9	14	2	1	15	6	12	11	5	8
	1	9	0	15	10	12	6	5	3	14	7	1	13	11	8	2	4
	2	13	10	3	9	0	7	14	4	8	6	5	12	11	1	2	15
	3	10	3	12	6	5	9	0	15	4	8	11	1	14	7	13	2
S_8		0	1	2	3	4	5	6	7	8	9	10	11	12	13	14	15
x_1x_6	0	1	10	2	12	15	9	4	7	14	3	5	0	8	6	11	13
	1	14	13	7	11	2	4	1	8	0	10	9	6	5	15	12	3
	2	10	15	12	1	9	2	7	4	13	0	6	11	3	5	8	14
	3	4	8	1	2	7	11	13	14	10	5	15	12	0	6	3	9

Table C.15 The 8 × 8 AES S-box. For $0 \leq i, j \leq 15$, the $S[(i \ll 4)|j]$ entry is in the i-th row and j-th column. Values in hexadecimal.

i\j	0	1	2	3	4	5	6	7	8	9	10	11	12	13	14	15
0	63_x	$7c_x$	77_x	$7b_x$	$f2_x$	$6b_x$	$6f_x$	$c5_x$	30_x	01_x	67_x	$2b_x$	fe_x	$d7_x$	ab_x	76_x
1	ca_x	82_x	$c9_x$	$7d_x$	fa_x	59_x	47_x	$f0_x$	ad_x	$d4_x$	$a2_x$	af_x	$9c_x$	$a4_x$	72_x	$c0_x$
2	$b7_x$	fd_x	93_x	26_x	36_x	$3f_x$	$f7_x$	cc_x	34_x	$a5_x$	$e5_x$	$f1_x$	71_x	$d8_x$	31_x	15_x
3	04_x	$c7_x$	23_x	$c3_x$	18_x	96_x	05_x	$9a_x$	07_x	12_x	80_x	$e2_x$	eb_x	27_x	$b2_x$	75_x
4	09_x	83_x	$2c_x$	$1a_x$	$1b_x$	$6e_x$	$5a_x$	$a0_x$	52_x	$3b_x$	$d6_x$	$b3_x$	29_x	$e3_x$	$2f_x$	84_x
5	53_x	$d1_x$	00_x	ed_x	20_x	fc_x	$b1_x$	$5b_x$	$6a_x$	cb_x	be_x	39_x	$4a_x$	$4c_x$	58_x	cf_x
6	$d0_x$	ef_x	aa_x	fb_x	43_x	$4d_x$	33_x	85_x	45_x	$f9_x$	02_x	$7f_x$	50_x	$3c_x$	$9f_x$	$a8_x$
7	51_x	$a3_x$	40_x	$8f_x$	92_x	$9d_x$	38_x	$f5_x$	bc_x	$b6_x$	da_x	21_x	10_x	ff_x	$f3_x$	$d2_x$
8	cd_x	$0c_x$	13_x	ec_x	$5f_x$	97_x	44_x	17_x	$c4_x$	$a7_x$	$7e_x$	$3d_x$	64_x	$5d_x$	19_x	73_x
9	60_x	81_x	$4f_x$	dc_x	22_x	$2a_x$	90_x	88_x	46_x	ee_x	$b8_x$	14_x	de_x	$5e_x$	$0b_x$	db_x
10	$e0_x$	32_x	$3a_x$	$0a_x$	49_x	06_x	24_x	$5c_x$	$c2_x$	$d3_x$	ac_x	62_x	91_x	95_x	$e4_x$	79_x
11	$e7_x$	$c8_x$	37_x	$6d_x$	$8d_x$	$d5_x$	$4e_x$	$a9_x$	$6c_x$	56_x	$f4_x$	ea_x	65_x	$7a_x$	ae_x	08_x
12	ba_x	78_x	25_x	$2e_x$	$1c_x$	$a6_x$	$b4_x$	$c6_x$	$e8_x$	dd_x	74_x	$1f_x$	$4b_x$	bd_x	$8b_x$	$8a_x$
13	70_x	$3e_x$	$b5_x$	66_x	48_x	03_x	$f6_x$	$0e_x$	61_x	35_x	57_x	$b9_x$	86_x	$c1_x$	$1d_x$	$9e_x$
14	$e1_x$	$f8_x$	98_x	11_x	69_x	$d9_x$	$8e_x$	94_x	$9b_x$	$1e_x$	87_x	$e9_x$	ce_x	55_x	28_x	df_x
15	$8c_x$	$a1_x$	89_x	$0d_x$	bf_x	$e6_x$	42_x	68_x	41_x	99_x	$2d_x$	$0f_x$	$b0_x$	54_x	bb_x	16_x

Table C.16 The 8 × 8 ARIA S-box 1. For $0 \leq i, j \leq 15$, the $S[(i \ll 4)|j]$ entry is in the i-th row and j-th column. Values in hexadecimal.

i\j	0	1	2	3	4	5	6	7	8	9	10	11	12	13	14	15
0	63_x	$7c_x$	77_x	$7b_x$	$f2_x$	$6b_x$	$6f_x$	$c5_x$	30_x	01_x	67_x	$2b_x$	fe_x	$d7_x$	ab_x	76_x
1	ca_x	82_x	$c9_x$	$7d_x$	fa_x	59_x	47_x	$f0_x$	ad_x	$d4_x$	$a2_x$	af_x	$9c_x$	$a4_x$	72_x	$c0_x$
2	$b7_x$	fd_x	93_x	26_x	36_x	$3f_x$	$f7_x$	cc_x	34_x	$a5_x$	$e5_x$	$f1_x$	71_x	$d8_x$	31_x	15_x
3	04_x	$c7_x$	23_x	$c3_x$	18_x	96_x	05_x	$9a_x$	07_x	12_x	80_x	$e2_x$	eb_x	27_x	$b2_x$	75_x
4	09_x	83_x	$2c_x$	$1a_x$	$1b_x$	$6e_x$	$5a_x$	$a0_x$	52_x	$3b_x$	$d6_x$	$b3_x$	29_x	$e3_x$	$2f_x$	84_x
5	53_x	$d1_x$	00_x	ed_x	20_x	fc_x	$b1_x$	$5b_x$	$6a_x$	cb_x	be_x	39_x	$4a_x$	$4c_x$	58_x	cf_x
6	$d0_x$	ef_x	aa_x	fb_x	43_x	$4d_x$	33_x	85_x	45_x	$f9_x$	02_x	$7f_x$	50_x	$3c_x$	$9f_x$	$a8_x$
7	51_x	$a3_x$	40_x	$8f_x$	92_x	$9d_x$	38_x	$f5_x$	bc_x	$b6_x$	da_x	21_x	10_x	ff_x	$f3_x$	$d2_x$
8	cd_x	$0c_x$	13_x	ec_x	$5f_x$	97_x	44_x	17_x	$c4_x$	$a7_x$	$7e_x$	$3d_x$	64_x	$5d_x$	19_x	73_x
9	60_x	81_x	$4f_x$	dc_x	22_x	$2a_x$	90_x	88_x	46_x	ee_x	$b8_x$	14_x	de_x	$5e_x$	$0b_x$	db_x
10	$e0_x$	32_x	$3a_x$	$0a_x$	49_x	06_x	24_x	$5c_x$	$c2_x$	$d3_x$	ac_x	62_x	91_x	95_x	$e4_x$	79_x
11	$e7_x$	$c8_x$	37_x	$6d_x$	$8d_x$	$d5_x$	$4e_x$	$a9_x$	$6c_x$	56_x	$f4_x$	ea_x	65_x	$7a_x$	ae_x	08_x
12	ba_x	78_x	25_x	$2e_x$	$1c_x$	$a6_x$	$b4_x$	$c6_x$	$e8_x$	dd_x	74_x	$1f_x$	$4b_x$	bd_x	$8b_x$	$8a_x$
13	70_x	$3e_x$	$b5_x$	66_x	48_x	03_x	$f6_x$	$0e_x$	61_x	35_x	57_x	$b9_x$	86_x	$c1_x$	$1d_x$	$9e_x$
14	$e1_x$	$f8_x$	98_x	11_x	69_x	$d9_x$	$8e_x$	94_x	$9b_x$	$1e_x$	87_x	$e9_x$	ce_x	55_x	28_x	df_x
15	$8c_x$	$a1_x$	89_x	$0d_x$	bf_x	$e6_x$	42_x	68_x	41_x	99_x	$2d_x$	$0f_x$	$b0_x$	54_x	bb_x	16_x

Table C.17 The 8 × 8 ARIA S-box 2. For $0 \leq i, j \leq 15$, the $S[(i \ll 4)|j]$ entry is in the i-th row and j-th column. Values in hexadecimal.

i\j	0	1	2	3	4	5	6	7	8	9	10	11	12	13	14	15
0	$e2_x$	$4e_x$	54_x	fc_x	94_x	$c2_x$	$4a_x$	cc_x	62_x	$0d_x$	$6a_x$	46_x	$3c_x$	$4d_x$	$8b_x$	$d1_x$
1	$5e_x$	fa_x	64_x	cb_x	$b4_x$	97_x	be_x	$2b_x$	bc_x	77_x	$2e_x$	03_x	$d3_x$	19_x	59_x	$c1_x$
2	$1d_x$	06_x	41_x	$6b_x$	55_x	$f0_x$	99_x	69_x	ea_x	$9c_x$	18_x	ae_x	63_x	df_x	$e7_x$	bb_x
3	00_x	73_x	66_x	fb_x	96_x	$4c_x$	85_x	$e4_x$	$3a_x$	09_x	45_x	aa_x	$0f_x$	ee_x	10_x	eb_x
4	$2d_x$	$7f_x$	$f4_x$	29_x	ac_x	cf_x	ad_x	91_x	$8d_x$	78_x	$c8_x$	95_x	$f9_x$	$2f_x$	ce_x	cd_x
5	08_x	$7a_x$	88_x	38_x	$5c_x$	83_x	$2a_x$	28_x	47_x	db_x	$b8_x$	$c7_x$	93_x	$a4_x$	12_x	53_x
6	ff_x	87_x	$0e_x$	31_x	36_x	21_x	58_x	48_x	01_x	$8e_x$	37_x	74_x	32_x	ca_x	$e9_x$	$b1_x$
7	$b7_x$	ab_x	$0c_x$	$d7_x$	$c4_x$	56_x	42_x	26_x	07_x	98_x	60_x	$d9_x$	$b6_x$	$b9_x$	11_x	40_x
8	ec_x	20_x	$8c_x$	bd_x	$a0_x$	$c9_x$	84_x	04_x	49_x	23_x	$f1_x$	$4f_x$	50_x	$1f_x$	13_x	dc_x
9	$d8_x$	$c0_x$	$9e_x$	57_x	$e3_x$	$c3_x$	$7b_x$	65_x	$3b_x$	02_x	$8f_x$	$3e_x$	$e8_x$	25_x	92_x	$e5_x$
10	15_x	dd_x	fd_x	17_x	$a9_x$	bf_x	$d4_x$	$9a_x$	$7e_x$	$c5_x$	39_x	67_x	fe_x	76_x	$9d_x$	43_x
11	$a7_x$	$e1_x$	$d0_x$	$f5_x$	68_x	$f2_x$	$1b_x$	34_x	70_x	05_x	$a3_x$	$8a_x$	$d5_x$	79_x	86_x	$a8_x$
12	30_x	$c6_x$	51_x	$4b_x$	$1e_x$	$a6_x$	27_x	$f6_x$	35_x	$d2_x$	$6e_x$	24_x	16_x	82_x	$5f_x$	da_x
13	$e6_x$	75_x	$a2_x$	ef_x	$2c_x$	$b2_x$	$1c_x$	$9f_x$	$5d_x$	$6f_x$	80_x	$0a_x$	72_x	44_x	$9b_x$	$6c_x$
14	90_x	$0b_x$	$5b_x$	33_x	$7d_x$	$5a_x$	52_x	$f3_x$	61_x	$a1_x$	$f7_x$	$b0_x$	$d6_x$	$3f_x$	$7c_x$	$6d_x$
15	ed_x	14_x	$e0_x$	$a5_x$	$3d_x$	22_x	$b3_x$	$f8_x$	89_x	de_x	71_x	$1a_x$	af_x	ba_x	$b5_x$	81_x

Table C.18 The 8 × 8 CAMELLIA S-box. For $0 \leq i, j \leq 15$, the $S[(i \ll 4)|j]$ entry is in the i-th row and j-th column. Values in hexadecimal.

i\j	0	1	2	3	4	5	6	7	8	9	10	11	12	13	14	15
0	70_x	82_x	$2c_x$	ec_x	$b3_x$	27_x	$c0_x$	$e5_x$	$e4_x$	85_x	57_x	35_x	ea_x	$0c_x$	ae_x	41_x
1	23_x	ef_x	$6b_x$	93_x	45_x	19_x	$a5_x$	21_x	ed_x	$0e_x$	$4f_x$	$4e_x$	$1d_x$	65_x	92_x	bd_x
2	86_x	$b8_x$	af_x	$8f_x$	$7c_x$	eb_x	$1f_x$	ce_x	$3e_x$	30_x	dc_x	$5f_x$	$5e_x$	$c5_x$	$0b_x$	$1a_x$
3	$a6_x$	$e1_x$	39_x	ca_x	$d5_x$	47_x	$5d_x$	$3d_x$	$d9_x$	01_x	$5a_x$	$d6_x$	51_x	56_x	$6c_x$	$4d_x$
4	$8b_x$	$0d_x$	$9a_x$	66_x	fb_x	cc_x	$b0_x$	$2d_x$	74_x	12_x	$2b_x$	20_x	$f0_x$	$b1_x$	84_x	99_x
5	df_x	$4c_x$	cb_x	$c2_x$	34_x	$7e_x$	76_x	05_x	$6d_x$	$b7_x$	$a9_x$	31_x	$d1_x$	17_x	04_x	$d7_x$
6	14_x	58_x	$3a_x$	61_x	de_x	$1b_x$	11_x	$1c_x$	32_x	$0f_x$	$9c_x$	16_x	53_x	18_x	$f2_x$	22_x
7	fe_x	44_x	cf_x	$b2_x$	$c3_x$	$b5_x$	$7a_x$	91_x	24_x	08_x	$e8_x$	$a8_x$	60_x	fc_x	69_x	50_x
8	aa_x	$d0_x$	$a0_x$	$7d_x$	$a1_x$	89_x	62_x	97_x	54_x	$5b_x$	$1e_x$	95_x	$e0_x$	ff_x	64_x	$d2_x$
9	10_x	$c4_x$	00_x	48_x	$a3_x$	$f7_x$	75_x	db_x	$8a_x$	03_x	$e6_x$	da_x	09_x	$3f_x$	dd_x	94_x
10	87_x	$5c_x$	83_x	02_x	cd_x	$4a_x$	90_x	33_x	73_x	67_x	$f6_x$	$f3_x$	$9d_x$	$7f_x$	bf_x	$e2_x$
11	52_x	$9b_x$	$d8_x$	26_x	$c8_x$	37_x	$c6_x$	$3b_x$	81_x	96_x	$6f_x$	$4b_x$	13_x	be_x	63_x	$2e_x$
12	$e9_x$	79_x	$a7_x$	$8c_x$	$9f_x$	$6e_x$	bc_x	$8e_x$	29_x	$f5_x$	$f9_x$	$b6_x$	$2f_x$	fd_x	$b4_x$	59_x
13	78_x	98_x	06_x	$6a_x$	$e7_x$	46_x	71_x	ba_x	$d4_x$	25_x	ab_x	42_x	88_x	$a2_x$	$8d_x$	fa_x
14	72_x	07_x	$b9_x$	55_x	$f8_x$	ee_x	ac_x	$0a_x$	36_x	49_x	$2a_x$	68_x	$3c_x$	38_x	$f1_x$	$a4_x$
15	40_x	28_x	$d3_x$	$7b_x$	bb_x	$c9_x$	43_x	$c1_x$	15_x	$e3_x$	ad_x	$f4_x$	77_x	$c7_x$	80_x	$9e_x$

Table C.19 The 8×8 CS S-box. For $0 \leq i, j \leq 15$, the $S[(i \ll 4)|j]$ entry is in the i-th row and j-th column. Values in hexadecimal.

i\j	0	1	2	3	4	5	6	7	8	9	10	11	12	13	14	15
0	29_x	$0d_x$	61_x	40_x	$9c_x$	eb_x	$9e_x$	$8f_x$	$1f_x$	85_x	$5f_x$	58_x	$5b_x$	01_x	39_x	86_x
1	97_x	$2e_x$	$d7_x$	$d6_x$	35_x	ae_x	17_x	16_x	21_x	$b6_x$	69_x	$4e_x$	$a5_x$	72_x	87_x	08_x
2	$3c_x$	18_x	$e6_x$	$e7_x$	fa_x	ad_x	$b8_x$	89_x	$b7_x$	00_x	$f7_x$	$6f_x$	73_x	84_x	11_x	63_x
3	$3f_x$	96_x	$7f_x$	$6e_x$	bf_x	14_x	$9d_x$	ac_x	$a4_x$	$0e_x$	$7e_x$	$f6_x$	20_x	$4a_x$	62_x	30_x
4	03_x	$c5_x$	$4b_x$	$5a_x$	46_x	$a3_x$	44_x	65_x	$7d_x$	$4d_x$	$3d_x$	42_x	79_x	49_x	$1b_x$	$5c_x$
5	$f5_x$	$6c_x$	$b5_x$	94_x	54_x	ff_x	56_x	57_x	$0b_x$	$f4_x$	43_x	$0c_x$	$4f_x$	70_x	$6d_x$	$0a_x$
6	$e4_x$	02_x	$3e_x$	$2f_x$	$a2_x$	47_x	$e0_x$	$c1_x$	$d5_x$	$1a_x$	95_x	$a7_x$	51_x	$5e_x$	33_x	$2b_x$
7	$5d_x$	$d4_x$	$1d_x$	$2c_x$	ee_x	75_x	ec_x	dd_x	$7c_x$	$4c_x$	$a6_x$	$b4_x$	78_x	48_x	$3a_x$	32_x
8	98_x	af_x	$c0_x$	$e1_x$	$2d_x$	09_x	$0f_x$	$1e_x$	$b9_x$	27_x	$8a_x$	$e9_x$	bd_x	$e3_x$	$9f_x$	07_x
9	$b1_x$	ea_x	92_x	93_x	53_x	$6a_x$	31_x	10_x	80_x	$f2_x$	$d8_x$	$9b_x$	04_x	36_x	06_x	$8e_x$
10	be_x	$a9_x$	64_x	45_x	38_x	$1c_x$	$7a_x$	$6b_x$	$f3_x$	$a1_x$	$f0_x$	cd_x	37_x	25_x	15_x	81_x
11	fb_x	90_x	$e8_x$	$d9_x$	$7b_x$	52_x	19_x	28_x	26_x	88_x	fc_x	$d1_x$	$e2_x$	$8c_x$	$a0_x$	34_x
12	82_x	67_x	da_x	cb_x	$c7_x$	41_x	$e5_x$	$c4_x$	$c8_x$	ef_x	db_x	$c3_x$	cc_x	ab_x	ce_x	ed_x
13	$d0_x$	bb_x	$d3_x$	$d2_x$	71_x	68_x	13_x	12_x	$9a_x$	$b3_x$	$c2_x$	ca_x	de_x	77_x	dc_x	df_x
14	66_x	83_x	bc_x	$8d_x$	60_x	$c6_x$	22_x	23_x	$b2_x$	$8b_x$	91_x	05_x	76_x	cf_x	74_x	$c9_x$
15	aa_x	$f1_x$	99_x	$a8_x$	59_x	50_x	$3b_x$	$2a_x$	fe_x	$f9_x$	24_x	$b0_x$	ba_x	fd_x	$f8_x$	55_x

Table C.20 The 8×8 E2 S-box. For $0 \leq i, j \leq 15$, the $S[(i \ll 4)|j]$ entry is in the i-th row and j-th column. Values in hexadecimal.

i\j	0	1	2	3	4	5	6	7	8	9	10	11	12	13	14	15
0	$e1_x$	42_x	$3e_x$	81_x	$4e_x$	17_x	$9e_x$	fd_x	$b4_x$	$3f_x$	$2c_x$	da_x	31_x	$1e_x$	$e0_x$	41_x
1	cc_x	$f3_x$	82_x	$7d_x$	$7c_x$	12_x	$8e_x$	bb_x	$e4_x$	58_x	15_x	$d5_x$	$6f_x$	$e9_x$	$4c_x$	$4b_x$
2	35_x	$7b_x$	$5a_x$	$9a_x$	90_x	45_x	bc_x	$f8_x$	79_x	$d6_x$	$1b_x$	88_x	02_x	ab_x	cf_x	64_x
3	09_x	$0c_x$	$f0_x$	01_x	$a4_x$	$b0_x$	$f6_x$	93_x	43_x	63_x	86_x	dc_x	11_x	$a5_x$	83_x	$8b_x$
4	$c9_x$	$d0_x$	19_x	95_x	$6a_x$	$a1_x$	$5c_x$	24_x	$6e_x$	50_x	21_x	80_x	$2f_x$	$e7_x$	53_x	$0f_x$
5	91_x	22_x	04_x	ed_x	$a6_x$	48_x	49_x	67_x	ec_x	$f7_x$	$c0_x$	39_x	ce_x	$f2_x$	$2d_x$	be_x
6	$5d_x$	$1c_x$	$e3_x$	87_x	07_x	$0d_x$	$7a_x$	$f4_x$	fb_x	32_x	$f5_x$	$8c_x$	db_x	$8f_x$	25_x	96_x
7	$a8_x$	ea_x	cd_x	33_x	65_x	54_x	06_x	$8d_x$	89_x	$0a_x$	$5e_x$	$d9_x$	16_x	$0e_x$	71_x	$6c_x$
8	$0b_x$	ff_x	60_x	$d2_x$	$2e_x$	$d3_x$	$c8_x$	55_x	$c2_x$	23_x	$b7_x$	74_x	$e2_x$	$9b_x$	df_x	77_x
9	$2b_x$	$b9_x$	$3c_x$	62_x	13_x	$e5_x$	94_x	34_x	$b1_x$	27_x	84_x	$9f_x$	$d7_x$	51_x	00_x	61_x
10	ad_x	85_x	73_x	03_x	08_x	40_x	ef_x	68_x	fe_x	97_x	$1f_x$	de_x	af_x	66_x	$e8_x$	$b8_x$
11	ae_x	bd_x	$b3_x$	eb_x	$c6_x$	$6b_x$	47_x	$a9_x$	$d8_x$	$a7_x$	72_x	ee_x	$1d_x$	$7e_x$	aa_x	$b6_x$
12	75_x	cb_x	$d4_x$	30_x	69_x	20_x	$7f_x$	37_x	$5b_x$	$9d_x$	78_x	$a3_x$	$f1_x$	76_x	fa_x	05_x
13	$3d_x$	$3a_x$	44_x	57_x	$3b_x$	ca_x	$c7_x$	$8a_x$	18_x	46_x	$9c_x$	bf_x	ba_x	38_x	56_x	$1a_x$
14	92_x	$4d_x$	26_x	29_x	$a2_x$	98_x	10_x	99_x	70_x	$a0_x$	$c5_x$	28_x	$c1_x$	$6d_x$	14_x	ac_x
15	$f9_x$	$5f_x$	$4f_x$	$c4_x$	$c3_x$	$d1_x$	fc_x	dd_x	$b2_x$	59_x	$e6_x$	$b5_x$	36_x	52_x	$4a_x$	$2a_x$

Table C.21 The 8 × 8 FOX S-box. For $0 \leq i, j \leq 15$, the $S[(i \ll 4)|j]$ entry is in the i-th row and j-th column. Values in hexadecimal.

i\j	0	1	2	3	4	5	6	7	8	9	10	11	12	13	14	15
0	$5d_x$	de_x	00_x	$b7_x$	$d3_x$	ca_x	$3c_x$	$0d_x$	$c3_x$	$f8_x$	cb_x	$8d_x$	76_x	89_x	aa_x	12_x
1	88_x	22_x	$4f_x$	db_x	$6d_x$	47_x	$e4_x$	$4c_x$	78_x	$9a_x$	49_x	93_x	$c4_x$	$c0_x$	86_x	13_x
2	$a9_x$	20_x	53_x	$1c_x$	$4e_x$	cf_x	35_x	39_x	$b4_x$	$a1_x$	54_x	64_x	03_x	$c7_x$	85_x	$5c_x$
3	$5b_x$	cd_x	$d8_x$	72_x	96_x	42_x	$b8_x$	$e1_x$	$a2_x$	60_x	ef_x	bd_x	02_x	af_x	$8c_x$	73_x
4	$7c_x$	$7f_x$	$5e_x$	$f9_x$	65_x	$e6_x$	eb_x	ad_x	$5a_x$	$a5_x$	79_x	$8e_x$	15_x	30_x	ec_x	$a4_x$
5	$c2_x$	$3e_x$	$e0_x$	74_x	51_x	fb_x	$2d_x$	$6e_x$	94_x	$4d_x$	55_x	34_x	ae_x	52_x	$7e_x$	$9d_x$
6	$4a_x$	$f7_x$	80_x	$f0_x$	$d0_x$	90_x	$a7_x$	$e8_x$	$9f_x$	50_x	$d5_x$	$d1_x$	98_x	cc_x	$a0_x$	17_x
7	$f4_x$	$b6_x$	$c1_x$	28_x	$5f_x$	26_x	01_x	ab_x	25_x	38_x	82_x	$7d_x$	48_x	fc_x	$1b_x$	ce_x
8	$3f_x$	$6b_x$	$e2_x$	67_x	66_x	43_x	59_x	19_x	84_x	$3d_x$	$f5_x$	$2f_x$	$c9_x$	bc_x	$d9_x$	95_x
9	29_x	41_x	da_x	$1a_x$	$b0_x$	$e9_x$	69_x	$d2_x$	$7b_x$	$d7_x$	11_x	$9b_x$	33_x	$8a_x$	23_x	09_x
10	$d4_x$	71_x	44_x	68_x	$6f_x$	$f2_x$	$0e_x$	df_x	87_x	dc_x	83_x	18_x	$6a_x$	ee_x	99_x	81_x
11	62_x	36_x	$2e_x$	$7a_x$	fe_x	45_x	$9c_x$	75_x	91_x	$0c_x$	$0f_x$	$e7_x$	$f6_x$	14_x	63_x	$1d_x$
12	$0b_x$	$8b_x$	$b3_x$	$f3_x$	$b2_x$	$3b_x$	08_x	$4b_x$	10_x	$a6_x$	32_x	$b9_x$	$a8_x$	92_x	$f1_x$	56_x
13	dd_x	21_x	bf_x	04_x	be_x	$d6_x$	fd_x	77_x	ea_x	$3a_x$	$c8_x$	$8f_x$	57_x	$1e_x$	fa_x	$2b_x$
14	58_x	$c5_x$	27_x	ac_x	$e3_x$	ed_x	97_x	bb_x	46_x	05_x	40_x	31_x	$e5_x$	37_x	$2c_x$	$9e_x$
15	$0a_x$	$b1_x$	$b5_x$	06_x	$6c_x$	$1f_x$	$a3_x$	$2a_x$	70_x	ff_x	ba_x	07_x	24_x	16_x	$c6_x$	61_x

Table C.22 The 8 × 8 HIEROCRYPT-1/L3 S-box. For $0 \leq i, j \leq 15$, the $S[(i \ll 4)|j]$ entry is in the i-th row and j-th column. Values in hexadecimal.

i\j	0	1	2	3	4	5	6	7	8	9	10	11	12	13	14	15
0	07_x	fc_x	55_x	70_x	98_x	$8e_x$	84_x	$4e_x$	bc_x	75_x	ce_x	18_x	02_x	$e9_x$	$5d_x$	80_x
1	$1c_x$	60_x	78_x	42_x	$9d_x$	$2e_x$	$f5_x$	$e8_x$	$c6_x$	$7a_x$	$2f_x$	$a4_x$	$b2_x$	$5f_x$	19_x	87_x
2	$0b_x$	$9b_x$	$9c_x$	$d3_x$	$c3_x$	77_x	$3d_x$	$6f_x$	$b9_x$	$2d_x$	$4d_x$	$f7_x$	$8c_x$	$a7_x$	ac_x	17_x
3	$3c_x$	$5a_x$	41_x	$c9_x$	29_x	ed_x	de_x	27_x	69_x	30_x	72_x	$a8_x$	95_x	$3e_x$	$f9_x$	$d8_x$
4	21_x	$8b_x$	44_x	$d7_x$	11_x	$0d_x$	48_x	fd_x	$6a_x$	01_x	57_x	$e5_x$	bd_x	85_x	ec_x	$1e_x$
5	37_x	$9f_x$	$b5_x$	$9a_x$	$7c_x$	09_x	$f1_x$	$b1_x$	94_x	81_x	82_x	08_x	fb_x	$c0_x$	51_x	$0f_x$
6	61_x	$7f_x$	$1a_x$	56_x	96_x	13_x	$c1_x$	67_x	99_x	03_x	$5e_x$	$b6_x$	ca_x	fa_x	$9e_x$	df_x
7	$d6_x$	83_x	cc_x	$a2_x$	12_x	23_x	$b7_x$	65_x	$d0_x$	39_x	$7d_x$	$3b_x$	$d5_x$	$b0_x$	af_x	$1f_x$
8	06_x	$c8_x$	34_x	$c5_x$	$1b_x$	79_x	$4b_x$	66_x	bf_x	88_x	$4a_x$	$c4_x$	ef_x	58_x	$3f_x$	$0a_x$
9	$2c_x$	73_x	$d1_x$	$f8_x$	$6b_x$	$e6_x$	20_x	$b8_x$	22_x	43_x	$b3_x$	33_x	$e7_x$	$f0_x$	71_x	$7e_x$
10	52_x	89_x	47_x	63_x	$0e_x$	$6d_x$	$e3_x$	be_x	59_x	64_x	ee_x	$f6_x$	38_x	$5c_x$	$f4_x$	$5b_x$
11	49_x	$d4_x$	$e0_x$	$f3_x$	bb_x	54_x	26_x	$2b_x$	00_x	86_x	90_x	ff_x	fe_x	$a6_x$	$7b_x$	05_x
12	ad_x	68_x	$a1_x$	10_x	eb_x	$c7_x$	$e2_x$	$f2_x$	46_x	$8a_x$	$6c_x$	14_x	$6e_x$	cf_x	35_x	45_x
13	50_x	$d2_x$	92_x	74_x	93_x	$e1_x$	da_x	ae_x	$a9_x$	53_x	$e4_x$	40_x	cd_x	ba_x	97_x	$a3_x$
14	91_x	31_x	25_x	76_x	36_x	32_x	28_x	$3a_x$	24_x	$4c_x$	db_x	$d9_x$	$8d_x$	dc_x	62_x	$2a_x$
15	ea_x	15_x	dd_x	$c2_x$	$a5_x$	$0c_x$	04_x	$1d_x$	$8f_x$	cb_x	$b4_x$	$4f_x$	16_x	ab_x	aa_x	$a0_x$

Table C.23 The 8 × 8 SAFER K/SK/+/++ S-box. For $0 \leq i, j \leq 15$, the $S[(i \ll 4)|j]$ entry is in the i-th row and j-th column. Values in hexadecimal.

i\j	0	1	2	3	4	5	6	7	8	9	10	11	12	13	14	15
0	01_x	$2d_x$	$e2_x$	93_x	be_x	45_x	15_x	ae_x	78_x	03_x	87_x	$a4_x$	$b8_x$	38_x	cf_x	$3f_x$
1	08_x	67_x	09_x	94_x	eb_x	26_x	$a8_x$	$6b_x$	bd_x	18_x	34_x	$1b_x$	bb_x	bf_x	72_x	$f7_x$
2	40_x	35_x	48_x	$9c_x$	51_x	$2f_x$	$3b_x$	55_x	$e3_x$	$c0_x$	$9f_x$	$d8_x$	$d3_x$	$f3_x$	$8d_x$	$b1_x$
3	ff_x	$a7_x$	$3e_x$	dc_x	86_x	77_x	$d7_x$	$a6_x$	11_x	fb_x	$f4_x$	ba_x	92_x	91_x	64_x	83_x
4	$f1_x$	33_x	ef_x	da_x	$2c_x$	$b5_x$	$b2_x$	$2b_x$	88_x	$d1_x$	99_x	cb_x	$8c_x$	84_x	$1d_x$	14_x
5	81_x	97_x	71_x	ca_x	$5f_x$	$a3_x$	$8b_x$	57_x	$3c_x$	82_x	$c4_x$	52_x	$5c_x$	$1c_x$	$e8_x$	$a0_x$
6	04_x	$b4_x$	85_x	$4a_x$	$f6_x$	13_x	54_x	$b6_x$	df_x	$0c_x$	$1a_x$	$8e_x$	de_x	$e0_x$	39_x	fc_x
7	20_x	$9b_x$	24_x	$4e_x$	$a9_x$	98_x	$9e_x$	ab_x	$f2_x$	60_x	$d0_x$	$6c_x$	ea_x	fa_x	$c7_x$	$d9_x$
8	00_x	$d4_x$	$1f_x$	$6e_x$	43_x	bc_x	ec_x	53_x	89_x	fe_x	$7a_x$	$5d_x$	49_x	$c9_x$	32_x	$c2_x$
9	$f9_x$	$9a_x$	$f8_x$	$6d_x$	16_x	db_x	59_x	96_x	44_x	$e9_x$	cd_x	$e6_x$	46_x	42_x	$8f_x$	$0a_x$
10	$c1_x$	cc_x	$b9_x$	65_x	$b0_x$	$d2_x$	$c6_x$	ac_x	$1e_x$	41_x	62_x	29_x	$2e_x$	$0e_x$	74_x	50_x
11	02_x	$5a_x$	$c3_x$	25_x	$7b_x$	$8a_x$	$2a_x$	$5b_x$	$f0_x$	06_x	$0d_x$	47_x	$6f_x$	70_x	$9d_x$	$7e_x$
12	10_x	ce_x	12_x	27_x	$d5_x$	$4c_x$	$4f_x$	$d6_x$	79_x	30_x	68_x	36_x	75_x	$7d_x$	$e4_x$	ed_x
13	80_x	$6a_x$	90_x	37_x	$a2_x$	$5e_x$	76_x	aa_x	$c5_x$	$7f_x$	$3d_x$	af_x	$a5_x$	$e5_x$	19_x	61_x
14	fd_x	$4d_x$	$7c_x$	$b7_x$	$0b_x$	ee_x	ad_x	$4b_x$	22_x	$f5_x$	$e7_x$	73_x	23_x	21_x	$c8_x$	05_x
15	$e1_x$	66_x	dd_x	$b3_x$	58_x	69_x	63_x	56_x	$0f_x$	$a1_x$	31_x	95_x	17_x	07_x	$3a_x$	28_x

Table C.24 The 8 × 8 SHARK S-box. For $0 \leq i, j \leq 15$, the $S[(i \ll 4)|j]$ entry is in the i-th row and j-th column. Values in hexadecimal.

i\j	0	1	2	3	4	5	6	7	8	9	10	11	12	13	14	15
0	$b1_x$	ce_x	$c3_x$	95_x	$5a_x$	ad_x	$e7_x$	02_x	$4d_x$	44_x	fb_x	91_x	$0c_x$	87_x	$a1_x$	50_x
1	cb_x	67_x	54_x	dd_x	46_x	$8f_x$	$e1_x$	$4e_x$	$f0_x$	fd_x	fc_x	eb_x	$f9_x$	$c4_x$	$1a_x$	$6e_x$
2	$5e_x$	$f5_x$	cc_x	$8d_x$	$1c_x$	56_x	43_x	fe_x	07_x	61_x	$f8_x$	75_x	59_x	ff_x	03_x	22_x
3	$8a_x$	$d1_x$	13_x	ee_x	88_x	00_x	$0e_x$	34_x	15_x	80_x	94_x	$e3_x$	ed_x	$b5_x$	53_x	23_x
4	$4b_x$	47_x	17_x	$a7_x$	90_x	35_x	ab_x	$d8_x$	$b8_x$	df_x	$4f_x$	57_x	$9a_x$	92_x	db_x	$1b_x$
5	$3c_x$	$c8_x$	99_x	04_x	$8e_x$	$e0_x$	$d7_x$	$7d_x$	85_x	bb_x	40_x	$2c_x$	$3a_x$	45_x	$f1_x$	42_x
6	65_x	20_x	41_x	18_x	72_x	25_x	93_x	70_x	36_x	05_x	$f2_x$	$0b_x$	$a3_x$	79_x	ec_x	08_x
7	27_x	31_x	32_x	$b6_x$	$7c_x$	$b0_x$	$0a_x$	73_x	$5b_x$	$7b_x$	$b7_x$	81_x	$d2_x$	$0d_x$	$6a_x$	26_x
8	$9e_x$	58_x	$9c_x$	83_x	74_x	$b3_x$	ac_x	30_x	$7a_x$	69_x	77_x	$0f_x$	ae_x	21_x	de_x	$d0_x$
9	$2e_x$	97_x	10_x	$a4_x$	98_x	$a8_x$	$d4_x$	68_x	$2d_x$	62_x	29_x	$6d_x$	16_x	49_x	76_x	$c7_x$
10	$e8_x$	$c1_x$	96_x	37_x	$e5_x$	ca_x	$f4_x$	$e9_x$	63_x	12_x	$c2_x$	$a6_x$	14_x	bc_x	$d3_x$	28_x
11	af_x	$2f_x$	$e6_x$	24_x	52_x	$c6_x$	$a0_x$	09_x	bd_x	$8c_x$	cf_x	$5d_x$	11_x	$5f_x$	01_x	$c5_x$
12	$9f_x$	$3d_x$	$a2_x$	$9b_x$	$c9_x$	$3b_x$	be_x	51_x	19_x	$1f_x$	$3f_x$	$5c_x$	$b2_x$	ef_x	$4a_x$	cd_x
13	bf_x	ba_x	$6f_x$	64_x	$d9_x$	$f3_x$	$3e_x$	$b4_x$	aa_x	dc_x	$d5_x$	06_x	$c0_x$	$7e_x$	$f6_x$	66_x
14	$6c_x$	84_x	71_x	38_x	$b9_x$	$1d_x$	$7f_x$	$9d_x$	48_x	$8b_x$	$2a_x$	da_x	$a5_x$	33_x	82_x	39_x
15	$d6_x$	78_x	86_x	fa_x	$e4_x$	$2b_x$	$a9_x$	$1e_x$	89_x	60_x	$6b_x$	ea_x	55_x	$4c_x$	$f7_x$	$e2_x$

Table C.25 The 8×8 SMS4 S-box. For $0 \le i, j \le 15$, the $S[(i \ll 4)|j]$ entry is in the i-th row and j-th column. Values in hexadecimal.

i\j	0	1	2	3	4	5	6	7	8	9	10	11	12	13	14	15
0	$d6_x$	90_x	$e9_x$	fe_x	cc_x	$e1_x$	$3d_x$	$b7_x$	16_x	$b6_x$	14_x	$c2_x$	28_x	fb_x	$2c_x$	05_x
1	$2b_x$	67_x	$9a_x$	76_x	$2a_x$	be_x	04_x	$c3_x$	aa_x	44_x	13_x	26_x	49_x	86_x	06_x	99_x
2	$9c_x$	42_x	50_x	$f4_x$	91_x	ef_x	98_x	$7a_x$	33_x	54_x	$0b_x$	43_x	ed_x	cf_x	ac_x	62_x
3	$e4_x$	$b3_x$	$1c_x$	$a9_x$	$c9_x$	08_x	$e8_x$	95_x	80_x	df_x	94_x	fa_x	75_x	$8f_x$	$3f_x$	$a6_x$
4	47_x	07_x	$a7_x$	fc_x	$f3_x$	73_x	17_x	ba_x	83_x	59_x	$3c_x$	19_x	$e6_x$	85_x	$4f_x$	$a8_x$
5	68_x	$6b_x$	81_x	$b2_x$	71_x	64_x	da_x	$8b_x$	$f8_x$	eb_x	$0f_x$	$4b_x$	70_x	56_x	$9d_x$	35_x
6	$1e_x$	24_x	$0e_x$	$5e_x$	63_x	58_x	$d1_x$	$a2_x$	25_x	22_x	$7c_x$	$3b_x$	01_x	21_x	78_x	87_x
7	$d4_x$	00_x	46_x	57_x	$9f_x$	$d3_x$	27_x	52_x	$4c_x$	36_x	02_x	$e7_x$	$a0_x$	$c4_x$	$c8_x$	$9e_x$
8	ea_x	bf_x	$8a_x$	$d2_x$	40_x	$c7_x$	38_x	$b5_x$	$a3_x$	$f7_x$	$f2_x$	ce_x	$f9_x$	61_x	15_x	$a1_x$
9	$e0_x$	ae_x	$5d_x$	$a4_x$	$9b_x$	34_x	$1a_x$	55_x	ad_x	93_x	32_x	30_x	$f5_x$	$8c_x$	$b1_x$	$e3_x$
10	$1d_x$	$f6_x$	$e2_x$	$2e_x$	82_x	66_x	ca_x	60_x	$c0_x$	29_x	23_x	ab_x	$0d_x$	53_x	$4e_x$	$6f_x$
11	$d5_x$	db_x	37_x	45_x	de_x	fd_x	$8e_x$	$2f_x$	03_x	ff_x	$6a_x$	72_x	$6d_x$	$6c_x$	$5b_x$	51_x
12	$8d_x$	$1b_x$	af_x	92_x	bb_x	dd_x	bc_x	$7f_x$	11_x	$d9_x$	$5c_x$	41_x	$1f_x$	10_x	$5a_x$	$d8_x$
13	$0a_x$	$c1_x$	31_x	88_x	$a5_x$	cd_x	$7b_x$	bd_x	$2d_x$	74_x	$d0_x$	12_x	$b8_x$	$e5_x$	$b4_x$	$b0_x$
14	89_x	69_x	97_x	$4a_x$	$0c_x$	96_x	77_x	$7e_x$	65_x	$b9_x$	$f1_x$	09_x	$c5_x$	$6e_x$	$c6_x$	84_x
15	18_x	$f0_x$	$7d_x$	ec_x	$3a_x$	dc_x	$4d_x$	20_x	79_x	ee_x	$5f_x$	$3e_x$	$d7_x$	cb_x	39_x	48_x

Table C.26 The 8×8 KHAZAD (tweaked) S-box. For $0 \le i, j \le 15$, the $S[(i \ll 4)|j]$ entry is in the i-th row and j-th column. Values in hexadecimal.

i\j	0	1	2	3	4	5	6	7	8	9	10	11	12	13	14	15
0	ba_x	54_x	$2f_x$	74_x	53_x	$d3_x$	$d2_x$	$4d_x$	50_x	ac_x	$8d_x$	bf_x	70_x	52_x	$9a_x$	$4c_x$
1	ea_x	$d5_x$	97_x	$d1_x$	33_x	51_x	$5b_x$	$a6_x$	de_x	48_x	$a8_x$	99_x	db_x	32_x	$b7_x$	fc_x
2	$e3_x$	$9e_x$	91_x	$9b_x$	$e2_x$	bb_x	41_x	$6e_x$	$a5_x$	cb_x	$6b_x$	95_x	$a1_x$	$f3_x$	$b1_x$	02_x
3	cc_x	$c4_x$	$1d_x$	14_x	$c3_x$	63_x	da_x	$5d_x$	$5f_x$	dc_x	$7d_x$	cd_x	$7f_x$	$5a_x$	$6c_x$	$5c_x$
4	$f7_x$	26_x	ff_x	ed_x	$e8_x$	$9d_x$	$6f_x$	$8e_x$	19_x	$a0_x$	$f0_x$	89_x	$0f_x$	07_x	af_x	fb_x
5	08_x	15_x	$0d_x$	04_x	01_x	64_x	df_x	76_x	79_x	dd_x	$3d_x$	16_x	$3f_x$	37_x	$6d_x$	38_x
6	$b9_x$	73_x	$e9_x$	35_x	55_x	71_x	$7b_x$	$8c_x$	72_x	88_x	$f6_x$	$2a_x$	$3e_x$	$5e_x$	27_x	46_x
7	$0c_x$	65_x	68_x	61_x	03_x	$c1_x$	57_x	$d6_x$	$d9_x$	58_x	$d8_x$	66_x	$d7_x$	$3a_x$	$c8_x$	$3c_x$
8	fa_x	96_x	$a7_x$	98_x	ec_x	$b8_x$	$c7_x$	ae_x	69_x	$4b_x$	ab_x	$a9_x$	67_x	$0a_x$	47_x	$f2_x$
9	$b5_x$	22_x	$e5_x$	ee_x	be_x	$2b_x$	81_x	12_x	83_x	$1b_x$	$0e_x$	23_x	$f5_x$	45_x	21_x	ce_x
10	49_x	$2c_x$	$f9_x$	$e6_x$	$b6_x$	28_x	17_x	82_x	$1a_x$	$8b_x$	fe_x	$8a_x$	09_x	$c9_x$	87_x	$4e_x$
11	$e1_x$	$2e_x$	$e4_x$	$e0_x$	eb_x	90_x	$a4_x$	$1e_x$	85_x	60_x	00_x	25_x	$f4_x$	$f1_x$	94_x	$0b_x$
12	$e7_x$	75_x	ef_x	34_x	31_x	$d4_x$	$d0_x$	86_x	$7e_x$	ad_x	fd_x	29_x	30_x	$3b_x$	$9f_x$	$f8_x$
13	$c6_x$	13_x	06_x	05_x	$c5_x$	11_x	77_x	$7c_x$	$7a_x$	78_x	36_x	$1c_x$	39_x	59_x	18_x	56_x
14	$b3_x$	$b0_x$	24_x	20_x	$b2_x$	92_x	$a3_x$	$c0_x$	44_x	62_x	10_x	$b4_x$	84_x	43_x	93_x	$c2_x$
15	$4a_x$	bd_x	$8f_x$	$2d_x$	bc_x	$9c_x$	$6a_x$	40_x	cf_x	$a2_x$	80_x	$4f_x$	$1f_x$	ca_x	aa_x	42_x

Table C.27 The 8 × 8 RAINBOW S-box. For $0 \leq i, j \leq 15$, the $S[(i \ll 4)|j]$ entry is in the i-th row and j-th column. Values in hexadecimal.

i\j	0	1	2	3	4	5	6	7	8	9	10	11	12	13	14	15
0	00_x	$0e_x$	$1c_x$	08_x	38_x	$e5_x$	10_x	19_x	70_x	16_x	cb_x	42_x	20_x	$e7_x$	32_x	$d4_x$
1	$e0_x$	cc_x	$2c_x$	65_x	97_x	$a7_x$	84_x	$1f_x$	40_x	67_x	cf_x	78_x	64_x	$2d_x$	$a9_x$	be_x
2	$c1_x$	$c2_x$	99_x	ec_x	58_x	$d1_x$	ca_x	fb_x	$2f_x$	$8e_x$	$4f_x$	$6d_x$	09_x	50_x	$3e_x$	$2a_x$
3	80_x	56_x	ce_x	11_x	$9f_x$	$0c_x$	$f0_x$	$a4_x$	$c8_x$	df_x	$5a_x$	$b1_x$	53_x	73_x	$7d_x$	$6f_x$
4	83_x	79_x	85_x	$f9_x$	33_x	$e9_x$	$d9_x$	$4b_x$	$b0_x$	74_x	$a3_x$	14_x	95_x	03_x	$f7_x$	dc_x
5	$5e_x$	$7a_x$	$1d_x$	$c0_x$	$9e_x$	55_x	da_x	26_x	12_x	$6b_x$	$a0_x$	$d5_x$	$7c_x$	98_x	54_x	72_x
6	01_x	48_x	ac_x	$0f_x$	$9d_x$	ad_x	22_x	36_x	$3f_x$	82_x	18_x	ba_x	$e1_x$	57_x	49_x	$2e_x$
7	91_x	$f1_x$	bf_x	$4a_x$	$b4_x$	62_x	63_x	ee_x	$a6_x$	51_x	$e6_x$	71_x	fa_x	$c9_x$	de_x	43_x
8	07_x	04_x	$f2_x$	$8c_x$	$0b_x$	21_x	$f3_x$	$6a_x$	66_x	$b2_x$	$d3_x$	$8f_x$	$b3_x$	$3c_x$	96_x	$5f_x$
9	61_x	76_x	$e8_x$	fd_x	47_x	$b6_x$	28_x	15_x	$2b_x$	88_x	06_x	52_x	ef_x	$d8_x$	$b9_x$	$b7_x$
10	bc_x	fc_x	$f4_x$	$a5_x$	$3a_x$	$0a_x$	81_x	$6e_x$	$3d_x$	60_x	aa_x	13_x	$b5_x$	ea_x	$4c_x$	39_x
11	24_x	87_x	$d6_x$	$1b_x$	41_x	$5d_x$	ab_x	17_x	$f8_x$	25_x	31_x	77_x	$a8_x$	$b8_x$	$e4_x$	$a1_x$
12	02_x	46_x	90_x	35_x	59_x	$c7_x$	$1e_x$	af_x	$3b_x$	fe_x	$5b_x$	$8a_x$	44_x	29_x	$6c_x$	db_x
13	$7e_x$	$d2_x$	05_x	37_x	30_x	89_x	75_x	$9c_x$	$c3_x$	$8d_x$	ae_x	$8b_x$	92_x	bb_x	$5c_x$	$d0_x$
14	23_x	$9a_x$	$e3_x$	$d7_x$	$7f_x$	45_x	94_x	ed_x	69_x	$9b_x$	$c4_x$	$4e_x$	$c6_x$	$c5_x$	dd_x	68_x
15	$4d_x$	eb_x	$a2_x$	$f6_x$	cd_x	27_x	$e2_x$	34_x	$f5_x$	$7b_x$	93_x	$1a_x$	bd_x	$0d_x$	86_x	ff_x

Table C.28 The 8 × 8 WHIRLPOOL S-box. For $0 \leq i, j \leq 15$, the $S[(i \ll 4)|j]$ entry is in the i-th row and j-th column. Values in hexadecimal.

i\j	0	1	2	3	4	5	6	7	8	9	10	11	12	13	14	15
0	18_x	23_x	$c6_x$	$E8_x$	87_x	$B8_x$	01_x	$4F_x$	36_x	$A6_x$	$d2_x$	$F5_x$	79_x	$6F_x$	91_x	52_x
1	60_x	Bc_x	$9B_x$	$8E_x$	$A3_x$	$0c_x$	$7B_x$	35_x	$1d_x$	$E0_x$	$d7_x$	$c2_x$	$2E_x$	$4B_x$	FE_x	57_x
2	15_x	77_x	37_x	$E5_x$	$9F_x$	$F0_x$	$4A_x$	dA_x	58_x	$c9_x$	29_x	$0A_x$	$B1_x$	$A0_x$	$6B_x$	85_x
3	Bd_x	$5d_x$	10_x	$F4_x$	cB_x	$3E_x$	05_x	67_x	$E4_x$	27_x	41_x	$8B_x$	$A7_x$	$7d_x$	95_x	$d8_x$
4	FB_x	EE_x	$7c_x$	66_x	dd_x	17_x	47_x	$9E_x$	cA_x	$2d_x$	BF_x	07_x	Ad_x	$5A_x$	83_x	33_x
5	63_x	02_x	AA_x	71_x	$c8_x$	19_x	49_x	$d9_x$	$F2_x$	$E3_x$	$5B_x$	88_x	$9A_x$	26_x	32_x	$B0_x$
6	$E9_x$	$0F_x$	$d5_x$	80_x	BE_x	cd_x	34_x	48_x	FF_x	$7A_x$	90_x	$5F_x$	20_x	68_x	$1A_x$	AE_x
7	$B4_x$	54_x	93_x	22_x	64_x	$F1_x$	73_x	12_x	40_x	08_x	$c3_x$	Ec_x	dB_x	$A1_x$	$8d_x$	$3d_x$
8	97_x	00_x	cF_x	$2B_x$	76_x	82_x	$d6_x$	$1B_x$	$B5_x$	AF_x	$6A_x$	50_x	45_x	$F3_x$	30_x	EF_x
9	$3F_x$	55_x	$A2_x$	EA_x	65_x	BA_x	$2F_x$	$c0_x$	dE_x	$1c_x$	Fd_x	$4d_x$	92_x	75_x	06_x	$8A_x$
10	$B2_x$	$E6_x$	$0E_x$	$1F_x$	62_x	$d4_x$	$A8_x$	96_x	$F9_x$	$c5_x$	25_x	59_x	84_x	72_x	39_x	$4c_x$
11	$5E_x$	78_x	38_x	$8c_x$	$d1_x$	$A5_x$	$E2_x$	61_x	$B3_x$	21_x	$9c_x$	$1E_x$	43_x	$c7_x$	Fc_x	04_x
12	51_x	99_x	$6d_x$	$0d_x$	FA_x	dF_x	$7E_x$	24_x	$3B_x$	AB_x	cE_x	11_x	$8F_x$	$4E_x$	$B7_x$	EB_x
13	$3c_x$	81_x	94_x	$F7_x$	$B9_x$	13_x	$2c_x$	$d3_x$	$E7_x$	$6E_x$	$c4_x$	03_x	56_x	44_x	$7F_x$	$A9_x$
14	$2A_x$	BB_x	$c1_x$	53_x	dc_x	$0B_x$	$9d_x$	$6c_x$	31_x	74_x	$F6_x$	46_x	Ac_x	89_x	14_x	$E1_x$
15	16_x	$3A_x$	69_x	09_x	70_x	$B6_x$	$d0_x$	Ed_x	cc_x	42_x	98_x	$A4_x$	28_x	$5c_x$	$F8_x$	86_x

Table C.29 The 8 × 8 ZODIAC S-box 1. For $0 \leq i, j \leq 15$, the $S[(i \ll 4)|j]$ entry is in the i-th row and j-th column. Values in hexadecimal.

i\j	0	1	2	3	4	5	6	7	8	9	10	11	12	13	14	15
0	$2d_x$	$f3_x$	$7c_x$	$6d_x$	$9d_x$	$b5_x$	26_x	74_x	$f2_x$	93_x	53_x	$b0_x$	$f0_x$	11_x	ed_x	83_x
1	78_x	$b6_x$	03_x	16_x	73_x	$3b_x$	$1e_x$	$8e_x$	70_x	bd_x	86_x	$1b_x$	47_x	$7e_x$	24_x	56_x
2	$f1_x$	77_x	88_x	46_x	97_x	$b1_x$	ba_x	$a3_x$	$b7_x$	10_x	$0a_x$	$c5_x$	37_x	$b3_x$	$c9_x$	$5a_x$
3	28_x	ac_x	64_x	$a5_x$	ec_x	ab_x	aa_x	$c6_x$	67_x	95_x	58_x	$0d_x$	$f8_x$	$9a_x$	$f6_x$	$6e_x$
4	66_x	dc_x	05_x	$3d_x$	$d3_x$	$8a_x$	$c3_x$	$d8_x$	89_x	$6a_x$	$e9_x$	36_x	49_x	43_x	bf_x	eb_x
5	$d4_x$	96_x	$9b_x$	68_x	$a0_x$	65_x	$5d_x$	57_x	92_x	$1f_x$	$d5_x$	71_x	$5c_x$	bb_x	22_x	$c1_x$
6	be_x	$7b_x$	bc_x	99_x	63_x	94_x	$5f_x$	$2a_x$	61_x	$b8_x$	34_x	32_x	19_x	fd_x	fb_x	17_x
7	40_x	$e6_x$	51_x	$1d_x$	41_x	44_x	$8f_x$	29_x	dd_x	04_x	80_x	de_x	$e7_x$	31_x	$d6_x$	$7f_x$
8	01_x	$a2_x$	$f7_x$	39_x	da_x	$6f_x$	23_x	ca_x	fe_x	$3a_x$	$d0_x$	$1c_x$	$d1_x$	30_x	$3e_x$	12_x
9	$a1_x$	cd_x	$0f_x$	$e0_x$	$a8_x$	af_x	82_x	59_x	$2c_x$	$f5_x$	$7d_x$	ad_x	$b2_x$	ef_x	$c2_x$	87_x
10	ce_x	75_x	06_x	13_x	02_x	90_x	$4f_x$	$2e_x$	72_x	33_x	85_x	$c0_x$	$8d_x$	cf_x	$a9_x$	81_x
11	$e2_x$	$c4_x$	27_x	$2f_x$	$6c_x$	$7a_x$	$9f_x$	52_x	$e1_x$	15_x	38_x	$2b_x$	fc_x	20_x	42_x	$c7_x$
12	08_x	$e4_x$	09_x	55_x	$5e_x$	$8c_x$	14_x	76_x	60_x	ff_x	df_x	$d7_x$	98_x	fa_x	$0b_x$	21_x
13	00_x	$1a_x$	$f9_x$	$a6_x$	$b9_x$	$e8_x$	$9e_x$	62_x	$4c_x$	$d9_x$	91_x	50_x	$d2_x$	ee_x	18_x	$b4_x$
14	07_x	84_x	ea_x	$5b_x$	$a4_x$	$c8_x$	$0e_x$	cb_x	48_x	69_x	$4b_x$	$4e_x$	$9c_x$	35_x	79_x	45_x
15	$4d_x$	54_x	$e5_x$	25_x	$3c_x$	$0c_x$	$4a_x$	$8b_x$	$3f_x$	cc_x	$a7_x$	db_x	$6b_x$	ae_x	$f4_x$	$e3_x$

Table C.30 The 8 × 8 ZODIAC S-box 2. For $0 \leq i, j \leq 15$, the $S[(i \ll 4)|j]$ entry is in the i-th row and j-th column. Values in hexadecimal.

i\j	0	1	2	3	4	5	6	7	8	9	10	11	12	13	14	15
0	00_x	$4a_x$	ce_x	$e7_x$	$d2_x$	62_x	$0c_x$	$e0_x$	$1f_x$	ef_x	11_x	75_x	78_x	71_x	$a5_x$	$8e_x$
1	76_x	$3d_x$	bd_x	bc_x	86_x	57_x	$0b_x$	28_x	$2f_x$	$a3_x$	da_x	$d4_x$	$e4_x$	$0f_x$	$a9_x$	27_x
2	53_x	04_x	$1b_x$	fc_x	ac_x	$e6_x$	$7a_x$	07_x	ae_x	63_x	$c5_x$	db_x	$e2_x$	ea_x	94_x	$8b_x$
3	$c4_x$	$d5_x$	$9d_x$	$f8_x$	90_x	$6b_x$	$b1_x$	$0d_x$	$d6_x$	eb_x	$c6_x$	$0e_x$	cf_x	ad_x	08_x	$4e_x$
4	$d7_x$	$e3_x$	$5d_x$	50_x	$1e_x$	$b3_x$	$5b_x$	23_x	38_x	34_x	68_x	46_x	03_x	$8c_x$	dd_x	$9c_x$
5	$7d_x$	$a0_x$	cd_x	$1a_x$	41_x	01_x	01_x	$8d_x$	$f6_x$	cb_x	52_x	$7b_x$	$d1_x$	$e8_x$	$4f_x$	29_x
6	$c0_x$	$b0_x$	$e1_x$	$e5_x$	$c7_x$	74_x	$b4_x$	aa_x	$4b_x$	99_x	$2b_x$	60_x	$5f_x$	58_x	$3f_x$	fd_x
7	cc_x	ff_x	40_x	ee_x	$b2_x$	$3a_x$	$6e_x$	$5a_x$	$f1_x$	55_x	$4d_x$	$a8_x$	$c9_x$	$c1_x$	$0a_x$	98_x
8	15_x	30_x	44_x	$a2_x$	$c2_x$	$2c_x$	45_x	92_x	$6c_x$	$f3_x$	39_x	66_x	42_x	$f2_x$	35_x	20_x
9	$6f_x$	77_x	bb_x	59_x	19_x	$1d_x$	fe_x	37_x	67_x	$2d_x$	31_x	$f5_x$	69_x	$a7_x$	64_x	ab_x
10	13_x	54_x	25_x	$e9_x$	09_x	ed_x	$5c_x$	05_x	ca_x	$4c_x$	24_x	87_x	bf_x	18_x	$3e_x$	22_x
11	$f0_x$	51_x	ec_x	61_x	17_x	16_x	$5e_x$	af_x	$d3_x$	49_x	$a6_x$	36_x	43_x	$f4_x$	47_x	91_x
12	df_x	33_x	93_x	21_x	$3b_x$	79_x	$b7_x$	97_x	85_x	10_x	$b5_x$	ba_x	$3c_x$	$b6_x$	70_x	$d0_x$
13	06_x	$a1_x$	fa_x	81_x	82_x	83_x	$7e_x$	$7f_x$	80_x	96_x	73_x	be_x	56_x	$9b_x$	$9e_x$	95_x
14	$d9_x$	$f7_x$	02_x	$b9_x$	$a4_x$	de_x	$6a_x$	32_x	$6d_x$	$d8_x$	$8a_x$	84_x	72_x	$2a_x$	14_x	$9f_x$
15	88_x	$f9_x$	dc_x	89_x	$9a_x$	fb_x	$7c_x$	$2e_x$	$c3_x$	$8f_x$	$b8_x$	65_x	48_x	26_x	$c8_x$	12_x

References

1. 3GPP: 3GPP TS 35.202 V5.0.0 Third Generation Partnership Project. Technical Specification Group Services and System Aspects; 3G Security; Specification of the 3GPP Confidentiality and Integrity Algorithms. Document 2: KASUMI Specification (Release 5) (2002)
2. Adams, C.M.: The CAST-256 Encryption Algorithm. First AES Conference, USA (1998)
3. Anderson, R.J., Biham, E., Knudsen, L.R.: Serpent and Smartcards. In: J.J. Quisquater, B. Schneier (eds.) Smart Card Research and Application Conference (CARDIS), LNCS 2000, pp. 246–253. Springer (1998)
4. Barreto, P.S.L.M., Nikov, V., Nikova, S., Rijmen, V., Tischhauser, E.: Whirlwind: a New Cryptographic Hash Function. Designs, Codes and Cryptography **56**, 141–162 (2010)
5. Barreto, P.S.L.M., Rijmen, V.: The ANUBIS Block Cipher. First NESSIE Workshop, Heverlee, Belgium (2000)
6. Barreto, P.S.L.M., Rijmen, V.: The KHAZAD Legacy-Level Block Cipher. First NESSIE Workshop, Heverlee, Belgium (2000)
7. Barreto, P.S.L.M., Rijmen, V.: The Whirlpool Hash Function. Submission to NESSIE (2003)
8. Barreto, P.S.L.M., Simplicio Jr, M.: Curupira, a Block Cipher for Constrained Platforms. 25th Brazilian Symposium on Computer Networks and Distributed Systems (2007)
9. Bertoni, G., Daemen, J., Peeters, M., Van Assche, G.: RadioGatún, a Belt-and-Mill Hash Function. Second Cryptographic Hash Function Workshop (2006)
10. Bertoni, G., Daemen, J., Peeters, M., Van Assche, G.: The Keccak Reference. http://keccak.noekeon.org (2011)
11. Biham, E., Anderson, R.J., Knudsen, L.R.: Serpent: a New Block Cipher Proposal. In: S. Vaudenay (ed.) Fast Software Encryption (FSE), LNCS 1372, pp. 222–238. Springer (1998)
12. Biham, E., Shamir, A.: Differential Cryptanalysis of the Data Encryption Standard. Springer (1993)
13. Bilgin, B., Nikova, S., Rijmen, V., Nikov, V., Stutz, G.: Threshold Implementations of All 3×3 and 4×4 S-boxes. IACR ePrint archive 2012/300 (2012)
14. Biryukov, A., Shamir, A.: Structural Cryptanalysis of SASAS. In: B. Pfitzmann (ed.) Advances in Cryptology, Eurocrypt, LNCS 2045, pp. 394–405. Springer (2001)
15. Bogdanov, A., Knudsen, L.R., Leander, G., Paar, C., Poschmann, A., Robshaw, M.J.B., Seurin, Y., Vikkelsoe, C.: Present: an Utra-Lightweight Block Cipher. In: P. Paillier, I. Verbauwhede (eds.) Cryptographic Hardware and Embedded Systems (CHES), LNCS 4727, pp. 450–466. Springer (2007)
16. Borst, J.: The Block Cipher: GRAND CRU. First NESSIE Workshop, Heverlee, Belgium (2000)
17. Brown, L., Kwan, M., Pieprzyk, J., Seberry, J.: Improving Resistance to Differential Cryptanalysis and the Redesign of LOKI. In: H. Imai, R.L. Rivest, T. Matsumoto (eds.) Advances in Cryptology, Asiacrypt, LNCS 739, pp. 36–50. Springer (1991)
18. Brown, L., Pieprzyk, J.: Introducing the New LOKI97 Block Cipher. First AES Conference, California, USA (1998)
19. Brown, L., Pieprzyk, J., Seberry, J.: LOKI - a Cryptographic Primitive for Authentication and Secrecy Applications. In: J. Seberry, J. Pieprzyk (eds.) Advances in Cryptology, Auscrypt, LNCS 453, pp. 229–236. Springer (1990)
20. Burwick, C., Coppersmith, D., D'Avignon, E., Genario, R., Halevi, S., Jutla, C., Matyas Jr, S.M., O'Connor, L., Peyravian, M., Safford, D., Zunic, N.: MARS – a Candidate Cipher for AES. First AES Conference, California, USA (1998)

21. Courtois, N., Pieprzyk, J.: Cryptanalysis of Block Ciphers with Overdefined Systems of Equations. In: Y. Zheng (ed.) Advances in Cryptology, Asiacrypt, LNCS 2501, pp. 267–287. Springer (2002)
22. Courtois, N.T.: How Fast can be Algebraic Attacks on Block Ciphers? IACR ePrint archive, 2006/168 (2006)
23. Daemen, J.: Cipher and Hash Function Design – Strategies based on Linear and Differential Cryptanalysis. Ph.D. thesis, Dept. Elektrotechniek, ESAT, Katholieke Universiteit Leuven, Belgium (1995)
24. Daemen, J., Govaerts, R., Vandewalle, J.: A New Approach to Block Cipher Design. In: R. Anderson (ed.) Fast Software Encryption (FSE), LNCS 809, pp. 18–32. Springer (1993)
25. Daemen, J., Knudsen, L.R., Rijmen, V.: The Block Cipher SQUARE. In: E. Biham (ed.) Fast Software Encryption (FSE), LNCS 1267, pp. 149–165. Springer (1997)
26. Daemen, J., Peeters, M., Van Assche, G., Rijmen, V.: NESSIE Proposal: NOEKEON. First NESSIE Workshop, Heverlee, Belgium (2000)
27. Daemen, J., Rijmen, V.: The Block Cipher BKSQ. In: J.J. Quisquater, B. Schneier (eds.) Smart Card Research and Applications (CARDIS), LNCS 1820, pp. 236–245. Springer (2000)
28. Davies, D., Murphy, S.: Pairs and Triplets of DES S-boxes. Journal of Cryptology **8**(1), 1–25 (1993)
29. Dawson, E., Chen, K., Henricksen, M., Millan, W., Simpson, L., Lee, H., Moon, S.J.: Dragon: a Fast Word-Based Stream Cipher. Submission to eStream Project (2006)
30. Diffie, W., Ledin, G.: SMS4 Encryption Algorithm for Wireless Networks, version 10.3 (2008)
31. ETSI: SAGE: Specification of the 3GPP Confidentiality and Integrity Algorithms 128-EEA3 and 128-EIA3. Document 1: 128-EEA3 and 128-EIA3 specification. ver. 1.6 (2011)
32. Filho, G.D., Barreto, P.S.L.M., Rijmen, V.: The Maelstrom-0 Hash Function. Proceedings of the 6th Brazilian Symposium on Information and Computer Systems Security (2006)
33. FIPS197: Advanced Encryption Standard (AES). FIPS PUB 197 Federal Information Processing Standard Publication 197, United States Department of Commerce (2001)
34. Gauravaram, P., Knudsen, L.R., Matusiewicz, K., Mendel, F., Rechberger, C., Schlaffer, M., Thomsen, S.: Grostl, a SHA-3 Candidate. Submission to NIST, Secure Hash Standard 3 (2008)
35. Guo, J., Peyrin, T., Poschmann, A., Robshaw, M.J.B.: The LED Block Cipher. In: B. Preneel, T. Takagi (eds.) Cryptographic Hardware and Embedded Systems (CHES), LNCS 6917, pp. 326–341. Springer (2011)
36. Gupta, K.C., Ray, I.G.: On Constructions of Involutory MDS Matrices. In: A.M. Youssef, A. Nitaj, A.E. Hassanien (eds.) AfricaCrypt, LNCS 7918, pp. 43–60. Springer (2013)
37. Isobe, T.: A Single-Key Attack on the Full GOST Block Cipher. In: A. Joux (ed.) Fast Software Encryption (FSE), LNCS 6733, pp. 290–305. Springer (2011)
38. Jacobson Jr, M.J., Huber, K.: The MAGENTA Block Cipher Algorithm. First AES Conference, California, USA (1998)
39. Junod, P., Macchetti, M.: Revisiting the IDEA Philosophy. In: O. Dunkelman (ed.) Fast Software Encryption (FSE), LNCS 5665, pp. 277–295. Springer (2009)
40. Junod, P., Vaudenay, S.: FOX: a New Family of Block Ciphers. In: H. Handschuh, M.A. Hasan (eds.) Selected Areas in Cryptography (SAC), LNCS 3357, pp. 114–129. Springer (2004)
41. Junod, P., Vaudenay, S.: Perfect Diffusion Primitives for Block Ciphers Building Efficient MDS Matrices. In: H. Handschuh, M.A. Hasan (eds.) Selected Areas in Cryptography (SAC), LNCS 3357, pp. 84–99. Springer (2004)

42. Kim, K.: Construction of DES-like S-boxes based on Boolean Functions Satisfying the SAC. In: H. Imai, R.L. Rivest, T. Matsumoto (eds.) Advances in Cryptology, Asiacrypt, LNCS 739, pp. 59–72. Springer (1991)
43. Kim, K., Lee, S., Park, S., Lee, D.: How to Strengthen DES Against Two Robust Attacks. Proceedings of 1995 Korea–Japan Joint Workshop on Info. Security and Cryptology, JW-ISC'95 (1995)
44. Kim, K., Park, S., Lee, S.: Reconstruction of s^2-DES S-boxes and Their Immunity to Differential Cryptanalysis. JW-ISC, 1993 Korean-Japan Joint Workshop on Information Security and Cryptology (1993)
45. Knudsen, L.R., Leander, G., Poschmann, A., Robsaw, M.J.B.: PRINTcipher: a Block Cipher for IC-Printing. In: S. Mangard, F.X. Standaert (eds.) Cryptographic Hardware and Embedded Systems (CHES), LNCS 6225, pp. 16–32. Springer (2010)
46. Küçük, O.: The Hash Function Hamsi. Submission to NIST, SHA-3 Competition (2009)
47. Kwon, D., Kim, J., Park, S., Sung, S.H., Sohn, Y., Song, J.H., Yeom, Y., Yoon, E.J., Lee, S., Lee, J., Chee, S., Han, D., Hong, J.: New Block Cipher: ARIA. In: J.I. Lim, D.H. Lee (eds.) Information Security and Cryptology, ICISC, LNCS 2971, pp. 432–445. Springer (2003)
48. Lacan, J., Fimes, J.: Systematic MDS Erasure Codes based on Vandermonde Matrices. IEEE Transactions on Communications Letters **8**(9), 570–572 (2004)
49. Lai, X., Massey, J.L.: A Proposal for a New Block Encryption Standard. In: I.B. Damgård (ed.) Advances in Cryptology, Eurocrypt, LNCS 473, pp. 389–404. Springer (1990)
50. Lai, X., Massey, J.L., Murphy, S.: Markov Ciphers and Differential Cryptanalysis. In: D.W. Davies (ed.) Advances in Cryptology, Eurocrypt, LNCS 547, pp. 17–38. Springer (1991)
51. Lee, C., Jun, K., Jung, M., Park, S., Kim, J.: Zodiac version 1.0 (revised) Architecture and Specification. Standardization Workshop on Information Security Technology, Korean contribution on MP18033, ISO/IEC JTC1/SC27 N2563 (2000)
52. Lee, C.H., Kim, J.S.: The New Block Cipher Rainbow. Samsung Advanced Institute of Technology (1997)
53. Lee, H.J., Lee, S.J., Yoon, J.H., Cheon, D.H., Lee, J.I.: The SEED Encryption Algorithm. RFC 4269 (2005)
54. Lim, C.H.: A Revised Version of Crypton - CRYPTON version 1.0. In: L.R. Knudsen (ed.) Fast Software Encryption (FSE), LNCS 1636, p. 31. Springer (1999)
55. MacWilliams, F.J., Sloane, N.J.A.: The Theory of Error-Correcting Codes. North-Holland Mathematical Library. North-Holland Publishing Co. (1977)
56. Massey, J.L.: SAFER K–64: a Byte-Oriented Block-Ciphering Algorithm. In: R. Anderson (ed.) Fast Software Encryption (FSE), LNCS 809, pp. 1–17. Springer (1994)
57. Matsui, M.: Linear Cryptanalysis Method for DES Cipher. In: T. Helleseth (ed.) Advances in Cryptology, Eurocrypt, LNCS 765, pp. 386–397. Springer (1993)
58. Matsui, M.: On Correlation between the Order of S-boxes and the Strength of DES. In: A. De Santis (ed.) Advances in Cryptology, Eurocrypt, LNCS 950, pp. 366–375. Springer (1994)
59. Matsui, M.: Block Encryption Algorithm MISTY. Technical report of IEICE, isec96-11, in Japanese, Mitsubishi Co. (1996)
60. Matsui, M.: New Block Encryption Algorithm MISTY. In: E. Biham (ed.) Fast Software Encryption (FSE), LNCS 1267, pp. 54–68. Springer (1997)
61. Matsui, M., Nakajima, J., Moriai, S.: A Description of the Camellia Encryption Algorithm. Request for comments, RFC3713 (2004)
62. Menezes, A.J., Van Oorschot, P., Vanstone, S.A.: Handbook of Applied Cryptography. CRC Press (1997)
63. Merkle, R.C.: Fast Software Encryption Functions. In: A.J. Menezes, S.A. Vanstone (eds.) Advances in Cryptology, Crypto, LNCS 537, pp. 476–501. Springer (1991)

64. Mileva, A.: Multipermutations in Crypto World: Different Faces of the Same Perfect Diffusion Layer. IACR ePrint, 2014/085 (2014)
65. Nakahara Jr, J.: A Linear Analysis of Blowfish and Khufu. In: E. Dawson, S. Duncan (eds.) Information Security Practice and Experience Conference (ISPEC), LNCS 4464, pp. 20–32. Springer (2007)
66. Nakahara Jr, J.: Analysis of Venkaiah et al.'s AES Design. International Journal of Network Security (IJNS) **9**, 285–289 (2009)
67. Nakahara Jr, J., Rijmen, V., Preneel, B., Vandewalle, J.: The MESH Block Ciphers. In: K. Chae, M. Yung (eds.) Information Security Applications (WISA), LNCS 2908, pp. 458–473. Springer (2003)
68. NBS: Data Encryption Standard (DES). FIPS PUB 46–3, Federal Information Processing Standards Publication 46–3 (1999)
69. NIST: Skipjack and KEA Specification, version 2.0. http://csrc.nist.gov/ (1998)
70. NTT: Specification of E2 – a 128-bit Block Cipher. First AES Conference, California, USA (1998)
71. Nyberg, K.: Differentially Uniform Mappings for Cryptography. In: T. Helleseth (ed.) Advances in Cryptology, Eurocrypt, LNCS 765, pp. 55–64. Springer (1993)
72. Pommerening, K.: Fourier Analysis of Boolean Maps – a Tutorial. available at http://www.staff.uni-mainz.de/pommeren/Kryptologie (2005)
73. Preneel, B., Leekwijck, W.V., Linden, L.V., Govaerts, R., Vandewalle, J.: Propagation Characteristics of Boolean Functions. In: I.B. Damgaard (ed.) Advances in Cryptology, Eurocrypt, LNCS 473, pp. 161–173. Springer (1990)
74. Rijmen, V., Daemen, J., Preneel, B., Bosselaers, A., De Win, E.: The Cipher SHARK. In: D. Gollmann (ed.) Fast Software Encryption (FSE), LNCS 1039, pp. 99–112. Springer (1996)
75. Rijmen, V., Preneel, B., De Win, E.: On Weaknesses of Non-Surjective Round Functions. Designs, Codes and Cryptography **12**(3), 253–266 (1997)
76. Saarinen, M.J.O.: Cryptographic Analysis of All 4×4-bit S-boxes. In: A. Miri, S. Vaudenay (eds.) Selected Areas in Cryptography (SAC), LNCS 7118, pp. 118–133. Springer (2011)
77. Sajadieh, M., Dakhilalian, M., Mala, H., Omoomi, B.: On Construction of Involutory MDS Matrices from Vandermonde Matrices in GF(2^q). Designs, Codes and Cryptography **64**(3), 287–308 (2012)
78. Schneier, B.: Description of a New Variable-Length Key, 64-bit Block Cipher (Blowfish). In: R. Anderson (ed.) Fast Software Encryption (FSE), LNCS 809, pp. 191–204. Springer (1994)
79. Schneier, B., Kelsey, J., Whiting, D., Wagner, D., Hall, C., Ferguson, N.: Twofish: a 128-bit Block Cipher. First AES Conference, California, USA (1998)
80. Schnorr, C.P., Vaudenay, S.: Black-Box Cryptanalysis of Hash Networks based on Multipermutations. In: A. De Santis (ed.) Advances in Cryptology, Eurocrypt, LNCS 950, pp. 47–57. Springer (1995)
81. Shannon, C.E.: Communication Theory of Secrecy Systems. Bell System Technical Journal **28**(4), 656–715 (1949)
82. Shimoyama, T., Kaneko, T.: Quadratic Relation of S-box and its Application to the Linear Attack of Full Round DES. In: H. Krawczyk (ed.) Advances in Cryptology, Crypto, LNCS 1462, pp. 200–211. Springer (1998)
83. Shimoyama, T., Yanami, H., Yokoyama, K., Takenaka, M., Itoh, K., Yajima, J., Torii, N., Tanaka, H.: The Block Cipher SC2000. In: M. Matsui (ed.) Fast Software Encryption (FSE), LNCS 2355, pp. 312–327. Springer (2002)
84. Shirai, T., Shibutani, K.: On the Diffusion Matrix Employed in the Whirlpool Hashing Function. The NESSIE project (2003)
85. Shirai, T., Shibutani, K., Akishita, T., Moriai, S., Iwata, T.: The 128-bit Blockcipher CLEFIA (extended abstract). In: A. Biryukov (ed.) Fast Software Encryption (FSE), LNCS 4593, pp. 181–195. Springer (2007)

86. Sony: CLEFIA. https://www.sony.net/Products/cryptography/clefia (2007)
87. Stern, J., Vaudenay, S.: CS-Cipher. In: S. Vaudenay (ed.) Fast Software Encryption (FSE), LNCS 1372, pp. 189–205. Springer (1998)
88. Toshiba: Specification of Hierocrypt-3. First NESSIE Workshop, Heverlee, Belgium (2000)
89. Toshiba: Specification of Hierocrypt-L1. First NESSIE Workshop, Heverlee, Belgium (2000)
90. Tsunoo, Y., Kubo, H., Miyauchi, H., Nakamura, K.: A Secure Cipher Evaluated by Statistical Methods. SCIS'98-4.2.B, The 1998 Symposium on Cryptography and Information Security, p.28–31, The Institute of Electronics, Information and Communication Engineers (1998)
91. Tsunoo, Y., Kubo, H., Miyauchi, H., Nakamura, K.: A New 128-bit Block Cipher CIPHERUNICORN-A. The Institute of Electronics, Information and Communication Engineers, Technical Report of IEICE, ISEC2000-5 (2000)
92. Vaudenay, S.: On the Need for Multipermutations: Cryptanalysis of MD4 and SAFER. In: B. Preneel (ed.) Fast Software Encryption (FSE), LNCS 1008, pp. 286–297. Springer (1995)
93. Vaudenay, S.: Provable Security for Block Cipher by Decorrelation. In: M. Morvan, C. Meinel, D. Krob (eds.) Proceedings of the 15th Annual Symposium on Theoretical Aspects of Computer Science (STACS), LNCS 1373, pp. 249–275. Springer (1998)
94. Watanabe, D., Furuya, S., Yoshida, H., Takaragi, K., Preneel, B.: A New Keystream Generator MUGI. In: J. Daemen, V. Rijmen (eds.) Fast Software Encryption (FSE), LNCS 2365, pp. 179–194. Springer (2002)
95. Webster, A.F., Tavares, S.E.: On the Design of S-boxes. In: H.C. Williams (ed.) Advances in Cryptology, Crypto, LNCS 219, pp. 523–534. Springer (1985)
96. Wu, H.: The Hash Function JH. Submission to NIST, SHA-3 competition (2011)
97. Youssef, A.M., Mister, S., Tavares, S.E.: On the Design of Linear Transformations for Substitution Permutation Encryption Networks. In: C. Adams, M. Just (eds.) Selected Areas in Cryptography (SAC), pp. 40–48. Springer (1997)
98. Zabotin, I.A., Glazkov, G.P., Isaeva, V.B.: GOST 28147-89, Cryptographic Protection for Data Processing Systems, Cryptographic Transformation Algorithm. Government Standard of the U.S.S.R., Inv. No. 3583, UDC 681.325.6:006.354 (1989)

Index

Λ-set, 485
3-dimensional cipher design, 85

Abelian group, 659
ACPC, 589
active S-box, 139, 376
active word, 486
adaptively chosen-ciphertext attack, 17
Advanced Encryption Standard, 7
AES, 7, 27
amplified boomerang, 339
AMX, 25, 106
ANF, 226, 681
ARX, 25
ASR attack, 575
asymmetric, 2

Babbage-Golic TMTO attack, 124
balanced word, 486
BDK attacks, 559
Bel-T, 109
biclique technique, 327
bijective mappings, 51
birthday paradox, 121, 123
Biryukov-Demirci attack, 529
Biryukov-Demirci relation, 49
bit-flipping errors, 20
block cipher, 2, 3
block size, 4
Boolean satisfiability, 225
boomerang attack, 336
boomerang distinguisher, 344
bottom-up analysis, 132, 163
branch rule for bitmasks, 373
branch-and-bound algorithm, 379
brute force, 117
byte oriented, 91

Cauchy matrix, 667
CBC, 18
CFB, 18
chosen-ciphertext attack, 16
chosen-key attack, 123
chosen-plaintext attack, 15
cipher state, 87
ciphertext, 2
ciphertext space, 11
ciphertext-only attack, 15
circulant matrix, 664
classical occupancy problem, 121
CNF, 226, 681
codebook, 2, 4, 271
collision, 122, 124, 339
commutative group, 659
commutative ring, 660
complete diffusion, 46
complete text diffusion, 52, 606
compression function, 104
computational graph, 42
confusion, 11, 14, 39, 675
CRYPTREC, 7
CTR, 18
cycle, 3

data complexity, 24
DDT, 133, 136
Demirci attack, 524
Demirci relation, 524
DES, 5, 104
design criteria, 38
DFR attacks, 643
dictionary attack, 120, 124
difference, 131
difference distribution, 133
Difference Distribution Table, 133

J. Nakahara Jr., *Lai-Massey Cipher Designs*, https://doi.org/10.1007/978-3-319-68273-0

difference operator, 132
differential, 142
differential branch number, 147, 205, 661
differential characteristic, 140
differential cryptanalysis, 131
differential distinguisher, 143
differential profile, 133, 676
differential trail, 137
differential uniformity, 676
differential-linear cryptanalysis, 474
differential-linear distinguisher, 479
differentially active, 218
differentially passive, 218
differentially uniform, 136
diffusion, 11, 14, 39
diffusion layer, 275
diffusion power, 151, 163
distinguisher, 265
divide-and-conquer attacks, 101
DNF, 681
dual code, 661

ECB, 18
EKS, 645
entropy, 13
equivalent keys, 13, 119
exhaustive key search, 50
exhaustive search, 117

false alarm, 128
Feistel Network, 7
Feistel Network cipher, 639
Fermat prime, 49
field, 660
finite fields, 49
fixed point, 220
fixed-point relation, 450
FOX, 90
full diffusion, 27
full text diffusion, 101
functional composition, 5, 47, 53, 61, 66, 73, 79

gap, 344
garbled word, 486
group, 659
Grover's algorithm, 120

Hadamard matrix, 668
half-round, 44
Hamming distance, 661
Hamming Weight, 119, 661
Higher-Order Differential-Linear attack, 480
higher-order multisets, 488
hybrid design, 109
Hypothesis of Stochastic Equivalence, 142

IDEA*, 100
IDEA-NXT, 90
IDEA-X, 44
IDEA-X/2, 45
ideal primitive, 221
impossible-boomerang distinguisher, 342
Impossible-Differential technique, 282
input difference, 132
integral cryptanalysis, 485
integrals, 485
interleaved, 53
interleaved layers, 50
internal permutation, 85
invariant, 46, 93
involution, 41
involutory, 53
involutory mappings, 88
involutory permutation, 617
involutory transformation, 61, 66, 292
IPES, 44
irreducible polynomial, 86
iteration, 5
iterative characteristic, 139
iterative trails, 141, 378

key avalanche, 51, 54, 104
key diffusion, 62, 68, 75, 99, 101, 104, 619
key invariant, 133
key schedule, 5
key space, 11
key-dependent distribution attack, 550
keyed MA-box, 96
keyed permutation, 2, 4, 45, 65
keyless-BD attack, 563
KM, 45
known-plaintext attack, 15
known-plaintext setting, 118, 120
KSA, 5, 41

LAT, 373
length-preserving, 3
lexicographic order, 4
Linear Approximation Table, 373
linear branch number, 383, 662
linear code, 661
linear cryptanalysis, 370
linear hull, 379
linear profile, 678
linear trail, 371, 613
linear transformation, 42

linear uniformity, 678
Low-High algorithm, 39, 103

MA, 45
MA-box, 39
MAD-box, 85
Markov cipher, 141
Maximum Distance Separable code, 85
MDS, 85, 91
MDS code, 662
MDS matrix, 86, 206
membership test, 479
memory complexity, 24
MESH, 25
MESH ciphers, 27
MESH-128, 64
MESH-128(8), 77
MESH-64(8), 71
MESH-96, 58
mini-cipher versions, 45
miss-in-the-middle technique, 282
MITM, 563
MITM attack, 564
Miyaguchi-Preneel, 76
modes of operation, 13, 18
modular division, 318
monoid, 659
multi-collision, 124
multipermutation, 663
multiple encryption, 6
multiplicative differential, 38, 280
multiplicative inverse, 37
multiset, 485
multiset attack, 485
multiset distinguisher, 487
multiset trail, 487

narrow differential trail, 137, 275
narrow linear trail, 371, 529
NESSIE, 7
NIST, 7
NLFSR, 7, 55, 61, 63, 68, 73, 90, 103, 619
non-Abelian group, 659
non-commutative operator, 318
non-involutory, 107
nonlinear feedback shift register, 55
nontrivial difference, 132, 137
nontrivial fixed point, 222

OFB, 18
online complexity, 120
orthomorphism mapping, 91
OT, 41
output difference, 132

passive S-box, 139, 376
passive word, 486
permutation, 2
permutation mappings, 126
PES, 37
PGP, 25
piling-up lemma, 378
plaintext, 2
plaintext block, 45
plaintext space, 11
primitive polynomial, 55, 68
propagation of differences, 132
PRP, 50, 153, 253, 384, 387, 453, 470
PseudoRandom Permutations, 17
public-key, 2

quantum computers, 120
quartet, 338

random permutation, 4
REESSE3+, 96
related-key attack, 16, 583
related-key boomerang attack, 593
related-key differential, 587
RIDEA, 83
right pair, 142
right quartet, 338
Rijndael, 7
ring, 49, 660
RKDL distinguisher, 584
round, 5

S-box, 25, 675
SAT solvers, 225
secret-key, 2
secret-key cryptosystem, 11
self-similarity condition, 319
self-synchronous stream cipher, 20
signal-to-noise ratio, 144, 266, 274
slide attack, 319
SPN, 7
SPN cipher, 639
Square attack, 484
square-root attacks, 122
standard form, 661
state-recovery attack, 118
stream ciphers, 4
Strong PseudoRandom Permutation, 17
substitution box, 374
success probability, 24
swap, 46
symmetric, 2
symmetric matrix, 664

T-functions, 39
Table-Lookup attack, 120
text diffusion, 10
time complexity, 24
TMDTO attack, 130
TMTO, 124, 538, 576
top-down boomerang distinguisher, 339
trivial bitmask, 375
trivial difference, 132
truncated differential, 256
truth table, 675

uncertainty, 13
unconditional security, 23
unicity distance, 118
unkeyed division, 101
unkeyed multiplication, 51, 60, 101
unkeyed permutation mapping, 97
unkeyed-$\boxdot$, 103
unkeyed-$\odot$, 103
user key, 42

Vandermonde matrix, 667
vector transposition, 86

weak key, 135, 224
weak subkeys, 65
weak-key assumptions, 145, 381
weak-key class, 147, 386
wide differential trail, 137
wide linear trail, 371
wide trail, 217, 251
WIDEA-n, 85
word swap, 41
wrong pair, 142

YBC, 106

Printed in the United States
By Bookmasters